ライフサイエンス

英語表現 使い分け辞典

第2版

編集／河本　健，大武　博
監修／ライフサイエンス辞書プロジェクト

羊土社
YODOSHA

羊土社
YODOSHA

【注意事項】本書の情報について

本書に記載されている内容は，発行時点における最新の情報に基づき，正確を期するよう，執筆者，監修・編者ならびに出版社はそれぞれ最善の努力を払っております．しかし科学・医学・医療の進歩により，定義や概念，技術の操作方法や診療の方針が変更となり，本書をご使用になる時点においては記載された内容が正確かつ完全ではなくなる場合がございます．また，本書に記載されている企業名や商品名，URL等の情報が予告なく変更される場合もございますのでご了承ください．

まえがき

本書は，生命科学分野の学術英語論文や学会抄録を書くための参考書「ライフサイエンス英語シリーズ」（『ライフサイエンス英語動詞使い分け辞典』，『ライフサイエンス論文作成のための英文法』等）の一冊である．本書『ライフサイエンス英語表現使い分け辞典』の初版を出版してから９年が経過した．初版は，３年半の歳月をかけて，なんとか完成にこぎつけた労作であった．改訂版を出すことなどは絶対にないと思っていたのだが，幸いにも好評を博し，今回，改訂版を出す運びとなった．

この９年の間に，元になるデータベースの大きさが３倍に増え，解析の精度が格段に向上した．精度が向上したことによって，初版に比べてはるかに系統立てた分かりやすい内容にまとめることができたと思う．これが改訂版を作った目的の１つである．さらに，この改訂版の最大の特徴は，名詞の単数・複数や冠詞の問題を全面的に取り上げたことである．日本人には難しいこれらの問題に対して，本書を参照すれば，英語の間違いをかなり減らすことができるはずである．

今回の改訂にあたって，ライフサイエンス辞書（LSD）プロジェクトのデータベースを再吟味して検討し，内容の充実に努めた．単語の使い方は実際の文から学ばなければなかなか習得できるものではないが，かといって１つの例文を見ただけでそれを学ぶことも不可能であろう．通常，１つの単語には様々な使い方があるので，相当数の用例を調べてもなかなか全貌を知ることはできない．しかし，このような個々の単語の用法を体系的に知る方法として共起表現を調べる方法がある．共起表現とは，ある単語の前後にどのような単語がよく用いられるかということだが，ここに単語の用法のエッセンスがあると言ってもよいであろう．もちろん直前直後だけが重要であるわけでもなく，また，特定の決まったパターンだけが使われるわけでもない．そこで，いろいろな表現をできるだけ収集して使用回数を示すことによって，単語の使い方を包括的に理解できるように執筆したものが本書である．

そもそも本書の特徴は，論文でよく使われる単語の共起表現の使用頻度を調べて用例と共に収録した点にある．たとえば「～の変化」という場合，日本人はchange ofを考えがちだが，実際にはchanges inの用例の方が圧倒的に多い．このような名詞＋前置詞，動詞（他動詞過去分詞／自動詞）＋前置詞の組み合わせには，決まったパターンが用いられることも多い．involved in，associated with，related to，required for，derived from，lead to，depend onなどの動詞＋前置詞のパターンがそれに相当する．このような論文でよく使われる共起表現を攻略することが，科学論文執筆のための大きなポイントとなるはずである．

共起表現検索で抽出できる単語の組み合わせは無限にも思えるが，よく使われるパターンはそれぞれの単語ごとにだいたい決まっており，しかも単語ごとにユニークである．本書では，このような組み合わせを個々の単語ごとに頻度情報に従ってパターン化し，特に日本人が苦手とする前置詞との共起表現に焦点を当てた．ある単語の次にどのような前置詞を用いるべきか迷ったときに参照すると大いに役に立つであろう．また，動詞＋前置詞だけでなく，effect on，treatment with，evidence for，patients withのような名詞＋前置詞，さらに副詞＋動詞，形容詞＋名詞など様々なパターンが，よく使われるものから順に並べてある．一般の辞書でこのようなことを調べることは困難であるので，本書は生命科学分野の論文や学会抄録を執筆するうえで，他にはない極めて有用なツールになるであろう．

本書を十分に活用して洗練された英語論文を仕上げていただけることを切に願っている．

2016年4月

編著者を代表して

河本　健

編集／河本　健

広島大学ライティングセンター特任教授

大武　博

福井県立大学名誉教授

監修／ライフサイエンス辞書プロジェクト

金子周司

京都大学大学院薬学研究科教授

鵜川義弘

宮城教育大学環境教育実践研究センター教授

大武　博

福井県立大学名誉教授

河本　健

広島大学ライティングセンター特任教授

竹内浩昭

静岡大学理学部生物科学科教授

竹腰正隆

東海大学医学部基礎医学系分子生命科学非常勤講師

藤田信之

製品評価技術基盤機構バイオテクノロジーセンター

本書の特徴および使い方

本書に収録した用語の収集は，ライフサイエンス辞書プロジェクトのLSDコーパス（PubMed論文抄録からの総語数約1億語）における単語の使用回数に基づいて行った．このコーパスに含まれる単語の種類は二十数万語に及んだが，そのうちの2,600語でコーパス全体の約85%が書かれていた．そこで，この頻出単語ランキングの上位3,000語の中から論文執筆の際に重要である約1,100語を選択して本書の見出し語とした．さらにそれらの関連語などで使い方に注意が必要な単語を上位5,000語の中から約300語を選んで追加した．総計1,430語の見出し語からなる本書は，個々の単語の使い方を示す活用辞典である．

本書に収録してある単語の使い方の情報を利用し，さらに個々の専門分野のキーワード，冠詞，代名詞，前置詞などを加えれば論文のほとんどの部分をカバーするであろう．

❊本書の特徴

本書には次のような特徴がある．

- 収集したすべての単語にその語形変化ごとの用例数（1億語のコーパス中での使用回数）を示した．
- 各単語の前後にどのような単語がよく用いられるかという共起表現を，その用例数と共に示した．
- PubMed論文抄録から典型的な例文を引用し，それを日本語訳と共に示した．

❇本書の構成と活用法

本書は，以下のように構成されている．

1 見出し語

見出し語は，論文執筆に重要な1,430語からなる．その品詞と意味とは，論文でよく用いられるものに絞って示してある（一般的な用法での意味は必ずしも考慮されていない）．また，重要度（使用回数）に応じて，星印を付してある．

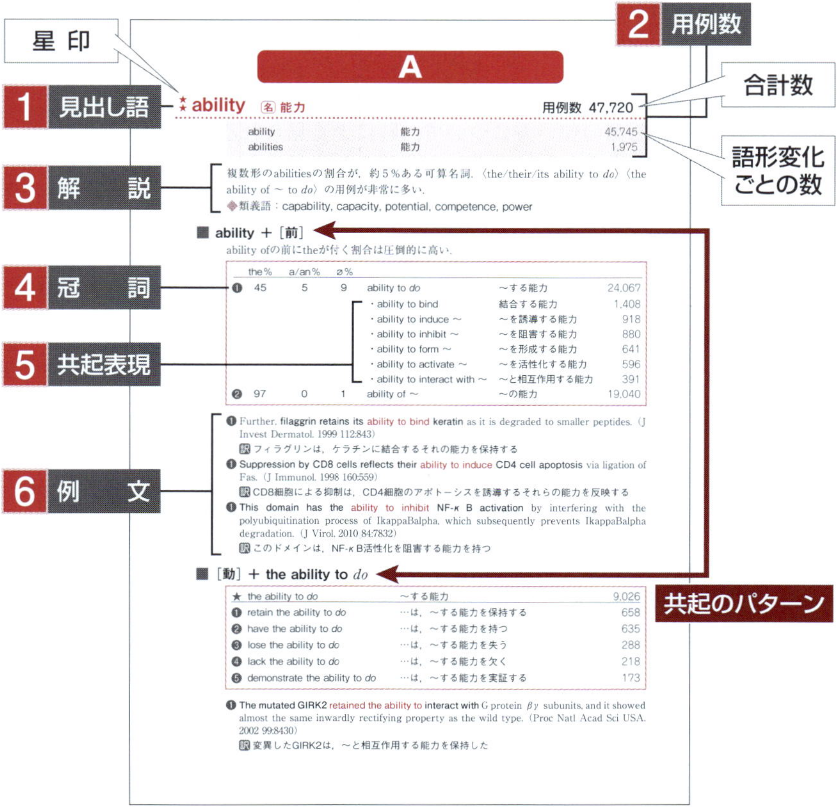

A

★★ ability 名 能力　　用例数 47,720

ability	能力	45,745
abilities	能力	1,975

複数形のabilitiesの割合が，約5％ある可算名詞．〈the/their/its ability to *do*〉〈the ability of ～ to *do*〉の用例が非常に多い．
◆類義語：capability, capacity, potential, competence, power

■ ability ＋［前］

ability ofの前にtheが付く割合は圧倒的に高い．

	the％	a/an％	ø％			
❶	45	5	9	ability to *do*	～する能力	24,067
				・ability to bind	結合する能力	1,408
				・ability to induce ～	～を誘導する能力	918
				・ability to inhibit ～	～を阻害する能力	880
				・ability to form ～	～を形成する能力	641
				・ability to activate ～	～を活性化する能力	596
				・ability to interact with ～	～と相互作用する能力	391
❷	97	0	1	ability of ～	～の能力	19,040

❶ Further, filaggrin retains its ability to bind keratin as it is degraded to smaller peptides.（J Invest Dermatol. 1999 112:843）
訳 フィラグリンは，ケラチンに結合するそれの能力を保持する

❶ Suppression by CD8 cells reflects their ability to induce CD4 cell apoptosis via ligation of Fas.（J Immunol. 1998 160:559）
訳 CD8細胞による抑制は，CD4細胞のアポトーシスを誘導するそれらの能力を反映する

❶ This domain has the ability to inhibit NF-κB activation by interfering with the polyubiquitination process of IkappaBalpha, which subsequently prevents IkappaBalpha degradation.（J Virol. 2010 84:7832）
訳 このドメインは，NF-κB活性化を阻害する能力を持つ

■［動］＋ the ability to *do*

★	the ability to *do*	～する能力	9,026
❶	retain the ability to *do*	…は，～する能力を保持する	658
❷	have the ability to *do*	…は，～する能力を持つ	635
❸	lose the ability to *do*	…は，～する能力を失う	288
❹	lack the ability to *do*	…は，～する能力を欠く	218
❺	demonstrate the ability to *do*	…は，～する能力を実証する	173

❶ The mutated GIRK2 retained the ability to interact with G protein βγ subunits, and it showed almost the same inwardly rectifying property as the wild type.（Proc Natl Acad Sci USA. 2002 99:8430）
訳 変異したGIRK2は，～と相互作用する能力を保持した

2 用例数

すべての見出し語に対してその語形変化ごとの用例数と合計数を示した．ただし動詞のing形については，名詞としての用例が多い場合は別項目として扱った．これによって，生命科学分野の論文でそれぞれの単語がどれくらいの頻度で用いられるかを知ることができる．頻度の高い単語を優先的に使い，さらに同じ表現の繰り返しを避けるためにやや頻度の低い単語を織りまぜながら論文を執筆するとよいであろう．語形変化ごとの用例数も示してあるので，名詞であればその単語が可算名詞であるか不可算名詞であるか，あるいは複数形で用いられることが多いかどうかなどの情報を得るためである．他動詞であれば，能動態で用いられることが多いのか受動態が多いのかなどを知ることができる．

3 解説

見出し語の使い方について，特によく見られる用法や注意すべき点などを示した．また，よく使われる類義語・反意語もできるだけ収録した．

4 冠詞

名詞の見出し語について，「名詞＋前置詞（同格 that 節）」のパターンごとに定冠詞 the が付く割合，不定冠詞 a あるいは an が付く割合，無冠詞の割合を示した．この3つの合計に，代名詞・数字など冠詞と同時に使うことができない単語が付く割合を加えると100％になる．

5 共起表現

本書ではすべての見出し語に対して，代表的な共起表現を使用回数と共に収録した．それらをパターンによって分類し，使用回数順に示した．パターンとしては，「過去分詞＋前置詞」「自動詞＋前置詞」「過去分詞＋to 不定詞」「副詞＋過去分詞」「名詞＋前置詞」「形容詞＋前置詞」「形容詞＋名詞」などがある．

6 例文

代表的な例文と日本語訳が収録してあるので，実際に論文においてどのように使われるかを知ることができる．ゴシック体の部分は，日本語訳と対応している．

❋本書の活用法（補足）

今回の改訂で新たに名詞＋前置詞の組み合わせごとの，冠詞の使い分けについて示したので補足しておく．一般に「名詞＋ of」の前には，the が付くことが非常に多い．とは言っても，必ず the が付くという訳でもない．たとえば，aim of の前に定冠詞 the が付く割合は99％である．また，ability of の前に the が付く割合は 97％である．つまり，これらの前に通常は the を付けるはずである．もし，the ではなく，不定冠詞 an を付けると判断するなら，その場合には相当に慎重に吟味しなければならない．もちろん，複数の目的があってそのうちの一つという意味なら，an aim of にするのが適切であろう．the aim of は，その目的が一つであることを意味するからだ．

これほど極端ではないが，abundance of の前に the が付く割合は71％，accuracy of の前に the が付く割合は 77％と，これらも非常に高い．したがって，ここでも the を付けることを基本に考えるのがよいであろう．一方，ablation of の前に the が付く割合は 7％しかなく，activation of の前に the が付く割合は 18％しかない．これらの場合には，前の例とは逆に，the を付けないことを基本に考える方がよいだろう．the が付かない理由は，これらの意味が「～を切除すること」「～を活性化すること」で，何かの性質を表すわけではないからであろう．

では，of 以外の前置詞の場合はどうだろうか．名詞によって事情は異なるが，一般的には the が付く割合がぐっと低くなる．たとえば同じ単語でも，abundance in の前に the が付く割合は 5％，accuracy in の前に the が付く割合は 12％とかなり低い．したがって，これらの前には the は付けないことを基本に考えるべきである．加えて，不定冠詞が付く割合もそれぞれ 0％と 5％と低く，無冠詞の割合がそれぞれ 65％と 61％と高いので，どちらも無冠詞で使うことを基本に考えるべきである．このように，後ろに続く前置詞によっても，状況が大きく変わる点に注意しなければならない．

名詞の使い方を考える場合，冠詞の問題だけでなく，可算・不可算および単数・複数の問題についても合わせて考えることが必須である．可算・不可算の区別は，複数形の割合で概ね判定できる．複数形の割合が低ければ低いほど，自信を持って不可算名詞であると言える．可算名詞と不可算名詞の境目は，おそらく1～2%ぐらいであるが，それ以外にも様々な問題があるので，コメントを参照していただきたい．例えば，abundanceやaccuracyを無冠詞で使う場合には，文中でこれらを不可算名詞として扱うことになる．しかし，abundanceは，複数形の用例もかなりある（約10%）ので，可算名詞としても使われることがわかる．その場合には，単数形に不定冠詞が付くこともあり得るわけである．同じ単語でも，状況によっては可算・不可算が変わってくることにも注意しよう．

最後に，複数形の割合が非常に高い名詞にも注目していただきたい．たとえばabnormalityの場合は，複数形のabnormalitiesの割合が85%と圧倒的に高い．このような単語は，複数形で使うことを基本に考えるのがよいであろう．

以上のように英語の名詞の使い方は，日本語の名詞と比べると非常に複雑である．適切に判断するのが難しいと感じるときには，よく使われるパターンを基本に考えてみるとよいだろう．本書をこのように活用することによって，英語の間違いを減らすことができるはずである．

LSDコーパスの1億語の範囲に限っても，共起の組み合わせは非常にたくさんあり，残念ながらそのすべてを限られた紙面に収録することはできない．また，あまり収録する分量が多くても，かえって単語の用法のエッセンスを掴むことができずに逆効果であろう．そこで本書では，論文でよく用いられる個々の単語の使い方を分かりやすくコンパクトにまとめるように心がけた．さらに詳しいことを知りたい場合は，ライフサイエンス辞書プロジェクトのホームページ（http://lsd-project.jp/）で公開されているWebLSDの「英語共起表現」を利用すれば調べることができる．

ライフサイエンス英語コーパスについて

❋コーパスとは

本書のもとになっているのは，ライフサイエンス辞書プロジェクトが独自に構築したライフサイエンス分野の専門英語のコーパスである．コーパスとは言語研究などのために一定の基準に従って収集された言語データのことをいうが，今日では「コンピュータで扱えるように体系化された大量の言語テキスト」すなわち「言葉のデータベース」の意味で使われることが多い．規模の大きな汎用の英語コーパスとしては British National Corpus（BNC）や Bank of English（Cobuild corpus）がある．これら大規模コーパスのコンピュータ分析をもとにした数量的な視点を取り入れることによって，辞書の編纂方法（見出し語の選択，意味の記載順，例文の選択など）が大きく様変わりしたと言われている．すでに国内外の多くの英語辞書でコーパスの活用を前面に打ち出している．

❋ライフサイエンス英語コーパスについて

ライフサイエンス分野では PubMed と呼ばれる無料の文献データベースが利用できることから，われわれは PubMed に収録されている学術論文の抄録を主な言語資料として，ライフサイエンス分野に特化した英語コーパスを構築した．生化学，分子生物学などの基礎的な分野から臨床医学などの応用分野に至るまで，ライフサイエンスのさまざまな分野を網羅する主要な学術誌（約 100 誌）を選び，2000 年から 20012 年までの 13 年間にアメリカまたはイギリスの研究機関から出された論文の抄録を収集して，1 億語からなるライフサイエンス英語コーパスを作成した．すべての文章には，付帯情報として PubMed の登録番号をもたせており，分析結果から容易に元の論文抄録を参照できるように工夫されている．

❇ライフサイエンス辞書とライフサイエンス英語コーパス

本書で利用したライフサイエンス英語コーパスには総語数にして約1億語の情報が含まれている．名詞や動詞の語尾活用を考慮すると，ユニークな単語の数は約20万語と見積もられるが，そのうち出現頻度の高い7万語でコーパス全体の99%をカバーしている．こうして構築されたライフサイエンス英語コーパスは，ライフサイエンス辞書のオンライン検索サービス WebLSD（http://lsd.pharm.kyoto-u.ac.jp/）において英語共起検索の形で実装されている．任意の検索語に対してその場でライフサイエンス英語コーパスを検索し，前後の隣接語を数量的に捉えることができるようになっている．それだけにとどまらず，英和辞書における見出し語の選択，複合語の抽出，用法・用例の抽出などにもコーパスが活用されており，ライフサイエンス辞書のすべてがこのコーパスをベースにしていると言っても過言ではない．

❇コーパスをもとに活きた英語を提示

科学論文の執筆を目的とした例文集や活用辞典は多数出版されているが，本書がそれらと決定的に異なるのは，コーパスのコンピュータ分析によって得られた数量的なデータを基礎に置いている点である．すなわち，見出し語の選択のみならず，解説する用法や用例の選択にあたっても，頻度情報を最大限に考慮して編纂を行った．これによって，実際の学術論文で好んで使用される「活きた英語」を提示できているものと思う．WebLSD とあわせて，ぜひ論文執筆等に活用していただきたい．

（藤田信之）

目　次

A

★★ ability ㊔ 能力 　　用例数 47,720

ability	能力	45,745
abilities	能力	1,975

複数形のabilitiesの割合が，約5％ある可算名詞．〈the/their/its ability to *do*〉〈the ability of ～ to *do*〉の用例が非常に多い．

◆類義語：capability, capacity, potential, competence, power

■ ability ＋［前］

ability ofの前にtheが付く割合は圧倒的に高い．

	the%	a/an%	ø%			
❶	45	5	9	ability to *do*	～する能力	24,067
				・ability to bind	結合する能力	1,408
				・ability to induce ～	～を誘導する能力	918
				・ability to inhibit ～	～を阻害する能力	880
				・ability to form ～	～を形成する能力	641
				・ability to activate ～	～を活性化する能力	596
				・ability to interact with ～	～と相互作用する能力	391
❷	97	0	1	ability of ～	～の能力	19,040

❶ Further, **filaggrin retains its ability to bind keratin** as it is degraded to smaller peptides.（J Invest Dermatol. 1999 112:843）

訳 フィラグリンは，ケラチンに結合するそれの能力を保持する

❶ **Suppression by CD8 cells reflects their ability to induce CD4 cell apoptosis** via ligation of Fas.（J Immunol. 1998 160:559）

訳 CD8細胞による抑制は，CD4細胞のアポトーシスを誘導するそれらの能力を反映する

❶ **This domain has the ability to inhibit NF-κB activation** by interfering with the polyubiquitination process of IkappaBalpha, which subsequently prevents IkappaBalpha degradation.（J Virol. 2010 84:7832）

訳 このドメインは，NF-κB活性化を阻害する能力を持つ

■［動］＋ the ability to *do*

★	the ability to *do*	～する能力	9,026
❶	retain the ability to *do*	…は，～する能力を保持する	658
❷	have the ability to *do*	…は，～する能力を持つ	635
❸	lose the ability to *do*	…は，～する能力を失う	288
❹	lack the ability to *do*	…は，～する能力を欠く	218
❺	demonstrate the ability to *do*	…は，～する能力を実証する	173

❶ **The mutated GIRK2 retained the ability to interact with** G protein βγ subunits, and it showed almost the same inwardly rectifying property as the wild type.（Proc Natl Acad Sci USA. 2002 99:8430）

訳 変異したGIRK2は，～と相互作用する能力を保持した

A

■ [過分] + ability to *do*

❶	reduced ability to *do*	低下した〜する能力	367
❷	decreased ability to *do*	低下した〜する能力	177
❸	impaired ability to *do*	損なわれた〜する能力	174

■ the ability of + [名] + to *do*

★	the ability of 〜 to *do*	…する〜の能力	10,306
❶	the ability of 〜 to induce …	…を誘導する〜の能力	606
❷	the ability of 〜 to bind	結合する〜の能力	559
❸	the ability of 〜 to inhibit …	…を抑制する〜の能力	479

❶ The ability of p53 to induce cell cycle arrest is linked to its ability to induce transcription of genes such as the cyclin-dependent kinase inhibitor p21.（Oncogene. 2001 20:659）

訳 細胞周期停止を誘導するp53の能力は，〜を誘導するそれの能力と関連する

■ [動] + the ability of + [名] + to *do*

❶	reduce the ability of 〜 to *do*	‥は，…する〜の能力を低下させる	259
❷	affect the ability of 〜 to *do*	‥は，…する〜の能力に影響する	235
❸	block the ability of 〜 to *do*	‥は，…する〜の能力を遮断する	211
❹	abolish the ability of 〜 to *do*	‥は，…する〜の能力を消失させる	191
❺	examine the ability of 〜 to *do*	‥は，…する〜の能力を調べる	180

❶ The negative regulation of pilus expression by FasX reduces the ability of GAS to adhere to human keratinocytes.（Mol Microbiol. 2012 86:140）

訳 〜は，ヒト角化細胞に接着するGASの能力を低下させる

ablation 名 消失／切除　　用例数 5,356

ablation	消失／切除	5,180
ablations	消失／切除	176

複数形の割合は3％あるが，単数形は無冠詞で使われることが圧倒的に多い．

◆類義語：disappearance, elimination, excision, resection, amputation

■ ablation + [前]

	the%	a/an%	ø%			
❶	7	4	86	ablation of 〜	〜の消失／〜の切除	1,947
❷	0	0	64	ablation in 〜	〜における消失／〜における切除	175

❶ Genetic ablation of these neurons abolished the increase in visual sensitivity during arousal without affecting baseline visual function or locomotor activity.（J Neurosci. 2012 32:15205）

訳 これらのニューロンの遺伝的消失

■ [形／名] + ablation

❶ genetic ablation	遺伝的消失	304
❷ catheter ablation	カテーテルアブレーション	233
❸ radiofrequency ablation	高周波アブレーション	196

★ able 形 できる　用例数 15,410

◆類義語：capable　◆反意語：unable

■ be able to *do*

★ be able to *do*	…は，～することができる	12,510
❶ be able to bind	…は，結合することができる	502
❷ be able to induce ～	…は，～を誘導することができる	409
❸ be able to detect ～	…は，～を検出することができる	251
❹ be able to identify ～	…は，～を同定することができる	242
❺ be able to inhibit ～	…は，～を阻害することができる	238

❷ Further studies found that **arsenite exposure was able to induce cyclin D1 expression.** (Cancer Res. 2005 65:9287)
訳 亜砒酸塩曝露は，サイクリンD1発現を誘導することができた

★ abnormal 形 異常な　用例数 10,173

◆類義語：aberrant, unusual

■ abnormal + [前]

❶ abnormal in ～	～において異常な	321
・be abnormal in ～	…は，～において異常である	233

❶ **Pulmonary function is frequently abnormal in patients with** congestive heart failure (CHF), the mechanism of which has not been completely characterized. (Circulation. 2002 106:1794)
訳 肺機能は，～の患者においてしばしば異常である

★ abnormality 名 異常／異常性　用例数 12,104

abnormalities	異常／異常性	10,519
abnormality	異常／異常性	1,585

複数形のabnormalitiesの割合は，約85%と圧倒的に高い．abnormalities inの用例が多い．

◆類義語：aberration, deviation

■ abnormalities ＋［前］

「～の異常」の意味では，abnormalities ofよりもabnormalities inの用例が多い．

	the%	a/an%	ø%			
❶	3	-	91	abnormalities in ～	～の異常	2,449
❷	11	-	88	abnormalities of ～	～の異常	920

❶ Congenital abnormalities in the coronary system can have major deleterious effects on heart function. (Circ Res. 2002 91:761)

訳 冠状動脈系の先天的な異常は，～に対する大きな有害な影響を持ちうる

❷ Constitutive inactivation of KIF3A produces abnormalities of left-right axis determination and embryonic lethality. (Proc Natl Acad Sci USA. 2003 100:5286)

訳 KIF3Aの構成的な不活性化は，左右軸決定の異常を生み出す

★ abolish 動 消失させる／破壊する　用例数 11,068

abolished	消失する／～を消失させた	8,115
abolishes	～を消失させる	1,559
abolish	～を消失させる	1,170
abolishing	～を消失させる	224

他動詞．

◆類義語：eliminate, lose, delete, remove, degrade, deplete, ablate, disappear

■ be abolished ＋［前］

★	be abolished	…は，消失する	2,442
❶	be abolished by ～	…は，～によって消失する	1,237
❷	be abolished in ～	…は，～において消失する	452

❶ The transforming and antiapoptotic activities of v-Rel were abolished by defined Ser-to-Ala mutations and restored by most Ser-to-Asp substitutions. (Mol Cell Biol. 1999 19:307)

訳 v-Relの抗アポトーシス活性は，限定されたセリンからアラニンへの変異によって消失した

❷ Inhibition was maximal at 16 hr and abolished in the presence of SR141716A, a selective CB_1 receptor antagonist. (J Neurosci. 2001 21:2425)

訳 抑制は，～の存在下において16時間で最大になりそして消失した

■［副］＋ abolished

❶	completely abolished	完全に消失する	781
❷	nearly abolished	ほとんど消失する	121

❶ The increased proton leak in the transgenic mitochondria was completely abolished by the addition of GTP. (J Biol Chem. 2003 278:22298)

訳 ～は，GTPの添加によって完全に消失した

■ abolish ＋［名句］

❶ abolish the ability of ～	…は，～の能力を消失させる	239
❷ abolish binding	…は，結合を消失させる	127
❸ abolish the effect of ～	…は，～の効果を消失させる	126

❶ By contrast, cellular depletion of STAT-5 with antisense ODNs completely **abolished the ability of** GH to promote differentiation.（J Biol Chem. 1999 274:8662）

訳 ～は，分化を促進する成長ホルモンの能力を完全に消失させた

abrogate 動 消失させる／抑止する　　用例数 5,953

abrogated	消失する／～を消失させた	3,869
abrogates	～を消失させる	1,052
abrogate	～を消失させる	795
abrogating	～を消失させる	237

他動詞．

◆類義語：abolish, block, silence, prevent, hinder, occlude, hamper, arrest, suppress, repress, inhibit

■ be abrogated ＋［前］

★ be abrogated	…は，消失する	1,188
❶ be abrogated by ～	…は，～によって消失する	683
❷ be abrogated in ～	…は，～において消失する	218

❶ Furthermore, ras- and myc-mediated transformation of mouse embryonic fibroblasts **was abrogated by** expression of HEY1 in a p53-dependent manner.（Proc Natl Acad Sci USA. 2004 101:3456）

訳 マウス胚線維芽細胞のrasおよびmycに仲介される形質転換は，HEY1の発現によって消失した

■［副］＋ abrogated

❶ completely abrogated	完全に消失する	366

❶ Corepression by ETO was **completely abrogated** by histone deacetylase inhibitors.（Mol Cell Biol. 2000 20:2075）

訳 ～は，ヒストン脱アセチル化酵素阻害剤によって完全に消失した

■ abrogate ＋［名句］

❶ abrogate the ability of ～	…は，～の能力を消失させる	130
❷ abrogate the effect of ～	…は，～の効果を消失させる	81

❶ Point mutations of conserved cysteine residues or a histidine in the RING finger domain, which are required for zinc binding, **abrogated the ability of** Cbl to negatively regulate Syk in COS-7 cells and Ramos B lymphocytic cells.（J Biol Chem. 2000 275:414）

訳 ～は，Sykを負に調節するCblの能力を消失させた

★★ absence 名 存在しないこと／非存在／不在 用例数 33,941

absence	存在しないこと／非存在	33,909
absences	存在しないこと／非存在	32

複数形のabsencesの割合は0.1%しかなく，原則，不可算名詞として使われる．in the absence ofの用例が圧倒的に多い．

◆類義語：lack, deficiency ◆反意語：presence

■ absence ＋ [前]

absence ofの前にtheが付く割合は圧倒的に高い．

	the %	a/an %	ø %			
❶	86	2	12	absence of ～	～の非存在	31,620
				・in the absence of ～	～が存在しない場合に	20,058
				・In the absence of ～	～が存在しない場合に	2,989
				・even in the absence of ～	～が存在しない場合でも	1,101
				・occur in the absence of ～	…は，～の存在なしに起こる	681
				・observed in the absence of ～	～の存在なしに観察される	203
				・complete absence of ～	～の完全な非存在	361
				・despite the absence of ～	～が存在しないにもかかわらず	298

❶ In the absence of stromal cells, primary epithelial cells were unable to proliferate. (Cancer Res. 2002 62:58)

訳 間質細胞が存在しない場合に，

❶ In cells infected with these mutants, protein synthesis continues even in the absence of the γ 134.5 gene. (J Virol. 1998 72:8620)

訳 タンパク質合成は，～の存在がなくても続く

❶ Interestingly, in contrast to native STAT6, activation of STAT6:ER* occurs in the absence of detectable tyrosine phosphorylation of the fusion protein. (J Immunol. 1998 161:1074)

訳 STAT6:ER*の活性化は，検出可能な～のチロシンリン酸化なしに起こる

❶ One caveat to this study is that it was performed in the complete absence of c-Myc. (Oncogene. 2010 29:2585)

訳 それは，c-Mycの完全な非存在下で行われた

■ the presence or/and absence of／the absence and/or presence of

❶	in the presence or absence of ～	～の存在下あるいは非存在下において	1,048
❷	in the presence and absence of ～	～の存在下および非存在下において	896
❸	in the absence and presence of ～	～の非存在下および存在下において	386
❹	in the absence or presence of ～	～の非存在下あるいは存在下において	304

❶ Tumour development was compared in Lmo2 transgenic mice in the presence or absence of the Rag1 gene. (Oncogene. 2001 20:4412)
訳 Rag1遺伝子の存在下あるいは非存在下において

absent 形 欠けている／存在しない 用例数 6,759

◆類義語：deficient, defective ◆反意語：present

■ be absent ＋［前］

★ be absent	…は，欠けている	3,934
❶ be absent in ～	…は，～において欠けている	1,726
❷ be absent from ～	…は，～から欠けている	773

❶ WT was the predominant morphotype in 26 (81%) of these samples and was absent in only 2 samples. (J Infect Dis. 2001 184:1480)
訳 ～は，2つのサンプルにおいてのみ欠けていた

❷ We have also determined that H4Ac16 is absent from a region of the X chromosome that includes a gene known to be dosage-compensated by a MSL-independent mechanism. (J Biol Chem. 2001 276:31483)
訳 H4Ac16は，～の領域から欠けている

■［副］＋ absent

❶ completely absent	完全に欠けている	173
❷ virtually absent	実質的に欠けている	111

❶ This response was completely absent in mice deficient in PDE4B but not PDE4D. (Proc Natl Acad Sci USA. 2002 99:7628)
訳 この反応は，PDE4Bの欠損したマウスにおいて完全に欠けていた

■ or absent

❶ reduced or absent	低下するかあるいは欠けている	138

❶ In particular, Bmp4 mRNA is reduced or absent in Nkx2.1$^{-/-}$ lungs. (Gene. 2004 327:25)
訳 Bmp4メッセンジャーRNAは，Nkx2.1$^{-/-}$の肺において低下するかあるいは欠けている

absolute 形 絶対の／完全な 用例数 4,418

◆類義語：obligate, complete

■ absolute ＋［名］

❶ absolute requirement	絶対的要求性	317
❷ absolute risk	絶対リスク	246

❶ The TRM5 enzyme does not have an absolute requirement for magnesium ions, whereas TrmD requires magnesium to express activity. (Biochemistry. 2004 43:9243)
訳 TRM5酵素は，マグネシウムイオンに対する絶対的要求性を持たない

absolutely 副 絶対的に／完全に 用例数 1,056

◆類義語：completely, entirely, fully, quite, thoroughly, totally

■ be absolutely ＋［過分／形］

★	be absolutely ～	…は，絶対的に～である	629
❶	be absolutely required for ～	…は，～のために絶対的に必要とされる	314
❷	be absolutely dependent on ～	…は，～に完全に依存する	92

❶ Gol is absolutely required for Lit activation.（J Biol Chem. 2005 280:112）
訳 Golは，Lit活性化のために絶対的に必要とされる

abundance 名 存在量 用例数 6,958

abundance	存在量	6,394
abundances	存在量	564

複数形の割合は約10%あるが，単数形は無冠詞で使われることが多い．

■ abundance ＋［前］

abundance ofの前にtheが付く割合は非常に高い．

	the%	a/an%	ø%			
❶	71	11	18	abundance of ～	～の存在量	2,428
❷	5	0	65	abundance in ～	～における存在量	457

❶ Liver-enriched inhibitory protein and liver-enriched activating protein were further shown to be an independent repressor and activator, respectively, for Hsf1 gene transcription, and the relative abundance of these two C/EBPβ isoforms was demonstrated to determine Hsf1 transcription.（J Biol Chem. 2012 287:40400）
訳 これら２つのC/EBPβアイソフォームの相対的存在量が，～を決定するために示された

■［形／名］＋ abundance

❶	relative abundance	相対的存在量	443
❷	protein abundance	タンパク質存在量	391
❸	mRNA abundance	メッセンジャーRNA存在量	356

abundant 形 大量の／豊富な 用例数 6,340

◆類義語：massive, rich

■ abundant ＋［前］

❶	abundant in ～	～において豊富な	1,411
	・be abundant in ～	…は，～において豊富である	556

❶ CRF-IR cell bodies were more abundant in the submucosal plexus（29.9-38.0%）than in the

myenteric plexus. (J Comp Neurol. 2006 494:63)
訳 CRF-IR細胞体は，筋層間神経叢においてより粘膜下神経叢（29.9-38.0%）において豊富であった

■ [副] + abundant

❶ most abundant	最も豊富な	1,341
❷ more abundant	より豊富な	579
・more abundant in ～	～においてより豊富な	238
・more abundant than ～	～より豊富な	124
❸ less abundant	より豊富でない	198

❶ Astrocytes are the most abundant and functionally diverse glial population in the vertebrate central nervous system (CNS). (Nature. 2005 438:360)
訳 アストロサイトは，～において最も豊富で機能的に多様なグリアの集団である

accelerate 動 加速する／促進する　用例数 9,159

accelerated	促進される／～を加速した／～を促進した	5,234
accelerate	～を加速する／～を促進する	1,695
accelerates	～を加速する／～を促進する	1,271
accelerating	～を加速する／～を促進する	959

主に他動詞として使われる.
◆類義語：facilitate, promote

■ accelerate + [名句]

❶ accelerate the rate of ～	…は，～の速度を加速させる	161
❷ accelerate the development of ～	…は，～の開発（発症）を加速させる	87
❸ accelerate the onset of ～	…は，～の開始を加速させる	45
❹ accelerate the progression of ～	…は，～の進行を加速させる	39

❶ Myc accelerates the rate of cell proliferation, at least in part, through its ability to down-regulate the expression of the cell cycle inhibitor $p27^{Kip1}$. (Am J Pathol. 2011 178:2470)
訳 Mycは，細胞増殖の速度を加速させる

■ be accelerated + [前]

★ be accelerated	…は，促進される	686
❶ be accelerated by ～	…は，～によって促進される	273
❷ be accelerated in ～	…は，～において促進される	141

❶ Tumor growth may be accelerated by in vivo passage, thus making these tumors more sensitive to some therapies than the original tumors. (Cancer Res. 2003 63:747)
訳 腫瘍の増殖は，生体内での継代によって促進されるかもしれない

accept ㊍ 認める／受け入れる　用例数 3,207

accepted	認められている／～を受け入れた	2,206
accept	～を受け入れる	424
accepting	～を受け入れる	395
accepts	～を受け入れる	182

他動詞．that節を伴う受動態の用例が多い．

◆類義語：assume, consider, believe, appreciate, regard, know, learn, think, view, understand

■ accepted ＋［that節］

★ accepted that ～	～ということは認められている	398
❶ it is widely accepted that ～	～ということは広く認められている	109
❷ it is generally accepted that ～	～ということは一般に認められている	105

❷ It is generally accepted that the internalization and desensitization of mu-opioid receptor (MOR) involves receptor phosphorylation and β-arrestin recruitment. (J Biol Chem. 2003 278:36733)

訳 ～ということは一般に認められている

■ accepted ＋［前］

❶ accepted as ～	～として認められている	149
・widely accepted as ～	～として広く認められている	46

❶ Surgery is widely accepted as an effective therapy for selected individuals with medically refractory epilepsy. (Lancet Neurol. 2008 7:525)

訳 外科手術は，～にとって効果的な治療であると広く認められている

★accompany ㊍ 伴う　用例数 11,422

accompanied	伴われる／～を伴った	8,662
accompanying	～を伴う	1,347
accompanies	～を伴う	758
accompany	～を伴う	655

他動詞．accompanied byの用例が圧倒的に多い．

◆類義語：associate, follow

■ be accompanied ＋［前］

★ be accompanied	…は，伴う	5,810
❶ be accompanied by ～	…は，～を伴う	5,721
・be accompanied by increased ～	…は，増大した～を伴う	361
・be accompanied by an increase in ～	…は，～の増大を伴う	292
・be accompanied by a decrease in ～	…は，～の減少を伴う	115

❶ A transient rise in cellular inositol 1,4,5,6-tetrakisphosphate was also observed and was accompanied by increased chloride channel activity. (Proc Natl Acad Sci USA. 2001 98:875)
訳 ～は，増大した塩化物イオンチャネル活性を伴った

❶ This reduction in the lesion volume has been accompanied by an increase in the intensity in the ^{18}F-FDG signal per voxel. (J Nucl Med. 2002 43:876)
訳 …は，～の強度の増大を伴っている

accomplish 動 達成する 用例数 3,101

accomplished	達成される／～を達成した	2,362
accomplish	～を達成する	563
accomplishes	～を達成する	112
accomplishing	～を達成する	64

他動詞.

◆類義語：achieve, attain

■ be accomplished ＋［前］

★ be accomplished	…は，達成される	5,810
❶ be accomplished by ～	…は，～によって達成される	824
❷ be accomplished in ～	…は，～において達成される	239
❸ be accomplished through ～	…は，～によって達成される	160

❶ Whether regeneration is accomplished by pluripotent cells or by the collective activity of multiple lineage-restricted cell types is unknown. (Science. 2011 332:811)
訳 再生は，多能性細胞によって達成される

■ to accomplish

★ to accomplish ～	～を達成する…	413
❶ To accomplish ～	～を達成するために	170

❶ To accomplish this goal, it is critical to understand the native mechanisms involved in the specification of HSCs during embryonic development. (Nature. 2011 474:220)
訳 このゴールを達成するために

accord 名 一致 用例数 647

accord	一致	623
accords	一致	24

in accord withの用例が圧倒的に多い.

■ in accord with

❶ in accord with ～	～と一致して	488

❶ These findings are in accord with known differences in tumorigenesis between uveal and cutaneous melanomas. (Invest Ophthalmol Vis Sci. 2003 44:2876)
訳 これらの知見は，～の間の腫瘍発生の既知の違いと一致している

accordance ㊔ 一致 用例数 586

in accordance withの用例が圧倒的に多い．

■ in accordance with

❶	in accordance with ～	～と一致して	530

❶ The human sequence also shows significant similarity to the cyclin-dependent kinases, in accordance with evidence that yeast Cdc7 is related to the cdks. (Gene. 1998 211:133)
訳 ～という証拠と一致して

according ㊍ 従って 用例数 6,052

according toの形で用いられる．

■ according to

★	according to ～	～に従って／～によれば	6,045
❶	According to ～	～によれば／～に従って	754
❷	classified according to ～	～に従って分類される	165
❸	stratified according to ～	～に従って層別化される	122
❹	into … groups according to ～	～に従って…グループに	81

❶ According to the results of dominant negative and reciprocal coimmunoprecipitation experiments, the Shaker and eag proteins do not interact. (Biophys J. 1998 75:1263)
訳 ドミナントネガティブおよび相互の免疫共沈降の実験の結果によれば

❷ Fistulas were classified according to Parks' criteria, and a consensus gold standard was determined for each patient. (Gastroenterology. 2001 121:1064)
訳 瘻孔は，Parksの基準に従って分類された

❹ These known genes were categorized into 15 groups according to their biological functions. (Gene. 2000 261:373)
訳 これらの既知の遺伝子は，それらの生物学的機能に従って15のグループに分類された

accordingly ㊖ したがって 用例数 2,431

文頭に用いられることが圧倒的に多い．
◆類義語：consequently, therefore, hence, thus

■ accordingly

❶	Accordingly, ～	したがって，～	2,168

❶ Accordingly, we hypothesized that the absence of a functional ATM protein might involve perturbations to the ubiquitin pathway as well. (Oncogene. 2002 21:4363)

訳 したがって，われわれは〜であると仮定した

★ account ㊖ 説明する／占める／明らかにする，㊔ 考慮／計算

用例数 15,976

account	説明する／占める／考慮	8,740
accounted	説明される／説明した	2,966
accounts	説明する／占める	2,501
accounting	説明する／明らかにする／占める	1,769

動詞としては自動詞だが，account forの形で他動詞的に使われることが圧倒的に多い．

◆類義語：explain, illustrate, consideration

■ ㊖ account for ＋［名句］

★ account for 〜	…は，〜を説明する／〜を占める	13,651
❶ account for approximately 〜 of …	…は，…のおよそ〜を占める	327
❷ account for the observed 〜	…は，観察された〜を説明する	264
❸ account for most 〜	…は，ほとんどの〜を説明する	238
❹ account for the majority of 〜	…は，〜の大部分を占める	162

❶ EDHF was found to account for approximately 80% of acetylcholine-mediated vasorelaxation.（Circulation. 2001 103:1702）

訳 EDHFは，〜のおよそ80%を占めることが見つけられた

❷ Yet the rotation angle of Tyr103 (134 degrees) is too large to account for the observed EPR spectrum in the wild type.（Biophys J. 2002 83:2845）

訳 〜は，観察されたEPRスペクトラムを説明するには大きすぎる

■ ㊖［形／動］＋ to account for

★ to account for 〜	〜を説明する…	1,788
❶ sufficient to account for 〜	〜を説明するのに十分な	170
❷ be proposed to account for 〜	…が，〜を説明するために提案される	115

❶ Binding to the N-terminal fragment is sufficient to account for the peptide binding activity of the entire molecule.（J Biol Chem. 2002 277:40742）

訳 N末端断片への結合は，ペプチドの結合活性を説明するのに十分である

❷ A model is proposed to account for human XOR regulation.（J Biol Chem. 2000 275:5918）

訳 ひとつのモデルが，ヒトのXOR調節を説明するために提案される

■ ㊖ be accounted for ＋［前］

★ be accounted for	…は，説明される	880
❶ be accounted for by 〜	…は，〜によって説明される	616

❶ The evolutionary history of serine proteases can be accounted for by highly conserved amino acids that form crucial structural and chemical elements of the catalytic apparatus.（EMBO J. 2001 20:3036）

訳 …は，～を形成する高度に保存されたアミノ酸によって説明されうる

■ 名［前］＋ account

❶ take into account ～	…は，～を考慮に入れる	714
❷ be taken into account	…は，考慮に入れられる	323

❶ An unbiased line of equivalence, taking into account the imprecision of both tests, was used to compare results. (Diabetes. 1999 48:1779)

訳 両方のテストの不正確性を考慮に入れて

❷ If programs in general practice to address dietary inequalities are to succeed, both patients' views and GPs' views must be taken into account. (Am J Clin Nutr. 2003 77:1043S)

訳 ～は，考慮に入れられなければならない

★ accumulate 動 蓄積する　用例数 10,778

accumulate	蓄積する	4,044
accumulated	蓄積した／蓄積される	3,569
accumulates	蓄積する	1,907
accumulating	蓄積する	1,258

他動詞として用いられることもあるが，自動詞の用例が非常に多い．

◆類義語：deposit

■ accumulate ＋［前］

❶ accumulate in ～	…は，～に蓄積する	3,220
❷ accumulate at ～	…は，～に蓄積する	564
❸ accumulate to ～	…は，～に蓄積する	391
❹ accumulate during ～	…は，～の間に蓄積する	213

❶ Moreover, virion-associated α TIF does not accumulate in the nucleus of cells infected in the presence of CTC-96. (J Virol. 2001 75:4117)

訳 ～は，細胞の核に蓄積しない

■［名］＋ accumulate

❶ cells accumulate	細胞が，蓄積する	394
❷ protein accumulates	タンパク質が，蓄積する	92

■ accumulating ＋［名］

❶ accumulating evidence	蓄積する証拠	519

❶ Accumulating evidence suggests that postmenopausal hormone use may decrease the risk for colorectal cancer. (Ann Intern Med. 1998 128:705)

訳 蓄積する証拠は，～ということを示唆する

★ accumulation ㊔ 蓄積 用例数 22,776

accumulation	蓄積	22,458
accumulations	蓄積	318

複数形の割合は約1.5%しかなく，不可算名詞として使われることが多い．

◆類義語：deposition, deposit, reservoir

■ accumulation ＋［前］

	the%	a/an%	ø%		
❶	42	8	49	accumulation of ～ ～の蓄積	12,288
				・result in the accumulation of ～ …は，～の蓄積という結果になる	311
				・lead to the accumulation of ～ …は，～の蓄積につながる	198
❷	1	1	72	accumulation in ～ ～における蓄積	2,192

❶ The absence or inactivation of the vWF-cleaving protease **results in the accumulation of large multimers**, which may cause thrombotic thrombocytopenic purpura.（Blood. 2001 98:1662）

訳 ～は，大きな多量体の蓄積という結果になる

❶ Loss of mismatch repair（MMR）function **leads to the accumulation of errors that normally occur during DNA replication**, resulting in genetic instability.（Cancer Res. 2001 61:4112）

訳 ～は，通常DNA複製の間に起こるエラーの蓄積につながる

❷ **Starch accumulation in sse1 suggests that** starch formation is a default storage deposition pathway.（Science. 1999 284:328）

訳 sse1におけるデンプンの蓄積は，～ということを示唆する

■［形／名］＋ accumulation

❶	nuclear accumulation	核蓄積	831
❷	protein accumulation	タンパク質の蓄積	358
❸	mRNA accumulation	メッセンジャーRNAの蓄積	338
❹	lipid accumulation	脂質の蓄積	328
❺	cAMP accumulation	cAMPの蓄積	307
❻	intracellular accumulation	細胞内蓄積	235

accuracy ㊔ 精度／正確さ 用例数 7,575

accuracy	精度	7,391
accuracies	精度	184

複数形の割合は約2.5%しかなく，不可算名詞として使われることが多い．

◆類義語：precision

■ accuracy ＋［前］

accuracy ofの前にtheが付く割合は非常に高い.

	the%	a/an%	ø%			
❶	77	12	11	accuracy of ～	～の精度	2,649
❷	12	5	61	accuracy in ～	～の際の精度	354
❸	22	3	63	accuracy for ～	～の精度	247

❶ Contrast injected with the steroid was used to assess the accuracy of the joint injection. (Arthritis Rheum. 2010 62:1862)

訳 ～が，関節内注入の精度を評価するために使われた

■［形］＋ accuracy

❶	diagnostic accuracy	診断精度	429
❷	high accuracy	高い精度	255

❶ PiB PET and ^{18}F-FDG PET have similar diagnostic accuracy in early cognitive impairment. (J Nucl Med. 2009 50:878)

訳 ～は，早期の認知障害において同等の診断精度をもつ

accurate 形 正確な　用例数 6,968

◆類義語：correct, precise, exact

■ accurate ＋［名］

❶	accurate diagnosis	正確な診断	153
❷	accurate assessment	正確な評価	132
❸	accurate identification	正確な同定	125
❹	accurate determination	正確な決定	125
❺	accurate prediction	正確な予測	111

❶ Improved testing should allow more accurate diagnosis of food allergy. (Curr Opin Pediatr. 2012 24:615)

訳 改善された検査は，食物アレルギーのより正確な診断を可能にするはずである

■［副］＋ accurate

❶	more accurate	より正確な	931
❷	highly accurate	高度に正確な	219
❸	most accurate	最も正確な	168

accurately 副 正確に　用例数 3,693

◆類義語：correctly, precisely, exactly

■ accurately + [動]

❶	accurately predict ～	…は，～を正確に予測する	231
❷	accurately reflect ～	…は，～を正確に反映する	104
❸	accurately identify ～	…は，～を正確に同定する	79
❹	accurately measure ～	…は，～を正確に測定する	77

❶ **The ability to accurately predict postoperative mortality** is expected to improve preoperative decisions for elderly patients considered for colorectal surgery. (Ann Surg. 2013 257:905)
訳 術後死亡率を正確に予測する能力

★ achieve 動 達成する　　用例数 19,607

achieved	達成される／～を達成した	12,075
achieve	～を達成する	5,282
achieving	～を達成する	1,642
achieves	～を達成する	608

他動詞.
◆類義語：accomplish, attain

■ be achieved + [前]/using

★	be achieved	…は，達成される／なされる	7,734
❶	be achieved by ～	…は，～によって達成される／なされる	2,190
	・be achieved by using ～	…は，～を使うことによって達成される／なされる	113
❷	be achieved in ～	…は，～において達成される／なされる	1,171
❸	be achieved with ～	…は，～によって達成される／なされる	705
❹	be achieved through ～	…は，～によって達成される／なされる	475
❺	be achieved using ～	…は，～を使って達成される／なされる	338

❶ Here we show that **the inducible regulation of RNAi also can be achieved by using a Cre-LoxP approach.** (Nucleic Acids Res. 2004 32:e85)
訳 RNAiの誘導可能な調節は，また，Cre-LoxPアプローチを使うことによってなされうる

❷ PTC-tube independence **was achieved in 51 of 76 (67%) patients** using the combined approach of PTBC and surgery for PTBC failures. (Transplantation. 2004 77:110)
訳 ～が，76名中51名（67%）の患者において達成された

❸ **Selective binding to DR4 or DR5 was achieved with** three to six-ligand amino acid substitutions. (J Biol Chem. 2005 280:2205)
訳 DR4あるいはDR5への選択的結合が，～によって達成された

❹ **The latter is achieved through a combination of** local and distal factors in the protein substrate. (Biochemistry. 2002 41:10002)
訳 後者は，～の組み合わせによって達成される

A

■ to achieve

★ to achieve ～	～を達成する…	4,082
❶ To achieve ～	～を達成するために	328
❷ be required to achieve ～	…は，～を達成するために必要とされる	145
❸ be difficult to achieve	…は，達成することが難しい	89
❹ be necessary to achieve ～	…は，～を達成するために必要な	83

❶ To achieve this goal, a different type of relationship is needed between academia and industry, and also within industry, to promote collaboration in the precompetitive space. (Trends Immunol. 2012 33:238)

訳 この目標を達成するために

❷ Intrinsic expression of protective Idd9 alleles in $CD4^+$ T-cells and nonlymphoid cells is required to achieve an optimal level of tolerance. (Diabetes. 2010 59:1478)

訳 ～は，最適レベルの耐性を達成するために必要とされる

❸ These results indicate that accessibility for a Nox4-specific peptide inhibitor might be difficult to achieve in vivo. (J Biol Chem. 2012 287:8737)

訳 ～は，生体内で達成することが難しいかもしれない

■ achieve ＋［名句］

❶ achieve this goal	…は，この目標を達成する	129
❷ achieve a complete response	…は，完全な応答を達成する	79

acquire　㊫ 得る／獲得する　用例数 9,822

acquired	得られる／獲得される／～を獲得した	7,038
acquire	～を獲得する／～を得る	2,001
acquiring	～を獲得する／～を得る	478
acquires	～を獲得する／～を得る	305

他動詞.

◆類義語：obtain, gain, procure

■ be acquired ＋［前］

★ be acquired	…は，獲得される／得られる	1,305
❶ be acquired by ～	…は，～によって獲得される／得られる	149
❷ be acquired in ～	…は，～において獲得される／得られる	142
❸ be acquired from ～	…は，～から獲得される／得られる	114

❶ These properties are acquired by partitioning the molecular electron density into quantum topological atoms. (J Am Chem Soc. 2003 125:1284)

訳 これらの性質は，～によって獲得される

❷ Data were acquired in the right lung during breath-holds at RV, FRC and TLC. (J Physiol. 2010 588:4759)

訳 データは，右肺において得られた

■ acquire ＋［名句］

❶	acquire the ability to *do*	…は，～する能力を獲得する	147

❶ However, while **HIV-1 can acquire the ability to use CXCR4**, SIVs that utilize CXCR4 have rarely been reported.（J Virol. 2009 83:9911）
訳 HIV-1は，CXCR4を使う能力を獲得できる

acquisition ㊔ 獲得　　用例数 5,857

acquisition	獲得	5,711
acquisitions	獲得	146

複数形の割合は2％しかなく，原則，**不可算名詞**として使われる．
◆類義語： gain

■ acquisition ＋［前］

	the％	a/an％	ø％			
❶	43	1	54	acquisition of ～	～の獲得	2,762
❷	4	5	91	acquisition in ～	～における獲得	143

❶ **The acquisition of invasiveness in ovarian cancer（OC）** is accompanied by the process of epithelial-to-mesenchymal transition（EMT）.（Oncogene. 2010 29:5741）
訳 卵巣癌（OC）の浸潤性の獲得

★★ act ㊫ 作用する／働く　　用例数 32,794

act	作用する／働く	13,265
acts	作用する／働く	10,310
acting	作用する／働く	8,116
acted	作用した／働いた	1,103

自動詞．act asの用例が非常に多い．
◆類義語：function, serve, behave, operate, work, play

■ act ＋［前］

❶	act as ～	…は，～として作用する	11,002
❷	act in ～	…は，～において作用する	2,831
	・act in concert	…は，協調して作用する	423
❸	act on ～	…は，～に対して作用する	2,034
❹	act to *do*	…は，～するように作用する	1,814
❺	act through ～	…は，～を通して作用する	1,236
❻	act at ～	…は，～において作用する	1,033
❼	act by ～	…は，～によって作用する	901

❶ Our previous studies suggest that **surfactant protein A (SP-A) can act as a ligand** in the attachment of M. tuberculosis to AMs. (J Clin Invest. 1999 103:483)

訳 サーファクタントタンパク質A (SP-A) は，リガンドとして作用できる

❷ Thus, **R-Ras and Ras may act in concert to regulate** integrin affinity via the activation of distinct downstream effectors. (Mol Biol Cell. 1999 10:1799)

訳 R-RasおよびRasは，～を調節するために協調して作用するかもしれない

❸ Therefore, **the inflammatory cytokines probably act on T cells** in vivo via an intermediary factor. (Proc Natl Acad Sci USA. 1998 95:3810)

訳 炎症性サイトカインは，おそらくT細胞に対して作用する

❹ The protein hormone **leptin acts to regulate body fat** and energy expenditure. (J Am Chem Soc. 2008 130:9106)

訳 レプチンは，体脂肪を調節するように作用する

■ act + [副]

❶	act directly	…は，直接作用する	473
	・act directly on ～	…は，～に対して直接作用する	303
❷	act synergistically	…は，相乗的に作用する	468
	・act synergistically with ～	…は，～と相乗的に作用する	193

❶ Further, **regulatory T cells may act directly on the innate immune system** to reduce or prevent disease. (Am J Pathol. 2003 162:691)

訳 調節性T細胞は，自然免疫系に対して直接作用するかもしれない

★ action ㊔ 作用　　用例数 28,914

action	作用	21,818
actions	作用	7,096

複数形の割合は25%あるが，単数形は無冠詞で使われることが多い.

◆類義語：function, operation, activity

■ action + [前]

action ofの前にtheが付く割合は非常に高い.

	the %	a/an %	ø %			
❶	71	8	21	action of ～	～の作用	6,892
				・by the action of ～	～の作用によって	323
				・through the action of ～	～の作用によって	290
❷	1	5	59	action in ～	～における作用	969
❸	1	10	51	action on ～	～に対する作用	690

❶ Compound 5e was converted to β-glucuronic acid conjugate 6e **by the action of pig liver esterase (PLE)**. (J Med Chem. 2002 45:937)

訳 ブタ肝エステラーゼ (PLE) の作用によって

❷ **Type 2 diabetes is characterized by abnormalities of insulin action in muscle, adipose tissue, and liver** and by altered β-cell function. (J Clin Invest. 2000 105:199)

訳 2型糖尿病は，筋，脂肪組織および肝臓におけるインスリン作用の異常によって特徴づけ

られる

❸ However, the mechanism of bisphosphonate action on bone is not fully understood. (Cancer Res. 2000 60:6001)

訳 骨に対するビスホスホネート作用の機構は，十分には理解されていない

■ [前] + action

❶ mechanism of action	作用の機構	2,113
❷ mode of action	作用のモード	587
❸ site of action	作用の部位	340

❶ However, little is known about the mechanism of action of these drugs in prostate cancer. (Cancer Res. 2001 61:7179)

訳 これらの薬剤の作用の機構については，ほとんど知られていない

❷ Both calcium and membrane binding affect the structure and the mode of action of L-1. (Biochemistry. 1998 37:15481)

訳 ～は，L-1の構造と作用のモードに影響を与える

■ [名／形] + action

❶ insulin action	インスリン作用	669
❷ inhibitory action	抑制性作用	394
❸ direct action	直接作用	226

★★ activate 動 活性化する　用例数 92,756

activated	活性化される／活性化した	55,881
activate	活性化する	14,944
activating	活性化する	11,698
activates	活性化する	10,233

他動詞の用例がほとんどだが，自動詞としても用いられる．

◆類義語：potentiate, enhance, augment

■ be activated + [前]

★ be activated	…は，活性化される	8,954
❶ be activated by ～	…は，～によって活性化される	4,106
❷ be activated in ～	…は，～において活性化されている	1,463
・be activated in response to ～	…は，～に応答して活性化される	254
❸ be activated during ～	…は，～の間に活性化されている	292
❹ be activated to *do*	…は，活性化されて～する	222
❺ be activated at ～	…は，～において活性化される	171

❶ p53 tumor suppressor is activated by phosphorylation and acetylation on DNA damage. (Cancer Res. 2002 62:2913)

訳 p53癌抑制因子は，リン酸化とアセチル化によって活性化される

❷ Stat3 is constitutively activated in many primary tumors and tumor cell lines, suggesting that signaling by this molecule may be important for cell transformation. (Oncogene. 2002 21:217)

訳 Stat3は，多くの原発腫瘍において構成的に活性化されている

❸ Transcription of the lactase gene is activated during enterocyte differentiation. (Gastroenterology. 2000 118:115)

訳 ラクターゼ遺伝子の転写は，腸細胞分化の間に活性化されている

■ activate ＋［名句］

❶	activate the expression of ～	…は，～の発現を活性化する	248
❷	activate the transcription of ～	…は，～の転写を活性化する	167

❶ Hoxa6 and Hoxa7 activate the expression of genes involved in germ layer specification during mESC differentiation in a cooperative and redundant fashion. (Mol Cell. 2011 43:1040)

訳 Hoxa6およびHoxa7は，～に関与する遺伝子の発現を活性化する

■［名／動／形］＋ to activate

★	to activate ～	～を活性化する…	6,113
❶	ability to activate ～	～を活性化する能力	596
❷	fail to activate ～	…は，～を活性化することができない	432
❸	be sufficient to activate ～	…は，～を活性化するのに十分である	225
❹	be shown to activate ～	…は，～を活性化することが示される	222

❶ In contrast, platelets from mice that lack $G\alpha_q$ show no decrease in the ability to activate Rap1 in response to epinephrine but show a partial reduction in ADP-stimulated Rap1 activation. (J Biol Chem. 2002 277:23382)

訳 …は，～に応答してRap1を活性化する能力の低下を示さない

❷ ANF treatment caused MEK phosphorylation and activation but failed to activate any of the Raf isoforms. (J Biol Chem. 1999 274:24858)

訳 しかし，Rafアイソフォームのどれも活性化することができなかった

★★ activation ㊔ 活性化　　用例数 134,988

activation	活性化	134,481
activations	活性化	507

複数形のactivationsの割合は0.4%しかなく，原則，不可算名詞として使われる.

◆類義語：transactivation, potentiation, enhancement

■ activation ＋［前］

	the%	a/an%	ø%			
❶	18	1	81	activation of ～	～の活性化	53,760
				・activation of ～ by …	…による～の活性化	2,939
				・activation of ～ in …	…における～の活性化	2,639
❷	1	0	97	activation in ～	～における活性化	6,421
❸	1	0	96	activation by ～	～による活性化	5,541

❶ Expression of chick Rap1GAP in PC-12 cells **inhibited activation of Rap1 by forskolin.** (J Biol Chem. 1999 274:21507)

訳 ～は，フォルスコリンによるRap1の活性化を抑制した

❷ **Studies of Wnt activation in gastric cancer have yielded conflicting results.** (Cancer Res. 2002 62:3503)

訳 胃癌におけるWnt活性化の研究は，相反する結果を生じている

❸ P (invF-1) contains a HilA binding site, termed a HilA box, **that is necessary and sufficient for activation by HilA.** (J Bacteriol. 2001 183:4876)

訳 HilAによる活性化に必要かつ十分である～

■ [前] ＋ activation

❶ lead to activation	…は，活性化につながる	624
❷ required for activation	活性化のために必要とされる	455
❸ mechanism of activation	活性化の機構	219
❹ mediated by activation	活性化によって仲介される	155

❶ We show that **NMDA receptor stimulation leads to activation of p21^{ras} (Ras)** through generation of nitric oxide (NO) via neuronal NO synthase. (Proc Natl Acad Sci USA. 1998 95:5773)

訳 NMDA受容体刺激は，p21^{ras}（Ras）の活性化につながる

❷ Our findings demonstrate that the cytoskeleton does not seem to regulate calcium influx and that **functional InsP3 receptors are not required for activation of I_{CRAC}.** (J Physiol. 2001 532:55)

訳 機能的なInsP3受容体は，I_{CRAC}の活性化のためには必要とされない

❸ However, **the mechanism of activation of transglutaminase by TIG3 is not known.** (Oncogene. 2005 24:2963)

訳 TIG3によるトランスグルタミナーゼの活性化の機構は，知られていない

■ activation ＋ [動]

❶ activation requires ～	活性化は，～を必要とする	298
❷ activation occurs	活性化が，起こる	243

❶ Escape from these vacuoles is mediated in part by a bacterial phospholipase C (PC-PLC), **whose activation requires cleavage of an N-terminal peptide.** (Mol Microbiol. 2000 35:289)

訳（そして）その活性化はN末端ペプチドの切断を必要とする

■ [形／過分] ＋ activation

❶ transcriptional activation	転写活性化	4,101
❷ ～-induced activation	～に誘導される活性化	1,907
❸ ～-dependent activation	～依存的な活性化	1,659
❹ ～-mediated activation	～に仲介される活性化	1,621
❺ constitutive activation	構成的な活性化	859
❻ complement activation	補完的な活性化	827

❶ The right promoter proximal arm **also plays a role in transcriptional activation** that is distinct from its role in AphB binding. (Mol Microbiol. 2002 44:533)

訳 ～は，また，転写活性化において役割を果たす

❷ Reticulocyte-enriched RBCs derived from sickle-cell disease (SCD) patients are most responsive to IAP-induced activation. (J Clin Invest. 2001 107:1555)

訳 ～は，IAPに誘導される活性化に最も応答性である

★★ active 形 活性のある，名 活性のあるもの 用例数 53,902

active	活性のある／活性のあるもの	53,871
actives	活性のあるもの	31

名詞としても使われるが，形容詞の用例が圧倒的に多い．

◆類義語：effective, efficacious, efficient

■ active ＋［前］

❶ active in ～	～において活性がある	2,432
・be active in ～	…は，～において活性がある	1,000
❷ active as ～	～として活性がある	363
❸ active against ～	～に対して活性がある	457
❹ active at ～	～において活性がある	345
❺ active during ～	～の間に活性がある	271

❶ The CTD phosphatase was found to be active in ternary elongation complexes. (Genes Dev. 1999 13:1540)

訳 CTDホスファターゼは，三成分伸長複合体において活性があることが見つけられた

❷ To be active as isomerases, DsbC and DsbG must be kept reduced. (J Biol Chem. 2002 277: 26886)

訳 イソメラーゼとして活性であるために

❸ Quinolones were active against the bacterial enteropathogens in the 3 sites. (J Infect Dis. 2002 185:497)

訳 キノロンは，細菌の腸管病原体に対して活性があった

■［副］＋ active

❶ constitutively active	構成的に活性がある	3,359
❷ highly active	高度に活性がある	1,130
❸ biologically active	生物学的に活性がある	1,067
❹ more active	より活性が高い	880
❺ catalytically active	触媒的に活性がある	815
❻ transcriptionally active	転写的に活性がある	674
❼ most active	最も活性が高い	545

❶ Mutation of the ZV repressor sequence greatly increased the induction of the promoter but did not make it constitutively active. (J Virol. 2002 76:10282)

訳 しかし，それを構成的に活性化しなかった

❷ Bacterially expressed P69 was inactive whereas the same protein expressed in insect cells was highly active. (J Biol Chem. 1999 274:1848)

訳 昆虫細胞において発現された同じタンパク質は，高い活性があった

❸ This approach provides a way to modulate the potency and specificity of biologically active compounds. (Proc Natl Acad Sci USA. 1999 96:1953)

訳 このアプローチは，生物学的に活性のある化合物の効力と特異性を調節する方法を提供する

❻ However, this vector-induced p53 is transcriptionally active and, therefore, p53 function is not inactivated by viral proteins. (Cancer Res. 1999 59:4247)

訳 このベクターに誘導されるp53は，転写的に活性がある

■ active ＋［名］

❶ active site	活性部位	12,979
❷ active form	活性型	1,230
❸ active conformation	活性立体構造	349
❹ active enzyme	活性酵素	340
❺ active role	積極的役割	338
❻ active state	活性状態	338

❶ The 3-amidopyrrolidin-4-one inhibitors were bound in the active site of the enzyme in two alternate directions. (J Med Chem. 2001 44:725)

訳 〜阻害剤は，酵素の活性部位に結合していた

actively 副 活発に／能動的に 用例数 2,770

◆類義語：vigorously, effectively, efficiently, positively

■ actively ＋［過分／現分］

❶ actively transcribed	活発に転写される	180
❷ actively growing	活発に増殖している	101
❸ actively involved	活発に関与している	78
❹ actively dividing	活発に分裂している	71

❶ Using reverse transcriptase PCR we were able to demonstrate that pspC is actively transcribed in vivo, when the bacteria are growing in the nasal cavity and in the lungs. (Infect Immun. 2002 70:2526)

訳 pspCは，生体内で活発に転写される

★★ activity 名 活性／活動 用例数 210,613

activity	活性／活動	189,908
activities	活性／活動	20,705

複数形のactivities割合は約10%あるが，単数形は無冠詞で使われることが断然多い．

◆類義語：action, ability

■ activity ＋［前］

activity ofの前にtheが付く割合は非常に高い．

	the%	a/an%	ø%			
❶	78	1	18	activity of ～	～の活性	30,284
❷	1	2	92	activity in ～	～における活性	19,265
❸	2	0	94	activity by ～	～による活性／～によって活性を	3,433
❹	3	1	85	activity against ～	～に対する活性	2,067

❶ **These results provide direct biochemical evidence for the catalytic activity of the hairpin ribozyme in a cellular environment**, and indicate that self-processing ribozyme transcripts may be well suited for cellular RNA-inactivation experiments.（Nucleic Acids Res. 1998 26: 3494）

訳 これらの結果は，細胞環境におけるヘアピンリボザイムの触媒活性に対する直接の生化学的証拠を提供する

❹ **The rproRIP1 had no detectable enzymatic activity against ribosomes from** any of the species assayed.（Eur J Biochem. 2000 267:1966）

訳 rproRIP1は，～からのリボソームに対する検出可能な酵素活性を持たなかった

■［前］＋ activity

❶	loss of activity	活性の喪失	316
❷	increase in activity	活性の増大	245

❶ In humans, **loss of activity of a lysosomal enzyme leads to** an inherited metabolic defect known as a lysosomal storage disorder.（J Mol Biol. 2012 423:736）

訳 リソソーム酵素の活性の喪失は，～につながる

■［動］＋ @ ＋ activity

❶	increase ～ activity	…は，～活性を増大させる	3,031
❷	regulate ～ activity	…は，～活性を調節する	1,998
❸	inhibit ～ activity	…は，～活性を抑制する	1,926
❹	reduce ～ activity	…は，～活性を減少させる	1,501
❺	enhance ～ activity	…は，～活性を増強する	1,174
❻	modulate ～ activity	…は，～活性を調節する	1,121
❼	decrease ～ activity	…は，～活性を減少させる	985
❽	show ～ activity	…は，～活性を示す	831

❹ **Mutation of the HNF3 element significantly reduced promoter activity in HepG2 cells**, whereas this element in isolation conferred HNF3β responsiveness to a heterologous promoter.（Gene. 2000 246:311）

訳 HNF3エレメントの変異は，HepG2細胞におけるプロモーター活性を有意に減少させた

❼ However, the effects of roscovitine appear to be distinct from those of LY294002, since roscovitine did not affect Akt activity while LY294002 **significantly decreased the activity of Akt**.（Oncogene. 2000 19:3059）

訳 ～は，Aktの活性を有意に低下させた

■ activity ＋［動］

❶ activity is required for ～	活性は，～のために必要とされる	480
❷ activity was observed	活性が，観察された	460
❸ activity is regulated	活性が，調節される	353
❹ activity was detected	活性が，検出された	304
❺ activity was measured	活性が，測定された	281
❻ activity was inhibited	活性が，抑制された	259

❶ These data suggest that **TAF$_{II}$250 acetyltransferase activity is required for cell cycle progression** and regulates the expression of essential proliferative control genes.（Mol Cell Biol. 2000 20:1134）

訳 TAF$_{II}$250アセチルトランスフェラーゼ活性は，細胞周期進行のために必要とされる

❹ **GUS activity was detected only in the phloem cells** but not in any other cell types of vegetative tissues.（Proc Natl Acad Sci USA. 1990 87:4144）

訳 GUS活性が，師部細胞においてのみ検出された

■［名／形］＋ activity

❶ kinase activity	キナーゼ活性	6,787
❷ promoter activity	プロモーター活性	4,442
❸ binding activity	結合活性	3,930
❹ transcriptional activity	転写活性	3,753
❺ catalytic activity	触媒活性	3,047
❻ enzymatic activity	酵素活性	2,568
❼ enzyme activity	酵素活性	2,434
❽ physical activity	身体活動性	2,045

★★ acute　形 急性の　　用例数 30,943

◆反意語：chronic

■［前］＋ acute

❶ patients with acute ～	急性～の患者	1,438
❷ treatment of acute ～	急性～の治療	289
❸ incidence of acute ～	急性～の発生率	231

❶ The effect of CD38 ligation on cell growth was also evaluated **in freshly isolated leukemic cells from patients with acute myelogenous leukemia（AML）**.（J Immunol. 1998 161:4702）

訳 急性骨髄性白血病（AML）の患者から新たに分離された白血病細胞において

■ acute and ＋［形］

❶ acute and chronic ～	急性および慢性の～	989

❶ Thus, prevention of the activation of calpain I **reduces the development of acute and chronic**

inflammation. (Am J Pathol. 2000 157:2065)
訳 ～は，急性および慢性炎症の発症を低下させる

■ acute ＋［名］

❶	acute rejection	急性拒絶反応	1,544
❷	acute phase	急性期	1,079
❸	acute infection	急性感染（症）	676

acutely 副 急性に　用例数 1,858

◆類義語：suddenly, abruptly　◆反意語：chronically

■ acutely ＋［過分］

❶	acutely infected	急性に感染した	195
❷	acutely isolated	急性単離した	116
❸	acutely dissociated	急性分離した	112

❶ **A total of 23 acutely infected persons were identified** only with the use of the nucleic acid amplification algorithm. (N Engl J Med. 2005 352:1873)
訳 合計23名の急性に感染した人々が同定された

adapt 動 適応する／適応させる／順応させる　用例数 4,870

adapted	適応される／適応した	3,098
adapt	適応する／～を適応させる	1,106
adapting	適応する／～を適応させる	456
adapts	適応する／～を適応させる	210

他動詞の用例が多いが，自動詞としても用いられる．

◆類義語：accommodate

■ be adapted ＋［前］

★	be adapted	…は，適応される	628
❶	be adapted to *do*	…は，～するために適応される	383
❷	be adapted for ～	…は，～のために適応される	167

❶ In conclusion, **ELISpot assays can be adapted to study B-cell as well as T-cell responses** to HCV. (Hepatology. 2006 43:91)
訳 ELISpotアッセイは，T細胞応答と同様にB細胞応答を研究するために適応されうる

❷ **This assay can be adapted for high-throughput screening for** potential prenyltransferase substrates and inhibitors. (Anal Biochem. 2005 345:302)
訳 このアッセイは，～のハイスループットスクリーニングのために適応されうる

■ adapt to

★	adapt to ～	～に適応する…	656
❶	to adapt to ～	～に適応するための	288

❶ The ability of biological systems to adapt to genetic and environmental perturbations is a fundamental but poorly understood process at the molecular level. (J Biol Chem. 2006 281: 8024)

訳 遺伝的および環境的撹乱に適応するための生物学的システムの能力

■ [形]-adapted

❶	dark-adapted ～	暗順応した～	345
❷	light-adapted ～	明順応した～	141

adaptation 名 適応／順応 用例数 6,642

adaptation	適応／順応	5,651
adaptations	適応／順応	991

複数形のadaptationsの割合は約15%あるが，単数形は無冠詞で使われることが非常に多い．

◆類義語：accommodation

■ adaptation ＋ [前]

	the%	a/an%	ø%			
❶	7	11	76	adaptation to ～	～への適応	1,227
❷	36	11	51	adaptation of ～	～の適応	793
❸	1	4	89	adaptation in ～	～における適応	311

❶ These phenotypes are clearly the result of adaptation to this environment, but their genetic basis remains unknown. (Science. 2010 329:72)

訳 これらの表現型は，明らかにこの環境へ適応の結果である

❷ The sequence changes identified here may be important in the adaptation of influenza viruses to humans. (Nature. 2005 437:889)

訳 ～は，インフルエンザウイルスのヒトへの適応において重要であるかもしれない

★ add 動 加える／添加する 用例数 11,218

added	加えられる／添加される／～を加えた	6,922
adding	～を加える／～を添加する	2,097
add	～を加える／～を添加する	1,373
adds	～を加える／～を添加する	826

他動詞．added toの用例が非常に多い．

◆類義語：supplement, append

■ be added ＋［前］

★	be added	…が，加えられる	2,657
❶	be added to ～	…が，～に加えられる	1,537
❷	be added in ～	…が，～において加えられる	83

❶ When Nun was added to a paused transcription elongation complex, it cross-linked to the DNA template.（Science. 1999 286:2337）

訳 Nunが，～に加えられたとき

❷ These new neurons, which are continually added in adulthood, may play a role in the functions of association neocortex.（Science. 1999 286:548）

訳 ～が，成人期において継続的に加えられる

★★ addition 名 添加／加えること　用例数 63,316

addition	添加	62,703
additions	添加	613

複数形の割合は1％しかなく，原則，不可算名詞として使われる．in additionの用例が非常に多い．

◆類義語：supplement, supplementation

■ addition ＋［前］

addition ofの前にtheが付く割合はかなり高い．

	the％	a/an％	ø％			
❶	61	2	37	addition of ～	～の添加	14,664
				・upon addition of ～	～を添加するとすぐに	460
				・reversed by the addition of ～	～の添加によって逆転される	109
				・inhibited by the addition of ～	～の添加によって抑制される	80

❶ Upon addition of exogenous N-formylated peptides, M3 trafficks rapidly to the cell surface.（J Immunol. 2001 167:1507）

訳 外来性のN-ホルミル化ペプチドを添加するとすぐに

❶ When aggregation was inhibited by the addition of an E-cadherin-blocking antibody, apoptosis increased synergistically.（J Biol Chem. 1999 274:9656）

訳 凝集がE-カドヘリン阻止抗体の添加によって抑制されたとき

■ in addition

★	in addition	加えて	44,650
❶	In addition,	加えて，	31,851
❷	in addition to ～	～に加えて	12,426
	・in addition to its role in ～	～におけるそれの役割に加えて	211
	・in addition to being ～	～であることに加えて	217

・in addition to providing ～	～を提供することに加えて	189

❶ In addition, we present further evidence for the suggestion that the binding of linker histone causes a subtle but global change in core histone-DNA interactions within the nucleosome. (Biochemistry. 1998 37:8622)

訳 加えて，われわれは～のさらなる証拠を提示する

❷ In addition to its traditional role in mediating protein targeting, the signal was found to play a surprising role in determining orientation of the PrP N terminus. (J Biol Chem. 2001 276:26132)

訳 ～におけるそれの伝統的な役割に加えて

★ additional 形 付加的な／追加の 用例数 27,077

■ additional ＋［名］

❶ additional studies	追加の研究／付加的な研究	747
❷ additional evidence	追加の証拠	486
❸ additional experiments	追加の実験	352
❹ additional information	追加の情報	319

❶ Additional studies are needed to further characterize the properties and capabilities of both ligands in health and disease. (J Nucl Med. 2002 43:678)

訳 追加の研究が，～をさらに特徴づけるために必要とされる

■［形］＋ additional

❶ two additional ～	2つの付加的な～	1,008
❷ three additional ～	3つの付加的な～	391
❸ no additional ～	追加の～のない	366

❶ Recently, two additional mitochondrial carriers with high similarity to UCP1 were molecularly cloned. (Diabetes. 2000 49:143)

訳 UCP1に高い類似性を持つ2つの付加的なミトコンドリア担体

❸ Crude receptor extract, which requires no additional purification, is used in the assay. (Anal Biochem. 2002 300:15)

訳 ～は，追加の精製を必要としない

■［動］＋ additional

❶ provide additional ～	…は，付加的な～を提供する	1,157
❷ identify additional ～	…は，付加的な～を同定する	525

❶ Our results provide additional in vivo evidence of a metabolic grid in maize (i.e. pathway convergence). (Plant Physiol. 1999 121:1037)

訳 われわれの結果は，～の付加的な生体内の証拠を提供する

A

★ additionally ㊖ そのうえ 用例数 8,390

文頭で用いられることが圧倒的に多い.

◆類義語：further, furthermore, moreover, in addition

■ additionally

❶ Additionally,	そのうえ	7,300

❶ Additionally, we cultured dissociated trigeminal ganglion cells in the presence of NGF, NT-3, or NGF-NT-3.（J Comp Neurol. 2000 425:202）

訳 そのうえ，われわれは～を培養した

★ address ㊍ 取り組む／立ち向かう 用例数 12,661

address	～に取り組む	7,692
addressed	～に取り組んだ／取り組まれた	3,140
addressing	～に取り組む	964
addresses	～に取り組む	865

他動詞.

■ address ＋［名句］

❶ address this issue	…は，この問題に取り組む	971
❷ address this question	…は，この問題に取り組む	850
❸ address the role of ～	…は，～の役割に取り組む	500
❹ address this problem	…は，この問題に取り組む	307
❺ address the mechanism	…は，その機構に取り組む	112
❻ address the hypothesis that ～	…は，～という仮説に取り組む	91

❶ To address this issue, we have recorded from identified motoneurons and compared their current-evoked firing patterns to network-driven ones in the larval zebrafish (Danio rerio).（J Neurosci. 2012 32:10925）

訳 この問題に取り組むために，

■ address ＋［名節］

❶ address whether ～	～かどうかに取り組む	376
❷ address how ～	どのように～かに取り組む	149

❶ We addressed whether parasites can directly induce T reg cells.（J Exp Med. 2010 207:2331）

訳 われわれは，～かどうかに取り組んだ

■ to address

★ to address ～	～に取り組む…	5,560
❶ To address ～	～に取り組むために	3,384
❷ used to address ～	～に取り組むために使われる	88

❶ To address this question, we generated mice lacking NMDA receptors (NMDARs) on either AgRP or POMC neurons. (Neuron. 2012 73:511)
訳 この問題に取り組むために,

■ [代名／名] ＋ address

❶ we address ～	われわれは, ～に取り組む	746
❷ study addresses ～	研究は, ～に取り組む	155
❸ review addresses ～	総説は, ～に取り組む	137

■ be addressed

★ be addressed	…は, 取り組まれる	1,264
❶ be addressed by ～	…は, ～によって取り組まれる	207
❷ be addressed in ～	…は, ～において取り組まれる	185

❶ However, the contribution of tissue-resident commensals to immunity and inflammation at other barrier sites has not been addressed. (Science. 2012 337:1115)
訳 ～は, 取り組まれていない

adequate　形 適切な／適当な／十分な　用例数 2,472

◆類義語：appropriate, proper, reasonable, sufficient

■ adequate ＋ [前]

❶ adequate to *do*	～するのに十分な	140
❷ adequate for ～	～にとって適切な	130

❶ The current national endoscopic capacity, as recently estimated, may be adequate to support widespread use of screening colonoscopy in the steady state. (Gastroenterology. 2005 129:1151)
訳 ～は, スクリーニング大腸内視鏡検査の広範な使用を支持するのに十分であるかもしれない

❷ Although this approach is adequate for many tachyarrhythmias, it has limitations. (Circulation. 1999 99:1906)
訳 このアプローチは, 多くの頻脈性不整脈にとって適切である

adhere　動 接着する／付着する　用例数 1,514

adhere	接着する	760
adhered	接着した	487
adhering	接着する	189
adheres	接着する	78

自動詞. adhere toの用例が圧倒的に多い.
◆同意語：attach, bind

A

■ adhere to

★ adhere to ～	…は，～に接着する	1,050
❶ to adhere to ～	～に接着する…	246

❶ However, **these cells were completely unable to adhere to surfaces** via β_2-integrins.（J Biol Chem. 2002 277:4285）
訳 これらの細胞は，表面にまったく接着できなかった

★ adhesion 名 接着　用例数 22,021

adhesion	接着	20,692
adhesions	接着	1,329

複数形のadhesionsの割合は約5％あるが，単数形は無冠詞で使われることが圧倒的に多い．

◆類義語：attachment, contact, binding

■ adhesion ＋［前］

	the％	a/an％	ø％			
❶	0	0	94	adhesion to ～	～への接着	1,897
❷	34	0	65	adhesion of ～	～の接着	1,256
				・adhesion of ～ to …	～の…への接着	491
❸	0	0	95	adhesion in ～	～における接着	412

❶ These active peptides reversed **the influence of CD9 expression on CHO cell adhesion to fibronectin**.（Blood. 2002 100:4502）
訳 フィブロネクチンへのCHO細胞接着に対するCD9発現の影響

❷ Preincubation of cells with CRP **also significantly increased the adhesion of monocytes to HAECs**.（Circulation. 2002 106:1439）
訳 ～は，また，単球のHAECへの接着を有意に増大させた

■ adhesion and ＋［名］

❶ adhesion and migration	接着と遊走	265

❶ These findings demonstrate for the first time that tuberin activates Rho and **regulates cell adhesion and migration**.（Oncogene. 2002 21:8470）
訳 ～は，細胞接着と遊走を調節する

■ ［名／形］＋ adhesion

❶ cell adhesion	細胞接着	5,748
❷ focal adhesions	接着斑	1,696
❸ intercellular adhesions	細胞間接着	1,040

★ adjacent　形 隣接する／近くの　用例数 10,118

◆類義語：proximate, vicinal, contiguous, close, nearby

■ adjacent ＋［前］

❶ adjacent to ～	～に隣接する	3,745
・immediately adjacent to ～	～にすぐ隣接する	294
・be adjacent to ～	…は，～にすぐ隣接する	286

❶ Furthermore, in contrast to phosphorylated general H3, H3.3 S31P **is localized in distinct chromosomal regions immediately adjacent to centromeres.** (Proc Natl Acad Sci USA. 2005 102:6344)
訳 ～は，セントロメアにすぐ隣接する別個の染色体領域に局在する

★ adjust　動 補正する／調整する　用例数 12,463

adjusted	補正される／補正した	9,352
adjusting	補正する	2,235
adjust	補正する	751
adjusts	補正する	125

研究・調査の条件を揃えるために「(結果の) 補正を行う」という意味で用いられることが多い．他動詞および自動詞の両方で用いられる．〈adjusted for〉〈adjusting for〉の用例が特に多い．

◆類義語：modulate, correct, control, regulate

■ adjusted ＋［前］

❶ adjusted for ～	～に対して補正される	1,669
・be adjusted for ～	…は，～に対して補正される	350
❷ adjusted to ～	～に対して補正される／～するように補正される	238

❶ **Relative risks were adjusted for age**, sex, cancer stage, the number of colonoscopic examinations, and the time to a first colonoscopy. (N Engl J Med. 2003 348:883)
訳 相対リスクは，年齢に対して補正された

❷ **Rates were directly adjusted to the age distribution of** the 2000 U.S. population. (Ann Intern Med. 2002 136:341)
訳 比率は，～の年齢分布に対して直接補正された

■ ［名］-adjusted

❶ age-adjusted ～	年齢補正された～	624
❷ risk-adjusted ～	リスク補正された～	380

❶ **The age-adjusted incidence of asthma** was 4.26/100 person-years during 2005-2007, higher than reported in other populations. (Am J Epidemiol. 2012 176:744)
訳 喘息の年齢補正された発生率

■ adjusted ＋［名］

❶	adjusted odds ratio	補正オッズ比	928
❷	adjusted hazard ratio	補正ハザード比	586
❸	adjusted models	補正されたモデル	139
❹	adjusted incidence	補正された発生率	136

■ to adjust

★	to adjust	補正する…	488
❶	to adjust for ～	～に対して補正するために	228

❶ This study demonstrates the feasibility of using CT **to adjust for** actual organ volumes in calculating organ-specific absorbed dose estimates.（J Nucl Med. 2004 45:1059）

訳 この研究は，～を計算する際に実際の臓器の体積に対して補正するためにCTを使う可能性を実証する

■ adjusting for

★	adjusting for ～	～に対して補正する	1,841
❶	after adjusting for ～	～に対して補正したあと	1,099

❶ **After adjusting for** age, the odds ratio for adverse birth outcome after receiving at least 1 dose of anthrax vaccination was 0.9 (95% CI, 0.4-2.4; P =.88).（JAMA. 2002 287:1556）

訳 年齢に対して補正したあと

adjustment 名 調整／適応 用例数 5,687

adjustment	調整／適応	5,052
adjustments	調整／適応	635

複数形のadjustmentsの割合は約10%あるが，単数形は無冠詞で使われることが非常に多い．adjustment forの用例が圧倒的に多い．

◆類義語：adaptation, modulation, regulation, accommodation

■ adjustment ＋［前］

	the%	a/an%	ø%			
❶	0	1	99	adjustment for ～	～に対する調整	3,609
				・after adjustment for ～	～に対する調整のあと	2,518
				・with adjustment for ～	～に対する調整をして	264
❷	15	14	68	adjustment of ～	～の調整	268

❶ **After adjustment for** age, smoking, and alcohol intake, coffee consumption was inversely associated with Parkinson's disease mortality in men (p_{trend} = 0.01) but not in women (p = 0.6).（Am J Epidemiol. 2004 160:977）

訳 年齢に対する調整のあと

❶ The data were analyzed by Poisson maximum-likelihood techniques **with adjustment for**

overdispersion. (J Infect Dis. 2004 189:1932)
訳 過分散に対する調整をして

administer 動 投与する／投薬する 用例数 9,958

administered	投与される／～を投与した	9,278
administering	～を投与する	447
administer	～を投与する	227
administers	～を投与する	6

他動詞．受動態の用例が圧倒的に多い．第4文型（S+V+O_1+O_2）の受動態としての用例も多い．

◆類義語：treat, cure, remedy, give

■ be administered ＋［前］

★ be administered	…は，投与される	3,542
❶ be administered to ～	…は，～に投与される	305
❷ be administered in ～	…は，～において投与される	106

❶ When LAM was administered to mice previously given a LAM-binding immunoglobulin M (IgM), LAM was very rapidly cleared from circulation. (Infect Immun. 2000 68:335)
訳 LAMがマウスに投与された

■ be administered ＋［副］

❶ be administered intravenously	…は，静脈内に投与される	138
❷ be administered orally	…は，経口的に投与される	90

❶ In some animals, myoinositol was administered intravenously over the same time period to raise plasma myoinositol levels by 5 to 10 mM. (J Am Soc Nephrol. 2002 13:1255)
訳 ミオイノシトールが静脈内に投与された

■［副］＋ administered

❶ orally administered ～	経口的に投与された～	214
❷ systemically administered ～	全身的に投与された～	115
❸ intravenously administered ～	静脈内に投与された～	93

❶ Orally administered LSF significantly inhibited EAE in both cases, decreasing peak clinical scores by >70% and >80%, respectively. (J Immunol. 1998 161:7015)
訳 経口的に投与されたLSFは，EAEを有意に抑制した

■ when administered

❶ when administered	投与されたとき	905

❶ However, MIP-$1^{-/-}$ splenocytes easily induced GVHD when administered to bm12 mice. (Blood. 1999 93:43)

訳 bm12マウスに投与されたとき

★ administration 名 投与／投薬　用例数 18,008

administration	投与	17,875
administrations	投与	133

複数形のadministrationsの割合は0.7%しかなく，原則，**不可算名詞**として使われる．

◆類義語：therapy, treatment

■ administration ＋［前］

	the%	a/an%	ø%			
❶	2	2	95	administration of ～	～の投与	3,609
				・after administration of ～	～の投与のあと	163
				・by administration of ～	～の投与によって	125

❶ **After administration of TTX**, the mfERG is further modified by the addition of NMDA.（Invest Ophthalmol Vis Sci. 2002 43:1673）

訳 TTXの投与のあと

■［形／名］＋ administration

❶	drug administration	薬物投与	1,024
❷	systemic administration	全身投与	642
❸	oral administration	経口投与	593
❹	intravenous administration	静脈内投与	411

❸ Standard EOGs were recorded **after oral administration of alcohol** in a group of patients with retinitis pigmentosa (RP).（Invest Ophthalmol Vis Sci. 2000 41:2730）

訳 アルコールの経口投与のあと

★ advance 名 進歩，動 進行させる　用例数 18,245

advanced	進行した／～を進行させた	8,019
advances	進歩／～を進行させる	7,408
advance	進歩／～を進行させる	2,076
advancing	～を進行させる	742

他動詞として使われるが，名詞の用例も多い．名詞としては複数形のadvancesの割合が，約85%と圧倒的に高い．advances inの用例が非常に多い．

◆類義語：progress, proceed, improve, advancement, progression

■ 名 advances ＋［前］

	the%	a/an%	ø%			
❶	9	-	86	advances in ～	～についての進歩	4,825

・advances in understanding ～	～を理解する際の進歩	295

❶ Recent advances in carbohydrate metabolism during pregnancy suggest that preventive measures should be aimed at improving insulin sensitivity in women predisposed to GDM. (Am J Clin Nutr. 2000 71:1256S))

訳 妊娠中の糖代謝についての最近の進歩は，～ということを示唆する

■ 名［形］＋ advances

❶ recent advances	最近の進歩	2,450
❷ significant advances	顕著な進歩	279
❸ major advances	大きな進歩	231
❹ technological advances	科学技術の進歩	189

■ 名 in advance of

❶ in advance of ～	～に先立って	103

❶ Neuronal populations in individuals with degenerative disorders show elevated levels of Par-4 protein in advance of cellular and functional loss.

訳 ～は，細胞および機能の喪失に先立って，上昇したレベルのPar-4タンパク質を示す

■ 動 advanced ＋［名］

❶ advanced disease	進行した疾患	456
❷ advanced stage	進行期	420
❸ advanced age	高齢	206

■ 動［前］＋ advanced

❶ patients with advanced ～	進行した～の患者	1,216
❷ treatment of advanced ～	進行した～の治療	119

❶ We therefore investigated the value of bone-targeted consolidation therapy in selected patients with advanced androgen-independent carcinoma of the prostate. (Lancet. 2001 357: 336)

訳 進行したアンドロゲン非依存性癌腫の患者

advantage 名 利点／優位性 用例数 7,577

advantage	利点／優位性	4,841
advantages	利点／優位性	2,736

複数形のadvantagesの割合が，約35%ある可算名詞．take advantage ofの用例も非常に多い．

◆類義語：benefit, merit

■ advantage ＋［前］

	the %	a/an %	ø %			
❶	16	36	39	advantage of ～	～の利点	2,201
❷	6	81	7	advantage in ～	～における利点	345
❸	7	74	5	advantage over ～	～より優位	343
❹	5	82	7	advantage for ～	～に対する利点	312
❺	3	81	8	advantage to ～	～にとっての利点	287

❷ Paradoxically, these alleles **may have a selective advantage in protection against** intracellular pathogens and occur at particularly high frequencies in sub-Saharan Africa (C variant) and South America (B variant). (Hum Mol Genet. 2000 9:1481)

訳 …は，～に対する保護において選択的な優位性を持つかもしれない

❸ Multi-photon laser scanning microscopes **have many advantages over single-photon systems**. (Biophys J. 2002 83:2292)

訳 ～は，単一光子システムよりも多くの利点を持つ

■ take advantage of

★	take advantage of ～	…は，～を利用する	1,311
❶	we take advantage of ～	われわれは，～を利用する	258
❷	Taking advantage of ～	～を利用して	154
❸	take advantage of the fact that ～	…は，～という事実を利用する	121
❹	take advantage of the ability of ～	…は，～の能力を利用する	89
❺	method takes advantage of ～	方法は，～を利用する	31

❶ In this report, **we take advantage of the fact that** a high percentage of microorganisms have both carbohydrate and lectin binding pockets at their surface. (Anal Chem. 2007 79:2312)

訳 われわれは，～という事実を利用する

❷ **Taking advantage of lasting hPLAP activity** after transcription of the reporter gene has ceased, we could show that SryhPLAP was expressed exclusively in all cells fated to become Sertoli cells. (Dev Biol. 2004 274:271)

訳 持続するhPLAP活性を利用して

adverse 形 有害な／反対の 用例数 8,881

◆類義語：aversive, deleterious, detrimental, harmful, hazardous, noxious

■ adverse ＋［名］

❶	adverse events	有害な事象	2,916
❷	adverse effects	有害な影響	1,537
❸	adverse outcomes	有害な結果	481

❶ Over the entire study period, **less than 20% of the patients stopped treatment because of adverse events**. (Am J Psychiatry. 2002 159:88)

訳 有害な事象ゆえに，20%未満の患者が治療を中止した

❷ Rituximab was well tolerated in this patient population, with most experiencing no significant adverse effects. (Arthritis Rheum. 2004 50:2580)
訳 ほとんどの患者が有意な有害な影響を経験することなく

adversely 副 不利に／逆に　用例数 885

■ adversely ＋［動／過分］

❶ adversely affect ～	…は，～に不利に影響する	647
❷ be adversely affected	…は，不利に影響を受ける	75

❶ The presence of ERACs does not adversely affect secretory protein traffic through the ER and does not lead to induction of the unfolded protein response. (Mol Biol Cell. 2004 15:908)
訳 ERACの存在は，分泌タンパク質の輸送に不利に影響しない

★★ affect 動 影響を与える　用例数 54,480

affect	～に影響を与える	21,028
affected	影響を受ける／冒される／～に影響を与えた	18,459
affects	～に影響を与える	7,690
affecting	～に影響を与える	7,303

他動詞．否定文の用例が多い．
◆類義語：influence, impact

■ affect ＋［名句］

❶ affect the expression of ～	…は，～の発現に影響を与える	347
❷ affect the ability of ～	…は，～の能力に影響を与える	316
❸ affect the rate of ～	…は，～の速度に影響を与える	204
❹ affect the binding	…は，結合に影響を与える	179
❺ affect the activity of ～	…は，～の活性に影響を与える	170
❻ affect the function of ～	…は，～の機能に影響を与える	159

❷ Deletion of interdomain B did not affect the ability of ZAP-70 to bind to the receptor. (Mol Cell Biol. 1999 19:948)
訳 ～は，受容体に結合するZAP-70の能力に影響を与えなかった

■［副］＋ affect

❶ significantly affect ～	…は，～に有意に影響を与える	1,408
❷ adversely affect ～	…は，～に不利に影響する	647
❸ directly affect ～	…は，～に直接影響を与える	425

❶ DHT itself did not significantly affect sexual receptivity. (Brain Res. 2002 948:102)
訳 DHT自身は，性的な受容性に有意に影響を与えなかった
❷ The lack of TNF receptors did not adversely affect the development of a type 1 IFN-gamma

response. (J Immunol. 1998 160:1340)
訳 TNF受容体の欠損は，～の発生に不利に影響しなかった

■ [動] + to affect

★ to affect ～	～に影響を与える…	2,013
❶ appear to affect ～	…は，～に影響を与えるように思われる	233
❷ fail to affect ～	…は，～に影響を与えることができない	142
❸ be shown to affect ～	…は，～に影響を与えることが示される	178
❹ be known to affect ～	…は，～に影響を与えることが知られている	108

❷ Pretreatment with the dopamine D2 receptor antagonist raclopride (1mg/kg, ip) failed to affect this phenomenon. (Brain Res. 2011 1399:33)
訳 ～は，この現象に影響を与えることができなかった

❸ Selective attention has previously been shown to affect neural activity in both extrastriate and striate visual cortex. (Nat Neurosci. 2002 5:1203)
訳 …は，～における神経活動に影響を与えることが以前に示されている

■ affected + [前]

❶ affected by ～	～によって影響を受ける	6,172
・be affected by ～	…は，～によって影響を受ける	1,858
❷ affected in ～	～において影響を受ける	913
❸ affected with ～	～によって冒された	263

❶ Manganese balance was only affected by the amount of manganese in the diet. (Am J Clin Nutr. 1999 70:37)
訳 マンガンバランスは，食餌中のマンガン含量によってのみ影響を受けた

❷ Sensorimotor gating is a neural filtering process that allows attention to be focused on a given stimulus, and is affected in patients with neuropsychiatric disorders. (Nat Genet. 1999 21:434)
訳 ～は，神経精神的な異常を持つ患者において影響を受ける

■ [副] + affected

❶ significantly affected	有意に（顕著に）影響を受ける	613
❷ severely affected	（病気などに）重度に冒される	312
❸ strongly affected	強く影響を受ける	193
❹ adversely affected	不利に影響を受ける	177
❺ minimally affected	最小限の影響を受ける	167
❻ differentially affected	特異的な影響を受ける	163

❶ Neither the surface fraction at steady state nor the stability of Shaker is significantly affected by glycosylation in COS cells. (Biochemistry. 2002 41:11351)
訳 ～は，COS細胞におけるグリコシル化によって顕著に影響を受ける

❷ There have been no studies reported to date, however, that have attempted to correlate severely affected pedigrees with a particular genotype. (Am J Hum Genet. 1999 65:1561)
訳 重度に冒された家系を特定の遺伝子型に相関させようと試みた～

■ affected ＋［名］

❶ affected individuals	（病気に）冒された個人	803
❷ affected patients	（病気に）冒された患者	222

■［前］＋ affecting

❶ without affecting ～	～に影響を与えることなしに	2,228
❷ by affecting ～	～に影響を与えることによって	505

❶ In NtBHA-treated rats, the age-dependent decline in food consumption and **ambulatory activity was reversed without affecting body weight.** (FASEB J. 2001 15:2196)
訳 歩行の活動は，体重に影響することなく逆行させられた

■［名］＋ affecting

❶ mutations affecting ～	～に影響する変異	466
❷ factors affecting ～	～に影響する因子	300
❸ genes affecting ～	～に影響する遺伝子	112

★★ affinity　名 親和性　　用例数 39,079

affinity	親和性	35,162
affinities	親和性	3,917

複数形のaffinitiesの割合は約10%あるが，単数形は無冠詞で使われることが多い.
◆類義語：connectivity, binding, bonding

■ affinity ＋［前］

affinity ofの前にtheが付く割合は圧倒的に高い.

	the%	a/an%	ø%			
❶	10	12	65	affinity for ～	～に対する親和性	5,370
❷	87	7	6	affinity of ～	～の親和性	4,196
				・affinity of ～ for …	～の…に対する親和性	1,475
❸	10	8	69	affinity to ～	～への親和性	1,267

❶ Ibogaine and noribogaine **were shown to have affinity for the serotonin transporter**, and inhibition of serotonin reuptake has been proposed to be involved in their anti-addictive actions. (Brain Res. 1998 800:260)
訳 ～は，セロトニン輸送体に対する親和性を持つことが示された

❷ Mutation of residues predicted to form part of this hydrophobic pocket **either abolished or significantly diminished the affinity of PDK1 for PIF.** (EMBO J. 2000 19:979)
訳 ～は，PDK1のPIFに対する親和性を消滅させるかあるいは顕著に低下させた

■ affinity-purified

❶ affinity-purified ～	アフィニティー精製された～	622

❶ Kinase activity of the affinity-purified proteins was assayed as autophosphorylation at amino acid Thr307 or against an Ugp1p-derived peptide.（J Biol Chem. 2011 286:44005）
訳 アフィニティー精製されたタンパク質のキナーゼ活性がアッセイされた

■［形／過分］＋ affinity

❶ high affinity	高親和性	9,455
❷ binding affinity	結合親和性	4,034
❸ low affinity	低親和性	2,120
❹ higher affinity	より高い親和性	1,169
❺ lower affinity	より低い親和性	706
❻ apparent affinity	見かけの親和性	380
❼ reduced affinity	低下した親和性	338
❽ increased affinity	増大した親和性	316

❶ GDNF binds with high affinity to the GDNF family receptor α-1（GFR α-1）, which is highly expressed in the midbrain.（J Comp Neurol. 2001 441:106）
訳 GDNF は，高親和性でGDNFファミリー受容体α-1に結合する

❷ Of all of the compounds synthesized, the 3-n-propyl derivative（−）-9 was found to be the most potent with a binding affinity of 3 nM.（J Med Chem. 1998 41:1962）
訳 ～が，3 nMの結合親和性を持つ最も強力なものであることが見つけられた

★★ age　名 年齢／歳，動 加齢する　　用例数 65,300

age	年齢／歳／加齢する	47,691
aged	加齢した	8,146
aging	加齢する	5,822
ages	年齢／歳／加齢する	3,641

名詞の用例が多いが，動詞としても用いられる．複数形のagesの割合が，約５％ある可算名詞．

■ 名 age ＋［前］

age ofの前にtheが付く割合はかなり高い．

	the%	a/an%	ø%			
❶	63	18	15	age of ～	～の年齢	2,688
				・age of onset	発症の年齢	348
❷	9	14	74	age at ～	～時年齢／～歳	2,308
				・age at onset	発症時年齢	446
				・age at diagnosis	診断時年齢	361
❸	2	4	92	age in ～	～における年齢	743

❶ The age of onset was highly variable, from early adulthood, and including a mild phenotype in advanced age.（Brain. 2012 135:1695）

訳 発症の年齢

❷ Age at diagnosis is a highly significant prognostic factor for survival of children with RTK.（J Clin Oncol. 2005 23:7641）

訳 診断時年齢は，～の生存にとって非常に重要な予後因子である

■ 名［前］＋ age

❶ ～ years of age	～の年齢／～歳	2,848
❷ ～ months of age	～ヶ月齢	1,329
❸ increase with age	…は，年齢とともに増大する	525
❹ after adjustment for age	年齢に関する補正のあと	546
❺ adjusted for age	年齢に関して補正される	436

❶ Respondents were 1024 women ≧ 25 years of age, including an oversampling of racial/ethnic minorities（68% white, 12% black, 12% Hispanic）.（Circulation. 2005 111:1321）

訳 回答者は，25歳以上の1024名の女性であった

❷ Although case-fatality rates increased with age, 736（47%）of 1561 non-fatal events occurred at age 75 years or older.（Lancet. 2005 366:1773）

訳 症例致死率は年齢とともに増大したけれども，

❸ After adjustment for age, gender, smoking, diabetes, hypertension, obesity, and HDL-C, these individuals had a 28% lower risk of recurrent myocardial infarction or vascular death（relative risk = 0.72; 95% confidence interval 0.52 to 0.99）.（J Am Coll Cardiol. 2005 45:1644）

訳 年齢，性別，喫煙，糖尿病，高血圧，肥満，およびHDL-Cに関する補正のあと

■ 名 age ＋［過分／形／名］

❶ age-related ～	年齢と関連する～	2,943
・age-related changes	年齢と関連する変化	297
❷ age-matched ～	年齢の一致する～	1,621
・age-matched controls	年齢の一致する対照群	395
❸ age-dependent ～	年齢に依存する～	899
❹ age groups	年齢群	708
❺ age range	年齢範囲	683
❻ age-adjusted ～	年齢を補正した～	624

❶ Age-related changes in brain steroids did not mirror associated changes in circulating P and T.（Brain Res. 2006 1067:115）

訳 脳のステロイドの年齢と関連する変化

■ 名 age and ＋［名］

❶ age and sex	年齢と性別	374

❶ After adjusting for age and sex, data are reported for all exposures with odds ratios >2.0 against either control group or for any allergic propensity.（Arthritis Rheum. 2004 51:656）

訳 年齢と性別に関して補正したあと，

A

■ ㊔［形］＋ age

❶	mean age	平均年齢	2,860
	・mean age of ～	～の平均年齢	479
❷	median age	年齢中央値	1,154

❶ The mean age of the participants was 78.2 years. (Invest Ophthalmol Vis Sci. 2006 47:65)
訳 参加者の平均年齢は，78.2歳であった

■ ㊍［名／形］＋ aged

❶	women aged ～	～歳の女性	874
❷	children aged ～	～歳の子供	514
❸	middle-aged ～	中年の～	656

❷ Participants were 2,010 children aged 9-10 years from 89 schools around Amsterdam Schiphol, Madrid Barajas, and London Heathrow airports. (Am J Epidemiol. 2006 163:27)
訳 参加者は，9-10歳の 2 ,010名の子供達であった

★★ agent ㊔ 薬剤／病原体 用例数 31,112

agents	薬剤／病原体	19,547
agent	薬剤／病原体	11,565

複数形のagentsの割合は，約65%と圧倒的に高い．

◆類義語：drug, pathogen

■ agent(s) ＋［前］

agent ofの前にtheが付く割合は圧倒的に高い．

	the %	a/an %	ø %			
❶	86	14	0	agent of ～	～の病原体	1,589
❷	3	-	80	agents in ～	～における薬剤	1,012
❸	6	-	88	agents for ～	～に対する薬剤	858

❶ Vibrio cholerae is the causative agent of the diarrheal disease cholera. (Proc Natl Acad Sci U S A. 2010 107:5581)
訳 コレラ菌は，下痢性疾患コレラの原因病原体である

❷ These studies suggest GSIs as potential therapeutic agents in the treatment of T-ALL. (Blood. 2009 113:6172)
訳 これらの研究は，T細胞性急性リンパ性白血病の治療においてGSIsを潜在的な治療薬として示唆している

■［形］＋ agent(s)

❶	therapeutic agents	治療薬	1,059
❷	causative agent	原因病原体	827

❸ chemotherapeutic agents	化学療法薬	735
❹ DNA-damaging agents	DNA傷害剤	641
❺ anticancer agents	抗がん剤	406

aggregate 名 凝集体，動 凝集する／集まる　用例数 7,501

aggregates	凝集体／凝集する／集まる	4,059
aggregate	凝集体／凝集する／集まる	2,170
aggregated	凝集した／集まった	1,081
aggregating	凝集する／集まる	191

名詞の用例が多いが，動詞としても用いられる．複数形のaggregatesの割合は，約70%と圧倒的に高い．

◆類義語：agglutinate, assemble, form

■ aggregates ＋［前］

	the%	a/an%	ø%			
❶	4	-	95	aggregates of ～	～の凝集体	424

❶ In contrast, fibrillar **aggregates of** α-synuclein exhibit a distinct domain organization.（J Biol Chem. 2003 278:37530）

訳 α-シヌクレインの原繊維凝集体は，～を示す

■ in aggregate

★ in aggregate	全体として	159
❶ In aggregate,	全体として，	75

❶ **In aggregate,** the results of this study indicate that reactive oxygen-nitrogen species may play a role in the pathogenesis of human ALI.（Am J Respir Crit Care Med. 2001 163:166）

訳 全体として，

aggregation 名 凝集　用例数 7,956

aggregation	凝集	7,867
aggregations	凝集	89

複数形のaggregationsの割合は1％しかなく，原則，**不可算名詞**として使われる．

◆類義語：agglutination, assembly

■ aggregation ＋［前］

	the%	a/an%	ø%			
❶	36	6	57	aggregation of ～	～の凝集	1,465
❷	0	0	88	aggregation in ～	～における凝集／～の凝集	416

❶ In our investigation, acid-induced **aggregation of** monoclonal IgG1 and IgG2 antibodies was studied at pH 3.5 as a function of salt concentration and buffer type. (Biochemistry. 2010 49: 9328)

訳 単クローンのIgG1およびIgG2抗体の酸に誘導される凝集

❷ The first of these is a generic model for competing cell populations, and **the second concerns aggregation in cell populations moving in response to chemical gradients**. (J Theor Biol. 2003 225:327)

訳 2番目は，化学勾配に応答して動く細胞集団の凝集に関することである

■ [名] + aggregation

❶ platelet aggregation	血小板凝集	1,033
❷ protein aggregation	タンパク質凝集	430

agree 動 一致する／同意する　用例数 1,939

agree	一致する／同意する	814
agreed	一致した／同意した	747
agrees	一致する／同意する	378

自動詞．agree withの用例が多い．

◆類義語：coincide, correspond, fit, match, represent, approve

■ agree + [前／that節]

❶ agree with ～	…は，～に一致する	802
・agree well with ～	…は，～によく一致する	350
❷ agree to *do*	…は，～することに同意する	112
❸ agree on ～	…は，～に関して一致する	77
❹ agree that ～	…は，～ということで一致する	145

❶ **These results agree with the previously described secretion-blocking activity of LcrG** and demonstrate that the interaction of LcrV with LcrG is necessary for controlling Yops secretion. (J Bacteriol. 2001 183:5082)

訳 これらの結果は，以前に述べられたLcrGの分泌阻止活性に一致する

❶ SUVs of the entire axial slices of liver **agree well with subjective visual evaluations**. (J Nucl Med. 2004 45:1892)

訳 ～は，主観的視覚評価によく一致する

❷ **A total of 1,440 children (96%) agreed to participate in the survey** at baseline, and of those eligible, 1,081 (88%) participated at followup. (Arthritis Rheum. 2003 48:2615)

訳 総計1,440名の子供（96%）が，調査に参加することに同意した

❹ **All methods agreed that** 26 pairs were indistinguishable and four pairs were different. (J Clin Microbiol. 2013 51:224)

訳 すべての方法が，～ということで一致した

agreement 名 一致　　用例数 6,546

agreement	一致	6,438
agreements	一致	108

複数形のagreementsの割合は約1.5%しかなく，原則，**不可算名詞**として使われる．in agreement withの用例が非常に多い．

◆類義語：accordance, accord, concert, concordance, correspondence, coincidence

■ agreement ＋［前］

	the%	a/an%	ø%			
❶	2	1	97	agreement with ～	～との一致	3,560
❷	34	7	57	agreement between ～	～の間の一致	846

❷ Overall, **there was 97.5% agreement between** the results obtained with the Copan Dacron swabs and those obtained with the Culturette swabs.（J Clin Microbiol. 2003 41:2686）

訳 ～の間に97.5%の一致があった

■ in agreement with

★	in agreement with ～	～との一致	1,939
❶	In agreement with ～	～と一致して	589
❷	be in agreement with ～	…は，～と一致している	496
❸	in good agreement with ～	～とよく一致して	536
❹	in excellent agreement with ～	～と非常によく一致して	250

❶ **In agreement with previously published reports**, an absence of homology was found between the α and β families of the PLP-dependent enzyme superfamily.（Biochemistry. 2000 39: 15242）

訳 以前に発表された報告と一致して

❷ The regression models obtained are interpretable and **the biological implications are in agreement with the known results**.（Bioinformatics. 2004 20:2799）

訳 生物学的意味は，既知の結果と一致している

❸ **These structural features are in good agreement with** the experimentally determined membrane structure of hemagglutinin fusion peptide from influenza virus.（Bioinformatics. 2004 20:970）

訳 これらの構造的特徴は，～とよく一致している

■［形］＋ agreement

❶	good agreement	よい一致	1,023
❷	excellent agreement	非常によい一致	475
❸	interobserver agreement	観察者間の一致	145

A

aim ㊔目的，㊍目的とする／～しようとする 用例数 9,648

aim	目的／～を目的とする	4,877
aimed	～を目的とした／目的とされる	3,551
aims	目的／～を目的とする	1,055
aiming	～を目的とする	165

名詞および動詞として使われる．the aim of ～ studyで始まる文は，通常，過去形で用いられる．

◆類義語：purpose, objective, goal, object, end, attempt, seek

■ ㊔ aim ＋［前］

aim ofの前にtheが付く割合は，99%で圧倒的に高い．ただし目的が複数あって，そのうちの一つという意味なら不定冠詞が用いられることもあり得る．

	the%	a/an%	⌀%		
❶	99	1	0	aim of ～　～の目的	3,540
				・the aim of this study was to ～ この研究の目的は，～することであった	2,020
				・the aim of this study was to determine ～ この研究の目的は，～を決定することであった	541
				・the aim of this study was to investigate ～ この研究の目的は，～を精査することであった	246
				・the aim of the present study was to ～ 現在の研究の目的は，～することであった	262
				・with the aim of ～　～の目的で	244

❶ The aim of this study was to determine if acetic acid evokes this wiping response by decreasing subepidermal pH.（Brain Res. 2000 862:217）

訳 この研究の目的は，～を決定することであった

❶ The aim of the present study was to evaluate systemic ofloxacin therapy as adjunct to flap surgery.（J Periodontol. 2000 71:202）

訳 現在の研究の目的は，～を評価することであった

❶ We are analyzing highly conserved heat shock genes of unknown or unclear function with the aim of determining their cellular role.（EMBO J. 2000 19:741）

訳 それらの細胞の役割を決定する目的で

■ ㊔ aim was to *do*

★	aim was to *do*	目的は，～することであった	638
❶	our aim was to *do*	われわれの目的は，～することであった	478

❶ Our aim was to determine the affinity of the human BSEP for bile salts and identify inhibitors.（Gastroenterology. 2002 123:1649）

訳 われわれの目的は，～を決定することであった

■ 動 aim to *do*

★ aim to *do*	…は，～することを目的とする	2,550
❶ we aimed to *do*	われわれは，～することを目的とした	1,098
❷ study aimed to *do*	研究は，～することを目的とした	359

❶ **We aimed to assess whether** gene products of human cytomegalovirus could be detected in colorectal cancers. (Lancet. 2002 360:1557)
訳 われわれは，～かどうかを評価することを目的とした

■ 動 aimed at

★ aimed at ～	～を目的とする…	1,830
❶ aimed at ～ing	～することを目的とする…	1,429
・studies aimed at ～ing	～することを目的とする研究	190
・strategies aimed at ～ing	～することを目的とする戦略	145
❷ be aimed at ～	…は，～を目的とする	216

❶ **We have initiated studies aimed at elucidating** the chemical nature of protein carbonyls. (Proc Natl Acad Sci U S A. 2001 98:69)
訳 われわれは，～を解明することを目的とした研究を開始した

★★ allow 動 許す／可能にする　用例数 35,429

allows	～を許す／～を可能にする	11,411
allow	～を許す／～を可能にする	10,552
allowed	許される／～を許した／～を可能にした	7,297
allowing	～を許す／～を可能にする	6,169

他動詞の用例が多い．第5文型のallow us to *do*のパターンも多い．allow forの形で自動詞としても使われる．

◆類義語：permit, enable, prompt, lead, tolerate

■ allow ＋［名句］

❶ allow the identification of ～	…は，～の同定を可能にする	241
❷ allow the development of ～	…は，～の開発を可能にする	130
❸ allow the detection of ～	…は，～の検出を可能にする	108

❶ **Mass spectrometry allowed the identification of 43 spots** on the cytotrophoblast map. (Biochemistry. 2001 40:4077)
訳 質量分析法は，43個のスポットの同定を可能にした

■ allow us to *do*

★ allow us to *do*	…は，われわれが～するのを可能にする	3,109
❶ allow us to identify ～	…は，われわれが～を同定するのを可能にする	240

❷ allow us to determine ～	…は，われわれが～を決定するのを可能にする	136
❸ allow us to propose ～	…は，われわれが～を提案するのを可能にする	100

❶ CS-BLI-based screening allowed us to identify agents that enhance or inhibit innate antitumor cytotoxicity. (Blood. 2012 119:e131)
訳 CS-BLIに基づくスクリーニングは，われわれが薬剤を同定するのを可能にした

■ [名] ＋ allow

❶ method allows ～	方法は，～を可能にする	232
❷ approach allows ～	アプローチは，～を可能にする	202
❸ system allows ～	システムは，～を可能にする	145
❹ data allow ～	データは，～を可能にする	97

❷ Our approach allows us to directly compare break-up magmatism generated at different locations and so isolate the key controlling factors. (Nature. 2010 465:913)
訳 われわれのアプローチは，われわれが～を直接比較するのを可能にする

■ [形] ＋ to allow

★ to allow ～	～を可能にする…	2,777
❶ sufficient to allow ～	～を可能にするのに十分な	137
❷ enough to allow ～	～を可能にするのに十分な	87

❶ Mutation of Phe350 and Phe351 in TM7 of the beta1-AR to Ala and Leu found in the beta3-AR was sufficient to allow activation by prototypic beta3-AR agonists. (Mol Pharmacol. 1998 53:856)
訳 …は，～による活性化を可能にするのに十分であった

■ allowing

❶ by allowing ～	～させることによって／～を許すことによって	392
❷ , thereby allowing ～	それによって，～を可能にする	294
❸ , thus allowing ～	このように，～を可能にする	285

❶ Genetic studies also suggest that fa (swb) compromises the functional autonomy of Notch by allowing the locus to become sensitive to chromosomal position effects emanating from distal sequences. (Genetics. 2000 155:1297)
訳 その遺伝子座を～に対して感受性にさせることによって

❷ We propose that lpf phase variation is a mechanism to evade cross-immunity between Salmonella serotypes, thereby allowing their coexistence in a host population. (Proc Natl Acad Sci USA. 1999 96:13393)
訳 それによって，それらの共存を可能にする

■ be allowed to *do*

★ be allowed	…は，可能にされる	658
❶ be allowed to *do*	…は，～することが可能にされる	501

❶ For the study of long-term survival, rAAV was administered by subretinal injection at P15, and the rats were allowed to live up to 8 months of age. (Proc Natl Acad Sci USA. 2000 97:11488)

訳 ラットは，8ヶ月齢まで生きることを可能にされた

■ allow for

❶ allow for ～	…は，～を計算に入れる	5,111

❶ This new method allows for the study of gene functions in neuroblast proliferation, axon guidance, and dendritic elaboration in the complex central nervous system. (Neuron. 1999 22:451)

訳 この新しい方法は，遺伝子機能の研究を計算に入れる

★ almost 副 ほとんど　用例数 7,748

◈類義語：nearly, virtually, mostly, largely, near

■ almost +［形］

❶ almost all	ほとんどすべて	1,207
・in almost all ～	ほとんどすべての～において	269
❷ almost identical ～	ほとんど同一の～	375
❸ almost complete ～	ほとんど完全な～	321
❹ almost no ～	ほとんど～のない	239

❶ The t (9; 22) chromosomal translocation is found in almost all patients with chronic myelogenous leukemia. (Proc Natl Acad Sci USA. 2000 97:2093)

訳 t(9;22)染色体転座は，慢性骨髄性白血病のほとんどすべての患者において見つけられる

❷ An almost identical expression pattern was observed in cells treated with either Taxol or EpoB. (Cancer Res. 2003 63:7891)

訳 ほとんど同一の発現パターンが，～で処理された細胞において観察された

■ almost +［副］

❶ almost exclusively	ほとんど排他的に	799
❷ almost completely	ほとんど完全に	724
・be almost completely ～	…は，ほとんど完全に～である	301
❸ almost entirely	ほとんど完全に	311

❷ Treatment with PN401 almost completely prevented the neuronal damage due to 3NP and completely prevented mortality. (Brain Res. 2003 994:44)

訳 PN401による処置は，ニューロンの損傷をほとんど完全に防いだ

❸ In contrast, aspartate residues within the CDRs were almost entirely excluded from the binding interface. (Proc Natl Acad Sci USA. 2004 101:12467)

訳 ～は，結合界面からほとんど完全に排除された

★★ alone 副 単独で　用例数 21,286

◆類義語：solely

■ alone or

❶	alone or in ～	単独であるいは～において	1,123
	・either alone or in combination with ～	単独であるいは～との組み合わせのどちらか	118

❶ Approximately 50% of cancer patients receive radiation treatment, either alone or in combination with other therapies. (Proc Natl Acad Sci USA. 2009 106:14391)
訳 単独であるいは他の治療法との組合せのどちらかで

■ [過分] + alone

❶	expressed alone	単独で発現される	188
	・when expressed alone	単独で発現されたとき	90
❷	used alone	単独で使われる	134

❶ However, this pUL38 domain was unable to activate mTORC1 when expressed alone. (J Virol. 2011 85:9103)
訳 単独で発現されたとき，このpUL38ドメインはmTORC1を活性化することができなかった

★★ also 副 ～もまた　用例数 182,176

◆類義語： too

■ be also + [過分／形]

★	be also	…もまた～	39,531
❶	be also observed	…もまた観察される	2,853
❷	be also found	…もまた見つけられる	2,192
❸	be also required	…もまた必要とされる	1,340
❹	be also associated	…もまた関連する	1,291
❺	be also present	…もまた存在する	1,063
❻	be also detected	…もまた検出される	1,007

❶ These effects were also observed in wild-type but not IL-17RA-deficient cells. (J Immunol. 2009 183:865)
訳 これらの効果は，野生型細胞においてだけでなく，L-17RA欠損細胞においても観察された

■ [代名／名] + also + [動]

❶	we also ～	われわれは，また～する	17,894
	・we also show ～	われわれは，また～を示す	3,070
❷	results also ～	結果は，また～する	1,890

・results also suggest ～	結果は，また～を示唆する	538
❸ data also ～	データ，また～する	1,429
❹ mice also ～	マウスは，また～する	1,099

❶ We also show that SK channel dendritic distribution is dynamic and under the control of protein kinase A.（J Neurosci. 2012 32:11435）
訳 われわれは，また，～ということを示す

★ alter 動 変化させる／改変する　用例数 36,307

altered	変えられる／変化した／改変される／～を変化させた	18,674
alter	～を変化させる	10,137
altering	～を変化させる	3,935
alters	～を変化させる	3,561

他動詞.
◆類義語：change, shift, modify, convert, vary

■ alter ＋［名句］

❶ alter the expression of ～	…は，～の発現を変化させる	287
❷ alter the ability of ～	…は，～の能力を変化させる	161
❸ alter the conformation of ～	…は，～の立体構造を変化させる	123
❹ alter the activity of ～	…は，～の活性を変化させる	121
❺ alter the structure of ～	…は，～の構造を変化させる	113

❶ Introduction of Bcl6 did not alter the expression of IL-21 or IL-4, the primary cytokines of human Tfh cells.（J Immunol. 2012 188:3734）
訳 Bcl6の導入は，IL-21あるいはIL-4の発現を変化させなかった

❷ Phenotypic differences between microvascular and macrovascular EC may alter the ability of these cells to support HCMV replication.（J Virol. 1998 72:5661）
訳 …は，～を支持するこれらの細胞の能力を変化させるかもしれない

■［動／名］＋ to alter

★ to alter ～	～を変化させる…	2,320
❶ fail to alter ～	…は，～を変化させることができない	163
❷ be shown to alter ～	…は，～を変化させることが示される	80
❸ ability to alter ～	～を変化させる能力	78

❶ However, soluble interleukin-1 receptor failed to alter the increase in CTpr levels.（J Infect Dis. 2001 184:373）
訳 可溶性のインターロイキン-1受容体は，～の増大を変化させることができなかった

■ be altered ＋［前］

★ be altered	…は，変えられる	3,181
❶ be altered in ～	…は，～において変化する	1,043

❷ be altered by ～	…は，～によって変えられる	722

❶ Expression of the nitric oxide synthases is altered in the retinal vasculature in the early stages of diabetic retinopathy. (FASEB J. 1999 13:1825)

訳 一酸化窒素合成酵素の発現は，網膜脈管構造において変化する

❷ Thus, responses of the basilar artery to important vasoactive agonists are not altered by diabetes mellitus. (Brain Res. 1998 783:326)

訳 ～は，糖尿病によって変えられない

■ ［副］＋ altered

❶ significantly altered	有意に変化した	644
❷ genetically altered	遺伝的に改変された	184
❸ dramatically altered	劇的に変化した	181

❶ Response amplitudes and kinetics were not significantly altered compared with cells dialyzed with cGMP alone. (J Neurosci. 1999 19:2938)

訳 反応の振幅と動力学は，～に比較して有意には変化しなかった

■ ［前］＋ altered

❶ associated with altered ～	変化した～と関連した	265
❷ mutants with altered ～	変化した～を持つ変異体	97

❷ Mutants with altered patterns of lignification have been identified in a population of mutagenised Arabidopsis seedlings. (Development. 2000 127:3395)

訳 変化したパターンの木化を持つ変異体が，～において同定されている

■ ［前］＋ altering

❶ by altering ～	～を変化させることによって	1,290
❷ without altering ～	～を変化させることなしに	691

❷ The Gly26→Arg mutation abolishes single-stranded DNA binding without altering the overall fold of the protein. (J Mol Biol. 1999 289:949)

訳 タンパク質の全体の折り畳みを変化させることなしに

★ alteration ㊔ 変化／変更　　用例数 15,414

alterations	変化／変更	11,657
alteration	変化／変更	3,757

複数形のalterationsの割合は約75%と非常に高いが，単数形は無冠詞で使われることが多い．

◆類義語：change, shift, conversion, modification, variation

■ alteration(s) ＋［前］

「～の変化」の意味では，alterations ofよりalterations inが使われることが圧倒的に多い．

	the%	a/an%	ø%			
❶	5	-	89	alterations in ～	～の変化	6,647
				・…-induced alterations in ～	…に誘導される～の変化	230
				・significant alterations in ～	～の有意な変化	172
				・alterations in gene expression	遺伝子発現の変化	137
				・be associated with alterations in ～	…は，～の変化と関連する	95
❷	17	10	69	alteration of ～	～の変化	1,663

❶ Thus, **immortalization of HTLV-1-infected cells appears to be independent of Tax-induced alterations in CBP/p300 function.** (J Virol. 2000 74:11988)
訳 HTLV-1に感染した細胞の不死化は，Taxに誘導されるCBP/p300機能の変化とは無関係であるように思われる

❶ **Human immunodeficiency virus (HIV) therapies have been associated with alterations in fat metabolism** and bone mineral density. (J Biol Chem. 2002 277:19247)
訳 ヒト免疫不全ウイルス（HIV）の治療は，脂質代謝の変化と関連する

❷ Our results suggest that **alteration of the affinity of the vinblastine binding site** involves only one nucleotide binding domain per transport cycle. (Biochemistry. 2001 40:15733)
訳 ビンブラスチン結合部位の親和性の変化

■［形］＋ alterations

❶	genetic alterations	遺伝的な変化	654
❷	structural alterations	構造的な変化	243

★ alternative　形 代替の，名 代替物／代わるもの　用例数 16,118

alternative	代替の／代わるもの	15,300
alternatives	代わるもの	818

形容詞の用例が多いが，可算名詞としても使われる．

■ 形 alternative ＋［名］

❶	alternative splicing	選択的スプライシング	2,480
❷	alternative pathway	代替経路	585
❸	alternative approach	代替アプローチ	302
❹	alternative mechanism	代替機構	236

A

■ ㊔ alternative ＋［前］

	the%	a/an%	⌀%			
❶	1	95	1	alternative to ～	～に代わるもの	1,475
				・alternative to conventional ～	従来の～に代わるもの	78
❷	2	97	0	alternative for ～	～に代わるもの	197

❶ Immunotherapy offers an alternative to conventional approaches.（J Org Chem. 2001 66:4115）
訳 免疫療法は，従来のアプローチに代わるものを提供する

★ amount ㊔ 量　用例数 17,420

amount	量	9,497
amounts	量	7,923

複数形のamountsの割合は約45%とかなり高い.

◆類義語：volume

■ amount ＋［前］

amount ofの前にtheが付く割合はかなり多い.

	the%	a/an%	⌀%			
❶	65	30	4	amount of ～	…量の～／～の量	8,422
				・small amount of ～	少量の～	340
				・large amount of ～	大量の～	265
				・significant amount of ～	かなりの量の～	255
				・total amount of ～	～の総量	240

❶ This method requires only a small amount of protein and, in many cases, may be used with unpurified proteins in cell lysates.（J Mol Biol. 2011 406:545）
訳 この方法は，ごく少量のタンパク質を必要とする

■ ［動］＋ the amount of

❶	increase the amount of ～	…は，～の量を増大させる	346
❷	reduce the amount of ～	…は，～の量を低下させる	272
❸	decrease the amount of ～	…は，～の量を低下させる	128

❶ Overexpression of Skp2 caused poly-ubiquitination of FOXO3 and degradation, whereas knockdown of Skp2 increased the amount of FOXO3 protein.（Oncogene. 2012 31:1546）
訳 Skp2のノックダウンは，FOXO3タンパク質の量を増大させた

amplification ㊔ 増幅　用例数 7,679

amplification	増幅	7,388
amplifications	増幅	291

複数形のamplificationsの割合は約５％あるが，単数形は無冠詞で使われることが多い．
◆類義語：replication, proliferation, growth

■ amplification ＋［前］

	the％	a/an％	ø％			
❶	23	6	68	amplification of ～	～の増幅	2,110
❷	1	4	75	amplification in ～	～における増幅	279

❶ This reduction in DNA synthesis, however, **did not prevent amplification of viral DNA** in the differentiated cellular compartment.（J Virol. 2003 77:2832）
訳 ～は，ウイルスDNAの増幅を阻止しなかった

❷ Finally, **Plk1 depletion significantly reduces centrosome amplification in** hydroxyurea-treated U2OS cells.（Proc Natl Acad Sci USA. 2002 99:8672）
訳 Plk1の欠乏は，～における中心体の増幅を有意に低下させる

■［名］＋ amplification

❶	PCR amplification	PCR増幅	707
❷	gene amplification	遺伝子増幅	493
❸	signal amplification	シグナル増幅	223

amplify 動 増幅する 用例数 5,996

amplified	～を増幅した／増幅される	3,895
amplify	～を増幅する	1,173
amplifying	～を増幅する	567
amplifies	～を増幅する	361

他動詞．
◆類義語：replicate, proliferate, propagate, grow

■ be amplified ＋［前］

★	be amplified	…は，増幅される	1,202
❶	be amplified by ～	…は，～によって増幅される	328
❷	be amplified in ～	…は，～において増幅される	200
❸	be amplified from ～	…は，～から増幅される	162

❶ **Extracted DNA was amplified by** the LightCycler (Roche) PCR assay.（J Clin Microbiol. 2002 40:3922）
訳 抽出されたDNAは，～によって増幅された

❷ AP-2γ at 20q13.2 encodes a transcription factor and **is frequently amplified in breast carcinoma**.（Cancer Res. 2004 64:8256）
訳 ～は，乳癌においてしばしば増幅される

❸ **GRα cDNA was also amplified from bovine lens.**（Invest Ophthalmol Vis Sci. 2003 44:5269）
訳 GRα cDNAは，また，ウシの水晶体から増幅された

A

amplitude ㊔ 振幅／大きさ 用例数 7,848

amplitude	振幅／大きさ	6,523
amplitudes	振幅／大きさ	1,325

複数形のamplitudesの割合が，約15%ある可算名詞．
◆類義語：magnitude, size, scale

■ amplitude ＋［前］

amplitude ofの前にtheが付く割合は圧倒的に高い．

	the%	a/an%	ø%			
❶	83	1	16	amplitude of ～	～の大きさ／～の振幅	1,926
				・reduce the amplitude of ～	…は，～の大きさを低下させる	138
				・increase the amplitude of ～	…は，～の大きさを増大させる	130

❶ Blocking presynaptic inhibition in vivo **increased the amplitude of** odorant-evoked input to glomeruli but had little effect on spatial patterns of glomerular input.（Neuron. 2005 48:1039）
訳 生体内でのシナプス前抑制のブロッキングは，匂い物質に惹起される入力の大きさを増大させた

analogous ㊒ 類似の／類似性の 用例数 3,828

analogous toの用例が非常に多い．
◆類義語：similar, equivalent, comparable, homologous

■ analogous ＋［前］

❶	analogous to ～	～に類似の	1,912
	・be analogous to ～	…は，～に類似する	470
	・in a manner analogous to ～	～に類似する様式で	136

❶ This model **is analogous to** one suggested previously for the relationship between oncogene function and apoptosis in carcinogenesis.（Cancer Res. 2003 63:1000）
訳 このモデルは，～に対して以前に示唆されたものに類似する

★★ analysis ㊔ 解析／分析 用例数 144,197

analysis	解析／分析	115,826
analyses	解析／分析	28,371

複数形のanalysesの割合は約20%あるが，単数形は無冠詞で使われることが多い．
◆類義語：assay, test, examination, study

■ analysis ＋［前］

	the%	a/an%	ø%			
❶	17	11	68	analysis of ～	～の解析	1,926
				・analysis of gene expression	遺伝子発現の解析	494
				・analysis of ～ in …	…における～の解析	2,244
				・analysis of ～ with …	…による～の解析	575
				・analysis of ～ by …	…による～の解析	509
❷	3	7	86	analysis in ～	～における解析	1,620
❸	4	10	77	analysis with ～	～による解析	1,284
❹	5	5	86	analysis by ～	～による解析	978

❶ **Tissue analysis of gene expression with cDNA microarrays provides** a measure of transcriptional or posttranscriptional regulation and cellular recruitment. (Circulation. 2000 101:1990)

訳 cDNAマイクロアレイによる遺伝子発現の組織解析は，～を提供する

❶ **Recent analysis of genetic alterations in human cancer points to** a major role for selection in neoplastic development but provides few details about the dynamics of the process. (Cancer Res. 2001 61:799)

訳 ヒト癌における遺伝的変化の最近の解析は，～を示す

❷ **DNase I footprinting analysis in a long DNA sequence provided** supporting evidence that f-ImImIm binds selectively to T.G mismatch sites. (Nucleic Acids Res. 2004 32:2000)

訳 長いDNA配列におけるデオキシリボヌクレアーゼＩフットプリンティング解析は，～を提供した

❸ **Analysis with a mathematical model suggests that** diluting cDNA libraries with other plasmids without the SV40 origin should improve the detection of COS cells expressing target cDNAs. (Anal Biochem. 2000 278:74)

訳 数学的モデルによる解析は，～ということを示唆する

❹ **Analysis by mass spectrometry suggested that** the inactive protein was a biosynthetic intermediate with only one oxygen atom incorporated into $Trp^{\beta 57}$ and no cross-link with residue β108. (Biochemistry. 2003 42:3224)

訳 質量分析法による解析は，～ということを示唆した

■ analysis ＋［動］

❶	analysis revealed ～	解析は，～を明らかにした	4,792
❷	analysis showed ～	解析は，～を示した	3,423
❹	analysis demonstrated ～	解析は，～を実証した	1,648
❸	analysis indicated ～	解析は，～を示した	1,507
❺	analysis identified ～	解析は，～を同定した	1,202
❻	analysis was performed	解析が，行われた	1,163

❶ **This analysis revealed that** the expression of a subset of ventral telencephalic markers, including Dlx2 and Gsh2, although greatly diminished, persist in $Shh^{-/-}$ mutants, and that these same markers were expanded in $Gli3^{-/-}$ mutants. (Development. 2002 129:4963)

訳 この解析は，～ということを明らかにした

❷ Moreover, **protein immunoblotting analysis showed that** CPT-I, as well as the inner CPT-II,

was localized in the mitoplast fraction. (J Biol Chem. 1998 273:23495)
訳 タンパク質免疫ブロット解析は，～ということを示した

❻ A retrospective analysis was performed on 54 patients who underwent surgery for squamous cell cancers of the larynx. (Cancer Res. 2000 60:3599)
訳 遡及的な解析が，～に対して行われた

■ [名／形] + analysis

❶	blot analysis	ブロット解析	4,582
❷	sequence analysis	シークエンス解析	3,748
❸	multivariate analysis	多変量解析	1,913
❹	regression analysis	回帰分析	1,841
❺	microarray analysis	マイクロアレイ解析	1,753
❻	genetic analysis	遺伝的解析	1,574
❼	mutational analysis	変異解析	1,566
❽	further analysis	さらなる解析	1,336
❾	PCR analysis	PCR解析	1,237

❼ We have investigated the regulation of σ^E through a transcriptional and mutational analysis of sigE and the surrounding genes. (Mol Microbiol. 1999 33:97)
訳 われわれは，sigEの転写および変異解析によって，σ^Eの調節を精査した

★★ analyze 動 解析する／分析する　用例数 30,247

analyzed	解析される／～を解析した	22,283
analyze	～を解析する	4,939
analyzing	～を解析する	2,794
analyzes	～を解析する	231

他動詞.

◆類義語：test, dissect, screen, examine, look at

■ analyze + [名句]

❶	analyze the role of ～	…は，～の役割を解析する	333
❷	analyze the effects of ～	…は，～の影響を解析する	282
❸	analyze data from ～	…は，～からのデータを解析する	231
❹	analyze the function of ～	…は，～の機能を解析する	148
❺	analyze the expression of ～	…は，～の発現を解析する	137

❶ We analyzed the role of CD11a, CD11b, and CD54 in a cell-to-cell adhesion assay using antibodies against these molecules. (Transplantation. 2001 72:1563)
訳 われわれは，～の役割を解析した

■ [代名／名] + analyze

❶	we analyzed ～	われわれは，～を解析した	4,612

❷ the authors analyzed ～	著者らは，～を解析した	166
❸ study analyzed ～	研究は，～を解析した	76

❶ In the current study, we analyzed the effects of nerve growth factor (NGF) treatment on LRP expression, distribution, and function within neurons in two neuronal cell lines. (J Biol Chem. 1998 273:13359)

訳 現在の研究において，われわれは～の影響を解析した

to analyze

★ to analyze ～	～を解析するために	3,428
❶ be used to analyze ～	…は，～を解析するために使われる	587
❶ To analyze ～	～を解析するために	429

be analyzed ＋［前］/using

★ be analyzed	…は，解析される	10,047
❶ be analyzed by ～	…は，～によって解析される	2,278
・be analyzed by using ～	…は，～を使うことによって解析される	174
❷ be analyzed for ～	…は，～に関して解析される	1,338
❸ be analyzed in ～	…は，～において解析される	1,096
❹ be analyzed with ～	…は，～を使って解析される	477
❺ be analyzed to *do*	…は，～するために解析される	433
❻ be analyzed using ～	…は，～を使って解析される	852

❶ Data were analyzed by using t tests, analysis of covariance, and multiple regression. (Am J Clin Nutr. 2000 71:725)

訳 データは，t検定を使うことによって解析された

❷ Selected strains of oral bacteria were analyzed for their ability to degrade wheat starch, maltose, maltotriose, and maltoheptaose. (J Dent Res. 1988 67:75)

訳 …は，～するそれらの能力に関して解析された

❻ Dose-response trends were analyzed using generalized logistic regression techniques. (Arthritis Rheum. 2002 46:1451)

訳 ～が，一般化ロジスティック回帰分析テクニックを使って解析された

［名］＋ be analyzed

❶ data were analyzed	データは，解析された	553
❷ samples were analyzed	サンプルは，解析された	193
❸ expression was analyzed	発現は，解析された	114

❶ Data were analyzed with parametric and nonparametric statistical tests. (Radiology. 2004 232:347)

訳 データは，パラメトリックおよびノンパラメトリックな統計検定を使って解析された

anchor 動 アンカーする／つなぎ止める，名 アンカー 用例数 4,304

anchored	アンカーされる／～をアンカーした	2,111
anchor	～をアンカーする／アンカー	1,713
anchors	～をアンカーする／アンカー	480

名詞としても使われるが，動詞の用例が多い．

◆類義語：attach, adhere, bind, associate

■ be anchored ＋［前］

★ be anchored	…は，アンカーされる	350
❶ be anchored to ～	…は，～にアンカーされる	173

❶ As an example of the utility of the high-resolution map, 22 cattle BAC fingerprint contigs **were directly anchored to cattle chromosome 19** [Bos taurus, (BTA) 19].（Proc Natl Acad Sci USA. 2005 102:18526）

訳 ～は，ウシの第19染色体に直接アンカーされた

■［名］-anchored

❶ GPI-anchored ～	GPIアンカーされた～	399
❷ membrane-anchored ～	膜にアンカーされた～	373

❷ **ADAM9 is a membrane-anchored metalloprotease** that is markedly up-regulated in several human carcinomas.（Cancer Res. 2005 65:9312）

訳 ADAM9は，膜にアンカーされたメタロプロテアーゼである

★ another 形 もう1つの，代名 もう1つのもの 用例数 13,449

◆類義語：other

■ 形 another ＋［名］

❶ another member	もう1つのメンバー	193
❷ another protein	もう1つのタンパク質	185
❸ another group	もう1つのグループ	161

■ 代名［前］＋ one another

one anotherは，each otherとほぼ同じ意味である．

★ one another	お互いに	1,699
❶ interact with one another	…は，お互いに相互作用する	97
❷ relative to one another	相互に関連している	54

❶ Here, we show that **CgrA and CgrC interact with one another directly**.（J Bacteriol. 2011 193:6152）

訳 CgrAとCgrCは，直接お互いに相互作用する

★ antagonist ㊔ 拮抗薬　用例数 17,303

antagonist	拮抗薬	10,961
antagonists	拮抗薬	6,342

複数形のantagonistsの割合が，約35%ある可算名詞.

◆反意語：agonist

■ antagonist ＋［前］

	the%	a/an%	ø%			
❶	2	96	2	antagonist of ～	～の拮抗薬	1,926

❶ We conclude that **Ang2 is both an agonist and an antagonist of Tie2**.（Mol Cell Biol. 2009 29: 2011）

訳 Ang2は，Tie2の作動薬と拮抗薬の両方である

★★ apoptosis ㊔ アポトーシス　用例数 47,840

不可算名詞. apoptosis inの用例が多い.

◆類義語：programmed cell death

■ apoptosis ＋［前］

	the%	a/an%	ø%			
❶	0	0	98	apoptosis in ～	～のアポトーシス／～におけるアポトーシス	6,923
				・apoptosis in ～ cells	～細胞のアポトーシス	1,781
❷	9	1	89	apoptosis of ～	～のアポトーシス	2,401
				・apoptosis of ～ cells	～細胞のアポトーシス	1,066

❶ These data identify PDE4 as a family of enzymes whose inhibition **induces apoptosis in CLL cells**.（Blood. 1998 92:2484）

訳 ～は，CLL細胞のアポトーシスを誘導する

❷ They induce low thymidine uptake of allogeneic T cells in MLR **due to extensive apoptosis of activated T cells**.（J Immunol. 2001 166:7042）

訳 活性化されたT細胞の広範なアポトーシスのせいで

■ ［名］-［過分／形］＋ apoptosis

❶	～-induced apoptosis	～に誘導されるアポトーシス	6,258
❷	～-mediated apoptosis	～に仲介されるアポトーシス	1,716
❸	～-dependent apoptosis	～依存的なアポトーシス	634

❶ To examine the mechanism of Fas resistance, **we studied Fas-induced apoptosis in human medial vascular smooth muscle cells (VSMCs)** from healthy coronary arteries.（Circ Res. 2000 86:1038）

訳 われわれは，Fasに誘導されるヒト内側血管平滑筋細胞（VSMC）のアポトーシスを研究した

■［前］＋ apoptosis

❶ induction of apoptosis	アポトーシスの誘導	1,378
❷ inhibitor of apoptosis	アポトーシスの阻害剤	549
❸ resistance to apoptosis	アポトーシスに対する耐性	293
❹ increase in apoptosis	アポトーシスの増大	199
❺ protection from apoptosis	アポトーシスからの保護	127

❶ These results suggest that **there is a partial but significant contribution of JNP to the induction of apoptosis in RGCs by** ON transection.（Invest Ophthalmol Vis Sci. 2002 43:1631）

訳 ～によるRGCのアポトーシスの誘導へのJNPの部分的だが有意な寄与がある

❸ Transformation occurs **after the selection of cells that have acquired resistance to apoptosis** that is triggered by these oncogenes, and a key mediator of this cell death process is the p53 tumor suppressor.（Oncogene. 2001 20:6983）

訳 アポトーシスに対する耐性を獲得した細胞の選択のあと

❺ In addition, we determined that c-Jun and AP1 activities **correlated with EPO-induced proliferation and/or protection from apoptosis**.（Mol Cell Biol. 1998 18:3699）

訳 ～は，EPOに誘導される増殖および/あるいはアポトーシスからの保護と相関した

■［動］＋ apoptosis

❶ induce apoptosis	…は，アポトーシスを誘導する	2,795
❷ undergo apoptosis	…は，アポトーシスを起こす	1,203
❸ inhibit apoptosis	…は，アポトーシスを抑制する	613
❹ block apoptosis	…は，アポトーシスを阻止する	298

❷ When virus is cleared, **the vast majority of these effector CD8 T cells undergo apoptosis**.（J Immunol. 2001 167:1333）

訳 これらのエフェクターCD8 T細胞の大部分は，アポトーシスを起こす

★ apparent ㊎ 明らかな／見かけの　用例数 11,935

◆類義語：evident, clear, obvious, pronounced

■［動］＋ apparent

❶ be apparent	…は，明らかである	1,121
・be apparent in ～	…は，～において明らかである	296
❷ become apparent	…が，明らかになる	337
・become apparent that ～	～ということが，明らかになる	120

❶ **Active angiogenesis was apparent in areas of** myeloma cell infiltration; the new endothelial cells were of human origin.（Blood. 1998 92:2908）

訳 活発な血管新生は，～の領域において明らかであった

❷ **It has become apparent that** chromatin modification plays a critical role in the regulation of cell-type-specific gene expression.（Proc Natl Acad Sci USA. 2004 101:16659）

訳 ～ということが明らかになっている

■ apparent ＋［名］

❶	apparent affinity	見かけの親和性	380
❷	apparent molecular mass	見かけの分子量	262
❸	apparent effect	明らかな影響	198
	・no apparent effect on ～	～に対する明らかな影響のない	149

❷ One abundant **protein with an apparent molecular mass of 50 kDa** was isolated, and the N-terminal sequence was determined.（J Bacteriol. 1999 181:7161）

訳 50キロダルトンの見かけの分子量を持つタンパク質

❸ **Cdk2-wt had no apparent effect on the cell division cycle**, whereas Cdk2-dn inhibited progression through several distinct stages.（Mol Cell Biol. 2001 21:2755）

訳 Cdk2-wtは，細胞分裂周期に対する明らかな影響を持たなかった

apparently 副 見かけ上／明らかに　用例数 4,530

◆類義語：clearly, distinctly, evidently, obviously, unambiguously

■ apparently ＋［形］

❶	apparently normal ～	見かけ上正常な～	190
❷	apparently healthy ～	見かけ上健康な～	144
❸	apparently due to ～	見かけ上～のせいで	112

❶ SOS induction and filamentation commenced **after an apparently normal cell division**, which sheared unresolved dimer chromosomes.（Mol Microbiol. 2000 36:973）

訳 見かけ上正常な細胞分裂のあと

❷ **In this trial of apparently healthy men**, supplementation with beta-carotene for an average of 12 years had no effect on the risk of subsequent type 2 DM.（JAMA. 1999 282:1073）

訳 この見かけ上健康な男性の治験において

★★ appear 動 思われる／見える／現れる　用例数 41,676

appears	思われる／現れる	19,687
appear	思われる／現れる	13,875
appeared	思われた／現れた	7,289
appearing	見える	825

自動詞．appear to *do*の用例が非常に多い．第2文型の自動詞（S+V+C）の用例も多い．

◆類義語：seem, think, believe, consider, regard, suspect, know, assume

■ appear to *do*

★	appear to *do*	…は，～するように思われる	31,009
❶	appear to be ～	…は，～であるように思われる	16,144
	・appear to be involved in ～	…は，～に関与しているように思われる	311

・appear to be associated with ～	…は，～と関連しているように思われる	311
・appear to be important	…は，重要あるように思われる	300
・appear to be due to ～	…は，～のせいであるように思われる	298
・appear to be mediated by ～	…は，～によって仲介されるように思われる	295
・there appears to be ～	～があるように思われる	223
❷ appear to have ～	…は，～を持つように思われる	1,961
❸ appear to play ～	…は，～を果たすように思われる	879
❹ appears to involve ～	…は，～に関わるように思われる	446

❶ Furthermore, **this decrease in AP-1 DNA binding appears to be due to a decrease in** the redox active protein, redox factor (Ref)-1. (J Immunol. 2002 168:5675)

訳 このAP-1のDNA結合の低下は，～の低下のせいであるように思われる

❶ **There appears to be no correlation between** diagnosed diseases (immunological versus idiopathic) and expression of ER in CDG gingiva. (J Periodontol. 2000 71:482)

訳 ～の間に相関はないように思われる

❸ **Mgf appears to play a role in** the survival of some hematopoietic cells in vitro by modulating the activity of p53. (Dev Biol. 1999 215:78)

訳 Mgfは，～において役割を果たすように思われる

■ appear ＋［that節］

★ appear that ～	～であるように思われる	1,480
❶ it appears that ～	～であるように思われる	1,238

❶ **It appears that** physical interactions of other cellular proteins (p53 and par-4) with WT1 can modulate the function of WT1. (J Biol Chem. 1998 273:10880)

訳 ～であるように思われる

■ appear ＋［形］

❶ appear normal	…は，正常であるように思われる	447
❷ appear similar	…は，似ているように思われる	120

❶ In contrast, **inflammation-mediated VP appeared normal in** Src-deficient mice. (Mol Cell. 1999 4:915)

訳 炎症に仲介されるVPは，～において正常であるように思われた

■ appear ＋［前］

❶ appear in ～	…は，～において現れる	1,221
❷ appear as ～	…は，～として現れる	511
❸ appear at ～	…は，～において現れる	365

❶ **The VP5 first appeared in the proximal axons** at 4 days, about 48 h after the appearance of gD. (J Virol. 2003 77:6117)

訳 VP5は，最初，軸索基部において現れた

appearance 名 出現／外見 　　用例数 4,847

appearance	出現／外見	4,757
appearances	出現／外見	90

複数形のappearancesの割合は2％しかなく，原則，**不可算名詞**として使われる．appearance ofの用例が非常に多い．

◆類義語：occurrence, emergence, advent, manifestation

■ appearance ＋［前］

appearance ofの前にtheが付く割合は圧倒的に高い．

	the％	a/an％	ø％			
❶	87	0	0	appearance of ～	～の出現	3,357
❷	10	7	44	appearance in ～	～における出現	148

❶ The mechanism underlying the appearance of these chromosomal alterations is poorly understood. (Cancer Res. 2002 62:2791)

訳 これらの染色体の変化の出現の根底にある機構は，あまり理解されていない

applicable 形 適用できる 　　用例数 6,759

■ applicable ＋［前］

❶ applicable to ～	～に適用できる	1,734
・be applicable to ～	…は，～に適用できる	895
❷ applicable for ～	～に適用できる	168

❶ Although developed using a resequencing microarray, the approach is applicable to any assay method that produces base call sequence information. (Nucleic Acids Res. 2006 34:5300)

訳 そのアプローチは，どのアッセイ法にも適用できる

■［副］＋ applicable

❶ generally applicable	一般的に適用できる	306
❷ broadly applicable	広く適用できる	240
❸ widely applicable	広く適用できる	163

★ application 名 適用／応用 　　用例数 20,035

application	適用／応用	11,749
applications	適用／応用	8,286

複数形のapplicationsが使われる割合は約40%とかなり高い．しかし，単数形は無冠詞で使われることが多い．

◆類義語：utilization, use

A

■ application ＋［前］

the%	a/an%	ø%			
❶ 42	6	51	application of ～	～の適用	7,839
			・application of ～ to …	～の…への適用	311
❷ 5	8	37	application to ～	～への適用	715
❸ 1	4	58	application in ～	～における適用	548
❹ 3	46	32	application for ～	～に対する適用	213

❶ Here we describe the application of chemically induced dimerization of FKBP to create nearly instantaneous high-affinity bivalent ligands capable of sequestering cellular targets from their endogenous partners. (Proc Natl Acad Sci USA. 2010 107:3493)

訳 われわれは，化学的に誘導されるFKBPの二量体形成の適用を述べる

❶ Importantly, bath application of glutamate to SCN slices rapidly and transiently increases PSA levels during both the subjective day and night. (J Neurosci. 2003 23:652)

訳 グルタミン酸のSCN切片へのバス適用は，急速で一過性にPSAレベルを増大させる

❷ A novel concept of two-dimensional fragment correlation mass spectrometry and its application to peptide sequencing is described. (Anal Chem. 2000 72:2337)

訳 ペプチドシークエンシングへのそれの適用が述べられる

■［形／名］＋ application

❶ clinical application	臨床応用	430
❷ topical application	局所適用	288
❸ bath application	バス適用（溶液槽適用）	218

★ apply 動 適用する／あてはまる 用例数 16,721

applied	適用される／～を適用した	12,363
apply	～を適用する／あてはまる	2,169
applying	～を適用する	1,735
applies	～を適用する	454

applied toの用例が非常に多い．他動詞として使われることが多いが，自動詞の用例もある．

◆類義語：use, employ, utilize

■ be applied ＋［前］

★ be applied	…が，適用される	5,258
❶ be applied to ～	…が，～に適用される	3,631
❷ be applied in ～	…が，～において適用される	309
❸ be applied for ～	…が，～に対して適用される	166

❶ This method was applied to the identification of substrates for the protein kinase Akt, which specifically phosphorylates the RXRXXS/T motif. (J Biol Chem. 2002 277:22115)

訳 この方法が，～の同定に適用された

❷ If this approach can **be applied in** acute central nervous system (CNS) infection, the long-term morbidity, which occurs in CNS infection, might be reduced. (Brain Res. 1998 802: 175)

訳 このアプローチは，急性の中枢神経系（CNS）感染において適用できる

❸ Capillary LC external accumulation interface with FTICR **was successfully applied for** the study of whole-proteome mouse tryptic digests. (Anal Chem. 2001 73:253)

訳 …が，～の研究に対してうまく適用された

■ [副] ＋ applied

❶ successfully applied	うまく適用される	225
❷ exogenously applied	外因的に適用される	91

■ [代名] ＋ apply

★ we apply ～	われわれは，～を適用する	1,881
❶ we apply ～ to …	われわれは，～を…に適用する	756

❶ For the first time, **we applied** the two-layer method **to** clinical whole-pancreas transplantation. (Transplantation. 2000 70:771)

訳 われわれは，2層法を～に適用した

■ apply to

❶ apply to ～	～にあてはまる	669

❶ Thus, mechanisms by which RecQ helicases and topoisomerase III proteins cooperate to maintain genomic stability in model organisms likely **apply to** humans. (Cancer Res. 2000 60: 1162)

訳 ～は，ヒトにあてはまりそうである

★★ approach 名 アプローチ／方法，動 接近する　用例数 45,687

approach	アプローチ／接近する	31,178
approaches	アプローチ／接近する	13,263
approaching	接近する	663
approached	接近した	583

動詞の用例もあるが，名詞として使われることが圧倒的に多い．複数形のapproachesの割合が，約30%ある可算名詞．

◈類義語：method, strategy, tool, technology, technique, procedure, way

■ approach ＋ [前]

	the%	a/an%	ø%			
❶	12	81	1	approach to ～	～するためにアプローチを…／～へのアプローチ	7,702
				・approach to ～ing	～するアプローチ	1,128
❷	3	89	1	approach for ～	～のためのアプローチ	3,108

・approach for ～ing	～するのためのアプローチ	1,233

❶ We used a molecular approach to identify a critical region of the X chromosome for neurocognitive aspects of TS. (Am J Hum Genet. 2000 67:672)

訳 われわれは，～の決定的に重要な領域を同定するために分子的アプローチを用いた

❶ Here we describe a novel therapeutic approach to thrombosis treatment. (Blood. 2010 116:4684)

訳 われわれは，血栓症治療への新規の治療的アプローチについて述べる

❷ ER α and ER β agonists are a promising new approach for treating specific conditions associated with menopause. (Curr Opin Pharmacol. 2010 10:629)

訳 ER α およびER β アゴニストは，特異的な状態を治療するための有望な新しいアプローチである

❷ IGRP inhibitors may be an attractive new approach for the treatment of insulin secretion defects in type 2 diabetes. (J Biol Chem. 2004 279:13976)

訳 IGRP抑制剤は，～の治療のための魅力的な新しいアプローチであるかもしれない

■ ［形］＋ approach

❶ new approach	新しいアプローチ	1,035
❷ novel approach	新規のアプローチ	1,017
❸ therapeutic approach	治療上のアプローチ	808
❹ genetic approach	遺伝的アプローチ	429
❺ alternative approach	代替のアプローチ	302
❻ promising approach	有望なアプローチ	288

■ approach ＋［動］

❶ approach provides ～	アプローチは，～を提供する	279
❷ approach allows ～	アプローチは，～を許す	202
❸ approach was used	アプローチが，使われた	290

❶ This approach provides a sensitive assay for V_2 receptor agonist ligands and may be amenable to many other G_s alph-coupled receptors. (Anal Biochem. 2002 300:212)

訳 このアプローチは，～に対する感受性の高いアッセイを提供する

❸ A proteomics approach was used to study DEP-induced responses in the macrophage cell line, RAW 264.7. (J Biol Chem. 2003 278:50781)

訳 プロテオミクスアプローチが，～を研究するために使われた

appropriate 形 適切な／妥当な 用例数 9,204

◆類義語：adequate, proper, reasonable

■ appropriate ＋［前］

❶ appropriate for ～	～のために適切な	566
・be appropriate for ～	…は，～のために適切である	218
❷ appropriate to ～	～に適切な／～するのに適切な	194

❶ The geometry of the hydrogen bonds is well accepted to be appropriate for intramolecular hydrogen-bond formation. (J Am Chem Soc. 2004 126:3488)

訳 ～は，分子内水素結合形成のために適切であるとよく認められている

❷ The purpose of this study was to test if similar bone stimulation could be induced by 2 single-dose drug delivery systems appropriate to periodontal therapy. (J Periodontol. 2002 73:1141)

訳 歯周療法に適切な２つの単一用量薬物送達システム

■ [副] ＋ appropriate

❶ most appropriate	最も適切な	210
❷ more appropriate	より適切な	181

❶ The most appropriate treatment for patients with ischemic mitral regurgitation (IMR) is often debated. (Circulation. 2003 108:II103)

訳 虚血性僧帽弁逆流症（IMR）の患者の最も適切な治療が，しばしば議論される

★★ approximately 副 およそ／おおよそ 用例数 53,399

◆類義語：about, ca., roughly

■ approximately ＋ [形]

❶ approximately 50%	およそ50%	2,679
・approximately 50% of ～	～のおよそ50%	672
❷ approximately 2-fold	およそ2倍	477
❸ approximately equal	おおよそ等しい	449
・approximately equal to ～	～におおよそ等しい	193

❶ H. pylori infects approximately 50% of the world's population, and infections can persist throughout the lifetime of the host. (Infect Immun. 2005 73:803)

訳 ピロリ菌は，世界の人口のおよそ50%に感染する

❸ The length of each peptide is approximately equal to 2 nm, similar to that of biological phospholipids. (Proc Natl Acad Sci USA. 2002 99:5355)

訳 おのおののペプチドの長さは，おおよそ2nmに等しい

■ [前] ＋ approximately

❶ be reduced by approximately ～	…は，およそ～ほど低下する	231
❷ value of approximately ～	およそ～の値	197
❸ rate of approximately ～	およそ～の速度	151

❶ The binding affinity with the -112A oligonucleotide was reduced by approximately one half, as compared with the -112G oligonucleotide. (Am J Hum Genet. 2002 70:718)

訳 ～が，およそ半分低下した

architecture 名 構築／構造 用例数 5,844

architecture	構築／構造	5,144
architectures	構築／構造	700

複数形のarchitecturesの割合が，約10%ある可算名詞.

◆類義語：assembly, structure

■ architecture ＋［前］

architecture ofの前にtheが付く割合は圧倒的に高い.

	the%	a/an%	ø%			
❶	95	1	4	architecture of ～	～の構築	1,390

❶ Determining the genetic architecture of late onset Alzheimer's disease remains an important research objective.（Brain. 2010 133:1155）

訳 遅発性のアルツハイマー病の遺伝的構築を決定すること

★ area 名 領域／面積 用例数 32,837

area	領域／面積	18,784
areas	領域／面積	14,053

複数形のareasの割合は約40%とかなり高い.

◆類義語：region, territory, site, locus

■ area ＋［前］

	the%	a/an%	ø%			
❶	48	39	3	area of ～	～の領域	2,845
❷	23	34	26	area in ～	～における領域	536
				・in the area of ～	～の領域において	322

❶ Through increased knowledge in the area of pharmacogenomics, it is hoped that that treatment of pain will move into the realm of personalized medicine.（Curr Opin Anaesthesiol. 2012 25:444）

訳 薬理ゲノム学の領域における増大した知識を通して

■ area under

★	area under ～	～下面積	1,310
❶	area under the curve	曲線下面積	521

❶ The area under the curve for FU correlated significantly with DLT（P = .006）and grade 3 to 4 diarrhea（P = .004）.（J Clin Oncol. 2005 23:6957）

訳 FUの曲線下面積が，DLTと有意に相関した

■ [名／形] + area

❶ surface area	表面積	1,693
❷ cortical areas	皮質領	743
❸ brain areas	脳領域	667
❹ tegmental area	被蓋野	591

argue 動 主張する／論じる 用例数 3,296

argue	主張する／論じる	1,962
argued	論じられる／論じた／主張した	476
argues	主張する／論じる	465
arguing	主張する／論じる	393

他動詞の用例が多いが，自動詞としても用いられる．argue thatの用例が非常に多い．

◆類義語：claim, discuss, debate

■ argue + [that節]

★ argue that ～	…は，～であると主張する	1,374
❶ we argue that ～	われわれは，～であると主張する	614
❷ results argue that ～	結果は，～であることを示す	149
❸ data argue that ～	データは，～であることを示す	139

❶ We argue that the majority of autosomal genes that function in male fertility in Drosophila are represented by one or more alleles in the ms collection.（Genetics. 2004 167:207）

訳 われわれは，～であると主張する

❷ The results argue that CRD2 is required for gD binding mainly to provide structural support for a gD binding site in CRD1.（J Virol. 2002 76:10894）

訳 それらの結果は，～であることを示す

■ it is argued + [that節]

❶ it is argued that ～	～ということが論じられる	121

❶ It is argued that decentralized democracy may be the best way to improve the chances for successful democracy.（Proc Natl Acad Sci USA. 2011 108:21297）

訳 ～ということが論じられる

■ argue + [前]

❶ argue against ～	…は，～に反対論を唱える	529
・results argue against ～	結果は，～に反対論を唱える	56
❷ argue for ～	…は，～に賛成論を唱える	393

❶ These results argue against recent reports suggesting that mismatched dNTP incorporations follow a conformational path distinctly different from that of matched dNTP incorporation, or that its conformational closing is a major contributor to fidelity.（Nucleic Acids Res. 2008 36:

2948)
訳 これらの結果は，最近の報告に反対論を唱える

★ arise 動 生じる／起こる　用例数 11,380

arise	生じる	4,752
arising	生じる	2,575
arises	生じる	2,225
arose	生じた	1,262
arisen	生じた	566

自動詞．arise fromの用例が非常に多い．
◆類義語：occur, take place, derive, originate, come from, emerge

■ arise ＋［前］

❶ arise from ～	…は，～から生じる	5,646
❷ arise in ～	…は，～において生じる	1,294
❸ arise by ～	…は，～によって生じる	397
❹ arise as ～	…は，～として生じる	332
❺ arise through ～	…は，～を経て生じる	314

❶ Characteristic δ^{13}C values arise from biases in the biosynthetic origins of the C18:0 fatty acids in milk and adipose fat.（Science. 1998 282:1478）
訳 特徴的なδ^{13}C値は，～から生じる

❷ Spinal cord oligodendrocyte precursors arise in the ventral ventricular zone as a result of local signals.（Development. 1999 126:2419）
訳 脊髄乏突起膠細胞前駆体は，～において生じる

■ ［動］＋ to arise

★ to arise	生じる…	610
❶ be thought to arise	…は，生じると考えられる	132
❷ appear to arise	…は，生じるように思われる	105

❶ Within this region, HSCs are thought to arise from hemangioblast precursors located in the ventral wall of the dorsal aorta.（Blood. 2000 96:1591）
訳 造血幹細胞は，血管芽細胞前駆体から生じると考えられる

arrangement 名 配置　用例数 3,141

arrangement	配置	2,409
arrangements	配置	732

複数形のarrangementsの割合が，約25%ある可算名詞．

■ arrangement ＋［前］

arrangement ofの前にtheが付く割合はかなり高い．

	the％	a/an％	ø％			
❶	68	21	6	arrangement of ～	～の配置	1,212

❶ In this article, **the spatial** arrangement of **the ORC subunits** within the ORC structure is described.（Proc Natl Acad Sci USA. 2008 105:10326）
訳 ORCサブユニットの空間的配置

★ array 名 アレイ／整列，動 整列させる　用例数 12,925

array	アレイ／整列	8,054
arrays	アレイ／整列	4,539
arrayed	整列した	315
arraying	整列する	17

動詞としても使われるが，名詞の用例が圧倒的に多い．複数形のarraysの割合が，約35%ある可算名詞．〈an array of ～〉の用例が多い．

◈類義語：series, range

■ array ＋［前］

array ofの前には，a/anが付く割合が非常に高い．

	the％	a/an％	ø％			
❶	8	92	0	array of ～	一連の～	3,020
				・an array of ～	一連の～	975
				・a diverse array of ～	多様な一連の～	345
				・a wide array of ～	広く一連の～	315
				・a broad array of ～	広く一連の～	133

❶ However, a wide array of **phenotypic differences has emerged between** mice and humans carrying biallelic null alleles of JAK3, TYK2, STAT1, or STAT5B.（Immunity. 2012 36:515）
訳 広く一連の表現型上の違いが，～の間に現れた

★ arrest 名 停止，動 停止させる　用例数 14,713

arrest	停止／～を停止させる	11,757
arrested	～を停止させた／停止した	2,114
arrests	停止／～を停止させる	604
arresting	／～を停止させる	238

名詞として使われることが多いが，動詞の用例もある．複数形のarrestsの割合は3％あるが，単数形は無冠詞で使われることも多い．

◈類義語：stop, stall, halt, offset

■ 名 arrest ＋［前］

	the %	a/an %	ø %			
❶	4	14	79	arrest in ～	～における停止	1,334
❷	26	23	48	arrest of ～	～の停止	649

❶ In addition, the Ack1 inhibitor AIM-100 not only inhibited Ack1 activation but also suppressed AKT tyrosine phosphorylation, **leading to cell cycle arrest in the G1 phase**. (Am J Pathol. 2012 180:1386)

訳 G1期における細胞周期停止につながる

■ 名［名／形］＋ arrest

❶	cell cycle arrest	細胞周期停止	2,656
❷	growth arrest	増殖停止	1,901
❸	cardiac arrest	心停止	1,085

■ 動 arrest ＋［名句］

❶	arrest the cell cycle	…は，細胞周期を停止させる	82

❶ **All four compounds arrest the cell cycle at G2/M**, though in addition high concentrations of aphidicolin arrest in G1. (J Virol. 2005 79:5695)

訳 4つの複合体のすべては，G2/M期で細胞周期を停止させる

■ 動 be arrested

★	be arrested	…は，停止する	397
❶	be arrested in ～	…は，～において停止する	114
❷	be arrested at ～	…は，～において停止する	111

❷ Development of the glands in these mice **was arrested at around day 13 of pregnancy**. (Dev Biol. 2001 229:163)

訳 ～は，妊娠のおよそ13日で停止した

■ 動［名］＋ arrested

❶	growth-arrested ～	増殖停止した～	233
❷	cells arrested	停止した細胞	178

❷ Consistent with a role of p202 in cell cycle regulation, **levels of p202 increase in cells arrested in the G0/G1 phase of cell cycle** after withdrawal of serum growth factors. (Oncogene. 2003 22:4775)

訳 p202のレベルは，細胞周期のG0/G1期で停止した細胞において増大する

article 名 記事／論文 用例数 7,988

article	記事／論文	6,247
articles	記事／論文	1,741

複数形のarticlesの割合が，約20%ある可算名詞.
◆類義語：paper, thesis

■ article ＋［動］

❶ article reviews ～	記事は，～を概説する	699
❷ article describes ～	記事は，～を述べる	330
❸ article provides ～	記事は，～を提供する	147
❹ article presents ～	記事は，～を提示する	136
❺ article discusses ～	記事は，～を議論する	134
❻ article summarizes ～	記事は，～を要約する	110

❶ This **article reviews** **examples of viruses and bacteria** known or thought to induce epigenetic changes in host cells, and how this might contribute to disease.（Trends Microbiol. 2010 18: 439）
訳 この記事は，ウイルスと細菌の例を概説する

■［代名／形］＋ article

❶ this article	この記事	5,153
・this article	この記事において	2,128
❷ the present article	現在の記事	114

ascertain 動 確かめる／確認する　用例数 1,878

ascertained	確かめられる／～を確かめた	955
ascertain	～を確かめる	853
ascertaining	～を確かめる	62
ascertains	～を確かめる	8

他動詞.
◆類義語：confirm, verify, validate

■ be ascertained ＋［前］

★ be ascertained	…が，確かめられる	506
❶ be ascertained by ～	…が，～によって確かめられる	110
❷ be ascertained from ～	…が，～から確かめられる	76

❶ **Protein localization** **was ascertained by** **immunolabeling**.（Biochemistry. 2010 49:7023）
訳 タンパク質局在が，免疫標識によって確かめられた

■ to ascertain

★ to ascertain ～	～を確かめる…	799
❶ To ascertain ～	～を確かめるために	201

❶ **To ascertain** **whether** young and old GCs perform distinct memory functions, we created a

transgenic mouse in which output of old GCs was specifically inhibited while leaving a substantial portion of young GCs intact. (Cell. 2012 149:188)
訳 ～かどうかを確かめるために

■ ascertain ＋［名節］

❶	ascertain whether ～	…は，～かどうかを確かめる	217

aspect 名 面／局面 用例数 9,694

aspects	面	7,706
aspect	面	1,988

複数形のaspectsの割合は，約80%と圧倒的に高い．aspects ofの用例が圧倒的に多い．
◆類義語：situation, condition, circumstance, profile, side

■ aspects ＋［前］

	the%	a/an%	ø%			
❶	6	-	55	aspects of ～	～の面	7,197
				・many aspects of ～	～の多くの面	736
				・in many aspects of ～	～の多くの面において	138
				・other aspects of ～	～の他の面	383
				・various aspects of ～	～のさまざまな面	345
				・different aspects of ～	～の異なる面	329

❶ Grass species differ in many aspects of inflorescence architecture, but in most cases the genetic basis of the morphological difference is unknown. (Genetics. 2005 169:1659)
訳 草の種は，花序の構造の多くの面において異なっている

❶ Auxin regulates various aspects of plant growth and development. (Plant Cell. 2005 17:3282)
訳 オーキシンは，植物の成長と発達のさまざまな面を調節する

★★ assay 名 アッセイ／試験，動 アッセイする 用例数 60,003

assay	アッセイ／～をアッセイする	30,283
assays	アッセイ／～をアッセイする	26,243
assayed	アッセイされる／～をアッセイした	3,117
assaying	～をアッセイする	360

名詞の用例が多いが，他動詞としても用いられる．複数形のassaysの割合は約45%とかなり高い．
◆類義語：method, analysis, test, analyze

■ 名 assay(s) ＋［前］

	the%	a/an%	ø%			
❶	19	77	1	assay for ～	～のためのアッセイ	1,787

❷	14	79	3	assay to *do*	～するためのアッセイ	1,467
❸	1	-	95	assays of ～	～のアッセイ	928

❶ We describe here the development of a bacteriophage replication **assay for** the detection of Mycobacterium tuberculosis by using mycobacteriophage D29. (J Clin Microbiol. 2004 42: 2115)
訳 われわれは，ここに～の検出のためのバクテリオファージ複製アッセイの開発について述べる

❷ Here we present a novel **assay to** detect and isolate DNA variants using stabile nanostructures formed directly on duplex DNA. (Genome Res. 2004 14:116)
訳 われわれは，～を検出するための新規のアッセイを示す

❸ Transcriptional clonality **assays of** the engrafted human IM cells demonstrated their clonal origin. (Blood. 2005 105:1699)
訳 移植されたヒトのIM細胞のアッセイは，～を実証した

■ 名 assays ＋［動］

❶	assays showed ～	アッセイは，～を示した	893
❷	assays demonstrated ～	アッセイは，～を実証した	747
❸	assays revealed ～	アッセイは，～を明らかにした	694
❹	assays indicated ～	アッセイは，～を示した	370
❺	assays were performed	アッセイが，行われた	307

❶ In vitro **assays showed** that the N-terminus is necessary for both processing steps. (Nucleic Acids Res. 2003 31:1744)
訳 試験管内アッセイは，～ということを示した

❷ Binding **assays demonstrated** that PEA-15 interfered with the ability of ERK2 to bind to nucleoporins. (J Biol Chem. 2004 279:12840)
訳 結合アッセイは，～ということを実証した

■ 名［形／名］＋ assays

❶	binding assays	結合アッセイ	1,392
❷	in vitro assays	試験管内アッセイ	727
❸	transfection assays	トランスフェクションアッセイ	726
❹	functional assays	機能的アッセイ	585
❺	reporter assay	レポーターアッセイ	558

■ 動 be assayed ＋［前］

★	be assayed	…は，アッセイされる	1,442
❶	be assayed for ～	…は，～に関してアッセイされる	456
❷	be assayed by ～	…は，～によってアッセイされる	311
❸	be assayed in ～	…は，～においてアッセイされる	211

❶ Forms of Fj that are constitutively secreted or anchored in the Golgi **were assayed for** function in vivo. (Development. 2004 131:881)
訳 ～が，機能に関してアッセイされた

❷ Secreted proMMP-3 was quantified by ELISA, and **MMP-3 activity** was assayed by **casein zymography**. (Invest Ophthalmol Vis Sci. 2003 44:3494)
訳 MMP-3活性が，カゼインザイモグラフィによってアッセイされた

❸ The transcript levels of seven Polycomb group genes were assayed in **embryos** mutant for various other genes in the family. (Mol Cell Biol. 2004 24:7737)
訳 ～が，胚においてアッセイされた

assemble 動 構築する／組み立てる／集合する 用例数 9,491

assembled	構築される／～を構築した／集合した	4,959
assemble	～を構築する／集合する	2,841
assembles	～を構築する／集合する	922
assembling	～を構築する／集合する	769

他動詞・自動詞の両方の用例がある．
◆類義語：organize, build, form, construct, aggregate

■ assemble ＋［前］

❶ assemble into ～	…は，集合して～を構築する	710
❷ assemble in ～	…は，～において集合する	172

❶ However, it is not known how these proteins interact and assemble into **a complex**. (J Biol Chem. 2001 276:25903)
訳 ～は，集合して複合体を構築する

■ be assembled ＋［前］

★ be assembled	…は，構築される	1,474
❶ be assembled into ～	…は，～に構築される	236
❷ be assembled from ～	…は，～から構築される	176
❸ be assembled in ～	…は，～において構築される	164
❹ be assembled by ～	…は，～によって構築される	122

❶ **The two genes** were assembled into **a mini-operon** that was induced to give high level expression of both enzymes. (J Biol Chem. 2001 276:29864)
訳 2つの遺伝子が，ミニオペロンに構築された

❷ **This structure** is assembled from **a host of proteins**, including a variety of disease gene products. (Curr Biol. 2010 20:R816)
訳 この構造は，多数のタンパク質から構築される

❸ In contrast, **the COX I and COX II subunits** were assembled in **cells with** 97% mutated mtDNA. (J Biol Chem. 2000 275:13994)
訳 COX I およびCOX II サブユニットが，～を持つ細胞において構築された

■ self-assemble

★ self-assemble	自己集合する	1,430
❶ self-assemble into ～	自己集合して～を構築する	270

❶ VacA monomers self-assemble into water-soluble oligomeric structures and can form anion-selective membrane channels. (J Biol Chem. 2004 279:2324)
訳 VacA単量体は，自己集合して～を構築する

★ assembly 名 構築／集合 用例数 26,487

assembly	構築／集合	24,823
assemblies	構築／集合	1,664

複数形のassembliesの割合は約5％あるが，単数形は無冠詞で使われることが非常に多い．

◆類義語：formation, construction, aggregation ◆反意語：disassembly

■ assembly ＋［前］

	the％	a/an％	ø％			
❶	53	3	43	assembly of ～	～の構築	7,213
				・required for the assembly of ～	～の構築のために必要とされる	112
❷	2	2	79	assembly in ～	～における構築	896
❸	2	0	56	assembly into ～	～への構築	328

❶ We found that formins were required for the assembly of one of the three budding yeast actin structures: polarized arrays of actin cables. (Nat Cell Biol. 2002 4:42)
訳 …が，～の構築のために必要とされた

❷ Bop1 is a conserved nucleolar protein involved in rRNA processing and ribosome assembly in eukaryotes. (J Biol Chem. 2002 277:29617)
訳 真核生物におけるリボソームの構築

❸ The important anticancer drug Taxol (paclitaxel) binds to tubulin in a stoichiometric ratio and promotes its assembly into microtubules. (Proc Natl Acad Sci USA. 2004 101:10006)
訳 ～は，それの微小管への構築を促進する

■ self/［名］＋ assembly

❶	self-assembly	自己組織化	1,412
❷	spindle assembly	紡錘体構築	623
❸	complex assembly	複合体構築	511

■ assembly and ＋［名］

❶	assembly and disassembly	集合と解体	206
❷	assembly and function	構築と機能	198

❶ Cycles of actin assembly and disassembly could occur repeatedly on the same phagosome. (Mol Biol Cell. 2004 15:5647)
訳 アクチンの集合と解体のサイクル

★★ assess ㊍ 評価する／査定する 用例数 37,219

assessed	評価される／～を評価した	21,220
assess	～を評価する	13,012
assessing	～を評価する	2,717
assesses	～を評価する	270

他動詞.

◆類義語：evaluate, estimate, value

■ assess ＋［名句／whether節］

❶	assess the effect of ～	…は，～の影響を評価する	1,144
❷	assess the role of ～	…は，～の役割を評価する	806
❸	assess the impact of ～	…は，～の影響を評価する	349
❹	assess the contribution of ～	…は，～の寄与を評価する	250
❺	assess the ability of ～	…は，～の能力を評価する	222
❻	assess the efficacy of ～	…は，～の有効性を評価する	179
❼	assess the association between ～	…は，～の間の関連を評価する	169
❽	assess the relationship between ～	…は，～の間の関連性を評価する	160
❾	assess whether ～	…は，～かどうかを評価する	1,185

❶ We assessed the effect of stunting, diarrhoeal disease, and parasitic infections during infancy on cognitive function in late childhood.（Lancet. 2002 359:564）

訳 われわれは，～の影響を評価した

❷ This study assessed the role of OPN in the K/BxN serum-transfer model of autoantibody-induced arthritis.（Arthritis Rheum. 2004 50:2685）

訳 この研究は，～におけるOPNの役割を評価した

❾ We aimed to assess whether tooth loss is associated with specific CVD mortality endpoints in a national population sample adjusting for potential confounders.（PLoS One. 2012 7:e30797）

訳 われわれは，～かどうかを評価することを目的とした

■［代名／名］＋ assess

❶	we assessed ～	われわれは，～を評価した	2,723
❷	study assessed ～	研究は，～を評価した	357
❸	the authors assessed ～	著者らは，～を評価した	193

■ to assess

★	to assess ～	～を評価する…	11,362
❶	be used to assess ～	…は，～を評価するために使われる	1,468
❷	be to assess ～	…は，～を評価することである	883
❸	aim to assess ～	…は，～を評価することを目的とする	265
❹	be performed to assess ～	…は，～を評価するために実行される	238
❺	To assess ～	～を評価するために	2,045

❶ Multiple logistic regression was used to assess the use of recommended therapies. (J Clin Oncol. 2004 22:3261)

訳 多重ロジスティック回帰分析が，～を評価するために使われた

❷ The objective was to assess the effect of smoking on the biotin status of women. (Am J Clin Nutr. 2004 80:932)

訳 目的は，～の影響を評価することであった

■ be assessed + [前]/using

★ be assessed	…が，評価される	11,054
❶ be assessed by ～	…が，～によって評価される	3,276
・be assessed by using ～	…が，～を使うことによって評価される	265
・be assessed by measuring ～	…が，～を測定することによって評価される	183
❷ be assessed in ～	…が，～において評価される	1,594
❸ be assessed using ～	…が，～を使って評価される	937
❹ be assessed with ～	…が，～を使って評価される	664
❺ be assessed for ～	…が，～に関して評価される	611
❻ be assessed at ～	…が，～において評価される	458

❶ Responsiveness to an acute stressor was assessed by measuring salivary cortisol levels before and after venipuncture. (Arthritis Rheum. 2003 48:2923)

訳 ～は，唾液のコルチゾールレベルを測定することによって評価された

❷ Expression of pERK, pAkt, and Ki-67 was assessed in archived tumor specimens by quantitative immunohistochemistry. (J Clin Oncol. 2004 22:4456)

訳 pERK，pAktおよびKi-67の発現が，～において評価された

❸ Classification criteria were assessed using immunohistochemical criteria. (J Neurosci. 2003 23:11178)

訳 分類の基準が，～を使って評価された

❹ The patients were assessed with a battery of neurocognitive tests at baseline and 12 weeks after beginning treatment. (Am J Psychiatry. 2004 161:985)

訳 患者が，～を使って評価された

■ [名] + be assessed

❶ function was assessed	機能が，評価された	237
❷ expression was assessed	発現が，評価された	133
❸ activity was assessed	活性が，評価された	122

❶ Recipient kidney function was assessed by calculating the estimated glomerular filtration rate (eGFR) using the Chronic Kidney Disease Epidemiology Collaboration formula. (Transplantation. 2012 94:1124)

訳 移植患者の腎機能が，～を算出することによって評価された

★assessment 名 評価　　用例数 10,644

assessment	評価	8,896
assessments	評価	1,748

複数形のassessmentsの割合は約15%あるが，単数形は無冠詞の用例が非常に多い．
◆類義語：evaluation, estimation

■ assessment ＋［前］

	the%	a/an%	ø%			
❶	27	12	60	assessment of ～	～の評価	4,702
❷	7	5	86	assessment in ～	～における評価	162

❶ New technologies for nondestructive quantitative assessment of human articular cartilage degeneration may facilitate the development of strategies to delay or prevent the onset of OA.（Arthritis Rheum. 2010 62:1412）
訳 ヒトの関節軟骨変性の非破壊定量的評価のための新しいテクノロジー

■［形／名］＋ assessment

❶	risk assessment	リスク評価	490
❷	clinical assessment	臨床評価	210
❸	quantitative assessment	定量的評価	186

assist 動 助ける／補助する，名 援助　用例数 5,190

assisted	助けられる／助けた	3,040
assist	助ける／援助	1,710
assists	助ける／援助	249
assisting	助ける	191

自動詞の用例が多いが，他動詞や名詞としても用いられる．
◆類義語：support, aid, help

■ assist in

★	assist in ～	…は，～を助ける	891
❶	assist in ～ing	…は，～するのを助ける	220
❷	to assist in ～	～を助ける…	199

❷ The formula and look-up tables based on the formula, can be used to assist in the design of microarray experiments.（Bioinformatics. 2004 20:2821）
訳 ～は，マイクロアレイ実験の設計を助けるために使われうる

★★ associate 動 関連する／結合する　用例数 150,641

associated	関連する／相関する／付随する／結合する	141,693
associate	結合する	4,279
associates	結合する	4,086
associating	結合する	583

他動詞の用例が多いが，自動詞としても使われる．他動詞の場合は，通常，受動態で用

いられる．associated withの用例が断然多い．

◆類義語：relate, correlate, couple, link, connect, concern, engage, bind, attach, accompany, follow

■ ［名］＋ be associated with

★ be associated with ～	…は，～と関連する	51,412
❶ expression was associated with ～	発現は，～と関連した	229
❷ therapy was associated with ～	治療は，～と関連した	204
❸ levels were associated with ～	レベルは，～と関連した	187
❹ infection is associated with ～	感染は，～と関連する	185
❺ activity is associated with ～	活性は，～と関連する	175

❺ These data indicate that even light-to-moderate activity is associated with lower CHD rates in women.（JAMA. 2001 285:1447）

訳 ～活性は，より低いCHD率と関連する

■ be associated with ＋［過分／形／名］

❶ be associated with increased ～	…は，増大した～と関連する	3,198
❷ be associated with reduced ～	…は，低下した～と関連する	1,007
❸ be associated with decreased ～	…は，低下した～と関連する	907
❹ be associated with lower ～	…は，より低い～と関連する	757
❺ be associated with higher ～	…は，より高い～と関連する	753
❻ be associated with an increase in ～	…は，～の増大と関連する	521
❼ be associated with the development of ～	…は，～の発症と関連する	329

❶ A variant of the XRCC3 gene, which is involved in DSB repair, has been associated with increased risk of malignant skin melanoma and bladder cancer.（Oncogene. 2002 21:4176）

訳 ～は，悪性皮膚黒色腫の増大したリスクと関連している

❻ Suppression of viremia was not associated with an increase in T cell proliferative responses.（J Infect Dis. 2000 181:1249）

訳 ウイルス血症の抑制は，T細胞の増殖応答の増大と関連しなかった

❼ These findings strongly suggest that loss of imprinting or switching of allelic expression of the p73 gene is associated with the development of renal cell carcinoma.（Oncogene. 1998 17:1739）

訳 ～は，腎細胞癌の発症と関連する

■ ［副］＋ associated with

★ associated with ～	～と関連する	101,812
❶ significantly associated with ～	有意に～と関連する	2,417
❷ strongly associated with ～	強く～と関連する	1,368
❸ independently associated with ～	独立に～と関連する	1,234
❹ inversely associated with ～	逆に～と相関する	635
❺ positively associated with ～	正に～と相関する	601

❶ The CR rate was significantly associated with higher levels of asparaginase（P =.012）.

(Blood. 2000 96:1709)

訳 CR比は，より高いレベルのアスパラギナーゼと有意に関連した

❷ Decreasing renal function was **strongly associated with** adverse outcome, increasing the risk for ischemic complications at all time points examined (all P < 0.002). (Circulation. 2002 105: 2361)

訳 低下している腎機能が，有害な結果と強く関連した

❸ Lymphatic invasion, tumor size, and age were **independently associated with** lymph node metastases. (Ann Surg. 1999 230:692)

訳 リンパ球浸潤，腫瘍サイズおよび年齢が，リンパ節転移と独立に関連した

■ ［名］＋ associated with

❶	factors associated with ～	～と関連する因子	1,194
❷	genes associated with ～	～と関連する遺伝子	848
❸	changes associated with ～	～と関連する変化	661

❶ Risk **factors associated with** rehospitalization were examined. (Am J Psychiatry. 1999 156: 863)

訳 ～と関連する危険因子

■ associate with

★	associate with ～	～と結合する	2,955
❶	to associate with ～	～と結合する…	903

❶ Hsp70 was found **to associate with** Mos ectopically expressed in COS-1 cells. (Oncogene. 1999 18:3461)

訳 Hsp70は，異所性発現したMosと結合することが見つけられた

★★ association ㊔ 関連／結合／会合　用例数 47,713

association	関連／結合／会合	39,455
associations	関連／結合／会合	8,258

複数形のassociationsの割合が，約15%ある可算名詞．

◆類義語：relation, relationship, relevance, link, connection, correlation, implication, binding

■ association ＋［前］

	the%	a/an%	ø%			
❶	55	11	29	association of ～	～の関連（結合）	8,683
				・association of ～ with …	～の…との関連（結合）	3,683
❷	8	13	58	association with ～	～との結合（関連）	8,380
				・in association with ～	～と関連して	1,938
				・its association with ～	～とそれの結合（関連）	985
❸	51	31	11	association between ～	～の間の関連（結合）	7,092

❶ These results provide a framework for future analysis of possible **association of human hereditary disorders with mutations** in LHX5. (Gene. 2000 260:95)
訳 ヒト遺伝性疾患の変異との関連

❷ Despite the subtype-specific immunogenicity of three of the four epitopes, **all four epitope peptides were found in association with** each of the three different HLA-B27 subtypes. (J Immunol. 2000 164:6120)
訳 4つのエピトープペプチドのすべてが，～と関連して見つけられた

❷ Thus, CARM1 functions as a secondary coactivator **through its association with p160 coactivators.** (Science. 1999 284:2174)
訳 p160コアクチベーターとそれの結合によって

❸ **These findings support an association between** obstetric complications and increased risk for early-onset schizophrenia. (Am J Psychiatry. 2000 157:801)
訳 これらの知見は，～の間の関連を支持する

■ association was ＋［過分］

❶	association was found	関連が，見つけられた	239
❷	association was observed	関連が，観察された	187

❶ **No significant association was found between** dietary vitamin K intake and BMD in men. (Am J Clin Nutr. 2003 77:512)
訳 ～の間に有意な関連は見つけられなかった

■［形］＋ association

❶	significant association	有意な関連	946
❷	no association	関連のない	835
❸	self-association	自己会合	827
❹	inverse association	逆相関	451
❺	strong association	強い関連（結合）	426

❷ **There was no association between IGF-I concentrations and breast-cancer risk** among the whole study group. (Lancet. 1998 351:1393)
訳 IGF-I濃度と乳癌リスクとの間に関連はなかった

■ association and ＋［名］

❶	association and dissociation	結合と解離	268

❶ We conclude that **monomeric P-selectin binds to PSGL-1 with fast association and dissociation rates** and relatively high affinity. (J Biol Chem. 1998 273:32506)
訳 単量体のP-セレクチンは，PSGL-1に速い結合と解離の速度で結合する

assume 動 推定する／仮定する　用例数 4,999

assumed	推定される／～を推定した	2,439
assuming	～を推定する	995
assume	～を推定する	982
assumes	～を推定する	583

他動詞. 〈be assumed to *do*〉〈be assumed that〉〈assume that〉の用例が多い.

◆類義語：presume, estimate, speculate, predict, expect, infer, postulate, deduce

■ be assumed to *do*

★	be assumed to *do*	…は，～すると推定される	581
❶	be assumed to be ～	…は，～であると推定される	309

❶ Affinity is often thought to be determined by the ligand unbinding rate, whereas **the binding rate is assumed to be diffusion-limited.** (J Neurosci. 1998 18:8590)

訳 結合速度が，拡散律速であると推定される

■ be assumed ＋ [that節]

★	be assumed that ～	～ということが推定される	317
❶	it has been assumed that ～	～ということが推定されてきた	114
❷	it is assumed that ～	～ということが推定される	93

❶ **It has been assumed that** all G_i-coupled receptors trigger the protective action of preconditioning by means of an identical intracellular signaling pathway. (Circ Res. 2001 89:273)

訳 ～ということが推定されてきた

■ assume ＋ [that節]

★	assume that ～	…は，～ということを推定する	310
❶	we assume that ～	われわれは，～ということを推定する	89

❶ **We assume that** latent infection in vaccinated individuals is completely undetectable. (J Theor Biol. 2009 261:548)

訳 われわれは，～ということを推定する

assumption 名 仮定／推定 用例数 2,855

assumption	仮定／推定	1,566
assumptions	仮定／推定	1,289

複数形のassumptionsの割合は約45%とかなり高い.

◆類義語：hypothesis, concept, notion, idea, view

■ assumption(s) ＋ [同格that節／前]

assumption thatの前にtheが付く割合は圧倒的に高い. assumption ofの前にtheが付く割合も非常に高い.

	the%	a/an%	ø%			
❶	90	8	0	assumption that ～	～という仮定	642
❷	74	20	5	assumption of ～	～の仮定	290
❸	3	-	82	assumptions about ～	～についての推定	224

❶ We present a method based on the assumption that proteins that function together in a pathway or structural complex are likely to evolve in a correlated fashion.（Proc Natl Acad Sci USA. 1999 96:4285）

訳 われわれは，～という仮定に基づく方法を提示する

❷ Arguments about natural selection often make use of the assumption of common ancestry, whereas arguments for common ancestry do not require the assumption that natural selection has been at work.（Proc Natl Acad Sci USA. 2009 106 S1:10048）

訳 自然淘汰に関する議論は，しばしば共通の祖先の仮定を利用する

❸ However, it is uncertain whether vaccine administration should be based on the current assumptions about the common mucosal immune system.（Infect Immun. 2001 69:1832）

訳 ワクチン投与は，～についての現在の推定に基づいているべきである

attach　動 結合させる／付着する　用例数 5,510

attached	結合している／～を結合させた／付着した	4,330
attach	～を結合させる／結合する／付着する	577
attaching	～を結合させる／結合する／付着する	409
attaches	～を結合させる／結合する／付着する	194

他動詞として用いられることが多いが，自動詞の用例もある．attach to／be attached toの用例が非常に多い．

◆類義語：adhere, bind, associate, anchor

■ be attached ＋［前］

★ be attached	…は，結合（付着）している	749
❶ be attached to ～	…は，～に結合（付着）している	545
・be covalently attached to ～	…は，～に共有結合している	135

❶ Both molecules were attached to the surfaces of different polystyrene microspheres trapped by optical tweezers.（Biophys J. 2002 83:1965）

訳 両方の分子が，異なるポリスチレン微粒子の表面に付着していた

❶ The resulting threonine carboxyl end of this protein is covalently attached to a pentaglycine cross-bridge of peptidoglycan.（Biochemistry. 2003 42:11307）

訳 このタンパク質のスレオニンカルボキシル末端は，～に共有結合している

■ attach to

★ attach to ～	…は，～に結合（付着）する	437
❶ to attach to ～	～に結合（付着）する…	90

❶ Stable cell lines were generated and tested for their ability to attach to α5β1-selective ligands.（J Biol Chem. 2000 275:20324）

訳 安定細胞株が作製され，α5β1選択的リガンドに結合するそれらの能力についてテストされた

attachment ㊔ 付着／結合　用例数 7,418

attachment	付着／結合	7,009
attachments	付着／結合	409

複数形のattachmentsの割合は約5％あるが，単数形は無冠詞で使われることが圧倒的に多い．

◆類義語：adhesion, binding

■ attachment ＋［前］

	the％	a/an％	ø％			
❶	53	2	45	attachment of ～	～の付着	1,495
				・attachment of ～ to …	～の…への付着	563
❷	0	2	84	attachment to ～	～への付着	996

❶ However, **attachment of live norovirus strains to** histo-blood group antigens has not been investigated to date.（J Virol. 2004 78:3035）

訳 生きたノロウイルス株の～への付着

❷ **Attachment to the pole**, however, is required to cause net kinetochore fibre shortening to generate polewards forces during anaphase.（Nat Cell Biol. 2004 6:227）

訳 極への付着

■ ［形／名］＋ attachment

❶	cell attachment	細胞付着	425
❷	covalent attachment	共有結合	334

attempt ㊔ 試み，㊍ 試みる　用例数 6,153

attempt	試み／試みる	2,262
attempts	試み／試みる	2,188
attempted	試みた／試みられた	1,386
attempting	試みる	317

名詞および動詞として用いられる．attempt to *do*の用例が非常に多い．複数形のattemptsの割合は，約60％と非常に高い．

◆類義語：trial, effort, purpose, aim, seek, undertake, address

■ ㊔ attempt to *do*

★	attempt to *do*	～しようとする試み	1,827
❶	in an attempt to *do*	～しようとして	1,054
	・in an attempt to identify ～	～を同定しようとして	139
	・in an attempt to understand ～	～を理解しようとして	64

❶ **In an attempt to identify upstream mediators**, we investigated Cav-2 tyrosine phosphorylation in an endogenous setting.（Biochemistry. 2004 43:13694）

訳 上流のメディエーターを同定しようとして

■ 名 [名／形] + attempts

❶ previous attempts to *do*	～しようとする以前の試み	327
❷ suicide attempts	自殺企図	194

❷ **Data on prior suicide attempts were obtained retrospectively from** interviews with the NIMH-Life-Chart method. (Am J Psychiatry. 2002 159:1160)

訳 以前の自殺企図に関するデータは，～から遡及的に得られた

■ 動 [代名] + attempt to *do*

❶ we attempted to *do*	われわれは，～しようと試みた	300

❶ In this study **we attempted to determine the important elements** within polyarginines that contribute to effective inhibition. (J Biol Chem. 2004 279:36788)

訳 われわれは，重要なエレメントを決定しようと試みた

attention 名 注意 用例数 6,803

attention	注意	6,801
attentions	注意	2

複数形のattentionsの用例はほとんどなく，原則，**不可算名詞**として使われる.

◆類義語：care

■ attention + [前]

	the%	a/an%	ø%			
❶	1	0	96	attention to ～	～への注意	977
				・draw attention to ～	…は，～への注意を促す	95
				・particular attention to ～	～への特別の注意	82
				・pay … attention to ～	‥は，～へ…の注意を払う	64
❷	1	0	84	attention in ～	～における注意	307
❸	1	0	87	attention on ～	～に対する注意	267
				・focus attention on ～	…は，～に対する注意を集める	123

❶ **Several recent studies have drawn attention to the importance of** Alx homeobox transcription factors during craniofacial development. (Hum Mol Genet. 2013 22:239)

訳 いくつかの最近の研究は，～の重要性への注意を促してきた

❶ In this review, we summarize the results of recent clinical trials of helminthic therapy, **with particular attention to mechanisms of action.** (Trends Parasitol. 2012 28:187)

訳 作用機序への特別の注意をもって

❸ Taken together, **these results focus attention on integral membrane lipid phosphatases** as regulators of isoprenoid phosphate metabolism and suggest that PDP1/PPAPDC2 is a functional isoprenoid diphosphate phosphatase. (J Biol Chem. 2010 285:13918)

訳 これらの結果は，膜内在性脂質ホスファターゼに対する注意を集める

■ [動] + ([形]) + attention

❶ have received ～ attention	…は，～な注意を受けてきた	383
・have received little attention	…は，ほとんど注意を受けてこなかった	112
❷ have attracted ～ attention	…は，～な注意を引きつけてきた	99

❶ However, **the roles of the other Cblns have received little attention.** (J Comp Neurol. 2011 519: 2225)

訳 他のCblnsの役割は，ほとんど注意を受けてこなかった

■ attention + [動]

❶ attention has been paid to ～	～へ注意が払われてきた	112
・little attention has been paid to ～	～へはほとんど注意が払われてこなかった	53

❶ However, **little attention has been paid to the cross-talk between** these two seemingly orthogonal signaling pathways. (Proc Natl Acad Sci USA. 2012 109:9095)

訳 ～の間のクロストークへはほとんど注意が払われてこなかった

■ [形] + attention

❶ little attention	ほとんど注意のない	298
❷ spatial attention	空間的注意	246
❸ much attention	たくさんの注意	227
❹ visual attention	視覚的注意	205

★ attenuate 動 減弱させる／低下させる／弱める 用例数 15,244

attenuated	～を減弱させた／減弱する	11,020
attenuates	～を減弱させる	1,828
attenuate	～を減弱させる	1,604
attenuating	～を減弱させる	792

他動詞.

◈類義語：weaken, relieve, alleviate, interfere, inhibit, decrease, reduce, diminish

■ be attenuated + [前]

★ be attenuated	…は，減弱される	2,394
❶ be attenuated by ～	…は，～によって減弱される	847
❷ be attenuated in ～	…は，～において減弱する	642

❶ **This effect was attenuated by inhibition of autophagy.** (Diabetes. 2013 62:1270)

訳 この効果は，オートファジーの阻害によって減弱された

❷ **The induction of all four mRNAs was greatly attenuated in peritoneal macrophages** derived from LXR α/β null mice. (J Biol Chem. 2002 277:31900)

訳 4つすべてのメッセンジャーRNAの誘導が，腹腔マクロファージにおいて大きく減弱した

■［副］＋ attenuated

❶	significantly attenuated ～	有意に減弱する／～を有意に減弱させた	1,059
❷	markedly attenuated ～	顕著に減弱する／～を顕著に減弱させた	341
❸	highly attenuated ～	大きく減弱する／～を大きく減弱させた	179

❶ These differences were significantly attenuated after HDL clustering was promoted using antibody against apolipoprotein A-I.（J Biol Chem. 2003 278:34331）

訳 これらの違いは，～のあと有意に減弱した

❷ Inhibition of PKC δ activation by rottlerin also markedly attenuated IL-13-induced Stat3 DNA binding activity.（J Biol Chem. 2004 279:15954）

訳 ～は，また，IL-13に誘導されるStat3のDNA結合活性を顕著に減弱させた

■ attenuate ＋［名句］

❶	attenuate the effect of ～	…は，～の効果を減弱させる	105
❷	attenuate the increase in ～	…は，～の増大を減弱させる	71
❸	attenuate the ability of ～	…は，～の能力を減弱させる	57
❹	attenuate the development of ～	…は，～の発生を減弱させる	53

❶ Mutation of LKB1 lysine residue 97 reduced HNE adduct formation and attenuated the effect of HNE on LKB1 activity.（J Biol Chem. 2012 287:42400）

訳 ～は，LKB1活性に対するHNEの効果を減弱させた

attenuation ㊔ 減弱／減衰　用例数 3,923

attenuation	減弱	3,902
attenuations	減弱	21

複数形のattenuationsの割合は0.5%しかなく，不可算名詞として使われることが多い．

◆類義語：decrease, decline, loss, reduction

■ attenuation ＋［前］

	the %	a/an %	ø %			
❶	26	8	64	attenuation of ～	～の減弱	1,851
❷	14	14	61	attenuation in ～	～における減弱	225

❶ Expression of WT-Hsp20 resulted in significant attenuation of apoptosis compared with the GFP control.（J Biol Chem. 2008 283:33465）

訳 WT-Hsp20の発現は，アポトーシスの顕著な減弱という結果になった

attributable ㊊ 起因する／起因しうる　用例数 3,544

attributable toの用例が圧倒的に多い．

A

■ attributable ＋［前］

❶ attributable to ～	～に起因する	3,016
・be attributable to ～	…は，～に起因する	1,286

❶ These defects are attributable to enhanced sensitivity to background signals, prolonged chemoattractant receptor signaling, and inappropriate CXCR2 downregulation. (Mol Cell Biol. 2012 32:4561)

訳 これらの欠損は，～への増強された感受性に起因する

attribute 動 起因すると考える／帰する 用例数 7,171

attributed	起因すると考えられる／起因すると考えた	5,402
attributes	起因すると考える	1,087
attribute	起因すると考える	621
attributing	起因すると考える	61

他動詞．attributed ～ to …（～を…に起因すると考える）という構文で用いられる．受動態は，be attributed toの用例が圧倒的に多い．

◈類義語：ascribe, result from, derive

■ be attributed

★ be attributed	…は，起因すると考えられる	3,917
❶ be attributed to ～	…は，～に起因すると考えられる	3,557

❶ A substantial proportion of endothelial dysfunction can be attributed to an effect of the abnormal lipid profile seen in such patients. (Circulation. 2002 106:3037)

訳 …は，～の影響に起因すると考えられうる

■ attribute ＋［名］＋ to

★ attribute ～ to …	～を…に起因すると考える	254
❶ we attribute ～ to …	われわれは，～を…に起因すると考える	161

❶ In this study, we attribute the enhancing effect of PT to the cell-binding B subunit (PT-B). (J Immunol. 2003 171:2314)

訳 われわれは，PTの増強する効果を細胞結合Bサブユニットに起因すると考える

augment 動 増大させる／増強する 用例数 6,218

augmented	増大される／増大する／～を増大した	3,219
augment	～を増大させる	1,458
augments	～を増大させる	946
augmenting	～を増大させる	595

他動詞．

◈類義語：enhance, potentiate, increase, up-regulate, activate

■ be augmented ＋［前］

★ be augmented	…は，増大される／増大する	653
❶ be augmented by ～	…は，～によって増大される	324
❷ be augmented in ～	…は，～において増大する	105

❶ Furthermore, p53-mediated activation of the BK2 promoter **is augmented by** the transcriptional co-activators, CBP/p300.（J Biol Chem. 2000 275:15557）
訳 ～は，転写のコアクチベーターCBP/p300によって増大される

❷ Killing of the latter **was** greatly **augmented in** the presence of αGalCer.（J Immunol. 2001 167:3114）
訳 後者の死滅は，～の存在下で大きく増大した

■ augment ＋［名句］

❶ augment the effect of ～	…は，～の効果を増大させる	40
❷ augment the ability of ～	…は，～の能力を増大させる	30

❷ This motif greatly **augments the ability of** TCF binding sites to respond to Wg signaling.（Curr Biol. 2008 18:1877）
訳 このモチーフは，Wgシグナル伝達に応答するTCF結合部位の能力を大きく増大させる

availability 名 利用できること 用例数 5,200

availability	利用できること	5,195
availabilities	利用できること	5

複数形のavailabilitiesの割合は0.1%しかなく，原則，**不可算名詞**として使われる．
◆類義語：usefulness, effectiveness

■ availability ＋［前］

availability ofの前にtheが付く割合は圧倒的に高い．

	the%	a/an%	ø%			
❶	84	0	16	availability of ～	～が利用できること	2,919

❶ Protein synthesis in reticulocytes depends on the **availability of** heme.（Blood. 2000 96:3241）
訳 網状赤血球におけるタンパク質合成は，ヘムが利用できることに依存する

★ available 形 利用できる／入手可能な／役に立つ 用例数 21,090

◆類義語：useful, accessible

■ be available ＋［前］

★ be available	…は，利用できる	6,613
❶ be available for ～	…は，～について利用できる	1,313

A

❷ be available at ～	…は，～において利用できる	939
❸ be available on ～	…は，～に関して利用できる	471
❹ be available to *do*	…は，～するために利用できる	447
❺ be available from ～	…は，～から入手できる	422
❻ be available in ～	…は，～において利用できる	357

❶ Complete follow-up data were available for all patients.（Hepatology. 2002 35:1501）
訳 完全な経過観察のデータが，すべての患者について利用できた

❸ However, little information is available on the acute response in the retina to detachment.（Invest Ophthalmol Vis Sci. 2000 41:2779）
訳 ～における急性応答に関して利用できる情報はほとんどない

❹ Several different sets of reference body mass index（BMI）values are available to define overweight in children.（Am J Clin Nutr. 2001 73:1086）
訳 ～が，子供の過体重を定義するために利用できる

❺ The predicted gene pairs are available from our World Wide Web site http://www.tigr.org/tigr-scripts/operons/operons.cgi.（Nucleic Acids Res. 2001 29:1216）
訳 予想される遺伝子対は，われわれのワールドワイドウェブサイトから入手できる

❻ Late follow-up（average 91 months）was available in 145 patients（86%）.（Ann Surg. 2001 234:454）
訳 ～が，145人の患者において利用できた

■［副］＋ available

❶ commercially available	市販の	1,170
❷ currently available	現在利用できる	1,080
❸ freely available	自由に利用できる	786
❹ publicly available	公的に利用できる	631
❺ readily available	容易に利用できる	550

❶ We have adapted a commercially available set of L. monocytogenes antisera to an enzyme-linked immunosorbent assay（ELISA）format for high-throughput, low-cost serotype determination.（J Clin Microbiol. 2003 41:564）
訳 われわれは，市販のリステリア菌抗血清のセットを酵素結合免疫吸着測定法（ELISA）に適応させてきた

❷ No analytical method is currently available to analyze EPA-regulated HAAs in biological samples at environmentally relevant low concentrations.（Anal Chem. 2003 75:4065）
訳 ～を分析するために現在利用できる分析方法はない

■ available ＋［名］

❶ available data	利用できるデータ	597
❷ available evidence	利用できる証拠	296

avenue 名 (目的を達するための) 道／手段 用例数 1,089

avenues	道／手段	670
avenue	道／手段	419

複数形のavenuesの割合は約60%と非常に高い.

◆類義語：means, tool, method, procedure, way, approach

■ avenues ＋ [前]

	the%	a/an%	ø%			
❶	1	-	98	avenues for ～	～のための道	409
				・new avenues for ～	～のための新しい道	228
				・open new avenues for ～	…は，～のための新しい道を切り開く	100
❷	1	-	92	avenues of ～	～の道	104

❶ These new findings may **open new avenues for** the treatment of dyslipidemias. (Curr Opin Lipidol. 2011 22:86)

訳 これらの新しい知見は，脂質異常症の治療のための新しい道を切り開くかもしれない

avoid 動 避ける／回避する 用例数 4,010

avoid	～を回避する	2,195
avoiding	～を回避する	168
avoided	～を回避した／回避される	167
avoids	～を回避する	130

他動詞.

◆類義語：circumvent, escape

■ avoid ＋ [名句]

❶	avoid the need for ～	…は，～の必要性を回避する	42
❷	avoid the use of ～	…は，～の使用を回避する	41

❶ The peptides can be radiolabeled to a high specific activity in a facile manner that **avoids the need for** purification. (Cancer Res. 2003 63:354)

訳 ～は，精製の必要性を回避する

■ to avoid

★	to avoid ～	～を回避する…	1,532
❶	To avoid ～	～を回避するために	170

❶ **To avoid** this problem, the metathesis was conducted with EtZnCl, which enabled the salt metathesis to proceed at low temperatures. (J Am Chem Soc. 2009 131:12483)

訳 この問題を回避するために，

B

balance　名 バランス／平衡，動 釣り合う　　用例数 8,194

balance	バランス／平衡／～を釣り合わせる	6,210
balanced	釣り合った／～を釣り合わせた	1,319
balancing	釣り合う／～を釣り合わせる	472
balances	バランス／平衡／～を釣り合わせる	193

他動詞としても使われるが，名詞の用例が非常に多い．複数形のbalancesの割合は約1%しかないが，単数形には不定冠詞が使われることがかなり多い．

◆類義語：equilibrium, equilibration, equilibrate

■ 名 balance ＋［前］

balance between/ofの前にtheが付く割合はかなり高い．

	the%	a/an%	ø%			
❶	68	30	2	balance between ～	～の間のバランス	371
❷	66	30	4	balance of ～	～のバランス	296
❸	33	12	48	balance in ～	～におけるバランス	79

❶ Here, we show that high mobility group box 1 (HMGB1) and p53 form a complex that **regulates the balance between tumor cell death and survival.** (Cancer Res. 2012 72:1996)

訳 ～は，腫瘍細胞の死と生存の間のバランスを調節する

❷ **It is unclear whether the balance of M1 and M2 macrophages can be altered** and whether this affects disease outcome. (J Immunol. 2012 189:551)

訳 M1とM2マクロファージのバランスが変えられ得るかどうかは不明である

■ 名［名］＋ balance

❶	energy balance	エネルギー収支	622

■ 動 be balanced

★	be balanced	…は，釣り合わされる	257
❶	be balanced by ～	…は，～によって釣り合わされる	106

★ barrier　名 障害／障壁／関門／バリア　　用例数 10,757

barrier	障害／障壁／関門／バリア	8,080
barriers	障害／障壁／関門／バリア	2,677

複数形のbarriersの割合が，約25%ある可算名詞．

◆類義語：obstacle, disturbance, difficulty, problem

■ barrier ＋［前］

barrier forの前にtheが付く割合は非常に高い．

	the%	a/an%	ø%			
❶	31	66	0	barrier to ～	～への障害	1,057
❷	72	27	0	barrier for ～	～に対する障壁	381
❸	53	47	0	barrier of ～	～の障害	254

❶ Tumor antigen-specific T-cell tolerance **imposes a significant barrier to the development of** effective therapeutic cancer vaccines. (Blood. 2005 105:1135)

訳 …は，～の開発への大きな障害を課す

❷ In contrast, **the activation barrier for C-H activation increases with decreasing ligand pK_a.** (J Am Chem Soc. 2009 131:11098)

訳 C-H活性化に対する活性化障壁は，リガンドのpK_aの低下に伴って上昇する

■ [名／形] + barrier

❶	blood-brain barrier	血液脳関門	1,160
❷	epithelial barrier	上皮バリア	311
❸	free energy barrier	自由エネルギー障壁	174

■ barrier + [名]

❶	barrier function	障壁機能	862

★ basal 形 基底の／基礎の／基本の　用例数 15,918

◆類義語：ground, basic, fundamental

■ basal + [名]

❶	basal ganglia	基底核	1,396
❷	basal levels	基礎レベル	761
❸	basal transcription	基礎転写	511
❹	under basal conditions	基礎条件下で	383

❹ **No lipase activities were detected under basal conditions**, but treatment with cytokines significantly stimulated the expression of both activities. (Circ Res. 2003 92:644)

訳 リパーゼ活性は，基礎条件下では検出されなかった

★★ base 動 基礎を置く，名 塩基／基礎　用例数 100,323

based	基づく	74,871
base	基礎／塩基	20,935
bases	基礎／塩基	4,498
basing	～の基礎を置く	19

名詞としても使われるが，動詞の用例が多い．特に，過去分詞のbased用例が圧倒的に多い．

■ based ＋［前］

❶ based on ～	～に基づいて	34,205
・Based on ～	～に基づいて	6,597
・based on the results of ～	～の結果に基づいて	129
・based on the use of ～	～の使用に基づいて	115
・based on the crystal structure of ～	～の結晶構造に基づいて	109
・based on the presence of ～	～の存在に基づいて	101
❷ based upon ～	～に基づいて	1,271

❶ **Based on the results of this study**, a model of PicoGreen/DNA complex formation is proposed. (Biophys J. 2010 99:3010)

訳 この研究の結果に基づいて，

■［名］＋ be based on

★ be based on ～	…は，～に基づいている	5,023
❶ method is based on ～	方法は，～に基づいている	264
❷ approach is based on ～	アプローチは，～に基づいている	135
❸ model is based on ～	モデルは，～に基づいている	99

❶ **This method is based on experimental evidence that** ligand binding sites also bind small organic molecules of various shapes and polarity. (Bioinformatics. 2012 28:286)

訳 この方法は，～という実験的証拠に基づいている

■［名］＋ based on

❶ model based on ～	～に基づくモデル	524
❷ method based on ～	～に基づく方法	356
❸ approach based on ～	～に基づくアプローチ	256
❹ system based on ～	～に基づくシステム	245
❺ analysis based on ～	～に基づく分析	237
❻ assay based on ～	～に基づくアッセイ	198

■［名］-based ＋［名］

❶ population-based study	集団ベース研究	253
❷ cell-based assays	細胞アッセイ	233
❸ structure-based design	構造に基づく設計	166

★ baseline 名 ベースライン／基線／起点　用例数 17,556

baseline	ベースライン／基線／起点	17,508
baselines	ベースライン／基線／起点	48

複数形のbaselinesの割合は0.3%しかなく，原則，**不可算名詞**として使われる．

■ [前] + baseline

❶	at baseline	最初に／ベースラインで	4,846
	・at baseline and at ～	最初および～で	340
	・at baseline and after ～	最初および～後	234
❷	return to baseline ～	…は，ベースライン～へ戻る	435
❸	to baseline levels	ベースラインレベルへ	157
❹	from baseline to ～	ベースラインから～へ	381
❺	compared with baseline	ベースラインと比べて／最初と比べて	295

❶ SF-AP radiographs were obtained at baseline and at 16 months and 30 months thereafter from subjects enrolled in a 6-center DMOAD trial.（Arthritis Rheum. 2004 50:2508）
訳 SF-APのX線写真が，最初およびその後16ヶ月と30ヶ月で得られた

❸ In fact, body temperature returned to baseline levels faster in the β-endorphin deficient mice.（Brain Res. 2003 978:169）
訳 体温は，ベースラインレベルに戻った

❹ Total testosterone increased significantly from baseline to the low normal range after 1 wk, and to upper normal range after two injections（p <.001).（Crit Care Med. 2001 29:1936）
訳 全テストステロン量は，ベースラインから低い正常レンジに有意に増大した

❺ Each intervention improved the targeted cognitive ability compared with baseline, durable to 2 years（P <.001 for all).（JAMA. 2002 288:2271）
訳 ベースラインと比べて

★basis 名 基盤／基礎／根拠 用例数 28,345

basis	基盤／基礎／根拠	23,847
bases	基盤／基礎／根拠	4,498

複数形のbasesの割合が，約15%ある可算名詞．on the basis ofの用例が非常に多い．
◆類義語：ground, reason

■ basis + [前]

basis ofの前にtheが付く割合は，99%と圧倒的に高い．また，basis forの前にtheが付く割合もかなり高い．

	the%	a/an%	ø%			
❶	99	1	0	basis of ～	～の基盤	13,128
				・on the basis of ～	～に基づいて	7,745
				・On the basis of ～	～に基づいて	3,149
				・identified on the basis of ～	～に基づいて同定される	156
				・selected on the basis of ～	～に基づいて選択される	110
❷	62	37	0	basis for ～	～のための基盤	8,065
				・basis for understanding ～	～を理解するための基盤	296

・provide a basis for ～ …は，～のための基盤を提供する		778
・serve as the basis for ～ …は，～のための基盤として役立つ		92

❶ On the basis of these results, we propose a mechanism for cPLA2 activation by calcium and phosphorylation.（J Biol Chem. 2003 278:41431）
訳 これらの結果に基づいて，

❶ These elements, many highly fragmented, were identified on the basis of sequence homology and structural characteristics.（Genome Res. 2004 14:860）
訳 ～は，シークエンスホモロジーに基づいて同定された

❷ This recognition mechanism provides a basis for understanding the effects of mutations that cause WAS.（Cell. 2002 111:565）
訳 この認識機構は，変異の影響を理解するための基盤を提供する

■［形］＋ basis

❶ molecular basis	分子基盤	3,806
・molecular basis of ～	～の分子基盤	1,882
・molecular basis for ～	～のための分子基盤	1,643
❷ structural basis	構造基盤	1,377
❸ genetic basis	遺伝的基盤	1,188
❹ mechanistic basis	機構基盤	543
❺ neural basis	神経基盤	268
❻ cellular basis	細胞基盤	267
❼ biochemical basis	生化学的基盤	183

❶ Understanding the structure and function of PKS will provide clues to the molecular basis of polyketide biosynthesis specificity.（Biochemistry. 2004 43:14529）
訳 …は，～の分子基盤への手がかりを提供するだろう

★ bear 動 有する／運ぶ／産む　用例数 12,534

bearing	～を有する	7,105
born	生まれた	2,423
borne	媒介される	1,450
bear	～を有する／生む	751
bears	～を有する／生む	611
bore	～を有した／生んだ	194

他動詞．現在分詞bearingの用例が非常に多い．bornとborneの２種類の過去分詞があり使い方が異なる．

◈類義語：have, carry

■［名］＋ bearing

❶ mice bearing ～	～を有するマウス	904

❷ cells bearing ～	～を有する細胞	349
❸ tumor-bearing ～	腫瘍を有する～	876
・tumor-bearing mice	腫瘍を有するマウス	469

❶ In **mice bearing immunogenic tumors**, the presence of pre-existing tumor-sensitized T cells is demonstrated by adoptive cell transfer experiments using purified spleen T cells from these mice. (J Immunol. 2001 167:6765)

訳 免疫原性の腫瘍を有するマウスにおいて，

❸ In this model, **tumor-bearing mice exhibit features similar to** those associated with human metastatic bone disease such as osteolytic bone destruction. (Cancer Res. 2002 62:5571)

訳 腫瘍を有するマウスは，～に似た特徴を示す

■ [名]-borne

❶ blood-borne ～	血液に媒介される～	242
❷ tick-borne ～	マダニに媒介される～	220
❸ food-borne ～	食物に媒介される～	186
❹ plasmid-borne ～	プラスミドに媒介される～	174

❷ Our findings show that the newly discovered species is a causal agent of **tick-borne relapsing fever**. (Lancet. 2003 362:1283)

訳 マダニに媒介される回帰熱

■ born ＋ [前]

❶ born in ～	～において生まれた	320
❷ born to ～	～に生まれた	288
❸ born at ～	～で生まれた	178

❶ Thirteen subjects exhibited deep subcortical white matter lesions, of whom nine (69.2%) **were born in the winter months (January to March)**. (Am J Psychiatry. 2001 158:1521)

訳 ～は，冬の月（１月から３月）に生まれた

❷ We followed **a group of infants born to HIV-infected women** from birth to five years of age with echocardiographic studies every four to six months. (N Engl J Med. 2000 343:759)

訳 HIVに感染した女性に生まれた新生児のグループ

❸ Subdural haematomas are thought to be uncommon in **babies born at term**. (Lancet. 2004 363:846)

訳 満期で生まれた赤ん坊

■ [名] ＋ born

❶ infants born	生まれた新生児	224
❷ children born	生まれた子供	143

★ become 動 なる／状態になる 用例数 20,765

become	～になる／～になった	10,199
became	～になった	4,343

becomes	～になる	4,277
becoming	～になる	1,946

第2文型（S+V+C）の自動詞.

B

■ become ＋［形／過分］

❶ become available	…が，入手可能になる	501
❷ become activated	…が，活性化された状態になる	343
❸ become apparent	…が，明らかになる	337
❹ become clear	…が，明らかになる	286
❺ become infected	…が，感染した状態になる	236
❻ become phosphorylated	…が，リン酸化状態になる	212
❼ become evident	…が，明らかになる	177
❽ become resistant	…が，抵抗性になる	167
❾ become active	…が，活性化状態になる	143

❶ Rodent models and genetically altered mice have recently become available to study many human diseases.（J Nucl Med. 2004 45:665）

訳 げっ歯類のモデルと遺伝的に改変されたマウスが，多くのヒトの疾患を研究するために最近利用可能になった

■ become ＋［副］

❶ become more ～	…は，もっと～になる	1,322
❷ become increasingly ～	…は，ますます～になる	871
・become increasingly important	…は，ますます重要になる	172
❸ become less ～	…は，より～でなくなる	282
❹ become progressively ～	…は，次第に～になる	168
❺ become highly ～	…は，高度に～になる	167

❶ With experience, human listeners can become more proficient at inferring positive-affect states from cat meows.（J Comp Psychol. 2003 117:44）

訳 ヒトの聞き手は，～を推測するのにもっと熟達できる

❷ Multivariate analysis has become increasingly common in the analysis of multidimensional spectral data.（Anal Chem. 2002 74:1824）

訳 多変量解析は，ますます一般的になってきた

■［副］＋ become

❶ have recently become ～	…は，最近～になった	121

■［名］＋ become

❶ cells become ～	細胞は，～になる	312

❶ Recent studies have provided important insight into the mechanisms by which iNKT cells become activated in response to diverse inflammatory stimuli.（Trends Immunol. 2013 34:50）

訳 iNKT細胞が，～に応答して活性化状態になる

■ it has become

❶ it has become ～ that …	…ということが，～になった	274
・it has become clear that …	…ということが，明らかになった	119

❶ Recently, it has become clear that the frequency of regulatory T cells (Treg) significantly increases in aged mice and humans. (Curr Opin Immunol. 2012 24:482)

訳 最近，～ということが明らかになった

★ begin 動 始める／始まる　用例数 10,141

beginning	～を始める／始まる	3,806
begin	～を始める／～に取りかかる／始まる	1,958
began	～を始めた／～に取りかかった／始まった	1,613
begins	～を始める／～に取りかかる／始まる	1,481
begun	～を始めた／始められる／始まった	1,283

他動詞と自動詞の両方で用いられる．他動詞としては，begin to *do*の用例が非常に多い．

◆類義語：start, launch

■ begin to *do*

★ begin to *do*	…は，～することに取りかかる	3,689
❶ begin to be ～	…は，～され始める	460
❷ begin to understand ～	…は，～を理解することに取りかかる	200
❸ begin to emerge	…は，現れ始める	150
❹ begin to elucidate ～	…は，～を解明することに取りかかる	135
❺ begin to define ～	…は，～を定義することに取りかかる	122
❻ begin to reveal ～	…は，～を明らかにすることに取りかかる	117
❼ begin to address ～	…は，～に取り組み始める	114
❽ begin to identify ～	…は，～を同定することに取りかかる	104
❾ begin to provide ～	…は，～を提供することに取りかかる	101

❶ To begin to understand the cooperative functions of Nkx2.2 and Arx in the development of endocrine cell lineages, we generated progenitor cell-specific deletions of Arx on the Nkx2.2 null background. (Dev Biol. 2011 359:1)

訳 ～の発生におけるNkx2.2とArxの協同的機能を理解することに取りかかるために

■ to begin

★ to begin	～に取りかかる…	837
❶ to begin to *do*	～することに取りかかる…	522
❷ To begin ～	～に取りかかるために	473

■ begin ＋［前］

❶ begin with ～	…は，～から始まる	920
❷ begin at ～	…は，～において始まる	823
❸ begin in ～	…は，～において始まる	557

❶ The two-step synthesis begins with the formation of alpha-chloro ketones by the coupling of a Weinreb amide (2-chloro-N-methoxy-N-methylacetamide) and an appropriate Grignard reagent. (Nat Protoc. 2012 7:1184)
訳 2段階合成は，α-クロロケトンの形成から始まる

behave 動 挙動する／行動する　用例数 2,210

behave	挙動する	768
behaves	挙動する	571
behaved	挙動した	475
behaving	挙動する	396

自動詞.
◆類義語：function, act, serve, operate, work

■ behave ＋［前］

❶ behave as ～	…は，～として挙動する	841
❷ behave like ～	…は，～のように挙動する	262
❸ behave in ～	…は，～において挙動する	134

❶ Thus, the 129 strain of mouse behaves as a functional CP49 knockout. (Invest Ophthalmol Vis Sci. 2004 45:884)
訳 129系のマウスは，機能的なCP49ノックアウトとして挙動する

★ behavior 名 行動／挙動　用例数 23,820

behavior	行動／挙動	18,739
behaviors	行動／挙動	5,081

複数形のbehaviorsが使われる割合は約20%あるが，単数形は無冠詞で使われることが多い.

■ behavior ＋［前］

behavior ofの前にtheが付く割合は圧倒的に高い.

	the%	a/an%	ø%			
❶	87	4	8	behavior of ～	～の行動	3,481
❷	6	5	80	behavior in ～	～における行動	1,488

❶ We propose that differences in the behavior of spindle checkpoint proteins in animal cells and budding yeast result primarily from evolutionary divergence in spindle assembly pathways. (J Cell Biol. 2004 164:535)

訳 動物細胞と出芽酵母における紡錘体チェックポイントタンパク質の挙動の違いは，主に～に由来する

❷ Low serotonin$_{1A}$ receptor (5-HT$_{1A}$R) binding is a risk factor for anxiety and depression, and deletion of the 5-HT$_{1A}$R results in anxiety-like behavior in mice. (Proc Natl Acad Sci USA. 2010 107:7592)

訳 5-HT$_{1A}$Rの欠損は，マウスにおける不安様行動という結果になる

■ [形／名] + behavior

❶ sexual behavior	性行動	370
❷ cell behavior	細胞挙動	340
❸ social behavior	社会的行動	282
❹ feeding behavior	食行動	277
❺ dynamic behavior	動的挙動	241
❻ aggressive behavior	攻撃行動	227
❼ seeking behavior	探索行動	194
❽ suicidal behavior	自殺行動	185

❶ Repeated exposure to female-related stimuli causes functional alterations in the neural circuitry mediating male rat sexual behavior. (Behav Neurosci. 2004 118:1473)

訳 ～は，雄ラットの性行動を仲介する神経回路網の機能的な変化を引き起こす

believe 動 信じる　用例数 6,234

believed	信じられる／～を信じた	4,801
believe	～を信じる	1,388
believes	～を信じる	24
believing	～を信じる	21

他動詞．believed to *do*の用例が非常に多い．

◆類義語：accept, appreciate, know, consider, regard, think, assume

■ be believed to *do*

★ be believed	…は，信じられている	3,391
❶ be believed to *do*	…は，～すると信じられている	3,106
・be believed to be ～	…は，～であると信じられている	1,251
・be believed to play ～	…は，～を果たすと信じられている	304
・be believed to contribute to ～	…は，～に寄与すると信じられている	89

❶ Sleep is believed to play an important role in memory consolidation. (Science. 2011 332:1571)

訳 睡眠は，～において重要な役割を果たすと信じられている

B

■ believed ＋［that節］

★ believed that ～	～ということが信じられている	583
❶ it is believed that ～	～ということが信じられている	195

❶ It is believed that these long-lasting changes contribute to learning and memory, drug tolerance, and ischemic preconditioning.（Proc Natl Acad Sci USA. 2004 101:2145）
訳 ～ということが信じられている

■［副］＋ believed

❶ widely believed	広く信じられている	196
❷ generally believed	一般的に信じられている	180
❸ previously believed	以前に信じられていた	151

❶ Although TfR is widely believed to be important for iron acquisition by all mammalian cells, direct experimental evidence is lacking.（Blood. 2003 102:3711）
訳 TfRは，～にとって重要であると広く信じられている

❷ It is generally believed that a slippery sequence and a downstream RNA structure are required for the programmed -1 ribosomal frameshifting.（J Virol. 2003 77:10280）
訳 ～ということが一般的に信じられている

■［代名］＋ believe

★ we believe ～	われわれは，～を信じている	1,105
❶ we believe that ～	われわれは，～ということを信じている	539
❷ what we believe to be ～	われわれが～であると信じているもの	177

❶ We believe that this is the first evidence that KGF can induce new alveolar formation in mature lungs.（Circulation. 2002 106:1120）
訳 われわれは，～ということを信じている

❷ These data reveal what we believe to be a novel role of Hsp90 chaperones in the regulation of the protein-folding environment in mitochondria of tumor cells.（J Clin Invest. 2011 121:1349）
訳 これらのデータは，われわれが～の新規の役割であると信じているものを明らかにする

belong 動 属する　　用例数 4,217

belongs	属する	1,517
belonging	属する	1,280
belong	属する	1,202
belonged	属する	218

自動詞．belong toの用例が圧倒的に多い．
◈類義語：pertain

■ belong to

❶ belong to ～	…は，～に属する	4,116

・belong to the ～ family of …	…は，～ファミリーの…に属する	281
・belong to a family of ～	…は，～のファミリーに属する	243

❶ Guanine nucleotide exchange factors (GEFs) belonging to the Dbl family of proteins represent one major class of proteins that regulate the activity of Rho GTPases. (J Biol Chem. 2003 278:43541)

訳 Dblファミリーのタンパク質に属するグアニンヌクレオチド交換因子（GEF）

❶ LMO4 belongs to a family of transcriptional regulators that comprises two zinc-binding LIM domains. (Mol Cell Biol. 2004 24:2074)

訳 LMO4は，～する転写調節因子のファミリーに属する

beneficial 形 有益な　　用例数 4,674

◈類義語：useful, valuable, available, helpful, convenient

■ be beneficial ＋［前］

★ be beneficial	…は，有益である	1,221
❶ be beneficial in ～	…は，～において有益である	430
❷ be beneficial for ～	…は，～にとって有益である	237
❸ be beneficial to ～	…は，～にとって有益である	148

❶ In conclusion, treatment with amniotic fluid stem cells may be beneficial in kidney diseases characterized by progressive renal fibrosis. (J Am Soc Nephrol. 2012 23:661)

訳 羊水幹細胞による治療は，腎疾患において有益であるかもしれない

❷ Cardiac resynchronization therapy has been beneficial for adult patients with poor left ventricular function and intraventricular conduction delay. (J Am Coll Cardiol. 2005 46:2277)

訳 心臓再同期療法は，～の成人患者にとって有益である

■ beneficial ＋［名］

❶ beneficial effects	有益な効果	1,052
❷ beneficial mutations	有益な変異	172

❶ The beneficial effects of statins are usually assumed to stem from their ability to reduce cholesterol biosynthesis. (FASEB J. 2005 19:1845)

訳 スタチンの有益な効果は，通常，コレステロールの生合成を低下させるそれらの能力に由来すると推定される

★ benefit 名 利点／利益，動 利益を得る　　用例数 12,991

benefit	利点／利益を得る	8,064
benefits	利点／利益を得る	4,671
benefited	利益を得た	199
benefiting	利益を得る	57

動詞としても使われるが，名詞の用例が非常に多い．複数形のbenefitsの割合は約40%とかなり高い．動詞としては，benefit fromの用例が非常に多い．

◆類義語：advantage, merit

■ 名 benefit ＋［前］

benefit ofの前にtheが付く割合はかなり高い.

	the %	a/an %	ø %			
❶	60	14	16	benefit of ～	～の利点	1,277
❷	4	21	65	benefit in ～	～における利点	677
❸	9	30	44	benefit for ～	～のための利点	351

❶ In conclusion, an impaired endothelial NO response could lessen the benefit of RAS inhibition in diabetic renal disease. (Am J Pathol. 2010 176:619)
訳 ～は，RAS阻害の利点を減弱させうる

❷ Despite this, antioxidant therapy has failed to show benefit in clinical trials. (Circulation. 2004 109:2448)
訳 抗酸化剤療法は，臨床治験における利点を示すことができないでいる

■ 名［形］＋ benefit

❶	clinical benefit	臨床的な利点	589
❷	survival benefit	生存優位性	509
❸	therapeutic benefit	治療上の利点	373
❹	significant benefit	著しい利点	154

❷ Participation in DM resulted in a significant survival benefit, most notably in symptomatic systolic HF patients. (Circulation. 2004 110:3518)
訳 ～は，有意な生存優位性という結果になった

■ 動 benefit from

❶	benefit from ～	…は，～から利益を得る	1,736
	・likely to benefit from ～	～からおそらく利益を得る	142

❶ We suggest that patients with late PTLDs and limited disease may benefit from local treatment. (Transplantation. 2002 74:1095)
訳 ～は，局所の治療から利益を得るかもしれない

★best 副 最もよく，名 最もよいもの，形 最もよい／最高の 用例数 8,287

well, goodの最上級．the bestの用例が非常に多い.
◆類義語：highest

■ 副 be best ＋［過分］

★	be best ～	…は，最もよく～である	1,221
❶	be best described	…は，最もよく記述されている	138
❷	be best known	…は，最もよく知られている	127

❸ be best explained	…は，最もよく説明されている	126

❶ In mammals, TOR is best known to regulate translation through the ribosomal protein S6 kinases (S6Ks) and the eukaryotic translation initiation factor 4E-binding proteins. (Oncogene. 2004 23:3151)

訳 TORは，翻訳を調節することが最もよく知られている

■ 副 the best ＋［過分］

❶ the best characterized ～	最もよく特徴づけられた～	269
❷ the best studied ～	最もよく研究された～	175
❸ the best fit ～	最も合う～	137

❶ Bcl-2 is the best characterized member of a large family of proteins that regulate apoptosis. (J Biol Chem. 2006 281:40493)

訳 Bcl-2は，アポトーシスを調節するタンパク質の大きなファミリーの最もよく特徴づけられたメンバーである

■ 名 the best ＋［前］

★ the best of ～	～の最もよいもの	272
❶ to the best of our knowledge	われわれの知る限りでは	200

❶ DMP is, to the best of our knowledge, the first non-SIM based prediction method to have been tested directly on new data. (Bioinformatics. 2004 20:1110)

訳 われわれの知る限りでは，

★ better 副 よりよく，形 よりよい　用例数 15,704

well, goodの比較級．

■ 副 to better ＋［動］

★ to better ～	よりよく～するために	3,382
❶ to better understand ～	～をよりよく理解するために	2,187
・to better understand how ～	どのように～かをよりよく理解するために	183
・to better understand the role of ～	～の役割をよりよく理解するために	182
❷ to better define ～	～をよりよく明らかにするために	285
❸ to better characterize ～	～をよりよく特徴づけるために	145

❶ We determined the intracellular localization of Rsp5p and the determinants necessary for localization, in order to better understand how Rsp5p activities are coordinated. (Mol Cell Biol. 2001 21:3564)

訳 どのようにRsp5p活性が同調されるかをよりよく理解するために

❷ Residual dipolar couplings (RDCs) are employed here to better define the global structure of the IRE RNA in solution. (J Mol Biol. 2003 326:1037)

訳 ～の全体構造をよりよく明らかにするために

B

■ 形 better ＋［名］

❶ better understanding	よりよい理解	1,863
・to gain a better understanding of ～	～についてよりよい理解を得るために	181
・lead to a better understanding of ～ …は，～についてのよりよい理解につながる		161
❷ better outcomes	よりよい結果	205
❸ better survival	よりよい生存	198

❶ The kinetics of unfolding of a collagen-like peptide, (Pro-Pro-Gly)$_{10}$, has been studied under isothermal conditions **to gain a better understanding** of the stabilization of the collagen triple helix. (J Mol Biol. 2004 337:917)

訳 コラーゲンの三重らせん体の安定化に対するよりよい理解を得るために

■ 形 better than

★ better than ～	～よりよい	1,875
❶ be better than ～	…は，～よりよい	288
❷ significantly better than ～	～より有意によい	285

❶ However, the protection conferred by strain 49237SOD was **significantly better than** that induced by the parental strain, 49237. (Infect Immun. 2002 70:2535)

訳 …は，～によって誘導されるそれより有意によかった

★★ bind 動 結合する／結合させる　用例数 83,192

bound	結合した／～を結合させた	34,117
bind	結合する／～を結合させる	25,199
binds	結合する／～を結合させる	23,876

他動詞・自動詞の両方の用例がある．自動詞は，bind toの用例が圧倒的に多い．他動詞には，「～を結合させる」と「～に結合する」の２つの用法がある．受動態のbound toは，「～に結合した」という意味で用いられる．bindingは別項参照．

◆類義語：associate, bond, attach, couple, link, connect, join, conjugate

■ 動 bound ＋［前］

❶ bound to ～	～に結合した	10,177
・be bound to ～	…は，～に結合している	1,070
・when bound to ～	～に結合したとき	518
❷ bound by ～	～と結合した（～によって結合される）	1,078
❸ bound in ～	～に結合した	691
❹ bound with ～	～で結合した	403
❺ bound at ～	～において結合した	403

❶ Plasminogen activator inhibitor type-1 (PAI-1) **is bound to** vitronectin (VN) in plasma and in the extracellular matrix. (Anal Biochem. 2001 296:245)

訳 ～は，ビトロネクチンに結合している

❶ When bound to an activating RNA and ATP, PKR undergoes autophosphorylation reactions at multiple serine and threonine residues. (Nucleic Acids Res. 2001 29:3020)

訳 ～に結合したとき

❷ The enhancer is bound by a heterodimer of the bHLH-Zip protein USF-1 and -2 and a cell-specific factor from thyroid C cell lines. (J Biol Chem. 2004 279:49948)

訳 エンハンサーは，～のヘテロダイマーと結合する

❹ Human SRP RNA bound with high affinity to a 63 amino acid residue region near the C terminus of SRP72. (J Mol Biol. 2005 345:659)

訳 …は，～に高い親和性で結合した

■ 動 bound ＋［副］＋ to

❶ bound specifically to ～	～に特異的に結合した	226
❷ bound directly to ～	～に直接結合した	135

■ 動［副］＋ bound

❶ covalently bound	共有結合した	446
・covalently bound to ～	～に共有結合した	183
❷ tightly bound	しっかり結合した	377
❸ specifically bound	特異的に結合した	348
❹ strongly bound	強くに結合した	156

❶ A15 was found to be highly specific for the active site of FXIIIa and was covalently bound to fibrin. (Circulation. 2004 110:170)

訳 ～は，フィブリンに共有結合していた

❷ The NOEs observed indicate that fluoroalcohol and water molecules are both tightly bound to the peptide in the vicinity of the interhelix bend. (Biophys J. 2004 86:3166)

訳 …は，共に～にしっかり結合している

■ 動［名］-bound

❶ membrane-bound ～	膜に結合した～	2,560
❷ GTP-bound ～	GTPに結合した～	522

■ 動 bind ＋［名］

❶ bind DNA	…は，DNAに結合する	848
❷ bind ligand	…は，リガンドに結合する	105

❶ Her1 binds DNA only as a homodimer but will also dimerize with Hes6. (Development. 2012 139:940)

訳 Her1は，ホモダイマーとしてのみDNAに結合する

■ 動 bind ＋［前］

❶ bind to ～	…は，～に結合する	19,565
・bind to DNA	…は，DNAに結合する	410

❷ bind with ～	…は，～で結合する	1,006
・bind with high affinity	…は，高親和性で結合する	388
・bind with high affinity to ～	…は，～に高親和性で結合する	282
❸ bind in ～	…は，～において結合する	943

B

❶ We find that **Su(H)-H complexes can bind to DNA with high efficiency** in vitro.（Genes Dev. 2002 16:1964）

訳 Su(H)-H複合体は，高い効率でDNAに結合できる

❷ We found that **NgR1 and NgR3 bind with high affinity to the glycosaminoglycan moiety of proteoglycans** and participate in CSPG inhibition in cultured neurons.（Nat Neurosci. 2012 15:703）

訳 NgR1とNgR3は，プロテオグリカンのグリコサミノグリカン部分に高い親和性で結合する

■ 動 bind ＋［副］＋ to

❶ bind directly to ～	…は，～に直接結合する	889
❷ bind specifically to ～	…は，～に特異的に結合する	793
❸ bind tightly to ～	…は，～にしっかり結合する	251
❹ bind preferentially to ～	…は，～に優先的に結合する	216

❶ **CC3 binds directly to the karyopherins** of the importin β family in a RanGTP-insensitive manner and associates with nucleoporins in vivo.（J Biol Chem. 2004 279:46046）

訳 CC3は，カリオフェリンに直接結合する

★★ binding 名 結合，-ing 結合する　用例数 203,324

binding	結合／結合する	203,258
bindings	結合	66

名詞の複数形のbindingsの用例はほとんどなく，原則，不可算名詞として使われる．

◆類義語：bond, binding, association, connection, adhesion, attachment

■ binding ＋［前］

	the%	a/an%	ø%			
❶	40	1	54	binding of ～	～の結合	22,304
				・binding of ～ to …	～の…への結合	7,833
❷	4	1	59	binding to ～	～へ結合する	21,694
				・capable of binding to ～	～へ結合することができる	225

❶ Studies using fibulin-$5^{-/-}$ mice indicated that fibulin-5 **is required for binding of ecSOD to vascular tissue**.（Circ Res. 2004 95:1067）

訳 ～は，ecSODの血管組織への結合のために必要とされる

❷ **Cocaine initiates its euphoric effects by binding to the dopamine transporter（DAT）**, blocking uptake of synaptic dopamine.（Mol Pharmacol. 2003 64:430）

訳 コカインは，ドパミントランスポーター（DAT）へ結合することによってそれの多幸性の効果を惹起する

■ binding ＋［名］

❶ binding site	結合部位	15,501
❷ binding protein	結合タンパク質	14,224
❸ binding domain	結合ドメイン	8,476
❹ binding affinity	結合親和性	4,034
❺ binding activity	結合活性	3,930
❻ binding pocket	結合ポケット	2,115
❼ binding motif	結合モチーフ	1,802
❽ binding assays	結合アッセイ	1,392
❾ binding properties	結合特性	1,378

❶ A potential binding site for influenza hemagglutinin is located near the interdomain hinge, a region that mediates ligand binding in KIR.（J Biol Chem. 2003 278:46081）

訳 インフルエンザ赤血球凝集素に対する潜在的な結合部位は，～の近くに位置する

■［名／形］＋ binding

❶ DNA binding	DNA結合	15,974
❷ ligand binding	リガンド結合	5,632
❸ RNA binding	RNA結合	3,777
❹ substrate binding	基質結合	2,676
❺ specific binding	特異的結合	1,547
❻ high affinity binding	高親和性結合	1,103

❺ Specific binding of HS to different ligand proteins depends on the sulfation pattern of HS.（J Cell Biol. 2004 166:1069）

訳 HSの～への特異的結合

★ biological 形 生物学的な 用例数 23,163

◆類義語：biologic

■ biological ＋［名］

❶ biological processes	生物学的過程	1,833
❷ biological activity	生物学的活性	1,483
❸ biological functions	生物学的機能	1,139
❹ biological systems	生物学的システム	928
❺ biological effects	生物学的影響	685
❻ biological significance	生物学的重要性	482
❼ biological role	生物学的役割	411
❽ biological properties	生物学的特性	391
❾ biological samples	生物学的試料	379

❶ This may have important implications for the biological activity of leptin in disease states associated with abnormal leptin levels (e.g., obesity and anorexia nervosa). (Diabetes. 2002 51:2105)

訳 これは，レプチンの生物学的活性に対する重要な意味を持つかもしれない

biologically 副 生物学的に 用例数 3,441

■ biologically ＋［形］

❶ biologically active	生物学的に活性がある	1,067
・be biologically active	…は，生物学的に活性がある	158
❷ biologically relevant	生物学的に関連する	665
❸ biologically important	生物学的に重要な	274

❶ This mechanism of SSHR translocation is suitable for facile delivery of biologically active peptides for cell-based and animal-based functional proteomic studies. (J Biol Chem. 2004 279:11425)

訳 このSSHR移行の機構は，生物学的に活性のあるペプチドの容易な運搬に適している

★★ block 動 ブロックする，名 遮断 用例数 48,430

blocked	ブロックされる／～をブロックした	19,167
block	～をブロックする／遮断	13,436
blocking	～をブロックする	8,126
blocks	～をブロックする／遮断	7,701

名詞としても使われるが，他動詞の用例の方が多い．名詞としては，複数形のblocksの割合が約30%ある可算名詞．

◈類義語：abrogate, silence, prevent, hinder, occlude, hamper, arrest, suppress, repress, blockade, blockage, suppression, depression

■ 動 be blocked ＋［前］

★ be blocked	…は，ブロックされる	6,950
❶ be blocked by ～	…は，～によってブロックされる	4,899
・be blocked by inhibitors of ～	…は，～の阻害剤によってブロックされる	105
・be blocked by pretreatment with ～	…は，～による前処置によってブロックされる	80
❷ be blocked in ～	…は，～においてブロックされる	422
❸ be blocked with ～	…は，～によってブロックされる	251
❹ be blocked at ～	…は，～においてブロックされる	119

❶ These changes were also blocked by inhibitors of protein kinase C activity. (Biochemistry. 2004 43:2578)

訳 これらの変化は，また，プロテインキナーゼC活性の阻害剤によってブロックされた

❷ We show vacuolar localization of Ams1 is blocked in mutants that are defective in the Cvt

and autophagy pathways.（J Biol Chem. 2001 276:20491）
訳 ～は，欠陥がある変異体においてブロックされる

■ 動［副］＋ blocked

❶	completely blocked ～	完全にブロックされる／～を完全にブロックした	1,062
❷	partially blocked ～	部分的にブロックされる／～を部分的にブロックした	377
❸	effectively blocked ～	効果的にブロックされる／～を効果的にブロックした	195

❶ This action of cGMP **was completely blocked by inhibitors of** cGMP-dependent kinase.（J Neurosci. 2004 24:6621）
訳 …は，～の阻害剤によって完全にブロックされた

❷ The induction of bTREK-1 mRNA by corticotropin **was partially blocked by the A-kinase antagonist H-89**.（Mol Pharmacol. 2003 64:132）
訳 ～は，Aキナーゼ拮抗剤H-89によって部分的にブロックされた

■ 動 block ＋［名句］

❶	block apoptosis	…は，アポトーシスをブロックする	298
❷	block the ability of ～	…は，～の能力をブロックする	285
❸	block activation of ～	…は，～の活性化をブロックする	206
❹	block the binding of ～	…は，～の結合をブロックする	194
❺	block the interaction	…は，相互作用をブロックする	179
❻	block the induction of ～	…は，～の誘導をブロックする	156
❼	block the effects of ～	…は，～の効果をブロックする	156
❽	block the formation of ～	…は，～の形成をブロックする	118
❾	block the increase in ～	…は，～の増大をブロックする	107

❶ **Snail2 knockdown completely blocks the ability of Twist1 to suppress** E-cadherin transcription.（Cancer Res. 2011 71:245）
訳 Snail2のノックダウンは，～を抑制するTwist1の能力を完全にブロックする

■ 動［名／動／形］＋ to block

★	to block ～	～をブロックする…	2,799
❶	ability to block ～	～をブロックする能力	285
❷	fail to block ～	…は，～をブロックできない	191
❸	be shown to block ～	…は，～をブロックすることが示される	115
❹	be sufficient to block ～	…は，～をブロックするのに十分である	91

❷ Unexpectedly, **JSI-124 failed to block STAT3 phosphorylation**, and CGNs were not protected from ToxB by other known STAT3 inhibitors.（J Biol Chem. 2012 287:16835）
訳 JSI-124は，STAT3のリン酸化をブロックできなかった

■ 名 block ＋［前］

	the%	a/an%	ø%			
❶	15	36	44	block of ～	～の遮断	940

❷	18	74	4	block in ～	～の遮断	860
❸	29	67	0	block to ～	～に対する遮断	365
❹	11	5	73	block by ～	～による遮断	301

❷ Unexpectedly, the introduced gene created a block in anthocyanin biosynthesis. (Plant Cell. 1990 2:279)

訳 誘導された遺伝子は，アントシアニン生合成の遮断を引き起こした

B

blockade 名 遮断／封鎖 用例数 7,178

blockade	遮断	7,139
blockades	遮断	39

複数形のblockadesの割合は0.5%しかなく，原則，不可算名詞として使われる.

◈類義語：block, blockage, suppression, depression

■ blockade ＋［前］

	the%	a/an%	ø%			
❶	4	3	92	blockade of ～	～の遮断	3,426

❶ Finally, the increase in FMRP expression was inhibited by blockade of NMDA receptors. (J Neurosci. 2004 24:10579)

訳 FMRP発現の増大は，NMDA受容体の遮断によって抑制された

★ bond 名 結合，動 結合する 用例数 21,430

bond	結合	13,066
bonds	結合	6,785
bonded	結合した	1,579

動詞としても用いられるが，名詞の用例が断然多い．複数形のbondsの割合が約35%ある可算名詞.

◈類義語：bonding, binding, association, bind, associate

■ bond ＋［前］

bond ofの前にtheが付く割合は圧倒的に高い．また，bond inの前にtheが付く割合も非常に高い.

	the%	a/an%	ø%			
❶	32	65	1	bond between ～	～の間の結合	620
❷	71	21	3	bond in ～	～の結合	530
❸	89	10	1	bond of ～	～の結合	448

❶ Previous research has identified a disulfide bond between cysteines 94 and 100 in the holo state. (Biochemistry. 2009 48:8603)

訳 以前の研究は，システイン94と100の間のジスルフィド結合を同定している

❷ We found that **the double bond in the acyl chain causes** a 5% reduction in drift time. (Anal Chem. 2009 81:8289)
訳 アシル鎖の二重結合は，～を引き起こす

■ ［名］＋ bond

❶	hydrogen bond	水素結合	2,458
❷	disulfide bond	ジスルフィド結合	1,625

break 名 切断，動 切断する　用例数 7,582

breaks	切断／～を切断する	3,677
break	切断／～を切断する	2,452
broken	切断される	744
breaking	～を切断する	665
broke	～を切断した	44

動詞としても用いられるが，名詞の用例が圧倒的に多い．複数形のbreaksの割合が，約65%と圧倒的に高い．

◆類義語：cleavage, breakage, truncation, ablation, scission, cut, cleave, truncate, ablate

■ breaks ＋［前］

	the%	a/an%	ø%			
❶	2	-	81	breaks in ～	～における切断	343

❶ **The expression of mito-PstI induces double-strand breaks in the mitochondrial DNA** (mtDNA), leading to OXPHOS deficiency, mostly due to mtDNA depletion. (J Neurosci. 2011 31:9895)
訳 mito-PstIの発現は，ミトコンドリアDNAにおける二重鎖切断を誘導する

■ ［名］＋ breaks

❶	strand breaks	鎖切断	2,119
	・double-strand breaks	二重鎖切断	1,626
	・DNA double-strand breaks	DNA二重鎖切断	860
	・DNA strand breaks	DNA鎖切断	237
❷	DNA breaks	DNA切断	584

bridge 名 ブリッジ／橋／架橋，動 架橋する　用例数 7,020

bridge	ブリッジ／橋／架橋	3,154
bridges	ブリッジ／橋／架橋	1,607
bridging	～を架橋する	1,469
bridged	架橋される／～を架橋した	790

名詞および動詞として使われる．複数形のbridgesの割合が，約30%ある可算名詞．
◆類義語：linker, link

■ bridge ＋［前］

	the%	a/an%	⌀%			
❶	15	84	0	bridge between ～	～の間のブリッジ	337
❷	14	78	7	bridge to ～	～へのブリッジ	174

❶ This domain is stabilized in part by a salt bridge between Asp^{65} and Arg^{101}. (J Biol Chem. 2007 282:37529)
訳 このドメインは，一部は，Asp^{65}とArg^{101}の間の塩橋によって安定化される

❷ TIPS can be successfully used as a bridge to surgical portosystemic shunting, as well as liver transplantation, but may cause technical difficulties when performing transplantation. (Ann Surg. 2005 241:978)
訳 TIPSは，外科的な門脈体静脈短絡へのブリッジとしてうまく使われうる

■［名］＋ bridge

❶	salt bridge	塩橋	599
❷	disulfide bridge	ジスルフィド架橋	171

broad 形 広い／広範な 用例数 8,436

◆類義語：wide, extensive, widespread

■ broad ＋［名］

❶	a broad range of ～	広範囲の～	1,406
❷	broad spectrum	広域性	1,290
	・a broad spectrum of ～	広域の～	471
❸	broad implications	広範な意味	212
❹	a broad array of ～	広く一連の～	133
❺	broad substrate specificity	広基質特異性	115

❶ Immobilized enzyme systems are important in a broad range of applications, from biological sensing to the industrial-scale biocatalytic synthesis of chiral products. (Anal Chem. 2004 76:6915)
訳 固定化酵素システムは，広範囲の適用において重要である

broadly 副 広く／広範に 用例数 2,576

◆類義語：widely, extensively, universally, ubiquitously

■ broadly ＋［現分／形／過分］

❶	broadly neutralizing	広く中和する	250
❷	broadly applicable	広く適用できる	240

・be broadly applicable	…は，広く適用できる	123
❸ broadly expressed	広範囲に発現する	203
❹ broadly distributed	広範囲に分布する	140

❷ The new one-step protocol **is broadly applicable** to any sensitive, laser-induced fluorescence method for detection of nucleic acids.（Anal Chem. 1999 71:5470）

訳 新しい一段階プロトコールは，どんな敏感なレーザーで誘導される蛍光法にも広く適用できる

❸ By contrast, Gli2 is expressed uniformly in all cells in the developing cerebellum except Purkinje cells and Gli3 is **broadly expressed** along the anteroposterior axis.（Development. 2004 131:5581）

訳 Gli3は，前後軸にそって広範囲に発現している

bulk　形 大量の，名 大部分　用例数 3,625

bulk	大量の／大部分	3,622
bulks	大部分	3

名詞としても使われるが，形容詞の用例が非常に多い．複数形のbulksの用例はほとんどなく，原則，**不可算名詞**として使われる．

◆類義語：abundant, massive, large amount of

■ 形 bulk ＋［名］

❶ bulk solution	大量の溶液	151
❷ bulk water	大量の水	141

❷ The potential model is able to reproduce the interactions between water and the metal ions regardless of whether they are at the mineral surface or in **bulk water**.（J Am Chem Soc. 2004 126:10152）

訳 それらがミネラル表面上にあるか大量の水の中にあるかどうかに関わらず

■ 名 bulk ＋［前］

	the%	a/an%	ø%			
❶	99	0	1	bulk of ～	～の大部分	563

❶ The **bulk of** these mutations were individually rare point mutations, 60% of which changed an amino acid.（Hum Mol Genet. 2002 11:133）

訳 これらの変異の大部分は，個々にまれな点突然変異であった

★ calculate ㊍ 計算する／算出する　　用例数 10,815

calculated	計算される／～を計算した	8,511
calculate	～を計算する	1,666
calculating	～を計算する	541
calculates	～を計算する	97

他動詞.

◆類義語：compute

■ be calculated ＋［前］/using

★ be calculated	…が，計算される	4,354
❶ be calculated for ～	…が，～に対して計算される	626
❷ be calculated from ～	…が，～から計算される	475
❸ be calculated by ～	…が，～によって計算される	413
・be calculated by using ～	…が，～を使うことによって計算される	102
❹ be calculated as ～	…が，～として計算される	291
❺ be calculated using ～	…が，～を使って計算される	414

❶ The local recurrence rate and the Mantel-Haenszel statistic for survival curves **were calculated for each group**. (Radiology. 2002 225:707)

訳 ～が，おのおののグループに対して計算された

❷ **Both of these parameters may be calculated from intergenic sequences**. (Proc Natl Acad Sci USA. 2004 101:3480)

訳 これらのパラメータの両方が，遺伝子間の配列から計算されるかもしれない

■［代名］＋ calculate

❶ we calculated ～	我々は，～を計算した	396

❶ **We calculated the proportion of patients** whose therapy adhered to each QI and to all applicable indicators (overall physician adherence). (Arthritis Rheum. 2007 57:822)

訳 われわれは，患者の比率を計算した

■ to calculate

★ to calculate ～	～を計算する…	1,301
❶ be used to calculate ～	…が，～を計算するために使われる	518

❶ Moreover, the relative orientations of pairs of methyl groups **were used to calculate effective torsional angles between** different planes of unsaturation of the retinal chromophore. (Biochemistry. 2004 43:12819)

訳 …が，～の間の効果的なねじれ角を計算するために使われた

★ call 動 呼ぶ／称する 用例数 11,008

called	呼ばれる／～を呼んだ	8,704
call	～を呼ぶ	1,247
calls	～を呼ぶ	708
calling	～を呼ぶ	349

他動詞の用例が多い．call A B（AをBと呼ぶ）の第5文型で用いられるが，受動態の用例が圧倒的に多い．

◆類義語：designate, term, name

■［名］＋ called

❶ protein called ～	～と呼ばれるタンパク質	283
❷ process called ～	～と呼ばれる過程	231
❸ method called ～	～と呼ばれる方法	162
❹ structures called ～	～と呼ばれる構造	103

❶ We found that **growing Dictyostelium cells secrete a 60 kDa protein called AprA** for autocrine proliferation repressor.（Development. 2005 132:4553）

訳 増殖するタマホコリカビ細胞は，AprAと呼ばれる60 kDaのタンパク質を分泌する

■ also called

❶ also called ～	～とも呼ばれる	787

❶ The cellular ATP-binding protein ABCE1（**also called HP68 or RNase L inhibitor**）appears to be critical for proper assembly of the HIV-1 capsid.（J Biol Chem. 2006 281:3773）

訳 HP68あるいはRNase L阻害剤とも呼ばれる

■［代名］＋ call

❶ we call ～	われわれは，～と呼ぶ	539
・, which we call ～	（それを,）われわれは～と呼ぶ	239
・that we call ～	われわれが～と呼ぶ…	96

❶ **The new fluorescent assay, which we call a protein beacon assay,** will be instrumental in quantitative dissection of fine details of RNAP interactions with promoters.（J Biol Chem. 2011 286:270）

訳 新しい蛍光アッセイ，それをわれわれはタンパク質ビーコンアッセイと呼ぶ

★ candidate 名 候補 用例数 13,121

candidate	候補	9,504
candidates	候補	3,617

複数形のcandidatesの割合が，約30%ある可算名詞．

■ 名 candidate ＋［前］

	the%	a/an%	⌀%			
❶	6	89	0	candidate for ～	～のための候補	1,157

❶ Our findings suggest that **FASN may be a unique candidate for molecular targeted therapy against** PEL and other B-NHL.（Proc Natl Acad Sci USA. 2012 109:11818）

訳 FASNは，～に対する分子標的治療のためのユニークな候補であるかもしれない

■ candidate ＋［名］

❶	candidate genes	候補遺伝子	1,573
	・identify candidate genes	…は，候補遺伝子を同定する	102

❶ **We tested this hypothesis and developed an approach to identify candidate genes associated with** disease development by focusing on cancer incidence since it varies greatly across human organs.（Bioinformatics. 2011 27:3300）

訳 われわれは，この仮説をテストし，～と関連する候補遺伝子を同定するアプローチを開発した

capability 名 能力　　用例数 3,578

capability	能力	1,845
capabilities	能力	1,733

複数形のcapabilitiesの割合は約50%と非常に高い．

◆類義語：ability, capacity, competence, potential, power

■ capability ＋［前］

capability ofの前にtheが付く割合は圧倒的に高い．

	the%	a/an%	⌀%			
❶	92	4	2	capability of ～	～の能力	621
❷	57	11	11	capability to *do*	～する能力	292
❸	30	36	20	capability for ～	～ための能力	98

❶ **Our results provide a demonstration of the capability of coronaviruses to evolve new gene functions** through recombination.（J Virol. 2010 84:12872）

訳 我々の結果は，新しい遺伝子機能を進化させるコロナウイルスの能力の実証を提供する

❷ However, **most Gram-negative bacteria have the capability to use phosphonates** as a nutritional source of phosphorus under conditions of phosphate starvation.（Nature. 2011 480:570）

訳 ほとんどのグラム陰性菌は，ホスホン酸塩を使う能力を持つ

★ capable 形 できる／能力がある 用例数 12,574

capable of ～ingの用例が圧倒的に多い.

◆類義語：able ◆反意語：incapable

■ be capable of

★ be capable of ～	…は，～できる	6,740
❶ be capable of binding	…は，結合できる	340
❷ be capable of inducing ～	…は，～を誘導できる	335
❸ be capable of forming ～	…は，～を形成できる	201
❹ be capable of activating ～	…は，～を活性化できる	176
❺ be capable of producing ～	…は，～を産生できる	150

❶ These peptides were capable of binding one or multiple A2, A3, and B7 supertype molecules with affinities typical of viral determinants. (J Immunol. 2005 175:5504)

訳 これらのペプチドは，～に結合できた

■ [名] ＋ capable of

★ capable of ～	～できる	12,446
❶ cells capable of ～	～できる細胞	298

❶ Dendritic cells are professional antigen-presenting cells capable of inducing and regulating innate and antigen-specific immune responses. (Cancer Res. 2005 65:10059)

訳 樹状細胞は，～を誘導しそして調節できる専門の抗原提示細胞である

★ capacity 名 能力／容量／収容能力 用例数 14,607

capacity	能力／容量	13,872
capacities	能力／容量	735

複数形のcapacitiesの割合が，約５％ある可算名詞.

◆類義語：ability, capability, competence, potential, power

■ capacity ＋ [前]

capacity ofの前にtheが付く割合は圧倒的に高い.

	the%	a/an%	ø%			
❶	91	3	6	capacity of ～	～の能力	3,679
				・capacity of ～ to …	～の…する能力	1,000
❷	49	13	6	capacity to *do*	～する能力	3,394
				・capacity to bind	結合する能力	142
				・capacity to induce ～	～を誘導する能力	139
				・capacity to produce ～	～を産生する能力	115
				・capacity to form ～	～を形成する能力	98
❸	43	25	13	capacity for ～	～の能力	1,250

❹	10	7	77	capacity in ～	～における能力	516

❶ Importantly, **CAS expression resulted in a striking enhancement of the capacity of Src-transformed cells to invade through Matrigel.** (Oncogene. 2004 23:7406)

訳 CAS発現は，Srcで形質転換された細胞のマトリゲルを通って浸潤する能力の著しい増強という結果になった

❷ The genomic sequences predict that miRNAs are likely to be derived from **larger precursors that have the capacity to form stem-loop structures.** (Genes Dev. 2002 16:720)

訳 ステムループ構造を形成する能力を持つより大きな前駆体

❸ Yet we have a limited understanding of **the signaling pathways that regulate the capacity for axon growth** during either development or regeneration. (Development. 2001 128:1175)

訳 軸索伸長のための能力を調節する情報伝達経路

C

■ ［名／形］＋ capacity

❶	exercise capacity	運動能力	427
❷	binding capacity	結合能	426
❸	heat capacity	熱容量	357
❹	proliferative capacity	増殖能力	313

★ care 名 治療／ケア，動 治療する／世話をする 用例数 22,401

care	治療／ケア／治療する	22,055
caring	治療する	213
cared	治療した	122
cares	治療する	11

名詞の用例が断然多いが，動詞としても用いられる．複数形のcaresの用例はほとんどなく，原則，**不可算名詞**として使われる．

◆類義語：treatment, therapy, intervention, practice, cure

■ care ＋［前］

	the%	a/an%	ø%			
❶	2	0	98	care for ～	～のための治療	921
				・care for patients	患者の治療／患者を治療する	152
❷	32	0	68	care of ～	～の治療	724
				・care of patients	患者の治療	178
❸	1	0	96	care in ～	～における治療	633

❶ **Temozolomide along with radiation has become the standard of care for patients with** newly diagnosed glioblastoma. (Curr Opin Neurol. 2005 18:632)

訳 放射線照射を伴うテモゾロミド投与は，～の患者の治療の標準になっている

❸ **Critical care in obstetrics has many similarities in pathophysiology to the care of nonpregnant women.** (Crit Care Med. 2005 33:S354)

訳 産科における救命医療は，病態生理において非妊娠女性の治療への多くの類似点を持つ

■［前］＋ care

❶ quality of care	治療の質	496
❷ standard of care	治療の標準	413

❶ In this study, **we assessed changes over time in the overall quality of care** and in the magnitude of racial disparities in nine measures of clinical performance.（N Engl J Med. 2005 353:692）

訳 われわれは，全体の治療の質の時を経た変化を評価した

■［名／形］＋ care

❶ health care	医療／健康管理	3,329
❷ intensive care unit	集中治療室	2,108
❸ primary care	初期治療	1,549
❹ critical care	救命医療	908
❺ medical care	医療	605

★carry 動 保有する／輸送する／保因する　用例数 16,425

carried	保持される／～を保有した	6,371
carrying	～を運ぶ／～を保有する	5,712
carry	～を運ぶ／～を保有する	2,956
carries	～を運ぶ／～を保有する	1,386

carry out（行う／実行する）の用例が圧倒的に多い.

◆類義語：have, bear, transport, perform, conduct, do

■ be carried out ＋［前］

★ be carried out	…が，行われる	2,936
❶ be carried out to *do*	…が，～するために行われる	481
・be carried out to determine ～	…が，～を決定するために行われる	96
❷ be carried out in ～	…が，～において行われる	459
❸ be carried out by ～	…が，～によって行われる	360
❹ be carried out on ～	…が，～に関して行われる	323

❶ **Functional experiments were carried out to determine whether** rs12746200 led to differences in mRNA expression.（Am J Clin Nutr. 2012 95:959）

訳 機能的実験が，～かどうかを決定するために行われた

❷ **The assay can be carried out in** single- as well as multiwell plate formats such as 96- and 384-well plates.（Anal Biochem. 2003 317:210）

訳 アッセイは，～において行われうる

■［名］＋ be carried out

❶ experiments were carried out	実験が，行われた	159

❷	studies were carried out	研究が，行われた	151

❶ Further experiments were carried out on a protein comprising the A, B, and C domains of rat synapsin I (rSynI-ABC). (J Biol Chem. 2004 279:11948)
訳 さらなる実験が，～に関して行われた

■ [代名] + carry out

★	carry out	…は，～を行う	3,100
❶	we carried out ～	われわれは，～を行った	621

❶ We carried out a number of in vitro and in vivo assays to determine the effect of this expanded RBE sequence on the Rep-RBE interaction and AAV targeted integration. (J Virol. 2003 77:1904)
訳 われわれは，いくつかの試験管内および生体内アッセイを行った

■ carry + [名句]

❶	carry a mutation	…は，変異を保有する	116

❶ Mice carrying a mutation in the Pms2 gene are predisposed to lymphomas and other tumors. (Oncogene. 1999 18:5325)
訳 Pms2遺伝子に変異を保有するマウスは，リンパ腫に罹りやすい

■ [名] + carrying

❶	mice carrying ～	～を保有するマウス	764

■ carried + [前]

❶	carried by ～	～に保持される	125

❶ Strains causing invasive disease could not be distinguished from strains carried by vaccinated children. (J Infect Dis. 2002 186:958)
訳 侵襲性の疾患を引き起こす菌株は，ワクチン接種された子供に保持された菌株と区別できなかった

★★ case 名 ケース／症例／場合　　用例数 42,716

cases	ケース／症例	27,153
case	ケース／症例	15,563

複数形のcasesの割合は，約65%と圧倒的に高い．

◆類義語：event, occasion, instance, circumstance, phenomenon, incident

■ cases + [前]/where

	the%	a/an%	ø%			
❶	5	-	42	cases of ～	～のケース／～の症例	4,996
❷	5	-	59	cases with ～	～を伴うケース／～を伴う症例	1,003

❸	3	-	57	cases in ～	～におけるケース／～における症例	709
❹	7	-	75	cases where～	～であるケース／～である症例	330

❶ In **cases of** advanced fibrosis, the pattern was generally that of type 2 NASH.（Hepatology. 2005 42:641）

訳 進行した線維症の症例において

■ in ＋［形／代名］＋ case(s)

❶	in some cases	いくつかのケースにおいて	1,734
❷	in all cases	すべてのケースにおいて	1,109
❸	in most cases	ほとんどのケースにおいて	847
❹	in both cases	両方のケースにおいて	774
❺	in each case	おのおののケースにおいて	715
❻	in many cases	多くのケースにおいて	623

❷ **In all cases**, the IP3R phosphorylation was diminished by the addition of LY294002, an inhibitor of phosphatidylinositol 3-kinase.（J Biol Chem. 2006 281:3731）

訳 すべてのケースにおいて,

■ case ＋［形／名］

❶	case-control study	症例対照研究	1,292
❷	case patients	症例患者	396

■ cases and ＋［名］

❶	cases and controls	症例と対照群	350

❶ **Cases and controls** were compared, using a matched multivariate analysis, to assess the impact of MMF as one component of triple-therapy adjusted for other drug therapies and known risk factors.（Transplantation. 2005 80:1174）

訳 症例と対照群が比較された

★ catalyze 動 触媒する　　用例数 16,753

catalyzed	～を触媒した／触媒される	7,565
catalyzes	～を触媒する	5,189
catalyze	～を触媒する	3,120
catalyzing	～を触媒する	879

他動詞.

■ catalyzed ＋［前］

❶	catalyzed by ～	～によって触媒される	2,573
	・be catalyzed by ～	…は，～によって触媒される	873
	・reaction catalyzed by ～	～によって触媒される反応	367

❶ This reaction is catalyzed by elongation factor-G (EF-G) and is associated with ribosome-dependent hydrolysis of GTP. (J Mol Biol. 2004 337:263)
訳 この反応は，伸長因子-Gによって触媒される

❶ The dehydrogenation reaction catalyzed by human glutaryl-CoA dehydrogenase was investigated using a series of alternate substrates. (Biochemistry. 2002 41:1274)
訳 ヒトのグルタリルCoA脱水素酵素によって触媒される脱水素反応が，～を使って精査された

C

■ [名]-catalyzed

❶ palladium-catalyzed ～	パラジウムに触媒される～	310
❷ enzyme-catalyzed ～	酵素に触媒される～	263
❸ acid-catalyzed ～	酸に触媒される～	271

■ [名]-catalyzed ＋ [名]

❶ ～-catalyzed reaction	～に触媒される反応	375
❷ ～-catalyzed hydrolysis	～に触媒される加水分解	166

❶ The synthesis of N-substituted benzimidazoles was possible from the palladium-catalyzed reaction of both classes of substrate with a variety of N-nucleophiles. (J Org Chem. 2012 77:9473)
訳 ～は，パラジウムに触媒される反応から可能であった

★★ cause 動 引き起こす，名 原因／理由 用例数 81,125

caused	引き起こされる／～を引き起こした	28,758
cause	～を引き起こす／原因	27,620
causes	～を引き起こす／原因	18,620
causing	～を引き起こす	6,127

他動詞として使われるが，名詞の用例も多い．複数形のcausesの割合は約40%とかなり高い．caused byの用例も非常に多い．

◆類義語：result in, lead to, give rise to, produce, yield, bring about, contribute, induce, generate, reason

■ 動 cause ＋ [名句]

❶ cause an increase in ～	…は，～の増大を引き起こす	670
❷ cause a significant ～	…は，有意な～を引き起こす	607
❸ cause disease	…は，疾患を引き起こす	600
❹ cause a decrease in ～	…は，～の低下を引き起こす	347
❺ cause loss of ～	…は，～の喪失を引き起こす	306
❻ cause apoptosis	…は，アポトーシスを引き起こす	299
❼ cause defects in ～	…は，～の欠損を引き起こす	259
❽ cause a reduction in ～	…は，～の低下を引き起こす	218

❶ Activation of thrombin receptors caused an increase in ligand affinity of thromboxane A2 receptors. (Proc Natl Acad Sci USA. 1998 95:10944)
訳 トロンビン受容体の活性化は，～の増大を引き起こした

■ 動 [形／過分] ＋ to cause

★ to cause ～	～を引き起こす…	3,574
❶ sufficient to cause ～	～を引き起こすのに十分な	473
❷ known to cause ～	～を引き起こすことが知られている	351
❸ shown to cause ～	～を引き起こすことが示される	299

❶ This rescue of weaver granule cells provides evidence that wvGIRK2 alone is not sufficient to cause granule cell death. (J Neurosci. 1999 19:7991)
訳 wvGIRK2だけでは，顆粒細胞死を引き起こすのに十分ではない

❸ Ceramide is a newly discovered second messenger that has been shown to cause cell growth arrest and apoptosis. (J Biol Chem. 1999 274:21121)
訳 細胞増殖停止とアポトーシスを引き起こすことが示されている二次メッセンジャー

■ 動 [名] ＋ be caused by

★ be caused by ～	…は，～によって引き起こされる	4,573
❶ disease is caused by ～	疾患は，～によって引き起こされる	136

❶ The disease is caused by mutations in an inositol polyphosphate 5-phosphatase designated OCRL. (Proc Natl Acad Sci USA. 1999 96:13342)
訳 その疾患は，～の変異によって引き起こされる

■ 動 caused by ＋ [名句]

★ caused by ～	～によって引き起こされる	14,662
❶ caused by mutations in ～	～の変異によって引き起こされる	889
❷ caused by loss of ～	～の喪失によって引き起こされる	176

■ 動 [名] ＋ caused by

❶ disease caused by ～	～によって引き起こされる疾患	602
❷ disorder caused by ～	～によって引き起こされる障害	390

■ 名 cause ＋ [前]

	the%	a/an%	ø%			
❶	40	56	3	cause of ～	～の原因	8,169
				・leading cause of ～	～の主要な原因	1,334
				・major cause of ～	～の主要な原因	1,147
				・common cause of ～	～の共通の原因	857
				・important cause of ～	～の重要な原因	456
				・cause of death	死の原因	878
				・cause of morbidity	罹患の原因	479

❶ Malaria is a leading cause of worldwide mortality from infectious disease. (Biochemistry. 1999 38:9872)

訳 マラリアは，感染症による世界的な大量死の主要な原因である

❶ Thrombotic diseases are a major cause of death and morbidity. (J Med Chem. 2000 43:4398)

訳 血栓性の疾患は，死亡と罹患の主要な原因である

★center 名 中心／中央，動 中央に置く／集中する　用例数 17,663

C

center	中心／中央に置く	10,288
centers	中央に置く	5,638
centered	中央に置く	1,687
centering	中央に置く	50

名詞の用例が多いが，動詞としても用いられる．複数形のcentersの割合が，約35%ある可算名詞．

◆類義語：middle, enrich, focus, concentrate, centralize

■ 名 center ＋［前］

center ofの前にtheが付く割合は圧倒的に高い．「～の中心」は通常唯一無二だからである．

	the%	a/an%	ø%			
❶	97	3	0	center of ～	～の中心	1,609
				・in the center of ～	～の中心において	107
				・at the center of ～	～の中心において	79
❷	56	36	0	center in ～	～における中心	377

❶ A C-terminal Pro-Cys motif is localized at the center of the tetramer and forms reversible enzyme disulfides that alter enzyme activity. (J Biol Chem. 2005 280:4639)

訳 C末端のPro-Cysモチーフは，その四量体の中心に局在する

■ 動 centered ＋［前］

❶	centered on ～	～に中心を置く	392
❷	centered at ～	～に中心を置く	316

❶ Competitive binding analysis and mutagenesis reveals a unique BTLA binding site centered on a critical lysine residue in cysteine-rich domain 1 of HVEM. (Proc Natl Acad Sci USA. 2005 102:13218)

訳 …は，～において決定的に重要なリジン残基に中心を置くユニークなBTLA結合部位を明らかにする

★central 形 中心的な／中枢の 用例数 28,090

◆類義語：pivotal, major

■ be central ＋［前］

★ be central	…は，中心をなす	1,505
❶ be central to ～	…は，～の中心をなす	1,220
・be central to the pathogenesis of ～	…は，～の病因の中心をなす	193
・be central to understanding ～	…は，～を理解するのに中心をなす	60

❶ Erythrocyte invasion by Plasmodium falciparum is central to the pathogenesis of malaria. (Nature. 2011 480:534)

訳 ～は，マラリアの病因の中心をなす

■ central ＋［名］

❶ central nervous system	中枢神経系	5,280
❷ central role	中心的な役割	3,154
❸ central region	中心領域	557

❷ Inflammation plays a central role in atherogenesis. (J Infect Dis. 2001 184:1109)

訳 炎症は，アテローム発生において中心的な役割を果たす

■ central and ＋［形］

❶ central and peripheral ～	中枢および末梢の～	438

❶ Nervous system tumors represent unique neoplasms that arise within the central and peripheral nervous system. (Cancer Res. 2003 63:3001)

訳 中枢および末梢神経系内で生じるユニークな新生物

★certain 形 ある／確信して 用例数 11,517

■ under certain ＋［名］

❶ under certain conditions	ある条件下で	273
❷ under certain circumstances	ある状況下で	102

❶ Under certain conditions, antagonism of anti-apoptotic BCL-2 family proteins can unleash pro-death molecules in cancer cells. (Oncogene. 2008 27 :S149)

訳 ある条件下で，抗アポトーシス性BCL-2ファミリータンパク質の拮抗作用は～を解放できる

■ certain ＋［名］＋ of

❶ certain types of ～	あるタイプの～	236
❷ certain forms of ～	ある型の～	126
❸ certain aspects of ～	～のある面	118

★challenge 名 攻撃／接種／曝露／挑戦／難題，動 接種する／曝露する／挑戦する／異議を唱える

用例数 20,849

challenge	攻撃／接種／挑戦／難題／～に接種する／～に異議を唱える	11,785
challenges	攻撃／接種／挑戦／難題／～に接種する／～に異議を唱える	3,705
challenged	接種される	3,006
challenging	～を接種する／～に異議を唱える	2,353

名詞の用例が多いが，他動詞としても使われる．複数形のchallengesの割合が，約20%ある可算名詞．

◈類義語：attack, inoculation, expose, inoculate

■ 名 challenge ＋［前］

	the%	a/an%	ø%			
❶	1	16	79	challenge with ～	～による攻撃	1,731
				・after challenge with ～	～による攻撃のあと	173
				・against challenge with ～	～による攻撃に対して	157
❷	5	71	17	challenge in ～	～における難題／挑戦	854
❸	51	20	28	challenge of ～	～の挑戦	569
❹	6	89	1	challenge for ～	～にとっての挑戦	494
❺	8	87	4	challenge to ～	～への挑戦	474

❶ Vaccination with WRSd1 conferred protection against challenge with each of three virulent S. dysenteriae 1 strains. (Infect Immun. 2002 70:2950)

訳 WRSd1のワクチン接種は，3つの病原性のあるS. dysenteriae1株のそれぞれによる攻撃に対する保護を与えた

❷ Successful treatment of molar furcation defects remains a challenge in clinical practice. (J Periodontol. 2004 75:824)

訳 臼歯根分岐部欠損の成功する治療は，臨床診療における難題のままである

■ 名［名／形］＋ challenge

❶	major challenge	大きな挑戦	502
❷	lethal challenge	致死的な攻撃	231
❸	allergen challenge	アレルゲン接種	191

❶ Severe sepsis is common and presents a major challenge for clinicians, managers, and healthcare policymakers. (Crit Care Med. 2003 31:2332)

訳 ～は，臨床医にとって大きな挑戦を提示する

❸ Biopsies were taken from atopic volunteers at 1, 3, 6, 24, 48, and 72 h after intradermal allergen challenge and were examined by immunohistochemistry. (J Immunol. 2002 169:4604)

訳 皮内アレルゲン接種のあと

■ 動 be challenged ＋［前］

★ be challenged	…は，接種される	855
❶ be challenged with ～	…は，～を接種される	417
・mice were challenged with ～	マウスは，～を接種された	95
❷ be challenged by ～	…は，～によって攻撃される	190

❶ Mice were challenged with LPS to produce acute endotoxemia and pulmonary vascular injury.（Am J Respir Crit Care Med. 2005 172:1119）

訳 マウスは，LPSを接種された

■ 動［名］＋ challenged with

★ challenged with ～	～を接種される	1,370
❶ mice challenged with ～	～を接種されたマウス	205

★★ change 名 変化，動 変化する／変化させる　用例数 107,833

changes	変化／変化する／～を変化させる	68,995
change	変化／変化する／～を変化させる	31,209
changed	変えられる／変化した／～を変化させた	4,106
changing	変化する／～を変化させる	3,523

名詞の用例が多いが，他動詞あるいは自動詞としても用いられる．複数形のchangesの割合は，約70%と圧倒的に高い．

◆類義語：alteration, modification, shift, conversion, replacement, variation, alter, convert, vary

■ 名 changes ＋［前］

「～の変化」の意味では，changes ofよりchanges inが使われることが圧倒的に多い．

	the%	a/an%	ø%			
❶	9	-	82	changes in ～	～の変化	38,915
				・changes in gene expression	遺伝子発現の変化	966
				・associated with changes in ～	～の変化と関連する	369
				・in response to changes in ～	～の変化に応答して	287
❷	34	-	63	changes of ～	～の変化	1,706
❸	7	-	89	changes to ～	～への変化	988

❶ Such actions often require changes in gene expression（Proc Natl Acad Sci USA. 2000 97:10424）

訳 そのような作用は，しばしば遺伝子発現の変化を必要とする

❶ Changes in antibody specificity were associated with changes in biological activity.（Infect Immun. 2004 72:13）

訳 抗体特異性の変化は，生物学的活性の変化と関連した

■ 名 changes ＋［過分］

❶	changes associated with ～	～と関連する変化	659
❷	changes induced by ～	～によって誘導される変化	312

❶ Hence, we set forth to determine whether **hormonal and metabolic changes associated with sexual maturation**, reproduction, aging, and calorie restriction affect Acrp30.（Diabetes. 2003 52:268）

訳 性成熟と関連するホルモンおよび代謝の変化

C

■ 名 changes ＋［動］

❶	changes occur	変化が，起こる	345
❷	changes were observed	変化が，観察された	321

❶ **Dramatic changes occur** in nuclear organization and function during the critical developmental transition from meiosis to mitosis.（Mol Biol Cell. 2002 13:558）

訳 劇的な変化が起こる

■ 名［形／過分］＋ changes

❶	conformational changes	立体構造的変化	3,696
❷	structural changes	構造的変化	1,805
❸	significant changes	有意な変化	1,402
❹	～-induced changes	～に誘導された変化	1,229

❶ Using fluorescence stopped-flow kinetic studies, **we have discerned the conformational changes in the Rho protein** that occur upon nucleotide and nucleic acid binding.（J Biol Chem. 2004 279:18370）

訳 われわれは，Rhoタンパク質の高次構造的変化を識別した

❸ **No significant changes in heart rate, blood pressure, serum potassium, or renal function were observed**.（Circulation. 2003 107:2690）

訳 心拍数，血圧，血清カリウム，あるいは腎機能の有意な変化は，観察されなかった

■ 動 be changed ＋［前］

★	be changed	…は，変えられる	923
❶	be changed to ～	…は，～に変えられる	392
❷	be changed from ～	…は，～から変えられる	90

❶ To investigate this possibility, we constructed two HIV-1 clones, in which **Trp23 or Phe40 was changed to Ala**.（J Virol. 2001 75:9357）

訳 Trp23あるいはPhe40が，アラニンに変えられた

❷ **The oligomeric state and quaternary structure of the mutant protein are drastically changed from the wild type protein**.（Nat Struct Biol. 2002 9:877）

訳 変異タンパク質のオリゴマー状態と四次構造は，野生型タンパク質とは劇的に変化している

character 名 特徴／特性 用例数 2,084

character	特徴／特性	1,609
characters	特徴／特性	475

複数形のcharactersの割合が，約20%ある可算名詞.

◆類義語：feature, hallmark, property, profile, unique

■ 名 character ＋［前］

character ofの前にtheが付く割合は圧倒的に高い.

	the%	a/an%	ø%			
❶	96	0	4	character of ～	～の特徴	544

❶ Here, we map the hydrophobic **character of** the binding surface of the IQN17 peptide, a soluble analogue of the N-peptide coiled coil. (Biochemistry. 2002 41:2956)

訳 われわれは，IQN17ペプチドの結合表面の疎水性特性をマップする

★ characteristic 名 特徴／特性，形 特徴的な 用例数 24,943

characteristics	特徴／特性	14,684
characteristic	特徴／特性／特徴的な／特有の	10,259

名詞および形容詞として用いられる．名詞としては，複数形のcharacteristicsの割合は，約95%と圧倒的に高い.

◆類義語：feature, character, hallmark, property, profile, unique

■ 名 characteristics ＋［前］

	the%	a/an%	ø%			
❶	52	-	36	characteristics of ～	～の特徴	5,532
❷	9	-	83	characteristics in ～	～における特徴	302

❶ Many of the **characteristics of** the in vitro replication closely mimicked those of in vivo replication. (J Biol Chem. 2004 279:17404)

訳 試験管内複製の特徴の多くは，生体内複製のそれらを綿密に模倣した

■ 名 characteristics ＋［現分／過分／形］

❶	characteristics including ～	～を含む特徴	182
❷	characteristics associated with ～	～と関連する特徴	161
❸	characteristics similar to ～	～に似た特徴	160

❸ The resistant cells had growth and morphological **characteristics similar to** those of untreated cells. (J Virol. 2002 76:8864)

訳 耐性細胞は，未処置の細胞のそれらに似た成長および形態学的特徴を持っていた

C

■ 名 [形／名] + characteristics

❶	clinical characteristics	臨床的特徴	610
❷	patient characteristics	患者の特徴	456
❸	baseline characteristics	ベースライン特性	349
❹	functional characteristics	機能的特徴	305
❺	demographic characteristics	人口統計的特徴	302

❶ The purpose of this study was to identify the clinical characteristics of children with LVNC. (Circulation. 2003 108:2672)
訳 この研究の目的は，LVNCの子供の臨床的特徴を同定することであった

■ 形 characteristic + [前]

❶	characteristic of ～	～の特徴である	4,708
	・be characteristic of ～	…は，～の特徴である	1,058

❶ The accumulation of gross chromosomal rearrangements (GCRs) is characteristic of cancer cells. (Proc Natl Acad Sci USA. 2004 101:15980)
訳 ～は，癌細胞の特徴である

★ characterization 名 特徴づけ　用例数 11,882

characterization	特徴づけ	11,652
characterizations	特徴づけ	230

複数形のcharacterizationsの割合は2％しかなく，不可算名詞として使われることが多い．characterization ofの用例が圧倒的に多い．

■ characterization + [前]

	the%	a/an%	ø%			
❶	50	8	42	characterization of ～	～の特徴づけ	9,669
				・functional characterization of ～	～の機能的な特徴づけ	392
				・biochemical characterization of ～	～の生化学的な特徴づけ	345
				・further characterization of ～	～のさらなる特徴づけ	345
				・molecular characterization of ～	～の分子的な特徴づけ	312

❶ Further characterization of this protein shows that it localizes to the cytoplasm of fly cells and is expressed through all stages of fly embryonic development. (Gene. 2004 342:49)
訳 このタンパク質のさらなる特徴づけは，～ということを示す

■ [動] + the characterization of

★ the characterization of ～	～の特徴づけ	1,420
❶ report the characterization of ～	…は，～の特徴づけを報告する	255
❷ describe the characterization of ～	…は，～の特徴づけを述べる	101

■ [名] + and characterization of

❶ identification and characterization of ～	～の同定および特徴づけ	609
・we report the identification and characterization of ～	われわれは，～の同定および特徴づけを報告する	153
❷ isolation and characterization of ～	～の単離および特徴づけ	391
❸ cloning and characterization of ～	～のクローニングおよび特徴づけ	325

❶ We report the identification and characterization of the major phase I metabolites of this drug candidate. (Mol Pharmacol. 2004 65:56)

訳 われわれは，この薬物候補の主要な第一相代謝産物の同定と特徴づけを報告する

★★ characterize 動 特徴づける　用例数 47,150

characterized	特徴づけられる／～を特徴づけた	35,263
characterize	～を特徴づける	9,373
characterizing	～を特徴づける	1,837
characterizes	～を特徴づける	677

他動詞．受動態の用例が多い．characterized byの用例が非常に多い．

■ be characterized + [前]

★ be characterized	…は，特徴づけられる	11,664
❶ be characterized by ～	…は，～によって特徴づけられる	7,484
・be characterized by the presence of ～	…は，～の存在によって特徴づけられる	166
❷ be characterized in ～	…は，～において特徴づけられる	713
❸ be characterized as ～	…は，～として特徴づけられる	448

❶ These peptides are characterized by the presence of an amino acid with a 6-membered cyclic guanidine side chain (capreomycidine) and two or more 2, 3-diaminopropionate residues. (Gene. 2003 312:215)

訳 これらのペプチドは，～を持つアミノ酸の存在によって特徴づけられる

❸ The orphan nuclear receptor CAR (NR1I3) has been characterized as a central component in the coordinate response to xenobiotic and endobiotic stress. (J Biol Chem. 2004 279:19832)

訳 オーファン核受容体CAR (NR1I3) は，～における中心的な成分として特徴づけられてきた

■ [名] + characterized by

★ characterized by ～	～によって特徴づけられる	14,521
❶ disorder characterized by ～	～によって特徴づけられる疾患	886
❷ disease characterized by ～	～によって特徴づけられる疾患	583

❶ Spinal muscular atrophy (SMA) is an autosomal recessive disorder characterized by a loss of α motoneurons in the spinal cord. (J Cell Biol. 2003 162:919)

訳 脊髄性筋萎縮症（SMA）は，α運動ニューロンの喪失によって特徴づけられる常染色体劣性遺伝疾患である

C

■ [副] + characterized

❶ well-characterized ～	よく特徴づけられた～	3,179
❷ previously characterized	以前に特徴づけられた	1,231
❸ further characterized	さらに特徴づけられた	487
❹ poorly characterized	あまり特徴づけられていない	483
❺ best characterized	最もよく特徴づけられた	347
❻ fully characterized	完全に特徴づけられた	325

❶ Escherichia coli RNase HI is a well-characterized model system for protein folding and stability. (J Mol Biol. 2001 314:863)

訳 ～は，タンパク質フォールディングのよく特徴づけられたモデルシステムである

❷ The regulation of STAT2 nuclear trafficking is distinct from the previously characterized STAT1 factor. (J Biol Chem. 2004 279:39199)

訳 STAT2の核輸送の調節は，以前に特徴づけられたSTAT1因子と異なっている

❸ Cleavage of M3R was further characterized by assays performed on the M3R molecule generated by in vitro translation. (Arthritis Rheum. 2001 44:2376)

訳 M3Rの切断は，～に対して行われたアッセイによってさらに特徴づけられた

■ [過去／過分] + and characterized

❶ identified and characterized ～	同定されそして特徴づけられる／～を同定しそして特徴づけた	514
❷ isolated and characterized ～	単離されそして特徴づけられる／～を単離しそして特徴づけた	489
❸ cloned and characterized ～	クローンされそして特徴づけられる／～をクローンしそして特徴づけた	400

❶ To this end, specific small molecule inhibitors of human B7.1 were identified and characterized. (J Biol Chem. 2002 277:7363)

訳 ～が，同定されそして特徴づけられた

■ [代名] + characterize

❶ we characterized ～	われわれは，～を特徴づけた	2,101

❶ In this study, we characterized an allelic mutation, Dsg3bal-Pas, with clinical features similar to those in Dsg3bal-2J. (J Invest Dermatol. 2002 119:1237)

訳 われわれは，対立遺伝子の変異Dsg3bal-Pasを特徴づけた

■ characterize ＋［名句］

❶	characterize the role of ～	…は，～の役割を特徴づける	326
❷	characterize the effects of ～	…は，～の効果を特徴づける	200

■ to characterize

★	to characterize ～	～を特徴づける…	5,194
❶	To characterize ～	～を特徴づけるために	920
❷	be used to characterize ～	…は，～を特徴づけるために使われる	630
❸	to further characterize ～	～をさらに特徴づけるために	488
❹	be to characterize ～	…は，～を特徴づけることである	416
❺	to better characterize ～	～をよりよく特徴づけるために	145

❷ Electrospray mass spectrometry was used to characterize the nonmetallated BBN conjugates. (Cancer Res. 2003 63:4082)

訳 エレクトロスプレー質量分析が，～を特徴づけるために使われた

❸ To further characterize the physiological function of InvA, the gene expression pattern during various stages of rickettsial intracellular growth was investigated. (Infect Immun. 2002 70:6346)

訳 InvAの生理的機能をさらに特徴づけるために

❹ The purpose of this study was to characterize an antigen that was recognized by antibodies present in sera of challenge-exposed pigs. (Infect Immun. 2000 68:4559)

訳 この研究の目的は，～を特徴づけることであった

choice 名 選択　　用例数 5,041

choice	選択	4,125
choices	選択	916

複数形のchoicesの割合が，20%ある可算名詞.

◆類義語：selection, option

■ choice ＋［前］

choice ofの前にtheが付く割合は非常に高い.

	the%	a/an%	ø%			
❶	76	6	16	choice of ～	～の選択	1,150
❷	56	35	8	choice between ～	～の間の選択	172

❶ Simulations are presented that demonstrate how the choice of airflow rate can affect quantitation. (Anal Chem. 2000 72:3949)

訳 気流速度の選択がどのように定量化に影響するかを実証するシミュレーションが示される

■ of choice

★ of choice	最適な…	695
❶ of choice for ～	～のために最適な…	361
❷ treatment of choice	最適な治療	153
❸ method of choice	最適な方法	102

❷ Surgery remains the treatment of choice for operable tumors, whereas chemotherapy is standard in locally advanced and metastatic disease. (J Clin Oncol. 2011 29:4820)

訳 外科手術は，手術可能な腫瘍のために最適な治療のままである

choose 動 選択する／選ぶ　用例数 3,134

chosen	選択される	1,688
choose	～を選択する	558
choosing	～を選択する	439
chose	～を選択した	386
chooses	～を選択する	63

他動詞.

◈類義語：select

■ be chosen ＋［前］

★ be chosen	…が，選択される	840
❶ be chosen for ～	…が，～のために選択される	215
❷ be chosen to *do*	…が，～するように選択される	146
❸ be chosen as ～	…が，～として選択される	142

❶ Five genes shown to have different transcript levels in response to either mild or severe stress were chosen for further analysis using real-time polymerase chain reaction. (Plant Physiol. 2003 133:1702)

訳 ～が，さらなる解析のために選択された

❷ Clinical measures were chosen to reflect the gingival health and tooth whiteness in an intent-to-treat study design. (J Periodontol. 2004 75:57)

訳 臨床的な尺度が，～を反映するように選択された

★ chronic 形 慢性の／長期にわたる　用例数 26,915

◈反意語：acute

■ chronic ＋［(形＋)名］

❶ chronic hepatitis	慢性肝炎	963
❷ chronic inflammation	慢性炎症	842
❸ chronic infection	慢性感染症	761

❹	chronic rejection	慢性拒絶	697
❺	chronic lymphocytic leukemia	慢性リンパ性白血病	661
❻	chronic obstructive pulmonary disease	慢性閉塞性肺疾患	616
❼	chronic kidney disease	慢性腎臓疾患	584
❽	chronic disease	慢性疾患	535

❹ Immune-mediated injury to the graft **has been implicated in the pathogenesis of chronic rejection**. (Transplantation. 2002 74:1053)
訳 ～は，慢性拒絶の病因に関与している

■ [前] ＋ chronic

❶	patients with chronic ～	慢性～の患者	1,627
❷	treatment of chronic ～	慢性～の治療	312
❸	associated with chronic ～	慢性～と関連した	294

❶ Reagents that regulate ZIP14 activity might be developed as **therapeutics to promote liver regeneration in patients with chronic liver disease**. (Gastroenterology. 2012 142:1536)
訳 慢性肝疾患の患者における肝再生を促進するための治療法

■ [形] ＋ and chronic

❶	acute and chronic ～	急性および慢性の～	989

❶ **Staphylococcus aureus can cause a variety of acute and chronic diseases.** (J Infect Dis. 2004 190:571)
訳 黄色ブドウ球菌は，さまざまな急性および慢性疾患を引き起こしうる

chronically 副 慢性的に／慢性に 用例数 1,950

◆反意語：acutely

■ chronically ＋ [過分]

❶	chronically infected	慢性的に感染した	531
❷	chronically instrumented	慢性的装置装着した	103

❶ However, treatment with HAART generally does not restore HIV-1-specific $CD4^+$ T cell responses **in chronically infected patients**. (J Infect Dis. 2001 183:657)
訳 慢性的に感染した患者において

circuit 名 回路 用例数 7,224

circuits	回路	3,812
circuit	回路	3,412

複数形のcircuitsの割合は約55%と非常に高い.
◆類義語：circuitry

■ circuits ＋［前］

	the％	a/an％	⌀％			
❶	10	-	84	circuits in ～	～における回路	303

❶ Axon branching plays a critical role in establishing the accurate patterning of neuronal circuits in the brain.（J Neurosci. 2010 30:16766）

訳 ～は，脳における神経回路の正確なパターン形成を確立する際に決定的に重要な役割を果たす

C

■［形］＋ circuits

❶	neural circuits	神経回路	640
❷	neuronal circuits	神経回路	255

circumstance 名 状況／環境 用例数 1,256

circumstances	状況	1,193
circumstance	状況	63

複数形のcircumstancesの割合は，約95％と圧倒的に高い．

◆類義語：case, occasion, instance, situation, condition, state, context

■［形］＋ circumstances

❶	certain circumstances	ある状況	134
	・under certain circumstances	ある状況下で	102
❷	some circumstances	ある状況	122

❶ Although nitric oxide（NO）is a crucial mediator in preconditioning under certain circumstances, the role of NO in CSD-induced neuroprotection is unclear.（Brain Res. 2005 1039:84）

訳 一酸化窒素（NO）は，ある状況下でのプレコンディショニングにおいて決定的に重要なメディエーターである

clarify 動 明らかにする／明白にする 用例数 2,450

clarify	～を明らかにする／～を明白にする	1,709
clarified	～を明らかにした／明らかにされた	466
clarifying	～を明らかにする／～を明白にする	169
clarifies	～を明らかにする／～を明白にする	106

他動詞．to clarifyの用例が非常に多い．

◆類義語：elucidate, reveal, disclose, uncover, manifest, define

■ to clarify

★ to clarify ～	～を明らかにするために	1,080
❶ To clarify ～	～を明らかにするために	384
❷ to clarify the role of ～	～の役割を明らかにするために	178

❷ To clarify the role of Notch in Th cell differentiation, we generated mice that conditionally inactivate Notch signaling in mature T cells. (J Exp Med. 2005 202:1037)

訳 Th細胞分化におけるNotchの役割を明らかにするために

★★ class 名 クラス／分類, 動 分類する　用例数 38,404

class	クラス	31,637
classes	クラス	6,735
classed	分類された	32

動詞としても使われるが，名詞の用例が圧倒的に多い．複数形のclassesの割合が，約20%ある可算名詞．

◆類義語：group, type, series, set

■ class ＋［前］

	the%	a/an%	ø%			
❶	14	72	0	class of ～	クラスの～	7,630
				・class of compounds	…クラスの化合物	239
				・class of proteins	…クラスのタンパク質	213

❶ This class of compounds binds to the active site via two metal ions that are coordinated by catalytic site residues, D443, E478, D498, and D549. (J Virol. 2010 84:7625)

訳 このクラスの化合物は，活性部位に結合する

■［冠／形／代名］＋ class(es) of

❶ a class of ～	あるクラスの～	1,355
❷ a new class of ～	新しいクラスの～	1,119
❸ this class of ～	このクラスの～	884
❹ two classes of ～	2つのクラスの～	717
❺ a novel class of ～	新規のクラスの～	575
❻ different classes of ～	異なるクラスの～	424
❼ distinct classes of ～	別個のクラスの～	255
❽ one class of ～	ひとつのクラスの～	211
❾ an important class of ～	重要なクラスの～	148

❶ This offers the possibility of a new class of sensor devices with unique capabilities. (Nat Commun. 2012 3:829)

訳 これは，新しいクラスのセンサーデバイスの可能性を提供する

classification 名 分類　　用例数 4,375

classification	分類	4,041
classifications	分類	334

複数形のclassificationsの割合は約8％あるが，単数形は無冠詞で使われることが非常に多い．

◈類義語：circuit, network

■ classification ＋［前］

	the％	a/an％	ø％			
❶	43	10	46	classification of ～	～の分類	1,221

❶ Discriminant analysis **is an effective tool for the classification of experimental units** into groups.（Bioinformatics. 2011 27:495）

訳 ～は，実験単位の分類のための効果的な手段である

classify 動 分類する　　用例数 4,459

classified	分類される／～を分類した	3,440
classify	～を分類する	648
classifying	～を分類する	295
classifies	～を分類する	76

他動詞．

◈類義語：categorize, sort

■ be classified ＋［前］/according to

★	be classified	…は，分類される	2,077
❶	be classified as ～	…は，～として分類される	1,161
	・be classified as having ～	…は，～を持つとして分類される	102
❷	be classified into ～	…は，～に分類される	410
❸	be classified by ～	…は，～によって分類される	110
❹	be classified according to ～	…は，～に従って分類される	113

❶ Of these subjects, **956（15%）were classified as having widespread pain**, 3061（48%）as having regional pain, and 2314（37%）as having no pain.（Arthritis Rheum. 2003 48:1686）

訳 956名（15%）が，広範な痛みを持つとして分類された

❷ **Patients were classified into two groups**: those with ≧0.5 mm in the MIT in any matched site（group 1）and those with MIT <0.5 mm（group 2）.（J Am Coll Cardiol. 2005 45:1532）

訳 患者は，2つのグループに分類された

■［名］＋ be classified

❶	patients were classified	患者は，分類された	138

❶ Patients were classified according to the region of onset and the physical signs in the hands. (Brain. 2003 126:2558)

訳 患者は，発症の領域に従って分類された

■ to classify

★ to classify ～	～を分類する…	402
❶ be used to classify ～	…は，～を分類するために使われる	91

❶ Clinical risk factors were used to classify hypothetical patients into cardiovascular and bleeding risk groups on the basis of published data. (Ann Intern Med. 2005 143:241)

訳 臨床上のリスク因子が，仮定上の患者を心血管および出血リスク群に分類するために使われた

★clear 形 明らかな，動 除去する 用例数 10,426

clear	明らかな／～を除去する	8,848
cleared	除去される／～を除去した	1,079
clearing	～を除去する	420
clears	～を除去する	79

形容詞の用例が多いが，他動詞としても用いられる．

◆類義語：apparent, evident, obvious, pronounced, distinct, scavenge, clean

■ 形 clear ＋［名節］

❶ it is clear that ～	～ということは明らかである	392
・it is now clear that ～	～ということは，今，明らかである	160
・it has become clear that ～	～ということが明らかになっている	119
❷ it is not clear whether ～	～かどうかは明らかではない	344
❸ it is not clear how ～	どのように～かは明らかではない	198

❶ It is clear that one of the roles of the Brca1 protein is to facilitate cellular responses to DNA damage. (Cancer Res. 2002 62:4588)

訳 ～ということは明らかである

❷ It is not clear whether these cellular changes are synonymous, sequential, or distinct responses to the protein. (J Biol Chem. 2004 279:17667)

訳 ～かどうかは明らかでない

❸ However, it is not clear how critical this priming is for immune responses or how it is normally induced in vivo. (Science. 2004 304:1808)

訳 どのように～かは明らかではない

■ 形［副／形］＋ clear

❶ be less clear	～は，より明らかでない	464
❷ no clear ～	明らかな～はない	134

❶ In higher eukaryotes, however, the mechanism by which DNA synthesis is reduced is less clear. (J Biol Chem. 2004 279:20067)

訳 DNA合成が減弱される機構は，より明らかでない

❷ There is no clear benefit of a short course of parent-initiated oral prednisolone for viral wheeze in children aged 1-5 years even in those with above-average eosinophil priming. (Lancet. 2003 362:1433)

訳 ～の明らかな利点はない

■ 形 clear ＋［名］

❶ clear evidence	明らかな証拠	393
❷ clear differences	明らかな違い	107

■ 動 cleared ＋［前］

❶ cleared from ～	～から除去される	245
・be cleared from ～	…は，～から除去される	122
❷ cleared by ～	～によって除去される	98

❶ Despite this physical entrapment within the subretinal space, IRBP is rapidly cleared from the IPM by an unknown mechanism. (J Comp Neurol. 2003 466:331)

訳 IRBPは，未知の機構によってIPMから急速に除去される

clearance 名 排除／クリアランス　用例数 7,428

clearance	排除／クリアランス	7,342
clearances	排除／クリアランス	86

複数形のclearancesの割合は1％しかなく，原則，不可算名詞として使われる．

◆類義語：exclusion, elimination

■ clearance ＋［前］

	the％	a/an％	ø％			
❶	21	3	74	clearance of ～	～の排除	2,423
				・clearance of apoptotic cells	アポトーシス細胞の排除	90
❷	1	1	71	clearance from ～	～からの排除	244

❶ Efficient clearance of apoptotic cells by phagocytes (efferocytosis) is critical for normal tissue homeostasis and regulation of the immune system. (J Immunol. 2011 187:5783)

訳 貪食細胞によるアポトーシス細胞の効率的な排除（貪食除去）は，正常な組織ホメオスタシスのために決定的に重要である

■［形／名］＋ clearance

❶ viral clearance	ウイルスの排除	441
❷ creatinine clearance	クレアチニンクリアランス	317
❸ bacterial clearance	細菌の排除	315

❶ $CD8^+$ cytotoxic T cells are critical for viral clearance from the lungs upon influenza virus

infection.（J Clin Invest. 2012 122:4037）
訳 CD8$^+$細胞傷害性T細胞は，肺からのウイルスの排除のために決定的に重要である

★clearly　副 明確に／明らかに　用例数 5,530

◆類義語：apparently, distinctly, evidently, obviously, unambiguously

■ clearly ＋［過分］

❶ clearly defined	明確に明らかにされる	418
・be clearly defined	…は，明確に明らかにされる	191
❷ clearly demonstrated	明確に実証される	206
❸ clearly established	明確に確立される	156
❹ not clearly understood	明確には理解されていない	132

❶ However, **the pathway of apoptosis is not clearly defined**.（J Biol Chem. 2002 277:37820）
訳 アポトーシスの経路は，明確には明らかにされていない

❹ The effect of road vehicle traffic pollution on asthma **is still not clearly understood**.（Am J Respir Crit Care Med. 2001 164:2177）
訳 ～は，まだ，明確には理解されていない

■ clearly ＋［動］

❶ clearly demonstrate ～	…は，～を明確に実証する	318
❷ clearly show ～	…は，～を明確に示す	214
❸ clearly indicate ～	…は，～を明確に示す	141

❶ **Our results clearly demonstrate that** target cell CypA, and not producer cell CypA, is important for HIV-1 CA-mediated function.（J Virol. 2004 78:12800）
訳 われわれの結果は，～ということを明確に実証する

❷ **These results clearly show that** iNOS is not required for control of T. cruzi infection in mice.（Infect Immun. 2004 72:4081）
訳 これらの結果は，～ということを明確に示す

★cleavage　名 切断　用例数 22,699

cleavage	切断	22,090
cleavages	切断	609

複数形のcleavagesの割合は約３％あるが，単数形は無冠詞で使われることが圧倒的に多い．

◆類義語：truncation, ablation, break, breakage, scission, deletion, cut

■ cleavage ＋［前］

	the%	a/an%	ø%			
❶	29	3	61	cleavage of ～	～の切断	5,311

❷	6	1	64	cleavage by ～	～による切断	910
❸	2	6	80	cleavage at ～	～における切断	826
❹	6	2	72	cleavage in ～	～における切断	512

❶ Cleavage of the A1-A2 domain junction by factor Xa R240A was not blocked by the 337-372 peptide.（J Biol Chem. 2004 279:33104）

訳 Xa因子R240AによるA1-A2ドメインジャンクションの切断は，337-372ペプチドによって阻止されなかった

❷ The cleavage by γ-secretase also results in the cytoplasmic release of a 59- or 57-residue-long C-terminal fragment（Cγ）.（Proc Natl Acad Sci USA. 2001 98:14979）

訳 γ-分泌酵素による切断は，また，～の細胞質内放出という結果になる

❸ Modified RNA/DNA hybrids were evaluated for cleavage at the PPT/U3 junction.（J Biol Chem. 2004 279:37095）

訳 PPT/U3ジャンクションにおける切断

C

■ cleavage ＋［動］

❶	cleavage occurs	切断が，起こる	160

❶ Cleavage occurs at specific sites consisting of alternating IU and UI base pairs.（EMBO J. 2001 20:4243）

訳 切断が，～からなる特異的部位で起こる

■［名／形／過分］＋ cleavage

❶	DNA cleavage	DNA切断	964
❷	proteolytic cleavage	タンパク質分解性切断	791
❸	bond cleavage	結合切断	465
❹	～-mediated cleavage	～に仲介される切断	371

■ cleavage ＋［名］

❶	cleavage site	切断部位	1,629
❷	cleavage products	切断産物	348
❸	cleavage activity	切断活性	332

cleave 動 切断する　用例数 8,240

cleaved	切断される／～を切断した	4,121
cleave	～を切断する	1,573
cleaves	～を切断する	1,554
cleaving	～を切断する	992

他動詞.

◆類義語：truncate, break, ablate, cut

■ be cleaved ＋［前］

★ be cleaved	…は，切断される	1,487
❶ be cleaved by ～	…は，～によって切断される	410
❷ be cleaved in ～	…は，～において切断される	128
❸ be cleaved at ～	…は，～において切断される	125

❶ The crystallin fragments obtained from aged bovine lenses were cleaved by the purified enzyme.（Invest Ophthalmol Vis Sci. 2004 45:1214）
訳 ～が，精製された酵素によって切断された

❷ We show here that BLM, but not the related RECQ-like helicase WRN, is rapidly cleaved in cells undergoing apoptosis.（J Biol Chem. 2001 276:12068）
訳 ～は，アポトーシスを起こしている細胞において急速に切断される

■［副］＋ cleaved

❶ proteolytically cleaved	タンパク分解性に切断される	125

★★ clinical 形 臨床の 用例数 58,558

■ clinical ＋［名］

❶ clinical trials	臨床治験	5,735
❷ clinical outcomes	臨床成績	1,560
❸ clinical practice	診療	1,286
❹ clinical studies	臨床研究	1,244
❺ clinical isolates	臨床分離株	1,046
❻ clinical features	臨床的特徴	981

❶ Key parameters can be determined via in vitro bench testing defining delivery standards for clinical trials of drugs with narrow therapeutic/toxicity ratios.（Am J Respir Crit Care Med. 2003 168:1205）
訳 ～を持つ薬剤の臨床治験

★ clinically 副 臨床的に 用例数 8,517

■ clinically ＋［形］

❶ clinically relevant	臨床的に関連する（重要な）	1,440
❷ clinically significant	臨床的に有意な	837
❸ clinically important	臨床的に重要な	626
❹ clinically useful	臨床的に有用な	351
・be clinically useful	…は，臨床的に有用である	110
❺ clinically meaningful	臨床的に有意義な	186

❶ The results of our recent in vitro studies of clinically relevant CsA concentrations demonstrated the modulation of macrophage scavenger receptors (MSRs) involved in atherogenesis. (Transplantation. 2004 77:1281)

訳 臨床的に重要なシクロスポリンA濃度のわれわれの最近の試験管内での研究の結果は，～を実証した

❷ Combination treatment produced a clinically significant improvement in health status and the greatest reduction in daily symptoms. (Lancet. 2003 361:449)

訳 併用療法は，健康状態の臨床的に有意な改善を生み出した

★close 形 近い／密接な，動 閉じる　　用例数 14,371

close	近い／密接な／～を閉じる	8,772
closed	～を閉じた／閉じた	4,323
closing	～を閉じる	1,072
closes	～を閉じる	204

他動詞としても使われるが，形容詞の用例が多い．close toの用例が非常に多い．closedも形容詞的に使われることが多い．

◆類義語：adjacent, near, nearby

■ 形 close ＋［前］

❶ close to ～	～に近い	3,368
・be close to ～	…は，～に近い	709
・located close to ～	～の近くに位置する	143
・very close to ～	～に非常に近い	166

❶ The equilibrium dissociation constant for the ternary complex was 18 ± 4 nm, which is close to that observed in the prokaryotic system. (J Biol Chem. 2000 275:20308)

訳 そして（それは，）原核生物システムにおいて観察されるそれに近い

❶ The lack of cleavage at sites located close to the glycosylated asparagine residue may result from steric blocking by the glycan. (Anal Chem. 2001 73:4530)

訳 グリコシル化されたアスパラギン残基の近くに位置する部位での切断の欠如は，～に由来するかもしれない

■ 形 close ＋［名］

❶ close proximity	近接	1,212
・in close proximity to ～	～に近接して	624
❷ close association	密接な関連／密接な結合	243
❸ close contact	密接な関連／密接な結合	186

❶ MKK4, located in close proximity to p53 gene, is thought to be a tumor suppressor and a metastasis suppressor gene. (Oncogene. 2004 23:5978)

訳 p53遺伝子に近接して位置する

❷ Approximately half of the LPTL motifs are in close association with proteins. (J Mol Biol. 2003 325:65)

訳 LPTLモチーフのおよそ半分は，タンパク質と密接に結合している

■ 動 closed + [名]

❶ closed conformation	閉構造	461
❷ closed state	閉状態	419

★ closely 副 密接に／近くに 用例数 10,340

closely relatedの用例が多い.

◆類義語：tightly, near, nearby

■ closely + [過分]

❶ closely related	密接に関連する	4,813
・closely related to ～	～に密接に関連する	1,554
・be closely related	…は，密接に関連する	890
❷ closely linked	密接に関連する	567
❸ closely associated with ～	～と密接に関連する	527

❶ All opsins from both ostracod species examined **are more closely related to each other than** to any other known opsin sequences. (J Mol Evol. 2004 59:239)

訳 …は，お互いに～より密接に関連する

■ closely + [動]

❶ closely resemble ～	…は，～によく似ている	975

❶ **The overall structure of the enzyme closely resembles that of M.RsrI.** (Nucleic Acids Res. 2003 31:5440)

訳 その酵素の全体の構造は，M.RsrIのそれによく似ている

■ [動] + closely + [前]

❶ correlate closely with ～	…は，～と密接に関連している	169

❶ **Peak levels of MKP-1 correlated closely with the decline in** p38 MAPK and ERK1/2 phosphorylation. (J Immunol. 2010 185:7562)

訳 MKP-1のピークレベルは，～の低下と密接に関連していた

■ [副] + closely

❶ more closely	もっと密接に	853
❷ most closely	最も密接に	663

❷ Based on protein sequence alignment, **TLP is most closely related to** the cysteine-rich proteins, a subclass of the family of LIM-only proteins. (Mol Cell Biol. 2001 21:8592)

訳 TLPは，～に最も密接に関連する

★cluster 名 クラスター／集団, 動 クラスター形成する 用例数 23,256

cluster	クラスター／集団／クラスター形成する	12,006
clusters	クラスター／集団／クラスター形成する	8,890
clustered	クラスター化された／クラスター形成した	2,360

動詞としても用いられるが，名詞の用例が非常に多い．複数形のclustersの割合は約45%とかなり高い．動詞としては，他動詞と自動詞の両方の用例がある．clusteringは別項参照．

◆類義語：group, cohort

C

■ 名 cluster ＋［前］

	the%	a/an%	ø%			
❶	22	71	5	cluster of ～	集団の～／～のクラスター	1,623
				・a cluster of ～	一群の～	825
❷	54	37	2	cluster in ～	～におけるクラスター	646

❶ We demonstrate a parallel shift in longitudinal gene expression that occurs in a cluster of genes related to the immune response. (Proc Natl Acad Sci USA. 2010 107:14799)
訳 免疫応答に関連する一群の遺伝子において起こる

■ 動 clustered ＋［前］

❶	clustered in ～	～においてクラスター化される／～においてクラスター形成する	378
❷	clustered into ～	～にクラスター化される／～にクラスター形成する	138

❶ Most LGMD2H mutations in TRIM32 are clustered in the NHL β-propeller domain at the C-terminus and are predicted to interfere with homodimerization. (Hum Mol Genet. 2011 20:3925)
訳 TRIM32におけるほとんどのLGMD2H変異は，NHL β-プロペラドメインにおいてクラスター化される

❶ Mutations clustered in the phosphatase domain and impaired 5-phosphatase activity, resulting in altered cellular PtdIns ratios. (Nat Genet. 2009 41:1032)
訳 変異は，ホスファターゼドメインにおいてクラスター形成した

clustering 名 クラスター化／クラスター形成 用例数 4,451

clustering	クラスター化／クラスター形成	4,435
clusterings	クラスター化／クラスター形成	16

名詞の用例が圧倒的に多い．複数形のclusteringsの割合は0.4%しかなく，原則，不可算名詞として使われる．

◆類義語：grouping

■ clustering ＋［前］

	the％	a/an％	ø％			
❶	33	4	63	clustering of ～	～のクラスター化	1,276

❶ Hierarchical clustering of compounds was carried out based on their bioactivity profiles. (Bioinformatics. 2010 26:2881)
訳 化合物の階層的クラスター化が行われた

★code 動 コードする，名 コード　用例数 15,514

coding	コードする／コーディング	10,271
code	コードする／コード	3,040
codes	コードする／コード	1,467
coded	コードされる／コードした	736

動詞の用例が多いが，名詞としても用いられる．
◆類義語：encode

■ 動 code for ＋［名句］

★ code for ～	…は，～をコードする	2,522
❶ code for a protein	…は，タンパク質をコードする	91

❶ The 1.5 kb cDNA codes for a protein of 495 amino acids. (Biochemistry. 2006 45:7323)
訳 その1.5 kbのcDNAは，495アミノ酸のタンパク質をコードする

■ 動［名］＋ coding for

★ coding for ～	～をコードする	1,135
❶ genes coding for ～	～をコードする遺伝子	273

❶ Most unexpectedly, genes coding for similar proteins were found in the genomes of eukaryotes, including humans. (J Biol Chem. 2005 280:15325)
訳 類似のタンパク質をコードする遺伝子が，真核生物のゲノムに見つけられた

■ 動 coding ＋［名］

❶ coding region	コード領域	1,887
❷ coding sequence	コード配列	1,196
❸ protein-coding genes	タンパク質をコードする遺伝子	540

■ 名 code ＋［前］

	the％	a/an％	ø％			
❶	56	35	5	code of ～	～のコード	116

❶ The source code of GPU-BLAST is freely available at http://archimedes.cheme.cmu.edu/biosoftware.html. (Bioinformatics. 2011 27:182)

訳 GPU-BLASTのソースコードは，～において無料で入手できる

■ 名［名／形］＋ code

❶	source code	ソースコード	417
❷	genetic code	遺伝コード	382

★ cohort 名 コホート／集団 用例数 13,113

cohort	コホート／集団	10,685
cohorts	コホート／集団	2,428

複数形のcohortsの割合が，約20%ある可算名詞.

◆類義語：group, cluster

■ 名 cohort ＋［前］

cohort ofの前には，a/anが付く割合が非常に高い.

	the%	a/an%	⌀%			
❶	3	88	3	cohort of ～	集団の～／～のコホート	3,203
				・a cohort of ～	一群の～	1,372
				・a large cohort of ～	大きな集団の～	240
				・cohort of patients	患者の集団	392

❶ We describe clinical, biochemical, and molecular findings in a cohort of patients with this treatable condition.（Ann Neurol. 2012 71:520）

訳 われわれは，この治療可能な状態の一群の患者における臨床的，生化学的および分子的知見を述べる

■ cohort ＋［名］

❶	cohort study	コホート研究	1,934
	・prospective cohort study	前向きコホート研究	502
	・retrospective cohort study	後向きコホート研究	315

coincide 動 一致する 用例数 2,401

coincided	一致した	934
coincides	一致する	788
coincide	一致する	411
coinciding	一致する	268

自動詞．coincide withの用例が圧倒的に多い.

◆類義語：correspond, agree, fit, match, represent

■ coincide with

❶ coincide with ～	…は，～と一致する	2,233

❶ Furthermore, RANK transcript expression **coincides with** the presence of multinuclear osteoclast-like cells. (J Exp Med. 2000 192:463)
訳 RANK転写物発現は，多核の破骨細胞様細胞の存在と一致する

coincident 形 一致する／同時に起こる　用例数 1,722

coincident withの用例が圧倒的に多い.
◆類義語：consistent, concurrent, congruent, corresponding, simultaneous, concomitant, synchronous, parallel

■ coincident ＋［前］

❶ coincident with ～	～に一致する	1,290
・be coincident with ～	…は，～に一致する	285

❶ Our results demonstrate that the onset of faradaic current **is coincident with** the onset of concentration enrichment. (Anal Chem. 2009 81:10149)
訳 ファラデー電流の開始は，濃度の濃縮の開始に一致する

collect 動 集める　用例数 8,452

collected	集められる／～を集めた	7,002
collecting	～を集める	987
collect	～を集める	394
collects	～を集める	69

他動詞.
◆類義語：harvest, gather, assemble, recover

■ be collected ＋［前］

★ be collected	…が，集められる	3,232
❶ be collected from ～	…が，～から集められる	719
❷ be collected at ～	…が，～において集められる	340
❸ be collected for ～	…が，～のために集められる	256
❹ be collected on ～	～に関する…が，集められる	193

❶ Formalin-fixed, paraffin-embedded tissue samples **were collected from** the United States and Paraguay. (Am J Pathol. 2001 159:1211)
訳 ホルマリンで固定され，パラフィンで包埋された組織サンプルが，アメリカ合衆国とパラグアイから集められた

❷ Specimens **were collected at** weekly intervals from birth to week 12 postpartum from 40 female Holstein calf-dam pairs in a dairy herd. (J Clin Microbiol. 2004 42:5664)

訳 検体が，1 週間おきに集められた

❹ Data were collected on risk factors, clinical breast examination, mammography, and cytology results. (Ann Surg. 2003 238:728)

訳 リスク因子に関するデータが集められた

■ ［名］＋ be collected

❶	data were collected	データが，集められた	615
❷	samples were collected	サンプルが，集められた	442

collection 名 コレクション／収集 用例数 4,362

collection	コレクション／収集	3,647
collections	コレクション／収集	715

複数形のcollectionsの割合が，約15%ある可算名詞.

◆類義語：acquisition

■ collection ＋［前］

	the %	a/an %	ø %			
❶	17	70	8	collection of ～	～のコレクション	1,849
				・a collection of ～	～のコレクション	772
				・a large collection of ～	～の大きなコレクション	105

❶ To translate such in vitro data to the intact heart, we used a collection of zebrafish cardiac mutants and transgenics to investigate whether cardiac conduction could influence in vivo cardiac morphogenesis independent of contractile forces. (Proc Natl Acad Sci USA. 2010 107:14662)

訳 われわれは，ゼブラフィッシュの心臓突然変異体のコレクションを使用した

■ ［名］＋ collection

❶	data collection	データ収集	395

collectively 副 まとめると 用例数 5,588

文頭で用いられることが断然多い.

◆類義語：taken together, in summary, in conclusion

■ Collectively

★	Collectively, ～	まとめると，～	4,301
❶	Collectively, these data ～	まとめると，これらのデータは～	934

❶ Collectively, these data suggest that MMSET has a role in PCa pathogenesis and progression through epigenetic regulation of metastasis-related genes. (Oncogene. 2013 32:2882)

訳 まとめると，これらのデータは～ということを示唆する

★ combination 名 組み合わせ／併用 用例数 26,638

combination	組み合わせ／併用	21,968
combinations	組み合わせ／併用	4,670

複数形のcombinationsの割合が，約20%ある可算名詞．〈combination of A and B〉〈in combination with〉の用例が多い．

◆類義語：conjunction, pair

■ combination ＋［前］

	the%	a/an%	ø%			
❶	35	59	5	combination of ～	～の組み合わせ	11,236
				・using a combination of ～	～の組み合わせを使って	1,254
				・by a combination of ～	～の組み合わせによって	841
				・different combinations of ～	～の異なる組み合わせ	292
				・various combinations of ～	～のさまざまな組み合わせ	213

❶ **Using a combination of evidence and expert opinion**, 10 indicators for quality of gout care were developed.（Arthritis Rheum. 2004 50:937）

訳 証拠と専門家の意見の組み合わせを使って

❶ **Different combinations of the regions are present in different mouse strains.**（Nucleic Acids Res. 2002 30:1394）

訳 それらの領域の異なる組み合わせが，異なるマウスの系統に存在する

■ in combination with

★	in combination with ～	～と組み合わせて	4,897
❶	used in combination with ～	～と組み合わせて使われる	242

❶ Synergy was observed **when SN38 was used in combination with** doxorubicin, bortezomib, as well as poly (ADP-ribose) polymerase inhibitor NU1025 and Fas-activator CH11.（Cancer Res. 2004 64:8746）

訳 SN38が～と組み合わせて使われたとき

■ combination ＋［名］

❶	combination therapy	併用療法	1,020
❷	combination treatment	併用治療	245

★ combine 動 組み合わせる／結合する／併用する 用例数 24,365

combined	組み合わされる／併用される／～を組み合わせた／結合した	18,038
combining	～を組み合わせる／結合する	3,659
combine	～を組み合わせる／結合する	1,452

combines	～を組み合わせる／結合する	1,216

他動詞受動態，特に，combined withの用例が多いが，自動詞としても用いられる．
◆分類：merge

■ combined ＋［前］

❶ combined with ～	～と組み合わされる	5,919
・be combined with ～	…は，～と組み合わされる	977
・when combined with ～	～と組み合わされると	241
・data combined with ～	～と組み合わされたデータ	94
・results combined with ～	～と組み合わされた結果	91
❷ combined to *do*	～するために組み合わされる	228
❸ combined in ～	～において組み合わされる	218

❶ The data **were combined with** the results obtained in 20 additional patients in a previously published pilot study.（Radiology. 1997 204:47）
訳 それらのデータは，～において得られた結果と組み合わされた

❶ **When combined with** parallel processes for metabolic and gene expression profiling, these platforms constitute a core technology in the high throughput determination of gene function.（Plant Cell. 2001 13:1499）
訳 代謝と遺伝子発現プロファイリングの平行プロセスと組み合わされると

■［代名］＋ combine

❶ we combined ～	われわれは，～を組み合わせた	483

❶ **We combined** whole-cell measurements with biophysical modeling to characterize the intrinsic stochastic and electrical properties of single neurons as observed at the soma.（J Neurosci. 2004 24:9723）
訳 われわれは，ホールセル測定を生物物理学的モデリングと組み合わせた

■ combined ＋［名］

❶ combined treatment	併用療法	346
❷ combined effects of ～	～の併用効果	256
❸ combined data	組み合わされたデータ	246

❶ The **combined treatment** with Tat and METH led to significantly greater 70-78% decreases in striatal DA overflow and content.（Brain Res. 2003 984:133）
訳 TatとMETHによる併用療法は，～の有意により大きな70-78%の低下につながった

■ combine ＋［前］

❶ combine to *do*	…は，結合して～する	279
❷ combine with ～	…は，～と結合する	164

❶ Anthrax lethal toxin consists of two distinct proteins that **combine to** form the active toxin.（Proc Natl Acad Sci USA. 1997 94:12059）
訳 結合して活性がある毒素を形成する２つの異なるタンパク質

C

☆ common 形 共通の／よくある，名 共通点 用例数 36,257

common	共通の／共通点	36,219
commons	共通点	38

形容詞として用いられることがほとんどだが，名詞の用例もある．
◆類義語：general, conventional, usual, normal

■ 形 be common ＋［前］

★ be common	…は，共通である／よくある	3,834
❶ be common in ～	…は，～において共通である／～によくある	1,256
・be common in patients	…は，患者において共通である	111
❷ be common to ～	…は，～に共通である	508
❸ be common among ～	…は，～の間で共通である	224

❶ Autoantibodies to prothrombin are common in patients with systemic lupus erythematosus. (Biochemistry. 2004 43:4047)
訳 プロトロンビンに対する自己抗体は，全身性エリテマトーデスの患者に共通である

❸ Serious bleeding was more common among patients who received higher doses of aspirin (>162 mg/d), with or without lotrafiban. (Circulation. 2003 108:399)
訳 重症の出血は，～を受けた患者の間によりよくあった

■ 形［副］＋ common

❶ the most common ～	最もよくある～	4,954
❷ more common	よりよくある	1,757
❸ less common	よりありふれていない	351

❶ Mutations in spastin are the most common cause of the condition. (Hum Mol Genet. 2005 14:19)
訳 ～は，その状態の最もよくある原因である

❷ The inferred haplotype harboring the [194]Trp allele was more common in controls than in cases (6.6 versus 5.3%, P = 0.07). (Cancer Res. 2003 63:8536)
訳 ～は，症例群においてより対照群においてよくあった

■ 形 common ＋［名］

❶ common cause of ～	～のよくある原因	857
❷ common ancestor	共通の祖先	757
❸ a common mechanism	共通の機構	499
❹ a common feature of ～	～の共通の特徴	498
❺ common pathway	共通の経路	342

■ 形［動］＋ a common ＋［名］

❶ be a common ～	…は，共通の～である	2,957
❷ share a common ～	…は，共通の～を共有する	773

❷ Structural predictions show that amassin and other OLF domain-containing vertebrate proteins share a common architecture. (J Cell Biol. 2003 160:597)
訳 amassinと他のOLFドメインを含む脊椎動物タンパク質は，共通の構造を共有する

■ 名［前］＋ common

★ in common	共通に	826
❶ in common with ～	～と共通に／～と同様に	310

❶ In common with other vertebrate species, the shark RAG2 coding region lacks introns and is closely linked in opposite orientation to the RAG1 gene. (FASEB J. 2003 17:470)
訳 他の脊椎動物種と同様に，

C

★ commonly 副 一般に／通常 用例数 8,210

◆類義語：generally, in general, usually, normally, universally

■ commonly ＋［過分］

❶ commonly used	一般に使われる	2,202
・be commonly used	…は，一般に使われる	431
・commonly used to *do*	一般に～するために使われる	297
・commonly used in ～	一般に～において使われる	233
❷ commonly associated with ～	一般に～と関連する	351
❸ commonly observed	一般に観察される	339
❹ commonly found in ～	一般に～において見つけられる	281

❶ Coenzyme Q10 is commonly used to treat congestive heart failure on the basis of data from several unblinded, subjective studies. (Ann Intern Med. 2000 132:636)
訳 コエンザイムQ10は，一般にうっ血性心不全を治療するために使われる

❷ Intra-amniotic infection (IAI) is commonly associated with preterm birth and adverse neonatal sequelae. (JAMA. 2004 292:462)
訳 ～は，一般に早産と関連する

❹ Low values of serum proteins and loss of lean body mass are commonly found in patients with chronic renal insufficiency (CRI) and especially in dialysis patients. (J Am Soc Nephrol. 2002 13:S22)
訳 ～は，通常慢性腎不全の患者に見つけられる

■［副］＋ commonly

❶ most commonly ～	最も一般に～	314
❷ more commonly ～	より一般に～	97

❶ The most commonly reported events were psychosis, severe depression, mania or agitation, hallucinations, sleep disturbance, and suicidal ideation. (Am J Psychiatry. 2005 162:189)
訳 最も一般に報告される出来事

communication ㊔ 情報交換／連絡／コミュニケーション

用例数 4,609

communication	情報交換	4,391
communications	情報交換	218

複数形のcommunicationsの割合は約５%あるが，単数形は無冠詞の用例が圧倒的に多い．

◆類義語：connection

■ communication ＋［前］

	the%	a/an%	ø%			
❶	5	6	87	communication between ～	～の間の情報交換	759
❷	4	2	94	communication in ～	～における情報交換	230
❸	2	2	89	communication with ～	～との情報交換	141

❶ **Communication between** the ATPase and substrate binding domains of Hsp70 is critical for regulated interaction between this molecular chaperone and its client proteins.（Mol Cell. 2005 20:493）

訳 Hsp70のATPアーゼおよび基質結合ドメインの間の情報交換は，～にとって決定的に重要である

★ comparable ㊓ 匹敵する／同等の

用例数 10,224

comparable to/withの用例が多い．

◆類義語：equivalent, equal, identical, same, consistent

■ comparable ＋［前］

❶ comparable to ～	～に匹敵する／～と同等である	4,083
・be comparable to ～	…は，～に匹敵する／～と同等である	1,758
・comparable to that	それに匹敵する／それと同等である	1,292
・levels comparable to ～	～に匹敵するレベル	223
❷ comparable with ～	～に匹敵する／～と同等である	1,173
❸ comparable in ～	～において同等である	538

❶ The phenotype of the Slc6a8-/y mouse **was comparable to** that of human patients.（J Clin Invest. 2012 122:2837）

訳 Slc6a8-/yマウスの表現型は，ヒトの患者のそれに匹敵した

❶ We found that most of the mutants were expressed well and were transported to the cell surface at **levels comparable to** that of the wild-type SER F protein.（J Virol. 2004 78:8513）

訳 ～は，野生型SER Fタンパク質のそれに匹敵するレベルで細胞表面へ運ばれた

❷ Wnt-1 stimulates 3T3-L1 preadipocytes to secrete factors that increase PKB/Akt phosphorylation at levels **comparable with** treatment with 10% serum.（J Biol Chem. 2002 277:38239）

訳 10%血清による処理に匹敵するレベルでPKB/Aktのリン酸化を増大させる因子

❸ Opioid growth factor and opioid growth factor receptor distribution were comparable in diabetic and control animals. (Diabetes. 2002 51:3055)

訳 オピオイド増殖因子およびオピオイド増殖因子受容体の分布は，糖尿病と対照群の動物において同等であった

★ compare 動 比較する　用例数 100,542

compared	比較される／～を比較した	88,140
compare	～を比較する	6,594
comparing	～を比較する	5,125
compares	～を比較する	683

他動詞．compare＋名＋with/toのパターンで使われる．特に，受動態でcompared with/toの用例が多い．as compared with/toの形でも用いられるが，asは省略されることが多い．

■ compared with/to ＋［名句］

❶ compared with ～	～と比較して／～と比較される	55,085
・compared with those	それらと比較して	3,032
・compared with wild-type ～	野生型～と比較して	2,496
・compared with control ～	コントロール～と比較して	1,964
・compared with normal ～	正常な～と比較して	1,060
・compared with patients	患者と比較して	843
・compared with placebo	偽薬と比較して	822
❷ compared to ～	～と比較して／～と比較される／～に匹敵する	19,273
・compared to those	それらと比較して	1,056

❶ These findings suggest that in the primate dentate gyrus, recurrent excitation is enhanced and inhibition is reduced compared with rodents. (J Comp Neurol. 2004 476:205)

訳 抑制は，げっ歯類と比較して低下する

❶ In conclusion, HIV-infected veterans in the HAART era are at a higher risk for diabetes compared with those in the pre-HAART era. (Hepatology. 2004 40:115)

訳 高活性抗レトロウイルス剤療法時代にHIVに感染した復員軍人は，高活性抗レトロウイルス剤療法時代前のそれらと比較して糖尿病のより高いリスクがある

❶ Compared with wild-type mice, WAT of ob/ob mice expressed substantially higher levels of many genes related to antioxidant response, inflammation, adipogenesis, lipogenesis, glucose uptake, and lipid transport. (Diabetes. 2013 62:845)

訳 野生型マウスと比較して，

❷ By immunofluorescence assay and real-time PCR, all mutant viruses demonstrated a modest delay in viral spread compared to that of reference HIV. (J Virol. 2004 78:5835)

訳 すべての変異ウイルスが，リファレンスのHIVのそれと比較してウイルス伝播のわずかな遅れを示した

■［接］＋ compared with/to

❶ as compared with ～	～と比較して	4,786

❷ when compared with ～	～と比較すると	3,269
❸ as compared to ～	～と比較して	1,816
❹ when compared to ～	～と比較すると	1,195

❶ The prepatent period of the resulting blood infection is 1 day longer as compared with wild type.（J Biol Chem. 2005 280:6752）

訳 ～は，野生型と比較して 1 日長い

❷ In contrast, adult male CRH-BP-deficient mice show significantly reduced body weight when compared with wild-type controls.（Proc Natl Acad Sci USA. 1999 96:11595）

訳 野生型対照群と比較すると

■ be compared ＋［前］

★ be compared	…が，比較される	11,021
❶ be compared with ～	…が，～と比較される	4,684
❷ be compared to ～	…が，～と比較される	1,818
❸ be compared in ～	…が，～において比較される	708
❹ be compared between ～	…が，～の間で比較される	572
❺ be compared for ～	…が，～に関して比較される	382
❻ be compared by ～	…が，～によって比較される	381
・be compared by using ～	…が，～を使うことによって比較される	138
❼ be compared using ～	…が，～を使って比較される	280

❶ Results were compared with in vivo intracellular recordings from aPCX layer II/III neurons and field recordings in urethane-anesthetized rats stimulated with odorants.（J Neurosci. 2004 24:652）

訳 結果が，生体の細胞内記録と比較された

❸ IFN-γ mechanisms were compared in embryonic fibroblasts（MEFs）and bone marrow Mφ（BMMφ）.（J Exp Med. 2001 193:483）

訳 IFN-γ機構が，胚性線維芽細胞（MEF）および骨髄Mφ（BMMφ）において比較された

❹ HRV was compared between CRT-on and CRT-off groups.（Circulation. 2003 108:266）

訳 HRVが，CRTオンとCRTオフのグループの間で比較された

■［名］＋ be compared

❶ results were compared	結果が，比較された	496
❷ data were compared	データが，比較された	162

❶ The results are compared to those that use a finite-difference method and to forward diffusion simulations.（Genetics. 2012 192:619）

訳 それらの結果は，有限差分法を使うそれらと比較される

■ compare ＋［名句］

❶ compare the effects of ～	…は，～の効果を比較する	461
❷ compare the results	…は，その結果を比較する	267
❸ compare the ability of ～	…は，～の能力を比較する	220
❹ compare the performance of ～	…は，～の性能を比較する	130

❺ compare the efficacy of ～	…は，～の効率を比較する	119

❶ We also compared the results to those of a recent study we conducted with a large set of antibiotics. (21357484J Neurosci. 2004 24:652)
訳 われわれは，また，その結果を最近の研究のそれらと比較した

■ [代名] ＋ compare

❶ we compared ～	われわれは，～を比較した	4,550

❶ Accordingly, we compared the effects of nonadrenergic vasopressin with those of epinephrine on cerebral cortical microvascular flow together with cortical tissue Po2 and Pco2 as indicators of cortical tissue ischemia. (Crit Care Med. 2007 35:2145)
訳 われわれは，非アドレナリン性バソプレシンの影響をエピネフリンのそれらと比較した

■ to compare

★ to compare ～	～を比較する…	4,062
❶ be to compare ～	…は，～を比較することである	691
・the purpose of this study was to compare ～	この研究の目的は，～を比較することであった	162
❷ be used to compare ～	…は，～を比較するために使われる	689
❸ To compare ～	～を比較するために	262

❶ The purpose of this study was to compare the nuclear transport of EGFR family proteins with that of FGFR-1. (J Biol Chem. 2012 287:16869)
訳 この研究の目的は，～の核移行を比較することであった

★comparison ㊔ 比較 用例数 21,997

comparison	比較	17,045
comparisons	比較	4,952

複数形のcomparisonの割合は約25%あるが，単数形は無冠詞の用例も多い．comparison withとcomparison toは，ほぼ同じ意味である．

■ comparison ＋ [前]

	the%	a/an%	ø%			
❶	10	31	57	comparison of ～	～の比較	6,818
				・comparison of ～ with …	～と…との比較	758
❷	0	5	95	comparison with ～	～との比較	3,085
❸	1	5	93	comparison to ～	～との比較	1,853
❹	15	40	43	comparison between ～	～の間の比較	431

❶ Comparison of our data with previous reports uncovered evidence that many inactivated structures display nonreactive conformations. (Proc Natl Acad Sci USA. 2012 109:2308)
訳 われわれのデータと以前の報告との比較は，～を明らかにした

❷ In addition, the effects of potassium and calcium channel blockers were also tested for

C

comparison with the results of previous studies. (Brain Res. 2001 900:119)

訳 ～は，また，以前の研究の結果との比較のためにテストされた

■ [前] + comparison

❶ in comparison	比較して	3,460
・in comparison with ～	～と比較して	1,592
・in comparison to ～	～と比較して	1,425
❷ by comparison	比較によって／比べると	837
❸ for comparison	比較のために	774

❶ Principal component analysis identified 103 SAGE tags as specific for invasive breast carcinomas, in comparison with in situ duct carcinomas or normal breast epithelium. (Cancer Res. 2002 62:5351)

訳 上皮内乳管癌と比較して

❶ In comparison to the parental cells, infection in these cells now progressed at a rapid rate with expanding plaque formation. (J Virol. 2004 78:8641)

訳 親細胞と比較して

❷ By comparison, the receptor affinities of the MeAla-containing peptides were significantly diminished. (J Med Chem. 2002 45:5280)

訳 比べると，

❸ For comparison, we used mice deficient in the antimicrobial molecule, neutrophil elastase (NE). (J Immunol. 2005 174:1557)

訳 比較のために，

■ [形／名] + comparison(s)

❶ sequence comparisons	配列比較	473
❷ direct comparison	直接比較	394
❸ healthy comparison subjects	健康な比較対象	314

❶ Sequence comparisons between human and rodents are increasingly being used for the identification of gene regulatory regions. (Bioinformatics. 2005 21:835)

訳 ヒトとげっ歯類の間の配列比較は，～のためにますます使われている

■ comparison + [名]

❶ comparison subjects	対照被検者	1,115
❷ comparison group	対照群	346

compatible 形 適合性の／矛盾しない／互換性の　用例数 2,283

compatible withの用例が圧倒的に多い．

◆類義語：suitable, fit, appropriate, adequate

■ be compatible + [前]

★ be compatible	…は，適合する	998

❶ be compatible with ～	…は，～に適合する	959

❶ These data **are compatible with** the hypothesis that the rotation of the TCRs may alter the downstream signaling. (Proc Natl Acad Sci USA. 2008 105:15523)
訳 これらのデータは，～という仮説に適合する

compete 動 競合する　用例数 4,995

competing	競合する	2,005
compete	競合する	1,685
competes	競合する	699
competed	競合した	606

主に自動詞として使われるが，他動詞の用例もある.

■ compete ＋［前］

❶ compete with ～	…は，～と競合する	1,831
・compete with ～ for …	…は，…に対して～と競合する	807
・compete with ～ for binding	…は，結合に対して～と競合する	373
❷ compete for ～	…は，～に対して競合する	906
・compete for binding	…は，結合に対して競合する	183

❶ The NTK domain of RSK1 **competed with** PKAc **for binding** to the pseudosubstrate region (amino acids 93-99) of PKARIα. (J Biol Chem. 2010 285:6970)
訳 RSK1のNTKドメインは，～への結合に対してPKAcと競合する

competition 名 競合／競争／競合作用　用例数 4,522

competition	競合／競争	4,466
competitions	競合／競争	56

複数形のcompetitionsの割合は1％しかなく，**不可算名詞**として使われることが多い.

■ competition ＋［前］

	the%	a/an%	ø%			
❶	27	16	54	competition between ～	～の間の競合（競争）	729
❷	5	3	88	competition for ～	～に対する競合（競争）	440
❸	3	2	89	competition with ～	～との競合（競争）	339

❶ This result suggests that **competition between** type IIa and type III RPTPs can regulate motor axon outgrowth, consistent with findings in Drosophila. (J Neurosci. 2005 25:3813)
訳 Ⅱa型とⅢ型のRPTPの間の競合は，～を調節しうる

❷ Development rates vary among individuals, often as a result of direct **competition for** food. (Behav Neurosci. 2005 119:1368)
訳 食物に対する直接競争の結果として

❸ The other Sp family members may be involved in photoreceptor-specific transcription directly or through their competition with Sp4. (J Biol Chem. 2005 280:20642)

訳 他のSpファミリーメンバーは，光受容体特異的転写に直接あるいはSp4とそれらの競合によって関与しているかもしれない

★complete 形 完全な，動 完了する 用例数 24,332

complete	完全な／～を完了する	19,324
completed	完了される／～を完了した	4,448
completing	～を完了する	387
completes	～を完了する	173

形容詞の用例が多いが，動詞としても用いられる．

◆類義語：full, perfect, finish

■ 形 complete ＋［名］

❶	complete loss of ～	～の完全な喪失	866
❷	complete response	完全な寛解／著効	864
❸	complete remission	完全な寛解	650
❹	complete absence of ～	～の完全な欠如	361
❺	complete inhibition of ～	～の完全な抑制	267

❶ Point mutation of a key predicted binding site residue (T317A) resulted in a complete loss of high affinity cAMP binding. (J Biol Chem. 2005 280:3771)

訳 鍵となる予想結合部位残基の点突然変異（T317A）は，高親和性cAMP結合の完全な喪失という結果になった

■ 形［副］＋ complete

❶	more complete ～	もっと完全な～	462
❷	nearly complete ～	ほとんど完全な～	458
❸	almost complete ～	ほとんど完全な～	321

❶ Thus, the E4032A mutation caused a nearly complete suppression of activation of RyR1 by Ca^{2+}. (Biophys J. 2002 82:2428)

訳 E4032A変異は，RyR1の活性のほとんど完全な抑制を引き起こした

■ 動［名］＋ complete

❶	patients completed ～	患者が，～を完了した	271
❷	participants completed ～	参加者が，～を完了した	137
❸	subjects completed ～	対象が，～を完了した	102

❶ One hundred six patients were enrolled; 101 patients completed 72 weeks of treatment. (Hepatology. 2012 56:2018)

訳 101名の患者が，72週の治療を完了した

■ 動 complete ＋［名句］

❶ complete the study	～が，研究を完了する	202
❷ complete questionnaires	～が，アンケートを完了する	70

★ completely 副 完全に　　用例数 12,981

◆類義語：entirely, fully, perfectly, totally, sufficiently, enough

■ completely ＋［過分／過去］

❶ completely blocked ～	完全に遮断される／～を完全に遮断した	1,062
・be completely blocked	…は，完全に遮断される	349
❷ completely inhibited ～	完全に抑制される／～を完全に抑制した	784
❸ completely abolished ～	完全に消失した／～を完全に消滅させた	781
❹ completely prevented ～	完全に阻止される／～を完全に阻止した	389
❺ completely abrogated ～	完全に抑止される／～を完全に抑止した	366
❻ not completely understood	完全には理解されていない	344

❶ Surprisingly, the anti-apoptotic effect of PTN **was completely blocked by the MAP kinase inhibitor UO126**, but was not affected by the PI 3-kinase inhibitor LY294002. (J Biol Chem. 2002 277:35862)

訳 ～が，MAPキナーゼ阻害剤UO126によって完全に遮断された

❷ Furthermore, processing and activation of different caspases by MX2870-1 **was completely inhibited by** increasing concentrations of PD169316. (Cancer Res. 2001 61:8504)

訳 …が，～によって完全に抑制された

❻ However, **the mechanisms responsible for activation and subsequent disease induction are not completely understood**. (J Exp Med. 2004 199:1631)

訳 活性化とそれに引き続く疾患誘導の原因である機構は，完全には理解されていない

■［副］＋ completely

❶ almost completely ～	ほとんど完全に～	724
❷ nearly completely ～	ほとんど完全に～	109

❶ Slow anion-channel **activity was almost completely abolished by removal of cytosolic ATP** or by the kinase inhibitors K-252a and H7. (Proc Natl Acad Sci USA. 1995 92:9535)

訳 ～活性が，細胞質ATPの除去によってほとんど完全に消失した

★★ complex 名 複合体，形 複雑な，動 複合体を形成する　　用例数 134,180

complex	複合体／複雑な／複合体を形成する	98,685
complexes	複合体／複雑な／複合体を形成する	33,272
complexed	複合体を形成した	2,075
complexing	複合体を形成する	148

名詞の用例が多いが，形容詞や動詞としても用いられる．complex/complexed withの用例が多い．

◆類義語：complicated

■ 名 complex ＋［前］

	the%	a/an%	ø%			
❶	6	44	46	complex with ～	～との複合体	6,186
				・in complex with ～	～と複合した…	508
❷	49	48	1	complex of ～	～の複合体	2,377
				・complex of ～ with …	～の…との複合体	208

❶ We also found that **RIP forms a complex with IKK in response to DNA damage**.（Genes Dev. 2003 17:873）

訳 RIP は，DNA損傷に応答してIKKとの複合体を形成する

❶ **This paper reports the 2.2 A resolution crystal structure of ssKex2 in complex with** an Ac-Arg-Glu-Lys-Arg peptidyl boronic acid inhibitor (R = 19.7, R_{free}= 23.4).（Biochemistry. 2004 43:2412）

訳 この論文は，～と複合したssKex2の2.2オングストローム分解能の結晶構造を報告する

❷ **The complex of TAR with argininamide serves as a model for** the RNA conformation in the tat-TAR complex.（J Mol Biol. 2002 317:263）

訳 TARのアルギニンアミドとの複合体は，～のモデルとして働く

■ 名［動］＋ a complex

❶	form a complex	…は，複合体を形成する	1,268

❶ **The ability of RET to form a complex with EGFR** was not dependent on recruitment of Shc or on their respective kinase activities.（Cancer Res. 2008 68:4183）

訳 EGFRと複合体を形成するRETの能力

■ 名［形／名］＋ complex

❶	protein complex	タンパク質複合体	2,387
❷	ternary complex	三元複合体／三重複合系	1,541
❸	～-DNA complex	～-DNA複合体	1,042
❹	receptor complex	レセプター複合体	940

■ 名 complex ＋［名］

❶	complex formation	複合体形成	2,843
❷	complex assembly	複合体構築	511

■ 名 complex ＋［現分］

❶	complex containing ～	～を含む複合体	811
❷	complex consisting of ～	～から成る複合体	261

形 [副] + complex

❶ more complex	より複雑な	1,625

❶ However, the existence of long-lived, low-activity Alu lineages in the human genome **suggests a more complex propagation mechanism**. (Genome Res. 2005 15:655)

訳 ～は，より複雑な伝播機構を示唆している

形 complex + [名]

❶ complex formation	複合形質	418
❷ complex process	複雑な過程／複合過程	391
❸ complex interactions	複雑な相互作用	339
❹ complex disease	複合疾患	311
❺ complex interplay	複雑な相互作用	257

動 complexed + [前]

❶ complexed with ～	～と複合体を形成した	1,436
❷ complexed to ～	～と複合体を形成した	276

❶ Furthermore, we determined that **HDAC1 complexed with Sp1 in PDAC cells** and that TSA treatment interfered with this association. (Cancer Res. 2003 63:2624)

訳 HDAC1は，PDAC細胞においてSp1と複合体を形成した

❷ When **purified native toxin was compared with toxins complexed to neurotoxin-associated proteins**, no significant differences in assay response were noted for serotypes A, B, and F. (Anal Biochem. 2006 353:248)

訳 精製された天然の毒素が，神経毒結合タンパク質と複合体を形成した毒素と比較された

complexity 名 複雑さ／複雑性 用例数 4,795

complexity	複雑さ／複雑性	4,457
complexities	複雑さ／複雑性	338

複数形のcomplexitiesの割合は7％あるが，単数形は無冠詞で使われることが非常に多い．

complexity + [前]

complexity ofの前にtheが付く割合は圧倒的に高い．

	the%	a/an%	ø%			
❶	93	1	6	complexity of ～	～の複雑さ	1,832
❷	19	9	67	complexity in ～	～における複雑さ	282

❶ **These findings underscore the complexity of cytokine interactions** while also providing insight into the multifaceted regulatory network controlling virus-specific $CD8^+$ T-cell functions. (Proc Natl Acad Sci USA. 2012 109:9971)

訳 これらの知見は，サイトカイン相互作用の複雑さを強調する

★ complication ㊔ 合併症 用例数 10,169

complications	合併症	8,018
complication	合併症	2,151

複数形のcomplicationsの割合は，約80%と圧倒的に高い．

■ complications ＋［前］

	the%	a/an%	ø%			
❶	42	-	54	complications of ～	～の合併症	965
❷	1	-	88	complications in ～	～における合併症	421
❸	4	-	92	complications after ～	～のあとの合併症	165

❶ Intestinal fibrosis causes many complications of Crohn's disease (CD). (Gastroenterology. 2011 141:819)

訳 腸管線維症は，クローン病の多くの合併症を引き起こす

■［形］＋ complications

❶ postoperative complications	術後合併症	224
❷ major complications	主な合併症	207
❸ infectious complications	感染性合併症	186

★★ component ㊔ 構成成分／成分／コンポーネント 用例数 48,126

component	構成成分	24,124
components	構成成分	24,002

複数形のcomponentsの割合は約50%と非常に高い．

◆類義語：composition, constituent, element

■ component ＋［前］

	the%	a/an%	ø%			
❶	14	85	1	component of ～	～の構成成分	12,863
				・essential component of ～	～の必須の構成成分	781
				・important component of ～	～の重要な構成成分	772
				・major component of ～	～の主要な構成成分	710
				・key component of ～	～の鍵となる構成成分	602
				・critical component of ～	～の決定的に重要な構成成分	526
❷	12	82	1	component in ～	～における構成成分	954

❶ Laminin α4 chain is a component of extracellular matrix (ECM) laminin-8 and -9 and serves dual roles as a structure protein and as a signaling molecule. (J Biol Chem. 2006 281:213)

訳 ラミニンα4鎖は，細胞外基質（ECM）ラミニン8および9の構成成分である

❷ Focal adhesion kinase (FAK) is a critical component in transducing signals downstream of both integrins and growth factor receptors. (Oncogene. 2006 25:1081)
訳 接着斑キナーゼ（FAK）は，～の下流にシグナルを伝達する際に決定的に重要な構成成分である

compose 動 構成する　用例数 8,672

composed	構成される／～を構成した	8,354
compose	～を構成する	202
composing	～を構成する	104
composes	～を構成する	12

他動詞．composed ofの用例が圧倒的に多い．
◆類義語：comprise, constitute, consist

■ composed ＋［前］

❶ composed of ～	～で構成されている	7,800
・be composed of ～	…は，～で構成されている	3,368
・composed of two ～	2つの～で構成されている	665

❶ Similar to the F-type ATP synthase (FATPase), the V-ATPase is composed of two subcomplexes, V_1 and V_0. (J Biol Chem. 2004 279:39856)
訳 ～は，２つの部分複合体V_1とV_0で構成されている

■［名］＋ composed of

❶ complex composed of ～	～で構成されている複合体	234
❷ protein composed of ～	～で構成されているタンパク質	166

❶ The human ASGP-R is a hetero-oligomeric complex composed of two subunits designated H_1 and H_2. (J Biol Chem. 2002 277:47305)
訳 H_1およびH_2と命名された２つのサブユニットで構成されている複合体

★composition 名 組成／成分／構成　用例数 10,078

composition	組成	9,088
compositions	組成	990

複数形のcompositionsの割合は約10%あるが，単数形は無冠詞で使われることが非常に多い．
◆類義語：component, constituent, element

■ composition ＋［前］

composition ofの前にtheが付く割合は圧倒的に高い．

	the%	a/an%	ø%			
❶	92	2	6	composition of ～	～の組成	2,782

❷ 21	0	70	composition in ～	～における組成	307

❶ Changes in the composition of the extracellular matrix may affect electrical coupling in cardiac myocytes. (Circ Res. 2005 96:558)
訳 細胞外基質の組成の変化は，～に影響を与えるかもしれない

■ [名] + composition

❶ body composition	体組成	825
❷ subunit composition	サブユニット組成	409
❸ lipid composition	脂質組成	378

★ comprise 動 構成する／～から成る／含む 用例数 10,028

comprising	～から成る	3,077
comprised	構成される／～から成った	2,829
comprise	～を構成する／～から成る	2,407
comprises	～を構成する／～から成る	1,715

他動詞．本来〈Xs comprise Y ＝ Y is comprised of Xs〉だが，comprised of Xsとcomprising Xsがほぼ同様の意味で用いられることが多い．
◆類義語：compose, constitute, consist

■ comprised + [前]

❶ comprised of ～	～で構成されている	1,840
・be comprised of ～	…は，～で構成されている	824
・comprised of two ～	2つの～で構成されている	136
・complex comprised of ～	～で構成されている複合体	49

❶ The 2.6-A resolution structure of human fatty acid synthase thioesterase domain reported here is comprised of two dissimilar subdomains, A and B. (Proc Natl Acad Sci USA. 2004 101: 15567)
訳 ～は，2つの異なるサブドメインで構成されている

❶ Here, we show that MRB1 contains a core complex comprised of six proteins and maintained by numerous direct interactions. (Nucleic Acids Res. 2012 40:5637)
訳 6つのタンパク質で構成されているコア複合体

■ comprise + [名句]

❶ comprise a family of ～	…は，～の一つのファミリーを構成する	109

❶ Bone morphogenetic proteins comprise a family of secreted ligands implicated in numerous aspects of organogenesis, including heart and neural crest development. (Development. 2004 131:3481)
訳 骨形成タンパク質は，分泌リガンドの一つのファミリーを構成する

■ [名] + comprising

❶ complex comprising ～	～から成る複合体	92

❶ We now describe another binding complex comprising mammalian Ufd1 and Npl4. (EMBO J. 2000 19:2181)
訳 哺乳類Ufd1とNpl4から成るもう一つの結合複合体

■ comprising + [名]

❶ comprising residues ～	残基～から成る…	100

❶ Antibody to BPV1 L2 peptides comprising residues 115 to 135 binds to intact BPV1 virions, but fails to neutralize at a 1:10 dilution. (J Viol. 2003 77:3531)
訳 残基115-135から成るBPV1 L2ペプチドに対する抗体は，～に結合する

C

compromise　動 損なう，名 折衷したもの　用例数 5,182

compromised	損なわれる／～を損なった	2,960
compromise	～を損なう／折衷したもの	1,197
compromising	～を損なう	547
compromises	～を損なう	478

動詞の用例が多いが，名詞としても用いられる．
◈類義語：impair, damage, lesion

■ be compromised + [前]

★ be compromised	…は，損なわれる	1,103
❶ be compromised in ～	…は，～において損なわれる	320
❷ be compromised by ～	…は，～によって損なわれる	231

❶ In parallel, jasmonic acid-regulated signaling is compromised in the ssi2 mutant. (Plant Cell. 2003 15:2383)
訳 ～は，ssi2変異体において損なわれている

❷ However, the activity of DP can be compromised by binding to the acute phase serum protein, α_1-acid glycoprotein (AGP). (J Med Chem. 2004 47:4905)
訳 DPの活性は，～への結合によって損なわれうる

■ [副] + compromised

❶ severely compromised	激しく損なわれる	258

❶ In contrast, Hsp40 interaction is severely compromised. (J Mol Biol. 2011 411:1099)
訳 Hsp40の相互作用は，激しく損なわれている

■ compromise + [名句]

❶ compromise ～ function	…は，～機能を損なう	196

❷ compromise the ability of ～	…は，～の能力を損なう	86

❷ This in turn compromises the ability of the glioma cell to migrate and proliferate. (J Biol Chem. 2006 281:19220)

訳 ～は，遊走し増殖するグリオーマ細胞の能力を損なう

■［前］＋ compromising

❶ without compromising ～	～を損なうことなしに	330

❶ SRL and reduced-dose TAC may achieve adequate immunosuppression without compromising renal function or enhancing EBV viremia significantly. (Transplantation. 2001 72:851)

訳 ～は，腎機能を損なうことなしに十分な免疫抑制を達成するかもしれない

compute 動 計算する／算出する 用例数 5,045

computed	算出される／～を計算した	3,759
computing	～を計算する／～を算出する	597
compute	～を計算する／～を算出する	581
computes	～を計算する／～を算出する	108

他動詞.

◆類義語：calculate

■ computed ＋［前］

❶ computed from ～	～から算出される	156
❷ computed for ～	～に対して算出される	138

❶ Prior to linkage analysis, standardized residuals were computed from regression analysis of each phenotype on age. (Arthritis Rheum. 2004 50:2489)

訳 標準化残差が，～の回帰分析から算出された

❷ Descriptive statistics were computed for all factors. (Hepatology. 2010 51:1972)

訳 記述統計量は，すべての因子に対して算出された

■ to compute

★ to compute ～	～を計算する…	388
❶ used to compute ～	～を計算するために使われる	119

❶ Logistic regression was used to compute odds ratios for the associations. (Am J Epidemiol. 2005 161:213)

訳 ロジスティック回帰分析が，～に対するオッズ比を計算するために使われた

concentrate 動 濃縮する／集中する，名 濃縮物 用例数 3,567

concentrated	濃縮される／集中した	2,348
concentrate	濃縮する／集中する／濃縮物	496

concentrating	濃縮する／集中する	422
concentrates	濃縮する／集中する／濃縮物	301

他動詞の用例が多いが，自動詞や名詞としても用いられる．
◆類義語：enrich, focus, center

■ concentrated ＋［前］

❶ concentrated in ～	～に集中している／～において濃縮される	899
・be concentrated in ～	…は，～に集中している／～において濃縮される	496
❷ concentrated at ～	～において濃縮される	334

❶ In several brain areas, **Notch1 and Notch2 mRNAs are relatively concentrated in white matter**, whereas Notch3 mRNA is not. (J Comp Neurol. 2001 436:167)
訳 Notch1とNotch2のメッセンジャーRNAは，白質に比較的集中している

❷ During the anaphase-telophase transition, **lamin B1 begins to become concentrated at the surface of the chromosomes**. (J Cell Biol. 2000 151:1155)
訳 ラミンB1は，染色体の表面において濃縮された状態になり始める

■ concentrate on

❶ concentrate on ～	…は，～に集中する	180

❶ Here **we concentrate on the relatively recent data** that document primarily the channel properties of transporters. (Annu Rev Physiol. 2007 69:87)
訳 われわれは，比較的最近のデータに集中する

★★ concentration 名 濃度　用例数 59,399

concentrations	濃度	32,195
concentration	濃度	27,204

複数形のconcentrationsの割合は約55%と非常に高い．

■ concentrations ＋［前］

	the %	a/an %	ø %			
❶	17	-	80	concentrations of ～	～の濃度	10,959
❷	6	-	82	concentrations in ～	～における濃度	1,973

❶ **At physiological concentrations of nucleotides**, human Pol epsilon readily inserts and extends from incorporated ribonucleotides. (J Biol Chem. 2012 287:42675)
訳 生理学的濃度のヌクレオチドにおいて

❷ Here **we report argpyrimidine concentrations in human lens** and serum proteins as determined by HPLC. (Anal Biochem. 2001 290:353)
訳 われわれは，ヒトの水晶体におけるアルグピリミジン濃度を報告する

C

■ [形] + concentrations

❶ high concentrations	高い濃度	1,849
・high concentrations of ～	高い濃度の～	405
❷ low concentrations	低い濃度	1,588
❸ higher concentrations	より高い濃度	981
❹ lower concentrations	より低い濃度	432
❺ increasing concentrations of ～	上昇する～の濃度／～の濃度の上昇	418
❻ physiological concentrations	生理学的濃度	405
❼ nanomolar concentrations	ナノモル濃度	390
❽ micromolar concentrations	マイクロモル濃度	389

❶ Very high concentrations of either 8Br-cAMP or 8Br-cGMP (≧100 μmol/L) were required to cross-activate both PKA and PKG. (Am J Pathol. 2003 163:1157)

訳 非常に高い濃度（100 μmol/L以上）の8Br-cAMPか8Br-cGMPのどちらかが，～するために必要とされた

■ [名] + concentrations

❶ plasma concentrations	血漿濃度	737
❷ glucose concentrations	グルコース濃度	480
❸ serum concentrations	血清濃度	395

■ at concentrations

★ at concentrations	濃度において	1,614
❶ at concentrations of ～	～の濃度において	321
❷ at concentrations that ～	～する濃度において	230

❶ The amino acid L-alanine was the only independent germinant in B. anthracis and then only at concentrations of >10 mM. (J Bacteriol. 2002 184:1296)

訳 >10 mMの濃度においてのみ

■ concentration as + [形] + as

❶ concentrations as low as ～	わずか～の低い濃度	199

❶ Effects on signaling were observed at cytokine concentrations as low as 0.3 ng/ml. (Am J Respir Crit Care Med. 2001 164:141)

訳 ～が，わずか0.3 ng/mlの低いサイトカイン濃度で観察された

■ concentrations + [動]

❶ concentrations were measured	濃度が，測定された	239
❷ concentrations increased	濃度が，増大した	174

❶ Plasma 17DMAG concentrations were measured by liquid chromatography/mass spectrometry. (J Clin Oncol. 2010 28:1520)

訳 血漿17DMAG濃度が測定された

■ concentration-dependent

★	concentration-dependent ～	濃度依存的な～	2,257
❶	in a concentration-dependent manner	濃度依存的様式で	718
❷	time- and concentration-dependent	時間および濃度依存的な	228

❶ These responses were inhibited in a concentration-dependent manner by celecoxib. (PLoS One. 2012 7:e43376)

訳 これらの反応は，濃度依存的様式で阻害された

C

concept 名 概念　　用例数 5,359

concept	概念	4,060
concepts	概念	1,299

複数形のconceptsの割合が，約25%ある可算名詞.

◈類義語：notion, idea, view, hypothesis

■ concept ＋［前／同格that節］

concept of/thatの前にtheが付く割合は圧倒的に高い.

	the%	a/an%	ø%			
❶	90	3	3	concept of ～	～の概念	1,287
❷	90	7	3	concept that ～	～という概念	1,201
				・support the concept that ～	…は，～という概念を支持する	157

❶ While the first of these options involves direct transannular C-C bond formation, the other two embody the concept of larger ring construction as a prelude to ring contraction. (J Org Chem. 2005 70:514)

訳 他の２つは，より大きな環構造の概念を具体化する

❷ These in vitro observations support the concept that opioid abuse favors HCV persistence in hepatic cells by suppressing IFN-α-mediated intracellular innate immunity and contributes to the development of chronic HCV infection. (Am J Pathol. 2005 167:1333)

訳 これらの試験管内の観察は，～という概念を支持する

■［前］＋ concept

❶	proof of concept	概念の証明	495
	・proof of concept for ～	～に対する概念の証明	108

❶ These data provide proof of concept for an HIV-1 vaccine that aims to elicit bnAbs of multiple specificities. (J Virol. 2012 86:4688)

訳 これらのデータは，HIV-1ワクチンに対する概念の証明を提供する

concert 名 協調／一致，動 協調する　用例数 2,912

concert	協調／一致	1,544
concerted	協調した／共同した	1,366
concerts	協調／一致	2

名詞あるいは動詞として使われる．名詞としてはin concert with/toの用例が圧倒的に多い．

◆類義語：agreement, accordance, accord, concordance, correspondence, coincidence

■ in concert

★ in concert	協調して	1,534
❶ in concert with ～	～と協調して	1,011
❷ in concert to *do*	～するために協調して	298

❶ The prevalence of hypertension among children in the US is increasing in concert with rising obesity rates.（Curr Opin Pediatr. 2005 17:642）

訳 合衆国における子供の間の高血圧症の発症率は，上昇している肥満率と協調して増大している

❷ This result indicated that both DNA methylation and histone deacetylase could act in concert to inhibit WTH3 and consequently stimulate MDR1 expression.（Cancer Res. 2005 65:10024）

訳 WTH3を抑制するために協調して働く

■ ［動］＋ in concert

❶ act in concert	…は，協調して働く	423
❷ work in concert	…は，協調して働く	176
❸ function in concert	…は，協調して機能する	106

★conclude 動 結論する　用例数 18,410

conclude	～を結論する	15,989
concluded	～を結論した／結論される	2,217
concludes	～を結論する	133
concluding	～を結論する	71

他動詞．conclude thatの用例が非常に多い．

■ conclude ＋［that節］

★ conclude that ～	…は，～であると結論する	17,540
❶ we conclude that ～	われわれは，～であると結論する	14,928
❷ the authors conclude that ～	著者らは，～であると結論する	201
❸ lead us to conclude that ～	…は，われわれが～であると結論するように仕向ける	135

❶ We conclude that the coreceptor SLAM plays a central role at the interface of acquired and innate immune responses. (J Exp Med. 2004 199:1255)
訳 われわれは，～であると結論する

■ be concluded ＋［that節］

★ be concluded that ～	～であると結論される	1,196
❶ it is concluded that ～	～であると結論される	868

❶ It is concluded that transcriptional regulation of the KChIP2 gene is a primary determinant of Ito expression in heart. (J Physiol. 2003 548:815)
訳 ～であると結論される

C

conclusion 名 結論 用例数 8,245

conclusion	結論	6,651
conclusions	結論	1,594

複数形のconclusionsの割合が，約20%ある可算名詞．文頭のIn conclusionやconclusion thatの用例が非常に多い．

■ in conclusion

❶ In conclusion,	結論として	4,101

❶ In conclusion, we demonstrate that MDR3 expression is directly up-regulated by FXR. (J Biol Chem. 2003 278:51085)
訳 結論として，われわれは～ということを実証する

■ conclusion ＋［同格that節］

conclusion thatの前にtheが付く割合は圧倒的に高い．

	the%	a/an%	ø%			
❶	89	7	0	conclusion that ～	～という結論	1,241
				・support the conclusion that ～	…は，～という結論を支持する	548

❶ The results support the conclusion that individual SR Ca^{2+} release units function similarly in slow-twitch and fast-twitch mammalian fibres. (J Physiol. 2003 551:125)
訳 それらの結果は，～という結論を支持する

concomitant 形 同時の／随伴性の，名 付随物 用例数 4,987

concomitant	同時の／随伴性の／付随物	4,973
concomitants	付随物	14

名詞としても使われるが，形容詞の用例が圧倒的に多い．
◈類義語：concurrent, simultaneous, coincident, synchronous, parallel

■ concomitant + [前]

❶ concomitant with ～	～と同時の	1,124

❶ The G1 cell cycle arrest was concomitant with an increase in cyclin-dependent kinase inhibitor p27Kip1. (Oncogene. 2005 24:5606)
訳 G1細胞周期停止は，サイクリン依存性キナーゼ阻害因子p27Kip1の増大と同時であった

■ concomitant + [名]

❶ concomitant increase in ～	～の同時増大	360
❷ concomitant decrease in ～	～の同時低下	196
❸ concomitant loss of ～	～の同時喪失	140

❶ MRTs expressed the T_{121} protein with a concomitant increase in mitotic activity. (Cancer Res. 2007 67:3002)
訳 有糸分裂活性の同時増大を伴って

■ [前] + concomitant

❶ with concomitant ～	同時～を伴う	570
❷ without concomitant ～	同時～のない	142

❶ Shutdown of CaCSE4 expression resulted in a sharp decline of CaCse4p levels with concomitant loss of cell viability. (Proc Natl Acad Sci USA. 2002 99:12969)
訳 CaCSE4発現のシャットダウンは，細胞生存率の同時喪失を伴うCaCse4pレベルの鋭い低下という結果になった

concomitantly 副 同時に　用例数 1,127

◆類義語：concurrently, simultaneously, coincidentally, synchronously

■ [動] + concomitantly

❶ occur concomitantly	…は，同時に起こる	89

❶ ERK activation occurred concomitantly with the induction of activator protein-1 (AP-1) binding, a nuclear factor required for activation of multiple genes involved in fibrosis. (J Biol Chem. 2000 275:27650)
訳 ERK活性化が，アクチベータータンパク質1（AP-1）結合の誘導と同時に起こった

concurrent 形 同時の／同時発生的な／併用の　用例数 2,756

◆類義語：concomitant, simultaneous, coincident, synchronous, parallel

■ concurrent + [前]

❶ concurrent with ～	～と同時発生的な／～と同時の	382

❶ Most interestingly, MKP-1 expression was potently induced by peptidoglycan, and this

induction was concurrent with MAP kinase dephosphorylation. (J Biol Chem. 2004 279:54023)
訳 この誘導は，MAPキナーゼ脱リン酸化と同時発生的であった

concurrently 副 同時に／同時発生的に 用例数 1,137

◈類義語：concomitantly, simultaneously, coincidentally, synchronously

■ [動] + concurrently

❶ occur concurrently	…が，同時に起こる	74
・occur concurrently with ～	…が，～と同時に起こる	52

❶ This change occurred concurrently with the acquisition of two new potential glycosylation site motifs in HA. (J Virol. 2007 81:11170)
訳 この変化は，～の獲得と同時に起こった

★★ condition 名 条件／状態，動 条件づける 用例数 53,831

conditions	条件／状態	40,591
condition	条件／状態	5,559
conditioning	条件づける	3,887
conditioned	条件づけられる	3,794

動詞としても用いられるが，名詞の用例の方が多い．複数形のconditionsの割合は，約90%と圧倒的に高い．under ～ conditionsの用例が非常に多い．

◈類義語：state, status, situation, aspect

■ 名 under + (@) + conditions

❶ under ～ conditions	～条件下で	18,947
❷ under conditions	条件下で	2,968

❶ Under these conditions, recombinant N-SHH protein promotes the proliferation of mesenchyme cells and the expression of noggin. (Dev Biol. 2002 245:280)
訳 これらの条件下で，

■ 名 [形／名] + conditions

❶ physiological conditions	生理学的条件	1,126
・under physiological conditions	生理学的条件下で	655
❷ growth conditions	生育条件	913
❸ experimental conditions	実験条件	839
❹ environmental conditions	環境条件	801
❺ pathological conditions	病的状況	614
❻ stress conditions	応力条件	579
❼ reaction conditions	反応条件	568
❽ culture conditions	培養条件	477

C

⑨ same conditions	同じ条件	469

■ ㊔ conditions ＋［前］

	the％	a/an％	ø％			
❶	19	-	80	conditions of ～	～の条件／～の状態	2,324
				・under conditions of ～	～の条件下で	1,365
❷	8	-	84	conditions in ～	～の条件	1,663
				・conditions in which ～	～である条件	538
❸	19	-	77	conditions for ～	～のための条件	831

❶ Dorsolateral PFC（DLPFC）**is thought to guide response selection under conditions of** response conflict or, alternatively, may refresh recently active representations within working memory.（Neuron. 2004 41:473）

訳 …は，～の条件下で反応選択を導くと考えられる

❷ **Under conditions in which E. coli was agglutinated by SIgA**, the binding of SIgA to E. coli was not increased by the presence of the pili, with or without adhesin.（Infect Immun. 2004 72: 1929）

訳 大腸菌がSIgAによって凝集させられた条件下で

❸ Here we demonstrate that purified EpsE is an Mg^{2+}-dependent ATPase and **define optimal conditions for the hydrolysis reaction.**（J Bacteriol. 2005 187:249）

訳 ～は，加水分解反応のための最適な条件を規定する

■ ㊔ conditions ＋［関］/including

❶ conditions that ～	～する条件	1,710
❷ conditions where ～	～である条件	542
❸ conditions including ～	～を含む条件	447

❶ Both csgD and csgBA transcription, required for expression of curli, **were inactive in an mlrA mutant grown under conditions that promote curli production.**（Mol Microbiol. 2001 41:349）

訳 ～は，curli産生を促進する条件下で増殖させられたmlrA変異体において不活性であった

❷ In the second approach, the efflux of cellular FC to high concentrations of cyclodextrins **was monitored under conditions where desorption of FC from the plasma membrane is rate limiting for efflux.**（Biochemistry. 2000 39:221）

訳 ～が，細胞膜からのFCの脱離が流出に対する律速である条件下でモニターされた

■ ㊍ conditioned ＋［名］

❶ conditioned medium	培養上清／ならし培地	914
❷ conditioned stimulus	条件づけた刺激	286

★ conduct ㊍ 行う　用例数 11,941

conducted	行われる／～を行った	9,660
conducting	～を行う	1,164
conduct	～を行う	1,029
conducts	～を行う	88

他動詞.
◈類義語：perform, carry out, do

■ conduct ＋［名句］

❶ conduct a prospective ～	…は，前向き～を行う	189
❷ conduct a randomized ～	…は，ランダム化された～を行う	181
❸ conduct a retrospective ～	…は，後向き～を行う	154
❹ conduct a case-control study	…は，症例対照研究を行う	91

❷ We conducted a randomized, double-blind trial to determine the effect of aspirin on the incidence of colorectal adenomas.（N Engl J Med. 2003 348:883）
訳 われわれは，アスピリンの影響を決定するために無作為化された二重盲検試験を行った

■［代名／名］＋ conduct

❶ we conducted ～	われわれは，～を行った	2,461
❷ the authors conducted ～	著者らは，～を行った	289

■ be conducted ＋［前］

★ be conducted	…が，行われる	4,679
❶ be conducted to *do*	…が，～するために行われる	1,332
・be conducted to determine ～	…が，～を決定するために行われる	357
・be conducted to evaluate ～	…が，～を評価するために行われる	132
❷ be conducted in ～	…が，～において行われる	844
❸ be conducted with ～	…が，～によって行われる	324
❹ be conducted on ～	…が，～に対して行われる	280

❶ A phase II trial was conducted to determine if high-dose radiation with concurrent hepatic arterial floxuridine would improve survival in patients with unresectable intrahepatic malignancies.（J Clin Oncol. 2005 23:8739）
訳 第二相の治験が，～かどうかを決定するために行われた

❷ Experiments were conducted in isolated canine renal tubules and in a canine autotransplant model of hypothermic preservation injury.（Transplantation. 2005 80:1455）
訳 実験が，単離されたイヌの腎尿細管において行われた

❸ All central cannula placements should be conducted with ultrasound assistance.（Crit Care Med. 2005 33:1764）
訳 ～は，超音波アシストによって行われるべきである

■［名］＋ be conducted

❶ study was conducted	研究が，行われた	893
❷ experiments were conducted	実験が，行われた	221
❸ analyses were conducted	分析が，行われた	219
❹ trial was conducted	治験が，行われた	153

❸ Analyses were conducted on data from the New York Social Environment Study (n = 4,000), a representative study of residents of New York, New York, conducted in 2005.（Am J

C

Epidemiol. 2011 173:1453)
訳 分析が，～からのデータに対して行われた

★ confer 動 与える　　用例数 11,710

confer	～を与える	4,313
conferred	与えられる／～を与えた	3,217
confers	～を与える	3,023
conferring	～を与える	1,157

他動詞．confer resistance toの用例が多い．
◆類義語：give, render, supply, provide

■ confer ＋［名句］

❶ confer resistance to ～	…は，～に対する抵抗性を与える	976
❷ confer protection against ～	…は，～に対する保護を与える	220
❸ confer susceptibility to ～	…は，～に対する感受性を与える	171
❹ confer sensitivity to ～	…は，～に対する感受性を与える	108
❺ confer specificity	…は，特異性を与える	102

❶ These data illustrate that a single molecule **can confer resistance to humoral and cellular immune attack**.（Transplantation. 2003 75:542）
訳 ～が，液性および細胞性の免疫攻撃に対する抵抗性を与えうる

■ to confer

★ to confer ～	～を与える…	1,458
❶ sufficient to confer ～	～を与えるのに十分である	400

❶ This finding supports the hypothesis that a single allele **may be sufficient to confer protection against cervical neoplasia**.（J Infect Dis. 2002 186:598）
訳 ～は，子宮頸部異常増殖に対する保護を与えるのに十分であるかもしれない

■ conferred ＋［前］

❶ conferred by ～	～によって与えられる	986
・be conferred by ～	…は，～によって与えられる	309

❶ **Target specificity is conferred by the F-box protein subunit of the SCF**（TIR1 in the case of Aux/IAAs）and there are multiple F-box protein genes in all eukaryotic genomes examined so far.（Nature. 2005 435:446）
訳 標的特異性は，SCFのF-boxタンパク質サブユニットによって与えられる

confine 動 限局する／制限する／限定する　　用例数 2,728

confined	限局される／～を限局した／制限される	2,416
confines	～を限局する／～を制限する	111

confining	～を限局する／～を制限する	107
confine	～を限局する／～を制限する	94

他動詞．confined toの用例が非常に多い．

◈類義語：restrict, limit, restrain

■ be confined ＋［前］

★ be confined	…は，限局される	1,034
❶ be confined to ～	…は，～に限局される	881

❶ PTTG1 protein expression **was confined to** the nucleus in HEEpiC cells but present in both the cytoplasm and nucleus in ESCC cells.（Cancer Res. 2008 68:3214）

訳 PTTG1タンパク質発現は，HEEpiC細胞の核に限局された

★★ confirm 動 確認する　用例数 30,033

confirmed	～を確認した／確認される	19,546
confirm	～を確認する	6,068
confirming	～を確認する	2,786
confirms	～を確認する	1,633

他動詞．

◈類義語：verify, validate, ascertain

■ confirm ＋［名節／名句］

❶ confirm that ～	…は，～ということを確認する	6,986
❷ confirm the presence of ～	…は，～の存在を確認する	662
❸ confirm the importance of ～	…は，～の重要性を確認する	298
❹ confirm the role of ～	…は，～の役割を確認する	217
❺ confirm the existence of ～	…は，～の存在を確認する	154

❶ Biochemical analysis **confirmed that** several of the new mutations result in increased Shp2 activity.（Cancer Res. 2004 64:8816）

訳 生化学的分析は，～ということを確認した

❷ Laboratory analysis **confirmed the presence of** circulating vaccine-derived poliovirus（cVDPV）type 1 in stool samples obtained from patients.（J Infect Dis. 2004 189:1168）

訳 検査室の分析が，～の存在を確認した

■［代名／名］＋ confirm

❶ we confirmed ～	われわれは，～を確認した	1,031
❷ results confirm ～	結果は，～を確認する	797
❸ analysis confirmed ～	分析は，～を確認した	695
❹ studies confirmed ～	研究は，～を確認した	387
❺ experiments confirmed ～	実験は，～を確認した	270

❻ assays confirmed ～	アッセイは，～を確認した	245

❷ These results confirm the binding of biliverdin reductase to hHO-1 and define binding sites of the two reductases.（J Biol Chem. 2003 278:20069）
訳 これらの結果は，ビリベルジン還元酵素の～への結合を確認する

■ to confirm

★ to confirm ～	～を確認するために	1,842
❶ be used to confirm ～	…は，～を確認するために使われる	246
❷ be needed to confirm ～	…は，～を確認するために必要とされる	137
❸ To confirm ～	～を確認するために	377

❶ Single-crystal X-ray analysis was used to confirm the structure of the fluorescent species.（J Am Chem Soc. 2004 126:16179）
訳 単結晶X線解析が，～の構造を確認するために使われた

■ be confirmed ＋［前］

★ be confirmed	…が，確認される	6,934
❶ be confirmed by ～	…が，～によって確認される	3,822
❷ be confirmed in ～	…が，～において確認される	1,032

❶ BLCA-4 expression was confirmed by both PCR and Western blot analysis.（Cancer Res. 2005 65:7145）
訳 BLCA-4発現が，PCRとウエスタンブロット解析の両方によって確認された

❷ The interaction was confirmed in cells using chromatin immunoprecipitation (ChIP).（J Biol Chem. 2011 286:42123）
訳 相互作用が，細胞において確認された

■［名］＋ be confirmed

❶ results were confirmed	結果が，確認された	188
❷ expression was confirmed	発現が，確認された	99
❸ interaction was confirmed	相互作用が，確認された	95
❹ findings were confirmed	知見が，確認された	93

❹ Findings were confirmed using all available follow-up data.（J Nucl Med. 2011 52:1408）
訳 知見は，すべての利用可能な追跡データを使って確認された

★ conformation　名 高次構造／立体構造　用例数 18,511

conformation	高次構造／立体構造	13,465
conformations	高次構造／立体構造	5,046

複数形のconformationsの割合が，約25%ある可算名詞.

◆類義語：structure, architecture

■ 名 conformation ＋［前］

conformation ofの前にtheが付く割合は圧倒的に高い．

	the％	a/an％	ø％			
❶	82	13	4	conformation of ～	～の高次構造	2,992
❷	26	55	12	conformation in ～	～における高次構造	772

❶ Specifically, we ask if the conformation of this region is important to attain optimal inhibitory interactions with the Ca-ATPase.（Biochemistry. 2005 44:16181）

訳 われわれは，この領域の高次構造が～を達成するのに重要であるかどうかを問う

❷ Small-angle neutron scattering（SANS）measurements were used to provide detailed information of the protein conformation in solution.（Biochemistry. 2005 44:15139）

訳 ～が，溶液中でのタンパク質高次構造の詳細な情報を提供するために使われた

■［過分／形］＋ conformation

❶	closed conformation	閉じた高次構造	461
❷	open conformation	開いた高次構造	427
❸	active conformation	活性のある高次構造	349
❹	helical conformation	らせん形高次構造	336

★ conformational 形 立体構造上の／高次構造上の 用例数 14,699

■ conformational ＋［名］

❶	conformational changes	立体構造上の変化	3,696
❷	conformational states	立体構造上の状態	418
❸	conformational flexibility	立体構造上の柔軟性	365

❶ These results provide the first insight into the dynamics of nucleotide binding and substrate-induced conformational changes at the active site of a protein histidine kinase.（Biochemistry. 2005 44:4375）

訳 これらの結果は，ヌクレオチド結合のダイナミクスと基質に誘導される立体構造上の変化への最初の洞察を与える

conjugate 動 結合する／接合する／抱合する，名 複合物／抱合体 用例数 8,235

conjugated	結合した／抱合した／～を結合させた	3,043
conjugate	～を結合させる／複合物／抱合体	2,367
conjugates	～を結合させる／複合物／抱合体	2,115
conjugating	結合する	710

名詞としても用いられるが，動詞の用例が非常に多い．

◆類義語：bind, associate, couple

■ conjugated ＋［前］

❶ conjugated to ～	～に結合させられる	761
❷ conjugated with ～	～と結合させられる	193

❶ Cholera toxin B conjugated to horseradish peroxidase（CTB-HRP）was injected into the genioglossus muscle on the right side of four isoflurane-anesthetized cats.（Brain Res. 2005 1032:23）

訳 西洋わさびペルオキシダーゼに結合させたコレラ毒素B（CTB-HRP）が，～に注射された

❷ To target dendritic cells, the most potent antigen-presenting cells, the particle carriers, were further conjugated with monoclonal antibodies.（Proc Natl Acad Sci USA. 2005 102:18264）

訳 ～は，さらにモノクローナル抗体と結合させられた

conjunction 名 組み合わせ／協同　用例数 3,420

conjunction	組み合わせ／協同	3,381
conjunctions	組み合わせ／協同	39

in conjunction withの用例が圧倒的に多い.

◆類義語：combination, connection, cooperation

■ in conjunction with

★ in conjunction with ～	～と組み合わせて	3,278
❶ be used in conjunction with ～	…は，～と組み合わせて使われる	200

❶ TIRF microscopy can be used in conjunction with CFP/YFP FRET to detect movements of the cytoplasmic tails of GIRK channels.（Neuron. 2003 38:145）

訳 TIRF顕微鏡法は，CFP/YFP FRETと組み合わせて使われうる

connect 動 連結する／つなぐ／関連づける　用例数 4,610

connected	連結される／関連する／～を連結した	2,186
connecting	～を連結する／～をつなぐ	1,272
connect	～を連結する／～をつなぐ	647
connects	～を連結する／～をつなぐ	505

他動詞として用いられることが多いが，自動詞の用例もある．connected toの用例が多い.

◆類義語：link, join, associate, relate, couple, bind, bond, attach, correlate, engage

■ connected ＋［前］

❶ connected to ～	～に連結されている／～に関連する	657
❷ connected by ～	～によって連結される	495
・domains connected by ～	～によって連結されたドメイン	68
❸ connected with ～	～と連結される	175

❶ Notably, the lumen was connected to the external cytoplasm through a small opening. (23093403 Crit Care Med. 1998 26:1096)

訳 その内腔は，外側の細胞質に連結されていた

❷ GlnBP consists of two globular domains connected by a hinge. (J Am Chem Soc. 2009 131:9532)

訳 GlnBPは，ヒンジによって連結された２つの球状ドメインからなる

■ connect ＋［名句］

❶	connect ～ to …	～を…に連結する（関連づける）	363

❶ Integrin cytoplasmic domains connect these receptors to the cytoskeleton. (J Biol Chem. 1998 273:6104)

訳 インテグリンの細胞質ドメインは，これらの受容体を細胞骨格につなぐ

connection 名 関連／結合／連絡 用例数 5,426

connections	関連／結合／連絡	3,707
connection	関連／結合／連絡	1,719

複数形のconnectionsの割合は，約70%と非常に高い．

◈類義語：link, relation, relationship, association, correlation, binding, communication

■ connection(s) ＋［前］

	the%	a/an%	ø%			
❶	43	55	2	connection between ～	～の間の関連（結合）	825
❷	3	-	80	connections with ～	～との関連（結合）	332
❸	6	-	89	connections in ～	～における関連（結合）	251
❹	69	-	31	connections of ～	～の関連（結合）	233
❺	10	-	79	connections to ～	～への関連（結合）	191

❶ The ability of inflammatory cytokines to decrease aquaporin expression may help explain the connection between inflammation and edema. (J Biol Chem. 2001 276:18657)

訳 ～は，炎症と浮腫の間の関連を説明するのに役立つかもしれない

❷ V1 has connections with at least 12 subdivisions of visual cortex, with half of the connections involving V2 and 20% V3. (J Comp Neurol. 2002 449:281)

訳 V1は，視覚野の少なくとも12の小部分との連絡を持つ

★ consequence 名 影響／結果 用例数 12,709

consequences	影響／結果	7,194
consequence	影響／結果	5,515

複数形のconsequencesの割合は約55%と非常に高い．consequences ofの用例が非常に多い．

◈類義語：result, outcome

■ consequences ＋［前］

consequences ofの前にtheが付く割合は圧倒的に高い.

	the%	a/an%	ø%			
❶	80	-	18	consequences of ～	～の影響／～の結果	4,283
❷	7	-	88	consequences for ～	～に対する影響	762
❸	10	-	67	consequences in ～	～における影響	198

❶ The consequences of this reaction for the enzyme included the hydroxylation at Cβ of two amino acid side chains in the vicinity of the cofactor, Trp and Leu. (J Am Chem Soc. 2004 126: 2006)

訳 その酵素に対するこの反応の結果は，２つのアミノ酸側鎖のCβの水酸化を含んでいた

❷ This has significant consequences for patient management and counseling. (Am J Hum Genet. 2004 75:1136)

訳 これは，患者の管理とカウンセリングに対して大きな影響を持つ

■ as consequence

★	as a consequence	結果として	2,003
❶	as a consequence of ～	～の結果として	1,347
❷	As a consequence,	結果として，	460

❶ The production of mitochondrial reactive oxygen species occurs as a consequence of aerobic metabolism. (J Biol Chem. 2012 287:4434)

訳 ミトコンドリアの活性酸素種の産生は，好気的代謝の結果として起こる

■［形］＋ consequences

❶	functional consequences	機能的影響	902
❷	important consequences	重要な影響	216

❶ Transgenic expression of GFI1 and GFI1B in T cells allowed us to determine the functional consequences of constitutive expression. (J Immunol. 2003 170:2356)

訳 構成的な発現の機能的影響を決定する

consequently 副 したがって／その結果として 用例数 4,064

◆類義語：accordingly, therefore, thus

■ consequently

❶	Consequently,	したがって，	2,310
❷	and consequently	そしてその結果として	1,286

❶ Consequently, we sought to determine whether the hypocretinergic system modulates the electrical activity of motoneurons. (J Neurosci. 2004 24:5336)

訳 したがって，われわれは～かどうかを決定しようと努めた

conservation 名 保存　　用例数 4,699

◈類義語：preservation

conservation	保存	4,690
conservations	保存	9

複数形のconservationsの割合は0.2%しかなく，原則，**不可算名詞**として使われる．

C

■ conservation ＋［前］

	the%	a/an%	ø%			
❶	38	5	51	conservation of ～	～の保存	1,883
❷	9	4	74	conservation in ～	～における保存	287
❸	18	2	69	conservation between ～	～の間の保存	149

❶ Nevertheless, the **conservation of** the catalytic site in all poxvirus orthologs suggested an important role in vivo.（J Virol. 2003 77:159）

訳 すべてのポックスウイルスのオーソログにおける触媒部位の保存は，生体内での重要な役割を示唆した

❷ Their strong **conservation in** all orthologues supports the possibility of developing broad-spectrum inhibitors of this enzyme.（J Biol Chem. 2003 278:19176）

訳 すべてのオーソログにおけるそれらの強い保存は，広域性の阻害剤を開発する可能性を支持する

■［名／形］＋ conservation

❶	sequence conservation	配列保存	552
❷	evolutionary conservation	進化上の保存	379

★★ conserve 動 保存する　　用例数 34,254

conserved	保存されている／～を保存した	33,732
conserving	～を保存する	277
conserve	～を保存する	196
conserves	～を保存する	49

他動詞．過去分詞の用例が圧倒的に多い．

◈類義語：preserve

■ be conserved ＋［前］

★	be conserved	…は，保存されている	4,289
❶	be conserved in ～	…は，～において保存されている	1,645
	・be conserved in all ～	…は，すべての～において保存されている	247
❷	be conserved among ～	…は，～の間で保存されている	530
❸	be conserved between ～	…は，～の間で保存されている	355
❹	be conserved across ～	…は，～に渡って保存されている	253

❺ be conserved from ～ to …	…は，～から…まで保存されている	205
❻ be conserved throughout ～	…は，～中で保存されている	139

❶ This glutamate (or aspartate) residue is conserved in all members of the Kir family. (Proc Natl Acad Sci USA. 2002 99:8430)

訳 ～は，Kirファミリーのすべてのメンバーにおいて保存されている

❷ Binding of IgG Fc is associated with a sequence that is highly conserved among all Eib proteins but otherwise unique. (Infect Immun. 2001 69:7293)

訳 すべてのEibタンパク質の間で高度に保存されている配列

❸ Our results suggest that hepcidin plays a key role in the antimicrobial defenses of bass and that its functions are potentially conserved between fish and human. (J Biol Chem. 2005 280: 9272)

訳 それの機能は，魚とヒトの間で潜在的に保存されている

❹ This structural motif binds and hydrolyzes GTP and is conserved across HCV isolates. (J Virol. 2004 78:11288)

訳 ～は，HCV単離株全体にわたって保存されている

❺ This Akt phosphorylation site in Foxa-2 is highly conserved from mammals to insects. (Proc Natl Acad Sci USA. 2003 100:11624)

訳 ～は，哺乳類から昆虫まで高度に保存されている

■ [副] ＋ conserved

❶ highly conserved	高度に保存されている	6,516
❷ evolutionarily conserved	進化的に保存されている	2,209
❸ well conserved	よく保存されている	478
❹ phylogenetically conserved	系統発生的に保存されている	187
❺ universally conserved	普遍的に保存されている	186

❶ The vault is a highly conserved ribonucleoprotein particle found in all higher eukaryotes. (J Mol Biol. 2004 344:91)

訳 ～は，すべての高等真核生物において見つけられる高度に保存されたリボ核タンパク質粒子である

❷ These data confirm an evolutionarily conserved role for Notch signaling in vertebrate liver development, and support the zebrafish as a model system for diseases of the human biliary system. (Development. 2004 131:5753)

訳 これらのデータは，Notchシグナル伝達の進化的に保存された役割を確認する

■ conserved ＋ [名]

❶ conserved residues	保存された残基	1,113
❷ conserved region	保存された領域	599
❸ conserved sequence	保存された配列	487
❹ conserved protein	保存されたタンパク質	438
❺ conserved role	保存された役割	290
❻ conserved domain	保存されたドメイン	281
❼ conserved motifs	保存されたモチーフ	269

❶ Finally, amino acid substitutions of conserved residues in this region of TERT were found to

impair telomerase activity and processivity. (J Biol Chem. 2002 277:36174)
訳 TERTのこの領域において保存されている残基のアミノ酸置換が，テロメラーゼ活性を損なうことが見つけられた

★consider 動 考える／考慮する／みなす 用例数 15,632

considered	考えられる／考慮される／みなされる／～を考慮した	10,991
consider	～を考慮する／～と考える／～とみなす	2,558
considering	～を考慮する／～と考える／～とみなす	1,654
considers	～を考慮する／～と考える／～とみなす	429

他動詞．第5文型受動態（be considered＋補語）の用例も多い．〈be considered＋名句〉〈be considered as＋名句〉〈be considered to be＋名句〉は，ほぼ同じ意味で用いられる．

◆類義語：regard, think, appear, believe, take into account

■ be considered ＋［前／名句］

★ be considered	…は，考えられる	7,881
❶ be considered to *do*	…は，～すると考えられる	1,255
・be considered to be ～	…は，～であると考えられる	849
❷ be considered in ～	…は，～において考えられる	763
❸ be considered a ～	…は，～であると考えられる	747
❹ be considered as ～	…は，～であると考えられる	678
❺ be considered for ～	…は，～に対して考慮される	491

❶ Subunit 1 (CcO I) is considered to be the most critical of the 13 CcO component subunits. (J Immunol. 2002 168:4721)
訳 ～は，13のCcO構成成分サブユニットの最も重要なものであると考えられる

❷ These findings are considered in the context of their possible relation to the sensitized vanilloid receptor mechanism unique to nociceptors. (J Comp Neurol. 2003 463:197)
訳 これらの知見は，～と関連して考えられる

❸ Oral contraceptives (OC) have historically been considered a risk factor for gingival diseases. (J Dent Res. 2001 80:2011)
訳 経口避妊薬(OC)は，歴史的に歯肉疾患に対するリスク因子であると考えられてきた

❹ Thus, Nodal is considered as a major inducer of mesendoderm during gastrulation. (Dev Biol. 2004 275:403)
訳 Nodalは，中内胚葉の主要な誘導物質であると考えられる

■［助］＋ be considered

❶ should be considered	…は，考えられるべきである	1,450

❶ PRDM16 should be considered a candidate for mutation in human clefting disorders, especially NSCP and PRS-like CP. (Hum Mol Genet. 2010 19:774)
訳 PRDM16は，変異に対する候補であると考えられるべきである

■ [副] + considered

❶ generally considered	一般に考えられる	237
❷ previously considered	以前に考えられた	162

❶ Radiation is **generally considered** to be an immunosuppressive agent that acts by killing radiosensitive lymphocytes. (J Immunol. 2004 173:2462)
訳 放射線照射は，一般に免疫抑制剤であると考えられる

■ consider + [名節]

❶ consider that ～	…は，～ということを考慮する	274
❷ consider how ～	…は，どのように～かを考慮する	153

❶ We **considered that** rapid AGAb uptake at the motor nerve terminal membrane might attenuate complement-mediated injury. (J Clin Invest. 2012 122:1037)
訳 われわれは，～ということを考慮した

■ [代名／名] + consider

❶ we consider ～	われわれは，～を考慮する	801
❷ review considers ～	総説は，～を考慮する	84

❶ **We consider** the possibility that high-precision monitoring of pulse profiles could lead to the formation of highly stable pulsar clocks. (Science. 2010 329:408)
訳 われわれは，～という可能性を考慮する

considerable 形 かなりの　用例数 4,786

◆類義語：substantial, appreciable

■ considerable + [名]

❶ considerable interest	かなりの関心	404
❷ considerable evidence	かなりの証拠	202

❶ Thus, there is **considerable interest** in the development of novel methods for the analysis of lipoprotein complexes. (Anal Chem. 2001 73:1084)
訳 ～のための新規の方法の開発にかなりの関心がある

considerably 副 かなり／相当　用例数 2,470

◆類義語：substantially, appreciably, fairly, quite

■ considerably + [副／形]

❶ considerably more ～	相当より～な	334
❷ considerably less ～	相当より～でない	185
❸ considerably higher	相当より高い	146

❹ considerably lower	相当より低い	126

❶ This molecular elevator is considerably more complex and better organized than previously reported artificial molecular machines. (Science. 2004 303:1845)
訳 …は，～より相当複雑でよく組織化されている

★ consist 動 成る／成り立つ 用例数 15,394

consisting	成る	5,983
consists	成る	5,357
consisted	成った	2,512
consist	成る	1,542

自動詞．consist ofの用例が圧倒的に多い．
◆類義語：compose, comprise, constitute

C

■ consist of ＋［名句］

★ consist of ～	…は，～から成る	14,772
❶ consist of two ～	…は，2つの～から成る	1,191
❷ consist of ～ exons	…は，～エクソンから成る	230
❸ consist of ～ patients	…は，～患者から成る	119
❹ consist of ～ domains	…は，～ドメインから成る	116
❺ consist of ～ subunits	…は，～サブユニットから成る	110

❶ The hairpin ribozyme in its natural context consists of two loops in RNA duplexes that are connected as arms of a four-way helical junction. (Biochemistry. 2001 40:2291)
訳 ～は，RNA二本鎖の２つのループから成る

❷ The results demonstrated that AKR1B10 consists of 10 exons and 9 introns, stretching approximately 13.8 kb. (Gene. 2009 437:39)
訳 AKR1B10は，10個のエクソンと９個のイントロンから成る

■［名］＋ consisting of

★ consisting of ～	～から成る…	5,787
❶ complex consisting of ～	～から成る複合体	261
❷ protein consisting of ～	～から成るタンパク質	208
❸ system consisting of ～	～から成るシステム	181

❶ The functional transcription factor exists as a heterodimeric complex consisting of HIF-1α and the aryl hydrocarbon receptor nuclear translocator (ARNT). (J Biol Chem. 2004 279:16128)
訳 機能的な転写因子は，HIF-1 αおよびアリール炭化水素受容体核輸送体（ARNT）からなるヘテロ二量体の複合体として存在する

⁑ consistent ㊢ 一致する　用例数 39,424

consistent withの用例が圧倒的に多い.

◈類義語：congruent, corresponding, concurrent, fit, equal, identical

■ consistent with ＋［名句］

★ consistent with ～	～と一致する	35,620
❶ Consistent with ～	～と一致して	7,191
・Consistent with this hypothesis,	この仮説と一致して,	265
・Consistent with this observation,	この観察と一致して,	223
・Consistent with these findings,	これらの知見と一致して,	175
❷ consistent with the hypothesis that ～	～という仮説と一致する	1,103
❸ consistent with a role for ～	～の役割と一致する	562
❹ consistent with a model in which ～	～であるモデルと一致する	557
❺ consistent with the idea that ～	～という考えと一致する	423
❻ consistent with the notion that ～	～という概念と一致する	296
❼ consistent with the presence of ～	～の存在と一致する	288

❶ Consistent with this observation, Arabidopsis growth was inhibited following inoculation with Bgh, suggesting a shift in resource allocation from growth to defence. (Plant J. 2004 40: 633)

訳 この観察と一致して

❷ The patterns of progression for the early and late progressors are consistent with the hypothesis that knee OA progression is episodic or phasic. (Arthritis Rheum. 2004 50:2479)

訳 …は, ～という仮説と一致する

❹ The data are consistent with a model in which UvsY plays bipartite roles in presynaptic filament assembly. (J Biol Chem. 2004 279:6077)

訳 それらのデータは, UvsYが～において二部からなる役割を果たすモデルと一致する

❺ The latter result is consistent with the idea that a negative charge at position 181 contributes to protonated Schiff base stability in the later intermediates. (Biochemistry. 2004 43:12614)

訳 後者の結果は, ～という考えと一致する

■［名］＋ be consistent with

★ be consistent with ～	…は, ～と一致する	15,130
❶ results are consistent with ～	結果は, ～と一致する	2,223
❷ data are consistent with ～	データは, ～と一致する	1,484
❸ findings are consistent with ～	知見は, ～と一致する	697
❹ observations are consistent with ～	観察は, ～と一致する	303

❶ These results are consistent with HLFY functioning as a signal that is necessary for the release of the assembled functional NMDA receptor complex from the ER. (J Biol Chem. 2004 279:28903)

訳 これらの結果は, HLFY機能と一致する

■ ［名］＋ consistent with

❶ manner consistent with ～	～と一致する様式	296
❷ changes consistent with ～	～と一致する変化	135

❶ In particular, CTP pools fluctuate in a manner consistent with a role in regulating fis expression.（J Biol Chem. 2004 279:50818）

訳 ～は，fisの発現を調節する際の役割と一致する様式で変動する

consistently 副 一貫して　用例数 3,999

■ consistently ＋［過分］

❶ consistently associated with ～	一貫して～と関連する	169
・be consistently associated with ～	…は，一貫して～と関連する	125
❷ consistently observed	一貫して観察される	116

❶ In contrast, obesity has been consistently associated with an increased risk of prostate cancer aggressiveness and mortality.（Am J Clin Nutr. 2007 86:s843）

訳 肥満は，～の増大したリスクと一貫して関連してきた

constitute 動 構成する　用例数 5,661

constitute	～を構成する	3,338
constitutes	～を構成する	1,618
constituted	～を構成した／構成される	383
constituting	～を構成する	322

他動詞．受動態が使われることはほとんどない．

◆類義語：compose, comprise, consist

■ constitute ＋［名句］

❶ constitute a ～ mechanism	…は，～機構を構成する	87
❷ constitute a family of ～	…は，～のファミリーを構成する	84

❷ SR proteins constitute a family of splicing factors that play key roles in both constitutive and regulated splicing in metazoan organisms.（Mol Cell Biol. 2003 23:4139）

訳 SRタンパク質は，スプライシング因子のファミリーを構成する

■ ［名］＋ constitute

❶ proteins constitute ～	タンパク質は，～を構成する	92

constitutive ㊎ 構成的な／恒常的な　用例数 7,144

■ constitutive ＋［名］

❶ constitutive activation	構成的な活性化	859
❷ constitutive expression	構成的な発現	833
❸ constitutive activity	構成的な活性	272

❶ However, constitutive activation of NF-κB is observed in a number of cancers including breast cancer. (Oncogene. 2002 21:2066)
訳 NF-κBの構成的な活性化が，いくつかの癌において観察される

❷ While constitutive expression of TBX2 is toxic to 293 cells, clones expressing TBX2EcR are viable in the absence of an EcR ligand. (Gene. 2004 342:67)
訳 TBX2の構成的な発現は，293細胞に毒性がある

★constitutively ㊖ 構成的に／恒常的に　用例数 8,098

constitutively activeの用例が非常に多い.

■ constitutively active

★ constitutively active	構成的に活性のある	3,359
❶ constitutively active form	構成的活性型	303
❷ constitutively active mutant	構成的に活性のある変異体	198
❸ be constitutively active	…は，構成的に活性がある	227

❶ Expression of the constitutively active form of FoxO3a blocks IL-2-mediated reversal of T-cell tolerance by retaining sirt1 expression. (Proc Natl Acad Sci USA. 2012 109:899)
訳 構成的活性型のFoxO3aの発現

■ constitutively ＋［過分］

❶ constitutively expressed	構成的に発現する	1,004
・be constitutively expressed	…は，構成的に発現する	506
❷ constitutively activated	構成的に活性化される	740
❸ constitutively phosphorylated	構成的にリン酸化される	157

❶ We report here that connective tissue growth factor mRNA is constitutively expressed in normal human skin. (J Invest Dermatol. 2002 118:402)
訳 ～は，正常なヒトの皮膚において構成的に発現している

❷ Here, we show that ERK is constitutively activated in melanoma cells expressing oncogenic B-RAF and that this activity is required for proliferation. (Oncogene. 2004 23:6292)
訳 ERKは，メラノーマ細胞において構成的に活性化される

■ constitutively ＋［動］

❶ constitutively express ～	…は，～を構成的に発現する	261

❶ The cells constitutively express high levels of both p53 proteins and also Mdm2.（Oncogene. 1998 16:1429）
訳 それらの細胞は，高レベルの～を構成的に発現している

★ construct 動 作製する，名 コンストラクト／作製物 用例数 21,702

constructed	作製される／～を作製した	8,028
construct	～を作製する／コンストラクト／作製物	6,733
constructs	～を作製する／コンストラクト／作製物	6,201
constructing	～を作製する	740

C

他動詞・名詞の両方で用いられる．複数形のconstructsの割合は約60%と非常に高い．
◆類義語：create, generate, assemble

■ 動 be constructed ＋［前］

★ be constructed	…が，作製される	4,008
❶ be constructed by ～	…が，～によって作製される	471
❷ be constructed to *do*	…が，～するために作製される	400
❸ be constructed in ～	…が，～において作製される	317
❹ be constructed from ～	…が，～から作製される	313

❶ A mutant virus was constructed by inserting the lacZ gene into the coding region of mtase1.（J Virol. 2003 77:3430）
訳 変異ウイルスが，lacZ遺伝子を～に挿入することによって作製された

❷ Multivariable logistic regression models were constructed to identify the most parsimonious model that could explain the variation in the log odds of obtaining consent.（Transplantation. 2012 94:873）
訳 多変量ロジスティック回帰モデルが，～を同定するために作製された

❹ Cardiac models were constructed from the MRI images at end-diastole, isovolumic systole, peak-systole and end-systole.（Circulation. 2002 106:I168）
訳 ～が，MRI画像から作製された

■ 動［名］＋ be constructed

❶ mutants were constructed	変異体が，作製された	130

❶ Site-directed mutants were constructed to probe the role of these residues in ligand binding and/or catalysis.（Biochemistry. 2003 42:10569）
訳 部位特異的な変異体が，これらの残基の役割を精査するために作製された

■ 動 construct ＋［名句］

❶ construct a series of ～	…は，一連の～を作製する	133
❷ construct a model	…は，モデルを作製する	76

❶ To better understand the mechanism by which DC-SIGN and DC-SIGNR selectively bind HIV-1 gp120, we constructed a series of deletion mutations in the repeat regions of both receptors.（J Virol. 2005 79:4589）

訳 われわれは，一連の欠失変異体を作製した

■ 動 [代名] + construct

❶	we constructed ～	われわれは，～を作製した	1,868

■ 動 to construct

★	to construct ～	～を作製する…	1,211
❶	be used to construct ～	…は，～を作製するために使われる	293

❶ This method **was used to construct** soluble receptors that bind trinitrotoluene, l-lactate or serotonin with high selectivity and affinity.（Nature. 2003 423:185）

訳 この方法が，可溶性の受容体を作製するために使われた

■ 名 constructs + [前]

	the%	a/an%	ø%			
❶	6	-	77	constructs in ～	～におけるコンストラクト	332
❷	1	-	85	constructs of ～	～のコンストラクト	302
❸	8	-	71	constructs with ～	～を持つコンストラクト	251

❶ **Constructs in** which amino acids 130-145 were exchanged between the antizymes confirmed the critical nature of this region.（J Biol Chem. 2002 277:45957）

訳 アミノ酸130-145がそれらのアンチザイム間で置換されているコンストラクトは，～を確認した

■ 名 constructs containing

❶	constructs containing ～	～を含むコンストラクト	488

❶ To investigate transcriptional regulation of MMP-9 gene expression, we utilized retroviral-based reporter **constructs containing** different lengths of the human MMP-9 promoter.（J Biol Chem. 2003 278:32994）

訳 異なる長さのヒトMMP-9プロモーターを含むレポーターコンストラクト

■ 名 [名] + constructs

❶	reporter constructs	レポーターコンストラクト	600
❷	deletion constructs	欠失コンストラクト	247

consumption 名 摂取／消費　用例数 5,876

複数形の用例はなく，原則，**不可算名詞**として使われる．

■ consumption + [前]

	the%	a/an%	ø%			
❶	23	2	74	consumption of ～	～の摂取	1,437
❷	1	0	97	consumption in ～	～における摂取	243

	the%	a/an%	ø%			
❸	5	0	86	consumption by ～	～による摂取	131

❶ Low to moderate consumption of alcohol 1-2 d/wk is independently associated with a reduced risk of MI. (Am J Clin Nutr. 2005 82:1336)

訳 アルコールの低から中程度の摂取1-2日/週は，MIの低下したリスクと独立に関連する

■ [名] ＋ consumption

❶	oxygen consumption	酸素消費	876
❷	alcohol consumption	アルコール摂取	846

C

★ contact 名 接触／連絡，動 接触する／連絡する　用例数 17,603

contact	接触／連絡／～と接触する	10,466
contacts	接触／連絡／～と接触する	6,152
contacted	～と接触した／接触される	500
contacting	～と接触する	485

動詞としても用いられるが，名詞の用例が非常に多い．複数形のcontactsの割合は約40%と高いが，単数形は無冠詞で使われることが多い．

◈類義語：communication, connection

■ contact(s) ＋ [前]

	the%	a/an%	ø%			
❶	2	0	94	contact with ～	～との接触	2,242
				・in contact with ～	～と接触して／～と連絡して	502
❷	9	12	77	contact between ～	～の間の接触	496
❸	10	-	89	contacts in ～	～における接触	377

❶ The cells are in contact with the neurons via axons that penetrate the Teflon barrier. (J Virol. 2005 79:10875)

訳 それらの細胞は，軸索を経てニューロンに連絡している

❷ Direct contact between the electrode and a vesicle lipid bilayer membrane shows a response that correlates with vesicle membrane cholesterol content. (Anal Chem. 2005 77:7393)

訳 電極とベジクル脂質二重層膜の間の直接接触は，～を示す

■ [名／形] ＋ contact

❶	cell-cell contact	細胞－細胞接触	467
❷	direct contact	直接接触	450

★★ contain 動 含む　用例数 102,475

containing	～を含む	54,178
contains	～を含む	20,771
contain	～を含む	16,338

contained	含まれる／～を含んだ	11,188

他動詞．containingの用例が多い．

◆類義語：include, involve

■ contain ＋［名句］

❶ contain two ～	…は，2つの～を含む	2,949
❷ contain both ～	…は，両方の～を含む	1,256
❸ contain only ～	…は，～だけを含む	1,231
❹ contain three ～	…は，3つの～を含む	1,056
❺ contain multiple ～	…は，複数の～を含む	934
❻ contain either ～	…は，どちらかの～を含む	801

❶ The present study shows that TbPDE2B, **which contains two tandem GAF domains**, binds cAMP with high affinity through its GAF-A domain.（J Biol Chem. 2005 280:3771）

訳 そして，（それは）2つの直列のGAFドメインを含む

■［名］＋ contain

❶ region contains ～	領域は，～を含む	447
❷ gene contains ～	遺伝子は，～を含む	374

❷ Here we present evidence that **every gap gene contains multiple enhancers** with overlapping activities to produce authentic patterns of gene expression.（Proc Natl Acad Sci USA. 2011 108:13570）

訳 すべてのgap遺伝子は，複数のエンハンサーを含む

■［名］＋ containing

❶ domain containing ～	～を含むドメイン	1,999
❷ protein containing ～	～を含むタンパク質	950
❸ complex containing ～	～を含む複合体	811
❹ cells containing ～	～を含む細胞	761
❺ DNA containing ～	～を含むDNA	600
❻ region containing ～	～を含む領域	582
❼ constructs containing ～	～を含むコンストラクト	488
❽ medium containing ～	～を含む培地	470

❷ **Using a GST-fusion protein containing the cytoplasmic domain of LRP-1**, we show that LRP-1 is a direct substrate for v-Src in vitro.（Oncogene. 2003 22:3589）

訳 LRP-1の細胞質ドメインを含むGST融合タンパク質を使って

❸ **M-RIP can assemble a complex containing both RhoA and MBS**, suggesting that M-RIP may play a role in myosin phosphatase regulation by RhoA.（J Biol Chem. 2003 278:51484）

訳 M-RIP は，RhoAとMBSの両方を含む複合体を構築できる

■ [名]-containing + [名]

❶	～-containing protein	～を含むタンパク質	1,046
❷	～-containing neurons	～を含むニューロン	267
❸	～-containing complexes	～を含む複合体	284
❹	～-containing peptides	～を含むペプチド	265

❹ **Several RGD-containing peptides were found to promote** invasion independently of M1 expression.（Infect Immun. 1998 66:4593）

訳 いくつかのRGDを含むペプチドは，～を促進することが見つけられた

■ contained + [前]

❶	contained in ～	～に含まれる	786
❷	contained within ～	～内に含まれる	742
	・be contained within ～	…は，～内に含まれる	303

❷ We found that **297 genes contained within 184 operons were regulated by** FNR and/or by O2 levels.（J Bacteriol. 2005 187:1135）

訳 184オペロン内に含まれる297遺伝子が，～によって制御された

★content 名 含有量／内容　用例数 13,834

content	含有量／内容	12,393
contents	含有量／内容	1,441

複数形のcontentsの割合は約10%あるが，単数形は無冠詞で使われることが非常に多い.

◈類義語：amount, quantity

■ content + [前]

content ofの前にtheが付く割合は非常に高い.

	the %	a/an %	ø %			
❶	73	10	15	content of ～	～の含有量	2,680
❷	18	2	69	content in ～	～における含有量	1,022

❶ **The O-acetyl content of W135P was low**, and its removal had no adverse effect upon the conjugate's immunogenicity.（Infect Immun. 2005 73:7887）

訳 W135PのO-アセチル含有量は低かった

❷ **Lyn content in B cell subpopulations was determined by FACS**.（Arthritis Rheum. 2005 52:3955）

訳 B細胞亜集団におけるLyn含有量が，FACSによって決定された

■ [名] + content

❶	DNA content	DNA含有量	472
❷	protein content	タンパク質含有量	433

❸	GC content	GC含有量	294

★ context ㊔ 脈絡／状況／関係 用例数 12,897

context	脈絡／状況／関係	11,440
contexts	脈絡／状況／関係	1,457

複数形のcontextsの割合が，約10%ある可算名詞.

◆類義語：condition, state, status, situation, aspect

■ context ＋［前］

context ofの前にtheが付く割合は圧倒的に高い.

	the%	a/an%	ø%			
❶	98	1	1	context of ～	～という脈絡／～の状況	6,244
				・in the context of ～	～と関連して	5,136
				・be discussed in the context of ～ …は，	～と関連して議論される	353
				・within the context of ～	～という脈絡の中で	519
❷	61	16	21	context in ～	～における状況	278

❶ These various results are discussed in the context of singlet oxygen-induced responses and lactofen's potential as a disease resistance-inducing agent.（Plant Physiol. 2005 139:1784）

訳 これらのさまざまな結果は，一重項酸素に誘導される反応と関連して議論される

❶ Within the context of MSP domain proteins（MDPs）, evidence suggests that this structure functions as a protein-protein interaction domain and a signaling element.（Trends Parasitol. 2005 21:224）

訳 MSPドメインタンパク質という脈絡の中で

■ in this context

❶	in this context	この状況において	527

❶ In this context, a dominant effect of EBNA3C is to decrease Rb protein levels.（Proc Natl Acad Sci USA. 2005 102:18562）

訳 この状況において，

■［名／形］＋ context

❶	sequence context	配列関係	415
❷	cellular context	細胞の状況	234

■ context-［形］

❶	context-dependent ～	状況依存的な～	539
❷	context-specific ～	状況特異的な～	155

❶ Thus, Wise can activate or inhibit Wnt signalling in a context-dependent manner.（Development. 2003 130:4295）

訳 ～は，状況依存的な様式でWntシグナル伝達を抑制する

★ continue 動 続ける　用例数 11,542

continued	続けられる／続けた	4,942
continue	続ける	3,008
continues	続ける	2,424
continuing	続ける	1,168

自動詞の用例が多いが，他動詞としても用いられる．continue to *do*の用例が非常に多い．

◆類義語：persist, last, sustain, keep

■ continue ＋［前］

❶ continue to *do*	…は，～し続ける	5,535
・continue to be ～	…は，～され続ける／～であり続ける	1,291
・continue to increase	…は，増大し続ける	315
・continue to grow	…は，成長し続ける	189
・continue to evolve	…は，進化し続ける	174
❷ continue for ～	…は，～の間続く	242

❶ Patients receiving dexrazoxane should continue to be monitored for cardiac toxicity.（J Clin Oncol. 1999 17:3333）

訳 デクスラゾキサン投与を受けている患者は，～に関してモニターされ続けるべきである

❷ Treatment continued for 1 year or until disease progression or unacceptable toxicity.（J Clin Oncol. 2004 22:2101）

訳 治療は，1 年間続いた

continuous 形 連続的な／連続の／継続的な　用例数 7,172

◆類義語：serial, successive

■ continuous ＋［名］

❶ continuous infusion	連続注入	367
❷ continuous flow	連続的な流れ	242

❶ Propylene glycol accumulated significantly in patients receiving continuous infusion of lorazepam.（Crit Care Med. 2002 30:2752）

訳 ロラゼパムの連続注入を受けている患者

★★ contrast 名 対照，動 対照をなす　用例数 54,885

contrast	対照／対照をなす	52,813
contrasts	対照／対照をなす	910
contrasting	対照をなす	722

C

contrasted	対照をなした	440

名詞の用例が多いが，動詞として用いられることもある．文頭の〈In contrast,〉あるいは〈in contrast to〉の用例が非常に多い．

■ 名 [前] ＋ ([形] ＋) contrast

❶ in contrast	対照的に	41,919
・In contrast,	対照的に,	29,335
・In marked contrast,	著しく対照的に,	142
・In sharp contrast,	著しく対照的に,	127
❷ by contrast	対照的に	3,705
・By contrast,	対照的に,	3,294

❶ In contrast, the pkASgrp78 cells had less ERK and more JNK phosphorylation upon H_2O_2 exposure.（J Biol Chem. 2003 278:29317）

訳 対照的に,

❷ By contrast, activation of CTLA-4-negative T cells, especially $CD4^+$ cells, in these cultures was profoundly suppressed.（J Immunol. 2004 172:4184）

訳 対照的に,

■ 名 contrast ＋ [前]

❶ contrast to ～	～との対照	11,559
・in contrast to ～	～とは対照的に	10,791
・in marked contrast to ～	～とは著しく対照的に	163
・in sharp contrast to ～	～とは著しく対照的に	137
❷ contrast with ～	～との対照	773
・in contrast with ～	～とは対照的に	495

❶ In contrast to **previous reports**, we found that Ca^{2+} does not bind directly to Na^+ channel C termini.（J Biol Chem. 2004 279:45004）

訳 以前の報告とは対照的に

❷ In contrast with **reports in the literature**, we have observed migration of the Tat-peptide conjugates primarily to the cytoplasm rather than the nucleus.（Transplantation. 2003 76:1043）

訳 文献における報告とは対照的に

■ 動 contrasted ＋ [前]

❶ contrasted with ～	～と対比される	235

❶ **The results are compared and** contrasted with **previous findings** from T-jump experiments on fast fibers.（Biophys J. 1996 71:1905）

訳 それらの結果が，以前の知見と比較されそして対比された

★★ contribute 動 寄与する／一因となる／もたらす 用例数 43,488

contribute	寄与する／一因となる	24,907
contributes	寄与する／一因となる	11,135
contributing	寄与する／一因となる	4,095
contributed	寄与した／一因となった／もたらされる	3,351

contribute toの用例が圧倒的に多い．現在形の自動詞として使われることがほとんどだが，他動詞の用例もある．

◆類義語：cause, result in, lead to, produce

■ contribute to ＋［名句］

★ contribute to ～	…は，～に寄与する／～の一因となる	37,668
❶ contribute to the development of ～	…は，～の発生に寄与する	850
❷ contribute to the pathogenesis of ～	…は，～の病因に寄与する	700
❸ contribute to the regulation of ～	…は，～の調節に寄与する	284
❹ contribute to the formation of ～	…は，～の形成に寄与する	190

❶ Genetic variants of IL4 **could contribute to the development of CDC in patients with acute leukemia.** (J Infect Dis. 2003 187:1153)

訳 …は，急性白血病の患者におけるCDCの発生に寄与しうる

❷ These findings indicate that **telomere dysfunction may contribute to the pathogenesis of** Werner syndrome and Bloom syndrome. (Mol Cell Biol. 2004 24:8437)

訳 テロメアの機能異常が，～の病因に寄与するかもしれない

■［副］＋ contribute to

❶ likely contribute to ～	…は，～に寄与しそうである	492

❶ **Increased oxidant production likely contributes to effective bacterial killing** in the lungs of SP-D$^{-/-}$ mice. (J Immunol. 2000 165:3934)

訳 増大したオキシダント産生は，効果的な細菌死滅に寄与しそうである

■［動／形］＋ to contribute to

★ to contribute to ～	～に寄与する…	2,287
❶ be thought to contribute to ～	…は，～に寄与すると考えられる	266
❷ appear to contribute to ～	…は，～に寄与するように思われる	224
❸ be likely to contribute to ～	…は，～に寄与しそうである	202
❹ be shown to contribute to ～	…は，～に寄与することが示される	135

❶ **The cell death response to oxidative stress is thought to contribute to** aging, neurological degeneration, and other disorders. (Mol Pharmacol. 2003 63:276)

訳 酸化ストレスへの細胞死応答は，～に寄与すると考えられる

■ contribute ＋［副］＋ to

❶ contribute significantly to ～	…は，～に著しく寄与する	914

❷ contribute substantially to ～	…は，～にかなり寄与する	229

❶ Although EDSBs are usually repaired with high fidelity, **errors in their repair contribute significantly to the rate of cancer** in humans. (Proc Natl Acad Sci USA. 2003 100:12871)
訳 それらの修復のエラーは，癌の発生率に著しく寄与する

■ [名] + contributing to

★ contributing to ～	～に寄与する…	3,162
❶ factors contributing to ～	～に寄与する因子	235
❷ mechanisms contributing to ～	～に寄与する機構	130

❶ To date, **only a few factors contributing to the specificity of this symbiosis have been identified.** (J Bacteriol. 2001 183:835)
訳 この共生の特異性に寄与するほんのわずかの因子だけが同定されている

★contribution 名 寄与／貢献　　用例数 14,295

contribution	寄与／貢献	9,264
contributions	寄与／貢献	5,031

複数形のcontributionsの割合が，約35%ある可算名詞.

■ contribution(s) + [前]

contribution ofの前にtheが付く割合は圧倒的に高い.

	the%	a/an%	ø%			
❶	91	6	3	contribution of ～	～の寄与	5,639
				・the contribution of ～ to …	…への～の寄与	1,770
				・the relative contributions of ～	～の相対的な寄与	484
❷	8	37	6	contribution to ～	～への寄与	2,246
❸	15	-	83	contributions from ～	～からの寄与	471

❶ **We systematically analyzed the contribution of the various inputs to the observed kinetic response** of channel activation. (J Biol Chem. 2003 278:10851)
訳 われわれは，観察された動力学的応答へのさまざまな入力の寄与を体系的に分析した

❶ **Mutational analysis was carried out to probe the relative contributions of the two domains to DNA binding.** (Biochemistry. 2003 42:12909)
訳 変異解析が，DNA結合への２つのドメインの相対的な寄与を精査するために実行された

❷ The role of different tissues in insulin action and **their contribution to the pathogenesis of diabetes remain unclear.** (J Clin Invest. 2004 114:214)
訳 糖尿病の病因へのそれらの寄与は，不明のままである

⁑ control ㊔ コントロール／制御／対照群, ㊓ 制御する

用例数 136,534

control	コントロール／制御／対照群／～を制御する	82,527
controls	コントロール／制御／対照群／～を制御する	27,863
controlled	制御される／～を制御した	16,759
controlling	～を制御する	9,385

名詞の用例が多いが，他動詞としても用いられる．複数形のcontrolsの割合は約30%あるが，単数形は無冠詞で使われることが圧倒的に多い．

◈類義語：regulation, modulation, adjustment, regulate, modulate, adjust, drive

■ ㊔ control ＋［前］

	the%	a/an%	ø%			
❶	40	0	60	control of ～	～の制御	14,170
				・in the control of ～	～の制御において	2,022
				・under the control of ～	～の制御下で	1,799
❷	2	0	97	control in ～	～における制御	1,116

❶ Inward rectifier potassium (Kir) channels **play important roles in the maintenance and control of cell excitability.** (J Biol Chem. 2004 279:22331)

訳 ～は，細胞興奮性の維持と制御において重要な役割を果たす

❷ **The findings emphasise the importance of prompt blood-pressure control in hypertensive patients** at high cardiovascular risk. (Lancet. 2004 363:2022)

訳 それらの知見は，高血圧の患者における迅速な血圧の制御の重要性を強調する

■ ㊔［過分／形／名］＋ control(s)

❶	healthy controls	健康な対照群	1,171
❷	matched controls	一致する対照群	1,006
❸	transcriptional control	転写の制御	837
❹	normal controls	正常な対照群	675
❺	wild-type controls	野生型の対照群	665

❶ **Serum from 22 patients with active RP and an equal number of age- and sex-matched healthy controls and RA patients were available for analysis.** (Arthritis Rheum. 2004 50:3663)

訳 活動的なRPの22名の患者および同じ数の年齢や性別の一致する健康な対照群およびRA患者からの血清が，解析に利用できた

■ ㊔ control ＋［名］

❶	control subjects	対照群被験者	4,180
❷	control group	対照群	3,225
❸	control mice	対照群マウス	1,984

❶ The mean distal ileum wall thickness at the first examination **was 3.0 mm in patients with rotavirus infection and 2.0 mm in control subjects** (P = .037). (J Infect Dis. 2004 189:1382)

訳 ～が，ロタウイルス感染症の患者において3.0 mmであり対照群被験者において2.0 mmであった

■ 名 [名] + and controls

❶ cases and controls	症例と対照群	350
❷ patients and controls	患者と対照群	315

■ 動 control + [名句]

❶ control the expression of ～	…は，～の発現を制御する	520
❷ control the activity of ～	…は，～の活性を制御する	204
❸ control the development of ～	…は，～の発達を制御する	108
❹ control the rate of ～	…は，～の割合を制御する	106
❺ control the timing of ～	…は，～のタイミングを制御する	96

❶ Multiple factors control the expression of the outer membrane porins OmpF and OmpC in Escherichia coli.（J Bacteriol. 2011 193:2252）

訳 複数の因子が，～の発現を制御する

■ 動 be controlled

★ be controlled	…は，制御される	4,966
❶ be controlled by ～	…は，～によって制御される	3,326
・expression is controlled by ～	発現は，～によって制御される	111
❷ be controlled in ～	…は，～において制御される	287

❶ In zebrafish, cdx4 expression is controlled by the Wnt pathway, but the molecular mechanism of this regulation is not fully understood.（EMBO J. 2011 30:2894）

訳 cdx4の発現は，Wnt経路によって制御される

■ 動 [名／過分] + controlled

❶ placebo-controlled ～	プラセボ対照～	351
❷ randomized controlled ～	無作為化対照～	249

❶ This pilot study was a 12-week randomized, double-blind, placebo-controlled trial of etanercept, with 14 subjects in each group.（Arthritis Rheum. 2004 50:2240）

訳 このパイロット研究は，エタネルセプトの12週の無作為化された二重盲検のプラセボ対照治験であった

■ 動 [名] + controlling

❶ mechanisms controlling ～	～を制御する機構	540

controversial 形 議論の余地のある 用例数 3,090

■ [動] + controversial

❶ be controversial	～は，議論の余地がある	1,206
❷ remain controversial	～は，議論の余地があるままである	1,157

❶ Tracheostomy practice in the setting of critical illness is controversial because evidence demonstrating unequivocal benefit is lacking. (Crit Care Med. 2005 33:2513)

訳 明確な利益を実証する証拠が欠けているので，～は議論の余地がある

❷ The role of complement component C5 in asthma remains controversial. (J Clin Invest. 2005 115:1590)

訳 喘息における補体成分C5の役割は，議論の余地があるままである

★ conventional 形 通常の／従来の 用例数 10,517

◈類義語：general, normal, usual, common

■ conventional + [名]

❶ conventional methods	通常の方法／従来の方法	244
❷ conventional therapy	通常の治療／従来の治療	236
❸ conventional chemotherapy	通常の化学療法／従来の化学療法	136

❶ Since conventional methods to measure cell volume, such as microscopy, are complex and time-consuming, cell volume has not been used as the basis for cell-based screening. (Anal Chem. 2005 77:1290)

訳 細胞の体積を測定するための通常の方法

■ [前] + conventional

❶ compared with conventional ～	通常の～と比べて／従来の～と比べて	237

❶ Compared with conventional radiotherapy, intensity-modulated radiotherapy (IMRT) can reduce irradiation of the parotid glands. (Lancet Oncol. 2011 12:127)

訳 従来の放射線療法と比べて

conversely 副 逆に 用例数 5,474

文頭に用いられる場合がほとんどである.

◈類義語：inversely, oppositely, in contrast

■ Conversely

❶ Conversely,	逆に，	4,886

❶ Conversely, overexpression of constitutively active calcineurin was sufficient to induce ECM protein expression. (J Biol Chem. 2004 279:15561)

訳 逆に，構成的に活性のあるカルシニュリンの過剰発現は，～を誘導するのに十分であった

conversion 名 変換／置換 用例数 8,972

conversion	変換	8,662
conversions	変換	310

複数形のconversionsの割合が約3％あるが，単数形は無冠詞で使われることが非常に多い．

◆類義語：shift, change, alteration, interconversion, switch, substitution, replacement

■ conversion ＋［前］

	the%	a/an%	ø%			
❶	53	0	40	conversion of ～	～の変換	4,080
				・conversion of ～ to …	～の…への変換	1,721
				・conversion of ～ into …	～の…への変換	346
				・catalyze the conversion of ～	…は，～の変換を触媒する	365
❷	7	2	81	conversion to ～	～への変換	1,002
❸	15	7	69	conversion from ～	～からの変換	213
				・conversion from ～ to …	～から…への変換	134

❶ The CysE enzyme, serine acetyltransferase, **catalyzes the conversion of serine to O-acetyl-L-serine** (OAS). (J Bacteriol. 2004 186:7610)

訳 ～は，セリンのO-アセチル-L-セリンへの変換を触媒する

❷ **Conversion to the abnormal isoform results in the formation and accumulation of** prion protein aggregates. (J Biol Chem. 2005 280:2455)

訳 異常なアイソフォームへの変換は，～の形成と蓄積という結果になる

convert 動 変える／変換する／変わる 用例数 9,345

converted	変わった／～を変えた	3,774
converting	～を変える／変わる	2,271
convert	～を変える／変わる	1,776
converts	～を変える／変わる	1,524

他動詞の用例が多いが，自動詞としても使われる．converted toの用例が多い．

◆類義語：change, alter, shift, switch

■ be converted ＋［前］

★	be converted	…は，変わる	2,131
❶	be converted to ～	…は，～に変わる	1,366
❷	be converted into ～	…は，～に変わる	465

❶ **About one-third of apoE2 was converted to apoE3**, and the repair was stable through 12 passages. (J Biol Chem. 2001 276:13226)

訳 apoE2のおよそ1/3が，apoE3に変わった

❷ If as few as three inappropriate genes are turned off, a normal cell can be converted into a cancer cell. (Oncogene. 2001 20:2988)

訳 正常細胞は，癌細胞に変わりうる

■ convert + [名句] + to/into

❶	convert ～ to …	‥は，～を…に変える	1,489
❷	convert ～ into …	‥は，～を…に変える	753

C

❶ ME2 is a genome-coded mitochondrial enzyme that converts malate to pyruvate and is involved in neuronal synthesis of the neurotransmitter γ-aminobutyric acid (GABA). (Am J Hum Genet. 2005 76:139)

訳 リンゴ酸をピルビン酸に変える酵素

cooperation 名 協同／協同作用／協力 用例数 1,492

複数形の用例はなく，原則，不可算名詞として使われる．

◆類義語：collaboration, cooperativity, conjunction

■ cooperation + [前]

	the%	a/an%	ø%			
❶	11	13	73	cooperation between ～	～の間の協同作用	307
❷	2	0	95	cooperation with ～	～との協同作用	271
				・in cooperation with ～	～と協同して	172
❸	56	8	36	cooperation of ～	～の協同作用	151

❶ Moreover, it provides genetic evidence for the cooperation between integrin and cytokine signaling pathways. (J Cell Biol. 2005 171:717)

訳 それは，インテグリンとサイトカインのシグナル伝達経路の間の協同作用の遺伝的証拠を提供する

❷ Regulation of gene transcription by the progesterone receptor (PR) in cooperation with coactivator/corepressor complexes coordinates crucial processes in female reproduction. (Mol Cell Biol. 2005 25:8150)

訳 コアクチベーター/コリプレッサー複合体と協同したプロゲステロン受容体（PR）による遺伝子転写の調節は，～における決定的な過程を調整する

cooperative 形 協同的な 用例数 3,857

◆類義語：collaborative

■ cooperative + [名]

❶	cooperative binding	協同的な結合	376
❷	cooperative interactions	協同的な相互作用	276

■ [副] + cooperative

❶ highly cooperative	高度に協同的な	190

❶ The data indicate that **the minimal DBD dimer binds both ATP and ADP at two equivalent but highly cooperative binding sites.** (Biochemistry. 2005 44:9645)

訳 最小のDBD二量体は，２つの同等だが高度に協同的な結合部位でATPおよびADPの両方に結合する

cooperativity ㊔ 協同性／協同作用 用例数 2,115

cooperativity	協同性／協同作用	2,104
cooperativities	協同性／協同作用	11

複数形のcooperativitiesの割合は約0.5%しかなく，原則，**不可算名詞**として使われる．

◆類義語：cooperation, collaboration, conjunction

■ cooperativity + [前]

cooperativity ofの前にtheが付く割合はかなり高い．

	the%	a/an%	ø%			
❶	64	3	31	cooperativity of ～	～の協同性	266
❷	10	1	83	cooperativity in ～	～における協同作用	240
❸	16	5	74	cooperativity between ～	～の間の協同作用	187

❸ **This model can explain the cooperativity between all binding partners of eIF4A** (eIF4G, RNA, ATP) and stimulation of eIF4A activity in the eIF4F complex. (Genes Dev. 2005 19:2212)

訳 このモデルは，eIF4Aのすべての結合パートナーの間の協同作用を説明できる

■ [形] + cooperativity

❶ positive cooperativity	正の協同性	197
❷ negative cooperativity	負の協同性	194

★coordinate ㊍ 協調させる／調整する／配位結合させる／協調する，㊌ 協調した／同等の，㊔ 座標 用例数 11,153

coordinated	協調した／～を協調させた／～を配位結合させた	4,423
coordinate	～を協調させる／～を配位結合させる／協調する／協調した／同等の	3,880
coordinates	～を協調させる／～を配位結合させる／協調する	1,601
coordinating	協調する	1,249

動詞の用例が多いが，形容詞や名詞としても用いられる．

◆類義語：synchronize, entrain

■ 動 be coordinated ＋［前］

★	be coordinated	…は，協調させられる	830
❶	be coordinated by ～	…は，～によって協調させられる	212
❷	be coordinated with ～	…は，～と協調する	166
❸	be coordinated to ～	…は，～に配位結合する	132

❶ RNA synthesis and processing are coordinated by proteins that associate with RNA polymerase II (pol II) during transcription elongation. (Mol Cell. 2005 20:225)

訳 RNA合成とプロセシングは，～と結合するタンパク質によって協調させられる

❷ Together these data suggest that ELMO1-mediated cytoskeletal changes may be coordinated with ERM protein crosslinking activity during dynamic cellular functions. (J Biol Chem. 2006 281:5928)

訳 ELMO1に仲介される細胞骨格の変化は，ERMタンパク質架橋結合活性と協調するかもしれない

❸ A magnesium ion is found to be coordinated to the phosphate tail of ATP as well as to a pyrophosphate group. (J Mol Biol. 2005 354:289)

訳 マグネシウムイオンは，ATPのリン酸末端に配位結合することが見つけられる

coordination 名 協調／配位　　用例数 4,208

coordination	協調	4,193
coordinations	協調	15

複数形のcoordinationsの割合は0.4%しかなく，原則，不可算名詞として使われる．

◆類義語：cooperation

■ coordination ＋［前］

	the%	a/an%	ø%			
❶	50	3	46	coordination of ～	～の協調	1,217
❷	24	12	61	coordination between ～	～の間の協調	173
❸	6	4	83	coordination to ～	～への協調	134
❹	1	4	80	coordination with ～	～との協調	119
				・in coordination with ～	～と協調して	47
❺	11	6	72	coordination in ～	～における協調	110

❶ The coordination of primase function within the replisome is an essential but poorly understood feature of lagging strand synthesis. (Mol Cell. 2005 20:391)

訳 レプリソーム内でのプライマーゼ機能の協調は，～の必須だがよく理解されていない特徴である

❷ Coordination between cell proliferation and cell death is essential to maintain homeostasis in multicellular organisms. (Cell. 2005 122:421)

訳 細胞増殖と細胞死の間の協調は，～を維持するために必須である

correct 動 補正する／訂正する，形 正確な 用例数 8,523

correct	～を補正する／～を訂正する／正確な	4,835
corrected	補正される／～を補正した／～を訂正した	2,902
correcting	～を補正する／～を訂正する	494
corrects	～を補正する／～を訂正する	292

形容詞としても用いられるが，動詞の用例が多い．

◆同義語：accurate, precise, exact

■ corrected ＋［前］

❶ corrected for ～	～に関して補正される	398
❷ corrected by ～	～によって補正される	366
・be corrected by ～	…は，～によって補正される	246

❶ All data were corrected for percentage recovery.（J Physiol. 1996 493:877）

訳 すべてのデータが，回収率に関して補正された

❷ This deficiency is corrected by the expression of wild-type（wt）protein L27 from a plasmid.（Mol Cell. 2005 20:427）

訳 この欠損は，～の発現によって補正される

correctly 副 正確に／正しく 用例数 2,197

◆類義語：accurately, precisely, exactly

■ correctly ＋［過分／過去］

❶ correctly identified	正確に同定される／～を正確に同定した	300
❷ correctly predicted	正確に予想される／～を正確に予想した	135
❸ correctly folded	正確に畳まれる／～を正確に折り畳んだ	100

❶ All T. gondii-infected animals were correctly identified by this technique.（J Clin Microbiol. 2001 39:2065）

訳 すべてのトキソプラズマ感染した動物が，この技術によって正確に同定された

★★ correlate 動 相関させる／相関する，名 相関関係にあるもの 用例数 36,596

correlated	相関する／相関した	23,060
correlates	相関する／相関関係にあるもの	6,694
correlate	相関する／相関関係にあるもの	5,826
correlating	相関する	1,016

自動詞・他動詞の両方で用いられる．自動詞の数字は現在形のみの例数を示す．〈correlate with〉〈be correlated with〉の用例が非常に多い．副詞を置く位置にも注意が必要である．自動詞のcorrelateでは，直前に来る場合と直後に来る場合の両方があ

る（ただし，wellは後に置く）．be correlatedの場合は，通常，be動詞とcorrelatedの間に置かれる．名詞として用いられる場合もある．

◈類義語：relate, associate, couple, link, connect, concern, attach

■ correlate with ＋［名句］

★ correlate with ～	…は，～と相関する	8,240
❶ correlate with increased ～	…は，増大した～と相関する	245
❷ correlate with changes in ～	…は，～の変化と相関する	97
❸ correlate with the degree of ～	…は，～の程度と相関する	73
❹ correlate with the ability of ～	…は，～の能力と相関する	72
❺ correlate with the presence of ～	…は，～の存在と相関する	71

❷ Importantly, the changes in quaternary structure correlate with changes in function.（Curr Opin Pharmacol. 2010 10:67）

訳 四次構造の変化は，機能の変化と相関する

■［名］＋ correlate with

❶ expression correlates with ～	発現は，～と相関する	199
❷ activity correlates with ～	活性は，～と相関する	104
❸ levels correlate with ～	レベルは，～と相関する	90

❶ Increased USP46 expression correlates with decreased ubiquitination and upregulation of PHLPP proteins in colon cancer cells, whereas knockdown of USP46 has the opposite effect.（Oncogene. 2013 32:471）

訳 増大したUSP46発現は，低下したユビキチン化と相関する

■［副］＋ correlate with

❶ directly correlate with ～	…は，～と直接相関する	123
❷ inversely correlate with ～	…は，～と逆相関する	97
❸ strongly correlate with ～	…は，～と強く相関する	91
❹ positively correlate with ～	…は，～と正に相関する	73

❶ Modulation of BDNF levels directly correlates with Akt phosphorylation and inhibitors of PI 3-kinase abrogate the BDNF responses.（J Biol Chem. 2004 279:33538）

訳 BDNFレベルの調節は，Aktのリン酸化と直接相関する

■ correlate ＋［副］＋ with

❶ correlate well with ～	…は，～とよく相関する	454
❷ correlate strongly with ～	…は，～と強く相関する	117
❸ correlate directly with ～	…は，～と直接相関する	89
❹ correlate inversely with ～	…は，～と逆相関する	73
❺ correlate positively with ～	…は，～と正に相関する	64
❻ correlate significantly with ～	…は，～と有意に相関する	63
❼ correlate closely with ～	…は，～と密接に相関する	63

❶ These results correlate well with our previous observations that morphine treatment causes CEM x174 cells to be more susceptible to SIV infection. (J Biol Chem. 2000 275:31305)
訳 これらの結果は，われわれの以前の観察とよく相関する

■ be correlated ＋［前］

★ be correlated	…は，相関する	4,828
❶ be correlated with ～	…は，～と相関する	4,123
・be correlated with increased ～	…は，増大した～と相関する	117
・be correlated with changes in ～	…は，～の変化と相関する	50
❷ be correlated to ～	…は，～と相関する	261

❶ We also demonstrate that the OM-induced morphological changes are correlated with increased cell motility in a STAT3-dependent manner. (Oncogene. 2003 22:894)
訳 OMに誘導される形態的変化は，増大した細胞運動と相関する

■［名］＋ be correlated with

❶ results were correlated with ～	結果が，～と相関した	91
❷ findings were correlated with ～	知見が，～と相関した	84

❶ Results were correlated with treatment outcome. (J Clin Oncol. 2012 30:3625)
訳 結果は，治療成績と相関した

■ be ＋［副］＋ correlated with

❶ be positively correlated with ～	…は，～と正に相関する	470
❷ be inversely correlated with ～	…は，～と逆相関する	420
❺ be highly correlated with ～	…は，～と高度に相関する	409
❸ be significantly correlated with ～	…は，～と有意に相関する	405
❹ be strongly correlated with ～	…は，～と強く相関する	304
❻ be negatively correlated with ～	…は，～と負に相関する	222

❶ Factor Ⅶc was positively correlated with total cholesterol ($p < 0.001$). (Am J Epidemiol. 1999 150:727)
訳 第7c因子は，総コレステロール量と正に相関した

❷ Furthermore, taurine-like immunoreactivity is inversely correlated with odor responsiveness. (J Neurosci. 2000 20:3282)
訳 タウリン様免疫反応性は，匂いの反応性と逆相関する

★ correlation ㊔ 相関　　用例数 17,167

correlation	相関	13,496
correlations	相関	3,671

複数形のcorrelationsの割合が，約20％ある可算名詞．correlation betweenの用例が多い．

◆類義語：relation, relationship, association, relevance, link, connection, implication

■ correlation ＋［前］

	the%	a/an%	ø%			
❶	24	58	10	correlation between ～	～の間の相関	5,262
❷	6	37	45	correlation with ～	～との相関	1,207
❸	42	26	29	correlation of ～	～の相関	1,064
				・correlation of ～ with …	～の…との相関	246

❶ This study investigates the correlation between the formation of reactive oxygen species (ROS) and auditory damage in noise-induced hearing loss.（Brain Res. 2000 878:163）

訳 この研究は，活性酸素種（ROS）の形成と聴覚性の障害の間の相関を精査する

❷ These measures were evaluated for their correlation with nucleotide sequence diversity.（J Clin Microbiol. 1998 36:2982）

訳 これらの計測が，ヌクレオチド配列の多様性とそれらとの相関に関して評価された

❸ Correlation of BG with clinical outcome was assessed in each patient.（J Clin Microbiol. 2012 50:2104）

訳 BGと臨床結果との相関が，評価された

■［形］＋ correlation

❶	significant correlation	有意な相関	734
❷	no correlation	相関のない	684
❸	strong correlation	強い相関	603
❹	positive correlation	正の相関	562
❺	inverse correlation	逆相関	434
❻	direct correlation	直接の相関	310

❶ We find a statistically significant correlation between the development of a virally induced tumor (sarcoid) and heterozygosity for the equine SCID allele.（Gene. 2002 283:263）

訳 われわれは，ウイルスに誘導された腫瘍（サルコイド）の発生と～のヘテロ接合性の間の統計的に有意な相関を見つける

❷ There was no correlation between mutation rate and recombination rate.（Am J Hum Genet. 2002 70:625）

訳 変異率と組換え率の間の相関はなかった

■ correlation ＋［動］

❶	correlation was found	相関が，見つけられた	281
❷	correlation was observed	相関が，観察された	200

❶ However, no significant correlation was found between surfactant protein concentrations and disease severity measured by arterial alveolar oxygen ratio.（Am J Respir Crit Care Med. 1999 159:1115）

訳 ～の間に有意な相関は見つけられなかった

correspond 動 一致する／相当する　用例数 4,740

corresponds	一致する／相当する	1,789
correspond	一致する／相当する	1,724
corresponded	一致した／相当した	1,227

自動詞．correspond toの用例が非常に多い．correspondingは別項参照．

◆類義語：coincide, agree, fit, match, represent, resemble

■ correspond +［前］

❶ correspond to ～	…は，～に一致する／～に相当する	4,038
❷ correspond with ～	…は，～と一致する	348

❶ This peptide sequence **corresponds to** a portion of the MV binding site on the human MV receptor CD46.

訳 このペプチド配列は，MV結合部位の一部に一致する

★ corresponding 形 -ing 対応する／一致する　用例数 16,700

corresponding toの用例が非常に多い．

◆類義語：consistent, congruent, concurrent, equivalent, fit

■ corresponding to

★ corresponding to ～	～に対応する…	5,235
❶ peptide corresponding to ～	～に対応するペプチド	407
❸ corresponding to residues ～	残基～に対応する…	129

❶ Ig-PLP1 is an Ig chimera expressing proteolipid protein-1 (PLP1) **peptide corresponding to** aa residues 139-151 of PLP. (J Immunol. 2003 171:1801)

訳 PLPのアミノ酸残基139-151に対応するプロテオリピッドタンパク-1（PLP1）のペプチド

■ corresponding +［名］

❶ a corresponding increase in ～	一致する～の増大	214
❷ the corresponding region	対応する領域	182

❶ Expression of exogenous K protein augmented the insulin-induced mitochondrial level of UCP2 protein that was not accompanied by **a corresponding increase in** ucp2 mRNA. (J Biol Chem. 2004 279:54599)

訳 外来性のKタンパク質の発現が，ucp2メッセンジャーRNAの一致する増大を伴わないUCP2タンパク質のインスリン誘導性のミトコンドリアレベルを増強した

★ couple ㊍ 共役させる／結合させる／関連づける，㊔ 共役／関連

用例数 19,177

coupled	共役する／結合する／関連する	16,213
couple	～を共役させる／～を結合させる／～を関連づける／結合する／一対	1,651
couples	～を共役させる／～を結合させる／～を関連づける／結合する／カップル	1,313

他動詞の用例が多いが，自動詞としても使われる．coupled toは物理的な結合や具体的な現象との関連を意味し，coupled withは結果など抽象的な概念との関連を表すことが多い．couplingは別項参照．

◈類義語：bind, associate, engage, link, connect, join, conjugate, relate, correlate

C

■ be coupled ＋［前］

★ be coupled	…は，共役（関連）する	1,957
❶ be coupled to ～	…は，～と共役（関連）する	1,227
❷ be coupled with ～	…は，～と共役（関連）する	355

❶ In addition, at physiological pH values, electron transfer **is coupled to** proton transfer for the P2＋/1＋ couple.（Biochemistry. 1998 37:11376）

訳 電子伝達は，プロトン伝達と共役する

❷ The insertion of the peptide into the membrane **is coupled with** work for creation of a vacancy for the peptide in the membrane.（Biophys J. 2001 81:285）

訳 ペプチドの膜への挿入は，～に対する作用と共役する

■［副］＋ coupled

❶ tightly coupled	堅固に共役（関連）する	245
❷ inductively coupled	誘導的に共役（関連）する	172

❶ We review here the evidence for the concept that the topology of the microloop formed by such activators is **tightly coupled** to the structural transitions in DNA mediated by RNA polymerase.（J Mol Biol. 1998 279:1027）

訳 ～は，構造的転移に堅固に共役する

■ couple ＋［名句］＋［前］

❶ couple ～ to …	‥は，～を…に結合（共役）させる	559
❷ couple ～ with …	‥は，～を…に結合（共役）させる	134

❶ Adhesion is achieved by homophilic interaction of the extracellular domains of cadherins on adjacent cells, with the cytoplasmic regions serving to **couple** the complex **to** the cytoskeleton.（J Biol Chem. 1999 274:37885）

訳 細胞質領域が複合体を細胞骨格に結合させるように働く状態で

■ couple ＋［前］

❶	couple to ～	…は，～に結合する	441
❷	couple with ～	…は，～に結合する	113

❶ The ErbB2/ErbB3 heregulin co-receptor has been shown to couple to phosphoinositide (PI) 3-kinase in a heregulin-dependent manner. (J Biol Chem. 2001 276:42153)

訳 ErbB2/ErbB3へレグリン・コレセプターが，ホスホイノシチド（PI）3-キナーゼに結合することが示された

★coupling 名 共役／カップリング，-ing 共役する　用例数 11,988

coupling	共役／共役する	11,132
couplings	共役	856

名詞の用例が多い．複数形のcouplings割合は約７％あるが，単数形は無冠詞で使われることが非常に多い．

◆類義語：conjugation, linkage, association, link

■ coupling ＋［前］

	the％	a/an％	ø％			
❶	43	6	47	coupling of ～	～の共役	1,901
				・coupling of ～ to …	～の…への共役	362
❷	32	13	55	coupling between ～	～の間の共役	1,110
❸	10	6	64	coupling to ～	～への共役	612

❶ This functional coupling of CD47 to heterotrimeric G proteins provides a mechanistic explanation for the biological effects of CD47 in a wide variety of systems. (J Biol Chem. 1999 274:8554)

訳 CD47のヘテロ三量体Gタンパク質へのこの機能的共役は，～に対する機構的説明を提供する

❷ Expression of a nonfunctional mutant Shal also induces a large increase in I_h, demonstrating a novel activity-independent coupling between the Shal protein and I_h enhancement. (Neuron. 2003 37:109)

訳 Shalタンパク質と～の間の新規の活性非依存的共役を実証する

❸ The results are consistent with aryloxy coupling to diDOPA followed by reoxidation to diDOPA quinone. (Biochemistry. 2000 39:11147)

訳 それらの結果は，～へのアリールオキシ共役と一致する

★course 名 経過／過程，動 経路をとる　用例数 10,953

course	経過／過程／経路をとる	9,737
courses	経過／過程／経路をとる	1,169
coursed	経路をとった	24
courseing	経路をとる	23

動詞の用例もあるが，ほとんどが名詞である．複数形のcoursesの割合が，約10%ある可算名詞．

■ course ＋［前］

course ofの前にtheが付く割合は非常に高い．

	the%	a/an%	ø%			
❶	79	18	3	course of ～	～の経過	1,901
				・during the course of ～	～の間に	929
				・in the course of ～	～の間に	834
				・early in the course of ～	～の間の早期に	203
				・over the course of ～	～の間に	676

❶ Inhibition of arginase activity during the course of infection has a clear therapeutic effect, as evidenced by markedly reduced pathology and efficient control of parasite replication. (FASEB J. 2005 19:1000)

訳 感染の間のアルギナーゼ活性の阻害は，明らかな治療効果を持つ

❶ Early in the course of a patient's illness, it is often impossible to determine whether the disease will be cured with cancer-directed treatment. (JAMA. 2004 292:2141)

訳 患者の疾病の間の早期に，

■［形／名］＋ course

❶	time course	時間経過	2,933
❷	clinical course	臨床経過	653
❸	disease course	疾患経過	319

★create 動 作製する／作る 用例数 12,575

created	作製される／～を作製した	5,346
create	～を作製する	3,871
creating	～を作製する	2,034
creates	～を作製する	1,324

他動詞．

◆類義語：generate, produce, construct, yield

■ create ＋［名句］

❶	create mice	…は，マウスを作製する	93
❷	create ～ model	…は，～モデルを作製する	81
❸	create a series of ～	…は，一連の～を作製する	72

❶ Here we created mice deficient in Kif3a in cartilage and focused on the cranial base and synchondroses. (Development. 2007 134:2159)

訳 われわれは，Kif3aを欠損したマウスを作製した

■ [代名] + create

❶ we created ～	われわれは，～を作製した	1,003

❶ In addition, we created an E4orf6 mutant that is selectively defective in rAAV augmentation of transduction. (J Virol. 1999 73:10010)
訳 われわれは，E4orf6変異体を作製した

■ to create

★ to create	～を作製する…	2,690
❶ be used to create ～	…は，～を作製するために使われる	285

❶ We demonstrate that the technique can be used to create large libraries of random fusions between two genes. (Nucleic Acids Res. 2002 30:e119)
訳 その技術は，2つの遺伝子の間のランダム融合の大きなライブラリーを作製するために使われうる

■ be created + [前]

★ be created	…が，作製される	1,888
❶ be created by ～	…が，～によって作製される	422
❷ be created in ～	…が，～において作製される	266
❸ be created to *do*	…が，～するために作製される	148

❶ A hybrid mouse was created by crossing a Wnt-reporter animal (BAT-LacZ) with a model of colorectal cancer (Apc1638N). (FASEB J. 2011 25:3136)
訳 ハイブリッドマウスが，～を交雑させることによって作製された

■ [名] + be created

❶ mice were created	マウスが，作製された	422

criterion ㊔ 判断基準／基準／クライテリア 用例数 9,720

criteria	判断基準／基準	8,534
criterion	判断基準／基準	1,186

複数形のcriteriaの割合が，約90%と圧倒的に高い．
◆類義語：standard

■ criteria + [前]

criteria ofの前にtheが付く割合はかなり高い．

	the%	a/an%	ø%			
❶	21	-	72	criteria for ～	～の判断基準	2,253
				・meet criteria for ～	…は，～の判断基準を満たす	263
❷	65	-	30	criteria of ～	～の判断基準	255

❶ One-hundred and seventy patients met criteria for acute lung injury (incidence, 27%; 95% confidence interval, 24–31%). (Crit Care Med. 2006 34:196)
訳 170名の患者が，急性肺障害の判断基準を満たした

■ [形／名] + criteria

❶	diagnostic criteria	診断上の基準	500
❷	inclusion criteria	組み入れ基準	460
❸	clinical criteria	臨床基準	178
❹	selection criteria	選択基準	152

❶ Diagnostic criteria for schwannomatosis, a recently described condition, are being developed. (Curr Opin Neurol. 2005 18:604)
訳 シュワン細胞腫に対する診断上の基準

★ critical 形 決定的に重要な　用例数 42,902

critical forの用例が多い.
◆類義語：crucial, definitive, conclusive, important

■ be critical + [前]

★	be critical	…は，決定的に重要である	13,146
❶	be critical for ～	…は，～のために決定的に重要である	8,822
	・be critical for the development of ～	…は，～の発達のために決定的に重要である	137
	・be critical for maintaining ～	…は，～を維持するために決定的に重要である	153
	・be critical for understanding ～	…は，～を理解するために決定的に重要である	119
❷	be critical to ～	…は，～に決定的に重要である	1,951
	・be critical to understanding ～	…は，～を理解するのに決定的に重要である	119
❸	be critical in ～	…は，～の際に決定的に重要である	1,021

❶ Organization of the genome is critical for maintaining cell-specific gene expression, ensuring proper cell function. (J Biol Chem. 2012 287:22080)
訳 そのゲノムの構成は，細胞特異的遺伝子発現を維持するために決定的に重要である

❷ We conclude that the regulation of serum IL-18 expression is critical to the outcome of innate immune responses to LPS. (J Immunol. 2002 169:2536)
訳 血清IL-18発現の調節は，～の結果に決定的に重要である

❸ This localization of grk RNA restricts the distribution of Gurken protein and is critical in defining both the anterior-posterior and dorsal-ventral axes of the egg. (Dev Biol. 2000 221:435)
訳 ～は，前後および背腹軸の両方を規定する際に決定的に重要である

■ critical + [名]

❶	critical role	決定的に重要な役割	7,053
❷	critical care	救急救命	908

❸ critical component	決定的に重要な構成成分	648
❹ critical step	決定的に重要な段階	620
❺ critical regulator	決定的に重要な制御因子	541
❻ critical period	決定的に重要な期間	480
❼ critical determinant	決定的に重要な決定要因	413
❽ critical region	決定的に重要な領域	392

❶ Here we show that **IKK α plays a critical role in regulating cyclin D1** during the cell cycle.（J Biol Chem. 2005 280:33945）

訳 IKK αは，サイクリンD1を調節する際に決定的に重要な役割を果たす

critically 副 決定的に／危篤状態に 用例数 3,790

◆類義語：crucially, conclusively, definitively

■ critically ＋［形］

❶ critically ill ～	危篤状態の～	1,035
・critically ill patients	危篤状態の患者	586

❶ **Similar loss of lymphocytes may be occurring in critically ill patients with other disorders.**（J Immunol. 2001 166:6952）

訳 リンパ球の同じような喪失が，危篤状態の他の疾患の患者において起こっているかもしれない

■ critically ＋［過分／形］

❶ critically involved in ～	～に決定的に関与している	361
❷ critically important	決定的に重要な	358
❸ critically dependent on ～	～に決定的に依存している	337
・be critically dependent on ～	…は，～に決定的に依存している	307

❶ This study also suggests that **oxidative stress is critically involved in** the DOX-induced chronic cardiotoxicity.（Cancer Res. 2001 61:3382）

訳 酸化ストレスは，～に決定的に関与している

cross-talk 名 クロストーク 用例数 1,541

cross-talk	クロストーク	1,519
cross-talks	クロストーク	22

cross-talk betweenの用例が非常に多い．複数形のcross-talksの割合は1％しかなく，不可算名詞として使われることが多い．

■ cross-talk ＋［前］

	the％	a/an％	ø％			
❶	18	14	65	cross-talk between ～	～の間のクロストーク	814

❶ Our study reveals novel cross-talk between HSCs and erythroblasts, and sheds a new light on the regulatory mechanisms regulating the balance between HSC self-renewal and differentiation. (23134786 J Biol Chem. 2005 280:38898)

訳 われわれの研究は，造血幹細胞と赤芽球の間の新規のクロストークを明らかにする

crucial 形 決定的に重要な 用例数 9,770

crucial forの用例が非常に多い.

◆類義語：critical, definitive, conclusive, important

■ be crucial ＋［前］

★ be crucial	…は，決定的に重要である	4,240
❶ be crucial for ～	…は，～のために決定的に重要である	2,716
・be crucial for understanding ～	…は，～を理解するために決定的に重要である	76
❷ be crucial to ～	…は，～に決定的に重要である	772
・be crucial to understanding ～	…は，～を理解するのに決定的に重要である	55
❸ be crucial in ～	…は，～において決定的に重要である	373

❶ Knowledge of the subcellular location of a protein is crucial for understanding its functions. (Bioinformatics. 2012 28:i32)

訳 タンパク質の細胞内局在に関する知識は，その機能を理解するために決定的に重要である

❷ Such skills are crucial to those surgeons intent on the future management of carotid occlusive disease. (Ann Surg. 2005 242:431)

訳 そのような技能は，～に懸命になっているそれらの外科医に決定的に重要である

■ crucial ＋［名］

❶ crucial role	決定的に重要な役割	1,971
❷ crucial step	決定的に重要なステップ	193
❸ crucial component	決定的に重要な構成成分	108
❹ crucial regulator	決定的に重要な制御因子	70

❶ Nitric oxide (NO) is a ubiquitous signaling molecule that plays a crucial role in oocyte maturation and embryo development. (Biochemistry. 2005 44:11361)

訳 一酸化窒素（NO）は，卵母細胞の成熟において決定的に重要な役割を果たす遍在性のシグナル伝達分子である

★★ culture 名 培養／培養物，動 培養する 用例数 47,411

culture	培養／培養物／～を培養する	19,247
cultured	培養される／～を培養した	14,906
cultures	培養／培養物	12,785
culturing	～を培養する	473

名詞および他動詞として用いられる．複数形のculturesの割合は約40%とかなり高い．

◆類義語：cultivation, incubation, cultivate, incubate, grow

■ 名 in culture

❶	in culture	培養した…／培養において	3,787

❶ In addition, **in neural cells in culture**, this lipid messenger also inhibited both interleukin 1-β -induced NF κ B activation and cyclooxygenase-2 expression. (J Biol Chem. 2003 278:43807)
訳 培養した神経細胞において

■ 名 cultures ＋［前］

	the%	a/an%	ø%			
❶	1	-	98	cultures of ～	～の培養	2,196

❶ In this study, **we use primary cultures of human keratinocytes** to demonstrate Ag-specific binding and internalization of BP IgE. (J Immunol. 2011 187:553)
訳 われわれは，ヒト角化細胞の初代培養を使用する

■ 名［名／形］＋ culture(s)

❶	cell culture	細胞培養	3,102
❷	tissue culture	組織培養	1,353
❸	primary cultures	初代培養	889
❹	organ culture	器官培養	426

■ 名 culture ＋［名］

❶	culture medium	培養液	964
❷	culture system	培養系	692
❸	culture conditions	培養条件	477
❹	culture supernatants	培養上清	476

■ 動 cultured ＋［前］

❶	cultured in ～	～において培養される	1,345
	・cells cultured in ～	～において培養された細胞	270
❷	cultured with ～	～とともに培養される／～を使って培養される	653
❸	cultured from ～	～から培養される	415

❸ **Primary human MSC were cultured from bone marrow aspirates** and then passaged at least three times before use in assays. (J Immunol. 2003 171:3426)
訳 初代ヒトMSCが，骨髄吸引液から培養された

■ 動［名］＋ be cultured

★	be cultured	～が，培養される	4,240
❶	cells were cultured	細胞が，培養された	325

❶ **A6 cells were cultured in media supplemented with** 1.5 microm aldosterone. (J Biol Chem. 2002 277:11965)

訳 A6細胞が，～を添加された培地で培養された

■ 動 cultured ＋［名］

❶ cultured cells	培養された細胞	2,062
・in cultured cells	培養された細胞において	1,087

★★ current 形 現在の，名 電流 用例数 41,835

current	現在の／電流	32,959
currents	電流	8,876

形容詞および名詞として用いられる．名詞としては，複数形のcurrentsの割合が約60%と非常に高い．

◈類義語：present, modern

■ 形 current ＋［名］

❶ the current study	現在の研究	2,719
・In the current study	現在の研究において	1,337
❷ current understanding	現在の理解	737
❸ current knowledge	現在の知識	514
❹ current models	現在のモデル	513

❶ In the current study, we investigated the involvement of TNF-R1 in H_2O_2-induced JNK activation.（J Biol Chem. 2003 278:44091）

訳 現在の研究において，

■ 名［形］＋ current(s)

❶ inward current	内向き電流	461
❷ postsynaptic currents	シナプス後電流	441
❸ synaptic currents	シナプス電流	337
❹ outward current	外向き電流	336
❺ sodium current	ナトリウム電流	280
❻ potassium current	カリウム電流	273

■ 名 currents ＋［前］

	the%	a/an%	ø%			
❶	2	-	74	currents in ～	～における電流	2,196

★ currently ㊫ 現在 用例数 8,730

◆類義語：presently, at present, now, today

■ currently ＋［形／過分］

❶ currently available	現在利用できる	1,080
❷ currently used	現在使われている	488
❸ currently unknown	現在知られていない	400
・be currently unknown	…は，現在知られていない	367
❹ currently known	現在知られている	245

❶ However, **only a limited number of tools are currently available for the analysis of** multiple genomic sequences. (Genome Res. 2004 14:313)
訳 限られた数の道具だけが，～の解析に現在利用できる

❷ **The currently used smallpox vaccine is associated with a high incidence of adverse events,** and there is a serious need for a safe and effective alternative vaccine. (J Virol. 2004 78:3811)
訳 現在使われている天然痘ワクチンは，有害反応の高頻度と関連する

❸ However, **the function of this domain is currently unknown.** (J Bacteriol. 2004 186:6595)
訳 このドメインの機能は，現在知られていない

D

★ daily 副 1日に，形 1日の／日々の　　用例数 7,906

副詞および形容詞として使われる．

◆類義語：everyday

■ 副 ［副／名］＋ daily

❶ twice daily	1日に2回	1,090
・～ mg twice daily	1日に2回～mg	346
❷ once daily	1日に1回	634
❸ ～ mg daily	1日に～mg	385
❹ ～ times daily	1日に～回	174

❶ The first 38 patients received MMF 15 mg/kg twice daily; the next 47 patients received MMF 3 times daily.（Blood. 2005 106:4381）

訳 最初の38名の患者は，MMFを１日に２回15 mg/kg受けた

❸ Female pregnant Sprague-Dawley rats were subjected to stress three times daily from day 15 to day 21 of gestation.（J Physiol. 2004 557:273）

訳 メスの妊娠SDラットが，１日に３回ストレスにさらされた

■ 副 daily for

★ daily for ～	～の間1日に…	812
❶ twice daily for ～	～の間1日に2回	176

❶ HSV-1 DNA copy number and frequency of shedding were determined by real-time polymerase chain reaction (PCR) analysis of tear and saliva samples collected twice daily for 30 consecutive days.（Invest Ophthalmol Vis Sci. 2005 46:241）

訳 ～は，連続30日間１日に２回集められた涙と唾液のサンプルのリアルタイムポリメラーゼ連鎖反応法（PCR）分析によって決定された

■ 形 daily ＋［名］

❶ activities of daily living	日常生活動作	254
❷ daily injections	毎日の注射	134
❸ daily doses	1日量	129

❸ We conducted a randomized clinical trial among 128 healthy subjects allocated to placebo or to 81-, 325-, or 650-mg daily doses of aspirin for 8 weeks.（Proc Natl Acad Sci USA. 2004 101:15178）

訳 プラセボあるいは，81 mg，325 mg，650 mgの１日量のアスピリンに割り当てられた被検者

★ damage 名 損傷／傷害，動 損傷する　　用例数 28,676

damage	損傷／傷害／～を損傷する	24,005
damaged	～を損傷した／損傷された	2,889

damaging	～を損傷する	1,511
damages	損傷／傷害／～を損傷する	271

動詞としても使われるが，名詞の用例が圧倒的に多い．複数形のdamagesの割合は0.5%しかなく，原則，不可算名詞として使われる．DNA damageの用例が多い．

◆類義語：injury, impairment, dysfunction, disturbance, injure, impair, compromise, lesion

■ 名 damage ＋［前］

	the%	a/an%	ø%			
❶	1	0	99	damage to ～	～への損傷	1,711
❷	2	0	96	damage in ～	～における損傷	1,594
❸	2	0	92	damage by ～	～による損傷	418

❶ At critical junctures, **oxidative damage to DNA requires** the base excision repair (BER) pathway. (Nucleic Acids Res. 2011 39:3156)

訳 DNAへの酸化的損傷は，～を必要とする

❷ However, a significant difference is that **$P53^{7KR}$ is activated more easily by DNA damage in thymus than WT P53.** (Proc Natl Acad Sci USA. 2005 102:10188)

訳 $P53^{7KR}$は，胸腺においてDNA損傷によって野生型P53より容易に活性化される

■ 名 damage ＋［過分／形］

❶	damage-induced ～	損傷に誘導される～	639
❷	damage caused by ～	～によって引き起こされる損傷	221
❸	damage induced by ～	～によって誘導される損傷	219
❹	damage-inducible ～	損傷に誘導される～	145

❸ **Mus81 plays a relatively minor role in the Rhp51-dependent repair of DNA damage induced by ultraviolet light.** (Nucleic Acids Res. 2004 32:5570)

訳 Mus81は，紫外線によって誘導されるDNA損傷のRhp51依存性の修復において比較的小さな役割を果たす

■ 名［名／形］＋ damage

❶	DNA damage	DNA損傷	9,612
❷	oxidative damage	酸化的損傷	1,198
❸	tissue damage	組織損傷	979
❹	brain damage	脳の損傷	331
❺	neuronal damage	ニューロンの損傷	299

■ 名［前］＋ DNA damage

❶	in response to DNA damage	DNA損傷に応答して	599
❷	after DNA damage	DNA損傷のあと	398

❶ The results demonstrate that **the Tip60 HAT plays a key role in the activation of ATM's**

kinase activity in response to DNA damage. (Proc Natl Acad Sci USA. 2005 102:13182)
訳 Tip60 HATは，DNA損傷に応答するATMのキナーゼ活性の活性化において重要な役割を果たす

❷ While TSA alone reduced cell survival after DNA damage, the combination of EX-527 and TSA had no further effect on cell viability and growth. (Mol Cell Biol. 2006 26:28)
訳 TSAは，単独でDNA損傷のあと細胞生存を低下させた

■ 名 [過分／形] ＋ DNA damage

❶ ～-induced DNA damage	～に誘導されたDNA損傷	435
❷ oxidative DNA damage	酸化的DNA損傷	303

❶ Furthermore, WRN RNAi-induced DNA damage was suppressed by overexpression of the telomere-binding protein TRF2. (Mol Cell Biol. 2005 25:10492)
訳 WRN RNAiに誘導されるDNA損傷が，テロメア結合タンパク質TRF2の過剰発現によって抑制された

■ 動 damaged ＋ [名]

❶ damaged DNA	損傷したDNA	672
❷ damaged cells	損傷した細胞	145

D

★★ data 名 データ 用例数 128,859

data	データ	128,853
datum	データ	6

dataは名詞の複数形．単数形はdatumだが，使われることはほとんどない．
◆類義語：result

■ data ＋ [動]

❶ data suggest ～	データは，～を示唆する	18,799
❷ data indicate ～	データは，～を示す	9,658
❸ data demonstrate ～	データは，～を実証する	6,034
❹ data show ～	データは，～を示す	4,918
❺ data support ～	データは，～を支持する	3,682
・data support the hypothesis that ～	データは，～という仮説を支持する	604

❶ These data suggest that the addition of potent FLT3 inhibitors such as SU11248 to AML chemotherapy regimens could result in improved treatment results. (Blood. 2004 104:4202)
訳 これらのデータは，～ということを示唆する

❷ These data indicate that rFIX is a safe and effective treatment for PUPs with hemophilia B. (Blood. 2005 105:518)
訳 これらのデータは，～ということを示す

❺ These data support the hypothesis that Plk1 activates the APC by directing the SCF-dependent destruction of Emi1 in prophase. (Mol Biol Cell. 2004 15:5623)
訳 これらのデータは，～という仮説を支持する

■ data ＋［過分］

❶	data obtained	得られたデータ	960
❷	data presented	提示されたデータ	811
❸	data collected	集められたデータ	500

❶ Data obtained from multicolor ratiometric microarrays correlate well with data obtained by using traditional approaches, but the arrays are faster and simpler to use.（Proc Natl Acad Sci USA. 2003 100:9330）

訳 多色レシオメトリックマイクロアレイから得られたデータは，～を使うことによって得られたデータとよく相関する

■ data ＋［前］

	the%	a/an%	ø%			
❶	2	-	97	data from ～	～からのデータ	7,716
				・using data from ～	～からのデータを使って	597
❷	3	-	91	data on ～	～に関するデータ	3,927
❸	7	-	91	data for ～	～に関するデータ	2,804

❶ Using data from our epidemiological study in Western Australia, we investigated the possibility of an increased rate of autism in twins.（Am J Hum Genet. 2002 71:941）

訳 ～におけるわれわれの疫学的な研究からのデータを使って

❸ The data for each tooth of each patient, including occlusal status and gingival width, were placed in a database and analyzed for a relationship between occlusal discrepancies and changes in gingival width.（J Periodontol. 2004 75:98）

訳 おのおのの患者のおのおのの歯に関するデータ

■［形／名］＋ data

❶	experimental data	実験データ	2,057
❷	expression data	発現データ	1,241
❸	sequence data	配列データ	1,174
❹	recent data	最近のデータ	1,137
❺	microarray data	マイクロアレイデータ	925

date ㊔ 日付，㊫ 日付をつける　用例数 6,719

date	日付／日付をつける	5,934
dates	日付／日付をつける	316
dating	日付をつける	294
dated	日付をつけた／日付をつけられる	175

動詞としても使われるが，名詞の用例，特に，to dateの用例が圧倒的に多い．

■ to date

★ to date	今日まで	5,175
❶ To date,	今日まで,	1,503

❶ To date, little is known about functional differences between the isoforms due to their indistinguishable activities in most in vitro assays. (FASEB J. 2012 26:1892)
訳 今日まで，～の間の機能的違いについてほとんど知られていない

★★ death 名 死 用例数 45,089

death	死	40,194
deaths	死	4,895

複数形のdeathsの割合は約10%あるが，単数形は無冠詞の用例が圧倒的に多い.

■ death ＋［前］

	the %	a/an %	ø %			
❶	2	0	98	death in ～	～における死	3,774
❷	32	1	66	death of ～	～の死	1,727
❸	0	0	95	death from ～	～による死	1,034

❶ Neuronal cell death in a number of neurological disorders is associated with aberrant mitochondrial dynamics and mitochondrial degeneration. (Mol Biol Cell. 2011 22:256)
訳 いくつかの神経疾患における神経細胞死は，～と関連している

❸ We investigated whether use of ACs was associated with the risk of death from prostate cancer. (22927523Mol Biol Cell. 2011 22:256)
訳 ～は，前立腺癌による死のリスクと関連していた

■［名／形］＋ death

❶ cell death	細胞死	15,753
❷ neuronal death	神経細胞死	931
❸ cardiac death	心臓死	910
❹ sudden death	突然死	773

decay 名 崩壊／減衰，動 崩壊する 用例数 5,724

decay	崩壊／減衰／崩壊する	4,842
decays	崩壊／減衰／崩壊する	411
decayed	崩壊した	319
decaying	崩壊する	152

動詞としても用いられるが，名詞の用例が多い．複数形のdecaysの割合は約5%あるが，単数形は無冠詞で使われることが多い.

◆類義語：breakdown, collapse, disruption

D

■ decay ＋［前］

	the％	a/an％	ø％			
❶	45	1	51	decay of ～	～の崩壊	997
❷	18	20	50	decay in ～	～における崩壊	225

❶ The decay of the peroxide intermediate is markedly affected by the D94E mutation, confirming the involvement of D94 in this reaction.（Biochemistry. 2005 44:6081）

訳 過酸化物中間体の崩壊は，～によって顕著に影響を受ける

❷ The mechanisms responsible for mRNA decay in mammalian cells, and how specific sequence elements accelerate decay, are unknown.（Curr Biol. 2002 12:R285）

訳 哺乳類細胞におけるmRNA崩壊に責任のある機構

decision 名 決定／決断 用例数 6,962

decision	決定／決断	4,165
decisions	決定／決断	2,797

複数形のdecisionsの割合は約40%とかなり高い．

◆類義語：determination

■ decision(s) ＋［前］

decision toの前にtheが付く割合は非常に高い．

	the％	a/an％	ø％			
❶	79	10	5	decision to *do*	～する決定	325
❷	0	-	98	decisions in ～	～における決定	241
❸	0	-	96	decisions about ～	～に関する決定	232

❶ The levels of IE and E genes are, in turn, thought to regulate the decision to enter the lytic cycle or latency.（J Virol. 2006 80:38）

訳 溶菌サイクルか潜伏期に入る決定を調節する

❷ Here, we address whether Hedgehog（Hh）signals independently regulate progenitor proliferation and neuronal fate decisions in the embryonic mouse retina.（J Neurosci. 2009 29:6932）

訳 ～は，胚マウス網膜における前駆細胞の増殖とニューロンの運命決定を制御する

❸ Results from this approach can assist in making decisions about drug dosing and frequency in the design of larger and longer clinical trials for diseases such as Alzheimer's disease, and may accelerate effective drug validation.（Ann Neurol. 2009 66:48）

訳 このアプローチの結果は，薬物の投与量と頻度に関する決定を行う際に役立ちうる

■ decision ＋［名］

❶	decision making	意思決定	1,774
❷	decision makers	意志決定者	115

❶ Assessments of the impact on clinical decision making use methods developed for evaluating adherence to practice guidelines.（Am J Psychiatry. 2004 161:946）

訳 臨床的意思決定に対する影響の評価は，～を評価するために開発された方法を使用する

■ [名] + decisions

❶	fate decisions	運命決定	428
❷	treatment decisions	治療の決定	246

★ decline 名 低下，動 低下する　用例数 10,982

decline	低下／低下する	6,177
declined	低下した	2,709
declines	低下／低下する	1,461
declining	低下する	635

名詞および自動詞として使われる．decline inの用例が多い．複数形のdeclines用例が，約20%ある可算名詞．

◆類義語：decrease, reduction, fall, reduce, lower

■ 名 decline + [前]

「～の低下」の意味では，decline ofよりもdecline inの用例が断然多い．

	the %	a/an %	ø %			
❶	24	43	29	decline in ～	～の低下	3,145
❷	46	29	25	decline of ～	～の低下	452

❶ Women may experience a decline in physical function during menopause.（Am J Epidemiol. 2004 160:484）

訳 女性は，～の間に身体的な機能の低下を経験するかもしれない

❷ The decline of this capacity is believed to be a major factor in the impairment of vision in age-related macular degeneration.（Proc Natl Acad Sci USA. 2004 101:10446）

訳 この能力の低下は，視覚障害の主要な因子であると信じられている

■ 名 [形] + decline

❶	cognitive decline	認識能の低下	575
❷	significant decline	有意な低下	169

■ 動 declined + [前]

❶	declined to ～	～に低下した	244
❷	declined in ～	～において低下した	218
❸	declined by ～	～だけ低下した	207
❹	declined with ～	～と共に低下した	200
❺	declined from ～ to …	～から…に低下した	97

❶ Interocular differences in contour-detection thresholds declined to normal levels in most of the patients within 8 weeks of the initiation of treatment.（Invest Ophthalmol Vis Sci. 2004 45:4016）

D

訳 ～は，8週間以内にほとんどの患者において正常なレベルに低下した

❷ Last, lamin B1 protein and mRNA declined in mouse tissue after senescence was induced by irradiation.（Mol Biol Cell. 2012 23:2066）

訳 ラミンB1タンパク質およびmRNAは，マウス組織において低下した

❸ Number and volume of Gd-enhancing lesions declined by 44%,（p<0.0001）and 41%（p=0.0018）, respectively.（Lancet. 2004 363:1607）

訳 ～は，44%低下した

❹ Interestingly, cell cycle control of IPCs by Sox2-mediated expression of p27（Kip1）gradually declined with age.（Neurosci. 2012 32:10530）

訳 ～は，年齢と共に徐々に低下した

★★ decrease　動 低下させる／減少させる／低下する，名 低下／減少

用例数 86,075

decreased	～を減少させた／低下した	46,145
decrease	～を減少させる／減少する／減少	25,605
decreases	～を減少させる／減少する／減少	10,003
decreasing	～を減少させる／減少する	4,322

動詞および名詞として用いられる．動詞としては，他動詞と自動詞の両方の用例がある．また，他動詞受動態が自動詞とほぼ同じ意味になることもある．名詞としては，複数形のdecreasesの割合が，約30%ある可算名詞．

◆類義語：reduce, diminish, lower, down-regulate, decline, fall, drop, reduction, attenuation, down-regulation

■ 動 decrease ＋［前］

❶ decrease with ～	…は，～と共に低下する	1,457
・decrease with increasing ～	…は，増大する～と共に低下する	456
・decrease with age	…は，年齢と共に低下する	141
❷ decrease from ～	…は，～から低下する	1,180
・decrease from ～ to …	‥は，～から…へ低下する	487
❸ decreased to ～	…は，～へ低下する	720

❶ However, the luminescence-decay rate decreases with increasing pH over a biologically relevant range of 4-8.（J Am Chem Soc. 2012 134:17372）

訳 発光の崩壊速度は，pHの上昇とともに低下する

❷ The false-positive blood culture rate decreased from 9.1% to 2.8%（P<.001）.（JAMA. 2003 289:726）

訳 ～が，9.1%から2.8%へ低下した

■ 動［名］＋ decrease

❶ levels decreased	レベルが，低下した	300
❷ expression decreased	発現が，低下した	189
❸ activity decreased	活性が，低下した	156

❹ rate decreased	速度が，低下した	115

■ 動 be decreased ＋［前］

★ be decreased	…は，低下する	4,370
❶ be decreased in ～	…は，～において低下する	1,312
❷ be decreased by ～	…は，～によって低下する／～だけ低下する	862
❸ be decreased to ～	…は，～へ低下する	124

❶ Expression of gp91phox and p67phox **was decreased in** HC pigs after antioxidant intervention, and SOD improved. (J Am Soc Nephrol. 2004 15:1816)

訳 ～は，抗酸化剤処置のあとHCブタにおいて低下した

❷ The expression of protein kinase B **was decreased by** approximately 60% in calpastatin transgenic muscle. (J Biol Chem. 2004 279:20915)

訳 プロテインキナーゼBの発現は，～においておよそ60%低下した

D

■ 動［名］＋ be decreased

❶ levels were decreased	レベルは，低下した	152
❷ expression was decreased	発現は，低下した	106

■ 動［副］＋ decreased

❶ significantly decreased	有意に低下した	2,923
❷ markedly decreased	顕著に低下した	628
❸ dramatically decreased	劇的に低下した	289

❶ However, the ratio of SF to LF was **significantly decreased** in 76% of the breast tumors and distributed evenly among the groups. (Cancer Res. 2004 64:5677)

訳 SF対LFの比は，乳癌の76%において有意に低下した

❷ Conversely, expression was **markedly decreased** or lost in invasive SCC and spindle cell carcinoma cell lines. (Cancer Res. 2003 63:4997)

訳 発現は，～において顕著に低下したかあるいは失われた

■ 動 decreased ＋［名］

❶ decreased expression	低下した発現	1,203
❷ decreased levels	低下したレベル	788
❸ decreased risk	低下したリスク	434
❹ decreased activity	低下した活性	262
❺ decreased binding	低下した結合	249

❶ Natural aging of wild-type mice is marked by **decreased expression of** BubR1 in multiple tissues, including testis and ovary. (Nat Genet. 2004 36:744)

訳 野生型マウスの自然な加齢は，BubR1の低下した発現によって特徴づけられる

❷ FFM accounted for 19% of body weight in those who gained weight, even in the presence of **decreased levels of** SRA. (Am J Clin Nutr. 2002 76:473)

訳 低下したレベルのSRAの存在下でさえ

■ 動 decrease ＋［名句］

❶	decrease the number of ～	…は，～の数を低下させる	308
❷	decrease the rate of ～	…は，～の速度を低下させる	303
❸	decrease the expression of ～	…は，～の発現を低下させる	227
❹	decrease the risk of ～	…は，～のリスクを低下させる	185
❺	decrease the incidence of ～	…は，～の頻度を低下させる	160

❶ Loss of FBLP-1 significantly impaired the growth and survival of BMSCs in vitro and **decreased the number of osteoblast (OB) progenitors** in bone marrow and OB differentiation in vivo. (J Biol Chem. 2012 287:21450)

訳 ～は，骨芽細胞(OB)前駆体の数を低下させた

■ 名 decrease ＋［前］

「～の低下」の意味では，decrease ofよりもdecrease inの用例が断然多い．

	the %	a/an %	ø %			
❶	9	77	9	decrease in ～	～の低下	17,081
				・result in a decrease in ～	…は，～の低下という結果になる	505
				・associated with a decrease in ～	～の低下と関連する	365
				・decrease in the number of ～	～の数の低下	342
				・lead to a decrease in ～	…は，～の低下につながる	324
				・accompanied by a decrease in ～	～の低下を伴った	151
❷	23	62	12	decrease of ～	～の低下	3,145

❶ **Despite the decrease in the number of new HCV infections**, the prevalence of advanced HCV-related liver disease is steadily increasing. (Transplantation. 2004 78:955)

訳 新たなHCV感染の数の低下にもかかわらず

❶ **This LTP was NMDA-dependent and associated with a decrease in** paired-pulse facilitation in both pathways. (J Neurosci. 2004 24:9507)

訳 このLTPはNMDA依存的で，～の低下と関連した

■ 名［形／過分］＋ decrease

❶	significant decrease	有意な低下	1,776
❷	～-fold decrease	～倍の低下	851
❸	marked decrease	顕著な低下	465
❹	～-dependent decrease	～に依存した低下	405
❺	dramatic decrease	劇的な低下	303

❶ Complementary experiments using CXCR3$^{-/-}$ mice as SCT donors **also resulted in a significant decrease in IPS**. (J Immunol. 2004 173:2050)

訳 ～は，また，IPSの顕著な低下という結果になった

❷ A C248S mutation caused a 5-fold decrease in K_d (NADPH), shifted the pK_a of K_d (NADPH) from 6.9 to 7.4, and decreased the ionic strength dependence of NADPH binding. (J Biol Chem. 2001 276:11812)

訳 C248S変異は，～の５倍の低下を引き起こした

★★ defect 名 欠陥／欠損 用例数 30,904

defects	欠陥／欠損	19,851
defect	欠陥／欠損	11,053

複数形のdefectsの割合は，約65%と圧倒的に高い．

◆類義語：deficit, deficiency, deletion, loss

D

■ defects ＋［前］

	the%	a/an%	⌀%			
❶	0	-	100	defects in ～	～の欠陥	8,147
				・have defects in ～	…は，～の欠陥を持つ	263
				・exhibit defects in ～	…は，～の欠陥を示す	229
				・show defects in ～	…は，～の欠陥を示す	188
				・lead to defects in ～	…は，～の欠陥につながる	144
				・due to defects in ～	～の欠陥のせいである	110
				・mutants with defects in ～	～の欠陥を持つ変異体	110
❷	46	-	51	defects of ～	～の欠陥	835

❶ Most renal cancers have defects in the von Hippel-Lindau tumor suppressor pVHL. (Proc Natl Acad Sci USA. 2005 102:11035)

訳 ほとんどの腎癌は，～の欠陥を持つ

❶ In mutants with defects in the chemotaxis signaling pathway, the agar plates remain dry and the cells' flagella are short. (Curr Biol. 2005 15:R599)

訳 走化性シグナル伝達経路の欠陥を持つ変異体において

■ defects ＋［現分／過分］

❶	defects including ～	～を含む欠陥	283
❷	defects associated with ～	～と関連する欠陥	190
❸	defects observed in ～	～において観察される欠陥	184
❹	defects caused by ～	～によって引き起こされる欠陥	148

❶ Of the embryos ablated at HH14, 76% demonstrated cardiac defects including overriding aorta and pulmonary atresia, while none of the sham-operated controls were affected. (Dev Biol. 2005 284:72)

訳 76%は，大動脈騎乗と肺動脈閉鎖症を含む心臓の欠陥をはっきり示した

■［形／名］＋ defects

❶	developmental defects	発達欠陥	447
❷	growth defect	発育欠陥	377

❸ birth defects	先天性欠損	322
❹ severe defects	～の重篤な欠陥	324
❺ genetic defects	遺伝的欠陥	261
❻ heart defects	心臓欠陥	257

❶ In all cases, **these developmental defects were associated with dysregulation of apoptosis and cellular proliferation.** (Mol Cell Biol. 2005 25:10329)

訳 これらの発育欠陥は，アポトーシスと細胞増殖の調節不全と関連した

❹ Deletion of the gene for protein L27 from the E. coli chromosome **results in severe defects in cell growth.** (Mol Cell. 2005 20:427)

訳 ～は，細胞増殖における重篤な欠陥という結果になる

★ defective 形 欠陥のある 用例数 11,977

◆類義語：deficient, imperfect

■ defective ＋［前］

❶ defective in ～	～に欠陥がある	3,885
・be defective in ～	…は，～に欠陥がある	2,028
・mutants defective in ～	～に欠陥のある変異体	442
・defective in ～ing	～するのに欠陥がある	402
❷ defective for ～	～に対して欠陥がある	757

❶ In contrast, **tumor cells expressing the EphA2 mutants are defective in RhoA GTPase activation** and cell migration. (Oncogene. 2005 24:7859)

訳 EphA2変異体を発現している腫瘍細胞は，RhoA GTP加水分解酵素の活性化に欠陥がある

❶ **Mutants defective in the cAMP/PKA pathway also exhibited resistance to** avicin G. (Proc Natl Acad Sci USA. 2005 102:12771)

訳 cAMP/PKA経路に欠陥のある変異体は，また，～に対する抵抗性を示した

❷ Furthermore, an S. flexneri ydeP mutant **was defective for both glutamate-independent and glutamate-dependent acid resistance.** (Mol Microbiol. 2005 58:1354)

訳 ～は，グルタミン酸非依存的およびグルタミン酸依存的な酸抵抗性の両方に対する欠陥があった

■ ［名］-defective ＋［名］

❶ replication-defective ～	複製欠陥のある	422
・replication-defective adenovirus	複製欠陥のあるアデノウイルス	90
❷ ～-defective mutant	～欠陥のある変異体	249

❶ **We used a replication-defective adenovirus vector** to deliver the ferritin transgenes. (Nat Med. 2005 11:450)

訳 われわれは，複製欠陥のあるアデノウイルスベクターを使用した

defense 名 防御

用例数 6,742

defense	防御	5,636
defenses	防御	1,106

複数形のdefensesの割合は約15%あるが，単数形は無冠詞の用例が非常に多い．

■ defense ＋［前］

	the%	a/an%	ø%			
❶	15	5	80	defense against ～	～に対する防御	1,298

❶ $CD8^+$ T cells are crucial for host **defense against** invading pathogens and malignancies. (J Immunol. 2005 175:5126)

訳 $CD8^+$ T細胞は，侵入する病原体に対する宿主防御にとって決定的に重要である

■［名］＋ defense

❶ host defense	宿主防御	1,930
・host defense against ～	～に対する宿主防御	532
❷ plant defense	植物防御	282

❶ To further characterize the role of MCH in **host defense** mechanisms against intestinal pathogens, Salmonella enterocolitis (using Salmonella enterica serovar Typhimurium) was induced in MCH-deficient mice and their wild-type littermates. (Infect Immun. 2013 81:166)

訳 腸内病原菌に対する宿主防御機構におけるMCHの役割をさらに特徴づけるために

■ defense ＋［名］

❶ defense responses	防御反応	331
❷ defense mechanisms	防御機構	330

★ deficiency 名 欠損／欠乏

用例数 13,653

deficiency	欠損／欠乏	12,189
deficiencies	欠損／欠乏	1,464

複数形のdeficienciesの割合が，約10%ある可算名詞．

◆類義語：deficit, defect, depletion, deprivation, lack, paucity, deletion, loss

■ deficiency ＋［前］

	the%	a/an%	ø%			
❶	7	18	40	deficiency in ～	～の欠損	1,773
❷	16	21	60	deficiency of ～	～の欠損	1,444

❶ This imbalance could result from a **deficiency in** suppression by regulatory T cells or strong activation signals could overcome such regulation. (Lancet. 2004 363:608)

訳 この不均衡は，抑制の欠損に起因しうる

❷ Pompe's disease **is caused by a deficiency of the lysosomal enzyme** acid α-glucosidase (GAA). (J Biol Chem. 2005 280:6780)

訳 〜は，リソソーム酵素の欠損によって引き起こされる

■ [名] + deficiency

❶ iron deficiency	鉄欠乏	481
❷ immune deficiency	免疫不全	264

❶ We used in vitro iron chelation to isolate **the effects of iron deficiency on Fgf23 expression.** (Proc Natl Acad Sci USA. 2011 108:E1146)

訳 Fgf23発現に対する鉄欠乏の影響

★★ deficient ㊋ 欠損した／欠乏した　用例数 31,139

◆類義語：defective, absent, imperfect, free

■ deficient + [前]

❶ deficient in 〜	〜を欠損した	5,026
・mice deficient in 〜	〜を欠損したマウス	1,571
・be deficient in 〜	…は，〜を欠損している	1,118
❷ deficient for 〜	〜を欠損した	728

❶ Finally, **neonatal AOD was ameliorated in mice deficient in Fcγ RIII** and was enhanced in Fcγ RIIB-deficient mice. (J Immunol. 2004 173:1051)

訳 新生児のAODは，FcγRⅢを欠損したマウスにおいて緩解した

■ deficient + [名]

❶ 〜-deficient mice	〜欠損マウス	7,901
❷ 〜-deficient cells	〜欠損細胞	1,856
❸ 〜-deficient mutant	〜欠損変異体	567
❹ 〜-deficient animals	〜欠損動物	385

❶ **IL-13-deficient mice failed to develop** an increased response to EFS or goblet cell hyperplasia after the 10-day OVA challenge. (J Immunol. 2004 172:6398)

訳 IL-13欠損マウスは，〜を発達させることができなかった

deficit ㊔ 欠陥／欠損　用例数 7,862

deficits	欠陥／欠損	5,282
deficit	欠陥／欠損	2,580

複数形のdeficitsの割合は，約65%と圧倒的に高い．

◆類義語：deficiency, defect, lack, deletion, loss

■ deficits ＋［前］

	the %	a/an %	ø %			
❶	16	-	81	deficits in ～	～における欠陥	1,939
❷	38	-	60	deficits of ～	～の欠陥	109

❶ The latter are the best predictor of functional outcome, though largely untreated by current pharmacotherapy; thus a better understanding of **the mechanisms underlying cognitive deficits in schizophrenia** is crucial. (J Physiol. 2012 590:715)

訳 統合失調症における認知障害の根底にある機構

■［名／形］＋ deficit(s)

❶	cognitive deficits	認知障害	514
❷	attention deficit	注意欠陥	390
❸	memory deficits	記憶欠陥	254

D

★★ define 動 明らかにする／定義する／規定する　用例数 39,586

defined	明らかにされる／定義される／～を明らかにした	23,422
define	～を明らかにする／～を定義する	10,946
defining	～を明らかにする／～を定義する	3,339
defines	～を明らかにする／～を定義する	1,879

他動詞.

◆類義語：reveal, disclose, elucidate, clarify, uncover, manifest, confine, determine

■ defined ＋［前］

❶	defined as ～	～として定義される	3,832
	・be defined as ～	…は，～として定義される	2,160
	・defined as ～ing	～すると定義される	176
❷	defined by ～	～によって明らかにされる／～によって定義される	3,374
	・as defined by ～	～によって明らかにされたように	551
❸	defined in ～	～において明らかにされる	579

❶ **The regulator phenotype was defined as the ability to suppress a DTH response to** a recall antigen in the presence of donor antigen. (Transplantation. 2001 72:571)

訳 調節性の表現型は，～へのDTH応答を抑制する能力として定義された

❷ CDK4 and CDK6 form a subfamily among the CDKs in mammalian cells, **as defined by sequence similarities.**

訳 シークエンス類似性によって明らかにされたように

❸ **The regulatory model of neurogenic gene expression defined in this study** permitted the identification of a neurogenic enhancer in the distant Anopheles genome. (Development. 2004 131:2387)

訳 この研究において明らかにされた神経原性遺伝子発現の調節モデル

■ [副] + defined

❶	well defined	よく明らかにされる	2,883
❷	poorly defined	あまり明らかにされていない	919
❸	clearly defined	はっきりと明らかにされる	418
❹	previously defined	以前に明らかにされた	365

❶ However, **the signaling pathways involved in VEGF-induced angiogenesis are not well defined**. (J Biol Chem. 2005 280:33262)

訳 VEGFに誘導される血管新生に関与するシグナル伝達経路は，よく明らかにされていない

❷ **The endogenous mechanisms limiting such reactions remain poorly defined**. (Proc Natl Acad Sci USA. 2003 100:2730)

訳 そのような反応を制限する内在性の機構は，あまり明らかにされないままである

❸ However, **the pathway of apoptosis is not clearly defined**. (J Biol Chem. 2002 277:37820)

訳 アポトーシスの経路は，はっきりとは明らかになっていない

❹ Two non-MHC QTL (chromosomes 1 and 4) were inherited from the BXSB and the rest were NZW-derived, **including two similar to previously defined loci**. (J Immunol. 2003 171:6442)

訳 以前に明らかにされた座位に似た２つを含んでいる

■ to be defined

★	to be defined	明らかにされる…	398
❶	remain to be defined	…は，明らかにされないままである	193

■ define + [名句]

❶	define the role of ～	…は，～の役割を明らかにする	519
❷	define a novel ～	…は，新規の～を明らかにする	490
❸	define a new ～	…は，新しい～を明らかにする	399
❹	define the mechanism	…は，機構を明らかにする	169
❺	define a role for ～	…は，～の役割を明らかにする	109
❻	define the molecular basis	…は，分子基盤を明らかにする	76
❼	define the function of ～	…は，～の機能を明らかにする	75
❽	define the relationship between ～	…は，～の間の関連性を明らかにする	58

❶ **To further define the role of ERG in regulating EC function**, we evaluated the effect of ERG knock-down on EC lumen formation in 3D collagen matrices. (Blood. 2011 118:1145)

訳 EC機能を調節する際のERGの役割をさらに明らかにするために

❷ Accordingly, **we define a novel signaling pathway that controls** EDG-1 up-regulation following stimulation of T cells by CCR7/CCL19. (22334704Blood. 2011 118:1145)

訳 われわれは，～を制御する新規のシグナル伝達経路を明らかにする

❺ In this report **we define a role for Gravin** as a temporal organizer of phosphorylation-dependent protein-protein interactions during mitosis. (Mol Cell. 2012 48:547)

訳 われわれは，Gravinの役割を明らかにする

■ to define

★ to define ～	～を明らかにする…	5,028
❶ be used to define ～	…は，～を明らかにするために使われる	374
❷ be to define ～	…は，～を明らかにすることである	268
❸ sought to define ～	…は，～を明らかにしようと努めた	139
❹ To define ～	～を明らかにするために	1,030

■ [代名] ＋ define

❶ we define ～	われわれは，～を明らかにする／～を定義する	1,282

❶ Here, **we define the mechanisms underlying** the requirement of Suds3 in pre/peri-implantation development. (Dev Biol. 2013 373:359)
訳 われわれは，～の根底にある機構を明らかにする

D

definition ㊔ 定義　　用例数 2,932

definition	定義	2,078
definitions	定義	854

複数形のdefinitionsの割合が，約30%ある可算名詞.

■ definition ＋ [前]

	the%	a/an%	ø%			
❶	50	26	17	definition of ～	～の定義	1,139
❷	35	44	9	definition for ～	～に対する定義	75

❶ Thus, **the definition of success must include** subjective symptom-based outcome in addition to anatomic outcome. (Curr Opin Urol. 2012 22:265)
訳 成功の定義は，～を含まなければならない

degeneration ㊔ 変性／退化　　用例数 6,137

degeneration	変性／退化	6,028
degenerations	変性／退化	109

複数形のdegenerationsの割合は2%しかなく，原則，**不可算名詞**として使われる.
◈類義語：denaturation

■ degeneration ＋ [前]

	the%	a/an%	ø%			
❶	24	0	76	degeneration of ～	～の変性	894
❷	6	0	83	degeneration in ～	～の変性	630

❶ The XOPS-mCFP transgene causes selective degeneration of rods without secondary loss of cones in animals up to 7 months of age.（Invest Ophthalmol Vis Sci. 2005 46:4762）

訳 XOPS-mCFP導入遺伝子は，桿体の選択的な変性を引き起こす

❷ The ER stress response is involved in retinal degeneration in hT17M Rho mice.（Invest Ophthalmol Vis Sci. 2012 53:3792）

訳 ERストレス応答は，hT17M Rhoマウスの網膜変性に関与している

■［形］＋ degeneration

❶	macular degeneration	黄斑変性症	800
❷	retinal degeneration	網膜変性症	787
❸	neuronal degeneration	ニューロン変性	337

★ degradation 名 分解 用例数 21,596

degradation	分解	21,577
degradations	分解	19

複数形のdegradationsの割合は0.1%しかなく，原則，**不可算名詞**として使われる.

◆類義語：breakdown, decomposition, disassembly

■ degradation ＋［前］

	the%	a/an%	ø%			
❶	29	3	67	degradation of ～	～の分解	7,093
❷	1	0	62	degradation by ～	～による分解	1,019
❸	2	0	65	degradation in ～	～における分解	964

❶ Auxin acts by promoting the degradation of transcriptional regulators called Aux/IAA proteins.（Dev Cell. 2005 9:109）

訳 オーキシンは，Aux/IAAタンパク質と呼ばれる転写調節因子の分解を促進することによって作用する

❷ These three STAT proteins are targeted for proteasome-mediated degradation by RNA viruses in the Rubulavirus genus of the Paramyxoviridae.（J Virol. 2005 79:10180）

訳 これら３つのSTATタンパク質は，RNAウイルスによるプロテアソームに仲介される分解を標的にしている

❸ Poly (A) tails stimulate RNA degradation in bacteria, suggesting that this is their ancestral function.（Cell. 2005 121:713）

訳 ポリ(A) 尾部は，細菌におけるRNA分解を刺激する

■［前］＋ degradation

❶	for degradation by ～	～による分解のために	158
❷	targeted for degradation	分解をターゲットにした	99

❶ We found that wild-type α-synuclein was selectively translocated into lysosomes for degradation by the chaperone-mediated autophagy pathway.（Science. 2004 305:1292）

訳 ～は，シャペロンに仲介される自己貪食経路による分解のために選択的にリソソームに移行した

■ [名] ＋ and degradation

❶ ubiquitination and degradation	ユビキチン化と分解	380
❷ phosphorylation and degradation	リン酸化と分解	210
❸ synthesis and degradation	合成と分解	166

❶ Furthermore, expression of SLAP and c-Cbl together **induced TCR ζ ubiquitination and degradation**, preventing the accumulation of fully assembled recycling TCR complexes. (Nat Immunol. 2006 7:57)

訳 ～は，TCRζのユビキチン化と分解を誘導した

■ [名／形／過分] ＋ degradation

❶ protein degradation	タンパク質分解	1,002
❷ proteasomal degradation	プロテアソームの分解	848
❸ ～-mediated degradation	～に仲介される分解	668
❹ ～-dependent degradation	～依存的分解	532
❺ rapid degradation	急速な分解	372

❸ We hypothesize that **LIN-23-mediated degradation of BAR-1 β-catenin regulates the transcription of Wnt target genes**, which in turn alter postsynaptic properties. (Neuron. 2005 46:51)

訳 BAR-1 β-カテニンのLIN-23に仲介される分解は，Wnt標的遺伝子の転写を調節する

❹ In contrast to p53, **FIP200 decreased cyclin D1 protein half-life by promoting proteasome-dependent degradation of cyclin D1**. (Cancer Res. 2005 65:6676)

訳 FIP200は，サイクリンD1のプロテアソーム依存的分解を促進することによって，サイクリンD1タンパク質の半減期を低下させた

D

degrade 動 分解する　　用例数 5,814

degraded	分解される／～を分解した	3,008
degrade	～を分解する	1,174
degrading	～を分解する	997
degrades	～を分解する	635

他動詞.

◈類義語：decompose, disassemble, break

■ degraded ＋ [前]

❶ degraded by ～	～によって分解される	720
・be degraded by ～	…は，～によって分解される	389
❷ degraded in ～	～において分解される	365

❶ Finally, pulse-chase labeling reveals that **ataxin-3 is degraded by the proteasome**, with expanded ataxin-3 being as efficiently degraded as normal ataxin-3. (J Biol Chem. 2005 280:32026)

訳 ataxin-3は，プロテアソームによって分解される

❷ Previous work defined two major pathways by which **normal transcripts are degraded in eukaryotes.** (Mol Biol Cell. 2005 16:5880)
訳 正常な転写物は，真核生物において分解される

■［副］＋ degraded

❶	rapidly degraded	急速に分解される	353

❶ Like many transcription factors, **α2 is rapidly degraded *in vivo*** by the ubiquitin-proteasome pathway. (Mol Cell Biol. 2006 26:371)
訳 α2は，生体内で急速に分解される

★★ degree ㊔ 程度／度　用例数 32,958

degrees	度／程度	21,089
degree	程度／度	11,869

温度などの尺度として使われる場合は，複数形の用例が圧倒的に多い．
◆類義語：extent, grade

■ degree ＋［前］

degree to whichの前にtheが付く割合は，100%と圧倒的に高い．

	the%	a/an%	ø%			
❶	50	32	11	degree of ～	～の程度	8,673
				・high degree of ～	高い程度の～	1,852
				・correlate with the degree of ～	…は，～の程度と相関する	208
				・relate to the degree of ～	…は，～の程度と関連する	51
❷	100	0	0	degree to which ～	～である程度	894

❶ Late-relapse GCT is uncommon and is associated with **a poor prognosis resulting from a high degree of resistance to chemotherapy.** (J Clin Oncol. 2005 23:6999)
訳 化学療法への高い程度の抵抗性に由来する予後不良

❶ Furthermore, the rated strength of the illusion **correlated with the degree of** premotor and cerebellar activity. (J Neurosci. 2005 25:10564)
訳 …は，～の程度と相関した

❷ The ratio between unwinding and annealing and **the degree to which both activities are ATP- and ADP-modulated are strongly influenced by** structured as well as unstructured regions in the RNA substrate. (Biochemistry. 2005 44:13591)
訳 両方の活性がATPおよびADPに仲介される程度は，～によって強く影響される

■［形］＋ degree(s)

❶	varying degrees	様々な程度	592
❷	to a lesser degree	より小さい程度で	342
❸	greater degree	より大きな程度	294

❹ some degree	ある程度	288
❺ different degrees	異なる程度	285
❻ various degrees	様々な程度	227
❼ higher degree	より高い程度	192
❽ the same degree	同じ程度	188
❾ a similar degree	類似の程度	163

❶ Electron paramagnetic resonance spectroscopy of the Cu-containing molecules **have varying degrees of electronic interaction** based on their charge and supramolecular structure. (J Am Chem Soc. 2005 127:9546)

訳 ～は，さまざまな程度の電気的相互作用を持つ

❷ Examination of NKT- and T-cell responses demonstrated that cross-linking of NKG2D and CD3 resulted in potent synergy **when combined with IL-12 and, to a lesser degree, with IL-18.** (Blood. 2006 107:1468)

訳 IL-12および，より小さな程度でIL-18と結合したとき

D

★ delay 動 遅延させる，名 遅延　　用例数 17,101

delayed	～を遅延させた／遅延した	9,495
delay	遅延／～を遅延させる	5,370
delays	遅延／～を遅延させる	1,749
delaying	～を遅延させる	487

他動詞の用例が多いが，名詞としても用いられる．複数形のdelaysの割合が，約20%ある可算名詞．

◆類義語：retard, retardation

■ 動 be delayed ＋［前］

★ be delayed	…は，遅延する	1,371
❶ be delayed in ～	…は，～において遅延する	322
❷ be delayed by ～	…は，～だけ遅延する／～によって遅延させられる	145
❸ be delayed until ～	…は，～まで遅延する	101

❶ Nonetheless, **cell cycle progression in tam was delayed in** both pachytene and meiosis Ⅱ. (Plant Physiol. 2004 136:4127)

訳 tamにおける細胞周期進行は，～において遅延した

❷ Similar findings were also observed in the urinary bladder, although **the inflammatory infiltrate was delayed by approximately a week relative to** that in the urethra. (Infect Immun. 2004 72:4210)

訳 炎症性浸潤は，～に比べておよそ1週間遅延した

■ 動［副］＋ delayed

❶ significantly delayed	有意に遅延した	349
❷ markedly delayed	顕著に遅延した	63

❶ The incidence of diabetes was significantly delayed by a single injection of the engineered NOD DCs into syngeneic recipients.（J Immunol. 2004 173:4331）
訳 糖尿病の発生は，～によって有意に遅延した

■ 動 delay ＋［名句］

❶	delay the onset of ～	…は，～の開始を遅延させる	216
❷	delay the development of ～	…は，～の発症を遅延させる	48
❸	delay the progression of ～	…は，～の進行を遅延させる	38

❶ Administration of the IgM NAA to MRL-lpr mice also delayed the onset of nephritis.（22407922 Plant Physiol. 2004 136:4127）
訳 ～は，腎炎の開始を遅延させた

■ 名 delay ＋［前］

	the％	a/an％	ø％			
❶	13	77	8	delay in ～	～の遅延	1,362
❷	25	39	35	delay of ～	～の遅延	299
❸	52	36	7	delay between ～	～の間の遅延	118

❶ This suggests that the decrease in the number of differentiated oligodendrocytes is attributable to a delay in the timing of their differentiation process.（J Neurosci. 2003 23:883）
訳 ～は，それらの分化過程のタイミングの遅延に起因する

delete 動 欠失させる／除去する 用例数 6,121

deleted	欠失した／欠落した／～を欠失させた	5,019
deleting	～を欠失させる	723
delete	～を欠失させる	287
deletes	～を欠失させる	92

◆類義語：remove, eliminate, abolish
他動詞.

■ deleted ＋［前］

❶	deleted in ～	～において欠失した／～を欠失した	754
❷	deleted for ～	～の欠失した	268
❸	deleted from ～	～から欠失した	232
	・be deleted from ～	…が，～から欠失する	143

❶ Mutants deleted in GAS1 and GAS2 had no defect in vegetative growth, conidiation, or appressoria formation, but they were reduced in appressorial penetration and lesion development.（Plant Cell. 2002 14:2107）
訳 GAS1およびGAS2を欠失した変異体は，栄養生長における欠損を持たなかった

❷ In a strain deleted for MSN2/4 and the PKA catalytic subunits, expression of HSP12 and HSP26 depends on HSF1 expression.（Genetics. 2005 169:1203）
訳 MSN2/ 4 とPKA触媒サブユニットを欠失した株において

❸ When this domain is deleted from the full-length toxin gene, actin cross-linking, but not cell rounding, is eliminated, indicating that this toxin carries multiple dissociable activities. (Proc Natl Acad Sci USA. 2004 101:9798)

訳 このドメインが完全長の毒素遺伝子から欠失すると

deleterious 形 有害な 用例数 2,494

◆類義語：adverse, aversive, detrimental, harmful, hazardous, noxious

■ deleterious ＋［名］

❶	deleterious effects	有害な影響	592
❷	deleterious mutations	有害な変異	353
❸	deleterious consequences	有害な結果	73

❶ The hypothalamic-pituitary-adrenal (HPA) axis may mediate the deleterious effects of stress on health. (Arch Gen Psychiatry. 2005 62:668)

訳 ～は，健康に対するストレスの有害な影響を仲介するかもしれない

★★ deletion 名 欠失 用例数 31,899

deletion	欠失	25,422
deletions	欠失	6,477

複数形のdeletionsの割合が，約20%ある可算名詞.

◆類義語：deficiency, deficit, defect, loss ◆反意語：insertion

■ deletion ＋［前］

	the%	a/an%	ø%			
❶	8	12	80	deletion of ～	～の欠失	10,888
				・by deletion of ～	～の欠失によって	435
❷	5	40	32	deletion in ～	～における欠失	1,212

❶ Deletion of the entire auf gene cluster had no effect on the ability of CFT073 to colonize the kidney, bladder, or urine of mice. (Infect Immun. 2004 72:3890)

訳 auf遺伝子クラスター全体の欠失は，～の能力に対して影響を持たなかった

❶ The rate of mat1 switching is also affected by deletion of repeats. (Genetics. 2002 162:591)

訳 ～は，また，リピートの欠失によって影響を受ける

❷ The only coding DNA sequence variant identified in these four genes was a 5-bp deletion in exon 6 of ELOVL4. (Invest Ophthalmol Vis Sci. 2001 42:2652)

訳 ～は，ELOVL4のエクソン６における５塩基対の欠失であった

■［名／形／過分］＋ deletion

❶	gene deletion	遺伝子欠失	514
❷	targeted deletion of ～	～の狙いを定めた欠失	488
❸	genetic deletion	遺伝的欠失	425

❹ specific deletion	特異的欠失	352
❺ ～-bp deletion	～塩基対の欠失	348
❻ ～-terminal deletion	～末端の欠失	344

❷ Mice with a **targeted deletion of** β3 integrin were used to examine the process by which tumor cells metastasize and destroy bone. (Proc Natl Acad Sci USA. 2003 100:14205)

訳 β3インテグリンの狙いを定めた欠失を持つマウスは，～を調べるために使われた

■ deletion ＋［名］

❶ deletion mutant	欠失変異体	1,349
❷ deletion analysis	欠失解析	782
❸ deletion mutations	欠失変異	302

delineate 動 描写する 用例数 2,803

delineate	～を描写した／描写される	1,455
delineated	～を描写する	863
delineating	～を描写する	330
delineates	～を描写する	155

他動詞.

◆類義語：describe

■ delineate ＋［名句］

❶ delineate the mechanism	…は，機構を描写する	95
❷ delineate the role of ～	…は，～の役割を描写する	89

❷ Here, we **delineate the role of** autocrine production of fibroblast growth factor 4 (Fgf4) and associated activation of the Erk1/2 (Mapk3/1) signalling cascade. (Development. 2007 134: 2895)

訳 われわれは，～のオートクリン産生の役割を描写する

deliver 動 送達する／届ける 用例数 6,832

delivered	～を送達した／送達される	3,772
deliver	～を送達する	1,862
delivering	～を送達する	713
delivers	～を送達する	485

他動詞.

■ delivered ＋［前］

❶ delivered to ～	～に送達される	857
・be delivered to ～	…は，～に送達される	455

❷ delivered by ～	～によって送達される	458
❸ delivered in ～	～において送達される	197
❹ delivered into ～	～に送達される	162

❶ Plasmid-encoding human IDO was delivered to donor lungs *in vivo* using a nonviral gene-transfer vector, polyethylenimine. (Am J Respir Crit Care Med. 2006 173:566)

訳 ヒトIDOをコードするプラスミドは，ドナーの肺に送達された

❷ When delivered by bilateral stereotaxic injection into the ventral hypothalamus (encompassing the arcuate nucleus) of mice, Ad-cMCD increases food intake and body weight. (J Biol Chem. 2005 280:39681)

訳 両側性の定位的な注入によって腹側視床下部へ送達されたとき

■ to deliver

★ to deliver ～	～を送達する…	1,130
❶ be used to deliver ～	…は，～を送達するために使われる	103

❶ Neural precursor cells (NPCs) are capable of tracking glioma tumors and thus could be used to deliver therapeutic molecules. (Ann Neurol. 2005 57:34)

訳 ～は，このように治療用の分子を送達するために使われうる

D

★ delivery ㊔ デリバリー／送達／出産　用例数 10,904

delivery	デリバリー／送達／出産	10,759
deliveries	デリバリー／送達／出産	145

複数形のdeliveriesの割合は1％しかなく，原則，不可算名詞として使われる．

◆類義語：transport

■ delivery ＋［前］

	the%	a/an%	ø%			
❶	31	1	65	delivery of ～	～のデリバリー	4,013
❷	1	0	78	delivery to ～	～へのデリバリー	790
❸	1	0	94	delivery in ～	～におけるデリバリー	222

❶ However, immune cell migration is also critically important for the delivery of protective immune responses to tissues. (Nat Immunol. 2005 6:1182)

訳 …は，また，～のデリバリーにとって決定的に重要である

❷ Detection and prediction of drug delivery to the tumor interstitium are of critical importance in cancer chemotherapy. (Cancer Res. 2001 61:3039)

訳 腫瘍間質への薬物デリバリーの検出と予想は，癌の化学療法において決定的に重要である

■［名］＋ delivery

❶ gene delivery	遺伝子送達	780
❷ drug delivery	薬物デリバリー	669
❸ preterm delivery	早産	248

❹ oxygen delivery	酸素送達	230

★★ demonstrate 動 実証する／明らかに示す　用例数 136,590

demonstrate	～を実証する／～を明らかに示す	66,068
demonstrated	実証される／～を実証した／～を明らかに示した	52,946
demonstrates	～を実証する／～を明らかに示す	9,172
demonstrating	～を実証する／～を明らかに示す	8,404

他動詞．demonstrate thatの用例が非常に多い．

◆類義語：document, corroborate, substantiate, prove, evidence, verify, show

■ ［代名／名］＋ demonstrate

❶ we demonstrate ～	われわれは，～を実証する	23,873
・Here we demonstrate ～	ここにわれわれは，～を実証する	3,087
・Finally, we demonstrate ～	最後に，われわれは～を実証する	574
❷ results demonstrate ～	結果は，～を実証する	11,005
❸ data demonstrate ～	データは，～を実証する	5,547
❹ findings demonstrate ～	知見は，～を実証する	3,120
❺ studies demonstrate ～	研究は，～を実証する	2,839
❻ analysis demonstrated ～	分析は，～を実証した	1,648
❼ experiments demonstrated ～	実験は，～を実証した	823
❽ assays demonstrated ～	アッセイは，～を実証した	747
❾ work demonstrates ～	研究は，～を実証する	527

❶ **Here, we demonstrate** that the tandem PHD1/2 fingers of MORF recognize the N-terminal tail of histone H3.（J Mol Biol. 2012 424:328）

訳 ここにわれわれは，～ということを実証する

■ demonstrate ＋ ［名節］

❶ demonstrate that ～	…は，～ということを実証する	76,332
❷ demonstrate how ～	…は，どのように～かを実証する	692

❷ These results **demonstrate that** α_2M-derived peptides target the receptor-binding sequence in TGF-β.（Biochemistry. 2003 42:6121）

訳 これらの結果は，～ということを実証する

■ ［副］＋ demonstrate ＋ ［that節］

❶ previously demonstrated that ～	…は，以前に，～ということを実証した	1,273
・we previously demonstrated that ～	われわれは，以前に，～ということを実証した	643
・we have previously demonstrated that ～	われわれは，以前に，～ということを実証した	511
❷ we further demonstrate that ～	われわれは，さらに，～ということを実証する	932

❸ we now demonstrate that ～	われわれは，今，～ということを実証する	614
❹ recently demonstrated that ～	…は，最近，～ということを実証した	502

❶ We have previously demonstrated that unspliced intron-containing CTE RNA is efficiently exported to the cytoplasm in mammalian cells. (Genes Dev. 2003 17:3075)
訳 われわれは，以前に，～ということを実証した

■ demonstrate ＋［副／副句］＋［that節］

❶ demonstrate for the first time that ～	…は，～ということを初めて実証する	840
❷ we demonstrate here that ～	われわれは，ここで，～ということを実証する	810

❶ We demonstrate for the first time that RSV activates EGFR in lung epithelial cells. (J Biol Chem. 2005 280:2147)
訳 われわれは，～ということを初めて実証する

■ demonstrate ＋［名句］

❶ demonstrate the importance of ～	…は，～の重要性を実証する	812
❷ demonstrate the utility of ～	…は，～の有用性を実証する	734
❸ demonstrate the presence of ～	…は，～の存在を実証する	730
❹ demonstrate a novel ～	…は，新規の～を実証する	652
❺ demonstrate the ability	…は，能力を実証する	550
❻ demonstrate a role for ～	…は，～の役割を実証する	527
❼ demonstrate the potential	…は，潜在能を実証する	517
❽ demonstrate the feasibility of ～	…は，～の実現可能性を実証する	480
❾ demonstrate the existence of ～	…は，～の存在を実証する	381

❶ These results demonstrate the importance of both carcinogen exposure and hormone stimulation on the induction of neoplasia in the prostate of Wistar-Unilever rats. (Cancer Res. 1998 58:3282)
訳 これらの結果は，～の重要性を実証する

❷ These data demonstrate the utility of the luxCt gene as a versatile and sensitive reporter of chloroplast gene expression in living cells. (Plant J. 2004 37:449)
訳 これらのデータは，luxCt遺伝子の有用性を実証する

❸ Blocking assays demonstrated the presence of unique B-cell epitopes in MSA-2a1, -2b, and -2c. (Infect Immun. 2002 70:3566)
訳 ブロッキングアッセイは，ユニークなB細胞エピトープの存在を実証した

■ to demonstrate

★ to demonstrate ～	～を実証する…	4,006
❶ be used to demonstrate ～	…が，～を実証するために使われる	1,618
❷ be the first to demonstrate ～	…は，～を実証する最初である	266
❸ fail to demonstrate ～	…は，～を実証することができない	200
❹ To demonstrate ～	～を実証するために	437

❶ In vitro kinase assay was used to demonstrate direct occludin phosphorylation by PKCβ.

D

(Diabetes. 2012 61:1573)

訳 試験管内キナーゼアッセイが，～を実証するために使われた

■ demonstrated ＋［前］

❶ demonstrated by ～	～によって実証される	3,654
・as demonstrated by ～	～によって実証されたように	1,304
❷ demonstrated in ～	～において実証される	2,220
・be demonstrated in ～	…が，～において実証される	1,564
❸ demonstrated to *do*	～することが実証される	1,809

❶ Poor absorption explained 4 of 5 false-negative results, as demonstrated by microinjection. (Circulation. 2003 107:1355)

訳 微量注入によって実証されたように

❷ Adherence to prednisolone was demonstrated in 17 of these 21. (Am J Respir Crit Care Med. 2001 164:1376)

訳 ～が，これらの21のうち17において実証された

❸ Aberrant Wnt signaling has been demonstrated to be important in colorectal carcinogenesis. (J Biol Chem. 2004 279:26707)

訳 異常なWntシグナル伝達が，結腸直腸発癌において重要であることが実証されている

demonstration ㊔ 実証　用例数 3,148

demonstration	実証	3,001
demonstrations	実証	147

複数形のdemonstrationsの割合が，約５％ある可算名詞.

◆類義語：evidence, indication, representation

■ demonstration ＋［前／同格that節］

demonstration thatの前にtheが付く割合は非常に高い.

	the%	a/an%	ø%			
❶	50	20	17	demonstration of ～	～の実証	1,687
❷	77	10	7	demonstration that ～	～という実証	1,062

❶ This is the first demonstration of a prognostic significance for STAT proteins in a malignancy. (Blood. 2002 99:252)

訳 これは，悪性腫瘍におけるSTATタンパク質の予後に対する重要性の最初の実証である

❷ These results led to the unexpected demonstration that BAPTA was a general inhibitor of cellular RNA synthesis in dermal fibroblasts. (Biochemistry. 2004 43:9576)

訳 これらの結果は，～という予想外の実証につながった

★depend ㊗ 依存する　用例数 21,498

depends	依存する	9,138
depend	依存する	5,372

depending	依存する	5,177
depended	依存した	1,811

自動詞．depend onの用例が圧倒的に多い．

◆類義語：rely

■ depend ＋［前］

❶ depend on ～	…は，～に依存する	17,844
❷ depend upon ～	…は，～に依存する	1,633

❶ These observations suggest that SB203580-mediated protection depends on the inhibition of p38α MAPK.（Circ Res. 2001 89:750）

訳 SB203580に仲介される保護は，p38α MAPKの阻害に依存する

❷ BCR1 expression depends upon the hyphal regulator Tec1p.（Curr Biol. 2005 15:1150）

訳 BCR1の発現は，菌糸の調節因子Tec1pに依存する

■ depend ＋［副］＋ on

❶ depend critically on ～	…は，～に決定的に依存する	222
❷ depend strongly on ～	…は，～に強く依存する	188

❶ The activity of Kir channels depends critically on the phospholipid PIP_2.（J Biol Chem. 2012 287:42278）

訳 Kirチャネルの活性は，リン脂質PIP_2に決定的に依存する

★ dependence 名 依存性／依存　用例数 10,546

dependence	依存性／依存	10,418
dependences	依存性／依存	128

複数形のdependencesの割合は１％しかないが，単数形に不定冠詞が付く用例は非常に多い．

■ dependence ＋［前］

dependence ofの前にtheが付く割合は圧倒的に高い．

	the%	a/an%	ø%			
❶	82	10	5	dependence of ～	～の依存性	4,620
				・the dependence of ～ on …	～の…への依存性	622
❷	17	39	19	dependence on ～	～への依存性	1,603

❶ Functional similarity to linker histones may explain the dependence of Rap30 binding on the bent DNA environment induced by the TATA box-binding protein.（Proc Natl Acad Sci USA. 1998 95:9117）

訳 リンカーヒストンへの機能的な類似性は，Rap30結合の～への依存性を説明するかもしれない

❷ It also accelerates the microneedle fabrication process and reduces its dependence on the

use of surfactants. (Proc Natl Acad Sci USA. 2009 106:18936)
訳 ～は，界面活性剤の使用へのそれの依存を低下させる

■ [名] + dependence

❶ voltage dependence	電圧依存性	814
❷ pH dependence	pH依存性	735
❸ temperature dependence	温度依存性	705

⁂ dependent 形 依存している／依存的な　用例数 116,339

dependent onの用例が非常に多い.

■ be dependent + [前]

★ be dependent	…は，依存している	11,443
❶ be dependent on ～	…は，～に依存している	9,288
・be highly dependent on ～	…は，～に高度に依存している	370
・be critically dependent on ～	…は，～に決定的に依存している	307
・be dependent on the presence of ～	…は，～の存在に依存している	253
・be strongly dependent on ～	…は，～に強く依存している	244
・be dependent on activation of ～	…は，～の活性化に依存している	81
❷ be dependent upon ～	…は，～に依存している	1,870

❶ The success of dental implants **is highly dependent on** integration between the implant and intraoral hard/soft tissue. (J Periodontol. 2002 73:322)
訳 歯科インプラントの成功は，～の間の統合に高度に依存している

❶ Strong engagement of the oxidative burst **was dependent on the presence of** functional Pst hrpS and hrpA gene products. (Plant J. 2000 24:569)
訳 …は，～の存在に依存していた

❶ Migration of endothelial and glioma cells toward TN-C or FN occurred in a dose-dependent manner and **was strongly dependent on** cell adhesion. (Cancer Res. 2002 62:2660)
訳 ～は，細胞接着に強く依存していた

❷ This mitogenic effect **is dependent upon** activation of a protein-tyrosine kinase cascade that results in activation of mitogen-activated protein kinase and phosphatidylinositol 3-kinase. (J Biol Chem. 2000 275:13789)
訳 この分裂促進的な効果は，～の活性化に依存している

■ [名] + be dependent on

❶ activity is dependent on ～	活性は，～に依存している	104
❷ expression is dependent on ～	発現は，～に依存している	82
❸ activation is dependent on ～	活性化は，～に依存している	72
❹ effect was dependent on ～	効果は，～に依存していた	70

❶ P-TEFb **activity is dependent on** phosphorylation of Thr186 in the CDK9 T loop. (J Virol. 2013 87:1211)
訳 P-TEFb活性は，Thr186のリン酸化に依存している

■ [名]-dependent + [名]

❶	～-dependent manner	～依存的な様式	9,332
	・in a dose-dependent manner	用量依存的な様式で	2,004
	・in a concentration-dependent manner	濃度依存的な様式で	718
	・in a time-dependent manner	時間依存的な様式で	313
	・in a dose- and time-dependent manner	用量および時間依存的な様式で	207
❷	～-dependent activation	～依存的な活性	1,659
❸	～-dependent transcription	～依存的な転写	1,491
❹	～-dependent mechanism	～依存的な機構	1,471
❺	～-dependent fashion	～依存的な様式	1,405
	・in a dose-dependent fashion	用量依存的な様式で～	346
❻	～-dependent pathway	～依存的な経路	1,262
❼	～-dependent signaling	～依存的なシグナル伝達	1,182
❽	…-dependent increase in ～	…依存的な～の増大	1,083
	・dose-dependent increase in ～	用量依存的な～の増大	353

❶ Fenoldopam increased renal blood flow in a dose-dependent manner compared with placebo, and, at the lowest dose, significantly increased renal blood flow occurred without changes in systemic blood pressure or heart rate. (Crit Care Med. 1999 27:1832)

訳 フェノルドパムは，用量依存的な様式で腎血流量を増大させた

❸ Skn7p modulates CDRE-dependent transcription by affecting Crz1p protein levels. (EMBO J. 2001 20:3473)

訳 Skn7pは，CDRE依存的な転写を調節する

❺ Activity of calpain purified from rabbit muscle was greatly inhibited in a dose-dependent fashion by MP. (Brain Res. 1997 748:205)

訳 ～は，MPによって用量依存的な様式で大きく抑制された

■ -dependent and -independent + [名]

★	～-dependent and -independent	～依存的および非依存的な	864
❶	～-dependent and -independent mechanisms	～依存的および非依存的な機構	184

❶ MDM2 is an oncoprotein that controls tumorigenesis through both p53-dependent and -independent mechanisms. (J Biol Chem. 2004 279:29841)

訳 MDM2 は，p53依存的および非依存的機構の両方によって腫瘍形成を調節するオンコプロテインである

dependently 副 依存的に 用例数 883

dose-dependentlyの用例が圧倒的に多い.

■ dose-dependently + [動]

★	dose-dependently	用量依存的に	767
❶	dose-dependently inhibited	用量依存的に抑制した	138
❷	dose-dependently increased	用量依存的に増大した	98

❶ In contrast to its effect on NO synthesis, **hsp90 dose-dependently inhibited** $O_2^{-\cdot}$ **generation from nNOS** with an IC_{50} of 658 nM. (Biochemistry. 2002 41:10616)
訳 hsp90は，nNOSからの$O_2^{-\cdot}$産生を用量依存的に抑制した

deplete 動 枯渇させる 用例数 6,910

depleted	欠けている／～を枯渇させた	5,429
depleting	～を枯渇させる	732
deplete	～を枯渇させる	499
depletes	～を枯渇させる	250

他動詞．過去分詞の用例が非常に多い．特に，depleted ofの用例が多い．
◆類義語：deprive, starve, lack

■ be depleted ＋［前］

★ be depleted	…は，欠けている	925
❶ be depleted of ～	…は，～が欠けている	212
❷ be depleted from ～	…は，～から欠けている	105
❸ be depleted in ～	…は，～において欠けている	104
❹ be depleted by ～	…は，～によって枯渇させられる	94

❶ **Genomic loci of piRNA biogenesis are depleted of protein-coding genes** and tend to overlap the start and end of transposons in sense and antisense, respectively. (Science. 2012 337:574)
訳 piRNA生合成の遺伝子座は，タンパク質をコードする遺伝子が欠けている

❷ In contrast, single-binding-site superantigens **were greatly depleted from the surface by 4 h.** (Infect Immun. 2005 73:5358)
訳 ～は，4時間までに表面から大きく欠けていた

❸ Like that of MMP13, **MMP9 expression is nearly depleted in Runx2 mutant mice.** (Mol Cell Biol. 2005 25:8581)
訳 MMP9の発現は，Runx2変異マウスにおいてほとんど欠けている

■［名］＋ depleted of

★ depleted of ～	～が枯渇した	1,012
❶ cells depleted of ～	～が枯渇した細胞	209
❷ mice depleted of ～	～が枯渇したマウス	126

❷ Moreover, **80% of C3H mice depleted of CD8 and CD4 cells died of E. muris infection** compared with only 44% of CD4 cell-depleted mice. (Infect Immun. 2004 72:966)
訳 CD8とCD4細胞が枯渇したC3Hマウスの80%が，E. muris感染で死亡した

■［名］-depleted ＋［名］

❶ ～-depleted cells	～が枯渇した細胞	550
❷ ～-depleted mice	～が枯渇したマウス	225

❶ **Only the CDC27- and APC1-depleted cells were enriched in the G2/M phase** with inhibited growth. (J Biol Chem. 2005 280:31783)

訳 CDC27およびAPC1が枯渇した細胞だけが，G2/M期に増加した

■ [名]-depleted

❶	cell-depleted ～	細胞が枯渇した～	407

❶ In contrast, the plasma cell-depleted bone marrow cells from 6 patients did not grow or produce myeloma in SCID-hu hosts.（Blood. 1999 94:3576）

訳 6人の患者からのプラズマ細胞が枯渇した骨髄細胞は，増殖しなかった

★ depletion ㊔ 枯渇／欠乏　用例数 12,464

depletion	枯渇／欠乏	12,353
depletions	枯渇／欠乏	111

複数形のdepletionsの割合は1％しかなく，原則，不可算名詞として使われる．

◆類義語：deficiency, depletion, deprivation, deficit, lack, paucity

■ depletion ＋ [前]

	the％	a/an％	ø％			
❶	7	1	91	depletion of ～	～の枯渇	5,924
				・～-mediated depletion of …	～に仲介された…の枯渇	221
❷	0	3	48	depletion in ～	～における枯渇	470

❶ With conventional culturing techniques, however, cell proliferation and ultimate density are limited by depletion of nutrients and accumulation of metabolites in the medium.（Nat Methods. 2005 2:685）

訳 ～は，栄養分の枯渇によって制限される

❷ Our results indicate that estrogen depletion in the brain may be a significant risk factor for developing AD neuropathology.（Proc Natl Acad Sci USA. 2005 102:19198）

訳 脳におけるエストロゲンの枯渇は，重要な危険因子であるかもしれない

■ [名] ＋ depletion

❶	cell depletion	細胞の枯渇	884
❷	ATP depletion	ATPの枯渇	322
❸	store depletion	貯蔵の枯渇	261

deposit ㊔ 沈着物，㊍ 沈着する　用例数 3,462

deposits	沈着物／沈着する	1,717
deposited	沈着した	1,342
deposit	沈着物／沈着する	320
depositing	沈着する	83

名詞および動詞として用いられる．名詞としては，複数形のdepositsの割合が約90％と

圧倒的に高い.
◆類義語：accumulate

■ 名 deposits ＋［前］

	the％	a/an％	ø％			
❶	7	-	74	deposits in ～	～における沈着物	269
❷	11	-	85	deposits of ～	～の沈着物	226

❶ The distribution of β-amyloid deposits in tissue was based on measurement of Eu- and Ni-coupled antibodies. (Anal Biochem. 2005 346:225)
訳 組織におけるβ-アミロイド沈着物の分布は，～に基づいていた

❷ Alzheimer's disease (AD) is characterized by extracellular deposits of fibrillar β-amyloid (Aβ) in the brain, a fulminant microglial-mediated inflammatory reaction, and neuronal death. (J Neurosci. 2005 25:299)
訳 アルツハイマー病(AD)は，線維性β-アミロイドの細胞外沈着物によって特徴づけられる

■ 名［名］＋ deposits

❶	amyloid deposits	アミロイド沈着物	277
❷	Aβ deposits	Aβ沈着物	118

■ 動 be deposited ＋［前］

★	be deposited	…は，沈着する	442
❶	be deposited in ～	…は，～に沈着する	126
❷	be deposited on ～	…は，～に沈着する	60

❶ In S. cerevisiae, histone variant H2A.Z is deposited in euchromatin at the flanks of silent heterochromatin to prevent its ectopic spread. (Cell. 2005 123:233)
訳 ヒストン変異体H2A.Zは，ユークロマチンに沈着する

deposition 名 沈着　　用例数 5,789

deposition	沈着	5,742
depositions	沈着	47

複数形のdepositionsの割合は0.8%しかなく，原則，不可算名詞として使われる.
◆類義語：accumulation

■ deposition ＋［前］

	the％	a/an％	ø％			
❶	34	4	61	deposition of ～	～の沈着	1,508
❷	4	0	68	deposition in ～	～における沈着	614
❸	3	0	71	deposition on ～	～における沈着	147

❶ Alzheimer's disease (AD) is characterized by the deposition of amyloid plaques in the parenchyma and vasculature of the brain. (Biochemistry. 2003 42:2768)

訳 アルツハイマー病（AD）は，アミロイド斑の沈着によって特徴づけられる

❷ These regions were active in default states in young adults and also showed amyloid deposition in older adults with AD.（J Neurosci. 2005 25:7709）

訳 ～は，また，ADのより高齢の成人におけるアミロイド沈着を示した

■ ［名］＋ deposition

❶ collagen deposition	コラーゲン沈着	288
❷ amyloid deposition	アミロイド沈着	265
❸ Aβ deposition	Aβ沈着	229

D

depress 動 弱める／抑制する 用例数 2,889

depressed	抑制される／弱める／うつ病の／～を抑制した	2,478
depress	～を弱める	196
depresses	～を弱める	134
depressing	～を弱める	81

他動詞．受動態の用例が非常に多い．

◈類義語：attenuate, weaken, suppress, inhibit, decrease, reduce

■ be depressed ＋［前］

★ be depressed	…は，弱められる	309
❶ be depressed in ～	…は，～において弱められる	77
❷ be depressed by ～	…は，～だけ弱められる／～によって抑制される	53

❶ Left ventricular systolic function was depressed in 30 patients (83%), with a median ejection fraction of 30% (range, 15% to 66%) at diagnosis.（Circulation. 2003 108:2672）

訳 左心室収縮機能が，30名の患者において弱められた

❷ Hypothalamic TRH is depressed by at least 75% compared with wild-type controls.

訳 視床下部TRHは，少なくとも75%弱められる

■ depressed ＋［名］

❶ depressed patients	うつ病患者	346
❷ depressed mood	抑うつ気分	92

★ depression 名 抑制／抑圧／うつ病 用例数 10,110

depression	抑制／抑圧／うつ病	10,021
depressions	抑制／抑圧／うつ病	89

複数形のdepressionsの割合は１%しかなく，不可算名詞として使われることが多い．

◈類義語：suppression, repression, inhibition, attenuation, decrease, reduction

■ depression ＋［前］

	the％	a/an％	ø％			
❶	2	6	92	depression in ～	～におけるうつ病	578
❷	16	16	67	depression of ～	～の抑制	550
❸	6	4	89	depression at ～	～における抑制	184

❶ **Depression in middle and old age is associated with** non-psychiatric hospitalisation, longer length of stay and higher mortality in clinical settings.（PLoS One. 2012 7:e34821）

訳 中高年におけるうつ病は，～と関連する

❷ **Selective depression of glutamate release may provide** an adaptive mechanism by which neurons limit excitotoxicity.（Neuron. 2004 42:423）

訳 グルタミン酸放出の選択的抑制は，～を提供するかもしれない

■［形／名］＋ depression

❶	major depression	大うつ病	1,266
❷	long-term depression	長期抑制	537
❸	synaptic depression	シナプス性の抑制	269

deprive 動 取り除く／欠乏させる　用例数 1,319

deprived	除かれる／～を取り除いた	1,240
depriving	～を取り除く	49
deprive	～を取り除く	18
deprives	～を取り除く	12

他動詞．deprived ofの用例が多い．

◆類義語：remove, eliminate, exclude, obviate, delete, withdraw

■ deprived ＋［前］

❶	deprived of ～	～が除かれる	288
	・be deprived of ～	…は，～が除かれる	81

❶ **When cell culture medium is deprived of threonine,** ES cells rapidly discontinue DNA synthesis, arrest cell division, and eventually die.（Proc Natl Acad Sci USA. 2011 108:15828）

訳 細胞培養液はスレオニンが除かれたとき

★★ derive 動 引き出す／由来する　用例数 47,016

derived	由来する／～を引き出した	44,207
derive	～を引き出す／由来する	1,994
derives	～を引き出す／由来する	579
deriving	由来する	236

他動詞として使われることが多いが，自動詞の用例もある．derived fromの用例が非

常に多い.

◈類義語：originate, stem, come from, arise, draw

■ derived ＋［前］

❶ derived from ～	～に由来する	17,341
・be derived from ～	…は，～に由来する	3,064
・cells derived from ～	～に由来する細胞	1,163
・peptides derived from ～	～に由来するペプチド	597
・cell lines derived from ～	～に由来する細胞株	456
❷ derived by ～	～によって引き出される	416

❶ Although **HRS cells are derived from mature B cells**, they have largely lost their B cell phenotype and show a very unusual co-expression of markers of various hematopoietic cell types.（J Clin Invest. 2012 122:3439）

訳 HRS細胞は，成熟したB細胞に由来する

❶ **Primary cultures of cells derived from the cysts of patients with ADPKD were used.**（J Am Soc Nephrol. 2002 13:2619）

訳 ADPKDの患者の嚢胞に由来する細胞の初代培養が使われた

❷ Liver-infiltrating lymphocytes were expanded polyclonally in bulk cultures, and **multiple clones were derived by limiting dilution.**（J Immunol. 1998 160:1479）

訳 複数のクローンが，限界希釈によって引き出された

■［名］-derived ＋［名］

❶ ～-derived cells	～由来細胞	1,086
❷ ～-derived macrophages	～由来マクロファージ	729
・bone marrow-derived macrophages	骨髄由来マクロファージ	317
❸ ～-derived cell lines	～由来細胞株	382
❹ ～-derived peptides	～由来ペプチド	360

❶ **Bone marrow-derived macrophages from mice lacking Abl and Arg kinases exhibit** inefficient phagocytosis of sheep erythrocytes and zymosan particles.（J Immunol. 2012 189:5382）

訳 AblおよびArgキナーゼを欠くマウスからの骨髄由来マクロファージは，～を示す

■［代名］＋ derive

❶ we derive ～	われわれは，～を引き出す	321

❶ Here **we derive a general result** that holds for any number of strategies, for a large class of population structures under weak selection.（Proc Natl Acad Sci USA. 2011 108:2334）

訳 われわれは，一般的な結果を引き出す

■ to derive

★ to derive ～	～を引き出すために	823
❶ be used to derive ～	…は，～を引き出すために使われる	178

❶ Magnetoencephalography data acquired throughout training **was used to derive functional networks.**（Brain. 2012 135:596）

訳 ～は，機能的ネットワークを引き出すために使われた

■ derive ＋［前］

❶ derive from ～	…は，～に由来する	569

❶ Disagreements probably **derive from** the manner of conduct of the study and the populations studied.（Hepatology. 2002 36:S35）

訳 不一致は，おそらく～に由来する

★★ describe 動 述べる／記述する／表す 用例数 44,839

described	述べられる／～を述べた／～を記述した	21,353
describe	～を述べる／～を記述する	18,090
describes	～を述べる／～を記述する	3,849
describing	～を述べる／～を記述する	1,547

他動詞.

◆類義語：state, mention, note, report, document

■ describe ＋［名句／how節］

❶ describe a novel ～	…は，新規の～を述べる	982
❷ describe how ～	…は，どのように～かを述べる	692
❸ describe a new ～	…は，新しい～を述べる	646
❹ describe a method	…は，方法を述べる	327
❺ describe the identification	…は，同定を述べる	317
❻ describe the development	…は，開発を述べる	315
❼ describe the use of ～	…は，～の使用を述べる	300
❽ describe the isolation	…は，単離を述べる	250
❾ describe the cloning	…は，クローニングを述べる	221

❶ Here, **we describe a novel form of** virus-induced perturbation of host cell nuclear structures.（J Virol. 2004 78:6360）

訳 われわれは，新規の型の～を述べる

❹ **We describe a method for** estimating growth parameters in various regions of a developing organ undergoing cell divisions, along with the corresponding changes in organ shape.（J Theor Biol. 2005 232:157）

訳 われわれは，～のための方法を述べる

■［代名／名］＋ describe

❶ we describe ～	われわれは，～を述べる	13,647
❷ paper describes ～	論文は，～を述べる	586
❸ report describes ～	報告は，～を述べる	521
❹ study describes ～	研究は，～を述べる	429
❺ review describes ～	総説は，～を述べる	344

❻ article describes ～	記事は，～を述べる	336

❶ Here we describe the isolation and characterization of HDAC10, a novel class II histone deacetylase. (J Biol Chem. 2002 277:187)

訳 ここにわれわれは，HDAC10の単離と特徴づけを述べる

❷ This paper describes a technique that controls the location for cell growth in vitro and demonstrates that the technique has minimal (if any) impact on intercellular signaling. (Anal Chem. 2002 74:4640)

訳 この論文は，～する技術を述べる

■ ［動］＋ to describe

★ to describe ～	～を述べる…	1,748
❶ be to describe ～	…は，～を述べることである	247
❷ be used to describe ～	…は，～を述べるために使われる	158

D

❷ The epaxial/hypaxial terminology is also used to describe regions of the embryonic somites based on fate mapping of somitic derivatives. (Dev Cell. 2003 4:159)

訳 …は，また，～の領域を述べるために使われる

■ described ＋［前］

❶ described in ～	～において述べられる	2,595
・be described in ～	…は，～において述べられる	1,148
❷ described by ～	～によって表される	1,185
❸ described for ～	～に対して述べられる	1,121
❹ described as ～	～として記述される	1,098
・described as ～ing	～すると記述される	104

❶ The principle described in this paper could be also applied to many other ECL analyses, such as immunoassays. (Anal Chem. 2004 76:5379)

訳 この論文で述べられる原理は，また，多くの他のECL分析にも応用できる

❸ The direct differential method is described for the systematic evaluation of scaled sensitivity coefficients in reaction networks. (Bioinformatics. 2005 21:1194)

訳 直接の差動的な方法が，～の系統的評価に対して述べられる

❹ Distamycin A has been described as an inhibitor of the cellular pathogenesis of vaccinia virus in culture. (J Virol. 2004 78:2137)

訳 ジスタマイシンAは，～の抑制剤として記述されてきた

■ ［副］＋ described

❶ previously described	以前に述べられた	3,114
❷ recently described	最近述べられた	1,179
❸ well described	よく表される	692

❶ Thus, yFACT functions in establishing transcription initiation complexes in addition to the previously described role in elongation.

訳 伸長における以前に述べられた役割に加えて

❷ In addition to the CD4 and CD8 coreceptors, a recently described adaptor, Unc119, may link

SFKs to the TCR.（Oncogene. 2004 23:7990）
訳 最近述べられたアダプターUnc119は，SFKをTCRにつなぐかもしれない

❸ All steps are well described by single exponential kinetics.（Biophys J. 2005 89:2024）
訳 すべてのステップが，ひとつの指数関数速度式によってよく表される

■ described ＋［副］

❶ described here	ここで述べられる	1,373
❷ described herein	ここで述べられる	340
❸ described previously	以前に述べられた	474

❶ The experiments described here show that Homo sapiens（Hs）Rad52 and yeast Rad52 proteins promote strand exchange as well.（Proc Natl Acad Sci USA. 2004 101:9568）
訳 ここで述べられる実験は，～ということを示す

description 名 記述 用例数 2,822

description	記述	2,196
descriptions	記述	626

複数形のdescriptionsの割合が，約20%ある可算名詞.

■ description ＋［前］

	the%	a/an%	ø%			
❶	32	56	7	description of ～	～の記述	1,773

❶ A complete portrait of a cell requires a detailed description of its molecular topography: proteins must be linked to particular organelles.（Nat Methods. 2011 8:80）
訳 細胞の完全なポートレイトは，それの分子トポグラフィーの詳細な記述を必要とする

■［形］＋ description

❶ the first description	最初の記述	271
❷ detailed description	詳細な記述	147

★design 動 設計する／計画する，名 設計／計画 用例数 26,763

design	～を設計する／～を計画する／設計／計画	12,972
designed	設計される／計画される／～を計画した	11,373
designing	～を設計する／～を計画する	1,224
designs	～を設計する／～を計画する／設計／計画	1,194

他動詞および名詞として用いられる．複数形のdesignsの割合は約10%あるが，単数形は無冠詞で使われることも多い.

◆類義語：plan

■ 動 designed ＋ ［前］

❶ designed to *do*	～するために計画される	6,261
・be designed to *do*	…は，～するために計画される	3,288
・study was designed to *do*	この研究は，～するために計画された	1,253
・designed to determine ～	～を決定するために計画される	478
・designed to test ～	～をテストするために計画される	279
・designed to evaluate ～	～を評価するために計画される	242
・designed to investigate ～	～を精査するために計画される	222
・designed to assess ～	～を評価するために計画される	209
・designed to identify ～	～を同定するために計画される	200
❷ designed for ～	～のために設計される	564

❶ This study was designed to determine the inhibitory effect of zinc on alcohol-induced endotoxemia and whether the inhibition is mediated by metallothionein (MT) or is independent of MT. (Am J Pathol. 2004 164:1959)

訳 この研究は，～の抑制効果を決定するために計画された

❷ Microarrays designed for this purpose use relatively few probes for each gene and are biased toward known and predicted gene structures. (Genomics. 2005 85:1)

訳 この目的のために設計されたマイクロアレイは，それぞれの遺伝子に対する比較的少数のプローブを用いる

■ 動 designed and ＋ ［過分］

❶ designed and synthesized	設計され，そして合成される	273

❶ A series of unique indazoles and pyridoindolones have been rationally designed and synthesized as novel classes of cannabinoid ligands based on a proposed bioactive amide conformation. (J Med Chem. 2003 46:2110)

訳 ～は，新規のクラスのカンナビノイドのリガンドとして合理的に設計されそして合成された

■ 動 ［代名］＋（have）designed

❶ we designed ～	われわれは，～を計画した	819
・we have designed ～	われわれは，～を計画してきた	351

❶ To study the association, we designed a case-control study. (N Engl J Med. 2000 343:1826)

訳 われわれは，症例対照研究を計画した

■ 動 to design

★ to design ～	～を設計するために	975
❶ be used to design ～	…は，～を設計するために使われる	163

❶ Two sources of information were used to design relevant mutations. (Hum Mol Genet. 2005 14:411)

訳 2つのソースの情報が，関連する変異を設計するために使われた

D

■ ㊔ design ＋［前］

design ofの前にtheが付く割合は圧倒的に高い.

	the %	a/an %	ø %			
❶	82	1	17	design of ～	～の設計	3,828
❷	7	44	48	design to *do*	～するための設計	197

❶ These results provide an explanation of CD8$^+$ T cell avidity maturation and **may contribute to the design of novel vaccines.** (Proc Natl Acad Sci USA. 2004 101:15154)
訳 ～は，新規のワクチンの設計に寄与するかもしれない

■ ㊔［名／形］＋ design

❶	drug design	薬の設計	597
❷	rational design	合理的な設計	566
❸	study design	研究計画	522

designate ㊍ 命名する　用例数 5,408

designated	命名された／～を命名した	5,149
designate	～を命名する	207
designating	～を命名する	31
designates	～を命名する	21

〈designate A B = designate A as B（AをBと命名する)〉の文型で用いられ，asは省略されることが多い．受動態の用例が圧倒的に多い.

◆類義語：term, name, call

■［代名］＋ designate

❶	we designate ～	われわれは，～を命名する	235

❶ **We designate this approach the LD decay (LDD) test.** (Proc Natl Acad Sci USA. 2006 103:135)
訳 われわれは，このアプローチをLD崩壊（LDD）テストと命名する

■［名］, designated

❶	gene, designated ～	遺伝子，～と命名された	269
❷	protein, designated ～	タンパク質，～と命名された	205

❶ **A novel gene, designated ML-1, was identified** from a human genomic DNA clone and human T cell cDNA sequences. (J Immunol. 2001 167:4430)
訳 新規の遺伝子，ML-1と命名された，が同定された

■ designated ＋［前］

❶	designated as ～	～と命名される	557

❶ Many eukaryotic proteins share a sequence designated as the zona pellucida (ZP) domain. (Annu Rev Biochem. 2005 74:83)

訳 多くの真核生物タンパク質は，透明帯（ZP）ドメインと命名された配列を共有している

destruction 名 破壊 用例数 3,489

複数形の用例はなく，原則，不可算名詞として使われる．

◆類義語：disruption, ablation, lesion

destruction ＋［前］

	the %	a/an %	ø %			
❶	39	6	54	destruction of ～	～の破壊	1,217
❷	1	0	98	destruction in ～	～における破壊	259
❸	3	0	83	destruction by ～	～による破壊	148

❶ Destruction of the lateral parabrachial nucleus produced a 44 % inhibition of peptone-induced pancreatic section. (J Physiol. 2003 552:571)

訳 外側傍小脳脚核の破壊は，～の44%抑制を引き起こした

❷ Cysteine proteases are postulated to play a role in tissue destruction in the joints of animals with arthritis. (Arthritis Rheum. 2001 44:703)

訳 システインプロテアーゼは，関節炎の動物の関節の組織破壊において役割を果たすと推論される

detail 名 詳細，動 詳しく述べる 用例数 5,458

detail	詳細／詳しく述べる	2,740
details	詳細／詳しく述べる	2,616
detailing	詳しく述べる	102

動詞としても用いられるが，名詞の用例が非常に多い．in detailの用例が多い．その他の用例では，複数形のdetailsが非常に多い．

in detail

★	in detail	詳細に	1,447
❶	in more detail	より詳細に	244
❷	be studied in detail	…は，詳細に研究される	206
❸	in greater detail	ずっと詳細に	116
❹	be examined in detail	…は，詳細に調べられる	91

❶ In this study we have analyzed the antibody responses and mechanism of protection induced by PiuA and PiaA in more detail. (Infect Immun. 2005 73:6852)

訳 われわれは，抗体の反応とPiuAとPiaAによって誘導される保護の機構をより詳細に解析した

❷ The fliL mutation was studied in detail. (J Bacteriol. 2005 187:6789)

訳 fliL変異が，詳細に研究された

■ details ＋［前］

	the%	a/an%	ø%			
❶	51	-	38	details of ～	～の詳細	1,561

❶ Therefore, a basic understanding of the structural details of lipid-free apoA-I **will be useful for elucidating the molecular details of the pathway.** (Cell. 2005 122:605)
訳 ～は，経路の分子的詳細を明らかにするために役に立つであろう

■ ［形］＋ details

❶ molecular details	分子的詳細	314
❷ mechanistic details	機構的詳細	161

detailed 形 詳細な　用例数 7,865

■ detailed ＋［名］

❶ detailed analysis	詳細な分析	906
❷ detailed information	詳細な情報	273
❸ detailed understanding of ～	～の詳細な理解	300
❹ detailed characterization of ～	～の詳細な特徴づけ	231

❶ In the first part of this study, **a detailed analysis of TGF-β isoform distribution was performed through** immunolocalization. (Am J Pathol. 2005 167:1005)
訳 TGF-βアイソフォームの分布の詳細な分析が，～によって行われた

■ ［副］＋ detailed

❶ more detailed ～	もっと詳細な～	531
❷ the first detailed ～	最初の詳細な～	172

❶ **More detailed studies revealed that** the mechanisms of heterotopic gene expression in Plg° mice required fibrin (ogen). (Proc Natl Acad Sci USA. 2005 102:10182)
訳 もっと詳細な研究が，～ということを明らかにした

★★ detect 動 検出する　用例数 46,089

detected	検出される／～を検出した	33,208
detect	～を検出する	8,863
detecting	～を検出する	3,153
detects	～を検出する	865

他動詞.

◈類義語：find, identify, discover, observe, recognize

■ be detected ＋［前］

★ be detected	…は，検出される	19,414
❶ be detected in ～	…は，～において検出される	8,885
・be detected in all ～	…は，すべての～において検出される	287
・be detected only in ～	…は，～においてのみ検出される	288
❷ be detected by ～	…は，～によって検出される	2,112
❸ be detected at ～	…は，～において検出される	816
❹ be detected with ～	…は，～によって検出される	531
❺ be detected on ～	…は，～において検出される	474

❶ GABA$_A$- receptor-mediated currents were detected in all neurons tested, with an average EC$_{50}$ of 22.2 μM.（Brain Res. 2001 921:183）

訳 GABA$_A$-受容体に仲介される電流は，テストされたすべてのニューロンにおいて検出された

❶ GUS activity was detected only in the phloem cells but not in any other cell types of vegetative tissues.（Proc Natl Acad Sci USA. 1990 87:4144-8）

訳 GUS活性は，師部細胞においてのみ検出された

❷ No AdV was detected by culture or PCR in throat swabs from healthy recruits, suggesting the absence of latency or asymptomatic shedding.（J Clin Microbiol. 2003 41:810）

訳 AdVは，培養やPCRによっては検出されなかった

❸ No antirasburicase antibodies were detected at day 14.（Blood. 2001 97:2998）

訳 抗rasburicase抗体は，14日目には検出されなかった

■［副］＋ detected

❶ readily detected	容易に検出される	350
❷ first detected	最初に検出される	316

❷ This mutation was first detected in clinical samples in Europe during the 2007-2008 influenza season.（J Clin Microbiol. 2010 48:3517）

訳 この変異は，臨床サンプルにおいて最初に検出された

■ as detected

★ as detected	検出されたように	650
❶ as detected by ～	～によって検出されたように	574

❶ Adherens junctions, as detected by cadherin-GFP expression, were distributed in the cell perimeter as high- or low-density segments.（Nat Commun. 2012 3:1099）

訳 カドヘリン-GFP発現によって検出されたように

■［代名／名］＋ detect

❶ we detected ～	われわれは，～を検出した	1,348
❷ analysis detected ～	分析は，～を検出した	129

❶ We detected no significant elevation of proinflammatory cytokines or activation of macrophages in the spleens of these animals, despite clear PrPSc deposition.（J Virol. 2005 79:5174）

訳 われわれは，炎症誘発性サイトカインの有意な上昇を検出しなかった

■ detect ＋［名句］

❶ detect the presence of ～	…は，～の存在を検出する	230

❷ This technique is also able to detect the presence of a refolding intermediate whose formation is otherwise prevented by cantilever recoil.（Biophys J. 2011 100:1800）

訳 この技術は，また，～の存在を検出できる

■ ［動／名］＋ to detect

★ to detect ～	～を検出する…	6,299
❶ be used to detect ～	…は，～を検出するために使われる	682
❷ fail to detect ～	…は，～を検出することができない	256
❸ ability to detect ～	～を検出する能力	387
❹ power to detect ～	～を検出する力	299

❶ Flow cytometric analysis was used to detect donor cells in tolerant mice.（Transplantation. 1998 65:1036）

訳 フローサイトメトリー分析が，ドナー細胞を検出するために使われた

❷ Without induction, LA failed to detect 50% of mecA-positive strains grown on two different media.（J Clin Microbiol. 2002 40:2251）

訳 LAは，２つの異なる培地で増殖されたmecA陽性株の50%を検出することができなかった

❸ The presence of endocervical cells is a valid and convenient surrogate for the ability to detect dyskaryosis.（Lancet. 1999 354:1763）

訳 核異型を検出する能力

■ ［前］＋ detecting

❶ capable of detecting ～	～を検出できる	219
❷ method for detecting ～	～を検出するための方法	163

❷ In this paper we present a new method for detecting block duplications in a genome.（J Mol Evol. 2003 56:28）

訳 われわれは，～を検出するための新しい方法を提示する

detectable 形 検出できる／検出可能な　用例数 9,427

■ be detectable ＋［前］

★ be detectable	…は，検出できる	1,578
❶ be detectable in ～	…は，～において検出できる	651
❷ be detectable by ～	…は，～によって検出できる	157
❸ be detectable at ～	…は，～において検出できる	115

❶ Among the α, β, and γ transcripts of the substance P gene, only the β and γ transcripts are detectable in these cells.（J Immunol. 1997 159:5654）

訳 βおよびγ転写物だけが，これらの細胞において検出できる

❷ Eotaxin mRNA **was detectable by** northern analysis in BAL cells exclusively from allergen-challenged segments.（Am J Respir Crit Care Med. 2001 163:1669）

訳 エオタキシンのメッセンジャーRNAは，ノーザン解析によって検出できた

■［副］＋ detectable

❶	barely detectable	かろうじて検出できる	262
❷	readily detectable	容易に検出できる	248

❶ However, expression of LipC is **barely detectable** in a wild-type background.（Mol Microbiol. 1999 34:317）

訳 LipCの発現は，野生型バックグラウンドにおいてかろうじて検出できる

❷ Enzyme activity appears to be suppressed in T. brucei, although the polypeptide is **readily detectable**.（J Biol Chem. 2000 275:19334）

訳 そのポリペプチドは容易に検出できるけれども

D

■ no/any detectable

❶	no detectable ～	検出可能な～のない	1,881
	・have no detectable ～	…は，検出可能な～を持たない	451
	・with no detectable ～	検出可能な～のない	195
❷	any detectable ～	どの検出可能な～も	222

❶ Overexpression of GTU1 **had no detectable** effect on cell growth or morphology.（J Cell Biol. 2002 158:1195）

訳 GTU1の過剰発現は，細胞増殖や形態に対する検出可能な影響を持たなかった

■ detectable ＋［名］

❶	detectable levels	検出可能なレベル	465
❷	no detectable effect	検出可能な影響のない	231

★ detection ㊔ 検出　　用例数 21,167

detection	検出	21,069
detections	検出	98

複数形のdetectionsの割合は0.5%しかなく，原則，**不可算名詞**として使われる．

◈類義語：identification, discovery

■ detection ＋［前］

	the%	a/an%	ø%			
❶	42	1	57	detection of ～	～の検出	9,493
				・assay for the detection of ～	～の検出のためのアッセイ	91

				・method for the detection of ～	～の検出のための方法	88
❷	1	0	93	detection in ～	～における検出	493
❸	3	0	84	detection by ～	～による検出	363

❶ In 1997, we developed a PCR assay for the detection of herpes simplex virus (HSV) DNA. (J Clin Microbiol. 2007 45:1618)

訳 われわれは，単純ヘルペスウイルス（HSV）のDNAの検出のためのPCRアッセイを開発した

■ [形] ＋ detection

❶ early detection	早期の検出	592
❷ sensitive detection	鋭敏な検出	211

❶ Long-term longitudinal studies are needed to evaluate this approach for early detection of head and neck cancer in at-risk populations. (Cancer Res. 2001 61:939)

訳 長期の縦断的な研究は，頭と首の癌の早期の検出のためのこのアプローチを評価するために必要とされる

■ detection ＋ [名]

❶ detection limit	検出限界	1,207
❷ detection method	検出方法	379

★ determinant 名 決定要因／決定基 用例数 11,071

determinants	決定要因／決定基	5,967
determinant	決定要因／決定基	5,104

複数形のdeterminantsの割合は約55%と非常に高い．

■ determinant ＋ [前]

	the%	a/an%	ø%			
❶	15	85	0	determinant of ～	～の決定要因	6,116
❷	30	68	1	determinant for ～	～の決定要因	850
❸	12	86	1	determinant in ～	～の決定要因	627

❶ Myocardial infarct size is a major determinant of prognosis. (Lancet. 2013 381:166)

訳 心筋梗塞サイズは，予後の主な決定因子である

■ [形／名] ＋ determinant(s)

❶ important determinant	重要な決定要因	1,058
❷ major determinant	主な決定要因	805
❸ critical determinant	決定的に重要な決定要因	569
❹ key determinant	鍵となる決定要因	507

❺ molecular determinants	分子的な決定基	410
❻ virulence determinant	病原性決定基	379
❼ genetic determinant	遺伝的な決定要因	339
❽ primary determinant	主要な決定要因	315
❾ structural determinants	構造的な決定基	302

determination 名 決定／定量 用例数 7,015

determination	決定／定量	6,389
determinations	決定／定量	626

複数形のdeterminationsの割合は約10%あるが，単数形は無冠詞で使われることが圧倒的に多い．

◆類義語：decision, quantitation, quantification

■ determination ＋［前］

	the%	a/an%	ø%			
❶	45	5	49	determination of ～	～の決定／～の定量	3,805
❷	1	0	95	determination in ～	～における決定	229

❶ Total RNA was isolated from HGF for determination of iNOS mRNA levels.（J Periodontol. 2002 73:392）

訳 全RNAが，iNOSメッセンジャーRNAレベルの定量のためにHGFから単離された

■［名］＋ determination

❶ sex determination	性決定	592
❷ structure determination	構造決定	497
❸ fate determination	運命決定	368

★ determine 動 決定する／測定する 用例数 91,382

determined	決定される／～を決定した	43,140
determine	～を決定する	37,762
determining	～を決定する	8,130
determines	～を決定する	2,350

他動詞．

◆類義語：define, decide, characterise

■ determine ＋［名句］

❶ determine the effect of ～	…は，～の効果を決定する	1,621
❷ determine the role of ～	…は，～の役割を決定する	1,342
❸ determine the structure of ～	…は，～の構造を決定する	483

D

❹ determine the extent	…は，範囲を決定する	418
❺ determine the mechanism	…は，機構を決定する	380
❻ determine the contribution of ～	…は，～の寄与を決定する	262
❼ determine the relationship between ～	…は，～の間の関連性を決定する	233
❽ determine the prevalence of ～	…は，～の罹患率を決定する	206
❾ determine the impact of ～	…は，～の影響を決定する	203

❷ Experiments were performed to determine the role of Rho/Rho kinase signaling in this process.（Circ Res. 2004 94:1367）

訳 実験が，～の役割を決定するために実行された

■ determine ＋［名節］

❶ determine whether ～	…は，～かどうかを決定する	11,924
❷ determine if ～	…は，～かどうかを決定する	2,831
❸ determine how ～	…は，どのように～かを決定する	1,144
❹ determine which ～	…は，どの～かを決定する	1,006

❶ Further studies are required to determine whether RYK modulates the signaling of EphB2 and EphB3.（J Biol Chem. 2002 277:23037）

訳 さらなる研究が，～かどうかを決定するために必要とされる

■ to determine

★ to determine ～	～を決定するために／～を決定すること	33,008
❶ To determine ～	～を決定するために	8,453
❷ be to determine ～	…は，～を決定することである	4,039
・the purpose of this study was to determine ～	この研究の目的は，～を決定することであった	912
❸ be used to determine ～	…は，～を決定するために使われる	2,307
❹ sought to determine ～	…は，～を決定しようとした	1,865
・we sought to determine ～	われわれは，～を決定しようとした	1,407
❺ be performed to determine ～	…は，～を決定するために実行される	556
❻ be needed to determine ～	…は，～を決定するために必要とされる	432
❼ be designed to determine ～	…は，～を決定するために設計される	424
❽ be undertaken to determine ～	…は，～を決定するために着手される	383
❾ be conducted to determine ～	…は，～を決定するために行われる	357

❶ To determine whether L9 plays a role in virus assembly, small interfering RNA（siRNA）-mediated knockdown was performed.（J Virol. 2013 87:1069）

訳 ～かどうかを決定するために

❷ The purpose of this study was to determine the role of the NFATC3 isoform in skeletal muscle development.（Dev Biol. 2001 232:115）

訳 この研究の目的は，～の役割を決定することであった

❸ Southern blot analysis was used to determine the rNKp30 gene copy number.（Transplantation. 2004 77:121）

訳 サザンブロット解析が，～を決定するために使われた

❹ In this study, we sought to determine the intracellular mechanisms of the magnitude-dependent actions of mechanical signals.（Arthritis Rheum. 2004 50:3541）

訳 われわれは，～の細胞内機構を決定しようとした

❼ This study was designed to determine the effects and downstream mechanisms of CO on cold I/R injury in a clinically relevant isolated perfusion rat liver model.（Hepatology. 2002 35:815）

訳 この研究は，影響を決定するために設計された

■ ［代名／名］＋（have）determined

❶ we determined ～	われわれは，～を決定した	5,150
・we determined that ～	われわれは，～ということを決定した	1,485
・we determined whether ～	われわれは，～かどうかを決定した	557
❷ we have determined ～	われわれは，～を決定した	1,970
・we have determined the crystal structure of ～	われわれは，～の結晶構造を決定した	275
❸ study determined ～	研究は，～を決定した	211

D

❶ Using this correlation, we determined that electrostatics and hydrogen bond polarizability play key roles in controlling redox-modulated molecular recognition.（J Am Chem Soc. 2003 125:7882）

訳 われわれは，～ということを決定した

■ be determined ＋［前］/using

★ be determined	…は，決定される	21,791
❶ be determined by ～	…は，～によって決定される	6,894
・be determined by using ～	…は，～を使うことによって決定される	228
・be determined by measuring ～	…は，～を測定することによって決定される	164
❷ be determined in ～	…は，～において決定される	1,624
❸ be determined using ～	…は，～を使って決定される	1,185
❹ be determined to be ～	…は，～であることが決定される	998
❺ be determined for ～	…は，～に対して決定される	985
❻ be determined at ～	…は，～において決定される	787
❼ be determined from ～	…は，～から決定される	705

❶ Drug susceptibility was determined by using two assays and demonstrated a biofilm-associated drug resistance phenotype.（Infect Immun. 2004 72:6023）

訳 薬物感受性が，2つのアッセイを使うことによって決定された

❷ The distribution of Kir7.1 protein was determined in frozen sections of bovine retina-RPE-choroid by indirect immunofluorescence analysis.（Invest Ophthalmol Vis Sci. 2003 44:3178）

訳 Kir7.1タンパク質の分布が，～の凍結切片において決定された

❹ The amino acid sequence of this peptide was determined to be identical to that of the diazepam-binding inhibitor (DBI).（J Clin Invest. 2000 105:351）

訳 このペプチドのアミノ酸配列が，～のそれと同一であることが決定された

❺ Values of K_i for the inhibitor trichostatin A were determined for each isozyme.（Biochemistry. 2004 43:11083）

訳 抑制剤トリコスタチンAのK_i値が，おのおののアイソザイムに対して決定された

❼ HSV1-sr39TK enzyme activity was determined from liver samples and compared with model parameter estimates.（J Nucl Med. 2004 45:1560）

訳 HSV1-sr39TK酵素活性が，肝臓サンプルから決定された

■ as determined

★ as determined	決定されたように	3,552
❶ as determined by ～	～によって決定されたように	3,204

❶ Analysis of IFN-treated cells demonstrated no defect in viral protein synthesis but did show a decrease in the level of released virus, as determined by immunoblot assays.（J Virol. 2003 77:13389）

訳 免疫ブロットアッセイによって決定されたように

■ [副] ＋ determined

❶ previously determined	以前に決定された	523
❷ experimentally determined	実験的に決定された	455

❶ Three hydrophobic residues are involved in docking, and the previously determined NMR structure indicates that these residues are clustered on the surface of the Ets-1 PNT domain.（Genes Dev. 2002 16:127）

訳 以前に決定されたNMR構造は，～ということを示す

■ to be determined

★ to be determined	決定される…	1,241
❶ remain to be determined	…は，決定されないままである	695
❷ have yet to be determined	…は，まだ，決定されねばならない	163

❶ Whether such an interaction is potentially beneficial or harmful remains to be determined.（Am J Clin Nutr. 2004 80:143）

訳 ～は，決定されないままである

❷ Although the NS1 protein is the most abundant viral protein synthesized in infected cells, its function has yet to be determined.（J Virol. 2004 78:6649）

訳 その機能は，まだ，決定されねばならない

■ [前] ＋ determining

❶ important in determining ～	～を決定する際に重要である	239
❷ method for determining ～	～を決定するための方法	180

★★ develop 動 開発する／発症する／発達する 用例数 67,847

developed	開発される／～を開発した／～を発症した／発達した	35,361
developing	～を開発する／～を発症する／発達する	15,834
develop	～を開発する／～を発症する／発達する	14,895
develops	～を開発する／～を発症する／発達する	1,757

「開発する/発症する」の意味では他動詞として,「発達する」の意味では自動詞として用いられる.

◆類義語：occur, arise, exploit

■ develop ＋［名句］

❶ develop a novel ～	…は，新規の～を開発する	666
❷ develop a new ～	…は，新しい～を開発する	566
❸ develop a method	…は，方法を開発する	513
❹ develop a model	…は，モデルを開発する	316

❶ We developed a novel approach to isolate CTCs from blood via immunomagnetic enrichment followed by fluorescence-activated cell sorting (IE-FACS). (Cancer Res. 2013 73:30)

訳 われわれは，～を単離するための新規のアプローチを開発した

❸ We have developed a method for turning on and off the expression of transgenes within Drosophila in both time and space. (Proc Natl Acad Sci USA. 2001 98:12602)

訳 われわれは，～のための方法を開発した

■ ［代名／名］＋ develop

❶ we developed ～	われわれは，～を開発した	4,620
❷ we have developed ～	われわれは，～を開発した	3,995
❸ mice developed ～	マウスが～を発症した／マウスが発達した	1,331
❹ patients developed ～	患者が，～を発症した	530

❶ C3H/He mice developed severe arthritis at all infectious doses, with 100% infection requiring 200 spirochetes. (Infect Immun. 1998 66:161)

訳 C3H/Heマウスは，重篤な関節炎を発症した

■ to develop

★ to develop ～	～を開発（発症）する…	1,241
❶ be to develop ～	…は，～を開発することである	459
❷ fail to develop ～	…は，～を開発することができない	369
❸ be used to develop ～	…は，～を開発するために使われる	332
❹ To develop ～	～を開発するために	310
❺ need to develop ～	～を開発する必要性	188
❻ efforts to develop ～	～を開発する努力	175
❼ be more likely to develop ～	…は，より～を発症しそうである	103

❶ The aim of this study is to develop an experimental system to evaluate the pharmacodynamics of dual phage-drug therapy. (PLoS One. 2012 7:e51017)

訳 この研究の目的は，～を評価する実験システムを開発することである

■ be developed ＋［前／that節］

★ be developed	…は，開発される	9,634

D

❶ be developed to *do*	…は，～するために開発される	2,272
❷ be developed for ～	…は，～のために開発される	1,463
❸ be developed that ～	～する…が開発される	605
❹ be developed as ～	…は，～として開発される	406
❺ be developed in ～	…は，～において開発される	343

❶ A rapid reversed-phase HPLC method was developed to directly measure log k'_w of eight local anesthetics.（Anal Biochem. 2001 292:102）
訳 急速逆相HPLC法が，～を直接測定するために開発された

❷ A new rat model of heterotopic heart and lung transplantation has been developed for MRI experiments.（Circulation. 2001 104:934）
訳 ～が，MRI実験のために開発された

❸ In 1997, a theoretical model was developed that predicted the existence of an internal, Na^+-driven fluid circulation from the poles to the equator of the lens.（Invest Ophthalmol Vis Sci. 2012 53:7087）
訳 ～の存在を予測する理論モデルが開発された

❺ A replication defective adenovirus vector（ΔE1）containing the human Bcl-2 gene（AdCMVhBcl-2）was developed in our laboratory.（Transplantation. 1999 67:775）
訳 ～が，われわれの研究室において開発された

■［副］＋ developed

❶ recently developed ～	最近開発された～	913
❷ newly developed ～	新しく開発された～	568
❸ previously developed ～	以前に開発された～	298

❶ Herein we report a comprehensive analysis of substitution rates in 50 RNA viruses using a recently developed maximum likelihood phylogenetic method.（J Mol Evol. 2002 54:156）
訳 最近開発された最尤系統学的手法を使って

■［名］＋ be developed

❶ assay was developed	アッセイが，開発された	282
❷ method was developed	方法が，開発された	262
❸ model was developed	モデルが，開発された	243

❶ A real-time PCR assay was developed to identify common staphylococcal species.（J Clin Microbiol. 2005 43:2876）
訳 リアルタイムPCRアッセイが，～を同定するために開発された

■ develop ＋［前］

❶ develop in ～	…は，～において発達する	625
❷ develop into ～	…は，～に発達する	304
❸ develop from ～	…は，～から発達する	212

❶ The majority of T cells develop in the thymus and exhibit well characterized phenotypic changes associated with their maturation.（Proc Natl Acad Sci USA. 1998 95:9459）
訳 T細胞の大部分は，胸腺において発達する

■ [前] + developing

❶	risk of developing ～	～を発症するリスク	1,101
❷	risk for developing ～	～を発症するリスク	309
❸	expressed in developing ～	発生中の～において発現する	155

❶ Our data suggest that long-erm exercise decreases the atherogenic activity of blood mononuclear cells in **persons at risk of developing ischemic heart disease.** (JAMA. 1999 281:1722)

訳 虚血性心疾患を発症するリスクのある人々

❸ Immunohistochemical analysis demonstrates that **Alk8 is expressed in developing zebrafish and mouse teeth.** (J Dent Res. 2001 80:1968)

訳 Alk8は，発生中のゼブラフィッシュとマウスの歯において発現する

D

★★ development 名 発生／発達／発症／開発 用例数 89,623

development	発生／発症／開発	87,775
developments	発生／発症／開発	1,848

複数形のdevelopmentsの割合は2％しかなく，原則，**不可算名詞**として使われる．development ofの用例が非常に多い．

◆類義語：onset, growth, exploitation

■ development + [前]

development ofの前にtheが付く割合は非常に高い．

	the%	a/an%	ø%			
❶	79	0	21	development of ～	～の発生／～の開発	35,913
				・the development of ～ in …	…における～の発生	1,547
				・the development of ～ for …	…のための～の開発	829
❷	0	0	100	development in ～	～における発生	3,658

❶ **The development of adhesions in the peritoneal and pelvic cavities,** which commonly form after surgery or infection, cause significant morbidity and mortality. (J Exp Med. 2002 195:1471)

訳 腹腔と骨盤腔における癒着の発生

❷ However, **the development of HSV vectors for the central nervous system** that exploit these properties has been problematical. (J Virol. 2001 75:4343)

訳 中枢神経系のためのHSVベクターの開発

❷ To directly address the role of the Pb99 protein in lymphoid development, Pb99-deficient mice were generated by gene targeting, and **lymphocyte development in these mice was analyzed.** (Mol Cell Biol. 2000 20:4405)

訳 これらのマウスにおけるリンパ球発生が分析された

■ [動] + the development of

❶ prevent the development of ～	…は，～の発生を防ぐ	451
❷ report the development of ～	…は，～の開発を報告する	369
❸ facilitate the development of ～	…は，～の発生（開発）を促進する	309
❹ describe the development of ～	…は，～の発生（開発）について述べる	244

❶ We report here the striking finding that **overexpression of this inhibitor** prevents the development of **cardiomyopathy** in this murine model of heart failure. (Proc Natl Acad Sci USA. 1998 95:7000)

訳 この阻害物質の過剰発現は，心筋症の発症を防ぐ

■ [前] + the development of

❶ contribute to the development of ～	…は，～の発生に寄与する	850
❷ lead to the development of ～	…は，～の発生につながる	848
❸ associated with the development of ～	～の発生と関連する	552
❹ required for the development of ～	～の発生のために必要とされる	339
❺ involved in the development of ～	～の発生に関与している	289
❻ risk factor for the development of ～	～の発生に対するリスク因子	262
❼ implicated in the development of ～	～の発生に関連づけられる	245
❽ target for the development of ～	～の開発のための標的	217
❾ essential for the development of ～	～の発生のために必須である	211

❶ The common Kd and/or Db alleles of NOD mice contribute to the development of **autoimmune diabetes**, but their respective contributions are unresolved. (Diabetes. 2000 49:131)

訳 ～は，自己免疫性糖尿病の発症に寄与する

■ [前] + development

❶ during development	発生の間に	3,806
❷ in development	発生において	2,126

❶ During development, CPECs differentiate from preneurogenic neuroepithelial cells and require bone morphogenetic protein (BMP) signaling, but whether BMPs suffice for CPEC induction is unknown. (J Neurosci. 2012 32:15934)

訳 発生の間に，

■ [形／名] + development

❶ embryonic development	胚発生	544
❷ normal development	正常な発生	334
❸ tumor development	腫瘍発生	229
❹ early development	初期発生	223

★ developmental 形 発生の／発生上の　用例数 14,040

■ developmental ＋［名］

❶ developmental processes	発生の過程	865
❷ developmental stages	発生のステージ	660
❸ developmental defects	発生上の欠陥	447

❶ **Developmental processes in multicellular animals depend on** an array of signal transduction pathways.（Dev Biol. 2002 248:199）

訳 多細胞動物における発生の過程は，～に依存する

❶ NtPSA1 RNA was found to accumulate to varying levels in different parts of the plant and **at different developmental stages**.（Plant Mol Biol. 1999 39:325）

訳 異なる発生のステージにおいて

D

diagnose 動 診断する　用例数 6,119

diagnosed	診断される／～を診断した	5,052
diagnose	～を診断する	605
diagnosing	～を診断する	462

他動詞．

■ diagnosed ＋［前］

❶ diagnosed with ～	～であると診断される	1,322
・be diagnosed with ～	…が，～であると診断される	447
❷ diagnosed in ～	～において診断される	446
❸ diagnosed as ～	～であると診断される	274
・diagnosed as having ～	～を持つと診断される	152

❶ **Twenty-one（5%）patients were diagnosed with cirrhosis**, and 13（3. 1%）developed liver-related complications, including 1 requiring transplantation and 2 developing hepatocellular carcinoma.（Gastroenterology. 2005 129:113）

訳 21人（5％）の患者が，肝硬変であると診断された

❷ **Severe sepsis was diagnosed in 34 of 343 patients**（9.9%）.（Crit Care Med. 2005 33:2521）

訳 重篤な敗血症が，343人の患者のうち34人において診断された

❸ **Adult patients diagnosed as having schizophrenia were enrolled** from 56 sites in the United States, including academic medical centers and community providers.（Arch Gen Psychiatry. 2006 63:490）

訳 統合失調症にかかっていると診断された大人の患者が登録された

■［副］＋ diagnosed

❶ newly diagnosed	新たに診断された	976
・with newly diagnosed ～	新たに診断された～の	423

❶ A total of 231 patients with newly diagnosed adenocarcinoma of the exocrine pancreas were compared with 388 general population controls. (Am J Epidemiol. 2005 162:222)
訳 合計231人の新たに診断された腺癌の患者

■ [名] ＋ diagnosed

❶ patients diagnosed	診断された患者	385
・patients diagnosed with ～	～であると診断された患者	248

❶ A total of 404 patients diagnosed with hematologic malignancies received a total body irradiation-based myeloablative conditioning regimen. (Blood. 2005 106:3308)
訳 血液悪性腫瘍であると診断された合計404人の患者

★ diagnosis 名 診断 用例数 15,883

diagnosis	診断	14,199
diagnoses	診断	1,684

複数形のdiagnosesの割合が，約10%ある可算名詞．
◆類義語：assessment

■ diagnosis ＋ [前]

	the%	a/an%	ø%			
❶	44	22	34	diagnosis of ～	～の診断	4,930
❷	26	17	47	diagnosis in ～	～における診断	303

❶ In contrast, we describe the diagnosis of PTDM by home glucometer monitoring. (Transplantation. 2005 80:775)
訳 われわれは，～によるPTDMの診断を述べる

■ [前] ＋ diagnosis

❶ age at diagnosis	診断時の年齢	361
❷ time of diagnosis	診断時	253

❶ Age at diagnosis is a highly significant prognostic factor for survival of children with RTK. (J Clin Oncol. 2005 23:7641)
訳 診断時の年齢は，高度に重要な予後因子である

❷ The mean trough drug concentrations at the time of diagnosis in affected recipients were within our putative target ranges. (Transplantation. 2005 80:749)
訳 移植患者における診断時の平均トラフ薬物濃度

■ [形] ＋ diagnosis

❶ clinical diagnosis	臨床診断	474
❷ early diagnosis	早期診断	352
❸ differential diagnosis	鑑別診断	293

■ diagnosis and ＋［名］

❶ diagnosis and treatment	診断と治療	669
❷ diagnosis and management	診断と管理	308

die 動 死亡する／死ぬ　用例数 7,272

died	死亡した	3,863
die	死亡する	2,350
dying	死亡する	991
dies	死亡する	68

自動詞.

◆類義語：kill

■ die ＋［前］

❶ die of ～	…は，～で死亡する	774
❷ die in ～	…は，～において死亡する	471
❸ die from ～	…は，～で死亡する	412
❹ die within ～	…は，～内に死亡する	384
❺ die at ～	…は，～において死亡する	384
❻ die during ～	…は，～の間に死亡する	315
❼ die by ～	…は，～によって死亡する	262

❶ Three patients died of liver failure while awaiting a transplant, and four patients died after the transplant.（Ann Surg. 2000 232:340）

訳 3人の患者が，肝不全で死亡した

❸ Two patients died from therapy-related complications during induction treatment.（Lancet Oncol. 2005 6:649）

訳 2人の患者が，治療に関連する合併症で死亡した

❹ Homozygous mice died within 12 h of birth.（Proc Natl Acad Sci USA. 2004 101:8485）

訳 ホモ接合体マウスは，出生の12時間以内に死亡した

❺ Tetraploid（4n）mouse embryos die at variable developmental stages.（Dev Biol. 2005 288:150）

訳 四倍体（4n）マウス胚は，さまざまな発生のステージにおいて死亡する

■［名］＋ die

❶ patients die	患者が，死亡する	544
❷ mice die	マウスが，死亡する	446
❸ cells die	細胞が，死亡する	213

■ to die

★ to die	死亡する…	379

❶ more likely to die	より死亡しそうである	78

❶ Among the subjects 1-20 years of age, **females were 60% more likely to die than males** (relative risk = 1.6, 95% confidence interval 1.4-1.8). (Am J Epidemiol. 1997 145:794)
訳 女性は男性より60%多く死亡しそうであった

★ differ 動 異なる　用例数 18,178

differ	異なる	9,715
differed	異なった	3,478
differs	異なる	2,857
differing	異なる	2,128

自動詞．副詞は，通常，differの後に置かれる．
◆類義語：vary

■ differ ＋［前］

❶ differ from ～	…は，～と異なる	3,976
❷ differ in ～	…は，～において異なる	3,730
❸ differ between ～	…は，～の間で異なる	1,628
❹ differ by ～	…は，～だけ異なる	1,172

❶ However, **the mechanism of Apc inactivation in AKR Min/+ mice often differs from that observed for B6 Min/+ mice.** (Proc Natl Acad Sci USA. 1998 95:10826)
訳 AKR Min/+マウスにおけるApc不活性化の機構は，B6 Min/+マウスに観察されるそれとしばしば異なる

❷ We have previously identified two broad electrophysiological classes of **spiral ganglion neuron that differ in their rate of accommodation.** (J Physiol. 2002 542:763)
訳 それらの適応の割合において異なるラセン神経節ニューロン

❸ Spermatophore and sperm numbers **did not differ between males presented with a previous mate and males paired with a new female.** (Proc Natl Acad Sci USA. 1999 96:10236)
訳 ～は，以前の連れ合いと一緒にいるオスと新しいメスとペアにしたオスの間で異ならなかった

❹ **The structures of two isoforms of Bcl-2 that differ by two amino acids** have been determined by NMR spectroscopy. (Proc Natl Acad Sci USA. 2001 98:3012)
訳 2アミノ酸だけ異なるBcl-2の２つのアイソフォームの構造

■ differ ＋［副］

❶ differ significantly	…は，～の間で有意に異なる	1,775
・did not differ significantly	…は，有意には異ならなかった	778
・differ significantly from ～	…は，～と有意に異なる	531
・differ significantly between ～	…は，～の間で有意に異なる	455
❷ differ markedly	…は，顕著に異なる	311
❸ differ substantially	…は，実質的に異なる	269

❶ The frequency of commonly reported adverse events **did not differ significantly between**

groups.（Gastroenterology. 2001 121:268）
訳 ～は，グループの間で有意には異ならなかった

★★ difference （名）違い／差 用例数 55,967

differences	違い／差	37,719
difference	違い／差	18,248

複数形のdifferencesの割合は，約65％と圧倒的に高い．

◆類義語：divergence, disparity, disagreement, distinction, variation, discrimination

■ differences ＋［前］

	the％	a/an％	ø％			
❶	8	-	80	differences in ～	～の違い（差）／～における違い（差）	19,982
				・significant differences in ～	～の有意な（顕著な）差	2,088
				・no significant differences in ～	～の有意な（顕著な）違いのない	958
				・statistically significant difference in ～	～の統計的に有意な違い	203
				・no difference in ～	～の違いのない	1,577
				・sex differences in ～	～の性差	462
				・…-specific differences in ～	～の…特異的な違い	457
❷	22	-	70	differences between ～	～の間の違い（差）	5,872
❸	9	-	79	differences among ～	～の間の違い（差）	729

❶ There were no significant differences in the incidence of rejection among patients stratified as with or without CCR5-Δ32 or by the CX3CR1-V249I or CX3CR1-T280M genotypes.（J Am Soc Nephrol. 2002 13:754）
訳 患者の間での拒絶の発生率の有意な差はなかった

❶ There was no difference in capillary blood volume between end diastole and end systole at baseline.（Circulation. 2002 105:218）
訳 ～の間に毛細血管血液量の差はなかった

❶ We have investigated the basis for the developmental stage-specific differences in the function of these two proteins.（Genetics. 2000 155:1643）
訳 これら２つのタンパク質の機能の発生ステージ特異的な違い

❷ There were no significant differences between vegetarians and nonvegetarians in mortality from cerebrovascular disease, stomach cancer, colorectal cancer, lung cancer, breast cancer, prostate cancer, or all other causes combined.（Am J Clin Nutr. 1999 70:516S）
訳 ～において菜食主義者と非菜食主義者の間に有意な差はなかった

❸ Precise assessment of these two traits allowed distinction of small but significant differences among genotypes.（Proc Natl Acad Sci USA. 1993 90:7869）
訳 遺伝子型の間の小さいが有意な違いの区別

D

■ differences ＋［動］

❶ differences were observed	違いが，観察された	582
❷ differences were found	違いが，見つけられた	582
❸ differences exist	違いが，存在する	261

❶ No significant **differences were observed** between PBS-treated and control groups.（Brain Res. 2001 913:133）
訳 ～の間に有意な差は観察されなかった

❷ No group **differences were found** in either the metabolism or the volume of the amygdala or the hippocampus.（Am J Psychiatry. 2000 157:1994）
訳 ～にグループ差は見つけられなかった

❸ However, **differences exist** in the location of two loops outside of the respective binding sites containing residues 114-125 and 222-227.（J Biol Chem. 1998 273:32818）
訳 違いが２つのループの場所に存在する

★★ different 形 異なる 用例数 79,733

◆類義語：distinct, discrete, dissimilar, disparate, separate

■ different ＋［前］

❶ different from ～	～と異なる	4,265
・be different from ～	…は，～と異なる	1,029
❷ different between ～	～の間で異なる	1,047
・be not different between ～	…は，～の間で異ならない	295
❸ different in ～	～において異なる	983
❹ different for ～	～に対して異なる	314

❶ The K_m of the W354F and W354A mutants were not significantly **different from** that of the wild-type.（Eur J Biochem. 1999 259:809）
訳 ～は，野生型のそれと有意には異ならなかった

❷ Blood-glycosylated hemoglobin concentrations were not significantly **different between** the two treatment groups.（Am J Pathol. 2000 157:2143）
訳 ～は，２つの治療群の間で有意には異ならなかった

❸ In these sensor kinases, the signal-sensing domains are vastly **different in** size and subdomain composition, with little apparent conservation between species, whereas the catalytic domains of these sensor kinases retain the high level of homology observed for the other phosphorelay proteins.（Mol Microbiol. 2002 46:297）
訳 ～は，大きさにおいて大いに異なる

■ ［副］＋ different

❶ significantly different	有意に（顕著に）異なる	2,910
・be not significantly different	…は，有意には異ならない	1,524
❷ very different	非常に異なる	1,055
❸ distinctly different	はっきりと異なる	407

❹ markedly different	著しく異なる	401

❷ This study shows that the Ly49 repertoire and its MHC-binding characteristics **can be very different among inbred mouse strains.** (J Immunol. 2001 166:5034)
訳 …は，マウス近交系の間で非常に異なりうる

❸ However, **this stimulation occurs in a manner distinctly different from** that observed with cognate E.coli SSB. (Nucleic Acids Res. 2002 30:2809)
訳 この刺激は，～とはっきりと異なる様式で起こる

■ [形] + different

❶ two different ～	2つの異なる～	4,311
❷ three different ～	3つの異なる～	2,308
❸ several different ～	いくつかの異なる～	1,178

D

❶ These results suggest that the activity of this casein kinase II protein kinase may be regulated by **the phosphorylation state of two different sites** in its multimeric structure. (Plant Physiol. 1993 103:955)
訳 2つの異なる部位のリン酸化状態

■ different + [名]

❶ different types of ～	異なるタイプの～	1,480
❷ different mechanisms	異なる機構	1,006
❸ different cell types	異なる細胞タイプ	688
❹ different species	異なる種	684
❺ different stages of ～	異なるステージの～	664
❻ different levels of ～	異なるレベルの～	599
❼ different regions of ～	異なる～の領域	596
❽ different classes of ～	異なるクラスの～	424

❶ **Eukaryotic nuclei contain three different types of RNA polymerases** (RNAPs), each consisting of 12-18 different subunits. (Proc Natl Acad Sci USA. 2000 97:6306)
訳 真核生物の核は，3つの異なるタイプのRNAポリメラーゼを含む

★ differential 形 差動的な／異なる，名 差　用例数 12,237

differential	差動的な／異なる／差	12,179
differentials	差	58

名詞としても用いられるが，形容詞の用例が圧倒的に多い．
◆類義語：different, distinct

■ differential + [名]

❶ differential expression	差動的な発現	1,349
❷ differential effects	差動的な効果	533
❸ differential regulation of ～	～の差動的な調節	426

❹ differential gene expression	差動的な遺伝子発現	307
❺ differential diagnosis	鑑別診断	293

❶ We also provide evidence that these differences in responsiveness **are modulated, at least in part, by** differential expression **of Hox genes** within the neural crest.
訳 ～は，Hox遺伝子の差動的な発現によって，少なくとも一部は，調節される

★ differentially 副 差動的に／異なって 用例数 7,200

◆類義語：differently

■ differentially ＋［過分］

❶ differentially expressed	差動的に発現される	2,325
❷ differentially regulated	差動的に調節される	783
・be differentially regulated	…は，差動的に調節される	560
❸ differentially methylated	差動的にメチル化される	211

❶ **The** differentially expressed **genes identified in this study** should be informative in studying oral epithelial cell carcinogenesis.（Oncogene. 1998 16:1921）
訳 この研究において同定された差動的に発現される遺伝子

❷ RNA transcript analysis determined that **these newly identified invasion loci** were differentially regulated **at the transcriptional level.**（Mol Microbiol. 2000 36:174）
訳 これらの新たに同定された侵入遺伝子座は，転写レベルで差動的に調節された

★ differentiate 動 分化する／分化させる／区別する 用例数 13,002

differentiated	分化した	6,522
differentiate	分化する／区別する	4,283
differentiating	分化する	1,863
differentiates	分化する／区別する	334

自動詞・他動詞の両方で使われるが，自動詞の用例が多い．他動詞の用例としては，過去分詞が多い．

◆類義語：distinguish, discriminate

■ differentiate ＋［前］

❶ differentiate into ～	…は，～へ分化する	2,055
❷ differentiate between ～	…は，～の間を区別する	479
❸ differentiate in ～	…は，～において分化する	340
❹ differentiate from ～	…は，～から分化する	289

❶ In the presence of exogenously supplied BDNF, **the grafted cells** differentiate into **both neurons and glia.**（Brain Res. 2000 874:87）
訳 移植された細胞は，ニューロンとグリアの両方に分化する

❷ **To** differentiate between **the physiological functions of these receptors,** selective antagonists are needed.（Biochemistry. 2002 41:6383）

訳 これらの受容体の間の生理学的な機能を区別するために

■ [副] + differentiated

❶	terminally differentiated ～	最終分化した～	528
❷	well differentiated ～	よく分化した～	316
❸	poorly differentiated	あまり分化していない	307
❹	fully differentiated ～	完全に分化した～	196

❶ Active genes are insulated from developmentally regulated **chromatin condensation in terminally differentiated cells.** (Mol Cell Biol. 2003 23:6455)

訳 最終分化した細胞におけるクロマチンの凝縮

❷ **A well differentiated human breast cancer** (BT20, n = 8) and a highly invasive metastatic human breast cancer (DU4475, n = 8) were implanted orthotopically in athymic nude mice. (Radiology. 2002 222:814)

訳 よく分化したヒト乳癌

❸ **Thirty percent of tumors were poorly differentiated.** (Ann Surg. 2002 235:814)

訳 腫瘍の30%は，あまり分化していなかった

■ differentiated + [名]

❶	differentiated cells	分化した細胞	594
❷	differentiated neurons	分化したニューロン	121

D

☆ differentiation 名 分化／識別 用例数 36,573

differentiation	分化	36,556
differentiations	分化	17

複数形のdifferentiationsの用例はほとんどなく，原則，**不可算名詞**として使われる．

◆類義語：specialization ◆反意語：dedifferentiation

■ differentiation + [前]

	the%	a/an%	ø%			
❶	42	0	56	differentiation of ～	～の分化	6,377
				・in the differentiation of ～	～の分化において	255
❷	0	0	98	differentiation in ～	～における分化	2,147
❸	2	0	77	differentiation into ～	～への分化	618

❶ Furthermore, **we demonstrate unique roles of Phox2 genes in the differentiation of specific motor neurons.** (EMBO J. 2005 24:4392)

訳 われわれは，特異的な運動ニューロンの分化におけるPhox2遺伝子のユニークな役割を実証する

❷ Both RC2.E10 cells and primary cortical precursor cells **undergo astroglial differentiation in response to cAMP stimulation** by treatment with 8-bromo-cAMP. (J Neurosci. 1999 19:9004)

訳 ～は，cAMP刺激に応答してアストログリアへの分化を起こす

❸ In PC12 cells, activated Raf and PI3 kinases **mediate Ras-induced cell cycle arrest and** differentiation into a neuronal phenotype.（Mol Cell Biol. 1999 19:1731）

訳 ～は，Rasに誘導される細胞周期停止とニューロン表現型への分化を仲介する

■［形／名］＋ differentiation

❶ cell differentiation	細胞分化	3,570
❷ terminal differentiation	最終分化	1,129
❸ neuronal differentiation	神経分化	944
❹ cellular differentiation	細胞分化	577

❷ Blimp-1 is a transcriptional repressor that is both required and sufficient to trigger terminal differentiation of B lymphocytes and monocyte/macrophages.（Nucleic Acids Res. 2000 28:4846）

訳 Blimp-1は，Bリンパ球と単球/マクロファージの最終分化を引き起こすのに必要でありかつ十分な転写抑制因子である

■［名］＋ and differentiation

❶ proliferation and differentiation	増殖と分化	1,174
❷ growth and differentiation	増殖と分化	839

differently 副 異なって／別々に　用例数 1,614

◆類義語：separately, independently, individually

■［動］＋ differently

❶ respond differently to ～	…は，～に異なって応答する	149
❷ be regulated differently	…は，異なって調節される	94
❸ behave differently	…は，異なって振る舞う	83

❶ This finding suggests that neurons in these structures are designed to respond differently to excitatory input.（J Comp Neurol. 1998 402:75）

訳 これらの構造におけるニューロンは，興奮性の入力に対して異なって応答するように設計されている

❷ Thus, induction of GL epsilon transcripts in mice and humans may be regulated differently.（J Immunol. 2001 166:411）

訳 ～は，異なって調節されるかもしれない

difficult 形 難しい／困難な　用例数 6,579

difficult to *do*の用例が非常に多い．to *do*はitで受けることが多く，その場合は「～することは困難である」という意味になる．

◆類義語：hard, intractable

■ difficult ＋［前］

❶ difficult to *do*	～することは困難である	4,056
・it is difficult to *do*	～することは困難である	323
・difficult to determine ～	～を決定することは困難である	194
・difficult to study ～	～を研究することは困難である	188
・difficult to identify ～	～を同定することは困難である	176
・difficult to detect ～	～を検出することは困難である	150
❷ difficult for ～	～にとって困難である	101
・difficult for ～ to *do*	～が…するのは困難である	41

❶ It is difficult to determine what fraction of these exposures represents caustic ingestions.（Curr Opin Pediatr. 2009 21:651）

訳 ～を決定することは困難である

❷ We found that it is more difficult for rats to detect LS than HS targets.（J Neurosci. 2012 32: 15577）

訳 ラットが～を検出するのはより困難である

■ ［動］＋ difficult

❶ be difficult	…は，困難である	3,121
❷ prove difficult	…は，困難であると判明する	242
❸ remain difficult	…は，困難なままである	131

❷ Because of its size, PC1 has proven difficult to handle biochemically, and structural information is consequently sparse.（Biochemistry. 2012 51:2879）

訳 PC1は，生化学的に扱うのが困難であると判明した

■ make it difficult to *do*

❶ make it difficult to *do*	…は，～することを困難にする	409

❶ Uncertainty of how HDACI-induced protein acetylation leads to cell death, however, makes it difficult to determine which tumors are likely to be responsive to these agents.（Proc Natl Acad Sci USA. 2005 102:4842）

訳 …は，～を決定することを困難にする

■ ［副］＋ difficult

❶ more difficult	より困難である	392
❷ often difficult	しばしば困難である	145
❸ very difficult	非常に困難である	119

difficulty 名 困難／障害 用例数 3,066

difficulty	困難／障害	1,651
difficulties	困難／障害	1,415

D

複数形のdifficultiesの割合は約45%と高いが，単数形は無冠詞で使われることも多い．
◆類義語：impairment, disturbance, distress, damage

■ difficulty ＋［前］

difficulty ofの前にtheが付く割合は圧倒的に高い．

	the%	a/an%	ø%			
❶	57	7	33	difficulty in ～	～の際の困難	485
				・difficulty in ～ing	～する際の困難	385
❷	96	0	3	difficulty of ～	～の困難	338

❶ The **difficulty in** creating this vaccine arises from the enormous genetic variation of the virus and the unusual importance of cytotoxic T lymphocytes (CTL) in controlling its spread. (Annu Rev Med. 2005 56:213)

訳 このワクチンを作る際の困難は，そのウイルスの莫大な遺伝的な変動から生じる

❷ These findings articulate the **difficulty of** eliminating or reducing the pathogen from plants. (Annu Rev Phytopathol. 2012;50:241)

訳 これらの知見は，～を除去するあるいは減らす困難を明確にする

diminish 動 減少させる／小さくする　用例数 9,589

diminished	減少させられる／減少した／～を減少させた	7,502
diminish	～を減少させる	848
diminishes	～を減少させる	847
diminishing	～を減少させる	392

他動詞．
◆類義語：decrease, reduce, lower, down-regulate, decline

■ be diminished ＋［前］

★	be diminished	…は，減少する／減らされる	1,236
❶	be diminished in ～	…は，～において減少する	399
❷	be diminished by ～	…は，～によって減らされる	246

❶ Conversely, CI1033 accumulation **was diminished in** cells expressing BCRP, suggesting that CI1033 is a substrate for this efflux pump. (Cancer Res. 2001 61:739)

訳 CI1033の蓄積が，BCRPを発現する細胞において減少した

❷ Increased iNOS protein expression observed in CCPA-treated hearts **was diminished by** DPCPX. (Circulation. 2000 102:902)

訳 ～が，DPCPXによって減らされた

■［副］＋ diminished

❶	significantly diminished	有意に減らされる	357
❷	greatly diminished	大きく減らされる	289
❸	markedly diminished	顕著に減らされる	264

❷ The induction of TIMP-1 mRNA by ATRA and bFGF **was greatly diminished by cycloheximide** and therefore required new protein synthesis.（Eur J Biochem. 2000 267:4150）
訳 〜は，シクロヘキシミドによって大きく減らされた

■ ［前］＋ diminished

❶ result in diminished 〜	…は，減少した〜という結果になる	224
❷ associated with diminished 〜	減少した〜と関連する	109

❷ Thus, **GS susceptibility is associated with diminished ER expression** in MCs.（Am J Pathol. 2002 160:1877）
訳 GS感受性は，減少したER発現と関連する

■ diminish ＋［名句］

❶ diminish the ability of 〜	…は，〜の能力を減少させる	812

❶ **Mutation of this binding site severely diminished the ability of Sp1 to activate** the Etsrp71 promoter.（Genomics. 2005 85:493）
訳 この結合部位の変異は，〜を活性化するSp1の能力を激しく減少させた

D

★★ direct ㊎ 直接の，㊍ 方向づける／向ける　用例数 52,349

direct	直接の／〜を方向づける	34,107
directed	向けられる／〜を方向づけた	15,406
directs	〜を方向づける	1,468
directing	〜を方向づける	1,368

形容詞の用例が多いが，他動詞として用いられることも多い．
◆類義語：orient

■ ㊎ direct ＋［名］

❶ direct evidence	直接の証拠	2,357
・provide direct evidence	…は，直接の証拠を提供する	1,110
❷ direct interaction	直接の相互作用	1,870
❸ direct effect	直接効果	1,182
❹ direct role	直接的役割	1,042
❺ direct binding	直接結合	876
❻ direct target	直接の標的	492
❼ direct contact	直接接触	450
❽ direct comparison	直接比較	394

❶ **The results provide, for the first time, direct evidence that** the virulence of CNA depends on its collagen-binding ability.（J Infect Dis. 2004 189:2323）
訳 それらの結果は，はじめて，〜という直接の証拠を提供する

❷ **We have identified a direct interaction between** cytoplasmic dynein and kinesin Ⅰ.（J Biol Chem. 2004 279:19201）

訳 われわれは，～の間の直接の相互作用を同定した

❹ Rather, studies revealed **STAT1 plays a direct role in** IFN-γ-inducible GILT expression.（J Immunol. 2004 173:731）

訳 STAT1は，～において直接的役割を果たす

■ 動 directed ＋［前］

❶ directed against ～	～に対して向けられる	1,898
・antibody directed against ～	～に対して向けられた抗体	669
❷ directed to ～	～へ向けられる	719
❸ directed at ～	～に向けられる	633
・directed at ～ing	～することに向けられる	166
❹ directed by ～	～によって方向づけられる	545
❺ directed toward ～	～へ向けられる	491

❶ In this study, **two monoclonal antibodies directed against different metal-chelate complexes** were expressed as recombinant Fab fragments.（Biochemistry. 2003 42:14173）

訳 異なる金属キレート複合体に対して向けられた２つのモノクローナル抗体

❷ **We developed a retroviral system for stable expression of siRNA directed to the unique fusion junction sequence of** TEL-PDGFβR in transformed hematopoietic cells.（J Clin Invest. 2004 113:1784）

訳 われわれは，～のユニークな融合ジャンクション配列へ向けられたsiRNAの安定発現のためのレトロウイルス系を開発した

❸ **Therapies directed at the vascular disease continue to** focus on the alleviation of vascular spasm.（Curr Opin Rheumatol. 2004 16:718）

訳 血管疾患に向けられた治療法は，～し続ける

❹ **This post-transcriptional modification is directed by sequence features present in** the 3'-untranslated region (3'-UTR).（Nucleic Acids Res. 2004 32:3392）

訳 この転写後の修飾は，～に存在する配列の特徴によって方向づけられる

■ 動 direct ＋［名句］

❶ direct the synthesis of ～	…は，～の合成を方向づける	82
❷ direct the expression of ～	…は，～の発現を方向づける	74
❸ direct the formation of ～	…は，～の形成を方向づける	65
❹ direct the assembly of ～	…は，～の集合を方向づける	63

❶ This work demonstrates that the hydrophobic effect **provides a powerful strategy to direct the synthesis of** entwined architectures.（Science. 2012 338:783）

訳 …は，～の合成を方向づける強力な戦略を提供する

direction 名 方向　　用例数 7,322

direction	方向	5,604
directions	方向	1,718

複数形のdirectionsの割合が，約25%ある可算名詞.

■ direction ＋［前］

direction ofの前にtheが付く割合は圧倒的に高い.

	the %	a/an %	ø %			
❶	88	0	10	direction of ～	～の方向	1,772
❷	4	-	92	directions for ～	～の方向性	243
❸	35	43	17	direction in ～	～における方向	153

❶ The effect also depended upon the rate of change in the **direction of** flow. (J Physiol. 2002 540:1023)

訳 その効果は，また，流れの方向の変化の速度に依存した

■ in ＋［形］＋ direction(s)

❶	in opposite directions	逆の方向に	184
❷	in the same direction	同じ方向に	140
❸	in both directions	両方の方向に	123

❶ Accordingly, we predicted a counterintuitive mode of control where the muscles and body must, on average, move **in opposite directions** (paradoxical movements). (J Physiol. 2004 556:683)

訳 ～は，逆の方向に動く

■［形］＋ directions

❶	future directions	将来の方向性	219
❷	new directions	新しい方向性	94

❶ In addition, we present the genomic resources currently available for Mimulus and discuss **future directions** for research. (J Physiol. 2004 556:683)

訳 ～は，研究の将来の方向性を議論する

★★ directly 副 直接に　　用例数 29,271

◆反意語：indirectly

■［動］＋ directly ＋［前］

❶	interact directly with ～	…は，～と直接相互作用する	1,028
❷	bind directly to ～	…は，～に直接結合する	974
❸	act directly on ～	…は，～に対して直接作用する	303
❹	contribute directly to ～	…は，～に直接寄与する	160
❺	participate directly in ～	…は，～に直接関与する	101
❻	correlate directly with ～	…は，～と直接関連する	91

❶ We demonstrate that SpoIIAA **interacts directly with** the σ^F. SpoIIAB complex, greatly decreasing the affinity of SpoIIAB for σ^F and thus causing the release of the latter. (J Mol Biol. 2004 340:203)

訳 SpoⅡAAは，～と直接相互作用する

❷ CC3 binds directly to the karyopherins of the importin β family in a RanGTP-insensitive manner and associates with nucleoporins in vivo. (Mol Cell Biol. 2004 24:7091)

訳 CC3は，～のカリオフェリンに直接結合する

■ directly ＋［動］

❶ directly interact with ～	…は，～と直接相互作用する	763
❷ directly bind	…は，直接結合する	758
・directly bind to ～	…は，～に直接結合する	417
❸ directly regulate ～	…は，～を直接制御する	585

❸ This perturbation in proliferation can be explained by the finding that HMGA2 directly regulates the RNA-binding protein IGF2BP2. (Dev Cell. 2012 23:1176)

訳 HMGA2は，RNA結合タンパク質IGF2BP2を直接制御する

■ be directly ＋［過分／形］

★ be directly ～	…は，直接～される	3,892
❶ be directly involved in ～	…は，～に直接関与している	399
❷ be directly related to ～	…は，～に直接関連する	381
❸ be directly proportional to ～	…は，～に正比例する	173
❹ be directly correlated with ～	…は，～に直接相関する	151

❶ Even for prolonged release, diffusion of FcRn into the plasma membrane can occur, indicating that FcRn is directly involved in IgG exocytosis. (Proc Natl Acad Sci USA. 2004 101:11076)

訳 ～は，IgGのエキソサイトーシスに直接関与している

■ to directly ＋［動］

★ to directly ～	直接～するために	1,760
❶ to directly test ～	～を直接テストするために	136
❷ to directly measure ～	～を直接測定するために	93

❶ To directly test the hypothesis that G_N plays a role in TSWV acquisition by thrips, we expressed and purified a soluble, recombinant form of the G_N protein (G_N-S). (J Virol. 2004 78:13197)

訳 ～という仮説を直接テストするために

■ directly or ＋［副］

❶ directly or indirectly	直接的にあるいは間接的に	657

❶ The absence of IFN-α in Pr4-infected macrophages suggests that MGF360/530 genes either directly or indirectly suppress a type I IFN response. (J Virol. 2004 78:1858)

訳 MGF360/530遺伝子は，直接的にあるいは間接的にtype I IFN応答を抑制する

disassembly 名 分解／解体　用例数 1,894

disassembly	分解／解体	1,893
disassemblies	分解／解体	1

複数形のdisassembliesの用例はほとんどなく，原則，**不可算名詞**として使われる．
◆類義語：finding, identification, observation

■ disassembly ＋［前］

	the%	a/an%	ø%			
❶	41	1	55	disassembly of ～	～の分解	598

❶ Human antisilencing factor 1 (Asf1) is one such factor which **is involved in both the assembly and disassembly of nucleosomes** in cellular systems. (22951827J Virol. 2004 78:1858)
訳 ～は，ヌクレオソームの構築と分解の両方に関与している

■［名］＋ and disassembly

❶	assembly and disassembly	構築と分解	399

D

discover 動 発見する　用例数 7,750

discovered	発見される／～を発見した	6,334
discover	～を発見する	1,023
discovering	～を発見する	365
discovers	～を発見する	28

他動詞．
◆類義語：find, identify

■ discover ＋［that節］

★	discover that ～	…は，～ということを発見した／～ということが発見された	1,694
❶	we discovered that ～	われわれは，～ということを発見した	849
❷	it was discovered that ～	～ということが発見された	103

❶ **We discovered that** the isomerase is a previously characterized protein called Rpe65. (Cell. 2005 122:449)
訳 われわれは，～ということを発見した

■ discover a novel ＋［名］

❶	discover a novel ～	…は，新規の～を発見する	109

❶ These results strongly argue that **we have discovered a novel signaling pathway**, controlled by Cx43, that enhances the generation of T(R) cells. (J Immunol. 2011 187:248)
訳 われわれは，新規のシグナル伝達経路を発見した

■ be discovered ＋［前］

★ be discovered	…は，発見される	1,778
❶ be discovered in ～	…は，～において発見される	338
❷ be discovered to *do*	…は，～することが発見される	105
❸ be discovered by ～	…は，～によって発見される	96

❶ This is the first PDE5 protein to be discovered in animal sperm.（Mol Biol Cell. 2006 17:114）
訳 これは，動物の精子において発見される最初のPDE5タンパク質である

■［副］＋ discovered

❶ recently discovered ～	最近発見された～	855
❷ newly discovered ～	新たに発見された～	493

❶ We also provide initial preclinical data on the antimetastatic efficacy of recently discovered small-molecule antagonists of XIAP.（Cancer Res. 2005 65:2378）
訳 最近発見されたXIAPの小分子拮抗剤

discovery 名 発見　　用例数 8,761

discovery	発見	7,869
discoveries	発見	892

複数形のdiscoveriesの割合は約10%あるが，単数形は無冠詞で使われることが多い．
◆類義語：finding, identification, observation

■ discovery ＋［前／同格that節］

discovery thatの前にtheが付く割合は圧倒的に高い．また，discovery ofの前にtheが付く割合も非常に高い．

	the%	a/an%	ø%			
❶	79	1	16	discovery of ～	～の発見	3,675
				・lead to the discovery of ～	…は，～の発見につながる	402
				・we report the discovery of ～	われわれは，～の発見を報告する	237
				・recent discovery of ～	～の最近の発見	185
❷	83	2	6	discovery that ～	～という発見	567
❸	6	3	66	discovery in ～	～における発見	173

❶ Studies in this model have led to the discovery of many genes that differentiate a resistant from a susceptible plant.（Heredity. 2005 94:507）
訳 このモデルにおける研究が，多くの遺伝子の発見につながってきた

❶ Here, we report the discovery of ExsE, a negative regulator of type Ⅲ secretion gene expression in Pseudomonas aeruginosa.（Proc Natl Acad Sci USA. 2005 102:8006）
訳 われわれは，ExsEの発見を報告する

❷ Significant findings include the discovery that hyperforin is an active ingredient of the herbal remedy St. (Annu Rev Nutr. 2005 25:297)
訳 重要な知見は，～という発見を含む

discriminate 動 識別する　用例数 3,141

discriminate	識別する	1,755
discriminating	識別する	652
discriminated	識別した	476
discriminates	識別する	258

自動詞・他動詞の両方の用例がある．to discriminateの用例が非常に多い．
◈類義語：distinguish, discern, sort, differentiate

■ discriminate ＋［前］

❶ discriminate between ～	…は，～の間を識別する	1,145
❷ discriminate among ～	…は，～の間を識別する	202

❶ At review, two observers attempted to discriminate between tumor recurrence and nonmalignant causes of symptoms. (Radiology. 2000 214:837)
訳 …が，腫瘍の再発と～の間を識別することを試みた

■ discriminate ＋［名］＋ from

❶ discriminate ～ from …	‥は，～を…と識別する	262

❶ A key feature of the immune system is its ability to discriminate self from nonself. (J Clin Invest. 2010 120:3641)
訳 自己を非自己と識別するそれの能力

■ to discriminate

★ to discriminate	識別する…	1,007
❶ ability to discriminate	識別する能力	177

discrimination 名 識別　用例数 4,003

discrimination	識別	3,845
discriminations	識別	158

複数形のdiscriminationsの割合が4％あるが，単数形は無冠詞で使われることが圧倒的に多い．
◈類義語：distinction, identification, difference

■ discrimination ＋［前］

	the%	a/an%	ø%			
❶	36	0	62	discrimination of ～	～の識別	534
❷	12	3	83	discrimination between ～	～の間の識別	398
❸	6	4	86	discrimination in ～	～における識別	150

❶ Behavioral studies indicate that prior experience can influence discrimination of subsequent stimuli.（Neuron. 2002 33:316）

訳 以前の経験が，後の刺激の識別に影響しうる

❷ Discrimination between edible and contaminated foods is crucial for the survival of animals.（Curr Biol. 2004 14:1065）

訳 食用と汚染された食物の間の識別は，動物の生存にとって決定的に重要である

★discuss 動 議論する　用例数 16,636

discussed	議論される／～を議論した	8,963
discuss	～を議論する	6,205
discusses	～を議論する	1,257
discussing	～を議論する	211

他動詞.

◆類義語：argue, debate

■ be discussed.

文末にbe discussed.が来る用例が非常に多い.

★	be discussed	…が，議論される	7,617
❶	be discussed.	…が，議論される	4,686

❶ Implications of this result for the validity of the simulated folding mechanism are discussed.（J Mol Biol. 2003 329:625）

訳 シミュレートされた折り畳み機構の妥当性に対するこの結果の意味が議論される

■ be discussed ＋［前］

❶	be discussed in ～	…が，～において議論される	1,512
	・be discussed in terms of ～	…が，～の点から議論される	421
	・be discussed in relation to ～	…が，～に関して議論される	222
	・be discussed in light of ～	…が，～を考慮に入れて議論される	139
❷	be discussed with ～	…が，～と議論される	299
	・be discussed with respect to ～	…が，～に関して議論される	119

❶ The mechanisms inherent in aconitase inactivation by $ONOO^-$ are discussed in terms of the mitochondrial matrix metabolic and thiol redox state.（Biochemistry. 2005 44:11986）

訳 …が，～の点から議論される

❷ These results are discussed with respect to their potential involvement in the pathogenesis of AD. (Brain Res. 2005 1044:206)

訳 これらの結果が，ADの病因へのそれらの潜在的な関与に関して議論される

■［名］＋ be discussed

❶	results are discussed	結果が，議論される	588
❷	findings are discussed	知見が，議論される	216
❸	data are discussed	データが，議論される	95

❶ Our findings are discussed in relation to the role of the Phe clamp in a Brownian ratchet model of translocation. (J Biol Chem. 2010 285:8130)

訳 われわれの知見が，～の役割に関して議論される

■［代名／名］＋ discuss

❶	we discuss ～	われわれは，～を議論する	3,446
❷	this review discusses ～	このレビューは，～を議論する	465

❶ We discuss these findings and the implications of this development for both systems and molecular neuroscience. (Neuron. 2005 45:183)

訳 われわれは，これらの知見を議論する

■ discuss ＋［名句／名節］

❶	discuss how ～	…は，どのように～かを議論する	620
❷	discuss the implications of ～	…は，～の意味を議論する	218
❸	discuss the role of ～	…は，～の役割を議論する	144
❹	discuss the possibility that ～	…は，～という可能性を議論する	81

❶ We discuss how such mechanical properties could be controlled at the microstructural level. (Ann Intern Med. 2005 142:627)

訳 われわれは，そのような力学的性質がどのように微細構造レベルで制御されうるかを議論する

❷ We discuss the implications of Tbx6-Mesp interactions for the evolution of cardiac mesoderm in invertebrates and vertebrates. (Development. 2005 132:4811)

訳 われわれは，Tbx6とMespの相互作用の意味を議論する

★★ disease ㊔ 疾患 用例数 134,840

disease	疾患	113,683
diseases	疾患	21,157

複数形のdiseasesの割合は約15%あるが，単数形は無冠詞で使われることが非常に多い．

◈類義語：illness, sickness, disorder, impairment, dysfunction

■ disease ＋［前］

	the%	a/an%	⌀%			
❶	7	8	81	disease in ～	～における疾患	4,797
				・disease in humans	ヒトにおける疾患	315
				・disease in patients with ～	～の患者における疾患	118

❶ African trypanosomes are ancient eukaryotes that cause lethal disease in humans and cattle.（Proc Natl Acad Sci USA. 2003 100:7539）

訳 アフリカトリパノソーマは，ヒトにおいて致死的な疾患を引き起こす古代の真核生物である

■［名／形］＋ disease

❶	heart disease	心疾患	4,036
❷	cardiovascular disease	循環器疾患	3,881
❸	liver disease	肝疾患	2,390
❹	artery disease	動脈疾患	2,356
❺	autoimmune disease	自己免疫疾患	1,576
❻	kidney disease	腎疾患	1,537

■ disease ＋［名］

❶	disease progression	疾患進行	2,991
❷	disease activity	疾患活動性	1,171
❸	disease severity	疾患重症度	1,126

★★ disorder ㊔ 障害／疾患　用例数 31,047

disorders	疾患／障害	15,789
disorder	疾患／障害	15,258

複数形のdisordersの割合は約50%と非常に高い．

◆類義語：disease, illness, deficiency, impairment, disturbance, dysfunction, deficit, damage

■ disorder ＋［前］

	the%	a/an%	⌀%			
❶	8	78	11	disorder of ～	～の疾患	690
❷	10	37	48	disorder in ～	～における疾患	681
❸	2	88	9	disorder with ～	～を持つ疾患	409

❶ Psoriasis is a common inflammatory disorder of the skin and other organs.（Am J Hum Genet. 2012 90:796）

訳 乾癬は，皮膚および他の器官のよくある炎症性疾患である．

❷ Bloom syndrome is a rare, autosomal recessive inherited disorder in humans.（Cancer Res. 2005 65:2526）

訳 ブルーム症候群は，ヒトにおける稀な常染色体劣性遺伝性疾患である

❸ Carney complex is a familial multiple neoplasia disorder with characteristic features such as cardiac and cutaneous myxomas and spotty pigmentation of the skin.（Lancet Oncol. 2005 6:501）

訳 …は，～のような特徴を持つ家族性の多発性異常増殖疾患である

■ disorder ＋［過分］

❶ disorder characterized by ～	～によって特徴づけられる疾患	886
❷ disorder caused by ～	～によって引き起こされる疾患	384

❶ Post-hypoxic myoclonus is a movement disorder characterized by brief, sudden involuntary muscle jerks.（Brain Res. 2005 1059:122）

訳 …は，～によって特徴づけられる運動障害である

❷ Huntington's disease（HD）is a neurodegenerative disorder caused by an expanded polyglutamine（polyQ）tract in the huntingtin（htt）protein.（J Cell Biol. 2005 169:647）

訳 ハンチントン病（HD）は，～によって引き起こされる神経変性疾患である

D

dispensable 形 必要でない 用例数 2,122

◆類義語：unnecessary

■ be dispensable ＋［前］

★ be dispensable	…は，必要でない	1,675
❶ be dispensable for ～	…は，～にとって必要でない	1,333

❶ Finally, we ablate Tbx18 in full knockout mice, but find no perturbations in hair follicle formation, suggesting that Tbx18 is dispensable for normal DP function.（J Invest Dermatol. 2013 133:344）

訳 Tbx18は，正常なDP機能にとって必要でない

displace 動（正しい位置から）移動させる／置換する 用例数 2,255

displaced	移動した／～を移動させた	1,155
displace	～を移動させる	499
displaces	～を移動させる	334
displacing	～を移動させる	267

他動詞.

◆類義語：shift, translocate, transfer, replace

■ be displaced ＋［前］

★ be displaced	…は，移動する	419
❶ be displaced by ～	…は，～によって移動する	124
❷ be displaced from ～	…は，～から移動する	101

❶ Germ cells are displaced by a repulsion mechanism from the posterior mesoderm into the endoderm. (Dev Cell. 2005 9:723)
訳 生殖細胞は，後部中胚葉から内胚葉に反発機構によって移動する

❷ It is poorly primed for catalytic hydrolysis because its ester carbonyl group is completely displaced from the enzyme's oxyanion hole. (Biochemistry. 2004 43:14111)
訳 それのエステルカルボニル基は，酵素のオキシアニオンホールから完全に移動する

displacement 名 置換　用例数 3,330

displacement	置換	2,972
displacements	置換	358

複数形のdisplacementsの割合は約10%あるが，単数形は無冠詞で使われることが非常に多い.

◆類義語：replacement, substitution

■ displacement ＋［前］

	the%	a/an%	ø%			
❶	25	9	64	displacement of ～	～の置換	1,100

❶ Displacement of SH3 from the linker is likely to influence efficient downregulation of c-Abl. (J Mol Biol. 2008 383:414)
訳 そのリンカーからのSH3の置換は，c-Ablの効率的な下方制御に影響しそうである

★ display 動 示す／提示する，名 提示　用例数 25,848

displayed	～を示した／提示される	10,652
display	～を示す／提示	10,194
displays	～を示す／提示	3,555
displaying	～を提示する	1,447

名詞として用いられることもあるが，他動詞の用例が非常に多い．特に，能動態の用例が多い.

◆類義語：exhibit, indicate, show, present, represent, exhibition, indication, representation

■ 動 display ＋［名句］

❶ display increased ～	…は，増大した～を示す	615
❷ display reduced ～	…は，低下した～を示す	403
❸ display significant ～	…は，有意な～を示す	316
❹ display enhanced ～	…は，増大した～を示す	312
❺ display high ～	…は，高い～を示す	289

❶ Furthermore, ethanol-fed A. baumannii displayed increased pathogenicity when confronted with a predator, Caenorhabditis elegans. (Mol Cell Biol. 2004 24:3874)

訳 ～が，増大した病原性を示した

■ 動［名］＋ display

❶	mice displayed ～	マウスは，～を示した	799
❷	cells displayed ～	細胞は，～を示した	520
❸	mutants displayed ～	変異体は，～を示した	226

❶ WT mice displayed concentration-dependent responding for ethanol.（J Neurosci. 2001 21:340）

訳 野生型マウスは，エタノールに対する濃度依存的な応答を示した

■ 動 displayed ＋［前］

❶	displayed by ～	～によって示される	460
❷	displayed on ～	～に提示される	244

❶ Based on the phenotype displayed by its mutation, we designated the gene corresponding to Cj0643 as cbrR（Campylobacter bile resistance regulator).（J Bacteriol. 2005 87:3662）

訳 それの変異によって示される表現型に基づいて

❷ This library of mutants was displayed on the surface of phage and challenged with protease Arg-c to select stably folded proteins.（J Bacteriol. 2005 187:3662）

訳 ～が，ファージの表面に提示された

■ 名 display ＋［前］

	the%	a/an%	ø%			
❶	54	9	33	display of ～	～の提示	411

❶ The display of self-antigens by thymic epithelial cells is key to inducing tolerance in the T lymphocyte compartment, a process enhanced by the Aire transcription factor.（J Exp Med. 2009 206:1245）

訳 胸腺上皮細胞による自己抗原の提示

★ disrupt 動 壊す／分裂させる／混乱させる 用例数 14,075

disrupted	壊される／～を壊した	5,969
disrupt	～を壊す	3,603
disrupts	～を壊す	2,525
disrupting	～を壊す	1,978

他動詞.

◆類義語：destroy, lesion, abolish, impair

■ disrupt ＋［名句］

❶	disrupt the interaction	…は，相互作用を壊す	219
❷	disrupt binding	…は，結合を壊す	78
❸	disrupt the formation of ～	…は，～の形成を混乱させる	71

❹ disrupt the function of ～	…は，～の機能を混乱させる	60

■ [代名／名] ＋ disrupt

❶ we disrupted ～	われわれは，～を壊した	198
❷ mutations disrupt ～	変異は，～を壊す	90

❶ To characterize the specific role of falcipain-2, **we disrupted** the falcipain-2 gene and assessed the effect of this alteration. (Proc Natl Acad Sci USA. 2004 101:4384)
訳 われわれは，falcipain-2遺伝子を壊した

■ be disrupted ＋ [前]

★ be disrupted	…は，壊される	2,157
❶ be disrupted by ～	…は，～によって壊される	686
❷ be disrupted in ～	…は，～において壊される	468

❶ This intramolecular interaction **is disrupted by** mutation of Tyr633 within the Mint1 autoinhibitory helix leading to enhanced APP binding and beta-amyloid production. (Proc Natl Acad Sci USA. 2012 109:3802)
訳 この分子内相互作用は，Tyr633の変異によって壊される

■ [副] ＋ disrupted

❶ severely disrupted	激しく壊される	108
❷ completely disrupted	完全に壊される	50
❸ specifically disrupted	特異的に壊される	46

■ [前] ＋ disrupting

❶ by disrupting ～	～を壊すことによって	494
❷ without disrupting ～	～を壊すことなしに	131

★disruption 名 破壊／分裂 用例数 11,767

disruption	破壊	11,061
disruptions	破壊	706

複数形のdisruptionsの割合は約5％あるが，単数形は無冠詞で使われることが多い.
◆類義語：disruption, ablation, lesion

■ disruption ＋ [前]

	the％	a/an％	ø％			
❶	10	4	84	disruption of ～	～の破壊	8,055
				・targeted disruption of ～	～に狙いを定めた破壊	640
❷	4	38	55	disruption in ～	～の破壊	447

❶ As early as 4 months of age, mice with **targeted disruption of** PPAR-γ in muscle showed glucose intolerance and progressive insulin resistance. (Nat Med. 2003 9:1491)

訳 筋肉のPPAR-γに狙いを定めた破壊を持つマウスは，グルコース不耐性および進行性のインスリン抵抗性を示した

❷ Disruption in one or more of the steps that control protein biosynthesis has been associated with alterations in the regulation of cell growth and cell cycle progression. (Oncogene. 2004 23:3138)

訳 ～するステップの１つかそれ以上の破壊

■［名］＋ disruption

❶ gene disruption	遺伝子破壊	507

dissect 動 精査する／解剖する　用例数 2,493

dissect	～を精査する	1,226
dissected	精査される／～を精査した	728
dissecting	～を精査する	523
dissects	～を精査する	16

他動詞.

◆類義語：analyze, examine, investigate, test

■ to dissect

★ to dissect ～	～を精査する…	958
❶ To dissect ～	～を精査するために	214

❶ To dissect the roles of functional domains in Fli1, we recently generated mutant Fli1 mice that express a truncated Fli1 protein ($Fli1^{\Delta CTA}$) that lacks the carboxy-terminal regulatory (CTA) domain. (Mol Cell Biol. 2010 30:5194)

訳 機能ドメインの役割を精査するために

■ dissect ＋［名句］

❶ dissect the role of ～	…は，～の役割を精査する	116

dissociate 動 解離する／分離する　用例数 3,546

dissociated	分離される／～を分離した／解離した	1,717
dissociate	解離する／～を分離する	854
dissociates	解離する／～を分離する	720
dissociating	解離する／～を分離する	255

他動詞・自動詞の両方の用例がある.

◆類義語：separate, divide, interrupt, resolve

■ dissociate ＋［前］

❶ dissociate from ～	…は，～から解離する	510

❶ Furthermore, we found that in response to DNA damage, **Chk1 rapidly dissociates from the chromatin.**（Curr Biol. 2006 16:150）
訳 Chk1は，クロマチンから急速に解離する

■ be dissociated ＋［前］

★ be dissociated	…は，分離される	383
❶ be dissociated from ～	…は，～から分離される	194
❷ be dissociated by ～	…は，～によって分離される	55

❶ In this study, **neurons were dissociated from brain slices** prepared from prepubertal female GnRH-EGFP mice.（J Neurosci. 2002 22:2313）
訳 ニューロンは，脳のスライスから分離された

dissociation 名 解離　　用例数 9,819

dissociation	解離	9,731
dissociations	解離	88

複数形のdissociationsの割合は１％しかなく，**不可算名詞**として使われることがかなり多い．
◆類義語：dissection, detachment, unbinding　◆反意語：association, binding

■ dissociation ＋［前］

	the％	a/an％	ø％			
❶	38	1	60	dissociation of ～	～の解離	2,500
❷	5	0	43	dissociation from ～	～からの解離	535
❸	10	66	18	dissociation between ～	～の間の解離	181

❶ Binding of CaM restores the elongated conformation and **facilitates dissociation of the C-terminal CaD fragment from F-actin.**（Biochemistry. 2003 42:2513）
訳 ～は，C末端CaD断片のF-アクチンからの解離を促進する

❷ **Upon dissociation from Rb**, the glycosylated YY1 is free to bind DNA.（J Biol Chem. 2003 278:14046）
訳 Rbから解離するとすぐに

■［過分／形］＋ dissociation

❶ ～-induced dissociation	～誘導性の解離	402
❷ apparent dissociation	見かけ上の解離	187

■ [名] + and dissociation

❶ association and dissociation	結合と解離	270

❶ The biophysical properties underlying the association and dissociation of these antibodies from cell surfaces is incompletely understood. (Transplantation. 2001 71:440)
訳 細胞表面からのこれらの抗体の結合と解離

distal 形 遠心の／末端の　用例数 9,014

◆類義語：distant, distally, efferent　◆反意語：proximal

■ distal + [前]

❶ distal to ～	～の遠心	757

❶ The 3' hairpin segment of the RNA binds distal to the active site and provides binding energy that contributes to enhanced catalytic efficiency. (Cell. 2005 120:599)
訳 ～は，活性部位の遠心に結合する

■ distal + [名]

❶ distal end	遠心端	211
❷ distal region	末端領域	143

★ distance 名 距離　用例数 10,338

distance	距離	7,510
distances	距離	2,828

複数形のdistancesの割合が，約25%ある可算名詞.

■ distance + [前]

distance betweenの前にtheが付く割合は非常に高い.

	the %	a/an %	ø %			
❶	77	4	17	distance between ～	～の間の距離	958
❷	35	52	11	distance of ～	～の距離	806
❸	24	25	40	distance from ～	～からの距離	683

❶ Consistently, exposure of intact cells to ouabain apparently increased the distance between the Na^+/K^+-ATPase and Src. (Mol Biol Cell. 2006 17:317)
訳 …は，～の間の距離を増大させた

❸ To date, there are no reports that have measured the distance from the contact point to the bony crest between implants. (J Periodontol. 2003 74:1785)
訳 接触点から骨稜までの距離を測定した報告はない

D

■ [形] + distance

❶ long distance	長距離	518
❷ genetic distance	遺伝的距離	131

★★ distinct 形 明らかに異なる／別個の　用例数 38,617

◆類義語：different, dissimilar, disparate, discrete, separate

■ be distinct + [前]

★ be distinct	…は，明らかに異なる	3,339
❶ be distinct from ～	…は，～と明らかに異なる	2,419
・be distinct from that	…は，それと明らかに異なる	339
❷ be distinct in ～	…は，～において明らかに異なる	107

❶ Its localization **was distinct from that** of elastin. (Invest Ophthalmol Vis Sci. 2002 43:1068)
訳 それの局在は，エラスチンのそれとは明らかに異なった

■ [副] + distinct

❶ functionally distinct	機能的に明らかに異なる	780
❷ structurally distinct	構造的に明らかに異なる	415
❸ genetically distinct	遺伝的に明らかに異なる	337
❹ mechanistically distinct	機械的に明らかに異なる	194
❺ morphologically distinct	形態的に明らかに異なる	190

❶ These findings suggest that the subdivisions of the cerebellar cortex produced by folding may create **functionally distinct** entities. (J Neurosci. 2004 24:3926)
訳 ～は，機能的に明らかに異なるものを作るかもしれない

❷ This Fc receptor is both functionally and **structurally distinct** from the classical Fcγ receptors and transports immunoglobulin G (IgG) within cells. (Mol Biol Cell. 2005 16:2028)
訳 このFcレセプターは，古典的なFcγレセプターとは機能的にも構造的にも明らかに異なる

■ distinct + [名]

❶ distinct mechanisms	別個の機構	755
❷ distinct roles	別個の役割	639
❸ distinct functions	別個の機能	517
❹ distinct regions	別個の領域	447
❺ distinct patterns	別個のパターン	439
❻ distinct pathways	別個の経路	387

❶ This study demonstrates that OGA and IAA act by **distinct mechanisms** and that OGA does not simply act by inhibition of IAA action. (Plant Physiol. 2002 130:895)
訳 OGAとIAAは，別個の機構によって働く

❸ Hepatitis delta virus (HDV) expresses two essential proteins with **distinct functions**. (J Virol.

2004 78:8120)

訳 デルタ型肝炎ウイルス（HDV）は，別個の機能を持つ２つの必須タンパク質を発現する

■ ［形］＋ distinct

❶	two distinct ～	2つの別個の～	4,466
❷	three distinct ～	3つの別個の～	1,294
❸	several distinct ～	いくつかの別個の～	389

❶ Ligand-independent ErbB2 activation **occurs principally by two distinct mechanisms**: overexpression and mutation. (J Biol Chem. 2002 277:28468)

訳 ～は，主に２つの別個の機構によって起こる

D

distinction ㊔ 区別／差異　用例数 1,203

distinction	区別／差異	865
distinctions	区別／差異	338

複数形のdistinctionsの割合が，約30%ある可算名詞．

◈類義語：discrimination, difference, divergence, disparity, disagreement, variation

■ distinction ＋［前］

	the%	a/an%	ø%			
❶	50	36	9	distinction between ～	～の間の区別	445

❶ **The distinction between positive and negative emotions** is fundamental in emotion models. (Science. 2012 338:1225)

訳 ポジティブな感情とネガティブな感情の間の区別

distinctly ㊖ 明らかに　用例数 757

◈類義語：apparently, clearly, evidently, obviously, unambiguously, definitely, explicitly

■ distinctly ＋［形］

❶	distinctly different	明らかに異なる	407

❶ In general, in the prokaryotic cyanobacterial cells, **the thylakoid membrane is distinctly different from the plasma membrane**. (Proc Natl Acad Sci USA. 2001 98:13443)

訳 チラコイド膜は，細胞膜とは明らかに異なっている

distinguish ㊍ 区別する　用例数 8,057

distinguish	区別する	3,801
distinguished	区別される／区別した	2,391
distinguishing	区別する	1,205
distinguishes	区別する	660

自動詞・他動詞の両方の用例がある.
◆類義語：discriminate, discern, sort, differentiate

■ distinguish ＋［前］

❶ distinguish between ～	…は，～の間を区別する	1,401

❶ While the A1 adenosine receptor does not distinguish between Gi1α and Gtα sequences, the 5-HT1A and 5-HT1B serotonin and M2 muscarinic receptors can couple with Gi1 but not Gt.（J Biol Chem. 2003 278:50530）
訳 A1アデノシン受容体は，Gi1α配列とGtα配列の間を区別しない

■ distinguish ＋［名］＋ from

❶ distinguish ～ from …	‥は，～を…と区別する	1,555

❶ A FLAG epitope was engineered on to the N terminus of the typeIII IP_3 R to distinguish the transfected from the endogenous isoform.（J Biol Chem. 2000 275:16084）
訳 形質移入されたアイソフォームを内在性のアイソフォームと区別する

■ be distinguished ＋［前］

★ be distinguished	…は，区別される	1,389
❶ be distinguished from ～	…は，～と区別される	515
❷ be distinguished by ～	…は，～によって区別される	499

❶ Finb/RREB-1 may be distinguished from coactivators, which increase transcription without sequence-specific DNA binding.（Mol Cell Biol. 2003 23:259）
訳 Finb/RREB-1は，コアクチベーターと区別されるかもしれない

■ to distinguish

★ to distinguish	区別する…	2,060
❶ be used to distinguish	…は，区別するために使われる	165

❶ The approach reported here can be used to distinguish between otherwise indistinguishable DNA trajectories in complex nucleoprotein machines.（Genes Dev. 2004 18:1898）
訳 ここで報告されるアプローチは，～の間を区別するために使われうる

distribute 動 分布させる　用例数 6,049

distributed	分布する／～を分布させた	5,689
distribute	～を分布させる	198
distributes	～を分布させる	95
distributing	～を分布させる	67

他動詞．受動態で用いられることが圧倒的に多い.
◆類義語：localize, locate, position

■ distributed ＋［前］

❶ distributed in ～	～に分布する	920
・be distributed in ～	…は，分布する	251
❷ distributed throughout ～	～全体に分布する	629
❸ distributed across ～	～じゅうに分布する	264
❹ distributed among ～	～の間に分布する	234
❺ distributed over ～	～一面に分布する	201

❶ In nonpolarized cells, mature apical proteins were uniformly distributed in the PM.（Mol Biol Cell. 2002 13:3400）

訳 ～は，PMに一様に分布した

D

■［副］＋ distributed

❶ widely distributed	広く分布する	755
❷ randomly distributed	ランダムに分布する	200
❸ uniformly distributed	一様に分布する	191
❹ evenly distributed	均等に分布する	179
❺ broadly distributed	広く分布する	140

❶ Cells expressing the orexins are restricted to a discrete region of the hypothalamus, but their terminal projections are widely distributed throughout the brain.（J Comp Neurol. 2003 465:593）

訳 ～は，脳全体に広く分布する

❷ These genes are randomly distributed in the genome and under-represented both for genes that are co-expressed in operons and for multiple members of gene families.（Nature. 2002 418:331）

訳 これらの遺伝子は，ゲノムにランダムに分布する

★ distribution 名 分布　　用例数 25,698

distribution	分布	21,766
distributions	分布	3,932

複数形のdistributionsの割合が，約15%ある可算名詞.

◆類義語：localization, partition, allocation

■ distribution ＋［前］

distribution ofの前にtheが付く割合は非常に高い.

	the%	a/an%	ø%			
❶	75	15	9	distribution of ～	～の分布	11,020
				・distribution of ～ in …	…における～の分布	1,313
				・differences in the distribution of ～	～の分布の違い	59
				・analysis of the distribution of ～		

					～の分布の分析	57
				・changes in the distribution of ～		
					～の分布の変化	43
❷	14	18	42	distribution in ～	～における分布	1,065

❶ We have recently shown that lysine mutations in p53's putative C-terminal acetylation sites result in increased stability and **cytoplasmic distribution of the p53 protein in a human lung cancer cell line.** (Oncogene. 2002 21:2605)

訳 ヒトの肺癌細胞株におけるp53タンパク質の細胞質内分布

❶ Here, **we focus on changes in the distribution of particular proteins,** mRNAs, and patterns of polyadenylation as essential prerequisites for cell fate determination and gametogenesis. (Dev Biol. 2004 269:319)

訳 われわれは，特定のタンパク質の分布の変化に焦点を合わせる

❷ **Type XV collagen has a widespread distribution in human tissues,** but a nearly restricted localization in basement membrane zones. (J Biol Chem. 2000 275:22339)

訳 15型コラーゲンは，ヒトの組織において広範な分布を持つ

■ ［形／名］＋ distribution

❶	subcellular distribution	細胞下分布	612
❷	spatial distribution	空間的分布	571
❸	tissue distribution	組織分布	520
❹	size distribution	大きさ分布	323

❶ First, **we examined the subcellular distribution of sodium channels** using electron microscopic observations of immunoperoxidase and immunogold labeling. (J Neurosci. 2004 24:329)

訳 われわれは，ナトリウムチャネルの細胞下分布を調べた

❸ **Some mice were sacrificed at predetermined times for quantitative tissue distribution of** ^{111}In. (J Nucl Med. 2003 44:1293)

訳 数匹のマウスは，～の定量的な組織分布のために予定された時間に屠殺された

divergence 名 分岐／相違　　用例数 3,594

divergence	分岐／相違	3,520
divergences	分岐／相違	74

複数形のdivergencesの割合は2％しかなく，**不可算名詞**として使われることが多い．

◆類義語：bifurcation, difference, disparity, disagreement, distinction

■ divergence ＋［前］

divergence ofの前にtheが付く割合はかなり高い．

	the%	a/an%	ø%			
❶	62	5	32	divergence of ～	～の分岐	827
❷	12	16	71	divergence in ～	～における相違	344
❸	25	2	71	divergence between ～	～の間の相違	294

❹	3	4	67	divergence from ～	～からの分岐	164

❶ Molecular clocks have been used to date the divergence of humans and chimpanzees for nearly four decades. (Proc Natl Acad Sci USA. 2005 102:18842)
訳 分子時計は，ヒトとチンパンジーの分岐の年代を定めるために使われている

❷ Divergence in gene content among these closely related strains was approximately 140 times greater than divergence at the nucleotide sequence level. (J Bacteriol. 2005 187:1783)
訳 これらの密接に関連する株の間の遺伝子含有量の相違は，～よりおよそ140倍大きかった

★ diverse 形 多様な／分岐した　用例数 13,929

◆類義語：various

■ diverse ＋［形／名］

❶ diverse cellular ～	多様な細胞の～	464
・diverse cellular processes	多様な細胞過程	174
❷ diverse biological ～	多様な生物学的～	395
❸ diverse array of ～	多様な～	392
❹ diverse set of ～	多様なセットの～	392
❺ diverse group of ～	多様なグループの～	317
❻ diverse functions	多様な機能	308
❼ a diverse range of ～	広範囲の～	252

❶ Sphingosine-1-phosphate (S1P), an important sphingolipid metabolite, regulates diverse cellular processes, including cell survival, growth, and differentiation. (Mol Cell Biol. 2005 25:11113)
訳 ～は，多様な細胞過程を調節する

❸ Thus, NOD-LRR proteins appear to be involved in a diverse array of processes required for host immune reactions against pathogens. (Annu Rev Biochem. 2005 74:355)
訳 宿主の免疫反応のために必要とされる多様な過程

❹ These genomes come from a phylogenetically diverse set of organisms, and range in size from 870 kb to more than 6Mb. (Curr Opin Microbiol. 2005 8:595)
訳 これらのゲノムは，系統学的に多様なセットの生物に由来する

■［副］＋ diverse

❶ structurally diverse	構造的に多様な	349
❷ more diverse	より多様な	222
❸ genetically diverse	遺伝的に多様な	191

❶ Our results were obtained on a test set of 44 structurally diverse protein targets. (J Med Chem. 2005 48:6821)
訳 われわれの結果は，44の構造的に多様なタンパク質標的のテストセットに関して得られた

diversity ㊔ 多様性 用例数 9,281

diversity	多様性	9,204
diversities	多様性	77

複数形のdiversitiesの割合は0.8%しかなく，不可算名詞として使われることが多い．
◆類義語：variety, divergence

■ diversity ＋［前］

diversity ofの前にtheが付く割合は非常に高い．

	the％	a/an％	ø％			
❶	72	16	9	diversity of ～	～の多様性	2,341
❷	13	2	82	diversity in ～	～における多様性	1,116
❸	14	3	80	diversity within ～	～内の多様性	221
❹	17	2	72	diversity among ～	～の間の多様性	177

❶ **The diversity of phenotypes suggests that** specific modifications of FrzCD act to differentially regulate motility and developmental aggregation in M. xanthus. (Mol Microbiol. 2006 59:45)
訳 表現型の多様性は，～ということを示唆する

❷ Selection in which fitnesses vary with the changing genetic composition of the population **may facilitate the maintenance of genetic diversity in a wide range of organisms**. (Genetics. 2004 167:499)
訳 ～は，広い範囲の生物における遺伝的な多様性の維持を促進するかもしれない

■［形／名］＋ diversity

❶	genetic diversity	遺伝的多様性	948
❷	sequence diversity	配列多様性	261
❸	functional diversity	機能的多様性	252
❹	structural diversity	構造的多様性	188
❺	nucleotide diversity	ヌクレオチド多様性	180

❶ Analysis of SNPs reveals long-range haplotypes across the entire dog genome, and **defines the nature of genetic diversity within and across breeds**. (Nature. 2005 438:803)
訳 ～は，品種内および品種全体の遺伝的多様性の性質を定義する

divide ㊍ 分ける／分裂する 用例数 5,246

divided	分けられる／～を分けた／分裂した	2,792
dividing	分裂する	1,369
divide	～を分ける／分裂する	903
divides	～を分ける／分裂する	182

他動詞の用例が多いが，自動詞として使われることもある．
◆類義語：separate, interrupt, dissociate, segregate, resolve

■ be divided ＋［前］

★ be divided	…は，分けられる	1,695
❶ be divided into ～	…は，～に分けられる	1,503
・be divided into two ～	…は，2つの～に分けられる	383
・patients were divided into ～	患者は，～に分けられた	153

❶ Thirty females with TMJ disc displacement with reduction were divided into two groups based on the presence or absence of the self-report of TMJ pain.（J Dent Res. 2001 80:1935）
訳 …が，～に基づいて２つのグループに分けられた

■ divide ＋［名］＋ into

❶ divide ～ into …	…は，～を…に分ける	224

■ dividing ＋［名］

❶ dividing cells	分裂細胞	510

D

document 動 記述する／述べる／実証する，名 文書 用例数 7,434

documented	記述された／記述した	5,014
document	記述する／文書	1,603
documents	記述する／文書	507
documenting	記述する	310

名詞の用例もあるが，動詞として用いられることが多い．
◆類義語：describe, report, demonstrate, prove, evidence, show

■ be documented ＋［前］

★ be documented	…が，記述される	1,335
❶ be documented in ～	…が，～において記述される	402
❷ be documented by ～	…が，～によって記述される	127

❶ Deregulation of Pim-2 expression has been documented in several human malignancies, including leukemia, lymphoma, and multiple myeloma.（Cancer Res. 2004 64:8341）
訳 Pim-2発現の調節解除が，いくつかのヒトの悪性腫瘍において述べられてきた

■［副］＋ documented

❶ well documented	よく記述される	1,084
・be well documented	…は，よく記述される	672
❷ previously documented	以前に述べられた	136

❶ Although the ability of IL-27 to promote T cell responses is well documented, the anti-inflammatory properties of this cytokine remain poorly understood.（J Immunol. 2006 176:237）

訳 T細胞応答を促進するIL-27の能力はよく記述されているけれども，

■ document ＋［that節］

❶	document that ～	…は，～ということを記述する	626

❶ Mouse studies **documented that** CD103 expression is required for efficient destruction of the graft renal tubules by CD8 effectors directed to donor MHC I alloantigens.（J Immunol. 2005 175:2868）

訳 マウスの研究が，～ということを記述した

■［代名／名］＋ document

❶	we document ～	われわれは，～を記述する	325
❷	results document ～	結果は，～を記述する	129
❸	study documents ～	研究は，～を記述する	90
❹	data document ～	データは，～を記述する	87
❺	findings document ～	知見は，～を記述する	68

★★ domain 名 ドメイン　用例数 132,929

domain	ドメイン	95,088
domains	ドメイン	37,841

複数形のdomainsの割合が，約30%ある可算名詞.

■ domain ＋［前］

domain ofの前にtheが付く割合は圧倒的に高い.

	the%	a/an%	ø%			
❶	95	5	0	domain of ～	～のドメイン	18,791
❷	69	17	1	domain in ～	～におけるドメイン	2,574

❶ The COOH-terminal **domain of** ZO-1 was required for its association with myosin 1C.（FASEB J. 2011 25:144）

訳 ZO-1のカルボキシ末端ドメインは，ミオシン1Cとそれの結合のために必要とされた

■［形／名］＋ domain

❶	binding domain	結合ドメイン	8,476
❷	terminal domain	末端ドメイン	7,275
❸	cytoplasmic domain	細胞質ドメイン	2,448
❹	transmembrane domain	膜貫通ドメイン	2,166
❺	catalytic domain	触媒ドメイン	2,096
❻	extracellular domain	細胞外ドメイン	1,723

★ dominant 形 優性の／ドミナントな／優位な 用例数 17,520

dominant negativeの用例が非常に多い.

◆類義語：predominant, superior

■ dominant negative ＋［名］

★ dominant negative ～	ドミナントネガティブな～	7,992
❶ dominant negative mutant	ドミナントネガティブ変異体	1,307
❷ dominant negative form of ～	ドミナントネガティブ型の～	639
❸ dominant negative effect	ドミナントネガティブ効果	406

❶ Expression of a dominant negative mutant of myocardin in Xenopus embryos interferes with myocardial cell differentiation.（Cell. 2001 105:851）

訳 ～のドミナントネガティブ変異体の発現

■［名］＋ of a dominant negative

★ of a dominant negative ～	ドミナントネガティブな～の	930
❶ expression of a dominant negative ～	ドミナントネガティブな～の発現	539
❷ overexpression of a dominant negative ～	ドミナントネガティブな～の過剰発現	196

❶ In macrophages infected with M. tuberculosis, SP-A reduced expression of a dominant negative isoform of C/EBPβ.（Infect Immun. 2004 72:645）

訳 SP-Aは，C/EBPβのドミナントネガティブアイソフォームの発現を低下させた

★ donor 名 ドナー 用例数 24,079

donor	ドナー	17,033
donors	ドナー	7,046

◆反意語：recipient

複数形のdonorsの割合が，約30%ある可算名詞.

■ donor(s) ＋［前］

	the%	a/an%	ø%			
❶	2	-	82	donors with ～	～のドナー	340
❷	41	48	9	donor in ～	～におけるドナー	174

❶ There may be an increased risk of primary nonfunction in livers procured from donors with hypernatremia.（Transplantation. 2010 90:438）

訳 高ナトリウム血症のドナーから得られた肝臓

■ donor ＋［名］

❶ donor cells	ドナー細胞	525
❷ donor age	ドナー年齢	309

■ [名／形] ＋ donor(s)

❶ electron donor	電子供与体	399
❷ healthy donors	健康なドナー	386
❸ blood donors	血液ドナー	274
❹ normal donors	正常なドナー	210

★★ dose ㊔ 用量／投与量, ㊍ 投薬する 用例数 45,404

dose	用量	33,280
doses	用量	10,326
dosing	投薬	1,529
dosed	投薬された	269

動詞としても用いられるが，名詞の用例が圧倒的に多い．複数形のdosesの割合が，約25%ある可算名詞．

◆類義語：amount

■ dose ＋ [前]

	the%	a/an%	ø%			
❶	18	68	3	dose of ～	…用量の～／～の用量	18,791

❶ One-day-old female hypercholesterolemic mice were administered a single dose of either vehicle or testosterone. (Circ Res. 2012 110:e73)

訳 1日齢の雌の高コレステロールマウスは，単一用量の溶媒かテストステロンのどちらかを投与された

■ [形／名／過分] ＋ dose

❶ high dose	高用量	2,883
❷ low dose	低用量	2,429
❸ single dose	単一用量	938
❹ radiation dose	放射線量	519
❺ maximum tolerated dose	最大耐量	392

down-regulate／downregulate ㊍ 下方制御する 用例数 8,364

down-regulated	下方制御される／～を下方制御した	3,119
down-regulate	～を下方制御する	802
down-regulates	～を下方制御する	700
down-regulating	～を下方制御する	441
downregulated	下方制御される／～を下方制御した	1,989
downregulate	～を下方制御する	566

downregulates	～を下方制御する	436
downregulating	～を下方制御する	311

他動詞．down-regulateおよびdownregulateのいずれでも用いられる．
◆類義語：decrease, reduce, diminish, lower, decline

■ down-regulated ＋［前］

★	be down-regulated	…は，下方制御される	1,328
❶	be down-regulated in ～	…は，～において下方制御される	473
❷	be down-regulated by ～	…は，～によって下方制御される	289

❶ Both forms are down-regulated in prostate and liver carcinomas relative to normal tissues.（Cancer Res. 2005 65:11010）
訳 両方の型が，前立腺癌および肝臓癌において下方制御される

❷ PAF-induced MMP-2 activation was also down-regulated by p38 MAPK and protein kinase A inhibitors.（J Biol Chem. 2006 281:2911）
訳 PAFに誘導されるMMP-2の活性化は，また，p38 MAPKによって下方制御された

down-regulation／downregulation ㊔ 下方制御

用例数 9,389

down-regulation	下方制御	5,845
down-regulations	下方制御	8
downregulation	下方制御	3,535
downregulations	下方制御	1

down-regulationおよびdownregulationのいずれでも用いられる．複数形の用例はほとんどなく，原則，不可算名詞として使われる．
◆類義語：decrease, reduction, decline, attenuation

■ down-regulation ＋［前］

	the％	a/an％	ø％			
❶	18	1	80	down-regulation of ～	～の下方制御	4,267
❷	4	14	30	down-regulation in ～	～における下方制御	164

❶ In addition, in VA13 cells, induction of E2F97 resulted in down-regulation of the tumor suppressor protein p53.（Oncogene. 1997 14:1191）
訳 E2F97の誘導は，癌抑制タンパク質p53の下方制御という結果になった

■ ［過分］＋ down-regulation

❶	～-mediated down-regulation	～に仲介される下方制御	203
❷	～-induced down-regulation	～に誘導される下方制御	180

❶ Rapamycin-induced down-regulation of cyclin D1 is inhibited by the GSK3β inhibitors lithium chloride, SB216763, and SB415286.（Cancer Res. 2005 65:1961）

D

訳 ラパマイシンに誘導されるサイクリンD1の下方制御は，～によって阻害される

★ downstream 形 副 下流の　用例数 16,986

◆反意語：upstream

■ 副 downstream ＋［前］

❶ downstream of ～	～の下流で	5,251
・act downstream of ～	…は，～の下流で働く	483
・function downstream of ～	…は，～の下流で機能する	334
・be downstream of ～	…は，～の下流にある	244
・immediately downstream of ～	～のすぐ下流に	234
❷ downstream from ～	～の下流に	625

❶ **FGF acts downstream of TGFβ signaling** in regulating CNC cell proliferation, and exogenous FGF2 rescues the cell proliferation defect in the frontal primordium of Tgfbr2 mutant. (Development. 2006 133:371)
訳 FGFは，TGFβシグナル伝達の下流で働く

❶ The kps gene cluster required for the biosynthesis of polysialic acid capsule **was mapped to a location immediately downstream of this PAI.** (Infect Immun. 2006 74:744)
訳 ～は，このPAIのすぐ下流の位置にマップされた

❷ Because **this gene is located downstream from vnfK** in an arrangement similar to the relative organization of the nifK and nifY genes, it was designated vnfY. (J Bacteriol. 2003 185:2383)
訳 この遺伝子は，vnfKの下流に位置する

■ 形 downstream ＋［名］

❶ downstream signaling	下流のシグナル伝達	1,086
・downstream signaling pathways	下流のシグナル伝達経路	242
❷ downstream target	下流の標的	957
・downstream target of ～	～の下流の標的	440
❸ downstream effector	下流のエフェクター	654

❶ Thus, **FAK may act in glioblastoma as a downstream target of growth factor signaling**, with integrins enhancing the impact of such signaling in the tumor microenvironment. (Am J Pathol. 2005 167:1379)
訳 FAKは，増殖因子シグナル伝達の下流の標的として神経膠芽腫において働くかもしれない

dramatic 形 劇的な　用例数 5,757

◆類義語：drastic, marked, striking, prominent, pronounced, remarkable, manifest, notable, significant

■ dramatic ＋［名］

❶ dramatic increase	劇的な増大	714
❷ dramatic reduction	劇的な低下	433

❸ dramatic changes	劇的な変化	319
❹ dramatic decrease	劇的な低下	303

❶ Akt expression was associated with a dramatic increase in tumor size, despite the absence of HIF-1. (Cancer Res. 2004 64:3500)

訳 Aktの発現は，腫瘍の大きさの劇的な増大と関連した

■ ［動（＋前）］＋ a dramatic

❶ result in a dramatic ～	…は，劇的な～という結果になる	378
❷ cause a dramatic ～	…は，劇的な～を引き起こす	214
❸ lead to a dramatic ～	…は，劇的な～につながる	187

❶ The transient expression of hiwi in the human leukemia cell line KG1 resulted in a dramatic reduction in cellular proliferation. (Blood. 2001 97:426)

訳 ～は，細胞増殖の劇的な低下という結果になった

D

★ dramatically 副 劇的に 用例数 7,957

◈類義語：drastically, markedly, remarkably, strikingly, extremely, unusually

■ dramatically ＋ ［過分／動／形］

❶ dramatically reduced	劇的に低下する	991
・be dramatically reduced	…は，劇的に低下する	397
❷ dramatically increased	劇的に増大する	660
❸ dramatically decreased	劇的に低下する	289
❹ dramatically enhanced ～	～を劇的に増強した／劇的に増強される	259
❺ dramatically different	劇的に異なる	237

❶ Transduction efficiency and binding of nuclear protein to RBS was dramatically reduced in male mice by castration. (Blood. 2003 102:480)

訳 ～は，オスのマウスにおいて劇的に低下した

❷ We also show that histone H1 hyperphosphorylation is dramatically increased in both ash2 and sktl mutant polytene chromosomes. (Genetics. 2004 167:1213)

訳 ～は，ash2およびsktlの両方の変異多糸染色体において劇的に増大する

❺ The ^{2}H NMR spectra of polyunsaturated bilayers are dramatically different from those of less unsaturated phospholipid bilayers. (J Am Chem Soc. 2002 124:298)

訳 …は，～のそれらと劇的に異なる

■ ［過分］＋ dramatically

❶ increased dramatically	劇的に増大した	305

❶ The activity of ILK was increased dramatically in these cells. (Cancer Res. 2004 64:1987)

訳 ILKの活性は，これらの細胞において劇的に増大した

drastically 副 強烈に／徹底的に 用例数 996

◆類義語：dramatically, markedly, remarkably, strikingly, extremely, unusually

■ drastically ＋［過分／動］

❶ drastically reduced ～	強烈に低下する／～を強烈に低下させた	284

❶ Each mutation drastically reduced CAT activity when inserted into a reporter construct.（Am J Hum Genet. 2001 69:712）
訳 おのおのの変異は，CAT活性を強烈に低下させた

★ drive 動 駆動する／推進する／運転する，名 駆動 用例数 20,123

driven	駆動される	10,263
drive	～を駆動する	4,717
driving	～を駆動する／～を推進する	3,074
drives	～を駆動する	1,856
drove	～を駆動した	213

名詞としても用いられるが，動詞の用例がほとんどである．
◆類義語：control, operate

■ driven ＋［前］

❶ driven by ～	～によって駆動される	3,853
・be driven by ～	…は，～によって駆動される	2,084
・gene driven by ～	～によって駆動される遺伝子	142

❶ Positive selection during thymocyte development is driven by the affinity and avidity of the TCR for MHC-peptide complexes expressed in the thymus.（J Immunol. 2005 175:7372）
訳 ～は，TCRの親和性と結合力によって駆動される

■［名］-driven

❶ promoter-driven ～	プロモーターに駆動される～	286
❷ light-driven ～	光に駆動される～	191

❶ Overexpression of AP-2 α into highly metastatic breast cell lines did not alter KiSS-1 promoter-driven luciferase gene activity.（J Biol Chem. 2006 281:51）
訳 ～は，KiSS-1プロモーターに駆動されるルシフェラーゼ遺伝子活性を変化させなかった

■ driving ＋［名］

❶ driving force	推進力	754

❶ Miniature inverted repeat transposable elements（MITEs）are thought to be a driving force for genome evolution.（Plant Cell. 2005 17:1559）
訳 ～は，ゲノム進化に対する推進力であると考えられる

drop ㊔ 低下，㊍ 低下する　用例数 1,914

drop	低下／低下する	1,045
dropped	低下した	454
drops	低下／低下する	350
dropping	低下する	65

名詞の用例が多いが，自動詞としても用いられる．複数形のdropsの割合が，約25%ある可算名詞．

◈類義語： decrease, decline, fall, reduction, attenuation, down-regulation

■ ㊔ drop ＋［前］

「～の低下」の意味では，drop ofよりもdrop inの用例が圧倒的に多い．

	the%	a/an%	ø%			
❶	12	79	3	drop in ～	～の低下	515
❷	19	71	0	drop of ～	～の低下	83

❶ However, **STN cells also exhibited a drop in spike threshold triggered by larger EPSPs,** allowing them to fire time-locked spikes in response to coincident input.（J Neurosci. 2010 30:13180）

訳 STN細胞は，また，より大きな興奮性シナプス後電位によって引き起こされるスパイク閾値の低下を示した

■ ㊍ drop ＋［前］

❶	drop to ～	～へ低下する	160

❶ However, **the survival rate drops to only 23%** for women with distant metastases.（J Neurosci. 2010 30:13180）

訳 生存率は，わずか23%に低下する

★★ due ㊎ 帰すべき　用例数 32,825

due toの形で，「～のせいで/～の理由で」という意味に用いられることが圧倒的に多い．

◈類義語：because of, owing to, responsible for

■ due to ＋［名句］

★	due to ～	～のせいで	30,962
❶	due to increased ～	増大した～のせいで	589
❷	due to the presence of ～	～の存在のせいで	397
❸	due to the lack of ～	～の欠如のせいで	295
❹	due to differences in ～	～の違いのせいで	271
❺	due to decreased ～	低下した～のせいで	245
❻	due to reduced ～	低下した～のせいで	240

D

❼ due to an increase in ～	～の増大のせいで	212
❽ due to loss of ～	～の喪失のせいで	190
❾ due to changes in ～	～の変化のせいで	180

❶ The cells treated with MDI plus 7-oxo-DHEA have a significantly increased level of total fat, **primarily due to increased levels of** Δ^9-16:1 and palmitic (16:0) acids. (Biochemistry. 2002 41: 5473)

訳 主には増大したレベルの～のせいで

❷ Upon cessation of chemotherapy, these cells will re-enter cell cycling, and experience extensive CIN **due to the presence of amplified centrosomes**. (Oncogene. 2004 23:6823)

訳 増幅された中心体の存在のせいで

■ [名] ＋ be due to

★ be due to ～	…は，～のせいである	1,717
❶ effect is due to ～	効果は，～のせいである	75

❶ Genetic analysis suggested that **the inhibitory effect is due to hydrolysis of the PhrA peptide** in a form as small as the pentapeptide (ARNQT). (J Bacteriol. 2002 184:43)

訳 その抑制性効果は，PhrAペプチドの加水分解のせいである

■ [副／副句] ＋ due to

❶ likely due to ～	おそらく～のせいで	851
・most likely due to ～	最も～のせいでありそうな	226
❷ primarily due to ～	主に～のせいで	380
❸ probably due to ～	おそらく～のせいで	376
❹ in part due to ～	部分的には～のせいで	352
❺ largely due to ～	主に～のせいで	308
❻ possibly due to ～	～のせいかもしれない	294
❼ presumably due to ～	おそらく～のせいで	283

❶ We conclude that previous reports linking SOCS1 to TLR signaling **are most likely due to effects on IFN-α/β receptor signaling**. (J Biol Chem. 2004 279:54702)

訳 ～は，最もIFN-α/β受容体シグナル伝達に対する影響のせいでありそうである

❷ The abnormal development of IL-1ra-deficient mice **is probably due to chronic overstimulation of the proinflammatory pathway** via IL-1, but a clear single pathological defect is not apparent. (J Immunol. 2002 169:393)

訳 ～は，おそらく炎症誘発性経路の慢性的過剰刺激のせいである

■ due ＋ [副句／副] ＋ to

❶ due in part to ～	部分的には～のせいで	695
❷ due primarily to ～	主に～のせいで	177

❶ **Enhanced signaling is due in part to the ability of** the CD19/CD21/CD81 complex to stabilize the BCR in sphingolipid- and cholesterol-rich membrane microdomains termed lipid rafts. (J Biol Chem. 2004 279:31973)

訳 増強されたシグナル伝達は，部分的には～の能力のせいである

duplication ㊔ 重複／複製　用例数 4,808

duplication	重複／複製	3,518
duplications	重複／複製	1,290

複数形のduplicationsの割合が，約25%ある可算名詞.

◆類義語：overlap, replication

■ duplication ＋［前］

	the%	a/an%	ø%			
❶	19	16	64	duplication of ～	～の重複	594
❷	8	27	59	duplication in ～	～における重複	172

❶ In addition, **tandem duplication of genes in an OG tends to be highly asymmetric.** (Plant Physiol. 2008 148:993)

訳 OGにおける遺伝子の縦列重複は，高度に非対称性になりがちである

■［名／形］＋ duplication

❶	gene duplication	遺伝子重複	692
❷	centrosome duplication	中心体複製	232
❸	tandem duplication	縦列重複	208

★duration ㊔ 持続時間　用例数 11,872

duration	持続時間	11,083
durations	持続時間	789

複数形のdurationsの割合は７%あるが，単数形は無冠詞で使われることが多い.

◆類義語：period

■ duration ＋［前］

	the%	a/an%	ø%			
❶	51	6	42	duration of ～	～の持続時間	5,798
				・median duration of ～	～の持続時間の中央値	341
				・mean duration of ～	～の平均持続時間	200
				・duration of response	応答の持続時間	160
				・duration of action	作用の持続時間	150
				・duration of treatment	処置の持続時間	135

❶ **The median duration of therapy was 12 weeks** (range, 2 to 111). (N Engl J Med. 2012 366:2455)

訳 治療の持続期間の中央値は，12週であった

D

★ dysfunction 名 機能障害／機能不全／障害 用例数 12,375

dysfunction	機能障害／機能不全	12,157
dysfunctions	機能障害／機能不全	218

複数形のdysfunctionsの割合は２％しかなく，原則，**不可算名詞**として使われる．

◆類義語：impairment, disturbance, disorder, deficit, deficiency, damage

■ dysfunction ＋［前］

	the％	a/an％	ø％			
❶	5	2	92	dysfunction in ～	～における機能障害	1,452
❷	15	2	83	dysfunction of ～	～の機能障害	619

❶ Although **the causes of neurological dysfunction in epilepsy are multifactorial**, accumulating evidence indicates that seizures in themselves may directly cause brain injury.（Ann Neurol. 2005 58:888）

訳 てんかんにおける神経性機能障害の原因は多因子性である

■［形／名］＋ dysfunction

❶	endothelial dysfunction	内皮障害	843
❷	mitochondrial dysfunction	ミトコンドリア障害	831
❸	ventricular dysfunction	心室機能障害	495
❹	renal dysfunction	腎機能障害	412
❺	cardiac dysfunction	心機能不全	382
❻	diastolic dysfunction	拡張機能障害	366
❼	systolic dysfunction	収縮不全	354
❽	organ dysfunction	臓器機能不全	345

E

★★ each　代名 おのおの／それぞれ，　形 おのおのの／それぞれの　用例数 59,448

■ 代名 each ＋［名］

each otherは，副詞ではなく名詞相当語句である．each otherは，one anotherとほぼ同じ意味になる．

❶ each of ～	～のおのおの	9,342
❷ each other	お互い	4,682
・interact with each other	…は，お互いに相互作用する	382

❶ Here, **we evaluate the contribution of each of the HAS proteins to outflow facility** in anterior segment perfusion culture.（Invest Ophthalmol Vis Sci. 2012 53:4616）

訳 われわれは，HASタンパク質のおのおのの流出能への寄与を評価する

❷ It is also unclear whether **the two NS5 domains interact with each other** to form a stable structure in which the relative orientation of the two domains is fixed.（Biochemistry. 2012 51: 5921）

訳 2つのNS5ドメインは，お互いに相互作用する

■ 形 each ＋［名］

❶ each patient	おのおのの患者	913
❷ each case	おのおののケース	856
❸ each group	おのおのの群	849

★★ early　副 早期に，形 早期の　用例数 53,865

◈類義語：initially, originally, initial, original, first　◈反意語：late

■ 副 early ＋［前］

❶ early in ～	～の早期に	3,178
・early in development	発生の早期に	288
・early in life	人生の早期に	278
・early in infection	感染の早期に	175
❷ as early as ～	早くも～	1,350
❸ early after ～	～のあと早期に	606
❹ early during ～	～の間の早期に	443

❶ **The majority of the human population becomes infected early in life by the gammaherpesvirus EBV.**（J Immunol. 2012 189:2965）

訳 人間集団の大多数は，人生の早期にガンマヘルペスウイルスEBVに感染する

❷ Increases in the Rel A subunit of NF-κ B **were observed as early as 30-min after administration of TNF.**（Brain Res. 2005 1048:24）

訳 ～は，TNF の投与のあと早くも30分で観察された

❸ Transient elevation of liver enzymes **was observed in most patients early after** islet infusion.（Transplantation. 2005 80:1718）

訳 …は，～のあと早期にほとんどの患者において観察された

■ 形 early ＋［名］

❶ early stages of ～	～の早期のステージ	1,799
❷ early onset	早期の発症	1,504
❸ early development	発生初期	801
❹ early phase	初期相	621
❺ early detection	早期の検出	592
❻ early events	早期の現象	544

❷ **Duchenne muscular dystrophy（DMD）is an inherited disease characterized by early onset of skeletal muscle degeneration** and progressive weakness.（Circulation. 2005 112:2462）

訳 デュシェンヌ型筋ジストロフィー（DMD）は，骨格筋の変性の早期の発症によって特徴づけられる遺伝的疾患である

easily 副 容易に　　用例数 3,185

◆類義語：readily

■［副］＋ easily

❶ more easily	より容易に	219

❶ However, a significant difference is that P53（7KR）**is activated more easily by DNA damage** in thymus than WT P53.（Proc Natl Acad Sci USA. 2005 102:10188）

訳 ～は，DNA傷害によってより容易に活性化される

■ easily ＋［形／過分］

❶ easily detected	容易に検出される	101
❷ easily accessible	容易に到達できる	37

❶ **Deletions and duplications can be easily detected**（a 2x decrease or a 1.5x increase in gene dosage）.（Anal Biochem. 2007 371:37）

訳 欠失と重複は，容易に検出されうる

★★ effect 名 影響／効果，動 引き起こす　　用例数 173,826

effects	影響／効果／～を引き起こす	89,929
effect	影響／効果／～を引き起こす	83,267
effected	引き起こされる／～を引き起こす	438
effecting	～を引き起こす	192

動詞としても使われるが，名詞の用例がほとんどである．複数形のeffectsの割合は約50%と非常に高い．〈effect(s) of〉〈effect(s) on〉の用例が多い．

◆類義語：influence, impact, efficacy, potency, efficiency

■ effect(s) ＋［前］

effects ofの前にtheが付く割合は圧倒的に高い．

	the％	a/an％	ø％			
❶	82	-	15	effects of ～	～の影響／～の効果	42,249
				・the effects of ～ on …	…に対する～の影響	6,178
❷	7	34	8	effect on ～	～に対する影響／～に対する効果	22,068
				・no effect on ～	～に対する影響のない	7,665
				・little effect on ～	～に対する影響のほとんどない	1,740
❸	2	-	83	effects in ～	～における影響／～における効果	2,999

❶ To explore the mechanisms involved, **we examined the effects of proteasome inhibition on** the AP-1 pathway. (J Immunol. 2001 167:1145)

訳 われわれは，～に対するプロテアソーム阻害の影響を調べた

❷ In contrast, HTR2A 102 T/C genotype **had no effect on mirtazapine side effects**. (Am J Psychiatry. 2003 160:1830)

訳 ～は，ミルトラザピン副作用に対する影響を持たなかった

❷ The not2::R165G also abrogated NOT2 ability to interact with ADA2 **but had little effect on** the integrity of the CCR4-NOT complex. (J Mol Biol. 2002 322:27)

訳 しかし，～に対する影響はほとんど持たなかった

❸ The morphine doses used in this study decrease clinical signs of pain but **can cause significant adverse effects in ventilated preterm neonates**. (Lancet. 2004 363:1673)

訳 ～は，人工呼吸の早産の新生児において著しく有害な影響を引き起こしうる

■［動］＋ the effects of

❶	examine the effects of ～	…は，～の影響を調べる	2,231
❷	investigate the effects of ～	…は，～の影響を精査する	1,475
❸	study the effects of ～	…は，～の影響を研究する	1,067
❹	determine the effects of ～	…は，～の影響を決定する	745
❺	evaluate the effects of ～	…は，～の影響を評価する	544
❻	assess the effects of ～	…は，～の影響を評価する	540

❶ **We therefore investigated the effects of ephedra alkaloids in patients with** autonomic impairment and explored their potential interaction with water ingestion. (Circulation. 2004 109:1823)

訳 われわれは，それゆえ，～の患者におけるマオウアルカロイドの効果を精査した

■［形／名］＋ effect(s)

❶	inhibitory effect	抑制的な影響	2,641
❷	side effects	副作用	2,161
❸	little effect	ほとんど影響のない	2,121
❹	significant effect	有意な（顕著な）影響	1,764
❺	adverse effects	有害作用	1,537

E

❻ protective effect	保護作用	1,385
❼ beneficial effects	有益な効果	1,052
❽ isotope effects	同位体効果	879
❾ similar effects	直接影響	708

❶ Moreover, **neomycin reversed the inhibitory effect of wortmannin** but not LY294002 on insulin stimulation of Akt kinase activity. (J Biol Chem. 2004 279:55277)
訳 ネオマイシンは，ワートマニンの抑制的な影響を逆転させた

❹ However, pretreatment with a general caspase inhibitor **had no significant effect on the amount of** cortical damage observed at 7 days post-injury. (Brain Res. 2002 949:88)
訳 …は，～の量に対する有意な影響を持たなかった

■ effect(s) ＋ [動]

❶ effects were observed	影響が，観察された	359
❷ effect is mediated	影響が，仲介される	226
❸ effect was blocked	影響が，遮断された	173
❹ effects were seen	影響が，見られた	145

★ effective 形 効果的な 用例数 26,437

◆類義語：efficacious, efficient, active

■ effective ＋ [前]

❶ effective in ～	～の際に効果的な	3,707
・be effective in ～	…は，～の際に効果的である	1,698
・effective in reducing ～	～を減らす際に効果的な	272
・effective in preventing ～	～を防ぐ際に効果的な	220
❷ effective for ～	～にとって効果的な	847
❸ effective at ～	～において効果的な	739
・effective at ～ing	～するのに効果的な	598
❹ as effective as ～	～と同じぐらい効果的な	458
❺ effective against ～	～に対して効果的な	385

❶ Under conditions of intense exposure, **varicella vaccine was highly effective in preventing moderate and severe disease** and about 80% effective in preventing all disease. (JAMA. 2004 292:704)
訳 水痘ワクチンは，中等度および重篤な疾患を予防する際に高度に効果的であった

❷ The results of this study demonstrated that β-radiation using 90Sr/90Y **is both safe and effective for preventing recurrence** in patients with in-stent restenosis. (Circulation. 2002 106:1090)
訳 ～は，再発を予防するために安全でありかつ効果的である

❸ Furthermore, **GFA-116 is also effective at inhibiting tumor growth** and metastasis to the lung of B16-F10 melanoma cells injected into immunocompetent mice. (Cancer Res. 2004 64:3586)
訳 GFA-116は，また，腫瘍の増殖を抑制するのに効果的である

❹ **Gentamicin cream was as effective as mupirocin in preventing S. aureus infections.** (J Am

Soc Nephrol. 2005 16:539)
訳 ゲンタマイシンクリームは，黄色ブドウ球菌感染を予防する際にムピロシンと同じぐらい効果的であった

■［副］＋ effective

❶ more effective	より効果的な	2,781
・more effective than ～	～より効果的な	948
❷ most effective	最も効果的な	1,057
❸ less effective	より効果的でない	814
❹ highly effective	高度に効果的な	762
❺ equally effective	同等に効果的な	289

❶ These results indicate that Lm-LLO-E7 is more effective than Lm-E7 at inducing DC maturation.（J Immunol. 2004 172:6030）
訳 Lm-LLO-E7は，～を誘導するのにLm-E7よりも効果的である

❷ Smallpox vaccination was historically the most effective defence measure against wild smallpox virus.（Lancet. 2003 362:1378）
訳 種痘は，野生型天然痘ウイルスに対する歴史的に最も効果的な防御手段であった

❸ When compared with curcumin, dexamethasone was less effective in suppression of NF-κB activation and induction of apoptosis in myeloma cells.（Blood. 2004 103:3175）
訳 クルクミンと比べると，デキサメサゾンは～の抑制においてより効果的でなかった

■ effective ＋［名］

❶ effective treatment	効果的な治療	905
❷ effective therapy	効果的な治療	392

■［形］＋ and effective

❶ safe and effective	安全で効果的な	647

❶ Rosiglitazone is a safe and effective oral agent for the treatment of diabetes mellitus after solid organ transplantation.（Transplantation. 2004 77:1009）
訳 ロシグリタゾンは，糖尿病の治療のための安全で効果的な経口剤である

effectively 副 効果的に　　用例数 6,783

◈類義語：efficiently, actively

■［副／前］＋ effectively

❶ more effectively	より効果的に	572
・more effectively than ～	～より効果的に	226
❷ as effectively as ～	～と同じぐらい効果的に	181

❶ Peripheral injection of BT seems to increase food intake more effectively than intracerebroventricular administration.（Brain Res. 2001 907:125）
訳 BTの末梢注射は，脳室内投与より効果的に食物摂取を増大させるように思われる

E

■ effectively ＋［動］

❶ effectively inhibited ～	効果的に抑制される／～を効果的に抑制した	238
❷ effectively blocked ～	効果的に遮断される／～を効果的に遮断した	195

❶ MCMV immune evasion genes effectively inhibited antigen presentation.（J Virol. 2003 77: 301）

訳 MCMV免疫回避遺伝子は，抗原提示を効果的に抑制した

effectiveness 名 有効性　　用例数 5,338

複数形の用例はなく，原則，不可算名詞として使われる．

◆類義語：efficacy, availability

■ effectiveness ＋［前］

effectiveness ofの前にtheが付く割合は圧倒的に高い．

	the%	a/an%	ø%			
❶	92	0	8	effectiveness of ～	～の有効性	2,991
❷	11	1	47	effectiveness in ～	～における有効性	219

❶ These results help to explain the effectiveness of cannabinoids in blocking the malaise generated by TNF-releasing disease processes by opposing effects on ryanodine channels.（J Neurosci. 2012 32:5237）

訳 これらの結果は，～を遮断する際のカンナビノイドの有効性を説明するのに役立つ

efficacious 形 有効な／効果的な　　用例数 1,610

◆類義語：effective, efficient

■ efficacious ＋［前］

❶ efficacious in ～	～において有効な	343
・be efficacious in ～	…は，～において有効である	196
・efficacious in ～ing	～する際に有効である	124
❷ efficacious for ～	～に対して有効な	86

❶ These results suggest that rapamycin in combination with etoposide-based chemotherapy may be efficacious in the treatment of AML.（Blood. 2005 106:4261）

訳 ～は，急性骨髄性白血病の治療において有効であるかもしれない

■ ［副］＋ efficacious

❶ more efficacious	より有効な	284
・more efficacious than ～	～より有効な	134
❷ highly efficacious	高度に有効な	97
❸ most efficacious	最も有効な	92

❶ PGE2 was more efficacious than β-agonist in activating PKA and inhibiting cytokine production. (Blood. 2006 107:2052)

訳 PGE2は，PKAを活性化させる際にβ-アゴニストより有効であった

★ efficacy 名 有効性 用例数 15,744

efficacy	有効性	15,464
efficacies	有効性	280

複数形のefficaciesの割合は2％しかなく，原則，不可算名詞として使われる.

◈類義語：efficiency, effectiveness, potency, effect

■ efficacy ＋［前］

efficacy ofの前にtheが付く割合は圧倒的に高い.

	the%	a/an%	ø%			
❶	86	0	14	efficacy of ～	～の有効性	6,413
❷	1	1	86	efficacy in ～	～における有効性	1,328
				・efficacy in ～ing	～する際の有効性	205

❶ To determine the efficacy of this compound in breast cancer cells, we first measured protein expression of three IAPs: XIAP, cIAP-1, and cIAP-2 in nine independent breast cancer cell lines. (Oncogene. 2005 24:7381)

訳 乳癌細胞におけるこの化合物の有効性を決定するために

❷ Although family psychoeducation has demonstrated efficacy in improving outcomes in schizophrenia, empirical support for its ability to enhance medication adherence is scarce. (Arch Gen Psychiatry. 2012 69:265)

訳 家族心理教育は，統合失調症における予後を改善する際に有効性を示してきた

■［形／名］＋ efficacy

❶ therapeutic efficacy	治療上の有効性	610
❷ clinical efficacy	臨床的有効性	422
❸ vaccine efficacy	ワクチンの有効性	347

❶ These results indicate that, as an adjuvant therapy, 3G4 could enhance the therapeutic efficacy of docetaxel in breast cancer patients. (Cancer Res. 2005 65:4408)

訳 3G4は，乳癌患者におけるドセタキセルの治療上の有効性を増強しうる

■［動］＋ the efficacy

❶ evaluate the efficacy	…は，有効性を評価する	414
❷ assess the efficacy	…は，有効性を査定する	276
❸ determine the efficacy	…は，有効性を決定する	168
❹ demonstrate efficacy	…は，有効性を示す	145

❶ Clinical trials are underway to evaluate the efficacy of inhibitors for class I and II histone deacetylases to treat malignancies. (Oncogene. 2006 25:176)

訳 阻害剤の有効性を評価するために臨床治験が進行中である

■ [名] ＋ and efficacy／efficacy and ＋ [名]

❶ safety and efficacy	安全性と有効性	797
❷ efficacy and safety	有効性と安全性	694

❶ The integration sites of viral vectors used in human gene therapy **can have important consequences for safety and efficacy**. (J Virol. 2005 79:11434)

訳 ～は，安全性と有効性に対する重要な結果を持ちうる

★ efficiency 名 効率　用例数 13,514

efficiency	効率	12,015
efficiencies	効率	1,499

複数形のefficienciesの割合は約10%あるが，単数形は無冠詞で使われることがかなり多い．

◆類義語：efficacy, potency, effect

■ efficiency ＋ [前]

efficiency ofの前にtheが付く割合は非常に高い．

	the%	a/an%	ø%			
❶	72	11	15	efficiency of ～	～の効率	4,403
❷	11	7	71	efficiency in ～	～における効率	495
❸	62	2	30	efficiency with ～	～を持つ効率	258
				・efficiency with which ～	～である効率	150

❶ **The efficiency of translocation is influenced by** the type of γ subunit present in the G protein. (J Biol Chem. 2004 279:51541)

訳 移行の効率は，～によって影響される

❷ **This decrease in efficiency in utilizing energy derived from** PSII for CO_2 fixation is due to an increase in photorespiration. (Plant Physiol. 1993 101:507)

訳 ～に由来するエネルギーを利用する際の効率のこの低下

❸ Genetic studies indicate that **the efficiency with which reovirus strains induce apoptosis is determined by the viral S1 gene**, which encodes attachment protein sigma1. (J Virol. 2001 75:4029)

訳 レオウイルス株がアポトーシスを誘導する効率は，ウイルスのS1遺伝子によって決定される

■ [形] ＋ efficiency

❶ catalytic efficiency	触媒効率	821
❷ high efficiency	高い効率	560

❶ **The catalytic efficiency of CPA 4-hydroxylation by rat P450 2B1** was 10- to 35-fold higher than that of rabbit P450 2B4 or 2B5. (Mol Pharmacol. 2004 65:1278)

訳 ラットP450 2B1によるCPA 4-水酸化の触媒効率

★ efficient 形 効率的な 用例数 15,993

◆類義語：effective, efficacious, active

■ efficient ＋［前］

❶ efficient in ～	～の際に効率的な／～において効率的な	551
・efficient in ～ing	～する際に効率的な	243
❷ efficient at ～	～において効率的な	287
・efficient at ～ing	～するのに効率的な	204

❶ The PP genotype may be more **efficient in** delivering B12 to tissues, resulting in enhanced B12 functional status. (Blood. 2002 100:718)

訳 PP遺伝子型は，B12を組織に送達する際により効率的であるかもしれない

❷ Moreover, effector memory T cells were more **efficient at** subsequently establishing a second generation of memory T cells. (J Immunol. 2004 172:6533)

訳 エフェクターメモリーT細胞は，引き続いて第２世代のメモリーT細胞を確立するのにより効率的であった

■［副］＋ efficient

❶ more efficient	より効率的な	1,498
・more efficient than ～	～より効率的な	323
・more efficient in ～	～においてより効率的な	165
❷ highly efficient	高度に効率的な	683
❸ less efficient	より効率的でない	540
❹ most efficient	最も効率的な	356

❶ Moreover, ICI and raloxifene are **more efficient than** tamoxifen in promoting ER α binding to the corepressor N-CoR in vivo and in vitro. (J Biol Chem. 2003 278:6912)

訳 ICIとラロキシフェンは，～を促進する際にタモキシフェンより効率的である

■ efficient ＋［名］

❶ efficient method	効率的な方法	284
❷ efficient synthesis	効率的な合成	151

■ for efficient

★ for efficient ～	効率的な～にとって…	2,824
❶ be required for efficient ～	…は，効率的な～に必要とされる	830

❶ Previous studies have suggested that cytoplasmic dynein **is required for efficient** neurofilament transport. (Mol Biol Cell. 2004 15:5092)

訳 ～は，効率的な神経フィラメント輸送に必要とされる

★efficiently 副 効率的に 用例数 9,222

◆類義語：effectively, actively

■［副］＋ efficiently

❶	more efficiently	より効率的に	1,498
	・more efficiently than ～	～より効率的に	524
❷	less efficiently	より効率的でない	1,498

❶ **Uninfected monocyte/macrophages crossed endothelial cell barriers six times more efficiently than neutrophils.** (Infect Immun. 2003 71:6728)

訳 非感染性の単球/マクロファージは，好中球より６倍効率的に内皮細胞バリアを通過した

■ efficiently ＋［前］

❶	efficiently in ～	～において効率的に	463
❷	efficiently to ～	～に効率的に	258
❸	as efficiently as ～	～と同じぐらい効率的に	317

❶ We found that **both Jurkat and PBMCs arrested efficiently in mitosis** when treated with nocodazole. (J Biol Chem. 2002 277:5187)

訳 ジャーカットおよびPBMCの両方は，有糸分裂において効率的に停止した

❸ This virus formed wild-type plaques and **replicated as efficiently as the wild-type KOS virus** in Vero cells. (J Virol. 2001 75:12431)

訳 ～は，野生型のKOSウイルスと同じぐらい効率的に複製した

effort 名 努力 用例数 8,275

efforts	努力	4,996
effort	努力	3,279

複数形のeffortsの割合は約60%と非常に高い．in an effort to *do*の用例も多い．

◆類義語：attempt, trial

■ efforts ＋［前］

	the%	a/an%	ø%			
❶	1	-	91	efforts to *do*	～しようとする努力	2,108
				・efforts to develop ～	～を開発しようとする努力	175
				・efforts to identify ～	～を同定しようとする努力	172
❷	5	-	89	efforts in ～	～における努力	204

❶ **Efforts to develop small molecule inhibitors of PARG activity** have until recently been hampered by a lack of structural information on PARG. (PLoS One. 2012 7:e50889)

訳 PARG活性の小分子阻害剤を開発しようとする努力

■ in an effort to *do*

★	in an effort to *do*	～しようとして	561
❶	in an effort to identify ～	～を同定しようとして	195
❷	in an effort to understand ～	～を理解しようとして	168

❶ In an effort to understand this negative regulation, we developed a novel genetic selection that detects altered expression from the HSP26 promoter. (Genetics. 2005 169:1203)
訳 この負の調節を理解しようとして

elderly 形 高齢の／年配の　用例数 3,425

◆類義語：old, aged

■ elderly ＋［名］

❶	elderly patients	高齢の患者	609
❷	elderly persons	高齢者	170
❸	elderly subjects	高齢の対象者	166

❶ The aim of this study was to quantify risks and benefits of different screening strategies in elderly patients with varying life expectancies. (Gastroenterology. 2005 129:1163)
訳 この研究の目的は，様々な平均余命の高齢の患者における異なる選別戦略のリスクと利点を定量化することであった

■ the elderly

★	the elderly	高齢者	1,030
❶	in the elderly	高齢者において	637

❶ Nocturnal polyuria is common in the elderly. (Lancet. 2000 355:486)
訳 夜間の多尿は，高齢者においてよくある

★★ element 名 エレメント　用例数 39,337

elements	エレメント	21,393
element	エレメント	17,944

複数形のelementsの割合は約55%と非常に高い.

■ elements ＋［前］

	the%	a/an%	⌀%			
❶	8	-	69	elements in ～	～におけるエレメント	2,105
❷	26	-	64	elements of ～	～のエレメント	1,987
❸	11	-	73	elements within ～	～内のエレメント	506

❶ Hence, we speculate that these novel regulatory elements in the IL-10 family gene locus function via an intermediate regulatory RNA.（J Immunol. 2005 175:7437）

訳 IL-10ファミリー遺伝子座におけるこれらの新規の調節性エレメント

❷ We explored the regulation of several elements of IgE-mediated signaling by a short priming with IL-3.（J Immunol. 2005 175:3006）

訳 われわれは，IgEに仲介されるシグナル伝達のいくつかのエレメントの調節を探索した

■ element ＋［過分／現分］

❶ element located	位置するエレメント	361
❷ elements required for ～	～のために必要とされるエレメント	204

❶ Here we show that U11 can bind the NRS as a mono-snRNP in vitro and that a G-rich element located downstream of the U11 site is required for efficient binding.（J Biol Chem. 2004 279:38201）

訳 U11部位の下流に位置するGに富むエレメントは，効率的な結合のために必要とされる

■［形／名］＋ element

❶ response element	応答エレメント	4,260
❷ regulatory element	調節性エレメント	3,264
❸ responsive element	応答性エレメント	1,230
❹ structural element	構造上のエレメント	932
❺ promoter element	プロモーターエレメント	913

❷ Transcriptional regulation of the VEGFR-2 is complex and may involve multiple putative upstream regulatory elements including E boxes.（J Biol Chem. 2005 280:29856）

訳 ～は，Eボックスを含む複数の推定上の上流調節性エレメントを含むかもしれない

★ elevate 動 上昇させる　用例数 23,025

elevated	上昇させられる／上昇した／～を上昇させた	21,558
elevate	～を上昇させる	571
elevating	～を上昇させる	514
elevates	～を上昇させる	382

他動詞．受動態の用例が非常に多い．

◈類義語：increase, multiply, raise, rise

■ be elevated ＋［前］

★ be elevated	…が，上昇する	3,366
❶ be elevated in ～	…が，～において上昇する	1,695
・levels were elevated in ～	レベルが，～において上昇した	134
❷ be elevated by ～	…が，～によって上昇する	161

❶ Interestingly, ARF6 protein levels were elevated in H-Ras- but not K-Ra-transformed cells.（Mol Cell Biol. 2003 23:645）

訳 ARF6タンパク質レベルが，～において上昇した

❷ Free radical production was elevated by ischemia-reperfusion but was significantly lower in SOD-treated animals. (J Am Soc Nephrol. 2001 12:2691)
訳 フリーラジカル産生は，虚血再灌流によって上昇した

■ [副] + elevated

❶	significantly elevated	有意に上昇した	1,110
❷	markedly elevated	顕著に上昇した	286

❶ Blood glucose levels were significantly elevated in the diabetic group (370 ± 8 mg/dl) compared to control group (104 ± 3 mg/dl). (Brain Res. 2002 956:268)
訳 血中グルコースレベルは，糖尿病群において有意に上昇した

■ elevated + [名]

❶	elevated levels of ～	上昇したレベルの～	2,213
❷	elevated expression of ～	上昇した～の発現	501
❸	elevated risk	上昇したリスク	337
❹	elevated temperatures	上昇した温度	310

❶ In an effort to establish a causal role of TGF-β_1 in this process, transgenic mice with elevated levels of active myocardial TGF-β_1 were generated. (Circ Res. 2000 86:571)
訳 活性のある心筋のTGF-β_1のレベルの上昇したトランスジェニックマウスが産生された

❷ Transformation is associated with elevated expression of the Wilm's tumor gene encoded transcription factor (WT1). (Blood. 2000 95:2198)
訳 形質転換は，ウイルムス腫瘍遺伝子の上昇した発現と関連する

E

elevation ㊔ 上昇 用例数 6,332

elevation	上昇	4,805
elevations	上昇	1,527

複数形のelevationsの割合が，約25%ある可算名詞.

◆類義語：increase, augmentation, rise

■ elevation + [前]

	the%	a/an%	ø%			
❶	18	17	62	elevation of ～	～の上昇	1,986
❷	16	52	27	elevation in ～	～の上昇	827

❶ Elevation of intracellular Ca^{2+} does not evoke ATP release but potentiates mechanosensitive ATP release. (J Biol Chem. 2001 276:23867)
訳 細胞内Ca^{2+}の上昇は，ATPの遊離を惹起しない

❷ Furthermore, light pulses known to induce phase delays caused significant elevation in mTim mRNA. (J Neurosci. 1999 19:RC15)
訳 ～は，mTimメッセンジャーRNAの有意な上昇を引き起こした

■ [形／過分] + elevation

❶ significant elevation	有意な上昇	200
❷ ～-induced elevation	～で誘導される上昇	134
❸ sustained elevation	持続性の上昇	105
❹ ～-fold elevation	～倍の上昇	100

★ elicit 動 誘発する／引き出す 用例数 12,529

elicited	誘発される／～を誘発した	6,525
elicit	～を誘発する	3,423
elicits	～を誘発する	1,673
eliciting	～を誘発する	908

他動詞.

◆類義語：evoke, provoke, induce

■ elicit + [名句]

❶ elicit ～ response	…は，～応答を誘発する	1,101
❷ elicit ～ effect	…は，～効果を誘発する	231
❸ elicit ～ antibodies	…は，～抗体を誘発する	164
❹ elicit ～ immunity	…は，～免疫を誘発する	144

❶ Tumor-secreted GRP94 **has been shown to elicit antitumor immune responses** when used as antitumor vaccines.（Oncogene. 2013 32:805）

訳 ～は，抗腫瘍免疫応答を誘発することが示されてきた

■ to elicit

★ to elicit ～	～を誘発する…	1,559
❶ fail to elicit ～	…は，～を誘発することができない	168
❷ ability to elicit ～	～を誘発する能力	90
❸ sufficient to elicit ～	～を誘発するのに十分な	86

❶ A fragment comprising amino acids 75 to 305 **failed to elicit significant protection.**（Infect Immun. 2003 71:1033）

訳 ～は，有意な保護を誘発することができなかった

■ elicited + [前]

❶ elicited by ～	～によって誘発される	2,121
・responses elicited by ～	～によって誘発される応答	203
❷ elicited in ～	～において誘発される	302

❶ **Many similarities exist in the cellular responses elicited by VEGF** and governed by integrins.（Mol Cell. 2000 6:851）

訳 多くの類似性が，VEGFによって誘発される細胞応答に存在する

❷ These data indicate that a biologically active, specific immune response to CD20 **can be elicited in mice vaccinated** with CD20 peptide conjugates.（Blood. 2002 99:3748）

訳 ～は，ワクチン接種されたマウスにおいて誘発されうる

eliminate 動 消失させる／除去する　用例数 9,538

eliminated	～を消失させた／消失した	4,483
eliminate	～を消失させる	2,350
eliminates	～を消失させる	1,371
eliminating	～を消失させる	1,334

他動詞.

◆類義語：remove, exclude, delete, abolish

■ eliminate ＋［名句］

❶ eliminate ～ activity	…は，～活性を消失させる	244
❷ eliminate the need for ～	…は，～に対する必要性を消失させる	226
❸ eliminate the ability of ～	…は，～の能力を消失させる	84

❶ This approach minimizes artifacts from fluorescence interference and **eliminates the need for antibody specificity** to methylated lysines.（Anal Biochem. 2011 419:217）

訳 ～は，抗体特異性に対する必要性を消失させる

■ be eliminated ＋［前］

★ be eliminated	…は，消失する	1,563
❶ be eliminated by ～	…は，～によって消失する	518
❷ be eliminated in ～	…は，～において消失する	158
❸ be eliminated from ～	…は，～から消失する	107

❶ **The transgene was eliminated by** back-crossing with wild-type Arabidopsis.（Plant Mol Biol. 2004 56:111）

訳 その導入遺伝子は，～によって消失した

❸ **Cholesterol is eliminated from neurons by** oxidization, which generates oxysterols.（J Biol Chem. 2004 279:34674）

訳 コレステロールは，～によってニューロンから消失した

■［副］＋ eliminated

❶ completely eliminated ～	～を完全に消失させた／完全に消失した	227
❷ virtually eliminated ～	～を実質的に消失させた／実質的に消失した	227

❶ Transection at the thoracic spinal level **completely eliminated PAG-induced DRRs**.（Brain Res. 2003 976:217）

訳 ～は，PAGに誘導されるDRRを完全に消失させた

E

elimination ㊔ 除去　用例数 4,623

elimination	除去	4,599
eliminations	除去	24

複数形のeliminationsの割合は0.5%しかなく，原則，**不可算名詞**として使われる．
◆類義語：removal, exclusion

■ elimination ＋［前］

	the %	a/an %	ø %			
❶	37	0	63	elimination of ～	～の除去	2,492

❶ **Elimination of E-Syt1 expression drastically reduced invasiveness** both in vitro and in vivo without modifying the oncogenic activity of CD74-ROS. (Cancer Res. 2012 72:3764)
訳 E-Syt1発現の除去は，侵襲性を劇的に低下させた

elongation ㊔ 伸長　用例数 7,178

elongation	伸長	7,165
elongations	伸長	13

複数形のelongationsの割合は0.2%しかなく，原則，**不可算名詞**として使われる．
◆類義語：extension, outgrowth

■ elongation ＋［前］

	the %	a/an %	ø %			
❶	36	4	58	elongation of ～	～の伸長	664
❷	0	0	97	elongation in ～	～における伸長	286
❸	1	0	96	elongation by ～	～による伸長	252

❶ **Mcm10 is an essential protein that participates in both the initiation and the elongation of DNA replication.** (Genetics. 2005 171:503)
訳 Mcm10は，DNA複製の開始と伸長の両方に関与する必須タンパク質である

❷ Moreover, overexpression of Fbx4 reduces endogenous Pin2/TRF1 protein levels and **causes progressive telomere elongation in human cells.** (J Biol Chem. 2006 281:759)
訳 ～は，ヒト細胞における進行性のテロメア伸長を引き起こす

■［名／形］＋ elongation

❶	transcription elongation	転写の伸長	750
❷	cell elongation	細胞伸長	261
❸	transcriptional elongation	転写の伸長	232
❹	chain elongation	鎖伸長	230

elucidate 動 解明する／明らかにする 用例数 9,707

elucidate	～を解明する／～を明らかにする	5,097
elucidated	解明される／～を解明した／～を明らかにした	3,045
elucidating	～を解明する／～を明らかにする	1,395
elucidates	～を解明する／～を明らかにする	170

他動詞.

◆類義語：reveal, clarify, disclose, uncover, manifest, define

■ elucidate ＋［名句］

❶ elucidate the mechanism	…は，機構を解明する	767
❷ elucidate the role of ～	…は，～の役割を解明する	518
❸ elucidate the function of ～	…は，～の機能を解明する	140

❷ The goal of this study was to elucidate the role of the EGF receptor (EGFR) as a component of the uPAR-signaling machinery. (J Biol Chem. 2003 278:1642)

訳 この研究の目的は，EGF受容体（EGFR）の役割を解明することであった

■ to elucidate

★ to elucidate ～	～を解明するために／～を解明すること	3,724
❶ To elucidate ～	～を解明するために	1,352
❷ be to elucidate ～	…は，～を解明することである	181
❸ help to elucidate ～	…は，～を解明するのに役立つ	123
❹ be used to elucidate ～	…は，～を解明するために使われる	114

❶ To elucidate the mechanism of cADPR action, we performed confocal Ca^{2+} imaging in saponin-permeabilized rat ventricular myocytes. (Circ Res. 2001 89:614)

訳 cADPR作用の機構を解明するために

■［副／動］＋ elucidate

❶ further elucidate ～	～をさらに解明する	393
・to further elucidate ～	～をさらに解明するために	258
❷ help elucidate ～	～を解明するのに役立つ	203

❶ To further elucidate the biosynthesis of this and related macrolides, we cloned and sequenced an 80kb region encompassing the FK520 gene cluster. (Gene. 2000 251:81)

訳 ～の生合成をさらに解明するために

❷ These results may help elucidate the ability of mu agonists to regulate the number and responsiveness of their receptors. (Brain Res. 2004 1028:121)

訳 これらの結果は，～の能力を解明するのに役立つかもしれない

■ be elucidated ＋［前］

★ be elucidated	…は，解明される	1,852
❶ be elucidated by ～	…は，～によって解明される	113

❷ be elucidated in ～	…は，～において解明される	78

❶ Molecular mechanisms were elucidated by expression profiling and in vitro analyses. (Gastroenterology. 2010 139:893)
訳 分子機構は，発現プロファイリングと試験管内分析によって解明された

■ to be elucidated

★ to be elucidated	解明される…	729
❶ remain to be elucidated	…は，解明されないままである	473

❶ The polymorphic gene (s) encoded within the sst1 locus that controls macrophage interactions with the two intracellular pathogens remains to be elucidated. (J Immunol. 2004 173:5112)
訳 ～は，解明されないままである

■ [副] ＋ elucidated

❶ fully elucidated	完全に解明される	377
・not been fully elucidated	完全には解明されていない	192

❶ Specific mechanisms recruiting neutrophils to the lung during hyperoxia-induced lung injury have not been fully elucidated. (J Immunol. 2004 172:3860)
訳 ～は，完全には解明されていない

★ emerge 動 現れる 用例数 10,543

emerging	現れる	5,124
emerged	現れた	3,363
emerge	現れる	1,385
emerges	現れる	671

自動詞.
◆類義語：appear, arise

■ emerge ＋ [前]

❶ emerge as ～	～として現れる	2,480
・have emerged as ～	…が，～として現れた	1,211
❷ emerge from ～	～から現れる	993
❸ emerge in ～	～において現れる	504

❶ The Wnt signaling pathway has emerged as a potential regulator of self-renewal for HSCs. (Curr Opin Hematol. 2004 11:88)
訳 Wntシグナル伝達経路が，～の潜在的な調節因子として現れた

❷ Squirrels also emerged from their burrows earlier and returned to them later over the measurement period. (PLoS One. 2012 7:e36053)
訳 リスは，また，そららの潜り穴から現れた

■ emerging ＋［名］

❶ emerging evidence	新出の証拠	455
❷ emerging data	新出のデータ	176

❶ Emerging evidence suggests a possible role of lycopene in the primary prevention of cardiovascular disease (CVD). (Am J Clin Nutr. 2005 81:990)
訳 新出の証拠は，～の可能な役割を示唆する

emergence 名 出現 用例数 2,915

emergence	出現	2,911
emergences	出現	4

複数形のemergencesの用例はほとんどなく，原則，不可算名詞として使われる．
◆類義語：occurrence, appearance

■ emergence ＋［前］

emergence ofの前にtheが付く割合は圧倒的に高い．

	the%	a/an%	ø%			
❶	87	2	11	emergence of ～	～の出現	2,332

❶ A crucial step in plant organogenesis is the emergence of organ primordia from the apical meristems. (Curr Biol. 2012 22:1739)
訳 植物の器官形成における決定的に重要な段階は，器官原基の出現である

emphasis 名 重点／強調 用例数 1,620

emphasis	重点	1,614
emphases	重点	6

複数形のemphasesの割合は約0.5%しかないが，単数形に不定冠詞が使われる用例は非常に多い．
◆類義語：stress

■ emphasis ＋［前］

	the%	a/an%	ø%			
❶	3	51	45	emphasis on ～	～に重点	1,174
				・with an emphasis on ～	～に重点を置いて	297
				・with emphasis on ～	～に重点を置いて	253
				・with particular emphasis on ～	～に特別に重点を置いて	142
				・with special emphasis on ～	～に特別に重点を置いて	68

E

❶ We review the recent literature on anogenital neoplasms in AIDS, with emphasis on cancers associated with HPV infection. (Curr Opin Oncol. 2004 16:455)
訳 HPV感染と関連する癌に重点を置いて

emphasize 動 強調する／力説する　用例数 3,005

emphasize	～を強調する	1,321
emphasizes	～を強調する	659
emphasizing	～を強調する	552
emphasized	強調される／～を強調した	473

他動詞.

◆類義語：highlight, underscore, stress, accentuate

■ emphasize ＋［名句／that節］

❶ emphasize the importance of ～	…は，～の重要性を強調する	555
❷ emphasize that ～	…は，～ということを強調する	281
❸ emphasize the need	…は，必要性を強調する	258
❹ emphasize the role of ～	…は，～の役割を強調する	92

❶ Our results emphasize the importance of SPHK2 expression in both lymphocytes and other tissues for immune modulation and drug metabolism. (J Biol Chem. 2005 280:36865)
訳 われわれの結果は，～におけるSPHK2発現の重要性を強調する

■［名］＋ emphasize

❶ results emphasize ～	結果は，～を強調する	236
❷ findings emphasize ～	知見は，～を強調する	113
❸ data emphasize ～	データは，～を強調する	79

★ employ 動 用いる／使用する　用例数 10,044

employed	用いられる／～を用いた	5,912
employing	～を用いる	2,076
employ	～を用いる	1,200
employs	～を用いる	856

他動詞.

◆類義語：use, utilize, exploit, apply

■ be employed ＋［前］

★ be employed	…は，用いられる	2,642
❶ be employed to *do*	…は，～するために用いられる	1,311
・be employed to identify ～	…は，～を同定するために用いられる	92
・be employed to determine ～	…は，～を決定するために用いられる	81

・be employed to study ～	…は，～を研究するために用いられる	69
・be employed to investigate ～	…は，～を精査するために用いられる	64
❷ be employed in ～	…は，～において用いられる	283
❸ be employed for ～	…は，～のために用いられる	205
❹ be employed as ～	…は，～として用いられる	204

❶ Site-specific cross-linking was successfully employed to investigate the DNA-protein interface of MutY.（Biochemistry. 2004 43:651）

訳 部位特異的なクロスリンキングが，MutY のDNA-タンパク質インターフェイスを精査するためにうまく用いられた

❷ Maintenance therapy was employed in 68%（54/79）of the patients who survived >3 weeks.（Transplantation. 2005 80:1033）

訳 維持療法が，３週間以上生存した患者の68%（54/79）において用いられた

■［代名］＋ employ

❶ we employed ～	われわれは，～を用いた	942
・we have employed ～	われわれは，～を用いている	308

❶ We employed microarray technology to identify SA target genes.（Infect Immun. 2005 73:5319）

訳 われわれは，SAの標的遺伝子を同定するためにマイクロアレイ・テクノロジーを用いた

★ enable 動 可能にする　用例数 12,502

enable	～を可能にする	4,010
enables	～を可能にする	3,513
enabled	～を可能にした	2,723
enabling	～を可能にする	2,256

他動詞．第５文型のS+V+O+to *do*のパターンで使われることが非常に多い．

◆類義語：allow, permit, lead, prompt

■ enable ＋［名／代名］＋ to *do*

★ enable ～ to *do*	…は，～が…することを可能にする	3,090
❶ enable us to *do*	…は，われわれが～することを可能にする	817
・enable us to identify ～	…は，われわれが～を同定することを可能にする	70

❶ Thus, the profiling approach enabled us to identify several novel functional domains in the RTA locus in the context of the herpesvirus genome.（J Virol. 2009 83:1811）

訳 プロファイリングアプローチは，われわれがいくつかの新規の機能ドメインを同定することを可能にした

■［名］＋ enable

❶ approach enables ～	アプローチは，～を可能にする	67
❷ method enables ～	方法は，～を可能にする	50

E

■ enable ＋［名句］

❶ enable the identification of ～	…は，～の同定を可能にする	114
❷ enable the development of ～	…は，～の開発を可能にする	96
❸ enable the use of ～	…は，～の使用を可能にする	73

❶ This approach enables the identification of smaller intragenic mutations including single-nucleotide variants that are not accessible even with high-resolution genomic array analysis.（Hum Mol Genet. 2012 21:R37）
訳 このアプローチは，より小さな遺伝子内変異の同定を可能にする

★★ encode 動 コードする／コード化する 用例数 55,562

encoding	～をコードする	21,143
encoded	コードされる／～をコードした	13,949
encodes	～をコードする	13,270
encode	～をコードする	7,200

他動詞．遺伝子などに対して用いられる．〈名詞＋encoding〉の用例が多い．
◆類義語：code ◆反意語：decode

■［名］＋ encoding

❶ gene encoding ～	～をコードする遺伝子	4,351
❷ cDNA encoding ～	～をコードするcDNA	824
❸ mRNA encoding ～	～をコードするmRNA	575
❹ transcripts encoding ～	～をコードする転写物	334

❶ These changes are associated with increased expression of genes encoding enzymes of fatty acid oxidation, ketogenesis and glycolysis.（Nature. 2004 432:1027）
訳 ～の酵素をコードする遺伝子の発現

■ encoded ＋［前］

❶ encoded by ～	～によってコードされる	6,160
・be encoded by ～	…は，～によってコードされる	1,586
・protein encoded by ～	～によってコードされるタンパク質	704
❷ encoded in ～	～においてコードされる	847

❶ The proteins encoded by clock genes regulate circadian variations of various cellular functions.（Cell. 2005 122:651）
訳 時計遺伝子によってコードされるタンパク質は，～の概日性変動を調節する

■ encode ＋［名句］

❶ encode a protein	…は，タンパク質をコードする	1,498
❷ encode a putative ～	…は，推定上の～をコードする	642
❸ encode a novel ～	…は，新規の～をコードする	558

❶ The mdm2 gene encodes a protein that is necessary for the negative regulation of p53 function in vivo.（Oncogene. 1996 13:1731）
訳 mdm2遺伝子は，～のために必要であるタンパク質をコードする

■ to encode

★ to encode ～	～をコードする…	1,609
❶ be predicted to encode ～	…は，～をコードすることが予想される	352
❷ be shown to encode ～	…は，～をコードすることが示される	100

❶ BRS1 is predicted to encode a secreted carboxypeptidase.（Proc Natl Acad Sci USA. 2001 98:5916）
訳 BRS1は，分泌型カルボキシペプチダーゼをコードすることが予想される

★★ end 名 端／終わり／目的，動 終わる　用例数 34,927

end	端／終わり／目的／終わる	27,686
ends	端／終わり／目的／終わる	6,428
ended	終わった	431
ending	終わる	382

動詞としても用いられるが，名詞の用例の方が圧倒的に多い．複数形のendsの割合が，約20%ある可算名詞．

◆類義語：terminal, purpose, aim, goal, objective, object

E

■ end ＋［前］

end ofの前にtheが付く割合は圧倒的に高い．

	the%	a/an%	ø%			
❶	91	0	4	end of ～	～の終わり／～の端	8,213
				・at the end of ～	～の終わりにおいて	1,829
				・5′ end of ～	～の5′端	1,120
				・one end of ～	～の片端	251
				・C-terminal end of ～	～のC末端	229
				・end of life care	終末医療	208
				・end of treatment	治療の終わり	154

❶ At the end of each developmental stage, insects perform a stereotypic behavioral sequence leading to ecdysis of the old cuticle.（Development. 2002 129:493）
訳 おのおのの発達のステージの終わりにおいて

■［前］＋（this/one）end

❶ To this end	この目的のために	725
❷ end-to-end	端と端	348
❸ at one end	片端において	206

❶ To this end, we identified three proteins that are stably associated with WRN in nuclear extracts.（J Biol Chem. 2004 279:13659）

訳 この目的のために，

❷ Evidently, end-to-end interactions are not required for high-affinity binding to acto-myosin S1.（Biochemistry. 2000 39:6891）

訳 端と端の相互作用は，～のためには必要とされない

■ end ＋［名］

❶ end point	終点／エンドポイント	2,493
❷ end stage	末期	1,445
❸ end joining	末端結合	1,036
❹ end labeling	末端標識化	719
❺ end products	最終産物	399

★ endogenous 形 内在性の 用例数 21,797

◆類義語：intrinsic ◆反意語：exogenous, extrinsic

■［名］＋ endogenous

❶ expression of endogenous ～	内在性の～の発現	283
❷ levels of endogenous ～	内在性の～のレベル	244

❶ Significantly, DNA binding-deficient ORF50 mutants are competent to autostimulate expression of endogenous ORF50 and to autoactivate ORF50 promoter reporters.（J Virol. 2005 79:8750）

訳 ～は，内在性のORF50の発現を自己刺激する能力がある

■ endogenous ＋［名］

❶ endogenous gene	内在性の遺伝子	247
❷ endogenous levels	内在性のレベル	246

■ endogenous and ＋［形］／［形］＋ and endogenous

❶ endogenous and exogenous ～	内在性および外来性の～	207
❷ exogenous and endogenous ～	外来性および内在性の～	137

engage 動 関与させる／結合させる 用例数 3,240

engaged	関与した／結合した／関与させた	1,228
engage	関与させる／関与する／結合させる	1,173
engages	関与させる／関与する／結合させる	451
engaging	関与させる／関与する／結合させる	388

他動詞および自動詞の両方で用いられる．

◆類義語：involve, participate, implicate, bind, associate

■ be engaged ＋［前］

★ be engaged	…は，関与する	351
❶ be engaged in ～	…は，～に関与する	159
❷ be engaged by ～	…は，～と結合する	60

❶ We determined whether PKC-θ is engaged in interferon（INF）signaling in T-cells.（J Biol Chem. 2004 279:29911）

訳 われわれは，PKC-θがT細胞におけるインターフェロン（INF）シグナル伝達に関与するかどうかを決定した

❷ Cyclin T1 is an unusually long cyclin and is engaged by cellular regulatory proteins.（J Biol Chem. 2005 280:13648）

訳 ～は，細胞の調節性タンパク質と結合する

■ engage in

❶ engage in ～	…は，～に関与する	185

❶ These acidic residues may engage in electrostatic interactions with basic residues on fractalkine that are necessary for receptor function but not for binding.（Mol Pharmacol. 2006 69:857）

訳 これらの酸性残基は，塩基性残基との静電的な相互作用に関与するかもしれない

E

engineer 動 操作する／構築する 用例数 5,759

engineered	操作される／～を操作した／～を構築した	5,248
engineer	～を操作する／～を構築する	439
engineers	～を操作する／～を構築する	72

他動詞．engineeringは別項参照．

◆類義語：manipulate

■ engineered ＋［前］

❶ engineered to ～	～するように操作される	1,050
・engineered to express ～	～を発現するように操作される	313
・cells engineered to ～	～するように操作された細胞	154
❷ engineered into ～	～へ操作される	163

❶ In cells engineered to express hemojuvelin, soluble hemojuvelin release was progressively inhibited by increasing iron concentrations.（Blood. 2005 106:2884）

訳 ～を発現するように操作された細胞において

■［副］＋ engineered

❶ genetically engineered	遺伝的に操作された	205

❶ In these previous experiments the tumor cells used were genetically engineered to express membrane FasL.（Invest Ophthalmol Vis Sci. 2005 46:2495）

訳 …は，～を発現するように遺伝的に操作された

■ [代名] ＋ engineer

❶	we engineered ～	われわれは，～を構築した	420

❶ Therefore, **we engineered** a green fluorescent protein (GFP) construct to measure N-glycosylation site occupancy. (FASEB J. 2012 26:4210)
訳 われわれは，緑色蛍光タンパク質(GFP)コンストラクトを構築した

engineering 名 工学／エンジニアリング　用例数 3,195

現在分詞としても使われるが，名詞の用例が多い．複数形の用例はなく，原則，**不可算名詞**として使われる．

■ engineering ＋ [前]

	the%	a/an%	ø%			
❶	54	2	44	engineering of ～	～の工学	519

❶ The development of zinc finger nuclease (ZFN) technology has enabled the genetic **engineering of** the rat genome. (Mol Pharmacol. 2012 81:220)
訳 ～は，ラットゲノムの遺伝子工学を可能にしてきた

■ [名／形] ＋ engineering

❶	tissue engineering	組織工学	449
❷	protein engineering	タンパク質工学	320
❸	genetic engineering	遺伝子工学	268

★★ enhance 動 増強する　用例数 58,983

enhanced	増強される／～を増強した	36,304
enhance	～を増強する	10,562
enhances	～を増強する	7,803
enhancing	～を増強する	4,314

他動詞．
◆類義語：potentiate, augment, facilitate, intensify, reinforce, increase, activate

■ be enhanced ＋ [前]/when

★	be enhanced	…は，増強される	4,558
❶	be enhanced by ～	…は，～によって増強される	1,967
❷	be enhanced in ～	…は，～において増強される	758
	・be enhanced in the presence of ～	…は，～の存在下において増強される	95
❸	be enhanced when ～	…は，～のとき増強される	163

❶ Interestingly, the ER α -MPG interaction was enhanced by the presence of estrogen response element (ERE)-containing DNA. (J Biol Chem. 2004 279:16875)

訳 …は，～の存在によって増強された

❷ Hydroxyapatite binding is enhanced in the presence of calcium. (J Biol Chem. 2003 278: 11843)

訳 ヒドロキシアパタイト結合は，カルシウム存在下において増強される

■ [副] + enhanced

❶ significantly enhanced ～	有意に増強される／～を有意に増強した	1,381
❷ greatly enhanced ～	大きく増強される／～を大きく増強した	608
❸ markedly enhanced ～	著しく増強される／～を著しく増強した	493
❹ further enhanced ～	さらに増強される／～をさらに増強した	396

❶ As demonstrated in this study, baseline apoptosis of eosinophils resulted from oxidant-mediated mitochondrial injury that was significantly enhanced by GC. (J Immunol. 2003 170:556)

訳 ～は，GCによって有意に増強された

❷ In agreement with this, current responses of SJ-RH30 cells to a pH 6.0 challenge were greatly enhanced by extracellular lactate. (J Physiol. 2005 562:759)

訳 ～は，細胞外乳酸によって大きく増強された

E

■ [前] + enhanced

❶ result in enhanced ～	…は，増強された～という結果になった	824
❷ lead to enhanced ～	…は，増強された～につながる	573
❸ be associated with enhanced ～	…は，増強された～と関連した	324

❷ Deletion of the membrane-binding domain leads to enhanced cytosolic interactions. (J Cell Biol. 2003 162:1233)

訳 膜結合ドメインの欠失は，増強されたサイトゾルの相互作用につながる

❸ This was associated with enhanced MEF2 DNA binding and transactivation activity. (Gastroenterology. 2004 127:1174)

訳 これは，増強されたMEF2のDNA結合と関連した

■ enhanced + [名]

❶ enhanced expression of ～	～の増強された発現	514
❷ enhanced sensitivity	増強された感受性	340
❸ enhanced binding	増強された結合	326

❶ $CD8^+$ T cells from poor responders also showed enhanced expression of cytokines. (J Am Soc Nephrol. 2003 14:1776)

訳 ～は，また，サイトカインの増強された発現を示した

■ enhance + [名句]

❶ enhance the ability of ～	…は，～の能力を増強する	348
❷ enhance the expression of ～	…は，～の発現を増強する	236
❸ enhance the activity of ～	…は，～の活性を増強する	222

❹	enhance the rate of ～	…は，～の速度を増強する	218
❺	enhance the efficacy of ～	…は，～の効率を増強する	152

❶ Moreover, **the SDR stressor** enhances the ability of **these macrophages to kill Escherichia coli** both in vitro and in vivo, through a Toll-like receptor 4-dependent mechanism. (Infect Immun. 2012 80:3429)

訳 SDRストレッサーは，大腸菌を殺すこれらのマクロファージの能力を増強する

enhancement 名 増強／亢進 用例数 8,603

enhancement	増強／亢進	8,056
enhancements	増強／亢進	547

複数形のenhancementsの割合が，約5％ある可算名詞.

◆類義語：potentiation, augmentation, reinforcement, facilitation, activation

■ enhancement ＋［前］

	the%	a/an%	ø%			
❶	29	18	44	enhancement of ～	～の増強	3,949
				・…-mediated enhancement of ～	…に仲介される～の増強	181
				・…-induced enhancement of ～	…に誘導される～の増強	140
❷	14	38	39	enhancement in ～	～における増強	617

❶ **NAC-**mediated enhancement of **apoptosis** was mimicked by incubating cells with GSH monoester, which increased intracellular GSH similarly to NAC. (J Biol Chem. 2004 279:50455)

訳 NACに仲介されるアポトーシスの亢進

❷ Specifically, Q425 binding of calcium **resulted in a 55,000-fold** enhancement in **affinity for CD4.** (Proc Natl Acad Sci USA. 2005 102:14575)

訳 ～は，CD4に対する親和性の55,000倍の増強という結果になった

■［名／形］＋ enhancement

❶	contrast enhancement	コントラストの増強	237
❷	significant enhancement	有意な増強	197
❸	～-fold enhancement	～倍の増強	188
❹	rate enhancement	速度の亢進	175

enough 副 十分に，形 十分な 用例数 1,938

enough toの用例が非常に多い.

◆類義語：sufficiently, fully, entirely, completely, quite

■ enough + [前]

❶ enough to *do*	～するのに十分に	1,203
・enough to allow ～	～を可能にするのに十分に	87
❷ enough for ～	～のために十分に	123

❶ More importantly, **expression levels were robust enough to allow direct visualization of the fusion protein** by fluorescence microscopy. (J Virol. 2004 78:7400)

訳 発現レベルは，融合タンパク質の直接の可視化を可能にするのに十分に強力であった

■ [形] + enough

❶ large enough	十分に大きい	159
❷ long enough	十分に長い	88

❶ The results show that the pool of pyridoxal 5'-phosphate that is not bound to proteins **is large enough to account for product inhibition of** both pyridoxal kinase and pyridoxine 5'-phosphate oxidase. (Anal Biochem. 2001 298:314)

訳 …は，～の生産物阻止を説明するのに十分大きい

enrich 動 濃縮する　用例数 8,055

enriched	～を濃縮した／濃縮される／強化される	7,584
enrich	～を濃縮する／～を強化する	305
enriching	～を濃縮する／～を強化する	86
enriches	～を濃縮する／～を強化する	80

他動詞.

◈類義語：concentrate

■ be enriched + [前]

★ be enriched	…は，濃縮される	2,093
❶ be enriched in ～	…は，～において濃縮される	1,127
❷ be enriched for ～	…は，～に対して濃縮される	347
❸ be enriched at ～	…は，～において濃縮される	164

❶ We find that **Ago1 protein is enriched in the oocytes** and is also highly expressed in the cytoplasm of follicle cells. (Dev Biol. 2012 365:384)

訳 Ago1タンパク質は，卵母細胞において濃縮される

❷ **These genes are enriched for functions related to biotic stress** and may reflect responses to the effects of inbreeding. (Proc Natl Acad Sci USA. 2012 109:11878)

訳 これらの遺伝子は，生物的ストレスに関連する機能に対して濃縮される

■ [副] + enriched

❶ highly enriched	高度に濃縮される	618
❷ significantly enriched	有意に濃縮される	145

❶ CaMKIV is highly enriched in the nucleus and thought to be critical for improved survival.（J Biol Chem. 2005 280:2165）
訳 CaMKIVは，核において高度に濃縮される

enroll 動 登録する　用例数 4,517

enrolled	登録される／登録した	4,307
enrolling	登録する	111
enroll	登録する	98
enrolls	登録する	1

他動詞の用例が多いが，自動詞としても使われる．
◆類義語：register

■ enrolled ＋［前］

❶ enrolled in ～	～に登録される	1,702
・be enrolled in ～	…は，～に登録される	653
❷ enrolled onto ～	～に登録される	245
❸ enrolled at ～	～に登録される	137
❹ enrolled between ～	～の間に登録される	109

❶ A total of 909 patients were enrolled in the trial.（N Engl J Med. 2012 366:991）
訳 合計909名の患者が，その治験に登録された

■［名］＋ be enrolled

★ be enrolled	…が，登録される	1,920
❶ patients were enrolled	患者が，登録された	531

❶ Thirty-one patients were enrolled onto the study.（J Clin Oncol. 2011 29:4351）
訳 31名の患者がその研究に登録された

■［名］＋ enrolled

❶ patients enrolled	登録された患者	581
・patients enrolled in ～	～に登録された患者	300
❷ women enrolled	登録された女性	112

❶ Of 539 patients enrolled in the initial 1-year study, 488 completed 1 year of the long-term extension（2% discontinued for lack of efficacy）.（Arthritis Rheum. 2008 58:953）
訳 最初の１年の研究に登録された539名の患者うち，488名が１年の長期延長を完了した

■ enrolled ＋［名］

❶ enrolled patients	登録された患者	133

■ [代名] + enroll

❶ we enrolled ～	われわれは，～を登録した	261

❶ We enrolled 184 patients, of whom 156 had metastatic melanoma. (Lancet. 2012 379:1893)
訳 われわれは，184名の患者を登録した

ensure 動 保証する／確実にする 用例数 4,414

ensure	～を保証する	2,693
ensures	～を保証する	803
ensuring	～を保証する	781
ensured	～を保証した／保証される	137

他動詞．受動態の用例は少ない．ensure thatの用例が非常に多い．
◆類義語：warrant, guarantee, assure, certify, insure

■ ensure + [that節／名句]

❶ ensure that ～	…は，～ということを保証する	1,303
・to ensure that ～	～ということを保証するために	541
❷ ensure ～ segregation	…は，～分離を保証する	95
❸ ensure ～ fidelity	…は，忠実度を保証する	62

❶ During cortical neurogenesis, cell proliferation and cell cycle exit are carefully regulated to ensure that the appropriate numbers of cells are produced. (J Neurosci. 2005 25:8627)
訳 …は，～ということを保証するために慎重に調節される

❶ The spindle checkpoint is a cell cycle surveillance mechanism that ensures the fidelity of chromosome segregation during mitosis and meiosis. (J Biol Chem. 2007 282:3672)
訳 紡錘体チェックポイントは，染色体分離の忠実度を保証する細胞周期監視機構である

■ to ensure

★ to ensure ～	～を保証するために	1,999
❶ required to ensure ～	～ということを保証するために必要とされる	57

enter 動 入る 用例数 5,986

enter	入る／入れる	2,657
entered	入った／入れた／入れられる	1,497
entering	入る／入れる	1,021
enters	入る／入れる	811

他動詞の用例が多いが，自動詞として用いられることもある．他動詞として，「入る」と「入れる」の両方の意味で使われるので注意が必要である．
◆類義語：translocate, migrate, move

■ enter ＋［名句］

❶ enter cells	…が，細胞に入る	223
❷ enter the nucleus	…が，核に入る	220
❸ enter the cell cycle	…が，細胞周期に入る	131
❹ enter mitosis	…が，有糸分裂に入る	128
❺ enter clinical trials	…が，臨床治験に入る	71

❷ The removal of I-κB, via the actions of inhibitor of κB (I-κB) kinase-2 (IKK-2), **allows NF-κB to enter the nucleus.** (Am J Respir Crit Care Med. 2005 172:962)
訳 ～は，NF-κBが核に入るのを可能にする

■ enter ＋［前］

❶ enter into ～	…が，～に入る	182

■［名］＋ enter

❶ cells enter	細胞が入る	223

❶ The compaction of chromatin that occurs **when cells enter mitosis** is probably the most iconic process of dividing cells. (Trends Genet. 2012 28:110)
訳 細胞が有糸分裂に入るとき

■ be entered

★ be entered	…が，入れられる	291
❶ be entered into ～	…が，～に入れられる	111
❷ patients were entered	患者が，入れられた	86
❸ be entered onto ～	…が，～に入れられる	82
・be entered onto ～ study	…が，研究に入れられる	43

❸ **Two thousand four hundred thirty-four patients were entered onto the study.** (20837956Am J Respir Crit Care Med. 2005 172:962)
訳 2434名の患者が，その研究に入れられた

entirely 副 完全に／全体に　用例数 2,404

◈類義語：completely, fully, perfectly, totally, enough, quite, grossly

■［副］＋ entirely

❶ almost entirely	ほとんど完全に	311

❶ Whereas lethal disease was observed in immunodeficient mice, **tumour development was almost entirely blocked in immunocompetent mice.** (Nature. 2005 434:904)
訳 腫瘍の発生は，免疫適格性のマウスにおいてほとんど完全に阻止された

■ entirely ＋［形］

❶	entirely different	完全に異なる	111
❷	entirely dependent on ～	完全に～に依存している	84
❸	entirely new	完全に新しい	80

❷ Further, this decrease in alteration frequency is entirely dependent on DNA mismatch repair.（Genetics. 2005 170:601）

訳 ～は，DNAミスマッチ修復に完全に依存している

■ entirely ＋［前］

❶	entirely of ～	完全に～から	127
	・composed entirely of ～	完全に～から成る	42
❷	entirely by ～	完全に～によって	125
❸	entirely on ～	完全に～に	100

❶ The peptide is composed entirely of natural amino acids and exhibits a marked（42%）change in fluorescence between its oxidized and its reduced states.（Anal Biochem. 2004 325:144）

訳 そのペプチドは，完全に天然のアミノ酸から成る

★entry 名 移行／移入　　用例数 12,389

entry	移行／移入	12,011
entries	移行／移入	378

複数形のentriesの割合は3％あるが，単数形は無冠詞で使われることが圧倒的に多い．

◆類義語：transition, transit, translocation, shift, transfer, entrance

■ entry ＋［前］

	the％	a/an％	ø％			
❶	0	2	96	entry into ～	～への移行（移入）	2,195
				・entry into mitosis	有糸分裂への移行	188
				・entry into cells	細胞への移入	176
				・entry into S phase	S期への移行	134
❷	46	0	54	entry of ～	～の移行	1,354
				・entry of ～ into …	～の…への移行	438
❸	2	1	91	entry in ～	～における移行	354

❶ Wee1 kinases delay entry into mitosis by phosphorylating and inactivating cyclin-dependent kinase 1（Cdk1）.（Curr Biol. 2005 15:1525）

訳 Wee1キナーゼは，有糸分裂への移行を遅らせる

❷ It is tempting to speculate that the absence of BBA74 interferes with the enhanced nutrient uptake that may be required for the entry of cells into log-phase growth.（Infect Immun. 2005 73:6791）

訳 ～は，細胞の対数期増殖への移行のために必要とされるかもしれない

E

❸ In contrast, Ca^{2+} released via ryanodine receptors (RYR) increases NFAT nuclear entry in myotubes. (Dev Biol. 2005 287:213)
訳 ～は，筋管におけるNFATの核移行を増大させる

■ [名／形] + entry

❶	Ca^{2+} entry	Ca^{2+}流入	636
❷	viral entry	ウイルスの移入	578
❸	virus entry	ウイルスの移入	491

★ environment 名 環境　　用例数 16,109

environment	環境	12,442
environments	環境	3,667

複数形のenvironmentsの割合が，約25%ある可算名詞.
◆類義語：milieu, circumstance, ecology

■ environment + [前]

environment ofの前にtheが付く割合は圧倒的に高い.

	the %	a/an %	ø %			
❶	92	5	2	environment of ～	～の環境	1,045
❷	58	26	9	environment in ～	～における環境	491
				・environment in which ～	～である環境	119
❸	19	76	3	environment for ～	～のための環境	382

❶ The local environment of delivered mesenchymal stem cells (MSCs) may affect their ultimate phenotype. (Circulation. 2003 108:2899)
訳 送達された間葉系幹細胞（MSC）の局所環境は，それらの最終の表現型に影響するかもしれない

❷ However, memory CD4 T cells lose the ability to reject skin grafts when transiently placed in an environment in which these low-level TCR stimulations are absent. (J Immunol. 2005 175:4829)
訳 これらの低レベルのTCR刺激が存在しない環境において

❸ The aim of this study was to develop a simulated environment for surgical trainees using similar principles. (Ann Surg. 2005 242:631)
訳 この研究の目的は，外科研修生のための模擬的環境を開発することであった

■ [形] + environment

❶	local environment	局所環境	316
❷	cellular environment	細胞環境	291

★ environmental 形 環境の 用例数 10,657

◈類義語：ambient

■ environmental ＋［名］

❶	environmental factors	環境因子	1,186
	・genetic and environmental factors	遺伝および環境因子	297
❷	environmental conditions	環境条件	801
❸	environmental cues	環境要因	437
❹	environmental stress	環境ストレス	361
❺	environmental signals	環境シグナル	357
❻	environmental stimuli	環境刺激	327

❶ The role of genetic and environmental factors on dental caries progression in young children was determined.（J Dent Res. 2005 84:1047）

訳 遺伝および環境因子の齲蝕進行に対する役割

■［前］＋ environmental

❶	in response to environmental ～	環境の～に応答して	336

❶ Taken together, these data suggest that OmrA and OmrB participate in the regulation of outer membrane composition in response to environmental conditions.（Mol Microbiol. 2006 59:231）

訳 ～は，環境条件に応答して外膜組成の調節に関与する

episode 名 発症 用例数 5,395

episodes	発症	3,647
episode	発症	1,748

複数形のepisodesの割合は，約70%と圧倒的に高い．

◈類義語：onset, event

■ episodes ＋［前］

	the %	a/an %	ø %			
❶	0	-	64	episodes of ～	～の発症	1,358
❷	5	-	67	episodes in ～	～における発症	188

❶ There were 31 episodes of infection after treatment; most cases were mild.（J Clin Oncol. 2000 18:317）

訳 治療のあと31の感染の発症があった

■［名／形］＋ episode(s)

❶	rejection episodes	拒絶の発症	360
❷	first episode	最初の発症	325

E

equal ㊊ 等しい／同じ，㊌ 匹敵する 用例数 4,189

equal	等しい／同じ／～に匹敵する	3,954
equals	～に匹敵する	141
equaled	～に匹敵した	80
equaling	～に匹敵する	14

動詞としても用いられるが，形容詞の用例がほとんどである．

◆類義語：equivalent, identical, same, comparable, consistent

■ equal ＋［前］

❶ equal to ～	～に等しい	1,355
・… than or equal to ～	～より…か等しい	346
・greater than or equal to ～	～より大きいか等しい	149
・less than or equal to ～	～より少ないか等しい	118
・be equal to ～	…は，～に等しい	246

❶ Indeed, the rate of wound healing in the CGS-21680-treated diabetic rats was greater than or equal to that observed in untreated normal rats.（J Exp Med. 1997 186:1615）

訳 ～は，未処置の正常ラットにおいて観察されたそれより大きいか等しかった

■［副］＋ equal

❶ approximately equal	ほぼ等しい	449
❷ nearly equal	ほとんど等しい	122

❶ The amplitude of the peak of Ca^{2+} sparks was approximately equal to 170 nmol/l.（Circ Res. 1997 81:462）

訳 Ca^{2+}スパークのピークの振幅は，170 nmol/lにほぼ等しかった

■ equal ＋［名］

❶ equal amounts	等しい量	166
❷ equal numbers	等しい数	129

equally ㊓ 同等に／同様に 用例数 3,144

◆類義語：equivalently, identically, similarly

■ equally ＋［副／形］

❶ equally well	同等によく	387
❷ equally effective	同等に効果的な	289
・be equally effective	…は，同等に効果的である	238
❸ equally important	同等に重要な	151

❶ In contrast, mu4 binds equally well to the GTP- and GDP-bound forms of ARF1 and is less dependent on switch I and switch II residues.（EMBO J. 2001 20:6265）

訳 mu4は，ARF1 のGTP結合型およびGDP結合型に同等によく結合する

❷ Pericardiectomy was equally effective in relieving symptoms regardless of the presence or absence of increased thickness. (Circulation. 2003 O108:1852)

訳 心膜切除術は，～において同等に効果的であった

■ equally ＋［前］

❶ equally to ～	～に同等に	118
・contribute equally to ～	…は，～に同等に寄与する	45
❷ equally in ～	～において同等に	107

equilibrium 名 平衡／平衡状態　用例数 8,262

equilibrium	平衡／平衡状態	7,775
equilibria	平衡／平衡状態	487

複数形のequilibriaの割合が，約５％ある可算名詞.

◆類義語：balance, equilibration

■ equilibrium ＋［前］

	the％	a/an％	ø％			
❶	43	32	24	equilibrium between ～	～の間の平衡	468
❷	0	9	88	equilibrium with ～	～との平衡	332
				・in equilibrium with ～	～と平衡状態にある	147
❸	74	15	10	equilibrium of ～	～の平衡	218

❶ A dynamic equilibrium between multiple sorting pathways maintains polarized distribution of plasma membrane proteins in epithelia. (J Biol Chem. 2005 280:14741)

訳 複数のソーティング経路の間の動的な平衡は，～を維持する

❷ In this scenario, only inactive Cdc2-cyclin B can form aggregates, and the aggregates are in equilibrium with inactive Cdc2-cyclin B in solution. (Mol Biol Cell. 2003 14:4695)

訳 その凝集物は，溶液中の不活性なCdc2-cyclin Bと平衡状態にある

■［前］＋ equilibrium

❶ at equilibrium	平衡状態で	244

❶ The extent of mono and interstrand cross-linking was compared with the level of binding at equilibrium. (Nucleic Acids Res. 1997 25:4123)

訳 ～が，平衡状態での結合のレベルと比較された

equivalent 形 同等の／等価な，名 等価物　用例数 8,050

equivalent	同等の／等価物	7,248
equivalents	同等の／等価物	802

名詞としても用いられるが，形容詞の用例が多い.

E

◆類義語：comparable, corresponding, equal, identical, same, consistent

■ equivalent ＋［前］

❶ equivalent to ～	～と同等である	2,092
・be equivalent to ～	…は，～と同等である	697
・equivalent to that	それと同等である	378
❷ equivalent in ～	～において同等である	322

❶ The growth rate of the wild type was equivalent to that of the mutant when citrate or phospholipid was employed as the sole carbon source.（J Bacteriol. 2000 182:4889）

訳 野生型の増殖速度は，変異体のそれと同等であった

❷ ELISpot reactivity to genotype 1-derived antigens was equivalent in patients of genotypes 1, 2, and 3.（Hepatology. 2006 43:91）

訳 ～は，遺伝子型１，２および３の患者において同等であった

■［副］＋ equivalent

❶ functionally equivalent	機能的に同等の	222
❷ nearly equivalent	ほとんど同等の	74

❶ This mutant protein is functionally equivalent to wild-type gp45.（Proc Natl Acad Sci USA. 1999 96:12448）

訳 この変異タンパク質は，野生型gp45と機能的に同等である

escape 名 回避，動 逃れる　用例数 3,730

escape	回避／逃れる	3,210
escaped	逃れた	221
escapes	逃れる／回避	190
escaping	逃れる	109

動詞としても用いられるが，名詞の用例が断然多い．複数形の用例はほとんどなく，原則，不可算名詞として使われる．動詞としては，他動詞と自動詞の両方の用例がある．

◆類義語：avoidance, evasion, avoid, circumvent

■ 名 escape ＋［前］

	the%	a/an%	ø%			
❶	3	4	86	escape from ～	～からの回避	642
❷	47	5	45	escape of ～	～の回避	170

❶ Escape from normal apoptotic controls is thought to be essential for the development of cancer.（Infect Immun. 2005 73:6311）

訳 正常なアポトーシスの制御からの回避は，癌の発生にとって必須であると考えられる

■ ㊍ to escape

❶ to escape ～	～を回避する…	410
❷ to escape from ～	～から逃れる…	103

❶ Such loss of Tax expression enables ATL cells to escape the host immune system. (Oncogene. 2005 24:6047)

訳 Tax発現のそのような喪失は，ATL細胞が宿主の免疫系を逃れることを可能にする

❷ Significantly, the transported proteins were able to escape from endosomes. (J Am Chem Soc. 2010 132:2642)

訳 輸送されたタンパク質は，エンドソームから逃れることができた

★ especially 副 特に 用例数 8,909

◆類義語：particularly, in particular, notably, specially

E

■ especially ＋［前］/when

❶ especially in ～	特に～において	1,878
❷ especially for ～	特に～に対して	535
❸ especially when ～	特に～のとき	441

❶ Elevated levels of plasminogen activator inhibitor-1 (PAI-1) are associated with myocardial infarction and stroke, especially in patients with diabetes. (Circulation. 2005 111:3261)

訳 特に糖尿病の患者において

❸ CAD has a substantial genetic basis, especially when it occurs early. (Am J Hum Genet. 2005 77:1011)

訳 特にそれが早期に起こるとき

■ especially ＋［形］

❶ especially important	特に重要である	201
・be especially important	…は，特に重要である	157
❷ especially useful	特に有用である	96

❶ A circadian translation program may be especially important in fly pacemaker cells. (Genetics. 2012 192:943)

訳 概日性の翻訳プログラムは，ハエのペースメーカー細胞において特に重要である

★★ essential 形 必須な／必要不可欠な 用例数 41,377

essential forの用例が多い．

◆類義語：mandatory, obligatory, necessary, fundamental, important

■ be essential ＋［前］

❶ be essential for ～	…は，～のために必須である	15,841
・be essential for normal ～	…は，正常な～のために必須である	494

・be essential for proper ～	…は，適当な～のために必須である	257
・be essential for maintaining ～	…は，～を維持するために必須である	187
・activity is essential for ～	活性は，～にとって必須である	186
・be essential for the development of ～	…は，～の発生のために必須である	177
・be essential for viability	…は，生存能のために必須である	163
・be essential for the formation of ～	…は，～の形成のために必須である	112
・be essential for understanding ～	…は，～を理解するために必須である	111
❷ be essential to ～	…は，～に必須である／～するために必須である	1,631
・be essential to maintain ～	…は，～を維持するために必須である	68
・be essential to understand ～	…は，～を理解するために必須である	60
❸ be essential in ～	…は，～において必須である	775
・be essential in ～ing	…は，～する際に必須である	224

❶ YY1 is essential for the development of mammalian embryos.（J Biol Chem. 2003 278:14046）
訳 YY1は，～の発生のために必須である

❶ Spt6 is essential for viability in Saccharomyces cerevisiae and regulates chromatin structure during pol Ⅱ transcription.（J Biol Chem. 2005 280:913）
訳 Spt6は，生存能のために必須である

❷ Smooth muscle cell proliferation around small pulmonary vessels is essential to the pathogenesis of pulmonary hypertension.（Proc Natl Acad Sci USA. 2003 100:12331）
訳 ～は，肺性高血圧症の病因に必須である

❷ This process is essential to maintain the proper activity of CRLs in cells.（J Biol Chem. 2012 287:29679）
訳 この過程は，～の適切な活性を維持するために必須である

❸ Selenium is essential in mammalian embryonic development.（J Biol Chem. 2004 279:8011）
訳 セレンは，哺乳類の胚発生において必須である

■ essential ＋［名］

❶ essential role	必須の役割	3,178
❷ essential component	必須の構成成分	909
❸ essential function	必須の機能	444
❹ essential step	必須のステップ	383

❶ Prolactin plays an essential role in the development of rodent mammary tumors and is a potent mitogen in human normal and cancerous breast tissues/cells.（Cancer Res. 2004 64:5677）
訳 プロラクチンは，～の発生において必須の役割を果たす

essentially 副 本質的に／基本的に 用例数 3,347

◆類義語：intrinsically, fundamentally, basically, virtually

■ essentially ＋［形］

❶ essentially identical	本質的に同一である	310
・be essentially identical	…は，本質的に同一である	197

❷ essentially all ～	基本的にすべての～	284
❸ essentially the same ～	本質的に同じ～	242
❹ essentially no ～	本質的に～のない	179
❺ essentially unchanged	本質的に変化のない	131

❶ The structure, characterized as a $T_3R_3^f$ zinc hexamer at 1.8 A resolution, is essentially identical to that of native insulin. (Biochemistry. 2004 43:16119)
訳 ～は，天然のインスリンのそれと本質的に同じである

❸ Both types of SIRT1 mutant mice and cells had essentially the same phenotypes. (Proc Natl Acad Sci USA. 2003 100:10794)
訳 ～は，本質的に同じ表現型を持っていた

★★ establish 動 確立する　用例数 34,715

established	確立される／～を確立した	19,838
establish	～を確立する	10,024
establishing	～を確立する	2,958
establishes	～を確立する	1,895

他動詞.

◆類義語：confirm, verify, validate, ascertain, demonstrate

■ establish ＋［名節］

❶ establish that ～	…は，～ということを確立する	4,638
❷ establish whether ～	…は，～かどうかを確立する	366
・to establish whether ～	～かどうかを確立するために	250

❶ These results establish that collagenase action regulates plaque collagen turnover and smooth muscle cell accumulation. (Circulation. 2004 110:1953)
訳 これらの結果は，～ということを確立する

❷ Differential screening was initially used to establish whether the cDNAs were differentially expressed. (Plant Mol Biol. 2003 51:619)
訳 …が，～かどうかを確立するために最初使われた

■ establish ＋［名句］

❶ establish a role for ～	…は，～の役割を確立する	244
❷ establish a novel ～	…は，新規の～を確立する	221
❸ establish a new ～	…は，新しい～を確立する	193
❹ establish the role of ～	…は，～の役割を確立する	138
❺ establish the importance of ～	…は，～の重要性を確立する	131

❶ Our results establish a role for HRG-3 as an intercellular heme-trafficking protein. (Cell. 2011 145:720)
訳 われわれの結果は，細胞間ヘム輸送タンパク質としてのHRG-3役割を確立する

❹ The aim of this study was to establish the role of GLP-1 on insulin secretion in patients with GB. (Diabetes. 2011 60:2308)

E

訳 この研究の目的は，GLP-1の役割を確立することであった

■ [代名／名] + establish

❶ we establish ～	われわれは，～を確立する	1,510
❷ results establish ～	結果は，～を確立する	1,228
❸ findings establish ～	知見は，～を確立する	661
❹ data establish ～	データは，～を確立する	554
❺ studies establish ～	研究は，～を確立する	513

❶ **We established a system for** propagation of Sindbis virus (SIN)-based replicons in tissue culture in the form of a tricomponent genome virus. (J Virol. 2005 79:637)

訳 われわれは，～のためのシステムを確立した

■ to establish

★ to establish ～	～を確立する…	3,994
❶ To establish ～	～を確立するために	416
❷ be to establish ～	…は，～を確立することである	190
❸ be used to establish ～	…は，～を確立するために使われる	183
❹ be required to establish ～	…は，～を確立するために必要とされる	107

❶ **To establish its importance**, we treated susceptible rats with a depleting anti-rat Vbeta13 monoclonal antibody and then exposed them to either polyinosinic:polycytidylic acid or a diabetogenic virus to induce diabetes. (Diabetes. 2012 61:1160)

訳 それの重要性を確立するために

■ be established + [前]

★ be established	…は，確立される	5,617
❶ be established by ～	…は，～によって確立される	676
❷ be established in ～	…は，～において確立される	551
❸ be established for ～	…は，～に対して確立される	226
❹ be established as ～	…は，～として確立される	203

❶ **The diagnosis was established by** serologic testing for IgM and IgG antibodies to La Crosse virus. (N Engl J Med. 2001 344:801)

訳 診断が，～によって確立された

❸ **A murine colony was established for** the CFTR ΔF508 (ΔF) mutation. (Gastroenterology. 2004 127:1162)

訳 マウスのコロニーが，～に対して確立された

■ to be established

★ to be established	確立される…	385
❶ remain to be established	…は，確立されないままである	224

❶ Enzymes with multiple catalytic sites are rare, and **their evolutionary significance remains to be established**. (Proc Natl Acad Sci USA. 2004 101:16555)

訳 それらの進化的重要性は，確立されないままである

■ ［副］＋ established

❶ well established	よく確立される	2,554
・a well-established ～	よく確立された～	412
・it is well established that ～	～ということは，よく確立されている	372
❷ previously established	以前に確立された	356

❶ It is well established that leukocyte chemotactic receptors, a subset of G protein-coupled receptors, undergo endocytosis after stimulation by ligand.（J Biol Chem. 2004 279:24372）

訳 ～ということがよく確立されている

❷ A tamoxifen-inducible Cre recombinase was introduced in the transgenic heart by breeding with previously established Mer-Cre-Mer transgenic mice.（FASEB J. 2003 17:749）

訳 以前に確立されたMer-Cre-Merトランスジェニックマウス

E

establishment 名 確立 用例数 3,921

establishment	確立	3,904
establishments	確立	17

複数形のestablishmentsの割合は0.4%しかなく，原則，不可算名詞として使われる．

■ establishment ＋［前］

establishment ofの前にtheが付く割合は非常に高い．

	the%	a/an%	ø%			
❶	75	0	25	establishment of ～	～の確立	3,052

❶ The results suggest a mechanism for the establishment of differences in transcription patterns during evolution.（Genome Res. 2012 22:2199）

訳 それらの結果は，転写パターンの違いの確立のための機構を示唆する

★ estimate 動 推定する／見積もる，名 推定／推定値／見積もり 用例数 24,060

estimated	推定される／～を推定した	9,859
estimates	推定／推定値／～を推定する	6,640
estimate	～を推定する／推定／推定値	6,225
estimating	～を推定する	1,336

他動詞の用例が多いが，名詞としても用いられる．複数形のestimatesの割合は，約80%と圧倒的に高い．

◆類義語：presume, assume, speculate, predict, expect, infer, postulate, deduce, estimation

■ 動 estimate ＋［名句］

❶ estimate the number of ～	…は，～の数を推定する	135
❷ estimate the effect of ～	…は，～の効果を推定する	98
❸ estimate the risk of ～	…は，～のリスクを推定する	64
❹ estimate the prevalence of ～	…は，～の有病率を推定する	63

❶ We sought to estimate the number of cancer patients diagnosed in the United States through 2030 by age and race.（J Clin Oncol. 2009 27:2758）

訳 われわれは，癌患者の数を推定しようと努めた

■ 動［代名］＋ estimate

❶ we estimate ～	われわれは，～を推定する	1,437
・we estimate that ～	われわれは，～であると推定する	535

❶ We estimate that exocytosis is accompanied by a transient change in synaptic cleft pH from 7.5 to approximately 6.9.（J Neurosci. 2003 23:11332）

訳 われわれは，～であると推定する

■ 動 to estimate

★ to estimate ～	～を推定するために	3,078
❶ be used to estimate ～	…は，～を推定するために使われる	870

❶ Coalescent methods were then used to estimate the in vivo recombination rate.（Genetics. 2004 167:1573）

訳 それから，～が，生体内組換え率を推定するために使われた

■ 動 be estimated ＋［前］/using

★ be estimated	…は，推定される	3,551
❶ be estimated to *do*	…は，～すると推定される	932
・be estimated to be ～	…は，～であると推定される	670
❷ be estimated by ～	…は，～によって推定される	570
・be estimated by using ～	…は，～を使うことによって推定される	109
❸ be estimated from ～	…は，～から推定される	411
❹ be estimated using ～	…は，～を使って推定される	334

❶ The number of TREK-2 channels in these cells was estimated to be approximately 500-1,000/cell.（Brain Res. 2002 931:56）

訳 これらの細胞におけるTREK-2チャネルの数は，細胞あたりおよそ500から1,000であると推定された

❷ Total collagen level was estimated by hydroxyproline quantification.（Gastroenterology. 2003 125:1750）

訳 総コラーゲンレベルは，ヒドロキシプロリンの定量によって推定された

❸ The effects of sex, geographic location, time, and deprivation category were estimated from a multifactorial Poisson regression model.（Gastroenterology. 2004 127:1051）

訳 ～は，多因子性ポアソン回帰モデルから推定された

■ 動 be estimated ＋［that節］

★	be estimated that ～	～であると推定される	151
❶	it is estimated that ～	～であると推定される	89

■ 名 estimates ＋［前］

	the%	a/an%	ø%			
❶	3	-	97	estimates of ～	～の推定（値）	2,835
❷	4	-	92	estimates for ～	～の推定（値）	508

❶ Estimates of the number of uninsured people in the United States usually exclude those with discontinuous coverage.（N Engl J Med. 2005 353:382）

訳 合衆国における無保険者の数の推定値は，通常，～を除外する

■ 名［名／形］＋ estimates

❶	risk estimates	リスクの推定	207
❷	previous estimates	以前の推定	140

E

etiology 名 病因／病因学 用例数 3,263

etiology	病因	2,923
etiologies	病因	340

複数形のetiologiesの割合が，約10%ある可算名詞.

◆類義語：occurrence, appearance

■ etiology ＋［前］

etiology ofの前にtheが付く割合は圧倒的に高い.

	the%	a/an%	ø%			
❶	94	2	4	etiology of ～	～の病因	1,528

❶ The etiology of intraventricular hemorrhage in preterm infants is multifactorial.（Ann Neurol. 2010 67:817）

訳 未熟児における脳室内出血の病因は，他因子性である

★★ evaluate 動 評価する／査定する 用例数 34,195

evaluated	評価される／～を評価した	20,770
evaluate	～を評価する／～を査定する	10,474
evaluating	～を評価する／～を査定する	2,561
evaluates	～を評価する／～を査定する	390

他動詞.

◆類義語：assess, estimate, value, examine

■ evaluate ＋［名句／whether節］

❶	evaluate the effect of ～	…は，～の影響を評価する	1,189
❷	evaluate whether ～	…は，～かどうかを評価する	855
❸	evaluate the role of ～	…は，～の役割を評価する	809
❹	evaluate the efficacy of ～	…は，～の有効性を評価する	283
❺	evaluate the ability of ～	…は，～の能力を評価する	279
❻	evaluate the association	…は，関連を評価する	275
❼	evaluate the impact of ～	…は，～の影響を評価する	224

❶ Therefore, **we evaluated the effect of stathmin expression on** the action of taxanes and Vinca alkaloids using a panel of human breast cancer cell lines. (Cancer Res. 2002 62:6864)

訳 われわれは，スタスミン発現の～に対する影響を評価した

❷ **To evaluate the role of polymorphisms in the eNOS gene in IHD**, we considered all available studies in a meta-analysis. (Circulation. 2004 109:1359)

訳 IHDにおけるeNOS遺伝子の多型性の役割を評価するために

■ ［代名／名］＋ evaluate

❶	we evaluated ～	われわれは，～を評価した	3,891
❷	study evaluated ～	研究は，～を評価した	617
❸	the authors evaluated ～	著者らは，～を評価した	204

❶ **We evaluated whether** men at risk for death from prostate cancer after radical prostatectomy can be identified using information available at diagnosis. (N Engl J Med. 2004 351:125)

訳 われわれは，～かどうかを評価した

■ to evaluate

★	to evaluate ～	～を評価する…	8,650
❶	To evaluate ～	～を評価するために	1,611
❷	be used to evaluate ～	…は，～を評価するために使われる	935
❸	the purpose of this study was to evaluate ～	この研究の目的は，～を評価することであった	310
❹	sought to evaluate ～	…は，～を評価しようと努めた	256
❺	be performed to evaluate ～	…は，～を評価するために実行される	198
❻	be designed to evaluate ～	…は，～を評価するために設計される	188

❷ **Small interfering RNA and antibodies were used to evaluate the involvement of RET.** (Cancer Res. 2005 65:11536)

訳 低分子干渉RNAと抗体が，RETの関与を評価するために使われた

■ be evaluated ＋［前］/using

★	be evaluated	…は，評価される	10,132
❶	be evaluated in ～	…は，～において評価される	1,815
❷	be evaluated by ～	…は，～によって評価される	1,674

	・be evaluated by using ～	…は，～を使うことによって評価される	147
❸	be evaluated for ～	…は，～に関して評価される	1,383
	・be evaluated for their ability to *do*	…は，～するそれらの能力に関して評価される	86
❹	be evaluated using ～	…は，～を使って評価される	639
❺	be evaluated with ～	…は，～を使って評価される	483
❻	be evaluated as ～	…は，～として評価される	352

❶ Knee cartilage was evaluated in 55 subjects who were categorized with radiography as healthy (n = 7) or as having mild OA (n = 20) or severe OA (n = 28).（Radiology. 2004 232: 592）

訳 膝軟骨が，55人の被検者において評価された

❷ Apoptotic cell death was evaluated by the TdT-mediated digoxigenin-dUTP nick-end labeling TUNEL assay, and by DNA laddering analysis.（Brain Res. 2004 1006:198）

訳 アポトーシス性の細胞死が，～によって評価された

❸ These analogs were evaluated for their ability to inhibit transcription in transiently transfected human lung epithelial A549 cells from either an AP-1 or NF-κB reporter.（Proc Natl Acad Sci USA. 2003 100:1169）

訳 …は，～における転写を抑制するそれらの能力に関して評価された

E

■ ［過分］＋ and evaluated

❶	be synthesized and evaluated	…は，合成され，そして評価される	282

❶ A number of 6-N-acyltriciribine analogues were synthesized and evaluated for antiviral activity and cytotoxicity.（J Med Chem. 2000 43:2457）

訳 ～が合成され，そして抗ウイルス活性と細胞傷害性に関して評価された

★ evaluation ㊔ 評価　用例数 10,306

evaluation	評価	9,429
evaluations	評価	877

複数形のevaluationsの割合は約10%あるが，単数形は無冠詞で使われることが多い．

◈類義語：assessment, estimation

■ evaluation ＋［前］

	the%	a/an%	ø%			
❶	21	13	65	evaluation of ～	～の評価	4,946
				・for evaluation of ～	～の評価のための	324
❷	5	5	85	evaluation in ～	～における評価	307
❸	6	10	82	evaluation for ～	～に対する評価	195

❶ This model system provides a powerful tool for evaluation of experimental therapeutics on hematopoietic stem cell function.（Blood. 1999 94:1）

訳 このモデル系は，造血幹細胞機能に対する実験的治療学の評価のための強力なツールを提供する

■ [形/名] + evaluation

❶ further evaluation	さらなる評価	453
❷ clinical evaluation	臨床的評価	440
❸ health evaluation	健康評価	235

❶ Further evaluation in diverse populations is necessary. (Am J Clin Nutr. 2005 81:990)
訳 多様な集団におけるさらなる評価が必要である

★ even 副 さえ/でも/なお 用例数 22,482

■ even + [前]

❶ even in ～	～においてでさえ	3,873
・even in the absence of ～	～の非存在下でさえ	1,101
・even in the presence of ～	～の存在下でさえ	795
❷ even after ～	～のあとでさえ	1,465
❸ even at ～	～においてでさえ	1,154

❶ Many of the mutants showed enhanced lipid mixing, calcein transfer, and syncytium formation even in the presence of the long SER F protein CT. (J Virol. 2004 78:8513)
訳 たとえ長いSER Fタンパク質の細胞質側末端の存在下でさえ

❸ Paramyxovirus nucleocapsids are flexible structures, a trait that has hitherto hampered structural analysis even at low resolution. (J Mol Biol. 2004 340:319)
訳 ～は，低い解像度でさえ今までに構造解析を妨害してきた

■ even + [接]

❶ even though ～	～であるけれども	2,921
❷ even when ～	～のときでさえ	2,619
❸ even if ～	たとえ～でも	620

❶ In the absence of ATM, telomere fusions occur even though telomere-specific Het-A sequences are still present. (Genes Dev. 2004 18:1850)
訳 テロメア特異的なHet-A配列はまだ存在するけれども

❷ Displacement results in autophosphorylation-independent activation of CaMKII which persists even when Ca^{2+} levels have gone down. (J Biol Chem. 2004 279:10206)
訳 Ca^{2+}レベルが低下したときでさえ

❸ Arvin therapy has no significant thrombolytic effect even if the thrombus is less than 24 hours old. (Circulation. 1970 42:729)
訳 たとえ血栓が24時間以内のものであっても

■ even + [副/形]

❶ even more ～	なおさら～/よりいっそう	788
❷ even greater ～	なおより大きな～	282

❶ Additionally, a loss of assembly competence was observed following longer term acid treatment, which was even more marked than that of the wild-type protein.（Biochemistry. 2004 43:1618）

訳 ～は，野生型のタンパク質のそれよりいっそう顕著であった

evenly ㊞ 均等に　用例数 395

◈類義語：uniformly

■ evenly ＋［過分］

❶	evenly distributed	均等に分布した	179
❷	evenly spaced	均等に配置された	48

❶ In immunogold electron microscopic studies, VZV gK was in enveloped virions and was evenly distributed in the cytoplasm in infected cells.（J Virol. 1999 73:4197）

訳 ～が，感染した細胞の細胞質において均等に分布した

■［過分］＋ evenly

❶	distributed evenly	均等に分布した	42

★★ event ㊔ 事象／現象／出来事　用例数 43,486

events	事象／現象	32,083
event	事象／現象	11,403

複数形のeventsの割合は，約75%と圧倒的に高い．

◈類義語：case, phenomenon, incident, occurrence, circumstance

■ events ＋［前］

	the %	a/an %	ø %			
❶	11	-	77	events in ～	～における事象	3,493
❷	46	-	50	events of ～	～の事象	771

❶ Right ventricular enlargement on the reconstructed CT 4-CH views predicts adverse clinical events in patients with acute PE.（Circulation. 2004 109:2401）

訳 …は，～の患者における有害な臨床事象を予想する

■ events ＋［現分／過分］

❶	events including ～	～を含む現象	436
❷	events leading to ～	～につながる事象	377
❸	events associated with ～	～に関連する事象	376

❶ Here, we describe the molecular events associated with activation of gene expression by urea-bound UreR.（J Biol Chem. 2002 277:37349）

訳 われわれは，～による遺伝子発現の活性化と関連する分子事象を述べる

E

■ events ＋［動］

❶ events occur	事象が，起こる	536

❶ Clinical cardiovascular **events occurred** in 3 patients treated with niacin (3.8%) and 7 patients treated with placebo (9.6%; P=0.20). (Circulation. 2004 110:3512)
訳 臨床的な心血管事象が，３人の患者において起こった

■［前］＋ events

❶ sequence of events	一連の事象	236
❷ cascade of events	一連の事象	170
❸ series of events	一連の事象	104

❶ The objective of this study is to better understand the **sequence of events** that occurs during repair from the time RNA polymerase II first encounters the lesion. (J Biol Chem. 2004 279:7751)
訳 この研究の目的は，一連の事象をよりよく理解することである

■［形］＋ events

❶ adverse events	有害事象	2,916
❷ signaling events	シグナル伝達事象	1,613
❸ cardiovascular events	心血管事象	1,115
❹ molecular events	分子的事象	866
❺ cardiac events	心臓事象	697
❻ early events	早期の事象	544

every 形 すべての／～ごとに　用例数 6,286

◆類義語：all

■ every ＋［名］

❶ every ～ weeks	～週間ごとに	824
❷ every ～ days	～日ごとに	350
❸ every ～ months	～ヶ月ごとに	289
❹ every ～ hours	～時間ごとに	180
❺ every other day	一日おきに	136
❻ every year	毎年	124

❶ After six cycles of chemotherapy, bevacizumab was given **every** 3 **weeks** for 17 additional treatments. (J Clin Oncol. 2011 29:4662)
訳 ベバシズマブが，３週間ごとに与えられた

★ evidence ㊔ 証拠，㊊ 立証する　用例数 62,327

evidence	証拠／形跡	59,966
evidenced	立証される	2,278
evidences	証拠／形跡	62
evidencing	立証する	21

主に名詞として使われるが，動詞の用例もある．複数形のevidencesの割合は0.1%しかなく，原則，**不可算名詞**として使われる．〈evidence＋that節〉〈evicence for〉〈evicence of〉の用例が非常に多い．

◆類義語：proof, demonstration, certification, hallmark, prove

■ ㊔ evidence ＋［同格that節／前］

	the%	a/an%	ø%			
❶	9	0	89	evidence that ～	～という証拠	16,956
❷	10	0	78	evidence for ～	～の証拠／～を支持する証拠	11,835
				・evidence for linkage	連鎖の証拠	420
				・evidence for the existence of ～	～の存在の証拠	260
				・evidence for the involvement of ～	～の関与の証拠	188
				・evidence for the role of ～	～の役割の証拠	186
				・evidence for the presence of ～	～の存在の証拠	167
❸	9	0	77	evidence of ～	～の証拠	10,843
				・with evidence of ～	～の証拠を持つ／～の証拠で	331
				・without evidence of ～	～の証拠なしに	317
				・for evidence of ～	～の証拠を求めて	306
				・evidence of linkage	連鎖の証拠	272
❹	7	0	92	evidence from ～	～からの証拠	1,120

❶ Finally, **we provide evidence that** Core-ISCOM could serve as an adjuvant for the HCV envelope protein E1E2. (J Immunol. 2001 166:3589)
訳 われわれは，～という証拠を提供する

❷ **We did not find any evidence for linkage between** type 2 diabetes and any other region on chromosome 20. (Diabetes. 2000 49:2212)
訳 われわれは，～の間の連鎖のどのような証拠も見つけなかった

❸ **Serial CT scans were retrospectively reviewed for evidence of local recurrence.** (Radiology. 1999 210:25)
訳 連続CTスキャンが，局所の再発の証拠を求めて遡及的に再吟味された

■ ㊔［動］＋ evidence

❶	provide evidence	…は，証拠を提供する	8,894
❷	present evidence	…は，証拠を提示する	2,296
❸	show evidence	…は，証拠を示す	948

E

❶ These results **provide evidence for** a role of K^+ uptake via IIR into astrocytes.（J Neurosci. 1998 18:4425）

訳 これらの結果は，～の証拠を提供する

❷ Here, **we present evidence that** $P2Y_6$ regulates chemokine production and release in monocytes.（J Biol Chem. 2001 276:26051）

訳 われわれは，～という証拠を提示する

❸ At 3 and 6 weeks after cell therapy, 92% (13 of 14) of MI^+cell hearts **showed evidence of myoblast graft survival**.（Circulation. 2001 103:1920）

訳 ～は，筋芽細胞移植生着の証拠を示した

■ 名 evidence ＋［動］

❶ evidence suggests that ～	証拠は，～ということを示唆する	2,770
❷ evidence indicates that ～	証拠は，～ということを示す	1,147

❶ Recent **evidence suggests that** endothelial DARC facilitates chemokine transcytosis to promote neutrophil recruitment.（PLoS One. 2011 6:e29624）

訳 最近の証拠は，～ということを示唆する

■ 名［形］＋ evidence

❶ no evidence	証拠のない	3,711
・there was no evidence	証拠はなかった	738
❷ direct evidence	直接の証拠	2,357
❸ recent evidence	最近の証拠	1,733
❹ the first evidence	最初の証拠	1,592
❺ strong evidence	強力な証拠	1,290
❻ experimental evidence	実験的な証拠	1,145
❼ further evidence	さらなる証拠	1,074
❽ genetic evidence	遺伝的な証拠	1,040

❶ **There was no evidence of mammary dysplasia or neoplasia during** the lifespan of multiparous transgenic mice.（Oncogene. 2002 21:198）

訳 ～の間の乳腺異形成あるいは新形成の証拠はなかった

❷ **These data provide direct evidence that** calcium influx through P2X2 receptors results in the activation of the MAP kinase cascade.（J Biol Chem. 1998 273:19965）

訳 これらのデータは，～という直接の証拠を提供する

■ 動 as evidenced by

❶ as evidenced by ～	～によって立証されたように	1,700

❶ RUFY proteins are localized predominantly to endosomes **as evidenced by their co-localization** with early endosome antigen marker (EEA1).（J Biol Chem. 2002 277:30219）

訳 それらの共局在によって立証されたように

evident ㊋ 明らかな／明白な　用例数 4,568

◆類義語：apparent, clear, obvious, pronounced

■ be evident ＋［前］

★	be evident	…は，明らかである	2,414
❶	be evident in ～	…は，～において明らかである	829
❷	be evident at ～	…は，～において明らかである	128
❸	be evident by ～	…は，～によって明らかである	120
❹	be evident from ～	…は，～から明らかである	117

❶ Changes were evident in patients receiving 16 or 24 mg/day of galantamine.（Am J Psychiatry. 2004 161:532）

訳 変化は，～を受けている患者において明らかであった

❷ Nuclear staining for KLF6 protein was evident at E15.5 in the corneal epithelium and the stroma.（Invest Ophthalmol Vis Sci. 2004 45:4327）

訳 KLF6タンパク質の核染色は，E15.5において明らかであった

■［動］＋ evident

❶	become evident	…は，～明らかになる	177

❶ Recently it has become evident that these structural models do not adequately explain the behavior of wild-type PAI-1 (wtPAI-1) in solution.（J Biol Chem. 2008 283:18147）

訳 ～ということが明らかになった

■［副］＋ evident

❶	most evident	最も明らかな	131
❷	more evident	もっと明らかな	96

evoke ㊌ 誘起する／惹起する／引き起こす　用例数 9,072

evoked	誘起される／～を誘起した	7,727
evoke	～を誘起する	749
evokes	～を誘起する	417
evoking	～を誘起する	179

他動詞.

◆類義語：elicit, provoke, induce

■ evoked ＋［前］

❶	evoked by ～	～によって誘起される	1,537
	・responses evoked by ～	～によって誘起される反応	118
❷	evoked in ～	～において誘起される	174

❶ In contrast, synaptic responses evoked by stimulation of CA3 pyramidal neurons are

mediated by calcium-impermeable AMPA receptors. (Nat Neurosci. 1998 1:572)
訳 CA3錐体ニューロンの刺激によって誘起されたシナプス性応答は，～によって仲介される

■ evoked ＋［名］

❶	evoked responses	誘起された反応／反応を誘起した	332
❷	evoked potentials	誘起された電位／電位を誘起した	241

★evolution 名 進化 用例数 15,405

evolution	進化	15,392
evolutions	進化	13

複数形のevolutionsの割合は0.1%しかなく，原則，不可算名詞として使われる．

■ evolution ＋［前］

evolution ofの前にtheが付く割合は圧倒的に高い．

	the %	a/an %	ø %			
❶	90	0	10	evolution of ～	～の進化	6,556
❷	4	3	86	evolution in ～	～における進化	702

❶ Retrotransposons play an important role in the evolution of genomic structure and function. (Genetics. 2002 162:1617)
訳 レトロトランスポゾンは，ゲノムの構造と機能の進化において重要な役割を果たす

❷ These results should form the basis of more realistic models of DNA and protein evolution in mitochondria. (Genetics. 2003 165:735)
訳 これらの結果は，ミトコンドリアにおけるDNAとタンパク質の進化のより現実的なモデルの基礎を形成するはずである

■［前］＋ evolution

❶	during evolution	進化の間に	396
❷	rate of evolution	進化の速度	80

❶ The elaboration of this mechanism may have had a significant role in the expansion of metazoan proteomes during evolution. (Nature. 2002 418:236)
訳 ～は，進化の間に後生動物のプロテオームの拡大において顕著な役割を担っていたかもしれない

evolve 動 進化する 用例数 8,701

evolved	進化した／～を進化させた	5,009
evolving	進化する／～を進化させる	1,733
evolve	進化する／～を進化させる	1,606
evolves	進化する／～を進化させる	353

自動詞の用例が多いが，他動詞としても使われる．

■ evolve ＋［前］

❶ evolve to *do*	…は，～するように進化する	982
❷ evolve from ～	…は，～から進化する	573
❸ evolve in ～	…は，～において進化する	567

❶ These findings indicate that **Rad54 and Rad51 have evolved to function with chromatin**, the natural substrate, rather than with naked DNA.（Genes Dev. 2002 16:2767）

訳 Rad54およびRad51は，クロマチンとともに機能するように進化してきた

❷ These results indicate that hDNase, though not an acid hydrolase, may enter the lysosomal trafficking pathway, and **may have evolved from a lysosomal enzyme**.（Biochemistry. 1998 37:15154）

訳 ～は，リソソーム酵素から進化したのかもしれない

★ examination 名 検査／検討／試験　用例数 10,159

examination	検査／検討	8,510
examinations	検査／検討	1,649

複数形のexaminationsの割合は約15%ある可算名詞だが，単数形は無冠詞の用例が断然多い．

◆類義語：test, investigation, trial

■ examination ＋［前］

	the%	a/an%	ø%			
❶	7	11	81	examination of ～	～の検討	4,946

❶ **An examination of the transcription start site verified that** the initiation site of the NECI-CAT mRNA in transgenic plants is identical with that of the native gene in vivo.（Plant Mol Biol. 2003 51:451）

訳 転写の開始部位の検討は，～ということを確認した

■ examination ＋［動］

❶ examination revealed ～	検査は，～を明らかにした	188
❷ examination showed ～	検査は，～を示した	102

❶ **Histological examination revealed that** the liver of mutant animals contained abnormal cells with enlarged nuclei.（Mol Cell Biol. 2004 24:1200）

訳 組織学的検査は，～ということを明らかにした

■［形］＋ examination

❶ physical examination	理学的検査／身体検査	447
❷ clinical examination	臨床検査	297
❸ histologic examination	組織学的検査	259
❹ microscopic examination	検鏡	249

❺ histological examination	組織学的検査	225

★★ examine 動 調べる 用例数 62,380

examined	～を調べた／調べられる	42,659
examine	～を調べる	15,387
examining	～を調べる	2,952
examines	～を調べる	1,382

他動詞.

◆類義語：test, investigate, study, survey, look at, explore, search, dissect, analyze

■ examine ＋［名句／whether節］

❶ examine the effect of ～	…は，～の効果を調べる	4,054
❷ examine whether ～	…は，～かどうかを調べる	2,957
❸ examine the role of ～	…は，～の役割を調べる	2,752
❹ examine the relationship between ～	…は，～の間の関係を調べる	650
❺ examine the ability of ～	…は，～の能力を調べる	541
❻ examine the expression of ～	…は，～の発現を調べる	450
❼ examine the association between ～	…は，～の間の関係を調べる	442

❶ To understand the way these changes contribute to cancer predisposition, **we examined the effects of defective mismatch repair** on the multistep process of pre-B-cell transformation by Abelson murine leukemia virus.（Mol Cell Biol. 2000 20:8373）

訳 われわれは，欠陥のあるミスマッチ修復の影響を調べた

❷ **We examined whether insulin and catecholamines share common pathways for** their stimulating effects on glucose uptake.（Circ Res. 1999 84:467）

訳 われわれは，インスリンとカテコールアミンが～に対する共通の経路を共有するかどうかを調べた

❸ **We further examined the role of** the most prominently induced site near the TCR-β enhancer（Eβ）in allelic exclusion by targeted mutagenesis.（J Immunol. 1998 160:1256）

訳 われわれは，～の役割をさらに調べた

■［代名／名］＋ examine

❶ we examined ～	われわれは，～を調べた	13,995
・in this study, we examined ～	この研究において，われわれは～を調べた	1,245
❷ study examined ～	研究は，～を調べた	1,605
❸ the authors examined ～	著者らは，～を調べた	815

❶ **In this study, we examined** an involvement of the mitochondria in oligodendrocyte apoptosis and the role of C5b-9 on this process.（J Immunol. 2001 167:2305）

訳 この研究において，われわれは～を調べた

■ to examine

★ to examine ～	～を調べる…	11,358
❶ To examine ～	～を調べるために	2,947
❷ be used to examine ～	…が，～を調べるために使われる	1,275
❸ be to examine ～	…は，～を調べることである	1,266
・the purpose of this study was to examine ～	この研究の目的は，～を調べることであった	302

❷ Finally, reverse transcription-PCR **was used to examine** the Akt3 expression in 27 primary breast carcinomas. (J Biol Chem. 1999 274:21528)

訳 ～が，Akt3発現を調べるために使われた

■ be examined ＋［前］

★ be examined	…が，調べられる	12,614
❶ be examined in ～	…が，～において調べられる	2,742
❷ be examined by ～	…が，～によって調べられる	1,896
・be examined by using ～	…が，～を使うことによって調べられる	168
❸ be examined for ～	…が，～に関して調べられる	1,056
❹ be examined using ～	…が，～を使って調べられる	881

❶ Here, their capacity to release eosinophil-activating cytokines **was examined in** cultured human airway smooth muscle. (Am J Respir Crit Care Med. 2002 165:1161)

訳 ～が，培養されたヒト気道平滑筋において調べられた

❷ The alteration of the phosphorylation levels of MAP1b and MAP2 **was examined by** Western blots using several phosphorylation-dependent antibodies to these proteins. (Brain Res. 2000 853:299)

訳 ～が，ウエスタンブロットによって調べられた

❸ A total of 14 known inhibitors of high-affinity L-glutamate transport **were examined for** their abilities to inhibit L-glutamate uptake by U373 cells. (Brain Res. 1999 839:235)

訳 …が，～を抑制するそれらの能力に関して調べられた

★example 名 例　　用例数 12,092

example	例	8,916
examples	例	3,175

複数形のexamplesの割合が，約25％ある可算名詞.

◆類義語：instance

■ example ＋［前］

	the%	a/an%	ø%			
❶	35	56	0	example of ～	～の例	2,691
				・the first example of ～	～の最初の例	750

E

・this is the first example of ～ これは，～の最初の例である	190
・provide an example of ～ …は，～の例を提供する	160

❶ This is the first example of a C-terminal mutation in SCN4A associated with human disease. (J Physiol. 2005 565:371)

訳 これは，ヒトの疾患と関連するSCN4AにおけるC末端の変異の最初の例である

■ ［前］＋（an）example

❶ for example	たとえば	4,825
❷ as an example	例として	432

❶ For example, distinct types of r-proteins S15a and P2 accumulate in ribosomes due to evolutionarily divergence of r-protein genes. (Plant Physiol. 2005 137:848)

訳 たとえば，

❷ As an example, the hybridization adsorption of a 16-mer single-stranded DNA (ssDNA) onto a two-component ssDNA array was monitored with LRSPR imaging. (Anal Chem. 2005 77:3904)

訳 例として，

excellent 形 非常によい／優れた／素晴らしい　用例数 5,643

◆類義語：good, superior, satisfactory

■ excellent ＋［名］

❶ excellent agreement	非常によい一致	475
・in excellent agreement with ～	～と非常によく一致して	250
❷ an excellent model	素晴らしいモデル	299
❸ excellent yields	非常によい収量	269
❹ excellent correlation	非常によい相関	160

❶ The results of experimental measurements made with a high-efficiency analytical column are in excellent agreement with these theoretical predictions. (Anal Chem. 2004 76:977)

訳 ～は，これらの理論的な予測と非常によく一致している

excess 名 過剰　用例数 6,886

excess	過剰	6,796
excesses	過剰	90

複数形のexcessesの割合は１％しかないが，単数形は不定冠詞を伴う用例が多い．

◆類義語：repletion

■ excess ＋［前］

	the％	a/an％	ø％			
❶	4	61	33	excess of ～	過剰の～	1,707
				・in excess of ～	～を上回って	427

❶ Blockade of LHRH receptors by an **excess of** LHRH agonist Decapeptyl suppressed the effects of AN-207. (Cancer Res. 2005 65:5857)

訳 過剰のLHRHのアゴニストDecapeptylによるLHRH受容体の遮断は，AN-207の効果を抑制した

❶ Treatment costs for ESRD are **in excess of** $14 billion annually (6.4% of Medicare budget). (J Am Soc Nephrol. 2005 16:S120)

訳 ESRDの治療のコストは，毎年140億ドルを上回っている

■［形］＋ excess

❶	～-fold excess	～倍過剰	188
❷	～ molar excess	～モル過剰	149

❷ Addition of a 10-fold **molar excess** of rat P450 reductase markedly increased the rates of metabolism by both fused and nonfused P450s 2C11. (Biochemistry. 2000 39:5196)

訳 10倍モル過剰のラットP450リダクターゼの添加は，～を顕著に増大させた

■ excess ＋［名］

❶	excess risk	過剰のリスク	244
❷	excess mortality	過剰の死亡率	178

★ exchange 名 交換，動 交換する　用例数 14,797

exchange	交換／交換する	13,320
exchanged	交換される／交換した	595
exchanges	交換／交換する	525
exchanging	交換する	357

動詞としても用いられるが，名詞の用例が圧倒的に多い．複数形のexchangesの割合は約５％あるが，単数形は無冠詞で使われることが圧倒的に多い．

◆類義語：replacement, substitution, displacement

■ exchange ＋［前］

	the％	a/an％	ø％			
❶	37	8	54	exchange of ～	～の交換	1,172
❷	4	0	94	exchange in ～	～における交換	419
❸	4	9	76	exchange between ～	～の間の交換	377

❶ In most tissues, gap junction cells facilitate the **exchange of** second messengers and metabolites between cells. (Biophys J. 2006 90:151)

E

訳 ～は，細胞間の二次メッセンジャーおよび代謝産物の交換を促進する

❷ However, the regional mechanism by which **the prone position improves gas exchange in acutely injured lungs** is still incompletely defined. (Am J Respir Crit Care Med. 2005 172:480)

訳 腹臥位は，急性に損傷した肺におけるガス交換を改善する

❸ A plausible scenario is that **information exchange between the ankyrin and transmembrane domains is involved in** activating defense signaling. (Plant J. 2005 44:798)

訳 アンキリンと膜貫通ドメインの間の情報交換は，～に関与している

■ exchange ＋［名］

❶	exchange factor	交換因子	1,044
❷	exchange activity	交換活性	320
❸	exchange rates	交換率	246

■［名］＋ exchange

❶	gas exchange	ガス交換	498
❷	hydrogen exchange	水素交換	359

excision ㊔ 切除　用例数 4,232

excision	切除	4,189
excisions	切除	43

複数形のexcisionsの割合は１％しかなく，不可算名詞として使われることが多い．

◆類義語：resection, ablation

■ excision ＋［前］

	the%	a/an%	ø%			
❶	28	7	65	excision of ～	～の切除	619

❶ **Surgical excision of the cervical lymph nodes** in healthy pre-EAE transgenic mice delayed the onset of EAE and resulted in a less severe disease. (J Immunol. 2008 181:4648)

訳 頸部リンパ節の外科切除

exclude ㊫ 除外する／排除する　用例数 5,129

excluded	除外される／～を除外した／～を排除した	2,929
excluding	～を除外する／～を排除する	1,086
exclude	～を除外する／～を排除する	853
excludes	～を除外する／～を排除する	261

他動詞．

◆類義語：rule out　◆反意語：include

■ be excluded ＋［前］

★ be excluded	…が，除外される	1,821
❶ be excluded from ～	…が，～から除外される	631
・be excluded from analysis	…が，分析から除外される	54
❷ be excluded as ～	…が，～として除外される	70
❸ be excluded by ～	…が，～によって除外される	60

❶ Four patients were excluded from analysis.（Radiology. 2002 223:645）
訳 4人の患者が分析から除外された

❷ Genetic admixture was excluded as a cause of the results.（Hum Mol Genet. 2003 12:625）
訳 遺伝的混合が，それらの結果の原因として除外された

■ exclude ＋［名句］

❶ exclude the possibility that ～	…は，～という可能性を排除する	94
❷ exclude patients with ～	…は，～の患者を除外する	56

❶ However, our findings do not exclude the possibility that VICE domain formation is required for viral replication in cells that are nonpermissive for ICP22 mutants.（J Virol. 2010 84:2384）
訳 われわれの知見は，～という可能性を排除しない

■ we exclude

❶ we excluded ～	われわれは，～を除外した	97

E

exclusively 副 もっぱら／独占的に／排他的に　用例数 5,511

◆類義語：solely, predominantly

■ exclusively ＋［前］

❶ exclusively in ～	もっぱら～において独占的に	1,272
・be expressed exclusively in ～	…は，もっぱら～において発現される	211
❷ exclusively to ～	もっぱら～に	445
❸ exclusively on ～	もっぱら～において	300
❹ exclusively by ～	もっぱら～によって	295

❶ We have recently identified a novel adaptor molecule, ALX, which is expressed exclusively in hematopoietic cells.（J Biol Chem. 2004 279:40647）
訳 ～は，もっぱら造血細胞において発現される

❷ In contrast, mNXF7 localizes exclusively to cytoplasmic granules and, despite its overall conserved sequence, lacks mRNA export activity.（Nucleic Acids Res. 2005 33:3855）
訳 mNXF7は，もっぱら細胞質顆粒に局在する

■ be exclusively ＋［過分］

★ be exclusively ～	…は，もっぱら～	660

❶ be exclusively expressed	…は，もっぱら発現される	108

■ almost exclusively

★ almost exclusively	ほとんどもっぱら	799
❶ almost exclusively in ～	ほとんどもっぱら～において	179

❶ In this report, we demonstrate that allurin mRNA is expressed almost exclusively in the oviduct and that its expression is increased 2.5-fold by human chorionic gonadotropin over a 12-h period. (Dev Biol. 2004 275:343)

訳 ～は，ほとんどもっぱら卵管において発現される

exert 動 発揮する／及ぼす 用例数 6,449

exert	～を発揮する	2,798
exerts	～を発揮する	2,184
exerted	発揮される／～を発揮した	1,192
exerting	～を発揮する	275

他動詞.

◆類義語：affect

■ exert ＋［名句］

❶ exert its effect	…は，それの効果を発揮する	324
❷ exert a dominant-negative effect	…は，ドミナントネガティブ効果を発揮する	44

❶ CAL has no effect on CFTR maturation, suggesting that it exerts its effects on mature CFTR. (J Biol Chem. 2004 279:1892)

訳 それは，成熟したCFTRに対して効果を発揮する

■ exerted ＋［前］

❶ exerted by ～	～によって発揮される	281
❷ exerted on ～	～に対して発揮される	85

❶ Removal of the forces exerted by the cytoplasmic microtubules had no effect on fragmentation. (Mol Biol Cell. 2005 16:141)

訳 細胞質微小管によって発揮される力の除去は，断片化に対して効果を持たなかった

★★ exhibit 動 示す 用例数 46,737

exhibited	～を示した／示される	20,237
exhibit	～を示す	16,809
exhibits	～を示す	7,102
exhibiting	～を示す	2,589

他動詞.

◆類義語：display, indicate, show, present, represent

■ exhibit ＋［名句］

❶ exhibit increased ～	…が，増大した～を示す	1,218
❷ exhibit reduced ～	…が，低下した～を示す	808
❸ exhibit significant ～	…が，有意な～を示す	773
❹ exhibit no ～	…が，～を示さない	694
❺ exhibit similar ～	…が，類似の～を示す	546
❻ exhibit enhanced ～	…が，増強された～を示す	544
❼ exhibit decreased ～	…が，低下した～を示す	487
❽ exhibit normal ～	…が，正常な～を示す	437
❾ exhibit defects in ～	…が，～の欠損を示す	229

❶ **163 genes exhibited increased expression** that was reduced 50% or more by calcineurin inhibition.（J Biol Chem. 2002 277:31079）

訳 163遺伝子が，増大した発現を示した

■［名］＋ exhibit

❶ mice exhibited ～	マウスは，～を示した	1,615
❷ cells exhibited ～	細胞は，～を示した	1,010

❶ **Both interleukin-4 and IFN-γ gene-deficient mice exhibited less severe colitis induction** by oxazolone.（J Exp Med. 2004 199:471）

訳 インターロイキン-4およびIFN-γの両方の遺伝子を欠損したマウスは，より重篤でない大腸炎誘導を示した

■ to exhibit

★ to exhibit ～	～を示す…	335
❶ be found to exhibit ～	…は，～を示すことが見つけられる	179
❷ be shown to exhibit ～	…は，～を示すことが示される	150

❶ **MED1 was found to exhibit thymine glycosylase activity** on O6-meG:T mismatches.（Proc Natl Acad Sci USA. 2003 100:15071）

訳 MED1は，チミングリコシラーゼ活性を示すことが見つけられた

■ exhibited ＋［前］

❶ exhibited by ～	～によって示される	261

❶ Natural or synthetic ER β-selective estrogens **may lack breast cancer promoting properties exhibited by estrogens** in hormone replacement regimens and may be useful for chemoprevention of breast cancer.（Cancer Res. 2004 64:423）

訳 ～は，エストロゲンによって示される乳癌を促進する性質を欠いているかもしれない

E

★exist 動 存在する　用例数 18,825

exist	存在する	6,619
existing	存在する／現存する	6,035
exists	存在する	5,472
existed	存在した	699

自動詞．現在形の用例が多い．

■ exist ＋［前］

❶ exist in ～	…が，～に存在する	3,936
❷ exists as ～	…が，～として存在する	1,315
❸ exists between ～	…が，～の間に存在する	1,107
❹ exist for ～	…が，～のために存在する	855

❶ **Sex differences exist in the effect of alcohol on** fibrosis as well as on the severity of hepatitis C.（Hepatology. 2002 36:S220）

訳 性差が，～に対するアルコールの影響に存在する

❷ **The protein exists as a 23.8 kDa homodimer** at pH 7 and unfolds with a T degrees of 122 degrees C.（Biochemistry. 2004 43:13026）

訳 そのタンパク質が，23.8キロダルトンのホモ二量体として存在する

❸ It is widely accepted that **a relationship exists between deafness and pigmentation** in the dog and also in other animals.（Genetics. 2004 166:1385）

訳 聴覚障害と色素沈着の間に関連性が存在する

■［名］＋ exist

❶ data exist	データが，存在する	270
❷ differences exist	違いが，存在する	261
❸ evidence exists	証拠が，存在する	235
❹ controversy exists	論争が，存在する	134
❺ relationship exists	関連性が，存在する	133
❻ information exists	情報が，存在する	129
❼ mechanisms exist	機構が，存在する	114
❽ correlation exists	相関が，存在する	91

■ there exist ＋［名句］

❶ there exists ～	～が存在する	261

❶ **There exists a robust day/night pattern** in the incidence of adverse cardiac events with a peak at approximately 10 a.m.（Proc Natl Acad Sci USA. 2004 101:18223）

訳 強力な昼/夜のパターンが存在する

■ to exist

★ to exist	存在すること	869

❶	known to exist	存在することが知られている	109
❷	shown to exist	存在することが示される	71

❶ An analysis based on retinal coverage indicates that this number of types could be contained within **the number of bipolar cells known to exist**. (J Comp Neurol. 2004 472:73)
訳 存在することが知られている双極細胞の数

existence 名 存在 用例数 5,920

複数形の用例はなく，原則，**不可算名詞**として使われる．the existence ofの用例が圧倒的に多い．
◆類義語：presence

■ existence ＋［前］

existence ofの前にtheが付く割合は圧倒的に高い．

	the%	a/an%	ø%			
❶	98	0	2	existence of ～	～の存在	5,499
				・suggest the existence of ～	…は，～の存在を示唆する	883
				・demonstrate the existence of ～	…は，～の存在を実証する	381
				・support the existence of ～	…は，～の存在を支持する	313
				・indicate the existence of ～	…は，～の存在を示す	290
				・reveal the existence of ～	…は，～の存在を明らかにする	268
				・evidence for the existence of ～	～の存在に対する証拠	260
				・consistent with the existence of ～	～の存在と一致している	101

❶ **The results suggest the existence of** multiple high-affinity binding sites within the cna promoter region. (J Bacteriol. 2003 185:4410)
訳 それらの結果は，～の存在を示唆する

❶ **We find strong evidence for the existence of** PM-class and CW-class GPI proteins. (Mol Microbiol. 2003 50:883)
訳 われわれは，～の存在の強力な証拠を見つける

exit 名 退出／終了／出口，動 出る 用例数 3,269

exit	終了／退出／出る	2,856
exiting	出る	164
exits	出る	144
exited	出た	105

名詞の用例が多いが，動詞としても用いられる．複数形のexitsの割合は2％しかなく，原則，**不可算名詞**として使われる．

◆類義語：end

■ ㊔ exit ＋［前］

	the%	a/an%	ø%			
❶	1	2	89	exit from ～	～からの離脱	518
				・exit from mitosis	有糸分裂からの離脱	183
				・exit from the cell cycle	細胞周期からの離脱	61

❶ We have identified the protein Xenopus nuclear factor 7（Xnf7）**as a novel APC inhibitor able to regulate the timing of exit from mitosis**.（J Cell Biol. 2005 169:61）

訳 有糸分裂からの離脱のタイミングを調節することができる新規のAPC抑制物質として

■ ㊔［形／名］＋ exit

❶	mitotic exit	有糸分裂終了	381
❷	cell cycle exit	細胞周期離脱	265

■ ㊍ exit ＋［名句］

❶	exit the cell cycle	…は，細胞周期を離脱する	121

❶ In the absence of growth signals, **cells exit the cell cycle** and enter into G0 or quiescence.（Genes Dev. 2011 25:801）

訳 細胞は，細胞周期を離脱する

exogenous ㊊ 外来性の／外因性の 用例数 8,555

◆類義語：extrinsic ◆反意語：endogenous, intrinsic

■ exogenous ＋［名］

❶	exogenous expression of ～	外来性の～の発現	200

❶ We have recently shown that **exogenous expression of c-myc promoter-binding protein 1（MBP-1）induces prostate cancer cell death**.（J Biol Chem. 2005 280:14325）

訳 外来性のc-mycプロモーター結合タンパク質1（MBP-1）の発現は，前立腺癌細胞死を誘導する

■［前］＋ exogenous

❶	addition of exogenous ～	外来性の～の添加	439
❷	in the absence of exogenous ～	外来性の～の非存在下で	242

❶ **Addition of exogenous PAPP-A also increased the myotube formation** and the activity of creatine kinase in C2C12 cultures.（J Biol Chem. 2005 280:37782）

訳 外来性のPAPP-Aの添加は，また，筋管形成を増大させた

■ [形] ＋ and exogenous

❶ endogenous and exogenous ～	内在性および外来性の～	207

❶ Both endogenous and exogenous antioxidants play an important and interdependent role in preventing clinically significant prostate cancer.（Cancer Res. 2005 65:2498）
訳 内在性および外来性の両方の抗酸化物は，～において重要で相互依存的な役割を果たす

expand 動 拡大させる／拡大する／広がる 用例数 9,488

expanded	～を拡大させた／拡大される／拡大した	5,075
expand	～を拡大させる／拡大する	2,231
expanding	～を拡大させる／拡大する	1,475
expands	～を拡大させる／拡大する	707

他動詞および自動詞の両方で用いられる．
◆類義語：enlarge, extend, spread, stretch, dilate, swell

E

■ expand ＋ [名句]

❶ expand our understanding of ～	…は，～についてのわれわれの理解を拡大する	168
❷ expand the range of ～	…は，～の範囲を拡大する	78
❸ expand the number of ～	…は，～の数を拡大する	72
❹ expand the repertoire of ～	…は，～のレパートリーを拡大する	68
❺ expand the scope of ～	…は，～の範囲を拡大する	65

❶ These results expand our understanding of the mechanisms underlying melanocyte loss in vitiligo and pathways linking environmental stressors and autoimmunity.（J Invest Dermatol. 2012 132:2601）
訳 これらの結果は，～の根底にある機構についてのわれわれの理解を拡大させる

❷ These findings expand the range of cell types infected in the blood.（Blood. 2010 116:4546）
訳 これらの知見は，感染した細胞タイプの範囲を拡大させる

■ [名] ＋ expand

❶ findings expand ～	知見は，～を拡大させる	65
❷ results expand ～	結果は，～を拡大させる	64

■ be expanded ＋ [前]

★ be expanded	…は，拡大される	656
❶ be expanded to ～	…は，～に拡大される	161
❷ be expanded in ～	…は，～において拡大される	131

❷ Endovascular stents were expanded in the aortae of obese insulin-resistant and type 2 diabetic Zucker rats, in streptozotocin-induced type 1 diabetic Sprague-Dawley rats, and in matched controls.（Circ Res. 2005 97:725）
訳 血管内ステントは，～の大動脈において拡大された

■ expand ＋［前］

❶ expand in ～	…は，～において拡大する	178
❷ expand to ～	…は，～に拡大するる	99

❶ During organogenesis, **tissues expand in size** and eventually acquire consistent ratios of cells with dazzling diversity in morphology and function.（Dev Biol. 2012 367:91）
訳 組織は，大きさが拡大する

■［副］＋ expanding/expanded

❶ rapidly expanding ～	急速に拡大する～	122
❷ greatly expanded	大きく拡大される	81
❸ clonally expanded	クローン性に拡大される	72

★ expansion　名 拡大　用例数 11,756

expansion	拡大	10,834
expansions	拡大	922

複数形のexpansionsの割合が，約10%ある可算名詞．
◆類義語：extension

■ expansion ＋［前］

	the%	a/an%	ø%			
❶	34	17	49	expansion of ～	～の拡大	4,442
❷	5	22	61	expansion in ～	～における拡大	786

❶ **This analysis detected extensive clonal expansion of hepatocytes**, as previously found in chronically infected chimpanzees and woodchucks.（J Virol. 2010 84:8308）
訳 この分析は，肝細胞の広範なクローン増殖を検出した

■［形］＋ expansion

❶ clonal expansion	クローン増殖	670
❷ in vivo expansion	生体内拡大	163
❸ rapid expansion	急速な拡大	125

★ expect　動 予想する／予期する　用例数 11,625

expected	予想される／～を予想した	11,004
expect	～を予想する	576
expects	～を予想する	23
expecting	～を予想する	22

他動詞．受動態の用例が非常に多い．

◆類義語：predict, presume, estimate, speculate, assume, infer, postulate, deduce

■ be expected ＋［前］

★ be expected	…は，予想される	3,295
❶ be expected to *do*	…は，～すると予想される	2,195
・be expected to be ～	…は，～であると予想される	495
・be expected to have ～	…は，～を持つと予想される	142
・be expected to increase	…は，増大すると予想される	95
❷ be expected from ～	…は，～から予想される	137
❸ be expected for ～	…は，～に対して予想される	108

❶ From the ubiquitous nature of the splicing machinery, **expression of U1A genes is expected to be constitutive**. (Plant Mol Biol. 2001 45:449)

訳 U1A遺伝子の発現は，構成的であると予想される

■ be expected that/if

❶ be expected that ～	…は，～ということが予想される	137
❷ be expected if ～	もし～なら，…が予想される	103

❶ It **is expected that** these assays could be applied to profile ER α isoforms in any estrogen-responsive tissues. (Anal Biochem. 2003 314:217)

訳 ～ということが予想される

❷ However, there was no increase in mPSC zinc sensitivity after binge ethanol **as would be expected if** a general arrest of synaptic maturation had occurred. (Brain Res. 2006 1089:101)

訳 もし～なら予想されるであろうように

■ as/than expected

❶ as expected	予想されるように	1,961
・as expected,	予想されるように,	1,242
・as expected for ～	～に対して予想されるように	197
・as expected from ～	～から予想されるように	172
❷ than expected	予想されるより…	945
・higher than expected	予想されるより高い	160
・lower than expected	予想されるより低い	114
・than expected from ～	～から予想されるより	98
・than expected by ～	～によって予想されるより	98

❶ **As expected,** Ty1 mutations with a positive fitness effect were in the minority. (Genetics. 2003 165:975)

訳 予想されたように，

❶ **As expected from the K_d value**, the enzyme is reversibly inhibited upon exposure to pathologically, and possibly physiologically, relevant concentrations of NO. (J Biol Chem. 2003 278:2341)

訳 K_d値から予想されるように

❷ The death rate compared with actuarial statistics **was significantly higher than expected** (P< 0.0001). (Circulation. 2000 101:2490)

訳 ～は，予想されたより有意に高かった

expectation 名 予想／見込み／期待 用例数 1,749

expectations	予想／見込み	975
expectation	予想／見込み	774

複数形のexpectationsの割合は約55%と非常に高い．

◆類義語：prediction, perspective, promise, estimate, prospect, likelihood, hope

■ expectation(s) ＋［同格that節／前］

expectation thatの前にtheが付く割合は圧倒的に高い．

	the%	a/an%	ø%			
❶	84	8	2	expectation that ～	～という予想	118
❷	54	-	38	expectations of ～	～の予想	108
❸	11	-	82	expectations for ～	～に対する予想	95

❶ The closed-loop model for initiation creates the **expectation that** sequences at the 3' end of eukaryotic mRNAs should regulate translation.（Gene. 2004 343:41）

訳 開始の閉ループモデルは，～という予想を生み出す

❷ Placebo analgesia has been shown to be driven by **expectations of** treatment effects.（Brain Res. 2010 1359:137）

訳 プラセボ鎮痛は，治療効果の期待によって駆り立てられることが示されてきた

■［前］＋ expectations

❶	contrary to expectations	予想に反して	147

❶ **Contrary to expectations**, murine Bapx1 does not affect the articulation of the malleus and incus.（Development. 2004 131:1235）

訳 予想に反して，

★ experience 名 経験，動 経験する 用例数 12,500

experience	経験／～を経験する	6,250
experienced	～を経験した	4,390
experiences	経験／～を経験する	1,124
experiencing	～を経験する	736

名詞および動詞として用いられる．複数形のexperiencesの割合が約20%あるが，単数形は無冠詞で使われることが非常に多い．

◆類義語：undergo

■ 名 experience ＋［前］

	the%	a/an%	ø%			
❶	9	0	65	experience with ～	～との経験	671

❷	56	10	22	experience of ～	～の経験	421
❸	8	2	76	experience in ～	～における経験	287

❶ We present our experience with patients who developed trastuzumab-related cardiotoxicity. (J Clin Oncol. 2005 23:7820)

訳 われわれは，～を発症した患者とのわれわれの経験を提示する

❷ The subjective experience of stress leads to reproductive dysfunction in many species, including rodents and humans. (Proc Natl Acad Sci USA. 2009 106:11324)

訳 ストレスの主観的経験は，～につながる

❸ The role of experience in the development of the cerebral cortex has long been controversial. (Science. 1998 279:566)

訳 大脳皮質の発達における経験の役割は，長い間議論の余地がある

■ 動 experienced ＋［前］

❶	experienced by ～	～によって経験される	284
❷	experienced in ～	～において経験される	85

❶ If replicated, these results may provide clues to the exceedingly high lung cancer incidence experienced by African Americans. (Cancer Res. 2005 65:9566)

訳 これらの結果は，アフリカ系アメリカ人によって経験される非常に高い肺癌の発生率への手がかりを与えるかもしれない

⁑ experiment 名 実験　　用例数 38,144

experiments	実験	32,446
experiment	実験	5,698

複数形のexperimentsの割合は，約85％と圧倒的に高い．

◈類義語：study, research

■ experiments ＋［前］/using

	the％	a/an％	ø％			
❶	4	-	91	experiments with ～	～を使った実験	2,110
❷	1	-	93	experiments in ～	～における実験	1,583
				・experiments in which ～	（それにおいて）～である実験	357
❸	1	-	90	experiments using～	～を使う実験	1,297
❹	3	-	82	experiments to *do*	～するための実験	671

❶ Through a series of coinfection experiments with heterologous viruses, the RTM1/RTM2-mediated restriction was shown to be highly specific for TEV. (Plant Cell. 2000 12:569)

訳 異種のウイルスを使った一連の同時感染実験によって

❷ Results from experiments in which posterior neural segments and/or paraxial mesoderm segments were placed at different axial levels suggest that signals setting Hoxd10 expression form a decreasing posterior-to-anterior gradient. (Development. 2001 128:2255)

訳 ～である実験からの結果

E

■ experiments ＋［動］

❶	experiments showed ～	実験は，～を示した	1,231
❷	experiments revealed ～	実験は，～を明らかにした	932
❸	experiments indicate ～	実験は，～を示す	918
❹	experiments demonstrated ～	実験は，～を実証した	823
❺	experiments suggest ～	実験は，～を示唆する	803
❻	experiments were performed	実験が，実行された	640

❶ Pulse-labeling **experiments showed** that most major late proteins failed to accumulate in the presence of the antibiotic. (J Virol. 2004 78:2137)

訳 パルスラベル実験は，～ということを示した

■［前］＋ experiments

❶	series of experiments	一連の実験	296
❷	set of experiments	セットの実験	120

❶ Based on the expression pattern, we performed a **series of experiments** to test the hypothesis that BMP-2 mediates myocardial regulation of cardiac cushion tissue formation in mice. (Dev Biol. 2004 269:505)

訳 われわれは，～をテストするために一連の実験を実行した

■［名／形］＋ experiments

❶	immunoprecipitation experiments	免疫沈降実験	560
❷	in vitro experiments	試験管内実験	528
❸	binding experiments	結合実験	449
❹	transfection experiments	トランスフェクション実験	410

★ experimental 形 実験の／実験的な　用例数 23,162

■ experimental ＋［名］

❶	experimental data	実験データ	2,057
❷	experimental evidence	実験的な証拠	1,145
❸	experimental results	実験結果	1,022
❹	experimental conditions	実験条件	839
❺	experimental studies	実験的研究	648
❻	experimental models	実験モデル	509

❶ The de novo receptor models appear to provide much better agreement with **experimental data**, particularly for receptor complexes with agonist ligands. (J Med Chem. 2003 46:4450)

訳 新規の受容体モデルは，実験データとのずっとよい一致を提供するように思われる

❻ Overexpression of eIF-4E in **experimental models** dramatically alters cellular morphology, enhances proliferation and induces cellular transformation, tumorigenesis and metastasis.

(Oncogene. 2004 23:3189)
訳 実験モデルにおけるeIF-4Eの過剰発現は，細胞の形態を劇的に変化させる

experimentally 副 実験的に 用例数 4,946

■ experimentally ＋［過分］

❶ experimentally determined	実験的に決定される	455
❷ experimentally observed	実験的に観察される	377
❸ experimentally infected	実験的に感染させられる	259
❹ experimentally induced	実験的に誘導される	216
❺ experimentally measured	実験的に測定される	159

❶ This decrease in the observed hydrogen bond frequency correlates with **a decrease in the experimentally determined thermal stability**.（Biochemistry. 2004 43:5314）
訳 実験的に決定された熱安定性の低下

■［過分］＋ experimentally

❶ observed experimentally	実験的に観察される	159

★explain 動 説明する 用例数 17,237

explain	～を説明する	9,276
explained	説明される／～を説明した	4,990
explains	～を説明する	1,808
explaining	～を説明する	1,163

他動詞．能動態現在形の用例が多い．
◈類義語：account for, illustrate, represent

■ explain ＋［名節］

❶ explain why ～	…は，なぜ～かを説明する	1,218
❷ explain how ～	…は，どのように～かを説明する	936

❶ Since GLI3 is a more effective repressor, **our results explain why** GLI3 is required only for anterior limb patterning and why GLI2 can compensate for GLI3A in posterior limb patterning.（Dev Biol. 2012 370:110）
訳 われわれの結果は，なぜ～かを説明する

❷ Because it is HG-CD147 that self-aggregates and stimulates MMP induction, **we now have a mechanism to explain how** caveolin-1 inhibits these processes.（Mol Biol Cell. 2004 15:4043）
訳 われわれは，今，どのように～かを説明するための機構を持っている

■ explain ＋［名句］

❶ explain the observed ～	…は，観察された～を説明する	269

E

❷ explain the lack of ～	…は，～の欠如を説明する	77
❸ explain the mechanism	…は，機構を説明する	76
❹ explain the differences	…は，違いを説明する	73
❺ explain the ability of ～	…は，～の能力を説明する	65

❶ **These results help to** explain the observed **differences in** collagen type I and type II fibrillar architecture and indicate the collagen type II cross-link organization, which is crucial for fibrillogenesis.（J Biol Chem. 2010 285:7087）

訳 これらの結果は，～の観察された違いを説明するのに役立つ

■ ［名／代名］＋ explain

❶ results explain ～	結果は，～を説明する	138
❷ we explain ～	われわれは，～を説明する	117
❸ model explains ～	モデルは，～を説明する	86

■ ［副］＋ explain

❶ partially explain	…は，～を部分的に説明する	125
❷ partly explain	…は，～を部分的に説明する	87

■ ［動］＋（to）explain

❶ to explain ～	～を説明する…	3,030
・help to explain ～	…は，～を説明するのに役立つ	449
・be proposed to explain ～	…は，～を説明するために提案される	258
❷ help explain ～	…は，～を説明するのに役立つ	621

❶ Self-fertilizing species often harbor less genetic variation than cross-fertilizing species, and **at least four different models have** been proposed to explain **this trend.**（Genetics. 2002 161: 99）

訳 少なくとも４つの異なるモデルが，この傾向を説明するために提案された

❶ **These findings may** help explain **why** those with high BMI have worse outcomes from their cancers.（Transplantation. 2012 94:539）

訳 これらの知見は，なぜ～かを説明するのに役立つかもしれない

■ be explained ＋［前］

★ be explained	…は，説明される	3,162
❶ be explained by ～	…は，～によって説明される	2,308
❷ be explained in ～	…が，～において説明される	220
・be explained in part	…が，部分的に説明される	89
・be explained in terms of ～	…が，～の点から説明される	86

❶ **Our results can** be explained by **assuming that** tER sites give rise to Golgi cisternae that continually mature.（Nat Cell Biol. 2002 4:750）

訳 われわれの結果は，～ということを仮定することによって説明されうる

❷ **The disagreement** is explained in terms of **individual metabolic properties** as opposed to those of the larger population.（J Theor Biol. 2005 233:1）

訳 不一致が，個々の代謝の性質の点から説明される

■［副］＋ explained

❶	best explained	最もよく説明される	145
❷	fully explained	十分に説明される	93

❶ Combined, these data are best explained by a direct role for DNA polymerase zeta in Ig hypermutation.（J Immunol. 2012 188:5528）
訳 これらのデータは，～の直接的役割によって最もよく説明される

explanation 名 説明 用例数 4,181

explanation	説明	3,418
explanations	説明	763

複数形のexplanationsの割合が，約20％ある可算名詞．explanation forの用例が非常に多い．

◆類義語：interpretation, report, description, conclusion

■ explanation ＋［前］

	the％	a/an％	ø％		
❶	8	81	0	explanation for ～　～に対する説明	2,394
				・provide an explanation for ～　…は，～に対する説明を提供する	487
				・possible explanation for ～　～に対する可能な説明	246
				・mechanistic explanation for ～　～に対する機構的説明	169
				・molecular explanation for ～　～に対する分子レベルの説明	165
❷	10	71	10	explanation of ～　～の説明	372

❶ These findings may provide an explanation for some of the racial differences in colon cancer incidence.（Am J Epidemiol. 2003 158:951）
訳 これらの知見は，～に対する説明を提供するかもしれない

exploit 動 利用する 用例数 4,107

exploited	～を利用した／利用される	1,947
exploit	～を利用する	1,059
exploiting	～を利用する	683
exploits	～を利用する	418

他動詞．

◆類義語：utilize, use, employ, take advantage of, apply

E

■ be exploited ＋［前］

★ be exploited	…は，利用される	1,254
❶ be exploited to *do*	…は，～するために利用される	494
❷ be exploited for ～	…は，～のために利用される	253
❸ be exploited in ～	…は，～において利用される	202
❸ be exploited as ～	…は，～として利用される	96

❶ Thus, these transformations **can be exploited to interconvert the two isomers** of each species with nanosecond switching speeds. (J Org Chem. 2005 70:8180)
訳 ～は，2つのアイソマーを相互変換するために利用されうる

❷ Moreover, the overlap between the DNA viral and tumor programs **can also be exploited for the development of** lytic cancer therapies. (Oncogene. 2005 24:7640)
訳 …は，また，～の開発のために利用されうる

■［代名］＋ exploit

❶ we exploited ～	われわれは，～を利用した	223

❶ Presently, **we exploited the fact that** L6 myotubes and 3T3/L1 adipocytes have substantially different (30% nonhomology) major aPKCs, viz. (J Biol Chem. 2006 281:17466)
訳 われわれは，～という事実を利用した

■ exploit ＋［名句］

❶ exploit the fact that ～	…は，～という事実を利用する	35

exploration 名 探索　　用例数 1,441

exploration	探索	1,365
explorations	探索	76

複数形のexplorationsの割合は約5％あるが，単数形は無冠詞で使われることが非常に多い．

◈類義語：search

■ exploration ＋［前］

	the%	a/an%	ø%			
❶	28	8	61	exploration of ～	～の探索	693
				・further exploration of ～	～のさらなる探索	97

❶ **These results encourage further exploration of the idea that** frontal systems linked with limbic circuits facilitate assessment of the relevance or personal significance in social contexts. (J Neurosci. 2010 30:13906)
訳 これらの結果は，～という考えのさらなる探索を奨励する

★ explore 動 探索する　用例数 13,404

explore	～を探索する	6,409
explored	～を探索した／探索される	5,477
exploring	～を探索する	1,008
explores	～を探索する	510

他動詞.

◆類義語：test, study, investigate, examine, survey, look at, search, dissect

■ explore ＋［名句］

❶	explore the role of ～	…は，～の役割を探索する	723
❷	explore the mechanism	…は，機構を探索する	377
❸	explore the effect of ～	…は，～の影響を探索する	330
❹	explore the possibility	…は，可能性を探索する	279
❺	explore the relationship between ～	…は，～の間の関連性を探索する	226
❻	explore the hypothesis that ～	…は，～という仮説を探索する	97

❶ The present study is designed to explore the role of G protein-coupled receptors (GPCRs) in the protection afforded by ischemic preconditioning (PC). (Circ Res. 2004 94:1133)

訳 現在の研究は，～の役割を探索するために計画されている

■ explore ＋［名節］

❶	explore whether ～	…は，～かどうかを探索する	428
❷	explore how ～	…は，どのように～かを探索する	301

❶ We explored whether the actin cytoskeleton is involved in Cx43 forward trafficking. (Circ Res. 2012 110:978)

訳 われわれは，～かどうかを探索した

■ ［代名／名］＋ explored

❶	we explored ～	われわれは，～を探索した	1,313
❷	study explored ～	研究は，～を探索した	136
❸	review explores ～	レビューは，～を探索する	84

■ to explore

★	to explore ～	～を探索するために	4,427
❶	To explore ～	～を探索するために	1,792
❷	to further explore ～	～をさらに探索するために	294
❸	be used to explore ～	…は，～を探索するために使われる	249
❹	be to explore ～	…は，～を探索することである	173

❶ To explore the mechanisms by which Ack1 promotes tumor progression, we investigated the role of AKT/PKB, an oncogene and Ack1-interacting protein. (Am J Pathol. 2012 180:1386)

E

訳 Ack1が腫瘍進行を促進する機構を探索するために

■ be explored ＋［前］

★ be explored	…は，探索される	2,342
❶ be explored in ～	…は，～において探索される	300
❷ be explored by ～	…は，～によって探索される	243

❶ The role of JC virus (JCV)-specific CTL **was explored in** the immunopathogenesis of progressive multifocal leukoencephalopathy (PML). (J Immunol. 2002 168:499)
訳 …が，～の免疫病原性において探索された

export 名 搬出，動 搬出する　用例数 6,346

export	搬出／～を搬出する	5,506
exported	～を搬出した／搬出される	695
exports	～を搬出する／搬出	83
exporting	搬出／～を搬出する	62

名詞の用例が多いが，動詞としても用いられる．複数形のexportsの割合は0.2%しかなく，原則，**不可算名詞**として使われる．

◆反意語：import

■ 名 export ＋［前］

	the%	a/an%	ø%			
❶	35	1	64	export of ～	～の搬出	1,265
				・export of ～ from …	～の…からの搬出	125
❷	4	0	78	export from ～	～からの搬出	208

❶ In addition, **export of** Gag **from** the nucleus was found to be a rate-limiting step in virus-like particle production. (J Virol. 2005 79:8732)
訳 Gagの核からの搬出が，～における律速段階であることが見つけられた

❷ Significantly, forced expression of CCR9 on mature SP thymocytes did not inhibit their **export from** the thymus, indicating that CCR9 down-regulation is not essential for thymocyte emigration. (J Immunol. 2006 176:75)
訳 ～は，胸腺からのそれらの搬出を抑制しなかった

■ 名［名］＋ export

❶ nuclear export	核外移行	1,695
❷ mRNA export	mRNAの搬出	458
❸ protein export	タンパク質の搬出	170

■ 動 be exported ＋［前］

★ be exported	…は，搬出される	342
❶ be exported from ～	…は，～から搬出される	96

❷ be exported to ～	…は，～へ搬出される	91

❶ Here we report that HB-EGF can be exported from the nucleus during stimulated processing and secretion of the growth factor.（Cancer Res. 2005 65:8242）
訳 HB-EGFは，核から搬出されうる

★ expose 動 曝露する 用例数 14,963

exposed	さらされる／曝露される／露出される／～をさらした	13,551
exposing	～をさらす／～を曝露する	662
expose	～をさらす／～を曝露する	440
exposes	～をさらす／～を曝露する	310

他動詞．exposed toの用例が非常に多い．
◈類義語：challenge

■ exposed ＋［前］

❶ exposed to ～	～にさらされる／～に曝露される	7,854
・be exposed to ～	…が，～にさらされる	2,611
・cells were exposed to ～	細胞が，～にさらされた	427
・mice were exposed to ～	マウスが，～にさらされた	139
・when exposed to ～	～にさらされたとき	549
・cells exposed to ～	～にさらされた細胞	308
❷ exposed in ～	～において曝露される	244
❸ exposed on ～	～において曝露される	242

❶ Human endothelial cells were exposed to direct vibration and rapid low-volume fluid oscillation.（J Physiol. 2004 555:565）
訳 ヒトの内皮細胞が，～にさらされた

❶ In cells exposed to high fluences, the G1 checkpoint is at least as extensive as in γ-irradiated cells.（Cancer Res. 2000 60:2623）
訳 高フルエンスにさらされた細胞において，

■［名］-exposed

❶ surface-exposed ～	表面が露出された～	563
❷ solvent-exposed ～	溶剤にさらされた～	514

❶ Other surface-exposed residues in the α-CTD contributed to acs transcription, suggesting that the α-CTD may interact with at least one protein other than CRP.（J Bacteriol. 2003 185:5148）
訳 α-CTDにおいて他の表面が露出された残基は，～に寄与した

■ by exposing

❶ by exposing ～ to …	～を…にさらすことによって	124

❶ Inhibition of total RNA synthesis achieved by exposing cells to a bFGF-neutralizing antibody

was similar in magnitude to that induced by ox-LDL. (Circulation. 2000 101:171)
訳 細胞をbFGF中和抗体にさらすことによって達成された全RNA合成の抑制

★ exposure 名 曝露／さらされること 用例数 28,603

exposure	曝露	26,819
exposures	曝露	1,784

複数形のexposuresの割合は約5％あるが，単数形は無冠詞で使われることが圧倒的に多い．

■ exposure ＋［前］

	the%	a/an%	ø%			
❶	3	1	93	exposure to ～	～への曝露	11,001
				・after exposure to ～	～への曝露のあと	1,234
				・by exposure to ～	～への曝露によって	625
				・of exposure to ～	～への曝露の	620
				・following exposure to ～	～への曝露に続いて	496
				・upon exposure to ～	～へ曝露するとすぐに	469
				・exposure to light	光への曝露	149
				・exposure to ionizing radiation	電離放射線への曝露	101
❷	17	1	82	exposure of ～	～の曝露	3,945
				・exposure of ～ to …	～の…への曝露	1,784
				・exposure of cells to ～	細胞の～への曝露	281

❶ Apoptosis was equally evident in cells deficient in caspase-9 or caspase-8 **after exposure to CDDO**, suggesting caspase-independent cell death. (Cancer Res. 2004 64:7927)
訳 CDDOへの曝露のあと

❶ PGE_2 production also increased significantly in the mesenteric lymph nodes **following exposure to viable Salmonella**, but not after exposure to killed bacteria. (J Immunol. 2004 172:2469)
訳 生きたサルモネラ菌への曝露に続いて

❶ **Upon exposure to drug**, cell death commences after a lag time, and the cell kill rate is dependent on the amount of drug in the critical intracellular compartment. (Cancer Res. 2004 64:711)
訳 薬剤にさらされるとすぐに，

❶ **Exposure to light decreased HLS1 protein levels** and evoked a concomitant increase in ARF2 accumulation. (Dev Cell. 2004 7:193)
訳 光への曝露は，HLS1タンパク質レベルを低下させた

❷ **Exposure of cells to 10 μM EM-1** for 2.5, 5 and 24 h resulted in a time-dependent down-regulation of mu receptors. (Brain Res. 2004 1028:121)
訳 細胞の10 μM EM-1への曝露

★★ express ㊍ 発現させる／発現する　用例数 118,803

expressed	発現される／～を発現した／～を発現させた	69,550
expressing	～を発現する	30,132
express	～を発現する／～を発現させる	16,910
expresses	～を発現する／～を発現させる	2,211

他動詞.

■ be expressed ＋［前］

★ be expressed	…が，発現される	23,576
❶ be expressed in ～	…が，～において発現される	11,542
・be expressed in ～ cells	…が，～細胞において発現される	1,324
・be expressed in ～ tissues	…が，～組織において発現される	487
・be expressed in Escherichia coli	…が，大腸菌において発現される	456
❷ be expressed at ～	…が，～において発現される	1,983
・be expressed at high levels	…が，高いレベルで発現される	439
❸ be expressed by ～	…が，～によって発現される	1,293
❹ be expressed on ～	…が，～において発現される	1,114
・be expressed on the surface of ～	…が，～の表面で発現される	124
❺ be expressed as ～	…が，～として発現される	840

❶ Wild-type PI31 and various truncation mutants **were expressed in Escherichia coli** and purified to homogeneity.（J Biol Chem. 2000 275:18557）
訳 ～が，大腸菌において発現された

❷ Although R5 viruses are present in the latent reservoir, **CCR5 was not expressed at high levels** on resting CD4$^+$ T cells.（J Virol. 2000 74:7824）
訳 CCR5は，高いレベルでは発現されなかった

❸ **Wit is expressed by a subset of neurons**, including motoneurons.（Neuron. 2002 33:545）
訳 Witは，ひとつのサブセットのニューロンによって発現される

❹ **Lu is expressed on the surface of** a subset of muscle and epithelial cells in diverse tissues and is thought to be involved in both normal and disease processes, including sickle cell disease and cancer.（J Biol Chem. 2002 277:44864）
訳 Luは，～の表面で発現される

❺ We identified two Xenopus RGS4 homologs, one of which, Xrgs4a, **was expressed as a Spemann organizer component**.（Development. 2000 127:2773）
訳 ～は，シュペーマンオーガナイザーの構成成分として発現された

■［名］＋ be expressed

❶ genes are expressed	遺伝子が，発現される	533
❷ protein is expressed	タンパク質が，発現される	333
❸ mRNA is expressed	mRNAが，発現される	278
❹ receptors are expressed	受容体が，発現される	186
❺ mutants were expressed	変異体が，発現された	101

E

■ [副] + expressed

❶ highly expressed	高度に発現される	2,721
❷ differentially expressed	差動的に発現される	2,325
❸ ubiquitously expressed	遍在性に発現される	1,037
❹ widely expressed	広く発現される	1,028
❺ constitutively expressed	構成的に発現される	1,004
❻ preferentially expressed	優先的に発現される	549
❼ ectopically expressed	異所性に発現される	512
❽ transiently expressed	一時的に発現される	466
❾ abundantly expressed	大量に発現される	465

❶ **MMP-9 was highly expressed in lungs** and supernatant of neutrophil cultures in severe pancreatitis, and, to a lesser degree, in mild pancreatitis. (Gastroenterology. 2002 122:188)
訳 MMP-9は，肺において高度に発現された

❷ We demonstrated previously that **E2F genes are differentially expressed in** developing epidermis. (J Biol Chem. 2001 276:23531)
訳 E2F遺伝子は，〜において差動的に発現される

■ expressed + [副] + in

❶ expressed only in 〜	〜においてのみ発現される	322
❷ expressed predominantly in 〜	主に〜において発現される	267
❸ expressed exclusively in 〜	もっぱら〜において発現される	250

❶ **MtPT4 is expressed only in mycorrhizal roots**, and the MtPT4 promoter directs expression exclusively in cells containing arbuscules. (Plant Cell. 2002 14:2413)
訳 MtPT4は，菌根においてのみ発現される

■ when expressed

★ when expressed	発現されるとき	1,922
❶ when expressed in 〜	〜において発現されるとき	1,202

❶ **When expressed in mammalian cells**, the carboxyl-terminal portion also localizes to perinuclear membranes similar to mammalian SREBPs. (J Biol Chem. 1998 273:16112)
訳 哺乳類細胞において発現されるとき，

■ express + [名句]

❶ express high levels of 〜	…は，高レベルの〜を発現する	1,164
❷ express 〜 receptors	…は，〜受容体を発現する	879

❶ Upon exposure to nicotine or NNK, **cells express high levels of IKBKE protein and mRNA**, which are largely abrogated by inhibition of STAT3. (Oncogene. 2013 32:151)
訳 細胞は，高レベルのIKBKEタンパク質とmRNAを発現する

■ [名／代名] ＋ express

❶ cells express ～	細胞が，～を発現する	1,777
❷ we expressed ～	われわれは，～を発現させた	806

❷ We expressed Cre recombinase under the transcription regulatory sequences of the human c-fes gene.（J Exp Med. 2001 194:581）

訳 われわれは，Creリコンビナーゼを発現させた

■ to express

★ to express ～	～を発現する…	3,118
❶ engineered to express ～	～を発現するように操作される	313
❷ found to express ～	～を発現することが見つけられる	175

❶ Keratinocytes engineered to express cdk4（R24C）and hTERT but not p53DD did not exhibit an extended life span.（Mol Cell Biol. 2002 22:5157）

訳 cdk4（R24C）とhTERTを発現するように操作されたケラチノサイト

E

■ [名／副] ＋ expressing

❶ cells expressing ～	～を発現する細胞	7,145
❷ mice expressing ～	～を発現するマウス	1,976
❸ cell lines expressing ～	～を発現する細胞株	492
❹ neurons expressing ～	～を発現するニューロン	377
❺ vector expressing ～	～を発現するベクター	319

❶ T cells expressing the Vγ4 T-cell receptor（TCR）promote myocarditis in coxsackievirus B3 (CVB3)-infected BALB/c mice.（J Virol. 2002 76:10785）

訳 Vγ4 T細胞受容体（TCR）を発現するT細胞

■ -expressing ＋ [名]

❶ ～-expressing cells	～を発現する細胞	2,837
❷ ～-expressing neurons	～を発現するニューロン	342

❶ These TLR2-expressing cells were also stimulated by living motile B. burgdorferi, suggesting that TLR2 recognition of lipoproteins is relevant to natural Borrelia infection.（J Biol Chem. 1999 274:33419）

訳 これらのTLR2発現細胞は，また，～によって刺激された

★★ expression 名 発現　　用例数 256,664

expression	発現	255,571
expressions	発現	1,093

複数形のexpressionsの割合は0.4%しかなく，原則，不可算名詞として使われる.

■ expression ＋［前］

	the%	a/an%	ø%			
❶	32	1	67	expression of ～	～の発現	93,469
				・expression of genes	遺伝子の発現	1,662
				・expression of a dominant negative ～	ドミナントネガティブな～の発現	539
				・expression of ～ in …	…における～の発現	9,180
❷	2	0	97	expression in ～	～における発現	25,217
❸	1	0	99	expression by ～	～による発現	4,379

❶ Aberrant embryos did not turn green, and **the expression of genes involved in photomorphogenesis was drastically attenuated.** (Plant Cell. 1998 10:383)
訳 光形態形成に関与する遺伝子の発現が，劇的に減弱した

❶ Hence, **human diabetes selectively alters the expression of Bax in the retina** and retinal vascular pericytes at the same time as it causes increased rates of apoptosis. (Am J Pathol. 2000 156:1025)
訳 ヒトの糖尿病は，網膜におけるBaxの発現を選択的に変える

❷ When examined by Northern blot, **expression in normal tissues was restricted to testis and heart**, and no expression was found in hematopoietic tissues. (Proc Natl Acad Sci USA. 2001 98:7492)
訳 正常な組織における発現は，精巣と心臓に限定された

❸ **Acute CD4 T cell-mediated rejection required MHC class II expression by the allograft**, indicating the importance of direct graft recognition. (J Clin Invest. 2000 106:1003)
訳 CD4 T細胞に仲介される急性の拒絶は，同種移植片によるMHCクラス２の発現を必要とした

■［前］＋ expression

❶	levels of expression	発現のレベル	730
❷	pattern of expression	発現のパターン	579
❸	changes in expression	発現の変化	335
❹	increase in expression	発現の増大	303

❶ Unlike with these in vitro-regulated proteins, **the levels of expression of OspA**, OspB, P72, flagellin, and BmpA remained unchanged throughout growth of the spirochetes in culture. (Infect Immun. 1998 66:5119)
訳 OspAの発現のレベル

■ expression ＋［動］

❶	expression increased	発現が，上昇した	500
❷	expression is regulated	発現が，調節される	497
❸	expression was observed	発現が，観察された	433
❹	expression was detected	発現が，検出された	406
❺	expression is induced	発現が，誘導される	336
❻	expression was increased	発現が，上昇した	316

❶ No mRNA expression was detected in heart, stomach, liver, small intestine, brain, or skin. (Blood. 2000 95:3125)

訳 メッセンジャーRNA発現は，心臓には検出されなかった

■ [名／過分／形] + expression

❶ gene expression	遺伝子発現	37,515
❷ protein expression	タンパク質発現	5,980
❸ mRNA expression	mRNA発現	4,952
❹ increased expression	増大した発現	4,076
❺ surface expression	表面発現	2,964
❻ ectopic expression	異所性の発現	2,651
❼ ～-specific expression	～特異的な発現	1,930

❶ Thus, YY-1 positively regulates IL-4 gene expression in lymphocytes. (J Biol Chem. 2001 276: 48871)

訳 YY-1は，リンパ球におけるIL-4遺伝子発現を正に調節する

❹ Increased expression of chemokine mRNA is observed in allogeneic but not syngeneic skin grafts 3-4 days after transplantation. (Transplantation. 2000 69:969)

訳 ケモカインのメッセンジャーRNAの増大した発現が，～において観察される

❻ Ectopic expression of ZF5 leads to an ITR-dependent repression of the autologous p5 promoter and reduces both AAV2 replication and the production of recombinant AAV2. (Proc Natl Acad Sci USA. 2001 98:14991)

訳 ZF5の異所性の発現は，～のITR依存性の抑制につながる

❼ In situ hybridization histochemistry and immunocytochemistry were used to examine lamina- and cell-specific expression of glutamate receptor (GluR) mRNAs and polypeptide subunits in motor and somatosensory cortex of macaque monkeys. (J Comp Neurol. 1999 407: 472)

訳 ～は，グルタミン酸受容体（GluR）メッセンジャーRNAのラミナ特異的および細胞特異的な発現を調べるために使われた

■ expression + [名]

❶ expression levels	発現レベル	3,558
❷ expression patterns	発現パターン	3,298
❸ expression profiles	発現プロフィール	2,019

❷ We report that during early limb development, expression patterns of HoxD genes in axolotls resemble those in amniotes and anuran amphibians. (Dev Biol. 1998 200:225)

訳 アホロートルにおけるHoxD遺伝子の発現パターンは，有羊膜類のそれらに似ている

★ extend 動 広げる／広がる／伸展する 用例数 18,302

extended	広げられる／伸展される／広がった／及んだ	9,776
extend	広げる／広がる／伸展する／及ぶ	4,169
extends	広げる／広がる／伸展する／及ぶ	2,217
extending	広げる／広がる／伸展する／及ぶ	2,140

他動詞および自動詞として用いられる.

◆類義語：enlarge, expand, spread, stretch

■ extend ＋［名句］

❶ extend our understanding of ～	…は，～についてのわれわれの理解を広げる	139
❷ extend these findings	…は，これらの知見を広げる	116
❸ extend these studies	…は，これらの研究を広げる	104
❹ extend the range of ～	…は，～の範囲を広げる	96

❶ Collectively, **these results extend our understanding of how** S. aureus avoids phagocyte-mediated clearance, and underscore LukAB as an important factor that contributes to staphylococcal pathogenesis. (Mol Microbiol. 2011 79:814)

訳 これらの結果は，どのように～かについてのわれわれの理解を広げる

■［代名／名］＋ extend

❶ we extend ～	われわれは，～を広げる	521
❷ results extend ～	結果は，～を広げる	161
❸ findings extend ～	知見は，～を広げる	119
❹ study extends ～	研究は，～を広げる	104

❶ Here, **we extend these findings to human cells.** (Nucleic Acids Res. 2010 38:1114)

訳 われわれは，これらの知見をヒト細胞に広げる

■ be extended ＋［前］

★ be extended	…は，～に広げられる	1,401
❶ be extended to ～	…は，～に広げられる／～に伸展される	801
❷ be extended by ～	…は，～によって広げられる	152

❶ **The methods developed here could quite easily be extended to protein crystallization**, phase diagram measurements, chemical reaction optimization, or multivariable experiments. (J Am Chem Soc. 2002 124:4432)

訳 ここで開発された方法は，極めて容易にタンパク質の結晶化に広げられうる

■ extended ＋［名］

❶ extended conformation	引き伸ばされた立体構造	257
❷ extended periods	長時間	204

■ extend ＋［前］

❶ extend to ～	…は，～に及ぶ／～に広がる	744
❷ extend from ～	…は，～から広がる／～から伸展する	666
❸ extend into ～	…は，～に及ぶ	335

❶ In such a region, **the effects of selection may extend to linked sites that are far away.** (Genetics. 2004 167:423)

訳 選択の効果は，ずっと離れたところにある関連した部位に及ぶかもしれない

extension ㊔ 伸長 用例数 7,368

extension	伸長	6,594
extensions	伸長	774

複数形のextensionsの割合が，約10%ある可算名詞．

◆類義語：expansion, stretch

■ extension ＋［前］

	the%	a/an%	ø%			
❶	30	31	38	extension of ～	～の伸長	1,578
❷	10	9	77	extension in ～	～における伸長	200
❸	8	47	33	extension to ～	～への伸長	168

❶ Human and other mammalian thymidylate synthase (TS) enzymes **have an N-terminal extension of approximately 27 amino acids** that is not present in bacterial TSs. (Biochemistry. 2010 49:2475)

訳 ～は，およそ27アミノ酸のN末端伸長を持つ

E

★ extensive ㊓ 広範な 用例数 10,317

◆類義語：wide, broad, widespread

■ extensive ＋［名］

❶	extensive studies	広範な研究	125
❷	extensive homology	広範なホモロジー	117
❸	extensive research	広範な研究	110
❹	extensive apoptosis	広範なアポトーシス	83

❶ **Despite extensive studies,** it is unclear how OCRL mutations result in a myriad of phenotypes found in Lowe syndrome. (Hum Mol Genet. 2012 21:3333)

訳 広範な研究にもかかわらず，

■［副］＋ extensive

❶	more extensive ～	より広範な～	679
❷	less extensive ～	より広範でない～	97

❶ **The 22A strain caused more extensive vacuolation in** the brains of SAMP8 and SAMR1 mice than in C57BL mice. (Brain Res. 2004 995:158)

訳 22A株は，～におけるより広範な空胞形成を引き起こした

extensively ㊐ 広範に 用例数 3,597

過去分詞の前で使われることも，後で使われることもある．

◆類義語：widely, broadly, universally, ubiquitously

■ be extensively ＋［過分］

★	be extensively ～	…は，広範に～される	1,366
❶	be extensively studied	…は，広範に研究される	510
	・be extensively studied in ～	…は，～において広範に研究される	111
❷	be extensively characterized	…は，広範に特徴づけられる	116
❸	be extensively used	…は，広範に使われる	100

❶ Neurogenic inflammation and the role of nerve growth factor（NGF）have been extensively studied in psoriasis.（J Invest Dermatol. 2004 122:812）

訳 ～は，乾癬において広範に研究されてきた

■ be ＋［過分］＋ extensively

❶	be studied extensively	…は，広範に研究される	389
	・be studied extensively in ～	…は，～において広範に研究される	90
❷	be used extensively	…は，広範に使われる	228

❶ The N-glycolylation of sialic acids is a unique carbohydrate modification that has been studied extensively in eukaryotes.（J Biol Chem. 2005 280:326）

訳 ～は，真核生物において広範に研究されてきた

★extent 名 程度／範囲　　用例数 14,682

extent	程度／範囲	14,241
extents	程度／範囲	441

複数形のextentsの割合が，3％ある可算名詞．

◆類義語：degree, grade, range

■ extent ＋［前］

extent ofやextent to whichの前にtheが付く割合は圧倒的に高い．

	the％	a/an％	ø％			
❶	90	0	9	extent of ～	～の程度／～の範囲	6,966
				・determine the extent of ～	…は，～の程度を決定する	214
				・reduce the extent of ～	…は，～の程度を低下させる	126
				・correlate with the extent of ～	…は，～の程度と相関する	124
❷	100	0	0	extent to which ～	～である程度	1,549
				・determine the extent to which ～	…は，～である程度を決定する	169
❸	11	81	4	extent in ～	～における程度	642

❶ To determine the extent of changes in differentiation, we used the Affymetrix GeneChip

Array system to observe global transcriptional changes of genes. (Cancer Res. 2004 64:1266)

訳 分化の変化の程度を決定するために

❶ Moreover, when PSI-BLAST is used, the magnitude of the E-value is found to be weakly correlated with the extent of enzyme function conservation in the third iteration of PS-BLAST. (J Mol Biol. 2003 333:863)

訳 ～は，酵素機能の保存の程度と弱く相関することが見つけられる

❷ The present research was designed to determine the extent to which the central nucleus of the amygdala (ACe) contributes to this effect. (J Neurosci. 2002 22:11026)

訳 現在の研究は，扁桃体（ACe）の中心核がこの効果に寄与する程度を決定するために設計された

■ to ＋［形／冠］＋ extent

❶ to a lesser extent	より低い程度で	1,554
❷ to a greater extent	より大きな程度で	530
・to a greater extent than ～	～より大きな程度で	279
・to a greater extent in ～	～においてより大きな程度で	120
❸ to the extent	程度まで	335
・to the extent of ～	～の程度まで	153
・to the extent that ～	～という程度まで	129
❹ to some extent	ある程度で	248
❺ to the same extent as ～	～と同じ程度で	244

❶ RGA and, to a lesser extent, GAI mRNAs were expressed ubiquitously in all tissues, whereas RGL1, 2, and 3 transcripts were present at high levels only in germinating seeds and/or flowers and siliques. (Plant Physiol. 2004 135:1008)

訳 そして，より低い程度で

❷ A non-ubiquitylatable Rac1 rescues the migration defect of Rac1-null cells to a greater extent than wild-type Rac1. (Oncogene. 2013 32:1735)

訳 ユビキチン化されないRac1 は，野生型のRac1より大きい程度でRac1欠損細胞の遊走欠陥を救出する

❸ Within African savannas, elephants often damage individual trees to the extent that they influence tree density. (Curr Biol. 2010 20:R854)

訳 それらが樹木の密度に影響を与える程度まで

❺ The selected replicons replicated to the same extent as those in parental cells. (Hepatology. 2003 38:769)

訳 選択されたレプリコンは，親細胞におけるそれらと同じ程度に複製した

★ extract 名 抽出物／抽出液，動 抽出する　用例数 14,432

extracts	抽出物／抽出する	8,149
extracted	抽出される／抽出した	2,997
extract	抽出物／抽出する	2,944
extracting	抽出する	342

動詞としても用いられるが，名詞の用例の方が多い．複数形のextractsの割合は，約80%と圧倒的に高い．

◆類義語：abstract

■ 名 extracts ＋［前］

	the%	a/an%	ø%			
❶	0	-	98	extracts from ～	～からの抽出物	1,189
❷	7	-	92	extracts of ～	～の抽出物	1,161

❶ **Viral nuclease activity was measured in extracts from cells infected with** well-defined viral mutants. (J Virol. 2005 79:15084)

訳 ウイルスのヌクレアーゼ活性が，～に感染した細胞からの抽出物において測定された

■ 名 extracts ＋［過分］

❶	extracts prepared from ～	～から調整された抽出物	236

■ 名［形／名］＋ extracts

❶	cell extracts	細胞抽出物	1,274
❷	nuclear extracts	核抽出物	1,264
	・nuclear extracts from ～	～からの核抽出物	321
❸	egg extracts	卵抽出液	480

■ 動 extracted ＋［前］

❶	extracted from ～	～から抽出される	1,464
❷	extracted by ～	～によって抽出される	129

❶ **Association constants extracted from the NMR data** are in line with those of other cyclic cis amides in chloroform solvent. (J Org Chem. 2005 70:8693)

訳 NMRデータから抽出された会合定数

■ 動 to extract

★	to extract ～	～を抽出する‥	563
❶	to extract ～ from …	～を…から抽出する‥	128

❶ We developed a fast, scalable and sensitive **method to extract TFBSs from ChIP-chip experiments** on genome tiling arrays. (Bioinformatics. 2005 21:i274)

訳 ChIP-chip実験からTFBSを抽出する方法

extraction 名 抽出　　用例数 3,685

extraction	抽出	3,557
extractions	抽出	128

複数形のextensionsの割合は３％あるが，単数形は無冠詞で使われることが圧倒的に多い．

■ extraction ＋［前］

	the%	a/an%	ø%			
❶	26	0	74	extraction of ～	～の抽出	1,578

❶ Extraction of seven drugs from raw urine is performed using specially designed SPME fibers coated uniformly with silica-C_{18} stationary phase.（Anal Chem. 2010 82:7502）
訳 未加工の尿からの７つの薬物の抽出が行われる

extremely 副 極端に／非常に 用例数 4,000

◈類義語：unusually, extraordinarily, very, exceedingly, markedly

■ extremely ＋［形］

❶	extremely low	極端に低い	475
❷	extremely high	極端に高い	411
❸	extremely sensitive to ～	～に極端に敏感な	148
	・be extremely sensitive to ～	…は，～に極端に敏感である	134
❹	extremely rare	極端にまれな	118
❺	extremely rapid	極端に急速な	101
❻	extremely stable	極端に安定な	98
❼	extremely useful	極端に有用な	93

❶ Patients with Tangier disease exhibit extremely low plasma HDL concentrations resulting from mutations in the ATP-binding cassette, sub-family A, member 1（ABCA1）protein.（J Clin Invest. 2005 115:1333）
訳 タンジール病の患者は，極端に低い血漿HDL濃度を示す

E

★ facilitate 動 促進する　用例数 20,602

facilitate	～を促進する	9,792
facilitates	～を促進する	4,604
facilitated	～を促進した／促進される	3,413
facilitating	～を促進する	2,793

他動詞．能動態の用例が多い．

◆類義語：accelerate, promote

■ facilitate ＋［名句］

❶ facilitate the development of ～	…は，～の開発を促進する	309
❷ facilitate the identification of ～	…は，～の同定を促進する	206
❸ facilitate the formation of ～	…は，～の形成を促進する	145
❹ facilitate the study of ～	…は，～の研究を促進する	108

❶ The availability of a large group of selective and nonselective CRF receptor peptide agonists will **facilitate the development of** CRF receptor selective drugs.（J Med Chem. 2004 47:3450）

訳 ～は，CRF受容体選択的薬剤の開発を促進するであろう

■ to facilitate

★ to facilitate ～	～を促進する…	3,792
❶ To facilitate ～	～を促進するために	522
❷ appear to facilitate ～	…は，～を促進すると思われる	65

❶ **To facilitate** this analysis, the entire accessory Sec system of S. gordonii was expressed in Escherichia coli.（J Biol Chem. 2012 287:24438）

訳 この分析を促進するために，

■ be facilitated ＋［前］

★ be facilitated	…は，促進される	1,065
❶ be facilitated by ～	…は，～によって促進される	903

❶ Heterodimer formation **is facilitated by** the presence of zinc but does not depend on copper loading of yCCS.（Biochemistry. 2000 39:14720）

訳 ヘテロ二量体形成は，亜鉛の存在によって促進される

■ thereby facilitating

❶ , thereby facilitating ～	それによって，～を促進する	213

❶ Biochemical analyses indicated that RTA interacts with RelA and promotes RelA ubiquitination, **thereby facilitating** RelA degradation.（J Virol. 2012 86:1930）

訳 それによって，RelAの分解を促進する

fact ㊔ 事実　用例数 5,804

fact	事実	5,605
facts	事実	199

複数形のfactsが使われる割合が，約３％ある可算名詞．the fact thatとin factの用例が非常に多い．

◆類義語：observation, discovery, finding

■ fact ＋［同格that節］

fact thatの前にtheが付く割合は，99％と圧倒的に高い．

	the％	a/an％	ø％			
❶	99	1	0	fact that ～	～という事実	3,325
				・despite the fact that ～	～という事実にもかかわらず	732
				・by the fact that ～	～という事実によって	465

❶ The binding between an enzyme and its substrate is highly specific, **despite the fact that** many different enzymes show significant sequence and structure similarity. (J Mol Biol. 2005 352:1105)

訳 多くの異なる酵素が顕著な配列と構造の類似性を示すという事実にもかかわらず

❶ Disruption of the skeletal muscle architecture and permeability also requires mast cell (MC) participation, as revealed **by the fact that** IR injury is markedly reduced in c-kit defective, MC-deficient mouse strains. (J Immunol. 2005 174:7285)

訳 ～という事実によって明らかにされたように

■ in fact

❶	in fact	実際に	1,857

❶ **In fact**, the opposite is true, in that these two cytokines appear to have profoundly different roles in regulating host immune responses. (Trends Immunol. 2006 27:17)

訳 実際に，

★★ factor ㊔ 因子／要因／ファクター　用例数 169,943

factor	因子	105,273
factors	因子	64,670

複数形のfactorsが使われる割合は約40％とかなり高い．

■ factor ＋［前］

	the％	a/an％	ø％			
❶	14	85	1	factor for ～	～の因子	4,174
				・risk factor for ～	～のリスク因子	2,426
❷	13	77	8	factor in ～	～における因子／～の際の因子	3,562

F

				・factor in determining ～	～を決定する因子	201
❸	19	65	16	factor of ～	～の因子	2,545
				・by a factor of ～	～倍だけ	833

❶ Cystatin C is a strong, independent **risk factor for** mortality in the elderly.（J Am Soc Nephrol. 2006 17:254）

訳 シスタチンCは，高齢者における死亡の強力で独立したリスク因子である

❷ Another important **factor in** tumor development seems to be the hypermethylation of CpG islands located within the promoter regions of tumor suppressor genes.（Cancer Res. 2005 65: 4673）

訳 腫瘍発症におけるもう１つの重要な因子は，～の過剰メチル化であるように思われる

■ [動] + factors

❶ to identify factors	因子を同定する…	310
・to identify factors that ～	～する因子を同定する…	104
・to identify factors associated with ～	～と関連する因子を同定する…	87

❶ Multinomial logistical regression analysis was used **to identify factors that** determine the UES response.（Gastroenterology. 2012 142:734）

訳 上部食道括約筋応答を決定する因子を同定するために，多項ロジスティック回帰分析が使われた

■ [名／形] + factor

❶ risk factor	リスク因子	3,714
❷ important factor	重要な因子	705
❸ major factor	主な因子	384
❹ key factor	鍵となる因子	374
❺ prognostic factor	予後因子	364
❻ survival factor	生存因子	335
❼ critical factor	決定的に重要な因子	286

★ fail 動 できない／失敗する　用例数 18,822

failed	できなかった	11,008
fail	できない	4,314
fails	できない	2,444
failing	できない	1,056

fail to *do*の用例が圧倒的に多い.

■ fail to *do*

★ fail to *do*	…は，～することができない	16,128
❶ fail to induce ～	…は，～を誘導することができない	841
❷ fail to activate ～	…は，～を活性化することができない	432

❸	fail to bind	…は，結合することができない	90
❹	fail to form ～	…は，～を形成することができない	87

❶ Ectopic expression of either Tax1 or Tax2 **failed to induce** apoptosis in T-cell lines. (J Virol. 2003 77:12152)
訳 Tax1あるいはTax2のいずれの異所性発現も，T細胞株のアポトーシスを誘導することができなかった

★ failure ㊔ 失敗／不全 用例数 21,485

failure	失敗／不全	20,740
failures	失敗／不全	745

複数形のfailuresが使われる割合は約3％あるが，単数形は無冠詞で使われることが多い．

◆類義語：abortion, insufficiency

■ failure ＋［前］

	the％	a/an％	ø％			
❶	34	19	47	failure of ～	～の失敗／～の不全	2,531
				・failure of ～ to *do*	～が…することの失敗	544
❷	15	22	59	failure to *do*	～することの失敗	2,199
❸	2	13	84	failure in ～	～における失敗／～における不全	1,062

❶ Here, we investigated the mechanism underlying **the failure of TGF-β to induce stress fibers** and inhibit cell migration in metastatic cells. (Oncogene. 2005 24:5043)
訳 TGF-βがストレスファイバーを誘導することの失敗

❷ A reduction of endodermal gene expression as well as **a failure to displace the visceral endoderm** occurs despite the formation of a normal foregut pocket. (Dev Biol. 2004 270:411)
訳 臓側内胚葉を置換することの失敗

❸ These data suggest that tamoxifen-stimulated secretion of TGFβ **might explain treatment failure in some patients.** (J Theor Biol. 2004 229:101)
訳 ～は，何人かの患者における治療の失敗を説明するかもしれない

■［名／形］＋ failure

❶	heart failure	心不全	7,187
❷	renal failure	腎不全	1,322
❸	treatment failure	治療失敗	620
❹	graft failure	移植失敗	610
❺	respiratory failure	呼吸不全	554
❻	liver failure	肝不全	476
❼	organ failure	臓器不全	444

fall ㊍ 低下する／収まる，㊔ 低下 用例数 4,374

fall	低下する／収まる／低下	1,938
fell	低下した／収まった	1,250
falls	低下する／収まる	698
falling	低下する／収まる	397
fallen	低下した／収まった	91

自動詞および名詞として用いられる.

◆類義語：decrease, reduce, diminish, lower, decline, reduction, attenuation

■ ㊍ fall ＋［前］

❶ fall into ～	…が，～に収まる	670
❷ fall to ～	…が，～に低下する	262
❸ fall from ～	…が，～から低下する	182
❹ fall by ～	…が，～だけ低下する	130

❶ **The mutants fell into two classes:** five that failed to respond to chi, and six that suggested a relaxed specificity for chi recognition.（Proc Natl Acad Sci USA. 2012 109:8901）

訳 それらの変異体は，2つのクラスに分かれた

❷ After statin therapy, **LDL cholesterol levels fell to a similar level**（101 ± 32 mg/dl）as in patients not receiving statins（98 ± 33 mg/dl）.（Circulation. 2005 112:357）

訳 LDLコレステロールレベルが，類似のレベルに低下した

❸ The rate of patients not receiving reperfusion **fell from 5.4% to 4.0%**（P=0.04）.（Circulation. 2012 126:189）

訳 ～が，5.4%から4.0%に低下した

■ ㊔ fall ＋［前］

「～の低下」の意味では，fall ofよりfall inが使われることが圧倒的に多い.

	the%	a/an%	ø%			
❶	32	52	13	fall in ～	～の低下	669
❷	59	30	11	fall of ～	～の低下	77

❶ A selective 84 pmol/l rise in arterial insulin **was thus associated with a fall in NHGO of approximately 50%**, which took 1 h to manifest.（Diabetes. 1996 45:1594）

訳 ～は，およそ50%のNHGOの低下とこのように関連付けられた

★★ family ㊔ ファミリー 用例数 63,239

family	ファミリー	51,776
families	ファミリー	11,463

複数形のfamiliesが使われる割合が，約20%ある可算名詞.

◆類義語：species, genus

■ family/families ＋［前］

	the%	a/an%	ø%			
❶	55	38	3	family of ～	ファミリーの～／～のファミリー	14,812
				・family of proteins	ファミリーのタンパク質／タンパク質のファミリー	1,658
				・family of transcription factors	ファミリーの転写因子／転写因子のファミリー	1,012
				・a large family of ～	～の大きなファミリー	396
❷	0	-	46	families with ～	～を持つファミリー	1,658

❶ The Daam family of proteins consists of Daam1 and Daam2.（Dev Cell. 2012 22:183）
訳 Daamファミリーのタンパク質は，Daam1 と Daam2からなる

❶ Claudin proteins form a large family of integral membrane proteins crucial for tight junction formation and function.（Cancer Res. 2005 65:7378）
訳 クローディンタンパク質は，複合的な膜タンパク質の大きなファミリーを形成する

F

■ family ＋［名］

❶	family members	ファミリーメンバー	4,756
❷	family history	家族歴	1,387
❸	family proteins	ファミリータンパク質	1,062
❹	family kinases	ファミリーキナーゼ	618

■［名］＋ family

❶	gene family	遺伝子ファミリー	2,703
❷	protein family	タンパク質ファミリー	1,623
❸	receptor family	受容体ファミリー	863
❹	kinase family	キナーゼファミリー	578

far 副 はるかに　　用例数 6,474

比較級の前で使われることが非常に多い．
◆類義語：much

■ far ＋［前］

❶	far from ～	～からほど遠い	523

❶ The etiology of primary biliary cirrhosis（PBC）is far from clear.（Hepatology. 2011 54:2099）
訳 原発性胆汁性肝硬変（PBC）の病因は，明らかからはほど遠い

■ far ＋［副／形］

❶ far more ～	はるかにより～	524
❷ far less ～	はるかにより～でない	249
❸ far greater	はるかにより大きい	161
❹ far-red	赤外の	387

❶ Placing these findings in the context of other studies, the effect of IFN-I on systemic autoimmunity **appears to be far more complex than originally perceived.** (Arthritis Rheum. 2005 52:3063)

訳 ～は，最初に認知されたよりはるかに複雑であるように思われる

■［副／前］＋ far

❶ so far	今までに	1,459
❷ thus far	ここまでは／これまでは	903
❸ by far	はるかに	260
❹ as far as ～	～限り／～まで	124

❶ **Controlled studies have so far provided conflicting data** on risk factors and comorbidity rates in PBC. (Hepatology. 2005 42:1194)

訳 制御された研究は，今までに相克的なデータを提供してきた

❷ However, **K345 variants have, thus far, failed to crystallize.** (J Mol Biol. 2006 355:422)

訳 K345変異体は，これまでは，結晶化できないでいる

❹ **As far as possible,** nomenclature for genes and proteins are standardized within genera and families. (Nucleic Acids Res. 2006 34:D382)

訳 可能な限り，

fashion ㊔ 様式　　用例数 5,589

fashion	様式	5,557
fashions	様式	32

複数形のfashionsの割合は0.6%しかないが，単数形には不定冠詞が付く用例が圧倒的に多い．

◆類義語：manner, pattern, mode, way

■［形］＋ fashion

❶ ～-dependent fashion	～依存的な様式	1,412
・in a dose-dependent fashion	用量依存的な様式で	346
❷ ～-specific fashion	～特異的な様式	256
❸ ～-independent fashion	～非依存的な様式	224

❶ Knockdown of PARP1 enhanced the lethality of CHK1 inhibitors **in a dose-dependent fashion.** (Mol Pharmacol. 2012 82:322)

訳 用量依存的な様式で

❷ Reverse transcription PCR results suggest that some SCPL genes **are expressed in a highly tissue-specific fashion**, whereas others are transcribed in a wide range of tissue types. (Plant Physiol. 2005 138:1136)
訳 ～は，高度に組織特異的な様式で発現する

■ fashion ＋［前］

	the%	a/an%	ø%			
❶	1	92	4	fashion with ～	～を伴う様式	239
❷	3	84	12	fashion in ～	～における様式	238

favor 動 支持する／促す／～に好都合である，名 支持 用例数 5,321

favor	～を支持する／～に好都合である／支持	1,976
favored	促される／～を支持した	1,575
favors	～を支持する／～に好都合である／支持	906
favoring	支持される／～を支持する／～に好都合である	864

主に動詞として用いられる．名詞としては，ほとんどin favor ofの形で使われる．
◆類義語：support, facilitate, accelerate, promote

F

■ 動 favor ＋［前］

❶	favor the formation of ～	…は，～の形成に好都合である	70

❶ **Expression of AKAP-Lbc in fibroblasts favors the formation of stress fibers** in a Rho-dependent manner. (J Biol Chem. 2001 276:44247)
訳 線維芽細胞におけるAKAP-Lbcの発現は，ストレスファイバーの形成に好都合である

■ 動 be favored ＋［前］

★	be favored	…は，促される／好都合である	525
❶	be favored by ～	…は，～によって促される	137
❷	be favored over ～	…は，～より好都合である	71

❶ Most trait differences in ancient hybrids could be recreated by complementary gene action in synthetic hybrids and **were favored by selection**. (Science. 2003 301:1211)
訳 ～は，選択によって促された

❷ **The short loop was favored over the longer loops**, particularly on supercoiled DNA. (J Mol Biol. 2004 339:53)
訳 短いループが，より長いループより好都合であった

■ 名 in favor of

❶	in favor of ～	～を支持する…	610

❶ However, **we find evidence in favor of both models 2 and 3** when the data is partitioned between groups of amino acids and between regions of the genes. (J Mol Evol. 2001 53:225)
訳 われわれは，モデル２および３の両方を支持する証拠を見つける

feasibility ㊔ 実現可能性　用例数 2,224

feasibility	実現可能性	2,222
feasibilities	実現可能性	2

複数形のfeasibilitiesの割合はほとんどなく，原則，**不可算名詞**として使われる．

◆類義語：possibility

■ feasibility ＋［前］

feasibility ofの前にtheが付く割合は圧倒的に高い．

	the%	a/an%	ø%			
❶	96	0	4	feasibility of ～	～の実現可能性	1,735
				・demonstrate the feasibility of ～	…は，～の実現可能性を実証する	480
				・the feasibility of using ～	～を使うことの実現可能性	289

❶ This study **demonstrates the feasibility of** using recombinant human adenoviral vectors to detect nodal metastases in a human prostate cancer model. (Nat Med. 2008 14:882)

訳 この研究は，組換え型ヒトアデノウイルスベクターを使うことの実現可能性を実証する

★ feature ㊔ 特徴，㊍ 特集する　用例数 28,897

features	特徴／特集する	20,817
feature	特徴／特集する	7,508
featuring	特集する	403
featured	特集される／特集した	169

動詞としても用いられるが，名詞の用例が非常に多い．複数形のfeaturesの割合は，約75%と圧倒的に高い．

◆類義語：characteristic, character, hallmark, property, profile

■ features ＋［前］

	the%	a/an%	ø%			
❶	37	-	46	features of ～	～の特徴	7,811
❷	9	-	72	features in ～	～における特徴	766

■ ［形］＋ feature(s)

❶	structural features	構造的な特徴	1,705
❷	clinical features	臨床的な特徴	981
❸	common feature	共通の特徴	561
❹	key features	鍵となる特徴	374
❺	unique features	ユニークな特徴	325

❻ many features	多くの特徴	255

❶ In summary, the loss of CfH accelerates the development of lupus nephritis and **recapitulates the functional and structural features of the human disease**.（J Am Soc Nephrol. 2011 22:285）
訳 ～は，そのヒト疾患の機能的および構造的特徴を再現する

feed 動（餌を）与える／養う／摂食する 用例数 5,468

fed	与えられる／与えた	4,483
feed	（餌を）与える	829
feeds	（餌を）与える	156

他動詞．主に第 4 文型（$S+V+O_1+O_2$）の受動態として用いられることが多い．動物が主語となる受動態の用例が圧倒的に多い．feedingは名詞の用例が多い．

◈類義語：give

■［名］＋ be fed

★ be fed ～	…は，～を与えられる	912
❶ mice were fed ～	マウスは，～を与えられた	209
❷ rats were fed ～	ラットは，～を与えられた	124

❶ The mutant **mice were fed a high-fat diet**（HFD）to induce insulin resistance and hyperglycemia.（Diabetes. 2012 61:566）
訳 マウスは高脂肪食を与えられた

■［名］＋ fed ／ fed ＋［名］

❶ mice fed ～	～を与えられたマウス	579
❷ ～-fed rats	～を与えられたラット	218

❶ **In diabetic mice fed a high-fat diet**, a 1-month chronic i.p. Ex9 treatment improved glucose tolerance and fasting glycemia.（J Clin Invest. 2005 115:3554）
訳 高脂肪食を与えられた糖尿病マウスにおいて

❷ **Mitochondria were isolated from the liver of both control and ethanol-fed rats** after 5 to 6 weeks of alcohol consumption.（Hepatology. 2003 38:141）
訳 ミトコンドリアが，コントロールおよびエタノールを与えられた両群のラットの肝臓から単離された

■ fed ＋［前］

❶ fed to ～	～に与えられた	121
❷ fed with ～	～を与えられた	86

❶ In the Langendorff model sulindac provided significant protection against cell death, when **the drug was fed to the animals** before the removal of the heart for the Langendorff procedure.（Proc Natl Acad Sci USA. 2009 106:19611）
訳 その薬剤が，動物に与えられた

F

★ few 形 ほとんどない／わずかな／少数の 用例数 12,846

通常，〈a few ～〉は「少数の～」という意味で用いられ，無冠詞の〈few ～〉は否定的に「～はほとんどない」の意味で使われる．

◆類義語：little, subtle, slight, insignificant

■ few ＋［名］

❶ few studies	研究はほとんどない	904
・few studies have examined ～	～を調べた研究はほとんどない	180
❷ few data	データはほとんどない	277
❸ the past few years	過去数年	413

❶ **Few studies have examined** the influence of diet on survival and chemotherapy-associated toxicities in patients with cancer.（J Clin Oncol. 2005 23:8453）

訳 ～を調べた研究はほとんどない

❷ **Few data are available regarding** satisfaction, psychological, and social function after CPM.（J Clin Oncol. 2005 23:7849）

訳 ～に関して利用できるデータはほとんどない

■［副／形］＋ few

❶ very few ～	非常にわずかな～	719
❷ relatively few ～	比較的わずかな～	463
❸ as few as ～	わずか～	405
❹ the first few ～	最初のいくつかの～	294

❶ Even in older mice, these follicle-like lesions grew no larger than the size of antral follicles and **contained very few proliferating cells.**（Cancer Res. 2005 65:9206）

訳 ～は，非常にわずかな増殖細胞を含んでいた

❸ **Our results showed that as few as 10 copies of stx_2 could be detected,** indicating that the RAM assay was as sensitive as conventional PCR.（J Clin Microbiol. 2005 43:6086）

訳 われわれの結果は，わずか10コピーのstx_2が検出されうることを示した

■ a few

★ a few ～	少数の～	4,586
❶ only a few ～	ほんの少数の～だけ	974

❶ In addition, **only a few studies examined** the newest CT colonography technology.（Ann Intern Med. 2005 142:635）

訳 ほんの少数の研究だけが，～を調べた

★ field 名 分野／場 用例数 20,052

field	分野／場	15,914
fields	分野／場	4,138

複数形のfieldsの割合が，約20%ある可算名詞．

◆類義語：area

■ field ＋［前］

field ofの前にtheが付く割合は圧倒的に高い（field of viewの場合を除く）．また，field inの前にtheが付く割合もかなり高い．

	the%	a/an%	⌀%			
❶	80	17	1	field of ～	～の分野	1,911
				・in field of ～	～の分野において	658
				・field of view	視野	168
❷	62	22	11	field in ～	～における場	213

❶ The past decade has seen remarkable advances in the field of stem cell biology.（Curr Opin Pediatr. 2012 24:577）

訳 過去10年は，幹細胞生物学の分野において注目すべき進歩が見られた

■［形／名］＋ field

❶	magnetic field	磁場	900
❷	electric field	電場	820
❸	visual field	視野	800
❹	receptive field	受容野	724
❺	force field	力場	385

F

final 形 最終の　用例数 5,414

前にtheが付くことが非常に多い．

◆類義語：eventual

■ the final ＋［名］

★	the final ～	最終～	3,208
❶	the final step	最終ステップ	287
❷	the final stages of ～	～の最終ステージ	119
❸	final product	最終産物	111

❶ This reaction is the final step in the specific GlcNAc utilization pathway and thus decides the metabolic fate of GlcNAc.（J Biol Chem. 2005 280:19649）

訳 この反応は，～における最終ステップである

★ finally 副 最後に／ついに　用例数 17,289

文頭で用いられることが圧倒的に多い．

◆類義語：eventually, ultimately, terminally, lastly, at last

■ Finally,

★ Finally,	最後に／ついに	16,257
❶ Finally, we show ～	最後に，われわれは～を示す	1,265
❷ Finally, we demonstrate ～	最後に，われわれは～を実証する	583
❸ Finally, we found ～	ついに，われわれは～を見つけた	335

❶ **Finally, we show that** the 50-kb Sgo1-binding domain is the chromosomal region where cohesins are protected from removal during meiosis I.（Genes Dev. 2005 19:3017）
訳 最後に，われわれは～ということを示す

❸ **Finally, we found that** multicopy expression of sypG resulted in robust biofilm formation.（Mol Microbiol. 2005 57:1485）
訳 ついに，われわれは～ということを見つけた

★★ find 動 見つける／分かる　用例数 123,117

found	見つけられる／～を見つけた	105,287
find	～を見つける	17,547
finds	～を見つける	283

他動詞．能動態はwe find/found that節の用例が圧倒的に多い．受動態は，〈be found to *do*〉〈be found in〉の用例が多い．findingは別項参照．

◆類義語：discover, detect, identify, observe

■ find ＋［that節］

★ find that ～	…は，～ということを見つける	45,737
❶ we found that ～	われわれは，～ということを見つけた	24,692
❷ we have found that ～	われわれは，～ということを見つけた	1,855
❸ we previously found that ～	われわれは，以前に～ということを見つけた	215
❹ the authors found that ～	著者らは，～ということを見つけた	147

❶ Consistent with these observations, **we found that** Fig4 physically associates with Vac14 in a common membrane-associated complex.（Mol Biol Cell. 2004 15:24）
訳 われわれは，～ということを見つけた

■ be found ＋［that節］

★ be found that ～	～ということが見つけられる	1,959
❶ it was found that ～	～ということが見つけられた	1,388

❶ In addition, **it was found that** two binding sites exist within this element of the fruA promoter.（Proc Natl Acad Sci USA. 2003 100:8782）
訳 ～ということが見つけられた

■ be found to *do*

★ be found to *do*	…が，～することが見つけられる	23,438
❶ be found to be ～	…が，～であることが見つけられる	11,435
・be found to be associated with ～	…が，～と関連することが見つけられる／～と結合していることが見つけられる	281
❷ be found to have ～	…が，～を持つことが見つけられる	1,697
❸ be found to bind	…が，結合することが見つけられる	481
❹ be found to contain ～	…が，～を含むことが見つけられる	474

❶ GFP-tagged DRP1A and DRP1C proteins **were found to be associated with** the cytoskeleton in G1 phase of the cell cycle.（Plant Mol Biol. 2003 53:297）

訳 ～が，細胞骨格と結合していることが見つけられた

❸ Human collagen type VI **was found to bind** both the recombinant A domain of SdrX and viable S. capitis expressing SdrX.（Infect Immun. 2004 72:6237）

訳 ヒトのVI型コラーゲンが，両方の～に結合することが見つけられた

■ be found ＋［前］

★ be found	…が，見つけられる	43,449
❶ be found in ～	…が，～において見つけられる	10,565
・mutations were found in ～	変異が，～において見つけられた	236
・differences were found in ～	違いが，～において見つけられた	166
❷ be found between ～	…が，～の間に見つけられる	1,258
・correlation was found between ～	相関が，～の間に見つけられた	249
❸ be found for ～	…が，～に対して見つけられる	1,085
❹ be found at ～	…が，～において見つけられる	845

❶ Multiple novel **mutations were found in** the OD4, UL9, UL36, and US1 genes, and we showed that S34 in the US1 gene is essential in ocular disease.（Invest Ophthalmol Vis Sci. 2003 44:2657）

訳 複数の新規の変異が，～において見つけられた

❷ A significant **correlation was found between** internal protein cavities and binding in a group of proteins with high resolution structures.（J Biol Chem. 2004 279:19628）

訳 有意な相関が，～の間に見つけられた

❸ A similar half-life **was found for** iNOS in several cell lines.（Proc Natl Acad Sci USA. 2004 101:18141）

訳 類似の半減期が，いくつかの細胞株におけるiNOSに対して見つけられた

■［代名／名］＋ found in

★ found in ～	～において見つけられる	21,694
❶ those found in ～	～において見つけられたそれら	733
・similar to those found in ～	～において見つけられたそれらに似ている	233
❷ proteins found in ～	～において見つけられたタンパク質	165
❸ motif found in ～	～において見つけられたモチーフ	159

F

❶ Chick Cv-2 has a conserved structure of five cysteine-rich repeats similar to those found in several BMP antagonists, and a C-terminal Von Willebrand type D domain. (Development. 2004 131:5309)

訳 ニワトリのCv-2は，いくつかのBMP拮抗剤において見つけられたそれらに類似した５つのシステインに富んだリピートの保存された構造を持つ

■ find no

★ find no ～	…は，～を見つけない	1,598
❶ we found no ～	われわれは，～を見つけなかった	942
・we found no evidence	われわれは，証拠を見つけなかった	364

❶ However, we found no evidence for significant differences in the patterns of crossover positioning between strains with different exchange frequencies. (Genetics. 2002 162:297)

訳 われわれは，有意な違いの証拠を見つけなかった

★★ finding ㊔ 知見／発見　用例数 64,746

findings	知見／発見	54,644
finding	知見／発見	10,102

複数形のfindingsの割合は，約85%と圧倒的に高い．

◆類義語：discovery, observation

■ findings ＋［動］

❶ findings suggest that ～	知見は，～ということを示唆する	8,163
❷ findings indicate that ～	知見は，～ということを示す	4,333
❸ findings demonstrate ～	知見は，～を実証する	3,120
❹ findings provide ～	知見は，～を提供する	2,621
・findings provide evidence	知見は，証拠を提供する	386
❺ findings support ～	知見は，～を支持する	1,923
・findings support the hypothesis that ～	知見は，～という仮説を支持する	321
❻ findings reveal ～	知見は，～を明らかにする	1,316
❼ findings show that ～	知見は，～ということを示す	946
❽ findings are consistent with ～	知見は，～に一致する	697
❾ findings have important implications for ～	知見は，～のための重要な意味を持つ	223

❶ Our findings suggest that frequent nut consumption is associated with a reduced risk of gallstone disease in men. (Am J Epidemiol. 2004 160:961)

訳 われわれの知見は，～ということを示唆する

❷ These findings indicate that nNOS can differentially regulate the ERK signal transduction pathway in a manner dependent on the presence of l-arginine and the production of NO*. (J Biol Chem. 2004 279:3933)

訳 これらの知見は，～ということを示す

❹ Together, these findings provide evidence that albumin is transported across the mammary epithelium by the same pathway as immunoglobulin. (J Physiol. 2004 560:267)

訳 これらの知見は，～という証拠を提供する

❽ These findings are consistent with an interaction between memory systems of the MTL and the striatum.（Behav Neurosci. 2004 118:438）
訳 これらの知見は，～の間の相互作用に一致する

■ findings ＋［前］

	the%	a/an%	ø%			
❶	11	-	51	findings in ～	～における知見	1,454
❷	55	-	40	findings of ～	～の知見	1,181
❸	23	-	71	findings from ～	～からの知見	727

❷ Overall, the findings of this study highlight the gender differences in neural responses associated with figurative language comprehension.（Brain Res. 2012 1467:18）
訳 この研究の知見は，～の性差を強調する

■［形］＋ findings

❶	recent findings	最近の知見	1,137
❷	previous findings	以前の知見	674
❸	new findings	新しい知見	396

F

★★ first ㊊ 最初の，㊓ 最初に，㊔ 最初のもの　用例数 67,792

形容詞，副詞および名詞として用いられる．
◆類義語：original, initial, early, originally, initially

■ ㊊ first ＋［名］

❶	for the first time	初めて	5,321
	・demonstrate for the first time that ～	…は，～ということを初めて実証する	615
	・show for the first time that ～	…は，～ということを初めて示す	542
❷	first step	最初のステップ	1,980
	・the first step in ～	～における最初のステップ	489
	・as a first step	最初のステップとして	372
❸	the first evidence	最初の証拠	1,591
	・the first evidence that ～	～という最初の証拠	865
	・provide the first evidence	…は，最初の証拠を提供する	839
❹	the first report	最初の報告	1,417
	・this is the first report	これは，最初の報告である	1,091
❺	the first demonstration	最初の実証	1,073

❶ In conclusion, these data demonstrate for the first time that PKCα stimulates NO production in endothelial cells and plays a role in regulation of blood flow in vivo.（Circ Res. 2005 97:482）
訳 これらのデータは，～ということを初めて実証する

❷ Nasopharyngeal colonization is the first step in the interaction between Streptococcus pneumonia（the pneumococcus）and its human host.（Infect Immun. 2005 73:7718）
訳 …は，～の間の相互作用における最初のステップである

❸ These results provide the first evidence that PDL1 is involved in fetomaternal tolerance.（J

Exp Med. 2005 202:231)
訳 これらの結果は，～という最初の証拠を提供する

■ 副 be first ＋［過分］

★ be first ～	…は，最初に～される	2,621
❶ be first detected	…は，最初に検出される	279
❷ be first identified	…は，最初に同定される	237

❶ Avian bornaviruses（ABV）were first detected and described in 2008.（J Virol. 2012 86:7023）
訳 鳥ボルナウイルス（ABV）は，最初に検出された

■ 名 the first ＋［前］

❶ the first to *do*	～する最初のもの	1,231
❷ the first of ～	～の最初のもの	365
❸ the first in ～	～における最初のもの	354

❶ These results are the first to demonstrate a critical role for the S4-S5 linker in the trafficking and/or function of IK and SK channels.（J Biol Chem. 2005 280:37257）
訳 これらの結果は，初めて～を実証するものである

❸ This report is the first in a series of three focused on establishing congruent strategies for carbohydrate sequencing.（Anal Chem. 2005 77:6250）
訳 この報告は，～における最初のものである

fit 動 適合させる，形 一致する／あてはまる，名 適合　用例数 5,909

fit	～を適合させる／適合する／一致する／適合	3,183
fitting	適合する／フィッティング	1,215
fitted	適合する／～を適合させた	832
fits	適合する／～を適合させる／適合	679

形容詞，動詞および名詞として用いられる．fitの過去・過去分詞としては，fittedとfitの両方が使われる．fitとfittingは名詞としても用いられる．

◆類義語：consistent, congruent, corresponding, concurrent, equal, match, correspond, coincide, represent

■ 動／形 be fit ＋［前］

★ be fit	…は，適合する	769
❶ be fit to ～	…は，～に適合する	200
❷ be fit with ～	…は，～と適合する	51

❶ The kinetic data can be fit to a three-state model involving a compact intermediate.（Biochemistry. 2005 44:627）
訳 動力学的データは，～に適合しうる

■ 動 be fitted ＋ [前]

★	be fitted	…は，適合する	395
❶	be fitted to ～	…は，～に適合する	179
❷	be fitted with ～	…は，～と適合する	75
❸	be fitted by ～	…は，～によって合わせられる	54

❶ A formal exemplar **model was fitted to the data** to track the changes in parameters to help interpret the fMRI results. (Proc Natl Acad Sci USA. 2012 109:333)

訳 モデルは，そのデータに適合した

❷ **Data were fitted with an exponential decay function** to evaluate PERG changes with time. (Invest Ophthalmol Vis Sci. 2005 46:1296)

訳 データは，指数減衰関数に適合した

■ 動 [名] ＋ fit

❶	model fits ～	モデルは，～に合う	36

❶ We show that the **GP model fits the data** much better than the traditional Poisson model. (Nucleic Acids Res. 2010 38:e170)

訳 GPモデルは，そのデータに合う

F

flexibility 名 可動性／柔軟性 用例数 3,412

flexibility	可動性／柔軟性	3,383
flexibilities	可動性／柔軟性	29

複数形のflexibilitiesが使われる割合は1％しかなく，原則，**不可算名詞**として使われる．

◆類義語：mobility, plasticity

■ flexibility ＋ [前]

flexibility ofの前にtheが付く割合は圧倒的に高い．

	the %	a/an %	ø %			
❶	88	2	8	flexibility of ～	～の可動性	869
❷	3	4	86	flexibility in ～	～の可動性	481

❶ A 5'-DNA-PNA-DNA-3' chimera was synthesized, thereby, conferring both a loss of charge and **altering the conformational flexibility of the oligonucleotide**. (Biochemistry. 2005 44:666)

訳 オリゴヌクレオチドの立体構造上の可動性を変化させる

❷ A number of torsion-angle transitions of the antiviral compound are involved, which suggests that **flexibility in antiviral compounds is important for binding**. (Proc Natl Acad Sci USA. 2005 102:7529)

訳 抗ウイルス性の化合物の可動性は，結合にとって重要である

■ [形] + flexibility

❶	conformational flexibility	立体構造上の可動性	365
❷	structural flexibility	構造上の可動性	99

fluctuation 名 ゆらぎ／変動　用例数 3,397

fluctuations	ゆらぎ／変動	2,861
fluctuation	ゆらぎ／変動	536

複数形のfluctuationsが使われる割合は，約85%と圧倒的に高い．

■ fluctuations + [前]

	the%	a/an%	ø%			
❶	13	-	84	fluctuations in ～	～のゆらぎ／～の変動	941
❷	24	-	76	fluctuations of ～	～のゆらぎ／～の変動	345

❶ We found that representative samples of the mRNAs induced by over-expression of WC-1 **show circadian fluctuations in their levels**. (Mol Microbiol. 2002 45:917)
訳 ～は，それらのレベルの概日性の変動を示す

❷ We show that the kinks **are due to thermal fluctuations of the molecules** at the crystal-solution interface. (J Mol Biol. 2000 303:667)
訳 ～は，分子の熱ゆらぎのせいである

★ focus 動 焦点を合わせる／注目する／集中する，名 焦点　用例数 14,644

focus	焦点を合わせる／注目する／焦点	5,889
focused	焦点を合わせた／注目した	5,294
focusing	焦点を合わせる／注目する	1,801
focuses	焦点を合わせる／注目する／焦点	1,660

動詞および名詞として用いられる．focus onの用例が非常に多い．
◆類義語：concentrate, center

■ 動 focus on

★	focus on ～	…は，～に集中する	9,192
❶	focus on ～ing	…は，～することに焦点を合わせる	897
❷	review focuses on ～	総説は，～に焦点を合わせる	768
❸	we focus on ～	われわれは，～に焦点を合わせる	704
❹	studies have focused on ～	研究は，～に焦点を合わせてきた	234

❷ **This review focuses on our current knowledge** obtained with tombusviruses and other plant viruses. (Curr Opin Virol. 2011 1:332)

訳 この総説は，我々の現在の知識に焦点を合わせる

❸ Here, **we focus on the function of Msx1 and Msx2**, homeobox genes implicated in several disorders affecting craniofacial development in humans. (Development. 2005 132:4937)

訳 われわれは，Msx1およびMsx2の機能に焦点を合わせる

❹ **Our studies have focused on understanding the role of H2A.Z** during cell cycle progression. (Mol Cell Biol. 2006 26:489)

訳 われわれの研究は，H2A.Zの役割を理解することに焦点を合わせてきた

■ 名 focus ＋［前］

focus ofの前にtheが付く割合はかなり高い．

	the%	a/an%	ø%			
❶	69	31	0	focus of ～	～の焦点	1,091
❷	8	81	8	focus on ～	～に対する集中	885
				・with a focus on ～	～に集中して	299
				・particular focus on ～	～に対する特別な集中	107

❶ **The focus of this study was to identify** the Tn5520 origin of conjugative transfer (oriT) and to study BmpH-oriT binding. (Mol Microbiol. 2006 59:288)

訳 この研究の焦点は，～を同定することであった

❷ We investigated hematopoietic cell-targeted deletion of the STAT3 gene in HSCs/HPCs **with a focus on mitochondrial function**. (Blood. 2012 120:2589)

訳 ミトコンドリア機能に集中して

F

★★ -fold 接尾 ～倍 用例数 43,698

◆類義語：-time, magnitude

■ ～-fold ＋［名］

❶	～-fold increase in …	…の～倍の増大	3,872
	・a ～-fold increase in …	…の～倍の増大	1,673
	・result in a ～-fold increase in …	‥は，…の～倍の増大という結果になる	234
	・approximately ～-fold increase in …	…のおよそ～倍の増大	272
❷	～-fold decrease in …	…の～倍の低下	747
❸	～-fold reduction in …	…の～倍の低下	579

❶ **Treatment of CHO cells with 100 μCi [^{3}H]dTTP resulted in a 14-fold increase in** bystander mutation incidence among neighboring A_L cells compared with controls. (Cancer Res. 2005 65:9876)

訳 100 μCi [^{3}H]dTTPによるCHO細胞の処理は，～の14倍の増大という結果になった

■ ～-fold ＋［形］

❶	～-fold higher	～倍より高い	2,943
	・～-fold higher than …	…より～倍高い	881
	・～-fold higher in …	…において～倍より高い	392
❷	～-fold more …	～倍より…	1,800

・～-fold more potent	～倍より強力な	301
❸ ～-fold lower	～倍より低い	1,369
❹ ～-fold greater	～倍より大きい	1,263
❺ ～-fold less …	～倍より…でない	776

❶ The estimated risk of developing SSPE **was 10-fold higher than the previous estimate** reported for the United States in 1982.（J Infect Dis. 2005 192:1686）
訳 ～は，以前の推定より10倍高かった

❶ In photorespiratory conditions, leaf glycine levels **were up to 46-fold higher in the mutant than in the wild type.**（J Biol Chem. 2005 280:26137）
訳 ～は，野生型においてより変異体において46倍まで高かった

❷ Blunt 27mer duplexes **can be up to 100-fold more potent than** traditional 21mer duplexes.（Nucleic Acids Res. 2005 33:4140）
訳 …は，～より100倍まで強力でありうる

❹ In acute i.c.v. studies, the bivalent ligands **functioned as agonists with potencies ranging from 1.6- to 45-fold greater than morphine.**（Proc Natl Acad Sci USA. 2005 102:19208）
訳 ～は，1.6から45倍の範囲でモルヒネより大きな効力を持つアゴニストとして機能した

■［動］＋～-fold

❶ increase ～-fold	…は，～倍増大する	650
❷ be increased ～-fold	…は，～倍増大する	286
❸ be reduced ～-fold	…は，～倍低下する	157

❷ **Expression of a HIF-1-dependent reporter gene was increased 3-fold** in cells subjected to IH.（J Biol Chem. 2005 280:4321）
訳 HIF-1依存性のレポーター遺伝子の発現は，３倍増大した

■［動］＋ a ～-fold

❶ show a ～-fold …	‥は，～倍の…を示す	272
❷ cause a ～-fold …	‥は，～倍の…を引き起こす	185

❶ Cells overexpressing MAGP-2 formed increased MAGP-2 matrix and **showed a 3-fold increase in intracellular type I procollagen.**（Arthritis Rheum. 2005 52:1812）
訳 ～は，細胞内Ｉ型プロコラーゲンの３倍の増大を示した

■［前］＋～-fold

❶ to ～-fold	～倍まで	5,559
・up to ～-fold	～倍まで	1,293
・2- to 3-fold	2～3倍	265
❷ by ～-fold	～倍だけ	2,290

❷ Analysis of pyridine nucleotides showed that both **NAD and NADH were increased by 2-fold** in CMS leaves.（Plant Physiol. 2005 139:64）
訳 NADとNADHは，２倍だけ増大した

■ than ～-fold

★ than ～-fold	～倍より…	1,350
❶ more than ～-fold	～倍以上	930
❷ greater than ～-fold	～倍以上	311

❶ Kinetic analysis demonstrated that **autophosphorylation results in a more than 9-fold increase in protein kinase activity.** (J Biol Chem. 2006 281:3954)

訳 自己リン酸化は，プロテインキナーゼ活性の９倍以上の増大という結果になる

❷ Hepatocellular carcinoma (HCC) is one of the most prevalent human malignancies, and individuals who are chronic hepatitis B virus (HBV) carriers **have a greater than 100-fold increased relative risk of developing HCC.** (Mol Cell. 2005 19:159)

訳 ～は，HCCを発症する100倍以上増大した相対リスクを持つ

■ ［副］＋～-fold

❶ approximately ～-fold	およそ～倍	3,640
・by approximately ～-fold	およそ～倍だけ	391
・approximately ～-fold higher	およそ～倍より高い	298
・increased approximately ～-fold	およそ～倍増大した	185
❷ about ～-fold	およそ～倍	589
❸ at least ～-fold	少なくとも～倍	558
❹ nearly ～-fold	ほぼ～倍	301
❺ over ～-fold	～倍以上	280

❶ Furthermore, **the catalytic efficiency of the mutant is reduced by approximately 350-fold** for cleavage of Z-Gly-Pro-7-amino-4-methylcoumarin. (J Biol Chem. 2005 280:19441)

訳 変異体の触媒効率は，およそ350倍だけ低下する

❷ **NOMV prepared from lpxL2 mutants was about 200-fold less active than wild-type NOMV** in rabbit pyrogen tests and in tumor necrosis factor α release assays. (Infect Immun. 2005 73:4070)

訳 lpxL2変異体から調製されたNOMVは，野生型NOMVよりおよそ200倍低い活性であった

❸ **Dimeric wild-type fVIIai was at least 75-fold more effective than monomeric fVIIai** in blocking f a association with TF. (Biochemistry. 2005 44:6321)

訳 二量体の野生型fVIIaiは，単量体のfVIIaiより少なくとも75倍効果的であった

★follow 動 追跡する／続く／伴う　用例数 26,211

followed	伴われる／追跡される／～を追跡した	21,235
follow	～を追跡する／～に続く	2,837
follows	～を追跡する／～に続く	2,139

他動詞．followingおよびfollow-upは，別項参照．

◆類義語：accompany, associate, continue, keep

F

■ 動 be followed ＋［前／副］

★ be followed	…は，追跡される／伴う	4,158
❶ be followed by ～	…は，あとに～を伴う	1,964
❷ be followed for ～	…は，～の間追跡される	703
・patients were followed for ～	患者は，～の間追跡された	140
❸ be followed up	…は，経過観察される	558
・patients were followed up	患者は，経過観察された	157

❶ The increase in VEGF mRNA was followed by an increase in immunoreactive VEGF protein. (J Biol Chem. 2002 277:18670)

訳 VEGF mRNAの増大は，あとに免疫反応性のVEGFタンパク質の増大を伴った

❸ Patients were followed up for a total of up to 10 years or until treatment failure. (Hepatology. 2004 39:915)

訳 患者は，合計10年までの間経過観察された

■ 動 as follows

★ as follows	以下の通り	1,282
❶ be as follows	…は，以下の通りである	754

❶ Stage subsets (tumor-node-metastasis system) in 83 eligible patients were as follows: T4N0/1, 31 patients (37%); T4N2, 22 patients (27%), and T1-3N3, 30 patients (36%). (J Clin Oncol. 2003 21:2004)

訳 ～は以下の通りであった

★★ following　形 次に述べる，名 次に述べること，前 ～のあと　用例数 33,773

followingは，前置詞，形容詞，名詞として使われることが多い．

◆類義語： after

■ 形 the following ＋［名］

★ the following ～	次に述べる～	4,269
❶ the following order	次に述べる順番	125
❷ the following observations	次に述べる観察	82

■ 名［動］＋ the following

❶ include the following	…は，次に述べることを含む	216
❷ reveal the following	…は，次に述べることを明らかにする	99
❸ suggest the following	…は，次に述べること示唆する	69

❶ Other major findings include the following: (1) There is a second element for late stripe expression adjacent to the traditional late element. (Dev Biol. 1999 211:39)

訳 他の主な知見は，次に述べることを含む

■ 前［名／副／動］＋ following

❶	… h following ～	～のあと…時間	337
❷	immediately following ～	～の直後に	289
❸	increased following ～	～のあとで増大した	219
❹	be observed following ～	…は，～のあとで観察される	193

❶ In normal embryonic kidney, a 3.2-kb podocan transcript was detected at low levels, and expression increased dramatically **within 24 h following birth**.（J Biol Chem. 2003 278:33248）

訳 出生後24時間以内に

❷ The late formation of free radicals may provide a window of opportunity for pharmacological rescue **immediately following exposure**, requiring both ROS and RNS scavengers.（Brain Res. 2004 1019:201）

訳 曝露の直後に

❹ Additionally, a loss of assembly competence **was observed following longer term acid treatment**, which was even more marked than that of the wild-type protein.（Biochemistry. 2004 43:1618）

訳 ～が，長期間の酸処理のあとで観察された

■ 前 following ＋［名］

❶	following treatment	処理のあと	682
❷	following infection	感染のあと	646
❸	following exposure to ～	～への曝露のあと	496
❹	following stimulation	刺激のあと	395
❺	following activation	活性化のあと	299

❶ These observations demonstrated that p21 was required for senescence development of HCT116 cells **following treatment with low concentrations of CPT**.（J Biol Chem. 2002 277:17154）

訳 低濃度のCPTによる処理のあと

★ follow-up 名 追跡／追跡調査　用例数 14,187

follow-up	追跡／追跡調査／追跡期間	14,126
follow-ups	追跡／追跡調査／追跡期間	61

複数形のfollow-upsの割合は0.4％しかないが，単数形には不定冠詞が使われることも多い．

■ follow-up ＋［前］

	the％	a/an％	ø％			
❶	8	72	18	follow-up of ～	～の追跡（期間）	1,834
❷	2	9	89	follow-up in ～	～における追跡（調査）	243
❸	11	4	85	follow-up for ～	～に対する追跡（調査）	228

❶ After a mean follow-up of 334 +/- 82 days, two desensitized patients (18%) received a kidney transplant compared with 14 patients (52%) in the nondesensitized group. (Transplantation. 2012 94:345)

訳 334±82日の平均追跡期間のあと

■ [名] + follow-up

❶	median follow-up	追跡期間中央値	1,647
❷	mean follow-up	平均追跡期間	750
❸	～-year follow-up	～年間の追跡調査	615
❹	long-term follow-up	長期間の追跡調査	428

■ follow-up + [名]

❶	follow-up period	追跡期間	732
❷	follow-up time	追跡時間	356
❸	follow-up study	追跡調査	282
❹	follow-up data	追跡データ	258

■ [前] + follow-up

❶	～ years of follow-up	～年の追跡調査	1,085
❷	during follow-up	追跡調査の間	780
❸	～ months of follow-up	～ヶ月の追跡調査	211
❹	duration of follow-up	追跡調査の期間	129
❺	be lost to follow-up	…は，追跡不能である	101

❶ Over a median of 13 years of follow-up, 71 incident, spontaneous IPH events occurred. (Ann Neurol. 2012 71:552)

訳 中央値で13年の追跡調査の間に

★ force 名 力，動 強制する　用例数 17,655

force	力／～を強制する	10,518
forces	力／～を強制する	4,387
forced	強制される／～を強制した	2,111
forcing	～を強制する	639

可算名詞の用例が多いが，動詞としても用いられる.

◆類義語：power, strength, capability, capacity, ability, potential, competence

■ 名 force + [前]

force for/ofの前にtheが付く割合はかなり高い.

	the%	a/an%	ø%			
❶	60	34	5	force for ～	～のための力	488
				・driving force for ～	～のための推進力	370

❷	13	52	32	force in ～	～における力	293
❸	64	31	5	force of ～	～の力	290

❶ Dynamic assembly and disassembly of actin filaments is a major driving force for cell movements.（Genes Dev. 2011 25:730）

訳 ～は，細胞運動のための主な推進力である

■ 名［形／名］＋ force

❶	atomic force	原子間力	1,177
❷	driving force	推進力	754
❸	task force	タスクフォース	236
❹	motive force	輸送力	210

■ 動 forced ＋［名］

❶	forced expression of ～	～の強制発現	600
❷	forced expiratory volume in ～	～における強制呼気量	171

❶ In addition, forced expression of TEL2 but not TEL blocks vitamin D3-induced differentiation of U937 and HL60 myeloid cells.（Blood. 2006 107:1124）

訳 TEL2の，しかしTELではない，強制発現はビタミンD3誘導性の分化をブロックする

F

★★ form 名 型，動 形成する　用例数 107,179

form	型／～を形成する	55,166
forms	型／～を形成する	24,735
formed	形成される／～を形成した	17,888
forming	～を形成する	9,390

名詞として用いられることが多いが，動詞としては他動詞の用例が多い．複数形のformsの割合は約40%とかなり高い．

◆類義語：configure, constitute, mode, type, shape

■ 名 form ＋［前］

	the%	a/an%	ø%			
❶	41	46	1	form of ～	～の型／～の形	16,616
				・in the form of ～	～の形で	1,388
				・active form of ～	活性型の～	803
				・mutant forms of ～	変異型の～	572
				・truncated form of ～	切断型の～	486
				・dominant negative form of ～	ドミナントネガティブ型の～	466
				・common form of ～	共通型の～	342
				・activated form of ～	活性型の～	340
				・soluble form of ～	可溶型の～	334

❶ At the N-terminus, a molecular tag **was introduced** in the form of **radioactive iodine** or biotin. (Biochemistry. 1999 38:3414)

訳 ～は，放射性ヨウ素の形で導入された

❶ **All of the** mutant forms of **Ste18p formed complexes with Ste4p**, as assessed by coimmunoprecipitation. (Mol Cell Biol. 1999 19:7705)

訳 変異型のSte18pのすべてが，Ste4pとの複合体を形成した

❶ **All mutants were assayed for changes in their interaction with a** truncated form of **the catalytic subunit of phosphorylase kinase**, γ(1-300). (Biochemistry. 2000 39:15887)

訳 すべての変異体が，ホスホリラーゼキナーゼの切断型の触媒サブユニットとそれらの相互作用の変化に関してアッセイされた

■ 動 form ＋［名句］

❶ form a complex	…は，複合体を形成する	575
❷ form the basis	…は，基礎を形成する	305
❸ form a hydrogen bond	…は，水素結合を形成する	140
❹ form a heterodimer	…は，ヘテロ二量体を形成する	122
❺ form a dimer	…は，二量体を形成する	122

❶ It is therefore possible that **these molecules** form a complex and thus are involved in the process of embryo implantation. (Proc Natl Acad Sci USA. 1998 95:5027)

訳 これらの分子は，複合体を形成する

■ 動［名／動］＋ to form

★ to form ～	～を形成するために…／～を形成する…	12,860
❶ ability to form ～	～を形成する能力	641
❷ fail to form ～	…は，～を形成することができない	174
❸ be shown to form ～	…は，～を形成することが示される	171
❹ be predicted to form ～	…は，～を形成することが予想される	123

❶ Even after replacement of this region with polyglutamine, **Sup35p retains its** ability to form **amyloids**. (Cell. 1998 93:1241)

訳 Sup35pは，アミロイドを形成する能力を保持する

❹ This 68-nt sequence was predicted to form **two stem-loops**, SL4 and SL5. (J Virol. 1998 72:4341)

訳 ～は，２つのステムループを形成することが予想された

■ 動 be formed ＋［前］

★ be formed	…は，形成される	4,433
❶ be formed by ～	…は，～によって形成される	934
❷ be formed in ～	…は，～において形成される	602
❸ be formed from ～	…は，～から形成される	313
❹ be formed at ～	…は，～において形成される	142
❺ be formed between ～	…は，～の間で形成される	138

❶ **The third ring** was formed by **intramolecular displacement** of a mesylate by the deprotected amine. (J Org Chem. 2012 77:3390)

訳 第３番目のリングは，分子内置換によって形成された

❸ The stator of the flagellar motor **is formed from** **the membrane proteins MotA and MotB**, which associate in complexes that contain multiple copies of each protein. (Biochemistry. 2001 40:13051)

訳 ～は，膜タンパク質MotAおよびMotBから形成される

★★ formation 名 形成　　用例数 70,903

formation	形成	70,758
formations	形成	145

複数形のformationsが使われる割合は0.2%しかなく，原則，**不可算名詞**として使われる．

◆類義語：assembly

■ 名 formation ＋［前］

formation ofの前にtheが付く割合はかなり高い．

	the%	a/an%	ø%		
❶	66	0	34	formation of ～　～の形成	28,073
				・result in the formation of ～　…は，～の形成という結果になる	1,043
				・lead to the formation of ～　…は，～の形成につながる	861
				・induce the formation of ～　…は，～の形成を誘導する	445
				・catalyze the formation of ～　…は，～の形成を触媒する	370
				・be required for the formation of ～　…は，～の形成に必要とされる	232
❷	3	0	93	formation in ～　～における形成	4,900

❶ Forced oxidation of membranes containing S molecules **also** **results in the formation of** **covalently linked dimers**. (J Bacteriol. 2000 182:6075)

訳 ～は，また，共有結合した二量体の形成という結果になる

❶ The generation of LCR HS sites in the absence of chromatin assembly **leads to the formation of** **S1- and $KMnO_4$-sensitive regions** in HS2 and HS3. (Mol Cell Biol. 2001 21:2629)

訳 ～は，S1および$KMnO_4$感受性の領域の形成につながる

❶ Analysis of SM α-actin null mice indicated that **SM α-actin** **is not** **required for the formation of** **the cardiovascular system**. (FASEB J. 2000 14:2213)

訳 平滑筋α-アクチンは，心血管系の形成に必要とされない

❶ **Factor XIII** **catalyzes the formation of** **isopeptide bonds between** noncovalently associated fibrin monomers in the final stages of the blood coagulation cascade. (Biochemistry. 2002 41:7947)

訳 第XIII因子は，～の間のイソペプチド結合の形成を触媒する

❷ Expression of the dominant negative form of MEKK1 had no influence on the ability of

F

retinoic acid to **induce either JNK activation or primitive endoderm formation in P19 stem cells.** (J Biol Chem. 2000 275:24032)
訳 P19幹細胞におけるJNKの活性化あるいは原始内胚葉形成のどちらかを誘導する

■ [名] + formation

❶	complex formation	複合体形成	2,848
	・complex formation with ～	～との複合体形成	235
	・complex formation between ～	～の間の複合体形成	205
❷	biofilm formation	バイオフィルム形成	1,273
❸	bond formation	結合形成	1,186
❹	tumor formation	腫瘍形成	1,118
❺	bone formation	骨形成	1,063

❶ Thus, **complex formation with p53** and ROS generation, rather than azurin redox activity, are important in the cytotoxic action of azurin towards macrophages. (Mol Microbiol. 2003 47:549)
訳 p53との複合体形成

foundation 名 基盤／根拠／基金／設立 用例数 1,675

foundation	基盤／根拠	1,542
foundations	基盤／根拠	133

複数形のfoundationsの割合が，約10%ある可算名詞.
◆類義語：basis, reason, establishment

■ foundation + [前]

	the%	a/an%	ø%			
❶	39	60	1	foundation for ～	～のための基盤／根拠	1,033
				・provide a foundation for ～	…は，～のための基盤を提供する	284
				・foundation for ～ing	～するための基盤	255

❶ **These results provide a foundation for understanding** tRNA maturation in organelles. (Proc Natl Acad Sci USA. 2012 109:16149)
訳 これらの結果は，～を理解するための基盤を提供する

★ fraction 名 部分／割合／画分 用例数 15,722

fraction	部分／割合／画分	11,592
fractions	部分／割合／画分	4,130

複数形のfractionsの割合が，約25%ある可算名詞．the fraction ofとa fraction ofの違いに注意.
◆類義語：portion, amount, ratio

■ fraction ＋［前］

	the%	a/an%	ø%			
❶	36	58	3	fraction of ～	わずかな～／～の割合／～の画分	5,420
				・the fraction of ～	～の割合	1,085
				・a fraction of ～	わずかな～	845
				・a small fraction of ～	極くわずかな～	502
				・a significant fraction of ～	かなりの部分の～	454
				・a large fraction of ～	大部分の～	387

❶ Therefore, **our principal aim was to determine the fraction of cirrhosis attributable to heavy alcohol use** among chronic HCV patients attending a liver clinic. (Hepatology. 2013 57:451)

訳 われわれの主目的は，大量の飲酒に起因する肝硬変の割合を決定することであった

❶ We report that **a fraction of NFAT protein constitutively localizes in the nucleus of** primary Tregs, where it selectively binds to Foxp3 target genes. (J Immunol. 2012 188:4268)

訳 わずかなNFATタンパク質が，～の核に構成的に局在する

■［名］＋ fraction

❶	ventricular ejection fraction	心室駆出分画	649
❷	membrane fraction	膜画分	388

framework 名 枠組み　用例数 6,107

framework	枠組み	5,680
frameworks	枠組み	427

複数形のframeworksの割合が，約５％ある可算名詞．

■ framework ＋［前］

framework ofの前にtheが付く割合は非常に高い．

	the%	a/an%	ø%			
❶	5	93	0	framework for ～	～のための枠組み	2,028
				・provide a framework for ～	…は，～のための枠組みを提供する	576
				・framework for understanding ～	～を理解するための枠組み	369
❷	71	27	0	framework of ～	～の枠組み	452
❸	8	79	2	framework to *do*	～するための枠組み	435

❶ **These results provide a framework for understanding** the tumor suppressor function of Merlin and indicate that Merlin mediates contact inhibition of growth by suppressing recruitment of Rac to matrix adhesions. (J Cell Biol. 2005 171:361)

訳 これらの結果は，～を理解するための枠組みを提供する

❷ **The framework of the MPT induction model includes** the potential states of the mitochondria (aggregated, orthodox and post-transition), their transitions from one state to another as

F

well as their interaction with the extra-mitochondrial environment. (J Theor Biol. 2010 265: 672)

訳 MPT誘導モデルの枠組みは，～を含んでいる

❸ Our analysis provides a **framework to** explain cofactor-independent dioxygenation within a protein architecture generally employed to catalyze hydrolytic reactions. (Proc Natl Acad Sci USA. 2010 107:657)

訳 我々の分析は，～を説明するための枠組みを提供する

★★ free 形 ～のない／フリーの／遊離の　用例数 33,302

◆類義語：deficient, absent

■ free ＋［名］

❶ free energy	自由エネルギー	3,542
❷ ～-free survival	～のない生存（率）	3,304
・progression-free survival	無進行生存（率）	981
・disease-free survival	無病生存（率）	832
・event-free survival	無再発生存（率）	612
❸ free radical	フリーラジカル	1,076
❹ ～-free medium	～のない培地	482

❷ For the 83 patients receiving a matched related donor bone marrow transplantation (BMT), the 3-year **disease-free survival** (DFS) is 67%. (Blood. 2006 107:1315)

訳 3年無病生存率（DFS）は67%である

■［名］-free

❶ cell-free ～	無細胞の～	1,999
・cell-free system	無細胞系	404
❷ serum-free ～	無血清の～	662

❶ We have developed a **cell-free system** capable of processing and joining noncompatible DNA ends. (EMBO J. 2005 24:849)

訳 われわれは，～できる無細胞系を開発してきた

❷ In **serum-free** medium, BMP-2 induced significantly less Smad-1/5 activation and Id-1 expression, and produced significant growth inhibition. (Oncogene. 2006 25:685)

訳 無血清培地において，

■ free ＋［前］

❶ free of ～	～のない	1,306

❶ At 16 months of follow-up, she remained **free of** recurrence. (J Clin Microbiol. 1999 37:807)

訳 彼女は，再発のないままであった

freely 副 自由に 用例数 1,750

◆類義語：ad libitum

■ freely ＋［形／現分］

❶	freely available	自由に利用できる	786
	・be freely available	…は，自由に利用できる	588
❷	freely moving	自由に動き回る	203
❸	freely accessible	自由に利用できる	107

❶ The source code implemented by MATLAB is freely available at: http://zhoulab.usc.edu/sMBPLS/. (Bioinformatics. 2012 28:2458)

訳 MATLABによって実行されるソースコードは，～において自由に利用できる

★★ frequency 名 頻度 用例数 31,429

frequency	頻度	25,236
frequencies	頻度	6,193

複数形のfrequenciesが使われる割合が，約20%ある可算名詞.

◆類義語：incidence, occurrence, rate

■ frequency ＋［前］

frequency ofの前にtheが付く割合はかなり高い.

	the%	a/an%	ø%			
❶	67	17	16	frequency of ～	～の頻度	8,807
				・increase the frequency of ～	…は，～の頻度を増大させる	319
				・reduce the frequency of ～	…は，～の頻度を低下させる	244
				・increase in the frequency of ～	～の頻度の増大	205
				・frequency of spontaneous ～	自然発症の～の頻度	203
❷	21	9	55	frequency in ～	～における頻度	949

❶ No age-related increase in the frequency of these mutations was observed in leukocytes. (Am J Hum Genet. 2003 73:939)

訳 これらの変異の頻度の年齢に関連した増大は，観察されなかった

❶ We compared the frequency of spontaneous TGCTs in single- and double-mutant mice to identify combinations that show evidence of enhancer or suppressor effects. (Genetics. 2004 166:925)

訳 われわれは，自然発症のTGCTの頻度を比較した

■［形］＋ frequency

❶	high frequency	高頻度	2,764
	・at high frequency	高頻度で	190

F

❷ low frequency	低頻度	1,432

❶ Ras mutations occur at high frequency in thyroid cancer. (Oncogene. 2001 20:7334)
訳 Rasの変異は，甲状腺癌において高頻度で起こる

frequent　形 高頻度の／頻繁な　用例数 5,488

◆類義語：frequent, continual

■ [副] ＋ frequent

❶ more frequent	より高頻度の	1,119
・more frequent in ～	～においてより高頻度の	399
❷ the most frequent ～	最も高頻度の～	757
❸ less frequent	より高頻度でない	375

❶ Methylation of SOCS-1, GSTP, APC, E-cadherin, and p15 was more frequent in HCC than in nontumor liver ($P < 0.05$). (Am J Pathol. 2003 163:1101)
訳 ～は，非腫瘍の肝臓においてより肝細胞癌において高頻度であった

★ frequently　副 しばしば／高頻度に／頻繁に　用例数 9,642

◆類義語：often, continually

■ frequently ＋ [前／過分]

❶ frequently in ～	～においてしばしば	855
❷ frequently used	しばしば使われる	364
❸ frequently associated with ～	～としばしば関連する	352
・be frequently associated with ～	…は，～としばしば関連する	237
❹ frequently observed	しばしば観察される	350

❶ The ErbB receptor family is implicated in the malignant transformation of several tumor types and is overexpressed frequently in breast, ovarian, and other tumors. (J Biol Chem. 2001 276:14842)
訳 …は，しばしば～において過剰発現する

❸ Enhanced activity of activator protein-1 (AP-1) is frequently associated with the promotion of skin tumorigenesis. (J Biol Chem. 2003 278:44265)
訳 ～は，皮膚の腫瘍形成の促進としばしば関連する

■ [副] ＋ frequently

❶ more frequently	より高頻度に	1,273
・more frequently in ～	～においてより高頻度に	326
・more frequently than ～	～より高頻度に	234
❷ most frequently	最も高頻度に	933
❸ less frequently	より高頻度でない	385

❶ In both groups of diabetic patients, **most abnormalities occurred** more frequently in the inferior retina.（Invest Ophthalmol Vis Sci. 2004 45:296）

訳 ほとんどの異常は，～においてより高頻度に起こった

❷ **Malignant gliomas are the** most frequently **occurring primary brain tumors** and are resistant to conventional therapy.（Oncogene. 2004 23:1821）

訳 悪性神経膠腫は，最も高頻度に起こる原発性脳腫瘍である

★ full 形 完全な／十分な　　用例数 19,480

◆類義語：complete, perfect

■ full ＋［名］／full-［名］

❶ full-length ～	全長の～／完全長～	9,709
・full-length protein	完全長のタンパク質	509
・full-length cDNA	完全長のcDNA	450
❷ full-thickness	完全な厚さ	279
❸ full activation	完全な活性化	253
❹ full range of ～	全範囲の～	231

❶ Furthermore, the intracellular localization of the protein **was different from that of the** full-length protein.（Oncogene. 2003 22:3441）

訳 ～は，完全長のタンパク質のそれとは異なっていた

★ fully 副 完全に　　用例数 12,756

否定文で使われることがかなり多い．

◆類義語：entirely, completely, perfectly, sufficiently, enough

■ fully ＋［過分／形］

❶ not fully understood	完全には理解されていない	991
❷ fully functional	完全に機能的な	503
❸ fully elucidated	完全に解明される	377
・be fully elucidated	…は，完全に解明される	326
❹ fully active	完全に活性のある	332
❺ fully characterized	完全に特徴づけられる	325

❶ **The molecular mechanisms involved in this regulation are** not fully understood.（J Immunol. 2004 172:4851）

訳 この調節に関与する分子機構は，完全には理解されていない

❷ Instead, **the catalytic site is** fully functional **in the complex**.（Biochemistry. 1997 36:3894）

訳 その触媒部位は，その複合体において完全に機能的である

F

⁑ function ㊔ 機能／関数，㊖ 機能する　用例数 169,515

function	機能／関数／機能する	124,870
functions	機能／関数／機能する	40,121
functioning	機能すること／機能する	3,843
functioned	機能した	681

主に名詞として使われるが，動詞として用いられることも多い．名詞では〈function of〉〈function in〉の用例が，動詞では〈function as〉〈function in〉の用例が多い．

◆類義語：action, act, serve, behave, operate, work

■ ㊔ function ＋［前］

function ofの前にtheが付く割合はかなり高い．a function ofは，「関数」を意味することが多い．

	the%	a/an%	ø%			
❶	62	28	7	function of ～	～の機能	25,610
				・investigate the function of ～	…は，～の機能を精査する	228
				・as a function of ～	～の関数として	3,687
				・as a function of time	時間の関数として	239
				・as a function of temperature	温度の関数として	213
				・be measured as a function of ～	…は，～の関数として測定される	114
❷	2	10	53	function in ～	～における機能	13,592
				function in ～ing	～する際の機能	697

❶ To investigate the function of the murine cystic fibrosis transmembrane conductance regulator (CFTR), a full-length cDNA encoding wild-type murine CFTR was assembled and stably expressed in Chinese hamster ovary (CHO) cells. (J Physiol. 1998 508:379)

訳 マウスの嚢胞性線維症膜コンダクタンス制御因子（CFTR）の機能を精査するために

❶ The oxidation reaction of piranha solutions with purified HiPco carbon nanotubes was measured as a function of temperature. (J Am Chem Soc. 2005 127:1541)

訳 ～が，温度の関数として測定された

❷ Here we investigated the role of GSK3β signaling in vascular biology by examining its function in endothelial cells (ECs). (J Biol Chem. 2002 277:41888)

訳 内皮細胞におけるそれの機能を調べることによって

■ ㊔ ［名］-of-function

❶ loss-of-function	機能喪失	3,655
・loss-of-function mutations	機能喪失変異	790
❷ gain-of-function	機能獲得	1,740

❶ Polycystic kidney disease (PKD) results from loss-of-function mutations in the PKD1 gene. (J Biol Chem. 2002 277:29577)

訳 ～は，PKD1遺伝子における機能喪失変異に由来する

■ 名 [形／名] + function

❶ renal function	腎機能	1,427
❷ unknown function	未知の機能	1,297
❸ protein function	タンパク質機能	1,130
❹ gene function	遺伝子機能	1,086
❺ cardiac function	心機能	1,027
❻ biological function	生物学的機能	991

❻ In contrast to the recent studies on GDF-9 and BMP-15, **nothing is known about the biological function of BMP-6 in the ovary.** (J Biol Chem. 2001 276:32889)
訳 卵巣におけるBMP-6の生物学的機能について何も知られていない

■ 動 function + [前]

❶ function as ～	～として機能する（～としての機能）	11,421
❷ function to ～	～する働きをする	3,259

❶ Therefore, **it may function as a therapeutic agent** based on its ability to disrupt critical signal transduction events in the intestinal cell necessary for perpetuation of the chronic inflammatory state. (Gastroenterology. 1999 116:602)
訳 それは，治療薬として機能するかもしれない

❷ Our study suggests that **PKD1 may function to regulate both pathways**, allowing cells to enter a differentiation pathway that results in tubule formation. (Mol Cell. 2000 6:1267)
訳 PKD1は，両方の経路を調節するように機能するかもしれない

■ 動 [動] + to function

★ to function	機能する～	3,062
❶ appear to function	…は，機能するように思われる	360
❷ be shown to function	…は，機能すると示される	227

❶ Thus, **mgU6-77 appears to function in the 2'-O-methylation of** two distinct classes of cellular RNA, snRNA, and rRNA. (Mol Cell. 1998 2:629)
訳 mgU6-77は，～の２'-O-メチル化において機能するように思われる

★★ functional 形 機能的な，名 機能的なもの　用例数 62,314

functional	機能的な	62,269
functionals	機能的なもの	45

形容詞として使われることが圧倒的に多い.

■ functional + [前]

❶ functional in ～	～において機能している	639
・be functional in ～	…は，～において機能している	375

F

❶ A PPAR response element was identified in the proximal promoter of **the human HMG-CoA synthase gene that is functional in** its native context. (J Biol Chem. 2001 276:27950)
訳 ～において機能しているヒトHMG-CoA合成酵素遺伝子

■ [副／形] + functional

❶	fully functional	十分に機能している	503
❷	different functional ～	異なる機能的な～	353
❸	important functional ～	重要な機能的な～	324
❹	distinct functional ～	別個の機能的な～	289

❶ However, SM mutants with deletions of this arginine-rich region localized normally in the nucleus and **were fully functional in gene activation.** (J Virol. 2001 75:6033)
訳 ～は，遺伝子活性化において十分に機能していた

❹ **The distinct functional properties of the two porins are likely to play** an integrated role with environmental regulation of their expression. (J Biol Chem. 2003 278:17539)
訳 2つのポリンの別個の機能的な性質は，おそらく～を果たすであろう

■ functional + [名]

❶	functional role	機能的な役割	1,829
❷	functional properties	機能的な性質	1,104
❸	functional groups	官能基	1,078
❹	functional studies	機能的な研究	1,051
❺	functional analysis	機能的な解析	964
❻	functional consequences	機能的影響	902
❼	the functional significance of ～	～の機能的な重要性	852
❽	functional domains	機能ドメイン	822

❶ However, **the functional role of NOS2 in** the pathogenesis of transplant arteriosclerosis remains unclear. (Circulation. 1998 97:2059)
訳 ～におけるNOS2の機能的な役割

■ [形] + and functional

❶	structural and functional ～	構造的および機能的～	1,459

❶ Here, we characterize the MscK channel activity and **compare it with the activity of its structural and functional homologue, MscS.** (EMBO J. 2002 21:5323)
訳 それを構造的および機能的ホモログMscSの活性と比較する

★ functionally 副 機能的に　用例数 9,987

■ functionally + [形]

❶	functionally important	機能的に重要な	869
	・be functionally important	…は，機能的に重要である	242

❷ functionally distinct	機能的に別個の	780
❸ functionally active	機能的に活性のある	308
❹ functionally relevant	機能的に関連する	276
❺ functionally redundant	機能的に重複する	237
❻ functionally equivalent	機能的に相当する	222
❼ functionally significant	機能的に重要な	220

❶ These findings indicate that sGC **is functionally important in cerebral arterioles.** (Brain Res. 1999 821:368)

訳 ～は，大脳の小動脈において機能的に重要である

❷ **This phenomenon is genetically and functionally distinct from** classical antigenic variation, which is mediated by the var multigene family of P. falciparum. (Nature. 1999 398:618)

訳 この現象は，遺伝的および機能的に～と別個である

❸ Furthermore, RNA levels for NF-κB-activated genes were analysed in order to determine if **NF-κB is functionally active in human breast cancer.** (Oncogene. 2000 19:1123)

訳 NF-κBは，ヒトの乳癌において機能的に活性がある

■ functionally ＋［過分］

❶ functionally related	機能的に関連する	393
❷ functionally characterized	機能的に特徴づけられる	181

❶ This protein, designated Blk, **is structurally and functionally related to** human Bik and localized to the mitochondrial membrane. (J Biol Chem. 1998 273:7783)

訳 …は，構造的および機能的に～と関連する

■ functionally ＋［動］

❶ functionally interact	…は，機能的に相互作用する	152
❷ functionally replace ～	…は，～を機能的に置換する	109

❶ Thus, our studies provide further support to the hypothesis that **XPD and p53 can functionally interact in a p53-mediated apoptotic pathway.** (Oncogene. 1999 18:4681)

訳 XPDとp53は，p53に仲介されるアポトーシス経路において機能的に相互作用しうる

■［副］＋ and functionally

❶ structurally and functionally ～	構造的および機能的に～	322
❷ physically and functionally ～	物理的および機能的に～	122

fundamental 形 基本的な／不可欠な，名 基礎　用例数 6,730

fundamental	基本的な／不可欠な／基礎	6,683
fundamentals	基礎	47

形容詞の用例が圧倒的に多いが，名詞としても用いられる．

◈類義語：ground, basal, basic, necessary, essential

F

■ fundamental ＋［前］

❶ fundamental to ～	～に不可欠である	492
・be fundamental to ～	…は，～に不可欠である	405

❶ Excitatory neurotransmission mediated by NMDA (N-methyl-D-aspartate) receptors is fundamental to the physiology of the mammalian central nervous system. (Nature. 2005 438: 185)

訳 ～は，哺乳類の中枢神経系の生理機能に不可欠である

■ fundamental ＋［名］

❶ fundamental role	基本的な役割	438
❷ fundamental question	基本的な疑問	196
❸ fundamental importance	基本的な重要性	193
❹ fundamental differences	基本的な違い	182

★★ further　副 さらに，形 さらに進んだ，動 発展させる／促進する

用例数 52,293

further	さらに／さらに進んだ／～を発展させる	52,136
furthering	～を発展させる	85
furthered	～を発展させた／発展させられる	39
furthers	～を発展させる	33

動詞としても用いられるが，副詞あるいは形容詞の用例が多い．

◆類 義 語：furthermore, moreover, additionally, in addition, additional, facilitate, promote, accelerate

■ 副 further ＋［動］＋［that節］

❶ we further show that ～	われわれは，～ということをさらに示す	1,185
❷ we further demonstrate that ～	われわれは，～ということをさらに実証する	714
❸ further suggest that ～	…は，～ということをさらに示唆する	563

❶ We further show that phospholipase C is an important signaling component for TRPA1 activation. (Neuron. 2004 41:849)

訳 われわれは，～ということをさらに示す

■ 副 to further ＋［動］

★ to further ～	さらに～するために	4,849
❶ to further investigate ～	～をさらに精査するために	544
❷ to further characterize ～	～をさらに特徴づけるために	488
❸ to further understand ～	～をさらに理解するために	379
❹ to further define ～	～をさらに定義するために	347

❺ to further explore ～	～をさらに探索するために	294
❻ to further elucidate ～	～をさらに解明するために	256
❼ to further examine ～	～をさらに調べるために	203

❶ To further investigate Kek1's function and evolution, we identified kek1 orthologs within dipterans.（Genetics. 2004 166:213）

訳 Kek1の機能と進化をさらに精査するために

■ 副 be further ＋［過分］

★ be further ～	さらに～される	5,270
❶ be further supported	…は，さらに支持される	370
❷ be further confirmed	…は，さらに確認される	365
❸ be further characterized	…は，さらに特徴づけられる	255
❹ be further enhanced	…は，さらに増強される	241

❸ The role of the A1AR in modulating inflammation, necrosis, and apoptosis in the kidney after IR renal injury was further characterized.（J Am Soc Nephrol. 2004 15:102）

訳 ～が，さらに特徴づけられた

F

■ 副 Further

★ Further,	さらに，	4,382
❶ Further, we show ～	さらに，われわれは～を示す	163
❷ Further, we demonstrate ～	さらに，われわれは～を実証する	95

❶ Further, we demonstrate that FHY3 and FAR1 are capable of homo- and hetero-interaction.（EMBO J. 2002 21:1339）

訳 さらに，われわれは～ということを実証する

■ 形 further ＋［名］

❶ further studies	さらに進んだ研究	1,838
・further studies are needed	さらに進んだ研究が，必要とされる	299
❷ further analysis	さらに進んだ分析	1,336
❸ further investigation	さらに進んだ研究	1,248
❹ further evidence	さらなる証拠	1,074

❶ Further studies are needed to study the role of hypoxia and other cytokines in relation to VEGF in this model.（Transplantation. 2001 72:164）

訳 さらに進んだ研究が，低酸素の役割を研究するために必要とされる

❷ Further analysis of follicular melanocytes in vivo and in primary cell culture demonstrated that pRB plays a cell-autonomous role in melanocyte survival.（Proc Natl Acad Sci USA. 2003 100:14881）

訳 濾胞性メラニン形成細胞のさらに進んだ分析

■ 動 further ＋［名句］

❶ further our understanding of ～	…は，～についてのわれわれの理解を発展させる	318

★★ furthermore 副 さらに／そのうえ 用例数 36,636

文頭に用いられることが圧倒的に多い.

◆類義語：further, moreover, additionally, in addition

■ Furthermore

★	Furthermore,	さらに,	35,704
❶	Furthermore, we show that ～	さらに, われわれは～ということを示す	1,026
❷	Furthermore, we found that ～	さらに, われわれは～ということを見つけた	479
❸	Furthermore, we demonstrate that ～	さらに, われわれは～ということを実証する	454

❶ **Furthermore, we show that** fear conditioning results in an increase of the autophosphorylated (active) form of αCaMKII in lateral amygdala (LA) spines. (J Neurosci. 2004 24:3281)

訳 さらに, われわれは～ということを示す

★ fusion 名 融合 用例数 25,959

fusion	融合	23,674
fusions	融合	2,285

複数形の割合は約10%あるが, 単数形は無冠詞の用例が多い.

◆類義語：union

■ fusion ＋［前］

	the%	a/an%	ø%			
❶	28	4	68	fusion of ～	～の融合	2,039
				・fusion of ～ with …	～の…との融合	229
				・fusion of ～ to …	～の…への融合／～の…する融合	208
❷	9	9	80	fusion in ～	～における融合	444
❸	4	8	81	fusion with ～	～との融合	442
❹	3	23	66	fusion to ～	～への融合／～する融合	349

❶ These oncoproteins act as aberrant transcription factors due to **the fusion of an ETS DNA binding domain to** a highly potent EWS (or FUS) transactivation domain. (Gene. 2005 363:1)

訳 ETSのDNA結合ドメインの～への融合

❸ Coronavirus spike (S) proteins **are responsible for binding and fusion with target cells** and thus play an essential role in virus infection. (J Virol. 2005 79:13209)

訳 ～は, 標的細胞との結合と融合に対して責任がある

■ fusion ＋［名］

❶	fusion protein	融合タンパク質	4,646
❷	fusion gene	融合遺伝子	422

■ [名] ＋ fusion

❶ membrane fusion	膜融合	1,587
❷ cell fusion	細胞融合	972

★ future 形 将来の，名 将来 　用例数 10,688

future	将来の／将来	10,679
futures	将来	9

形容詞の用例が非常に多いが，名詞としても使われる．複数形のfuturesの割合は１％しかなく，原則，**不可算名詞**として使われる．

■ 形 future ＋ [名]

❶ future studies	将来の研究	1,229
❷ future research	将来の研究	698

❶ This mutant will be used in **future studies** of interactions of T. denticola with host cells and tissue.（Infect Immun. 2001 69:6193）

訳 この変異体は，～の将来の研究において使われるであろう

■ 形 [前] ＋ future

❶ basis for future ～	将来の～に対する基礎	128
❷ foundation for future ～	将来の～に対する枠組み	95
❸ design of future ～	将来の～のデザイン	95

❶ These data form the **basis for future** studies to determine the role of Wnt signaling in the developing gastrointestinal tract.（Dev Biol. 2003 256:18）

訳 これらのデータは，将来の研究に対する基礎を形成する

■ 名 future ＋ [前]

future ofの前にtheが付く割合は，圧倒的に高い．

	the%	a/an%	ø%			
❶	98	2	0	future of ～	～の将来	180

■ 名 in the future

❶ in the future	将来において	906

❶ Pharmacotherapy to enhance recovery may be possible **in the future**.（Curr Opin Neurol. 1997 10:52）

訳 ～は，将来において可能かもしれない

G

★gain (名) 獲得／増加, (動) 得る　用例数 12,806

gain	獲得／増加／～を得る	9,396
gained	～を得た／得られた	1,729
gains	獲得／増加／～を得る	1,205
gaining	～を得る	476

名詞および動詞として用いられる．gain-of-functionの用例が非常に多い．

◆類義語：acquisition, obtain, acquire, procure, increase

■ (名) gain ＋［前］

	the%	a/an%	ø%			
❶	33	26	32	gain of ～	～の獲得	2,492
				・gain-of-function ～	機能獲得型の～	1,319
				・gain of function	機能の獲得	418
❷	17	23	52	gain in ～	～における増加	513

❶ **Gain of function in embryos leads to ectopic neurogenesis** and to the specification of ectopic olfactory sensory structures, including olfactory bulb and sensory epithelia.（Mol Cell Biol. 2005 25:3608）

訳 胚における機能の獲得は，異所性の神経形成につながる

❷ **L-NAME significantly reduced body weight gain in DIO** but not in chow-fed rats.（Brain Res. 2004 1016:222）

訳 L-NAMEは，DIOにおける体重増加を有意に減らした

■ (名)［名］＋ gain

❶ weight gain	体重増加	1,447

■ (動) gain ＋［名］

❶ to gain insight into ～	～への洞察を得るために	965
❷ gain access to ～	…は，～に到達する	303

❶ **To gain insights into this complex process**, we recorded myoelectric signals from multiple upper-limb muscles in subjects with cortical lesions.（Proc Natl Acad Sci USA. 2012 109:14652）

訳 この複雑な過程への洞察を得るために

■ (動) to gain

★ to gain ～	～を得る…	2,611
❶ To gain ～	～を得るために	1,338

■ (動) gained ＋［前］

❶ gained from ～	～から得られる	718

・insights gained from ～	～から得られる洞察	53

★★ gene 名 遺伝子　用例数 322,460

gene	遺伝子	197,306
genes	遺伝子	125,154

複数形のgenesの割合は約40%とかなり高い．通例，特定の遺伝子名にはtheを付け，遺伝子シンボル単独にはtheを付けない．

■ gene ＋［動］

❶ gene encodes ～	遺伝子は，～をコードする	1,570

❶ The aglZ gene encodes a 153-kDa protein that interacts with purified MglA in vitro.（J Bacteriol. 2004 186:6168）

訳 aglZ遺伝子は，～をコードする

■［動］＋ genes

❶ to identify genes	遺伝子を同定する…	902
・to identify genes that ～	～する遺伝子を同定する…	274
・to identify genes involved in ～	～に関与する遺伝子を同定する…	108

❶ A major focus of current cancer research is to identify genes that can be used as markers for prognosis and diagnosis, and as targets for therapy.（Bioinformatics. 2004 20:3166）

訳 ～は，予後と診断のためのマーカーとして使われうる遺伝子を同定することである

■ gene(s) ＋［現分／過分］

❶ gene encoding ～	～をコードする遺伝子	4,351
❷ genes involved in ～	～に関与する遺伝子	3,421
❸ genes including ～	～を含む遺伝子	1,743

❶ The described methodology can be applied to discover new genes encoding proteins from families with well-characterized structural and functional domains.（J Mol Biol. 2004 337:307）

訳 述べられた方法論は，～のファミリーのタンパク質をコードする新しい遺伝子を発見するために適用されうる

■［前］＋ genes

❶ expression of genes	遺伝子の発現	1,661

❶ Expression of genes involved in DNA replication was also down-regulated in resistant cell lines.（Cancer Res. 2004 64:8167）

訳 DNA複製に関与する遺伝子の発現は，また，下方制御された

G

■ gene ＋［前］

	the%	a/an%	ø%			
❶	60	38	0	gene for ～	～に対する遺伝子	2,288
				・candidate gene for ～	～に対する候補遺伝子	300
❷	59	37	1	gene of ～	～の遺伝子	1,945
				・target gene of ～	～の標的遺伝子	118

❶ A disintegrin-like metalloproteinase with thrombospondin motifs-16（Adamts16）is an important candidate gene for hypertension.（Proc Natl Acad Sci USA. 2012 109:20555）
訳 ～は，高血圧に対する重要な候補遺伝子である

■［名／形］＋ genes

❶	target genes	標的遺伝子	4,825
❷	tumor suppressor gene	癌抑制遺伝子	1,858
❸	specific genes	特異的遺伝子	1,770
❹	～-regulated genes	～に制御される制御遺伝子	1,631
❺	candidate genes	候補遺伝子	1,573

■ gene ＋［名］

❶	gene expression	遺伝子発現	37,513
❷	gene transcription	遺伝子転写	3,815
❸	gene transfer	遺伝子導入	3,254
❹	gene therapy	遺伝子治療	2,856
❺	gene product	遺伝子産物	2,802
❻	gene family	遺伝子ファミリー	2,703
❼	gene regulation	遺伝子制御	2,254

★ general 形 一般の／全身の，名 一般　用例数 18,945

形容詞の用例が多いが，名詞としても使われる．名詞としては，主にin generalの形で使われる．

◆類義語：common, conventional, ordinary, usual, normal, universal

■ 形 general ＋［名］

❶	general population	一般集団	1,415
❷	general mechanism	一般的な機構	524

■ 名 in general

❶	in general	一般に	2,996

❶ In general, the studies are confounded by a number of methodologic issues, including the selection of an appropriate control population.（Crit Care Med. 2001 29:N114）

訳 一般に，

★ generally 副 一般に／普通に 用例数 9,788

◆類義語：commonly, usually, normally, universally, in general

■ generally ＋［過分／形］

❶ generally accepted	一般に受け入れられる	308
・it is generally accepted that ～	～ということは一般に受け入れられている	105
❷ generally applicable	一般に適用できる	306
・be generally applicable	…は，一般に適用できる	206
❸ generally considered	一般に考えられる	237
❹ generally thought	一般に考えられる	234
❺ generally assumed	一般に推定される	196
❻ generally believed	一般に信じられる	180

❶ It is generally accepted that MMPs are secreted in a latent form (proMMP) and are activated only upon removal of their inhibitory propeptides. (FASEB J. 2004 18:734)

訳 ～ということは一般に受け入れられている

❷ This assay should be generally applicable to other AdoMet-dependent methyltransferases. (Anal Biochem. 2004 326:100)

訳 このアッセイは，一般に～に適用できるはずである

❹ Lineage commitment is generally thought to occur at the $CD4^{+}8^{+}$ (double positive) stage of differentiation and to result in silencing of the opposite coreceptor gene. (J Exp Med. 2003 197:1709)

訳 …は，～において起こると一般に考えられている

■ more generally

❶ more generally	もっと一般に	501

❶ More generally, the present work illustrates the power of segmental isotopic labeling for probing molecular interactions in large proteins by NMR. (Proc Natl Acad Sci USA. 2002 99: 8536)

訳 もっと一般に，

★★ generate 動 作製する／生成する／産生する 用例数 49,539

generated	生成される／～を産生した	27,873
generate	～を作製する／～を産生する	12,269
generating	～を作製する／～を産生する	5,526
generates	～を作製する／～を産生する	3,871

他動詞.

◆類義語：produce, make, elaborate, construct, create

G

■ we generate

❶ we generated ～	われわれは，～を作製した	4,367
・we generated mice	われわれは，マウスを作製した	532
・we generated transgenic mice	われわれは，トランスジェニックマウスを作製した	318

❶ To further examine the functional role of USF2 in vivo, **we generated mice with genetic deletion of USF2 gene.** (Oncogene. 2006 25:579)
訳 われわれは，USF2遺伝子の遺伝的欠失を持つマウスを作製した

■ ［動／名］＋ to generate

★ to generate ～	～を作製する…／～を産生する…	7,844
❶ be used to generate ～	…は，～を作製するために使われる	688
❷ ability to generate ～	～を産生する能力	293

❶ The C57BL/6, 129, and B6,129 mouse strains or stocks **have been commonly used to generate targeted mutant mice.** (Am J Pathol. 2001 158:323)
訳 ～は，狙いを定めた変異マウスを作製するために一般に使われている

■ generate and ＋［動］

❶ generated and characterized ～	～を作製し，そして特徴づけした	150

❶ To assess the role of Mlh3 in mammalian meiosis, **we have generated and characterized Mlh3$^{-/-}$ mice.** (Nat Genet. 2002 31:385)
訳 われわれは，Mlh3$^{-/-}$マウスを作製しそして特徴づけした

■ be generated ＋［前］

★ be generated	…は，生成される	8,068
❶ be generated by ～	…は，～によって生成される	1,971
❷ be generated in ～	…は，～において産生される	972
❸ be generated from ～	…は，～から生成される	708
❹ be generated to *do*	…は，～するために作製される	255
❺ be generated with ～	…は，～によって生成される	210
❻ be generated at ～	…は，～において生成される	209

❶ Variant mRpgr isoforms **are generated by alternative splicing** and by utilizing two in-frame initiation codons. (J Biol Chem. 1998 273:19656)
訳 ～は，選択的スプライシングによって生成される

❸ **Dendritic cells were generated from monocytes of 4 patients with MM** and used to present a recombinant Sp17 protein to autologous T cells. (Blood. 2002 100:961)
訳 樹状細胞が，多発性骨髄腫の４人の患者の単球から生成された

★ generation ㊔ 産生／生成／世代　用例数 20,131

generation	産生／生成／世代	18,780
generations	産生／生成／世代	1,351

複数形のgenerationsの割合は約5％あるが，単数形は無冠詞で使われることが圧倒的に多い．

◈類義語：production

■ generation ＋［前］

generation ofの前にtheが付く割合はかなり高い．

	the％	a/an％	ø％			
❶	64	6	30	generation of ～	～の産生（生成）	9,441
				・… generations of ～	…世代の～	288
				・result in the generation of ～	…は，～の生成という結果になる	268
				・lead to the generation of ～	…は，～の生成につながる	212
				・contribute to the generation of ～	…は，～の生成に寄与する	125
				・be required for the generation of ～	…は，～の生成ために必要とされる	104
❷	1	1	40	generation in ～	～における産生	620
❸	2	0	71	generation by ～	～による産生	285

❶ Cleavage of HA results in the generation of variably sized fragments that stimulate multiple angiogenic and inflammatory responses in a size-specific manner.（Am J Pathol. 2009 174:2254）

訳 HAの切断は，様々な大きさの断片の生成という結果になる

❶ Here we report that the transcription factor XBP-1 is required for the generation of plasma cells.（Nature. 2001 412:300）

訳 転写因子XBP-1は，形質細胞の生成のために必要とされる

❷ Tissue factor（TF）is a critical determinant of thrombin generation in normal hemostasis and in atherothrombotic disease.（Circulation. 2000 101:2144）

訳 組織因子（TF）は，正常な止血におけるトロンビン産生の重要な決定因子である

■［形／名］＋ generation

❶	second generation	第2世代	669
❷	first generation	第1世代	462
❸	ROS generation	活性酸素種産生	454
❹	force generation	力産生	369

★★ genetic 形 遺伝的な／遺伝学的な 用例数 53,242

◆類義語：heritable

■ genetic ＋［名］

❶ genetic variation	遺伝的変異	1,837
❷ genetic analysis	遺伝的解析	1,574
❸ genetic studies	遺伝的研究	1,533
❹ genetic basis	遺伝的基礎	1,188
❺ genetic factors	遺伝的因子	1,057
❻ genetic evidence	遺伝的証拠	1,040
❼ genetic background	遺伝的背景	975
❽ genetic diversity	遺伝的多様性	948
❾ genetic screen	遺伝子選別	825

❶ In this paper, **we present an analysis of genetic variation in** three wild populations of the barn swallow, Hirundo rustica.（Heredity. 2004 93:8）

訳 われわれは，～における遺伝的変異の分析を示す

■ genetic and ＋［形］

❶ genetic and biochemical ～	遺伝的および生化学的～	641
❷ genetic and environmental ～	遺伝的および環境的～	563

★ genetically 副 遺伝的に 用例数 8,104

■ genetically ＋［過分／形］

❶ genetically engineered	遺伝的に操作された	789
❷ genetically modified	遺伝的に改変された	577
❸ genetically distinct	遺伝的に異なる	337
❹ genetically heterogeneous	遺伝的に異種の	271
❺ genetically deficient	遺伝的に欠損した	249
❻ genetically determined	遺伝的に決定された	222
❼ genetically defined	遺伝的に定義された	196
❽ genetically diverse	遺伝的に多様な	191

❶ In this report, **we investigate arrhythmic susceptibility in a murine model genetically engineered to express** progressively decreasing levels of Cx43.（Circ Res. 2004 95:1035）

訳 われわれは，～を発現するように遺伝的に操作されたマウスモデルにおける不整脈の感受性を精査する

❷ **Pluripotent embryonic stem（ES）cells have been used to produce genetically modified mice as experimental models of** human genetic diseases.（Proc Natl Acad Sci USA. 2002 99:3586）

訳 多能性の胚性幹（ES）細胞は，～の実験モデルとして遺伝的に改変されたマウスを産生す

るために使われている

❺ Using genetically deficient recipients, we determined that $\alpha\beta^{+}CD4^{+}$ T cells were required to regulate cGVHD.（Blood. 2004 104:1565）

訳 遺伝的に欠損したレシピエントを使って

★★ give 動 与える 用例数 30,126

given	与えられる	16,583
give	～を与える	5,600
gave	～を与えた	3,769
gives	～を与える	2,563
giving	～を与える	1,611

他動詞.〈given＋名詞句／名詞節〉の第4文型受動態の形で使われることも多い. give rise toの用例も多い.

◆類義語：provide, confer, render, feed, administer

■ be given ＋［前］

★ be given	…が，与えられる	3,312
❶ be given to ～	…が，～に与えられる	500
❷ be given for ～	…が，～の間与えられる／～に対して与えられる	131
❸ be given in ～	…が，～において与えられる	128

❶ Postoperative iodine-131 was given to 85.4% of patients with papillary cancer and 79.3% of patients with follicular cancer.（Ann Intern Med. 1998 129:622）

訳 手術後のヨウ素-131が，乳頭癌の患者の85.4%に与えられた

■ given ＋［that節］

❶ given that ～	～ということを考慮に入れれば	1,182

❶ Given that the two cell lines are from the same species, the loss of function of both of these receptors in CHOK1 cells is surprising.（J Virol. 1999 73:2916）

訳 2つの細胞株が同じ種からであることを考慮に入れれば

■ given ＋［名句］

❶ given the importance of ～	～の重要性を考慮に入れれば	138
❷ given the role of ～	～の役割を考慮に入れれば	72

■［名］＋ given

❶ mice given ～	～を与えられたマウス	351
❷ patients given ～	～を与えられた患者	237

❶ However, mice given streptomycin in their drinking water developed long-term carriage of S. aureus, and they were colonized with inocula as low as 10^5 CFU.（Infect Immun. 1999 67:5001）

訳 それらの飲料水にストレプトマイシンを与えられたマウス

G

■ when/any ＋ given

❶ when given ～	～を与えられたとき	530
❷ any given ～	任意の～も	452
・at any given ～	任意の～においても	152

❶ As_2O_3 inhibits ΔPML-RARα formation and differentiation-induction **when given** in **combination with ATRA.**（Oncogene. 2003 22:4083）

訳 ATRAと組み合わせて与えられたとき

❷ **At any given time, approximately 30% of organelles/granules are in motion.**（Blood. 2005 106:4066）

訳 どの任意の時間においても，およそ30%のオルガネラ／顆粒が動いている

■ give rise to

❶ give rise to ～	…は，～を生じる	4,206

❶ **Bone marrow stem cells give rise to a variety of hematopoietic lineages** and repopulate the blood throughout adult life.（Science. 2000 290:1779）

訳 骨髄幹細胞は，さまざまな造血性の系列を生じる

goal ㊔ 目的 用例数 7,371

goal	目的	5,957
goals	目的	1,414

複数形のgoalsの割合が，約20%ある可算名詞.

◆類義語：purpose, aim, objective, object, end

■ goal ＋［前］

goal ofの前にtheが付く割合は非常に高い.

	the%	a/an%	ø%		
❶	76	20	2	goal of ～　～の目的	3,125
				・the goal of this study was to *do*　この研究の目的は，～することであった	922
				・the goal of this study was to determine ～　この研究の目的は，～を決定することであった	263
				・with the goal of ～ing　～するという目的を持って	328
				・the goal of the present study was to *do*　現在の研究の目的は，～することであった	172
❷	6	75	1	goal in ～　～における目的	242

❶ **The goal of this study was to determine** the extent to which the crystalline content of the HA surface affected osteoblast function in vitro.（J Periodontol. 2005 76:1697）

訳 この研究の目的は，～を決定することであった

❶ **With the goal of identifying a CETP inhibitor** with high in vitro potency and optimal in vivo efficacy, a conformationally constrained molecule was designed based on the highly potent

and flexible 13. (J Med Chem. 2009 52:1768)
訳 CETP阻害剤を同定するという目的を持って

❷ A major goal in genomics is to understand how genes are regulated in different tissues, stages of development, diseases, and species. (Genome Res. 2006 16:123)
訳 ゲノム科学における主な目的は，～を理解することである

■ goal was to *do*

★ goal was to *do*	目的は，～することであった	492
❶ our goal was to *do*	われわれの目的は，～することであった	248
❷ the authors' goal was to *do*	著者らの目的は，～することであった	102

❶ Our goal was to evaluate the role of the androgen receptor coactivator SRC-1 in prostate cancer progression. (Cancer Res. 2005 65:7959)
訳 われわれの目的は，～を評価することであった

good 形 よい　　用例数 8,225

◆類義語：excellent, superior, satisfactory

■ good ＋［名］

❶ good agreement	よい一致	1,023
・in good agreement with ～	～とよく一致して	536
❷ good yields	よい収率	307
❸ good correlation	よい相関	293
・good correlation between ～	～の間のよい相関	123
❹ good model	よいモデル	152

❶ The optical mapping data were in good agreement with theory. (Circ Res. 2005 97:277)
訳 光学的マッピングデータは，理論とよく一致していた

❸ Modeling studies demonstrated a good correlation between calculated relative binding energies and activity/resistance data. (J Med Chem. 2005 48:3736)
訳 モデリング研究は，計算された相対結合エネルギーと活性/抵抗性データの間のよい相関を実証した

govern 動 支配する　　用例数 4,787

governing	～を支配する	1,608
govern	～を支配する	1,370
governed	～を支配した／支配される	1,163
governs	～を支配する	646

他動詞.

◆類義語：dominate, control, manage

G

■ be governed ＋［前］

★ be governed	…は，支配される	830
❶ be governed by ～	…は，～によって支配される	727

❶ Circadian rhythms **are governed by** a highly coupled, complex network of genes.（Biophys J. 2011 101:2563）

訳 概日リズムは，遺伝子の高度に共役した複雑なネットワークによって支配される

■［名］＋ governing

❶ mechanisms governing ～	～を支配する機構	375
❷ factors governing ～	～を支配する因子	81

❶ These findings demonstrate that the **mechanisms governing** this form of sleep-dependent plasticity require cortical activity.（J Neurosci. 2005 25:9266）

訳 この型の睡眠依存的な可塑性を支配する機構は，～を必要とする

great 形 大きな／大きい　用例数 4,367

◆類義語：large, huge

■ great ＋［名］

❶ great promise	大きな見込み	423
・hold great promise for ～	…は，～のために大いに有望である	147
❷ great interest	大きな興味	375
・be of great interest	…は，大いに興味深い	171
❸ great potential	大きな潜在能	356
・have great potential	…は，大きな潜在能を持つ	171
❹ a great deal of ～	たくさんの～／大きな～	209
❺ the great majority of ～	大多数の～	140
❻ be of great importance	…は，非常に重要である	119

❶ Stem cell therapy **holds great promise for** the replacement of damaged or dysfunctional myocardium.（Proc Natl Acad Sci USA. 2004 101:12277）

訳 幹細胞治療は，～のために大いに有望である

❷ From this, it is clear that SNP and mutation discovery **is of great interest** in today's Life Sciences.（Bioinformatics. 19:i44）

訳 ～は，今日の生命科学において大いに興味深い

❹ There is currently **a great deal of** interest in proteins that fold in a single highly cooperative step.（J Mol Biol. 2001 312:569）

訳 現在，～するタンパク質に大いに関心がある

★★ greater 形 より大きい／より大きな　用例数 31,834

greatの比較級.

■ greater than

★ greater than ～	～よりも大きい	7,519
❶ be greater than ～	…は，～よりも大きい	1,352
❷ greater than that	それよりも大きい	1,167
・greater than that of ～	～のそれよりも大きい	539
・greater than that in ～	～におけるそれよりも大きい	137
❸ greater than in ～	～においてよりも大きい	272

❷ The hypotensive effect of inhaled Y-27632 on hypoxic PH **was greater than that of inhaled nitric oxide**, and the effect lasted for at least 5 hours.（Am J Respir Crit Care Med. 2005 171:494）

訳 ～は，吸入された一酸化窒素のそれよりも大きかった

❸ **Overestimation of PA in weight-reduced black women is greater than in weight-reduced white women** and never-overweight black and white women.（Am J Clin Nutr. 2004 79:1013）

訳 体重減少した黒人女性におけるPAの過剰評価は，体重減少した白人女性においてよりも大きい

■ greater ＋［前］

❶ greater in ～	～においてより大きい	2,587
・be greater in ～	…は，～においてより大きい	1,251
・greater in ～ than in …	…においてよりも～においてより大きい	395
❷ greater for ～	～にとってより大きい	635

❶ Premature errors **were, however, greater in males than in females**, suggesting greater impulsivity in males.（Behav Neurosci. 2003 117:76）

訳 ～は，しかし，女性においてよりも男性においてより大きかった

❷ Unlike intensive care unit length of stay, **mean hospital length of stay was greater for survivors than for nonsurvivors**（18.48 vs. 12.22 days, p <.001）.（Crit Care Med. 2002 30:2413）

訳 平均病院滞在期間は，非生存者に対してよりも生存者に対してより大きかった

■［副］＋ greater

❶ significantly greater	有意により大きい	3,144
・significantly greater in ～	～において有意により大きい	595
・significantly greater than ～	～より有意に大きい	568
❷ ～-fold greater	～倍大きい	1,272
❸ much greater	ずっとより大きい	736
❹ ～ times greater	～倍大きい	561

❶ Our data suggest that the prevalence of PSC in the United States, with its attendant medical burdens, **is significantly greater than previously estimated**.（Gastroenterology. 2003 125:1364）

訳 ～は，以前に見積もられたより有意に大きい

❷ Especially notable is the catalytic potency of the NDK activity, **which is 75-fold greater than**

that of poly P synthesis. (Proc Natl Acad Sci USA. 2002 99:16684)
訳 そして，（それは）ポリP合成のそれより75倍大きい

■ greater ＋［名］

❶ greater risk	より大きなリスク	560
・greater risk of ～	～のより大きなリスク	321
・at greater risk	より大きなリスクにある	160
❷ to a greater extent	より大きな程度で	530
・to a greater extent than ～	～よりも大きな程度で	279
❸ greater degree	より大きな程度	294
❹ a greater number of ～	より多くの～	272
❺ greater sensitivity	より大きな感受性	256
❻ greater reduction	より大きな低下	247
❼ greater increase	より大きな増大	246

❶ Men that are heterozygous with respect to the mutated allele have 50% greater risk of prostate cancer than non-carriers, and homozygotes have more than double the risk. (Nat Genet. 2002 32:581)
訳 変異したアレルに関してヘテロ接合性である男性は，非保因者より50%大きな前立腺癌のリスクを持つ

❷ Multiple substitutions reduce affinity to a greater extent than the loss of the major P1 anchor residue. (J Immunol. 2003 171:5683)
訳 複数の置換は，～の喪失よりも大きな程度で親和性を低下させる

❹ Tropical and subtropical inbreds possess a greater number of alleles and greater gene diversity than their temperate counterparts. (Genetics. 2003 165:2117)
訳 熱帯性および亜熱帯性の近交系は，より多くのアレルを持つ

★ greatly　副 大きく／非常に　用例数 8,388

◆類義語：extremely, unusually, extraordinarily, very, markedly, vastly

■ greatly ＋［過分／動］

❶ greatly reduced ～	大きく低下している／～を大きく低下させた	1,423
・be greatly reduced	…は，大きく低下している	557
❷ greatly increased ～	大きく増大している／～を大きく増大させた	671
❸ greatly enhanced ～	大きく増強される／～を大きく増強させた	608
❹ greatly diminished ～	大きく減弱される／～を大きく減弱させた	289

❶ Intracellular and secretion levels of SseF are greatly reduced in an sscB mutant strain compared to the wild-type strain. (J Bacteriol. 2004 186:5078)
訳 ～は，sscB変異株において大きく低下している

❷ Ephedra use is associated with a greatly increased risk for adverse reactions compared with other herbs, and its use should be restricted. (Ann Intern Med. 2003 138:468)
訳 マオウの使用は，副作用に対する大きく増大したリスクと関連する

★★ group ㊔ グループ／群／基，㊖ グループ化する 用例数 110,454

group	グループ／群／基／グループ化する	70,050
groups	グループ／群／基／グループ化する	39,115
grouped	グループ化した／グループ化される	927
grouping	グループ化する	362

動詞としても用いられるが，名詞の用例が圧倒的に多い．複数形のgroupsの割合は約40%とかなり高い．

◆類義語：class, type, set, series, cohort, cluster

■ group ＋［前］

	the%	a/an%	ø%			
❶	27	65	0	group of ～	～のグループ／…グループの～	3,125
				・group of patients	…グループの患者	600
				・group of proteins	…グループのタンパク質	208

❶ XBP1s is an independent prognostic marker and can be used with beta2 microglobulin and t (4;14) to **identify a group of patients with a poor outcome**. (Plant Mol Biol. 2010 73:569)

訳 予後の悪い一群の患者を同定する

■［冠／形／代名］＋ group of

❶	a group of ～	一群の～	2,476
❷	two groups of ～	二つのグループの～	525
❸	this group of ～	このグループの～	519
❹	both groups of ～	両方のグループの～	260
❺	a diverse group of ～	多様なグループの～	216
❻	a large group of ～	大きなグループの～	187
❼	one group of ～	一つのグループの～	185
❽	a heterogeneous group of ～	不均一なグループの～	172
❾	a small group of ～	小さなグループの～	138

■［形／名］＋ group

❶	control group	対照群	3,225
❷	placebo group	プラセボ群	1,771
❸	hydroxyl group	ヒドロキシル基	819
❹	methyl group	メチル基	805
❺	treatment group	治療群	761

★grow ㊀生育する／増殖する／成長する 用例数 17,296

growing	増殖する	6,570
grown	生育される／培養される	5,493
grow	増殖する／～を生育する	3,015
grew	増殖した／～を生育した	1,587
grows	増殖する／～を生育する	631

他動詞・自動詞の両方で用いられる．他動詞としては，受動態の用例が多い．

◆類義語：replicate, proliferate, propagate, amplify

■ grown ＋［前］

❶ grown in ～	～において生育（培養）される	1,870
・be grown in ～	…が，～において生育（培養）される	477
・cells grown in ～	～において培養された細胞	377
❷ grown on ～	～において生育（培養）される	682
❸ grown at ～	～において生育（培養）される	392
❹ grown under ～	～下で生育（培養）される	392

❶ In the present study, **the activities of these two enzymes were measured in cells grown in media containing** various concentrations of selenite, molybdate, and various purine substrates.（J Bacteriol. 2002 184:2039）

訳 これら２つの酵素の活性が，～を含む培地で培養された細胞において測定された

❷ Serum-starved **cells grown on a laminin matrix** exhibited integrin-dependent neurite outgrowth.（Mol Cell Biol. 2000 20:158）

訳 ラミニン基質上で培養された細胞

■［名］＋ be grown

★ be grown	…が，生育（培養）される	1,088
❶ cells were grown	細胞が，培養された	196

❶ GltC-dependent gltAB expression was drastically reduced when **cells were grown in media containing** arginine or ornithine or proline, all of which are inducers and substrates of the Roc catabolic pathway.（J Bacteriol. 2004 186:3399）

訳 細胞が，～を含む培地で培養された

■ grow ＋［前］

❶ grow in ～	…は，～において増殖する	981
❷ grow on ～	…は，～において増殖する	479

❶ **EBNA3AHT-infected LCLs were unable to grow in medium without 4HT.**（J Virol. 2003 77:10437）

訳 EBNA3AHTに感染したLCLは，4HTのない培地では増殖できなかった

■ [名] + grow

❶ cells grew	…は，増殖した	100

❶ Tumor **cells grew** more rapidly and had increased vascularization when coinjected with the RESDECs.（Blood. 2004 103:1325）
訳 腫瘍細胞は，より急速に増殖した

■ [名／形] + to grow

★ to grow ～	増殖する…／～を増やす…	1,468
❶ ability to grow ～	～を増やす能力	189
❷ be unable to grow	…は，増殖することができない	138

❶ Differences in an individual's **ability to grow** new blood vessels may influence the rate of progression of these diseases.（FASEB J. 2000 14:871）
訳 新しい血管を増やす個々の能力の違いは，これらの疾患の進行の速度に影響するかもしれない

■ growing + [名]

❶ growing body of ～	増大する一連の～	429
・a growing body of evidence	増大する一連の証拠	218
❷ growing evidence	増大する証拠	416
❸ growing number of ～	多数の～	363
❹ growing cells	増殖する細胞	320

❶ There is **a growing body of evidence** suggesting that apoptosis is involved in ischemic brain injury.（Brain Res. 1998 784:25）
訳 増大する一連の証拠がある

❸ Expanded trinucleotide repeats underlie a **growing number of** human diseases.（Mol Cell Biol. 1999 19:5675）
訳 拡大した三塩基反復が，多数のヒトの疾患の根底にある

G

★★ growth 名 増殖／成長　用例数 96,758

growth	増殖／成長	96,708
growths	増殖／成長	50

複数形のgrowthsの割合は0.05%しかなく，原則，**不可算名詞**として使われる．
◆分類：proliferation, replication

■ growth + [前]

	the%	a/an%	ø%			
❶	55	0	45	growth of ～	～の増殖	7,702
				・growth of ～ cells	～細胞の増殖	462
❷	1	1	95	growth in ～	～における増殖	4,926
❸	1	1	95	growth on ～	～における増殖	831

❶ **EPIs inhibit the growth of epithelial cells** but induce them to secrete the neutrophil attractant IL-8, while PEPI blocks neutrophil activation by tumor necrosis factor, preventing release of oxidants and proteases.（Cell. 2002 111:867）

訳 EPIは，上皮細胞の増殖を抑制する

❷ These results suggest that **m09 is dispensable for viral growth in these organs** and that the presence of the transposon sequence in the viral genome does not significantly affect viral replication in vivo.（J Virol. 2000 74:7411）

訳 m09は，これらの臓器におけるウイルス増殖にとって必要ではない

■［前］＋ growth

❶	required for growth	増殖のために必要とされる	218
❷	essential for growth	増殖のために必須である	192
❸	inhibition of growth	増殖の抑制	139

❶ We further show that **PEPCK is required for growth of Mtb** in isolated bone marrow-derived murine macrophages and in mice.（Proc Natl Acad Sci USA. 2010 107:9819）

訳 PEPCKは，Mtbの増殖のために必要とされる

■ growth ＋［名］

❶	growth factor	成長因子	24,124
❷	growth arrest	増殖停止	1,901
❸	growth inhibition	増殖阻害	1,878
❹	growth rate	増殖速度	1,527
❺	growth conditions	増殖条件	913
❻	growth defect	増殖欠陥	533
❼	growth phase	増殖相	472
❽	growth control	増殖制御	464
❾	growth suppression	増殖抑制	450

H

★ half ㊔ 半分，㊞ 半分の　　用例数 14,618

half	半分／半数／半分の	14,144
halves	半分／半数	474

名詞および形容詞として用いられる．複数形のhalvesの割合は約５％あるが，単数形は無冠詞で使われることが非常に多い．

■ ㊔ half ＋［前］

	the%	a/an%	ø%			
❶	20	0	78	half of ～	～の半分（半数）	4,791
				・half of all ～	すべての～の半数	254
				・half of patients	患者の半数	96

❶ Approximately half of all annotated canine genes contain SINEC_Cf repeats, and these elements are occasionally transcribed.（Genome Res. 2005 15:1798）

訳 すべてのアノテーションされたイヌの遺伝子のおよそ半数は，～を含む

■ ㊔［形／副］＋ half

❶	terminal half	末端半分	1,236
	・the N-terminal half of ～	～のN末端半分	286
❷	one-half	2分の1	761
❸	more than half	半分より多い	542
❹	approximately half	およそ半分	350
❺	about half	およそ半分	311
❻	nearly half	半分近く	194

❶ The binding site of the aglycone is located in the N-terminal half of the protein.（Mol Pharmacol. 2003 63:283）

訳 ～は，そのタンパク質のN末端半分に位置する

❷ Approximately one-half of the responding neurons develop rods within 6 h or with as little as 10 nM Aβ(1-42).（J Neurosci. 2005 25:11313）

訳 応答ニューロンのおよそ２分の１は，6時間以内に桿体を発達させる

❸ A single acquired mutation of JAK2 was noted in more than half of patients with a myeloproliferative disorder.（Lancet. 2005 365:1054）

訳 ～が，骨髄増殖性疾患の患者の半数以上において認められた

■ ㊞ half-［名／形］

❶	half-life	半減期	2,890
	・half-life of ～	～の半減期	1,387
❷	half-maximal ～	最大半量の～	604

❶ The half-life of FOXA1 was increased when coexpressed with BRCA1.（Oncogene. 2006 25:1391）

訳 FOXA1の半減期は，BRCA1と共発現されたとき増大した

hallmark 名 顕著な特徴 用例数 2,946

hallmark	顕著な特徴	2,062
hallmarks	顕著な特徴	884

複数形のhallmarksの割合が，約30%ある可算名詞．hallmark ofの用例が断然多い．

◆類義語：feature, characteristic, property

■ hallmark ＋［前］

	the％	a/an％	ø％			
❶	18	77	0	hallmark of ～	～の顕著な特徴	1,639
				・pathological hallmark of ～	病理学的な～の特徴	94

❶ High-frequency recombination is a hallmark of HIV-1 replication.（J Virol. 2010 84:7651）

訳 高頻度の組換えは，HIV-1複製の顕著な特徴である

hand 名 手／ハンド 用例数 6,657

hand	手	6,086
hands	手	571

複数形のhandsの割合が，約10%ある可算名詞．on the other handの用例が多い．on（the）one handと対で使われることもある．

■ on ＋ @ ＋ hand

❶	on the other hand,	他方では	2,586
❷	on the one hand	一方では	98
❸	on one hand	一方では	56

❶ On the other hand, no aromatase expression was detected in precancerous（n = 42）or normal cervical（n = 17）tissue samples.（Cancer Res. 2005 65:11164）

訳 他方では，

❷ Together, these findings may explain why arsenic can initiate oxidative stress and telomere erosion, leading to apoptosis and anti-tumor therapy on the one hand and chromosome instability and carcinogenesis on the other.（J Biol Chem. 2003 278:31998）

訳 そして，（それは）一方ではアポトーシスおよび抗腫瘍療法にそして他方では染色体不安定および発癌につながる

harbor 動 持つ／内部に持つ 用例数 4,999

harboring	～を持つ	2,240
harbor	～を持つ	1,415
harbors	～を持つ	747

harbored	～を持った	597

「(変異遺伝子や導入遺伝子を内部に) 持つ」場合に用いられることが多い.
◆類義語：have, carry

■ harbor ＋ [名句]

❶ harbor mutations	…は，変異を持つ	141

■ [名] ＋ harboring

❶ cells harboring ～	～を持つ細胞	263
❷ mice harboring ～	～を持つマウス	252

❷ To examine whether Shh regulates cortical interneuron specification, **we studied mice harboring conditional mutations in Shh** within the neural tube. (Development. 2005 132:4987)
訳 われわれは，Shhに条件的な変異を持つマウスを研究した

harvest 動 集める／採取する，名 収穫　用例数 2,280

harvested	集められる／採取される／～を集めた	1,235
harvesting	～を集める／～を採取する	699
harvest	～を集める／～を採取する	303
harvests	～を集める／～を採取する	43

名詞としても用いられるが，他動詞の用例が多い．細胞や臓器などに使われることが多い.
◆類義語：collect, gather, retrieve, recover, recovery

■ harvested ＋ [前]

❶ harvested from ～	～から集められる／～から採取される	424
・be harvested from ～	…が，～から集められる／～から採取される	127
❷ harvested at ～	～において集められる／～において採取される	133
❸ harvested for ～	～のために集められる／～のために採取される	100

❶ **Draining lymph node cells were harvested from HLA-DR*0401 transgenic mice** that had been immunized with HC gp-39. (Arthritis Rheum. 2003 48:2375)
訳 流入領域リンパ節細胞が，HLA-DR*0401トランスジェニックマウスから採取された

★ healthy 形 健康な　用例数 12,270

■ healthy ＋ [名]

❶ healthy subjects	健康な被検者	1,271
❷ healthy controls	健康な対照群	1,171
❸ healthy volunteers	健康なボランティア	952

H

❹	healthy individuals	健康な個々人	699
❺	healthy adults	健康な成人	425
❻	healthy donors	健康なドナー	386
❼	healthy comparison subjects	健康な比較対照群	314

❶ Serum 40-kilodalton ligand **was 10- to 30-fold higher in patients with colon cancer than in healthy subjects**. (Gastroenterology. 2004 127:741)
訳 ～は，健康な被検者においてより大腸癌患者において10-30倍高かった

★help 動 役立つ／助ける　用例数 12,034

help	役立つ／助ける	9,068
helps	役立つ／助ける	1,650
helped	役立った／助けた	732
helping	役立つ／助ける	584

他動詞および自動詞として用いられる．help to *do*の用例が非常に多い．ただし，toは省略されることの方が多い．
◆類義語：aid, support, assist

■ help ＋［動］

❶	help explain ～	…は，～を説明するのに役立つ	621
❷	help identify ～	…は，～を同定するのに役立つ	387
❸	help define ～	…は，～を明らかにするのに役立つ	242
❹	help elucidate ～	…は，～を解明するのに役立つ	235

❷ **In order to help identify candidate TSGs**, we have constructed a chromosome 6 tile path microarray. (Oncogene. 2006 25:1261)
訳 候補TSGを同定するのに役立てるために

❹ **These findings may help elucidate the mechanisms by which** chronic asymptomatic chlamydial infection contribute to atherogenesis. (J Immunol. 2002 168:1435)
訳 これらの知見は，～である機構を解明するのに役立つかもしれない

■ help ＋［前］

❶	help to *do*	…は，～するのに役立つ	3,206
	・help to explain ～	…は，～を説明するのに役立つ	449
	・help to identify ～	…は，～を同定するのに役立つ	172
❷	help in ～	…は，～の際に役立つ	492
	・help in ～ing	…は，～する際に役立つ	193

❶ **This finding may help to explain the reduced risk of breast cancer associated with** regular use of nonsteroidal anti-inflammatory drugs. (Cancer Res. 2005 65:10113)
訳 この発見は，～に関連する乳癌の低下したリスクを説明するのに役立つかもしれない

❷ **They should also help in studying the relationship between** oncogenes such as c-Myc and DNA damage. (Oncogene. 2005 24:8038)
訳 それらは，また，～の間の関連を研究する際に役立つはずだ

■ help us

★ help us	…は，われわれが～するのに役立つ	275
❶ help us to *do*	…は，われわれが～するのに役立つ	128
・help us to understand ～	…は，われわれが～を理解するのに役立つ	56
❷ help us understand ～	…は，われわれが～を理解するのに役立つ	66

❷ These results will help us understand the role of ASICs in exercise physiology and provide a molecular target for potential drug therapies to treat muscle pain.（FASEB J. 2013 27:793）

訳 これらの結果は，われわれがASICの役割を理解するのに役立つであろう

■ ［名］＋ help

❶ results help	結果は，役立つ	130
❷ findings help	知見は，役立つ	77

■ to help

★ to help ～	～に役立つ…	1,514
❶ To help ～	～に役立てるために	269

helpful ㊊ 役に立つ　　用例数 714

◆類義語：available, useful, valuable, beneficial, convenient

■ be helpful ＋［前］

★ be helpful	…は，役に立つ	498
❶ be helpful in ～	…は，～の際に役に立つ	303
❷ helpful in ～ing	…は，～する際に役に立つ	247

❶ Studies using rodent models may be helpful in improving the success of islet transplantation.（Transplantation. 2004 78:615）

訳 げっ歯類モデルを使う研究は，膵島移植の成功を改善する際に役に立つかもしれない

hence ㊖ それゆえ　　用例数 4,971

◆同義語： therefore, thus, consequently, accordingly

■ hence

❶ Hence,	それゆえ，	2,405
❷ and hence	そしてそれゆえ	1,963

❶ Hence, we conducted a retrospective study to examine the impact of HRT on the natural history of lung cancer.（J Clin Oncol. 2006 24:59）

訳 それゆえ，

H

★★ here　副 ここに　用例数 106,131

文頭の用例が断然多い．Hereのあとに，カンマはあってもなくてもよい．

◆類義語：herein

■ Here we ＋［動］

★ Here we ～	ここにわれわれは，～	44,157
❶ Here, we ～	ここに，われわれは～	34,788
❷ Here we show ～	ここにわれわれは，～を示す	12,300
❸ Here we report ～	ここにわれわれは，～を報告する	9,314
❹ Here we describe ～	ここにわれわれは，～を述べる	3,396
❺ Here we demonstrate ～	ここにわれわれは，～を実証する	3,087

❷ **Here we show that** isolated cardiac myocytes overexpressing MDM2 acquired resistance to hypoxia/reoxygenation-induced cell death.（J Biol Chem. 2006 281:3679）

訳 ここにわれわれは～ということを示す

■ we ＋［動］＋ here

❶ we report here ～	われわれは，ここに～を報告する	4,380
❷ we show here ～	われわれは，ここに～を示す	3,846

❶ **We report here that** a normal budding yeast chromosome（ChrVII）can undergo remarkable cycles of chromosome instability.（Genes Dev. 2006 20:159）

訳 われわれは，ここに～ということを報告する

■［過分］＋ here

❶ ～ presented here	ここに提示された～	2,028
・data presented here	ここに提示されたデータ	433
・results presented here	ここに提示された結果	425
❷ ～ reported here	ここに報告された～	1,401
・studies reported here	ここに報告された研究	155
❸ ～ described here	ここに述べられた～	1,373

❶ **The data presented here define a set of interactions between** proteins involved in dislocation of misfolded polypeptides from the ER.（Proc Natl Acad Sci USA. 2005 102:14296）

訳 ここに提示されたデータは，～の間の１組の相互作用を明らかにする

❷ In contrast, **the studies reported here point to a fundamentally different mode** for assembly of this transporter.（J Biol Chem. 2004 279:33290）

訳 ここに報告された研究は，本質的に異なるモードを指し示す

herein　副 ここに　用例数 4,895

文頭の用例が非常に多い．Hereinのあとには，カンマが入ることが多い．

◆類義語：here

■ Herein, we ＋［動］

★ Herein, we ～	ここに，われわれは～	1,757
❶ Herein, we report ～	ここに，われわれは～を報告する	412
❷ Herein, we show ～	ここに，われわれは～を示す	201

❶ **Herein, we report that** CK2 inhibits U1 snRNA gene transcription by RNA polymerase Ⅱ.（J Biol Chem. 2005 280:27697）

訳 ここに，われわれは～ということを報告する

■ we ＋［動］＋ herein

❶ we report herein ～	われわれは，ここに～を報告する	205
❷ we show herein ～	われわれは，ここに～を示す	82

■［過分／動］＋ herein

❶ described herein	ここに述べられた	340
❷ presented herein	ここに示された	209
❸ reported herein	ここに報告された	202

❶ On the other hand, **several of the compounds described herein**, such as (+)-trans-5c, show comparable activity at all three transporters.（J Med Chem. 2005 48:7970）

訳 ここに述べられた化合物のいくつか

❷ **The study presented herein reveals that** the heme of human CBS undergoes a coordination change upon reduction at elevated temperatures.（Biochemistry. 2005 44:16785）

訳 ここに示された研究は，～ということを明らかにする

H

heterodimer 名 ヘテロ二量体 用例数 4,836

heterodimer	ヘテロ二量体	2,938
heterodimers	ヘテロ二量体	1,898

複数形のheterodimersの割合は約40%とかなり高い.

◆反意語：homodimer

■ heterodimer ＋［前］

	the%	a/an%	ø%			
❶	4	95	1	heterodimer with ～	～とのヘテロ二量体	235
❷	12	86	0	heterodimer of ～	～のヘテロ二量体	183

❶ **The hMSH6 protein forms a heterodimer with hMSH2** that is capable of recognizing a DNA mismatch.（J Biol Chem. 2008 283:31641）

訳 hMSH6タンパク質は，hMSH2とのヘテロ二量体を形成する

heterozygous 形 ヘテロ接合性の 用例数 4,320

◆反意語：homozygous

■ heterozygous ＋［前］

★ heterozygous for ～	～に対してヘテロ接合性の	777
❶ mice heterozygous for ～	～に対してヘテロ接合性であるマウス	255
❷ be heterozygous for ～	…は，～に対してヘテロ接合性である	233

❶ **Mice heterozygous for** PINCH1 are viable and indistinguishable from wild-type littermates.（Mol Cell Biol. 2005 25:3056）

訳 PINCH1に対してヘテロ接合性であるマウスは生存可能であり，そして野生型の同腹仔と区別がつかない

■ heterozygous ＋［名］

❶ heterozygous mice	ヘテロ接合性のマウス	361
❷ heterozygous mutations	ヘテロ接合性の変異	237

★★ high 形 高い 用例数 119,096

◆反意語：low

■ high ＋［名］

❶ high affinity	高親和性	9,455
❷ high levels	高いレベル	8,590
❸ high risk	高いリスク	5,637
❹ high resolution	高解像度	4,838
❺ high throughput	ハイスループット	4,614
❻ high dose	高い用量	2,883
❼ high density	高密度	2,859
❽ high frequency	高頻度	2,764
❾ a high degree of ～	高度の～	1,385
❿ high concentrations of ～	高濃度の～	1,233

❶ Purified Hap_s binds with **high affinity** to fibronectin, laminin, and collagen IV but not to collagen II.（Infect Immun. 2002 70:4902）

訳 精製されたHapは，高い親和性でフィブロネクチンと結合する

❷ ARC is expressed at **high levels** in striated muscle and displays fiber-type restricted expression patterns.（Hum Mol Genet. 2004 13:213）

訳 ARCは，横紋筋において高いレベルで発現する

■［副／前］＋ high

❶ very high	非常に高い	1,691
❷ relatively high	比較的高い	1,147

❸ as high as ～	～ほども高い	789
❹ extremely high	極度に高い	411

❷ The dissociation rate constant for the formation of LPO-Fe-(II)-O_2 is relatively high, in contrast to hemoprotein model compounds. (J Biol Chem. 2004 279:39465)

訳 ～は，比較的高い

❸ The uptake of 2-deoxyglucose in MIN6 cells was similarly inhibited (IC_{50} of 2.0 μmol/l), whereas glucokinase activity was unaffected at drug levels as high as 1 mmol/l. (Diabetes. 2003 52:1695)

訳 グルコキナーゼ活性は，1mmol/lもの高い薬物レベルでも影響されなかった

★★ higher 形 より高い 用例数 54,807

highの比較級.

■ higher than

★ higher than ～	～より高い	5,579
❶ higher than that	それより高い	1,322
・higher than that of ～	～のそれより高い	673
❷ higher than in ～	～におけるより高い	607
❸ … orders of magnitude higher than ～	～より…桁高い	112

❶ The cost of the Wada test was 3.7 times higher than that of functional MR imaging. (Radiology. 2004 230:49)

訳 ～は，磁気共鳴機能画像法のそれより3.7倍高かった

❷ In KO11, expression levels and activities were threefold higher than in strain B for xylose isomerase (xylA) and twofold higher for xylulokinase (xylB). (J Bacteriol. 2001 183:2979)

訳 KO11において，発現レベルおよび活性は，B株においてより３倍高かった

❸ These affinities were 2-3 orders of magnitude higher than those previously derived by radiolabeled fatty acyl-CoA ligand binding assay. (J Biol Chem. 2002 277:23988)

訳 これらの親和性は，～より2-3桁高かった

■ higher ＋［前］

❶ higher in ～	～においてより高い	5,485
・higher in ～ than in …	…においてより～において高い	781
❷ higher for ～	～に対してより高い	870

❶ Leptin levels were higher in women than in men regardless of diabetic status. (Diabetes. 1996 45:822)

訳 レプチンのレベルは，男性においてより女性において高かった

❷ The EST density was significantly higher for the D genome than for the A or B. (Genetics. 2004 168:701)

訳 EST密度は，Aゲノムに対してよりDゲノムに対して有意に高かった

■ higher ＋［名］

❶	higher levels	より高いレベル	3,516
❷	higher risk	より高いリスク	1,401
❸	higher affinity	より高い親和性	1,169
❹	higher concentrations	より高い濃度	981
❺	higher rates of ～	より速い速度の～	718
❻	higher frequency	より高い頻度	547
❼	higher doses	より高い用量	483
❽	higher incidence of ～	～のより高い発生率	421

❶ These results suggest that the use of GFs during chemotherapy **may be detrimental in those cancers expressing higher levels of the specific receptor**.（Oncogene. 2004 23:981）
訳 ～は，より高いレベルの特異的受容体を発現しているそれらの癌において有害であるかもしれない

❹ Aminoguanidine was as effective at 1 μM as **at the higher concentrations**.（Invest Ophthalmol Vis Sci. 2000 41:1176）
訳 より高い濃度において

■［副］＋ higher

❶	significantly higher	有意により高い	6,393
❷	～-fold higher	～倍より高い	2,949
❸	much higher	ずっとより高い	1,262
❹	～ times higher	～倍より高い	820
❺	slightly higher	わずかにより高い	374
❻	substantially higher	実質的により高い	371

❶ **Serum creatinine levels were also significantly higher in** the two everolimus groups than in the azathioprine group.（N Engl J Med. 2003 349:847）
訳 血清クレアチニンレベルは，また，～において有意により高かった

❷ **The increased mutation frequencies were 10- to 500-fold higher than those observed for** wild-type Pol recombinants.（J Virol. 2003 77:2946）
訳 増大した変異の頻度は，～に対して観察されたそれらより10から500倍高かった

❸ **HBV rates in this population were much higher than those in the general population**, and vaccination levels were low.（J Infect Dis. 2003 188:571）
訳 この集団におけるHBVの割合は，一般の集団におけるそれらよりずっと高かった

highest ㊢ 最も高い／最高の　用例数 9,359

highの最上級．the highestの用例が非常に多い．
◆類義語：best

■ highest ＋［前］

❶	highest in ～	～において最も高い	793

❷ highest for ～	～にとって最も高い	173

❶ Hospitalizations for congestive heart failure and stroke **were highest in the southeastern United States.** (Circulation. 2005 111:1233)
訳 ～は，合衆国南東部において最も高かった

■ the highest ＋［名］

★ the highest ～	最も高い～	6,496
❶ the highest levels	最も高いレベル	451
❷ the highest affinity	最も高い親和性	236
❸ the highest dose	最も高い用量	207

❶ Lesional samples **exhibited the highest levels of viral gene expression**, with NAWM exhibiting an intermediate level between lesional and control tissues. (Brain. 2005 128:516)
訳 ～は，最も高いレベルのウイルス遺伝子発現を示した

highlight 動 強調する　用例数 8,794

highlight	～を強調する	4,360
highlights	～を強調する	2,234
highlighted	～を強調した／強調される	1,187
highlighting	～を強調する	1,013

他動詞．現在形能動態の用例が多い．
◆類義語：emphasize, underscore, stress, accentuate

■［名／代名］＋ highlight

❶ results highlight ～	結果は，～を強調する	745
❷ findings highlight ～	知見は，～を強調する	453
❸ we highlight ～	われわれは，～を強調する	438
❹ review highlights ～	レビューは，～を強調する	393
❺ data highlight ～	データは，～を強調する	284
❻ study highlights ～	研究は，～を強調する	220

❷ **These findings highlight the role of epigenetics in the regulation of** development and oncogenesis by Gfi1. (Mol Cell Biol. 2005 25:10338)
訳 これらの知見は，～の調節におけるエピジェネティクスの役割を強調する

■ highlight ＋［名句］

❶ highlight the importance of ～	…は，～の重要性を強調する	989
❷ highlight the potential	…は，潜在能を強調する	293
❸ highlight the need for ～	…は，～に対する必要性を強調する	237
❹ highlight the role of ～	…は，～の役割を強調する	199

❶ **Our results highlight the importance of** the i-AAA complex and proteolysis at the inner

membrane in cells lacking mitochondrial DNA. (Mol Biol Cell. 2006 17:213)
訳 われわれの結果は，～の重要性を強調する

★★ highly 副 高度に／非常に　　用例数 45,455

◆類義語：extremely, unusually, very, markedly

■ highly ＋［過分］

❶ highly conserved	高度に保存されている	6,517
❷ highly expressed in ～	～において高度に発現する	1,905
・be highly expressed in ～	…は，～において高度に発現する	1,476
❸ highly correlated	高度に相関する	815
❹ highly enriched	高度に濃縮される	618
❺ highly purified	高度に精製される	602
❻ highly regulated	高度に調節される	571

❶ The TOUSLED (TSL)-like nuclear protein kinase family is highly conserved in plants and animals. (Plant Physiol. 2004 134:1488)
訳 ～は，植物と動物において高度に保存されている

❷ SWAP-70, an unusual phosphatidylinositol-3-kinase-dependent protein that interacts with the RhoGTPase Rac, is highly expressed in mast cells. (Mol Cell Biol. 2004 24:10277)
訳 ～は，肥満細胞において高度に発現する

■ highly ＋［形］

❶ highly sensitive	高度に感受性の／非常に敏感な	1,379
❷ highly specific	高度に特異的な	1,293
❸ highly active	高度に活性がある	1,130
❹ highly selective	高度に選択的な	871
❺ highly homologous	高度に相同的である	807
❻ highly significant	高度に重要な	768
❼ highly effective	高度に効果的な	762
❽ highly variable	高度に変わりやすい	698
❾ highly efficient	高度に効率的な	683
❿ highly similar	高度に類似した	525

❶ Furthermore, yeast lacking Asf1p are highly sensitive to mutations in DNA polymerase α and to DNA replicational stresses. (Mol Cell Biol. 2004 24:10313)
訳 Asf1pを欠いている酵母菌は，DNAポリメラーゼαの変異に高度に感受性である

❷ A15 was found to be highly specific for the active site of FXIIIa and was covalently bound to fibrin. (Circulation. 2004 110:170)
訳 A15は，FXIIIaの活性部位に対して高度に特異的であることが見つけられた

★history ㊔ 歴史／病歴 用例数 11,566

history	歴史／病歴	10,733
histories	歴史／病歴	833

複数形のhistoriesの割合が，約５％ある可算名詞.

◆類義語：record

■ history ＋［前］

	the%	a/an%	ø%			
❶	26	42	27	history of ～	～の歴史／病歴	6,480
				・family history of ～	～の家族歴	842
				・natural history of ～	～の自然経過／自然史	655
				・evolutionary history of ～	～の進化史	310
				・history of diabetes	糖尿病の病歴	147
				・history of breast cancer	乳癌の病歴	123

❶ Young women with breast cancer who have a **family history of** breast cancer and who test negative for deleterious mutations in BRCA1 and BRCA2 are at significantly greater risk of CBC than other breast cancer survivors.（J Clin Oncol. 2013 31:433）

訳 乳癌の家族歴を持つ乳癌の若い女性たち

hold ㊗ 保持する 用例数 4,835

held	保持される／～を保持した	1,853
hold	～を保持する	1,324
holds	～を保持する	1,105
holding	～を保持する	553

他動詞の用例が多い．hold promiseの用例が多い.

◆類義語：retain, keep, maintain, sustain, have

■ hold ＋［名］

❶ hold promise	…は，有望である	545
・hold promise for ～	…は，～のために有望である	320
・hold great promise	…は，大いに有望である	215
・hold promise as ～	…は，～として有望である	134

❶ Thus, an HPV16 E7 immunogen **holds promise for** noninvasive treatment and prevention of human cervical cancer.（Cancer Res. 2005 65:2018）

訳 HPV16 E7免疫原は，ヒトの子宮頸癌の非侵襲的治療と予防のために有望である

■ held ＋［前］/ together

❶ held in ～	～に保持される	352
・be held in ～	…は，～に保持される	202

H

❷ held at ～	～において保持される	152
❸ held together by ～	～によってつながれる	111

❶ NC174 was nested against HCDR3 and was held in place by two tryptophan side-chains（L91 and L96）from LCDR3.（J Mol Biol. 2000 302:853）

訳 ～は，2つのトリプトファン側鎖によって正しい位置に保持された

❸ The extracellular domain is a trimer held together by collagen-like and coiled-coil domains adjacent to the CRD.（J Biol Chem. 2005 280:22993）

訳 細胞外ドメインは，コラーゲン様およびコイルドコイルドメインによってつながれた三量体である

★ homologous 形 相同的な／相同の　用例数 12,512

◆類義語：similar, analogous

■ homologous ＋［前］

❶ homologous to ～	～と相同的である	2,886
・be homologous to ～	…は，～と相同的である	1,056
❷ homologous with ～	～と相同的である	151

❶ TEL2/ETV7 is highly homologous to the ETS transcription factor TEL/ETV6, a frequent target of chromosome translocation in human leukemia.（Blood. 2006 107:1124）

訳 TEL2/ETV7は，ETS転写因子TEL/ETV6と高度に相同的である

■［副］＋ homologous

❶ highly homologous	高度に相同的である	807
❷ structurally homologous	構造的に相同的である	173

■ homologous ＋［名］

❶ homologous recombination	相同組換え	2,714
❷ homologous proteins	相同タンパク質	273
❸ homologous chromosomes	相同染色体	238
❹ homologous sequences	相同配列	202

★ homology 名 相同性　用例数 11,700

homology	相同性	11,293
homologies	相同性	407

複数形のhomologiesの割合は３%あるが，単数形は無冠詞の用例が非常に多い．

■ homology ＋［前］

	the%	a/an%	ø%			
❶	2	1	82	homology to ～	～への相同性	2,867

				・protein with homology to ～	～に相同性を持つタンパク質	98
❷	1	2	83	homology with ～	～との相同性	1,370
❸	15	7	69	homology between ～	～の間の相同性	339
❹	53	3	43	homology of ～	～の相同性	170

❶ Vibrio cholerae encodes a small RNA with homology to Escherichia coli RyhB.（Infect Immun. 2005 73:5706）

訳 ～は，大腸菌RyhBに相同性を持つ低分子RNAをコードする

❷ The fold shares regions of structural homology with the DNA-binding domain of homeodomain proteins.（J Mol Biol. 2005 350:964）

訳 …は，～のDNA結合ドメインと構造的相同性の領域を共有する

■［名／形］＋ homology

❶	sequence homology	配列相同性	1,395
❷	significant homology	有意な相同性	431
❸	structural homology	構造的相同性	302

❶ SRG shows no significant nucleotide sequence homology to any known genes in the Genbank.（Cancer Res. 2005 65:10716）

訳 SRGは，Genbankにおける既知のどの遺伝子にも有意なヌクレオチド配列相同性を示さない

homozygous 形 ホモ接合性の　用例数 5,743

homozygous forの用例が多い．

◆反意語：heterozygous, wild-type

■ homozygous ＋［前］

❶	homozygous for ～	～に対してホモ接合性である	1,592
	・mice homozygous for ～	～に対してホモ接合性であるマウス	443
	・be homozygous for ～	…は，～に対してホモ接合性である	419

❶ Mice homozygous for the Y522S mutation exhibit skeletal defects and die during embryonic development or soon after birth.（FASEB J. 2006 20:329）

訳 Y522S変異に対してホモ接合性であるマウス

■ homozygous ＋［名／形］

❶	homozygous mutant	ホモ接合性の変異体	380
❷	homozygous deletion	ホモ接合性の欠失	260

❶ Markers for L1 identity, ACR4 and ATML1, are not expressed in homozygous mutant embryos.（Plant J. 2005 44:114）

訳 ～は，ホモ接合性変異体胚において発現されない

⁑ however 副 しかし 用例数 111,896

文頭の用例が非常に多い.

◆類義語：but, nevertheless, nonetheless

■ however

❶	However,	しかし,	79,618
❷	; however,	しかし,	30,733

❶ However, sJ1, like sDll, induces a NIH 3T3 cell tranformed phenotype mediated by FGF signaling.（J Biol Chem. 2004 279:13285）

訳 しかし,

❷ Cortical interneurons are of particular interest for cell transplantation; however, only a limited subset of these neurons can be generated from ESCs.（Nat Commun. 2012 3:841）

訳 しかし，これらのニューロンの限られたサブセットだけが胚性幹細胞から生成されうる

hypertrophy 名 肥大 用例数 4,074

hypertrophy	肥大	4,069
hypertrophies	肥大	5

複数形のhypertrophiesの割合が0.1%しかなく，原則，**不可算名詞**として使われる.

◆類義語：hyperplasia

■ hypertrophy ＋［前］

	the%	a/an%	ø%			
❶	2	0	98	hypertrophy in ～	～における肥大	309
❷	16	0	83	hypertrophy of ～	～の肥大	147

❶ Two reports in this issue of the JCI, by Morimoto et al. and Li et al., suggest that curcumin may inhibit cardiac hypertrophy in rodent models and provide beneficial effects after myocardial infarction or in the setting of hypertension.（J Clin Invest. 2008 118:850）

訳 クルクミンは，げっ歯類モデルにおける心肥大を抑制するかもしれない

■［形］＋ hypertrophy

❶	cardiac hypertrophy	心肥大	998
❷	ventricular hypertrophy	心室肥大	422

★ hypothesis 名 仮説 用例数 24,843

hypothesis	仮説	22,456
hypotheses	仮説	2,387

複数形のhypothesesの割合が，約10%ある可算名詞．the hypothesis thatの用例が断然多い.

◆類義語：assumption, notion, concept, idea

■ hypothesis ＋［同格that節／前］

hypothesis that/ofの前にtheが付く割合は圧倒的に高い．

	the%	a/an%	ø%			
❶	95	2	0	hypothesis that ～	～という仮説	12,937
❷	86	10	0	hypothesis of ～	～の仮説	616

■［動／前］＋ the hypothesis ＋［同格that節］

★	the hypothesis that ～	～という仮説	12,007
❶	test the hypothesis that ～	…は，～という仮説をテストする	4,769
	・we tested the hypothesis that ～	われわれは，～という仮説をテストした	1,865
	・To test the hypothesis that ～	～という仮説をテストするために	714
❷	support the hypothesis that ～	…は，～という仮説を支持する	3,338
	・results support the hypothesis that ～	結果は，～という仮説を支持する	649
	・data support the hypothesis that ～	データは，～という仮説を支持する	604
❸	consistent with the hypothesis that ～	～という仮説と一致している	1,103
	・results are consistent with the hypothesis that ～	結果は，～という仮説と一致している	267
❹	lead to the hypothesis that ～	…は，～という仮説につながる	306
❺	support for the hypothesis that ～	～という仮説に対する支持	216
❻	we investigated the hypothesis that ～	われわれは，～という仮説を精査した	108
❼	we examined the hypothesis that ～	われわれは，～という仮説を調べた	92

❶ Thus, **we tested the hypothesis that** CDK4 may be one of the critical downstream genes involved in Myc carcinogenesis.（Mol Cell Biol. 2004 24:7538）

訳 われわれは，～という仮説をテストした

❷ All these **results support the hypothesis that** myopodin functions as a tumor suppressor gene to limit the growth and to inhibit the metastasis of cancer cells.（Am J Pathol. 2004 164:1799）

訳 これらの結果すべてが，～という仮説を支持する

❸ These **results are consistent with the hypothesis that** past, but not recent, multivitamin use may be associated with modestly reduced risk of colorectal cancer.（Am J Epidemiol. 2003 158:621）

訳 これらの結果は，～という仮説と一致している

❹ This finding **led to the hypothesis that** these regions constitute a network supporting a default mode of brain function.（Proc Natl Acad Sci USA. 2003 100:253）

訳 この知見は，～という仮説につながった

■ this hypothesis

★	this hypothesis	この仮説	4,780
❶	to test this hypothesis	この仮説をテストするために	1,686
❷	consistent with this hypothesis	この仮説と一致している	301

❶ **To test this hypothesis**, we created mice that lack the three known β ARs（β-less mice).（Science. 2002 297:843）

H

訳 この仮説をテストするために，

★ hypothesize 動 仮定する　用例数 11,903

hypothesized	～を仮定した／仮定される	9,183
hypothesize	～を仮定する	2,632
hypothesizing	～を仮定する	60
hypothesizes	～を仮定する	28

他動詞.

◆類義語：postulate, presume, assume

■ hypothesize ＋［that節］

★ hypothesize that ～	～ということを仮定する	8,920
❶ we hypothesized that ～	われわれは，～ということを仮定した	5,764
❷ the authors hypothesized that ～	著者らは，～ということを仮定した	125

❶ **We hypothesized that** a TLR2 deficiency would attenuate S aureus-induced cardiac proinflammatory mediator production and the development of cardiac dysfunction. (Circulation. 2004 110:3693)

訳 われわれは，～ということを仮定した

■ be hypothesized ＋［that節］

★ be hypothesized that ～	～ということが仮定される	924
❶ it has been hypothesized that ～	～ということが仮定されている	349
❷ it was hypothesized that ～	～ということが仮定された	271

❷ **It was hypothesized that** cortisol infusion would fully mimic developmental changes in myocardial responsiveness to adrenergic and cholinergic stimulation. (J Physiol. 2005 562: 493)

訳 ～ということが仮定された

■ be hypothesized to *do*

★ be hypothesized to ～	…は，～すると仮定される	1,011
❶ be hypothesized to be ～	…は，～であると仮定される	274
❷ be hypothesized to play ～	…は，～を果たすと仮定される	94

❶ The heteromodal association cortex has **been hypothesized to be** selectively involved in the pathophysiology of schizophrenia. (Am J Psychiatry. 2004 161:322)

訳 …は，～に選択的に関与していると仮定されている

I

idea ㊔ 考え　　用例数 3,748

idea	考え	3,174
ideas	考え	574

the idea thatの用例が非常に多い．複数形のideasの割合が，約15％ある可算名詞．

◈類義語：notion, view, concept, hypothesis

■ idea ＋［前］

idea that/ofの前にtheが付く割合は圧倒的に高い．

	the％	a/an％	ø％			
❶	98	2	0	idea that ～	～という考え	2,069
				・support the idea that ～	…は，～という考えを支持する	1,002
				・be consistent with the idea that ～	…は，～という考えと一致する	270
❷	95	5	0	idea of ～	～の考え	251
				・support the idea of ～	…は，～の考えを支持する	63

❶ These results **support the idea that** the binding of the MjK2 RCK domain to membranes takes place via an electrostatic interaction with anionic lipid surfaces.（Biochemistry. 2005 44:62）

訳 これらの結果は，～という考えを支持する

❶ These results **are consistent with the idea that** the CEm plays an active role in fear conditioning.（J Neurosci. 2005 25:1847）

訳 これらの結果は，～という考えと一致する

★ identical ㊒ 同一の／同じ／等しい　　用例数 12,754

identical toの用例が多い．

◈類義語：equal, same, equivalent, comparable, consistent

■ identical ＋［前］

❶ identical to ～	～と同一である／～に等しい	3,930
・be identical to ～	…は，～と同一である	1,358
・identical to that of ～	～のそれと同一である	591
❷ identical in ～	～において同一である	670
❸ identical with ～	～と同一である	381

❶ In contrast, the Texcoco strain retains a gp45 gene, encoding an open reading frame **identical to that of** JG-29.（Infect Immun. 2001 69:3782）

訳 JG-29のそれと同一であるオープンリーディングフレームをコードしている

❷ All Japanese isolates, defined as the Japanese（J）genotype, were **identical in** the respective genomic region and proved the most divergent from the E strains, carrying four distinct variations.（J Virol. 2004 78:8349）

訳 ～は，それぞれのゲノム領域において同一であった

■ ［副］＋ identical

❶ nearly identical	ほとんど同じである	1,000
・nearly identical to ～	～とほとんど同じである	301
❷ virtually identical	実質的に同じである	393
❸ almost identical	ほとんど同じである	375
❹ essentially identical	本質的に同じである	310

❶ The lipid binding activity of the reduced mutant, as determined by its ability to form discoidal lipoproteins, was nearly identical to that of the wild type protein. (J Biol Chem. 2001 276: 34162)

訳 ～は，野生型タンパク質のそれとほとんど同じであった

■ not identical

❶ not identical	同じではない	439
・similar but not identical	似ているが同じではない	143

❶ We have also identified a FLA1-related gene (FLA2) whose sequence is similar but not identical to FLA1. (J Biol Chem. 2002 277:17580)

訳 その配列がFLA1に似ているが同じではないFLA1関連遺伝子（FLA2）

★ identification 名 同定／識別 用例数 18,854

identification	同定	18,559
identifications	同定	295

identification ofの用例が非常に多い．複数形のidentificationsの割合は1.5%しかなく，原則，不可算名詞として使われる．

◆類義語：detection, discovery, discrimination, observation

■ identification ＋［前］

	the%	a/an%	ø%			
❶	59	0	41	identification of ～	～の同定	13,986
				・lead to the identification of ～	…は，～の同定につながる	887
				・we report the identification of ～	われわれは，～の同定を報告する	759
				・result in the identification of ～	…は，～の同定という結果になる	290
				・identification of genes	遺伝子の同定	277
				・we describe the identification of ～	われわれは，～の同定について述べる	149

❶ An analysis of the growth rates of 17 point mutants led to the identification of VP1 amino acids that are critical in virus-host cell receptor interactions. (J Biol Chem. 2004 279:49172)

訳 …は，～において決定的に重要であるVP1アミノ酸の同定につながった

❶ This analysis resulted in the identification of approximately 40 distinct protein families within the P-loop kinase class. (J Mol Biol. 2003 333:781)

訳 この分析は，～の同定という結果になった

❶ Identification of genes needed for a particular trait can be accomplished by a comparative genomics approach using three or more organisms. (Bioinformatics. 2005 21:1693)

訳 ～のために必要とされる遺伝子の同定

■ identification and ＋［名］／［名］＋ and identification

❶	identification and characterization of ～	～の同定と特徴づけ	599
❷	detection and identification	検出と同定	113

❶ Here we report the identification and characterization of two endogenous peptide ligands of Methuselah, designated Stunted A and B. (Nat Cell Biol. 2004 6:540)

訳 われわれは，～の同定と特徴づけを報告する

★★ identify 動 同定する 用例数 131,348

identified	同定される／～を同定した	83,679
identify	～を同定する	37,533
identifying	～を同定する	6,968
identifies	～を同定する	3,168

他動詞.

◆類義語：find, discover, detect, observe, recognize, confirm

■ identify ＋［名句］

❶	identify a novel ～	…は，新規の～を同定する	1,987
❷	identify genes	…は，遺伝子を同定する	1,194
❸	identify patients	…は，患者を同定する	797
❹	identify proteins	…は，タンパク質を同定する	405
❺	identify factors	…は，因子を同定する	388
❻	identify a ～ mechanism	…は，～機構を同定する	329
❼	identify residues	…は，残基を同定する	233
❽	identify regions	…は，領域を同定する	206

❶ We also identified a novel gain-of-function sly1-d mutation that increased GA signaling by reducing the levels of the DELLA protein in plants. (Plant Cell. 2004 16:1392)

訳 われわれは，また，新規の機能獲得sly1-d変異を同定した

❷ We have studied the gene expression profile of cells expressing tumor-derived p53-D281G to identify genes transactivated by mutant p53. (Oncogene. 2004 23:4430)

訳 変異したp53によってトランス活性化される遺伝子を同定するために

■ [代名／名] + identify

❶	we identified ～	われわれは，～を同定した	11,086
❷	results identify ～	結果は，～を同定する	1,500
❸	analysis identified ～	解析は，～を同定した	1,202
❹	findings identify ～	知見は，～を同定する	809
❺	data identify ～	データは，～を同定する	679

❸ Southwestern **analysis identified** two GluRE binding proteins of 34 and 36 kDa in glucose-treated extracts. (Mol Biol Cell. 2004 15:4347)
訳 サウスウェスタン分析は，2つのGluRE結合タンパク質を同定した

■ to identify

★	to identify ～	～を同定する…	22,904
❶	To identify ～	～を同定するために	3,653
❷	be used to identify ～	…は，～を同定するために使われる	2,360
❸	be to identify ～	…は，～を同定することである	951
❹	sought to identify ～	…は，～を同定しようと努めた	509
❺	approach to identify ～	…は，～を同定するためのアプローチ	474
❻	screen to identify ～	…は，～を同定するための選別	358

❷ A polymerase chain reaction (PCR)-amplified subtractive hybridization technique **was used to identify** genes specific to the BPF clonal group. (Mol Microbiol. 2003 47:1101)
訳 …が，～に特異的な遺伝子を同定するために使われた

■ be identified + [前]

★	be identified	…が，同定される	30,411
❶	be identified in ～	…が，～において同定される	5,826
❷	be identified as ～	…が，～として同定される	4,724
	・be identified as being ～	…が，～であるとして同定される	115
❸	be identified by ～	…が，～によって同定される	2,714
	・be identified by using ～	…が，～を使うことによって同定される	143
❹	be identified from ～	…が，～から同定される	706
❺	be identified with ～	…が，～を使って同定される	551
❻	be identified for ～	…が，～に対して同定される	544
❼	be identified using ～	…が，～を使って同定される	508

❶ Mutations in this gene **were identified in** three independent gli1 alleles. (Plant J. 2004 37:617)
訳 この遺伝子の変異が，～において同定された

❷ Additionally, HopPtoG **was identified as** a suppressor on the basis of an enhanced HR produced by a hopPtoG mutant. (Plant J. 2004 37:554)
訳 HopPtoGが，抑制因子として同定された

❸ G- and F-actin **were identified by** DNase and rhodamine phalloidin, respectively. (Invest Ophthalmol Vis Sci. 2005 46:96)
訳 G-アクチンとF-アクチンが，～によって同定された

■ [名] + be identified

❶ genes were identified	遺伝子, が同定された	375
❷ mutations were identified	変異が, 同定された	355
❸ proteins were identified	タンパク質が, 同定された	283
❹ sites were identified	部位が, 同定された	230
❺ patients were identified	患者が, 同定された	203

❶ A total of 969 provisional orthologous genes were identified as preferentially expressed genes (PEGs) in various chicken tissues/organs (LOD>3.0). (Gene. 2004 340:213)
訳 合計969の暫定的なオルソロガス遺伝子が, ～として同定された

■ be identified + [that節]

❶ be identified that ～	～する…が同定される	1,203
・genes have been identified that ～	～する遺伝子が同定された	43

❶ Using a combination of global and candidate-specific two hybrid screens, eight proteins were identified that interact with Sen1p. (Nucleic Acids Res. 2004 32:2441)
訳 Sen1pと相互作用する８つのタンパク質が同定された

■ [副] + identified

❶ previously identified	以前に同定された／以前に同定した	3,200
❷ recently identified	最近同定された	1,914
❸ newly identified	新たに同定された	1,026
❹ first identified	最初に同定された	569
❺ originally identified	最初に同定された	473
❻ correctly identified	性格に同定された	300

❶ TBX3, but not TBX3 + 2a, is able to bind to the previously identified T-box binding site in a gel shift assay. (Cancer Res. 2004 64:5132)
訳 ～は, 以前に同定されたT-ボックス結合部位に結合できる

❷ Here, we have studied a recently identified hematopoietic-specific Rho GTPase, RhoH. (Blood. 2005 105:1467)
訳 われわれは, 最近同定された造血性特異的なRho GTP加水分解酵素, RhoH, を研究した

❸ Here we report a newly identified form of inhibitory synaptic plasticity, termed depolarization-induced potentiation of inhibition, in rodents. (Nat Neurosci. 2004 7:525)
訳 われわれは, 新たに同定された型の抑制性シナプスの可塑性を報告する

■ to be identified

★ to be identified	同定される…	720
❶ remain to be identified	…は, 同定されないままである	172
❷ have yet to be identified	…は, また, 同定されねばならない	132

I

■ identified and ＋［過分］

❶ identified and characterized	同定され，そして特徴づけられる	514

❶ While the main structural components of the COPII coat **have been** identified and characterized, the regulatory event(s) promoting COPII vesicle biogenesis and cargo selection still remains largely unknown.（Mol Cell. 2004 14:413）

訳 ～が同定されそして特徴づけられた

■［前］＋ identifying

❶ method for identifying ～	～を同定するための方法	175
❷ tool for identifying ～	～を同定するための道具	117
❸ useful for identifying ～	～を同定するために有用である	113
❹ useful in identifying ～	～を同定する際に有用である	110

❶ Here **we developed a** method for identifying novel cancer targets via negative-selection RNAi screening using a human breast cancer xenograft model at an orthotopic site in the mouse.（Nature. 2011 476:346）

訳 われわれは，～を同定するための方法を開発した

❷ **This assay could prove** useful in identifying region-selective NAD(H) catabolism that may contribute to neurodegeneration.（Brain Res. 2010 1316:112）

訳 このアッセイは，～を同定する際に有用であると判明しうる

identity 名 一致／同一性／正体　　用例数 9,918

identity	一致／同一性／正体	8,970
identities	一致／同一性／正体	948

複数形のidentitiesの割合は約10%あるが，単数形は無冠詞で使われることが断然多い.

◆類義語：coincidence, accordance, correspondence, similarity, analogy

■ identity ＋［前］

identity ofの前にtheが付く割合は圧倒的に高い.

	the%	a/an%	ø%			
❶	97	0	3	identity of ～	～の正体	2,248
❷	1	5	88	identity with ～	～との一致	1,044
❸	2	5	92	identity to ～	～との一致	876
❹	1	3	80	identity in ～	～における一致	358
❺	26	1	62	identity between ～	～の間の一致	199

❶ However, **the** identity of **the AP subunits that recognize these signals remains controversial.**（J Cell Biol. 2003 163:1281）

訳 これらのシグナルを認識するAPサブユニットの正体は，議論の余地が残るままである

❷ **SCD4 encodes a 352-amino acid protein that shares 79% sequence** identity with **the SCD1, SCD2, and SCD3 isoforms.**（J Biol Chem. 2003 278:33904）

訳 SCD4 は，SCD1，SCD2およびSCD3アイソフォームと79%の配列の一致を共有する

352アミノ酸のタンパク質をコードする

❸ CP-1 was a G3 rotavirus as its VP7 had 92 to 96% deduced amino acid identity to those of G3 rotaviruses. (J Clin Microbiol. 2002 40:937)

訳 ～は，G3ロタウイルスのそれらと92～96%の推定アミノ酸の一致を持っていた

■ ［名／形］＋ identity

❶	sequence identity	配列の一致／配列同一性	1,572
❷	amino acid identity	アミノ酸の一致	500
❸	molecular identity	分子的な同一性	212

illness 名 疾患／病気　用例数 5,317

illness	疾患／病気	4,411
illnesses	疾患／病気	906

複数形のillnessesの割合は約15%あるが，単数形は無冠詞で使われることが断然多い．

◆類義語：disease, disorder, sickness

■ illness ＋［前］

	the %	a/an %	ø %			
❶	8	5	82	illness in ～	～における疾患	310

❶ Rates of mental illness in children are increasing throughout the world. (N Engl J Med. 2005 352:1749)

訳 小児における精神疾患の割合は，増加している

■ ［形］＋ illness

❶	mental illness	精神疾患	332
❷	critical illness	重篤疾患	303
❸	respiratory illness	呼吸器疾患	205

■ ［前］＋ illness

❶	severity of illness	疾患の重症度	358

❶ Severity of illness correlated positively with both blood EBV load (P = .015) and $CD8^+$ lymphocytosis (P = .0003). (J Infect Dis. 2013 207:80)

訳 疾患の重症度は，～と正に相関した

illustrate 動 例証する　用例数 5,667

illustrate	～を例証する	3,034
illustrated	～を例証した／例証される	1,097
illustrates	～を例証する	1,081
illustrating	～を例証する	455

他動詞．能動態現在形の用例が多い．

◆類義語：account for, explain, represent, demonstrate

■ illustrate ＋［名句］

❶ illustrate the importance of ～	…は，～の重要性を例証する	178
❷ illustrate the potential	…は，潜在能を例証する	176
❸ illustrate the utility of ～	…は，～の有用性を例証する	123
❹ illustrate the power of ～	…は，～の力を例証する	70

❶ **Previous studies have** illustrated the importance of **accumulation of KRas mutations** in Myc-mediated tumor formation.（Oncogene. 2013 32:1296）

訳 以前の研究は，KRas変異の蓄積の重要性を例証してきた

■ illustrate ＋［名節］

❶ illustrate that ～	…は，～ということを例証する	690
❷ illustrate how ～	…は，どのように～かを例証する	542

❶ **These results** illustrate that expression gene profiling can contribute to the elucidation of important β-cell biological functions.（Diabetes. 2004 53:1496）

訳 これらの結果は，～ということを例証する

❷ In this work, **we** illustrate how **the proteasomal subunit S5a regulates hHR23a protein structure.**（Proc Natl Acad Sci USA. 2003 100:12694）

訳 われわれは，どのようにプロテアソームサブユニットS5aがhHR23aタンパク質構造を調節するかを例証する

■［代名／名］＋ illustrate

❶ we illustrate ～	われわれは，～を例証する	494
❷ results illustrate ～	結果は，～を例証する	399
❸ data illustrate ～	データは，～を例証する	180
❹ findings illustrate ～	知見は，～を例証する	177
❺ study illustrates ～	研究は，～を例証する	155

■ to illustrate

★ to illustrate ～	～を例証する…	543
❶ To illustrate ～	～を例証するために	154

❶ To illustrate **the utility of this method,** we investigate lineages of cells expressing one of 3 naturally regulated proteins, each with a different representative expression behavior.（Proc Natl Acad Sci USA. 2009 106:18149）

訳 この方法の有用性を例証するために，

★ image 名 イメージ／画像, 動 画像化する 用例数 16,097

images	イメージ／画像／画像化する	9,017
image	イメージ／画像／画像化する	5,675
imaged	画像化した／画像化された	1,405

名詞の用例が多いが，動詞としても使われる．複数形のimagesの割合は約60%と非常に高い．imagingは別項参照．

◆類義語：picture

■ 名 images ＋［前］

	the%	a/an%	ø%			
❶	5	-	82	images of ～	～の図	1,457
❷	10	-	69	images in ～	～における図	274
❸	5	-	74	images from ～	～からの図	241

❶ **Fluorescence images of the ganglion cells** were obtained in vivo with an adaptive optics scanning laser ophthalmoscope. (Invest Ophthalmol Vis Sci. 2008 49:467)

訳 神経節細胞の蛍光像

■ 動 imaged ＋［前］

❶	imaged with ～	～を使って画像化される	225
	・be imaged with ～	…は，～を使って画像化される	154
❷	imaged by ～	～によって画像化される	163
❸	imaged in ～	～において画像化される	134

❶ **All rabbits were imaged with a Micro PET-CT scanner** on days 0, 2, 5, 7, 14, 21, 28, and 35. (Invest Ophthalmol Vis Sci. 2011 52:5899)

訳 すべてのウサギが，マイクロPET-CTスキャナーを使って画像化された

I

★ imaging 名 イメージング／造影 用例数 25,873

imageの現在分詞でもあるが，ほとんどの場合で不可算名詞として使われる．

■ imaging ＋［前］

	the%	a/an%	ø%			
❶	7	0	93	imaging of ～	～のイメージング	2,472
❷	0	0	100	imaging in ～	～におけるイメージング	801
❸	0	0	85	imaging with ～	～によるイメージング	632

❶ An ultra-high resolution spectral-domain optical coherence tomography (SD-OCT) **system was built for in vivo imaging of retinas of birds of prey**. (Invest Ophthalmol Vis Sci. 2010 51:5789)

訳 システムは，猛禽類の網膜の生体内イメージングのために作られた

■ [形/名] ＋ imaging

❶ magnetic resonance imaging	磁気共鳴画像法	3,507
❷ molecular imaging	分子イメージング	471
❸ optical imaging	光学イメージング	395
❹ fluorescence imaging	蛍光イメージング	394
❺ functional imaging	機能イメージング	387
❻ in vivo imaging	生体内イメージング	354
❼ live cell imaging	生細胞イメージング	343

immediately 副 直ちに／すぐに　用例数 5,249

◆類義語：readily, soon, shortly

■ immediately ＋ [前]

❶ immediately after ～	～の直後	1,496
❷ immediately before ～	～の直前	350
❸ immediately following ～	～に続いてすぐ	289

❶ Similar results were obtained using normal donors when **neutralizing antibodies to IFN-γ were administered immediately after the BMT**. (J Clin Invest. 1998 102:1742)

訳 IFN-γに対する中和抗体が，骨髄移植の直後に投与された

❸ Increased nuclear staining with the membrane impermeable dye propidium iodide **was observed immediately following kainate treatment**, indicating a loss of plasma membrane integrity. (Brain Res. 1999 828:27)

訳 ～が，カイニン酸処理に続いてすぐに観察された

■ immediately ＋ [形]

❶ immediately upstream of ～	～のすぐ上流	329
❷ immediately adjacent to ～	～の直近	294
❸ immediately downstream of ～	～のすぐ下流	234

❶ **Several groups have identified a region immediately upstream of the β domain** that is important for outer membrane translocation, the so-called linker region. (J Biol Chem. 2004 279:31495)

訳 いくつかのグループが，βドメインのすぐ上流の領域を同定した

❷ **The CAAA VNTR is located immediately adjacent to the transcriptional promoters** for flanking open reading frames and may affect their activity. (J Clin Microbiol. 2000 38:1516)

訳 CAAA VNTRは，転写のプロモーターの直近に位置する

immobilize 動 固定化する／不動化する　用例数 3,186

immobilized	固定化される／～を固定化した	2,925
immobilize	～を固定化する	118

immobilizing	～を固定化する	111
immobilizes	～を固定化する	32

他動詞．受動態の用例が非常に多い．

◆類義語：fix

■ immobilized ＋［前］

❶	immobilized on ～	～上に固定化される	418
	・be immobilized on ～	…は，～上に固定化される	129
❷	immobilized in ～	～において固定化される	102
❸	immobilized onto ～	～上に固定化される	84

❶ Individual BAC molecules **were immobilized on** glass slides coated with Poly-L-lysine. (Plant J. 1999 17:581)

訳 個々のBAC分子が，ガラススライド上に固定された

★ immunity 名 免疫　　用例数 12,537

immunity	免疫	12,517
immunities	免疫	20

複数形のimmunitiesの割合は0.2%しかなく，原則，**不可算名詞**として使われる．

◆類義語：immunization

■ immunity ＋［前］

	the%	a/an%	ø%			
❶	1	0	99	immunity to ～	～に対する免疫	1,415
❷	3	0	89	immunity in ～	～における免疫	1,016
❸	1	1	94	immunity against ～	～に対する免疫	759

❶ Acquired **immunity to** Streptococcus pneumoniae (pneumococcus) has long been assumed to depend on the presence of anticapsular antibodies. (Proc Natl Acad Sci USA. 2005 102:4848)

訳 肺炎レンサ球菌に対する獲得免疫は，～に依存すると長い間考えられている

❸ An intact T cell compartment and IFN-γ signaling are required for protective **immunity against** Chlamydia. (J Immunol. 2005 174:5729)

訳 ～は，クラミジアに対する保護性の免疫のために必要とされる

■［形／名］＋ immunity

❶	innate immunity	自然免疫	1,662
❷	protective immunity	防御免疫	1,234
❸	adaptive immunity	適応免疫	913
❹	～-mediated immunity	～仲介性免疫	633
❺	cell immunity	細胞免疫	554

I

❻	antitumor immunity	抗腫瘍性免疫	460
❼	humoral immunity	液性免疫	446
❽	cellular immunity	細胞性免疫	407

immunization 名 免疫化／免疫 用例数 6,130

immunization	免疫化	5,753
immunizations	免疫化	377

複数形のimmunizationsの割合は約５%あるが，単数形は無冠詞で使われることが圧倒的に多い．

◆類義語：immunity

■ immunization ＋［前］

	the%	a/an%	ø%			
❶	0	4	96	immunization with ～	～による免疫化	1,822
❷	3	2	92	immunization of ～	～の免疫化	707
				・immunization of ～ with …	…による～の免疫化	313

❶ NTN was induced by **immunization with** rabbit IgG followed by rabbit anti-mouse glomerular basement membrane.（Arthritis Rheum. 2005 52:642）

訳 NTNが，ウサギIgGによる免疫化によって誘導された

❷ **Immunization of** mice **with** the AS15 peptide provided significant protection against subsequent parasite challenge, resulting in a lower parasite burden in the brain.（Infect Immun. 2012 80:3279）

訳 AS15ペプチドによるマウスの免疫化

■［前］＋ immunization

❶	after immunization	免疫化のあと	459
❷	following immunization	免疫化に続いて	181

❶ **After immunization** with PE, we could detect $B220^{+}$ and, as reported previously, $B220^{-}$ antigen-binding cells.（J Exp Med. 2003 197:1233）

訳 PEによる免疫化のあと

immunize 動 免疫する 用例数 4,422

immunized	免疫される／～を免役した	4,047
immunizing	～を免疫する	250
immunize	～を免疫する	124
immunizes	～を免疫する	1

他動詞．immunized withの用例が非常に多い．

■ immunized ＋［前］

❶ immunized with ～	～によって免疫される	1,848
・mice immunized with ～	～によって免疫されたマウス	755
・mice were immunized with ～	マウスが，～によって免疫された	187

❶ In contrast, **spleen cells from mice immunized with p53 and Flt3L exhibited** a higher Ag-specific proliferative response than mice immunized with p53 alone.（Cancer Res. 2001 61: 8227）

訳 p53およびFlt3Lによって免疫されたマウスからの脾臓細胞は，～を示した

■ immunized ＋［名］

❶ immunized mice	免役されたマウス	656
❷ immunized animals	免役された動物	185

★ impact　名 影響／衝撃，動 影響する　用例数 15,916

impact	影響／衝撃／影響する	13,628
impacts	影響／衝撃／影響する	1,625
impacted	影響した／影響される	417
impacting	影響する	246

動詞として用いられることもあるが，名詞の用例がほとんどである．複数形のimpactsの割合が，約10％ある可算名詞．

◆類義語：effect, impulse, influence, affect, impinge

■ impact ＋［前］

impact ofの前にtheが付く割合は圧倒的に高い．

	the％	a/an％	ø％			
❶	97	2	1	impact of ～	～の影響	5,973
				・impact of ～ on …	…に対する～の影響	1,905
❷	7	52	13	impact on ～	～に対する影響	3,878
				・significant impact on ～	～に対する有意な影響	438
				・no impact on ～	～に対する影響のない	241
				・major impact on ～	～に対する大きな影響	188
				・negative impact on ～	～に対する負の影響	139
				・little impact on ～	～に対してほとんど影響のない	136

❶ **Our objective was to determine the impact of these changes on transplant outcome.**（Blood. 1997 90:858）

訳 われわれの目的は，移植の結果に対するこれらの変化の影響を決定することであった

❷ A diacylglycerol analog **had no impact on permeability.**（Am J Respir Crit Care Med. 2005 172:1153）

訳 ～は，透過性に対する影響を持たなかった

I

★ impair ㊍ 損なう 用例数 21,070

impaired	損なわれる／〜を損なった	16,135
impairs	〜を損なう	2,231
impair	〜を損なう	2,149
impairing	〜を損なう	555

他動詞．受動態の用例が多い．

◆類義語：compromise, damage, lesion, disrupt

■ be impaired ＋［前］

★ be impaired	…は，損なわれる	2,932
❶ be impaired in 〜	…は，〜において損なわれる	1,351
・be impaired in their ability to *do*	…は，するそれらの能力において損なわれる	67
❷ be impaired by 〜	…は，〜によって損なわれる	254

❶ The fsr mutants were not impaired in their ability to colonize the nematode intestine.（Infect Immun. 2002 70:5647）

訳 fsr変異体は，〜をコロニー形成するそれらの能力において損なわれなかった

■［副］＋ impaired

❶ severely impaired	ひどく損なわれる	682
❷ significantly impaired	有意に損なわれる	568
❸ markedly impaired	顕著に損なわれる	237

❶ These results demonstrate that porcine T-cell function is severely impaired in the xenogeneic murine microenvironment.（Transplantation. 2004 78:1609）

訳 ブタのT細胞機能は，〜においてひどく損なわれる

■ impaired ＋［名］

❶ impaired glucose tolerance	損なわれたグルコース耐性	242
❷ impaired ability to *do*	損なわれた〜する能力	176

❷ Following HU-induced replication arrest, NBS and ATR-Seckel cells show similarly impaired G2/M checkpoint arrest and an impaired ability to restart DNA synthesis at stalled replication forks.（EMBO J. 2005 24:199）

訳 損なわれた〜を再開する能力

■ impaire ＋［名句］

❶ impaire the ability of 〜	…は，〜の能力を損なう	100

❶ Disruption of migR and fevR also impaired the ability of F. tularensis to prevent neutrophil oxidant production.（Infect Immun. 2009 77:2517）

訳 migRとfevRの破壊は，また，〜する野兎病菌の能力を損なった

impairment ㊔ 障害　　用例数 7,650

impairment	障害	6,193
impairments	障害	1,457

複数形のimpairmentsの割合が，約20%ある可算名詞．

◆類義語：dysfunction, disturbance, disorder, deficit, deficiency, damage

■ impairment ＋［前］

	the%	a/an%	ø%			
❶	21	11	63	impairment of ～	～の障害	1,563
❷	10	24	61	impairment in ～	～における障害	1,081

❶ Myocardial energetics **may play a key role in the impairment of diastolic function in obesity.** (Circulation. 2012 125:1511)

訳 ～は，肥満における拡張機能の障害において鍵となる役割を果たすかもしれない

❷ **Oxidative stress has been implicated in cognitive impairment in both old experimental animals and aged humans.** (Proc Natl Acad Sci USA. 2003 100:8526)

訳 酸化ストレスは，高齢の実験動物および高齢のヒトの両方における認知障害に関連づけられてきた

■［形／名］＋ impairment

❶	cognitive impairment	認知障害	1,055
❷	functional impairment	機能障害	335
❸	memory impairment	記憶障害	292

implant ㊍ 移植する，㊔ 移植片／インプラント　　用例数 5,313

implanted	移植される／～を移植した	2,189
implant	～を移植する／移植片／インプラント	1,521
implants	～を移植する／移植片／インプラント	1,507
implanting	～を移植する	96

名詞としても用いられるが，他動詞の用例が多い．

◆類義語：transplant, graft, engraft

■ implanted ＋［前］

❶	implanted in ～	～において移植される	364
	・be implanted in ～	…は，～において移植される	149
❷	implanted with ～	～を移植される	332
❸	implanted into ～	～に移植される	266

❷ To test this hypothesis, **male Sprague-Dawley rats were implanted with guide cannulae** directed to the medial portion of the AccSh. (Brain Res. 2005 1050:156)

訳 雄のSDラットは，ガイドカニューレを移植された

I

❸ When implanted into severe combined immunodeficient (SCID) mice, the growth of these tumors was dramatically reduced on the order of 5- to 10-fold as compared with controls. (J Biol Chem. 2005 280:40066)

訳 重症複合免疫不全（SCID）マウスに移植されたとき

implement 動 実行する　用例数 3,469

implemented	実行される／～を実行した	2,126
implement	～を実行する	632
implementing	～を実行する	510
implements	～を実行する	201

他動詞.

◆類義語：carry out, perform, execute

■ be implemented ＋［前］

★ be implemented	…は，実行される	1,117
❶ be implemented in ～	…は，～において実行される	372
❷ be implemented to *do*	…は，～するために実行される	113

❶ To date, only relatively simple methods have been implemented in available software. (Bioinformatics. 2012 28:1045)

訳 比較的単純な方法だけが，利用できるソフトウェアにおいて実行されてきた

★ implicate 動 関連づける／関与させる／意味づける　用例数 24,116

implicated	関連づけられる／～を関連づけた	17,481
implicate	～を関連づける	3,410
implicating	～を関連づける	2,173
implicates	～を関連づける	1,052

他動詞. implicated inの用例が非常に多い.

◆類義語：involve, engage, participate, associate, relate

■ be implicated ＋［前］

★ be implicated	…は，関連づけられる	11,260
❶ be implicated in ～	…は，～に関連づけられる	9,626
・have been implicated in ～	…は，～に関連づけられてきた	6,329
・be implicated in regulating ～	…は，～を調節することに関連づけられる	173
❷ be implicated as ～	…は，～として関連づけられる	1,041

❶ Keratinocyte growth factor (KGF) is an angiogenic and mitogenic polypeptide that has been implicated in cancer growth and tissue development and repair. (Am J Pathol. 1998 153:213)

訳 ケラチノサイト増殖因子（KGF）は，癌の増殖に関連づけられてきた血管新生および分裂促進的なポリペプチドである

❷ Oxidative stress has been implicated as a causal factor in the aging process of the heart and other tissues. (FASEB J. 2001 15:700)

訳 酸化ストレスは，原因因子として心臓および他の組織の老化過程に関連づけられてきた

■ be implicated in ＋［名句］

❶ be implicated in the pathogenesis of ～	…は，～の病因に関連づけられる	636
❷ be implicated in the regulation of ～	…は，～の調節に関連づけられる	442
❸ be implicated in the development of ～	…は，～の発症に関連づけられる	204
❹ be implicated in a variety of ～	…は，さまざまな～に関連づけられる	156
❺ be implicated in the control of ～	…は，～の制御に関連づけられる	145
❻ be implicated in the pathophysiology of ～	…は，～の病態生理に関連づけられる	95

❶ Immune-mediated injury to the graft has been implicated in the pathogenesis of chronic rejection. (Transplantation. 2002 74:1053)

訳 ～は，慢性拒絶の病因に関連づけられてきた

❷ Serotonin systems have been implicated in the regulation of hippocampal function. (Proc Natl Acad Sci USA. 1998 95:15026)

訳 セロトニン系は，海馬の機能の調節に関連づけられてきた

■ ［副］＋ implicated in

★ implicated in ～	～に関連づけられる	13,820
❶ previously implicated in ～	～に以前に関連づけられた	543
❷ strongly implicated in ～	～に強く関連づけられる	131

❶ In addition to the karyopherins, we identified Rai1p, a protein previously implicated in rRNA processing. (Mol Cell Biol. 2003 23:2042)

訳 リボソームRNAプロセシングに以前に関連づけられたタンパク質

■ ［名］＋ implicated in

❶ protein implicated in ～	～に関連づけられたタンパク質	222
❷ genes implicated in ～	～に関連づけられた遺伝子	206

■ implicate ＋［名］＋［前］

❶ implicate ～ in …	‥は，～を…に関連づける	1,917
・results implicate ～ in …	結果は，～を…に関連づける	135
❷ implicate ～ as …	‥は，～を…として関連づける	1,281
・results implicate ～ as …	結果は，～を…として関連づける	113
❸ implicate a role for ～	…は，～の役割を関連づける	190

❶ These results implicate BARD1 phosphorylation in the cellular response to DNA damage. (J Biol Chem. 2005 280:24669)

訳 これらの結果は，BARD1のリン酸化をDNA損傷への細胞応答に関連づける

❷ These data implicate stathmin as a regulator of the microtubule network during T cell activation. (J Immunol. 2012 188:5421)

訳 これらのデータは，スタスミンを～の制御因子として関連づける

★ implication ㊔ 意味／影響 用例数 11,732

implications	意味	11,150
implication	意味	582

複数形のimplicationsの割合は，約95%と圧倒的に高い．ここでいう「意味／影響」とは，「潜在的重要性」「予想される結果」「密接な関係」などを指している．

◆類義語：importance, significance, association

■ implications ＋［前］

implications ofの前にtheが付く割合は圧倒的に高い．

	the%	a/an%	ø%			
❶	3	-	93	implications for ～	～のための意味／～に対する意味	6,642
				・have implications for ～	…は，～のための意味を持つ	2,170
				・have important implications for ～	…は，～のための重要な意味を持つ	1,749
				・findings have implications for ～	知見は，～のための意味を持つ	250
				・implications for understanding ～	～を理解するための意味	506
				・implications for the development of ～	～の開発のための意味	183
				・implications for our understanding of ～	～についてのわれわれの理解のための意味	156
				・implications for the design of ～	～の設計のための意味	135
❷	83	-	16	implications of ～	～の意味	2,294
				・implications of these findings	これらの知見の意味	329
				・implications of these results	これらの結果の意味	250
				・discuss the implications of ～	…は，～の意味を議論する	218
				・implications of ～ are discussed	～の意味が，議論される	189
❸	2	-	89	implications in ～	～における意味	627

❶ These data **have implications for understanding** the fundamental link between I/R injury and alloimmunity.（Transplantation. 2002 74:916）

訳 これらのデータは，～を理解するための意味を持つ

❶ This neonatal immune bias **has important implications for the development of** vaccine strategies, particularly against viral infections.（J Immunol. 2000 164:3698）

訳 …は，～の開発のための重要な意味を持つ

❷ We discuss the implications of these findings for current models of β-globin regulation. (Blood. 2000 95:3600)

訳 われわれは，～の現在のモデルに対するこれらの知見の意味を議論する

❷ The biological implications of these oxidative processes are discussed. (J Biol Chem. 2001 276:24621)

訳 これらの酸化的過程の生物学的意味が議論される

❸ The effects are independent of aldosterone and blood pressure and have important implications in renin-dependent hypertension and chronic cardiac failure when circulating Ang II is elevated. (Circulation. 1998 98:2765)

訳 ～は，レニン依存性の高血圧において重要な意味を持つ

■ [形] + implications

❶ important implications	重要な意味	2,261
❷ clinical implications	臨床的な意味	440
❸ significant implications	重要な意味	365
❹ therapeutic implications	治療上の意味	297
❺ broad implications	広範な意味	212

imply ㊍ 意味する／暗示する／示唆する　用例数 8,253

imply	～を意味する／～を示唆する	2,836
implying	～を意味する／～を示唆する	2,745
implies	～を意味する／～を示唆する	2,203
implied	示唆される／～を示唆した	469

他動詞．imply thatの用例が非常に多い．「意味する」という訳語を使うが，「暗に意味する」という意味合いがある．

◆類義語：suggest, mean, indicate

■ imply + [that節]

❶ imply that ～	…は，～ということを意味する	5,425

❶ These results imply that Doc1p/Apc10 may play a role to regulate the binding of specific substrates, similar to that of the coactivators. (EMBO J. 2003 22:786)

訳 これらの結果は，～ということを意味する

■ imply + [名句]

❶ imply the existence of ～	…は，～の存在を意味する	127
❷ imply a role for ～	…は，～の役割を意味する	105

❶ In addition, our findings imply the existence of a negative modifier of HEN1 activity in the Columbia genetic background. (Nucleic Acids Res. 2010 38:5844)

訳 われわれの知見は，HEN1活性の負の修飾因子の存在を意味する

■ [名] + imply

❶	results imply ～	結果は，～を意味する	793
❷	data imply ～	データは，～を意味する	397
❸	findings imply ～	知見は，～を意味する	259

■ implying + [that節]

❶	, implying that ～	そして，（それは）～ということを意味している	1,704

❶ Overexpression of Cse4p suppresses this defect in the H4 mutant, implying that the two proteins act together in centromere structure.（Cell. 1998 94:607）
訳 そして，（それは）～ということを意味している

import 名 移行／移入，動 移入する　用例数 3,994

import	移行／移入／移入する	3,364
imported	移入した	544
imports	移行／移入／移入する	51
importing	移入する	35

動詞としても使われるが，名詞の用例が圧倒的に多い．複数形のimportsの割合は0.7%しかなく，原則，不可算名詞として使われる．
◆類義語：entry

■ import + [前]

	the%	a/an%	ø%			
❶	53	1	45	import of ～	～の移行	756
❷	2	0	89	import into ～	～への移行	151

❶ These findings, together with the observation that Sec13 and Nup93 could interact directly with Msk, suggest their direct involvement in the nuclear import of MAD.（Mol Cell Biol. 2010 30:4022）
訳 ～は，MADの核内移行への直接的関与を示唆する

■ [形] + import

❶	nuclear import	核内移行	1,217
❷	protein import	タンパク質移行	371

★ importance 名 重要性　用例数 16,266

importance	重要性	16,263
importances	重要性	3

複数形のimportancesの用例はほとんどなく，原則，不可算名詞として使われる．

◆類義語：significance, implication

■ importance ＋［前］

importance ofの前にtheが付く割合は圧倒的に高い．

	the%	a/an%	ø%			
❶	97	3	0	importance of ～	～の重要性	11,950
				・importance of ～ing	～することの重要性	1,130
				・highlight the importance of ～	…は，～の重要性を強調する	989
				・demonstrate the importance of ～	…は，～の重要性を実証する	812
				・underscore the importance of ～	…は，～の重要性を強調する	600
				・emphasize the importance of ～	…は，～の重要性を強調する	555
				・the relative importance of ～	～の相対的な重要性	519
				・the functional importance of ～	～の機能的な重要性	361
				・the potential importance of ～	～の潜在的な重要性	187
❷	5	1	48	importance in ～	～における重要性	1,190
				・importance in ～ing	～する際の重要性	230

❶ These results underscore the importance of the class II-mediated immune response in recovery from HBV infection.（J Infect Dis. 1999 179:1004）

訳 これらの結果は，～の重要性を強調する

❶ The relative importance of the two mechanisms has not been investigated except for a limited study, which suggested that the role of duplicate genes in compensation is negligible.（Nature. 2003 421:63）

訳 2つの機構の相対的な重要性

❷ Regulation of this protein may be of critical importance in modulating the role of Ang II in vascular disease.（Mol Pharmacol. 2000 57:460）

訳 このタンパク質の調節は，～の役割を調節する際に決定的に重要かもしれない

■ be ＋ of importance

★	of importance	重要である	343
❶	be of importance	…は，重要である	153
	・be of importance in ～	…は，～において重要である	63
	・be of importance for ～	…は，～にとって重要である	41

❶ We conclude that MBL may be of importance in first-line immune defense against several important pathogens.（Infect Immun. 2000 68:688）

訳 MBLは，～において重要であるかもしれない

I

■ be + of + [形] + importance

❶ be of critical importance	…は，決定的に重要である	134
❷ be of great importance	…は，とても重要である	119
❸ be of particular importance	…は，特に重要である	116
❹ be of fundamental importance	…は，機能的に重要である	112

❶ The stability of human embryonic stem cells (hESCs) is of critical importance for both experimental and clinical applications. (22802633Infect Immun. 2000 68:688)

訳 ヒト胚性幹細胞（hESCs）の安定性は，～にとって決定的に重要である

★ important 形 重要な　用例数 81,453

◆類義語：key, vital, crucial, critical, significant

■ be important + [前]

★ be important	…は，重要である	20,285
❶ be important for ～	…は，～にとって重要である／～のために重要である	9,687
・be important for ～ function	…は，～機能にとって重要である	477
・be important for understanding ～	…は，～を理解するために重要である	230
・be important for maintaining ～	…は，～を維持するために重要である	137
・be important for the development of ～	…は，～の発達のために重要である	109
・be important for binding	…は，結合のために重要である	104
❷ be important in ～	…は，～において重要である／～の際に重要である	4,574
・be important in regulating ～	…は，～を調節する際に重要である	173
・be important in determining ～	…は，～を決定する際に重要である	172
・be important in the pathogenesis of ～	…は，～の病因において重要である	162
・be important in the regulation of ～	…は，～の調節において重要である	101
❸ it is important to *do*	～することは重要である	785
・it is important to understand ～	～を理解することは重要である	145

❶ The second helical region, and the specific amino acids that the helix exposes to solvent, may be particularly important for binding and selectivity. (Biochemistry. 1999 38:3874)

訳 ～は，結合と選択性にとって特に重要であるかもしれない

❷ Variable flanking regions, such as the loop connecting β-strands 2 and 3, are thought to be important in determining the RNA-binding specificities of individual RRMs. (J Biol Chem. 2002 277:33267)

訳 …は，～のRNA結合特異性を決定する際に重要であると考えられる

❸ In the context of development, it is important to understand the mechanisms that coordinate growth and patterning of tissues. (Proc Natl Acad Sci USA. 2005 102:3318)

訳 ～する機構を理解することは重要である

■ important + [名]

❶ important role	重要な役割	12,899
❷ important implications for ～	～のための重要な意味	1,910

❸ important regulator of ～	～の重要な制御因子	1,130
❹ important mechanism	重要な機構	979
❺ important component of ～	～の重要な構成要素	965
❻ important determinant of ～	～の重要な決定要因	919

❶ Thus, constitutively increased expression of initiation factors 4E and 2α **may play an important role in the development of lymphomas** and is correlated with their biological aggressiveness. (Am J Pathol. 1999 155:247)

訳 ～は，リンパ腫の発生において重要な役割を果たすかもしれない

❷ **These observations have important implications for understanding the mechanism of** the conversion between the normal (PrP^{C}) and pathogenic (PrP^{Sc}) forms of prion protein. (J Biol Chem. 1999 274:36859)

訳 これらの観察は，～の機構を理解するための重要な意味を持つ

■ ［副］＋ important

❶ most important	最も重要な	1,797
❷ functionally important	機能的に重要な	869
❸ more important	より重要な	856
❹ potentially important	潜在的に重要な	690
❺ clinically important	臨床的に重要な	626

❶ According to BD results, **the most important factor for aldolase binding to actin** is the quaternary structure of aldolase and actin. (Biophys J. 1999 76:17)

訳 アルドラーゼのアクチンへの結合のための最も重要な因子

❹ The wide epithelial distribution and the conservation across species **suggests a potentially important role for occludin 1B** in the structure and function of the tight junction. (Mol Biol Cell. 2000 11:627)

訳 ～は，オクルディン1Bの潜在的に重要な役割を示唆する

★ importantly 副 重要なことに／顕著に　用例数 8,422

I

文頭の用例が非常に多い.

■ importantly,

❶ Importantly,	重要なことに,	5,947
❷ More importantly,	もっと重要なことに,	772
❸ Most importantly,	最も重要なことに,	658

❶ **Importantly, we show that** TSAd can act as a potent transcriptional activator in T cells. (J Exp Med. 2001 193:1425)

訳 重要なことに，われわれは～ということを示す

❸ **Most importantly,** TOC with or without ascorbate pretreatment significantly improved survival in Ob rats following ischemia in a dose-dependent manner. (Hepatology. 2001 34:13)

訳 最も重要なことに,

★★ improve ㊍ 改善する／改良する 用例数 35,197

improved	改善される／改善した	17,922
improve	改善する	10,946
improves	改善する	3,242
improving	改善する	3,087

他動詞の用例が多い．

◆類義語：advance

■ improve ＋［名句］

❶ improve ～ function	…は，～機能を改善する	856
❷ improve ～ outcomes	…は，～結果を改善する	613
❸ improve survival	…は，生存率を改善する	574
❹ improve our understanding of ～	…は，～についてのわれわれの理解を改善する	396
❺ improve the quality of ～	…は，～の質を改善する	232
❻ improve the accuracy of ～	…は，～の正確性を改善する	161
❼ improve the efficacy of ～	…は，～の有効性を改善する	138
❽ improve the efficiency of ～	…は，～の効率を改善する	120

❹ These findings could **improve our understanding of** the development and progression of these premalignant lesions.（Gastroenterology. 2012 142:730）
㊞ これらの知見は，～についてのわれわれの理解を改善しうる

■［動／名］＋ to improve

★ to improve ～	～を改善する…	5,029
❶ be shown to improve ～	…は，～を改善することが示される	202
❷ potential to improve ～	～を改善する潜在能	202
❸ be used to improve ～	…は，～を改善するために使われる	151
❹ efforts to improve ～	～を改善する努力	140
❺ strategies to improve ～	～を改善する戦略	137

❸ ECM1 and TMPRSS4 are excellent diagnostic markers of malignant thyroid nodules and **may be used to improve the diagnostic accuracy of** FNA biopsy.（Ann Surg. 2005 242:353）
㊞ …は，～の診断上の正確性を改善するために使われるかもしれない

❹ **Recent efforts to improve medical education include** adopting a new framework based on 6 broad competencies defined by the Accreditation Council for Graduate Medical Education.（Ann Intern Med. 2010 153:751）
㊞ 医学教育を改善する最近の努力は，～を含んでいる

■ be improved ＋［前］

★ be improved	…は，改善される	1,275
❶ be improved by ～	…は，～によって改善される	436
❷ be improved in ～	…は，～において改善される	116

❶ The capacity of RecA to compete with RdgC is improved by the DinI protein. (J Biol Chem. 2006 281:4708)
訳 ～は，DinIタンパク質によって改善される

■ ［副］＋ improved

❶ significantly improved	有意に改善される	1,328
❷ greatly improved	大きく改善される	196

❶ At 12 months, exercise capacity was significantly improved in the treatment group. (Circulation. 2004 110:II213)
訳 運動能力が，治療群において有意に改善された

■ improved ＋［名］

❶ improved survival	改善された生存率	871
❷ improved understanding	改善された理解	371
❸ improved outcomes	改善された結果	239

❶ However, in the 45 patients with acute lung injury, improved survival correlated with a higher colony count (p<0.04). (Am J Respir Crit Care Med. 2005 172:854)
訳 改善された生存率は，より高いコロニー数と相関した

★ improvement ㊔ 改善　用例数 11,822

improvement	改善	8,303
improvements	改善	3,519

複数形のimprovementsの割合が，約30%ある可算名詞．improvement inの用例が非常に多い．

◆類義語：refinement

■ improvement ＋［前］

	the %	a/an %	ø %			
❶	13	23	57	improvement in ～	～の改善	3,822
❷	26	22	49	improvement of ～	～の改善	868
❸	2	80	13	improvement over ～	～を超える改善	274

❶ Treatment with gefitinib was not associated with significant improvement in survival in either coprimary population. (Lancet. 2005 366:1527)
訳 ゲフィチニブによる治療は，生存率の有意な改善と関連しなかった

❸ Our novel approach, which offers an improvement over previous methodologies, is of value for quantitative work of this nature. (Brain Res. 2008 1236:73)
訳 ～は，以前の方法論を超える改善を提供する

■ ［形／名］＋ improvement

❶ significant improvement	有意な改善	1,306

I

❷ quality improvement	質の改善	433
❸ greater improvement	より大きな改善	265
❹ clinical improvement	臨床的な改善	227

inactivate ㊐ 不活性化する／失活させる 用例数 7,878

inactivated	不活性化される／～を不活化した	4,159
inactivating	～を不活性化する／～を失活させる	1,591
inactivate	～を不活性化する／～を失活させる	1,261
inactivates	～を不活性化する／～を失活させる	867

他動詞.

◆類義語：deactivate ◆反意語：activate

■ be inactivated ＋［前］

★ be inactivated	…は，不活性化される	1,116
❶ be inactivated by ～	…は，～によって不活性化される	447
❷ be inactivated in ～	…は，～において不活性化される	186

❶ The enzyme requires Fe^{2+} as a cofactor and is inactivated by 4-chloro-3-hydroxyanthranilate. (Biochemistry. 2005 44:7632)

訳 …は，4‐クロロ‐3‐ヒドロキシアントラニル酸塩によって不活性化される

❷ To determine the function of Runx1 in skeletal muscle, we generated mice in which Runx1 was selectively inactivated in muscle. (Genes Dev. 2005 19:1715)

訳 Runx1は，筋肉において選択的に不活性化された

■ inactivating ＋［名］

❶ inactivating mutations	不活性化変異	280

★ inactivation ㊔ 不活性化／不活化 用例数 14,220

inactivation	不活性化／不活化	14,184
inactivations	不活性化／不活化	36

複数形のinactivationsの割合は0.3％しかなく，原則，不可算名詞として使われる.

◆反意語：activation

■ inactivation ＋［前］

	the％	a/an％	ø％			
❶	17	2	80	inactivation of ～	～の不活性化	6,242
				・inactivation of p53	p53の不活性化	128
❷	1	1	46	inactivation in ～	～における不活性化	578
❸	2	0	69	inactivation by ～	～による不活性化	483

❶ Overexpression of c-Myc and inactivation of p53 are hallmarks of human Burkitt's lymphomas. (Cancer Res. 2005 65:5454)

訳 c-Mycの過剰発現とp53の不活性化は，～の特徴である

❷ Repression of BRCA1 expression by hypoxia represents an intriguing mechanism of functional BRCA1 inactivation in the absence of genetic mutation. (Cancer Res. 2005 65:11597)

訳 遺伝的変異の非存在下における機能的なBRCA1不活性化の興味深い機構

■ [前] ＋ inactivation

❶	recovery from inactivation	不活性化から回復	162
❷	rate of inactivation	不活性化の速度	94

❶ Recovery from inactivation was observed, and its extent depended on the pH_i and the amount of time that the channel was inactive. (Proc Natl Acad Sci USA. 2005 102:17630)

訳 不活化からの回復が観察された

★ incidence ㊔ 発生率／頻度　用例数 15,156

incidence	発生率／頻度	14,823
incidences	発生率／頻度	333

複数形のincidencesの割合は2％しかないが，単数形に不定冠詞が用いられる例は多い．

◆類義語：occurrence, frequency, rate

■ incidence ＋ [前]

incidence ofの前にtheが付く割合はかなり高い．

	the%	a/an%	ø%			
❶	63	22	15	incidence of ～	～の発生率	8,998
				・increased incidence of ～	～の増大した発生率	570
				・high incidence of ～	～の高い発生率	538
				・reduce the incidence of ～	…は，～の発生率を低下させる	537
				・higher incidence of ～	～のより高い発生率	420
				・cumulative incidence of ～	～の累積的発生率	369
				・lower incidence of ～	～のより低い発生率	258
				・increase in the incidence of ～	～の発生率の増大	111
❷	14	12	57	incidence in ～	～における発生率	393

❶ Angiotensin-converting enzyme inhibitors reduce the incidence of both types of arrhythmia. (J Am Coll Cardiol. 2005 45:525)

訳 ～は，両方のタイプの不整脈の発生率を低下させる

❶ An increased incidence of LM subtypes should direct melanoma screening to heavily sun-exposed sites, where these subtypes predominate. (J Invest Dermatol. 2005 125:685)

訳 LMサブタイプの増大した発生率

I

■ [名] + incidence

❶ cancer incidence	癌の発生率	474
❷ tumor incidence	腫瘍の発生率	229

■ incidence and + [名]

❶ incidence and severity	発生率と重症度	239

❶ Fasudil treatment resulted in a dose-dependent reduction in both the incidence and severity of AAA. (Circulation. 2005 111:2219)

訳 ～は，AAAの発生率と重症度の両方の用量依存的な低下という結果になった

★★ include 動 含む　用例数 108,680

including	～を含む	69,559
included	含まれる／～を含んだ	16,613
include	～を含む	14,958
includes	～を含む	7,550

他動詞．〈, including〉の用例が非常に多い．

◆類義語：contain, involve

■ [名] + include

❶ genes include ～	遺伝子は，～を含む	144
❷ features include ～	特徴は，～を含む	128
❸ factors include ～	因子は，～を含む	124
❹ examples include ～	例は，～を含む	121

❷ Additional clinical features include oral leukokeratosis, follicular keratosis, and cysts (steatocysts and pilosebaceous cysts). (J Invest Dermatol. 2011 131:1018)

訳 付加的な臨床的特徴は，口腔白色角化症，毛包性角化症および嚢胞を含む

■ include + [名句]

❶ include the use of ～	…は，～の使用を含む	139

❶ Conventional treatments for infantile hemangioma include the use of corticosteroids, laser, surgery, and immunomodulator therapy. (Curr Opin Ophthalmol. 2011 22:419)

訳 小児血管腫のための従来の治療は，副腎皮質ステロイドの使用，レーザー，外科手術，免疫調節薬療法を含む

■ be included + [前]

★ be included	…が，含まれる	2,967
❶ be included in ～	…が，～に含まれる	1,487
・be included in the analysis	…が，その分析に含まれる	159

・be included in the study	…が，その研究に含まれる	152
❷ be included as ～	…が，～として含まれる	103

❶ A total of 66,303 patients were included in the analysis, 63,171 (95.3%) underwent a primary gastric bypass procedure and 3132 patients (4.7%) underwent a gastric band-related reoperation. (Ann Surg. 2013 257:279)

訳 合計で66,303名の患者がその分析に含まれた

■ [名] ＋ be included

❶ patients were included	患者が，含まれた	146
❷ studies were included	研究が，含まれた	98

■ [名], including

★ , including ～	，そして（それらには）～が含まれる	54,134
❶ proteins, including ～	タンパク質，そして（それらには）～が含まれる	1,449
❷ genes, including ～	遺伝子，そして（それらには）～が含まれる	1,370
❸ factors, including ～	因子，そして（それらには）～が含まれる	1,037

❶ The L-complex contains approximately 16 proteins, including the two RNA-editing ligases (RELs), REL1 and REL2. (Proc Natl Acad Sci USA. 2005 102:4712)

訳 L-複合体はおよそ16個のタンパク質を含む，そして（それらには）2つのRNA編集リガーゼ（REL），REL1およびREL2，が含まれる

incorporate 動 取り込む／組み入れる 用例数 9,552

incorporated	取り込まれる／～を取り込んだ	4,694
incorporating	～を取り込む／～を組み入れる	2,060
incorporate	～を取り込む／～を組み入れる	1,694
incorporates	～を取り込む／～を組み入れる	1,104

他動詞.

◆類義語：integrate

■ be incorporated ＋ [前]

★ be incorporated	…が，取り込まれる	2,133
❶ be incorporated into ～	…が，～に取り込まれる	1,466
❷ be incorporated in ～	…が，～において取り込まれる	148

❶ Organic cations of increasing size were used as current carriers through the PC2 channel after PC2 was incorporated into lipid bilayers. (J Biol Chem. 2005 280:29488)

訳 PC2が脂質二重層に取り込まれたあと

I

incorporation 名 取り込み　用例数 7,527

incorporation	取り込み	7,477
incorporations	取り込み	50

複数形のincorporationsの割合は0.7%しかなく，原則，不可算名詞として使われる．

◆類義語：uptake

■ incorporation ＋［前］

	the%	a/an%	ø%			
❶	41	2	57	incorporation of ～	～の取り込み	3,647
				・incorporation of ～ into …	…への～の取り込み	748
❷	1	0	76	incorporation into ～	～への取り込み	1,072
❸	4	0	68	incorporation in ～	～における取り込み	229

❶ Incorporation of this mutant pneumolysin into current pneumococcal vaccines may increase their efficacy.（Infect Immun. 2006 74:586）

訳 現在の肺炎球菌ワクチンへのこの変異ニューモリシンの取り込みは，それらの効力を増大させるかもしれない

❷ Our findings suggest that inefficient incorporation into myofilament is responsible for the instability of mutant slow TnT in ANM muscle.（J Biol Chem. 2005 280:13241）

訳 筋フィラメントへの非効率的な取り込みは，～の原因である

■［名］＋ incorporation

❶	thymidine incorporation	チミジンの取り込み	464
❷	nucleotide incorporation	ヌクレオチドの取り込み	238
❸	BrdU incorporation	ブロモデオキシウリジンの取り込み	201

★★ increase 動 増大させる／増大する，名 増大　用例数 249,388

increased	～を増大させた／増大した	128,967
increase	～を増大させる／増大する／増大	68,443
increases	～を増大させる／増大する／増大	30,956
increasing	増大する	21,022

動詞の用例が多いが，名詞としても用いられる．動詞としては，他動詞と自動詞の両方の用例がある．また，他動詞受動態が自動詞とほぼ同じ意味になることもある．名詞としては，複数形のincreasesの割合が，約25%ある可算名詞．

◆類義語：elevate, up-regulate, raise, augment, elevation, augmentation, rise, increment, up-regulation

■ 動 increase ＋［前／副］

❶	increase with ～	…は，～と共に増大する	3,778
	・increase with increasing ～	…は，増大する～と共に増大する	599

・increase with age	…は，年齢と共に増大する	519
・increase with time	…は，時間と共に増大する	122
❷ increase from ～	…は，～から増大する	2,475
・increase from ～ to …	…は，～から…へ増大する	1,063
❸ increase to ～	…は，～へ増大する	1,308
❹ increase ～-fold	…は，～倍増大する	650

❶ The defective ERK signaling was caused by the dual specific phosphatase 6 (DUSP6), whose protein expression increased with age due to a decline in repression by miR-181a. (Nat Med. 2012 18:1518)

訳 タンパク質発現は，年齢と共に増大した

❹ Epithelial CCL20 mRNA increased 19-fold at 3 h, and protein increased approximately 16-fold at 6 h after injury. (FASEB J. 2011 25:2659)

訳 上皮CCL20 メッセンジャーRNAは，19倍増大した

■ 動 [名] ＋ increase

❶ levels increased	レベルは，増大した	619
❷ expression increased	発現は，増大した	500
❸ activity increased	活性は，増大した	439
❹ rate increased	速度は，増大した	324

❶ Her serum prolactin levels increased from 7 to 133 ng/mL (normal < 20 ng/mL) and hematocrit from 17% to 22% to 35%. (Blood. 2002 100:2687)

訳 彼女の血清プロラクチンのレベルは，7 ng/mLから133 ng/mLへ増大した

■ 動 be increased ＋ [前／副]

★ be increased	…は，増大する	11,198
❶ be increased in ～	…は，～において増大する	3,427
・be increased in patients	…は，患者において増大する	117
❷ be increased by ～	…は，～によって増大する／～だけ増大する	1,781
❸ be increased to ～	…は，～へ増大する	304
❹ be increased ～-fold	…は，～倍増大する	286
❺ be increased from ～	…は，～から増大する	241

❶ Colonic IL-16 protein levels were increased in patients with Crohn's disease (P<0.05) but not ulcerative colitis. (Gastroenterology. 2000 119:972)

訳 結腸のIL-16タンパク質レベルは，クローン病の患者において増大した

❷ BEST1 promoter activity was increased by SOX9 overexpression and decreased by siRNA-mediated SOX9 knockdown. (J Biol Chem. 2010 285:26933)

訳 BEST1プロモーター活性は，SOX9過剰発現によって増大した

■ 動 [名] ＋ be increased

❶ levels were increased	レベルは，増大した	328
❷ expression was increased	発現は，増大した	316
❸ activity was increased	活性は，増大した	229

■ 動 [副] + increased

❶ significantly increased	有意に増大した	5,820
❷ markedly increased	顕著に増大した	1,206
❸ greatly increased	大きく増大した	671
❹ dramatically increased	劇的に増大した	660

❶ The number of SR-A-positive cells was significantly increased in PIN as compared with normal prostatic tissue (P = 0.0176). (Cancer Res. 2004 64:2076)

訳 SR-A陽性細胞の数は，PINにおいて有意に増大した

■ 動 increased + [名]

❶ increased risk	増大したリスク	6,016
❷ increased expression	増大した発現	4,076
❸ increased levels	増大したレベル	2,571
❹ increased sensitivity	増大した感受性	1,038
❺ increased susceptibility	増大した感受性	926

❶ Travelers with cardiovascular disease may be at increased risk for venous thrombosis as a result of depressed ejection fraction or immobility. (Ann Intern Med. 2004 141:148)

訳 循環器疾患を持つ旅行者は，静脈血栓症の増大したリスクにあるかもしれない

❸ Increased levels of pro-NGF mRNA and protein were observed in the RCS rat model of retinal dystrophy. (J Biol Chem. 2004 279:41839)

訳 増大したレベルのプロNGFメッセンジャーRNAおよびタンパク質が，～において観察された

■ 動 increase + [名句]

❶ increase the risk of ～	…は，～のリスクを増大させる	1,144
❷ increase the number of ～	…は，～の数を増大させる	1,139
❸ increase the rate of ～	…は，～の速度を増大させる	825
❹ increase the expression of ～	…は，～の発現を増大させる	648

❷ HIV-induced immune impairment may increase the risk of neurosyphilis. (J Infect Dis. 2004 189:369)

訳 ～は，神経梅毒のリスクを増大させるかもしれない

■ 動 increasing + [名]

❶ increasing number of ～	増大する数の～	795
❷ increasing evidence	増大する証拠	754
❸ increasing concentrations of ～	～の増大する濃度	418

❶ The rate of substrate hydrolysis by AVP-pVIc increased with increasing concentrations of polyE. (Biochemistry. 2005 44:8721)

訳 AVP-pVIcによる基質の加水分解の速度は，ポリEの増大する濃度とともに増大した

■ 名 increase ＋［前］

「～の増大」の意味では，increase ofよりもincrease inの用例が圧倒的に多い．

	the%	a/an%	ø%		
❶	13	78	5	increase in ～　～の増大	42,920
				・increase in the number of ～　～の数の増大	1,060
				・result in an increase in ～　…は，～の増大という結果になる	814
				・lead to an increase in ～　…は，～の増大につながる	712
				・cause an increase in ～　…は，～の増大を引き起こす	670
				・be associated with an increase in ～　…は，～の増大と関連する	521
				・show an increase in ～　…は，～の増大を示す	435
				・increase in the rate of ～　～の速度の増大	326
				・increase in the expression of ～　～の発現の増大	313
				・be accompanied by an increase in ～　…は，～の増大を伴う	292
				・increase in the level of ～　～のレベルの増大	283
❷	19	70	9	increase of ～　～の増大	2,833

❶ Affected animals demonstrated a dramatic increase in the number of megakaryocytes in the bone marrow and the spleen. (Blood. 2003 102:3363)

訳 冒された動物は，～における巨核球の数の劇的な増大を示した

❶ TGF-β1 treatment leads to an increase in MLK3 activity. (J Biol Chem. 2004 279:29478)

訳 TGF-β1処理は，MLK3活性の増大につながる

❶ S-phase arrest was also associated with an increase in Chk1 kinase activity. (Mol Pharmacol. 2002 62:680)

訳 S期停止は，また，Chk1キナーゼ活性の増大と関連した

I

■ 名［形］＋ increase

❶	～-fold increase	～倍の増大	4,480
❷	significant increase	有意な増大	3,328
❸	～-dependent increase	～依存性の増大	1,182
❹	～-induced increase	～に誘導される増大	1,101
❺	marked increase	顕著な増大	970

❶ Six-month-old IR/IRS-1 and LIRKO mice both showed up to a 10-fold increase in β cell mass, which involved epithelial-to-mesenchymal transition. (Am J Psychiatry. 2003 160:290)

訳 …は，～の10倍までの増大を示した

❷ The dephosphorylation at Tyr-397 in FAK triggered by wild-type α-actinin and PTP 1B caused a significant increase in cell migration. (J Biol Chem. 2006 281:1746)

訳 ～は，細胞遊走の有意な増大を引き起こした

increasingly 副 ますます　用例数 4,198

■ increasingly ＋［形／過分］

❶ increasingly important	ますます重要な	392
❷ increasingly recognized	ますます認識される	334
・be increasingly recognized	…は，ますます認識される	260
❸ increasingly used	ますます使われる	222

❸ Lung transplantation is increasingly used as the treatment for many end-stage pulmonary diseases.（Transplantation. 2000 70:1599）

訳 肺移植が，～のための治療としてますます使われる

■［動］＋ increasingly

❶ be increasingly ～	…は，ますます～である	1,588
❷ become increasingly ～	…は，ますます～になる	871

❷ Haplotype analysis has become increasingly important for the study of human disease as well as for reconstruction of human population histories.（Am J Hum Genet. 2000 67:518）

訳 ハプロタイプ分析は，ヒト疾患の研究にとってますます重要になっている

incubate 動 インキュベートする／保温する　用例数 3,477

incubated	インキュベートされる／～をインキュベートした	3,114
incubating	～をインキュベートする	346
incubate	～をインキュベートする	14
incubates	～をインキュベートする	3

他動詞．受動態の用例が非常に多い．

◆類義語：culture, cultivate

■ be incubated ＋［前］

★ be incubated	…は，インキュベートされる	1,463
❶ be incubated with ～	…は，～とインキュベートされる	969
・cells were incubated with ～	細胞は，～とインキュベートされた	164
❷ be incubated in ～	…は，～においてインキュベートされる	241

❶ Prostate cancer, lymphoma, and leukemic cells were incubated with the drug for 4, 16, or 24 hours.（Cancer Res. 2011 71:4968）

訳 細胞は，その薬剤とインキュベートされた

incubation 名 インキュベーション／培養　用例数 5,402

incubation	インキュベーション	5,164
incubations	インキュベーション	238

複数形のincubationsの割合は約5％あるが，単数形は無冠詞で使われることが断然多い．

◆類義語：cultivation, culture

■ incubation ＋［前］

	the％	a/an％	⌀％			
❶	7	3	84	incubation of ～	～のインキュベーション	1,811
				・incubation of ～ with …	…と～のインキュベーション	827
				・incubation of cells	細胞のインキュベーション	139
❷	0	2	92	incubation with ～	～とのインキュベーション	1,233
❸	0	4	91	incubation in ～	～におけるインキュベーション	278
❹	0	5	89	incubation at ～	～におけるインキュベーション	226

❶ Incubation of cells with LPS and L-4F（1 to 50 μg/ml）reduced THP-1 adhesion in a concentration-dependent manner.（Circ Res. 2005 97:236）

訳 LPSおよびL-4F（1～50 μg/ml）と細胞のインキュベーションは，～を低下させた

❷ HCV replicon RNA could be fully cleared from replicon cells after prolonged incubation with R1479.（J Biol Chem. 2006 281:3793）

訳 R1479との長時間のインキュベーションのあと

■［前］＋ incubation

❶	by incubation with ～	～とのインキュベーションによって	205
❷	～ h of incubation	～時間のインキュベーション	168

❶ This activity was eliminated by incubation with the RANKL decoy receptor osteoprotegerin fusion protein.（J Immunol. 2006 176:625）

訳 この活性は，RANKLデコイ受容体オステオプロテジェリン融合タンパク質とのインキュベーションによって除去された

❷ These folate-QDs tend to accumulate in multi-vescicular bodies of KB cells after 6 h of incubation.（J Am Chem Soc. 2005 127:11364）

訳 6時間のインキュベーションのあと

I

indeed 副 実際に／確かに　用例数 4,579

◆同義語： actually, practically, really, in fact

■ indeed

❶	Indeed,	実際に，	2,538
❷	be indeed ～	…は，確かに～である	869

❶ Indeed, a single shuffled sequence could give rise to several prion variants.（Proc Natl Acad Sci USA. 2005 102:12825）

訳 実際に，

❷ Here we provide evidence that elimination of intracellular staphopain B activity is indeed the

function of SspC. (J Bacteriol. 2005 187:1751)
訳 細胞内staphopain B活性の除去は，確かにSspCの機能である

★★ independent 形 依存しない／非依存性の／独立した

用例数 43,432

independent ofの用例が非常に多い．
◆類義語：irrespective, unrelated, irrelevant

■ ［動］＋ independent ＋［前］

★ be independent	…は，依存しない	7,421
❶ be independent of ～	…は，～に依存しない	6,600
・be largely independent of ～	…は，概ね～に依存しない	202
❷ occur independent of ～	…は，～に依存せずに起こる	125
❸ be independent from ～	…は，～に依存しない（《非標準用法》：ofを使うのが一般的）	96

❶ This effect was independent of the presence of Trk because K252a did not prevent GM1-mediated release of neurotrophin-3. (J Biol Chem. 2002 277:49466)
訳 この効果は，～の存在に依存しなかった

■ independent ＋［名］

❶ ～-independent manner	～非依存性の様式	1,249
❷ ～-independent growth	～非依存性の増殖	1,044
❸ ～-independent mechanism	～非依存性の機構	789
❹ independent predictor of ～	～の独立予測因子	729

❶ The HDAC1 complex binds MDM2 in a p53-independent manner and deacetylates p53 at all known acetylated lysines in vivo. (EMBO J. 2002 21:6236)
訳 HDAC1複合体は，p53非依存性の様式でMDM2に結合する

■ ［名］-independent

❶ anchorage-independent ～	足場非依存性の～	1,034
・anchorage-independent growth	足場非依存性増殖	792
❷ ligand-independent ～	リガンド非依存性の～	526
❸ androgen-independent ～	アンドロゲン非依存性の～	492

❶ Many ligand-independent receptor tyrosine kinases are tumorigenic. (J Biol Chem. 2000 275:35328)
訳 多くのリガンド非依存性の受容体チロシンキナーゼは，腫瘍原性である

■ -dependent and -independent ＋［名］

★ ～-dependent and -independent …	～依存性および非依存性の…	868
❶ ～-dependent and -independent mechanisms	～依存性および非依存性の機構	219

❷ ～-dependent and -independent pathways	～依存性および非依存性の経路	135

❶ Thus, the Nodal pathway **regulates ventral forebrain patterning through both Hedgehog signaling-dependent and -independent mechanisms.** (Neuron. 2001 29:341)
訳 ～は，ヘッジホッグシグナル伝達依存性および非依存性の両方の機構によって腹側の前脳パターン形成を調節する

★ independently 副 独立して／無関係に　用例数 10,702

■［動］＋ independently of

★ independently of ～	～とは無関係に／～とは独立して	3,480
❶ occur independently of ～	…は，～とは無関係に起こる	524
❷ function independently of ～	…は，～とは独立して機能する	193
❸ act independently of ～	…は，～とは独立して作用する	150

❶ In particular, as demonstrated for the β_2AR, **this occurs independently of changes in GPCR kinase phosphorylation.** (J Biol Chem. 2006 281:2932)
訳 これは，GPCRキナーゼリン酸化の変化とは無関係に起こる

■ be independently ＋［過分］

★ be independently ～	…は，独立して～である	1,790
❶ be independently associated with ～	…は，独立して～と関連する	899
❷ be independently regulated	…は，独立して調節される	79

❶ **Lymphatic invasion, tumor size, and age were independently associated with lymph node metastases.** (Ann Surg. 1999 230:692)
訳 リンパ性浸潤，腫瘍サイズ，および年齢は，独立してリンパ節転移と関連した

■ independently ＋［動］

❶ independently predict ～	…は，独立して～を予測する	274

❶ **Body mass index (BMI) independently predicts mortality** in studies of HIV infected patients initiating antiretroviral therapy (ART). (PLoS One. 2011 6:e29625)
訳 ボディ・マス・インデックス(BMI)は，独立して死亡率を予測する

★★ indicate 動 示す　用例数 112,539

indicate	～を示す	50,388
indicating	～を示す／～を示している	28,079
indicated	示される／～を示した	20,129
indicates	～を示す	13,943

他動詞．indicate thatの用例が圧倒的に多い．
◆類義語：show, present, exhibit, display, represent, point

I

■ indicate ＋［名節］

❶ indicate that ～	…は，～ということを示す	87,094
❷ indicate how ～	…は，どのように～かを示す	101

❶ These results indicate that reduced expression and increased phosphorylation of phospholamban provides compensation for decreased SERCA2 protein levels in heterozygous heart.（J Biol Chem. 2000 275:38073）

訳 これらの結果は，～ということを示す

■ indicate ＋［名句］

❶ indicate the presence of ～	…は，～の存在を示す	866
❷ indicate a role for ～	…は，～の役割を示す	510
❸ indicate the importance of ～	…は，～の重要性を示す	312
❹ indicate the existence of ～	…は，～の存在を示す	290
❺ indicate the involvement of ～	…は，～の関与を示す	181
❻ indicate a requirement for ～	…は，～の必要性を示す	130

❶ Together, these results indicate the presence of a delicate balance between Wnt/beta-catenin and Shh signaling mechanisms in the progression from progenitors to dopamine neurons.（J Neurosci. 2010 30:9280）

訳 これらの結果は，～の間のデリケートなバランスの存在を示す

■［名］＋ indicate

❶ results indicate ～	結果は，～を示す	16,677
❷ data indicate ～	データは，～を示す	8,989
❸ findings indicate ～	知見は，～を示す	4,810
❹ studies indicate ～	研究は，～を示す	3,425
❺ analysis indicated ～	分析は，～を示した	1,507
❻ evidence indicates ～	証拠は，～を示す	1,258
❼ observations indicate ～	観察は，～を示す	950
❽ experiments indicate ～	実験は，～を示す	918

❷ These data indicate that Merkel cells are poised to release glutamate and neuropeptides.（Proc Natl Acad Sci USA. 2004 101:14503）

訳 これらのデータは，～ということを示す

■ indicated ＋［前］

❶ indicated by ～	～によって示される	2,080
・as indicated by ～	～によって示されるように	1,433

❶ The ashen strain only has a defect in secretion, as indicated by retention of melanosomes in melanocytes.（J Invest Dermatol. 2004 122:452）

訳 メラノサイトにおけるメラノソームの貯留によって示されるように

■ , indicating

★	, indicating ～	そして，（それは）～を示している	24,836
❶	, indicating that ～	そして，（それは）～ということを示している	17,305

❶ This enhancement is reduced by the addition of a large excess of Ca(II), **indicating that** these ions bind in the active site.（Biochemistry. 2004 43:15286）
訳 そして，（それは）～ということを示している

indication 名 徴候／適応症／指示 用例数 2,440

indications	徴候／適応症／指示	1,294
indication	徴候／適応症／指示	1,146

複数形のindicationsの割合は約55%と非常に高い.
◆類義語：evidence, demonstration, representation, symptom, sign

■ indication ＋［前／同格that節］

	the %	a/an %	ø %			
❶	15	55	5	indication of ～	～の徴候	414
❷	50	26	15	indication for ～	～の適応症	294
❸	22	54	3	indication that ～	～という徴候	208

❶ To obtain an independent, physiological **indication of** the stress produced by several tests, we measured changes in heart rate using telemetry.（J Neurosci. 2003 23:6255）
訳 いくつかのテストによって生み出されたストレスの独立した生理的な兆候を得るために

❷ Symptomatic osteoarthritis is the **indication for** surgery in more than 90% of patients, and its incidence is increasing because of an ageing population and the obesity epidemic.（Lancet. 2012 380:1768）
訳 症候性変形性関節症は，手術の適応症である

❸ There was no **indication that** the T15-Id^- B cells either proliferated or differentiated into Ab-secreting cells following immunization.（J Immunol. 2002 168:1273）
訳 ～という徴候はなかった

indicative 形 示している 用例数 2,381

indicative ofの用例が圧倒的に多い.

■ indicative ＋［前］

❶	indicative of ～	～を示している	2,353
	・be indicative of ～	…は，～を示している	647

❶ This behavior **is indicative of** reversible hydrogen bonding between nitrobenzene radical anions and arylureas.（J Am Chem Soc. 2005 127:6423）
訳 この挙動は，～の間の可逆的水素結合を示している

indirect ㊎ 間接の／間接的な 用例数 4,543

◆反意語：direct

■ indirect ＋［名］

❶	indirect immunofluorescence	間接免疫蛍光	489
❷	indirect evidence	間接的な証拠	258
❸	indirect effects	間接的な効果	215
❹	indirect pathway	間接的な経路	190

■ direct and/or indirect

❶	direct and indirect ～	直接および間接の～	399
❷	direct or indirect ～	直接あるいは間接の～	194

❶ First, many angiogenic factors are now known to exert both direct and indirect effects on bone and cartilage formation. (Circ Res. 2005 96:930)

訳 多くの血管新生因子が，今，骨および軟骨形成に対して直接および間接の両方の効果を発揮することが知られている

❷ SPNs are subject to control from both supraspinal and spinal inputs that exert effects through activation of direct or indirect pathways. (J Neurosci. 2005 25:1063)

訳 直接あるいは間接の経路の活性化によって効果を発揮する

★★ individual ㊔ 個人／個々，㊎ 個々の 用例数 45,941

individual	個々の～／個人	25,048
individuals	個々人／人々	20,893

名詞および形容詞として用いられる．名詞としては，複数形のindividualsの割合は，約90%と圧倒的に高い．

◆類義語：each, respective, single, private

■ ㊔ individuals ＋［前］

	the%	a/an%	ø%			
❶	0	-	77	individuals with ～	～を持つ個々人	4,382
❷	2	-	64	individuals in ～	～における個々人	696
❸	2	-	41	individuals from ～	～からの個々人	583

❶ The correlation was slightly lower in individuals with decreased MMSE scores. (Invest Ophthalmol Vis Sci. 2002 43:2572)

訳 相関は，低下したMMSEスコアを持つ個々人においてわずかにより低かった

■ ㊔［過分／形］＋ individuals

❶	infected individuals	感染した個々人	1,136
❷	affected individuals	発症した個々人	803

❸ healthy individuals	健康な個々人	699

■ 形 individual ＋［名］

❶ individual cells	個々の細胞	731
❷ individual patients	個々の患者	451
❸ individual differences	個々の違い	441
❹ individual genes	個々の遺伝子	361

❸ Analysis of individual differences provides a new window into the neurobiology of emotion processing that complements traditional approaches.（Curr Opin Neurobiol. 2004 14:233）

訳 個々の違いの分析は，～を提供する

★ induce 動 誘導する／誘発する　用例数 190,482

induced	誘導される／～を誘導した	140,261
induce	～を誘導する	23,644
induces	～を誘導する	17,582
inducing	～を誘導する	8,995

他動詞.

◆類義語：evoke, provoke, elicit, cause, increase

■ induce ＋［名句］

❶ induce apoptosis	…は，アポトーシスを誘導する	2,795
❷ induce the expression of ～	…は，～の発現を誘導する	1,018
❸ induce the formation of ～	…は，～の形成を誘導する	445
❹ induce a conformational change	…は，立体構造の変化を誘導する	258
❺ induce an increase in ～	…は，～の増大を誘導する	244
❻ induce phosphorylation of ～	…は，～のリン酸化を誘導する	221
❼ induce the production of ～	…は，～の産生を誘導する	208

❶ The mechanism by which PDE4 inhibitors induce apoptosis in CLL remains unknown.（Blood. 2003 101:4122）

訳 PDE4 抑制剤がCLLのアポトーシスを誘導する機構は，未知のままである

❷ In turn, KLF2 induces the expression of CRABP-II and RARγ, further potentiating inhibition of adipocyte differentiation by RA.（Diabetes. 2012 61:1112）

訳 KLF2 は，CRABP-IIおよびRARγ の発現を誘導する

■［名／動］＋ to induce

★ to induce ～	～を誘導する…	11,087
❶ ability to induce ～	～を誘導する能力	918
❷ be sufficient to induce ～	…は，～を誘導するのに十分である	901
❸ fail to induce ～	…は，～を誘導できない	844
❹ be shown to induce ～	…は，～を誘導することが示される	508

❷ These experiments indicate that a period as short as one-third the period of gestation **is sufficient to induce protection against** mammary carcinogenesis. (Proc Natl Acad Sci USA. 2001 98:11755)

訳 …は，～からの保護を誘導するのに十分である

❸ **TNF-α failed to induce ERK1/2 activation in** the actinomycin D-treated cells. (J Immunol. 1999 162:3672)

訳 TNF-αは，～におけるERK1/2活性化を誘導できなかった

■ be induced ＋［前］

★ be induced	…は，誘導される	10,641
❶ be induced by ～	…は，～によって誘導される	4,376
❷ be induced in ～	…は，～において誘導される	2,445
❸ be induced to *do*	…は，～するように誘導される	625

❶ Whereas **spi2a was induced by either bacterial products or IFN-γ**, ISG15 was induced only by bacterial products. (J. Immunol. 2002 168:2415)

訳 spi2aは，細菌性の産物かIFN-γのどちらかによって誘導された

❷ **EAE/AU was induced in Lewis rats** with myelin basic protein in complete Freund's adjuvant (CFA). (Invest Ophthalmol Vis Sci. 2001 42:2894)

訳 EAE/AU は，Lewisラットにおいて誘導された

❸ As **C2C12 cells are induced to differentiate into permanently arrested myotubes**, the abundance of the p18(L) transcript decreases. (Mol Cell Biol. 1998 18:2334)

訳 C2C12細胞は，永久に停止した筋管に分化するように誘導される

■［名］＋ induced by

★ induced by ～	～によって誘導される	10,641
❶ apoptosis induced by ～	～によって誘導されるアポトーシス	1,233
❷ activation induced by ～	～によって誘導される活性化	369
❸ changes induced by ～	～によって誘導される変化	313
❹ cell death induced by ～	～によって誘導される細胞死	294

❶ C33 and mC5 cells **were also more resistant to apoptosis induced by H_2O_2** and to the loss of mitochondrial membrane potential induced by H_2O_2 and antimycin A. (J Biol Chem. 1999 274: 26217)

訳 ～は，また，H_2O_2によって誘導されるアポトーシスにもっと抵抗性であった

■［名］-induced

❶ stress-induced ～	ストレスに誘導される～	2,071
❷ radiation-induced ～	放射線に誘導される～	1,093
❸ agonist-induced ～	作用物質に誘導される～	1,009
❹ ligand-induced ～	リガンドに誘導される～	961
❺ virus-induced ～	ウイルスに誘導される～	909
❻ drug-induced ～	薬剤に誘導される～	887

❶ The MAPK kinase MKK6 selectively stimulates p38 MAPK and **confers protection against stress-induced apoptosis** in cardiac myocytes. (J Biol Chem. 2000 275:23825)

訳 ～は，ストレスに誘導されるアポトーシスに対する保護を与える

■ induced ＋［名］

❶ …-induced apoptosis	…に誘導されるアポトーシス	6,258
❷ …-induced increase in ～	…に誘導される～の増大	1,751
❸ …-induced activation of ～	…に誘導される～の活性化	1,714
❹ …-induced cell death	…に誘導される細胞死	1,507
❺ …-induced changes in ～	…に誘導される～の変化	1,029
❻ …-induced phosphorylation of ～	…に誘導される～のリン酸化	903

❶ IC486068 enhanced radiation-induced apoptosis in endothelial cells and reduced cell migration and tubule formation of endothelial cells in Matrigel following irradiation.（Cancer Res. 2004 64:4893）

訳 IC486068は，放射線に誘導される内皮細胞のアポトーシスを増強した

❸ Taken together, these data demonstrate that CD3/CD28-induced activation of IKKβ and expression of Bcl-xL promote the survival of primary human $CD4^+$ T lymphocytes.（J Immunol. 2000 165:1743）

訳 CD3/CD28に誘導されるIKKβの活性化およびBcl-xLの発現は，～を促進する

★ inducible 形 誘導性の／誘導できる　用例数 11,107

◆類義語：inductive, provocative, regulatory

■ inducible ＋［名］

❶ inducible expression	誘導性の発現	498
❷ inducible genes	誘導性の遺伝子	380
❸ inducible promoter	誘導性のプロモーター	283

❶ Class II transactivator（CIITA）is required for both constitutive and inducible expression of MHC class II genes.（Oncogene. 2001 20:4219）

訳 …は，～の構成的発現および誘導性発現の両方のために必要とされる

■［名］-inducible

❶ interferon-inducible ～	インターフェロン誘導性の～	229
❷ stress-inducible ～	ストレス誘導性の～	209

■ inducible ＋［前］

❶ inducible by ～	～によって誘導できる	293
・be inducible by ～	…は，～によって誘導できる	178
❷ inducible in ～	～において誘導できる	130

❶ The UGT1A1 gene has been shown to be inducible by nuclear receptors PXR and CAR.（J Biol Chem. 2003 278:15001）

訳 UGT1A1遺伝子は，核内受容体PXRおよびCARによって誘導できることが示されている

I

★ induction ㊔ 誘導／誘発 用例数 37,728

induction	誘導	37,629
inductions	誘導	99

複数形のinductionsの割合は0.3%しかなく，原則，**不可算名詞**として使われる．

◈類義語：derivation, elicitation, facilitation

■ induction ＋［前］

	the %	a/an %	ø %			
❶	34	4	62	induction of ～	～の誘導	23,325
				・induction of apoptosis	アポトーシスの誘導	1,378
				・…-mediated induction of ～	…に仲介される～の誘導	649
				・…-dependent induction of ～	…依存的な～の誘導	364
				・lead to the induction of ～	…は，～の誘導につながる	160
				・be required for the induction of ～	…は，～の誘導のために必要とされる	144
❷	2	0	38	induction by ～	～による誘導	1,421
❸	1	2	43	induction in ～	～における誘導	1,330

❶ Use of specific caspase inhibitors revealed that **the induction of apoptosis is caspase 8 dependent**, but caspase 3 independent. (J Immunol. 2000 165:2500)

訳 アポトーシスの誘導は，カスパーゼ8依存的である

❶ Transfection of antisense Bcl-2 into HDMECs **blocked VEGF-mediated induction of IL-8**. (Cancer Res. 2001 61:2183)

訳 ～は，IL-8のVEGFに仲介される誘導を阻止した

❶ Here we provide genetic evidence that **Dif is required for the induction of** only a subset of antimicrobial peptide genes. (Genes Dev. 1999 13:792)

訳 Difは，～の誘導のために必要とされる

■［名］＋ induction

❶	gene induction	遺伝子誘導	526
❷	tolerance induction	耐性誘導	505

★ infect ㊍ 感染させる 用例数 35,924

infected	感染した／～を感染させた	32,669
infect	～を感染させる	1,655
infecting	～を感染させる	863
infects	～を感染させる	737

他動詞．受動態，特にinfected withの用例が多い．

■ be infected ＋［前］

★ be infected	…が，感染させられる	6,803
❶ be infected with ～	…が，～を感染させられる	1,144
・mice were infected with ～	マウスが，～を感染させられた	155
・cells were infected with ～	細胞が，～を感染させられた	134
❷ be infected by ～	…が，～を感染させられる	150

❶ 293A **cells were infected with** adenovirus-doxycycline (Ad-Dox)-inducible hBVR cDNA. (J Biol Chem. 2004 279:19916)

訳 293A細胞が，～を感染させられた

■［副］＋ infected with

★ infected with ～	～に感染した	6,803
❶ chronically infected with ～	～に慢性的に感染した	156
❷ latently infected with ～	～に潜伏感染した	121
❸ experimentally infected with ～	～に実験的に感染した	97

❶ In this study, we have observed that sera from patients **chronically infected with** hepatitis C virus (HCV) displayed significantly lower C3 levels than sera from healthy individuals. (J Virol. 2012 86:2221)

訳 C型肝炎ウイルスに慢性的に感染した患者

■［名］＋ infected with

❶ cells infected with ～	～に感染した細胞	1,208
❷ mice infected with ～	～に感染したマウス	907
❸ patients infected with ～	～に感染した患者	321

❶ In **cells infected with** either the recombinant adenovirus-HBV or baculovirus-WHV, the replication level of the X-negative construct was about 10% of that of the wild-type virus. (J Virol. 2004 78:4566)

訳 ～に感染した細胞において

■ infected ＋［名］

❶ infected cells	感染した細胞	4,705
・latently infected cells	潜在的に感染した細胞	1,208
❷ infected mice	感染したマウス	1,551
❸ infected patients	感染した患者	1,358
❹ infected individuals	感染した個々人	1,137

■［名］-infected

❶ HIV-infected ～	HIV感染した～	2,674
❷ virus-infected ～	ウイルス感染した～	997

❶ Human immunodeficiency virus (HIV) RNA testing is the gold standard for monitoring antiretroviral therapy in HIV-infected patients.（J Clin Microbiol. 2005 43:5950）

訳 ～は，HIV感染した患者において抗レトロウイルス剤療法をモニターするための判断基準である

■ to infect

★	to infect ～	～を感染させる…	913
❶	ability to infect ～	～を感染させる能力	101

❶ Canine parvovirus（CPV）and feline panleukopenia virus（FPV）differ in their ability to infect dogs and dog cells.（J Virol. 2003 77:12211）

訳 …は，～を感染させるそれらの能力において異なる

★★ infection 名 感染　　用例数 79,581

infection	感染	66,021
infections	感染	13,560

複数形のinfectionsの割合は約15%あるが，単数形は無冠詞で使われることが断然多い．

◆類義語：transmission

■ infection ＋［前］

infection ofの後には宿主，infection withの後には病原体などが用いられることが多い．

	the %	a/an %	ø %			
❶	5	2	73	infection of ～	～の感染	5,264
				・infection of ～ with …	～の…による感染	657
❷	1	1	97	infection with ～	～による感染	4,979
❸	3	11	68	infection in ～	～における感染	4,587
❹	0	2	86	infection by ～	～による感染	2,152

❶ The increase in activity was induced by infection of myocytes with a recombinant adenovirus（AdPak1）containing cDNA for a constitutively active Pak1.（Circ Res. 2004 94:194）

訳 ～は，筋細胞の組換え型アデノウイルスによる感染によって誘導された

❷ Here we resolve this paradox by studying the early immune response in mice after infection with different doses of Leishmania major.（J Exp Med. 2004 199:1559）

訳 異なる用量の森林型熱帯リーシュマニアによる感染のあと

❸ Human β-defensins 2 and 3（HBD-2 and HBD-3）are inducible peptides present at sites of infection in the oral cavity.（J Clin Microbiol. 2004 42:1024）

訳 ～は，口腔における感染の部位に存在する誘導性ペプチドである

❹ When treated with dsRNA, shrimp showed increased resistance to infection by two unrelated viruses, white spot syndrome virus and Taura syndrome virus.（J Virol. 2004 78:10442）

訳 エビは，2つの無関係のウイルスによる感染への増大した抵抗性を示した

■［前］＋ infection

❶ response to infection	感染に対する応答	652
❷ site of infection	感染の部位	420
❸ course of infection	感染のコース	333
❹ ～ days after infection	感染の後～日目	263
❺ susceptibility to infection	感染に対する感受性	232
❻ susceptible to infection	感染しやすい	224

❸ The dynamics of hepatitis C virus（HCV）quasi species in the E2 region **may correlate with the course of infection** after orthotopic liver transplantation（OLT）.（J Infect Dis. 2004 189: 2037）

訳 ～は，感染のコースと相関するかもしれない

❹ Blood chemistry appeared to be normal with exception of creatine phosphokinase, which **peaked at 7 days after infection**.（Blood. 2010 115:1823）

訳（そして，それは）感染の後７日目にピークになった

❺ Thus, costimulatory molecule function **is critical in determining the initial susceptibility to infection** with P. carinii.（J Immunol. 2003 171:1969）

訳 ～は，感染に対する初期の感受性を決定する際に決定的に重要である

■［名／形］＋ infection

❶ HIV infection	HIV感染	2,565
❷ virus infection	ウイルス感染	2,124
❸ viral infection	ウイルス感染	1,968
❹ chronic infection	慢性感染	761

infer ㊍ 推論する／推測する　用例数 3,444

inferred	推論される／～を推論した	1,782
infer	～を推論する／～を推測する	1,276
inferring	～を推論する／～を推測する	329
infers	～を推論する／～を推測する	57

他動詞.

◆類義語：assume, presume, estimate, speculate, predict, expect, postulate, deduce

■ infer ＋［that節］

❶ infer that ～	…は，～ということを推論する	439
・we infer that ～	われわれは，～ということを推論する	331

❶ **We infer that** these inconsistencies result from the processing used to match gene products to reactions within KEGG's metabolic pathways.（Nucleic Acids Res. 2005 33:4035）

訳 われわれは，～ということを推論する

I

■ ［動］＋ to infer

★ to infer ～	～を推論する…	654
❶ be used to infer ～	…は，～を推論するために使われる	119

❶ Furthermore, **our model can be used to infer interactions** not only for single-domain pairs but also for multiple domain pairs. (Bioinformatics. 2005 21:4394)

訳 われわれのモデルは，相互作用を推論するために使われうる

■ be inferred ＋［前］

★ be inferred	…は，推論される	696
❶ be inferred from ～	…は，～から推論される	291
❷ be inferred to *do*	…は，～すると推測される	100

❶ Many structural characteristics of the receptor **can be inferred from the structural and dynamical features** identified in this study. (Biochemistry. 2005 44:8790)

訳 ～は，構造的および動力学的特徴から推論されうる

infiltration ㊔ 浸潤　　用例数 3,776

infiltration	浸潤	3,761
infiltrations	浸潤	15

複数形のinfiltrationsの割合は0.4%しかなく，原則，**不可算名詞**として使われる．

◆類義語：invasion

■ infiltration ＋［前］

	the%	a/an%	ø%			
❶	15	6	79	infiltration of ～	～の浸潤	1,006
				・infiltration of ～ into …	～の…への浸潤	98
❷	2	1	95	infiltration in ～	～における浸潤	299
❸	0	2	98	infiltration into ～	～への浸潤	253

❶ **Reperfusion after stroke leads to infiltration of inflammatory cells into the ischemic brain.** (Ann Neurol. 2011 70:606)

訳 脳卒中のあとの再灌流は，炎症細胞の虚血脳への浸潤につながる

❷ **Neutrophil infiltration in Mac-1 KO mice was severely impaired** at 24 hours. (Blood. 2006 107:1063)

訳 Mac-1ノックアウトマウスにおける好中球浸潤が，激しく損なわれた

❸ **Shh also induced keratinocyte infiltration into the underlying collagen matrix.** (J Invest Dermatol. 2005 124:457)

訳 Shhは，また，下にあるコラーゲン基質へのケラチノサイト浸潤を誘導した

■ ［名］＋ infiltration

❶ cell infiltration	細胞浸潤	583

❷ neutrophil infiltration	好中球浸潤	409
❸ macrophage infiltration	マクロファージ浸潤	298

★ inflammation 名 炎症　用例数 20,523

inflammation	炎症	20,507
inflammations	炎症	16

複数形のinflammationsの用例はほとんどなく，原則，不可算名詞として使われる．

■ inflammation ＋［前］

	the%	a/an%	ø%			
❶	1	0	99	inflammation in ～	～における炎症	1,867

❶ Our findings indicate that **TRPV1 has a role in acute and chronic inflammation in the mouse knee joint.** (Arthritis Rheum. 2005 52:3248)
訳 TRPV1は，マウスの膝関節における急性および慢性の炎症において役割を担う

■ ［前］＋ inflammation

❶ sites of inflammation	炎症の部位	384
❷ resolution of inflammation	炎症の消散	153
❸ markers of inflammation	炎症のマーカー	142

❶ Monocyte-macrophage chemoattractant protein 1 (MCP-1) **recruits macrophages to sites of inflammation.** (Infect Immun. 2005 73:5735)
訳 ～は，マクロファージを炎症の部位に動員する

■ ［名／形］＋ inflammation

❶ airway inflammation	気道炎症	738
❷ chronic inflammation	慢性炎症	842
❸ lung inflammation	肺の炎症	494
❹ systemic inflammation	全身性炎症	473

★★ influence 名 影響，動 影響する　用例数 30,326

influence	影響／～に影響する	17,779
influences	影響／～に影響する	5,604
influenced	影響される／～に影響した	5,112
influencing	～に影響する	1,831

名詞および他動詞として用いられる．〈influence of〉〈influence on〉の用例が多い．複数形のinfluencesの割合が，約30%ある可算名詞．

◆類義語：effect, affect, impact

I

■ ㊔ influence ＋［前］

influence ofの前にtheが付く割合は圧倒的に高い.

	the %	a/an %	ø %			
❶	93	3	3	influence of ～	～の影響	4,277
				・influence of ～ on …	…に対する～の影響	1,316
				・examine the influence of ～	…は，～の影響を調べる	296
				・investigate the influence of ～	…は，～の影響を精査する	236
				・determine the influence of ～	…は，～の影響を決定する	136
				・study the influence of ～	…は，～の影響を研究する	120
				・assess the influence of ～	…は，～の影響を評価する	112
❷	12	45	14	influence on ～	～に対する影響	1,890
				・no influence on ～	～に対する影響のない	143
				・significant influence on ～	～に対する有意な影響	106
				・little influence on ～	～に対する影響のほとんどない	93

❶ In the present investigation, **we examined the influence of glutamate application on glutamine, serine and aspartate release** from rat cortical glial cultures.（Brain Res. 2003 978: 213）

訳 われわれは，グルタミン，セリンおよびアスパラギン酸放出に対するグルタミン酸適用の影響を調べた

❷ **Paroxetine had no influence on fatigue in patients receiving chemotherapy.**（J Clin Oncol. 2003 21:4635）

訳 パロキセチンは，化学療法を受けている患者における疲労に対して影響を持たなかった

■ ㊍ influence ＋［名句］

❶	influence the development of ～	…は，～の発生に影響する	173
❷	influence the expression of ～	…は，～の発現に影響する	151
❸	influence the rate of ～	…は，～の速度に影響する	107
❹	influence the outcome of ～	…は，～の結果に影響する	100

❶ To develop a better mechanistic understanding of how viruses **can influence the development of autoimmune disease**, we exposed prediabetic mice to various viral infections.（J Clin Invest. 2004 113:74）

訳 ～は，自己免疫疾患の発症に影響しうる

■ ㊍［動］＋ to influence

★	to influence ～	～に影響する…	1,719
❶	appear to influence ～	…は，～に影響するようである	147

❷ be shown to influence ～	…は，～に影響することが示される	115
❸ be known to influence ～	…は，～に影響することが知られている	92

❸ Genetics and environmental conditions early in life **are known to influence** height.（Am J Clin Nutr. 2004 80:185）

訳 ～は，身長に影響することが知られている

■ 動 influenced ＋［前］

❶ influenced by ～	～によって影響される	3,708
・be influenced by ～	…は，～によって影響される	2,076
・strongly influenced by ～	～によって強く影響される	336
・significantly influenced by ～	～によって顕著に影響される	108

❶ The activities of PML-RAR α in early myeloid cells are therefore **strongly influenced by** the presence of neutrophil elastase.（Mol Cell Biol. 2005 25:23）

訳 ～は，それゆえ，好中球エラスターゼの存在によって強く影響される

influx 名 流入　　用例数 4,563

influx	流入	4,524
influxes	流入	39

複数形のinfluxesの割合は0.9%しかないが，単数形には不定冠詞が使われることもある．

◆類義語：inflow, entry

■ influx ＋［前］

	the%	a/an%	ø%			
❶	33	39	28	influx of ～	～の流入	755
❷	0	2	35	influx through ～	～を通した流入	393
❸	2	7	70	influx in ～	～における流入	300
❹	3	3	66	influx into ～	～への流入	217

❶ Osmotic down shock causes an immediate **influx of** Ca^{2+} in yeast, likely through a membrane stretch-sensitive channel.（FASEB J. 2007 21:1813）

訳 浸透圧ダウンショックは，Ca^{2+}の即時の流入を引き起こす

❷ Our hypothesis is that persistent increases in Ca^{2+} **influx through** the LTCC cause apoptosis if the excessive influx results in SR Ca^{2+} overload.（Circ Res. 2005 97:1009）

訳 LTCCを通したCa^{2+}流入の持続性の増大は，アポトーシスを引き起こす

❹ Reduced excitability was achieved by adaptation of all transporters to reduce Ca^{2+} **influx into** the cytosol.（J Biol Chem. 2004 279:41642）

訳 サイトゾルへのCa^{2+}流入を低下させる

■［名］＋ influx

❶ Ca^{2+} influx	Ca^{2+}の流入	950
❷ neutrophil influx	好中球の流入	196

I

★ information ㊔ 情報 用例数 29,634

複数形の用例は全くなく，原則，不可算名詞として使われる.

■ information ＋［前］

	the％	a/an％	⌀％			
❶	4	0	96	information about ～	～に関する情報	3,223
❷	1	0	96	information on ～	～に関する情報	3,127
❸	3	0	94	information from ～	～からの情報	1,498

❶ Transgenic mice have provided invaluable information about gene function and regulation. (Proc Natl Acad Sci U S A. 2001 98:10728)

訳 トランスジェニックマウスは，遺伝子機能に関する貴重な情報を提供してきた

❷ Structural information on the MotA/MotB complex is very limited. (Biochemistry. 2001 40:13051)

訳 MotA/MotB複合体に関する構造的な情報は，非常に限られている

❸ We obtained peptide sequence information from the 90-kDa subunit (GTC-90) that allowed us to identify a number of GTC-90 cDNAs. (J Biol Chem. 1998 273:29565)

訳 われわれは，90キロダルトンのサブユニットからペプチド配列情報を得た

■［形／名］＋ information

❶ structural information	構造的な情報	937
❷ new information	新しい情報	607
❸ sequence information	配列情報	550
❹ little information	ほとんど情報のない	499
・little information is available	利用できる情報はほとんどない	151
・there is little information	ほとんど情報はない	153
❺ genetic information	遺伝的情報	423
❻ sensory information	感覚情報	377
❼ positional information	位置情報	347
❽ visual information	視覚情報	322

❹ Little information is available concerning the presence of MDC in the kidney or their role in renal pathophysiology. (J Am Soc Nephrol. 2000 11:595)

訳 ～におけるMDCの存在に関して利用できる情報はほとんどない

■［動］＋ information

❶ provide information	…は，情報を提供する	888
❷ obtain information	…は，情報を得る	136

infuse ㊍ 注入する　用例数 1,996

infused	注入される／～を注入した	1,848
infusing	～を注入する	126
infuse	～を注入する	20
infuses	～を注入する	2

他動詞．受動態の用例が多い．

◆類義語：inject, transfuse

■ infused ＋［前］

❶ infused with ～	～を注入される	344
❷ infused into ～	～へ注入される	275

❹ Six-month-old apolipoprotein E-deficient mice were infused with Ang II (1.44 mg/kg/d) for 1 month. (Circulation. 2005 111:2219)

訳 6ヶ月齢のアポリポタンパクE欠損マウスが，Ang Ⅱを注入された

❹ This phenomenon may be of clinical relevance because cold, oxygen-rich solutions are often infused into the eye during intraocular surgery. (J Physiol. 2004 559:883)

訳 冷たい酸素に富んだ溶液が頻繁に目に注入される

inherent ㊎ 固有の／本来備わっている　用例数 2,113

◆類義語：intrinsic, endogenous, endemic

■ inherent ＋［前］

❶ inherent in ～	～における固有の	334
❷ inherent to ～	～に固有の	197

❶ Despite the intricacies inherent in these successive transformations, malaria parasites remain staggeringly successful at disseminating through their vertebrate host populations. (Trends Parasitol. 2005 21:573)

訳 これらの連続的な形質転換における固有の複雑さにもかかわらず

❷ We suggest that models of Era function should reflect the rapid exchange of nucleotides in addition to the GTPase activity inherent to Era. (J Bacteriol. 2000 182:3460)

訳 Eraに固有のGTP加水分解酵素活性に加えて

★★ inhibit ㊍ 抑制する／阻害する　用例数 80,518

inhibited	抑制される／～を抑制した	37,286
inhibit	～を抑制する／～を阻害する	18,804
inhibits	～を抑制する／～を阻害する	16,549
inhibiting	～を抑制する／～を阻害する	7,879

他動詞．

◆類義語：suppress, repress, block, abrogate, prevent, attenuate, interfere

■ inhibit ＋［名句］

❶	inhibit the growth of ～	…は，～の増殖を抑制する	693
❷	inhibit the activity of ～	…は，～の活性を抑制する	485
❸	inhibit apoptosis	…は，アポトーシスを抑制する	478
❹	inhibit the binding of ～	…は，～の結合を抑制する	408
❺	inhibit the expression of ～	…は，～の発現を抑制する	371
❻	inhibit the ability of ～	…は，～の能力を抑制する	339
❼	inhibit cell proliferation	…は，細胞増殖を抑制する	334
❽	inhibit transcription	…は，転写を抑制する	290
❾	inhibit the activation of ～	…は，～の活性化を抑制する	273

❶ **VES has been shown to inhibit the growth of a wide variety of tumor cells** in cell culture and animal models.（Cancer Res. 2001 61:6569）

訳 VESは，さまざまな腫瘍細胞の増殖を抑制することが示されている

❸ Our results indicate that **genetic changes that inhibit apoptosis** can cooperate with PMLRAR α to initiate APL.（J Exp Med. 2001 193:531）

訳 アポトーシスを抑制する遺伝的変化

■［名／動］＋ to inhibit

★	to inhibit ～	～を抑制する…	2,349
❶	ability to inhibit ～	～を抑制する能力	880
❷	be shown to inhibit ～	…は，～を抑制することが示される	499
❸	fail to inhibit ～	…は，～を抑制することができない	356
❹	be found to inhibit ～	…は，～を抑制することが見つけられる	262

❶ CcpA has different affinities for different cres, but **this does not correlate with its ability to inhibit transcription.**（Mol Microbiol. 2005 56:155）

訳 これは，転写を抑制するそれの能力と相関しない

❷ **Several antiangiogenic factors have been shown to inhibit tumor growth in animal models.**（Cancer Res. 2005 65:7984）

訳 いくつかの抗血管新生因子は，動物モデルにおける腫瘍の増殖を抑制することが示されている

■［副］＋ inhibit

❶	potently inhibit ～	…は，～を強力に抑制する	407
❷	selectively inhibit ～	…は，～を選択的に抑制する	397
❸	specifically inhibit ～	…は，～を特異的に抑制する	388
❹	significantly inhibit ～	…は，～を有意に抑制する	319
❺	directly inhibit ～	…は，～を直接抑制する	316

❸ In fact, overexpression of mCycT1, but not hCycT1, **specifically inhibits Tat-TAR function in human cells.**（EMBO J. 1998 17:7056）

訳 ～は，ヒト細胞におけるTat-TAR機能を特異的に抑制する

■ [名] + be inhibited

★ be inhibited	…が，抑制される／阻害される	9,782
❶ activity was inhibited	活性が，阻害された	259
❷ binding was inhibited	結合が，抑制された	87
❸ expression was inhibited	発現が，抑制された	81

❶ The catalytic activity was inhibited by an electrophilic phosphonate diester, consistent with a nucleophilic catalytic mechanism. (J Biol Chem. 2012 287:36096)

訳 触媒活性が，求電子性ホスホネートジエステルによって阻害された

■ be inhibited + [前]

❶ be inhibited by ～	…が，～によって抑制される	6,561
❷ be inhibited in ～	…が，～において抑制される	743
❸ be inhibited with ～	…が，～によって抑制される	194

❶ The excitatory effect of DAMGO was also inhibited by pretreatment with pertussis toxin. (Mol Pharmacol. 2003 63:89)

訳 …が，また，～による前処置によって抑制された

❷ Cell-cycle progression was inhibited in excision repair-defective rad1 mutants, but not in rad2 cells, indicating a role for Rad1 processing of the DSB ends. (Genetics. 1999 152:1513)

訳 細胞周期進行が，除去修復欠損rad1変異体において抑制された

■ [副] + inhibited

❶ significantly inhibited ～	有意に抑制される／～を有意に抑制した	1,851
❷ completely inhibited ～	完全に抑制される／～を完全に抑制した	975
❸ strongly inhibited ～	強く抑制される／～を強く抑制した	925
❹ potently inhibited ～	強力に抑制される／～を強力に抑制した	814
❺ specifically inhibited ～	特異的に抑制される／～を特異的に抑制した	782
❻ selectively inhibited ～	選択的に抑制される／～を選択的に抑制した	697
❼ partially inhibited ～	部分的に抑制される／～を部分的に抑制した	665

❶ Proliferation of human epithelial cells is significantly inhibited by a function-perturbing alpha3 integrin antibody. (Mol Biol Cell. 1999 10:259)

訳 ヒト上皮細胞の増殖は，～によって有意に抑制される

❷ Receptor internalization was completely inhibited by truncation of the C terminus. (Mol Biol Cell. 1999 10:1179)

訳 受容体のインターナリゼーションは，C末端の切断によって完全に抑制された

❼ Additionally, HDL-mediated cholesterol efflux was partially inhibited by filipin and sphingomyelinase treatment. (J Biol Chem. 2001 276:3158)

訳 ～は，フィリピンとスフィンゴミエリナーゼ処理によって部分的に抑制された

☆☆ inhibition ㊔ 抑制／阻害 用例数 61,633

inhibition	抑制／阻害	61,537
inhibitions	抑制／阻害	96

複数形のinhibitionsの割合は0.2%しかなく，原則，**不可算名詞**として使われる．

◆類義語：suppression, repression, depression, prevention, attenuation, reduction

■ inhibition ＋［前］

	the%	a/an%	ø%			
❶	14	0	81	inhibition of ～	～の抑制	35,067
				・lead to inhibition of ～	…は，～の抑制につながる	282
				・due to inhibition of ～	～の抑制のせいで	146
❷	7	0	62	inhibition by ～	～による抑制	3,062
❸	3	2	55	inhibition in ～	～における抑制	1,607

❶ Activation of PI3-kinase **leads to inhibition of** apoptosis, promotion of cell survival, and enhanced proliferative responses to G-CSF.（J Immunol. 1998 160:4979）

訳 PI3-キナーゼの活性化は，アポトーシスの抑制につながる

❷ However, only APC/C isolated from mitotic cells was sensitive to **inhibition by** MCC.（J Cell Biol. 2001 154:925）

訳 ～は，MCCによる抑制に感受性であった

■［名／過分／形］＋ inhibition

❶	growth inhibition	増殖抑制	1,878
❷	～-mediated inhibition	～に仲介される阻害（抑制）	1,450
❸	～-dependent inhibition	～依存的な阻害（抑制）	752
❹	～-induced inhibition	～に誘導される阻害（抑制）	633
❺	pharmacological inhibition	薬理学的な阻害（抑制）	627
❻	selective inhibition	選択的な阻害（抑制）	451
❼	significant inhibition	有意な阻害（抑制）	440
❽	feedback inhibition	フィードバック阻害（抑制）	394
❾	specific inhibition	特異的な阻害（抑制）	379

❷ Thus, the TGF-β/Smad signaling pathway is required for hypoxia-**mediated inhibition** of adipocyte differentiation in MSCs.（J Biol Chem. 2005 280:22688）

訳 TGF-β/Smadシグナル伝達経路は，間葉系幹細胞において低酸素に仲介される脂肪細胞分化の抑制のために必要とされる

❸ Pharmacodynamic analysis demonstrated significant dose-**dependent inhibition** of complement hemolytic activity for up to 14 hours at 2 mg/kg.（Circulation. 1999 100:2499）

訳 薬力学的な分析は，有意な用量依存的な補体溶血性活性の阻害を実証した

★ inhibitory 形 抑制性の　　用例数 20,063

◈類義語：suppressive　◈反意語：excitatory

■ inhibitory ＋［名］

❶ inhibitory effect	抑制性の影響	2,641
❷ inhibitory activity	抑制性の活性	1,425
❸ inhibitory action	抑制性の作用	271

❶ The **inhibitory effect** of Smad7 on adipocyte differentiation and its cooperation with TGF-β was associated with the C-domain of Smad7.（J Cell Biol. 2000 149:667）
訳 脂肪細胞分化に対するSmad7の抑制性の影響

❷ The **inhibitory activity** of IRI was prevented by pretreatment with aryl sulfatase, suggesting the presence of a critical sulfo ester in IRI.（Biochemistry. 2000 39:3452）
訳 IRIの抑制性の活性が，前処置によって防がれた

■［形］＋ and inhibitory

❶ excitatory and inhibitory ～	興奮性および抑制性の～	462

initially 副 最初に　　用例数 6,369

◈類義語：originally, first, early

■ Initially

❶ Initially,	最初に	425

❶ **Initially,** nascent chains were found to associate with the heat shock protein（Hsp）70 family member BiP.（Mol Biol Cell. 2005 16:3740）
訳 最初に，新生鎖が～と結合することが見つけられた

■ initially ＋［過分］

❶ initially identified	最初に同定される	279
・be initially identified	…が，最初に同定される	155
❷ initially formed	最初に形成される	122

❶ Five concave and six convex thrombi **were initially identified**.（Radiology. 2005 237:338）
訳 5つの凹面と６つの凸面の血栓が，最初に同定された

★ initiate 動 開始する　　用例数 18,065

initiated	開始される／～を開始した	7,427
initiate	～を開始する	5,120
initiating	～を開始する	2,931
initiates	～を開始する	2,587

他動詞.

◆類義語：start, commence

■ initiate ＋［名句］

❶ initiate transcription	…は，転写を開始する	166
❷ initiate ～ formation	…は，～形成を開始する	163
❸ initiate apoptosis	…は，アポトーシスを開始する	118
❹ initiate a cascade of ～	…は，～のカスケードを開始する	81

❹ In response to clastogens, the ataxia telangiectasia mutated (ATM) protein is rapidly activated, **which in turn initiates a cascade of DNA damage response**. (J Biol Chem. 2010 285: 12055)

訳 そしてそれは次にDNA損傷応答のカスケードを開始する

■［動／形］＋ to initiate

★ to initiate ～	～を開始する…	2,493
❶ be required to initiate ～	…は，～を開始するのに必要とされる	93
❷ be sufficient to initiate ～	…は，～を開始するのに十分である	87

❷ We now show that **the inhibition of cyclin-dependent kinase (Cdk) 1 is sufficient to initiate cytokinesis**. (J Biol Chem. 2005 280:36502)

訳 サイクリン依存性キナーゼ（Cdk）1 の抑制は，細胞質分裂を開始するのに十分である

■ be initiated ＋［前］

★ be initiated	…は，開始される	3,369
❶ be initiated by ～	…は，～によって開始される	1,348
❷ be initiated in ～	…は，～において開始される	297
❸ be initiated at ～	…は，～において開始される	196

❶ **Mismatch repair (MMR) is initiated by MutS family proteins (MSH)** that recognize DNA mismatches and recruit downstream repair factors. (Mol Cell. 2005 20:771)

訳 ミスマッチ修復（MMR）は，MutSファミリータンパク質（MSH）によって開始される

❸ Our results indicate that **spikes are initiated at multiple sites** within the dendritic arbors of DSGCs and that each dendritic spike initiates a somatic spike. (Neuron. 2005 47:739)

訳 棘波は，複数の部位で開始される

★ initiation 名 開始　　用例数 16,972

initiation	開始	16,940
initiations	開始	32

複数形のinitiationsの割合は0.2%しかなく，原則，**不可算名詞**として使われる.

◆類義語：onset, start

■ initiation ＋［前］

	the%	a/an%	ø%			
❶	50	0	48	initiation of ～	～の開始	5,723
				・initiation of DNA replication	DNA複製の開始	347
				・initiation of transcription	転写の開始	150
				・required for the initiation of ～	～の開始のために必要とされる	139
				・initiation of translation	翻訳の開始	134
				・initiation of treatment	処置の開始	114
❷	0	0	90	initiation in ～	～における開始	360

❶ The accumulation of Cdc6 promotes the initiation of DNA replication.（Cell. 2005 122:825）

訳 Cdc6の蓄積は，DNA複製の開始を促進する

❶ Cdc7, a protein kinase required for the initiation of eukaryotic DNA replication, is activated by a regulatory subunit, Dbf4.（Genes Dev. 2005 19:2295）

訳 真核生物DNA複製の開始のために必要とされるタンパク質キナーゼ

■ initiation and ＋［名］

❶	initiation and progression	開始と進行	363

❶ As a cluster of recent Nature papers now show, altered expression of specific miRNA genes contributes to the initiation and progression of cancer.（Cell. 2005 122:6）

訳 特異的なマイクロRNA遺伝子の変化した発現は，癌の開始と進行に寄与する

inject 動 注入する／注射する　用例数 8,704

injected	注入される／～を注入した	7,875
injecting	～を注入する／～を注射する	630
inject	～を注入する／～を注射する	145
injects	～を注入する／～を注射する	54

他動詞.

◆類義語：infuse, transfuse

■ injected ＋［前］

❶	injected with ～	～を注入される	1,741
	・be injected with ～	…は，～を注入される	591
	・mice injected with ～	～を注射されたマウス	333
❷	injected into ～	～へ注入される	1,591
	・be injected into ～	…が，～へ注入される	819
	・when injected into ～	～へ注入されたとき	238
❸	injected in ～	～において注入される	149

❶ Neonatal mice injected with rabbit anti-mouse BP180（mBP10）IgG develop a BP-like

I

disease. (Blood. 2006 107:1063)

訳 ウサギの抗マウスBP180 (mBP10) IgGを注入された新生仔マウス

❷ ICG was injected into the mouse tail vein, and images were taken by in vivo confocal microscopy. (Invest Ophthalmol Vis Sci. 2003 44:4489)

訳 ICGが，マウスの尾静脈に注入された

❷ Concentrated vitreous protein from injured eyes caused a 60% decrease in retinal neovascularization when injected into the vitreous cavity of OIR rats. (Invest Ophthalmol Vis Sci. 2006 47:405)

訳 OIRラットの硝子体腔へ注入されたとき

■ injected ＋［副］

❶	injected intravenously	静脈内に注入される	213
❷	injected intraperitoneally	腹腔内に注入される	117
❸	injected subcutaneously	皮下に注入される	114

❶ Twelve-week-old rabbits were injected intravenously with cell-associated HTLV-1 (ACH-transformed R49). (J Virol. 2010 84:5124)

訳 12週齢のウサギは，～を静脈内に注入された

■［副］＋ injected

❶	intravenously injected	静脈内に注入される	104

❶ All mice were intravenously injected with ^{64}Cu-bis-DOTA-hypericin (24 h after laser treatment in treated mice). (J Nucl Med. 2011 52:792)

訳 すべてのマウスは，～を静脈内に注入された

★ injection ㊔ 注入／注射　用例数 18,483

injection	注入／注射	14,190
injections	注入／注射	4,293

複数形のinjectionsの割合は約25%あるが，単数形は無冠詞で使われることが多い.

◆類義語：infusion, transfusion

■ injection ＋［前］

	the%	a/an%	ø%			
❶	7	14	75	injection of ～	～の注入	6,536
				・injection of ～ into …	…への～の注入	221
❷	1	7	82	injection into ～	～への注入	437
❸	2	9	67	injection in ～	～における注入	308
❹	0	7	86	injection with ～	～の注入	250

❶ Intracerebral injection of the clones into naive mice induced degeneration, not only in the brain, but also in the spinal cord. (J Virol. 2005 79:14640)

訳 未処置マウスへのそれらのクローンの脳内注入

■ after injection

★	after injection	注入のあと	1,290
❶	after injection of ～	～の注入のあと	417
❷	～ h after injection	注入のあと～時間	234
❸	～ min after injection	注入のあと～分	142

❶ After injection of the retrograde tracer Fluoro-Gold (FG) into SNc, the rats received pairings of a visual CS with food. (J Neurosci. 2005 25:3881)

訳 逆行性トレーサーFluoro-Gold (FG) のSNcへの注入のあと

❷ Nonbound 131I-TM-601 was eliminated by 48 h after injection with the remaining radiolabeled peptide bound to tumor for at least 6-8 d. (J Nucl Med. 2005 46:580)

訳 ～は，残りの放射標識されたペプチドの注入のあと48時間までに除去された

■ [形] ＋ injection

❶	intravenous injection	静脈内注射	528
❷	intraperitoneal injection	腹腔内注射	350
❸	single injection	単回投与	309
❹	subcutaneous injection	皮下注射	232

★ injury 名 損傷／傷害　　用例数 28,518

injury	損傷／傷害	26,849
injuries	損傷／傷害	1,669

複数形のinjuriesの割合は約５％あるが，単数形は無冠詞の用例が圧倒的に多い．

◆類義語：damage, impairment, dysfunction, disturbance, lesion

■ injury ＋ [前]

	the %	a/an %	ø %			
❶	1	0	96	injury in ～	～における損傷	1,968
❷	3	3	93	injury to ～	～への損傷	736

❶ The molecular mechanisms responsible for oxidant-induced neuronal injury in meningitis are explored in some depth. (J Infect Dis. 2002 186:S225)

訳 髄膜炎において酸化体に誘導されるニューロンの損傷の原因である分子機構

❷ Mechanical injury to the skin results in activation of the complement component C3 and release of the anaphylatoxin C3a. (J Clin Invest. 2004 114:399)

訳 皮膚への機械的損傷は，補体成分C3の活性化という結果になる

■ injury ＋ [過分]

❶	injury induced	誘導される損傷	468

❶ PXR coordinates hepatic responses to prevent liver injury induced by environmental toxins. (Mol Pharmacol. 2006 69:99)

訳 PXRは，環境の毒素によって誘導される肝傷害を防ぐために肝臓の応答と同調する

■ [前] ＋ injury

❶ response to injury	損傷への応答	381
❷ ～ days after injury	損傷のあと～日	161
❸ site of injury	損傷の部位	157

■ [名] ＋ injury

❶ lung injury	肺傷害	2,085
❷ brain injury	脳傷害	1,561
❸ liver injury	肝傷害	1,322
❹ tissue injury	組織傷害	867

★ input 名 入力 用例数 11,809

input	入力	7,068
inputs	入力	4,741

複数形のinputsの割合は約40%あるが，単数形は無冠詞の用例が断然多い．

■ input ＋ [前]

	the%	a/an%	ø%			
❶	16	8	76	input to ～	～への入力	970
❷	0	1	96	input from ～	～からの入力	956

❶ **Input to** the central nervous system from olfactory sensory neurons (OSNs) is modulated presynaptically. (Neuron. 2005 48:1039)

訳 嗅覚感覚神経から中枢神経系への入力

❷ Sensory **input from** various receptors in the periphery first becomes integrated in the spinal cord dorsal horn. (Brain Res. 2005 1045:72)

訳 末梢におけるさまざまな受容体からの感覚性入力

■ [形／名] ＋ input

❶ synaptic input	シナプス入力	450
❷ sensory input	感覚入力	354

insensitive 形 非感受性の 用例数 3,534

■ insensitive ＋ [前]

❶ insensitive to ～	～に対して非感受性の	1,758
・be insensitive to ～	…は，～に対して非感受性である	1,037

・be relatively insensitive to ～	…は，～に対して相対的に非感受性である	192

❶ Kv4.2 channels were relatively insensitive to all compounds tested.（Mol Pharmacol. 2006 69: 718）

訳 Kv4.2チャネルは，テストされたすべての化合物に対して相対的に非感受性であった

insert 動 挿入する，名 挿入断片 　用例数 5,909

inserted	挿入される／～を挿入した	2,830
insert	～を挿入する／挿入断片	1,756
inserts	～を挿入する／挿入断片	873
inserting	～を挿入する	450

名詞としても用いられるが，他動詞の用例が多い．

■ inserted ＋［前］

❶	inserted into ～	～へ挿入される	1,015
	・be inserted into ～	…は，～へ挿入される	522
❷	inserted in ～	～において挿入される	266
❸	inserted at ～	～において挿入される	160

❶ We conclude that AtPEX2 and AtPEX10 are inserted into peroxisome membranes directly from the cytosol.（Plant Physiol. 2005 139:690）

訳 ～は，ペルオキシソーム膜へ挿入される

★ insertion 名 挿入 　用例数 12,027

insertion	挿入	9,318
insertions	挿入	2,709

複数形のinsertionsの割合は約25%あるが，単数形が無冠詞で使われることも多い．

◆類義語：intercalation

■ insertion ＋［前］

	the%	a/an%	ø%			
❶	29	5	66	insertion of ～	～の挿入	2,366
				・insertion of ～ into …	…への～の挿入	102
❷	6	70	21	insertion in ～	～における挿入	585
❸	10	6	63	insertion into ～	～への挿入	486

❶ The mechanism(s) of insertion of new introns into genes remains unknown.（Curr Biol. 2004 14:1505）

訳 遺伝子への新しいイントロンの挿入の機構は，未知のままである

I

■ insertion ＋［名］

❶ insertion site	挿入部位	509
❷ insertion mutants	挿入変異体	212

■［名］＋ insertion

❶ transposon insertion	トランスポゾン挿入	438
❷ membrane insertion	膜挿入	376

★ insight ㊔ 洞察　　用例数 17,193

insights	洞察	8,779
insight	洞察	8,414

複数形のinsightsの割合は約50%あるが，単数形は無冠詞で使われることが圧倒的に多い．また，insight intoの用例が圧倒的に多い．

◆類義語：perception

■ insight ＋［前］

	the%	a/an%	ø%			
❶	1	3	95	insight into ～	～への洞察	7,529
				・insight into how ～	どのように～かへの洞察	490
				・insight into the mechanism	機構への洞察	468
				・insight into the role of ～	～の役割への洞察	229
				・insight into the pathogenesis of ～	～の病因への洞察	90
				・insight into the function of ～	～の機能への洞察	87

❶ Several approaches were employed to gain insight into the mechanism by which amastigotes avoid eliciting superoxide production.（Infect Immun. 2005 73:8322）

訳 いくつかのアプローチが，～である機構への洞察を得るために用いられた

■［動］＋ insight into

❶ provide insight into ～	…は，～への洞察を提供する	2,552
・results provide insight into ～	結果は，～への洞察を提供する	189
・findings provide insight into ～	知見は，～への洞察を提供する	116
❷ gain insight into ～	…は，～への洞察を得る	1,030
❸ give insight into ～	…は，～への洞察を与える	137
❹ offer insight into ～	…は，～への洞察を提供する	134

❶ Furthermore, these findings provide insight into the development of new therapeutic strategies for idiopathic pulmonary fibrosis patients.（J Immunol. 2005 175:1894）

訳 これらの知見は，～の開発への洞察を提供する

■ ［形］＋ insights into

❶	new insights into ～	～への新たな洞察	2,210
	・provide new insights into ～	…は，～への新たな洞察を提供する	1,102
❷	important insights into ～	～への重要な洞察	485
❸	novel insights into ～	～への新規の洞察	435
❹	mechanistic insights into ～	～への機構的な洞察	380
❺	further insights into ～	～へのさらなる洞察	194

❶ Our findings provide new insights into how integrins function as nanomachines to precisely control cell adhesion and signaling.（J Cell Biol. 2012 199:497）
訳 我々の知見は，どのように～かへの新しい洞察を提供する

instability 名 不安定性　用例数 5,574

instability	不安定性	5,409
instabilities	不安定性	165

複数形のinstabilitiesの割合は3％あるが，単数形は無冠詞の用例が断然多い．
◆類義語：lability

■ instability ＋［前］

	the％	a/an％	ø％			
❶	58	4	38	instability of ～	～の不安定性	541
❷	4	4	85	instability in ～	～における不安定性	522

❷ Here, we review the general causes of chromosomal instability in human tumors.（J Dent Res. 2005 84:107）
訳 われわれは，ヒトの腫瘍における染色体の不安定性の一般的な原因を概説する

■ ［形／名］＋ instability

❶	genomic instability	ゲノムの不安定性	1,034
❷	chromosomal instability	染色体の不安定性	407
❸	genetic instability	遺伝的な不安定性	407
❹	genome instability	ゲノム不安定性	234

instance 名 例／場合　用例数 1,980

instances	例／場合	1,175
instance	例／場合	805

複数形のinstancesの割合は約60％と非常に高い．
◆類義語：example, event, occasion, case, circumstance

■ instances ＋［前］

	the％	a/an％	ø％			
❶	5	-	62	instances of ～	～の例	541

❶ We found **instances of** discordance between HUMARA results and those obtained by pyrosequencing and qTCA methods, as well as by directly quantifying AR gene expression. (Blood. 2012 119:e100)

訳 我々は，～の間の不一致の例を発見した

■［前］＋（some）instance（s）

❶	for instance	たとえば	463
❷	in some instances	いくつかの場合では	223

❶ **For instance**, CO_2 is known to induce production of polysaccharide capsule virulence determinants in pathogenic bacteria and fungi via unknown mechanisms. (Curr Biol. 2005 15:2013)

訳 たとえば，

★ instead 副 代わりに　　用例数 8,132

◆類義語：alternatively, in contrast

■ instead

❶	Instead,	代わりに	2,908
❷	instead of ～	～の代わりに	1,811
	・instead of ～ing	～する代わりに	347

❶ **Instead,** our findings suggest that only transient interaction of Claspin with replication forks potentiates its Chk1-activating function. (Mol Biol Cell. 2005 16:5269)

訳 代わりに，

❷ One primary difference is the fact that wild-type (wt) AAV utilizes three capsid subunits **instead of** two to form the virion shell. (J Virol. 2005 79:9933)

訳 ～は，2つの代わりに3つのキャプシドのサブユニットを利用する

insufficient 形 不十分な　　用例数 2,248

◆類義語：short

■ be insufficient ＋［前］

★	be insufficient	…は，不十分である	1,225
❶	be insufficient to *do*	…は，～するには不十分である	768
❶	be insufficient for ～	…は，～にとって不十分である	224

❶ Human data **are insufficient to** support this claim for any specific pesticide, largely because of challenges in exposure assessment. (Am J Epidemiol. 2009 169:919)

訳 ヒトのデータは，この主張を支持するには不十分である

★ intake 名 摂取／摂取量 用例数 11,162

intake	摂取／摂取量	9,424
intakes	摂取／摂取量	1,738

複数形のintakesの割合は約15%あるが，単数形は無冠詞の用例が圧倒的に多い．

◆類義語：ingestion

■ intake ＋［前］

	the%	a/an%	ø%			
❶	19	9	71	intake of ～	～の摂取	1,304
❷	0	0	99	intake in ～	～における摂取量	434

❶ These fatty acids are required in the human diet, although **excess dietary intake of omega-6 fatty polyunsaturated fatty acids may have a negative influence on human health.** (Dev Biol. 2013 373:14)

訳 オメガ6脂肪多価不飽和脂肪酸の過剰な食事摂取は，ヒトの健康に負の影響を持つかもしれない

❷ **Cholecystokinin decreased food intake in saline-injected mice** but not in VS-treated mice. (Proc Natl Acad Sci USA. 2005 102:15184)

訳 コレシストキニンは，生理食塩水を注射されたマウスにおける食物摂取量を低下させた

■［名］＋ intake

❶	food intake	食物摂取量	1,718
❷	energy intake	エネルギー摂取量	805
❸	dietary intake	食事摂取量	474
❹	alcohol intake	アルコール摂取量	371

I

★ integrate 動 統合する／組み込む 用例数 10,156

integrated	統合される／組み込まれる／～を統合した	5,909
integrate	～を統合する／～を組み込む	1,821
integrating	～を統合する／～を組み込む	1,325
integrates	～を統合する／～を組み込む	1,101

他動詞．

◆類義語：unify, incorporate

■ integrated ＋［前］

❶	integrated into ～	～に組み込まれる	718
❷	integrated with ～	～と統合される	367

❶ However, most surviving transplanted cells **were integrated into the wall of the lateral**

ventricle and expressed vimentin, a marker also expressed by ependymocytes. (J Comp Neurol. 2005 491:96)
訳 ～が，側脳室の壁に組み込まれた

❷ These data were integrated with U. S. Census Bureau data on 1995 earnings by age, educational attainment, and gender. (Am J Psychiatry. 2000 157:940)
訳 これらのデータは，アメリカ商務省のデータと統合された

■ integrate ＋［名句］

❶	integrate information	…は，情報を統合する	126

❶ How human beings integrate information from external sources and internal cognition to produce a coherent experience is still not well understood. (J Neurosci. 2012 32:10649)
訳 人間は，外部ソースからの情報を統合する

integration ㊔ 組込み／統合　　用例数 7,714

integration	組込み／統合	7,516
integrations	組込み／統合	198

複数形のintegrationsの割合は2.5%しかなく，原則，不可算名詞として使われる．

■ integration ＋［前］

	the%	a/an%	ø%			
❶	50	6	44	integration of ～	～の組込み	2,664
				・integration of ～ into …	～の…への組込み	228
❷	1	0	90	integration in ～	～における組込み	339
❸	2	2	75	integration into ～	～への組込み	264

❶ Most significantly, Metnase promotes integration of exogenous DNA into the genomes of host cells. (Proc Natl Acad Sci USA. 2005 102:18075)
訳 ～は，外来性DNAの宿主細胞のゲノムへの組込みを促進する

integrity ㊔ 完全性／統合性　　用例数 5,646

integrity	完全性／統合性	5,644
integrities	完全性／統合性	2

複数形のintegritiesの用例はほとんどなく，原則，不可算名詞として使われる．
◆類義語：completeness

■ integrity ＋［前］

integrity ofの前にtheが付く割合は圧倒的に高い．

	the%	a/an%	ø%			
❶	86	2	12	integrity of ～	～の完全性	2,049
❷	3	0	93	integrity in ～	～における完全性	294

❶ The authors conclude that the integrity of the VMN is important for full expression of sociosexual behaviors in male rats. (Behav Neurosci. 2005 119:1227)
訳 VMNの完全性は，～にとって重要である

■ [形] + integrity

❶ structural integrity	構造的な完全性	537
❷ genomic integrity	遺伝的な統合性	321

⁑ interact 動 相互作用する 用例数 35,573

interact	相互作用する	14,268
interacts	相互作用する	11,181
interacting	相互作用する	8,139
interacted	相互作用した	1,985

自動詞．interact withの用例が圧倒的に多い．

■ interact + [前]

❶ interact with ～	…は，～と相互作用する	23,052
・interact with each other	…は，お互いに相互作用する	101
❷ interact in ～	…は，～において相互作用する	951
❸ interact to *do*	…は，～するために相互作用する	573

❶ OsDr1 and OsDrAp1 are nuclear proteins that interact with each other and with the TATA binding protein/DNA complex. (Plant Cell. 2002 14:181)
訳 OsDr1およびOsDrAp1は，お互いに相互作用する核タンパク質である

■ [名] + interact with

❶ proteins interact with ～	タンパク質は，～と相互作用する	237
❷ domain interacts with ～	ドメインは，～と相互作用する	131

❶ A number of regulatory proteins interact with FtsZ and modulate FtsZ assembly/disassembly processes, ensuring the spatiotemporal integrity of cytokinesis. (J Bacteriol. 2012 194:3189)
訳 いくつかの制御タンパク質は，FtsZと相互作用する

■ [副] + interact with

❶ directly interact with ～	…は，～と直接相互作用する	763
❷ physically interact with ～	…は，～と物理的に相互作用する	712
❸ specifically interact with ～	…は，～と特異的に相互作用する	414
❹ functionally interact with ～	…は，～と機能的に相互作用する	168
❺ genetically interact with ～	…は，～と遺伝的に相互作用する	101

❷ Prohibitin physically interacts with all three Rb family proteins in vitro and in vivo, and was very effective in repressing E2F-mediated transcription. (Oncogene. 1999 18:3501)
訳 プロヒビチンは，３つのRbファミリータンパク質すべてと物理的に相互作用する

■ interact ＋［副］＋ with

❶	interact directly with ～	…は，～と直接相互作用する	1,028
❷	interact specifically with ～	…は，～と特異的に相互作用する	376
❸	interact strongly with ～	…は，～と強く相互作用する	151
❹	interact genetically with ～	…は，～と遺伝的に相互作用する	126
❺	interact physically with ～	…は，～と物理的に相互作用する	104

❶ Gemin4 also interacts directly with several of the Sm core proteins.（J Cell Biol. 2000 148: 1177）

訳 Gemin4は，また，Smコアタンパク質のいくつかと直接相互作用する

■［名／動］＋ to interact with

★	to interact with ～	～と相互作用する…	3,195
❶	ability to interact with ～	～と相互作用する能力	391
❷	be shown to interact with ～	…は，～と相互作用することが示される	300
❸	be found to interact with ～	…は，～と相互作用することが見つけられる	162
❹	be known to interact with ～	…は，～と相互作用することが知られている	122

❶ Consistent with this activity, the R213Q mutant was found to have the ability to interact with DNA sequences located within the MDM2 promoter.（Oncogene. 2000 19:3095）

訳 R213Q変異体は，MDM2プロモーター内に位置するDNA配列と相互作用する能力を持つことが見つけられた

❷ FRPs have been shown to interact with Wnt proteins and antagonize Wnt signaling in a Xenopus developmental model.（J Biol Chem. 1999 274:16180）

訳 FRPは，Wntタンパク質と相互作用することが示されている

★★ interaction ㊔ 相互作用　　用例数 109,090

interaction	相互作用	55,658
interactions	相互作用	53,432

複数形のinteractionsの割合は約50％と非常に高い．

◆類義語：interplay

■ interaction ＋［前］

interaction ofの前にtheが付く割合はかなり高い．

	the％	a/an％	ø％			
❶	17	12	51	interaction with ～	～との相互作用	10,873
				・its interaction with ～	～とそれの相互作用	1,745
				・through interaction with ～	～との相互作用によって	404
				・by interaction with ～	～との相互作用によって	307
				・required for interaction with ～		

					～との相互作用のために必要とされる	155
❷ 46	33	16	interaction between ～		～の間の相互作用	10,145
❸ 64	10	25	interaction of ～		～の相互作用	9,195
			・interaction of ～ with …		～の…との相互作用	4,541

❶ This suggests that the conserved C-terminal region of FtsZ **is required for proper polymerization of FtsZ in Caulobacter and for its interaction with FtsA**. (Mol Microbiol. 1998 27:1051)

訳 ～は，カウロバクターにおけるFtsZの適切な重合およびFtsAとそれの相互作用のために必要とされる

❶ The repression of basal transcription by HSF-4a **occurs through interaction with the basal transcription factor TFⅡF**. (J Biol Chem. 2001 276:14685)

訳 ～は，基本転写因子TFⅡFとの相互作用によって起こる

❷ The alanine helix **provides a model system for studying the energetics of interaction between water and the helical peptide group**, a possible major factor in the energetics of protein folding. (Proc Natl Acad Sci USA. 2000 97:10786)

訳 ～は，水とらせん状ペプチド基との間の相互作用のエネルギー論を研究するためのモデルシステムを提供する

❸ **This study demonstrates for the first time the interaction of a GAS3/PMP22 family member with an integrin protein** and suggests that such interactions and their functional consequences are a physiologic role of GAS3/PMP22 proteins. (J Biol Chem. 2002 277:41094)

訳 この研究は，GAS3/PMP22ファミリーメンバーのインテグリンタンパク質との相互作用を初めて実証する

■ interaction ＋［現分／過分］

❶ interactions involving ～	～に関わる相互作用	447
❷ interactions mediated by ～	～によって仲介される相互作用	156
❸ interactions using ～	～を使う相互作用	137
❹ interactions including ～	～を含む相互作用	131

■［名／形］＋ interaction

❶ protein interactions	タンパク質相互作用	3,824
❷ direct interaction	直接の相互作用	1,480
❸ electrostatic interactions	静電的な相互作用	1,042
❹ specific interactions	特異的な相互作用	793
❺ physical interaction	物理的な相互作用	730
❻ molecular interactions	分子的な相互作用	683
❼ hydrophobic interactions	疎水性の相互作用	679

interest 名 興味／関心　　用例数 8,852

interest	興味／関心	8,596
interests	興味／関心	256

複数形のinterestsの割合は約３%あるが，単数形は無冠詞で使われることが断然多い．

I

of interestの用例が非常に多い.

■ ［名／動］＋ of interest

★ of interest	興味深い／関心の	3,803
❶ region of interest	関心領域	684
❷ be of interest	…は，興味深い	596
・be of interest to *do*	～することは興味深い	173
・it was of interest to *do*	～することは興味深かった	56
・be of interest for ～	…は，～にとって興味深い	83

❷ Given the relatively large number of induced genes (6% of the genome), **it was of interest to determine** which genes provide functions essential to transformation. (Mol Microbiol. 2004 51: 1051)

訳 ～を決定することは興味深かった

❷ **More potent ligands are of interest for hepatic delivery of therapeutic agents.** (J Am Chem Soc. 2012 134:1978)

訳 もっと強力なリガンドは，治療薬の肝臓送達にとって興味深い

■ of ＋［形］＋ interest

❶ of particular interest	特に興味深い	573
・be of particular interest	…は，特に興味深い	277
・Of particular interest be ～	～は，特に興味深い	167
・Of particular interest,	特に興味深いことには,	57
❷ of great interest	大変興味深い	245
❸ of considerable interest	かなり興味深い	124

❶ **This motif is of particular interest**, because it has also been linked to regulation of the Skp1/Cul1/F-box complex, SCF^{Skp2}. (Proc Natl Acad Sci USA. 2005 102:18562)

訳 このモチーフは特に興味深い

❶ **Of particular interest is the role of nutrient on the population dynamics.** (J Theor Biol. 2010 265:225)

訳 人口動態に対する栄養物の役割は，特に興味深い

■ interest ＋［前］

	the％	a/an％	ø％			
❶	5	9	80	interest in ～	～に対する関心	2,415
				・interest in ～ing	～することに対する関心	538
				・there is considerable interest in ～	～に対するかなりの関心がある	101
				・there is great interest in ～	～に対する大きなの関心がある	51
❷	0	1	98	interest for ～	～に対する関心	324

❶ **There is considerable interest in using this technology** as the basis for diagnostic assays. (Curr Opin Rheumatol. 2005 17:606)

訳 このテクノロジーを使うことにかなりの関心がある

interesting ㊊ 興味深い　用例数 2,064

■ be interesting

★ be interesting	…は，興味深い	211
❶ it is interesting that ～	～ということは興味深い	65

❶ It is interesting that this amplicon was preferentially detected in carcinomas carrying wild-type Rb. (Oncogene. 2010 29:5700)
訳 ～ということは興味深い

■ ［副］＋ interesting

❶ most interesting	最も興味深い	104
❷ particularly interesting	特に興味深い	88

❶ This is particularly interesting because CK2 is required for initiation at a subset of pol II promoters. (J Biol Chem. 2011 286:23160)
訳 これは，特に興味深い

★ interestingly ㊓ 興味深いことに　用例数 10,250

文頭で使われることが圧倒的に多い.
◆類義語：intriguingly

■ interestingly

❶ Interestingly,	興味深いことに	9,743
❷ Most interestingly,	最も興味深いことに	113

❶ Interestingly, we found that RAP interacts with oxidoreductase ERp57 and mediates its interaction with LRP. (Biochemistry. 2005 44:5794)
訳 興味深いことに，われわれは～ということを見つけた

★ interface ㊔ 界面／インターフェース　用例数 11,699

interface	界面／インターフェース	9,633
interfaces	界面／インターフェース	2,066

複数形のinterfacesの割合が，約20%ある可算名詞.

■ interface ＋［前］

interface between/ofの前にtheが付く割合は圧倒的に高い.

	the %	a/an %	ø %			
❶	84	15	0	interface between ～	～の間の界面	921
❷	91	7	1	interface of ～	～の界面	709
❸	66	27	1	interface in ～	～における界面	263

I

❶ We present a microscopic model of the interface between liquid water and a hydrophilic, solid surface, as obtained from ab initio molecular dynamics simulations. (J Am Chem Soc. 2005 127:6830)

訳 われわれは，液状の水と親水性で固形の表面の間の界面の微視的なモデルを提示する

❷ Both potentiators bind within the dimer interface of the nondesensitized receptor at a common site located on the twofold axis of molecular symmetry. (J Neurosci. 2005 25:9027)

訳 ～は，脱感作されていない受容体の二量体界面内に結合する

■ [名] + interface

❶ dimer interface	二量体界面	860
❷ binding interface	結合界面	319
❸ water interface	水界面	306
❹ subunit interface	サブユニット界面	258

★ interfere 動 干渉する／妨げる　　用例数 10,121

interfering	干渉する／妨げる	5,597
interfere	干渉する／妨げる	2,646
interferes	干渉する／妨げる	1,265
interfered	干渉した／妨げた	613

自動詞．interfere withの用例が非常に多い．

◆類義語：disturb, interrupt, impede, obstruct, hamper, intercept

■ interfere + [前]

❶ interfere with ～	…は，～に干渉する	5,480
・interfere with ～ binding	…は，～結合に干渉する	289
・interfere with ～ function	…は，～機能に干渉する	282
・interfere with the ability of ～	…は，～の能力に干渉する	116

❶ Transcriptional repressors are thought to inhibit gene expression by interfering with the binding or function of RNA Polymerase II, perhaps by promoting local chromatin condensation. (Proc Natl Acad Sci USA. 2012 109:9460)

訳 転写抑制因子は，～の結合あるいは機能に干渉することによって，遺伝子発現を抑制すると考えられる

❶ A point mutation in maf that interferes with Maf binding to DivIVA also interferes with the ability of Maf to inhibit cell division. (21564336J Biol Chem. 2003 278:39477)

訳 ～は，細胞分裂を抑制するMafの能力に干渉する

■ interfering RNA

❶ small interfering RNA	低分子干渉RNA	3,381
❷ short interfering RNA	短鎖干渉RNA	479

interference ㊔ 干渉　用例数 6,984

interference	干渉	6,840
interferences	干渉	144

複数形のinterferencesの割合は2%しかなく，原則，不可算名詞として使われる．

■ interference ＋［前］

	the%	a/an%	ø%			
❶	5	0	83	interference with ～	～による干渉	799
❷	19	0	43	interference of ～	～の干渉	279
❸	1	0	48	interference in ～	～における干渉	238
❹	4	0	84	interference from ～	～からの干渉	182

❶ In contrast to kif3b/kif17 double mutants, **simultaneous interference with kif3b and kif3c leads to** the complete loss of photoreceptor and hair cell cilia, revealing redundancy of function. (Proc Natl Acad Sci USA. 2012 109:2388)

訳 kif3bおよびkif3cによる同時干渉は，～につながる

■［名］＋ interference

❶	RNA interference	RNA干渉	3,245

★ intermediate ㊔ 中間体，㊓ 中間の／中等の　用例数 23,822

intermediate	中間体／中間の／中等の	17,016
intermediates	中間体	6,806

名詞および形容詞として用いられる．複数形のintermediatesの割合が，約30%ある可算名詞．

◆類義語：middle, intermediary, interstitial, moderate

■ ㊔ intermediate ＋［前］

	the%	a/an%	ø%			
❶	25	71	2	intermediate in ～	～における中間体	799

❶ These results strongly indicate that N-oleoylglycine **is an intermediate in oleamide biosynthesis** and provide further evidence that PAM does have a role in primary fatty acid amide production in vivo. (Biochemistry. 2004 43:12667)

訳 ～は，オレアミド生合成における中間体である

■ ㊓ be intermediate ＋［前］

★	be intermediate	…は，中間である	365
❶	be intermediate between ～	…は，～の間の中間である	164

❶ Membrane voltage responses of FS cells to Na^{+}-K^{+} ATPase blockade **were intermediate**

I

between the two PYR cell groups (P < 0.05). (J Physiol. 2010 588:4401)
訳 〜は，2つのPYR細胞群の間の中間であった

■ 形 intermediate ＋［名］

❶	intermediate state	中間の状態	271
❷	intermediate levels	中等レベル	159

interplay 名 相互作用／関係，動 相互作用する　用例数 1,992

interplay	相互作用／関係	1,971
interplays	相互作用／関係	17
interplaying	相互作用する	4

名詞の用例が多く，動詞として使われることはほとんどない．複数形のinterplaysの割合は0.9%しかないが，単数形には不定冠詞が付くことが多い．
◆類義語：interaction

■ interplay ＋［前］

interplay between/ofの前にtheが付く割合は非常に高い．

	the%	a/an%	ø%			
❶	71	18	6	interplay between 〜	〜の間の相互作用	1,242
❷	70	25	5	interplay of 〜	〜の相互作用	463

❶ However, the mechanisms that regulate the interplay between these cellular processes remain poorly understood. (J Biol Chem. 2012 287:36593)
訳 これらの細胞プロセスの間の相互作用を調節する機構

❷ Study of these flies will provide insight into the critical interplay of genetics and environment in PD. (Curr Biol. 2005 15:1572)
訳 〜は，PDにおける遺伝学と環境の決定的に重要な相互作用への洞察を提供するであろう

■［名／形］＋ interplay

❶	complex interplay	複合体相互作用	257
❷	dynamic interplay	動的な相互作用	111

interpret 動 解釈する　用例数 4,144

interpreted	解釈される／〜を解釈した	2,185
interpret	〜を解釈する	1,151
interpreting	〜を解釈する	757
interprets	〜を解釈する	51

他動詞．

■ be interpreted ＋［前］

★ be interpreted	…は，解釈される	1,537
❶ be interpreted as ～	…は，～として解釈される	503
❷ be interpreted in terms of ～	…は，～の点から解釈される	213
❸ be interpreted to *do*	…は，～すると解釈される	167

❶ These effects can be interpreted as evidence that structural distance impacts the establishment of agreement overall, consistent with sentence processing models which predict that hierarchical structure impacts the processing of syntactic dependencies.（Brain Res. 2012 1456:49）
訳 これらの効果は，～という証拠として解釈されうる

■［名］＋ be interpreted

❶ results are interpreted	結果は，解釈される	133

❶ The results are interpreted in terms of differences in cell-wall biosynthesis between plant species.（J Am Chem Soc. 2010 132:6335）
訳 それらの結果は，～の違いの点から解釈される

■［代名］＋ interpret

❶ we interpret ～	我々は，～を解釈する	318

❶ We interpret these results to indicate a role for the OBD central channel in binding and threading ssDNA during unwinding of circular SV40 DNA.（Biochemistry. 2010 49:2087）
訳 われわれは，これらの結果は～の役割を示していると解釈する

■ to interpret

★ to interpret ～	～を解釈する…	533
❶ difficult to interpret	解釈するのが困難である	94

❶ Evidence for this hypothesis is mixed though different methodologies have made these findings difficult to interpret.（Brain Res. 2011 1385:151）
訳 ～は，これらの知見を解釈するのを困難にしてきた

interpretation 名 解釈　　用例数 4,133

interpretation	解釈	3,570
interpretations	解釈	563

複数形のinterpretationsの割合が，約15%ある可算名詞．

■ interpretation ＋［前／同格that節］

interpretation thatの前にtheが付く割合は圧倒的に高い.

	the％	a/an％	ø％			
❶	52	12	33	interpretation of ～	～の解釈	2,362
❷	91	5	3	interpretation that ～	～という解釈	148

❶ Nevertheless, they have a number of limitations that **challenge the interpretation of the results.** (J Am Coll Cardiol. 2010 55:415)

訳 ～は，それらの結果の解釈に異を唱える

★ intervention　名 介入／インターベンション　用例数 14,333

intervention	介入／インターベンション	8,865
interventions	介入／インターベンション	5,468

複数形のinterventionsの割合は約40%とかなり高い.

◆類義語：treatment, therapy, manipulation, involvement

■ intervention(s) ＋［前］

	the％	a/an％	ø％			
❶	5	12	82	intervention in ～	～における介入	645
❷	1	-	97	interventions to *do*	～するための介入	507
❸	2	-	93	interventions for ～	～のための介入	351

❶ Therefore, TAL-specific CTL **may serve as a novel target for therapeutic intervention in patients with MS.** (J Immunol. 2005 175:8365)

訳 ～は，多発性硬化症の患者における治療介入のための新規の標的として働くかもしれない

❷ **Interventions to improve detection and prevention of glucocorticoid-induced osteoporosis are necessary.** (Arthritis Rheum. 2002 46:3136)

訳 グルココルチコイド誘導性骨粗鬆症の検出および予防を改善するための介入が必要である

■［形］＋ intervention

❶	therapeutic intervention	治療介入	1,076
❷	coronary intervention	冠動脈インターベンション	745
❸	surgical intervention	外科的処置	310

intrinsic　形 内因性の／内在性の　用例数 9,612

◆類義語：endogenous, inherent　◆反意語：extrinsic, exogenous

■ intrinsic ＋［前］

❶	intrinsic to ～	～に内因性である	464

・be intrinsic to ～	…は，～に内因性である	206

❶ Both these mechanisms are intrinsic to the myocyte.（J Physiol. 2004 559:205）
訳 これらの機構の両方は，筋細胞に内因性である

■ intrinsic and ＋［形］

❶ intrinsic and extrinsic ～	内因性および外因性の～	184

❶ The precise control of the cell cycle requires regulation by many intrinsic and extrinsic factors.（Dev Cell. 2005 9:843）
訳 ～は，多くの内因性および外因性の因子による調節を必要とする

★ introduce　動 導入する／紹介する　用例数 10,699

introduced	導入される／～を導入した／～を紹介した	6,363
introduce	～を導入する／～を紹介する	2,295
introducing	～を導入する／～を紹介する	1,400
introduces	～を導入する／～を紹介する	641

他動詞.
◆類義語：transfect, transform

■ be introduced ＋［前］

★ be introduced	…が，導入される	2,854
❶ be introduced into ～	…が，～に導入される	1,211
❷ be introduced in ～	…が，～において導入される	223
❸ be introduced at ～	…が，～において導入される	166

❶ Additional missense mutations were introduced into the mal61/HA-581NS allele, altering potential phosphorylation and ubiquitination sites.（Biochemistry. 2000 39:4518）
訳 付加的なミスセンス変異が，mal61/HA-581NSアレルに導入された

■［名］＋ be introduced

❶ mutations were introduced	変異が，導入された	188

■［代名］＋ introduce

❶ we introduce ～	われわれは，～を紹介する	1,106

❶ We introduce a method for quickly determining the rate of implicit learning.（PLoS One. 2012 7:e33400）
訳 われわれは，～のための方法を紹介する

introduction ㊔ 導入　用例数 4,671

introduction	導入	4,577
introductions	導入	94

複数形のintroductionsの割合は2％しかなく，原則，**不可算名詞**として使われる．

■ introduction ＋［前］

	the％	a/an％	ø％			
❶	48	0	52	introduction of ～	～の導入	3,987
				・introduction of ～ into …	～の…への導入	383
❷	13	6	57	introduction into ～	～への導入	112

❶ **Introduction of SNAP-29 into presynaptic superior cervical ganglion neurons** in culture significantly inhibited synaptic transmission in an activity-dependent manner. (Proc Natl Acad Sci USA. 2001 98:14038)
訳 SNAP-29のシナプス前上頸神経節ニューロンへの導入

invasion ㊔ 浸潤　用例数 8,932

invasion	浸潤	8,801
invasions	浸潤	131

複数形のinvasionsの割合は1.5％しかなく，原則，**不可算名詞**として使われる．
◆類義語：infiltration

■ invasion ＋［前］

	the％	a/an％	ø％			
❶	19	1	79	invasion of ～	～の浸潤	1,510
❷	1	0	99	invasion in ～	～における浸潤	402
❸	0	0	100	invasion by ～	～による浸潤	384

❶ Functionally, **FOXA1 silencing increases migration and invasion of luminal cancer cells**, both of which are characteristics of basal subtype cells. (Oncogene. 2013 32:554)
訳 FOXA1のサイレンシングは，管腔癌細胞の遊走と浸潤を増大させる

■［名］＋ invasion

❶	tumor invasion	腫瘍浸潤	310
❷	tumor cell invasion	腫瘍細胞浸潤	214

inverse ㊋ 逆の　用例数 3,046

◆類義語：reverse, opposite, backward

■ inverse ＋［名］

❶ inverse association	逆相関	451
❷ inverse correlation	逆相関	434
❸ inverse relationship	逆相関	410
❹ inverse agonist	逆作動薬	207

❶ We found a striking inverse correlation between hA3G mRNA levels and HIV viral loads（P≦0.00009）and a highly significant positive correlation between hA3G mRNA levels and CD4 cell counts（P≦0.00012）in these patients.（J Virol. 2005 79:11513）

訳 われわれは，hA3Gメッセンジャー RNAレベルとHIVウイルス量の間の顕著な逆相関を見つけた

inversely 副 逆に　用例数 2,998

◆類義語：conversely, oppositely, adversely

■ be inversely ＋［過分／形］

★ be inversely ～	…は，逆に～である	1,772
❶ be inversely associated with ～	…は，～と逆に相関する	527
❷ be inversely related to ～	…は，～に逆に相関する	456
❸ be inversely correlated with ～	…は，～と逆に相関する	420
❹ be inversely proportional to ～	…は，～に逆比例する	172

❷ Thus, the strength of the DBDs' specific association with DNA is inversely related to the stability of the free DBDs.（Biochemistry. 2005 44:14202）

訳 DNAとDBDの特異的な結合の強さは，遊離のDBDの安定性に逆に相関する

❸ In the vegetarians and the vegans, plasma DHA was inversely correlated with plasma LA.（Am J Clin Nutr. 2005 82:327）

訳 血漿DHAは，血漿LAと逆に相関した

❹ In cells infected with live C. pneumoniae, the increase was inversely proportional to the MOI.（Infect Immun. 2005 73:4323）

訳 その増大は，感染多重度に逆比例した

★★ investigate 動 精査する／調べる／研究する　用例数 51,722

investigated	～を精査した／精査される	30,769
investigate	～を精査する	18,328
investigating	～を精査する	2,128
investigates	～を精査する	497

主に他動詞として，「詳細に調査研究する」意味で用いられる．

◆類義語：study, work, research, test, explore, examine, survey, look at, search, dissect

■ investigate ＋［名句］

❶	investigate the role of ～	…は，～の役割を精査する	3,885
❷	investigate the effect of ～	…は，～の効果を精査する	2,784
❸	investigate the mechanism	…は，機構を精査する	1,438
❹	investigate the relationship between ～	…は，～との間の関連性を精査する	484
❺	investigate the function of ～	…は，～の機能を精査する	450
❻	investigate the ability of ～	…は，～の能力を精査する	361
❼	investigate the contribution of ～	…は，～の寄与を精査する	320
❽	investigate the impact of ～	…は，～の影響を精査する	246
❾	investigate the hypothesis that ～	…は，～という仮説を精査する	243
❿	investigate the influence of ～	…は，～の影響を精査する	236
⓫	investigate the possibility that ～	…は，～という可能性を精査する	225

❶ Here **we investigated the role of GSK3β signaling in** vascular biology by examining its function in endothelial cells (ECs). (J Biol Chem. 2002 277:41888)
訳 われわれは，～におけるGSK3βシグナル伝達の役割を精査した

❷ **We investigated the effects of OFQ on** contractions of muscle strips obtained from different regions of the gastrointestinal tract. (Gastroenterology. 1999 116:108)
訳 われわれは，～に対するOFQの効果を精査した

■ investigate ＋［名節］

❶	investigate whether ～	…は，～かどうかを精査する	4,363
❷	investigate how ～	…は，どのように～かを精査する	960

❶ **We investigated whether** this supine hypertension could be driven by residual sympathetic activity. (Circulation. 2000 101:2710)
訳 われわれは，～かどうかを精査した

■［代名／名］＋ investigate

❶	we investigated ～	われわれは，～を精査した	12,385
	・we have investigated ～	われわれは，～を精査した	3,082
❷	study investigated ～	研究は，～を精査した	1,026
❸	the authors investigated ～	著者らは，～を精査した	353

■ to investigate

★	to investigate ～	～を精査する…	14,899
❶	To investigate ～	～を精査するために	5,886
❷	be to investigate ～	…は，～を精査することである	1,177
❸	be used to investigate ～	…は，～を精査するために使われる	1,043
❹	model to investigate ～	～を精査するためのモデル	279
❺	be undertaken to investigate ～	…は，～を精査するために着手される	214
❻	be designed to investigate ～	…は，～を精査するために設計される	182

❷ The aim of this study was to investigate whether dietary glycine could reduce development of chronic rejection. (Transplantation. 2000 69:773)

訳 この研究の目的は，～かどうかを精査することであった

❸ An antibody raised against a peptide from human caspase 3 was used to investigate how extracellular signals controlled spatial patterning of cell death. (Development. 2002 129:3269)

訳 ～が，どのように細胞外シグナルが～を制御するかを精査するために使われた

■ be investigated ＋［前］/using

★	be investigated	…は，精査される	9,458
❶	be investigated in ～	…は，～において精査される	1,677
❷	be investigated by ～	…は，～によって精査される	1,579
	・be investigated by using ～	…は，～を使うことによって精査される	156
❸	be investigated using ～	…は，～を使って精査される	1,068
❹	be investigated for ～	…は，～のために精査される	314
❺	be investigated as ～	…は，～として精査される	299

❶ Repair of all 12 single-base mismatches in recombination intermediates was investigated in Chinese hamster ovary cells. (Genetics. 1998 149:1935)

訳 ～が，チャイニーズハムスター卵巣細胞において精査された

❷ Synthesis of Na,K-ATPase $\alpha1$ polypeptide was investigated by measuring ^{35}S-methionine incorporation. (Invest Ophthalmol Vis Sci. 2002 43:2714)

訳 ～が，^{35}S-メチオニンの取り込みを測定することによって精査された

❸ The cellular basis of TdP was investigated using a novel approach of transmural optical imaging in the canine wedge preparation (n=14). (Circulation. 2002 105:1247)

訳 …が，～の新規のアプローチを使って精査された

★ investigation 名 研究／調査　　用例数 11,295

investigation	研究	7,653
investigations	研究	3,642

I

複数形investigationsの割合が，約30％ある可算名詞.

◆類義語：study, test, examination, research, survey, work, trial

■ investigation ＋［前］

	the％	a/an％	ø％			
❶	23	23	52	investigation of ～	～の研究／～に関する研究	2,978
❷	2	22	67	investigation into ～	～についての研究	314
❸	23	23	52	investigation in ～	～における研究	211

❶ Thus, this model should prove useful for further investigation of disease pathogenesis as well as vaccine studies of visceral leishmaniasis. (Infect Immun. 2003 71:401)

訳 このモデルは，疾患の病因のさらなる研究のために役に立つと判明するはずである

■ [形／前] + investigation

❶ further investigation	さらなる研究	1,449
❷ clinical investigation	臨床研究	233
❸ present investigation	現在の研究	227
❹ recent investigation	最近の研究	216
❺ previous investigations	以前の研究	200
❻ under investigation	研究中である	373

★★ in vitro 形 試験管内の，副 試験管内で　用例数 64,848

in vitroの形で形容詞または副詞として用いられる．in vivo（生体内で）に対する表現として用いられる．

◆類義語：ex vivo ◆反意語：in vivo

■ 形 in vitro + [名]

❶ in vitro studies	試験管内の研究	1,673
❷ in vitro assays	試験管内アッセイ	727
❸ in vitro model	試験管内モデル	697
❹ in vitro transcription	試験管内転写	662
❺ in vitro experiments	試験管内実験	528

❶ **In vitro studies show that** p53 inhibits RPA-stimulated WRN helicase activity on an 849-bp M13 partial duplex substrate.（Cancer Res. 2005 65:1223）

訳 試験管内の研究は，～ということを示す

■ 形 [名] + in vitro

❶ activity in vitro	試験管内での活性	1,208
❷ growth in vitro	試験管内での成長	340

❶ **This study led to the successful discovery of novel compounds with increased antituberculosis activity in vitro** and a better understanding of the requisite pharmacological properties to advance this class.（J Med Chem. 2005 48:8261）

訳 この研究は，試験管内での増大した抗結核性活性を持つ新規の化合物の成功した発見につながった

■ 副 in vitro and + [副]／[副] + and in vitro

❶ in vitro and in vivo	試験管内および生体内で	6,196
・both in vitro and in vivo	試験管内および生体内の両方で	1,761
❷ in vivo and in vitro	生体内および試験管内で	2,327

❶ We have shown that **IP3R can be phosphorylated by Akt kinase in vitro and in vivo.**（J Biol Chem. 2006 281:3731）

訳 IP3Rは，試験管内および生体内でAktキナーゼによってリン酸化されうる

★★ in vivo ㊮ 生体内の，㊫ 生体内で　用例数 70,364

in vivoの形で形容詞または副詞として用いられる．
◆反意語：ex vivo, in vitro

■ ㊮ in vivo ＋［名］

❶ in vivo studies	生体内研究	911
❷ in vivo evidence	生体内の証拠	600
❸ in vivo model	生体内モデル	466
❹ in vivo function	生体内機能	421
❺ in vivo imaging	生体内イメージング	354

❷ Here, we provide in vivo evidence showing CD44 antagonism to breast cancer metastasis.（Cancer Res. 2005 65:6755）
訳 われわれは，～を示す生体内の証拠を提供する

■ ㊮［名］＋ in vivo

❶ function in vivo	生体内における機能	977
❷ activity in vivo	生体内における活性	825
❸ expression in vivo	生体内における発現	543
❹ growth in vivo	生体内における増殖	395

❶ In this study, the C-terminal domain (CTD) of the large subunit of fission yeast RFC is shown to be essential for its function in vivo.（Nucleic Acids Res. 2005 33:4078）
訳 ～は，生体内におけるそれの機能にとって必須であることが示される

★★ involve ㊍ 関与させる／関わる／巻き込む／含む　用例数 80,436

involved	関与している／～に関わった	47,751
involving	～に関わる	13,341
involves	～に関わる	12,473
involve	～に関わる	6,871

他動詞．受動態は，be involved inの用例が非常に多い．
◆類義語：implicate, engage, participate, contain, include

■ be involved ＋［前］

★ be involved	…は，関与している	18,714
❶ be involved in ～	…は，～に関与している	17,327
・be involved in regulation of ～	…は，～の調節に関与している	611
・be involved in regulating ～	…は，～を制御することに関与している	373
・be involved in the pathogenesis of ～	…は，～の病因に関与している	317
・be involved in the development of ～	…は，～の発生に関与している	162
・be involved in mediating ～	…は，～を仲介することに関与している	160
❷ be involved with ～	…は，～と関係する	126

❶ The dopamine D2 receptor (D2) is involved in the regulation of pituitary hormone secretion. (Brain Res. 2002 939:95)

訳 ～は，下垂体ホルモン分泌の調節に関与している

❶ The results suggest that Fas-FasL interaction associated with apoptosis is involved in the pathogenesis of acquired ocular toxoplasmosis in mice. (Infect Immun. 1999 67:928)

訳 ～は，後天性眼トキソプラズマ症の病因に関与している

❶ These data suggest that the TIM-1-TIM-4 interaction is involved in regulating T cell proliferation. (Nat Immunol. 2005 6:455)

訳 TIM-1-TIM-4相互作用は，T細胞増殖を制御することに関与している

■ [名] ＋ be involved in

❶	proteins are involved in ～	タンパク質は，～に関与している	208
❷	genes are involved in ～	遺伝子は，～に関与している	155
❸	pathway is involved in ～	経路は，～に関与している	151

■ [副] ＋ involved in

❶	directly involved in ～	直接～に関与している	618
❷	critically involved in ～	決定的に大きく～に関与している	361

❶ This glutamate (E12) is directly involved in the binding of cations in the complex interface. (Biochemistry. 2001 40:2419)

訳 ～は，複合体の界面における陽イオンの結合に直接関与している

■ [名] ＋ involved in

❶	genes involved in ～	～に関与する遺伝子	3,421
❷	proteins involved in ～	～に関与するタンパク質	1,715
❸	mechanisms involved in ～	～に関与する機構	1,156
❹	pathways involved in ～	～に関与する経路	677
❺	factors involved in ～	～に関与する因子	673

❶ Aberrant embryos did not turn green, and the expression of genes involved in photomorphogenesis was drastically attenuated. (Plant Cell. 1998 10:383)

訳 光形態形成に関与する遺伝子の発現が，強烈に減弱された

■ [動] ＋ to be involved in

★	to be involved in ～	～に関与している…	2,528
❶	appear to be involved in ～	…は，～に関与しているように思われる	311
❷	be shown to be involved in ～	…は，～に関与していることが示される	232
❸	be thought to be involved in ～	…は，～に関与していると考えられる	196
❹	be known to be involved in ～	…は，～に関与していることが知られている	180

❹ Other target nuclei are known to be involved in the regulation of sleep, including the lateral dorsal and pedunculopontine tegmentum. (J Comp Neurol. 2005 487:204)

訳 他の標的の核は，睡眠の調節に関与していることが知られている

■ involve ＋［名句］

❶	involve the formation of ～	…は，～の形成に関わる	235
❷	involve the activation of ～	…は，～の活性化に関わる	183
❸	involve the use of ～	…は，～の使用に関わる	168
❹	involve changes in ～	…は，～の変化に関わる	166

❶ Although the extent and mechanisms of packaging vary, **the process involves the formation of** nucleic-acid superstructures.（Nat Commun. 2012 3:901）

訳 その過程は，～の形成に関わる

■［動］＋ to involve

★	to involve ～	～に関わる…	1,504
❶	appear to involve ～	…は，～に関わるように思われる	447
❷	be thought to involve ～	…は，～に関わるように考えられる	194

❶ The in vivo response to radiotherapy is not well understood but **appears to involve the p53 tumor suppressor protein**.（Cancer Res. 2002 62:7316）

訳 ～は，p53癌抑制タンパク質に関わるように思われる

■［名］＋ involving

❶	mechanism involving ～	～に関わる機構	1,177
❷	pathway involving ～	～に関わる経路	615
❸	process involving ～	～に関わる過程	486
❹	interactions involving ～	～に関わる相互作用	447
❺	studies involving ～	～に関わる研究	345

❶ The results indicate that constant high plasma levels of progesterone **attenuate inflammatory hyperalgesia by a mechanism involving inhibition of** N-methyl-D-aspartate receptor activation at the spinal cord level.（Brain Res. 2000 865:272）

訳 …は，～の抑制に関わる機構によって炎症性の痛覚過敏を減弱させる

involvement ㊔ 関与　　用例数 9,009

involvement	関与	8,989
involvements	関与	20

複数形involvementsの割合は0.2％しかなく，原則，**不可算名詞**として使われる．

◈類義語：participation, engagement, intervention

■ involvement ＋［前］

involvement ofの前にtheが付く割合は非常に高い．

	the％	a/an％	⌀％			
❶	78	3	19	involvement of ～	～の関与	5,459

I

				・involvement of ～ in …	…への～の関与	1,672
				・evidence for the involvement of ～	～の関与の証拠	188
❷	2	1	43	involvement in ～	～への関与	1,630

❶ The results provide direct evidence for the involvement of unknown cellular factors in the membrane integration process of connexins. (J Biol Chem. 1998 273:7856)
訳 それらの結果は，コネキシンの膜組込み過程への未知の細胞因子の関与の直接の証拠を提供する

❷ Recent evidence suggests it has a pathological role in cerebral insults, but its involvement in intracerebral hemorrhage (ICH) is unknown. (Brain Res. 2001 901:38)
訳 脳内出血 (ICH) へのそれの関与は知られていない

■ [動] ＋ the involvement of

❶	suggest the involvement of ～	…は，～の関与を示唆する	501
❷	investigate the involvement of ～	…は，～の関与を精査する	166
❸	indicate the involvement of ～	…は，～の関与を示す	145

❶ Pervanadate reversed the gamma-T3-induced down-regulation of STAT3 activation, suggesting the involvement of a protein-tyrosine phosphatase. (J Biol Chem. 2010 285:33520)
訳 そして，(それは) プロテインチロシンホスファターゼの関与を示唆している

❸ The reduction in AMPAR amount is accompanied by receptor ubiquitination and can be blocked by suppression of proteasome activity, indicating the involvement of proteasome-mediated receptor degradation. (22396428 J Biol Chem. 2010 285:33520)
訳 そして，(それは) プロテアソームに仲介される受容体の分解の関与を示している

■ [形] ＋ involvement

❶	possible involvement	可能な関与	273
❷	potential involvement	可能な関与	193

irrespective 形 無関係の　用例数 1,461

irrespective ofの用例が圧倒的に多い.
◆類義語：independent, unrelated, irrelevant

■ irrespective ＋ [前]

❶	irrespective of ～	～とは無関係に	1,455
	・, irrespective of ～	, ～とは無関係に	726
	・irrespective of whether ～	～かどうかとは無関係に	120

❶ Patients were analysed in the treatment group to which they were randomised, irrespective of whether they received their allocated treatment. (Lancet Neurol. 2010 9:581)
訳 ～かどうかとは無関係に

★★ isolate 動 単離する／分離する／隔離する，名 単離体／分離株

用例数 49,519

isolated	単離される／～を単離した	30,529
isolates	～を単離する／単離体／分離株	13,946
isolate	～を単離する／単離体／分離株	4,474
isolating	～を単離する	570

他動詞の用例が多いが，名詞としても用いられる．複数形のisolatesの割合は，約90%と圧倒的に高い．

◆類義語：separate, dissociate, segregate, resolve

■ 動 be isolated ＋［前］

★ be isolated	…は，単離される	6,646
❶ be isolated from ～	…は，～から単離される	2,561
❷ be isolated by ～	…は，～によって単離される	570
❸ be isolated in ～	…は，～において単離される	350
❹ be isolated as ～	…は，～として単離される	222

❶ IECs **were isolated from** patients with Crohn's disease or ulcerative colitis and from normal controls.（Gastroenterology. 1998 115:1426）

訳 IECが，クローン病の患者から単離された

❷ Lipid raft and nonraft fractions from T cells **were isolated by** ultracentrifugation.（Arthritis Rheum. 2003 48:1343）

訳 ～が，超遠心によって単離された

■ 動［副］＋ isolated

❶ freshly isolated	新鮮に単離された	891
❷ previously isolated	以前に単離された	324

❶ Four weeks after constrictor placement, $CD31^+$ mononuclear cells（MNCs）were **freshly isolated** from the peripheral blood of each animal.（Circulation. 2003 107:461）

訳 ～が，おのおのの動物の末梢血から新鮮に単離された

■ 動［名］＋ isolated from

❶ cells isolated from ～	～から単離された細胞	930
❷ strains isolated from ～	～から単離された株	128

❶ T **cells isolated from** HCD4/DQ8 mice also responded well to modified（deamidated）versions of the gliadin peptides, whereas HCD4DQ6 mice did not.（J Immunol. 2002 169:5595）

訳 HCD4/DQ8マウスから単離されたT細胞

■ 動 isolated and ＋［過分］

❶ isolated and characterized	単離され，そして特徴づけられる	489

I

❶ A novel dominant inhibitory HLH factor, designated HHM (human homologue of maid), **was isolated and characterized**. (Hepatology. 2000 32:357)
訳 ～が，単離されそして特徴づけられた

■ 動 to isolate

★ to isolate ～	～を単離するために…	1,664
❶ be used to isolate ～	…は，～を単離するために使われる	290

❶ **Phage display was used to isolate peptides that bind** the paralog XRCC3. (Cancer Res. 2004 64:3002)
訳 ファージディスプレイが，～に結合するペプチドを単離するために使われた

■ 名 isolates + [前]

	the%	a/an%	ø%			
❶	3	-	45	isolates of ～	～の単離体（分離株）	1,646
❷	4	-	42	isolates from ～	～からの単離体（分離株）	1,432

❶ The results from this study suggest that probe SSA-2 **may serve as a species-specific DNA probe for the identification of clinical isolates of** S. sanguinis. (J Clin Microbiol. 2003 41:3481)
訳 …は，～の臨床分離株の同定のための種特異的なDNAプローブとして働くかもしれない

■ 名 [形] + isolates

❶ clinical isolates	臨床分離株	1,046
❷ primary isolates	一次分離株（単離体）	217

isolation 名 単離／分離／隔離／孤立 用例数 5,853

isolation	単離	5,805
isolations	単離	48

複数形のisolationsの割合は0.8%しかなく，原則，**不可算名詞**として使われる．
◆類義語：separation, segregation, purification

■ isolation + [前]

	the%	a/an%	ø%			
❶	57	0	43	isolation of ～	～の単離	2,362
				・we report the isolation of ～	われわれは，～の単離を報告する	70
❷	3	0	84	isolation from ～	～からの単離	142

❶ **We report the isolation of a novel reovirus strain** from a 6.5-week-old child with meningitis. (J Infect Dis. 2004 189:1664)
訳 われわれは，新規のレオウイルス株の単離を報告する

❷ Repopulation, **as measured by cell isolation from recipient livers** 1-7 months after transplantation, was on average 20%. (Transplantation. 2002 73:1818)

訳 レシピエントの肝臓からの細胞の単離によって測定されたとき

■ [形] ＋ isolation

❶ reproductive isolation	生殖隔離	210
❷ social isolation	社会的隔離	91

■ isolation and ＋ [名]

❶ the isolation and characterization of ～	～の単離と特徴づけ	109

❶ Here we report the isolation and characterization of such a secondary mutation.（J Bacteriol. 2002 184:5052）

訳 ここにわれわれは，そのような二次性の変異の単離と特徴づけを報告する

■ [前] ＋ isolation

❶ in isolation	孤立して／単独で	137

❶ The HR-C region in isolation formed a weakly stable trimeric coiled coil.（J Biol Chem. 2004 279:20836）

訳 HR-C領域は，単独で弱く安定な三量体のコイルドコイルを形成した

J

★ junction ㊔ 接合部　　用例数 12,395

junction	接合部	6,982
junctions	接合部	5,413

複数形のjunctionsの割合は約45%とかなり高い.

■ junction ＋［前］

junction of/betweenの前にtheが付く割合は圧倒的に高い.

	the%	a/an%	ø%			
❶	96	4	0	junction of ～	～の接合部	337
				・at the junction of ～	～の接合部において	130
❷	84	13	1	junction between ～	～の間の接合部	162

❶ In particular, UTRs of the M and F genes **enhanced levels of GFP expression** at the junction of **the P and M genes** without altering replication of NDV, suggesting that UTRs could be used for enhanced expression of a foreign gene by NDV.（J Virol. 2010 84:2629）

訳 ～は，P遺伝子とM遺伝子の接合部におけるGFP発現のレベルを増強させた

keep 動 維持する　　用例数 1,926

keeping	～を維持する	672
kept	維持される／～を維持した	605
keep	～を維持する	519
keeps	～を維持する	130

他動詞.

◈類義語：maintain, sustain, hold, retain

■ kept ＋［前］

❶ kept in ～	～において維持される	128
❷ kept at ～	～において維持される	59

❶ Control mice were **kept in** a normal environment (RH = 50%-80%, no AF, T = 21-23 ℃) for the same duration. (Invest Ophthalmol Vis Sci. 2005 46:2766)

訳 対照群マウスは，正常の環境において維持された

■ be kept ＋［形］

★ be kept	…は，維持される	357
❶ be kept constant	…は，一定に維持される	48

■ in keeping with

❶ in keeping with ～	～と一致して	290

❶ **In keeping with** the idea of the involvement of a signaling system, this growth effect by Sxl is not cell autonomous. (Development. 2005 132:4801)

訳 シグナル伝達系の関与の考えと一致して

★ key 形 鍵となる／重要な／不可欠な，名 鍵　　用例数 29,500

key	鍵となる／重要な／不可欠な／鍵	29,419
keys	鍵	81

主に形容詞として用いられが，名詞の用例もある．複数形のkeysの割合が，約20%ある可算名詞．名詞としては，the key toの用例が多い．

◈類義語：important, vital, crucial, critical, significant

■ 形 key ＋［名］

❶ key role	鍵となる役割	3,928
・play a key role in ～	…は，～における鍵となる役割を果たす	1,141
❷ key regulator	鍵となる調節因子	1,120
❸ key step	鍵となる段階	772

❹ key component	鍵となる構成成分	760
❺ key enzyme	鍵となる酵素	474
❻ key residues	鍵となる残基	394
❼ key factor	鍵となる因子	374
❽ key features	鍵となる特徴	374
❾ key determinant	鍵となる決定要因	361

❶ These results indicate that DAPK plays a key role in mediating ischemic neuronal injury. (J Biol Chem. 2005 280:42290)

訳 DAPKは，虚血性ニューロン障害を仲介する際に鍵となる役割を果たす

❷ The chromosomal passenger complex (CPC) is a key regulator of mitosis in many organisms, including yeast and mammals. (Development. 2005 132:4777)

訳 ～は，多くの生物において有糸分裂の鍵となる調節因子である

❸ Decapping is a key step in mRNA turnover. (Mol Cell. 2005 20:905)

訳 キャップ除去は，メッセンジャーRNAの代謝回転における鍵となる段階である

■ 形 be key ＋［前］

★ be key	…は，不可欠である／鍵となる～である	1,895
❶ be key to ～	…は，～に不可欠である	467
・be key to ～ing	…は，～を理解するのに不可欠である	69
❷ be key for ～	…は，～にとって不可欠である	91

❶ Lymphocyte proliferation is key to the regulation of the immune system. (Oncogene. 2006 25:2170)

訳 リンパ球の増殖は，免疫系の調節に不可欠である

■ 名 the key to

❶ the key to ～	～への鍵	266

❶ The key to the success of this method is generation of the nitrile anion in the presence of the heteroaryl halide. (J Org Chem. 2005 70:10186)

訳 この方法の成功への鍵は，～におけるニトリルアニオン生成である

★ kinetics 名 動力学／速度論　用例数 13,478

単複同形.

■ kinetics ＋［前］

kinetics ofの前にtheが付く割合は圧倒的に高い.

	the%	a/an%	ø%			
❶	83	0	15	kinetics of ～	～の動力学／速度論	5,259
❷	11	0	83	kinetics in ～	～における動力学／速度論	443

❶ The choices of linker and metal complex were also found to have significant impact on the kinetics of the reaction between the 1,10-phenanthroline-5,6-dione ligand and NADH. (J Am Chem Soc. 2012 134:18022)

訳 …は，また，～の間の反応の動力学に有意な影響を持つことも見つけられた

■ ［名］＋ kinetics

❶ binding kinetics	結合速度論	355
❷ steady-state kinetics	定常状態速度論	292

⁑ know 動 知る　　用例数 62,054

known	知られている／既知の	60,826
know	～を知っている	916
knowing	～を知っている	233
knew	～を知っていた	57
knows	～を知っている	22

他動詞．〈be known to *do*〉〈known as〉〈little is known about〉の用例が多い．

◆類義語：learn, appreciate, accept, assume, consider, believe, think, view, understand

■ be known to *do*

★ be known to *do*	…は，～すると知られている	8,081
❶ be known to be ～	…は，～であると知られている	2,099
❷ be known to play ～	…は，～を果たすと知られている	378
❸ be known to have ～	…は，～を持つと知られている	347
❹ be known to regulate ～	…は，～を調節すると知られている	267
❺ be known to cause ～	…は，～を引き起こすと知られている	229
❻ be known to induce ～	…は，～を誘導すると知られている	200

❶ Tyrosine kinase activity is known to be important in neuronal growth cone guidance.（J Cell Biol. 2001 155:427）

訳 チロシンキナーゼ活性は，～において重要であると知られている

■ known ＋［前］

❶ known as ～	～として知られている	6,696
・also known as ～	～としても知られている	2,609
❷ known for ～	～ゆえに知られている	877

❶ Normal cells do not divide indefinitely due to a process known as replicative senescence.（Mol Cell Biol. 2000 20:273）

訳 複製的老化として知られている過程ゆえに

❶ The rgg gene（also known as ropB）is required for the expression of streptococcal erythrogenic toxin B （SPE B）, an extracellular cysteine protease that contributes to virulence.（Infect Immun. 2001 69:822）

訳 ropBとしても知られている

■ [名] + is known about

★ is known about ～	～について…が知られている	7,777
❶ little is known about ～	～についてほとんど知られていない	6,473
❷ less is known about ～	～についてさらにわずかしか知られていない	610
❸ much is known about ～	～についてたくさん知られている	238
❹ nothing is known about ～	～について何も知られていない	182
❺ what is known about ～	～について知られていること	173

❷ Less is known about the neurobiology of $P2X_7$ receptors in the enteric nervous system (ENS). (J Comp Neurol. 2001 440:299)

訳 ～の神経生物学についてさらにわずかしか知られていない

■ little is known + [前]

★ little is known	ほとんど知られていない	7,617
❶ little is known about ～	～についてほとんど知られていない	6,473
・little is known about how ～	どのように～かについてほとんど知られていない	762
・very little is known about ～	～についてほんのわずかしか知られていない	408
・relatively little is known about ～	～について比較的わずかしか知られていない	400
・little is known about the mechanisms	機構についてほとんど知られていない	267
・little is known about the role of ～	～の役割についてほとんど知られていない	267
❷ little is known regarding ～	～に関してほとんど知られていない	393
❸ little is known of ～	～についてほとんど知られていない	367

❶ The 330-kbp viral genome has been sequenced, yet little is known about how viral mRNAs are synthesized and processed. (J Virol. 2001 75:1744)

訳 けれども，どのようにウイルスのメッセンジャーRNAが合成されそしてプロセスされるかについてほとんど知られていない

❶ Relatively little is known about the mechanism by which the signal is transmitted from the sensory site to the catalytic site. (J Biol Chem. 2003 278:13192)

訳 シグナルがセンサー部位から触媒部位へ伝えられる機構については比較的わずかしか知られていない

❸ Little is known of molecular mechanisms of human mechanosensation. (Neuron. 2003 37:731)

訳 ～の分子機構についてはほとんど知られていない

■ it is not known + [名節]

★ it is not known	知られていない	1,464
❶ it is not known whether ～	～かどうかは知られていない	818
❷ it is not known how ～	どのように～かは知られていない	245
❸ it is not known if ～	～かどうかは知られていない	148

❶ However, it is not known whether specific domains of FN can also regulate apoptosis. (J Biol Chem. 1999 274:30906)

訳 しかし，～かどうかは知られていない

■ known ＋［that節］

★	known that ～	～ということが知られている	997
❶	it is well known that ～	～ということがよく知られている	215
❷	it is known that ～	～ということが知られている	146

❷ It is known that albumin combines with cysteine in circulation to form albumin-Cys[34]-S-S-Cys.（J Biol Chem. 2001 276:30111）

訳 ～ということが知られている

knowledge ㊔ 知識 　用例数 9,100

複数形の用例はなく，原則，不可算名詞として使われる．

◆類義語：understanding, realization

■ knowledge ＋［前］

	the%	a/an%	ø%			
❶	13	4	57	knowledge of ～	～に関する知識	3,442
❷	13	0	67	knowledge about ～	～に関する知識	573

❶ Knowledge of their distribution is crucial for understanding the evolution of the inflammatory process.（Arthritis Rheum. 2005 52:3718）

訳 それらの分布に関する知識

❷ Growing knowledge about gene-disease associations will lead to new opportunities for genetic testing.（Am J Epidemiol. 2002 156:311）

訳 遺伝子ー疾患の関連に関する増加する知識

■ to our knowledge

❶	to our knowledge	われわれの知る限りでは	1,751
	・to our knowledge, this is the first ～	われわれの知る限りでは，これは最初の～である	552
	・to the best of our knowledge	われわれの最もよく知る限りでは	200

❶ To our knowledge, this is the first report of control of mdm2 at the post-transcriptional level and in a p53-independent manner.（J Biol Chem. 2000 275:26024）

訳 われわれの知る限りでは，これはmdm2の制御の最初の報告である

■［形］＋ knowledge

❶	current knowledge	現在の知識	514
❷	prior knowledge	予備知識	212
❸	detailed knowledge	詳細な知識	123

L

★ label 動 標識する，名 標識　　用例数 21,176

labeled	標識される／～を標識した	15,836
label	～を標識する／標識	3,596
labels	～を標識する／標識	979
labelled	標識される／～を標識した	765

名詞としても用いられるが，他動詞の用例が多い．labeledはアメリカ英語，labelledはイギリス英語で用いられる．labelingは別項参照．

◆類義語：mark, tag

■ labeled ＋［前］

❶ labeled with ～	～によって標識される	1,903
・be labeled with ～	…は，～によって標識される	611
❷ labeled by ～	～によって標識される	384
❸ labeled in ～	～において標識される	212
❹ labeled for ～	～に対して標識される	171

❶ Ganglion cells were labeled with biotinylated dextran amine in four embryos.（J Neurosci. 2001 21:8129）

訳 神経節細胞は，ビオチン化されたデキストランアミンによって標識された

❷ Kinetochores were labeled by expression of epitope-tagged CENP-A, which stably marks prekinetochore domains in human cells.（J Cell Biol. 2000 151:1113）

訳 動原体が，～の発現によって標識された

■［副］＋ labeled

❶ fluorescently labeled	蛍光性に標識された	752
❷ retrogradely labeled	逆行性に標識された	315
❸ isotopically labeled	同位体的に標識された	172
❹ doubly labeled	二重に標識された	148

❶ Inhibition has been measured through the monitoring of fluorescence resonance energy transfer between fluorescently labeled peptide and RNA components.（J Mol Biol. 2004 336:625）

訳 抑制は，蛍光性に標識されたペプチドとRNAコンポーネントの間の蛍光共鳴エネルギー転移のモニタリングによって測定されてきた

■［形／名］-labeled ＋［名］

❶ double-labeled	二重標識された	286
❷ spin-labeled	スピン標識された	250
❸ fluorescein-labeled	フルオレセイン標識された	236
❹ dye-labeled	色素標識された	183

■ labeled + [名]

❶	labeled cells	標識された細胞	697
❷	labeled neurons	標識されたニューロン	404
❸	labeled proteins	標識されたタンパク質	153

labeling / labelling ㊔ 標識／標識化　用例数 9,516

labeling	標識／標識化	9,035
labelings	標識／標識化	7
labelling	標識／標識化	481
labellings	標識／標識化	1

labellingはイギリス英語で用いられる．複数形の用例はほとんどなく，原則，**不可算名詞**として使われる．

■ labeling + [前]

	the%	a/an%	ø%			
❶	10	0	87	labeling of ～	～の標識	1,534
❷	0	1	97	labeling with ～	～による標識	539
❸	1	0	84	labeling in ～	～における標識	360

❶ **Labeling of δTyr-212 in the extracellular domain** was inhibited >90% by d-tubocurarine, whereas addition of either carbamylcholine or isoflurane had no effect. (Biochemistry. 2003 42:13457)

訳 細胞外ドメインにおけるδTyr-212の標識

■ labeling + [名]

❶	labeling experiments	標識実験	346
❷	labeling studies	標識研究	292

❶ **Double-labeling experiments suggest that** neuronal GLT-1 protein is primarily localized to the dendrites of excitatory neurons. (J Neurosci. 1998 18:4490)

訳 二重標識実験は，～ということを示唆する

■ [名／形] + labeling

❶	end labeling	末端標識	719
❷	double labeling	二重標識	386
❸	metabolic labeling	代謝標識	301
❹	spin labeling	スピン標識	243
❺	photoaffinity labeling	光親和性標識	230
❻	isotope labeling	同位体標識	208

★★ lack 動 欠く，名 欠如 用例数 40,352

lacking	～を欠く	17,102
lack	～を欠く／欠如	16,977
lacks	～を欠く／欠如	3,736
lacked	～を欠いた	2,537

他動詞の用例が多いが，名詞としても用いられる．動詞としては現在分詞の用例が多く，受動態で用いられることはない．名詞の複数形のlacksの用例はほとんどないが，単数形に不定冠詞が使われる例は非常に多い．

◆類義語：deficiency, deprivation, depletion, deficit, defect, paucity, insufficiency, shortage, shortness, deletion, loss, deprive, starve

■ 動［名］＋ lacking

❶	mice lacking ～	～を欠いているマウス	3,580
❷	cells lacking ～	～を欠いている細胞	1,908
❸	mutants lacking ～	～を欠いている変異体	889
❹	strains lacking ～	～を欠いている株	394
❺	embryos lacking ～	～を欠いている胚	272

❶ Knock-out (KO) mice lacking FMRP were compared with congenic C57BL/6J wild-type (WT) controls. (J Neurosci. 2005 25:9460)

訳 FMRPを欠いているノックアウト（KO）マウスが，コンジェニックなC57BL/6J野生型（WT）対照群と比較された

❷ Cells lacking PTEN possess elevated levels of PtdIns(3,4,5)P_3, PKB, and S6K activity and heterozygous PTEN$^{+/-}$ mice develop a variety of tumors. (Curr Biol. 2005 15:1839)

訳 PTENを欠いている細胞は，上昇したレベルの～を持つ

■ 動 lack ＋［名句］

❶	lack both ～	…は，両方の～を欠く	788
❷	lack functional ～	…は，機能的な～を欠く	404
❸	lack either ～	…は，どちらかの～を欠く	364
❹	lack expression of ～	…は，～の発現を欠く	223
❺	lack endogenous ～	…は，内在性の～を欠く	193

❶ Conversely, the lipid-rich necrotic core (LR-NC), lacking both vasculature and matrix, shows no or only slight enhancement. (Circulation. 2005 112:3437)

訳 脈管構造および基質の両方を欠いている

■ 名 lack ＋［前］

	the%	a/an%	ø%			
❶	43	29	27	lack of ～	～の欠如	10,899
				・due to the lack of ～	～の欠如のせいで	295
				・be hampered by the lack of ～		

…は，～の欠如によって妨害される		138
・lack of expression	発現の欠如	132
・lack of effect	効果の欠如	114
・be limited by the lack of ～ …は，～の欠如によって制限される		86

❶ Lack of expression of hTR and hTERT in ALT cell lines is associated with histone H3 and H4 hypoacetylation and methylation of Lys9 histone H3. (Cancer Res. 2005 65:7585)
訳 ALT 細胞株におけるhTRおよびhTERTの発現の欠如

★ large ㊫ 大きな　用例数 48,514

◆類義語：huge, great, massive

■ large ＋［名］

❶ large-scale ～	大規模な～	3,849
❷ a large number of ～	たくさんの～	2,444
❸ large amounts of ～	大量の～	765
❹ in large part	大部分は	462
❺ a large proportion of ～	大部分の～／～の大部分	397
❻ a large family of ～	大きなファミリーの～	396
❼ a large fraction of ～	大部分の～／～の大部分	387
❽ a large increase in ～	～の大幅な増大	345
❾ large quantities of ～	大量の～	277

❶ SNPs discovered by RRS also offer unique advantages for large-scale genotyping. (Nature. 2000 407:513)
訳 ～は，また，大規模な遺伝子型同定に対するユニークな利点を提供する

❷ Angiogenesis and enhanced microvascular permeability are hallmarks of a large number of inflammatory diseases. (Blood. 2002 99:538)
訳 ～は，たくさんの炎症性疾患の顕著な特徴である

❸ Large amounts of the FMC were generated in the first 3h of this incubation, and remained throughout the 9d incubation. (J Biol Chem. 2003 278:22144)
訳 大量のFMCが生成された

❹ Epithelial Na^+ transport is regulated in large part by mechanisms that control expression of the epithelial Na^+ channel (ENaC) at the cell surface. (J Biol Chem. 2004 279:5042)
訳 …は，大部分は～する機構によって調節される

■ be large ＋［副］

★ be large ～	…は，大きい	821
❶ be large enough	…は，十分大きい	104

❶ The groove is large enough to accommodate the loop between beta-strands beta4 and beta5 of the lectin domain of E-selectin that has been implicated in neutrophil adhesion. (J Biol Chem. 1998 273:11770)
訳 その溝は，～の間のループを適応させるのに十分大きい

■ [副] ＋ large

❶ very large	とても大きな	663
❷ relatively large	比較的大きな	483

★ largely 副 大部分は／主として／ほとんど 用例数 12,670

否定的な内容の文で使われることの方が多い.

◆類義語：predominantly, mostly, mainly, almost, nearly, quite

■ largely ＋ [形]

❶ largely unknown	ほとんど知られていない	2,535
・be largely unknown	…は，ほとんど知られていない	1,435
・remain largely unknown	…は，ほとんど知られていないままである	910
❷ largely due to ～	大部分は～のせいである	308
・be largely due to ～	…は，大部分は～のせいである	203
❸ largely unexplored	ほとんど探索されていない	303
・remain largely unexplored	…は，ほとんど探索されないままである	171
❹ largely independent of ～	ほとんど～に依存しない	241
❺ largely unaffected	ほとんど影響されていない	211
❻ largely responsible for ～	～に主として責任のある	205
❼ largely undefined	ほとんど定義されていない	169
❽ largely dependent on ～	大部分は～に依存する	160
❾ largely restricted to ～	ほとんど～に限られる	153

❶ However, the molecular mechanisms by which Sprouty exerts its antagonistic effect **remain largely unknown**. (Proc Natl Acad Sci USA. 2002 99:6041)

訳 ～は，ほとんど知られていないままである

❷ We think **this benefit is largely due to these drugs' ability** to inhibit angiogenesis. (Lancet. 2003 361:605)

訳 この利点は，大部分はこれらの薬剤の能力のせいである

❻ Furthermore, our data showed that the EC2 sequence ^{173}LETFTVKSCPDAIKEVFDNK^{192} **was largely responsible for the CD9-mediated CHO cell phenotype**. (Blood. 2002 100:4502)

訳 ～は，CD9に仲介されるCHO細胞の表現型に主として責任があった

★ larger 形 より大きな 用例数 12,113

largeの比較級.

■ larger ＋ than/[前]

❶ larger than ～	～より大きな	1,845
・be larger than ～	…は，～より大きい	465
・larger than that of ～	～のそれより大きな	121
❷ larger in ～	～においてより大きな	472

❸ larger for ～	～にとってより大きな	141

❶ Furthermore, a point mutant of Cx46, with leucine substituted by glycine in position 35, displayed a conductance much larger than that of the wild type.（Biophys J. 2006 90:140）
訳 ～は，野生型のそれよりずっと大きなコンダクタンスを示した

❷ The rate of decline for each decade was larger in men than in women from the 40s onward.（Circulation. 2005 112:674）
訳 ～は，女性においてより男性において大きかった

■［副］＋ larger

❶ much larger	ずっとより大きな	727
・much larger than ～	～よりずっと大きな	193
❷ significantly larger	有意により大きな	614
❸ ～ times larger	～倍大きな	192

❶ The murine PE1 open reading frame（ORF）is much larger than the previously reported human PE1 ORF.（Biochemistry. 2000 39:8917）
訳 マウスのPE1のオープンリーディングフレーム（ORF）は，以前報告されたヒトのPE1のORFよりずっと大きい

❷ Furthermore, microinjection of 8-bromo-cAMP（50μM）induced significantly larger phase shifts than vehicle.（Brain Res. 2005 1058:10）
訳 ～は，溶媒よりも有意に大きな位相シフトを誘導した

last 形 最後の／最近の，動 続く　用例数 6,627

last	最後の／最近の／続く	6,065
lasted	続いた	437
lasts	続く	125

形容詞の用例が多いが，別の意味の動詞としても用いられる．lateの最上級で，〈the last ～〉の用例が非常に多い．

■ the last ＋［名／形］

★ the last ～	最近の～／最後の～	4,586
❶ the last decade	最近の10年	376
・over the last decade	最近の10年の間に	147
・in the last decade	最近の10年のうちに	124
❷ the last two ～	最後の2つの～	245
❸ the last year	ここ1年	205
❹ the last step	最終段階	135

❶ In the last decade, continued efforts in pancreas cancer research have led to the development of new, more effective therapies.（Gastroenterology. 2005 128:1642）
訳 最近の10年のうちに，

★ late ㊊ 遅発の／後期の／遅い，㊖ 遅く　用例数 16,743

形容詞および副詞として用いられる．

◆類義語：slow, delayed　◆反意語：early, initially

■ ㊊ late ＋［名］

❶ late stage	後期	932
❷ late onset	遅発	850
❸ late phase	後期	500

❶ Inhibition of fungal growth on edr2 mutant leaves **occurred at a late stage of the infection process** and coincided with formation of necrotic lesions approximately 5 days after inoculation.（Plant J. 2005 44:245）

㊞ ～が，感染の過程の後期に起こった

■ ㊊［形］＋ and late

❶ early and late ～	早期および後期の～	918

❶ This allowed us to **study the early and late phases of spontaneous induction of the B cell autoimmune response.**（J Immunol. 2005 174:6872）

㊞ B細胞自己免疫応答の自発性誘導の初期相および遅延相を研究する

■ ㊖ late ＋［前］

❶ late in ～	～において遅く	908
・late in infection	感染の後期において	97

❶ In the absence of inducer, viral membrane formation was delayed and crescents and occasional immature forms **were detected only late in infection.**（J Virol. 2004 78:257）

㊞ ～は，感染の後期においてのみ検出された

★ later ㊊ より遅い，㊖ あとで／より遅く　用例数 10,232

lateの比較級．

■ ㊊ later ＋［名］

❶ later stages	より遅い段階／後期	801
・at later stages	より遅い段階で	336
❷ at later times	より遅い時点で	196
❸ in later life	晩年に	136

❶ **At later stages**, it is known that neuroblasts switch on expression of Grainyhead (Grh) and maintain it through many subsequent divisions.（Development. 2005 132:3835）

㊞ より遅い段階で，

■ 副 later ＋［前］

❶ later in ～	～におけるより遅い時期に	995
・later in life	晩年に	223
・later in development	発達のより遅い時期に	148

❶ Joint injury in young adults leads to an increased risk of developing osteoarthritis (OA) later in life. (Arthritis Rheum. 2005 52:2386)

訳 若年成人における関節傷害は，晩年に骨関節炎（OA）を発症する増大したリスクにつながる

■ 副［名］＋ later

❶ ～ days later	～日あとで	540
❷ ～ weeks later	～週間あとで	392
❸ ～ h later	～時間あとで	308

❶ Five days later, fetuses were subjected to 0.5 h hypoxaemia during either i.v. saline or a selective CGRP antagonist in randomised order. (J Physiol. 2005 566:587)

訳 5日あとで，

★ lead 動 つながる／導く／仕向ける，名 手がかり／鉛 用例数 71,229

leading	つながる／主な	18,924
leads	つながる／～を仕向ける／手がかり	18,544
lead	つながる／～を仕向ける／手がかり	17,371
led	つながった／～を仕向けた	16,390

自動詞の用例が多いが，他動詞としても用いられる．名詞の用例も見られる．lead to の用例が圧倒的に多い．第５文型のS+V+O+to *do*のパターンでも使われる．

◆類義語：bring, guide, cause, result, allow, permit, enable, prompt

■ lead to ＋［名句］

★ lead to ～	…は，～につながる	60,912
❶ lead to increased ～	…は，増大した～につながる	1,945
❷ lead to the identification of ～	…は，～の同定につながる	887
❸ lead to the formation of ～	…は，～の形成につながる	861
❹ lead to the development of ～	…は，～の発生につながる	848
❺ lead to an increase in ～	…は，～の増大につながる	712
❻ lead to activation of ～	…は，～の活性化につながる	581
❼ lead to the discovery of ～	…は，～の発見につながる	403
❽ lead to loss of ～	…は，～の喪失につながる	367
❾ lead to a decrease in ～	…は，～の低下につながる	324
❿ lead to the hypothesis that ～	…は，～という仮説につながる	306

❷ The application of phenotype-based statistical modeling approaches has led to the identification of new markers for the development of diabetic kidney disease.（Diabetes. 2004 53:784）

訳 ～は，糖尿病性腎臓疾患の発症の新しいマーカーの同定につながった

❸ Homodimerization of caspase-8 or caspase-9 leads to the formation of a stable dimeric complex.（J Biol Chem. 2002 277:50761）

訳 ～は，安定な二量体複合体の形成につながる

❻ Here we show that overexpression of AND-34 leads to activation of the Rho family GTPases Cdc42 and Rac.（Cancer Res. 2003 63:6802）

訳 AND-34の過剰発現は，～の活性化につながる

■［名］＋ leading to

★ leading to ～	～につながる	14,617
❶ pathway leading to ～	～につながる経路	418
❷ events leading to ～	～につながる事象	377
❸ mechanisms leading to ～	～につながる機構	311

❶ The purpose of this study was to elucidate the molecular pathway leading to autophagic cell death.（Oncogene. 2005 24:980）

訳 この研究の目的は，自己貪食性細胞死につながる分子経路を解明することであった

■ lead us to *do*

★ lead us to *do*	…は，われわれに～するように仕向ける	1,406
❶ lead us to propose ～	…は，われわれに～を提案するように仕向ける	268
❷ lead us to hypothesize that ～	…は，われわれに～と仮定するように仕向ける	163
❸ lead us to conclude that ～	…は，われわれに～と結論するように仕向ける	135
❹ lead us to investigate ～	…は，われわれに～を精査するように仕向ける	100

❶ Our results lead us to propose a model in which Arf-GEF recruitment is linked to Golgi maturation via Arf1 activation.（Dev Cell. 2012 22:799）

訳 われわれの結果は，われわれに～であるモデルを提案するように仕向ける

■ leading ＋［名］

❶ leading cause of ～	～の主な原因	1,334
❷ leading edge	先端	808

❶ Ischemic heart disease (IHD) is a leading cause of death in India.（Am J Clin Nutr. 2004 79:582）

訳 虚血性心疾患（IHD）は，インドにおける死亡の主な原因である

★ least 形 最も少ない，名 最小のもの　用例数 26,599

littleの最上級．at leastの用例が圧倒的に多い．

◆反意語：most

■ at least ＋［名／代名／形］

★ at least	少なくとも	24,590
❶ at least one	少なくともひとつ	3,529
・at least one of ～	少なくとも～のひとつ	490
❷ at least two ～	少なくとも2つの～	2,806
・at least two distinct ～	少なくとも2つの別個の～	294
❸ at least some ～	少なくともいくつかの～	635

❶ The enhanced invasive potential **required the presence of MMP-9 and at least one of** the cell surface receptors, CD44, $\alpha_V\beta_3$, or $\alpha_V\beta_5$ integrin.（Cancer Res. 2005 65:11545）

訳 …は，MMP-9および～の少なくとも１つの存在を必要とした

❷ **The yeast transcription factor Ste12 is required in at least two distinct signaling processes**, each regulated by many of the same protein kinases.（J Biol Chem. 2006 281:1964）

訳 酵母の転写因子Ste12は，少なくとも２つの別個のシグナル伝達過程において必要とされる

■ at least ＋［副／副句］

❶ at least in part	少なくとも部分的には	2,705
❷ at least partially	少なくとも部分的には	672
❸ at least partly	少なくとも部分的には	289

❶ Subsequent research in animal models has shown that **the vaccine-enhanced disease is mediated at least in part by** memory cells producing Th2 cytokines.（J Immunol. 2005 174:7234）

訳 ワクチンにより増強されるその疾患は，少なくとも部分的には～によって仲介される

■［前］＋ at least

❶ persisted for at least ～	少なくとも～の間持続した	191
❷ the presence of at least ～	少なくとも～の存在	166

❶ Thirty-six percent of episodes of clinical remission off medication **persisted for at least 2 years**, and only 6% of such episodes persisted for 5 years.（Arthritis Rheum. 2005 52:3554）

訳 ～は，少なくとも２年間持続した

★ length ㊔ 長さ 用例数 30,401

length	長さ	27,538
lengths	長さ	2,863

複数形のlengthsの割合は約10%あるが，単数形は無冠詞で使われることが非常に多い．

L

■ length ＋［前］

	the%	a/an%	ø%			
❶	52	10	37	length of ～	～の長さ	5,337

				・length of stay	滞在の長さ	1,092
				・hospital length of stay	入院期間	161
				・the entire length of ～	～の全長	156
				・mean length of ～	～の平均長	139
				・average length of ～	～の平均長	111
❷	14	3	77	length in ～	～における長さ	347

❶ One of these **is the central pathway spanning the entire length of the extracellular domain** and covering a distance of approximately 70 A.（Proc Natl Acad Sci USA. 2011 108:13800）

訳 ～は，細胞外ドメインの全長に渡る中心経路である

■［形／名］＋ length

❶	full-length ～	全長の～／完全長の～	9,709
❷	telomere length	テロメア長	835
❸	chain length	鎖長	746
❹	fragment length	断片長	611
❺	cycle length	周期長	404

★lesion ㊔ 損傷／病変，㊦ 傷害する／破壊する　用例数 28,947

lesions	損傷／病変／～を傷害する／～を破壊する	19,032
lesion	損傷／病変／～を傷害する／～を破壊する	8,793
lesioned	傷害される／～を破壊した	976
lesioning	～を傷害する／～を破壊する	146

動詞としても使われるが，名詞の用例が多い．複数形のlesionsの割合は，約70%と圧倒的に高い．

◆類義語：damage, injury, destruction, disruption, rupture, injure, destroy, disrupt, abolish, impair

■ ㊔ lesions ＋［前］

	the％	a/an％	ø％			
❶	7	-	80	lesions in ～	～における損傷	1,903
❷	14	-	85	lesions of ～	～の損傷	1,456

❶ **Several of these mutants have lesions in peroxisomal protein genes.**（Proc Natl Acad Sci USA. 2004 101:1786）

訳 これらの変異体のいくつかは，ペルオキシソームタンパク質遺伝子に損傷を持つ

❷ Surprisingly, **lesions of the SCN in mPer2（Luciferase）knockin mice** did not abolish circadian rhythms in peripheral tissues, but instead caused phase desynchrony among the tissues of individual animals and from animal to animal.（Proc Natl Acad Sci USA. 2004 101:5339）

訳 mPer2（ルシフェラーゼ）ノックインマウスにおけるSCNの損傷

■ 名 [形／名] + lesions

❶ atherosclerotic lesions	アテローム硬化型の病変	732
❷ DNA lesions	DNA損傷	630
❸ skin lesions	皮膚病変	343

■ 名 lesion + [名]

❶ lesion formation	病変形成	352
❷ lesion size	病変サイズ	291
❸ lesion site	病変部位	234

■ 動 lesioned + [動]

❶ lesioned rats	傷害を受けたラット	288
❷ lesioned animals	傷害を受けた動物	144

❶ However, firing rates of large aspiny-like neurons **were faster in both hemispheres of the 6-OHDA-lesioned rats as compared to normal animals.** (Brain Res. 1999 833:58)

訳 ～は，正常の動物に比べて6-ヒドロキシドパミンで傷害を受けたラットの両方の半球においてより速かった

★★ less 形 副 より少ない／より～でない，名 より少ししかないもの

用例数 38,024

littleの比較級.

◈反意語：more

■ less than

★ less than ～	～より少ない	8,881
❶ be less than ～	…は，～より少ない	1,325
❷ of less than ～	～より少ない…	1,014
❸ less than that	それより少ない	546

❶ **Their activity was less than that observed for the ER partial agonist, 4-hydroxytamoxifen (ZOHT).** (Mol Pharmacol. 2004 66:970)

訳 それらの活性は，ERの部分作用薬4-hydroxytamoxifen（ZOHT）に観察されたそれよりも少なかった

❷ In homeostatic COS1 cells, the Neh2 degron **confers on Nrf2 a half-life of less than 10 min.** (J Biol Chem. 2004 279:31556)

訳 ～は，Nrf2に10分未満の半減期を与える

■ [副] + less

❶ significantly less	有意により少ない	2,132
❷ much less	ずっとより少ない	1,624

❸ 〜-fold less	〜倍より少ない	778

❶ The mean chemotaxis index **of the diabetics was significantly less than that of the controls** (p < or = 0.02). (J Dent Res. 1998 77:1497)
訳 糖尿病患者の〜は，対照群のそれより有意に少なかった

❷ However, S-(−)-scopolamine (0.5–1.5 mg/kg) produced similar increases in DA release in the mPFC, but **the effect was much less than that of oxotremorine**. (Brain Res. 2002 958:176)
訳 その効果は，オキソトレモリンのそれよりもずっと少なかった

❸ The level of interleukin 12 in the serum of CD154-deficient T-cell recipients **was 5-fold less than that of wild-type cell recipients**. (Gastroenterology. 2000 119:715)
訳 〜は，野生型細胞レシピエントのそれよりも５倍少なかった

■ less ＋［形］

❶ less likely to *do*	より〜しそうにない	1,270
・be less likely to *do*	…は，より〜しそうにない	835
❷ less effective	より効果的でない	814
❸ less sensitive	より敏感でない	682
❹ less severe	より重篤でない	609
❺ less efficient	より効率的でない	540
❻ less clear	より明瞭でない	537
❼ less stable	より安定でない	532
❽ less potent	より強力でない	452
❾ less active	より活発でない	448

❶ Patients with obsessive-compulsive disorder **were less likely to respond to placebo than patients with** generalized social phobia or panic disorder. (Am J Psychiatry. 2004 161:1485)
訳 …は，〜の患者よりもプラセボに応答しそうになかった

❷ When compared with curcumin, dexamethasone **was less effective in suppression of** NF-κB activation and induction of apoptosis in myeloma cells. (Blood. 2004 103:3175)
訳 …は，〜の抑制においてより効果的でなかった

■ less ＋［副］

❶ less well 〜	よりよく〜ではない	890
・less well understood	よりよく理解されてはいない	265
❷ less efficiently	より効果的でなく	385
❸ less frequently	より頻繁でなく	367

❶ CD19 promotes the proliferation and survival of mature B cells, but **its role in early B cell development is less well understood**. (J Immunol. 2003 171:5921)
訳 初期のB細胞発生におけるそれの役割は，よりよく理解されてはいない

■ less ＋［前］

❶ less in 〜	〜においてより少ない	495

❶ **Suicidal behavior was significantly less in patients treated with** clozapine vs olanzapine (hazard ratio, 0.76; 95% confidence interval, 0.58–0.97; P =.03). (Arch Gen Psychiatry. 2003 60:

82)
訳 自殺行動は，～で治療された患者において有意により少なかった

■ less is known about

❶ less is known about ～	～についてより少ししか知られていない	610

❶ Less is known about the mechanism by which this protocol prolongs xenograft survival. (Transplantation. 2001 71:319)
訳 ～である機構については，より少ししか知られていない

lesser 形 より少ない　用例数 2,838

littleの二重比較級．〈to a lesser ～〉の用例が圧倒的に多い．

■ to a lesser ＋［名］

★ to a lesser ～	より少ない～で	1,908
❶ to a lesser extent	より少ない程度で	1,554
❷ to a lesser degree	より少ない程度で	342

❶ Blocking antibodies demonstrated that their formation was dependent upon expression of OPGL and, to a lesser extent, on tumor necrosis factor α. (Infect Immun. 2002 70:3143)
訳 それらの形成はOPGLの発現に，そしてより少ない程度で，腫瘍壊死因子αに依存していた

★★ level 名 レベル／水準　用例数 172,544

levels	レベル	123,215
level	レベル	49,329

複数形のlevelsの割合は，約70％と圧倒的に多い．

■ levels ＋［前］

	the%	a/an%	ø%			
❶	22	-	76	levels of ～	～のレベル	52,403
				・levels of ～ in …	…における～のレベル	3,885
				・levels of expression	発現のレベル	730
❷	4	-	85	levels in ～	～におけるレベル	11,012

❶ In this study we have used a novel fusion assay based on intracistronic complementation of lacZ, in combination with fluorescent X-gal histochemistry and immunocytochemistry to assess levels of NCAM expression in individual muscle cells. (Dev Biol. 2000 221:112)
訳 個々の筋肉細胞におけるNCAM発現のレベルを評価するために

❷ Retinal VEGF mRNA levels in 1-week diabetic animals were 3.2-fold higher than in nondiabetic controls ($P < 0.0001$). (Invest Ophthalmol Vis Sci. 2001 42:2408)
訳 1週の糖尿病の動物における網膜のVEGFメッセンジャーRNAレベルは，非糖尿病の対照群におけるより3.2倍高かった

■ at + @ + level(s)

★ at ～ level(s)	～レベルで	17,167
❶ at the level of ～	～のレベルで	2,849
❷ expressed at ～ levels	～レベルで発現される	1,628
❸ at high levels	～高いレベルで	1,146

❶ NFAT family members appear to be regulated primarily at the level of their subcellular localization. (J Biol Chem. 2001 276:3666)

訳 NFATファミリーのメンバーは，主にそれらの細胞内局在のレベルで調節されるように思われる

❷ In the embryo, slt-1 is expressed at high levels in anterior epidermis. (Neuron. 2001 32:25)

訳 slt-1は，～において高いレベルで発現される

■ ［形／過分］+ level(s)

❶ high levels	高いレベル	8,590
❷ low levels	低いレベル	4,040
❸ higher levels	より高いレベル	3,516
❹ increased levels	増大したレベル	2,571
❺ elevated levels	上昇したレベル	2,366
❻ lower levels	より低いレベル	1,708
❼ reduced levels	低下したレベル	1,533
❽ molecular level	分子レベル	1,370
❾ normal levels	正常レベル	1,130

❶ Cyclin B accumulates to high levels in all mitotic vih cells, particularly at the spindle poles. (Curr Biol. 2004 14:1723)

訳 サイクリンBは，すべての有糸分裂中のvih細胞において高レベルに蓄積する

❹ Increased levels of IL-1 were first detected 8 h after infection with gonococci, suggesting that the earlier IL-8 and IL-6 responses were not mediated through the IL-1 signaling pathway. (Infect Immun. 2001 69:5840)

訳 増大したレベルのIL-1が，感染のあと８時間ではじめて検出された

❽ Although several auxin biosynthetic pathways have been proposed, none of these pathways has been precisely defined at the molecular level. (Genes Dev. 2002 16:3100)

訳 これらの経路のどれも分子レベルで正確には定義されていない

■ ［名］+ levels

❶ protein levels	タンパク質レベル	5,423
❷ mRNA levels	メッセンジャーRNAレベル	4,989
❸ expression levels	発現レベル	3,558
❹ transcript levels	転写物レベル	1,162
❺ plasma levels	血漿レベル	1,162
❻ serum levels	血清レベル	1,117

■ levels ＋［動］

❶ levels increased	レベルが，増大した	818
❷ levels were measured	レベルが，測定された	431

❶ Nrf2 mRNA levels increased approximately 2-fold 6 h after D3T treatment.（Mol Cell Biol. 2002 22:2883）
訳 Nrf2メッセンジャーRNAレベルが，D3T処理のあと6時間でおよそ2倍増大した

lie 動 位置する　　用例数 3,800

lies	位置する	1,909
lie	位置する	1,434
lying	位置する	417
lay	位置した	40

自動詞．過去形のlayは，別の他動詞（「～を置く」）の現在形と同じ綴りであるから混同しないこと．
◆類義語：locate, localize, position

■ lie ＋［前］

❶ lie in ～	…は，～に位置する	996
❷ lie within ～	…は，～内に位置する	560
❸ lie at ～	…は，～に位置する	274
❹ lie between ～	…は，～の間に位置する	230
❺ lie on ～	…は，～に位置する	210

❶ The site of hydrolysis, Leu480, lies in the α chain on the N-terminal side of the cysteine linking the α and β chains.（J Biol Chem. 2006 281:1489）
訳 ～は，N末端側のα鎖に位置する

❷ The 13q peak lies within a region previously linked strongly to panic disorder.（Hum Mol Genet. 2005 14:3337）
訳 ～は，以前にパニック障害に強く連鎖した領域内に位置する

★ light 名 光，形 軽い　　用例数 26,039

light	光／軽い	25,899
lights	光	140

形容詞としても使われるが，名詞の用例が非常に多い．複数形のlightsの割合は1％しかなく，原則，**不可算名詞**として使われる．
◆類義語：ray

■ light ＋［前］

❶	shed light on ～	…は，～に光を当てる／～を解明する	1,167
	・shed new light on ～	…は，～に新たな光を当てる	300
❷	in light of ～	～の観点から	1,061

❶ **Recent research has shed light on the factors determining** the end of the critical period, and on how cortical plasticity might be re-established in adulthood.（Curr Biol. 2005 15:R1000）

訳 最近の研究は，～を決定する因子を解明した

❷ **These findings are discussed in light of** the genetic and antigenic variability observed among some ASFV isolates.（Biochemistry. 2005 44:8408）

訳 これらの知見は，～の観点から議論される

■ light-［過分／現分／形］

❶	light-induced ～	光誘導性の～	810
❷	light-harvesting ～	集光性の～	453
❸	light-dependent ～	光依存性の～	318

❶ **UV light-induced CREB phosphorylation** was absent in ATM-deficient cells, confirming that ATM is required for CREB phosphorylation in UV irradiation-damaged cells.（J Biol Chem. 2006 281:1692）

訳 紫外光に誘導されるCREBのリン酸化

likelihood　名 可能性／見込み　　用例数 4,075

likelihood	可能性／見込み	3,997
likelihoods	可能性／見込み	78

複数形のlikelihoodsの割合が，約２％ある可算名詞．

◆類義語：probability

■ likelihood ＋［前／同格that節］

likelihood thatの前にtheが付く割合は圧倒的に高い．likelihood ofの前にtheが付く割合もかなり高い．

	the％	a/an％	ø％			
❶	62	25	12	likelihood of ～	～の可能性	1,932
				・likelihood of ～ing	～する可能性	429
				・increase the likelihood of ～	…は，～の可能性を増大させる	215
				・increased likelihood of ～	～の増大した可能性	123
				・greater likelihood of ～	～のより大きな可能性	113
❷	96	2	2	likelihood that ～	～という可能性	263

❶ **Etomidate administration increased the likelihood of developing adrenal insufficiency** (pooled relative risk 1.33; 95% confidence interval 1.22-1.46; Q statistic, 10.7; I2 statistic, 43.9%).（Crit Care Med. 2012 40:2945）

訳 エトミデート投与は，副腎不全を発症する可能性を増大させた

❶ We hypothesized that a mother's untreated caries was associated with increased likelihood of her children's untreated caries, after controlling for other factors.（J Dent Res. 2010 89:954）

訳 母親の未処置の齲蝕は，子供の未処置の齲蝕の増大した可能性と関連していた

❷ These data increase the likelihood that ADAM33, CDKAL1, and PTPN22 are true psoriasis-risk genes.（J Invest Dermatol. 2009 129:629）

訳 これらのデータは，～という可能性を増大させる

★★ likely 形 ありそうな／しそうな，副 おそらく　用例数 32,120

◆類義語：probable, probably, presumably, perhaps, possibly

■ 形 be likely to *do*

★ be likely to *do*	…は，おそらく～する	6,788
❶ be likely to be ～	…は，おそらく～である	3,054
・be likely to be important	…は，おそらく重要である	204
・be likely to be involved in ～	…は，おそらく～に関与している	166
❷ be likely to have ～	…は，おそらく～を持っている	442
❸ be likely to play ～	…は，おそらく～を果たす	353
❹ be likely to contribute to ～	…は，おそらく～に寄与する	202

❶ These proteins are likely to be involved in viral DNA replication, among other functions.（J Virol. 2002 76:338）

訳 これらのタンパク質は，ウイルスDNA複製におそらく関与している

❸ This short-term form of presynaptic depression is triggered by postsynaptic depolarization and is likely to play an important role in information processing.（J Physiol. 2004 556:95）

訳 ～は，情報処理においておそらく重要な役割を果たす

■ 形 [副] ＋ likely

❶ more likely	よりありそうな	4,931
・be more likely to *do*	…は，より～しそうである	2,650
・be more likely to have ～	…は，より～を持ちそうである	634
・be more likely than ～ to *do*	…は，～より…しそうである	280
❷ most likely	最もありそうな	3,546
・be most likely to *do*	…は，最も～しそうである	315
・, most likely by ～	，おそらく～によって	166
・be most likely due to ～	…は，最も～のせいでありそうである	132
❸ less likely	よりありそうにない	1,587
・be less likely to *do*	…は，～しそうにない	835
・be less likely than ～ to *do*	…は，～ほど…しそうにない	100

❶ Our findings support the concept that SNPs altering the conserved amino acids are more likely to be associated with cancer susceptibility.（Cancer Res. 2004 64:2251）

訳 保存されたアミノ酸を変えるSNPは，より癌の感受性に関連していそうである

❶ Caucasians were more likely than African Americans to be treated with ECT（odds ratio=4.71; 95% confidence interval [CI]=3.77-5.90）.（Am J Psychiatry. 2004 161:1635）

訳 白人は，おそらくアフリカ系アメリカ人より電撃療法によって治癒しそうであった

❷ Neurones located in outer lamina II, particularly radial and vertical cells, **were most likely to respond to Hcrt-2.** (J Physiol. 2002 538:517)

訳 ～は，最もHcrt-2に応答しそうであった

❸ By contrast, a randomized controlled trial showed that **infants treated with sunflower oil are less likely to experience nosocomial infections than are control infants.** (Curr Opin Pediatr. 2005 17:481)

訳 ひまわり油で処置された乳児は，対照群の乳児よりも院内感染を経験しそうにない

■ 形 likely ＋［that節］

★ likely that ～	～ということはありそうな	1,330
❶ it is likely that ～	～ということはありそうである	950

❶ **It is likely that** the R249S mutation resulted in an ensemble of native and native-like conformations in a dynamic equilibrium. (J Mol Biol. 2004 336:187)

訳 ～ということはありそうである

■ 副 likely ＋［形／動／過分］

❶ likely due to ～	～のせいでありそうな	851
❷ likely contribute to ～	…は，～に貢献しそうである	492
❸ likely involved in ～	～に関与しそうな	204

❶ **This is likely due to the low-density gas** improving albuterol deposition in the distal airways. (Am J Respir Crit Care Med. 2002 165:1317)

訳 これは，低密度ガスのせいでありそうである

likewise 副 同様に　用例数 1,843

文頭の用例が非常に多い．

◆類義語： similarly

■ likewise

❶ Likewise,	同様に,	1,292

❶ **Likewise, Rad14p promotes transcription of** other non-GAL genes such as CUP1, CTT1, and STL1 after transcriptional induction. (J Biol Chem. 2013 288:793)

訳 同様に，Rad14pは～の転写を促進する

★★ limit 動 制限する／限定する，名 限界　用例数 38,996

limited	限られる／～を制限した	19,772
limiting	～を制限する	8,758
limit	～を制限する／限界	6,358
limits	～を制限する／限界	4,108

他動詞の用例が多いが，名詞としても用いられる．複数形のlimitsの割合は約40%とか

なり高い.

◆類義語：limitation, restrain, restrict, confine

■ 動 be limited ＋［前］

★ be limited	…は，制限される	6,306
❶ be limited by ～	…は，～によって制限される	1,825
・be limited by the lack of ～	…は，～の欠如によって制限される	86
❷ be limited to ～	…は，～に限られる	1,741
❸ be limited in ～	…は，～に限られる	326

❶ Cyclosporin A is effective in the treatment of asthma patients, but its chronic use is limited by toxicity.（J Med Chem. 2003 46:674）

訳 それの慢性的な使用は，毒性によって制限される

❷ Thus, like the mouse, active human SNF2L is limited to neurons and a few other tissues.（J Biol Chem. 2004 279:45130）

訳 活性のあるヒトのSNF2Lは，ニューロンと少数の他の組織に限られている

■ 動［副］＋ limited

❶ only limited ～	限られた～のみ	343
❷ very limited ～	非常に限られた～	315

❶ The device is presently limited in scope to occlusal surfaces, and only limited ECM data from clinical trials are available.（J Dent Res. 2004 83:C76）

訳 臨床治験からの限られたECMデータのみが利用できる

■ 動 limited ＋［名］

❶ a limited number of ～	限られた数の～	661
❷ limited proteolysis	限定されたタンパク質分解	355

❶ However, only a limited number of tools are currently available for the analysis of multiple genomic sequences.（Genome Res. 2004 14:313）

訳 限られた数の手段だけが，～の解析のために現在利用できる

■ 動［名］-limiting

❶ rate-limiting ～	律速の～	3,360
・rate-limiting step	律速段階	1,240
・rate-limiting enzyme	律速酵素	446
❷ dose-limiting ～	用量規定の～	737

■ 名 limit ＋［前］

limit forの前にtheが付く割合は非常に高い.

	the %	a/an %	ø %			
❶	56	35	5	limit of ～	～の限界	1,680
				・limit of detection	検出の限界	528

				・detection limit of ～	～の検出限界	326
❷	73	23	0	limit for ～	～に対する限界	217

❶ The 40 min assay has a detection limit of 5 zmol (approximately 3 000 molecules) of CIP. (Anal Chem. 2010 82:3803)

訳 40分のアッセイは，5ゼプトモルの検出限界を持つ

■ 名 [名／形] ＋ limit

❶	detection limit	検出限界	660
❷	upper limit	上限	355
❸	lower limit	下限	258

limitation 名 制限／限定　用例数 5,351

limitations	制限／限定	3,246
limitation	制限／限定	2,105

複数形のlimitationsの割合は約60%と非常に高い.

◆類義語：restriction

■ limitations ＋ [前]

limitations ofの前にtheが付く割合は非常に高い.

	the %	a/an %	ø %			
❶	74	-	21	limitations of ～	～の制限	1,114
❷	7	-	85	limitations in ～	～における制限	350

❶ Due to the limitations of currently available tests, approaches combining structural and functional tests are essential in order to provide reliable detection of endpoints. (Curr Opin Pharmacol. 2013 13:115)

訳 現在利用できるテストの制限のせいで，

■ [形／名] ＋ limitation

❶	major limitation	主な制限	135
❷	nutrient limitation	栄養制限	115

★ linear 形 直線の　用例数 11,021

◆類義語：straight, proportional

■ linear ＋ [名]

❶	linear regression	直線回帰	1,102
❷	linear relationship	直線関係	318
❸	linear correlation	直線相関	223

❷ Here we find a linear relationship between genetic and geographic distance in a worldwide sample of human populations, with major deviations from the fitted line explicable by admixture or extreme isolation.（Proc Natl Acad Sci USA. 2005 102:15942）
訳 われわれは，遺伝的および地理的距離の間の直線関係を見つける

★★ link　(動) 連結する／関連づける，(名) 関連／結合　用例数 40,605

linked	連結される／関連する／連鎖する／結合する	27,668
link	つなぐ／結合する／関連／結合	8,911
links	つなぐ／結合する／関連／結合	4,026

動詞および名詞として用いられる．動詞としては，他動詞の用例が多いが自動詞としても使われる．linked toの用例が非常に多い．名詞としては，複数形のlinksの割合が，約30%ある可算名詞．

◆類義語：connect, couple, relate, associate, correlate, bind, attach, relation, relationship, correlation, connection, association

■ (動) linked ＋［前］

❶ linked to ～	～に関連する／～に連結される	10,014
・be linked to ～	…は，～に関連する／～に連結される	4,880
❷ linked with ～	～と関連する／～と連結される	828
❸ linked by ～	～によって連結される	533

❶ Only D17Z1 has been linked to CENP-A chromatin assembly.（Proc Natl Acad Sci USA. 2012 109:13704）
訳 D17Z1だけが，CENP-Aクロマチン構築に関連してきた

■ (動)［副］＋ linked

❶ closely linked	密接に関連する	567
❷ covalently linked	共有結合する	501
❸ tightly linked	密接に関連する	346

❶ Apoptosis is closely linked to proliferation.（Oncogene. 2002 21:1750）
訳 アポトーシスは，増殖と密接に関連する

❷ Surface proteins of Staphylococcus aureus are covalently linked to the cell wall envelope by a mechanism requiring C-terminal sorting signals with an LPXTG motif.（J Biol Chem. 2004 279:31383）
訳 黄色ブドウ球菌の表面タンパク質は，細胞壁エンベロープに共有結合している

■ (動) link ＋［名句］＋ to

❶ link ～ to …	～を…へ結びつける	2,758

❶ Integrins link the extracellular matrix to the cytoskeleton via a complex of proteins: the integrin-cytoskeleton link.（Curr Biol. 2008 18:R389）
訳 インテグリンは，細胞外基質を細胞骨格へ結びつける

■ 動 [名]-linked

❶	cross-linked	架橋結合した	2,187
	・cross-linked to ～	～に架橋結合した	315
❷	X-linked ～	X染色体に連鎖する～	1,977

❶ We found that **several proteins were** cross-linked to **apolipoprotein B.** (J Biol Chem. 2002 277:22010)

訳 いくつかのタンパク質がアポリポタンパク質Bに架橋結合した

■ 名 link ＋ [前]

	the%	a/an%	ø%			
❶	31	69	0	link between ～	～の間の関連	4,218
				・direct link between	～の間の直接の関連	316
				・mechanistic link between	～の間の機構的な関連	251
				・functional link between	～の間の機能的な関連	214

❶ **These findings argue against a** direct link between **virulence traits and antimicrobial resistance** in E. coli. (J Infect Dis. 2004 190:1739)

訳 これらの知見は，病原性の形質と抗菌剤耐性との間の直接の関連に反対論を唱える

★ linkage 名 連鎖／連関／結合　　用例数 11,534

linkage	連鎖／連関／結合	10,320
linkages	連鎖／連関／結合	1,214

複数形のlinkagesの割合は約10%あるが，単数形は無冠詞で使われることが非常に多い．

◆類義語：link

■ linkage ＋ [前]

	the%	a/an%	ø%			
❶	2	10	80	linkage to ～	～への連鎖	1,001
				・linkage to chromosome ～	染色体～への連鎖	101
❷	41	4	53	linkage of ～	～の連鎖／～の結合	733
❸	40	33	26	linkage between ～	～の間の連関	635

❶ **We found evidence for** linkage to chromosome **8** in adolescent-onset IGE families in which JME was not present. (Am J Hum Genet. 1999 64:1411)

訳 われわれは，第８染色体への連鎖の証拠を見つけた

❷ **Biotin protein ligases catalyze specific covalent** linkage of **the coenzyme biotin to biotin-dependent carboxylases.** (J Biol Chem. 2011 286:13071)

訳 ビオチン・タンパク質リガーゼは，補酵素ビオチンのビオチン依存性カルボキシラーゼへの特異的な共有結合を触媒する

❸ **These data established for the first time a mechanistic** linkage between **the import and activation of exogenous fatty acids** in yeast. (J Biol Chem. 2001 276:37051)

訳 これらのデータは，外来性脂肪酸の移入と活性化の間の機構的な連関をはじめて確立した

■ ［前］＋ linkage

❶ evidence for linkage	連鎖の証拠	420
❷ evidence of linkage	連鎖の証拠	272

■ ［形］＋ linkage

❶ genetic linkage	遺伝的連鎖	441
❷ significant linkage	有意な連関	200

■ linkage ＋［名］

❶ linkage disequilibrium	連鎖不均衡	1,412
❷ linkage analysis	連鎖解析	1,030
❸ linkage studies	連鎖研究	248
❹ linkage map	連鎖マップ	223

linker 名 リンカー／つなぎ役　用例数 5,623

linker	リンカー	4,760
linkers	リンカー	863

複数形のlinkersの割合が，約15％ある可算名詞.

◆類義語：bridge

■ linker ＋［前］

linker betweenの前にtheが付く割合は非常に高い.

	the%	a/an%	ø%			
❶	72	28	0	linker between ～	～の間のリンカー	175

❶ However, a specific interaction is observed with individual DPC molecules at a site close to the linker between the two EF-hands.（Biochemistry. 2005 44:6502）

訳 特異的な相互作用が，2つのEFハンドの間のリンカーに近い位置の個々のDPC分子に観察される

■ linker ＋［名］

❶ linker region	リンカー領域	416
❷ linker histone	リンカーヒストン	210

literature 名 文献　用例数 6,405

literature	文献	6,362
literatures	文献	43

複数形のliteraturesの割合は0.7%しかなく，原則，不可算名詞として使われる．
◆類義語：paper, article

■ literature ＋［前］

literature on/regardingの前にtheが付く割合は非常に高い．

	the%	a/an%	ø%			
❶	75	4	21	literature on ～	～に関する文献	663
❷	79	0	21	literature regarding ～	～に関する文献	162

❶ We review the current literature on complications seen with the use of vaginal mesh for both stress urinary incontinence and POP.（22617058J Bacteriol. 2001 183:6740）
訳 われわれは，合併症に関する現在の文献を概説する

■［形／過分］＋ literature

❶	recent literature	最近の文献	492
❷	current literature	現在の文献	260
❸	published literature	出版された文献	193

★ little 名 わずかしないもの，形 小さい／わずかな，副 ほとんど～でない

用例数 21,212

◆類義語：few, subtle, slight, insignificant

■ 名 little is known ＋［前］

★	little is known	ほとんど知られていない	7,617
❶	little is known about ～	～についてはほとんど知られていない	6,473
❷	little is known regarding ～	～に関してはほとんど知られていない	393
❸	little is known of ～	～についてはほとんど知られていない	367

❶ Currently, little is known about how the organism adapts to environmental stresses and maintains its cellular integrity.（J Bacteriol. 2001 183:6740）
訳 生物がどのように環境のストレスに適応するかについてはほとんど知られていない

■ 形 little ＋［名］

❶	little effect on ～	～に対してほとんど効果のない	1,747
	・have little effect on ～	…は，～に対してほとんど効果を持たない	1,107
❷	little information	ほとんど情報のない	499
	・there is little information	ほとんど情報はない	153
❸	little evidence	ほとんど証拠のない	492
❹	little change	ほとんど変化のない	345
❺	little attention	ほとんど注意のない	298

❶ Sunflower oil had little effect on the fatty acid composition of lipid fractions.（Lancet. 2003 361:477）

訳 ひまわり油は，～に対してほとんど効果を持たなかった

❷ However, **there is little information** on the outcomes of ITP patients refractory to splenectomy.（Blood. 2004 104:956）

訳 ～の結果に関する情報はほとんどない

■ 形 little or no／little if any

❶ little or no ～	ほとんどあるいは全く～ない	1,902
・little or no effect on ～	～に対してほとんどあるいは全く影響のない	337
・have little or no effect on ～	…は，～に対してほとんどあるいは全く影響を持たない	225
・with little effect on ～	～に対する影響はほとんどなしに	112
❷ little, if any,	もしあったとしてもほとんどない	130
❸ little if any	もしあったとしてもほとんどない	129

❶ Contact with L7 or L11 **had little or no effect on** the distribution of two other mRNAs in the cytoplasm of SN cell bodies.（J Neurosci. 2002 22:2669）

訳 …は，～の分布に対してほとんどあるいは全く影響を持たなかった

❶ LPA stimulated the transcriptional activity of the IL-8 gene **with little effect on** IL-8 mRNA stability.（J Biol Chem. 2004 279:9653）

訳 LPAは，IL-8メッセンジャーRNAの安定性に対する影響はほとんどなしに，IL-8遺伝子の転写活性を刺激した

❸ A synthetic peptide 337-372 blocked the cleavage at Lys36 (IC_{50} = 230 mM) while showing **little if any** effect on cleavage at Arg336.（J Biol Chem. 2003 278:16502）

訳 だが一方，Arg336における切断に対して，もしあったとしてもほとんど影響を示さない

■ 形［副］＋ little

❶ very little	非常に少ない	1,247
❷ relatively little	比較的少ない	874

❶ Both techniques agree that there is **very little** energy transfer from BIDS to FM.（Biochemistry. 2004 43:11917）

訳 BIDSからFMへのエネルギー転移は非常に少ない

■ 副 as little as

❶ as little as ～	わずか～ほど	540

❶ This method readily detects **as little as** 0.1 pmol of each TAG molecular species from chloroform extracts and is linear over a 1000-fold dynamic range.（Anal Biochem. 2001 295:88）

訳 この方法は，クロロホルム抽出物からそれぞれわずか0.1 pmolのTAG分子種を容易に検出する

L

★ localization 名 局在／局在性／局在化　用例数 21,903

localization	局在／局在性	21,602
localizations	局在／局在性	301

複数形のlocalizationsの割合は１%しかなく，原則，**不可算名詞**として使われる．
◆類義語：position, location, site

■ localization ＋［前］

	the%	a/an%	ø%			
❶	34	2	62	localization of ～	～の局在	8,541
				・localization of ～ to …	～の…への局在	712
❷	2	2	48	localization to ～	～への局在	959
❸	3	1	69	localization in ～	～における局在	940

❶ The roX RNAs are required for the **localization of** MSL complexes **to** the X chromosome. (Genetics. 2004 166:1825)
訳 roX RNAは，MSL複合体のX染色体への局在のために必要とされる

■ ［形／名］＋ localization

❶	nuclear localization	核局在	3,066
❷	subcellular localization	細胞内局在	1,792
❸	membrane localization	膜局在	636
❹	cellular localization	細胞局在	532
❺	intracellular localization	細胞内局在	363
❻	protein localization	タンパク質局在	322
❼	cytoplasmic localization	細胞質局在	317

❶ Here, we show that **the nuclear localization of Lgs entirely depends on Pygo**, which itself is constitutively localized to the nucleus; thus, Pygo functions as a nuclear anchor. (Nat Cell Biol. 2004 6:626)
訳 Lgsの核局在は，完全にPygoに依存する

★ localize 動 局在させる／局在を突き止める／局在する

用例数 24,127

localized	局在する／～の局在を突き止めた	17,287
localize	～を局在させる／～の局在を突き止める／局在する	3,356
localizes	～を局在させる／～の局在を突き止める／局在する	2,905
localizing	～を局在させる／～の局在を突き止める／局在する	579

他動詞受動態の用例が多いが，自動詞としても使われる．副詞は，localizedと前置詞の間に置かれることも多いが，それは場所を強調する副詞だからである．ただし，局在することを強調したい場合，あるいは慣用的には，localizedの前に副詞が置かれることも多い．
◆類義語：locate, position, map

■ ［名］＋ be localized

★ be localized	…は，局在する	5,613
❶ protein is localized	タンパク質は，局在する	152

❶ The Scotin **protein is localized** to the ER and the nuclear membrane.（J Cell Biol. 2002 158:235）

訳 Scotinタンパク質は，小胞体と核膜に局在する

■ localized ＋［前］

❶ localized to ～	～に局在する	5,374
・localized to the nucleus	核に局在する	213
・localized to the plasma membrane	細胞膜に局在する	136
・localized to the cytoplasm	細胞質に局在する	103
・be localized to ～	…は，～に局在する	2,890
❷ localized in ～	～に局在する	1,951
・localized in the nucleus	核に局在する	149
・localized in the cytoplasm	細胞質に局在する	117
❸ localized at ～	～に局在する	467
❹ localized on ～	～に局在する	381
❺ localized within ～	～内に局在する	346

❶ Slm9 protein level is constant and Slm9 is **localized to the nucleus** throughout the cell cycle.（Genetics. 2000 155:623）

訳 Slm9は核に局在する

❶ Cell fractionation studies showed that active p35/CDK5 was mainly **localized to the plasma membrane**.（Am J Pathol. 2004 165:1175）

訳 活性のあるp35/CDK5は，主に細胞膜に局在した

❷ Interestingly, ectopically expressed PRAK was **localized in the nucleus** and can be redistributed by coexpression of p38α or p38β to the locations of p38α and p38β.（Mol Biol Cell. 2003 14:2603）

訳 異所性に発現されたPRAKは，核に局在した

■ ［副］＋ localized

❶ highly localized	高度に局在する	190
❷ predominantly localized	主に局在する	179
・predominantly localized to ～	主に～に局在する	83
❸ primarily localized	主に局在する	128

❶ Intracellularly, ASIC1a is **predominantly localized to** the endoplasmic reticulum in Chinese hamster ovary cells, and this intracellular localization is also observed in neurons.（J Biol Chem. 2010 285:13002）

訳 ASIC1aは，主に小胞体に局在する

■ localized ＋［副］

❶ localized predominantly	主に局在する	169
・localized predominantly to ～	主に～に局在する	88
❷ localized primarily	主に局在する	151
❸ localized exclusively	もっぱら局在する	111

❷ During lytic infection, ORF29p is localized primarily to infected cell nuclei.（J Virol. 2005 79: 13070）

訳 ORF29pは，主に感染した細胞の核に局在する

■ localize ＋［前］

❶ localize to ～	…は，～に局在する	3,129
・protein localizes to ～	タンパク質は，～に局在する	115
❷ localize in ～	…は，～に局在する	393

❶ We find that Orbit/MAST protein localizes to Drosophila growth cones.（Neuron. 2004 42: 913）

訳 Orbit/MASTタンパク質は，ショウジョウバエの成長円錐に局在する

■ to localize

★ to localize ～	～の局在を突き止める…／局在する…	1,503
❶ to localize to ～	～に局在する…	332
❷ be used to localize ～	…は，～の局在を突き止めるために使われる	135
❸ fail to localize	…は，局在することができない	116

❷ A genomic GYG2 clone was used to localize the gene to Xp22.3 by fluorescence in-situ hybridization.

訳 ～は，その遺伝子がXp22.3に局在することを突き止めるために使われた

❸ Increases in Bax immunoreactivity were quantitatively and temporally variable and Bax failed to localize to mitochondria.（Brain Res. 2003 994:146）

訳 Baxは，ミトコンドリアに局在することができなかった

locally 副 局所的に　　用例数 2,439

◆類義語：topically, focally, regionally

■ locally ＋［過分］

❶ locally advanced	局所的に進行した	371
・patients with locally advanced ～	局所的に進行した～の患者	142
・locally advanced ～ cancer	局所的に進行した～癌	131

❶ Fifty-one patients with locally advanced NSCLC were enrolled at 13 sites.（J Clin Oncol. 2005 23:5918）

訳 51名の局所的に進行したNSCLCの患者が登録された

★ locate ㊍ 位置づける　　用例数 18,604

located	位置する／～を位置づけた	17,691
locate	～を位置づける	620
locating	～を位置づける	194
locates	～を位置づける	99

他動詞．受動態の用例が非常に多い．

◆類義語：localize, position, map, lie

■［名］＋ be located

★ be located	…は，位置する	7,204
❶ gene is located	遺伝子は，位置する	209
❷ site is located	部位は，位置する	194

■ located ＋［前］

❶ located in ～	～に位置する	5,538
・be located in ～	…は，～に位置する	2,427
・located in ～ region	～領域に位置する	559
❷ located at ～	～に位置する	2,299
・located at ～ terminus	～末端に位置する	139
❸ located on ～	～に位置する	1,864
・located on chromosome ～	染色体～に位置する	310
❹ located within ～	～内に位置する	1,686
❺ located between ～	～の間に位置する	817
❻ located near ～	～の近くに位置する	592

❶ The M3 gene is located in a region of the genome that is transcribed during latency.（J Virol. 2000 74:6741）

訳 M3遺伝子は，潜伏期の間に転写されるゲノムのある領域に位置する

❷ The epitope for the polyol-responsive mAb NT73, which reacts with Escherichia coli RNA polymerase, was located at the C terminus of the β' subunit.（Anal Biochem. 2003 323:171）

訳 ～は，β′サブユニットのC末端に位置した

❸ Surprisingly, we find that these residues are located on the surface of the protein and not in the hydrophobic core.（J Mol Biol. 2003 334:515）

訳 これらの残基は，そのタンパク質の表面に位置する

❹ HLA-DRB1, a major genetic determinant of susceptibility to rheumatoid arthritis (RA), is located within 1,000 kb of the gene encoding tumor necrosis factor (TNF).（Arthritis Rheum. 2003 48:90）

訳 ～は，腫瘍壊死因子（TNF）をコードする遺伝子の1,000 kb内に位置する

L

■ located ＋［副］

❶ located upstream of ～	～の上流に位置する	280
❷ located immediately ～	すぐ～に位置する	189

❸	located close to ～	～の近くに位置する	143
❹	located downstream of ～	～の下流に位置する	138
❺	located adjacent to ～	～に隣接して位置する	136

❶ The element located upstream of the glmS gene in Gram-positive organisms functions as a metabolite-dependent ribozyme that responds to glucosamine-6-phosphate. (Proc Natl Acad Sci USA. 2004 101:6421)

訳 glmS遺伝子の上流に位置するエレメント

■ [名] + located

❶	residues located	位置する残基	346
❷	site located	位置する部位	311
❸	gene located	位置する遺伝子	271
❹	element located	位置するエレメント	216
❺	region located	位置する領域	205

❶ There are three cysteine residues located in the vicinity of the active site. (Biochemistry. 2000 39:13307)

訳 活性部位の近くに位置する3つのシステイン残基がある

★ location 名 位置／部位／局在 用例数 14,137

location	位置／部位／局在	9,306
locations	位置／部位／局在	4,831

複数形のlocationsの割合が，約35%ある可算名詞.

◆類義語：position, locus, site, region

■ location + [前]

location ofの前にtheが付く割合は圧倒的に高い.

	the %	a/an %	ø %			
❶	94	0	6	location of ～	～の位置／～の局在	3,666
❷	16	23	25	location in ～	～における位置	455

❶ We examined the location of tau in control and AD cortices using biochemical and morphologic methods. (Am J Pathol. 2012 181:1426)

訳 われわれは，対照群とアルツハイマー病の皮質におけるタウの局在を調べた

■ [形] + location

❶	subcellular location	細胞内局在	228
❷	chromosomal location	染色体上の位置	213

★ locus ㊔ 座位／部位　用例数 29,866

locus	座位／部位	17,074
loci	座位／部位	12,792

複数形のlociの割合は約45%とかなり高い.
◆類義語：region, site, area

■ locus ＋［前］

locus ofの前にtheが付く割合は圧倒的に高い．また，locus inの前にtheが付く割合もかなり高い.

	the%	a/an%	ø%			
❶	69	20	0	locus in ～	～における座位	938
❷	83	14	0	locus of ～	～の座位	731
❸	35	60	0	locus on ～	～上の座位	647
❹	26	69	1	locus for ～	～の座位	435

❶ Remarkably, **expression of the Mtbesx1 locus in Msesx1 mutants** restored polar localization of tagged proteins, indicating establishment of the MtbESX-1 apparatus in M. smegmatis. (Mol Microbiol. 2012 83:654)

訳 Msesx1変異体におけるMtbesx1座位の発現

❷ **These results reveal the locus of amplification** in cochlear traveling waves and connect the characteristics of normal hearing to molecular forces. (Neuron. 2012 76:989)

訳 これらの結果は，増幅の座位を明らかにする

❸ A gene that codes for the antigen, or a protein that regulates the antigen, **was previously mapped to a locus on chromosome 6.** (Diabetes. 2004 53:978)

訳 ～が，以前に第６染色体上の座位にマップされた

■［名／形］＋ locus/loci

❶	quantitative trait loci	量的形質遺伝子座	765
❷	genetic loci	遺伝子座	472
❸	susceptibility loci	感受性領域	438
❹	gene locus	遺伝子座	425
❺	microsatellite loci	マイクロサテライト領域	244

★★ long ㊒ 長い，㊓ 長く／長い間　用例数 42,956

■ ㊒ long-［名／現分／過分］

❶	long-term ～	長期間の～	18,499
	・long-term potentiation	長期間の増強	1,284
	・long-term survival	長期の生存	1,035
	・long-term memory	長期記憶	506

❷ long-range ～	長期にわたる／遠大な～	2,009
❸ long-lasting ～	持続性の～	1,320
❹ long-lived ～	長く生きた～	1,255
❺ long-standing ～	長年にわたる～	1,255

❶ Synaptic long-term potentiation is maintained through gene transcription, but how the nucleus is recruited remains controversial. (J Neurosci. 2005 25:7032)
訳 シナプスの長期間の増強は，～によって維持される

■ 副 have long been

★ have long been ～	…は，長い間～されている	1,570
❶ have long been known	…は，長い間知られている	226
❷ have long been recognized	…は，長い間認識されている	190
❸ have long been considered	…は，長い間考えられている	114

❷ Src kinase has long been recognized as a factor in the progression of colorectal cancer and seems to play a specific role in the development of the metastatic phenotype. (Cancer Res. 2005 65:1814)
訳 Srcキナーゼは，長い間～として認識されている

■ 副 as long as

❶ as long as ～	～である限り	625

❶ The strain in $BaTiO_3$ layers is fully maintained as long as the $BaTiO_3$ thickness does not exceed the combined thicknesses of the $CaTiO_3$ and $SrTiO_3$ layers. (Nature. 2005 433:395)
訳 $BaTiO_3$の厚さが～を超えない限り

lose 動 失う／減らす 用例数 7,557

lost	失われる／～を失った	5,915
lose	～を失う／～を減らす	1,054
loses	～を失う／～を減らす	303
losing	～を失う／～を減らす	285

他動詞.
◆類義語：eliminate, delete, remove, deplete, abolish, miss

■ be lost ＋［前］/when

★ be lost	…は，失われる	2,541
❶ be lost in ～	…は，～において失われる	612
❷ be lost from ～	…は，～から失われる	156
❸ be lost when ～	…は，～のときに失われる	138
❹ be lost to follow-up	…は，追跡不能である	101

❶ We demonstrate here that this residual lymphocyte population is lost in mice lacking both the γ-like chain and Jak3. (Mol Cell Biol. 2004 24:2584)

訳 …は，～を欠くマウスにおいて失われている

❷ Immunohistology indicated that hr44 is present in vesicular structures of the iris and ciliary body and **is lost from the epithelial layers of** inflamed eyes coincidentally with CD63. (Invest Ophthalmol Vis Sci. 2003 44:2650)

訳 …は，～の上皮層から失われている

■ lose ＋［名句］

❶	lose the ability to *do*	…は，～する能力を失う	288
❷	lose their ability to *do*	…は，～するそれらの能力を失う	104

❶ We observed that **TRAP mutant proteins that had lost the ability to bind RNA were no longer recognized by AT**. (J Biol Chem. 2002 277:10608)

訳 RNAに結合する能力を失ったTRAP変異タンパク質は，もはやATによって認識されなかった

★★ loss ㊔ 喪失／減少　　用例数 50,409

loss	喪失／減少	48,908
losses	喪失／減少	1,501

複数形のlossesの割合は３%あるが，単数形は無冠詞で使われることが多い.

◆類義語：deficit, defect, deletion, deficiency, lack, paucity decrease, reduction, decline, fall

■ loss ＋［前］

	the%	a/an%	ø%			
❶	15	10	71	loss of ～	～の喪失	34,161
				・loss of function	機能喪失	1,280
				・loss of heterozygosity	ヘテロ接合性の喪失	964
				・complete loss of ～	～の完全な喪失	866
				・progressive loss of ～	～の進行性の喪失	354
				・loss of activity	活性の喪失	316
				・partial loss of ～	～の部分的な喪失	313
				・significant loss of ～	～の有意な減少	301
❷	10	16	62	loss in ～	～における喪失／～の喪失	2,581

❶ **The defect is due to loss of function of AMPK**. (Circulation. 2005 111:21)

訳 その欠損は，AMPKの機能喪失のせいである

❶ A substitution of Leu-30 or Thr-36 with alanine **resulted in a complete loss of transport activities**. (J Biol Chem. 2004 279:31478)

訳 ～は，輸送活性の完全な喪失という結果になった

❷ **Calcium supplementation protected both men and women from bone loss in** the longitudinal study of whites. (Am J Clin Nutr. 2004 80:1066)

訳 カルシウムの補充は，男性および女性の両方を～における骨の喪失から保護した

■ [前] + loss of

❶ result in loss of ～	…は，～の喪失という結果になる	797
❷ lead to loss of ～	…は，～の喪失につながる	367
❸ associated with loss of ～	～の喪失と関連する	288
❹ due to loss of ～	～の喪失のせいで	190
❺ caused by loss of ～	～の喪失によって引き起こされる	176

❸ Human immunodeficiency virus (HIV) disease is associated with loss of $CD4^+$ T cells, chronic immune activation, and progressive immune dysfunction. (J Exp Med. 2004 200:331)

訳 ヒト免疫不全ウイルス（HIV）疾患は，$CD4^+$ T細胞の喪失と関連する

■ [名／形] + loss

❶ weight loss	体重減少	2,231
❷ bone loss	骨の減少	1,012
❸ hearing loss	聴力損失	992
❹ cell loss	細胞喪失	859
❺ graft loss	移植片喪失	607
❻ neuronal loss	神経細胞喪失	509
❼ progressive loss	進行性喪失	369

★★ low 形 低い　用例数 66,433

■ low + [前]

❶ as low as ～	～ほども低い	932
❷ low in ～	～において低い	892
・be low in ～	…は，～において低い	370
❸ low to *do*	～するには低い	567

❶ In contrast, synthesis of genomic HDV RNA was totally inhibited by α-amanitin at concentrations as low as 2.5 μg/ml. (J Virol. 2002 76:3920)

訳 ～は，2.5 μg/mlほども低い濃度でα-アマニチンによって完全に抑制された

❷ Thus, PDE1C expression is low in cultured human SMCs made quiescent by attaching to fibrillar collagen type I. (Circ Res. 2002 90:151)

訳 PDE1C発現は，培養されたヒトのSMCにおいて低い

❸ The number of pairs concordant for definite osteophytes in the sample was too low to assess this feature alone. (Arthritis Rheum. 2000 43:2410)

訳 ～は，この特徴を評価するには低すぎた

■ [副] + low

❶ very low	非常に低い	2,917
・very low levels	非常に低いレベル	459

❷ relatively low	相対的に低い	1,212
❸ extremely low	極端に低い	475

❶ In Arabidopsis thaliana, IMB1 is expressed at very low levels in dry seeds, but is markedly induced during seed imbibition.（Plant J. 2003 35:787）
訳 IMB1は，～において非常に低いレベルで発現する

■ low ＋［名］

❶ low levels	低いレベル	4,040
・at low levels	低いレベルで	698
・expressed at low levels	低いレベルで発現した	237
❷ low-dose	低い用量	2,429
❸ low affinity	低い親和性	2,120
❹ low concentrations	低い濃度	1,588
❺ low pH	低いpH	1,449
❻ low temperature	低い温度	1,312

❶ β-CM mRNA and protein are expressed at low levels in the mammalian retina and RPE-choroid.（Invest Ophthalmol Vis Sci. 2003 44:44）
訳 β-CMメッセンジャーRNAおよびタンパク質は，～において低いレベルで発現する

■［形］＋ and low／low and ＋［形］

❶ high and low	高いおよび低い	886
❷ low and high	低いおよび高い	688

❶ DNA conformation and E2-DNA contacts are similar in both high and low affinity complexes.（J Biol Chem. 2000 275:31245）
訳 ～は，高いおよび低い親和性の複合体の両方において類似している

★★ lower 形 より低い，動 低下させる　用例数 43,218

lower	より低い／～を低下させる	37,981
lowering	～を低下させる／低下	2,616
lowered	低下した／～を低下させた	1,917
lowers	～を低下させる	704

lowの比較級の用例が多いが，他動詞としても用いられる．
◆類義語：reduce, diminish, down-regulate, decrease, decline, fall

■ lower than

★ lower than ～	～より低い	4,209
❶ lower than that	それより低い	1,180
❷ lower than in ～	～においてより低い	453

❶ GFR in 9-month-old diabetic mice was significantly lower than that of 9-month-old control

L

mice. (Diabetes. 2004 53:3248)

訳 ～は，9ヶ月齢の対照群マウスのそれよりも有意に低かった

❷ In women with FM or GWI, **HRV was significantly** lower than in **men with** FM or GWI. (Arthritis Rheum. 2004 51:700)

訳 HRVは，～の男性においてよりも有意に低かった

■ lower ＋［前］

❶	lower in ～	～においてより低い	3,400
	・be lower in ～	…は，～においてより低い	1,437
	・lower in ～ than in …	…においてより～において低い	417
❷	lower for ～	～に対してより低い	512

❶ In addition, plasma levels of antidiuretic hormone **were significantly** lower in **trpv4$^{-/-}$ mice** than in **wild-type littermates** after a hyperosmotic challenge. (Proc Natl Acad Sci USA. 2003 100:13698)

訳 ～は，野生型の同腹仔においてよりtrpv4$^{-/-}$マウスにおいて有意に低かった

■［名］＋ be lower

★	be lower	…は，より低い	3,602
❶	levels were lower	レベルは，～においてより低かった	128

■［副］＋ lower

❶	significantly lower	有意により低い	4,420
❷	～-fold lower	～倍より低い	1,377
❸	much lower	ずっとより低い	1,070
❹	substantially lower	実質的により低い	310
❺	slightly lower	わずかにより低い	298

❷ CHO and Rat2 cell lines produced more infectious virus, but **this production was still 40**-fold lower **than production in human cells.** (J Virol. 2001 75:3141)

訳 この産生は，まだ，ヒトの細胞における産生より40倍低かった

■［前］＋ lower

❶	associated with lower ～	より低い～と関連する	154
❷	at lower levels	より低いレベルで	103

❶ High retinol and hemoglobin concentrations in late, but not in early, pregnancy **were strongly and independently** associated with lower **birth weight** and smaller placental size at birth. (Am J Clin Nutr. 2004 79:103)

訳 ～は，より低い出生時体重と強くそして独立に関連した

❷ PAK5 is highly expressed in the brain and **is expressed** at lower levels **in** several other tissues. (Mol Cell Biol. 2003 23:7134)

訳 …は，～においてより低いレベルで発現する

■ lower ＋［名］

❶	lower levels	より低いレベル	1,708
❷	lower risk	より低いリスク	1,075
❸	lower affinity	より低い親和性	706
❹	lower concentrations	より低い濃度	432

M

machinery ㊔ 装置／機構 用例数 5,961

machinery	装置／機構	5,750
machineries	装置／機構	211

複数形のmachineriesの割合は4%あるが，単数形は無冠詞で使われることも多い．
◆類義語：mechanism, machine, apparatus, device

■ machinery ＋［前］

machinery of/inの前にtheが付く割合は圧倒的に高い．machinery forの前にtheが付く割合も非常に高い．

	the%	a/an%	ø%			
❶	82	0	18	machinery in ～	～における機構	320
❷	95	1	4	machinery of ～	～の機構	242
❸	73	4	19	machinery for ～	～のための機構	200

❶ **The direct interactions between transcriptional repressors and the core transcriptional machinery in bacteria and archaea are sufficient to generate** a sophisticated suite of mechanisms that provide flexible control. (Curr Biol. 2010 20:R764)
訳 細菌と古細菌における転写抑制因子とコア転写装置の間の直接相互作用は，～を産生するには十分である

■ ［名／形］＋ machinery

❶	transcription machinery	転写装置	305
❷	transcriptional machinery	転写装置	263
❸	replication machinery	複製機構	214
❹	molecular machinery	分子装置	164

★ magnitude ㊔ 大きさ／規模／振幅 用例数 10,537

magnitude	大きさ	10,020
magnitudes	大きさ	517

複数形のmagnitudesの割合は約5%あるが，単数形は無冠詞で使われることも多い．
◆類義語：size, scale, amplitude

■ magnitude ＋［前］

magnitude ofの前にtheが付く割合は圧倒的に高い．

	the%	a/an%	ø%			
❶	87	6	7	magnitude of ～	～の大きさ	4,347

❶ Fourier analysis revealed that **the increase in the magnitude of the fluctuations was mediated by** the low-frequency components. (Invest Ophthalmol Vis Sci. 2003 44:1035)

M

訳 変動の大きさの増大は，～によって仲介された

■ ［前］＋ magnitude

❶ ～ orders of magnitude	～桁の大きさ	2,085
・～ orders of magnitude higher than …	…より～桁高い	112
❷ similar in magnitude	大きさにおいて似ている	105

❶ The affinity of RPA for damaged ssDNA was 5 orders of magnitude higher than that of the damage recognition protein XPA (xeroderma pigmentosum group A protein). (Biochemistry. 2000 39:850)

訳 損傷された一本鎖DNAに対するRPAの親和性は，損傷認識タンパク質XPAのそれより５桁高かった

❷ The up-regulation of GABA(A) receptors was similar in magnitude to that previously found for rats with bilateral transection of the inferior cerebellar peduncle. (Brain Res. 2007 1186:188)

訳 GABA(A)受容体の上方制御は，以前に見つけられたそれと大きさにおいて似ていた

mainly 副 主に　　用例数 5,562

前置詞の前に置かれることがかなり多い.

◆類義語：predominantly, mostly, primarily, principally, chiefly, largely

■ mainly ＋［前］

❶ mainly in ～	主に～において	691
・expressed mainly in ～	主に～において発現される	67
・occur mainly in ～	…は，主に～において起こる	64
❷ mainly by ～	主に～によって	445

❶ Apoptosis occurs mainly in cardiac myocytes, and is shown for the first time to be limited to hypoxic regions during acute infarction. (J Clin Invest. 1997 100:1363)

訳 アポトーシスは，主に心筋細胞において起こる

■ mainly ＋［形／副／過分］

❶ mainly due to ～	主に～のせいで	198
・be mainly due to ～	…は，主に～のせいである	126
❷ mainly because of ～	主に～のせいで	77
❸ mainly expressed in ～	主に～において発現される	48

❶ This is mainly due to a change in Schottky barrier height (SBH). (Adv Mater. 2012 24:3532)

訳 これは，主に～の変化のせいである

★ maintain 動 維持する　　用例数 24,619

maintained	維持される／～を維持した	8,552
maintain	～を維持する	8,009

maintaining	～を維持する	6,138
maintains	～を維持する	1,920

他動詞.

◆類義語：sustain, keep, hold, retain

■ be maintained ＋［前］

★ be maintained	…は，維持される	4,472
❶ be maintained in ～	…は，～において維持される	1,011
❷ be maintained by ～	…は，～によって維持される	542
❸ be maintained at ～	…は，～において維持される	376
❹ be maintained for ～	…は，～の間維持される	321

❶ After lens formation, **Pax6 expression is maintained in the lens epithelium**, whereas its level abruptly decreases in differentiated fiber cells.（Invest Ophthalmol Vis Sci. 2004 45:3589）

訳 Pax6の発現は，水晶体上皮において維持される

❷ Finally, once initiated, **Bar expression can be maintained by positive autoregulation.**（Development. 2004 131:5573）

訳 Bar発現は，正の自己調節によって維持されうる

■［動／名／形］＋ to maintain

★ to maintain ～	～を維持する…	4,641
❶ be required to maintain ～	…は，～を維持するために必要とされる	451
❷ ability to maintain ～	～を維持する能力	130
❸ be necessary to maintain ～	…は，～を維持するために必要である	124

❶ We also show that **endogenous Ras is required to maintain normal levels of dMyc**, but not dPI3K signaling during wing development.（Genes Dev. 2002 16:2286）

訳 内在性のRasが，正常なレベルのdMycを維持するために必要とされる

■［前］＋ maintaining

❶ be essential for maintaining ～	…は，～を維持するために必須である	187
❷ be critical for maintaining ～	…は，～を維持するために決定的に重要である	153
❸ be important for maintaining ～	…は，～を維持するために重要である	137
❹ be required for maintaining ～	…は，～を維持するために必要とされる	115

❶ **Potassium channels are essential for maintaining a normal ionic balance** across cell membranes.（Nature. 2004 431:830）

訳 カリウムチャネルは，正常なイオンのバランスを維持するために必須である

★ maintenance ㊔ 維持　　用例数 12,670

複数形の用例はなく，原則，不可算名詞として使われる．

◆類義語：retention

M

■ maintenance ＋［前］

maintenance ofの前にtheが付く割合はかなり高い.

the%	a/an%	ø%			
❶ 64	0	36	maintenance of ～	～の維持	7,580
			・maintenance of ～ integrity	～完全性の維持	244
			・maintenance of ～ stability	～安定性の維持	209
			・contribute to the maintenance of ～	…は，～の維持に寄与する	155
			・be required for the maintenance of ～	…は，～の維持のために必要とされる	153

❶ The Werner protein (WRN) belongs to the RecQ family of DNA helicases and is required for the maintenance of genomic stability in human cells. (Cancer Res. 2003 63:7136)

訳 ～は，遺伝的な安定性の維持のために必要とされる

❶ Consistent with our in vitro observations, we show that WRN contributes to the maintenance of DNA synthesis fidelity in vivo. (J Biol Chem. 2012 287:12480)

訳 WRNは，生体内でDNA合成の忠実度の維持に寄与する

■ ［名］＋ and maintenance of

❶ development and maintenance of ～	～の発生と維持	279
❷ establishment and maintenance of ～	～の確立と維持	226
❸ induction and maintenance of ～	～の誘導と維持	153
❹ formation and maintenance of ～	～の形成と維持	146

❷ We have developed a genetic approach to examine the role of spontaneous activity and synaptic release in the establishment and maintenance of an olfactory sensory map. (Neuron. 2004 42:553)

訳 われわれは，嗅覚系感覚地図の確立と維持における自発活性とシナプス性遊離の役割を調べるために遺伝的アプローチを開発した

★★ major 形 主要な／主な　用例数 50,463

◆類義語：main, predominant, primary, principal

■ major ＋［名］

❶ major role	主要な役割	2,304
・play a major role in ～	…は，～において主要な役割を果たす	1,639
❷ major cause	主要な原因	1,177
❸ major component	主要な要素	798
❹ major groove	主溝	718
❺ major source	主要な供給源	646
❻ major determinant	主要な決定要因	626
❼ major challenge	主要な課題	502

❶ Drug efflux systems play a major role in resistance to a wide range of noxious compounds in several Gram negative species. (Mol Microbiol. 2006 59:126)

訳 ～は，広範囲の有害化合物への抵抗性において主要な役割を果たす

❷ Papillomaviruses (HPVs) are a major cause of human disease, and are responsible for approximately half a million cases of cervical cancer each year. (J Mol Biol. 2006 355:360)

訳 パピローマウイルス（HPV）は，ヒトの疾患の主要な原因である

★ majority 名 大多数／大部分 用例数 10,764

majority	大多数	10,757
majorities	大多数	7

複数形のmajoritiesの用例はほとんどないが，単数形には不定冠詞が付くことも多い．majority ofの用例が圧倒的に多い．

◆類義語：bulk

■ majority ＋［前］

majority ofの前にtheが付く割合は圧倒的に高い．ただし，いくつかに分類した集団を比較する際，過半数を占める最大集団を表すときにa majority ofが使われることもある．

	the%	a/an%	ø%			
❶	86	12	2	majority of ～	～の大多数	9,766
				・the majority of ～	～の大多数	7,645
				・responsible for the majority of ～	～の大多数に責任のある	114
				・a majority of ～	～の大多数	864
				・the vast majority of ～	～の圧倒的多数	862
				・majority of patients	患者の大多数	674
				・majority of cases	症例の大多数	219
				・the great majority of ～	～の圧倒的多数	140

❶ The vast majority of these infections are caused by Candida species. (J Dent Res. 2005 84:966)

訳 これらの感染の圧倒的多数は，カンジダ種によって引き起こされる

❶ The majority of patients with PTDM may be safely treated with ROSI +/− a sulfonylurea. (Transplantation. 2005 80:1402)

訳 PTDMの患者の大多数は，～によって安全に治療されるかもしれない

★★ make 動 作る／なす／する／させる 用例数 32,539

made	なされる／作られる／～を作った／～にした	14,735
make	～を作る／～にする	7,511
making	～を作る／～にする	6,265
makes	～を作る／～にする	4,028

M

他動詞．第３文型だけでなく，第５文型で使われることも多い．
◆類義語：produce, create, achieve

■ be made ＋［前］

★ be made	…は，なされる／作られる	7,815
❶ be made in ～	…は，～においてなされる／～において作られる	1,655
・be made in understanding ～	…は，～の理解においてなされる	205
❷ be made to *do*	…は，～させられる	705
❸ be made by ～	…は，～によってなされる	459
❹ be made from ～	…は，～から作られる	321
❺ be made with ～	…は，～を使ってなされる	310
❻ be made for ～	…は，～に対してなされる	300

❶ Major advances have **been made in understanding** the mechanisms of some of the subtypes.（Lancet. 2005 366:1653）
訳 大きな進歩が，～の機構の理解においてなされてきた

❷ Double mutants **were made to** exchange the D597/M336 pair in nNOS with N368/V106 in eNOS.（Biochemistry. 2005 44:15222）
訳 二重変異体は，nNOSのD597/M336ペアをeNOSのN368/V106と交換させられた

■［名］＋ be made

❶ measurements were made	測定は，なされた	262

❶ Biochemical **measurements were made** by using standard procedures.（Am J Clin Nutr. 2009 90:686）
訳 生化学的測定は，標準的方法を使うことによってなされた

■ be made ＋［形］

❶ be made possible	…が，可能にされる	146
❷ be made available	…が，利用できるようにされる	118

■ make it ＋［形］＋ to *do*

❶ make it possible to *do*	…は，～することを可能にする	746
❷ make it difficult to *do*	…は，～することを困難にする	409

❶ The large size of COS cells **makes it possible to** test the effect of neuroligins presented over an extended surface area.（J Biol Chem. 2005 280:22365）
訳 ～は，ニューロリギンの効果をテストすることを可能にする

■ make ＋［形］

❶ make possible ～	…は，～を可能にする	190

■ make use of

❶	make use of ～	…は，～を利用する	561

❶ We make use of the fact that recombination is a very unlikely event among these markers.（Genetics. 2003 164:1161）
訳 われわれは，～という事実を利用する

■ make up

❶	make up ～	…は，～を作り上げる	588
❷	be mad up of ～	…は，～からなる	166

❷ WT HP67 is made up of two subdomains, which form a tightly packed interface.（J Mol Biol. 2006 355:1066）
訳 野生型HP67は，2つのサブドメインからなる

management 名 管理 用例数 9,644

management	管理	9,640
managements	管理	4

複数形のmanagementの用例はほとんどなく，原則，不可算名詞として使われる．

■ management ＋［前］

	the%	a/an%	ø%			
❶	59	0	41	management of ～	～の管理	4,241
				・management of patients	患者の管理	497
❷	3	1	85	management in ～	～における管理	237

❶ Management of patients with cholera involves aggressive fluid replacement; effective therapy can decrease mortality from more than 50% to less than 0.2%.（Lancet. 2012 379:2466）
訳 コレラの患者の管理は，積極的な補液に関わる

■［形］＋ management

❶	clinical management	臨床管理	463
❷	medical management	医療管理	267
❸	patient management	患者管理	224

manifest 動 顕在化させる／明らかにする／現す／現れる，形 著明な 用例数 3,435

manifested	顕在化する／明らかにされる／現れた／～を明らかにした	1,570
manifest	～を現す／現れる／著明な	1,262
manifests	～を現す／現れる	388

M

manifesting	～を現す／現れる	215

他動詞，自動詞および形容詞として用いられる．

◆類義語：reveal, elucidate, clarify, uncover, define, marked, striking, prominent

■ (動) be manifested ＋［前］

★ be manifested	…は，顕在化する	463
❶ be manifested by ～	…は，～によって顕在化する	144
❷ be manifested in ～	…は，～において顕在化する	120
❸ be manifested as ～	…は，～として顕在化する	84

❶ The effect **is manifested by** increases in the fraction of cells in a population that contains multiple γ-H2AX foci.（Oncogene. 2005 24:7257）

訳 その効果は，～の増大によって顕在化する

❸ Diabetes **is manifested as** hyperglycemia due to a relative deficiency of the production of insulin by the pancreatic β-cells.（Diabetes. 1999 48:2270）

訳 糖尿病は，高血糖として顕在化する

■ (形) be manifest ＋［前］

❶ be manifest in ～	…は，～において著明な	62

❶ Strikingly, these effects of Mnt knockdown **are** even **manifest in** cells lacking c-myc.（Mol Cell Biol. 2004 24:1560）

訳 これらのMntノックダウンの効果は，c-mycを欠く細胞において実に著明である

manipulate (動) 操る／操作する　用例数 2,880

manipulated	操られる／～を操った	1,091
manipulate	～を操る	918
manipulating	～を操る	806
manipulates	～を操る	65

他動詞．

◆類義語：engineer, operate, drive, handle

■ be manipulated ＋［前］

★ be manipulated	…は，操られる	501
❶ be manipulated by ～	…は，～によって操られる	121
❷ be manipulated to *do*	…は，～するように操られる	96

❶ This study indicates that the lineage differentiation of ES cells can **be manipulated by** the expression of GATA factors.（Dev Biol. 2005 286:574）

訳 ES細胞の系統分化は，GATA因子の発現によって操られうる

❷ This ancient cellular antiviral response can **be manipulated to** provide an effective research tool to knock down the level of expression of selected target genes, providing a very powerful new method for the analysis of cell signalling pathways.（Oncogene. 2004 23:8376）

訳 ～は，選択的標的遺伝子の発現のレベルをノックダウンする効果的なリサーチツールを提供するように操られうる

■ to manipulate

★	to manipulate ～	～を操る…	578
❶	ability to manipulate ～	～を操る能力	89

❶ The **ability to manipulate** biological cells and micrometre-scale particles plays an important role in many biological and colloidal science applications. (Nature. 2005 436:370)

訳 生物学的細胞およびマイクロメータースケールの粒子を操る能力は，～において重要な役割を果たす

manipulation 名 操作　　用例数 4,183

manipulation	操作	3,049
manipulations	操作	1,134

複数形のmanipulationsの割合は約25%あるが，単数形は無冠詞で使われることが圧倒的に多い．

◆類義語：operation, treatment

■ manipulation ＋［前］

	the%	a/an%	ø%			
❶	18	3	77	manipulation of ～	～の操作	1,684
				・genetic manipulation of ～	～の遺伝的操作	168

❶ Moreover, **genetic manipulation of** MRP expression results in concomitant changes in PDE activity and protein levels, thus affecting cAMP degradation in parallel with cAMP efflux. (Mol Pharmacol. 2011 80:281)

訳 MRP発現の遺伝的操作は，PDE活性とタンパク質レベルの同時変化という結果になる

★manner 名 様式　　用例数 22,409

manner	様式	22,279
manners	様式	130

複数形のmannersの割合は0.6%しかないが，単数形には不定冠詞が付くことが非常に多い．

◆類義語：fashion, pattern, mode, modality, way

■ in a ＋［形］＋ manner

★	in a ～ manner	～の様式で	14,173
❶	in a ～-dependent manner	～依存的な様式で	5,525
	・in a dose-dependent manner	用量依存的な様式で	755
	・in a concentration-dependent manner	濃度依存的な様式で	322

M

・in a time-dependent manner	時間依存的な様式で	113
❷ in a ～-specific manner	～特異的な様式で	1,019
・in a tissue-specific manner	組織特異的な様式で	192
・in a sequence-specific manner	配列特異的な様式で	165
❸ in a ～-independent manner	～非依存的な様式で	794
❹ in a similar manner	類似の様式で	297

❶ Rivastigmine inhibited acetylcholinesterase in all positive structures in a dose-dependent manner (10^{-6}–10^{-4} M). (Brain Res. 2005 1060:144)

訳 リバスチグミンは，用量依存的な様式ですべての陽性の構造においてアセチルコリンエステラーゼを抑制した

❷ Our studies provide the first direct evidence that Mdm2 can function in the absence of Mdm4 to regulate p53 activity in a tissue-specific manner. (Mol Cell Biol. 2006 26:192)

訳 Mdm2は，Mdm4非存在下において，組織特異的な様式でp53の活性を調節するように機能しうる

❹ The TPR1 mutant digested in a similar manner to wild type; however, TPR2a and TPR2b mutants each displayed greater resistance to chymotryptic digestion. (J Biol Chem. 2004 279: 16185)

訳 TPR1変異体は野生型に似た様式で消化した

■ in a/this manner

❶ in a manner	様式で	3,025
・in a manner that ～	～である様式で	993
・in a manner similar to ～	～に似た様式で	556
・in a manner consistent with ～	～と一致する様式で	250
・in a manner dependent on ～	～に依存する様式で	208
・in a manner analogous to ～	～に類似した様式で	136
❷ in this manner	この様式で	282

❶ Animal models have confirmed that brain lysates and purified protein can accelerate brain pathology in a manner similar to prions. (Curr Opin Neurol. 2012 25:721)

訳 ～は，プリオンに似た様式で脳の症状を加速させうる

★★ many 形 多くの，名 多くのもの　用例数 49,362

◆類義語：numerous, large number of

■ 形 many ＋［形／名］

❶ many other ～	多くの他の～	1,305
❷ many different ～	多くの異なる～	991
❸ many genes	多くの遺伝子	833
❹ many aspects of ～	～の多くの面	736

❷ Aggressive cancer phenotypes are a manifestation of many different genetic alterations that promote rapid proliferation and metastasis. (Cancer Res. 2005 65:10801)

訳 高悪性度の癌の表現型は，多くの異なる遺伝的変化の発現である

❹ Stat proteins are latent cytoplasmic transcription factors that are crucial in many aspects of mammalian development. (Blood. 2006 107:1085)

訳 Statタンパク質は，哺乳類の発生の多くの面において決定的に重要である不顕性の細胞質転写因子である

■ 形 many ＋［前］

❶ as many as ～	～ほども多くの	481

❶ In fact, miRNAs may regulate as many as one-third of human genes. (Development. 2005 132:4645)

訳 マイクロRNAは，ヒト遺伝子の３分の１もの多くを調節するかもしれない

■ 名 many ＋［前］

❶ many of ～	～の多くのもの	5,938
・, many of which ～	（そして，）それらの多くは～	698

❶ Human tumors frequently display recurrent chromosome aberrations, many of which are hallmarks of particular tumor subtypes. (Genome Res. 2005 15:1831)

訳（そして，）それらの多くは特定の腫瘍のサブタイプの特徴である

marked 形 顕著な／著しい　用例数 9,951

◆類義語：striking, prominent, profound, manifest, dramatic, remarkable, significant

■ marked ＋［名］

❶ a marked increase in ～	～の顕著な増大	902
・result in a marked increase in ～	…は，～の顕著な増大という結果になる	95
❷ a marked reduction in ～	～の顕著な低下	579
❸ a marked decrease in ～	～の顕著な低下	419
❹ in marked contrast	著しく対照的に	325

❶ Additionally, we observed a marked increase in immunoreactivity of cytochrome C although its release from mitochondria was not apparent. (Brain Res. 2003 994:146)

訳 われわれは，シトクロムCの免疫反応性の顕著な増大を観察した

❷ The Ad-uPAR-Cath B-infected cells revealed a marked reduction in tumor growth and invasiveness as compared with the parental and vector controls. (Cancer Res. 2004 64:4069)

訳 ～は，腫瘍増殖の顕著な低下を示した

❹ In marked contrast to PNS effects, the peptides inhibited proliferation of neural precursor cells of the E10.5 hindbrain. (Dev Biol. 2004 271:161)

訳 PNS効果と著しく対照的に

★ markedly 副 顕著に／著しく　用例数 11,751

◆類義語：remarkably, strikingly, dramatically, prominently, notably, extremely, unusually, significantly

■ markedly ＋［過分／形］

❶ markedly reduced	顕著に低下した／顕著に低下させた	1,798
・be markedly reduced	…は，顕著に低下する	691
❷ markedly increased	顕著に増大した／顕著に増大させた	1,206
❸ markedly decreased	顕著に低下した／顕著に低下させた	628
❹ markedly enhanced	顕著に増強される／顕著に増強させた	493
❺ markedly different	顕著に異なる	401
❻ markedly inhibited	顕著に抑制される／顕著に抑制した	395
❼ markedly attenuated	顕著に減弱した／顕著に減弱させた	341
❽ markedly elevated	顕著に上昇した／顕著に上昇させた	286
❾ markedly diminished	顕著に減少した／顕著に減少させた	264

❶ Immunostaining was markedly reduced in all tissues in both eyes from donor 5.（Invest Ophthalmol Vis Sci. 2000 41:1617）
訳 免疫染色は，すべての組織において顕著に低下した

❷ The half-life of the chimeric mRNA was markedly increased in cells stimulated with TNF-α and IL-4.（J Immunol. 2003 171:4369）
訳 キメラのメッセンジャーRNAの半減期は，～において顕著に増大した

❸ At 4 weeks after acute lesion, GFRα-2 mRNA was markedly decreased in SN bilaterally, whereas GFRα-1 mRNA in SN and ST was not affected.（Brain Res. 2004 1016:170）
訳 GFRα-2メッセンジャーRNAは，SNにおいて顕著に低下した

❺ In this study we show that the effects of Gads deficiency on murine $CD4^+$ and $CD8^+$ T cells are markedly different.（J Immunol. 2004 173:1711）
訳 ～は，顕著に異なる

■［動］＋ markedly

❶ differ markedly	…は，顕著に異なる	311
・differ markedly from ～	…は，～とは顕著に異なる	118
・differ markedly in ～	…は，～において顕著に異なる	96
❷ increase markedly	…は，顕著に増大する	237

❶ However, these two compounds differed markedly in their interactions with HIV-1 integrase.（Proc Natl Acad Sci USA. 2004 101:6894）
訳 これらの２つの化合物は，HIV-1インテグラーゼとそれらの相互作用において顕著に異なった

★ marker ㊔ マーカー　　用例数 29,891

markers	マーカー	17,578
marker	マーカー	12,313

複数形のmarkersの割合は約60%と非常に高い．

◆類義語：tag, label

■ marker(s) + [前]

	the%	a/an%	ø%			
❶	0	-	92	markers of ～	～のマーカー	2,779
❷	1	95	3	marker for ～	～のためのマーカー	1,806
❸	3	-	84	markers in ～	～におけるマーカー	965

❶ A number of transcripts were identified as novel markers of MSCs. (Arthritis Rheum. 2012 64: 2632)

訳 いくつかの転写物が，MSCの新規マーカーとして同定された

■ [形／名] + markers

❶	molecular markers	分子マーカー	631
❷	microsatellite markers	マイクロサテライトマーカー	578
❸	genetic markers	遺伝子マーカー	560
❹	cell markers	細胞マーカー	408
❺	specific markers	特異的マーカー	401
❻	surface markers	表面マーカー	353
❼	differentiation markers	分化マーカー	328

★ match 動 一致させる／匹敵させる，名 一致　用例数 14,091

matched	匹敵する／一致する	8,857
matching	匹敵する／一致する	2,270
match	～を匹敵させる／～を一致させる／一致	1,978
matches	～を匹敵させる／～を一致させる／一致	986

他動詞の用例が多いが，名詞としても用いられる．

◆類義語：fit, correspond, coincide

■ matched + [前]

❶	matched for ～	～に関して匹敵する	479
	・matched for age	年齢に関して匹敵する	234
❷	matched to ～	～に匹敵する	447
	・be matched to ～	…は，～に匹敵する	211
❸	matched by ～	～において匹敵する	213
	・matched by age	年齢において匹敵する	73
❹	matched with ～	～と匹敵する	190

❶ Fasting liver biopsy specimens were obtained from 42 HCV-infected subjects and 10 non-HCV-infected subjects matched for age and body mass index. (Hepatology. 2003 38:1384)

訳 ～が，42名のHCVに感染した被検者および年齢とボディマス指数に関して匹敵する10名のHCVに非感染の被検者から得られた

❷ Controls were matched to cases on age, ethnicity, and area of residence. (Am J Epidemiol.

1999 150:561)

訳 対照群は，年齢，民族性，および住居の地域に関して症例に匹敵した

■ matched ＋［名／形］

❶	matched controls	匹敵する対照群	1,005
❷	matched normal ～	匹敵する正常な～	316
❸	matched control subjects	匹敵する対照被験者	294
❹	matched healthy ～	匹敵する健康な～	277

❶ Fifteen patients aged 13 to 19 years were compared with 11 healthy, age-matched controls.（Circulation. 2002 105:2274）

訳 年齢が13から19歳の15名の患者が，11名の健康で年齢の一致する対照群と比較された

❹ Twenty age- and sex-matched healthy control subjects were also studied.（Circulation. 2004 109:2319）

訳 20名の年齢および性別の一致した健康な対照被検者も，また，研究された

■ ［名］-matched

❶	age-matched ～	年齢の匹敵する～	1,621
	・age- and sex-matched ～	年齢および性別の匹敵する～	252
❷	HLA-matched ～	HLAの一致する～	242

★ maturation 名 成熟　　用例数 10,175

複数形の用例はなく，原則，不可算名詞として使われる．

■ maturation ＋［前］

	the%	a/an%	ø%			
❶	48	1	51	maturation of ～	～の成熟	2,480
❷	3	0	83	maturation in ～	～における成熟	427

❶ The K7L gene product of the smallpox virus is a protease implicated in the maturation of viral proteins.（J Biol Chem. 2012 287:39470）

訳 ～は，ウイルスタンパク質の成熟に関連づけられるプロテアーゼである

■ ［形／名］＋ maturation

❶	cell maturation	細胞成熟	455
❷	oocyte maturation	卵成熟	350

maximal 形 最大の　　用例数 7,168

■ be maximal ＋［前］

★	be maximal	…は，最大である	553

❶	be maximal at ～	…は，～において最大である	211

❶ Cell proliferation **was maximal at** 2 wk, correlating with peak $\alpha_v\beta_3$ expression.（FASEB J. 2005 19:1857）

訳 細胞増殖は，2週間目において最大であった

■ maximal ＋［名］

❶	maximal activation	最大活性化	211
❷	maximal activity	最大活性	204
❸	maximal inhibition	最大阻害	154

■ half-maximal

❶	half-maximal ～	最大～の半分	604

maximum ㊔ 最大／最大値，㊋ 最大の 用例数 9,205

maximum	最大／最大値／最大の	8,743
maxima	最大／最大値	446
maximums	最大／最大値	16

名詞および形容詞として使われる．名詞としては複数形の割合が，約20%ある可算名詞．複数形にはmaximaとmaximumsの2種類がある．

◆類義語：peak

■ maximum ＋［前］

	the%	a/an%	ø%			
❶	11	86	3	maximum of ～	～の最大値	548
❷	7	76	11	maximum at ～	～における最大値	282

❶ In steady-state assays, the enzyme exhibits a pH **maximum of** 8.5 and is also found to function as a selenite oxidase.（J Am Chem Soc. 2005 127:16567）

訳 その酵素は，8.5のpHの最大値を示す

❷ The induction of htrA gradually reached a **maximum at** around 20 min, suggesting that HtrA may be involved in a late competence response.（J Bacteriol. 2005 187:3028）

訳 htrAの誘導は，およそ20分において徐々に最大値に到達した

■［動］＋ a maximum

❶	reach a maximum	…は，最大値に達する	271

❶ Immunostaining revealed that VEGFR-2（VEGF receptor）colocalized with CD31（endothelial cell marker）at all time points in the abscess rim, whereas F4/80（macrophage）immunostaining **reached a maximum** at day 7 and decreased by day 10.（J Nucl Med. 2009 50:2058）

訳 F4/80（マクロファージ）免疫染色は，7日目に最大値に達した

M

means ㊔ 手段／平均値 用例数 8,737

「手段」という意味では，単数扱い（単複同形）で使われることが多い．「平均値」の意味の複数形としても用いられる．

◆類義語：tool, method, procedure, way

■ means ＋［前］

	the%	a/an%	ø%			
❶	5	36	55	means of ～	～の手段	4,421
				・by means of ～	～によって	2,267
				・means of ～ing	～する手段	1,590
				・as a means of ～	～の手段として	412
❷	9	80	7	means to *do*	～する手段	1,550
				・as a means to *do*	～する手段として	320
❸	6	76	12	means for ～	～のための手段	832
				・means for ～ing	～するのための手段	459
❹	46	44	2	means by which ～	それによって～である手段	426

❶ This chemoselectivity difference is explored **by means of** molecular modeling.（J Am Chem Soc. 2001 123:10436）

訳 この官能基選択性の違いが，分子モデリングによって探索される

❷ In addition, **as a means to find more stable derivatives** and gain further insights into structure-activity relationships, several PatA derivatives were synthesized and assayed in the IL-2 reporter gene assay.（J Am Chem Soc. 2004 126:10582）

訳 もっと安定な誘導体を見つけるための手段として

■［動］＋ a means

❶ provide a means	…は，手段を提供する	744
・provide a means to *do*	…は，～する手段を提供する	303
・provide a means of ～	…は，～の手段を提供する	201
・provide a means for ～	…は，～のための手段を提供する	200

❶ This system could **provide a means for clathrin-mediated endocytosis** to quickly recycle vesicle proteins in highly excitable cells.（Nat Commun. 2012 3:1154）

訳 このシステムは，クラスリンに仲介されるエンドサイトーシスのための手段を提供しうる

★★ measure ㊫ 測定する，㊔ 尺度 用例数 58,852

measured	測定される／～を測定した	32,863
measure	～を測定する／尺度	10,876
measures	～を測定する／尺度	10,556
measuring	～を測定する	4,557

名詞としても用いられるが，他動詞の用例が多い．名詞としては，複数形の用例の方が多い．

◆類義語：determine, estimate, assay, quantify, measurement

■ 動 be measured ＋［前］/using

★ be measured	…が，測定される	14,254
❶ be measured by ～	…が，～によって測定される	2,962
・be measured by using ～	…が，～を使うことによって測定される	242
❷ be measured in ～	…が，～において測定される	2,745
❸ be measured using ～	…が，～を使って測定される	1,008
❹ be measured at ～	…が，～において測定される	879
❺ be measured with ～	…が，～によって測定される	864
❻ be measured for ～	…が，～に関して測定される	445
❼ be measured as ～	…が，～として測定される	430

❶ Ca^{2+} efflux from hepatocyte populations **was measured by using** extracellular fura-2.（Gastroenterology. 2002 123:1291）
訳 肝細胞集団からのCa^{2+}流出が，細胞外fura-2を使うことによって測定された

❷ Platelet procoagulant activity **was measured in** a prothrombinase assay.（Blood. 2002 100:2839）
訳 血小板凝固活性が，プロトロンビナーゼアッセイにおいて測定された

❹ Endogenous hormone concentrations **were measured at** baseline and at 3 and 6 mo.（Am J Clin Nutr. 2002 75:145）
訳 内在性のホルモン濃度が，ベースラインにおいて測定された

❺ Forearm blood flow **was measured with** venous occlusion plethysmography in the resting forearm.（J Physiol. 2000 525:253）
訳 前腕血流が，静脈性閉塞脈波検査によって測定された

■ 動 ［名］＋ be measured

❶ levels were measured	レベルが，測定された	431
❷ activity was measured	活性が，測定された	281
❸ concentrations were measured	濃度が，測定された	240

❶ Tissue iNOS **levels were measured** by immunohistochemistry.（J Nucl Med. 2012 53:994）
訳 組織のiNOSレベルが，免疫組織化学によって測定された

■ 動 ［副］＋ measured

❶ directly measured	直接測定される	182
❷ experimentally measured	実験的に測定される	159

■ 動 as measured by

❶ as measured by ～	～によって測定されたとき	3,349

❶ Env- and Gag-specific antibodies **as measured by** enzyme-linked immunosorbent assay were also low or undetectable at this time.（J Virol. 2001 75:10200）
訳 酵素結合免疫吸着測定法によって測定されたとき

■ 動 measure ＋［名句］

❶ measure the effects of ～	…は，～の影響を測定する	167
❷ measure the rate of ～	…は，～の速度を測定する	142

❶ We **measured the effects of** PHLPP overexpression or knockdown with small interfering RNAs on Akt activation and cell death.（Gastroenterology. 2012 142:377）

訳 われわれは，～の影響を測定した

■ 動［代名］＋ measure

❶ we measure ～	われわれは，～を測定する	3,096

■ 動［動／名］＋ to measure

★ to measure ～	～を測定する…	5,057
❶ be used to measure ～	…は，～を測定するために使われる	1,112
❷ method to measure ～	～を測定するための方法	147
❸ assay to measure ～	～を測定するためのアッセイ	140

❶ Size exclusion chromatography coupled to SF-ICPMS **was used to measure** the iron-containing biomolecules in the samples.（Anal Chem. 2001 73:4422）

訳 ～が，鉄を含む生体分子を測定するために使われた

■ 動［前］＋ measuring

❶ be assessed by measuring ～	…は，～を測定することによって査定される	183
❷ be determined by measuring ～	…は，～を測定することによって決定される	164
❸ method for measuring ～	～を測定するための方法	144

❶ Cell viability was assessed by tetrazolium salt（WST-1）assay, and apoptosis **was determined by measuring** DNA cleavage or phosphatidylserine exposure.（Invest Ophthalmol Vis Sci. 2002 43:2546）

訳 アポトーシスは，DNA切断を測定することによって決定された

■ 名 measure ＋［前］

	the%	a/an%	ø%			
❶	11	84	1	measure of ～	～の尺度／～の計測	2,737
				・as a measure of ～	～の尺度として	118

❶ Capillary filtration coefficient **as a measure of** microvascular fluid permeability（conductance）was analyzed before and after start of the synthetic colloid.（Crit Care Med. 2001 29:123）

訳 微小血管の液体透過性（コンダクタンス）の尺度として毛細管濾過係数が解析された

■ 名［名］＋ measures

❶ outcome measures	結果の計測	853
❷ control measures	対策	207

★ measurement 名 測定／測定値 用例数 23,585

measurements	測定／測定値	16,647
measurement	測定／測定値	6,938

複数形のmeasurementsの割合が，約70%と圧倒的に高い．

◆類義語：assay, quantification

■ measurements ＋［前］

	the%	a/an%	ø%			
❶	2	-	92	measurements of ～	～の測定	4,454
❷	3	-	85	measurements in ～	～における測定	618

❶ **Measurements of** the distinct phytoplankton phosphorus pools may be required to assess nutrient limitation accurately from elemental composition.（Nature. 2004 432:897）

訳 別個の植物プランクトンのリン貯蔵の測定は，～するために必要とされるかもしれない

❷ High levels of interference are observed, consistent with **measurements in** other fish species.（Genetics. 1998 148:1225）

訳 他の魚の種における測定値と一致して

■ measurements ＋［動］

❶	measurements were made	測定が，なされた	262
❷	measurements showed ～	測定値は，～を示した	237
❸	measurements indicate ～	測定値は，～を示す	237
❹	measurements were performed	測定が，実行された	168
❺	measurements revealed ～	測定値は，～を明らかにした	151

❶ Cytokine **measurements were made** on lymph node and spleen supernatants for interferon-γ, interleukin (IL)-4, IL-10, and IL-13.（J Infect Dis. 2002 186:428）

訳 サイトカインの測定が，リンパ節に対してなされた

❷ Direct blood pressure and heart rate **measurements showed** little or no hemodynamic effect.（Blood. 2004 103:1356）

訳 直接血圧および心拍数測定値は，ほとんどあるいは全く血行力学的な影響を示さなかった

■［形／名］＋ measurement(s)

❶	direct measurement	直接測定	273
❷	kinetic measurements	動態学的測定	246
❸	fluorescence measurements	蛍光測定	227
❹	experimental measurements	実験的測定	219

★★ mechanism 名 機構／機序 用例数 131,651

mechanism	機構／機序	75,029
mechanisms	機構／機序	56,622

複数形のmechanismsの割合は約45%とかなり高い.

◆類義語：machinery, mode

■ mechanism ＋［前］/whereby

mechanism ofの前にtheが付く割合はかなり高い.

	the %	a/an %	ø %		
❶	64	25	2	mechanism of ～ 　～の機構	18,436
				・mechanism of action 　作用の機序	2,113
				・insight into the mechanism of ～ 　～の機構への洞察	491
				・investigate the mechanism of ～ 　…は，～の機構を精査する	302
				・understand the mechanism of ～ 　…は，～の機構を理解する	265
				・understanding of the mechanism of ～ 　～の機構を理解すること	138
❷	14	78	0	mechanism for ～ 　～の機構／～のための機構	10,387
				・provide a mechanism for ～ 　…は，～の機構を提供する	327
				・mechanism for regulating ～ 　～を調節する機構	309
				・mechanism for controlling ～ 　～を調節する機構	134
❸	58	38	0	mechanism by which ～ 　（それによって）～である機構	7,546
❹	17	75	1	mechanism in ～ 　～における機構	2,431
				・mechanism in which ～ 　（そこで）～である機構	850
❺	4	90	1	mechanism to *do* 　～する機構／～するための機構	1,858
❻	22	75	0	mechanism whereby ～ 　（それによって）～である機構	963

❶ **These findings suggest a mechanism of action for metformin** and identify novel therapeutic targets in insulin-resistant states.（Nat Med. 2000 6:998）

訳 これらの知見は，メトホルミンの作用の機序を示唆する

❶ Our findings identified unique GEF specificity determinants in Rac1 and **provide important insights into the mechanism of DH/PH selection** of GTPase targets.（Nat Struct Biol. 2001 8:1037）

訳 ～は，DH/PH選択の機構への重要な洞察を提供する

❶ Here **we investigate the mechanism of TPA induction of FGF-BP gene expression** in the human ME-180 SCC cell line.（J Biol Chem. 1998 273:19130）

訳 われわれは，FGF-BP遺伝子発現のTPA誘導の機構を精査する

❷ **These findings provide a mechanism for** BRCA1-associated breast carcinogenesis.（Nat Genet. 2001 28:266）

訳 これらの知見は，～の機構を提供する

❷ Our data suggest that glypican binding to the releasable C-terminal portion of Slit **may serve as a mechanism for regulating the biological activity of** Slit and/or the proteoglycan.（J Biol Chem. 2001 276:29141）

訳 …は，～の生物学的活性を調節するための機構として役立つかもしれない

❸ However, the mechanism by which selenium suppresses tumor development remains unknown. (Cancer Res. 2001 61:2307)

訳 セレンが腫瘍の発生を抑制する機構は未知のままである

❹ The results are not readily explained by a mechanism in which caldesmon acts only by stabilizing an inactive state of actin-tropomyosin. (Biochemistry. 2001 40:5757)

訳 カルデスモンが～の不活性状態を安定化させることによってのみ作用する機構

■ mechanism(s) ＋ [現分／過分／形]

❶ mechanisms underlying ～	～の根底にある機構	3,966
❷ mechanism involving ～	～に関わる機構	1,177
❸ mechanisms involved in ～	～に関与する機構	1,156
❹ mechanisms responsible for ～	～の原因である機構	1,079
❺ mechanisms regulating ～	～を調節する機構	761
❻ mechanisms controlling ～	～を制御する機構	540
❼ mechanisms including ～	～を含む機構	403
❽ mechanisms governing ～	～を支配する機構	375
❾ mechanisms leading to ～	～につながる機構	311

❶ These studies provide new insights into the molecular mechanisms underlying the effect of glucocorticoids on gene expression. (Mol Cell Biol. 1999 19:6471)

訳 これらの研究は，遺伝子発現に対するグルココルチコイドの影響の根底にある分子機構への新しい洞察を提供する

❷ Inhibitors of NF-κB activation were observed to efficiently block the suppressive effect of LPS on SR-B1 and ABCA1, suggesting a mechanism involving NF-κB. (Infect Immun. 2002 70:2995)

訳 NF-κBに関わる機構を示唆している

■ mechanism ＋ [動]

❶ mechanism involves ～	機構は，～に関わる	253
❷ mechanism remains ～	機構は，～のままである	253
❸ mechanism is proposed	機構が，提案される	199

❸ An organometallic mechanism is proposed to explain these results. (J Am Chem Soc. 2002 124:1638)

訳 有機金属の機構が，これらの結果を説明するために提案される

■ [形] ＋ mechanism(s)

❶ molecular mechanisms	分子機構	5,710
❷ a novel mechanism	新規の機構	2,718
❸ ～-dependent mechanism	～依存的な機構	1,471
❹ underlying mechanisms	根底にある機構	1,230
❺ regulatory mechanism	調節機構	1,048
❻ different mechanisms	異なる機構	1,006

❶ Structural and kinetic studies have provided extensive information about the molecular mechanisms of kinase activation by phosphorylation. (J Mol Biol. 2005 353:600)
訳 ～は，リン酸化によるキナーゼ活性化の分子機構についての広範な情報を提供してきた

★★ mediate 動 仲介する　用例数 120,845

mediated	仲介される／～を仲介した	89,500
mediate	～を仲介する	14,286
mediates	～を仲介する	9,771
mediating	～を仲介する	7,288

他動詞.
◆類義語：modulate, control, regulate

■ be mediated ＋［前］

★ be mediated	…は，仲介される	14,095
❶ be mediated by ～	…は，～によって仲介される	9,662
・be mediated by activation of ～	…は，～の活性化によって仲介される	100
・be mediated in part by ～	…は，一部は～によって仲介される	548
❷ be mediated through ～	…は，～によって仲介される	1,573
❸ be mediated via ～	…は，～によって仲介される	616

❶ The inhibitory effect of TGF-β1 on sodium uptake and αENaC expression in ATII cells was mediated by activation of the MAPK, ERK1/2. (J Biol Chem. 2003 278:43939)
訳 ～は，MAPKの活性化によって仲介された

❷ Transcription activation of anaerobically induced genes in Escherichia coli is mediated through the action of the global anaerobic regulator FNR. (J Mol Biol. 2002 315:275)
訳 ～は，包括的な嫌気性調節因子FNRの作用によって仲介される

❸ Overexpression of StAR and CYP7A1 were mediated via infection with recombinant adenoviruses. (Hepatology. 2004 40:910)
訳 StARおよびCYP7A1の過剰発現は，組換えアデノウイルスによる感染によって仲介された

■［名］＋ be mediated

❶ effect is mediated	効果は，仲介される	226
❷ interaction is mediated	相互作用は，仲介される	169
❸ expression is mediated	発現は，仲介される	112
❹ response is mediated	応答は，仲介される	105
❺ activation is mediated	活性化は，仲介される	100

❶ Interestingly, this effect is mediated at least partially by insulin, suggesting that effects of octopamine on metabolism are independent of its effects on sleep. (J Biol Chem. 2012 287:32406)
訳 この効果は，少なくとも一部はインスリンによって仲介される

■ -mediated ＋［名］

❶	～-mediated apoptosis	～に仲介されるアポトーシス	1,716
❷	～-mediated signaling	～に仲介されるシグナル伝達	1,702
❸	～-mediated activation of …	～に仲介される…の活性化	1,334
❹	～-mediated inhibition of …	～に仲介される…の抑制	1,120
❺	～-mediated phosphorylation	～に仲介されるリン酸化	892
❻	～-mediated transcription	～に仲介される転写	882
❼	～-mediated endocytosis	～に仲介されるエンドサイトーシス	791
❽	～-mediated repression	～に仲介される抑制	718
❾	～-mediated induction	～に仲介される誘導	703

❶ We demonstrate here that **matrix attachment is an efficient regulator of Fas-mediated apoptosis in endothelial cells**. (J Cell Biol. 2001 152:633)

訳 基質付着は，内皮細胞におけるFasに仲介されるアポトーシスの効果的な調節因子である

■［名］-mediated

❶	receptor-mediated ～	受容体に仲介される～	3,117
❷	cell-mediated ～	細胞に仲介される～	3,008
❸	integrin-mediated ～	インテグリンに仲介される～	686

❶ Previous studies have suggested that **most papillomaviruses enter the host cell via clathrin-dependent receptor-mediated endocytosis** but have not addressed later steps in viral entry. (Proc Natl Acad Sci USA. 2004 101:14252)

訳 ほとんどのパピローマウイルスは，クラスリン依存性受容体に仲介されるエンドサイトーシスによって宿主細胞に入る

■ mediate ＋［名句］

❶	mediate the effects of ～	…は，～の効果を仲介する	326
❷	mediate the interaction	…は，誘導を仲介する	172
❸	mediate the formation of ～	…は，～の形成を仲介する	108
❹	mediate the activation of ～	…は，～の活性化を仲介する	101

❶ While **these receptors mediate the effects of EGL-20**, we find that the Frizzled receptor LIN-17 can antagonize MIG-1 signaling. (Dev Cell. 2006 10:367)

訳 これらの受容体は，EGL-20の効果を仲介する

■［名］＋ mediate

❶	receptors mediate ～	受容体は，～を仲介する	219
❷	cells mediate ～	細胞は，～を仲介する	180
❸	proteins mediate ～	タンパク質は，～を仲介する	165

M

■ ［動／形／名］＋ to mediate

★ to mediate ～	～を仲介する…	4,311
❶ be shown to mediate ～	…は，～を仲介することが示される	244
❷ be sufficient to mediate ～	…は，～を仲介するのに十分である	200
❸ ability to mediate ～	～を仲介する能力	180
❹ be thought to mediate ～	…は，～を仲介すると考えられる	152
❺ be known to mediate ～	…は，～を仲介することが知られている	133

❶ **Endocannabinoid signaling has been shown to mediate synaptic plasticity** by retrogradely inhibiting presynaptic transmitter release in several systems.（J Neurosci. 2012 32:13597）

訳 内在性カンナビノイド・シグナル伝達は，シナプス可塑性を仲介することが示されている

★★ member 名 メンバー　　用例数 31,778

members	メンバー	18,653
member	メンバー	13,125

複数形のmembersの割合は約60%と非常に高い．member(s) ofの用例が非常に多い．

◆類義語：class

■ member ＋［前］

member ofの前には，a/anが付く割合が圧倒的に高い．

	the%	a/an%	ø%			
❶	4	91	0	member of ～	～のメンバー	10,005
				・a member of ～	～のメンバー	6,305
				・a new member of ～	～の新しいメンバー	404
				・member of a family of ～	～のファミリーのメンバー	315
				・a novel member of ～	～の新規のメンバー	291

❶ **Liver X receptor β（LXRβ）is a member of the nuclear receptor supergene family expressed in the central nervous system**, where it is important for cortical layering during development and survival of dopaminergic neurons throughout life.（Proc Natl Acad Sci USA. 2012 109: 13112）

訳 肝臓X受容体β（LXRβ）は，中枢神経系において発現される核内受容体スーパー遺伝子ファミリーのメンバーである

■ ［名／形］＋ members

❶ family members	ファミリーメンバー	4,751
❷ other members	他のメンバー	1,162

metastasis ㊔ 転移 用例数 9,838

metastasis	転移	6,179
metastases	転移	3,659

複数形のmetastasesの割合は約35%あるが，単数形は無冠詞で使われることが圧倒的に多い．

■ metastasis ＋［前］

	the%	a/an%	ø%			
❶	0	0	98	metastasis in ～	～における転移	514
❷	31	0	69	metastasis of ～	～の転移	374
❸	0	0	100	metastasis to ～	～への転移	164

❶ CRC cells overexpressing miR-28 developed tumors more slowly in mice compared with control cells, but **miR-28 promoted tumor metastasis in mice.** (Gastroenterology. 2012 142: 886)

訳 miR-28は，マウスにおける腫瘍転移を促進した

■ ［名］＋ metastasis/metastases

❶	cancer metastasis	癌転移	384
❷	tumor metastasis	腫瘍転移	367
❸	liver metastases	肝転移	313
❹	bone metastases	骨転移	251

■ ［名］＋ and metastasis

❶	invasion and metastasis	浸潤と転移	434
❷	growth and metastasis	増殖と転移	381
❸	progression and metastasis	進行と転移	194

★★ method ㊔ 方法 用例数 60,610

method	方法	35,315
methods	方法	25,295

複数形のmethodsの割合は約40%とかなり高い．

◆類義語：procedure, means, mode, manner, technique, methodology

■ method ＋［前］

	the%	a/an%	ø%			
❶	6	93	0	method for ～	～のための方法	6,129
				・method for determining ～	～を決定するための方法	180

				・method for identifying ～	～を同定するための方法	175
				・method for detecting ～	～を検出するための方法	163
				・method for measuring ～	～を測定するための方法	144
❷	8	75	0	method to *do*	～する方法	3,779
				・method to identify ～	～を同定するための方法	198
				・method to measure ～	～を測定するための方法	143
				・method to detect ～	～を検出するための方法	140
				・method to determine ～	～を決定するための方法	126
❸	37	52	2	method of ～	～の方法	2,441
				・method of ～ing	～する方法	868
				・method of choice	選択の方法	102

❶ DNA microarrays represent **an important new method for determining the complete expression profile of a cell.** (Proc Natl Acad Sci USA. 2001 98:8944)
訳 細胞の完全な発現プロファイルを決定するための重要な新しい方法

❷ Then, **a method to identify the role of natural selection in molecular evolution** by comparing within- and between-species DNA sequence variation will be presented. (Gene. 1999 238:39)
訳 分子進化における自然淘汰の役割を同定するための方法

❸ **This method of analysis involved the use of highly polymorphic microsatellite markers** and statistics to identify regions of hemizygous deletion in unmatched melanoma cell line DNAs. (Am J Hum Genet. 2000 67:417)
訳 この分析の方法は，高度に多型のマイクロサテライトマーカーの使用を含んでいた

■ method ＋［動］

❶	methods were used to *do*	方法が，～するために使われた	349
❷	method provides ～	方法は，～を提供する	279
❸	method is based on ～	方法は，～に基づいている	264
❹	method was developed	方法が，開発された	262
❺	method allows ～	方法は，～を可能にする	232

❷ **This method provides rapid detection of coat protein in** the low-femtomole range, as estimated by titering plaque-forming units of MS2. (Anal Chem. 1998 70:3863)
訳 この方法は，～におけるコートタンパク質の迅速な検出を提供する

❸ **The method is based on the sequential treatment of cell walls with** specific hydrolytic enzymes followed by dialysis. (Anal Biochem. 2002 301:136)
訳 その方法は，～による細胞壁の連続的な処理に基づいている

❹ **A method was developed to assess human telomere lengths at the individual cell level** in tissue sections from standard formalin-fixed paraffin-embedded tissues. (Am J Pathol. 2002 160:1259)
訳 個々の細胞レベルでヒトのテロメア長を評価する方法が開発された

■［形／過分／現分］＋ method(s)

❶	new method	新しい方法	1,143
❷	novel method	新規の方法	651
❸	～-based method	～に基づく方法	517

❹ computational methods	計算方法	427
❺ statistical methods	統計手法	402
❻ current methods	現法	352
❼ different methods	異なる方法	332
❽ existing methods	現存する方法	308

methodology 名 方法論／手順 用例数 3,988

methodology	方法論	3,154
methodologies	方法論	834

複数形のmethodologiesの割合が，約20%ある可算名詞.

◆類義語：method, procedure, technique

■ methodology + [前]

	the%	a/an%	ø%			
❶	11	68	13	methodology for ～	～のための方法論	417
				・methodology for ～ing	～するための方法論	151
❷	11	49	18	methodology to *do*	～するための方法論／～するために方法論を	305

❶ Tandem mass spectrometry (MS/MS) **is the key experimental methodology for high-throughput glycan identification and characterization.** (Bioinformatics. 2005 21:i431)

訳 ～は，ハイスループットなグリカンの同定と特徴づけのための鍵となる実験的方法論である

❷ In this work, **we have developed a general methodology to synthesize** and screen one-bead-one-compound peptide libraries containing free C-termini. (Biochemistry. 2008 47:3061)

訳 われわれは，～を合成するための一般的な方法論を開発した

★ methylation 名 メチル化 用例数 13,879

methylation	メチル化	13,791
methylations	メチル化	88

複数形のmethylationsの割合は0.6%しかなく，原則，不可算名詞として使われる.

■ methylation + [前]

	the%	a/an%	ø%			
❶	14	0	71	methylation of ～	～のメチル化	2,358
				・methylation of ～ in …	…における～のメチル化	176
				・methylation of histone H3	ヒストンH3のメチル化	111
❷	3	0	53	methylation in ～	～におけるメチル化	935

❸	0	0	63	methylation at ～	～におけるメチル化	452

❶ Rpd3S recognizes methylation of histone H3 at lysine 36 (H3K36me), which is required for its deacetylation activity. (EMBO J. 2012 31:3564)

訳 Rpd3Sは，リジン36におけるヒストンH3のメチル化を認識する

■ ［名］＋ methylation

❶	DNA methylation	DNAメチル化	2,911
❷	promoter methylation	プロモーターメチル化	442
❸	cytosine methylation	シトシンメチル化	315

middle 形 中央の／中間の，名 中央／中間 用例数 5,249

middle	中央／中央の／中間の	5,247
middles	中央	2

形容詞の用例が多いが，名詞としても使われる．複数形のmiddlesの用例はほとんどなく，原則，不可算名詞として使われる．

◆類義語：center, intermediate

■ 名 middle ＋［前］

middle ofの前にtheが付く割合は99%と圧倒的に高い．

	the %	a/an %	ø %			
❶	99	0	1	middle of ～	～の中央	656
				・in the middle of ～	～の中央において	384

❶ S1 nuclease analysis reveals a single region of hypersensitivity in the middle of the repeat tract, whereas V1 digestion is consistent with a hydrogen bonded or well stacked structure. (J Biol Chem. 2005 280:29340)

訳 S1ヌクレアーゼ分析は，～の中央において過感受性の単一の領域を明らかにする

■ 形 middle ＋［過分／名］

❶	middle-aged ～	中年の～	656
❷	middle age	中年	151
❸	middle region	中間領域	110

migrate 動 遊走する／移動する 用例数 5,023

migrate	遊走する	2,161
migrating	遊走する	1,599
migrated	遊走した	943
migrates	遊走する	320

自動詞．主に細胞に対して用いられる．

◆類義語：move, transfer, shift, translocate

■ migrate ＋［前］

❶	migrate to ～	…は，～へ遊走する	694
❷	migrate into ～	…は，～の中へ遊走する	360
❸	migrate from ～	…は，～から遊走する	241
❹	migrate in ～	…は，～において遊走する	233
❺	migrate as ～	…は，～として遊走する	176
❻	migrate through ～	…は，～を経て遊走する	141
❼	migrate with ～	…は，～と共に遊走する	120
❽	migrate toward ～	…は，～の方へ遊走する	114

❶ In mouse embryos, **germ cells arise during gastrulation and migrate to the early gonad.** (Development. 2003 130:4279)

訳 生殖細胞は，原腸形成の間に生じ，そして初期の生殖腺へ遊走する

❷ **These cells are able to migrate into the inflamed skin** of wild-type mice. (J Immunol. 2002 169:4307)

訳 これらの細胞は，炎症を起こした皮膚の中へ遊走できる

■［名］＋ migrate

❶	cells migrate	細胞は，遊走する	231

❶ **Muscle progenitor cells migrate from the lateral somites into** the developing vertebrate limb, where they undergo patterning and differentiation in response to local signals. (Genes Dev. 2012 26:2088)

訳 筋肉前駆細胞は，側方体節から～の中へ遊走する

★ migration ㊔ 遊走／移動／転移　用例数 19,443

migration	遊走／移動／転移	19,112
migrations	遊走／移動／転移	331

複数形のmigrationsの割合が1.5%しかなく，原則，**不可算名詞**として使われる．

◆類義語：movement, transfer

■ migration ＋［前］

	the%	a/an%	ø%			
❶	40	1	59	migration of ～	～の遊走（転移）	2,863
❷	1	0	98	migration in ～	～における遊走	1,071
❸	0	2	82	migration to ～	～への遊走	473

❶ Here, we show that a novel autocrine loop of LPA **promotes the migration of ovarian cancer cells**, which is a critical step of tumor metastasis. (FASEB J. 2003 17:1570)

訳 ～は，卵巣癌細胞の転移を促進する

❷ **Invasive cell migration in both normal development and metastatic cancer is regulated by** various signaling pathways, transcription factors and cell-adhesion molecules. (Development. 2005 132:3493)

訳 正常な発生および転移性癌の両方における浸潤性の細胞遊走は，さまざまなシグナル伝達経路によって調節される

❸ The data suggest that, in HIV-1 infection, lymphocyte migration to the intestine may be promoted by increased MAdCAM-1 expression. (J Infect Dis. 2002 185:1043)

訳 腸へのリンパ球の遊走は，増大したMAdCAM-1発現によって促進されるかもしれない

■ ［名／形］＋ migration

❶ cell migration	細胞遊走	4,683
❷ neuronal migration	神経細胞移動	531
❸ branch migration	分岐点移動	282
❹ transendothelial migration	経内皮遊走	257

■ migration and ＋［名］／［名］＋ and migration

❶ migration and invasion	遊走と浸潤	439
❷ proliferation and migration	増殖と遊走	409
❸ adhesion and migration	接着と遊走	289
❹ migration and proliferation	遊走と増殖	215

mild 形 軽度の／穏やかな　用例数 6,321

◆類義語：moderate, intermediate

■ mild ＋［前］

❶ mild to ～	軽度から～	666
・mild to moderate ～	軽度から中程度の～	545

❶ Subcutaneous ghrelin administration enhances short-term food intake in dialysis patients with mild to moderate malnutrition. (J Am Soc Nephrol. 2005 16:2111)

訳 ～は，軽度から中程度の栄養不良の透析患者における短期間の食物摂取を亢進させる

■ ［前］＋ mild

❶ patients with mild ～	軽度の～の患者	294
❷ under mild conditions	穏やかな条件下で	220

mimic 動 模倣する，名 擬態　用例数 8,501

mimic	模倣する／擬態	3,316
mimics	～を模倣する／擬態	2,065
mimicked	模倣される／～を模倣した	1,946
mimicking	～を模倣する	1,174

主に動詞として用いられる．

■ mimic ＋［名句］

❶ mimic the effects of ～	…は，～の効果を模倣する	311

❶ EMSY is located within an amplicon in sporadic breast and ovarian cancers, suggesting that its overexpression may **mimic the effects of** BRCA2 inactivation. (Cell. 2003 115:507)
訳 それの過剰発現は，BRCA2不活性化の効果を模倣するかもしれない

■ to mimic

★ to mimic ～	～を模倣する…	938
❶ designed to mimic ～	～を模倣するように設計される	117

❶ The results of the present study demonstrate that ESI MS can be used to model metal ion release from transferrin under conditions that are **designed to mimic** the physiological environment. (Anal Chem. 2001 73:2565)
訳 生理的な環境を模倣するように設計された状態下で

■ be mimicked ＋［前］

★ be mimicked	…は，模倣される	835
❶ be mimicked by ～	…は，～によって模倣される	688
❷ be mimicked in ～	…は，～において模倣される	85

❶ This effect **was mimicked by** addition of the calcium/calmodulin-dependent kinase (CaMK) inhibitor KN93 to depolarizing medium. (J Biol Chem. 2003 278:41472)
訳 この効果は，カルシウム/カルモジュリン依存性キナーゼ（CaMK）阻害剤KN93の添加によって模倣された

mix ㊍ 混合する，㊔ 混合／混合物　用例数 8,620

mixed	混合される／混合する	7,727
mix	混合／混合物／混合する	852
mixes	混合／混合物／混合する	41

名詞としても用いられるが，動詞の用例が多い．
◆類義語：mixture

■ mixed ＋［前］

❶ mixed with ～	～と混合される	505
・be mixed with ～	…は，～と混合される	207

❶ When DCs **were mixed with** Matrigel and injected subcutaneously into mice, only immature DCs promoted the ingrowth of patent blood vessels. (FASEB J. 2010 24:1411)
訳 樹状細胞は，マトリゲルと混合された

mixture ㊔ 混合物　　用例数 8,063

mixture	混合物	4,834
mixtures	混合物	3,229

複数形のmixturesの割合は約40%とかなり高い.

◆類義語：mix

■ mixture +［前］

mixture ofの前には，a/anが付く割合が非常に高い.

	the%	a/an%	ø%			
❶	8	91	1	mixture of ～	～の混合物	2,656
				・a mixture of ～	～の混合物	1,778

❶ Mice were infected with a mixture of the three strains, the infection was allowed to proceed, and the strains colonizing the organs were identified. (Infect Immun. 2012 80:3189)

訳 マウスは，その３つの菌株の混合物を感染させられた

mode ㊔ 様式　　用例数 11,356

mode	様式	7,816
modes	様式	3,540

複数形のmodesの割合が，約30%ある可算名詞.

◆類義語：manner, fashion, mechanism, machinery, type, form

■ mode +［前］

	the%	a/an%	ø%			
❶	46	28	4	mode of ～	～の様式	3,467
				・mode of action	作用の様式	587
				・mode of binding	結合の様式	158
				・mode of inheritance	遺伝の様式	126
				・mode of interaction	相互作用の様式	110

❶ To study the mode of action of Plk3 in the Golgi fragmentation cascade, we examined functional as well as physical interactions between Plk3 and MEK1/ERKs. (Oncogene. 2004 23:3822)

訳 ゴルジ体の断片化カスケードにおけるPlk3の作用の様式を研究するために

■ ［名／形］+ mode

❶	binding mode	結合様式	601
❷	different modes	異なる様式	201
❸	normal mode	正常な様式	196

❶ The binding mode of these residues is compared with that of biotin and peptidic ligands. (J Mol Biol. 2006 356:738)

訳 これらの残基の結合様式が，～のそれと比較される

★★ model ㊔ モデル，㊫ モデルを作成する　用例数 133,710

model	モデル／～のモデルを作成する	95,491
models	モデル／～のモデルを作成する	35,996
modeled	～のモデルを作成した	2,003
modelled	～のモデルを作成した	220

動詞としても用いられるが，名詞の用例が圧倒的に多い．複数形のmodelsの割合が，約30％ある可算名詞．modelledはイギリス英語．modelingは別項参照．

■ model ＋［前］

	the%	a/an%	ø%			
❶	16	81	0	model of ～	～のモデル	18,390
				・support a model of ～	…は，～のモデルを支持する	187
				・propose a model of ～	…は，～のモデルを提案する	124
❷	10	88	1	model for ～	～のモデル／～のためのモデル	9,118
				・model for studying ～	～を研究するためのモデル	440
				・propose a model for ～	…は，～のモデルを提案する	353
				・provide a model for ～	…は，～のモデルを提供する	311
				・suggest a model for ～	…は，～のモデルを示唆する	238
				・serve as a model for ～	～のモデルとして働く	216
				・model for understanding ～	～を理解するためのモデル	184
				・support a model for ～	…は，～のモデルを支持する	141
❸	3	95	0	model in which ～	（そこで，）～であるモデル	4,443
				・support a model in which ～	…は，～であるモデルを支持する	1,233
				・be consistent with a model in which ～	…は，～であるモデルと一致している	497
				・propose a model in which ～	…は，～であるモデルを提案する	457
				・suggest a model in which ～	…は，～であるモデルを示唆する	446
❹	14	70	1	model to *do*	～するためのモデル／～するためにモデルを	3,294

M

・model to study ～	～を研究するためのモデル	511
・model to investigate ～	～を精査するするためのモデル	218
・model to explain ～	～を説明するためのモデル	161

❶ Thus, our findings support a model of adipogenesis in which SFRP5 inhibits WNT signaling to suppress oxidative metabolism and stimulate adipocyte growth during obesity. (J Clin Invest. 2012 122:2405)

訳 われわれの知見は，～である脂肪生成のモデルを支持する

❷ The human cytomegalovirus (HCMV) early UL4 promoter has served as a useful model for studying the activation of early viral gene expression. (J Virol. 2000 74:9845)

訳 ～は，初期のウイルス遺伝子発現の活性化を研究するための有用なモデルとして役立っている

❸ Our data support a model in which calcium gating of SK channels is mediated by binding of calcium to calmodulin and subsequent conformational alterations in the channel protein. (Nature. 1998 395:503)

訳 われわれのデータは，SKチャネルのカルシウムゲーティングがカルシウムのカルモジュリンへの結合によって仲介されるモデルを支持する

❹ These mice may serve as an animal model to study the role of ECM proteins in osteoporosis. (Nat Genet. 1998 20:78)

訳 これらのマウスは，骨粗鬆症におけるECMタンパク質の役割を研究するための動物モデルとして役立つかもしれない

■ model ＋［関］

❶ model whereby ～	（それによって）～であるモデル	553
❷ model where ～	（そこで）～であるモデル	532

❶ We propose a model whereby nicotinamide inhibits deacetylation by binding to a conserved pocket adjacent to NAD^+, thereby blocking NAD^+ hydrolysis. (J Biol Chem. 2002 277:45099)

訳 われわれは，ニコチンアミドが保存されたポケットに結合することによって脱アセチル化を抑制するモデルを提案する

■ model ＋［動］

❶ models were used	モデルが，使われた	601
❷ model predicts ～	モデルは，～を予測する	494
❸ model provides ～	モデルは，～を提供する	476
❹ model is proposed	モデルが，提案される	337
❺ model suggests that ～	モデルは，～ということを示唆する	280
❻ model was developed	モデルが，開発された	243
❼ model is presented	モデルが，提示される	208

❸ This model provides a possible explanation for the role of Cdc5 in DNA damage checkpoint adaptation. (Science. 1999 286:1166)

訳 このモデルは，～の可能な説明を提供する

❹ A model is proposed to account for human XOR regulation. (J Biol Chem. 2000 275:5918)

訳 ひとつのモデルが，ヒトのXOR調節を説明するために提案される

■ [名/形] + model

❶ mouse model	マウスモデル	5,858
❷ animal model	動物モデル	2,477
❸ murine model	マウスモデル	1,678
❹ rat model	ラットモデル	988
❺ mathematical model	数学的モデル	896
❻ regression model	回帰モデル	743
❼ in vitro model	試験管内モデル	697
❽ new model	新しいモデル	645
❾ structural model	構造モデル	542

■ model + [名]

❶ model system	モデルシステム	3,163
❷ model organisms	モデル生物	579

modeling/modelling 名 モデル化/モデリング 用例数 8,009

modeling	モデル化/モデリング	7,157
modelling	モデル化/モデリング	852

複数形の用例はなく，原則，**不可算名詞**として使われる．modellingはイギリス英語．

■ modeling + [前]

	the%	a/an%	ø%			
❶	17	0	80	modeling of ～	～のモデル化	1,218

❶ **Modeling of signal transduction pathways plays a major role in** understanding cells' function and predicting cellular response. (PLoS One. 2012 7:e50085)

訳 シグナル伝達経路のモデル化は，～において主要な役割を果たす

■ [形] + modeling

❶ molecular modeling	分子モデル化	1,225
❷ homology modeling	ホモロジーモデリング	381
❸ computational modeling	計算モデル化	276
❹ mathematical modeling	数学的モデル化	254
❺ computer modeling	コンピューターモデリング	209

moderate 形 中程度の/中等度の，動 緩和する 用例数 8,460

moderate	中程度の/中等度の/緩和する	8,255
moderated	緩和した/緩和される	96

M

moderating	緩和する	65
moderates	緩和する	44

形容詞の用例が圧倒的に多いが，動詞としても使われる．

◆類義語：intermediate, mild

■ moderate ＋［前］

❶ moderate to severe ～	中程度から重症の～	641
❷ moderate to high ～	中程度から高い～	154

❶ Patients with moderate to severe nonproliferative diabetic retinopathy (NPDR) had POBF 18% higher than the control (mean OBF, 943 μl/min). (Invest Ophthalmol Vis Sci. 2004 45: 4504)

訳 中程度から重症の非増殖性糖尿病性網膜症（NPDR）の患者は，対照群より18%高いPOBFを持っていた

■［前］＋ moderate

❶ mild to moderate ～	軽度から中程度の～	545
❷ patients with moderate ～	中程度の～を持つ患者	296
❸ low to moderate ～	低から中程度の～	156

❶ Inflammatory and prothrombotic markers are elevated in individuals with mild to moderate renal disease. (J Am Soc Nephrol. 2004 15:3184)

訳 ～は，軽度から中程度の腎臓疾患を持つ個々人において上昇する

■ moderate ＋［名］

❶ moderate levels	中程度のレベル	215
❷ moderate intensity	中程度の強度	121

❶ At moderate levels of MnSOD activity (two- to six-fold increase compared to parent cells), these accumulations were blocked. (Oncogene. 2005 24:8154)

訳 中程度のレベルのMnSOD活性において

moderately 副 中程度に／適度に　　用例数 2,153

◆類義語：partially, mildly, somewhat, reasonably

■ moderately ＋［過分］

❶ moderately reduced	中程度に低下した	88
❷ moderately increased	中程度に増大した	74

❷ The results suggest that alcohol consumption is associated with moderately increased severity of periodontal disease. (J Periodontol. 2001 72:183)

訳 飲酒量は，歯周疾患の中程度に増大した重症度と関連する

■ moderately ＋［形］

❶ moderately high	中程度に高い	97
❷ moderately severe	中程度に重症の	79

■［副］＋ moderately

❶ only moderately	ほんのわずかに	183

modest 形 わずかな　用例数 4,393

◆類義語：subtle, slight, insignificant, small, weak, minor

■ modest ＋［名］

❶ modest increase	わずかな増大	265
❷ modest effect	わずかな効果	192
❸ modest changes	わずかな変化	108
❹ modest reduction	わずかな低下	106

❶ Other B-cell lymphomas from Ung$^{-/-}$ mice exhibited a modest increase in mutation frequency. (Oncogene. 2005 24:3063)

訳 ～は，変異の頻度のわずかな増大を示した

■［副］＋（a）modest

❶ only modest ～	ほんのわずかな～	374
❷ only a modest ～	ほんのわずかな～	287
❸ more modest ～	よりわずかな～	159
❹ relatively modest ～	比較的わずかな～	118

❶ To date, only modest numbers of pediatric studies using magnetic resonance imaging have been conducted. (Curr Opin Pediatr. 2005 17:619)

訳 磁気共鳴画像法を使うほんのわずかな数の小児科の研究が行われてきた

modestly 副 わずかに　用例数 1,352

◆類義語：slightly, marginally, weakly

■ modestly ＋［過分］

❶ modestly increased	わずかに増大した	115
❷ modestly reduced	わずかに低下した	110

❷ Surprisingly, the NCX1 knockout mice live to adulthood with only modestly reduced cardiac function as assessed by echocardiography. (Circ Res. 2004 95:604)

訳 NCX1ノックアウトマウスは，ほんのわずかに低下した心機能を持った状態で成体期まで生きる

M

■ [副] + modestly

❶	only modestly	ほんのわずかに	329

★ modification ㊔ 修飾／改変　用例数 18,753

modification	修飾	11,765
modifications	修飾	6,988

複数形のmodificationsの割合は約35%あるが，単数形は無冠詞で使われることが多い．
◆類義語：alteration, change, shift, conversion

■ modification + [前]

	the%	a/an%	ø%			
❶	23	10	63	modification of ～	～の修飾	4,377
❷	5	22	57	modification in ～	～における修飾	437
❸	2	3	89	modification by ～	～による修飾	426

❶ Inactivation was accomplished by modification of proteins and lipids in the viral envelope using the hydrophobic photoinduced alkylating probe 1,5 iodonaphthylazide (INA). (J Virol. 2005 79:12394)
訳 不活性化が，タンパク質および脂質の修飾によって達成された

■ [形／名] + modification

❶	post-translational modification	翻訳後修飾	612
❷	posttranslational modification	翻訳後修飾	578
❸	chemical modification	化学修飾	435
❹	covalent modification	共有結合的修飾	409
❺	histone modification	ヒストン修飾	297
❻	protein modification	タンパク質修飾	220

★ modify ㊌ 修飾する／改変する　用例数 20,034

modified	修飾される／改変される／～を修飾した	14,199
modify	～を修飾する／～を改変する	2,640
modifying	～を修飾する／～を改変する	2,351
modifies	～を修飾する／～を改変する	844

他動詞．
◆類義語：change, alter, shift, convert, vary

■ be modified + [前]

★	be modified	…が，修飾される	2,172

❶ be modified by ～	…が，～によって修飾される	830
❷ be modified to *do*	…が，～するように改変される	344
❸ be modified with ～	…が，～を使って修飾される	178
❹ be modified in ～	…が，～において修飾される	165

❶ Using microarray analysis, **we detected 41 genes whose expression was modified by LIF** in Shp-2$^{\Delta 46-110}$ ES cells.（Blood. 2005 105:635）

訳 われわれは，発現がLIFによって修飾される41個の遺伝子を検出した

❷ To determine how hemodynamic forces might modulate TM expression, **the surgical protocol was modified to alter blood flow** and pressure-induced vessel distension.（Circ Res. 2003 92:41）

訳 外科的なプロトコールは，血流を変えるように改変された

❸ Because this protein does not contain cysteine, two cysteine mutants（Val60Cys and Phe62Cys）**have been prepared and covalently modified with fluorescein.**（Proc Natl Acad Sci USA. 2002 99:14171）

訳 ～が，調製されそしてフルオレセインによって共有結合的に修飾された

■［副］＋ modified

❶ genetically modified	遺伝的に改変される	577
❷ chemically modified	化学的に修飾される	274

❶ **Genetically modified cells** that harbor deletions in various members of the RB/E2F family confirm our data from the wild-type cells.（J Biol Chem. 2004 279:40511）

訳 遺伝的に改変された細胞

❷ **Chemically modified DNA oligonucleotides have been crucial to the development of antisense therapeutics.**（Biophys J. 2003 85:2525）

訳 化学的に修飾されたDNAオリゴヌクレオチドは，アンチセンス薬物療法の開発に決定的に重要である

★modulate 動 調節する／制御する 用例数 24,717

modulate	～を調節する	9,266
modulated	調節される／～を調節した	5,400
modulating	～を調節する	5,357
modulates	～を調節する	4,694

他動詞.

◆類義語：regulate, control, adjust, mediate, drive

■ modulate ＋［名句］

❶ modulate the activity of ～	…は，～の活性を調節する	458
❷ modulate the expression of ～	…は，～の発現を調節する	302
❸ modulate gene expression	…は，遺伝子発現を調節する	220
❹ modulate the function of ～	…は，～の機能を調節する	171
❺ modulate transcription	…は，転写を調節する	146

M

❸ Endocytosis and trafficking within the endocytosis pathway are known to modulate the activity of different signaling pathways. (Development. 2004 131:5807)
訳 ～は，異なるシグナル伝達経路の活性を調節することが知られている

■ ［名／動］＋ to modulate

★	to modulate ～	～を調節する…	2,916
❶	ability to modulate ～	～を調節する能力	177
❷	be shown to modulate ～	…は，～を調節することが示される	170
❸	be known to modulate ～	…は，～を調節することが知られている	80

❶ In theory, any cell expressing one or more donor alloantigens has the ability to modulate the subsequent immune response to an allograft expressing the same molecules. (Transplantation. 2002 73:S16)
訳 1つあるいはそれ以上のドナーの同種異系抗原を発現するどの細胞も，同じ分子を発現する同種移植片に対する引き続いた免疫応答を調節する能力を持つ

❷ Tax has been shown to modulate the activities of several cellular promoters. (J Virol. 2001 75:2161)
訳 Taxは，いくつかの細胞性プロモーターの活性を調節することが示されている

■ ［副］＋ modulate

❶	directly modulate ～	…は，～を直接調節する	173
❷	differentially modulate ～	…は，～を差動的に調節する	141
❸	negatively modulate ～	…は，～を負に調節する	109

❶ These same pathways also may directly modulate TGF-β1 signaling. (J Am Soc Nephrol. 2004 15:2032)
訳 これらの同じ経路は，また，TGF-β1シグナル伝達を直接調節するかもしれない

■ be modulated ＋ ［前］

★	be modulated	…は，調節される	2,808
❶	be modulated by ～	…は，～によって調節される	2,047
❷	be modulated in ～	…は，～において調節される	194

❶ Mammalian target of rapamycin (mTOR) is a central regulator of protein synthesis whose activity is modulated by a variety of signals. (Genes Dev. 2004 18:2893)
訳 ～は，その活性が種々のシグナルによって調節されるタンパク質合成の中心的な調節因子である

★ modulation 名 調節／制御　用例数 10,697

modulation	調節／制御	10,350
modulations	調節／制御	347

複数形のmodulationsの割合は3％あるが，不可算名詞として使われることが多い．

◆類義語：regulation, control, adjustment, contrast, accommodation

■ modulation ＋［前］

	the％	a/an％	ø％			
❶	26	3	68	modulation of ～	～の調節	7,023
				・…-dependent modulation of ～	…依存性の～の調節	186
				・down-modulation of ～	～の下方調節	179
❷	6	2	54	modulation by ～	～による調節	430
❸	6	9	64	modulation in ～	～における調節	256

❶ Individual variation in CYP3A4 may play a role in breast and prostate carcinogenesis **through modulation of sex hormone metabolite levels.**（Am J Epidemiol. 2004 160:825）
訳 性ホルモン代謝物レベルの調節によって

❶ This represents **a new pathway for integrin-dependent modulation of gene expression** and control of cellular differentiation.（J Clin Invest. 2004 114:408）
訳 遺伝子発現のインテグリン依存性の調節のための新しい経路

❷ In addition, **intracellular cGMP is necessary for the KCa channel modulation by NO.**（J Physiol. 2004 555:219）
訳 細胞内cGMP は，NOによるKCaチャネル調節のために必要である

moiety　名 部分／鎖　　用例数 5,746

moiety	部分／鎖	4,050
moieties	部分／鎖	1,696

複数形のmoietiesの割合が，約30%ある可算名詞．
◆類義語：part

■ moiety ＋［前］

moiety ofの前にtheが付く割合は圧倒的に高い．moiety inの前にtheが付く割合もかなり高い．

	the％	a/an％	ø％			
❶	93	5	2	moiety of ～	～の部分	978
❷	69	27	1	moiety in ～	～における部分	331

❶ We found that NgR1 and NgR3 **bind with high affinity to the glycosaminoglycan moiety of proteoglycans** and participate in CSPG inhibition in cultured neurons.（Nat Neurosci. 2012 15:703）
訳 …は，プロテオグリカンのグリコサミノグリカン部分に高親和性で結合する

★★ molecular　形 分子の　　用例数 66,294

■ molecular ＋［名］

❶	molecular mechanisms	分子機構	5,710

M

❷ molecular weight	分子量	4,045
❸ molecular basis	分子基盤	3,806
❹ molecular dynamics	分子動力学	3,105
❺ molecular mass	分子量	2,087
❻ molecular level	分子レベル	1,370
❼ molecular modeling	分子モデル化	1,225
❽ molecular events	分子現象	866
❾ molecular biology	分子生物学	799

❶ The molecular mechanisms underlying this process remain to be fully elucidated.（J Biol Chem. 2004 279:36982）

訳 この過程の根底にある分子機構は，十分には解明されないままである

❸ Here we investigate the molecular basis for cooperativity between adjacent spectrin domains 16 and 17 from chicken brain α-spectrin（R16 and R17）.

訳 われわれは，～の間の協同性に対する分子基盤を精査する

monitor 動 モニターする，名 モニター 用例数 9,087

monitored	モニターされる／～をモニターした	5,202
monitor	～をモニターする／モニター	3,494
monitors	～をモニターする	391

名詞としても用いられるが，他動詞の用例が多い．monitoringは別項参照．

◆類義語：observe

■ be monitored ＋［前］

★ be monitored	…が，モニターされる	3,019
❶ be monitored by ～	…が，～によってモニターされる	821
❷ be monitored for ～	…が，～に関してモニターされる／～の間モニターされる	324
❸ be monitored in ～	…が，～においてモニターされる	324
❹ be monitored with ～	…が，～によってモニターされる	174

❶ The fidelity of chromosomal duplication is monitored by cell cycle checkpoints operational during mitosis.（Mol Cell Biol. 2005 25:9232）

訳 ～が，細胞周期チェックポイントによってモニターされる

❷ Mice were monitored for the development and severity of arthritis.（Arthritis Rheum. 2005 52:620）

訳 マウスが，関節炎の発症と重症度に関してモニターされた

■ as monitored

★ as monitored	モニターされながら	238
❶ as monitored by ～	～によってモニターされながら	212

❶ The peptides were screened for cellular uptake efficiency as monitored by fluorescence microscopy.（Anal Biochem. 2005 341:290）

訳 蛍光顕微鏡によってモニターされながら

■ to monitor

★ to monitor ～	～をモニターする…	2,448
❶ be used to monitor ～	…は，～をモニターするために使われる	484

❶ They can also be used to monitor changes in gene expression in response to drug treatments.（Nat Genet. 1999 21:48）
訳 それらは，また，～に応答する遺伝子発現の変化をモニターするために使われうる

monitoring 名 モニタリング，-ing モニターする　用例数 6,469

monitoring	モニタリング／～をモニターする	6,467
monitorings	モニタリング	2

複数形の用例はほとんどなく，原則，不可算名詞として使われる．

■ monitoring ＋［前］

	the %	a/an %	ø %			
❶	16	3	80	monitoring of ～	～のモニタリング	1,301

❶ Monitoring of immune function revealed no consistent changes.（N Engl J Med. 2012 367:11）
訳 免疫機能のモニタリングは，～を明らかにした

morbidity 名 罹患／罹患率　用例数 5,975

morbidity	罹患／罹患率	5,820
morbidities	罹患率	155

複数形のmorbiditiesの割合は2.5％しかなく，原則，不可算名詞として使われる．
◆類義語：prevalence, susceptibility, onset

■ morbidity ＋［前］

	the %	a/an %	ø %			
❶	3	0	97	morbidity in ～	～における罹患	338
❷	16	3	80	morbidity of ～	～の罹患	131

■［前］＋ morbidity

❶ cause of morbidity	罹患の原因	479

■ morbidity and ＋［名］／［名］＋ and morbidity

❶ morbidity and mortality	罹患（率）と死亡（率）	2,611
❷ mortality and morbidity	死亡（率）と罹患（率）	364

❶ Colorectal cancer remains a major cause of morbidity and mortality in United States. (Curr Opin Gastroenterol. 2003 19:57)
訳 直腸結腸癌は，合衆国における罹患と死亡の主要な原因のままである

★★ more 形 副 もっと／より／それ以上　用例数 114,407

◈反意語：less

■ 形/副 more than

★ more than ～	～以上	15,086
❶ more than half	半分以上	542
❷ more than 50%	50%以上	330
❸ no more than ～	～しかない	281

❶ Still, more than half of the hotspots lie in noncoding regions of the mouse genome. (Genome Res. 2004 14:574)
訳 ホットスポットの半分以上は，～の非翻訳領域にある

❸ All QTL were of minor effect, explaining no more than 20% of the observed variation in phenotypic value. (Heredity (Edinb). 2010 105:562)
訳 観察された変動の20%しか説明しない

■ 形/副 [副] ＋ more

❶ significantly more ～	有意により～	3,765
❷ much more ～	ずっともっと～	2,351
❸ …-fold more ～	…倍より～	1,803
❹ … times more ～	…倍より～	1,338
❺ even more ～	さらにもっと～	788
❻ far more ～	はるかにより～	523

❷ UAP56 mRNA is more abundant than URH49 mRNA in many tissues, although in testes URH49 mRNA is much more abundant. (Nucleic Acids Res. 2004 32:1857)
訳 URH49メッセンジャーRNAは，ずっともっと豊富である

❺ The inhibition was even more pronounced when IFN-γ was included in the regimen. (Cancer Res. 2000 60:3200)
訳 IFN-γがその投与計画に含まれているときに，その抑制はさらにもっと明白であった

■ 副 more ＋ [形]

❶ more likely	よりありそうな	4,931
・be more likely	…は，よりありそうである	3,251
❷ more effective	より効果的な	2,781
❸ more sensitive	より敏感な	2,419
❹ more severe	より重症の	1,991
❺ more potent	より強力な	1,942
❻ more common	よりよくある	1,757

❼ more complex	より複雑な	1,625
❽ more efficient	より効率的な	1,498
❾ more stable	より安定な	1,211
❿ more rapid	より急速な	1,135

❶ Trisomy 8 cells were more likely to express activated caspase-3 than were normal cells. (Blood. 2002 100:4427)

訳 ～は，正常な細胞より活性化されたカスパーゼ-3を発現しそうであった

❷ The 5gP DNA vaccine appeared to be more effective than the corresponding protein subunit vaccine, regardless of adjuvant. (Invest Ophthalmol Vis Sci. 2004 45:506)

訳 5gP DNAワクチンは，～より効果的であるように思われた

❸ AC5 was up to 15-fold more sensitive to inhibitors than AC2. (J Biol Chem. 2004 279:19955)

訳 AC5は，AC2より15倍まで抑制剤に敏感であった

❺ Moreover, a truncated form of human BAD lacking the N-terminal 28 amino acids is more potent than wild-type BAD in inducing apoptosis. (Mol Cell Biol. 2001 21:3025)

訳 N末端28アミノ酸を欠く切断型のヒトBADは，野生型BADより強力である

❻ Env-specific responses were significantly more common in patients presenting at <30 DPI than in those presenting at 30 to 365 DPI (21 versus 0.5%, P = 0.001). (J Virol. 2003 77:2663)

訳 Env特異的な反応は，～を示している患者において有意によりよくあった

■ 副 more ＋［副］

❶ more rapidly	より急速に	1,315
❷ more frequently	より頻繁に	1,273
❸ more efficiently	より効率的に	1,068
❹ more importantly,	もっと重要なことに，	1,021
❺ more slowly	よりゆっくりと	956
❻ more recently	より最近に	894
❼ more closely	より緊密に	853
❽ more strongly	より強く	826
❾ more often	よりしばしば	821

❷ In addition, nonfunctional agr occurred more frequently among strains isolated from infections of joint prostheses. (J Infect Dis. 2004 190:1498)

訳 非機能性のagrが，～から単離された菌株の間により頻繁に生じた

❹ More importantly, the sustained responses were largely PKA-independent, and were sensitive to specific CaMKII inhibitors or adenoviral expression of a dominant-negative CaMKII mutant. (Circ Res. 2004 95:798)

訳 もっと重要なことに，

■ 形 or more

❶ one or more	1つあるいはそれ以上	2,951
❷ two or more	2つあるいはそれ以上	990

❶ These data suggest that one or more DAP12-pairing receptors negatively regulate signaling through TLRs. (Nat Immunol. 2005 6:579)

訳 1つあるいはそれ以上のDAP12とペアになる受容体は，シグナル伝達を負に制御する

M

★ moreover 副 さらに　　用例数 20,677

文頭で用いられることが非常に多い.

◆類義語：furthermore, further, additionally, in addition

■ Moreover,

★ Moreover,	さらに,	20,170
❶ Moreover, we show that ～	さらに，われわれは～ということを示す	372
❷ Moreover, we found that ～	さらに，われわれは～ということを見つけた	231

❶ Moreover, we show that Kid triggers apoptosis in human cells. (EMBO J. 2003 22:246)

訳 さらに，われわれは～ということを示す

morphological 形 形態学的な／形態的な　　用例数 4,809

◆類義語：morphologic, topological

■ morphological ＋［名］

❶ morphological changes	形態学的変化	723
❷ morphological features	形態学的特徴	221
❸ morphological defects	形態学的欠損	140

❶ These findings raise the possibility that infectious and autoimmune etiologies can lead to similar morphological changes in the nerves. (Ann Neurol. 2005 57:768)

訳 ～は，類似の形態学的変化につながりうる

■ morphological and ＋［形］

★ morphological and ～	形態学的および～	611
❶ morphological and biochemical ～	形態学的および生化学的な～	117
❷ morphological and functional ～	形態的および機能的な～	110

❶ Apoptosis is characterized by various cell morphological and biochemical features, one of which is the internucleosomal degradation of genomic DNA. (J Biol Chem. 2000 275:21302)

訳 アポトーシスは，さまざまな細胞の形態学的および生化学的特徴によって特徴づけられる

★ mortality 名 死亡率／死亡　　用例数 24,617

mortality	死亡率／死亡	24,528
mortalities	死亡率	89

複数形のmortalitiesの割合は0.5%しかなく，原則，不可算名詞として使われる.

◆類義語：lethality

■ mortality ＋［前］

	the %	a/an %	ø %			
❶	4	1	94	mortality in ～	～における死亡率	2,649
				・mortality in patients	患者における死亡率	425
❷	8	0	91	mortality from ～	～による死亡率／～からの死亡	593
❸	1	2	93	mortality after ～	～のあとの死亡率	482
❹	40	8	49	mortality of ～	～の死亡率	470
❺	6	1	91	mortality among ～	～の間の死亡率	422
❻	5	0	94	mortality for ～	～の死亡率	343
❼	1	0	96	mortality at ～	～における死亡率	223

❶ Increased mortality in patients with chronic pulmonary hypertension has been associated with elevated right atrial (RA) pressure. (Circulation. 2005 112:I212)
訳 慢性肺性高血圧の患者における増大した死亡率は，～と関連している

■［前］＋ mortality

❶	risk of mortality	死亡のリスク	314
❷	associated with mortality	死亡と関連した	219
❸	cause of mortality	死亡の原因	175
❹	predictor of mortality	死亡の予測因子	165
❺	reduction in mortality	死亡率の低下	142
❻	increase in mortality	死亡率の増大	138

❶ Moderate drinkers have a lower risk of mortality after myocardial infarction (MI). (Circulation. 2005 112:3839)
訳 中程度の飲酒者は，心筋梗塞のあと，より低い死亡のリスクを持つ

❻ Each 10-ml/min decrement in CrCl was associated with an increase in mortality (hazard ratio, 1.14; $P < 0.0001$). (J Am Soc Nephrol. 2003 14:2373)
訳 ～は，死亡率の増大と関連した

■ mortality ＋［名］

❶	mortality rates	死亡率	1,428
❷	mortality risk	死亡のリスク	477

❶ The AG-associated mortality rate was 2.5 deaths/100,000 children. (J Infect Dis. 2005 192:S80)
訳 AGに関連する死亡率は，～であった

■［名／形／過分］＋ mortality

❶	all-cause mortality	すべての原因の死亡率	1,180
❷	hospital mortality	病院での死亡率	1,070
❸	increased mortality	増大した死亡率	799
❹	～-day mortality	～日の死亡率	565
❺	～-related mortality	～に関連する死亡	509

M

❻ high mortality	高い死亡率	403
❼ cancer mortality	癌による死亡	398
❽ higher mortality	より高い死亡率	384
❾ overall mortality	全体の死亡率	376

❶ rPAF-AH was well tolerated and not antigenic, but did not decrease 28-day all-cause mortality in patients with severe sepsis. (Crit Care Med. 2004 32:332)
訳 ～は，重症の敗血症の患者における28日目のすべての原因の死亡率を低下させなかった

■ [名] + and mortality／mortality and + [名]

❶ morbidity and mortality	罹患（率）と死亡（率）	2,670
❷ mortality and morbidity	死亡（率）と罹患（率）	361
❸ incidence and mortality	発生（率）と死亡（率）	141

❶ Prevention and early detection are crucial in reducing morbidity and mortality from skin cancer. (Curr Opin Oncol. 1999 11:123)
訳 ～は，皮膚癌による罹患率と死亡率を低下させる際に決定的に重要である

★★ most 副 形 名 最も／ほとんど　　用例数 76,027

◆類義語：best ◆反意語：least

■ 副 most + [形]

❶ most common ～	最もよくある～	5,598
❷ most likely	最もありそうな	3,546
・be most likely	…は，最もありそうである	1,036
・most likely to *do*	最も～しそうな	569
・the most likely ～	最もありそうな～	448
・most likely due to ～	最も～のせいでありそうな	225
❸ most important ～	最も重要な～	1,797
❹ most potent ～	最も強力な～	1,376
❺ most abundant ～	最も豊富な～	1,341
❻ most effective ～	最も効果的な～	1,057
❼ most frequent ～	最も頻繁な～	884
❽ most recent ～	最も最近の～	827
❾ most significant ～	最も有意な～	762

❶ Alzheimer disease (AD), a progressive neurodegenerative disorder, is the most common cause of dementia in the elderly. (Radiology. 2003 226:315)
訳 ～は，高齢者における認知症の最もよくある原因である

❷ The most likely explanation for this finding is the metabolism of sulindac sulfide to inactive metabolites by the peroxidase activity of cyclooxygenase. (J Biol Chem. 2003 278:25790)
訳 この発見に対する最もありそうな説明は，～である

❸ One of the most important p53 functions is its ability to activate apoptosis, and disruption of this process can promote tumor progression and chemoresistance. (Oncogene. 2003 22:9030)

訳 最も重要なp53機能の１つは，～である

❹ The most potent antibacterials were simple hydroxybutyl and methoxybutyl derivatives, and hydrophobically substituted piperidinylbutyl derivatives.（J Med Chem. 2003 46:2731）

訳 最も強力な抗菌薬は，～であった

■ 副 most ＋［副］

❶ most commonly ～	最も一般に～	1,397
・the most commonly used ～	最も一般に使われる～	248
❷ most frequently ～	最も頻繁に～	933
❸ most importantly,	最も重要なことに,	849
❹ most notably ～	最も著しく～	686
❺ most closely ～	最も密接に～	663
❻ most highly ～	最も高く～	582
❼ most often ～	最も頻繁に～	562
❽ most widely ～	最も広く～	507
❾ most strongly ～	最も強く～	491

❶ Acetaminophen is one of the most commonly used drugs for the safe and effective treatment of pain and fever.（FASEB J. 2005 19:635）

訳 アセトアミノフェンは，痛みと発熱の安全で効果的な治療のために最も一般的に使われる薬の１つである

❷ Malignant gliomas are the most frequently occurring primary brain tumors and are resistant to conventional therapy.（Oncogene. 2004 23:1821）

訳 悪性グリオーマは，もっとも高頻度に起こる原発性の脳腫瘍である

❸ Most importantly, the G420H mutant was unable to deliver the HDL cholesteryl ester to a metabolically active membrane compartment for efficient hydrolysis.（J Biol Chem. 2004 279:24976）

訳 最も重要なことに，

❺ Brk（breast tumor kinase）is a nonreceptor tyrosine kinase that is most closely related to the Frk family of kinases, and more distantly to Src family kinases.（J Biol Chem. 2002 277:34634）

訳 ～は，Frkファミリーのキナーゼに最も密接に関連する非受容体型のチロシンキナーゼである

■ 形 most ＋［名］

❶ most patients	ほとんどの患者	1,293
❷ most cases	ほとんどの場合	1,135
・in most cases	ほとんどの場合において	847

❷ In most cases, high affinity at μ and κ receptors, and lower affinity at δ receptor was observed, resulting in good selectivity for μ and κ receptors.（J Med Chem. 2004 47:165）

訳 ほとんどの場合において，

■ 名 most ＋［前］

	the%	a/an%	ø%			
❶	1	0	99	most of ～	～のほとんど	7,112

・, most of which ～	（そして,）それのほとんどは～	401

❶ Mutations on presenilins are responsible for most of familial forms of Alzheimer's disease.（J Biol Chem. 2003 278:12064）

訳 プレセニリンの変異は，家族性のアルツハイマー病のほとんどの原因である

❶ Such therapies, most of which are generic and inexpensive today, seem to offer a marked survival advantage compared with patients in whom such therapies are omitted.（Circulation. 2004 109:745）

訳（そして,）それらのほとんどは今日ではジェネリックで安価である

★ motif 名 モチーフ　用例数 28,727

motif	モチーフ	18,283
motifs	モチーフ	10,444

複数形のmotifsの割合が，約35%ある可算名詞.

■ motif ＋［前］

motif ofの前にtheが付く割合は非常に高い.

	the %	a/an %	ø %			
❶	39	56	2	motif in ～	～におけるモチーフ	1,400
❷	78	20	1	motif of ～	～のモチーフ	1,022

❶ The sterol-sensing domain (SSD) is a conserved motif in membrane proteins responsible for sterol regulation.（J Biol Chem. 2011 286:26298）

訳 ステロールセンシングドメイン（SSD）は，～に責任のある膜タンパク質において保存されているモチーフである

■［名／形／過分］＋ motif

❶ binding motif	結合モチーフ	1,802
❷ sequence motif	配列モチーフ	531
❸ recognition motif	認識モチーフ	341
❹ structural motif	構造上のモチーフ	337
❺ conserved motif	保存されたモチーフ	242

move 動 移動する／移動させる／運動する　用例数 6,213

move	移動する／～を移動させる／運動する	2,205
moving	移動する／～を移動させる／運動する	2,040
moved	移動した／運動した	1,062
moves	移動する／～を移動させる／運動する	906

自動詞の用例が多いが，他動詞としても用いられる.

◆類義語：migrate, transfer, shift, translocate

■ move ＋［前］

❶ move to ～	…は，～へ移動する	420
❷ move from ～	…は，～から移動する	399
❸ move in ～	…は，～において運動する	370
❹ move into ～	…は，～へ移動する	231
❺ move toward ～	…は，～へ向かって移動する	185
❻ move through ～	…は，～を経て移動する	161
❼ move along ～	…は，～に沿って移動する	152

❷ Mutation of the signature motifs of capsid proteins of filamentous RNA viruses in p64 results in the formation of tailless virions, which **are unable to move from cell to cell**.（J Virol. 2003 77: 2377）

訳 ～は，細胞から細胞へ移動できない

★movement 名 運動／移動／動き 用例数 14,321

movement	運動／移動／動き	9,758
movements	運動／移動／動き	4,563

複数形のmovementsの割合は約30%あるが，単数形は無冠詞で使われることが非常に多い．

◆類義語：transfer, migration, shift, transition, translocation

■ movement ＋［前］

	the%	a/an%	ø%			
❶	56	3	41	movement of ～	～の運動／～の動き	2,391
❷	2	2	95	movement in ～	～における運動	407
❸	7	13	60	movement to ～	～への移動	169

❶ **Two models have been proposed to account for the movement of chromatids during** anaphase.（Nature. 2004 427:364）

訳 ～の間の染色分体の移動を説明するために２つのモデルが提案されている

❷ **Receptor tyrosine kinases have been shown to promote cell movement in a variety of systems**.（Development. 1996 122:409）

訳 受容体チロシンキナーゼは，さまざまなシステムにおいて細胞運動を促進することが示されている

★much 副 ずっと，形 多くの，名 多量／多くのこと 用例数 17,543

◆類義語：far, very, greatly

■ 副 much ＋［副／形］

❶ much more ～	ずっとより～	2,351

	・be much more ～	…は，ずっとより～である	1,080
❷	much less ～	ずっとより～でない	1,624
❸	much higher	ずっとより高い	1,262
❹	much lower	ずっとより低い	1,070
❺	much greater	ずっとより大きい	736
❻	much larger	ずっとより大きい	727
❼	much smaller	ずっとより小さい	499
❽	much slower	ずっとより遅い	334
❾	much faster	ずっとより速い	325

❶ Nasopharyngeal carcinoma is much more common in Asian countries than in Western countries. (Am J Epidemiol. 2005 162:1174)

訳 上咽頭癌は，西洋諸国においてよりアジア諸国においてずっとよくある

❷ Transfection with S2 increased the amounts of type I and type II TGF-β receptors (TβRI and TβRII), whereas S2δS was much less effective. (Proc Natl Acad Sci USA. 2004 101:3083)

訳 S2δSは，ずっとより効果的ではなかった

❸ The incidence of R117H mutations was much higher than expected. (Am J Hum Genet. 1996 58:823)

訳 R117Hの変異の発生率は，予想されるよりずっと高かった

■ 形 much ＋ [前]

❶	as much as ～	～ほども多い	926

❶ Mutation of two of these sites reduced reporter gene activity by as much as 69%. (J Biol Chem. 2004 279:19800)

訳 これらの部位の２つの変異は，69%もレポーター遺伝子の活性を低下させた

■ 名 much ＋ [前]

	the%	a/an%	ø%			
❶	1	0	98	much of ～	～の多く	1,893

❶ Much of this response comes from inflammatory cells within the lower respiratory tract. (J Infect Dis. 2005 191:1225)

訳 この応答の多くは，炎症細胞に起因する

★ mutagenesis 名 変異誘発／突然変異誘発 用例数 10,836

mutagenesis	変異誘発／突然変異誘発	10,829
mutageneses	変異誘発／突然変異誘発	7

複数形のmutagenesesの用例はほとんどなく，原則，**不可算名詞**として使われる．

■ mutagenesis ＋［前］

	the％	a/an％	⌀％			
❶	2	3	95	mutagenesis of ～	～の変異誘発	2,038

❶ **Mutagenesis of** these binding sites resulted in a loss of DNA binding in vitro, and mutation of one of these sites in vivo resulted in an increase in transcription of both the cif and cifR genes. (J Bacteriol. 2012 194:5315)

訳 これらの結合部位の変異誘発は，DNA結合の喪失という結果になった

■［形］＋ mutagenesis

❶	site-directed mutagenesis	部位特異的変異誘発	3,386
❷	insertional mutagenesis	挿入変異	321
❸	random mutagenesis	ランダム変異誘発	271
❹	site-specific mutagenesis	部位特異的変異誘発	233

■ mutagenesis ＋［名］

❶	mutagenesis studies	変異誘発研究	635
❷	mutagenesis experiments	変異誘発実験	260

☆ mutant 形 変異の，名 変異体／ミュータント　用例数 120,754

mutant	変異の／変異体／ミュータント	76,396
mutants	変異体／ミュータント	44,358

◆反意語：wild-type

形容詞および名詞として用いられる．複数形のmutantsの割合は約60%と非常に高い．

■ 形 mutant ＋［名］

❶	mutant mice	変異マウス	3,225
❷	mutant proteins	変異タンパク質	1,500
❸	mutant cells	変異細胞	1,468
❹	mutant strains	変異株	1,122
❺	mutant embryos	変異胚	942
❻	mutant forms	変異型	805
❼	mutant p53	変異p53	761
❽	mutant enzymes	変異酵素	731
❾	mutant alleles	変異誘発遺伝子	678

❶ In agreement with this correlation, the incidence of cells responding to bitter stimuli was reduced by 70% in **mutant mice** lacking α-gustducin. (J Neurosci. 2003 23:9947)

訳 ～が，α-gustducinを欠く変異マウスにおいて70%低下した

■ 形 mutant and ＋［形］

❶	mutant and wild-type ～	変異および野生型の～	415

❶ CCl4 caused similar acute liver injury in mutant and wild-type mice.（J Clin Invest. 2005 115:3072）
訳 CCl4は，変異および野生型マウスにおいて類似の急性肝傷害を引き起こした

■ 名 mutant(s) ＋［前］

	the%	a/an%	ø%			
❶	25	71	4	mutant of ～	～の変異体	3,399
❷	9	-	73	mutants in ～	～における変異体	1,813
❸	5	-	58	mutants with ～	～を持つ変異体	1,699

❶ In HeLa cells, LIGHT induces NF-κB and JNK activation, which can be blocked by the dominant negative mutant of TRAF2.（Mol Cell Biol. 2005 25:2130）
訳 そして，（それは）TRAF2のドミナントネガティブ変異体によってブロックされうる

❸ In the current study, several human IgG1 mutants with increased binding affinity to human FcRn at pH 6.0 were generated that retained pH-dependent release.（J Immunol. 2006 176:346）
訳 ～に対する増大した結合親和性を持つ変異体

■ 名 mutants ＋［動］

❶	mutants have ～	変異体は，～を持つ	789
❷	mutants showed ～	変異体は，～を示した	553
❸	mutants exhibit ～	変異体は，～を示す	400

❶ Biochemical analyses revealed that the Asp255 mutants showed no detectable in vitro DNA packaging activity.（J Biol Chem. 2006 281:518）
訳 Asp255変異体は，検出できる試験管内DNAパッケージング活性を示さなかった

■ 名 mutants ＋［現分／形］

❶	mutants lacking ～	～を欠く変異体	889
❷	mutants defective in ～	～に欠陥のある変異体	442

❶ Mutants lacking the transcription regulators Mga and CovR（CsrR）also failed to form biofilm.（Mol Microbiol. 2005 57:1545）
訳 転写調節因子MgaおよびCovR（CsrR）を欠く変異体は，また，バイオフィルムを形成することができなかった

■ 名［形］＋ mutant

❶	double mutant	二重変異体	2,253
❷	deletion mutant	欠失変異体	1,349
❸	null mutant	無発現変異体	1,135
❹	dominant negative mutant	ドミナントネガティブ変異体	949

❺ deficient mutant	欠損変異体	835

❶ The double mutant T628A,T635A did not respond to Nedd4-2 or K44A. (J Biol Chem. 2005 280:39161)

訳 二重変異体T628A,T635Aは，Nedd4-2にもK44Aにも応答しなかった

mutate 動 変異させる／変異導入する　用例数 8,947

mutated	変異した／変異させられる／～を変異させた	7,859
mutating	～を変異させる	887
mutate	～を変異させる	173
mutates	～を変異させる	28

他動詞．

■ be mutated ＋［前］

★ be mutated	…は，変異している	1,934
❶ be mutated to ～	…が，～に変異させられる／変異している	621
・be mutated to alanine	…が，アラニンに変異させられる	110
❷ be mutated in ～	…は，～において変異している	614

❶ Each residue of the peptide was mutated to alanine to assess its functional importance. (J Biol Chem. 2001 276:28503)

訳 そのペプチドのおのおのの残基が，アラニンに変異させられた

❷ Rb2/p130 is mutated in a human cell line of lung small cell carcinoma as well as in primary lung tumors. (Oncogene. 2003 22:6959)

訳 Rb2/p130は，ヒトの細胞株において変異している

■［副］＋ mutated

❶ frequently mutated	しばしば変異している	182

★★ mutation 名 変異／突然変異　用例数 110,016

mutations	変異／突然変異	65,388
mutation	変異／突然変異	44,628

複数形のmutationsの割合は約60%と非常に高い．

◆類義語：variation

■ mutation(s) ＋［前］

	the%	a/an%	ø%			
❶	0	-	95	mutations in ～	～の変異	22,014
				・caused by mutations in ～	～の変異によって引き起こされる	889

				・mutations in the gene	遺伝子の変異	522
				・associated with mutations in ～	～の変異と関連した	160
❷	5	2	92	mutation of ～	～の変異	8,010
❸	5	-	86	mutations at ～	～における変異	1,629
❹	2	-	84	mutations within ～	～内の変異	990
❺	11	-	54	mutations on ～	～における変異	953

❶ Cystic fibrosis is caused by mutations in the gene encoding the cystic fibrosis transmembrane conductance regulator (CFTR). (Science. 2004 304:600)

訳 囊胞性線維症は，～をコードする遺伝子の変異によって引き起こされる

❷ Electron transfer to ETF is not affected, to any large extent, by mutation of Val-344. (Biochemistry. 2000 39:9188)

訳 Val-344の変異によって

❸ Semidominant mutations at the IXR1 and IXR2 loci of Arabidopsis confer isoxaben and thiazolidinone resistance. (Proc Natl Acad Sci USA. 2001 98:10079)

訳 IXR1およびIXR2座位における半優性の変異

■ mutations ＋［動］

❶	mutations cause ～	変異は，～を引き起こす	479
❷	mutations were found in ～	変異が，～において見つけられた	372
❸	mutations were identified in ～	変異が，～において同定された	355
❹	mutation results in ～	変異は，～という結果になる	249
❺	mutations affect ～	変異は，～に影響する	224
❻	mutations were detected in ～	変異が，～において検出された	211

❶ Both mutations cause truncation of the expressed PTF1A protein C-terminal to the basic-helix-loop-helix domain. (Nat Genet. 2004 36:1301)

訳 両方の変異が，発現されたPTF1Aタンパク質C末端の切断を引き起こす

❷ AXIN2 mutations were found in two (2.7%) HCCs but not in HBs. (Oncogene. 2002 21:4863)

訳 AXIN2変異が，～において見つけられた

■ mutations ＋［現分／過分］

❶	mutations affecting ～	～に影響する変異	466
❷	mutations associated with ～	～と関連する変異	263

❶ XLP is caused by mutations affecting SAP, an adaptor that recruits Fyn to SLAM family receptors. (Immunity. 2004 21:693)

訳 XLPは，SAPに影響を与える変異によって引き起こされる

■［前］＋ mutations

❶	effects of mutations	変異の影響	248
❷	screen for mutations	変異に関する選別	113

❶ The effects of mutations on the bindings of cAMP and cGMP to CRP were also investigated. (Biochemistry. 2002 41:11857)

訳 cAMPおよびcGMPのCRPへの結合に対する変異の影響が，また，精査された

■［名／形］＋ mutations

❶	point mutations	点突然変異	2,488
❷	missense mutations	ミスセンス変異	1,375
❸	null mutation	無発現変異	819
❹	loss of function mutations	機能喪失変異	791
❺	somatic mutations	体細胞変異	694

N

name ㊍ 命名する，㊔ 名前　用例数 4,641

named	～を命名した／命名される	2,703
name	～を命名する／名前	1,604
names	～を命名する／名前	334

名詞としても用いられるが，他動詞の用例が非常に多い．name A B（AをBと命名する）の第5文型で用いられることが多い．

◆類義語：designate, term, call

■ ［代名］＋ name

❶ we have named ～	われわれは，～を命名した	270
❷ we named ～	われわれは，～を命名した	210

❶ **We have named this gene ZNRD1 for** zinc ribbon domain-containing 1 protein.（Genomics. 2000 63:139）

訳 われわれは，この遺伝子を～にちなんでZNRD1と命名した

■ name ＋［名句］＋［前］

❶ name ～ for …	…にちなんで～と命名する	248

■ ［名］＋ named

❶ protein named ～	～と命名されたタンパク質	140
❷ gene named ～	～と命名された遺伝子	93

❶ **Recently a secreted protein named Mac-1 was identified** and shown to enhance survival of the pathogen.（J Biol Chem. 2004 279:52789）

訳 最近，Mac-1と命名された分泌タンパク質が同定された

naturally ㊓ 自然に　用例数 4,669

◆類義語：spontaneously

■ naturally ＋［現分／過分］

❶ naturally occurring ～	自然に起こる	2,707
❷ naturally infected	自然感染した	169
❸ naturally processed	自然にプロセスされた	120

❶ Whereas **Bcl-x$_L$ plays a more important role during the period of naturally occurring neuronal death**, Bcl-w plays a more important role at later stages.（Development. 2001 128:447）

訳 Bcl-x$_L$は，自然に起こる神経細胞死の期間中に，より重要な役割を果たす

★ nature ㊔ 性質　　用例数 11,839

nature	性質	11,809
natures	性質	30

複数形のnaturesの割合は0.3%しかなく，原則，不可算名詞として使われる．
◆類義語：property, propensity, character

■ nature ＋［前］

nature ofの前にtheが付く割合は，99%と圧倒的に高い．

	the%	a/an%	ø%			
❶	99	1	0	nature of ～	～の性質	4,347
				・dynamic nature of ～	～の動的な性質	211
				・depend on the nature of ～	…は，～の性質に依存する	178
				・molecular nature of ～	～の分子的な性質	143
				・complex nature of ～	～の複雑な性質	103
				・insight into the nature of ～	～の性質への洞察	82

❶ In this report, we examined how the efficiency of dual RMCE catalyzed by Flp and Cre **depends on the nature of transcription units** that express the recombinases. (Nucleic Acids Res. 2012 40:e62)
訳 ～は，転写ユニットの性質に依存する

■ nature

❶	in nature	現実に／実際に	381

❶ Despite its simplicity, **our model can explain part of the huge difference between genotypic and phenotypic mutation rates that is observed in nature**. (Genetics. 2006 172:197)
訳 われわれのモデルは，現実に観察される遺伝子型および表現型の変異率の間の莫大な違いの一部を説明しうる

★ nearly ㊙ ほとんど／ほぼ　　用例数 9,397

◆類義語：almost, virtually, mostly, largely, near

■ nearly ＋［形／過分／副］

❶	nearly all	ほとんどすべて（の）	1,333
❷	nearly identical	ほぼ同一の	1,000
	・be nearly identical	…は，ほぼ同一である	434
❸	nearly complete	ほぼ完全な	458
❹	nearly half	ほぼ半分の	269
❺	nearly normal	ほぼ正常の	160
❻	nearly every	ほとんどすべての	146
❼	nearly equal	ほぼ等しい	122

❽ nearly abolished	ほとんど消失した	121
❾ nearly completely	ほとんど完全に	109

❶ Calmodulin (CaM) is a major effector for the intracellular actions of Ca^{2+} in nearly all cell types. (Science. 2004 306:698)

訳 カルモジュリン（CaM）は，ほとんどすべての細胞のタイプにおいてCa^{2+}の細胞内作用に対する主要な効果器である

❶ Nearly all of these segments are also conserved in the chicken and dog genomes, with an average of 95 and 99% identity, respectively. (Science. 2004 304:1321)

訳 これらのセグメントのほとんどすべては，また，〜において保存されている

❷ The lipid binding activity of the reduced mutant, as determined by its ability to form discoidal lipoproteins, was nearly identical to that of the wild type protein. (J Biol Chem. 2001 276:34162)

訳 〜は，野生型タンパク質のそれとほぼ同一であった

❸ With the availability of a nearly complete sequence of the human genome, aligning expressed sequence tags (EST) to the genomic sequence has become a practical and powerful strategy for gene prediction. (Genome Res. 2001 11:889)

訳 ヒトゲノムのほぼ完全な配列の利用が可能になって

necessarily 副 必ずしも／必然的に　用例数 1,203

not necessarilyの用例が圧倒的に多い．

■ not necessarily

★ not necessarily 〜	必ずしも〜ではない	987
❶ do not necessarily 〜	…は，必ずしも〜しない	410
❷ be not necessarily 〜	…は，必ずしも〜ではない	321

❶ However, models suggest that this does not necessarily accelerate adaptation by diploid populations. (Science. 2003 299:555)

訳 これは，必ずしも二倍体集団による適応を加速しない

❷ However, the mudrB gene, though important for Mu activity in maize, is not necessarily a component of Mu elements in other grasses. (Plant Physiol. 2001 125:1293)

訳 〜は，必ずしもMuエレメントの構成要素ではない

★ necessary 形 必要な　用例数 20,020

◆類義語：essential, fundamental

■ be necessary ＋［前］

★ be necessary	…は，必要である	12,041
❶ be necessary for 〜	…は，〜のために必要である	7,003
・activity is necessary for 〜	活性は，〜のために必要である	107
・domain is necessary for 〜	ドメインは，〜のために必要である	79
❷ be necessary to *do*	…は，〜するのに必要である	2,371
・it is necessary to *do*	〜することは，必要である	310

❶ To improve survival for bortezomib-resistant patients, it is necessary to develop new therapeutic strategies. (Blood. 2012 119:2568)
訳 新しい治療戦略を開発することは必要である

■ necessary for ＋［名句］

★ necessary for ～	～のために必要な	11,071
❶ necessary for ～ing	～するために必要な	726
❷ necessary for normal ～	正常な～のために必要な	263
❸ necessary for efficient ～	効果的な～のために必要な	226
❹ necessary for the development of ～	～の発生のために必要な	98
❺ necessary for the formation of ～	～の形成のために必要な	83
❻ necessary for the induction of ～	～の誘導のために必要な	75
❼ necessary for activation of ～	～の活性化のために必要な	74

❷ To our knowledge, this is the first demonstration that visual cortical processing is necessary for normal tactile perception. (Nature. 1999 401:587)
訳 視覚皮質プロセシングは，正常な触覚認知にとって必要である

❹ Human immunodeficiency virus type 1 (HIV-1) Nef is a key pathogenic factor necessary for the development of AIDS. (J Virol. 2003 77:3041)
訳 ～は，AIDSの発症に欠かせない鍵となる病原性因子である

■ necessary to *do*

★ necessary to *do*	～するのに必要な	3,448
❶ necessary to maintain ～	～を維持するのに必要な	166
❷ necessary to determine ～	～を決定するのに必要な	146
❸ necessary to achieve ～	～を達成するのに必要な	129

❶ One of the intriguing aspects of FTI biology is that continuous drug exposure is not necessary to maintain phenotypic reversion. (Cancer Res. 1999 59:2059)
訳 ～は，表現型の復帰を維持するのに必要ではない

■ necessary and sufficient／necessary but not sufficient

❶ necessary and sufficient	必要かつ十分な	2,040
・necessary and sufficient for ～	～のために必要かつ十分な	1,282
・necessary and sufficient to *do*	～するのに必要かつ十分な	682
・both necessary and sufficient	必要かつ十分な	641
❷ necessary but not sufficient	必要であるが十分ではない	394
・necessary but not sufficient for ～	～のために必要であるが十分ではない	273

❶ P (invF-1) contains a HilA binding site, termed a HilA box, that is necessary and sufficient for activation by HilA. (J Bacteriol. 2001 183:4876)
訳 ～は，HilAによる活性化のために必要かつ十分である

❷ These findings provide evidence that Mei3-RKDIII defines a Ran1/Pat1-binding site that is necessary but not sufficient for inhibition of the kinase. (Genetics. 1998 150:1007)
訳 そのキナーゼの抑制のために必要であるが十分ではないRan1/Pat1結合部位

★ need 名 必要性，動 助 必要とする 用例数 24,282

need	必要性／～を必要とする	11,538
needed	必要とされる／～を必要とした	10,022
needs	必要性／～を必要とする	2,609
needing	～を必要とする	113

名詞，動詞および助動詞として用いられるが，助動詞の用例は少ない．名詞としては，複数形のneedsの割合が，約15%ある可算名詞．

◆類義語：requirement, demand, request, claim, necessity, prerequisite, require, necessitate, entail

■ 名 need(s) ＋［前］

needs ofの前にtheが付く割合は圧倒的に高い．また，need for/toの前にtheが付く割合もかなり高い．

	the %	a/an %	ø %			
❶	70	25	5	need for ～	～の必要性	5,524
				・without the need for ～	～の必要性なしに	567
				・highlight the need for ～	…は，～の必要性を強調する	237
				・eliminate the need for ～	…は，～の必要性を除く	175
				・need for further ～	さらなる～の必要性	154
				・need for new ～	新しい～の必要性	126
				・need for additional ～	追加の～の必要性	123
				・urgent need for ～	～の緊急の必要性	122
❷	67	14	10	need to *do*	～する必要性	2,080
				・without the need to *do*	～する必要性なしに	97
❸	94	-	5	needs of ～	～の必要性	401

❶ The pE vectors have **an episomal design to allow long-term production of high-titer virus** without the need for **subcloning the producer line.** (Anal Biochem. 2001 297:86)

訳 産生株をサブクローンする必要性なしにタイターの高いウイルスの長期の産生を可能にするエピソームの設計

❷ **These results highlight** the need to **consider the age of** the donor and embryo quality when making embryo transfer decisions involving use of donor eggs. (Am J Epidemiol. 2001 154:1043)

訳 これらの結果は，～の年齢を考慮する必要性を強調する

■ 動 be needed ＋［前］

★	be needed	…は，必要とされる	6,402
❶	be needed to *do*	…は，～するために必要とされる	3,457
❷	be needed for ～	…は，～のために必要とされる	1,034
❸	be needed in ～	…は，～において必要とされる	229

❶ Additional study is needed to determine whether these regions of LOH harbor tumor suppressor genes and whether specific regions of LOH correlate with clinical characteristics. (Blood. 2001 98:1188)

訳 付加的な研究が，～かどうかを決定するために必要とされる

❷ The BSAP inhibitory domain (residues 358 to 385) is needed for this repression. (Mol Cell Biol. 2000 20:1911)

訳 ～が，この抑制のために必要とされる

■ 動 need to *do*

★ need to *do*	…は，～する必要がある	2,700
❶ we need to *do*	われわれは，～する必要がある	129

❶ We need to understand the consequences of these epigenetic effects. (Am J Clin Nutr. 2013 97:94)

訳 われわれは，これらのエピジェネティックな効果の結果を理解する必要がある

★★ negative 形 負の／陰性の／ネガティブの／否定的な，名 ネガティブなもの

用例数 40,210

negative	負の／陰性の／ネガティブの／ネガティブなもの	40,046
negatives	ネガティブなもの	164

名詞としても使われるが，形容詞の用例が圧倒的に多い．

◆反意語：positive

■ negative ＋［前］

❶ negative for ～	～に対して陰性である	801
・be negative for ～	…は，～に対して陰性である	379
❷ negative in ～	～において陰性である	236

❶ Of 48 blood samples tested, 42 tested positive by fliC PCR for serovar Typhi; 4 of these were negative for tviA and tviB. (J Clin Microbiol. 2005 43:4418)

訳 これらのうちの4つは，tviAおよびtviBに対して陰性であった

■ negative ＋［名／形］

❶ negative regulator	負の調節因子	2,173
❷ negative feedback	負のフィードバック	1,218
❸ negative regulation	負の制御	895
❹ negative selection	負の選択	825
❺ negative regulatory ～	負の調節性の～	723
❻ negative effect	負の効果	706
❼ negative predictive ～	負の予測的な～	686
❽ negative charge	負電荷	602

❶ We conclude that Gyp6 is a negative regulator of Nhx1-dependent trafficking out of the PVC.

(J Biol Chem. 2004 279:4498)
訳 Gyp6は，Nhx1依存性の輸送の負の調節因子である

■ ［形］＋ and negative／negative and ＋［形］

❶ positive and negative ～	正および負の～	1,731
・positive and negative predictive ～	正および負の予測的な～	177
❷ negative and positive ～	負および正の～	253

❶ DNA looping is often involved in positive and negative regulation of gene transcription in both prokaryotes and eukaryotes.（J Mol Biol. 2003 334:53）
訳 DNAループ形成は，遺伝子転写の正および負の調節にしばしば関与している

■ ［形］＋ negative

❶ dominant-negative ～	ドミナントネガティブの～	7,992
・dominant-negative mutant	ドミナントネガティブ変異体	952
・dominant-negative form	ドミナントネガティブ型	488
・dominant-negative effect	ドミナントネガティブ効果	329
❷ false-negative ～	偽陰性の～	489

❶ In contrast, expression of a dominant-negative mutant of KLF8 dramatically suppresses the transformed phenotypes induced by v-Src.（Oncogene. 2007 26:456）
訳 KLF8のドミナントネガティブ変異体の発現は，～を劇的に抑制する

★negatively 副 負に／ネガティブに 用例数 6,467

◈反意語：positively

■ negatively ＋［過分／動］

❶ negatively regulate ～	…は，～を負に調節する	2,174
❷ negatively charged ～	負に荷電した～	1,271
❸ negatively regulated by ～	～によって負に調節される	694
・be negatively regulated by ～	…は，～によって負に調節される	405
❹ negatively correlated with ～	～と負に相関する	335

❷ Site-directed mutagenesis of negatively charged glutamate (Glu^{274}) and aspartate (Asp^{253}) residues had no effect on divalent cation block.（Biophys J. 2004 86:1470）
訳 負に荷電したグルタミン酸（Glu^{274}）およびアスパラギン酸（Asp^{253}）残基の部位特異的な変異導入は，～に対する効果を持たなかった

❸ Here, we show that a novel pro-apoptotic pathway that is induced by RXR agonist is negatively regulated by casein kinase 1α (CK1α).（J Biol Chem. 2004 279:30844）
訳 ～は，カゼインキナーゼ1α（CK1α）によって負に調節される

❹ Dendritic arborization indices negatively correlated with NFT densities while no significant correlations were found with Aβ plaque densities.（Am J Pathol. 2003 163:1615）
訳 樹状突起分枝インデックスは，NFT密度と負に相関した

■ [副] ＋ and negatively

❶ positively and negatively	正および負に	208

❶ Recent data support the idea that **mycobacterial products can** positively and negatively **regulate the inflammatory response.** (Trends Microbiol. 2005 13:98)
訳 マイコバクテリアの産物が，正および負に炎症応答を調節しうる

★ network 名 ネットワーク 用例数 25,120

network	ネットワーク／回路網	16,615
networks	ネットワーク／回路網	8,455
networked	ネットワーク化された	50

動詞としても使われるが，名詞の用例が圧倒的に多い．複数形のnetworksの割合が，35%ある可算名詞．
◆類義語：circuitry

■ network ＋ [前]

	the%	a/an%	ø%			
❶	20	75	1	network of ～	～のネットワーク	2,250
				・complex network of ～	～の複雑なネットワーク	183
				・network of interactions	相互作用のネットワーク	114
❷	56	35	0	network in ～	～におけるネットワーク	551

❶ Our results show that **eukaryotic translation termination involves a** network of interactions **between the two release factors and the ribosome.** (Proc Natl Acad Sci USA. 2012 109:18413)
訳 真核生物の翻訳終結は，2つの終結因子とリボソームの間の相互作用のネットワークを含む

■ [形／名] ＋ network

❶ regulatory network	制御ネットワーク	707
❷ neural network	神経回路網	493
❸ signaling network	シグナル伝達ネットワーク	345

nevertheless 副 にもかかわらず 用例数 3,048

文頭の用例が非常に多い．
◆類義語： nonetheless, however, but

■ nevertheless

❶ Nevertheless,	にもかかわらず，	2,339

❶ Nevertheless, **the mechanism of SCN5A downregulation is unclear.** (Circulation. 2011 124:

1124)
訳 にもかかわらず，SCN5A下方制御の機構は不明である

★★ new ㊙形 新しい　用例数 66,943

◆類義語：novel

■ new ＋［名］

❶	new insights	新しい洞察	2,210
	・provide new insights into ～	…は，～への新しい洞察を提供する	1,102
❷	new method	新しい方法	1,143
❸	a new class of ～	新しいクラスの～	1,120
❹	new approach	新しいアプローチ	1,035
❺	new model	新しいモデル	645
❻	new mechanism	新しい機構	639
❼	new information	新しい情報	607
❽	new evidence	新しい証拠	525
❾	new data	新しいデータ	516

❶ The results **provide new insights into** the relationship between the function of a protein and its conformational flexibility as well as its global stability.（J Mol Biol. 2005 345:599）
訳 それらの結果は，～の間の関連性への新しい洞察を提供する

■［動］＋ a new ＋［名］

❶	provide a new ～	…は，新しい～を提供する	1,203
❷	represent a new ～	…は，新しい～に相当する	886
❸	identify a new ～	…は，新しい～を同定する	717
❹	describe a new ～	…は，新しい～を述べる	646
❺	develop a new ～	…は，新しい～を開発する	566
❻	suggest a new ～	…は，新しい～を示唆する	537
❼	define a new ～	…は，新しい～を定義する	399
❽	reveal a new ～	…は，新しい～を明らかにする	386
❾	present a new ～	…は，新しい～を提供する	380

❶ These compounds **represent a new** class of arginase inhibitors.（J Med Chem. 2010 53:4266）
訳 これらの化合物は，新しいクラスのアルギナーゼ阻害剤に相当する

★ newly ㊙副 新たに　用例数 8,450

■ newly ＋［過分］

❶	newly synthesized	新たに合成された	1,343
❷	newly identified	新たに同定された	1,026

N

❸ newly diagnosed	新たに診断された	976
・patients with newly diagnosed ～	新たに診断された～の患者	255
❹ newly developed	新たに開発された	568
❺ newly formed	新たに形成された	550
❻ newly discovered	新たに発見された	493
❼ newly described	新たに述べられた	339
❽ newly generated	新たに生成された	283

❶ These effects were associated with decreased intracellular degradation and **increased secretion of newly synthesized apoB** as VLDL.（J Biol Chem. 2004 279:44938）

訳 新たに合成されたapoBの増大した分泌

❷ CTGF, **a newly identified growth factor**, is highly expressed in dermal granulation tissue.（Invest Ophthalmol Vis Sci. 2001 42:2534）

訳 新たに同定された増殖因子

next 副 次に，形 次の　　用例数 4,669

◆類義語：subsequently, following

■ 副 next

❶ Next, ～	次に，～	864
・Next, we ～	次に，われわれは，～	382
❷ we next ～	われわれは，次に，～	455

❶ **Next, we used** two-dimensional polyacrylamide gel electrophoresis and mass spectrometry to identify 42 proteins that were overexpressed in the cancer cells relative to normal cells.（Oncogene. 2006 25:2628）

訳 次に，われわれは～を使った

❷ **We next examined whether** MHC-II-restricted $CD4^{+}8^{-}$ thymocytes remain competent to initiate CD8 lineage gene expression.（J Immunol. 2005 175:4465）

訳 われわれは，次に～かどうかを調べた

■ 形 next ＋［名］

❶ next generation	次世代	849
❷ next step	次の段階	119

★ normal 形 正常な，名 正常なもの　　用例数 66,971

normal	正常な／正常なもの	66,821
normals	正常なもの	150

形容詞の用例がほとんどだが，名詞として使われることもある.

◆類義語：usual, general, conventional, usual, common

■ [動] + normal

❶ be normal	…は，正常である	2,150
・be normal in ～	…は，～において正常である	670
❷ appear normal	…は，正常であるように思われる	447

❷ While fruiting-body aggregation **appears normal in the mutant**, it fails to sporulate ($<10^{-6}$ the wild-type number of viable spores). (J Bacteriol. 2000 182:2438)

訳 ～は，変異体において正常のように思われる

■ normal + [名]

❶ normal cells	正常な細胞	1,357
❷ normal levels	正常なレベル	1,130
❸ normal subjects	正常な被験者	1,128
❹ normal development	正常な発達	1,084
❺ normal tissues	正常な組織	893
❻ normal mice	正常なマウス	817
❼ normal controls	正常な対照群	675
❽ normal growth	正常な成長	534

■ [前] + normal

❶ required for normal ～	正常な～に必要とされる	1,215
❷ compared with normal ～	正常な～と比べて	1,075
❸ essential for normal ～	正常な～に必須の	606
❹ patients with normal ～	正常な～の患者	387

❶ Furthermore, analyses of gene knockout mice revealed that **Nfia is specifically required for normal expression of the GABRA6 gene** in cerebellar granule neurons. (J Biol Chem. 2004 279:53491)

訳 Nfiaは，GABRA6遺伝子の正常な発現のために特異的に必要とされる

❷ JD-1 cells from a patient with a DP COOH-terminal truncation **were also more weakly adherent compared with normal keratinocytes**. (J Cell Biol. 2002 159:1005)

訳 ～は，また，正常なケラチノサイトと比べてより弱く接着した

■ than/near normal

❶ than normal ～	正常な～より	821
❷ near normal ～	ほぼ正常な～	398

❶ Our results show that **despite the presence of higher than normal levels of DNA-PK kinase activity**, c-Abl fails to become activated after IR exposure in ATM-deficient cells. (J Biol Chem. 2000 275:30163)

訳 正常なレベルより高いDNA-PKキナーゼ活性の存在にもかかわらず

normalize ㊍ 標準化する／規準化する 用例数 2,725

normalized	標準化される／～を標準化した	2,107
normalize	～を標準化する	310
normalizing	～を標準化する	160
normalizes	～を標準化する	148

他動詞.

◆類義語：standardize

■ normalized ＋［前］

❶ normalized to ～	～に標準化される	257
❷ normalized by ～	～によって標準化される	186

❶ In addition, when **PV curves were** **normalized to** V_{30}, there were no differences between the infants with CF and healthy control subjects in the fractional volumes at any airway pressure. (Am J Respir Crit Care Med. 2004 170:505)

㊋ PV曲線は，V_{30}に標準化された

★normally ㊖ 正常に／通常 用例数 9,883

◆類義語：generally, in general, commonly, usually, universally

■［動］＋ normally

❶ develop normally	…は，正常に発達する	391
❷ occur normally	…は，通常起こる	214
・occur normally in ～	…は，通常～において起こる	112
❸ grow normally	…は，正常に成長する	165
❹ respond normally to ～	…は，～に正常に応答する	120
❺ proceed normally	…は，正常に進行する	118
❻ function normally	…は，正常に機能する	99

❶ **The XPG nuclease-deficient animals** **develop normally** and exhibit no obvious defect in class switch recombination. (Mol Cell Biol. 2004 24:2237)

㊋ XPGヌクレアーゼ欠損動物は，正常に発達する

❷ Oscillatory and negative flows **occur normally in** **the cardiovascular system**, which predispose those regions to atherosclerosis. (J Physiol. 2004 561:575)

㊋ ～は，通常心血管系において起こる

■ normally ＋［過分／動／形］

❶ normally expressed	通常発現する	336
・be normally expressed	…は，通常発現する	197
❷ normally occur	…は，通常起こる	169
❸ normally associated with ～	通常～に結合する	166
❹ normally present	通常存在する	140

❺ normally found	通常見つけられる	125

❶ Collagenase-3 is normally expressed in hypertrophic chondrocytes, periosteal cells, and osteoblasts during bone development.（J Mol Biol. 1999 292:837）

訳 コラゲナーゼ-3は，肥大軟骨細胞において通常発現する

❸ Ankyrin was not normally associated with the plasma membrane of these cells.（J Cell Biol. 1996 133:647）

訳 アンキリンは，通常これらの細胞の細胞膜に結合しなかった

N

notably 副 特に／著しく　用例数 4,178

文頭で使われることが多い.

◆類義語：particularly, in particular, especially, specially

■ notably

❶ Notably,	特に,	2,526
・Notably, we ～	特に，われわれは，～	112
❷ most notably	最も顕著に	686

❶ Notably, we found that cells expressing mutant Htt displayed reduced SIRT3 levels.（J Biol Chem. 2012 287:24460）

訳 特に，われわれは，～ということを見つけた

note 動 注目する／認める／述べる，名 注目／記録　用例数 5,770

noted	認められる／～に注目した／～を述べた／注目される	4,677
note	～に注目する／～を述べる／注目／記録	848
notes	記録／～に注目する	165
noting	～に注目する／～を述べる	80

他動詞の用例が多いが，名詞としても用いられる.

◆類義語：recognize, focus

■ 動 be noted ＋［前］

★ be noted	…は，認められる	3,026
❶ be noted in ～	…は，～において認められる	1,025
❷ be noted for ～	…は，～に対して認められる	220
❸ be noted between ～	…は，～の間で認められる	186

❶ Mild preeclampsia was noted in 25% of patients and severe preeclampsia in 10%.（JAMA. 2002 288:2320）

訳 軽度の子癇前症が，患者の25%において認められた

❷ Similar discrepancies were noted for liver and bladder.（J Nucl Med. 2004 45:559）

訳 類似の矛盾が，肝臓と膀胱に対して認められた

■ 動 [代名] + note + [that節]

★	noted that ～	～ということに注目した／～ということを述べた	329
❶	we noted that ～	われわれは，～ということに注目した	126

❶ We noted that patients with large hepatic adenomas had severe iron refractory anemia similar to that observed in anemia of chronic disease. (Blood. 2002 100:3776)
訳 われわれは，～ということに注目した

■ 名 of note

❶	Of note,	注目すべきことに，	274

❶ Of note, a higher level of the HIV-1 accessory protein Vpr has been detected in the cerebrospinal fluid of AIDS patients with neurological disorders. (J Virol. 2000 74:9717)
訳 注目すべきことに，

nothing 代名 何も～ない　用例数 412

■ nothing + [動]

❶	nothing is known	何も知られていない	231
	・nothing is known about ～	～については，何も知られていない	182

❶ Almost nothing is known about the mechanisms of dissimilatory metal reduction by Gram-positive bacteria, although they may be the dominant species in some environments. (Proc Natl Acad Sci USA. 2012 109:1702)
訳 ～の機構については，ほとんど何も知られていない

notion 名 考え／概念　用例数 2,679

notion	考え／概念	2,568
notions	考え／概念	111

the notion thatの用例が非常に多い.
◆類義語：concept, idea, view, hypothesis

■ notion + [同格that節／前]

notion thatやnotion ofの前にtheが付く割合は圧倒的に高い.

	the%	a/an%	ø%			
❶	98	2	0	notion that ～	～という考え	1,848
				・support the notion that ～	…は，～という考えを支持する	936
				・be consistent with the notion that ～	…は，～という考えと一致する	189
❷	95	1	0	notion of ～	～の考え	299

・support the notion of ～ …は，～の考えを支持する		85

❶ These results support the notion that the peripheral expansion of the CD4$^+$CD25$^+$ T cells is controlled in part by costimulation. (J Immunol. 2004 173:2428)
訳 これらの結果は，～という考えを支持する

N

★★ novel 形 新規の／新しい，名 小説 用例数 57,281

novel	新規の／新しい／小説	57,274
novels	小説	7

名詞としても使われるが，形容詞の用例が圧倒的に多い．
◆類義語：new

■ novel ＋［名］

❶ a novel mechanism	新規の機構	2,718
❷ a novel role	新規の役割	1,296
❸ a novel approach	新規のアプローチ	888
❹ a novel method	新規の方法	600
❺ a novel function	新規の機能	591
❻ a novel protein	新規のタンパク質	590
❼ a novel class of ～	新規のクラスの～	575
❽ a novel gene	新規の遺伝子	518
❾ a novel pathway	新規の経路	391

❶ These results suggest a novel mechanism for the progression of prostate cancer. (Cancer Res. 2004 64:8860)
訳 これらの結果は，前立腺癌の進行の新規の機構を示唆する

❼ Thus, XEDAR belongs to a novel class of death receptors that lack a discernible death domain but are capable of activating apoptosis in a caspase 8- and FADD-dependent fashion. (J Biol Chem. 2004 279:41873)
訳 XEDARは，新規のクラスの細胞死受容体に属する

■［動］＋ a novel ＋［名］

❶ identify a novel ～	…は，新規の～を同定する	1,987
❷ represent a novel ～	…は，新規の～を表す	1,418
❸ suggest a novel ～	…は，新規の～を示唆する	1,258
❹ provide a novel ～	…は，新規の～を提供する	1,241
❺ reveal a novel ～	…は，新規の～を明らかにする	1,138
❻ describe a novel ～	…は，新規の～を述べる	982
❼ develop a novel ～	…は，新規の～を開発する	666
❽ demonstrate a novel ～	…は，新規の～を実証する	652
❾ encode a novel ～	…は，新規の～をコードする	558

❶ We identified a novel protein, enkurin, that is expressed at high levels in the testis and vomeronasal organ and at lower levels in selected other tissues. (Dev Biol. 2004 274:426)
訳 われわれは，新規のタンパク質を同定した

★now 副 名 今　　用例数 17,179

◆類義語：presently, at present, currently, today, immediately

■ 副 we now ＋［動］／ we have now ＋［過分］

★ we now ～	われわれは，今，～	5,251
❶ we now report ～	われわれは，今，～を報告する	1,750
❷ we now show that ～	われわれは，今，～ということを示す	1,325
❸ we now demonstrate that ～	われわれは，今，～ということを実証する	615
❹ we have now identified ～	われわれは，今，～を同定した	192
❺ we now describe ～	われわれは，今，～を述べる	176

❶ We now report that threshold levels of neuregulin-1 (NRG1) type III on axons determine their ensheathment fate. (Neuron. 2005 47:681)
訳 われわれは，今，～ということを報告する

❷ We now show that large volumes of bone can be engineered in a predictable manner, without the need for cell transplantation and growth factor administration. (Proc Natl Acad Sci USA. 2005 102:11450)
訳 われわれは，今，～ということを示す

❹ We have now identified a second Tctex1-related protein (here termed LC9) in Chlamydomonas. (Mol Biol Cell. 2005 16:5661)
訳 われわれは，今，～を同定した

■ 副 be now ＋［過分／形］

★ be now ～	…は，今，～である	4,227
❶ be now known	…は，今，知られている	262
❷ be now available	…は，今，利用できる	235
❸ be now recognized	…は，今，認識されている	161

❶ TDP-43 is now known to play important roles in neuronal RNA metabolism. (Brain Res. 2012 1462:16)
訳 TDP-43は，今，～において重要な役割を果たすことが知られている

■ 副 it is now ＋［形］

★ it is now ～	…は，今，～である	739
❶ it is now clear that ～	～ということは，今，明らかである	160
❷ it is now possible to *do*	～することは，今，可能である	128

❶ It is now clear that tyrosine kinases represent attractive targets for therapeutic intervention in cancer. (Proc Natl Acad Sci USA. 2005 102:14344)
訳 ～ということは，今，明らかである

N

■ 名 until now

❶ until now	今まで	474

❶ However, until now there has been no satisfactory and rigorous means to define or to measure the efficiency of an antiviral CTL response. (Oncogene. 2005 24:6035)
訳 今まで

★ null 形 ヌルの／無発現の／無の，名 無 用例数 15,169

null	ヌルの／無発現の／無	15,075
nulls	無	94

名詞としても使われるが，形容詞の用例が圧倒的に多い．
◆類義語：deficient

■ null ＋［名］

❶ null mice	ヌルのマウス（ノックアウトマウス）	3,716
❷ null mutant	ヌルの変異体	1,135
❸ null cells	ヌルの細胞	874
❹ null mutation	ヌルの変異	819
❺ null allele	ヌルのアレル	452
❻ null embryos	ヌルの胚	274
❼ null animals	ヌルの動物	203
❽ null phenotype	ヌルの表現形	175
❾ null hypothesis	帰無仮説	153

❶ Null mice produced no detectable 1α-hydroxylase transcript. (Proc Natl Acad Sci USA. 2006 103:75)
訳 ヌルのマウスは，検出できる1α-水酸化酵素転写物を産生しなかった

■ null ＋［前］

❶ null for ～	～に対してヌルの	297

❶ In contrast, HCT116 cells null for the p53 alleles (HCT116 $p53^{-/-}$) exhibited centrosome amplification after irradiation. (Oncogene. 2005 24:4017)
訳 p53のアレルに対してヌルであるHCT116細胞（HCT116 $p53^{-/-}$）は，～を示した

★★ number 名 数，動 数える／番号をつける 用例数 67,789

number	数	55,441
numbers	数	11,982
numbering	数える	215
numbered	番号をつけられた	151

動詞としても使われるが，名詞の用例が圧倒的に多い．複数形のnumbers の割合が，約20%ある可算名詞．

◆類義語：count

■ number ＋［前］

the number ofとa number ofとでは，意味が大きく違うので注意が必要である．

	the %	a/an %	ø %			
❶	50	44	5	number of ～	～の数／数の～	44,286
				・the number of ～	～の数	15,633
				・increase in the number of ～	～の数の増大	1,137
				・increase the number of ～	…は，～の数を増大させる	1,072
				・reduce the number of ～	…は，～の数を低下させる	795
				・the number of ～ing	～する数	619
				・reduction in the number of ～	～の数の低下	442
				・decrease in the number of ～	～の数の低下	375
				・the number of cells	細胞の数	340
				・decrease the number of ～	…は，～の数を低下させる	293
				・the number of patients	患者の数	260
				・a number of ～	いくつかの～	11,915
				・a number of genes	いくつかの遺伝子	318
				・identify a number of ～	…は，いくつかの～を同定する	256
				・a number of studies	いくつかの研究	251
❷	14	3	79	number in ～	～における数	537

❶ Ectopic expression of TIM-1 during T cell differentiation **results in a significant** increase in the number of **cells producing IL-4** but not IFN-γ.（Proc Natl Acad Sci USA. 2005 102:17113）

訳 ～は，IL-4を産生する細胞の数の有意な増大という結果になる

❶ In vivo experiments showed that **inhibiting MLCK** increased the number of **apoptotic cells** and retarded the growth of mammary cancer cells in mice.（Mol Cell Biol. 2005 25:6259）

訳 MLCKを抑制することが，アポトーシス細胞の数を増大させた

❷ O-glycosylation and phosphorylation of Sp1 **are thought to modulate the expression of** a number of genes **in normal and diabetic state.**（J Biol Chem. 2006 281:3642）

訳 ～は，正常および糖尿病の状態においていくつかの遺伝子の発現を調節すると考えられる

■［形］＋ number of

❶	a large number of ～	多数の～	2,444
❷	a small number of ～	少数の～	1,153
❸	the total number of ～	～の合計数	843
❹	a limited number of ～	限られた数の～	661
❺	a significant number of ～	かなりの数の～	418

❻ an increasing number of ～	増大する数の～	375
❼ an increased number of ～	増大した数の～	357
❽ a greater number of ～	より多くの～	272
❾ the mean number of ～	～の平均数	262

❶ Interestingly, a large number of genes involved in RNA processing showed distinct down-regulation during the initial liver phase of infection. (J Infect Dis. 2005 191:400)

訳 RNAプロセシングに関与する多数の遺伝子が，～を示した

❷ A small number of mammalian signaling pathways mediate a myriad of distinct physiological responses to diverse cellular stimuli. (Science. 2005 309:1857)

訳 少数の哺乳類のシグナル伝達経路が，多数の別個の生理学的応答を仲介する

❸ The total number of CARTp-IR cells was significantly reduced (by five-fold) at P14, and this reduced number persisted into adulthood. (J Comp Neurol. 2005 489:501)

訳 CARTp-IR細胞の合計数が，有意に低下した

numerous 形 多数の　　用例数 8,456

◆類義語：many, large number of

■ numerous ＋［名］

❶ numerous studies	多数の研究	476
❷ numerous genes	多数の遺伝子	202
❸ numerous proteins	多数のタンパク質	120

❶ Numerous studies have revealed that P. aeruginosa biofilms are highly refractory to antibiotics. (J Immunol. 2005 175:7512)

訳 多数の研究が，～ということを明らかにしている

❷ This increase affects the expression of numerous genes on other chromosomes as well. (Cancer Res. 2004 64:6941)

訳 この増大は，多数の遺伝子の発現に影響を与える

O

objective ㊔ 目的，㊖ 客観的な／他覚的な　用例数 5,956

objective	目的／客観的な／他覚的な	5,246
objectives	目的	710

名詞の用例が多いが，形容詞としても使われる．複数形のobjectivesの割合が，約20%ある可算名詞．

◆類義語：purpose, aim, goal, object, end

■ objective ＋［前］

objective ofの前にtheが付く割合は圧倒的に高い．

	the%	a/an%	ø%			
❶	99	1	0	objective of ～	～の目的	4,347
				・the objective of this study was to ～	この研究の目的は，～することであった	976
				・the objective of this study was to determine ～	この研究の目的は，～を決定することであった	284
				・the objective of this study was to examine ～	この研究の目的は，～を調べることであった	98
				・the objective of the present study was to ～	現在の研究の目的は，～することであった	134

❶ The objective of this study was to determine **whether** preoperative antibiotics are associated with a reduced risk of death.（Ann Surg. 2005 242:107）

訳 この研究の目的は，～かどうかを決定することであった

■ objective was to *do*

★	objective was to *do*	目的は，～することであった	1,148
❶	objective was to determine ～	目的は，～を決定することであった	290
❷	objective was to examine ～	目的は，～を調べることであった	131
❸	objective was to evaluate ～	目的は，～を評価することであった	96
❹	objective was to assess ～	目的は，～を評価することであった	92

❶ **Our** objective was to determine **the effects of** Campylobacter rectus（C. rectus）infection on pregnancy outcomes in a mouse model.（J Periodontol. 2005 76:551）

訳 われわれの目的は，～の影響を決定することであった

★ observation ㊔ 観察　用例数 23,122

observations	観察	15,273
observation	観察	7,849

複数形のobservationsの割合は，約65%と非常に高い．

◆類義語：detection, discovery, discrimination, finding, identification

■ observations ＋［動］

❶	observations suggest that ～	観察は，～ということを示唆する	1,966
❷	observations indicate that ～	観察は，～ということを示す	860
❸	observations provide ～	観察は，～を提供する	481
❹	observations support ～	観察は，～を支持する	428
❺	observations demonstrate ～	観察は，～を実証する	402
❻	observations are consistent with ～	観察は，～と一致する	303
❼	observations show that ～	観察は，～ということを示す	176

❶ These observations suggest that the Inaba antigen should be maximized in cholera vaccine designs.（J Infect Dis. 2002 186:246）
訳 これらの観察は，～ということを示唆する

❷ These observations indicate that the early stage of diabetes mellitus provokes accelerated renal cortical superoxide anion production in a setting of normal or increased NO production.（J Am Soc Nephrol. 2001 12:1630）
訳 これらの観察は，～ということを示す

❸ These observations provide evidence for a possible association of JCV with human medulloblastomas.（Proc Natl Acad Sci USA. 1999 96:11519）
訳 これらの観察は，～の証拠を提供する

■ observation（s）＋［前／同格that節］

observation thatの前にtheが付く割合は圧倒的に高い．また，observation ofの前にtheが付く割合もかなり高い．

	the％	a/an％	ø％			
❶	89	3	0	observation that ～	～という観察	2,209
❷	65	3	24	observation of ～	～の観察	1,545
❸	11	-	83	observations in ～	～における観察	453

❶ Our findings with the NIKS cells support the observation that spontaneous immortalization is not linked to alterations in squamous differentiation or the ability to undergo apoptosis.（J Invest Dermatol. 2000 114:444）
訳 NIKS細胞に関するわれわれの知見は，～という観察を支持する

❷ The observation of such a continuum is noteworthy because, so far, this signature of fractional spin excitations has been observed only in one-dimensional systems.（Nature. 2012 492:406）
訳 そのような連続体の観察は，注目に値する

❸ We report observations in patients with visual extinction demonstrating that detection of visual events is gated by attention at the level of processing at which a stimulus is selected for action.（Proc Natl Acad Sci USA. 2002 99:16371）
訳 われわれは，～の患者における観察を報告する

■［形］＋ observation（s）

❶	previous observations	以前の観察	452
❷	experimental observations	実験上の観察	423
❸	recent observations	最近の観察	293

O

❹ direct observation	直接の観察	254
❺ clinical observations	臨床的観察	202

❶ This is consistent with previous observations that the CD2 LCR contains a T-cell-specific enhancer. (J Virol. 2001 75:4641)
訳 これは，～という以前の観察と一致する

■ [前] + these/this observation(s)

❶ consistent with this observation	この観察と一致して	232
❷ based on these observations	これらの観察に基づいて	224

❷ Based on these observations, we hypothesized that h-IAPP cytotoxicity is mediated by membrane damage induced by early h-IAPP aggregates. (Diabetes. 1999 48:491)
訳 これらの観察に基づいて，

★★ observe 動 観察する　用例数 84,850

observed	観察される／～を観察した	81,622
observe	～を観察する	2,706
observing	～を観察する	499
observes	～を観察する	23

他動詞.
◆類義語：discover, identify, detect, find, see

■ [名] + be observed

★ be observed	…が，観察される	35,884
❶ differences were observed	違いが，観察された	582
❷ activity was observed	活性が，観察された	460
❸ expression was observed	発現が，観察された	433
❹ effect was observed	効果が，観察された	369
❺ changes were observed	変化が，観察された	323
❻ responses were observed	応答が，観察された	261
❼ results were observed	結果が，観察された	236
❽ correlation was observed	相関が，観察された	200

❶ No differences were observed in either the levels of cardiac Ca^{2+}-handling proteins or the degree of cardiac regulatory protein phosphorylation between wild-type and transgenic mice. (Circ Res. 2000 86:1218)
訳 違いは，～には観察されなかった

■ be observed + [前]

❶ be observed in ～	…は，～において観察される	12,266
・be observed in both ～	…は，両方の～において観察される	339
・be observed in all ～	…は，すべての～において観察される	319

・be observed only in ～	…は，～においてのみ観察される	314
・be observed in the presence of ～	…は，～の存在下において観察される	168
❷ be observed for ～	…は，～に対して観察される	2,747
❸ be observed with ～	…は，～に観察される	1,825
❹ be observed between ～	…は，～の間に観察される	1,444
❺ be observed at ～	…は，～において観察される	1,345
❻ be observed to *do*	…は，～することが観察される	1,293
❼ be observed after ～	…は，～のあとに観察される	581
❽ be observed on ～	…は，～において観察される	500
❾ be observed by ～	…は，～によって観察される	482

O

❶ An opposite effect was observed in the presence of the nACh receptor antagonist d-tubocurarine.（Brain Res. 2003 967:235）
訳 逆の影響が，～の存在下において観察された

❷ Differing relative extents of guanine oxidation were observed for the different mismatches.（J Am Chem Soc. 2001 123:8649）
訳 ～が，異なるミスマッチに対して観察された

❸ Cells harvested after more than 3 days of infection showed extensive degradation of the cysteine-less α-subunit, which is not observed with the wild-type enzyme.（J Biol Chem. 2000 275:30734）
訳 ～は，野生型酵素には観察されない

■ be observed ＋［接］

❶ be observed when ～	…は，～のとき観察される	1,011
❷ be observed after ～	…は，～のあと観察される	581

❶ The same effect was observed when Cys^{154} was substituted with isoleucine.（Biochemistry. 2004 43:6475）
訳 同じ効果が，Cys^{154}がイソロイシンによって置き換えられたとき観察された

■［副］＋ observed

❶ previously observed	以前に観察された	947
❷ experimentally observed	実験的に観察される	377
❸ frequently observed	しばしば観察される	350
❹ commonly observed	通常観察される	339
❺ only observed	唯一観察される	288
❻ directly observed	直接観察される	173

❶ Similar changes were previously observed in structures of nucleotide-free Ras and Ef-Tu.（J Biol Chem. 2004 279:47352）
訳 類似の変化が，～の構造において以前に観察された

■ observed ＋［that節］

★ observed that	～ということを観察した／～ということが観察された	3,732
❶ we observed that ～	われわれは，～ということを観察した	2,392

❷ it was observed that ～	～ということが観察された	199

❶ Consistent with this finding, **we observed that** T antigen suppressed the death program engaged by Bin1. (Oncogene. 2000 19:4669)
訳 われわれは，～ということを観察した

⁑ obtain 動 得る　用例数 33,507

obtained	得られる／～を得た	28,317
obtain	～を得る	3,973
obtaining	～を得る	1,139
obtains	～を得る	78

他動詞．受動態，特にobtained fromの用例が多い．
◆類義語：acquire, gain

■ be obtained ＋ [前]/using

★ be obtained	…が，得られる	13,208
❶ be obtained from ～	…が，～から得られる	2,722
❷ be obtained by ～	…が，～によって得られる	1,428
・be obtained by using ～	…が，～を使うことによって得られる	116
❸ be obtained for ～	…が，～に対して得られる	1,268
❹ be obtained in ～	…が，～において得られる	1,231
❺ be obtained with ～	…が，～を使って得られる	986
❻ be obtained at ～	…が，～において得られる	556
❼ be obtained using ～	…が，～を使って得られる	502

❶ **Data were obtained from both male and female mice** and results demonstrate that the acute effect of leptin on LepR PMV neurons was identical for both sexes. (J Neurosci. 2011 31:13147)
訳 データが，雄と雌の両方のマウスから得られた

❷ Genomic sequences corresponding to the MDL1, MDL2, and MDL4 cDNAs **were obtained by polymerase chain reaction amplification of genomic DNA.** (Plant Physiol. 1999 119:1535)
訳 ～が，ゲノムDNAのポリメラーゼ連鎖反応増幅によって得られた

❸ **Similar results were obtained for three other ESR1 variants** including c.454-351A > G, in the same linkage disequilibrium block. (Hum Mol Genet. 2005 14:2405)
訳 類似の結果が，3つの他のESR1変異体に対して得られた

■ [名] ＋ were obtained

❶ results were obtained	結果が，得られた	916
❷ data were obtained	データが，得られた	458
❸ samples were obtained	試料が，得られた	402
❹ images were obtained	画像が，得られた	201
❺ informed consent was obtained	インフォームドコンセントが，得られた	163
❻ measurements were obtained	測定値が，得られた	141

❼	approval was obtained	承認が，得られた	125

■ [名] + obtained

❶	results obtained	得られた結果	1,245
❷	data obtained	得られたデータ	960
❸	samples obtained	得られた試料	416
❹	images obtained	得られた画像	306
❺	cells obtained	得られた細胞	304
❻	values obtained	得られた値	277
❼	specimens obtained	得られた検体	231

O

❶ The results obtained with RT-PCR correlated with the proteomic data. (Infect Immun. 2001 69:822)

訳 RT-PCRを使って得られた結果は，プロテオミクスのデータと相関した

❸ Moreover, they bind and kill malignant B cells from peripheral blood samples obtained from patients with hairy cell leukemia, marginal zone lymphoma, and chronic lymphocytic leukemia. (Blood. 2010 115:4778)

訳 ～の患者から得られた末梢血試料

■ to obtain

★	to obtain ～	～を得る…	3,410
❶	To obtain ～	～を得るために	351
❷	be used to obtain ～	…は，～を得るために使われる	335

❷ In the present study, the differential display-PCR technique was used to obtain a prostate-specific approximately 339-bp cDNA fragment. (Cancer Res. 2003 63:329)

訳 ～が，前立腺特異的なおよそ339 bpのcDNA断片を得るために使われた

■ [代名] + obtain

❶	we obtained ～	われわれは，～を得た	876

❶ We obtained information about mortality from all sites. (Lancet. 2012 380:50)

訳 われわれは，～に関する情報を得た

occlusion 名 閉塞／閉塞症 用例数 3,224

occlusion	閉塞	3,019
occlusions	閉塞	205

複数形のobstructionsの割合は約５%があるが，単数形は不可算名詞として使われることが非常に多い．

◆類義語：obstruction, blockade

■ occlusion ＋［前］

	the%	a/an%	ø%			
❶	10	10	79	occlusion of ～	～の閉塞	376
❷	3	3	87	occlusion in ～	～における閉塞	129

❶ **Spinal ischemia was induced by transient occlusion of the descending aorta** combined with systemic hypotension.（J Neurosci. 2007 27:11179）

訳 脊髄虚血は，下行大動脈の一過性閉塞によって誘導された

■［名／形］＋ occlusion

❶	artery occlusion	動脈閉塞	613
❷	coronary occlusion	冠動脈閉塞	200

★★ occur 動 起こる　用例数 67,801

occurs	起こる	22,761
occur	起こる	18,739
occurred	起こった	18,625
occurring	起こる	7,676

自動詞.

◆類義語：take place, arise, derive, emerge

■ occur ＋［前／接］

❶	occur in ～	…が，～において起こる	19,503
	・occur in the absence of ～	…が，～の非存在下において起こる	681
	・occur in the presence of ～	…が，～の存在下において起こる	251
	・occur in the context of ～	…が，～の状況において起こる	106
❷	occur at ～	…が，～において起こる	5,156
	・occur at the level of ～	…が，～のレベルで起こる	194
❸	occur during ～	…が，～の間に起こる	3,122
❹	occur with ～	…が，～で起こる	2,071
❺	occur through ～	…が，～によって起こる	1,670
❻	occur by ～	…が，～によって起こる	1,594
❼	occur within ～	…が，～内で起こる	1,585
❽	occur after ～	…が，～のあと起こる	1,463
❾	occur when ～	…が，～のとき起こる	1,369

❶ **An increase in CD4$^+$ T cells occurred in the gastric mucosa** during acute H. pylori infection as early as 1 week after infection.（Gastroenterology. 2000 118:307）

訳 CD4$^+$ T細胞の増大が，胃粘膜において起こった

❷ **Idx-1 expression occurs at developmental day e8.5**, and insulin expression occurs at e9.5, respectively.（J Biol Chem. 2002 277:16028）

訳 Idx-1発現が，発生の胎生8.5日目に起こる

❸ Medulloblastoma is the most common malignant brain tumour that occurs during childhood. (Lancet Oncol. 2004 5:209)

訳 髄芽腫は，小児期の間に起こる最もよくある悪性脳腫瘍である

■ [名] + occur

❶	events occurred	出来事が，起こった	364
❷	changes occur	変化が，起こる	345
❸	activation occurs	活性化が，起こる	243
❹	deaths occurred	死が，起こった	230
❺	expression occurs	発現が，起こる	207
❻	mutations occur	変異が，起こる	188

■ occur + [副]

❶	occur only	…が，唯一起こる	974
	・occur only in ～	…が，～においてのみ起こる	354
❷	occur independently of ～	…が，～とは独立に起こる	524
❸	occur early	…が，早期に起こる	423
❹	occur primarily	…が，主に起こる	362
❺	occur frequently	…が，頻繁に起こる	337
❻	occur predominantly	…が，主に起こる	249
❼	occur prior to ～	…が，～より前に起こる	230
❽	occur normally	…が，通常起こる	214
❾	occur rapidly	…が，急速に起こる	214

❷ Differentiation that occurs independently of TH and RA apparently involves a different pathway. (EMBO J. 2001 20:5261)

訳 THおよびRAとは独立に起こる分化

■ naturally occurring

❶	naturally occurring	自然に起こる	2,707

❶ Naturally occurring mutations in HIV-1-infected patients have important implications for therapy and the outcome of clinical studies. (Nat Med. 1996 2:753)

訳 HIV-1に感染した患者において自然に起こる変異は，治療のための重要な意味を持つ

occurrence 名 発生　　用例数 4,319

occurrence	発生	4,019
occurrences	発生	300

複数形のoccurrencesの割合が，約5%ある可算名詞.

◆類義語：appearance, emergence, advent, manifestation, development, onset, incidence

O

■ occurrence +［名］

occurrence ofの前にtheが付く割合は圧倒的に高い.

	the%	a/an%	ø%			
❶	81	3	13	occurrence of ～	～の発生	2,933
❷	6	24	28	occurrence in ～	～における発生	226

❶ Recent work has identified changes in gene expression occurring in metastatic and non-metastatic islet cell tumors, **which appear to correlate with the occurrence of lymph node and liver metastases.** (Curr Opin Oncol. 2006 18:23)

訳 そして,（それは）リンパ節および肝転移の発生と相関するように思われる

offer 動 提供する　用例数 9,741

offer	～を提供する	4,408
offers	～を提供する	3,596
offered	～を提供した／提供される	1,172
offering	～を提供する	565

他動詞.

◆類義語：provide, present

■ offer +［名句］

❶	offer insight into ～	…は,～への洞察を提供する	253
❷	offer the potential	…は,可能性を提供する	208
❸	offer a ～ approach	…は,～なアプローチを提供する	192
❹	offer the possibility	…は,可能性を提供する	129
❺	offer the opportunity	…は,機会を提供する	101

❷ **PET offers the potential to increase diagnostic accuracy.** (J Nucl Med. 2005 46:936)

訳 PETは,診断上の精度を増大させる可能性を提供する

■［名／代名］+ offer

❶	findings offer ～	知見は,～を提供する	174
❷	results offer ～	結果は,～を提供する	159
❸	we offer ～	われわれは,～を提供する	131

❷ **These results offer a possible explanation to** some experimental data on adsorption. (Biophys J. 2005 89:1516)

訳 これらの結果は,～への可能な説明を提供する

★ often 副 しばしば／たびたび　用例数 17,510

◆類義語：frequently

■ be often + [過分／形]

❶ be often associated with ～	…は，しばしば～と関連する	407
❷ be often used	…は，しばしば使われる	271
❸ be often found	…は，しばしば見つけられる	133
❹ be often difficult	…は，しばしば難しい	133
❺ be often observed	…は，しばしば観察される	105

❶ All of these scaffolding proteins **are often associated with** cell junctions.（Proc Natl Acad Sci USA. 2005 102:9814）

訳 ～は，しばしば細胞間結合と関連する

❷ Serotyping and other phenotypic methods **are often used** to characterize the capsular polysaccharide of group B streptococci (GBS).（J Clin Microbiol. 2005 43:6113）

訳 …は，～を特徴づけるためにしばしば使われる

■ often + [動]

❶ often have ～	…は，しばしば～を持つ	340
❷ often result in ～	…は，しばしば～という結果になる	260
❸ often require ～	…は，しばしば～を必要とする	240
❹ often lead to ～	…は，しばしば～につながる	234
❺ often contain ～	…は，しばしば～を含む	121

❷ Disruption of translation in mitochondria **often results in** ovule abortion before and immediately after fertilization.（Plant J. 2005 44:866）

訳 ミトコンドリアにおける翻訳の破壊は，しばしば～という結果になる

■ [副] + often

❶ more often ～	よりしばしば～	821
・more often than ～	～よりしばしば	203
・more often in ～	～においてよりしばしば	138
❷ most often	最もしばしば	562
❸ less often ～	よりしばしば～でない	160

❶ Hypertension and dyslipidemia are risk factors for atherosclerosis and occur together **more often than** expected by chance.（Science. 2004 306:1190）

訳 ～は，偶然に予想されるよりしばしば共に起こる

❷ Evolution of chromosome 7 abnormalities was seen **most often** in refractory patients who had failed to respond to therapy.（Blood. 2002 99:3129）

訳 ～が，治療抵抗性患者において最もしばしば見られた

old 形 歳の／齢の／年取った／古い　用例数 9,270

◆類義語：elderly, aged

■ [名]-old

❶ ～-year-old	～歳の	1,599
❷ ～-month-old	～月齢の	928
❸ ～-week-old	～週齢の	741
❹ ～-day-old	～日齢の	695

❶ **A 38-year-old male with advanced AIDS**, who had failed to respond to triple-drug antiretroviral therapy, underwent baboon BMT in 1995.（Transplantation. 2004 78:1582）

訳 38歳の進行したAIDSの男性

❸ **Three- to 4-week-old mice were infected with** C. parvum, and infection was monitored by quantifying fecal oocyst shedding.（Infect Immun. 2006 74:549）

訳 3から4週齢のマウスが，～に感染させられた

■ -old ＋［名］

❶ ～-old mice	～齢のマウス	561
❷ ～-old rats	～齢のラット	427
❸ ～-old male	～歳の男性	214

■ old ＋［名］

❶ old age	老齢	276
・in old age	老齢において	96

❶ Recently, we have reported that developmental exposure of rats to Pb **resulted in latent elevation of APP mRNA, APP, and Aβ in old age**.（FASEB J. 2005 19:2083）

訳 ～は，老齢においてAPPメッセンジャーRNA，APPおよびAβの潜在性の上昇という結果になった

older 形 より年上の／より高齢の／より古い　用例数 9,095

■ older ＋［名］

❶ older adults	より年上の成人	949
❷ older patients	より年上の患者	677
❸ older age	より年上	664
❹ older women	より年上の女性	304

❶ These findings suggest that weight loss, even with regain, **could accelerate sarcopenia in older adults**.（Am J Clin Nutr. 2005 82:872）

訳 ～は，より年上の成人における筋肉減少症を促進させうる

■ older than

★ older than ～	～より年上の	728
❶ patients older than ～	～より年上の患者	128

❶ This effect could not be observed in slices from rats older than postnatal day 5.（J Neurosci. 2005 25:2285）

訳 この効果は，出生後５日より高齢のラットからのスライスにおいて観察できなかった

■ ［名］＋ or/and older

❶ ～ years or older	～歳あるいはそれより年上の	641
❷ ～ years and older	～歳およびそれより年上の	348
❸ ～ years of age or older	～歳の年齢あるいはそれより年上の	239

❶ SARS-associated mortality may exceed 50% for persons aged 60 years or older.（J Virol. 2005 79:5833）

訳 SARSに関連した死亡率は，60歳あるいはそれより年上の人々に対して50%を超えるかもしれない

❸ The study population included women 18 years of age or older who underwent at least one diagnostic mammography examination between 1996 and 2001.（Radiology. 2005 235:775）

訳 …は，～の間に少なくとも１つの診断的マンモグラフィー検査を受けた18歳あるいはそれより年上の女性を含んでいた

once 副 一度　　用例数 5,239

■ once

❶ once daily	1日に一度	634
・～ mg once daily	1日に一度～ミリグラム	128
・once daily for ～	～の間1日に一度	91
❷ once a ～	～につき一度	322
・once a day	1日につき一度	76
❸ once per ～	～につき一度	228
・once per cell cycle	細胞周期につき一度	92

❶ Denileukin diftitox was administered intravenously at a dose of 18 μg/kg once daily for 5 days every 3 weeks, up to eight cycles.（J Clin Oncol. 2004 22:4095）

訳 ～は，３週ごとに５日間，１日に一度，18 μg/kgの用量で静脈内に投与された

❷ The pregnant Sprague-Dawley rats received 1 or 3 g/kg of alcohol or an isocaloric solution by intragastric intubation once a day from gestational day (GD) 5 to GD 20.（Brain Res. 2005 1042:125）

訳 妊娠したSDラットは，１日に一度胃内挿管によって１あるいは3g/kgのアルコールあるいは等カロリーの溶液を受けた

★★ only 副 のみ／わずかに／だけ　　用例数 89,668

◆類義語：solely, simply, merely, slightly

■ only ＋［前／接］

❶ only in ～	～においてのみ	7,334

・found only in ～	～においてのみ見つけられる	440
・observed only in ～	～においてのみ観察される	340
・expressed only in ～	～においてのみ発現する	322
・detected only in ～	～においてのみ検出される	306
・present only in ～	～においてのみ存在する	285
❷ only when ～	～のときのみ	1,777
❸ only to ～	～にのみ	1,363
❹ only by ～	～によってのみ	1,341
❺ only for ～	～に対してのみ	1,260
❻ only at ～	～においてのみ	1,207
❼ only after ～	～のあとでのみ	959
❽ only if ～	～の場合に限り	615
・but only if ～	しかし，～の場合に限り	130

❶ **Half of all transcribed sequences are found only in the nucleus** and for the most part are unannotated.（Science. 2005 308:1149）

訳 すべての転写される配列の半分は，核においてのみ見つけられる

❶ In fact, **abundant ErbB-3 expression is detected only in gefitinib-sensitive NSCLC cell lines.**（Proc Natl Acad Sci USA. 2005 102:3788）

訳 大量のErbB-3の発現が，ゲフィチニブ感受性のNSCLC細胞株においてのみ検出される

❼ PLK-1 also specifically inhibited particle binding by COS cells, **only after transfection with human MARCO cDNA.**（J Immunol. 2005 175:6058）

訳 ヒトのMARCO cDNAのトランスフェクションのあとでのみ

❽ Surprisingly, **this effect is observed with catalytically inactive forms of Cdc9p protein, but only if they possess a functional PCNA-binding site.**（Genetics. 2005 171:427）

訳 この効果が触媒的に不活性型のCdc9pタンパク質について観察されるが，（それは）それらが機能的なPCNA結合部位を持つ場合に限る

■ only ＋［過分］

❶ the only known ～	唯一の既知の～	437
❷ only limited ～	限られた～のみ	343
❸ only observed in ～	～においてのみ観察される	125
❹ only found in ～	～においてのみ見つけられる	72

■ only ＋［副］

❶ only partially	部分的にのみ	1,177
・be only partially ～	…は，部分的にのみ～である	518
❷ only slightly	ごくわずかにのみ	834
❸ only weakly	ごく弱くのみ	526

❶ In parabiosis experiments, **splenic DCs were only partially replaced by circulating precursors** over a 6 week period.（Immunity. 2005 22:439）

訳 脾臓のDCが，循環する前駆体によって部分的にのみ置換された

■ only ＋［名句］

❶ only one	ただ1つ	4,050
❷ only two	ただ2つ	1,243
❸ only a small ～	ほんの小さい～	1,049
・only a small fraction of ～	ほんの小さい割合の～	220
・only a small number of ～	ほんの少数の～	112
・only a small proportion of ～	ほんの小さい割合の～	98
❹ only a few ～	ほんの少数の～	974
❺ only a single ～	ただ1つの～	765

❶ Furthermore, on certain balanced translocations **only one** of the derivative chromosomes displays the phenotype.（Hum Mol Genet. 2005 14:2813）

訳 派生染色体のただ１つが，その表現型を示す

❺ However, **only a single** animal in this group demonstrated viral escape in the immunodominant Gag181-189 CM9 response.（J Virol. 2005 79:15556）

訳 このグループのただ１つの動物が，～を示した

■ not/but only

❶ not only ～	～だけでなく	9,557
・not only ～ but also …	～だけでなく…も	1,060
・not only in ～	～においてだけでなく	766
・not only for ～	～に対してだけでなく	444
・not only to ～	～にだけでなく	426
・not only by ～	～によってだけでなく	353
❷ but only ～	しかし，～のみ	3,717
・but only in ～	しかし，～においてのみ	355
・but only when ～	しかし，～のときのみ	247
・but only if ～	しかし，～の場合のみ	130

❶ Using several different methodologies, we found reduced mineralization **not only** in $Akp2^{-/-}$ **but also** in $Enpp1^{-/-}$ and [$Akp2^{-/-}$; $Enpp1^{-/-}$] femurs and tibias.（Am J Pathol. 2005 166:1711）

訳 われわれは，$Akp2^{-/-}$においてだけでなく$Enpp1^{-/-}$ および［$Akp2^{-/-}$; $Enpp1^{-/-}$］の大腿骨においても低下した石灰化を見つけた

★ onset 名 開始／発症　用例数 16,106

onset	開始／発症	16,017
onsets	開始／発症	89

複数形のonsetsの割合は0.5%しかなく，原則，**不可算名詞**として使われる．

◆類義語：initiation, start, development, occurrence, episode

■ onset ＋［前］

onset ofの前にtheが付く割合は圧倒的に高い．

	the％	a/an％	ø％			
❶	80	1	19	onset of ～	～の開始／～の発症	7,125
				・at the onset of ～	～の開始において／～の発症において	715
				・after the onset of ～	～の開始のあとに／～の発症のあとに	575
				・before the onset of ～	～の開始の前に／～の発症の前に	492
				・onset of symptoms	症状の発症	223
				・prior to the onset of ～	～の開始より前に／～の発症より前に	153
				・onset of diabetes	糖尿病の発症	124
				・onset of disease	疾患の発症	122
❷	4	7	81	onset in ～	～における発症	266

❶ Children who underwent transplantation after the onset of symptoms had minimal neurologic improvement.（N Engl J Med. 2005 352:2069）

訳 症状の発症のあと移植を受けた子供たち

❶ Expression of the transcription factor zic1 at the onset of gastrulation is one of the earliest molecular indicators of neural fate determination in Xenopus.（Dev Biol. 2006 289:517）

訳 原腸形成の開始における転写因子zic1の発現

■［形／名］＋ onset

❶	early-onset ～	早発性の～	1,504
❷	late-onset ～	遅発性の～	850
❸	disease onset	疾患の発症	476
❹	new-onset ～	初発の～	432
❺	adult-onset ～	成人発症の～	423
❻	symptom onset	症状発症	269
❼	childhood-onset ～	小児発症の～	259

❶ Mutations in torsinA cause dominantly inherited early-onset torsion dystonia in humans.（Neuron. 2005 48:875）

訳 torsinAにおける変異が，優性に遺伝する早発性捻転ジストニアを引き起こす

■［前］＋ onset

❶	age at onset	発症時年齢	446
❷	age of onset	発症の年齢	348

❶ Previous evidence suggests that the inheritance of bipolar disorder（BP）may vary depending on the age at onset（AAO）.（Am J Hum Genet. 2005 77:545）

訳 ～は，発症時年齢（AAO）に応じて異なるかもしれない

operate 動 作動する／手術する　用例数 4,278

operating	作動する／～を手術する	2,601
operated	手術される／作動した	1,805
operate	作動する／～を手術する	1,522
operates	作動する／～を手術する	951

自動詞として用いられることが多いが，他動詞の用例もある．

◆類義語：function, serve, work, drive

■ operate ＋［前］

❶	operate in ～	…は，～において作動する	985
❷	operate at ～	…は，～において作動する	387
❸	operate on ～	…は，～において作動する	295

❶ Ddc1 therefore **appears to operate in a positive feedback loop** that promotes Mek1 function. (Genes Dev. 2002 16:363)

訳 ～は，ポジティブフィードバックループにおいて作動するように思われる

■ ［名］-operated

❶	sham-operated ～	偽の手術をされた～	596
❷	store-operated ～	貯蔵量に応じて働く～	547

❶ **Compared with sham-operated rats**, the ligated rats had significantly decreased baseline Pmax and max dP/dt. (Circulation. 2004 110:㊙187)

訳 偽の手術をされたラットと比べると

opportunity 名 機会　用例数 4,727

opportunity	機会	2,593
opportunities	機会	2,134

複数形のopportunitiesの割合は約45%とかなり高い．

◆類義語：chance, occasion

■ opportunity/opportunities ＋［前］

	the%	a/an%	ø%			
❶	28	66	5	opportunity to *do*	～する機会	1,707
				・a unique opportunity to *do*	～するユニークな機会	295
				・opportunity to study ～	～を研究する機会	210
				・opportunity to examine ～	～を調べる機会	117
				・opportunity to investigate ～	～を精査する機会	96
❷	4	-	89	opportunities for ～	～のための機会	1,067
				・new opportunities for ～	～のための新しい機会	242

❶ Human artificial chromosomes (HACs) **provide** a unique opportunity to **study kinetochore formation** and to develop a new generation of vectors with potential in gene therapy. (Nucleic Acids Res. 2005 33:e130)

訳 ～は，動原体形成を研究するユニークな機会を提供する

❷ Learning how critical cell regulatory pathways are controlled **may lead to** new opportunities for **cancer treatment**. (Cancer Res. 2005 65:8575)

訳 ～は，癌の治療のための新しい機会につながるかもしれない

■ [動] + opportunity

❶ provide an opportunity	…は，機会を提供する	424
❷ offer the opportunity	…は，機会を提供する	101

oppose 動 反対する　用例数 1,846

opposed	～に反対した／反対される	1,456
oppose	～に反対する	204
opposes	～に反対する	186

他動詞． as opposed toの用例が多い．

◆類義語：disagree, counteract, contrast

■ as opposed to

❶ as opposed to ～	～とは対照的に	1,090

❶ However, **the role of the PDT effect on tumor cells** as opposed to **the host tissues has not been determined**. (Cancer Res. 2002 62:1604)

訳 宿主組織とは対照的に，腫瘍細胞に対するPDT効果の役割は決定されていない

optimal 形 最適な／至適な　用例数 9,367

■ be optimal + [前]

❶ be optimal for ～	…は，～にとって最適である	176

❶ **No single extraction method** was optimal for **all organisms**. (J Clin Microbiol. 2005 43:5122)

訳 どの単一の抽出法も，あらゆる生物にとって最適であるわけではなかった

■ [前] + optimal

❶ be required for optimal ～	…は，最適な～のために必要とされる	436
❷ be necessary for optimal ～	…は，最適な～のために必要である	104

❶ Additional studies with truncated soluble DR1 demonstrated that **transmembrane domains in DR1 molecules** are **also** required for optimal **activity**. (J Immunol. 2001 167:5167)

訳 DR1分子の膜貫通ドメインは，また，最適な活性のために必要とされる

■ optimal ＋［名］

❶ optimal activity	最適な活性	176
❷ optimal treatment	最適な治療	168
❸ optimal conditions	最適な条件	156

optimize ㊍ 最適化する／至適化する 用例数 5,197

optimized	最適化される／～を最適化した	2,765
optimize	～を最適化する	1,504
optimizing	～を最適化する	756
optimizes	～を最適化する	172

他動詞.

■ be optimized ＋［前］

❶ be optimized for ～	…は，～に対して最適化される	198
❷ be optimized to *do*	…は，～するように最適化される	164
❸ be optimized by ～	…は，～によって最適化される	86

❶ Three sites of diversity on the ligands **were optimized for** this reaction using a positional scanning approach.（J Org Chem. 2005 70:8835）

訳 ～は，この反応に対して最適化された

❷ Thus, combination chemotherapy must **be optimized to** increase tumor response and at the same time lower its toxicity.（Cancer Res. 2005 65:6934）

訳 併用化学療法は，腫瘍の反応を増大させるように最適化されねばならない

■ optimized ＋［名］

❶ optimized conditions	最適化された条件	198

❶ Under **optimized conditions**, 1 nmol of the substituted MoFe protein catalyzes the formation of 21 nmol of CH_4 within 20 min.（Proc Natl Acad Sci U S A. 2012 Nov 27;109(48):19644-）

訳 最適化された条件下で，

option ㊔ 選択肢／選択 用例数 3,548

options	選択肢／選択	2,193
option	選択肢／選択	1,355

複数形のoptionsの割合は約60%と非常に高い.

◆類義語：choice, selection

■ options ＋［前］

	the%	a/an%	ø%			
❶	21	-	72	options for ～	～に対する選択肢	654

❶ Treatment **options for** myelofibrosis are limited.（N Engl J Med. 2012 366:787）
訳 骨髄線維症に対する治療選択肢は限られている

■［名／形］＋ options

❶ treatment options	治療選択肢	721
❷ therapeutic options	治療選択肢	393

★ order 名 オーダー／順番／桁／次数，動 整理する　用例数 26,336

order	オーダー／順番／桁／～を整理する	19,133
ordered	整理された／～を整理した／規則正しい	3,568
orders	オーダー／順番／～を整理する	2,706
ordering	順番	929

名詞の用例が多いが，他動詞としても用いられる．複数形は，ほとんどorders of magnitudeの形で使われる．in order toの形でも使われる．

◆類義語：turn, arrange

■ 名 order(s) ＋［前］

	the%	a/an%	ø%			
❶	47	39	6	order of ～	～のオーダー	3,277
				・～ orders of magnitude	～桁	2,086
				・be on the order of ～	…は，およそ～である	121

❶ Benzene's permeability is about 2 **orders of magnitude** higher than in comparable gas-phase experiments.（J Am Chem Soc. 2005 127:15112）
訳 ベンゼンの透過性は，～においてよりおよそ２桁高い

❷ The Gibbs free energy for the transition to the unfolded form **is on the order of** -4 to -6 kJ/mol at physiological temperatures（37 ℃）.（Biochemistry. 2001 40:11660）
訳 ～は，生理的な温度においておよそ−4 から−6 kJ/molである

■ 名［形／名］＋ order

❶ higher-order	高次の	1,783
・higher-order structures	高次構造	485
❷ rank order	順位	366

❶ The N-terminal tail domains（NTDs）of histones play important roles in the formation of **higher-order** structures of chromatin and the regulation of gene functions.（J Am Chem Soc. 2009 131:15104）
訳 ～は，クロマチンの高次構造の形成において重要な役割を果たす

■ 名 in order to *do*

★ in order to *do*	～するために	6,716
❶ in order to determine ～	～を決定するために	485
❷ in order to identify ～	～を同定するために	383
❸ in order to understand ～	～を理解するために	331

❶ In order to determine whether DEK is a bona fide HPV oncogene target in primary cells, DEK expression was monitored in human keratinocytes transduced with HPV E6 and/or E7.（J Virol. 2005 79:14309）

訳 ～かどうかを決定するために

■ 動［副］＋ ordered

❶ highly ordered	非常に規則正しい	294
❷ well ordered	よく整理された	246

❶ The initiation of these phases follows a highly ordered pattern that appears important for the production of virus particles.（J Virol. 2002 76:10401）

訳 これらの相の開始は，非常に規則正しいパターンに従う

O

★ organization 名 構成／構築／組織化 用例数 11,437

organization	構成／構築	10,736
organizations	構成／構築	701

複数形のorganizationsの割合は約5％あるが，単数形は無冠詞で使われることが多い．

◆類義語：constitution, composition, formation

■ organization ＋［前］

organization ofの前にtheが付く割合は非常に高い．

	the％	a/an％	ø％			
❶	77	4	18	organization of ～	～の構成	4,135
❷	10	14	71	organization in ～	～における構成	546

❶ However, the spatial organization of these molecules remains unclear.（J Biol Chem. 2013 288:382）

訳 これらの分子の空間的構成は，不明なままである

■［形］＋ organization

❶ genomic organization	ゲノム構築	345
❷ cytoskeletal organization	細胞骨格系	320
❸ spatial organization	空間的構成	308
❹ structural organization	構造構成	294
❺ functional organization	機能的構成	275

organize ㊍ 組織化する／構築する　用例数 5,076

organized	組織化される／～を組織化した	3,040
organizing	～を組織化する	1,040
organize	～を組織化する	758
organizes	～を組織化する	238

他動詞.

◆類義語：assemble, build, orchestrate, construct, form

■ be organized ＋［前］

★ be organized	…は，組織化される	1,105
❶ be organized into ～	…は，～に組織化される	375
❷ be organized in ～	…は，～において組織化される	221
❸ be organized as ～	…は，～として組織化される	101
❹ be organized by ～	…は，～によって組織化される	74

❶ The alternative exons are organized into four clusters, and the exons within each cluster are spliced in a mutually exclusive manner.（Mol Cell Biol. 2005 25:10251）

訳 代わりのエクソンは，４つのクラスターに組織化される

■［副］＋ organized

❶ highly organized	高度に組織化される	190
❷ self-organized	自己構築した	108

orient ㊍ 方向づける／向かわせる　用例数 3,182

oriented	向けられる／方向づけた	2,282
orienting	方向づける	392
orient	方向づける	370
orients	方向づける	138

自動詞としても用いられるが，他動詞がほとんどである.

◆類義語：direct

■ oriented ＋［前］

❶ oriented in ～	～に向けられる	173
❷ oriented toward ～	～に向けられる	121

❶ Such nonshared genes usually appear to be truncated and form clusters in which they are oriented in the same direction.（Plant Cell. 2005 17:343）

訳 それらは，同じ方向に向けられる

❷ The apparatus was oriented toward the preferred direction of the recorded cell, or the 180 degrees opposite direction.（J Neurosci. 2005 25:2420）

訳 その装置が，～の優先方向に向けられた

orientation ㊔ 方向性／方向／配向 用例数 8,919

orientation	方向性／方向／配向	7,585
orientations	方向性／方向／配向	1,334

複数形のorientationsの割合が，約15%ある可算名詞.
◈類義語：direction

■ orientation ＋［前］

orientation ofの前にtheが付く割合は圧倒的に高い.

	the%	a/an%	ø%			
❶	82	3	12	orientation of ～	～の方向性	2,218
			1	・relative orientation of ～	～の相対的方向性	211
❷	20	19	47	orientation in ～	～における方向性	312

❶ Cre- and Flp-mediated recombination switches the orientation of the gene-trap cassette, permitting conditional rescue in one orientation and conditional knockout in the other.（Proc Natl Acad Sci USA. 2012 109:15389）
訳 CreおよびFlpに仲介される組換えは，ジーントラップカセットの方向性を切り替える

★ origin ㊔ 起源／開始点 用例数 13,507

origin	起源／開始点	10,468
origins	起源／開始点	3,039

複数形のoriginsの割合が，約20%ある可算名詞.
◈類義語：genesis

■ origin ＋［前］

origin ofの前にtheが付く割合は圧倒的に高い.

	the%	a/an%	ø%			
❶	89	7	4	origin of ～	～の起源	3,287
				・origin of replication	複製の起源	306
❷	8	56	36	origin for ～	～に対する起源	271

❶ We propose an equation that allows us to investigate the origin of evolution.（J Theor Biol. 2009 256:586）
訳 われわれは，われわれが進化の起源を精査することを可能にする式を提案する

■［名／形］＋ origin

❶	replication origin	複製開始点	293
❷	evolutionary origin	進化的起源	214
❸	cell origin	細胞の起源	121
❹	common origin	共通の起源	116

O

originally 副 最初に／本来 用例数 2,237

◆類義語：initially, early

■ originally ＋［過分］

❶ originally identified	最初に同定された	473
・originally identified as ～	～として最初に同定された	211
❷ originally described	最初に記述された	155
❸ originally isolated	最初に単離された	150
❹ originally proposed	最初に提唱された	102

❶ Negative cofactor 2 (NC2) is an evolutionarily conserved transcriptional regulator that was originally identified as an inhibitor of basal transcription. (Proc Natl Acad Sci USA. 2002 99:12727)

訳 ～は，基礎転写の抑制因子として最初に同定された進化的に保存された転写制御因子である

❷ Steroid receptor binding factor (RBF) was originally isolated from avian oviduct nuclear matrix. (Biochemistry. 2000 39:753)

訳 ステロイド受容体結合因子（RBF）は，～から最初に単離された

originate 動 生じる／始まる 用例数 3,688

originate	生じる	1,120
originating	生じる	1,091
originated	生じる	906
originates	生じる	571

自動詞.

◆類義語：derive, stem, arise, draw, take place

■ originate ＋［前］

❶ originate from ～	…は，～から生じる	2,117
❷ originate in ～	…は，～において生じる	695
❸ originate at ～	…は，～において生じる	115

❶ These axons originate from cells that have been identified as serial homologs of motor neuron-1 of other abdominal ganglia. (J Comp Neurol. 2001 440:245)

訳 これらの軸索は，～である細胞から生じる

❶ Bruton's tyrosine kinase (Btk) is a critical transducer of signals originating from the B cell antigen receptor (BCR). (Proc Natl Acad Sci USA. 1999 96:2221)

訳 Brutonチロシンキナーゼ（Btk）は，B細胞抗原受容体（BCR）から生じるシグナルの決定的に重要なトランスデューサーである

★★ other 形 他の，代名 もう一方のもの　用例数 127,697

other	他の／もう一方のもの	121,043
others	もう一方のもの	6,654

形容詞の用例が多いが，名詞としても使われる．複数形のothersの割合が，約５%ある可算名詞．

◆類義語：consequence, result, achievement, prognosis

■ 形 other ＋［名］

❶ on the other hand	他方では	2,594
❷ other proteins	他のタンパク質	1,807
❸ other factors	他の因子	1,244
❹ other genes	他の遺伝子	1,235
❺ other members	他のメンバー	1,162
❻ other tissues	他の組織	1,051
❼ other species	他の種	1,022

❶ On the other hand, evidence indicates that the globally distributed prototypic arenavirus lymphocytic choriomeningitis virus (LCMV) is a neglected human pathogen. (J Virol. 2012 86:4578)

訳 他方では，

■ 形［名］＋ other than

★ other than ～	～以外の	3,154
❶ factors other than ～	～以外の因子	261
❷ mechanisms other than ～	～以外の機構	111

❶ This suggests that factors other than CD45 levels regulate HIV-1 transcription within the ectocervix. (J Infect Dis. 2009 200:965)

訳 CD45レベル以外の因子が，HIV-1転写を制御する

■ 代名［前］＋ each other

each otherは，one anotherとほぼ同じ意味である．

★ each other	お互いに	3,154
❶ interact with each other	…は，お互いに相互作用する	382
❷ similar to each other	お互いに類似している	141
❸ related to each other	お互いに関連している	112
❹ relative to each other	お互いに関連している	100

❶ We confirmed reports that RPL26 and NCL interact with each other and then explored the potential role of this interaction in the translational control of p53 after stress. (J Biol Chem. 2012 287:16467)

訳 RPL26とNCLは，お互いに相互作用する

☆★ outcome ㊔ 結果／予後／成果 用例数 31,784

outcome	結果／予後	16,089
outcomes	結果／予後	15,695

複数形のoutcomesの割合は約50%と非常に高い.

◆類義語：consequence, result, achievement, prognosis

■ outcome(s) ＋［前］

outcome ofの前にtheが付く割合は圧倒的に高い.

	the%	a/an%	ø%			
❶	80	7	12	outcome of ～	～の結果（予後）	2,793
❷	3	-	93	outcomes in ～	～における結果（予後）	1,521
❸	5	-	94	outcomes for ～	～に対する結果	605
❹	11	8	80	outcome after ～	～のあとの結果	436

❶ The addition of oxaliplatin to fluorouracil (FU) and leucovorin (LV) **improves the outcome of patients with colorectal cancer (CRC)**. (J Clin Oncol. 2004 22:4753)

訳 ～は，直腸結腸癌（CRC）の患者の予後を改善する

❷ **A Markov model was developed to project outcomes in patients with testicular cancer** who were undergoing CT surveillance in the decade after orchiectomy. (Radiology. 2013 266:896)

訳 Markovモデルは，精巣癌の患者における予後を推定するために開発された

■［形］＋ outcome

❶	clinical outcomes	臨床結果	1,560
❷	the primary outcome	一次結果	952
❸	adverse outcomes	有害な結果	481
❹	poor outcome	良くない結果	478

❶ **We therefore assessed the long-term clinical outcomes of 59 POEMS patients treated with** ASCT at our institution. (Blood. 2012 120:56)

訳 われわれは，それゆえ，～で治療された59名のPOEMS患者の長期臨床結果を評価した

output ㊔ 出力 用例数 5,056

output	出力	4,488
outputs	出力	568

複数形のoutputsの割合は約10%あるが，単数形は無冠詞で使われることが非常に多い.

◆反意語：input

■ output ＋［前］

output ofの前にtheが付く割合は圧倒的に高い.

	the %	a/an %	ø %			
❶	83	10	6	output of ～	～の出力	606
❷	6	1	90	output in ～	～における出力	207
❸	38	3	57	output from ～	～からの出力	176

❶ The spike output of neural pathways can be regulated by modulating output neuron excitability and/or their synaptic inputs.（J Neurosci. 2009 29:15001）
訳 神経路のスパイク出力は，～によって調節されうる

overcome 動 克服する　用例数 5,120

overcome	～を克服する／克服される	3,841
overcoming	～を克服する	556
overcomes	～を克服する	480
overcame	～を克服した	243

他動詞.

■ overcome ＋［名句］

❶	overcome this problem	…は，この問題を克服する	134
❷	overcome this limitation	…は，この制限を克服する	119

❶ To overcome this problem, we sought to identify constrained peptidomimetic inhibitors that would provide potential new drug leads.（Biochemistry. 2005 44:12362）
訳 この問題を克服するために，

■ to overcome

★	to overcome ～	～を克服する…	1,785
❶	To overcome ～	～を克服するために	425

■ be overcome ＋［前］

❶	be overcome by ～	…は，～によって克服される	577

❶ This limitation can be overcome by using a synthetic system in which Spo0A activation is controlled by inducing expression of phosphorelay kinase KinA.（Proc Natl Acad Sci USA. 2012 109:E3513）
訳 この制限は，合成システムを使うことによって克服されうる

★ overexpress 動 過剰発現させる 用例数 11,762

overexpressed	過剰発現される／〜を過剰発現した	5,369
overexpressing	〜を過剰発現させる	4,961
overexpress	〜を過剰発現させる	1,299
overexpresses	〜を過剰発現させる	133

他動詞.

◆類義語：express

■ be overexpressed ＋［名］

★ be overexpressed	…は，過剰発現される	2,092
❶ be overexpressed in 〜	…は，〜において過剰発現される	1,468
・be overexpressed in Escherichia coli	…は，大腸菌において過剰発現される	98

❶ GPR variants were overexpressed in Escherichia coli, and the active form（P41）was assayed for activity against SASP and the zymogen form（P46）was assayed for the ability to autoprocess to P41.（J Bacteriol. 2005 187:7119）

訳 GPR変異体が，大腸菌において過剰発現された

■ when overexpressed

★ when overexpressed	過剰発現されたとき	568
❶ when overexpressed in 〜	〜において過剰発現されたとき	253

❶ The Agouti-related protein（AGRP）, an appetite modulator, induces hyperphagia when administered intracerebroventricularly or when overexpressed in transgenic mice.（Am J Clin Nutr. 2005 82:1097）

訳 トランスジェニックマウスにおいて過剰発現されたとき

■ -overexpressing ＋［名］

❶ 〜-overexpressing cells	〜過剰発現細胞	506
❷ 〜-overexpressing mice	〜過剰発現マウス	126

❶ This channel activity was never observed in Bcl-2-overexpressing cells.（Mol Biol Cell. 2005 16:2424）

訳 このチャネル活性は，Bcl-2過剰発現細胞においては決して観察されなかった

■［名／前］＋ overexpressing

❶ cells overexpressing 〜	〜を過剰発現する細胞	964
・in cells overexpressing 〜	〜を過剰発現する細胞において	207
❷ mice overexpressing 〜	〜を過剰発現するマウス	564
・transgenic mice overexpressing 〜	〜を過剰発現するトランスジェニックマウス	311

❶ Increased cytotoxicity occurred in cells overexpressing NBR and NPR, whereas overexpressed NQO1 had no effect.（Proc Natl Acad Sci USA. 2005 102:9282）

訳 増大した細胞傷害性が，NBRおよびNPRを過剰発現する細胞において起こった

★ overexpression 名 過剰発現　用例数 22,205

overexpression	過剰発現	22,202
overexpressions	過剰発現	3

複数形のoverexpressionsの用例はほとんどなく，原則，**不可算名詞**として使われる．
◈類義語：expression

■ overexpression ＋［前］

	the%	a/an%	ø%		
❶	4	1	95	overexpression of ～　～の過剰発現	15,008
				・overexpression of wild-type ～　野生型～の過剰発現	408
				・by overexpression of ～　～の過剰発現によって	372
				・overexpression of a dominant-negative ～　ドミナントネガティブな～の過剰発現	196
				・…-specific overexpression of ～　…特異的な～の過剰発現	187
				・…-mediated overexpression of ～　…に仲介される～の過剰発現	180
				・transient overexpression of ～　一時的な～の過剰発現	141
❷	3	1	27	overexpression in ～　～における過剰発現	982

❶ **Overexpression of a dominant-negative RAGE construct inhibited** the IL-7-mediated production of MMP-13.（Arthritis Rheum. 2009 60:792）
訳 ドミナントネガティブなRAGEコンストラクトの過剰発現は，～を阻害した

❶ To elucidate the specific roles of ACSL5 in fatty acid metabolism, **we used adenoviral-mediated overexpression of ACSL5** (Ad-ACSL5) in rat hepatoma McArdle-RH7777 cells.（J Biol Chem. 2006 281:945）
訳 われわれは，アデノウイルスに仲介されるACSL5の過剰発現を利用した

★ overlap 動 重複する，名 重複　用例数 10,064

overlapping	重複する	4,956
overlap	重複する／重複	3,577
overlaps	重複する／重複	1,019
overlapped	重複する	512

自動詞，他動詞および名詞として用いられる．複数形のoverlapsの割合が，約10%ある可算名詞．
◈類義語：duplicate, duplication, redundancy

■ 動 overlapping ＋［名］

❶	overlapping functions	重複する機能	236

❷ overlapping peptides	重複するペプチド	134

❶ Using double and triple mutant combinations we show that these three copine genes **have overlapping functions essential for the viability of plants.** (Plant J. 2006 45:166)
訳 ～は，植物の生存能に必須である重複する機能を持つ

■ 動［副］＋ overlapping

❶ partially overlapping ～	部分的に重複する～	229

■ 名 overlap ＋［前］

	the %	a/an %	ø %			
❶	22	10	61	overlap between ～	～の間の重複	492
❷	11	21	53	overlap in ～	～における重複	437
❸	4	8	60	overlap with ～	～との重複	410
❹	39	13	42	overlap of ～	～の重複	383

❶ **There was little clinical overlap between these two syndromes.** (J Clin Microbiol. 2012 50:3598)
訳 これら２つの症候群の間の臨床的重複はほとんどなかった

❸ Efforts at deciphering the molecular basis for this disease have been complicated by the clinical and genetic heterogeneity as well as **extensive phenotypic overlap with other syndromes.** (Hum Mol Genet. 2005 14:R235)
訳 他の症候群との広範な表現型の重複

owing to 形 ～のために／～のせいで 用例数 1,709

◆類義語：due to
owing toの用例が圧倒的に多い.

■ owing to

❶ owing to ～	～のために／～のせいで	1,709

❶ **Owing to the complexity of drug actions,** a broader genomics approach aims at finding new drug targets and optimizing therapy for the individual patient. (Hum Mol Genet. 2005 14:R207)
訳 薬物作用の複雑さのために

★ oxidation 名 酸化 用例数 12,598

oxidation	酸化	12,371
oxidations	酸化	227

複数形のoxidationsの割合は２％しかなく，原則，**不可算名詞**として使われる.

■ oxidation ＋［前］

	the%	a/an%	⌀%			
❶	37	2	60	oxidation of ～	～の酸化	3,673
❷	0	3	84	oxidation in ～	～における酸化	476

❶ In particular, there has been no method to detect the oxidation of methionine on polyacrylamide gels. (Anal Biochem. 2012 421:767)

訳 ポリアクリルアミドゲル上でメチオニンの酸化を検出する方法はなかった

P

★ pain ㊔ 痛み／疼痛 用例数 10,809

pain	痛み／疼痛	10,777
pains	痛み／疼痛	32

複数形のpainsの割合は0.3%しかなく，原則，不可算名詞として使われる．
◆類義語：ache

■ pain ＋［前］

	the %	a/an %	ø %			
❶	0	0	98	pain in ～	～の痛み	11,272

❶ Approximately 100 million people suffer from chronic pain in the USA. (Curr Opin Anaesthesiol. 2012 25:566)
訳 およそ1億人が，アメリカ合衆国において慢性痛を患っている

■［形／名］＋ pain

❶	neuropathic pain	神経因性疼痛	599
❷	chest pain	胸痛	505
❸	chronic pain	慢性痛	476
❹	back pain	背部痛	339
❺	knee pain	膝関節痛	291
❻	abdominal pain	腹痛	275

paper ㊔ 論文 用例数 7,627

paper	論文	6,820
papers	論文	807

複数形のpapersの割合が，約10%ある可算名詞．
◆類義語：article, literature

■ in this paper

★	in this paper	この論文	3,179
❶	In this paper,	この論文において，	2,003
	・In this paper, we report ～	この論文において，われわれは～を報告する	235
	・In this paper, we show ～	この論文において，われわれは～を示す	194
	・In this paper, we describe ～	この論文において，われわれは～を述べる	183
	・In this paper, we present ～	この論文において，われわれは～を提示する	158
❷	described in this paper	この論文において述べられる	117

❶ In this paper, we report that the desmosomal protein desmoplakin (DP) is not essential for cell adhesion in the intestinal epithelium. (Mol Biol Cell. 2012 23:792)

訳 この論文において，われわれは～ということを報告する

★ parallel 名 対比，副 形 動 平行した　用例数 11,751

parallel	平行した	9,346
paralleled	平行した	1,269
parallels	対比／平行する	922
paralleling	平行する	214

in parallelの用例が多い．

◆類義語：coincident, synchronous, simultaneous, concomitant, concurrent

■ 名 [前] + parallel

★ in parallel	同時に／平行して	2,566
❶ in parallel with ～	～と平行して／～と同時に	896
❷ in parallel,	同時に，	471
❸ in parallel to ～	～と平行して／～と同時に	292

❶ Association of Aurora A with TPX2 leads to activation of the kinase, in parallel with phosphorylation of TPX2.（J Biol Chem. 2004 279:9008）

訳 TPX2のリン酸化と同時に

❷ In parallel, the mice showed enhanced resistance to an established rapidly growing tumor and to viral infection at a mucosal site.（J Exp Med. 2004 199:815）

訳 同時に，

■ 副 形 parallel + [前]

❶ parallel to ～	～に平行に	650

❶ In vertical roots bundles of MFs in the elongation and maturation zone were oriented parallel to the longitudinal axis of cells.（Plant Physiol. 1997 113:1447）

訳 ～は，細胞の長軸に平行に向いていた

■ 動 be paralleled + [前]

★ be paralleled	…は，平行する	405
❶ be paralleled by ～	…は，～と平行する	375

❶ Reversal of HSC activation was paralleled by a marked decrease in MEF2 protein and activity.（Gastroenterology. 2004 127:1174）

訳 HSC活性化の反転は，MEF2タンパク質および活性の顕著な低下と平行した

★ part 名 部分／一部　用例数 27,390

part	部分／一部	24,706
parts	部分	2,684

〈in part〉〈part of〉の用例が非常に多い．複数形の割合が，約10%ある可算名詞．

◆類義語：portion, piece, moiety, division, region, role

■ part ＋［前］

	the%	a/an%	ø%			
❶	10	25	62	part of ～	～の一部	11,272
				・as part of ～	～の一部として…	2,801
				・be part of ～	…は，～の一部である	2,442
				・form part of ～	…は，～の一部を形成する	382
				・an integral part of ～	～の不可欠な部分	361
				・be an integral part of ～	…は，～の不可欠な部分である	220
				・terminal part of ～	～の末端部分	169
				・an important part of ～	～の重要な部分	163
				・an essential part of ～	～の必要不可欠な部分	129

❶ TCR and cytokine signals **induce naive T cells to undergo spontaneous divisions as part of a homeostatic response to** conditions of T cell deficiency. (J Immunol. 2002 169:3752)

訳 …は，～に対するホメオスタシス応答の一部としてナイーヴなT細胞が自発性の分裂を起こすのを誘導する

❶ **Staphylococcus aureus is part of the indigenous microbiota of humans.** (J Biol Chem. 2012 287:6693)

訳 黄色ブドウ球菌は，ヒトの常在菌の一部である

■ in part

★	in part	一部は	11,361
❶	in part by ～	一部は～によって	3,747
	・be mediated in part by ～	…は，一部は～によって仲介される	548
	・be regulated in part by ～	…は，一部は～によって調節される	211
❷	at least in part	少なくとも一部は	2,705
	・, at least in part, by ～	少なくとも一部は～によって	897
❸	in part to ～	一部は～に	1,529
	・be due in part to ～	…は，一部は～のせいである	456
❹	in part through ～	一部は～によって	1,052
❺	in part from ～	一部は～から	452

❶ Here we found that **SGK degradation is mediated in part by Nedd4-2.** (J Biol Chem. 2005 280:4518)

訳 SGKの分解は，一部はNedd4-2によって仲介される

❷ These data indicate that hypotension evoked by stimulation of IL **is mediated, at least in part, by direct or indirect projections to the LHA** and through the PAG. (Brain Res. 2000 859:83)

訳 ～は，LHAへの直接あるいは間接の投射によって少なくとも一部は仲介される

❸ The increase of adipose tissue mass associated with obesity **is due in part to an increase in the number of adipocytes.** (Proc Natl Acad Sci USA. 2004 101:9607)

訳 ～は，一部は脂肪細胞の数の増大のせいである

★ partially 副 部分的に

用例数 13,651

◈類義語：partly, in part

■ partially ＋［過分／現分］

❶ partially inhibited	部分的に抑制される／部分的に抑制した	544
・be partially inhibited	…が，部分的に抑制される	192
❷ partially purified	部分的に精製される／部分的に精製した	491
❸ partially restored	部分的に回復される／部分的に回復した	431
❹ partially blocked	部分的にブロックされる／部分的にブロックした	377
❺ partially reversed	部分的に逆行される／部分的に逆行した	342
❻ partially rescued	部分的に救出される／部分的に救出した	336
❼ partially folded	部分的に折り畳まれる	293
❽ partially overlapping	部分的に重複している	229

❶ Binding of hemolin to LPS **was partially inhibited by calcium and phosphate**. (Eur J Biochem. 2002 269:1827)

訳 ～が，カルシウムとリン酸によって部分的に抑制された

❷ In this study, native neuraminidase **was partially purified from cultures of S. pneumoniae** by serial chromatography with DEAE-Sepharose and Sephacryl S-200. (Infect Immun. 2004 72: 4309)

訳 ～が，肺炎レンサ球菌の培養液から部分精製された

❹ This stage could be reproduced by intradermal injection of IL-1α or IL-1β and **was partially blocked by injection of neutralizing Ab** against IL-1α but not IL-1β. (J Immunol. 2002 168: 3586)

訳 ～が，中和抗体の注入によって部分的にブロックされた

■ partially ＋［形］

❶ partially redundant	部分的に重複する	207
❷ partially dependent on ～	～に部分的に依存する	198
❸ partially unfolded	部分的に畳まれていない	184

❶ Lamellae **are initiated by parallel and partially redundant signaling pathways** involving Rac GTPases and the adaptor protein Nck, which stimulate SCAR, an Arp2/3 activator. (J Cell Biol. 2003 162:1079)

訳 ～は，平行で部分的に重複するシグナル伝達経路によって惹起される

■ ［副／名］＋ partially

❶ only partially	部分的にだけ	1,177
❷ at least partially	少なくとも部分的に	672

❶ We found that S315D retained a complete centrosome duplication activity, while **S315A only partially retained it**. (Oncogene. 2001 20:6851)

訳 S315Aは，それを部分的にだけ保持した

P

★ participate 動 関与する／参加する　用例数 10,409

participate	関与する／参加する	5,290
participates	関与する／参加する	2,571
participating	関与する／参加する	1,412
participated	関与した／参加した	1,136

自動詞．participate inの用例が圧倒的に多い．

◆類義語：involve, implicate, engage

■ participate ＋［前］

❶	participate in ～	…は，～に関与する／～に参加する	9,261
	・participate in the regulation of ～	…は，～の調節に関与する	230
	・participate in the study	…は，研究に関与する	110
	・participate in the formation of ～	…は，～の形成に関与する	92
	・participate in regulating ～	…は，～を調節することに関与する	66

❶ Epithelial Na^+ channels (EnaC) **participate in the regulation of** extracellular fluid volume homeostasis and blood pressure. (J Biol Chem. 2004 279:9743)

訳 ～は，細胞外液量ホメオスタシスの調節に関与する

■［動］＋ to participate in

★	to participate in ～	～に関与する／～に参加する…	1,116
❶	shown to participate in ～	～に関与することが示される	99
❷	known to participate in ～	～に関与することが知られている	89

❶ One of these clusters includes residues **known to participate in** the proteolytic cleavage of these viruses at high pH. (J Virol. 2005 79:554)

訳 …は，～のタンパク質分解性切断に関与することが知られている残基を含む

★ particular 形 特別な／特定の，名 特徴　用例数 14,111

particular	特別な／特徴	14,101
particulars	特別な	10

形容詞および名詞として使われる．名詞としては，in particularの用例が圧倒的に多い．複数形のparticularsの用例はほとんどなく，原則，**不可算名詞**として使われる．

◆類義語：special, specific

■ 名［前］＋ particular

❶	in particular	特に	6,300
	・In particular,	特に，	3,606

❶ **In particular,** we focus on adenylate kinase. (Biophys J. 2001 80:2439)

訳 特に，われわれは～に注目する

■ 形 particular ＋［名］

❶ of particular interest	特に興味深い	573
・be of particular interest	…は，特に興味深い	277
❷ particular emphasis on ～	～に特に重点	194
・with particular emphasis on ～	～に特に重点を置いて	142
❸ of particular importance	特に重要な	191
・be of particular importance	…は，特に重要である	116
❹ particular focus on ～	～に対する特別な注目	106
・with a particular focus on ～	～に特に注目して	54
・with particular focus on ～	～に特に注目して	47
❺ particular attention to ～	～への特別な注意	80
・with particular attention to ～	～に特に注意して	49

❶ Of particular interest are associations between common variations in the genes of factor XIII and altered risk profiles for thrombosis.（Blood. 2002 100:743）

訳 ～の間の関連は，特に興味深い

❷ In this article, we review the role of RCTs in achieving the goals of CER, with particular emphasis on the role of publicly funded clinical trials.（J Clin Oncol. 2012 30:4194）

訳 公的資金を受けた臨床治験の役割に特に重点を置いて

❸ Inferring positive selection at single amino acid sites is of particular importance for studying evolutionary mechanisms of a protein.（J Mol Evol. 2004 59:11）

訳 …は，～を研究するために特に重要である

★particularly 副 特に　用例数 13,445

◈類義語：especially, in particular, notably, specially

■ particularly ＋［前］

❶ particularly in ～	特に～において	2,336
・particularly in patients with ～	特に～の患者において	78

❶ We found that the expression of the Drg-1 protein was significantly reduced in breast tumor cells, particularly in patients with lymph node or bone metastasis as compared to those with localized breast cancer.（Oncogene. 2004 23:5675）

訳 特に～の患者において

■ particularly ＋［形］

❶ particularly important	特に重要な	544
・be particularly important	…は，特に重要である	419
❷ particularly useful	特に役に立つ	321
❸ particularly high	特に高い	203
❹ particularly sensitive	特に敏感な	178

❶ The carboxy terminus of RsbT proved to be particularly important for accumulation in Bacillus subtilis.（J Bacteriol. 2004 186:2789）

訳 …は，～の蓄積にとって特に重要であると判明した

■ particularly ＋［代名］

❶ , particularly those	特に～のそれら（人々）	587
・, particularly those with ～	特に～を持つそれら（人々）	113

❶ In summary, **elderly individuals with AKI**, particularly those with **previously diagnosed CKD, are at significantly increased risk for ESRD**, suggesting that episodes of AKI may accelerate progression of renal disease.（J Am Soc Nephrol. 2009 20:223）

訳 急性腎障害の高齢の個々人，特に以前に慢性腎疾患と診断された人々は，末期腎不全の有意に高いリスクがある

past ㊊ 過去の　　用例数 7,751

■ past ＋［名］

❶ the past decade	過去10年	979
・over the past decade	過去10年の間に	454
❷ the past year	過去1年	854
・in the past year	過去1年において	351
❸ the past few years	過去2，3年	413
❹ the past two decades	過去20年	230
❺ the past several years	過去数年	189

❶ **Lipid-lowering therapy with statins has revolutionized the management of this condition** over the past decade.（Curr Opin Cardiol. 2005 20:525）

訳 スタチンによる脂質降下療法は，過去10年の間にこの状態の管理に革命を起こしてきた

❷ **The most exciting developments** in the past year **have contributed to our understanding of the pathophysiology** behind this destructive process.（Curr Opin Hematol. 2005 12:390）

訳 過去１年において最も目覚ましい進展が，われわれの病態生理の理解に寄与してきた

■ ［前］＋ the past ＋［名］

❶ over the past ～	過去～の間に	2,101
❷ in the past ～	過去～において	1,828
❸ during the past ～	過去～の間に	775

❶ Over the past **year**, substantial new information has been published on the neurohormonal control of pancreatic exocrine function.（Curr Opin Gastroenterol. 2005 21:531）

訳 過去１年の間に，

❸ **This article reviews what is new in the field of gastrointestinal infections in the immunocompromised host** during the past **year**.（Curr Opin Gastroenterol. 2003 19:37）

訳 この記事は，過去１年の間に免疫不全宿主における消化器感染の分野において何が新しいかを概説する

★ pathogenesis 名 病因 　用例数 16,369

pathogenesis	病因	16,350
pathogeneses	病因	19

複数形のpathogenesesの割合が0.1%しかなく，原則，**不可算名詞**として使われる．
◆類義語：etiology

■ pathogenesis ＋［前］

pathogenesis ofの前にtheが付く割合は圧倒的に高い．

	the %	a/an %	ø %		
❶	96	0	4	pathogenesis of ～　～の病因	9,413
				・play … role in the pathogenesis of ～ ‥は，～の病因において…な役割を果たす	1,006
				・contribute to the pathogenesis of ～ …は，～の病因に寄与する	700
				・be implicated in the pathogenesis of ～ …は，～の病因に関連づけられる	636
				・be involved in the pathogenesis of ～ …は，～の病因に関与している	317
				・understanding of the pathogenesis of ～ ～の病因の理解	198
				・pathogenesis of Alzheimer's disease アルツハイマー病の病因	162
				・be important in the pathogenesis of ～ …は，～の病因において重要である	162
				・pathogenesis of atherosclerosis 粥状動脈硬化の病因	142
				・insights into the pathogenesis of ～ ～の病因への洞察	117
❷	3	1	68	pathogenesis in ～　～における病因	478

❶ Aberrations in p53-mediated terminal differentiation **may therefore play a role in the pathogenesis of nephron dysgenesis and dysfunction**. (J Clin Invest. 2002 109:1021)

訳 ～は，それゆえ，ネフロンの発育異常および機能障害の病因において役割を果たすかもしれない

❶ The increased level of cyclin E2 in breast tumors suggests that, similar to cyclin E1, **it may contribute to the pathogenesis of breast cancer**. (Oncogene. 2002 21:8529)

訳 それは，乳癌の病因に寄与するかもしれない

❶ Immune-mediated injury to the graft **has been implicated in the pathogenesis of chronic rejection**. (Transplantation. 2002 74:1053)

訳 ～は，慢性拒絶の病因に関連づけられてきた

P

⁑ pathway ㊔ 経路　　用例数 109,401

pathway	経路	68,302
pathways	経路	41,099

複数形のpathwaysの割合は約40%とかなり高い.

◆類義語：route

■ pathway ＋［前］

pathway ofの前にtheが付く割合は非常に高い．また，pathway to *do*の前にtheが付く割合もかなり高い.

	the %	a/an %	ø %			
❶	57	26	3	pathway in ～	～における経路	5,043
❷	71	26	0	pathway of ～	～の経路	2,619
❸	39	57	0	pathway for ～	～の経路	2,551
❹	60	28	1	pathway to *do*	～する経路	1,436

❶ Phosphatidylinositol 3-kinase [PI3K]/Akt signaling **is a critical pathway in cell survival.** (Proc Natl Acad Sci USA. 2005 102:10858)

訳 ～は，細胞生存における決定的に重要な経路である

❸ **The histidine kinase CheA is known to be an essential component in the signaling pathway for bacterial chemotaxis.** (Proc Natl Acad Sci USA. 2002 99:6169)

訳 ヒスチジンキナーゼCheAは，細菌の走化性のシグナル伝達経路における必須の構成要素であると知られている

❹ These findings suggest that **Gram-positive bacteria induce NO production using a PAFR signaling pathway to activate STAT1** via Jak2. (J Immunol. 2006 176:573)

訳 グラム陽性菌は，STAT1を活性化するPAFRシグナル伝達経路を使ってNO産生を誘導する

■ pathway ＋［前］＋ which

❶	pathway by which ～	（それによって）～である経路	355
❷	pathway in which ～	（そこで）～である経路	248

❶ **These findings define a pathway by which changes in endocytic trafficking can regulate tissue growth** in a non-cell-autonomous manner. (Dev Cell. 2005 9:699)

訳 これらの知見は，エンドサイトーシス輸送の変化が組織の増殖を調節しうる経路を定義する

■ pathway ＋［現分］

❶	pathway involving ～	～を含む経路	615
❷	pathway leading to ～	～につながる経路	346

❷ Cellular stress causes a transient increase in transcription of these heat shock operons through relief of HrcA-mediated repression, but **the pathway leading to derepression is unclear.** (J Bacteriol. 2005 187:7535)

訳 抑制解除につながる経路は不明である

■ [名／形] ＋ pathway

❶	signaling pathway	シグナル伝達経路	7,879
❷	signal transduction pathway	シグナル伝達経路	1,288
❸	～-dependent pathway	～依存性の経路	1,262
❹	secretory pathway	分泌経路	1,026
❺	biosynthetic pathway	生合成経路	1,009
❻	～-independent pathway	～非依存性の経路	668
❼	alternative pathway	代替経路	585
❽	apoptotic pathway	アポトーシス経路	567

❸ **These findings demonstrate a PPARγ-dependent pathway in regulation of** sodium transport in the CD that underlies TZD-induced fluid retention. (Proc Natl Acad Sci USA. 2005 102: 9406)

訳 これらの知見は，～の調節におけるPPARγ依存性の経路を実証する

P

★★ patient ㊔ 患者　　用例数 239,127

patients	患者	210,903
patient	患者	28,224

複数形のpatientsの割合が，90%と圧倒的に高い．

■ patients ＋ [前]

	the%	a/an%	ø%			
❶	0	-	84	patients with ～	～の患者	69,882
				・patients with ～ disease	～疾患の患者	4,331
				・patients with ～ cancer	～癌の患者	1,776
				・patients with chronic ～	慢性～の患者	1,627
				・patients with acute ～	急性～の患者	1,438
				・patients with advanced ～	進行した～の患者	1,216
				・patients with severe ～	重症～の患者	972
				・patients with schizophrenia	統合失調症の患者	805
❷	3	-	68	patients in ～	～における患者	5,173
				・patients in whom ～	～である患者	523
❸	0	-	80	patients without ～	～のない患者	1,835
				・compared with patients without ～	～のない患者と比べて	121
❹	2	-	74	patients at ～	～における患者	1,777
				・patients at risk	リスクのある患者	359

❶ **Patients with chronic hepatitis C virus (HCV) infection** frequently report fatigue, lassitude, depression, and a perceived inability to function effectively. (Hepatology. 2002 35:433)

訳 慢性のC型肝炎ウイルス（HCV）感染のある患者

❹ **Patients at risk for ARDS underwent bronchoalveolar lavage within 24 hrs of** being identified, then again 72 hrs later. (Crit Care Med. 2000 28:1)

訳 急性呼吸促迫症候群のリスクのある患者が，～の24時間以内に気管支肺胞洗浄を受けた

■ ［前］＋ patients with

❶	treatment of patients with ～	～の患者の治療	579
❷	management of patients with ～	～の患者の管理	398
❸	survival in patients with ～	～の患者の生存（率）	336
❹	mortality in patients with ～	～の患者の死亡率	336
❺	proportion of patients with ～	～の患者の比率	288
❻	cells from patients with ～	～の患者から細胞	284
❼	observed in patients with ～	～の患者において観察される	260
❽	therapy in patients with ～	～の患者の治療	252
❾	samples from patients with ～	～の患者から試料	242

❶ **Taxanes seem to be active agents in the treatment of patients with sex cord-stromal tumors** of the ovary. (J Clin Oncol. 2004 22:3517)

訳 タキサンは，性索間質性腫瘍の患者の治療における活性のある薬剤であるように思われる

❼ **Responses have been observed in patients with T-cell lymphomas**, despite prior treatment with multiple chemotherapeutic agents. (Blood. 2004 103:4636)

訳 応答が，T細胞リンパ腫の患者において観察されてきた

■ patients ＋［動］

❶	patients received ～	患者が，～を受けた	1,934
❷	patients underwent ～	患者が，～を受けた	1,326
❸	patients showed ～	患者が，～を示した	645
❹	patients were treated	患者が，治療された	627
❺	patients were enrolled	患者が，登録された	531
❻	patients were randomized	患者が，ランダム化された	486
❼	patients died	患者が，死んだ	451

❶ **Patients received T-1249 monotherapy by subcutaneous injection**, for 14 days, at doses ranging from 6.25 to 192 mg/day. (J Infect Dis. 2004 189:1075)

訳 患者が，皮下注射によるT-1249単剤治療を受けた

❷ Overall, **86 patients underwent surgery**, with one perioperative death. (Ann Surg. 2005 242:413)

訳 86名の患者が手術を受けた

■ patients ＋［現分／過分］

❶	patients undergoing ～	～を受けた患者	2,533
❷	patients receiving ～	～を受けた患者	2,465
❸	patients treated with ～	～で治療された患者	1,862

❶ **Changes were evident in patients receiving 16 or 24 mg/day of galantamine.** (Am J Psychiatry. 2004 161:532)

訳 変化が，16あるいは24 mg/日のガランタミンを受けた患者において明らかであった

❷ Closure of the abdomen in patients undergoing intestinal transplantation can be extremely difficult, if not impossible. (Lancet. 2003 361:2173)
訳 腸の移植を受ける患者の腹部の縫合

❸ Improvement in P50 gating appears to be greatest in patients treated with clozapine. (Am J Psychiatry. 2004 161:1822)
訳 ～は，クロザピンで治療された患者において最も大きいように思われる

■ ［形／名／過分］＋ patients

❶	all patients	すべての患者	4,499
❷	cancer patients	癌患者	2,298
❸	～-treated patients	～で治療された患者	1,562
❹	infected patients	感染した患者	1,359
❺	consecutive patients	治療継続患者	1,331
❻	most patients	ほとんどの患者	1,293
❼	diabetic patients	糖尿病患者	1,072
❽	～-positive patients	～陽性患者	749

P

★★ pattern 名 パターン，動 パターンを形成する　用例数 49,423

patterns	パターン／パターンを形成する	24,416
pattern	パターン／パターンを形成する	24,028
patterned	パターンを形成した	979

動詞としても用いられるが，名詞の用例が圧倒的に多い．複数形のpatternsの割合は約50%と非常に高い．

◈類義語：fashion, manner, mode, form, type

■ pattern ＋［前］

pattern ofの前にtheが付く割合はかなり高い．

	the%	a/an%	ø%			
❶	62	30	0	pattern of ～	～のパターン	9,991
				・pattern of expression	発現のパターン	579
				・pattern of gene expression	遺伝子発現のパターン	188
❷	8	-	79	patterns in ～	～におけるパターン	1,880

❶ AID mRNA is expressed in GC B cells and GC-derived lymphomas, but the pattern of expression of the AID protein is not known. (Blood. 2004 104:3318)
訳 AIDタンパク質の発現のパターンは，知られていない

■ ［名／形］＋ pattern

❶	expression pattern	発現パターン	2,050
❷	similar pattern	類似のパターン	459
❸	temporal pattern	時間パターン	274

❹ specific pattern	特異的なパターン	255
❺ spatial pattern	空間的パターン	228
❻ distribution pattern	分布パターン	199
❼ complex pattern	複雑なパターン	191
❽ the same pattern	同じパターン	188

❶ The **expression pattern** of the Pdlim2 gene was studied by Northern blot analysis and in situ hybridization. (Invest Ophthalmol Vis Sci. 2004 45:3955)
訳 Pdlim2遺伝子の発現パターンが，ノーザンブロット分析によって研究された

■ pattern ＋［名］

❶ pattern recognition	パターン認識	617
❷ pattern formation	パターン形成	458

❷ We have employed a pharmacological approach to explore the molecular mechanisms of **pattern formation**. (Dev Biol. 2005 287:416)
訳 われわれは，パターン形成の分子機構を探索するために薬理学的なアプローチを使用した

★ peak 名 ピーク，動 ピークになる　用例数 14,718

peak	ピーク／ピークになる	10,487
peaks	ピーク／ピークになる	2,691
peaked	ピークになった	1,176
peaking	ピークになる	364

名詞の用例が多いが，自動詞としても用いられる．名詞の形容詞的用法も多い．
◆類義語：maximum

■ 名 peak ＋［前］

peak ofの前にtheが付く割合は非常に高い．

	the%	a/an%	ø%			
❶	74	22	3	peak of ～	～のピーク	786
				・at the peak of ～	～のピークにおいて	157
❷	17	72	5	peak at ～	～におけるピーク	413

❶ On day 8, **at the peak of** the splenic response in vehicle-treated mice, virus-specific IgM and IgG antibody-secreting cells (ASC) were decreased 22- and 457-fold in MnTBAP-treated animals. (J Virol. 2013 87:2577)
訳 脾臓応答のピークにおいて

❷ The action spectrum exhibits a major **peak at** approximately 280 nm and a minor peak at approximately 360 nm. (Plant Physiol. 2000 122:99)
訳 作用スペクトルは，およそ280 nmに主要なピークを示す

■ 動 peaked ＋［前］

❶ peaked at ～	～においてピークになった	530

❶ Expression of p24 in the muscle cells peaked at day 7 and became undetectable after day 12. (J Virol. 2005 79:14688)

訳 筋肉細胞におけるp24の発現は，７日目でピークになった

★★ perform 動 行う／実行する 用例数 37,211

performed	行われる／～を行った／～を実行した	30,187
perform	～を行う／～を実行する	4,035
performing	～を行う／～を実行する	2,060
performs	～を行う／～を実行する	929

他動詞.

◈類義語：conduct, carry out, do, undertake

■ be performed ＋［前］/using

★ be performed	…が，行われる	17,852
❶ be performed in ～	…が，～において行われる	2,971
・be performed in ～ patients	…が，～の患者におけて行われる	450
❷ be performed to *do*	…が，～するために行われる	2,634
・be performed to determine ～	…が，～を決定するために行われる	556
・be performed to identify ～	…が，～を同定するために行われる	267
・be performed to assess ～	…が，～を評価するために行われる	238
・be performed to evaluate ～	…が，～を評価するために行われる	198
❸ be performed on ～	…が，～において行われる	2,167
❹ be performed with ～	…が，～を使って行われる	1,317
❺ be performed using ～	…が，～を使って行われる	1,286
❻ be performed by ～	…が，～によって行われる	918
❼ be performed at ～	…が，～において行われる	758
❽ be performed for ～	…が，～に対して行われる	604

❶ To evaluate the dose response, a second arm of the study was performed in the left eyes. (Invest Ophthalmol Vis Sci. 2004 45:635)

訳 ～が，左目において行われた

❷ Multiple regression analyses were performed to determine if each one of these measurements was affected by age and gender. (J Periodontol. 2004 75:1061)

訳 …が，～かどうかを決定するために行われた

❸ In vivo imaging was performed on mice with s.c. MDA-MB-468 and MDA-MB-435 tumors. (Cancer Res. 2003 63:7870)

訳 …が，～のマウスにおいて行われた

❹ Immunohistochemical analysis of neuronal, epithelial, and immune cells in the eye was performed with antibodies against established cell markers. (Invest Ophthalmol Vis Sci. 2003 44:2367)

訳 …が，～に対する抗体を使って行われた

P

■ [名] + be performed

❶ analysis was performed	分析が，行われた	1,165
❷ studies were performed	研究が，行われた	833
❸ experiments were performed	実験が，行われた	640
❹ assays were performed	アッセイが，行われた	308
❺ measurements were performed	測定が，行われた	168
❻ tests were performed	テストが，行われた	123
❼ procedures were performed	手続きが，行われた	120
❽ transplants were performed	移植が，行われた	117

❶ Northern blot analysis was performed to examine the level of insulin-like growth factor-I (IGF-I) gene expression. (Brain Res. 2004 1006:198)

訳 ノーザンブロット分析が，～を調べるために行われた

■ [代名／名] + perform

❶ we performed ～	われわれは，～を行った	4,482
❷ subjects performed ～	被験者は，～を行った	179
❸ the authors performed ～	著者らは，～を行った	105

❶ We performed continuous renal replacement therapy on patients. (Crit Care Med. 2004 32: 932)

訳 われわれは，患者に対して持続的腎代替療法を行った

■ perform + [名句]

❶ perform ～ analysis	…は，～分析を行う	1,022
❷ perform ～ study	…は，～研究を行う	546

★ performance 名 性能／能力／成績　用例数 15,209

performance	性能／能力／成績	14,994
performances	性能／能力／成績	215

複数形のperformancesの割合は１％しかなく，原則，不可算名詞として使われる．

◆類義語：ability, capability

■ performance + [前]

performance ofの前にtheが付く割合は圧倒的に高い．

	the%	a/an%	ø%			
❶	82	0	17	performance of ～	～の性能	2,919
				・evaluate the performance of ～	…は，～の性能を評価する	148
				・compare the performance of ～	…は，～の性能を比較する	130

				・improve the performance of ～	…は，～の性能を改善する	94
				・assess the performance of ～	…は，～の性能を評価する	79
❷	3	1	87	performance in ～	～における性能	1,007
❸	2	0	91	performance on ～	～における性能	710

❶ We evaluated the performance of all models using the area under the receiver operator characteristic curve.（Crit Care Med. 2012 40:2295）
訳 われわれは，すべてのモデルの性能を評価した

❷ Impaired cognitive performance in FM patients correlated with pain complaints, but not with depressive or anxiety symptoms.（Arthritis Rheum. 2001 44:2125）
訳 FMの患者の損なわれた認識能は，～と相関した

❸ Results were compared with performance on conventional tests of memory.（Brain. 2000 123:472）
訳 結果が，従来のテストにおける成績と比較された

P

■［形／名］＋ performance

❶ high performance	高性能	1,562
❷ cognitive performance	認知能力	319
❸ task performance	作業能力	260
❹ memory performance	記憶性能	239

perhaps 副 おそらく／多分 用例数 3,875

◆類義語：probably, presumably, likely, maybe, possibly

■ perhaps ＋［前］

❶ , perhaps by ～	おそらく～によって	264
❷ , perhaps in ～	おそらく～において	100

❶ These spectral changes are suggestive of a stabilization of the phenolate anion, perhaps by hydrogen bonding, rather than an increase in the microenvironment dielectric constant or dye immobilization.（Biochemistry. 2005 44:15937）
訳 おそらく水素結合によって

■ and perhaps

★ and perhaps ～	そして，おそらく～	1,228
❶ and perhaps other ～	そして，おそらく他の～	257

❶ Recent studies have revealed that elevated phospholipase D activity generates survival signals in breast and perhaps other human cancers.（J Biol Chem. 2005 280:35829）
訳 乳癌そしておそらく他のヒトの癌において

★ period ㊔ 期間 用例数 22,653

period	期間	18,223
periods	期間	4,430

複数形のperiodsの割合が，約20%ある可算名詞．
◆類義語：term, duration

■ period ＋［前］

	the %	a/an %	ø %			
❶	33	66	1	period of ～	～の期間	3,924
				・over a period of ～	～の期間に	472
				・during the period of ～	～の期間に	315
				・short period of ～	短い期間の～	110
❷	37	58	4	period in ～	～における期間	535

❶ Subsequently, the patient was placed on a strict 3-month maintenance protocol and **was evaluated over a period of 1 year**. (J Periodontol. 2005 76:1996)
訳 ～が，1 年の期間に評価された

❶ Thus, hundreds of unzipping events **can be tested in a short period of time** (few minutes), independently of the unzipping voltage amplitude. (Biophys J. 2004 87:3205)
訳 ～は，短い時間の間にテストされうる

■［名／形］＋ period

❶	～-year period	～年の期間	982
❷	study period	研究期間	900
	・during the study period	研究期間中に	379
❸	follow-up period	フォローアップ期間	732
❹	time period	時間期間	640
❺	～-month period	～ヶ月の期間	502
❻	critical period	危機的な時期	480

❶ Both the percentage of E. faecium among the enterococci and the proportion of vancomycin-resistant E. faecium **increased significantly over this 10-year period**. (J Clin Microbiol. 2005 43:462)
訳 ～が，この10年の期間に有意に増大した

❷ **During the study period**, 3,898 patients underwent 4416 surgical procedures. (Ann Surg. 2006 243:96)
訳 研究期間中に，

permit ㊀ 可能にする／認可する 用例数 6,385

permit	～を可能にする	2,278
permits	～を可能にする	2,075
permitted	可能にされる／～を可能にした	1,131

permitting	～を可能にする	901

S+V+O（第3文型）およびS+V+O+to *do*（第5文型）の他動詞として使われる.
◆類義語：allow, enable, prompt, lead

■ permit ＋［名句］

❶ permit identification of ～	…は，～の同定を可能にする	70
❷ permit the identification of ～	…は，～の同定を可能にする	62

❶ The strategy has permitted identification of several prospective essential genes.（Genetics. 2002 162:1573）
訳 その戦略は，いくつかの予想される必須遺伝子の同定を可能にしてきた

■ permit ＋ us ＋ to *do*

❶ permit us to *do*	…は，われわれが～することを可能にする	122

❶ The strong correlation permitted us to determine extensive consensus sequences at the donor and acceptor sites of longer introns.（Genome Res. 2004 14:67）
訳 その強い相関は，われわれが詳細なコンセンサス配列を決定することを可能にした

persist 動 持続する　用例数 6,321

persisted	持続した	2,540
persist	持続する	2,018
persists	持続する	1,346
persisting	持続する	417

自動詞.
◆類義語：last, sustain, continue

■ persist ＋［前］

❶ persist in ～	…は，～において持続する	1,538
❷ persist for ～	…は，～の間持続する	899
❸ persist after ～	…は，～のあと持続する	392
❹ persist at ～	…は，～において持続する	256
❺ persist throughout ～	…は，～の間中持続する	189

❶ Current enhancement persisted in the presence of LY294002, suggesting that c-Src is downstream of PI3K.（Circ Res. 2004 94:626）
訳 ～は，LY294002の存在下において持続した

❷ The loss of I_h and the increased neuronal excitability persisted for 1 week following seizures.（Neuron. 2004 44:495）
訳 ～は，発作のあと1週間持続した

P

■ [名] + persist

❶	expression persisted	発現は，持続した	44

❶ Gene expression persisted in muscle and heart, but diminished in tissues undergoing rapid cell division, such as neonatal liver. (Nat Biotechnol. 2005 23:321)
訳 遺伝子発現は，筋肉および心臓において持続した

perspective 名 展望／観点／見込み　用例数 2,298

perspective	展望／観点	1,813
perspectives	展望／観点	485

複数形のperspectivesの割合が，約20%ある可算名詞.
◆類義語：vision, prediction, expectation, promise, estimate, prospect, likelihood

■ perspective + [前]

perspective ofの前にtheが付く割合はかなり高い.

	the%	a/an%	ø%			
❶	1	79	10	perspective on ～	～に関する展望	314
❷	68	26	1	perspective of ～	～の観点	237

❷ From the perspective of a radiologist, all structures on the presented images should be assessed. (J Am Coll Cardiol. 2010 55:1566)
訳 放射線科医の観点から，示された画像に関するすべての構造が評価されるべきである

■ [動] + a + [形] + perspective

❶	provide a ～ perspective	…は，～展望を提供する	136

❶ These results provide a new perspective on the pathogenesis of cardiomyopathies and open new avenues for treatment of the disease. (Nat Commun. 2012 3:1238)
訳 これらの結果は，心筋症の病因に関する新しい展望を提供する

★★ phase 名 相／期　用例数 45,484

phase	相／期	40,544
phases	相／期	4,940

複数形のphasesの割合が，約10%ある可算名詞.

■ phase + [前]

phase ofの前にtheが付く割合は圧倒的に高い.

	the%	a/an%	ø%			
❶	86	11	0	phase of ～	～の相／～の期	5,168

❶ After SIV infection, unlike Th17 cells, Tc17 cells were not depleted during the acute phase of infection. (J Immunol. 2011 186:745)
訳 Tc17細胞は，感染の急性期の間に枯渇しなかった

■ ［形／名］＋ phase

❶	stationary phase	定常期	1,536
❷	solid phase	固相	1,266
❸	gas phase	気相	1,144
❹	acute phase	急性期	1,079
❺	early phase	初期相	621
❻	late phase	遅延相	500
❼	growth phase	増殖相	472
❽	chronic phase	慢性期	460
❾	second phase	第二相	401

P

phenomenon 名 現象 用例数 6,517

phenomenon	現象	4,527
phenomena	現象	1,990

複数形のphenomenaの割合が，約30%ある可算名詞.
◆類義語：event, case, incident, circumstance

■ phenomenon ＋［前］

phenomenon ofの前にtheが付く割合は圧倒的に高い.

	the %	a/an %	ø %			
❶	86	11	0	phenomenon of ～	～の現象	441
❷	7	70	0	phenomenon in ～	～における現象	313

❸ The phenomenon of multidrug resistance (MDR) has decreased the hope for successful cancer chemotherapy. (Cancer Res. 2011 71:3735)
訳 多剤耐性(MDR)の現象は，～に対する希望を低下させてきた

★★ phenotype 名 表現型 用例数 41,441

phenotype	表現型	29,087
phenotypes	表現型	12,354

複数形のphenotypesの割合が，約30%ある可算名詞.

■ phenotype ＋［前］

phenotype ofの前にtheが付く割合は圧倒的に高い.

	the %	a/an %	ø %			
❶	93	6	0	phenotype of ～	～の表現型	3,788
❷	43	37	14	phenotype in ～	～における表現型	1,985

❶ The **phenotype of** a ylyA mutant was ascribed to a defect in spore germination efficiency.（J Bacteriol. 2013 195:253）

訳 ylyA変異体の表現形は，～の欠損のせいであった

■［形／名］＋ phenotype

❶	mutant phenotype	変異表現型	660
❷	disease phenotype	疾患表現型	329
❸	clinical phenotype	臨床像	325

phenotypic 形 表現型の　用例数 6,816

■ phenotypic ＋［名］

❶	phenotypic changes	表現型の変化	388
❷	phenotypic variation	表現型のバリエーション	350
❸	phenotypic analysis	表現型の分析	263
❹	phenotypic differences	表現型の違い	240
❺	phenotypic effects	表現型の影響	203
❻	phenotypic characteristics	表現型の特徴	182

❶ BT549 clones expressing BAF57 demonstrated marked **phenotypic changes**, slow growth kinetics, and restoration of contact inhibition.（Mol Cell Biol. 2005 25:7953）

訳 ～は，顕著な表現型の変化を実証した

★ phosphorylate 動 リン酸化する　用例数 17,905

phosphorylated	リン酸化される／～をリン酸化した	13,424
phosphorylates	～をリン酸化する	2,032
phosphorylate	～をリン酸化する	1,774
phosphorylating	～をリン酸化する	675

他動詞．受動態の用例が多い．

■ be phosphorylated ＋［前］

★	be phosphorylated	…は，リン酸化される	3,271
❶	be phosphorylated by ～	…は，～によってリン酸化される	917

❷ be phosphorylated in ～	…は，～においてリン酸化される	671
・be phosphorylated in response to ～	…は，～に応答してリン酸化される	105
❸ be phosphorylated on ～	…は，～においてリン酸化される	353
・be phosphorylated on tyrosine	…は，チロシンにおいてリン酸化される	74
❹ be phosphorylated at ～	…は，～においてリン酸化される	294

❶ DOKL is phosphorylated by the Abl tyrosine kinase in vivo.（Mol Cell Biol. 1999 19:8314）
訳 DOKLは，Ablチロシンキナーゼによってリン酸化される

❷ Here we report that Nbs is specifically phosphorylated in response to γ-radiation, ultraviolet light and exposure to hydroxyurea.（Nature. 2000 405:477）
訳 Nbsは，γ線照射に応答して特異的にリン酸化される

❸ Several intracellular signaling proteins, such as CBL and VAV, were phosphorylated on tyrosine in response to CD80, CD86, and anti-CD28 mAb.（J Biol Chem. 1999 274:3116）
訳 …が，～に応答してチロシンにおいてリン酸化された

P

■［名／動／副］＋ phosphorylated

❶ tyrosine-phosphorylated	チロシンリン酸化される	1,300
❷ become phosphorylated	…は，リン酸化された状態になる	212
❸ constitutively phosphorylated	構成的にリン酸化される	157
❹ highly phosphorylated	高度にリン酸化される	143
❺ rapidly phosphorylated	急速にリン酸化される	119

❶ Both proteins are tyrosine-phosphorylated in response to insulin and IGF-1 in transfected cells, although the kinetics differ.（J Biol Chem. 2003 278:25323）
訳 両方のタンパク質は，インスリンおよびIGF-1に応答してチロシンリン酸化される

■ phosphorylated ＋［名］

❶ phosphorylated form	リン酸化型	326
❷ phosphorylated proteins	リン酸化タンパク質	208

❶ The phosphorylated form of Chk1 possessed higher intrinsic protein kinase activity and eluted more quickly on gel filtration columns.（Mol Cell Biol. 2001 21:4129）
訳 リン酸化型のChk1は，より高い内在性のタンパク質リン酸化活性を持っていた

■［副］＋ phosphorylate

❶ directly phosphorylate	…は，直接～をリン酸化する	241

★★ phosphorylation 名 リン酸化　用例数 55,489

phosphorylation	リン酸化	55,229
phosphorylations	リン酸化	260

複数形のphosphorylationsの割合は0.5％しかなく，原則，不可算名詞として使われる．

■ phosphorylation ＋［前］

	the％	a/an％	ø％		
❶	15	0	80	phosphorylation of ～　～のリン酸化	19,971
				・tyrosine phosphorylation of ～　～のチロシンリン酸化	2,713
				・…-induced phosphorylation of ～　…誘導性の～のリン酸化	898
				・…-dependent phosphorylation of ～　…依存性の～のリン酸化	610
				・…-mediated phosphorylation of ～　…に仲介される～のリン酸化	583
				・increased phosphorylation of ～　増大した～のリン酸化	497
				・phosphorylation of serine　セリンのリン酸化	227
				・lead to phosphorylation of ～　…は，～のリン酸化につながる	227
❷	2	0	51	phosphorylation in ～　～におけるリン酸化	2,329
				・phosphorylation in response to ～　～に応答したリン酸化	126
❸	2	0	75	phosphorylation at ～　～におけるリン酸化	1,548
				・phosphorylation at serine　セリンにおけるリン酸化	137
❹	0	0	62	phosphorylation by ～　～によるリン酸化	1,546

❶ We also observed that **VEGF stimulation resulted in increased tyrosine phosphorylation of** the focal adhesion components paxillin and p130cas. (J Immunol. 2000 164:1169)

訳 VEGF刺激は，増大した～のチロシンリン酸化という結果になった

❶ In vitro peptide binding experiments provide evidence that **phosphorylation of pS1 (Ser722) may play a role in modulating FAK binding to the SH3 domain** of the adapter protein p130^{Cas}. (Mol Biol Cell. 2001 12:1)

訳 pS1（Ser722）のリン酸化は，SH3ドメインへのFAKの結合を調節する際に役割を果たすかもしれない

❷ **Many protein kinases are regulated by phosphorylation in the activation loop**, which is required for enzymatic activity. (Mol Cell Biol. 1998 18:2923)

訳 多くのプロテインキナーゼは，活性化ループにおけるリン酸化によって調節される

❸ Phosphorylation of Akt, **a mediator of shear stress-induced eNOS phosphorylation at Ser-1177**, was decreased in the diabetic penis at baseline, but it was restored by ES. (Proc Natl Acad Sci USA. 2005 102:11870)

訳 ストレス誘発性のSer-1177におけるeNOSのリン酸化の仲介因子

❹ **Induction of eIF-2α phosphorylation by Gcn2p** during glucose limitation requires the function of the HisRS-related domain but is largely independent of the ribosome binding sequences of Gcn2p. (Mol Cell Biol. 2000 20:2706)

訳 Gcn2pによるeIF-2αのリン酸化の誘導

■ [前] + phosphorylation

❶ regulated by phosphorylation	リン酸化によって調節される	242

❶ We find that LSF binds the LTR as a tetramer and that **binding is regulated by phosphorylation mediated by** mitogen-activated protein kinases (MAPKs). (J Virol. 2005 79:5952)
訳 結合は，～に仲介されるリン酸化によって調節される

■ phosphorylation + [名／形]

❶ phosphorylation site	リン酸化部位	1,774
❷ phosphorylation-dependent ～	リン酸化依存性の～	552
❸ phosphorylation state	リン酸化状態	551
❹ phosphorylation status	リン酸化状態	343

P

★ physiological 形 生理的な／生理学的な　用例数 16,984

◆類義語：physiologic

■ physiological + [名]

❶ physiological conditions	生理的条件	1,126
・under physiological conditions	生理的条件下で	728
❷ physiological role	生理的役割	946
❸ physiological function	生理的機能	672
❹ physiological processes	生理的過程	621
❺ physiological concentrations	生理的濃度	405
❻ physiological significance	生理的重要性	320
❼ physiological responses	生理応答	320
❽ physiological relevance	生理関連性	268
❾ physiological levels of ～	生理的レベルの～	259

❶ These results do not support the idea that **mitochondria produce considerable amounts of reactive oxygen species under physiological conditions.** (J Biol Chem. 2002 277:44784)
訳 ミトコンドリアは，生理的条件下でかなりの量の活性酸素種を産生する

❷ BTB-kelch proteins can elicit diverse biological functions but **very little is known about the physiological role of these proteins in vivo.** (Mol Cell Biol. 2005 25:8531)
訳 生体内でのこれらのタンパク質の生理的役割については非常にわずかしか知られていない

physiologically 副 生理学的に／生理的に　用例数 2,531

■ physiologically + [形]

❶ physiologically relevant	生理学的に関連する	1,228

・be physiologically relevant	…は，生理学的に関連している	134
・physiologically relevant concentrations	生理学的に関連する濃度	104
❷ physiologically important	生理学的に重要な	196

❶ However, the role of MTA1 in tumorigenesis in a physiologically relevant animal system remains unknown. (Development. 2004 131:3469)
訳 生理学的に関連した動物システムの腫瘍形成におけるMTA1の役割は，未知のままである

pivotal 形 中心的な／きわめて重要な 用例数 2,696

◆類義語：central, important, key

■ pivotal ＋［名］

❶ pivotal role	中心的役割	1,429
・play a pivotal role in ～	…は，～において中心的役割を果たす	957

❶ Retinoic acid (RA) is a signaling molecule that plays a pivotal role in major cellular processes and vertebrate development. (Oncogene. 2006 25:1400)
訳 レチノイン酸（RA）は，主要な細胞過程において中心的役割を果たすシグナル伝達分子である

■ be pivotal ＋［前］

★ be pivotal	…は，きわめて重要である	314
❶ be pivotal in ～	…は，～においてきわめて重要である	123
❷ be pivotal for ～	…は，～にとってきわめて重要である	83

❶ T lymphocytes bearing $\alpha\beta$ T cell receptors are pivotal in the immune response of most vertebrates. (Nat Immunol. 2005 6:239)
訳 ～は，ほとんどの脊椎動物の免疫応答においてきわめて重要である

★ place 動 置く，名 場所 用例数 11,903

place	～を置く／場所	5,777
placed	置かれる／～を置いた	4,500
places	～を置く／場所	860
placing	～を置く	766

他動詞および名詞として用いられる．名詞としては，take placeとin place ofの用例が多い．

◆類義語：site, arrange

■ 動 be placed ＋［前］

★ be placed	…が，置かれる	2,149
❶ be placed in ～	…が，～に置かれる	672
❷ be placed on ～	…が，～に置かれる	540
❸ be placed into ～	…が，～へ置かれる	175

❶ Implants were placed in the area approximately 5 months following the regenerative procedure. (J Periodontol. 2004 75:322)
訳 インプラントが，その領域に置かれた

■ 名 in place of

❶	in place of ～	～の代わりに	750

❶ Anagrelide was used in place of hydroxyurea in two patients because of cytopenias caused by the latter agent. (Transplantation. 2002 74:1090)
訳 アナグレリドが，ヒドロキシウレアの代わりに使われた

■ 名 take place

★	take place	…が，起こる	2,244
❶	take place in ～	…が，～において起こる	596
❷	take place at ～	…が，～において起こる	181
❸	take place during ～	…が，～の間に起こる	130

❶ Most cholesterol turnover takes place in the liver and involves the conversion of cholesterol into soluble and readily excreted bile acids. (J Biol Chem. 2003 278:22980)
訳 ほとんどのコレステロールの代謝回転が，肝臓で起こる

plasticity 名 可塑性 用例数 8,264

plasticity	可塑性	8,251
plasticities	可塑性	13

複数形のplasticitiesの用例はほとんどなく，原則，不可算名詞として使われる．

■ plasticity ＋［前］

	the %	a/an %	ø %			
❶	1	0	95	plasticity in ～	～における可塑性	1,201
❷	49	1	48	plasticity of ～	～の可塑性	850

❶ Moreover, because the cAMP-PKA-CREB pathway is required for structural synaptic plasticity in learning and memory, DSK/CCKLR signaling may also contribute to these mechanisms. (J Cell Biol. 2012 196:529)
訳 cAMP-PKA-CREB 経路は，学習と記憶における構造的シナプス可塑性のために必要とされる

■［形／名］＋ plasticity

❶	synaptic plasticity	シナプス可塑性	2,366
❷	～-dependent plasticity	～依存性可塑性	316
❸	～-term plasticity	～期可塑性	222
❹	neuronal plasticity	神経可塑性	220

P

platform ㊔ プラットフォーム／基盤 用例数 4,200

platform	プラットフォーム／基盤	3,215
platforms	プラットフォーム／基盤	985

複数形のplatformsの割合が，約25%ある可算名詞.

◆類義語：basis, foundation

■ platform ＋［前］

	the%	a/an%	ø%			
❶	6	91	1	platform for ～	～のためのプラットフォーム	984
				・platform for ～ing	～するためのプラットフォーム	225
				・provide a platform for ～	…は，～のためのプラットフォームを提供する	151
❷	16	77	0	platform to *do*	～するためのプラットフォーム	251

❶ This mouse model **provides a platform for** more accurately dissecting the early events in APL pathogenesis.（J Clin Invest. 2011 121:1636）

訳 このマウスモデルは，～をより正確に分析するためのプラットフォームを提供する

★★ play ㊓ 果たす／働く 用例数 55,112

play	～を果たす	29,167
plays	～を果たす	23,156
played	～を果たした／果たされる	1,885
playing	～を果たす	904

他動詞．play ～ role(s) inの用例が圧倒的に多い.

◆類義語：achieve, attain

■ play a role in ＋［名句］

★ play a role in ～	…は，～において役割を果たす	9,875
❶ play a role in regulating ～	…は，～を調節する際に役割を果たす	380
❷ play a role in the pathogenesis of ～	…は，～の病因において役割を果たす	282
❸ play a role in the regulation of ～	…は，～の調節において役割を果たす	228
❹ play a role in the development of ～	…は，～の発生において役割を果たす	202

❷ This represents a potential mechanism by which IL-4/IL-13 could **play a role in the pathogenesis of** lung fibrosis.（J Immunol. 2004 173:3425）

訳 IL-4/IL-13は，肺の線維症の病因において役割を果たしうる

■ play ＋ a/an ＋［形］＋ role in

❶ play an important role in ～	…は，～において重要な役割を果たす	9,053
❷ play a critical role in ～	…は，～において決定的に重要な役割を果たす	4,276
❸ play a key role in ～	…は，～において鍵となる役割を果たす	2,792
❹ play a central role in ～	…は，～において中心的な役割を果たす	1,947
❺ play a major role in ～	…は，～において主要な役割を果たす	1,639
❻ play an essential role in ～	…は，～において必須な役割を果たす	1,594
❼ play a crucial role in ～	…は，～において決定的な役割を果たす	1,243
❽ play a significant role in ～	…は，～において重要な役割を果たす	1,061
❾ play a pivotal role in ～	…は，～において中心的な役割を果たす	957

❶ Histone deacetylases have recently been shown to play an important role in regulating gene expression.（J Biol Chem. 2001 276:45168）

訳 ヒストン脱アセチル化酵素は，遺伝子発現を調節する際に重要な役割を果たすことが最近示されている

❻ By contrast, G protein signaling plays an essential role in chemotaxis.（J Cell Biol. 2002 157:921）

訳 Gタンパク質シグナル伝達は，走化性において必須の役割を果たす

■ play in

★ play in ～	…が～において果たす‥	1,572
❶ role that … play in ～	…が～において果たす役割	361

❶ In colon cancer, the frequently mutated K-ras oncogene also can regulate VEGF expression, but the role that K-ras may play in hypoxia is unknown.（Cancer Res. 2004 64:1765）

訳 K-rasが低酸素において果たすかもしれない役割は知られていない

■［動／形］＋ to play

★ to play ～	～を果たす…	5,596
❶ appear to play ～	…は，～を果たすように思われる	880
❷ be thought to play ～	…は，～を果たすと考えられる	713
❸ be shown to play ～	…は，～を果たすことが示される	678
❹ be known to play ～	…は，～を果たすと知られている	378
❺ be likely to play ～	…は，おそらく～を果たす	353
❻ be believed to play ～	…は，～を果たすと信じられている	304
❼ be proposed to play ～	…は，～を果たすと提唱される	259

❶ FLT3 is a type III receptor tyrosine kinase that is thought to play a key role in hematopoiesis.（Mol Cell. 2004 13:169）

訳 FLT3は，造血において鍵となる役割を果たすと考えられるⅢ型受容体チロシンキナーゼである

■ [名] + play

❶	cells play ～	細胞は，～を果たす	622
❷	pathway plays ～	経路は，～を果たす	290
❸	proteins play ～	タンパク質は，～を果たす	246

❶ Ly49H^{+} NK cells play a critical role in innate antiviral immune responses to murine CMV (MCMV).（J Immunol. 2004 173:6312）

訳 Ly49H^{+} NK細胞は，～に対する抗ウイルス性自然免疫応答において決定的に重要な役割を果たす

■ played by

★	played by ～	～によって果たされる…	751
❶	role played by ～	～によって果たされる役割	483

❶ The role played by organic chemistry in the pharmaceutical industry continues to be one of the main drivers in the drug discovery process.（Science. 2004 303:1810）

訳 医薬品産業において有機化学によって果たされる役割は，～であり続ける

★point 名 点／ポイント，動 示す／指示する　用例数 27,938

point	点／ポイント／示す	18,379
points	点／ポイント／示す	8,497
pointed	示される／示した	532
pointing	点／ポイント／示す	530

名詞の用例が多いが，自動詞としても用いられる．複数形のpointsの割合が，約35%ある可算名詞．動詞としては，point toの用例が多い．

◆類義語：spot, dot, indicate, show

■ 名 point + [前]

point ofの前にtheが付く割合はかなり高い．

	the%	a/an%	ø%			
❶	61	29	1	point of ～	～の点	1,830
❷	19	76	1	point for ～	～のための点	726
❸	13	66	2	point in ～	～における点	661

❶ Large biochemical networks pose a unique challenge from the point of view of evaluating conservation laws.（Bioinformatics. 2006 22:346）

訳 大きな生化学的なネットワークは，保存則を評価する観点からのユニークな難題を提起する

❷ Pathguide is useful as a starting point for biological pathway analysis and for content aggregation in integrated biological information systems.（Nucleic Acids Res. 2006 34:D504）

訳 ～は，生物学的経路分析のための出発点として有用である

■ 名 [名／形] ＋ point(s)

❶	time points	タイムポイント	2,255
❷	end point	終点／エンドポイント	2,493
❸	single point	一点	534
❹	starting point	出発点	437
❺	set point	セットポイント	310

❶ The most highly induced gene at all time points was mclca3 (gob5), a putative calcium-activated chloride channel involved in the regulation of mucus production and/or secretion. (Am J Pathol. 2005 167:1243)

訳 すべてのタイムポイントにおいて最も高度に誘導された遺伝子は，mclca3 (gob5) であった

P

■ 名 point ＋ [名]

❶	point mutations	点突然変異	2,488
❷	point mutants	点突然変異体	423

■ 動 point to

★	point to ～	…は，～を示す	3,059
❶	results point to ～	結果は，～を示す	286
❷	findings point to ～	知見は，～を示す	191
❸	data point to ～	データは，～を示す	181
❹	evidence points to ～	証拠は，～を示す	135

❸ These data point to a novel role for AMPK in modulating endothelial cell NO bioactivity and HSP90 function. (Circulation. 2005 111:3473)

訳 これらのデータは，～の際のAMPKの新規の役割を示す

pool 名 プール／貯蔵所 用例数 7,720

pool	プール	5,482
pools	プール	2,238

複数形のpoolsの割合が，約30%ある可算名詞.

■ pool ＋ [前]

	the%	a/an%	ø%			
❶	41	53	0	pool of ～	～のプール	1,201

❶ Ribozymes could be selected from a pool of random sequences depending on the length of their stems. (Proc Natl Acad Sci USA. 2012 109:17972)

訳 リボザイムは，ランダムな配列のプールから選択されうる

poor 形 よくない／乏しい　用例数 9,250

◆類義語：bad

■ poor ＋［名］

❶	poor prognosis	予後不良	1,145
❷	poor outcome	よくない結果	478
❸	poor survival	よくない生存率	232

❷ Cytogenetic features may further refine this prognosis and **identify patients with a poor outcome**.（Blood. 2003 102:2756）

訳 ～は，よくない結果の患者を同定する

■ ［前］＋ poor

❶	be associated with poor ～	…は，よくない～と関連する	354
❷	correlate with poor ～	…は，よくない～と相関する	165
❸	patients with poor ～	よくない～の患者	121

❶ Recently, **elevated levels of β-catenin have been associated with poor prognosis in** human adenocarcinoma of the breast.（Oncogene. 2001 20:5093）

訳 β-カテニンの上昇したレベルは，～における予後不良と関連している

★ poorly 副 不十分に／あまり～でない　用例数 12,407

◆類義語：insufficiently

■ poorly ＋［過分］

❶	poorly understood	あまり理解されていない	7,239
	・be poorly understood	…は，あまり理解されていない	4,511
	・be still poorly understood	…は，まだあまり理解されていない	316
	・mechanisms are poorly understood	機構は，あまり理解されていない	138
	・remain poorly understood	…は，あまり理解されないままである	1,662
❷	poorly defined	あまり明らかにされていない	919
❸	poorly characterized	あまり特徴づけられていない	483
❹	poorly differentiated	あまり分化していない	307

❶ These networks are known to intersect with one another, but **the mechanisms are poorly understood**.（Genes Dev. 2004 18:1413）

訳 その機構は，あまり理解されていない

❷ **The endogenous mechanisms limiting such reactions remain poorly defined**.（Proc Natl Acad Sci USA. 2003 100:2730）

訳 そのような反応を制限する内在性の機構は，あまり明らかにされないままである

population ㊔ 集団／人口　用例数 7,720

population	集団／人口	5,482
populations	集団／人口	2,238

複数形のpopulationsの割合が，約35%ある可算名詞.

■ population ＋［前］

population ofの前にtheが付く割合は非常に高い.

	the%	a/an%	ø%			
❶	76	14	10	population of ～	～の集団	5,180
❷	51	35	11	population in ～	～における集団（人口）	781

❶ We have previously characterized a **population of** Trm cells that persists within the brain after acute virus infection.（J Immunol. 2012 189:3462）

訳 われわれは，以前にTrm細胞の集団の特徴づけを行った

■［形／名］＋ population

❶	general population	母集団	1,415
❷	patient population	患者集団	679
❸	study population	研究対象集団	589
❹	human population	人間集団	453
❺	large population	大きな集団	351

portion ㊔ 部分　用例数 8,164

portion	部分	6,303
portions	部分	1,861

複数形のportionsの割合が，約20%ある可算名詞.

◆類義語：region, part

■ portion ＋［前］

	the%	a/an%	ø%			
❶	43	53	1	portion of ～	～の部分	5,500
				・C-terminal portion of ～	～のC末端部	237
				・a significant portion of ～	～のかなりの部分	215
				・a large portion of ～	～の大部分	169
				・a small portion of ～	～の少しの部分	124

❶ Native ActRIIB has four isoforms that differ in the length of the **C-terminal portion of** their extracellular domains.（J Biol Chem. 2010 285:21037）

訳 細胞外ドメインのC末端部の長さが異なる４つのアイソフォーム

P

★★ position ㊔ 位置，㊍ 位置づける　用例数 33,052

position	位置／～を位置づける	18,582
positions	位置／～を位置づける	9,564
positioned	位置づけられる／位置する／～を位置づけた	2,498
positioning	位置決め／～を位置づける	2,408

名詞として用いられることが多いが，動詞としての用例もある．複数形のpositionsの割合が，約35%ある可算名詞．

◆類義語：location, locus, site, region, area, situation, locate, localize, map

■ ㊔ position ＋［前］

position ofの前にtheが付く割合は圧倒的に高い．

	the%	a/an%	ø%			
❶	92	2	3	position of ～	～の位置	3,517
❷	8	-	72	positions in ～	～における位置	924

❶ The recent structure of PfPGI has confirmed the hypothesis that the enzyme belongs to the cupin superfamily and **identified the position of the active site.** (J Mol Biol. 2004 343:649)

訳 ～は，活性部位の位置を同定した

■ ㊔［前］＋ position

❶ residue at position ～	～番目の残基	424
❷ substitution at position ～	～番目の置換	172

❶ A sixth mutation in the core, involving **a highly conserved Met residue at position 67,** appeared intolerant to substitution. (J Mol Biol. 2004 344:769)

訳 高度に保存されている67番目のメチオニン残基

■ ㊍ positioned ＋［前］

❶ positioned to *do*	～するように位置する	315
・be positioned to *do*	…は，～するように位置する	132
❷ positioned in ～	～に位置する	272
❸ positioned at ～	～に位置する	234

❶ Here we report ten crystal structures of the HDV ribozyme in its pre-cleaved state, showing that **cytidine is positioned to activate** the 2'-OH nucleophile in the precursor structure. (Nature. 2004 429:201)

訳 シチジンは，～を活性化するように位置する

★★ positive ㊒ 陽性の／正の，㊔ 陽性であること　用例数 43,657

positive	陽性の／正の	43,093
positives	陽性であること	564

形容詞の用例が多いが，名詞としても用いられる．

◆類義語：active, aggressive

■ positive ＋［前］

❶ positive for ～	～に対して陽性である	2,151
・be positive for ～	…は，～に対して陽性である	1,005

❶ All three PCR tests were positive for 17 specimens, all of which also tested positive by LCR. (J Clin Microbiol. 2003 41:2174)

訳 3つのPCRテストのすべてが，17の検体に対して陽性であった

■ positive and ＋［形］／［形］＋ and positive

❶ positive and negative ～	正および負の～	1,677
❷ negative and positive ～	負および正の～	222

❶ Through characterizing an additional 1000 bp of upstream DNA sequences of the murine p53 gene, we identified new positive and negative regulatory elements. (Oncogene. 2006 25:555)

訳 われわれは，新しい正および負の調節エレメントを同定した

P

■ positive ＋［名］

❶ ～-positive cells	～陽性細胞	2,100
❷ positive selection	正の選択	1,547
❸ positive feedback	正のフィードバック	891
❹ ～-positive patients	～陽性患者	749
❺ positive regulator	正の調節因子	576
❻ positive correlation	正の相関	562
❼ positive charge	正の荷電	543

❷ The cell surface protein CD34 is frequently used as a marker for positive selection of human hematopoietic stem/progenitor cells in research and in transplantation. (Blood. 2003 101:112)

訳 細胞表面タンパク質CD34は，～の正の選択のためのマーカーとしてしばしば使われる

positively 副 正に／陽性に　　用例数 6,018

◆類義語：actively

■ positively ＋［過分／動］

❶ positively charged	正に荷電した	1,294
❷ positively regulate ～	…は，～を正に調節する	609
❸ be positively associated with ～	…は，～と正に関連する	487
❹ be positively correlated with ～	…は，～と正に相関する	470

❹ In both Ca^{2+} concentrations, the conductance was positively correlated with hair cell CF. (Neuron. 2003 40:983)

訳 コンダクタンスは，～と正に相関した

■ positively and ＋［副］

❶ positively and negatively	正および負に	208

★possess　動 持つ／所有する　用例数 10,112

possess	～を持つ／～を所有する	4,654
possesses	～を持つ／～を所有する	2,829
possessing	～を持つ／～を所有する	1,432
possessed	～を持った／～を所有した	1,197

他動詞．通常，受動態では用いられない．

◆類義語：hold, have

■［名］＋ possess

❶ cells possess ～	細胞は，～を持つ	218

■［動］＋ to possess

★ to possess ～	～を持つ…	632
❶ be shown to possess ～	…は，～を持つことが示される	164
❷ be found to possess ～	…は，～を持つことが見つけられる	89

❶ ALD1 **was shown to possess** aminotransferase activity in vitro, suggesting it generates an amino acid-derived defense signal. (Plant J. 2004 40:200)

訳 ALD1は，アミノトランスフェラーゼ活性を持つことが示された

★possibility　名 可能性　用例数 10,025

possibility	可能性	9,134
possibilities	可能性	891

〈動詞＋the possibility that〉の用例が非常に多い．複数形のpossibilitiesの割合が，約10%ある可算名詞．

◆類義語：potential, probability, feasibility, promise

■ possibility ＋［前／同格that節］

possibility that/of/forの前にtheが付く割合は圧倒的に高い．

	the%	a/an%	ø%			
❶	98	2	0	possibility that ～	～という可能性	5,132
				・raise the possibility that ～	…は，～という可能性を示唆する	2,366
				・, raising the possibility that ～	そして，（それは）～という可能性を示唆している	592

				・results raise the possibility that ～	結果は，～という可能性を示唆する	181
				・findings raise the possibility that ～	知見は，～という可能性を示唆する	167
				・suggest the possibility that ～	…は，～という可能性を示唆する	280
				・investigate the possibility that ～	…は，～という可能性を精査する	225
❷	97	2	1	possibility of ～	～の可能性	2,460
				・raise the possibility of ～	…は，～の可能性を示唆する	377
				・suggest the possibility of ～	…は，～の可能性を示唆する	344
				・the possibility of using ～	～を使う可能性	166
❸	82	9	1	possibility for ～	～に対する可能性	144

❶ The post-translational modification by tTG reduced the RNA binding activity of the core protein, raising the possibility that tTG may regulate the biological functions of the HCV core protein. (J Biol Chem. 2001 276:47993)

訳 そして，（それは）～という可能性を示唆している

❶ These results raise the possibility that some forms of synaptic memory may be stored in a digital manner in the brain. (Proc Natl Acad Sci USA. 1998 95:4732)

訳 これらの結果は，～という可能性を示唆する

❶ We investigated the possibility that vascular endothelial growth factor (VEGF) treatment could regulate KDR/Flk-1 receptor expression in endothelial cells. (J Biol Chem. 1998 273:29979)

訳 われわれは，～という可能性を精査した

❷ This observation raises the possibility of redundancy in the phyA-signaling pathway, which could account for the incomplete block of phyA signaling observed in the far1 mutant. (Genes Dev. 1999 13:2017)

訳 この観察は，phyAシグナル伝達経路における重複性の可能性を示唆する

■ this/one possibility

❶	one possibility is that ～	1つの可能性は，～ということである	127
❷	to test this possibility	この可能性をテストするために	116
❸	to investigate this possibility	この可能性を精査するために	108

❶ One possibility is that these localized surrounds may provide a substrate for figure-ground segmentation of visual scenes. (J Neurosci. 1999 19:10536)

訳 1つの可能性は，～ということである

❷ To test this possibility, we generated a panel of reassortant viruses that expressed the NA genes of human H2N2 viruses isolated from 1957 to 1968 with all other genes from the avian virus A/duck/Hong Kong/278/78 (H9N2). (J Virol. 2001 75:11773)

訳 この可能性をテストするために，

★possible 形 可能な／あり得る 用例数 23,346

◆類義語：feasible, probable, likely, capable

■ possible ＋［前／that節］

❶ possible to *do*	～することは可能である	3,240
・it is possible to *do*	～することは可能である	880
・make it possible to *do*	…は，～することを可能にする	746
・not possible to *do*	～することは可能ではない	196
・possible to identify ～	～を同定することは可能である	147
・possible to use ～	～を使うことは可能である	117
・possible to determine ～	～を決定することは可能である	109
❷ possible in ～	～において可能である	356
❸ it is possible that ～	～ということはあり得る	699

❶ The doubly labeled water method now **makes it possible to quantitatively and objectively test a hypothesis proposed almost 50 y ago.** (Am J Clin Nutr. 1998 68:956S)
訳 ～は，ほとんど50年前に提唱された仮説を定量的および客観的にテストすることを可能にする

❶ If this is true, then **it should be possible to identify** other attentional mechanisms tied to other response modalities. (Nat Neurosci. 2001 4:656)
訳 ～を同定することは可能なはずである

❶ However, **it is usually not possible to determine** directly the contribution of specific phytonutrients to the total ORAC value. (Anal Biochem. 2000 287:226)
訳 ～を決定することは通常可能ではない

❸ Therefore, **it is possible that** CCK-induced activation of myenteric neurons also depends upon vagal activation. (Brain Res. 2000 878:155)
訳 ～ということはあり得る

■ made possible

★ made possible	可能にされる	285
❶ be made possible by ～	…は，～によって可能にされる	121

❶ Such statistically valid sampling of the kinetics **is made possible by the use of the discrete molecular dynamics method** with square-well interactions. (J Mol Biol. 1999 293:917)
訳 ～は，別個の分子動力学的方法の使用によって可能にされる

■ possible ＋［名］

❶ possible role	あり得る役割	1,335
❷ possible mechanism	あり得る機構	744
❸ possible explanation	あり得る説明	332
❹ possible involvement of ～	～のあり得る関与	235

❶ **The results suggest a possible role for** cytokine immunogenetics in the stability of peripheral tolerance. (Transplantation. 2000 69:1527)
訳 それらの結果は，～のあり得る役割を示唆する

❷ This provides empirical evidence in support of one possible mechanism for respiratory muscle failure in emphysema. (Am J Respir Crit Care Med. 2001 163:1081)

訳 これは，～のひとつのあり得る機構を支持する経験的な証拠を提供する

■ [形] + possible

❶ all possible ～	すべてのあり得る～	482
❷ one possible ～	1つのあり得る～	362

❶ Our results stress the importance of complete reference tables of all possible binding sites for comparing protein binding preferences for various DNA sequences. (Nucleic Acids Res. 2002 30:1255)

訳 われわれの結果は，すべてのあり得る結合部位の完全な参照テーブルの重要性を強調する

P

★ possibly 副 もしかしたら／多分 用例数 7,327

◆類義語：potentially, probably, presumably, conceivably, likely, maybe, perhaps

■ possibly + [前]

❶ , possibly by ～	もしかしたら～によって	528
❷ , possibly through ～	もしかしたら～によって	362

❶ Asp-151 is essential for maintaining this conformation, possibly by anchoring its side chain into the partially formed substrate pocket through interaction with Arg-180. (J Biol Chem. 1998 273:9695)

訳 もしかしたらそれの側鎖を～に固定することによって

■ possibly + [接／形]

❶ , possibly because ～	もしかしたら～の理由で	272
❷ , possibly due to ～	もしかしたら～のせいで	242

❷ Embryonic dopamine neurons survive poorly after transplant into models of Parkinson's disease, possibly due to programmed cell death (apoptosis). (Brain Res. 1998 786:96)

訳 もしかしたらプログラム細胞死（アポトーシス）のせいで

postulate 動 仮定する／推論する 用例数 3,737

postulated	仮定される／～を仮定した	2,589
postulate	～を仮定する	897
postulates	～を仮定する	205
postulating	～を仮定する	46

他動詞.

◆類義語：hypothesize, propose, assume, presume, estimate, speculate, predict, expect, infer, deduce, consider

■ postulate ＋［that節］

★ postulate that ～	…は，～ということを仮定する	851
❶ we postulate that ～	われわれは，～ということを仮定する	556

❶ We postulate that IrtAB is a transporter of Fe-carboxymycobactin.（J Bacteriol. 2006 188: 424）

訳 われわれは，～ということを仮定する

■ be postulated ＋［that節］

★ be postulated that ～	～ということが仮定される	324
❶ it has been postulated that ～	～ということが仮定されてきた	162
❷ it is postulated that ～	～ということが仮定されている	143

❶ It has been postulated that the induction of acute GVHD requires the presence of Peyer patches（PPs）.（Blood. 2006 107:410）

訳 ～ということが仮定されてきた

■ be postulated to *do*

★ be postulated to *do*	…は，～すると仮定される	904
❶ have been postulated to *do*	…は，～すると仮定されてきた	397
❷ be postulated to be ～	…は，～であると仮定される	251
❸ be postulated to play ～	…は，～を果たす仮定される	97

❶ RNAi has been postulated to function as an adaptive antiviral immune mechanism in the worm, but there is no experimental evidence for this.（Nature. 2005 436:1044）

訳 RNAiは，～として機能すると仮定されている

❷ Cyclin D1 is postulated to be a target of the canonical Wnt pathway and critical for intestinal adenoma development.（J Biol Chem. 2005 280:28463）

訳 サイクリンD1は，標準的なWnt経路の標的であると仮定される

potency ㊔ 効力／力価　用例数 5,987

potency	効力／力価／活性	5,348
potencies	効力／力価／活性	639

複数形のpotenciesの割合は約10%あるが，単数形は無冠詞で使われることが多い．

◆類義語：efficiency, efficacy, effect

■ potency ＋［前］

potency ofの前にtheが付く割合は非常に高い．

	the%	a/an%	ø%			
❶	79	3	16	potency of ～	～の効力	1,263
❷	3	4	77	potency in ～	～における効力	296

❸	10	8	75	potency for ～	～に対する効力	221
❹	9	2	81	potency against ～	～に対する効力	176

❶ The potency of FGF-2 in H-510 cells might reflect this additional MEK/S6K2 signalling. (Oncogene. 2001 20:7658)

訳 H-510細胞におけるFGF-2の効力は，～を反映するかもしれない

■ potency and ＋［名］

❶	potency and selectivity	力価と選択性	181

❶ The potency and selectivity of these inhibitors may affect the efficacy and toxicity of therapy. (J Biol Chem. 2002 277:1576)

訳 これらの抑制剤の力価と選択性は，～に影響を与えるかもしれない

P

■［形／過分］＋ potency

❶	inhibitory potency	阻害活性	215
❷	high potency	高い効力	206
❸	increased potency	増大した効力	103

★ potent 形 強力な　　用例数 17,979

◆類義語：strong, powerful, robust, intensive, intense

■［副］＋ potent

❶	more potent	より強力な	1,942
	・more potent than ～	～より強力な	705
❷	the most potent ～	最も強力な～	1,264
❸	less potent	より強力でない	452
❹	highly potent	高度に強力な	361

❶ In each assay, the sulfamoylated estrone derivatives were >10-fold more potent than their parent compounds. (Cancer Res. 2000 60:5441)

訳 ～は，それらの親化合物より10倍以上強力であった

❷ Adenophostin A is the most potent known agonist of D-myo-inositol 1, 4,5-trisphosphate [Ins (1,4,5) P3] receptors. (Mol Pharmacol. 1999 55:109)

訳 …は，～の最も強力な既知の作用薬である

■ potent ＋［前］

❶	potent in ～	～において強力な	405
	・potent in ～ing	～するのに強力な	155
❷	potent at ～	～において強力な	117

❶ CDD was more potent in killing cells than the full-length Bit1 protein when equivalent amounts of cDNA were transfected. (Cancer Res. 2013 73:1352)

訳 CDDは，細胞を殺すのに～よりも強力である

■ potent ＋［名］

❶	potent inhibitor	強力な抑制剤（阻害剤）	1,163
❷	potent inducer	強力な誘導因子	301
❸	potent activator	強力な活性化因子	210
❹	potent inhibition	強力な抑制（阻害）	202

❶ Laulimalide **is a** potent inhibitor **of cellular proliferation with IC50 values in** the low nanomolar range, whereas isolaulimalide is much less potent with IC50 values in the low micromolar range.（Cancer Res. 1999 59:653）

訳 …は，～において50%抑制濃度値を持つ細胞増殖の強力な抑制剤である

■ potent and ＋［形］

❶	potent and selective ～	強力かつ選択的な～	696
❷	potent and specific ～	強力かつ特異的な～	171

❶ Substitution at the 4- and 5-positions of the 2-iminopyrrolidine **yielded both** potent and selective **inhibitors of hiNOS.**（J Med Chem. 1998 41:3675）

訳 ～は，強力でかつ選択的なhiNOSの阻害剤を生じた

★★ potential　㊔ 潜在能／電位，㊏ 潜在的な／可能な　用例数 67,692

potential	潜在能／電位／潜在的な／可能な	61,393
potentials	潜在能／電位	6,299

名詞および形容詞として用いられる．複数形のpotentialsの割合は約10%あるが，単数形は無冠詞で使われることが多い．

◆類義語：capability, ability, competence, possible, feasible, capable

■ ㊔ potential ＋［前］

potential of/toの前にtheが付く割合は圧倒的に高い．

	the%	a/an%	ø%			
❶	89	2	9	potential of ～	～の潜在能	5,575
				・therapeutic potential of ～	～の治療上の潜在能	411
				・potential of ～ing	～する潜在能	329
				・demonstrate the potential of ～	…は，～の潜在能を実証する	230
❷	59	3	34	potential for ～	～のための潜在能	4,170
				・potential for ～ing	～するための潜在能	593
				・have potential for ～	…は，～のための潜在能を持つ	235
❸	80	3	13	potential to *do*	～する潜在能	3,991
				・have the potential to *do*	…は，～する潜在能を持つ	2,376

❶ These results demonstrate the potential of MHC class II gene transfer to permit tolerance to solid organ allografts.（J Clin Invest. 2001 107:65）

訳 これらの結果は，～の潜在能を実証する

❷ Detection sensitivities of 10^{-7} mol/l are demonstrated, with the potential for significant improvement.（Anal Chem. 2001 73:1007）

訳 顕著な改善の可能性を持って

❸ Human induced pluripotent stem cells have the potential to become an unlimited cell source for cell replacement therapy.（PLoS One. 2012 7:e50880）

訳 ヒトiPS細胞は，細胞補充療法のための無限の細胞源となる潜在能を持つ

■ ㊔［名／形］＋ potential

❶ membrane potential	膜電位	2,376
❷ action potential	活動電位	2,335
❸ therapeutic potential	治療上の潜在能	1,090

P

■ ㊒ potential ＋［名／形］

❶ potential role	潜在的な役割	2,143
❷ potential therapeutic ～	潜在的な治療上の～	1,807
❸ potential mechanism	潜在的な機構	879
❹ potential target	潜在的な標的	811
❺ potential use	可能な利用	526
❻ potential applications	可能な適用	483

❶ In this study, we investigated the potential role of AA as a proapoptotic agent in CML.（Cancer Res. 1999 59:5047）

訳 われわれは，AAの潜在的な役割を精査した

❷ Conceivably, SAF-1 could be a potential therapeutic target to control the overexpression of MMP-1 associated with the pathogenesis of OA.（Arthritis Rheum. 2003 48:134）

訳 SAF-1は，潜在的な治療上の標的でありうる

❸ These findings provide a potential mechanism for the ability of virus to persist in neurons and to escape immune surveillance.（J Immunol. 1999 162:4024）

訳 これらの知見は，ウイルスの能力の潜在的な機構を提供する

★ potentially ㊙ 潜在的に　用例数 11,793

◆類義語：latently

■ potentially ＋［形］

❶ potentially important	潜在的に重要な	690
❷ potentially useful	潜在的に有用な	417
・be potentially useful	…は，潜在的に有用である	151

❶ The findings highlight a potentially important role for imbalance between cysteine proteases and cystatin C in arterial wall remodeling and establish that cystatin C deficiency occurs in vascular disease.（J Clin Invest. 1999 104:1191）

訳 それらの知見は，～の間の不均衡に対する潜在的に重要な役割を強調する

potentiate 動 増強する 用例数 4,432

potentiated	増強される／～を増強した	1,952
potentiate	～を増強する	1,046
potentiates	～を増強する	977
potentiating	～を増強する	457

他動詞.

◆類義語：enhance, augment, facilitate, intensify, reinforce, activate

■ potentiate ＋［名句］

❶ potentiate the effect of ～	…は，～の効果を増強する	108

❶ Additionally, our data imply that **caffeine may potentiate the effects of marijuana** on hippocampal function.（Blood. 2002 100:982）

訳 カフェインは，マリファナの効果を増強するかもしれない

■ potentiated ＋［前］

❶ potentiated by ～	～によって増強される	426
・be potentiated by ～	…は，～によって増強される	272

❶ **Fas-mediated apoptosis was potentiated by inhibition of β-catenin nuclear signaling.**（Blood. 2002 100:982）

訳 Fasに仲介されるアポトーシスが，β-カテニン核シグナル伝達の抑制によって増強された

potently 副 強力に 用例数 2,438

◆類義語：strongly, robustly, intensively, intensely

■ potently ＋［動］

❶ potently inhibit ～	…は，～を強力に抑制する	724
❷ potently induce ～	…は，～を強力に誘導する	182
❸ potently activate ～	…は，～を強力に活性化する	134
❹ potently suppress ～	…は，～を強力に抑制する	120
❺ potently stimulate ～	…は，～を強力に刺激する	108

❶ Moreover, piperlongumine **potently inhibits the growth of** spontaneously formed malignant breast tumours and their associated metastases in mice.（Nature. 2011 475:231）

訳 …は，～の増殖を強力に抑制する

■ potently ＋［過分］

❶ potently inhibited	強力に抑制された／強力に抑制した	145

❶ Activity was potently inhibited by bestatin and apstatin in a slow binding competitive fashion, with K_i* values of 3 and 44 nm, respectively. (J Biol Chem. 2002 277:26057)
訳 活性は，～によって強力に抑制された

power 名 力／威力／能力，動 強化する　用例数 7,719

power	力／威力／～を強化する	6,966
powered	～を強化した／強化される	509
powers	力／威力／～を強化する	216
powering	～を強化する	28

名詞の用例が非常に多いが，動詞としても使われる．複数形のpowersの割合は2％しかなく，原則，不可算名詞として使われる．
◆類義語：capability, capacity, ability, potential, competence, force, strength

■ 名 power ＋［前］

power ofの前にtheが付く割合は圧倒的に高い．

	the％	a/an％	ø％			
❶	87	7	6	power of ～	～の力	1,500
❷	22	0	60	power to *do*	～する力	518
				・power to detect ～	～を検出する力	299

❶ Our results demonstrate the power of ERM to identify key genes involved in cell-survival signaling. (Curr Biol. 2000 10:1233)
訳 われわれの結果は，～に関与する鍵となる遺伝子を同定するためのERMの威力を実証する

❷ Such negative associations may only be in evidence transiently, because the statistical power to detect them diminishes as the mutations accumulate. (J Exp Med. 2005 201:891)
訳 それらを検出する検定力が，変異が蓄積するにつれて低下する

■ 動 powered ＋［前］

❶	powered by ～	～によって強化される	150

❶ The motors are powered by ATP binding and hydrolysis. (Biochemistry. 2010 49:2097)
訳 それらのモーターは，ATPの結合と加水分解によって強化される

powerful 形 強力な　用例数 5,419

◆類義語：potent, strong, robust, intensive, intense

■ powerful ＋［名］

❶	powerful tool	強力な手段	822
❷	powerful approach	強力なアプローチ	272
❸	powerful method	強力な方法	208

P

❶ These studies indicate that **BSMV-VIGS** is a **powerful tool** for dissecting the genetic pathways of disease resistance in hexaploid wheat. (Plant Physiol. 2005 138:2165)
訳 BSMV-VIGSは，～の遺伝的経路を精査するための強力な手段である

■ ［副］＋ powerful

❶	more powerful	より強力な	260
❷	most powerful	最も強力な	243

❶ Simulations indicate that **our proposed approach is more powerful than association tests** that are based on each separate sample. (Am J Hum Genet. 2005 76:592)
訳 われわれの提案するアプローチは，関連検定より強力である

practice 名 実行／実際／診療，動 実行する／開業する

用例数 7,350

practice	実行／実際／診療／～を実行する	5,217
practices	実行／実際／診療／～を実行する	1,705
practicing	開業する	245
practiced	実行された／～を実行した	183

名詞として用いられることが多いが，動詞としての用例もある．practicesの割合は約25%あるが，単数形は無冠詞で使われることが多い．

◆類義語：performance

■ practice ＋［前］

practice ofの前にtheが付く割合は圧倒的に高い．

	the%	a/an%	ø%			
❶	88	2	9	practice of ～	～の実際	444
				・practice of ～ing	～することの実際	104
❷	3	7	85	practice in ～	～における実際	211

❶ **The practice of pediatric cardiac intensive care has evolved considerably over the last several years.** (Crit Care Med. 2011 39:1974)
訳 小児の心臓集中治療の実際は，最近数年の間にかなり進化してきた

■ ［形］＋ practice

❶	clinical practice	診療	1,286
❷	current practice	現在の診療	148
❸	medical practice	医療行為	133

precise ㊖ 正確な　用例数 6,323

◆類義語：correct, accurate, exact

■ precise ＋［名］

❶ precise role	正確な役割	462
❷ precise mechanism	正確な機構	304
❸ precise control	正確な制御	208

❶ However, **the precise role of CNK in Ras signaling is not known**, and mammalian CNKs are proposed to have distinct functions. (Proc Natl Acad Sci USA. 2005 102:11757)
訳 Ras シグナル伝達におけるCNKの正確な役割は知られていない

P

precisely ㊙ 正確に　用例数 2,088

◆類義語：accurately, correctly, exactly

■［副］＋ precisely

❶ more precisely	より正確に	240

❶ **To more precisely define the molecular mechanisms** whereby FGFR1 causes angiogenesis in the prostate we exploited a transgenic mouse model, JOCK-1, in which activation of a conditional FGFR1 allele in the prostate epithelium caused rapid angiogenesis and progressive hyperplasia. (Oncogene. 2007 26:4897)
訳 分子機構をより正確に定義するために

★★ predict ㊍ 予測する／予想する　用例数 36,011

predicted	予測される／～を予測した	21,497
predict	～を予測する	8,088
predicts	～を予測する	3,463
predicting	～を予測する	2,963

他動詞.
◆類義語：expect, presume, estimate, speculate, assume, infer, postulate, deduce

■ be predicted ＋［前］

★ be predicted	…は，予測される	4,772
❶ be predicted to *do*	…は，～すると予測される	2,951
・be predicted to be ～	…は，～であると予測される	780
・be predicted to encode ～	…は，～をコードすると予測される	352
・be predicted to have ～	…は，～を持つと予測される	220
・be predicted to form ～	…は，～を形成すると予測される	123
・be predicted to contain ～	…は，～を含むと予測される	110
❷ be predicted by ～	…は，～によって予測される	543

❸ be predicted from ～	…は，～から予測される	228

❶ Arabidopsis contains 34 genes that **are predicted to encode** calcium-dependent protein kinases (CDPKs). (Plant Physiol. 2002 128:1008)

訳 アラビドプシスは，カルシウム依存性タンパク質キナーゼ（CDPK）をコードすると予測される34の遺伝子を含む

❸ Here I review what can **be predicted from** these models and how well these predictions match experiments. (Biochemistry. 2004 43:2141)

訳 私は，これらのモデルから何が予測できるかを概説する

■［副］＋ predicted

❶ independently predicted	独立に予測される	176
❷ accurately predicted	正確に予測される	175

■ as predicted

★ as predicted	予測されたように	832
❶ as predicted by ～	～によって予測されたように	357
❷ as predicted from ～	～から予測されたように	115

❶ **As predicted by** this model, formation of A2E was completely inhibited when abcr$^{-/-}$ mice were raised in total darkness. (Proc Natl Acad Sci USA. 2000 97:7154)

訳 このモデルによって予測されたように，

■ predict ＋［that節］

★ predict that ～	…は，～ということを予測する	2,441
❶ we predict that ～	われわれは，～ということを予測する	304
❷ model predicts that ～	モデルは，～ということを予測する	264
❸ theory predicts that ～	理論は，～ということを予測する	126

❷ First, this **model predicts that** Ds and Ft can bind. (Development. 2004 131:3785)

訳 このモデルは，～ということを予測する

■ predict ＋［名句］

❶ predict the presence of ～	…は，～の存在を予測する	100
❷ predict the risk of ～	…は，～のリスクを予測する	95

■［動／名］＋ to predict

★ to predict ～	～を予測する…	3,514
❶ be used to predict ～	…は，～を予測するために使われる	519
❷ ability to predict ～	～を予測する能力	206
❸ model to predict ～	～を予測するモデル	100

❶ These recombination patterns **were used to predict** the chromosome 21 exchange patterns established during meiosis I. (Am J Hum Genet. 2005 76:91)

訳 これらの組換えパターンが，第21番染色体交換パターンを予測するために使われた

★ prediction 名 予測 用例数 11,076

prediction	予測	6,405
predictions	予測	4,671

複数形のpredictionsの割合は約40％とかなり高いが，単数形は無冠詞で使われることも多い．

◆類義語：expectation, anticipation

■ prediction(s) ＋［前／同格that節］

prediction thatの前にtheが付く割合は非常に高い．

	the%	a/an%	ø%			
❶	34	8	55	prediction of ～	～の予測	1,977
❷	9	-	78	predictions for ～	～に対する予測	313
❸	75	10	9	prediction that ～	～という予測	243
❹	13	-	86	predictions from ～	～からの予測	220

❶ Secondary structure prediction of the selected sequences was completed to determine the motifs that each ligand binds, and the hairpin loop preferences for 1 and 2 were computed.（Biochemistry. 2008 47:12670）

訳 選択された配列の２次構造予測が完了した

❸ The multiple mechanisms favor the prediction that ATRA will induce FR-β expression in a broad spectrum of AML cells.（Blood. 2003 101:4551）

訳 複数の機構が，～という予測を支持する

■［名／形］＋ prediction(s)

❶	structure prediction	構造予測	546
❷	theoretical predictions	理論的予測	240
❸	model prediction	モデル予測	202
❹	risk prediction	リスク予測	201

predictive 形 予測的な／予測する／前兆となる 用例数 6,169

◆類義語：anticipatory

■ predictive ＋［前］

❶	predictive of ～	～の前兆となる	1,715
	・be predictive of ～	…は，～の前兆となる	804

❶ $GABA_A$ δ receptor subunit gene expression was predictive of a poor outcome among Evans stage IV-S patients.（J Clin Oncol. 2004 22:4127）

訳 ～が，予後不良の前兆となった

P

predispose 動 罹患の原因になる／罹患しやすくさせる

用例数 2,327

predispose	罹患の原因になる／～を罹患しやすくさせる	888
predisposing	罹患の原因になる	583
predisposed	罹患しやすい／罹患の原因になった	469
predisposes	罹患の原因になる／～を罹患しやすくさせる	387

predispose toの用例が多い．自動詞および他動詞の両方の用例がある．

■ predispose to／be predisposed to

❶ predispose to ～	…は，～の罹患の原因になる	814
❷ be predisposed to ～	…は，～に罹患しやすい	153

❶ The intensive search for genetic variants that predispose to type 2 diabetes was launched with optimism, but progress has been slower than was hoped.（Science. 2005 307:370）

訳 2型糖尿病の罹患の原因になる遺伝的な変異の集中的検索が開始された

❷ Patients with BSS are predisposed to multiple skin appendage tumors such as cylindroma, trichoepithelioma, and spiradenoma.（J Invest Dermatol. 2005 124:919）

訳 BSSの患者は，多発性皮膚付属器腫瘍に罹患しやすい

★ predominantly 副 主に／優勢に

用例数 8,764

過去分詞の前で使われることも，後で使われることもある．

◆類義語：mainly, mostly, primarily, principally, chiefly, preferentially

■ predominantly ＋［前］

❶ expressed predominantly in ～	主に～において発現する	267
❷ localized predominantly to ～	主に～に局在する	88

❶ The novel gene is expressed predominantly in adipose tissue, and its expression is induced early during 3T3-L1 adipocyte differentiation.（J Biol Chem. 2004 279:47066）

訳 その新規の遺伝子は，主に脂肪組織において発現する

❷ Virus antigen was localized predominantly to anterior horn cells in infected IFN-$\gamma^{-/-}$ H-2_q mice.（J Virol. 2003 77:12252）

訳 ウイルス抗原は，主に前角細胞に局在した

■ predominantly ＋［過分］

❶ predominantly expressed in ～	～において主に発現する	312
❷ predominantly localized to ～	～に主に局在する	83

❶ GC-GAP is predominantly expressed in the brain with low levels detected in other tissues.（J Biol Chem. 2003 278:34641）

訳 GC-GAPは，脳において主に発現する

prefer 動 好む　　用例数 4,353

preferred	好まれる／好んだ	3,399
prefer	好む	369
prefers	好む	313
preferring	好む	272

他動詞の用例が多いが，自動詞としても用いられる．

◈類義語：favor, select

■ preferred ＋［前］

❶ preferred over ～	～よりも好まれる	105
・be preferred over ～	…は，～よりも好まれる	81
❷ preferred to ～	～より好まれる	84

❶ In the context of the Cdk2-pTpY/CycA complex, phospho-threonine is preferred over phospho-tyrosine by more than 10-fold.（Anal Biochem. 2001 289:43）

訳 ホスホスレオニンは，ホスホチロシンより10倍以上好まれる

❷ Living donor transplantation is preferred to cadaveric transplantation.（Annu Rev Med. 50:193）

訳 生体ドナー移植は，死体移植より好まれる

■ prefer ＋［前］

❶ prefer to *do*	…は，～することを好む	155

❶ Most people prefer to use their right eye for viewing.（Curr Biol. 2001 11:R828）

訳 ほとんどの人々は，右目を使うことを好む

preference 名 選択性／好み／優先　　用例数 6,304

preference	選択性／好み／優先	4,421
preferences	選択性／好み／優先	1,883

複数形のpreferencesの割合が，約30%ある可算名詞．

◈類義語：priority

■ preference ＋［前］

preference ofの前にtheが付く割合は圧倒的に高い．

	the%	a/an%	ø%			
❶	7	67	18	preference for ～	～に対する選択性	1,999
				・strong preference for ～	～に対する強い選択性	197
❷	81	11	4	preference of ～	～の選択性	364
❸	6	18	70	preference to ～	～への選択性	192
				・in preference to ～	～に優先して	116

P

❶ Sulfolobus solfataricus contains a membrane-associated protein kinase activity that **displays a strong preference for threonine** as the phospho-acceptor amino acid residue. (J Bacteriol. 2004 186:463)

訳 ～は，スレオニンに対する強い選択性を示す

❸ **Warfarin is commonly used in preference to aspirin** for this disorder, but these therapies have not been compared in a randomized trial. (N Engl J Med. 2005 352:1305)

訳 ワルファリンは，一般にアスピリンに優先して使われる

★ preferentially 副 優先的に　用例数 7,014

◆類義語：predominantly

■ preferentially ＋［過分／動］

❶ preferentially expressed	優先的に発現する	549
・be preferentially expressed	…は，優先的に発現する	404
❷ preferentially bind	…は，優先的に結合する	324
・preferentially bind to ～	…は，～に優先的に結合する	144
❸ preferentially associated with ～	～と優先的に結合した	111

❶ **NF-Ya is preferentially expressed in HSC-enriched bone marrow subpopulations**, and NF-Ya mRNA rapidly declines with HSC differentiation. (Proc Natl Acad Sci USA. 2005 102:11728)

訳 NF-Yaは，造血幹細胞を濃縮された骨髄亜集団において優先的に発現する

❷ **NSC13778 preferentially binds C/EBPα** 1,000-fold better than it binds C/EBPβ. (Anal Biochem. 2005 340:259)

訳 NSC13778は，C/EBPαに優先的に結合する

■［動］＋ preferentially ＋［前］

❶ bind preferentially to ～	…は，～に優先的に結合する	231
❷ interact preferentially with ～	…は，～と優先的に相互作用する	83

❶ We conclude that **Siglec-8 binds preferentially to the sLex structure** bearing an additional sulfate ester on the galactose 6-hydroxyl. (J Biol Chem. 2005 280:4307)

訳 Siglec-8は，sLex構造に優先的に結合する

preparation 名 調製／準備／標本／画分　用例数 8,152

preparation	調製／準備／標本／画分	4,798
preparations	調製／準備／標本／画分	3,354

複数形のpreparationsの割合は約40%とかなり高い.

◆類義語：specimen, sample, formulation, arrangement

■ preparation (s) ＋ [前]

preparation ofの前にtheが付く割合はかなり高い.

	the %	a/an %	ø %			
❶	68	11	20	preparation of ～	～の調製	1,685
❷	9	3	87	preparation for ～	～のための準備	251
				・in preparation for ～	～に備えて	136
❸	2	-	95	preparations from ～	～からの画分	245

❶ An improved method for the preparation of a series of oxazole-containing dual PPARα/γ agonists is described. (J Org Chem. 2003 68:2623)

訳 一連のオキサゾールを含む二重のPPARα/γ作動薬の調製のために改良された方法が述べられる

❷ In preparation for future clinical trials, clinical studies have begun to provide more quantitative measures of disease onset and progression. (Curr Opin Neurol. 2002 15:483)

訳 将来の臨床治験に備えて,

❸ Plasma membrane preparations from activated monocytes also induced mesangial IL-6 and MCP-1 synthesis. (J Immunol. 1999 163:2168)

訳 活性化された単球からの細胞膜画分は, また, 糸球体間質のIL-6およびMCP-1合成を誘導した

■ [名] ＋ preparation

❶	sample preparation	試料調製	544
❷	slice preparation	切片標本	221
❸	membrane preparations	膜画分	171

prepare 動 調製する　　用例数 9,778

prepared	調製される／～を調製した	8,334
prepare	～を調製する	977
preparing	～を調製する	414
prepares	～を調製する	53

他動詞. 受動態の用例が非常に多い.

◆類義語：arrange, construct

■ prepared ＋ [前]

❶	prepared from ～	～から調製される	1,994
	・extracts prepared from ～	～から調製された抽出物	237
❷	prepared by ～	～によって調製される	1,233
	・be prepared by ～	…は, ～によって調製される	802
❸	prepared in ～	～において調製される	659
❹	prepared with ～	～によって調製される	327
❺	prepared for ～	～のために調製される	312

P

❶ DNA end-joining activity was also substantially reduced in extracts prepared from antisense mRNA-expressing cells. (Mol Cell Biol. 1999 19:3869)

訳 ～は，また，アンチセンスメッセンジャーRNA発現細胞から調製された抽出物において大幅に低下した

❷ Here, we describe an ADAR1 substrate RNA that can be prepared by a combination of chemical synthesis and enzymatic ligation. (Nucleic Acids Res. 2012 40:9825)

訳 ～は，化学合成と酵素的連結の組み合わせによって調製されうる

■ to prepare

★	to prepare ～	～を調製するために	812
❶	be used to prepare ～	…は，～を調製するために使われる	139

❶ The rat homolog of HRG4, RRG4 was expressed and used to prepare an antibody. (Invest Ophthalmol Vis Sci. 1998 39:690)

訳 RRG4は，抗体を調製するために発現され，使われた

★★ presence 名 存在 用例数 58,859

presence	存在	58,856
presences	存在	3

複数形のpresencesの用例はほとんどなく，原則，**不可算名詞**として使われる．the presence ofの用例が圧倒的に多い．

◆類義語：existence ◆反意語：absence

■ presence ＋［前］

presence ofの前にtheが付く割合は圧倒的に高い．

	the%	a/an%	ø%			
❶	95	1	4	presence of ～	～の存在	52,928

■［前］＋ the presence of

❶	In the presence of ～	～の存在下で	2,222
❷	even in the presence of ～	～の存在下でさえ	795
❸	dependent on the presence of ～	～の存在に依存して	685
❹	despite the presence of ～	～の存在にも関わらず	557
❺	only in the presence of ～	～の存在下においてのみ	545
❻	due to the presence of ～	～の存在が原因で	397
❼	depend on the presence of ～	…は，～の存在に依存する	361
❽	observed in the presence of ～	～の存在下で観察される	292
❾	consistent with the presence of ～	～の存在と一致して	288
❿	associated with the presence of ～	～の存在と関係する	280
⓫	characterized by the presence of ～	～の存在によって特徴づけられる	267

❷ It is suggested that this mechanism might protect cartilage from extensive degradation even in the presence of acute inflammation.（Arthritis Rheum. 2002 46:2207）

訳 この機構は，急性炎症の存在下でさえ，軟骨を広範な分解から保護するかもしれない

❸ The plasma membrane localization is dependent on the presence of an amino-terminal pleckstrin homology domain.（J Biol Chem. 2000 275:14295）

訳 細胞膜局在は，アミノ末端プレクストリン相同ドメインの存在に依存する

❻ These receptors have a unique structural composition due to the presence of multiple C-type lectin-like domains within a single polypeptide backbone.（J Biol Chem. 2002 277:32320）

訳 複数のC型レクチン様ドメインの存在のせいで

❽ An opposite effect was observed in the presence of the nACh receptor antagonist d-tubocurarine.（Brain Res. 2003 967:235）

訳 逆の効果が，nACh受容体拮抗剤d-ツボクラリンの存在下において観察された

❾ These results are consistent with the presence of an active Ca^{2+}/H^{+} antiport in the thylakoid membrane.（Plant Physiol. 1999 119:1379）

訳 これらの結果は，～の存在と一致する

P

■［動］＋ the presence of

❶ reveal the presence of ～	…は，～の存在を明らかにする	1,264
❷ suggest the presence of ～	…は，～の存在を示唆する	936
❸ indicate the presence of ～	…は，～の存在を示す	866
❹ require the presence of ～	…は，～の存在を必要とする	862
❺ demonstrate the presence of ～	…は，～の存在を実証する	730
❻ confirm the presence of ～	…は，～の存在を確認する	662
❼ show the presence of ～	…は，～の存在を示す	427
❽ detect the presence of ～	…は，～の存在を検出する	230

❶ Western blot analysis revealed the presence of C/EBPα and C/EBPβ in human granulosa-lutein cell nuclear extracts.（J Biol Chem. 274:26591）

訳 ウエスタンブロット分析は，C/EBPα およびC/EBPβ の存在を明らかにした

❺ We previously demonstrated the presence of estrogen receptor (ER) β in cells of the megakaryocytic lineage.（Circ Res. 2001 88:438）

訳 われわれは，以前にエストロゲン受容体（ER）β の存在を実証した

■ the presence or/and absence of ／ the absence and/or presence of

❶ in the presence or absence of ～	～の存在あるいは非存在下において	1,048
❷ in the presence and absence of ～	～の存在および非存在下において	896
❸ in the absence and presence of ～	～の非存在および存在下において	386
❹ in the absence or presence of ～	～の非存在あるいは存在下において	304

❶ This study examined peptic ulcer development in the presence or absence of gastric neutrophils in patients requiring long-term use of NSAIDs.（Gastroenterology. 1999 116:254）

訳 この研究は，胃の好中球の存在あるいは非存在下における消化性潰瘍の発症を調べた

★★ present ㊊ 存在する／現在の，㊍ 提示する　用例数 86,183

present	現在の／存在する／～を提示する	66,141
presented	提示される／～を提示した	13,235
presenting	～を提示する	4,131
presents	～を提示する	2,676

presentは形容詞として用いられることが多いが，違う意味の他動詞としても使われる．

◆類義語：current, modern, provide, show, indicate

■ ㊊ present ＋［前］

❶ present in ～	～において存在する	19,246
・be present in ～	…は，～において存在する	10,648
・present in all ～	すべての～において存在する	701
・present in both ～	両方の～において存在する	468
・present only in ～	～においてのみ存在する	285
❷ present at ～	～で存在する	2,190
・present at ～ in …	…において～で存在する	223
❸ present on ～	～において存在する	1,507
❹ present within ～	～内に存在する	551
❺ present as ～	～として存在する	491

❶ The ear gene product is **present in all** early embryonic cells, but becomes restricted to specific tissues in late embryogenesis. (Genetics. 2002 160:1051)

訳 ear遺伝子産物は，すべての初期胚細胞において存在する

❷ The αIIb/β3-integrin receptor is **present at** high levels only **in** megakaryocytes and platelets. (Blood. 2002 100:3588)

訳 ～は，巨核球および血小板においてのみ高いレベルで存在する

■ ㊊ present ＋［名］

❶ the present study	現在の研究	12,212
・in the present study	現在の研究において	6,469
・in the present study, we examined ～	現在の研究において，われわれは～を調べた	369
・in the present study, we investigated ～	現在の研究において，われわれは～を精査した	347
・… of the present study	現在の研究の…	1,316
・… of the present study was to ～	現在の研究の…は，～することであった	939
・the purpose of the present study was to ～	現在の研究の目的は，～することであった	344
・the aim of the present study was to ～	現在の研究の目的は，～することであった	262
・the present study examined ～	現在の研究は，～を調べた	425

・the present study was undertaken	現在の研究が，着手された	332
・the present study was designed	現在の研究が，計画された	284
❷ the present work	現在の研究	786
❸ the present results	現在の結果	467
❹ the present report	現在の報告	331

❶ **In the present study, we examined the properties of lens epithelial cells** derived from mice in which the αB-crystallin gene had been knocked out.（FASEB J. 2001 15:221）

訳 現在の研究において，われわれは水晶体上皮細胞の性質を調べた

❶ **The purpose of the present study was to determine whether** RPE65 is expressed at the protein level in mammalian cones, as well as in those of amphibians.（Invest Ophthalmol Vis Sci. 2002 43:1604）

訳 現在の研究の目的は，～かどうかを決定することであった

❶ **The present study examined whether** variation of c-Fos expression across the 24-hour light-dark cycle may also be different in these subdivisions.（J Comp Neurol. 2001 440:31）

訳 現在の研究は，～かどうかを調べた

P

■ 動 [代名／名] ＋ present

❶ we present ～	われわれは，～を提示する	11,089
・we present evidence that ～	われわれは，～という証拠を提示する	1,248
・we present a model	われわれは，モデルを提示する	322
・we present data	われわれは，データを提示する	307
❷ this paper presents ～	この論文は，～を提示する	203
❸ this study presents ～	この研究は，～を提示する	193

❶ Here **we present evidence that** Ypt31/32p, members of Rab family of GTPases, regulate Sec2p function.（J Cell Biol. 2002 157:1005）

訳 われわれは，～という証拠を提示する

■ 動 presented ＋ [前]

❶ presented in ～	～において提示される	1,108
❷ presented by ～	～によって提示される	838
❸ presented to *do*	～することが提示される	677
・be presented to *do*	…は，～することが提示される	417

❶ In vitro experimental **data presented in this report** suggest that SarA expression is autoregulated.（J Bacteriol. 2000 182:5893）

訳 このレポートにおいて提示されるデータ

■ 動 be presented ＋ [that節／前]

★ be presented	…が提示される	5,827
❶ be presented that ～	～する…が提示される	524
・evidence is presented that ～	～する証拠が提示される	232
❷ be presented for ～	～に関する…が提示される	487

❶ **Evidence is presented that indicates** a physiological significance for the interaction of hUBC9 with ATF2. (J Biol Chem. 1998 273:5892)

訳 ～を示す証拠が提示される

■ 動 [名] + presented

❶ data presented	提示されるデータ	811
・data presented here	ここで提示されるデータ	433
❷ results presented	提示される結果	633
・results presented here	ここで提示される結果	425

❶ The **data presented here** demonstrate down-regulation of p107 protein in response to various antiproliferative signals, and implicate calpain in p107 posttranslational regulation. (Oncogene. 1999 18:1789)

訳 ここで提示されるデータは，～を実証する

■ 動 present + [前]

❶ present with ～	…は，～を呈する	985

❶ All patients **presented with** respiratory distress, and three required mechanical ventilation. (Am J Respir Crit Care Med. 2005 172:768)

訳 すべての患者が，呼吸困難を呈した

presentation 名 提示／発表 用例数 7,185

presentation	提示／発表	6,506
presentations	提示／発表	679

複数形のpresentationsの割合は約10%あるが，単数形は無冠詞で使われることが非常に多い．

◆類義語：representation, manifestation

■ presentation + [前]

	the%	a/an%	ø%			
❶	38	4	55	presentation of ～	～の提示	1,826
❷	2	1	89	presentation by ～	～による提示	380
❸	2	1	87	presentation to ～	～への提示	352

❶ Furthermore, intestinal stress modifies the **presentation of** pancreatic self-antigens in PLNs. (Proc Natl Acad Sci USA. 2005 102:17729)

訳 腸のストレスは，膵臓の自己抗原の提示を変える

❷ We demonstrate that antigen **presentation by** these DC subsets is sufficient to control a subcutaneous L. major infection. (J Exp Med. 2004 199:725)

訳 これらのDCのサブセットによる抗原提示は，皮下の森林型熱帯リーシュマニア感染を制御するのに十分である

■ [前] + presentation

❶ at presentation	受診時に	480

❶ Patients were divided into five categories on the basis of symptoms at presentation (none, nonanginal chest pain, atypical angina, typical angina, and dyspnea). (N Engl J Med. 2005 353: 1889)

訳 患者は，受診時の症状に基づいて５つのカテゴリーに分類された

■ [名／形] + presentation

❶ antigen presentation	抗原提示	898
❷ clinical presentation	臨床像	481

P

preserve 動 保存する／維持する／保持する　用例数 4,993

preserved	保存される／～を保存した	2,717
preserve	～を保存する	933
preserving	～を保存する	858
preserves	～を保存する	485

他動詞.

◆類義語：conserve, store, reserve

■ be preserved + [前]

★ be preserved	…は，保存される	1,036
❶ be preserved in ～	…は，～において保存される	317

❶ Interestingly, although VN sites were preserved in all patients, AN site removal resulted in decline in both auditory and VN tasks. (Brain. 2005 128:2742)

訳 VN部位は，すべての患者において保存されていた

presumably 副 おそらく／多分　用例数 3,855

◆類義語：potentially, probably, conceivably, likely, maybe, perhaps, possibly

■ presumably + [前]

❶ , presumably by ～	おそらく～によって	354
❷ , presumably through ～	おそらく～を通して	145

❶ Organelle transport by myosin-V is down-regulated during mitosis, presumably by myosin-V phosphorylation. (Science. 2001 293:1317)

訳 おそらくミオシン-Vリン酸化によって

■ presumably because/due

❶ , presumably because ～	おそらく～ゆえに	291
・, presumably because of ～	おそらく～ゆえに	154
❷ , presumably due to ～	おそらく～ゆえに	224

❶ We were not able to obtain derivatives of SW480 cells that stably expressed these constitutively activated mutants, presumably because of toxicity.（Cancer Res. 2004 64:3966）
訳 おそらく毒性ゆえに

presume ㊍ 推定する　用例数 1,728

presumed	推定される／～を推定した	1,688
presume	～を推定する	22
presumes	～を推定する	12
presuming	～を推定する	6

他動詞.

◆類義語：estimate, speculate, assume, expect, predict, infer, postulate, deduce

■ be presumed ＋［前］

★ be presumed	…は，推定される	505
❶ be presumed to *do*	…は，～すると推定される	423
・be presumed to be ～	…は，～であると推定される	191

❶ DNA-binding is presumed to be essential for all nuclear actions of thyroid hormone.（J Clin Invest. 2003 112:588）
訳 DNA結合は，～に必須であると推定される

pretreatment ㊔ 前処置　用例数 5,797

pretreatment	前処置	5,751
pretreatments	前処置	46

複数形のpretreatmentsの割合は0.8%しかなく，原則，不可算名詞として使われる.

■ pretreatment ＋［前］

	the%	a/an%	ø%			
❶	1	1	97	pretreatment with ～	～による前処置	1,959
				・blocked by pretreatment with ～	～による前処置によってブロックされる	108
❷	4	2	94	pretreatment of ～	～の前処置	1,490
				・pretreatment of cells	細胞の前処置	209

❶ This effect of PDGF was blocked by pretreatment with wortmannin and attenuated in cells pretreated with cytochalasin D. (J Biol Chem. 2005 280:42300)

訳 このPDGFの効果は，ワートマニンによる前処置によってブロックされた

❷ These effects were blocked by pretreatment of cells with either HA1077 or Y-27632, which inhibit the kinases downstream of RhoA. (J Biol Chem. 2005 280:31172)

訳 これらの効果は，～による細胞の前処置によってブロックされた

★ prevalence 名 罹患率／有病率／流行　用例数 10,226

prevalence	罹患率／有病率／流行	9,988
prevalences	罹患率／有病率／流行	238

複数形のprevalencesの割合は２％しかなく，原則，**不可算名詞**として使われる．

◆類義語：morbidity, susceptibility, onset

■ prevalence ＋［前］

prevalence ofの前にtheが付く割合は非常に高い．

	the%	a/an%	ø%			
❶	75	19	6	prevalence of ～	～の罹患率	6,282
				・high prevalence of ～	～の高い罹患率	509
				・higher prevalence of ～	～のより高い罹患率	291
				・increased prevalence of ～	～の増大した罹患率	206
				・prevalence of obesity	肥満の罹患率	148
				・to determine the prevalence of ～	～の罹患率を決定する…	137
❷	13	11	46	prevalence in ～	～における罹患率	309

❶ African Americans (AAs) have a higher prevalence of obesity and type 2 diabetes than do whites. (Am J Clin Nutr. 2005 82:1210)

訳 アフリカ系アメリカ人（AA）は，白人より高い肥満および２型糖尿病の罹患率を持つ

❶ The aim of this study was to determine the prevalence of cardiovascular dysfunction and its predictors in children with acquired immunodeficiency syndrome (AIDS). (J Am Coll Cardiol. 2003 41:1598)

訳 この研究の目的は，心血管機能障害の罹患率を決定することであった

★★ prevent 動 阻止する／防ぐ　用例数 34,508

prevent	～を阻止する	11,940
prevented	阻止される／～を阻止した	10,811
prevents	～を阻止する	6,083
preventing	～を阻止する	5,674

他動詞．

◆類義語：hinder, occlude, hamper, abrogate, block, suppress, repress

P

■ prevent ＋［名句］

❶ prevent the development of ～	…は，～の発症を阻止する	451
❷ prevent apoptosis	…は，アポトーシスを阻止する	334
❸ prevent the formation of ～	…は，～の形成を阻止する	274
❹ prevent activation of ～	…は，～の活性化を阻止する	191
❺ prevent the induction of ～	…は，～の誘導を阻止する	150
❻ prevent the increase in ～	…は，～の増大を阻止する	148
❼ prevent the loss of ～	…は，～の喪失を阻止する	108

❶ Use of oral contraceptives (OCs) may prevent the development of rheumatoid arthritis, but the influence of OC use on disease outcome is unresolved. (Arthritis Rheum. 2011 63:2183)

訳 ～は，関節リウマチの発症を阻止するかもしれない

■［名／形／動］＋ to prevent

★ to prevent ～	～を阻止する…	5,726
❶ strategy to prevent ～	～を阻止する戦略	228
❷ sufficient to prevent ～	～を阻止するのに十分な	190
❸ required to prevent ～	～を阻止するために必要とされる	177
❹ fail to prevent ～	…は，～を阻止することができない	174
❺ ability to prevent ～	～を阻止する能力	130

❷ The presence of zeaxanthin in these strains was sufficient to prevent photooxidative damage in the npq1 lor1 background. (Plant Cell. 2003 15:992)

訳 ～は，光酸化ダメージを阻止するのに十分であった

❸ Fourth, a 100-fold excess of n-alkanethiol molecules is required to prevent aggregation of DENs during extraction. (J Am Chem Soc. 2004 126:16170)

訳 ～は，DENの凝集を阻止するために必要とされる

■ be prevented ＋［前］

★ be prevented	…は，阻止される	2,741
❶ be prevented by ～	…は，～によって阻止される	1,866
❷ be prevented in ～	…は，～において阻止される	172
❸ be prevented from ～	…は，～から妨げられる	106
・be prevented from ～ing	…は，～することから妨げられる	92

❶ The hyperoxia-induced cell cycle arrest was prevented by low adhesion conditions. (J Biol Chem. 2003 278:36099)

訳 ～が，低接着状態によって阻止された

■［副］＋ prevented

❶ completely prevented	完全に阻止される	389
❷ partially prevented	部分的に阻止される	100

❶ Obesity and insulin resistance developing in WT mice on an HF diet were completely

prevented in mice lacking PAI-1. (Diabetes. 2004 53:336)
訳 ～が，PAI-1を欠くマウスにおいて完全に阻止された

prevention 名 予防／阻止 用例数 7,740

複数形の用例はなく，原則，不可算名詞として使われる．
◆類義語：inhibition

■ prevention ＋［前］

	the%	a/an%	ø%			
❶	46	1	52	prevention of ～	～の予防	3,287
				・primary prevention of ～	～の1次予防	133

❶ This review highlights advances in the primary and secondary prevention of cardiovascular disease (CVD) in women in the preceding 12 months. (Curr Opin Cardiol. 2012 27:542)
訳 このレビューは，循環器疾患の1次および2次予防における進歩を強調する

★ previous 形 以前の 用例数 27,749

◆類義語：former, prior, recent

■ previous ＋［名］

❶	previous studies	以前の研究	9,082
	・previous studies have shown that ～	以前の研究は，～ということを示している	1,234
	・previous studies have demonstrated that ～	以前の研究は，～ということを実証している	361
	・previous studies have suggested that ～	以前の研究は，～ということを示唆している	331
	・consistent with previous studies	以前の研究と一致して	169
❷	previous work	以前の研究	2,492
❸	previous reports	以前の報告	1,279
	・in contrast to previous reports	以前の報告とは対照的に	116
❹	previous findings	以前の知見	674
❺	previous results	以前の結果	600
❻	previous observations	以前の観察	452
❼	previous data	以前のデータ	421
❽	previous research	以前の研究	415

❶ Previous studies have shown that chronic i.v. treatment with morphine or heroin decreased mu opioid receptor activation of G-proteins in specific brain regions. (Brain Res. 2001 895:1)
訳 以前の研究は，～ということを示している

❶ Consistent with previous studies, treatment of primary cortical cultures from ARE reporter mice revealed selective promoter activity in astrocytes. (J Neurosci. 2004 24:1101)
訳 以前の研究と一致して

P

❷ Previous work **has shown that** astrocytes function as the progenitors of these new neurons through immature intermediate D cells. (J Comp Neurol. 2004 478:359)

訳 以前の研究は，～ということを示している

❸ We conclude that, in contrast to previous reports, Syk does not play a major role in GPCR signaling. (Blood. 2003 101:4155)

訳 以前の報告とは対照的に

★★ previously 副 以前に　　用例数 57,523

過去分詞の前で使われることが多いが，後に来ることもある．主に，過去形や完了形の文で用いられ，現在形の文ではあまり使われない．

◆類義語：before, formerly, already

■ previously ＋［過分］

❶ previously reported	以前に報告された／以前に報告した	5,964
・previously reported that ～	～ということを以前に報告した	1,394
❷ previously shown	以前に示した／以前に示された	4,117
・we have previously shown that ～	われわれは，～ということを以前に示している	1,856
・previously shown to ～	～することが以前に示された	1,668
・be previously shown	…は，以前に示される	1,041
❸ previously identified	以前に同定された／以前に同定した	3,200
❹ previously described	以前に述べられた／以前に述べた	3,114
❺ previously demonstrated	以前に実証された／以前に実証した	1,846
❻ previously unrecognized	これまでに知られていない	1,339
❼ previously characterized	以前に特徴づけられた	1,231
❽ previously observed	以前に観察された	947
❾ previously published	以前に発表された	932

❶ We previously reported that LeCTR1 expression is up-regulated in response to ethylene. (Plant Mol Biol. 2004 54:387)

訳 われわれは，～ということを以前に報告した

❷ We have previously shown that CREB-binding protein (CBP/p300) act as an important SOX9 co-activator during chondrogenesis. (J Biol Chem. 2005 280:8343)

訳 われわれは，～ということを以前に示した

❷ Protein expression was particularly dense in **regions** previously shown to **contain known GABA$_B$ receptor subunits**. (Brain Res. 2003 989:135)

訳 既知のGABA$_B$受容体サブユニットを含むことが以前に示された領域

❸ We previously identified the Salmonella-specific regulatory protein RtsA, which induces expression of hilA and, thus, the SPI1 genes. (J Bacteriol. 2004 186:68)

訳 われわれは，以前に～を同定した

❹ This β-oxidation pathway **is distinct from the** previously described **aerobic fatty acid degradation pathway** and requires enzymes encoded by two operons, yfcYX and ydiQRSTD. (J Biol Chem. 2004 279:37324)

訳 ～は，以前に述べられた好気的脂肪酸分解経路とは別個である

■ [過去／過分] ＋ previously

❶ shown previously	以前に示した／以前に示された	1,460
・we have shown previously that ～	われわれは，～ということを以前に示している	614
・shown previously to ～	～することが以前に示された	547
❷ reported previously	以前に報告された／以前に報告した	1,157
・reported previously that ～	～ということを以前に報告した	337
❸ described previously	以前に述べられた	474
❹ demonstrated previously	以前に実証した／以前に実証された	396

❶ We have shown previously that mice lacking the Nrf2 are more susceptible to hyperoxia than are wild-type mice.（J Biol Chem. 2004 279:42302）

訳 われわれは，～ということを以前に示した

❷ We reported previously that the neuropeptide oxytocin attenuates stress-induced hypothalamo-pituitary-adrenal（HPA）activity and anxiety behavior.（J Neurosci. 2004 24:2974）

訳 われわれは，～ということを以前に報告した

■ than previously

★ than previously ～	以前に～より	1,758
❶ than previously thought	以前に考えられたより	415
❷ than previously reported	以前に報告されたより	294

❶ Both the primary and carrier states of infection with EBV are more complex than previously thought.（Virol. 2003 77:1840）

訳 ～は，以前に考えられたより複雑である

■ Previously,

★ Previously,	以前に，	2,989
❶ Previously, we showed that ～	以前に，われわれは～ということを示した	462
❷ Previously, we reported ～	以前に，われわれは～を報告した	363

❶ Previously, we showed that inactivation of Ku70 in Arabidopsis results in telomere lengthening.（Proc Natl Acad Sci USA. 2003 100:611）

訳 以前に，われわれは～ということを示した

★ primarily 副 主に／一次的に　　用例数 13,856

過去分詞の前で使われることも，後で使われることもある．

◆類義語：principally, mainly, mostly, predominantly, chiefly, exclusively

■ primarily ＋ [過分／形／接]

❶ primarily due to ～	主に～のせいで	380
❷ primarily responsible for ～	～に主に責任のある	373

P

❸ primarily because of ～	主に～の理由で	160
❹ primarily mediated by ～	～によって主に仲介される	116
❺ primarily expressed in ～	～において主に発現する	115
❻ primarily associated with ～	～と主に関連する	115

❶ Rather, the poor L1 expression is primarily due to inadequate transcriptional elongation. (Nature. 2004 429:268)
訳 ～は，主に不適当な転写の伸長のせいである

■ ［過分／形／動］＋ primarily ＋［前］

❶ expressed primarily in ～	主に～において発現する	232
❷ mediated primarily by ～	主に～によって仲介される	212
❸ due primarily to ～	主に～のせいで	177
❹ based primarily on ～	主に～に基づいて	113
❺ found primarily in ～	主に～において見つけられる	112
❻ result primarily from ～	…は，主に～に由来する	112
❼ determined primarily by ～	主に～によって決定される	100

❶ The cpk gene is expressed primarily in the kidney and liver and encodes a hydrophilic, 145-amino acid protein, which we term cystin. (J Clin Invest. 2002 109:533)
訳 cpk遺伝子は，主に腎臓および肝臓において発現する

❷ The nuclear import and export of macromolecular cargoes through nuclear pore complexes is mediated primarily by carriers such as importin-β. (Nature. 2004 432:872)
訳 ～は，主にインポーチン-βのような運搬体によって仲介される

❸ Acute coronary occlusions leading to ST-segment elevation myocardial infarctions (STEMIs) are due primarily to rupture of atherosclerotic plaques. (Circulation. 2004 110:278)
訳 …は，主に～の破裂のせいである

★★ primary 形 一次の／初代の／原発性の　用例数 51,028

◆類義語：principal, main, predominant, major

■ primary and ＋［形］

❶ primary and secondary ～	一次性および二次性の～	865
❷ primary and metastatic ～	原発性および転移性の～	206

❶ Controversy about the effectiveness of primary and secondary prevention in pregnant women is discussed. (Lancet. 2004 363:1965)
訳 妊婦における一次性および二次性の予防の有効性に関する論争が議論される

■ primary ＋［名］

❶ primary care	プライマリケア	1,549
❷ primary outcome	一次結果	1,171
❸ primary tumors	原発腫瘍	974
❹ primary cultures	初代培養	889

❺	primary sequence	一次配列	653
❻	primary infection	一次感染	477
❼	primary structure	一次構造	415
❽	primary prevention	一次予防	393

principle ㊔ 原理　　用例数 4,597

principle	原理	2,309
principles	原理	2,288

複数形のprinciplesの割合が約50%と非常に多い．

◆類義語：rationale

P

■ ［前］＋ principle

❶	in principle	原理的には	517
❷	proof of principle	原理の証明	485

❶ Tobacco rattle virus (TRV), a bipartite RNA virus, **has a wide host range and so in principle could serve as an efficient vector for** VIGS in a diverse array of plant species. (Plant J. 2005 44:334)

訳 …は，広い宿主レンジを持ち，そしてそれで原理的には～の効率的なベクターとして働きうる

❷ Initially, **as a proof of principle**, green fluorescent protein-siRNA was administered intratracheally into transgenic mice overexpressing green fluorescent protein. (Am J Pathol. 2005 167:1545)

訳 原理の証明として，

■ principles ＋ ［前］

	the %	a/an %	⌀ %			
❶	57	-	38	principles of ～	～の原理	617
❷	14	-	81	principles for ～	～のための原理	118

❶ **Recent studies of the fundamental principles of microbial cellulose utilization support** the feasibility of CBP. (Curr Opin Biotechnol. 2005 16:577)

訳 微生物のセルロース利用の基本的な原理の最近の研究は，～を支持する

■ ［形／名］＋ principles

❶	first principles	第一原理	239
❷	design principles	デザイン原理	144
❸	general principles	一般原理	118

prior 形 前の／事前の／以前の，副 前に　用例数 7,960

prior toの用例が非常に多い．

◆類義語：former, previous, recent

■ prior to

★ prior to ～	～の前	7,960
❶ prior to ～ing	～する前	535
❷ occur prior to ～	…は，～の前に起こる	230
❸ just prior to ～	～の直前	168
❹ prior to the onset of ～	～の開始の前	153
❺ … min prior to ～	～の前の…分間	124
❻ … days prior to ～	～の前の…日間	123

❸ Our results strongly suggest that this molecule is induced and maintained on T_{reg} **following or just prior to their arrival in tissues.** (J Immunol. 2005 174:5444)

訳 組織におけるそれらの出現のあとあるいは直前

❺ In addition, home cage behaviors **were recorded for 30 min prior to drug treatment** and for 2 h following drug treatment. (Brain Res. 2005 1044:176)

訳 ～が，薬物治療の前の30分間に記録された

■ prior ＋［名］

❶ prior studies	以前の研究	617
❷ prior knowledge	以前の知識	212

probability 名 確率／蓋然性　用例数 7,181

probability	確率／蓋然性	6,069
probabilities	確率／蓋然性	1,112

複数形のprobabilitiesの割合が，約15%ある可算名詞．probability ofの用例が非常に多い．

◆類義語：chance, likelihood, feasibility, possibility

■ probability ＋［前／同格that節］

probability thatの前にtheが付く割合は圧倒的に高い．また，probability ofの前にtheが付く割合もかなり高い．

	the%	a/an%	ø%			
❶	67	15	18	probability of ～	～の確率（蓋然性）	2,956
				・probability of ～ing	～する確率	637
❷	87	11	2	probability that ～	～という確率	376

❶ **The probability of death within 100 days was 11%** (95% CI, 8% to 14%). (Ann Intern Med. 2000 133:504)

訳 100日以内での死亡の確率は，11%であった

❷ Mutation of CIN genes increases the probability that whole chromosomes or large fractions of chromosomes are gained or lost during cell division.（Proc Natl Acad Sci USA. 2002 99:16226）

訳 CIN遺伝子の変異は，～という確率を増大させる

■ [形／名] + probability

❶ open probability	開確率	582
❷ release probability	放出確率	240
❸ high probability	高い確率	205

★ probably 副 おそらく／ありそうに　用例数 7,128

P

◆類義語：presumably, conceivably, likely, maybe, perhaps, possibly, potentially

■ probably due/because

❶ probably due to～	おそらく～のせいで	375
・be probably due to～	…は，おそらく～のせいである	172
❷ , probably because ～	おそらく～なので	187
・, probably because of ～	おそらく～のせいで	101

❶ This reduction is probably due to a downregulation of the Epidermal Growth Factor receptor (Egfr) pathway, since an activated form of the Egfr can rescue the phenotype of embryos mutant for the PS integrins.（Development. 2000 127:2607）

訳 この低下は，おそらく～の下方制御のせいである

❷ Nitrocefin turnover was significantly increased, probably because of an increased rate of breakdown of the intermediate species due to a lack of stabilizing forces.（J Biol Chem. 2002 277:24744）

訳 おそらく～の分解の増大した速度のせいで

■ probably + [動]

❶ probably reflect ～	…は，おそらく～を反映する	189
❷ probably contribute to ～	…は，おそらく～に寄与する	141
❸ probably represent ～	…は，おそらく～を示す	138
❹ probably result from ～	…は，おそらく～に由来する	105

❶ These differences probably reflect different functions of these cortical areas in mediating innate odour preference or associative memory.（Nature. 2011 472:191）

訳 これらの違いは，おそらく～の異なる機能を反映する

■ probably + [前]

❶ , probably by ～	おそらく～によって	219
❷ , probably through ～	おそらく～によって	119

❶ Candida albicans maintains a commensal relationship with human hosts, probably by adhering to mucosal tissue in a variety of physiological conditions.（Infect Immun. 1999 67:

6040)
訳 おそらく粘膜組織に接着することによって

■ most probably

❶	most probably	ほとんどおそらく／最もありそうに	239

❶ The POU domain **most probably** binds to Ets-2 directly, while the N-terminal domain inhibits transcription. (Mol Cell Biol. 2001 21:7883)
訳 POUドメインは，ほとんどおそらくEts-2に直接結合するであろう

★ probe ㊔ プローブ，㊅ 探索する　用例数 20,208

probe	プローブ／～を探索する	10,540
probes	プローブ／～を探索する	6,287
probing	～を探索する	1,851
probed	～を探索した／探索される	1,530

名詞の用例が非常に多いが，動詞としても使われる．複数形のprobesの割合は約50%と非常に高い．

◆類義語：explore, test, study, investigate, examine, survey, look at, search, dissect

■ ㊔ probes ＋［前］

	the%	a/an%	ø%			
❶	3	-	88	probes for ～	～のためのプローブ	525
❷	3	-	87	probes to *do*	～するプローブ	385

❶ Fluorescent **probes for** the direct imaging of cellular hypoxia could be useful tools that complement radiochemical imaging and immunohistochemical staining methods. (J Org Chem. 2012 77:3531)
訳 細胞の低酸素状態の直接イメージングのための蛍光プローブは，有用なツールでありうる

■ ㊅ probe ＋［名句］

❶	probe the role of ～	…は，～の役割を探索する	143

❶ To **probe the role of** this amino acid in KPC-2, we performed site-saturation mutagenesis. (J Biol Chem. 2012 287:31783)
訳 このアミノ酸の役割を探索するために

■ ㊅ to probe

★	to probe ～	～を探索する…	2,182
❶	used to probe ～	～を探索するために使われる	483
❷	To probe ～	～を探索するために	342

❶ This approach could also be **used to probe** associated metabolic diseases. (J Biol Chem. 2012 287:31929)
訳 このアプローチは，また，関連する代謝性疾患を探索するために使われうる

■ 動 be probed ＋［前］

★	be probed	…は，探索される	769
❶	be probed by ～	…は，～によって探索される	260
❷	be probed with ～	…は，～を使って探索される	148

❶ Although **protein properties can** be probed by **mutagenesis**, this approach has been limited by its low throughput.（Proc Natl Acad Sci USA. 2012 109:16858）
訳 タンパク質の性質は，変異誘発によって探索されうる

★ problem 名 問題 用例数 11,803

problem	問題	6,677
problems	問題	5,126

複数形のproblemsの割合は約45%とかなり高い．
◆類義語：question, matter, issue

■ problem ＋［前］

problem ofの前にtheが付く割合は圧倒的に高い．

	the%	a/an%	ø%			
❶	81	19	0	problem of ～	～の問題	917
				・problem of ～ing	～する際の問題	265
❷	9	83	1	problem in ～	～における問題	861
❸	18	75	1	problem for ～	～に対する問題	249

❶ **We address the** problem of **identify**ing **gene transcriptional modules** from gene expression data by proposing a new approach.（Bioinformatics. 2007 23:473）
訳 われわれは，遺伝子転写モジュールを同定する際の問題に取り組む

❷ **Measles remains an important** problem in **Africa**, where human immunodeficiency virus type 1（HIV-1）infection is prevalent.（J Infect Dis. 2005 192:1950）
訳 麻疹は，アフリカにおいて重要な問題のままである

■ problems ＋［過分］

❶	problems associated with ～	～に関連する問題	185

❶ Adeno-associated viral（AAV）vectors **overcome many of the** problems associated with **other vector systems.**（Hum Mol Genet. 2002 11:733）
訳 ～は，他のベクターシステムに関連する問題の多くを克服する

■［動］＋ this problem

❶	to address this problem	この問題に取り組むために	220
❷	to overcome this problem	この問題を克服するために	97

❶ To address this problem, we demonstrate a method based on graph theory that quantifies the

P

relative importance of viral variants. (Bioinformatics. 2012 28:1624)
訳 この問題に取り組むために

■ [名/形] + problem

❶	health problem	健康問題	433
❷	clinical problem	臨床問題	226
❸	major problem	主要な問題	153

★ procedure 名 方法/処置/手術/手順 用例数 14,392

procedure	方法/処置	8,207
procedures	方法/処置	6,185

複数形のproceduresの割合は約45%とかなり高い.
◆類義語：protocol, method

■ procedure + [前]

	the%	a/an%	ø%			
❶	10	82	4	procedure for ～	～のための方法	696
				・procedure for ～ing	～するための方法	209
❷	18	70	1	procedure to *do*	～するための方法	430
❸	31	58	1	procedure in ～	～における方法	234

❶ We describe a procedure for isolating agonists for mammalian G protein-coupled receptors of unknown function. (Nat Biotechnol. 1998 16:1334)
訳 われわれは，～の作動薬を単離するための方法を述べる

proceed 動 進行する/進む 用例数 4,216

proceeds	進行する	2,026
proceed	進行する	1,390
proceeded	進行した	568
proceeding	進行する	232

自動詞.
◆類義語：progress, advance, improve

■ proceed + [前]

❶	proceed through ～	…は，～を経て進行する	642
❷	proceed via ～	…は，～を経て進行する	582
❸	proceed in ～	…は，～において進行する	432
❹	proceed with ～	…は，～と共に進行する	370
❺	proceed by ～	…は，～によって進行する	327

❻	proceed to ～	…は，～に進行する	322

❶ The reaction **proceeds through** the formation of a carbodiimide, followed by a sequential addition-dehydration with acyl hydrazides. (J Org Chem. 2005 70:6362)
訳 その反応は，カルボジイミドの形成を経て進行する

❷ These findings indicate that decarboxylation **proceeds via** a dienolate intermediate. (Biochemistry. 2005 44:13932)
訳 脱炭酸反応は，～を経て進行する

■ ［名］＋ proceed

❶	reaction proceeds	反応が，進行する	217

P

★★ process 名 過程　　用例数 67,169

process	過程	38,990
processes	過程	28,179

複数形のprocessesの割合は約40%とかなり高い．
◆類義語：course, procedure

■ 名 process ＋［前］

process ofの前にtheが付く割合は圧倒的に高い．process by whichの前にtheが付く割合も非常に高い．

	the%	a/an%	ø%			
❶	83	14	1	process of ～	～の過程	4,525
				・in process of ～	～の過程において	767
				・process of ～ing	～する過程	488
				・during process of ～	～の過程の間に	245
❷	32	51	1	process in ～	～における過程	1,789
❸	77	23	0	process by which ～	（それによって）～である過程	395

❶ The hemagglutinin-neuraminidase (HN) protein of Newcastle disease virus (NDV) plays a crucial role in the **process of** infection. (J Virol. 2004 78:4176)
訳 ～は，感染の過程において決定的に重要な役割を果たす

❷ In addition to these shared pathways, it appears that developmental **processes in** Hydra make use of pathways involving a variety of peptides. (Dev Biol. 2002 248:199)
訳 ヒドラにおける発生過程は，さまざまなペプチドを含む経路を利用する

❸ DNA replication is the **process by which** cells make one complete copy of their genetic information before cell division. (Science. 2001 294:96)
訳 DNA複製は，細胞がそれらの遺伝情報のひとつの完全なコピーを作る過程である

■ 名 process ＋［動］

❶	process involves ～	過程は，～を含む	225
❷	process requires ～	過程は，～を必要とする	208

❸ process occurs ～	過程が，起こる	124

❷ This process requires the participation of the CD28 costimulatory receptor. (J Immunol. 2004 173:632)
訳 この過程は，～の関与を必要とする

■ 名 process(es) ＋［現分／過分］

❶ processes including ～	～を含む過程	1,512
❷ processes such as ～	～のような過程	1,124
❸ process involving ～	～を含む過程	486
❹ processes involved in ～	～に関与する過程	279
❺ process called ～	～と呼ばれる過程	231
❻ processes underlying ～	～の根底にある過程	208
❼ process termed ～	～と名付けられた過程	201
❽ process known as ～	～として知られている過程	197
❾ processes associated with ～	～と関連する過程	165

❶ Stat3 plays diverse roles in biological processes including cell proliferation, survival, apoptosis, and inflammation. (J Immunol. 2004 172:7703)
訳 Stat3は，細胞増殖を含む生物学的過程において多様な役割を果たす

❸ We concluded that spore coat assembly is a dynamic process involving diverse patterns of protein assembly and localization. (J Bacteriol. 2004 186:4441)
訳 ～は，多様なパターンのタンパク質アセンブリーを含む動的な過程である

❹ However, cellular processes involved in the regulation and expression of laccase remain largely unknown in C. neoformans. (Mol Microbiol. 2003 50:1271)
訳 ～の調節と発現に関与する細胞過程

❺ Many long-lasting memories require a process called consolidation, which involves the exchange of information between the cortex and hippocampus. (Nature. 2004 431:699)
訳 多くの長期記憶は，固定と呼ばれる過程を必要とする

❽ Loss of cell attachment to the ECM causes apoptosis, a process known as anoikis. (Cell. 2004 116:751)
訳 アノイキスとして知られている過程

■ 名［前］＋ this process

❶ in this process	この過程に	674
・involved in this process	この過程に関与する	115
・role in this process	この過程における役割	105

❶ The protease(s) involved in this process has not been fully characterized. (J Neurosci. 2002 22:4842)
訳 この過程に関与するプロテアーゼは，十分には特徴づけられていない

■ 名［形］＋ process(es)

❶ cellular processes	細胞過程	623
❷ biological processes	生物学的過程	391

❸	developmental processes	発生過程	251
❹	physiological processes	生理学的過程	163
❺	～-dependent process	～依存性の過程	140

★ processing ㊔ プロセシング／処理 用例数 19,878

複数形の用例はなく，**不可算名詞**として使われる．

◈類義語：treatment

■ processing ＋［前］

	the%	a/an%	ø%			
❶	38	0	60	processing of ～	～のプロセシング	4,217
❷	0	0	90	processing in ～	～におけるプロセシング	1,017
❸	1	0	85	processing by ～	～によるプロセシング	390

❶ **Aβ is produced by proteolytic processing of the amyloid precursor protein (APP).** (Brain Res. 2004 1018:86)

訳 Aβは，アミロイド前駆体タンパク質（APP）のタンパク質分解性プロセシングによって産生される

❷ **Information processing in the nervous system involves** the activity of large populations of neurons. (Trends Neurosci. 1998 21:259)

訳 神経系における情報処理は，～を含む

■［形／名］＋ processing

❶	information processing	情報処理	733
❷	proteolytic processing	タンパク質分解性プロセシング	665
❸	RNA processing	RNAプロセシング	627

★★ produce ㊎ 産生する／作製する／引き起こす 用例数 61,567

produced	産生される／～を産生した／～を引き起こした	29,406
produce	～を産生する	17,057
producing	～を産生する	8,316
produces	～を産生する	6,788

他動詞．

◈類義語：cause, result in, lead to, give rise to, yield, bring about, generate, make, create

■ produced ＋［前］

❶	produced by ～	～によって産生される	7,705
	・be produced by ～	…は，～によって産生される	1,752
❷	produced in ～	～において産生される	2,150

P

❸ produced from ～	～から産生される	758
❹ produced during ～	～の間に産生される	344
❺ produced at ～	～において産生される	303

❶ Leptin, the ob gene product, is produced by adipocytes, and it acts to decrease caloric intake and increase energy expenditure. (Diabetes. 1999 48:272)
訳 ～は，脂肪細胞によって産生される

❷ TPO is produced in the liver and levels are low in patients with cirrhosis. (Hepatology. 2003 37:558)
訳 TPOは，肝臓において産生される

❸ In mammals, NO is produced from Arg by the enzyme NO synthase. (Plant Cell. 2004 16:332)
訳 一酸化窒素は，アルギニンから産生される

■［名／代名］＋ produced

❶ cells produced ～	細胞は，～を産生した	535
❷ we produced ～	われわれは，～を作製した	428
❸ mice produced ～	マウスは，～を産生した	407

❶ NK T cells produced interleukin-4 in the spleen within 24 h of infection, and these cells were $CD4^-$. (Infect Immun. 2002 70:2215)
訳 NK T細胞は，感染の24時間以内に脾臓においてインターロイキン-4を産生した

■ produce ＋［名句］

❶ produce high levels of ～	…は，高いレベルの～を引き起こす	203
❷ produce an increase in ～	…は，～の増大を引き起こす	152
❸ produce a dose-dependent ～	…は，用量依存的な～を引き起こす	118

❷ Lateral striatal lesions produced an increase in late responses. (Behav Neurosci. 1999 113:253)
訳 ～は，後期応答の増大を引き起こした

■［名／動／形］＋ to produce

★ to produce ～	～を産生する…	7,995
❶ fail to produce ～	…は，～を産生することができない	386
❷ ability to produce ～	～を産生する能力	369
❸ be used to produce ～	…は，～を産生するために使われる	229
❹ be shown to produce ～	…は，～を産生すると示される	143
❺ be sufficient to produce ～	…は，～を産生するのに十分である	127

❶ This defect extended to NK cells, which also failed to produce IFN-γ when stimulated by IL-12. (J Immunol. 2001 166:5712)
訳 ～は，IFN-γを産生することができなかった

❷ These data suggest that strain 93246 is nonencapsulated N. meningitidis but has the ability to produce extracellular amylopectin from sucrose. (J Clin Microbiol. 2003 41:273)
訳 ～は，細胞外アミロペクチンを産生する能力を持つ

☆ product 名 産物　　用例数 37,557

product	産物	18,848
products	産物	18,709

複数形のproductsの割合は約50%と非常に高い.

■ product(s) + [前]

product ofの前にtheが付く割合は非常に高い.

	the%	a/an%	⌀%			
❶	71	27	1	product of ～	～の産物	3,872
❷	20	-	66	products in ～	～における産物	770

P

❶ The formation of estrogens from C19 steroids is catalyzed by aromatase cytochrome P450 (P450arom), **the product of the cyp19 gene**. (Proc Natl Acad Sci USA. 1998 95:6965)

訳 cyp19遺伝子の産物

❷ **Expression of EBV-encoded gene products in T lymphocytes could contribute to viral pathogenesis** during acute EBV infection as well as in individuals coinfected with EBV and HIV. (J Immunol. 1999 163:6261)

訳 Tリンパ球におけるEBVにコードされる遺伝子産物の発現は，ウイルスの病因に寄与しうる

■ [名／形] + product

❶	gene product	遺伝子産物	3,872
❷	natural product	天然物	859
❸	protein product	タンパク質産物	594

❶ Here **we identify Tsg gene products from human, mouse, Xenopus, zebrafish and chick.** (Nature. 2001 410:475)

訳 われわれは，ヒト，マウス，アフリカツメガエル，ゼブラフィッシュおよびニワトリからTsg遺伝子産物を同定する

❷ **Other natural products with specific biologic effects may modulate the activity of FXR** or other relatively promiscuous nuclear hormone receptors. (Science. 2002 296:1703)

訳 特異的な生物学的効果を持つ他の天然物は，FXRの活性を調節するかもしれない

☆ production 名 産生　　用例数 46,923

production	産生	46,907
productions	産生	16

複数形のproductionsの割合はほとんどなく，原則，不可算名詞として使われる.

◆類義語：generation

■ production ＋［前］

	the%	a/an%	ø%			
❶	45	2	48	production of ～	～の産生	15,548
				・production of ～ in …	…における～の産生	912
				・increased production of ～	増大した～の産生	592
				・result in the production of ～	…は，～の産生という結果になる	297
				・lead to the production of ～	…は，～の産生につながる	264
				・…-induced production of ～	…に誘導される～の産生	242
				・required for the production of ～	～の産生のために必要とされる	116
❷	1	0	97	production in ～	～における産生	3,912
❸	1	0	59	production by ～	～による産生	2,891

❶ Nutrient secretagogues **can increase the production of succinyl-CoA in rat pancreatic islets.** (Diabetes. 2002 51:2669)
訳 ～は，ラットの膵島におけるスクシニルCoAの産生を増大させうる

❶ Development of T cell defects in tumor-bearing hosts **are often associated with increased production of immature myeloid cells.** (J Immunol. 2001 166:5398)
訳 ～は，未成熟の骨髄細胞の増大した産生としばしば関連する

❶ Expression of either CoFADX-1 or CoFADX-2 in somatic soybean embryos **resulted in the production of calendic acid.** (J Biol Chem. 2001 Jan 276:2637)
訳 ～は，カレンジン酸の産生という結果になった

❶ Importantly, the signaling response elicited by β-amyloid and prion fibrils **leads to the production of neurotoxic products.** (J Neurosci. 1999 19:928)
訳 ～は，神経毒性のある産物の産生につながる

❷ We have previously shown that colonic epithelial cells **are a major site of MIP-α production in human colon** and that enterocyte MIP-3α protein levels are elevated in inflammatory bowel disease. (J Biol Chem. 2003 278:875)
訳 ～は，ヒトの大腸におけるMIP-α産生の主要な部位である

❸ We and others have demonstrated that the **Stat4 is critical for IFN-γ production by activated T cells** and Th1 cells. (J Immunol. 2000 165:1374)
訳 Stat4は，活性化されたT細胞によるIFN-γ産生にとって決定的に重要である

★profile ㊔ プロファイル／特性　　用例数 17,921

profiles	プロファイル／特性	9,248
profile	プロファイル／特性	8,673

複数形のprofilesの割合は約50%と非常に高い．

◆類義語：property, characteristic, character, hallmark, feature

■ profile(s) ＋ [前]

profile ofの前にtheが付く割合は圧倒的に高い．

	the%	a/an%	⌀%			
❶	81	17	2	profile of ～	～のプロファイル	2,357
❷	11	-	81	profiles in ～	～におけるプロファイル	579
❸	23	-	74	profiles for ～	～のプロファイル	381

❶ Here, we analyzed the gene expression profile of ET-1-stimulated mesangial cells to identify determinants of collagen accumulation. (J Biol Chem. 2011 286:11003)

訳 われわれは，ET-1刺激されたメサンギウム細胞の遺伝子発現プロファイルを分析した

❷ The IgG subclass profiles in SP are also similar to those in serum. (J Immunol. 2005 175:4127)

訳 SPにおけるIgGサブクラスのプロファイルは，また，血清におけるそれらと似ている

❸ Gene expression profiles for each tumor class were used to complete unsupervised hierarchical clustering analyses and identify differentially expressed genes contributing to these associations. (Cancer Res. 2005 65:10602)

訳 それぞれの腫瘍のクラスの遺伝子発現プロファイルが，～を完了するために使われた

■ [名] ＋ profile(s)

❶	expression profiles	発現プロファイル	2,019
❷	safety profile	安全プロファイル	355
❸	cytokine profile	サイトカインプロファイル	300

profoundly 副 大いに／深く 用例数 1,472

◈類義語：deeply, greatly, extremely, very, markedly, vastly

■ profoundly ＋ [動]

❶	profoundly affect ～	…は，～に大いに影響する	180

❶ The bone morphogenetic proteins (BMPs) profoundly affect embryonic development, differentiation and disease. (Nat Med. 2005 11:387)

訳 骨形成タンパク質（BMP）は，胚発生，分化および疾患に大いに影響する

prognosis 名 予後 用例数 4,118

prognosis	予後	4,006
prognoses	予後	112

複数形のprognosesの割合は3％あるが，単数形は無冠詞で使われることが多い．

■ prognosis ＋ [前]

	the%	a/an%	⌀%			
❶	3	12	85	prognosis in ～	～における予後	453

P

❷	41	2	57	prognosis of ～	～の予後	423
❸	43	15	42	prognosis for ～	～の予後	225

❶ **High serum LDH levels are associated with poor prognosis in patients with cancer**, including renal cell carcinoma (RCC). (J Clin Oncol. 2012 30:3402)

訳 高血清LDHレベルは，癌の患者におけるよくない予後と関連している

■ [形] + prognosis

❶ poor prognosis	よくない予後	1,145
❷ worse prognosis	より悪い予後	115
❸ good prognosis	よい予後	101

★ program 名 プログラム／計画，動 プログラムする 用例数 17,461

program	プログラム／計画／～をプログラムする	8,494
programs	プログラム／計画／～をプログラムする	4,873
programmed	プログラムされた／プログラムした	3,201
programming	～をプログラムする／プログラミング	893

名詞の用例が断然多いが，動詞としても用いられる．複数形のprogramsの割合が，約35%ある可算名詞．

◆類義語：plan, planning, project, design, schedule

■ 名 program + [前]

	the%	a/an%	ø%			
❶	48	50	0	program of ～	～のプログラム	722
❷	43	49	2	program in ～	～におけるプログラム	370
❸	24	69	1	program for ～	～のプログラム	354

❶ **Diseased skin often exhibits a deregulated program of the keratinocyte maturation** necessary for epidermal stratification and function. (J Invest Dermatol. 2005 125:294)

訳 患部皮膚は，しばしばケラチノサイト成熟の調節解除されたプログラムを示す

❷ We also present evidence that the ecdysone-induced Broad Complex of zinc finger transcription factor genes **is required for full activation of the myogenic program in these cells**. (Dev Biol. 2005 288:612)

訳 ～は，これらの細胞における筋原性プログラムの完全活性化のために必要とされる

❸ Because of the seriousness of smoking among childhood cancer survivors, this intervention model **may be appropriate as a multicomponent treatment program for survivors** who smoke. (J Clin Oncol. 2005 23:6516)

訳 ～は，生存者のための多成分治療プログラムとして適切であるかもしれない

■ 名 [名／形] + program

❶ differentiation program	分化プログラム	310
❷ transcriptional program	転写プログラム	275

❸	computer program	コンピュータープログラム	215
❹	developmental program	発生プログラム	204

■ 動 programmed

❶	programmed cell death	プログラム細胞死	1,801
❷	programmed to ～	～するようにプログラムされる	123

❷ Most tissues are patterned so that **progenitors in different locations are programmed to have different properties**.（Dev Biol. 2005 283:29）
訳 異なる部位の前駆体は，異なる性質を持つようにプログラムされている

P

progress 名 進歩／進行，動 進行する　用例数 7,843

progress	進歩／進行／進行する	5,646
progressed	進行した	1,167
progresses	進行する／進歩	605
progressing	進行する	425

名詞および自動詞として用いられる．複数形のprogressesの用例はほとんどなく，原則，**不可算名詞**として使われる．
◆類義語：progression, advance, proceed, improve

■ 名 progress ＋［前］

progress ofの前にtheが付く割合は圧倒的に高い．

	the％	a/an％	ø％			
❶	11	0	85	progress in ～	～における進歩	1,684
				・progress in understanding ～	～を理解する上での進歩	271
❷	82	0	17	progress of ～	～の進歩	221
❸	4	0	93	progress toward ～	～への進歩	172

❶ Here **we summarize recent progress in the biology of leptin**, concentrating on its central nervous system（CNS）actions.（Nat Neurosci. 1998 1:445）
訳 われわれは，レプチンの生物学における最近の進歩を要約する

■ 名 progress ＋［動］

❶	progress has been made in ～	進歩が，～においてなされてきた	543

❶ **Rapid progress has been made in understanding** the synaptic changes required for memory encoding.（Nat Neurosci. 2002 5:1035）
訳 急速な進歩が，～を理解する上でなされてきた

■ 名［形］＋ progress

❶	recent progress	最近の進歩	685

❷ significant progress	著しい進歩	300
❸ considerable progress	かなりの進歩	157

■ ㊫ progress ＋［前］

❶ progress to ～	…は，～に進行する	1,132
❷ progress through ～	…は，～を経て進行する	335
❸ progress from ～	…は，～から進行する	244

❶ In contrast, lesions induced by high-risk HPV types have the potential to progress to cancer. (J Virol. 2001 75:7564)
訳 ～は，癌に進行する潜在能を持つ

❷ These hyperproliferative lesions appeared to progress through two distinct stages. (Cancer Res. 1998 58:4314)
訳 これらの過剰増殖性の病変は，2つの別個のステージを経て進行するように思われた

★ progression ㊔ 進行　用例数 24,689

progression	進行	24,646
progressions	進行	43

複数形のprogressionsの割合は0.2%しかなく，原則，**不可算名詞**として使われる．
◆類義語：progress, advance

■ progression ＋［前］

progression ofの前にtheが付く割合はかなり高い．

	the %	a/an %	ø %			
❶	69	3	28	progression of ～	～の進行	5,596
				・progression of atherosclerosis	アテローム性動脈硬化の進行	185
				・progression of disease	疾患の進行	156
				・progression of prostate cancer	前立腺癌の進行	141
❷	1	0	92	progression in ～	～における進行	1,631
				・disease progression in ～	～における疾患進行	394
				・cell cycle progression in ～	～における細胞周期進行	317
❸	10	2	84	progression to ～	～への進行	1,299
				・progression to AIDS	エイズへの進行	155
❹	6	2	89	progression through ～	～を経た進行	692
				・progression through the cell cycle	細胞周期を経た進行	156

❶ A component of mycoplasma respiratory diseases is immunopathologic, suggesting that lymphocyte activation is a key event in the progression of these chronic inflammatory diseases. (J Immunol. 2002 168:3493)
訳 リンパ球活性化は，これらの慢性炎症性疾患の進行における鍵となる事象である

❷ We have recently reported that overrepresentation of 8q24 (c-myc) is associated with clinical progression in prostate cancer. (Am J Pathol. 2002 160:1799)

訳 ～は，前立腺癌における臨床的な進行と関連する

❸ Survival analyses with respect to the rate of progression to AIDS were performed to identify the effects of closely related HLA-B*35 subtypes with different peptide-binding specificities. (N Engl J Med. 2001 344:1668)

訳 AIDSへの進行の割合に関する生存率分析が，～を同定するために行われた

❹ Cyclin D1 is a critical regulator involved in cell cycle progression through the G1 phase into the S phase, thereby contributing to cell proliferation. (Cancer Res. 2005 65:9934)

訳 サイクリンD1は，G1期を経た細胞周期進行に関与する決定的に重要な調節因子である

■［前］＋ progression

❶ rate of progression	進行の速度	152
❷ risk of progression	進行のリスク	127
❸ associated with progression	進行と関連した	113

■［名／形］＋ progression

❶ cell cycle progression	細胞周期進行	3,120
❷ disease progression	疾患進行	2,991
❸ tumor progression	腫瘍進行	1,881
❹ cancer progression	癌進行	1,120

P

progressive 形 進行性の　　用例数 7,546

■ progressive ＋［名］

❶ progressive disease	進行性の疾患	479
❷ progressive loss of ～	～の進行性の喪失	354
❸ progressive increase in ～	～の進行性の増大	208

❷ Huntington's disease (HD) is characterized by a progressive loss of neurons in the striatum and cerebral cortex and is caused by a CAG repeat expansion in the gene encoding huntingtin. (J Neurosci. 2002 22:8266)

訳 ハンチントン病（HD）は，ニューロンの進行性の喪失によって特徴づけられる

❸ Induction of atherosclerosis in LDL receptor-null mice by feeding them a high-fat diet resulted in a progressive increase in Egr-1 expression in the aorta. (J Clin Invest. 2000 105:653)

訳 ～は，Egr-1発現の進行性の増大という結果になった

■［前］＋ progressive

❶ characterized by progressive ～	進行性～によって特徴付けられる	164
❷ patients with progressive ～	進行性～の患者	138
❸ lead to progressive ～	…は，進行性～につながる	132

❷ Altered bone turnover may be a diagnostic or therapeutic target in patients with progressive OA.（Arthritis Rheum. 2002 46:3178）

訳 ～は，進行性骨関節炎の患者における診断上あるいは治療上の標的であるかもしれない

progressively 副 進行性に／次第に 用例数 2,401

自動詞の後で使われることも，前で使われることもある．

◆類義語：continuously, successively

■ ［動］＋ progressively

❶	increase progressively	…は，進行性に増大する	180
❷	become progressively ～	…は，進行性に～になる	168

❶ Additionally, the proportion of revertant viruses increased progressively during the course of infection in these animals, and two of these animals developed fatal SAIDS.（J Virol. 1998 72:5820）

訳 ～は，感染の過程の間に進行性に増大した

❷ Interestingly, retinomas demonstrate a low level of genomic instability that becomes progressively more severe in retinoblastoma tumors.（Curr Opin Ophthalmol. 2009 20:351）

訳 ～は，網膜芽細胞腫腫瘍において進行性により重篤になる

■ progressively ＋［動］

❶	progressively increase	…は，進行性に増大する	259
❷	progressively decrease	…は，進行性に低下する	109

❶ The degree of morphological change progressively increased with successive rounds of confluence.（Cancer Res. 2002 62:4605）

訳 形態変化の程度は，～と共に進行性に増大した

project 動 投射する／計画する，名 プロジェクト 用例数 6,484

project	投射する／～を計画する／プロジェクト	3,359
projects	投射する／～を計画する／プロジェクト	1,306
projecting	投射する／～を計画する	971
projected	投射した／～を計画した／計画される	848

自動詞の用例が多いが，他動詞や名詞としても使われる．

■ project ＋［前］

❶	project to ～	…は，～へ投射する	1,505
	・neurons projecting to ～	～に投射するニューロン	116
❷	project from ～	…は，～から投射する	165

❶ These actions are believed to involve connections with dopamine (DA) neurons projecting to the nucleus accumbens (NAc).（J Comp Neurol. 2006 494:863）

訳 これらの作用は，側坐核（NAc）へ投射するドーパミン（DA）ニューロンとの連絡に関わ

ると信じられている

projection ㊔ 投射　用例数 6,401

projections	投射	3,596
projection	投射	2,805

複数形のprojectionsの割合は約55%と非常に高い.

■ projections ＋ [前]

	the%	a/an%	ø%			
❶	6	-	81	projections to ～	～への投射	705
❷	12	-	87	projections from ～	～からの投射	518
❸	6	-	81	projections of ～	～の投射	334

❶ The prefrontal cortex and **the hippocampus exhibit converging projections to the nucleus accumbens** and have functional reciprocal connections via indirect pathways. (Neuron. 2005 47:255)

訳 海馬は，側坐核への収束する投射を示す

❷ **We used retrograde tracing techniques to examine the projections from the inferior colliculus to the cochlear nucleus** in guinea pigs. (J Comp Neurol. 2001 429:206)

訳 われわれは，下丘から蝸牛神経核への投射を調べるために逆行性追跡技術を使用した

★★ proliferation ㊔ 増殖　用例数 34,347

proliferation	増殖	34,295
proliferations	増殖	52

複数形のproliferationsの割合は0.2%しかなく，原則，不可算名詞として使われる.

◆類義語：growth, replication

■ proliferation ＋ [前]

	the%	a/an%	ø%			
❶	41	3	56	proliferation of ～	～の増殖	4,792
❷	1	0	93	proliferation in ～	～における増殖	2,586

❶ Atrial natriuretic peptide (ANP) **inhibits the proliferation of many cells**, in part through interfering with signal transduction enacted by G protein-coupled growth factor receptors. (J Biol Chem. 2000 275:7365)

訳 ～は，多くの細胞の増殖を抑制する

❷ We now report that **human cell proliferation in NOD-scid mice increased** after in vivo depletion of NK cells. (J Immunol. 2000 165:518)

訳 NOD重症複合免疫不全マウスにおけるヒト細胞の増殖は増大した

P

■ [前] + proliferation

❶ inhibition of proliferation	増殖の抑制	198

■ [名／形] + proliferation

❶ cell proliferation	細胞増殖	11,259
❷ cellular proliferation	細胞増殖	1,214
❸ ～-induced proliferation	～誘導性増殖	560
❹ increased proliferation	増加した増殖	502

❸ In addition, the specific JAK3 tyrosine kinase inhibitor WHI-P131 significantly reduced IL-21-induced proliferation of BaF3/IL-21Rα cells. (Biochemistry. 2002 41:8725)

訳 特異的JAK3チロシンキナーゼ抑制剤WHI-P131は，BaF3/IL-21Rα細胞のIL-21誘導性増殖を有意に低下させた

■ proliferation and + [名]／[名] + and proliferation

❶ proliferation and differentiation	増殖と分化	1,048
❷ proliferation and survival	増殖と生存	555
❸ proliferation and migration	増殖と遊走	370
❹ proliferation and apoptosis	増殖とアポトーシス	338
❺ survival and proliferation	生存と増殖	333
❻ migration and proliferation	遊走と増殖	260
❼ growth and proliferation	成長と増殖	260

❶ Studies with p27 knockout mice have revealed abnormalities in proliferation and differentiation of multiple cell types. (J Clin Invest. 2001 108:383)

訳 ～は，複数の細胞タイプの増殖と分化における異常を明らかにしている

proliferative 形 増殖性の 用例数 5,365

◆類義語：productive, proliferating, multiplicative

■ proliferative + [名]

❶ proliferative responses	増殖応答	641
❷ proliferative capacity	増殖能	313
❸ proliferative potential	増殖能	207
❹ proliferative activity	増殖活性	179

❶ Positive proliferative responses to insulin were observed in 25% of type 1 diabetic patients and 10% of siblings. (Diabetes. 2004 53:1692)

訳 インスリンに対する正の増殖応答が，～において観察された

★ prolong 動 延長させる／遷延させる 用例数 10,143

prolonged	延長した／～を延長させた	8,651
prolong	～を延長させる	642
prolongs	～を延長させる	474
prolonging	～を延長させる	376

他動詞.

◆類義語：spread, extend, expand, stretch, dilate

■ prolong ＋［名句］

❶ prolong the survival of ～	…は，～の生存を延長させる	142

❶ We found that **PTA treatment significantly** **prolonged the survival of** **tumor-bearing mice**. (Cancer Res. 2011 71:6116)

訳 PTA処理は，腫瘍を持つマウスの生存を有意に延長させた

■［副］＋ prolonged

❶ significantly prolonged	有意に延長した／有意に延長させた	363
❷ more prolonged	より延長した	155

■ prolonged ＋［名］

❶ prolonged survival	延長した生存	459
❷ prolonged exposure	延長した曝露	327
❸ prolonged periods	延長した期間	212

❶ We also found that **CNF1 synthesis leads to** **prolonged survival** **of UPEC** in association with human neutrophils. (Infect Immun. 2005 73:5301)

訳 CNF1合成は，UPECの延長した生存につながる

❷ **Down-regulation of PKC by** **prolonged exposure** **to phorbol ester** revealed a second form of $Ca^{2+}{}_i$ oscillation at low agonist concentrations. (J Biol Chem. 2002 277:35947)

訳 ホルボールエステルへの延長した曝露によるPKCの下方制御

■［前］＋ prolonged

❶ after prolonged ～	延長した～のあと	336
❷ during prolonged ～	延長した～の間	254
❸ associated with prolonged ～	延長した～と関連した	116

❶ HCV replicon RNA could be fully cleared from replicon cells **after prolonged** **incubation with R1479**. (J Biol Chem. 2006 281:3793)

訳 R1479との延長したインキュベーションのあと

P

prominent 形 顕著な／卓越した 用例数 4,911

◆類義語：striking, marked, profound, manifest, dramatic, remarkable, significant

■ prominent ＋［前］

❶ prominent in ～	～において顕著な	586
・be prominent in ～	…は，～において顕著である	238

❶ This feature **was prominent in** three patients but rarely seen in only one patient recorded outside epileptogenic cortex.（Brain. 2009 132:3047）

訳 この特徴は，３人の患者において顕著であった

■［副］＋ prominent

❶ most prominent	最も顕著な	550
❷ more prominent	より顕著な	303

❶ This was **most prominent** in HIV-infected patients: resistance increased from 6.3% in 1988 to 53% in 1995.（J Infect Dis. 1999 180:1809）

訳 これは，HIVに感染した患者において最も顕著であった

■ prominent ＋［名］

❶ prominent role	顕著な役割	486
❷ prominent feature	顕著な特徴	212

❶ Heterobivalent tyrosine recombinases play a **prominent role** in numerous bacteriophage and transposon recombination systems.（J Mol Biol. 2005 347:11）

訳 ～は，多数のバクテリオファージにおいて顕著な役割を果たす

promise 動 見込みがある，名 見込み／有望／約束 用例数 8,769

promising	有望な／見込みがある	5,288
promise	見込み／有望／見込みがある	2,999
promises	見込み／有望／見込みがある	470
promised	見込みがあった	12

動詞および名詞として用いられる．複数形のpromisesの割合は約２％しかなく，原則，**不可算名詞**として使われる．

◆類 義 語：engagement, agreement, expectation, prospect, chance, likelihood, feasibility, possibility

■ 動 promising ＋［名］

❶ promising approach	有望なアプローチ	288
❷ promising results	有望な結果	241
❸ promising target	有望な標的	171
❹ promising strategy	有望な戦略	164

❶ Targeting signaling pathways activated by BCR-ABL is a promising approach for drug development.（Cancer Res. 2005 65:2047）
訳 ～は，薬剤開発のための有望なアプローチである

■ 動 promise to *do*

❶	promise to *do*	…は，～する見込みがある	620

❶ The approach promises to improve the accuracy of structures determined by NMR, and extend the size limit.（Science. 1997 278:1111）
訳 そのアプローチは，NMRによって決定される構造の正確性を改善する見込みがある

■ 名 promise ＋［前］

promise ofの前にtheが付く割合は圧倒的に高い．

	the%	a/an%	ø%			
❶	2	1	96	promise for ～	～のための見込み	999
				・promise for ～ing	～するための見込み	272
❷	0	0	100	promise as ～	～としての見込み	505
❸	2	2	96	promise in ～	～における見込み	363
				・promise in ～ing	～する際の見込み	120
❹	87	3	9	promise of ～	～の見込み	355

❶ This technology holds great promise for human gene therapy as a nonviral technology to deliver therapeutic genes.（Proc Natl Acad Sci USA. 2005 102:17059）
訳 この技術は，ヒトの遺伝子治療のために大いに有望である

❷ Adenovirus 5（Ad5）vectors show promise as human immunodeficiency virus vaccine candidates.（J Virol. 2005 79:15556）
訳 アデノウイルス5（Ad5）ベクターは，ヒト免疫不全ウイルスワクチン候補として有望である

❹ Regenerative medicine holds the promise of replacing damaged tissues largely by stem cell activation.（Proc Natl Acad Sci USA. 2010 107:9323）
訳 再生医療は，損傷した組織を置き換える見込みを持つ

■ 名［動］＋（［形］＋）promise

❶	show promise	…は，有望である	817
	・show great promise	…は，おおいに有望である	105
❷	hold promise	…は，有望である	549
	・hold great promise	…は，おおいに有望である	216

❶ Several novel tests show promise in the early detection of sepsis.（Curr Opin Pediatr. 2012 24:165）
訳 いくつかの新規のテストが，敗血症の早期の検出において有望である

P

★★ promote 動 促進する 用例数 39,662

promote	～を促進する	14,501
promotes	～を促進する	11,533
promoting	～を促進する	8,217
promoted	促進される／～を促進した	5,411

他動詞.

◆類義語：facilitate, accelerate

■ promote ＋ [名句]

❶ promote apoptosis	…は，アポトーシスを促進する	435
❷ promote the formation of ～	…は，～の形成を促進する	397
❸ promote cell survival	…は，細胞の生存を促進する	341
❹ promote survival	…は，生存を促進する	334
❺ promote the development of ～	…は，～の発生を促進する	236
❻ promote proliferation	…は，増殖を促進する	223
❼ promote differentiation	…は，分化を促進する	217
❽ promote angiogenesis	…は，血管新生を促進する	187
❾ promote tumorigenesis	…は，腫瘍形成を促進する	171

❶ Bax is a Bcl-2 family member that promotes apoptosis and counters the protective effect of Bcl-2. (Cancer Res. 2001 61:659)

訳 Baxは，アポトーシスを促進するBcl-2ファミリーメンバーである

❷ Rac activity promoted the formation of peripheral lamellae that mediated random migration. (J Cell Biol. 2005 170:793)

訳 Rac活性は，～の形成を促進した

■ [名] ＋ promote

❶ signaling promotes ～	シグナル伝達は，～を促進する	154
❷ activity promotes ～	活性は，～を促進する	90

❶ In these neoplasms, HH signaling promotes proliferation and survival, contributes to the maintenance of cancer stem cells, and enhances tolerance or resistance to chemotherapeutic agents. (Am J Pathol. 2012 180:2)

訳 HHシグナル伝達は，増殖と生存を促進する

■ [形／名／過分] ＋ to promote

★ to promote ～	～を促進する…	6,193
❶ sufficient to promote ～	～を促進するのに十分な	305
❷ ability to promote ～	～を促進する能力	299
❸ shown to promote ～	～を促進することが示される	241
❹ known to promote ～	～を促進することが知られる	193

❷ The ability of Myc to activate telomerase may contribute to its ability to promote tumor formation.（Genes Dev. 1998 12:1769）

訳 テロメラーゼを活性化するMycの能力は，腫瘍形成を促進するそれの能力に寄与するかもしれない

■ be promoted +［前］

★ be promoted	…は，促進される	630
❶ be promoted by ～	…は，～によって促進される	455

❶ This conversion **is promoted by** two factors: the strong Fe(III)-OOH bond, which inhibits Fe-O bond lysis, and the addition of protons, which facilitates O-O bond cleavage.（J Am Chem Soc. 2011 133:7256）

訳 この変換は，2つの因子によって促進される

P

prompt 動 促す／駆り立てる，形 迅速な 用例数 2,197

prompted	～を促した／促される	1,107
prompt	～を促す／迅速な	766
prompting	～を促す	217
prompts	～を促す	107

他動詞の用例が多いが，形容詞としても用いられる．第5文型のS+V+O+to *do*のパターンで使われることが多い．

◆類義語：permit, enable, allow, lead, rapid, fast

■ prompt us to *do*

❶ prompt us to *do*	…は，われわれに～するよう促す	445
・prompt us to investigate ～	…は，われわれに～を精査するよう促す	100
・prompt us to examine～	…は，われわれに～を調べるよう促す	70

❶ These observations **prompted us to examine** the phenotype following intestine-specific Mttp deletion because murine, like human enterocytes, secrete virtually exclusively apoB48.（J Biol Chem. 2006 281:4075）

訳 これらの観察は，われわれに～を調べるよう促した

prone 形 傾向の／しやすい 用例数 2,629

◆類義語：susceptible, apt

■ prone to

★ prone to ～	～しやすい／～に陥りやすい	814
❶ be prone to ～	…は，～しやすい	336
❷ more prone to ～	より～しやすい	126

❶ Finally, we show that mice lacking Sema3F **are prone to** seizures.（J Neurosci. 2005 25:3613）

訳 Sema3Fを欠くマウスは，けいれんを起こしやすい

pronounced ㊒ 顕著な／明白な 用例数 3,426

◆類義語：clear, apparent, evident, marked, striking, prominent

■ pronounced ＋［前］

❶ pronounced in ～	～において顕著な	592
❷ pronounced for ～	～に対して顕著な	134

❶ Decreased allograft survival **was most pronounced in patients** who were on hemodialysis before transplantation (RR 1.62, P =0.004). (Transplantation. 2004 77:1405)

訳 ～は，患者において最も顕著であった

■［副］＋ pronounced

❶ more pronounced	より顕著な	1,005
・be more pronounced	～は，より顕著である	501
❷ most pronounced	最も顕著な	421
❸ less pronounced	より顕著でない	234

❶ **This effect seemed to be more pronounced in women with a smoking history.** (J Clin Oncol. 2006 24:59)

訳 この効果は，喫煙歴のある女性においてより顕著であるように思われた

propagate ㊓ 増幅する／増殖する／伝播する／増殖させる 用例数 2,265

propagated	増幅した／増殖した／～を増殖させた	802
propagate	増幅する／増殖する／～を増殖させる	619
propagating	増幅する／増殖する／～を増殖させる	601
propagates	増幅する／増殖する／～を増殖させる	243

自動詞および他動詞として用いられる．

◆類義語：replicate, grow, proliferate, amplify

■ propagate ＋［前］

❶ propagate through ～	…は，～を経て伝播する	74
❷ propagate in ～	…は，～において増幅する	71

❶ We further demonstrate that **these intracellular signals propagate through the cytoplasm** and activate anchorage-dependent ERK signaling. (J Biol Chem. 2012 287:5211)

訳 これらの細胞内シグナルは，細胞質を経て伝播する

■ be propagated ＋［前］

★ be propagated	…は，増殖する	299
❶ be propagated in ～	…は，～において増幅する	56

❶ The shuttle vector is propagated in cultured cells, then recovered and analyzed in yeast using selection for reporter gene expression. (Nucleic Acids Res. 2005 33:5667)
訳 シャトルベクターは，培養細胞において増幅する

propensity 名 性質／性向　用例数 2,514

propensity	性質	2,243
propensities	性質	271

複数形のpropensitiesの用例が，約10%ある可算名詞.
◆類義語：property, nature, character

■ propensity ＋［前］

propensity ofの前にtheが付く割合は圧倒的に高い.

	the%	a/an%	ø%			
❶	23	42	16	propensity to *do*	～する性質	611
				・propensity to form ～	～を形成する性質	92
❷	40	33	16	propensity for ～	～の性質	405
❸	89	6	5	propensity of ～	～の性質	372

❶ The mutant, p53tet-R337H, had a significantly higher propensity to form amyloid-like fibrils. (J Mol Biol. 2003 327:699)
訳 ～は，アミロイド様原線維を形成する有意により高い性質を持っていた

❷ The propensity for symmetric distribution of SSE is especially evident for large proteins with the number of SSE ≧ 10. (J Mol Biol. 2005 353:1171)
訳 SSEの対称性分布の性質は，10以上のSSEの数を持つ大きなタンパク質にとって特に明らかである

❸ Epilepsy is characterised by the propensity of the brain to generate spontaneous recurrent bursts of excessive neuronal activity, seizures. (J Physiol. 2013 591:765)
訳 てんかんは，～する脳の性質によって特徴づけられる

proper 形 適切な／適当な　用例数 6,149

◆類義語：adequate, appropriate, reasonable, pertinent

■［動／形］＋ for proper

★	for proper ～	適切な～のために	1,821
❶	be required for proper ～	…は，適切な～のために必要とされる	510
❷	be essential for proper ～	…は，適切な～のために必須である	257
❸	be critical for proper ～	…は，適切な～のために決定的に重要である	115
❹	be necessary for proper ～	…は，適切な～のために必要である	98

❶ PR-Set7 is required for proper cell cycle progression and is subject to degradation by the CRL4(Cdt2) ubiquitin ligase complex as a function of the cell cycle and DNA damage. (Genes Dev. 2012 26:2580)

P

訳 PR-Set7は，適切な細胞周期進行のために必要とされる

■ proper ＋［名］

❶	proper folding	適切な折り畳み	202
❷	proper development	適切な発達	201
❸	proper regulation	適切な制御	158
❹	proper function	適切な機能	154
❺	proper localization	適切な局在化	144

❷ These mechanisms are essential for proper development, as evading them leads to tissue outgrowths composed of dividing but terminally differentiated cells.（J Cell Biol. 2010 189:981）

訳 これらの機構は，適切な発達のために必須である

★★ property 名 特性／性質　　用例数 37,894

properties	特性／性質	34,537
property	特性／性質	3,357

複数形のpropertiesの割合は，約90%と圧倒的に高い.

◆類義語：character, nature, propensity

■ properties ＋［前］

properties ofの前にtheが付く割合は非常に高い.

	the%	a/an%	ø%			
❶	75	-	23	properties of ～	～の性質／～の特性	16,112
				・binding properties of ～	～の結合特性	746
				・functional properties of ～	～の機能的特性	617
				・mechanical properties of ～	～の機械的特性	431
				・biochemical properties of ～	～の生化学的特性	351
				・physical properties of ～	～の身体的特性	304
				・kinetic properties of ～	～の動力学的特性	284
				・biophysical properties of ～	～の生物物理学的特性	259
				・structural properties of ～	～の構造的特性	245

❶ In this work, the properties of these raft-like domains are further explored and compared with properties thought to be central to raft function in plasma membranes.（Proc Natl Acad Sci USA. 2001 98:10642）

訳 これらのラフト様ドメインの性質が，さらに探索される

❶ To characterize the functional properties of ADH8A and ADH8B, recombinant proteins were purified from SF-9 insect cells.（J Biol Chem. 2004 279:38303）

訳 ADH8AおよびADH8Bの機能的特性を特徴づけるために

■ properties ＋［形／現分］

❶	properties similar to ～	～に似た性質	321

❷ properties including ～	～を含む性質	244

proportion ㊔ 割合／比率／比例　用例数 9,429

proportion	割合／比率／比例	7,900
proportions	割合／比率／比例	1,529

複数形のproportionsの割合が，約15%ある可算名詞.

◆類義語：rate, ratio

■ proportion ＋［前］

	the%	a/an%	ø%			
❶	50	44	4	proportion of ～	～の割合	6,674
				・proportion of patients	患者の割合	780
				・significant proportion of ～	～のかなりの割合	456
				・large proportion of ～	～の大きな割合	425
				・high proportion of ～	～の高い割合	371
				・higher proportion of ～	～のより高い割合	331
				・substantial proportion of ～	～のかなりの割合	291
				・small proportion of ～	～の低い割合	276
				・proportion of cells	細胞の割合	151

❶ Our primary outcome was the proportion of patients with a favourable clinical response (cure or improvement) on the day that intravenous antibiotic was discontinued. (Lancet. 2005 366: 1695)

訳 われわれの一次結果は，～の患者の割合であった

❶ Dementia is present in a significant proportion of patients admitted to general inpatient units. (Am J Psychiatry. 2000 157:704)

訳 認知症は，～に入院した患者のかなりの割合に存在する

■ in proportion to

❶ in proportion to ～	～に比例して	322

❶ The increase in activity was in proportion to serum glucose concentration. (Proc Natl Acad Sci USA. 2005 102:17952)

訳 活性の増大は，血清グルコース濃度に比例していた

proportional ㊏ 比例する　用例数 3,646

◆類義語：linear

■ proportional to

★ proportional to ～	～に比例する	1,446
❶ be proportional to ～	…は，～に比例する	705
❷ inversely proportional to ～	～に逆比例する	220

P

❸	directly proportional to ～	～に直接比例する	194

❶ In other organisms, **the rate of telomere shortening is proportional to the length of** the terminal 3' single-strand overhang. (Nucleic Acids Res. 2005 33:4536)
訳 テロメア短縮の割合は，～の長さに比例する

❷ The extent of induction of lung growth by mechanical strain **was inversely proportional to the number of** alveolar type Ⅱ cells remaining in the lung epithelium. (Am J Respir Crit Care Med. 2005 171:1395)
訳 …は，～の数に逆比例した

proposal 名 提唱／提案　　用例数 1,394

proposal	提唱／提案	1,040
proposals	提唱／提案	354

複数形のproposalsの割合が，約25%ある可算名詞．
◆類義語：proposition, suggestion

■ proposal ＋［同格that節］

proposal thatの前にtheが付く割合は圧倒的に高い．

	the%	a/an%	ø%			
❶	82	9	1	proposal that ～	～という提唱	549
				・support the proposal that ～	…は，～という提唱を支持する	100

❶ These findings **support the proposal that** common genetic variations in the eNOS gene contribute to atherosclerosis susceptibility, presumably by effects on endothelial NO availability. (Circulation. 2004 109:1359)
訳 これらの知見は，～という提唱を支持する

★★ propose 動 提唱する／提案する　　用例数 36,031

proposed	提唱される／～を提唱した	18,375
propose	～を提唱する	16,937
proposes	～を提唱する	589
proposing	～を提唱する	130

他動詞．we proposeの用例が圧倒的に多い．
◆類義語：advocate, suggest

■ we propose ＋［that節／名句］

★	we propose ～	われわれは，～を提唱する	15,204
❶	we propose that ～	われわれは，～ということを提唱する	10,717
❷	we propose a model	われわれは，モデルを提唱する	1,061
❸	we propose a mechanism	われわれは，機構を提唱する	155

❶ We propose that this latter shift correlates with a change in the timing of specification of the secondary embryonic axis.（Dev Biol. 2003 263:231）

訳 われわれは，～ということを提唱する

❷ We propose a model in which CDF protein binding facilitates Raf-1 activation.（J Biol Chem. 2004 279:27807）

訳 われわれは，～であるモデルを提唱する

■ to propose

★ to propose ～	～を提唱する…	590
❶ lead us to propose ～	われわれが～を提唱するように導く	268

❶ Our findings also lead us to propose a role for immortal DNA strands in tissue aging as well as cancer.（Cancer Res. 2002 62:6791）

訳 われわれの知見は，また，われわれが～の役割を提唱するように導く

■ be proposed ＋［前］

★ be proposed	…は，提唱される	10,383
❶ be proposed to *do*	…は，～することが提唱される／～するために提唱される	3,944
・be proposed to be ～	…は，～であることが提唱される	835
・be proposed to play ～	…は，～を果たすことが提唱される	259
・be proposed to explain ～	…は，～を説明するために提唱される	258
・be proposed to function	…は，機能することが提唱される	171
・be proposed to act	…は，働くことが提唱される	116
・be proposed to account for ～	…は，～を説明するために提唱される	115
・be proposed to mediate ～	…は，～を仲介することが提唱される	104
❷ be proposed as ～	…は，～として提唱される	1,126
❸ be proposed for ～	…は，～に対して提唱される	643
❹ be proposed in ～	…は，～において提唱される	326

❶ Therefore, vIL-6 has been proposed to play an important role in tumor progression.（J Virol. 2002 76:8252）

訳 vIL-6は，腫瘍の進行において重要な役割を果たすことが提唱されている

❶ A model is proposed to explain the interactions between photoperiod and the CLV genes.（Genetics. 2005 169:907）

訳 光周期とCLV遺伝子の間の相互作用を説明するため 1 つのモデルが提唱される

❷ Apoptosis（programmed cell death）has been proposed as a contributing pathophysiological mechanism.（Am J Psychiatry. 2004 161:109）

訳 アポトーシス（プログラム細胞死）は，寄与する病態生理学的機構として提唱されている

❸ Four different conditioning strategies are proposed for evaluation of effects at multiple, closely linked loci when family data are used.（Am J Hum Genet. 2002 70:124）

訳 4つの異なる条件づけの戦略が，効果の評価のために提唱される

■ proposed ＋［that節］

★ proposed that ～	～ということが提唱される	2,801
❶ it has been proposed that ～	～ということが提唱されている	321

❶ It has been proposed that the targeting of these processing enzymes to secretory granules involves their association with lipid rafts in granule membranes. (Biochemistry. 2003 42: 10445)
訳 ～ということが提唱されている

■ ［副］＋ proposed

❶ previously proposed	以前に提唱された／以前に提唱した	630
❷ recently proposed	最近提唱された／最近提唱した	342

❶ Our results support a previously proposed model of hypertrophy in which hypertrophy can precede satellite cell activation. (J Physiol. 2012 590:2151)
訳 われわれの結果は，～の以前に提唱されたモデルを支持する

■ proposed ＋［名］

❶ proposed mechanism	提唱された機能	326
❷ proposed role	提唱された役割	259
❸ proposed model	提唱されたモデル	228
❹ proposed method	提唱された方法	223

★ protect 動 保護する／防ぐ 用例数 16,425

protected	保護される／～を保護した	6,518
protect	～を保護する／防ぐ	5,056
protects	～を保護する／防ぐ	3,040
protecting	～を保護する／防ぐ	1,811

他動詞および自動詞の両方で用いられる．
◆類義語：prevent

■ protect ＋［名句］

❶ protect the host	…は，宿主を保護する	127

❶ The innate immune system protects the host from bacterial and viral invasion. (J Biol Chem. 2012 287:37406)
訳 自然免疫系は，細菌やウイルスの侵入から宿主を保護する

■ ［動］＋ to protect

★ to protect	保護する…	650
❶ fail to protect	…は，保護することができない	148
❷ be shown to protect	…は，保護することが示される	106

❷ NACA was shown to protect PC12 cells from glutamate (Glu) toxicity, as evaluated by LDH and MTS assays. (Brain Res. 2005 1056:132)
訳 NACAは，～からPC12細胞を保護することが示された

■ be protected ＋［前］

★ be protected	…は，保護される	1,589
❶ be protected from ～	…は，～から保護される	888
❷ be protected against ～	…は，～から保護される	299
❸ be protected by ～	…は，～によって保護される	145

❶ In sharp contrast, TNFR2-deficient mice were completely protected from glomerulonephritis at all time points, despite an intact systemic immune response.（J Clin Invest. 2005 115:1199）

訳 TNFR2欠損マウスは，糸球体腎炎から完全に保護された

❷ Immunized mice were partially protected against intranasal challenge with 235,000（10 50% lethal doses）Ames strain B. anthracis spores.（Infect Immun. 2006 74:794）

訳 免疫されたマウスは，鼻腔内の曝露から部分的に保護された

P

■［副］＋ protected

❶ completely protected	完全に保護される	181
❷ significantly protected	有意に保護される	138
❸ partially protected	部分的に保護される	127

■ protect ＋［前］

❶ protect against ～	～を防ぐ	1,841
・to protect against ～	～を防ぐ…	539
❷ protect from ～	～を防ぐ	216

❶ Estrogen has been shown to protect against the development of autoimmune disease, yet the mechanism is not known.（J Immunol. 2004 173:2227）

訳 エストロゲンは，自己免疫疾患の発症を防ぐことが示されている

★ protection ㊔ 保護　　用例数 13,573

protection	保護	13,492
protections	保護	81

複数形のprotectionsの割合は0.6%しかなく，原則，不可算名詞として使われる．

◆類義語：conservation, safeguard

■ protection ＋［前］

	the %	a/an %	ø %			
❶	4	1	95	protection against ～	～からの保護	3,221
❷	3	2	94	protection from ～	～からの保護	1,599
				・protection from apoptosis	アポトーシスからの保護	127
❸	36	1	62	protection of ～	～の保護	1,219
❹	1	1	89	protection in ～	～における保護	538

❶ Similarly, in guinea pigs, **immunization with the vaccine provided significant protection against genital HSV-2 disease** but did not prevent mucosal infection.（J Infect Dis. 2005 192:2117）

訳 ワクチンによる免疫化は，性器のHSV-2 疾患からの有意な保護を提供した

❷ Finally, functional studies reveal that **PGE2-mediated protection from apoptosis is completely inhibited by** a dominant-negative NR4A2 construct.（J Biol Chem. 2006 281:2676）

訳 PGE2に仲介されるアポトーシスからの保護は，～によって完全に抑制される

❸ **A primary role for chaperones appears to be protection of effectors from Lon-associated degradation** prior to secretion.（Mol Microbiol. 2005 55:941）

訳 シャペロンの主要な役割は，Lonに伴われる分解からの効果器の保護であるように思われる

■ [動] ＋ protection

❶ provide protection	…は，保護を与える	225
❷ confer protection	…は，保護を与える	185

❷ E235-mediated induction of senescence was not dependent on p21 or p53; however, **p21 conferred protection against the growth inhibitory effects of** E235.（Mol Pharmacol. 2013 83:594）

訳 p21は，～の増殖抑制効果からの保護を与える

■ [形／過分] ＋ protection

❶ significant protection	有意な保護	331
❷ ～-mediated protection	～に仲介される保護	300
❸ complete protection	完全な保護	251

★protective 形 保護的な　用例数 10,407

■ protective ＋ [名]

❶ protective effect	保護効果	1,385
❷ protective immunity	防御免疫	1,234
❸ protective role	保護的役割	613
❹ protective antigen	感染防御抗原	449
❺ protective efficacy	保護効率	265

❶ **The protective effect of alcohol did not vary by beverage type**, but did change with NHL subtype.（Lancet Oncol. 2005 6:469）

訳 アルコールの保護効果は，飲料のタイプによって異ならなかった

■ be protective ＋ [前]

★ be protective	…は，保護的である	819
❶ be protective against ～	～に対して保護的である	320

❶ Findings suggest that one-carbon nutrients, particularly vitamin B6 and methionine, **may be**

protective against NHL. (Am J Epidemiol. 2005 162:953)
訳 〜は，NHLに対して保護的であるかもしれない

protocol 名 プロトコル／手順 用例数 9,979

protocol	プロトコル／手順	7,070
protocols	プロトコル／手順	2,909

複数形のprotocolsの割合が，約30%ある可算名詞．
◆類義語：procedure, method

■ protocol ＋［前］

	the %	a/an %	ø %			
❶	11	82	3	protocol for 〜	〜のためのプロトコル	665
				・protocol for 〜ing	〜するためのプロトコル	176
❷	12	67	1	protocol to *do*	〜するプロトコル	251

❶ To investigate how meristem organization is self-perpetuated, we developed a protocol for the analysis of meristem growth in 3-D. (Plant J. 2002 31:229)
訳 われわれは，〜の分析のためのプロトコルを開発した

■［名／形］＋ protocol

❶ treatment protocol	治療プロトコル	179
❷ experimental protocol	実験プロトコル	126

prove 動 判明する／証明する 用例数 7,706

proven	判明した／〜を証明した	2,596
prove	判明する／〜を証明する	2,482
proved	判明した／〜を証明した	2,169
proving	判明する／〜を証明する	261
proves	判明する／〜を証明する	198

第２文型の自動詞（prove＋形容詞）の用例が多いが，他動詞（prove＋that節）の用例もかなりある．過去分詞としては，provedとprovenの両方が用いられる．
◆類義語：turn out, realize, demonstrate, document, evidence

■ prove ＋［形］

❶ prove useful	…は，役に立つと判明する	869
❷ prove difficult	…は，困難であると判明する	242
❸ prove effective	…は，効果的であると判明する	166
❹ prove valuable	…は，価値があると判明する	125

❶ Insights gained from glycerol binding may prove useful in the design of a peroxidase-specific ligand.（J Mol Biol. 2004 335:503）

訳 ～は，ペルオキシダーゼ特異的なリガンドの設計において役に立つと判明するかもしれない

■ prove to *do*

★ prove to *do*	…は，～すると判明する	2,199
❶ prove to be ～	…は，～であると判明する	2,001

❶ Thus 19F NMR has proved to be a useful spectral tool to probe molecular recognition in a hydrophobic cavity of a metal-assembled coiled coil.（J Am Chem Soc. 2004 126:4192）

訳 19F NMRは，有用なスペクトルツールであると判明した

■ prove ＋［that節］

❶ prove that ～	…は，～ということ証明する	492

❶ Our previous studies proved that SPL utilizes the 5'-dA* generated by the SAM cleavage reaction to abstract the H(6proR) atom to initiate the SP repair process.（Biochemistry. 2012 51:7173）

訳 われわれの以前の研究は，～ということを証明した

★★ provide 動 提供する／与える　用例数 107,463

provide	～を提供する	55,455
provides	～を提供する	26,914
provided	提供される／～を提供した	14,772
providing	～を提供する	10,322

他動詞として用いられることが圧倒的に多いが，自動詞の用例もある．

◆類義語：present, offer, supply, confer, render

■ provide ＋［名句］

❶ provide evidence	…は，証拠を提供する	8,894
・provide evidence that ～	…は，～という証拠を提供する	5,134
・provide evidence for ～	…は，～の証拠を提供する	2,484
・provide the first evidence	…は，最初の証拠を提供する	1,126
❷ provide insight into ～	…は，～への洞察を提供する	4,099
・provide new insights into ～	…は，～への新しい洞察を提供する	1,102
❸ provide a mechanism	…は，機構を提供する	1,321
❹ provide information	…は，情報を提供する	888
❺ provide a basis for ～	…は，～に対する基盤を提供する	778
❻ provide a means	…は，手段を提供する	744
❼ provide support for ～	…は，～に対する支持を提供する	607
❽ provide an explanation for ～	…は，～に対する説明を提供する	487
❾ provide a rationale for ～	…は，～に対する理論的根拠を提供する	418

❶ In conclusion, **these results** provide evidence that chronic liver disease can paradoxically result in cell death resistance in vivo. (Hepatology. 2004 39:433)

訳 これらの結果は，～という証拠を提供する

❷ **These findings** provide insights into genetic predisposition to oxidative stress and the relationship between OSEs and macrophage apoptosis that may explain advanced atherosclerosis in human Hp2-2 plaques. (J Am Coll Cardiol. 2012 60:112)

訳 これらの知見は，～への洞察を提供する

■ ［名／代名］＋ provide

❶	results provide ～	結果は，～を提供する	5,493
❷	we provide ～	われわれは，～を提供する	4,936
❸	data provide ～	データは，～を提供する	3,390
❹	findings provide ～	知見は，～を提供する	2,621
❺	study provides ～	研究は，～を提供する	2,021
❻	observations provide ～	観察は，～を提供する	481
❼	model provides ～	モデルは，～を提供する	476
❽	work provides ～	研究は，～を提供する	443
❾	method provides ～	方法は，～を提供する	279

❸ Our data provide **information on** the amino acids that affect preferences and point to a conserved loop as being of key importance. (Proc Natl Acad Sci USA. 2012 109:E3295)

訳 われわれのデータは，～に関する情報を提供する

■ to provide

★	to provide ～	～を提供する…	6,658
❶	To provide ～	～を提供するために	430
❷	used to provide ～	～を提供するために使われる	208

❶ To provide **insight into the pathogenesis of IRU**, the phenotype and specificity of intraocular T cells in a single patient were analyzed. (J Infect Dis. 2002 186:701)

訳 IRUの病原性への洞察を提供するために

■ provide ＋［名／代名］＋ with

❶	provide ～ with …	‥は，～に…を与える	1,370
	・provide us with ～	…は，われわれに～を提供する	144

❶ **This study** provides us with **the mechanistic basis** to develop alternative approaches to inhibit Gcn5 activity for cancer therapy. (J Biol Chem. 2011 286:41344)

訳 この研究は，われわれに機構的な基礎を提供する

■ providing

❶	, providing evidence	そして，（それは）証拠を提供する	449
❷	, thereby providing ～	それによって，～を提供する	435

❶ Three of the duplications were at 2.85 Mb of the E. coli chromosome, providing evidence **for the replicability of** the adaptation to high temperature. (Proc Natl Acad Sci USA. 2001 98:525)

P

訳 そして，（それは）～の複製能力の証拠を提供する

❷ S13 provides a direct link between the tRNA-binding site and the movements in the head of the small subunit seen during translocation, thereby providing a possible pathway of signal transduction.（J Mol Biol. 2005 349:47）

訳 それによって，シグナル伝達の可能な経路を提供する

■ provided ＋［前］

❶ provided by ～	～によって提供される	2,304
・be provided by ～	…は，～によって提供される	847
❷ provided to ～	～するために提供される	298
❸ provided in ～	～において提供される	75

❶ Proof that the Acd1 gene is an orthologue of ZmLls1 is provided by in vivo complementation of the acd1 mutant by the ZmLls1 gene.（Plant Mol Biol. 2004 54:175）

訳 …は，～の生体内補完によって提供される

■ provide for

❶ provide for ～	…は，～を与える／～に備える	452

❶ The likely redundant control of morphogenesis may provide for plasticity at this critical stage of early development.（Development. 2003 130:873）

訳 ～は，可塑性を与えるかもしれない

★proximal 形 近位の　用例数 10,060

◆反意語：distal

■ proximal to

❶ proximal to ～	～の近位にある	995
・be proximal to ～	…は，～の近位にある	142

❶ This region is proximal to the binding sites of known and potential natural transcription factors.（J Biol Chem. 2005 280:3707）

訳 この領域は，～の結合部位の近位にある

■ proximal ＋［名］

❶ proximal promoter	近位プロモーター	691
❷ proximal region	近位部	356

proximity 名 近接／近く　用例数 3,110

proximity	近接／近く	3,075
proximities	近接／近く	35

複数形のproximitiesの割合は1％しかなく，原則，不可算名詞として使われる．

◈類義語：vicinity

■ proximity ＋［前］

proximity ofの前にtheが付く割合はかなり高い．

	the %	a/an %	ø %			
❶	4	1	88	proximity to ～	～の近く	1,303
				・in close proximity to ～	～のすぐ近くに	624
				・in proximity to ～	～の近くに	228
❷	68	1	31	proximity of ～	～の近く	665
				・close proximity of ～	～のすぐ近く	118

❶ In PP1, the potential redox active site is located in close proximity to the phosphatase active site.（FASEB J. 1999 13:1866）

訳 潜在的なレドックス活性部位は，ホスファターゼ活性部位のすぐ近くに位置する

❷ Prominent HO-1 staining of neuronal cells in the proximity of blood vessels and circumventricular organs was also observed.（Brain Res. 2003 962:1）

訳 血管および脳室周囲器官の近くに神経細胞の顕著なHO-1染色が，また，観察された

publish 動 出版する／発表する　用例数 6,815

published	発表される／出版される／～を発表した	6,709
publishing	～を発表する／～を出版する	60
publish	～を発表する／～を出版する	42
publishes	～を発表する／～を出版する	4

他動詞．

■ published ＋［前］

❶ published in ～	～に発表される	562
・be published in ～	…は，～に発表される	129

❶ Numerous studies examining this disorder have been published in the past 2 years.（Curr Opin Neurol. 2005 18:645）

訳 ～が，過去２年間に発表されている

■ ［副］＋ published

❶ previously published	以前に発表された	932
❷ recently published	最近発表された	403

❶ These findings stand in contrast to previously published reports of 1,25(OH)$_2$D$_3$ production in keratinocytes.（Proc Natl Acad Sci USA. 2006 103:75）

訳 これらの知見は，以前に発表された報告と対照をなす

■ published ＋［名］

❶ published data	発表されたデータ	549

P

❷ published studies	発表された研究	345
❸ published reports	発表された報告	198
❹ published literature	発表された文献	193

❶ Analysis of published data indicated allelic loss of XPC in most human lung tumors and allelic loss of Gadd45a in some human lung and other cancer types.（Proc Natl Acad Sci USA. 2005 102:13200）

訳 発表されたデータの分析は，XPCのアレルの喪失を示した

purification 名 精製 用例数 3,623

purification	精製	3,577
purifications	精製	46

複数形のpurificationsの割合は１％しかなく，原則，不可算名詞として使われる．

■ purification ＋［前］

	the%	a/an%	ø%			
❶	38	3	59	purification of ～	～の精製	1,146

❶ We report the expression and purification of human Nod2 from insect cells.（J Am Chem Soc. 2012 134:13535）

訳 われわれは，ヒトNod2の発現と精製を報告する

■［名］＋ purification

❶ affinity purification	アフィニティー精製	431
❷ protein purification	タンパク質精製	114

★ purify 動 精製する 用例数 21,108

purified	精製される／～を精製した	19,999
purifying	～を精製する	509
purify	～を精製する	506
purifies	～を精製する	94

他動詞．

◆類義語：isolate

■ purified ＋［前］

❶ purified from ～	～から精製される	1,893
・be purified from ～	…は，～から精製される	581
・purified from Escherichia coli	大腸菌から精製される	134
❷ purified to ～	～にまで精製される	798
・purified to homogeneity	均一にまで精製される	490

❸ purified by ～	～によって精製される	672
❹ purified in ～	～において精製される	160

❶ The proteins were expressed and purified from Escherichia coli both individually and as a complex.（J Biol Chem. 2003 278:10649）

訳 それらのタンパク質は，大腸菌から発現されそして精製された

❷ The phosphoenolpyruvate carboxylase from this organism was purified to homogeneity.（J Bacteriol. 2004 186:5129）

訳 ～は，均一にまで精製された

❸ The protein was purified by affinity chromatography on an anti-SOD antibody-Sepharose column.（Anal Biochem. 2003 318:132）

訳 そのタンパク質は，アフィニティークロマトグラフィーによって精製された

P

■ purified and ＋［過分］／［過分］＋ and purified

❶ expressed and purified	発現されそして精製される	345
❷ purified and characterized	精製されそして特徴づけられる	257

❷ Seven mutant enzymes with defective phenotypes were purified and characterized.（Mol Microbiol. 2003 48:67）

訳 ～が，精製されそして特徴づけられた

■［副］＋ purified

❶ highly purified	高度に精製された	602
❷ partially purified	部分的に精製された	491

❶ Conversely, a highly purified archaeal RNase HII type 2 protein has a pronounced activity.（Proc Natl Acad Sci USA. 2002 99:16654）

訳 高度に精製された古細菌のリボヌクレアーゼHII2型タンパク質は，明白な活性を持つ

■［前］＋ purified

❶ addition of purified ～	精製された～の添加	173
❷ activity of purified ～	精製された～の活性	112

■ affinity-purified

❶ affinity-purified ～	アフィニティー精製された～	622

❶ Affinity-purified CYP2D6(193-212)-specific Ab inhibited the metabolic activity of CYP2D6.（J Immunol. 2003 170:1481）

訳 アフィニティー精製されたCYP2D6(193-212)特異的抗体は，～を抑制した

■ we（have）purified

❶ we purified ～	われわれは，～を精製した	408
・we have purified ～	われわれは，～を精製した	263

❶ To search for possible novel regulators of c-Cbl, we purified a number of c-Cbl-associated proteins by affinity chromatography and identified them by mass spectrometry.（Oncogene.

2004 23:4690)
訳 われわれは，アフィニティークロマトグラフィーによっていくつかのc-Cbl関連タンパク質を精製した

■ purified ＋［名］

❶	purified protein	精製されたタンパク質	625
❷	purified enzyme	精製された酵素	312
❸	purified components	精製された成分	152

purpose 名 目的　用例数 7,786

purpose	目的	6,753
purposes	目的	1,033

複数形のpurposesの割合が，約15%ある可算名詞.
◆類義語：aim, goal, objective, object, end

■ purpose ＋［前］

purpose ofの前にtheが付く割合は，99%と圧倒的に高い.

	the%	a/an%	ø%			
❶	99	0	1	purpose of ～	～の目的	5,652

■ the purpose of ＋［名］

★	the purpose of ～	～の目的	5,504
❶	the purpose of this study was to *do*	この研究の目的は，～することであった	3,024
	・the purpose of this study was to determine ～	この研究の目的は，～を決定することであった	912
	・the purpose of this study was to evaluate ～	この研究の目的は，～を評価することであった	310
	・the purpose of this study was to examine ～	この研究の目的は，～を調べることであった	302
	・the purpose of this study was to investigate ～	この研究の目的は，～を精査することであった	265
❷	the purpose of this review is to *do*	この総説の目的は，～することである	426
	・the purpose of this review is to summarize ～	この総説の目的は，～を要約することである	75
❸	the purpose of the present study was to *do*	現在の研究の目的は，～することであった	344
	・the purpose of the present study was to determine ～	現在の研究の目的は，～を決定することであった	110
❹	for the purpose of ～ing	～する目的のために	154

❶ The purpose of this study was to determine the role of matrix metalloproteinases (MMP) in Pseudomonas aeruginosa keratitis. (Invest Ophthalmol Vis Sci. 2006 47:256)

訳 この研究の目的は，～におけるマトリックスメタロプロテアーゼ（MMP）の役割を決定することであった

❷ The purpose of this review is to summarize recent advances in the diagnosis and treatment of several severe skin diseases seen in children.（Curr Opin Pediatr. 2011 23:403）

訳 この総説の目的は，最近の進歩を要約することである

Q

quantification ㊔ 定量化　　用例数 2,617

quantification	定量化	2,590
quantifications	定量化	27

複数形のquantificationsの割合は1％しかなく，原則，**不可算名詞**として使われる．

◆類義語：quantitation

■ quantification ＋［前］

	the％	a/an％	ø％			
❶	31	3	64	quantification of ～	～の定量化	1,349

❶ Accurate **quantification of** protein content and composition has been achieved using isotope-edited surface enhanced resonance Raman spectroscopy.（J Am Chem Soc. 2008 130: 9624）

訳 タンパク質の含量と組成の正確な定量化が達成されている

quantify ㊍ 定量化する　　用例数 7,846

quantified	定量化される／～を定量化した	3,662
quantify	～を定量化する	3,047
quantifying	～を定量化する	931
quantifies	～を定量化する	206

他動詞．

◆類義語：quantitate, measure

■ quantify ＋［名句］

❶ quantify the effect of ～	…は，～の効果を定量化する	117

■［代名］＋ quantify

❶ we quantified ～	われわれは，～を定量化した	397

❶ **We quantified** the expression of four photosynthesis-related genes in seven of these enhanced deetiolation (end) mutants and found that photosynthesis-related gene expression is attenuated.（Plant Physiol. 2012 159:366）

訳 われわれは，～の発現を定量化した

■ to quantify

★ to quantify ～	～を定量化する…	2,110
❶ be used to quantify ～	…が，～を定量化するために使われる	393

❶ Logistic regression models **were used to quantify** the effect that polyparasitism has on anemia (hemoglobin level <11 g/dl).（J Infect Dis. 2005 192:2160）

訳 ロジスティック回帰分析モデルが，～である効果を定量化するために使われた

■ be quantified ＋［前］/using

★ be quantified	…が，定量化される	2,052
❶ be quantified by ～	…が，～によって定量化される	684
❷ be quantified in ～	…が，～において定量化される	274
❸ be quantified using ～	…が，～を使って定量化される	237
❹ be quantified with ～	…が，～によって定量化される	128

❶ CEACAM6 expression was quantified by real-time polymerase chain reaction (PCR) and Western blot. (Ann Surg. 2004 240:667)

訳 CEACAM6の発現が，リアルタイムポリメラーゼ連鎖反応法（PCR）およびウエスタンブロットによって定量化された

❷ PDGF AA and BB were quantified in peritoneal fluid by ELISA. (Oncogene. 2006 25:2060)

訳 血小板由来成長因子AAおよびBBが，酵素結合免疫吸着検定法によって腹水において定量化された

Q

quantitate ㊍ 定量化する／定量する　用例数 1,322

quantitated	定量化される／～を定量化した	687
quantitate	～を定量化する	499
quantitating	～を定量化する	123
quantitates	～を定量化する	13

他動詞.

◆類義語：quantify, measure

■ be quantitated ＋［前］

★ be quantified	…が，定量化される	405
❶ be quantitated by ～	…が，～によって定量化される	148

❶ CO liberated into the headspace was quantitated by gas chromatography. (Anal Biochem. 2005 341:280)

訳 ～が，ガスクロマトグラフィーによって定量化された

■ to quantitate

★ to quantitate ～	～を定量化する…	343
❶ be used to quantitate ～	…が，～を定量化するために使われる	75

❶ Electron beam tomography (EBT) can be used to quantitate the amount of CAC and assist in prognostication of future cardiac events. (J Am Coll Cardiol. 2002 39:408)

訳 ～が，CACの量を定量化するために使われうる

quantitation ㊔ 定量化　用例数 1,430

quantitation	定量化	1,426
quantitations	定量化	4

複数形のquantitationsの割合は0.3%しかなく，原則，**不可算名詞**として使われる．
◆類義語：quantification

■ quantitation ＋［前］

	the%	a/an%	ø%			
❶	26	2	71	quantitation of ～	～の定量化	865

❶ This study sought to determine the utility of quantitation of right ventricular (RV) function in predicting RV failure in patients undergoing left ventricular assist device (LVAD) implantation. (J Am Coll Cardiol. 2012 60:521)
訳 この研究は，右心室(RV)機能の定量化の有用性を決定しようと努めた

★ question ㊔ 疑問／問題，動 疑問を持つ　用例数 11,275

question	疑問／問題／～に疑問を持つ	6,287
questions	疑問／問題／～に疑問を持つ	4,472
questioned	～に疑問を持った／疑問を持たれる	451
questioning	～に疑問を持つ	65

他動詞の用例もあるが，名詞として使われることが非常に多い．複数形のquestionsの割合は約40%とかなり高い．
◆類義語：problem, query, doubt, suspect

■ question(s) ＋［前］

question ofの前にtheが付く割合は圧倒的に高い．

	the%	a/an%	ø%			
❶	91	8	0	question of ～	～という問題	1,349
				・the question of whether ～	～かどうかという問題	152
				・the question of how ～	どのように～かという問題	62
❷	2	-	89	questions about ～	～に関する問題	799
❸	2	81	0	question in ～	～における問題	421
❹	4	-	78	questions regarding ～	～に関する問題	320

❶ We address the question of whether there exists an effective evolutionary model of amino-acid substitution that forms a metric-distance function. (Bioinformatics. 2004 20:1214)
訳 われわれは，～かどうかという問題に取り組む

❷ This raises questions about the biological mechanism(s) that produce new mutations and has implications for the study of male-driven evolution. (Genome Res. 2005 15:1086)
訳 これは，生物学的機構に関する問題を提起する

■ [動] + this/the question

❶ to address this question	この問題に取り組むために	617
❷ raise the question of ～	…は，という問題を提起する	445
❸ address the question of ～	…は，～という問題に取り組む	214

❶ To address this question, we performed chimera analysis using $ET_A^{-/-}$ embryonic stem cells. (Dev Biol. 2003 261:506)

訳 この問題に取り組むために，

❷ This raises the question of whether M protein is also involved in the induction of apoptosis. (J Virol. 2001 75:12169)

訳 これは，～かどうかという問題を提起する

■ [前] + question

❶ call into question ～	…は，～に異議を唱える	198

Q

❶ Our results call into question the identification of PQQ as a new vitamin. (Nature. 2005 433:E10)

訳 われわれの結果は，新しいビタミンとしてのPQQの同定に異議を唱える

■ question(s) + [動]

❶ questions remain	疑問が，残っている	334
❷ question is whether ～	疑問は，～かどうかである	143

❶ However, key questions remain regarding the interpretation of EX1 hydrogen exchange. (Biochemistry. 2004 43:587)

訳 ～の解釈に関する重要な疑問が残っている

quickly 副 急速に／速く 用例数 1,648

◆類義語：rapidly, briefly

■ more quickly

❶ more quickly	より急速に	245

❶ Therefore, we analyzed Ca^{2+} clearance from the cytosol of both cell types and found that Th2 cells extrude Ca^{2+} more quickly than Th1 cells. (J Immunol. 2000 164:1153)

訳 Th2細胞は，Th1細胞より急速にCa^{2+}を排除する

quite 副 全く／非常に／きわめて 用例数 2,121

◆類義語：totally, fully, thoroughly, enough, perfectly, grossly

■ quite + [形]

❶ quite different	全く異なる	347

・be quite different	…は，全く異なる	226
❷ quite similar	きわめて類似している	156

❶ The fish immune system is quite different from the mammalian system because the anterior kidney forms the main site for hematopoiesis in this species. (J Immunol. 2005 174:6608)
訳 魚の免疫システムは，哺乳類のシステムとは全く異なる

❷ The peptide folding pattern of the catalytic domain is quite similar to the patterns observed in many methyltransferases. (Biochemistry. 1999 38:8323)
訳 その触媒ドメインのペプチド折り畳みパターンは，多くのメチルトランスフェラーゼにおいて観察されたパターンときわめて類似している

R

raise 動 上昇させる／産生する／提起する　用例数 9,206

raised	上昇した／～を上昇させた／産生された	3,376
raise	～を上昇させる／～を産生する	2,242
raising	～を上昇させる／～を産生する	1,829
raises	～を上昇させる／～を産生する	1,759

他動詞.〈raise the possibility that〉〈raised against〉の用例が多い.

◆類義語：elevate, produce, increase

■ raise ＋［名句］

❶ raise the possibility	…は，可能性を示唆する	2,776
・raise the possibility that ～	…は，～という可能性を示唆する	2,375
・, raising the possibility	そして，（それは）可能性を示唆している	681
・raise the possibility of ～	…は，～の可能性を示唆する	379
❷ raise the question of ～	…は，～という疑問を提起する	445

❶ Studies of the genetic basis of type 2 diabetes suggest that variation in the calpain-10 gene affects susceptibility to this common disorder, **raising the possibility that** calpain-sensitive pathways may play a role in regulating insulin secretion and/or action.（Diabetes. 2001 50: 2013）

訳 そして，（それは）～という可能性を示唆している

❷ **This raises the question of how** an essentially disordered protein is transformed into highly organized fibrils.（J Biol Chem. 2001 276:10737）

訳 これは，どのように～かという疑問を提起する

■［名］＋ raise the possibility

❶ results raise the possibility	結果は，可能性を示唆する	195
❷ findings raise the possibility	知見は，可能性を示唆する	181
❸ data raise the possibility	データは，可能性を示唆する	107

❶ **These results raise the possibility that** pharmacological manipulation of ACC2 may lead to loss of body fat in the context of normal caloric intake.（Science. 2001 291:2613）

訳 これらの結果は，～という可能性を示唆する

■ raised ＋［前］

❶ raised against ～	～に対して産生される	770
・antibodies raised against ～	～に対して産生された抗体	131
❷ raised to ～	～に対して産生される	211
❸ raised in ～	～において上昇する	208
❹ raised by ～	～によって産生される	135

❶ **Antibodies raised against the two sequences** labeled bursicon-containing neurons in the central nervous systems of P. americana.（J Comp Neurol. 2002 452:163）

訳 2つの配列に対して産生された抗体

❷ Polyclonal antibodies raised to the N-terminal PMT region bound efficiently to full-length native toxin, suggesting that the N terminus is surface located. (Infect Immun. 2001 69:7839)

訳 N末端PMT領域に対して産生されたポリクローナル抗体は，～に効率的に結合した

❸ The risk was significantly raised in young seronegative recipients if the donor was older than the recipient. (Transplantation. 2000 69:897)

訳 そのリスクは，～において有意に上昇した

randomize 動 無作為化する／ランダム化する／無作為抽出する

用例数 9,690

randomized	無作為化される／～を無作為化した	9,600
randomizing	～を無作為化する	44
randomize	～を無作為化する	25
randomizes	～を無作為化する	21

他動詞．過去分詞の用例が圧倒的に多い．

■ be randomized ＋［前］

★ be randomized	…が，無作為化される	2,084
❶ be randomized to *do*	…が，～するように無作為化される	1,482
・be randomized to receive ～	…が，～を受けるように無作為化される	630
❷ be randomized into ～	…が，～に無作為化される	119

❶ After a 3-week placebo run-in, 37 patients were randomized to receive placebo (n=7) or ezetimibe 10 mg/d (n=30) for 8 weeks. (Circulation. 2004 109:966)

訳 37人の患者が，偽薬（n= 7）あるいはエゼチミベ10 mg/d（n=30）を受けるように無作為化された

■［名］＋ be randomized

❶ patients were randomized	患者が，無作為化された	486

■ randomized ＋［形／名］

❶ randomized trials	無作為化試験	745
❷ randomized controlled trials	無作為化対照試験	689
❸ randomized clinical trials	無作為化臨床試験	309
❹ randomized, double-blind, placebo-controlled trial	無作為化二重盲検偽薬対照試験	163
❺ randomized, placebo-controlled trial	無作為化偽薬対照試験	102

❹ We conducted a randomized, double-blind, placebo-controlled trial of oxandrolone therapy for surgical/trauma patients requiring >7 days of ventilation. (Ann Surg. 2004 240:472)

訳 われわれは，オキサンドロロン療法の無作為化二重盲検偽薬対照試験を行った

randomly 副 無作為に／ランダムに 用例数 6,142

■ randomly ＋［過分／動］

❶ randomly assigned	無作為に割り当てられる／無作為に割り当てた	3,319
・be randomly assigned	…は，無作為に割り当てられる	2,469
・be randomly assigned to ～	…は，～するように無作為に割り当てられる／～に無作為に割り当てられる	2,007
・be randomly assigned to receive ～	…は，～を受けるように無作為に割り当てられる	843
・patients were randomly assigned	患者は，無作為に割り当てられた	649
・participants were randomly assigned	参加者は，無作為に割り当てられた	140
・be randomly assigned in ～	…は，～において無作為に割り当てられる	104
・we randomly assigned ～	われわれは，～を無作為に割り当てた	379
❷ randomly selected	無作為に選択される	777
・randomly selected from ～	～から無作為に選択される	123
❸ randomly distributed	ランダムに分布する	200
❹ randomly allocated	無作為に割り当てられる	155
❺ randomly chosen	無作為に選ばれる	117

❶ Thirty-five patients were randomly assigned to receive aprepitant and 34 to receive placebo for the first course.（J Clin Oncol. 2012 30:3998）

訳 35名の患者が，～を受けるように無作為に割り当てられた

❷ Controls（n = 218）were randomly selected from the study population.（JAMA. 2003 290: 2677）

訳 対照群（n = 218）は，研究対象集団から無作為に選択された

★★ range 名 範囲／レンジ，動 範囲である／及ぶ 用例数 44,588

range	範囲／レンジ／範囲である	34,713
ranging	範囲である	5,253
ranged	範囲であった	3,129
ranges	範囲／レンジ／範囲である	1,493

名詞の用例が多いが，動詞としても用いられる．複数形のrangesの割合が，約5％ある可算名詞．

◆類義語：spectrum, array, extent, cover, reach, extend, span

■ 名 range ＋［前］

	the%	a/an%	ø%			
❶	23	75	1	range of ～	～の範囲／範囲の～	15,853
				・a wide range of ～	広範囲の～	4,183
				・a broad range of ～	広範囲の～	1,406
				・dynamic range of ～	～のダイナミックレンジ	251

R

・full range of ～	全範囲の～	231
・a diverse range of ～	広範囲の～	224
・a wider range of ～	より広範囲の～	146

❶ The few downstream target genes that have been identified indicate a wide range of downstream effectors. (Annu Rev Genet. 39:219)

訳 同定されている少数の下流の標的遺伝子は，広範囲の下流の効果器を示す

❶ Superoxide dismutases (SODs) catalyze the dismutation of superoxide radicals in a broad range of organisms, including plants. (Anal Biochem. 2004 332:314)

訳 スーパーオキシドジスムターゼ（SOD）は，広範囲の生物においてスーパーオキシドラジカルの不均化を触媒する

■ 名 ［形／名］＋ range

❶ long range	長い範囲／長距離	2,009
❷ age range	年齢範囲	683
❸ host range	宿主範囲	633
❹ interquartile range	四分位範囲	600
❺ short range	短い範囲	465
❻ concentration range	濃度範囲	462
❼ temperature range	温度範囲	416

❶ Such long range interactions are expected to be of functional significance for activation and signal transduction in heptahelical G-protein-coupled receptors. (J Biol Chem. 2000 275:13431)

訳 そのような長い範囲の相互作用は，活性化のために機能的に重要であると予測される

■ 動 ranging ＋［前］

❶ ranging from ～ to …	～から…の範囲である	3,174
・values ranging from ～ to …	～から…の範囲である値	125
・concentrations ranging from ～ to …	～から…の範囲である濃度	125
❷ ranging in ～	～において範囲である	427
・ranging in size from ～ to …	大きさにおいて～から…の範囲である	159
・ranging in age from ～ to …	年齢において～から…の範囲である	91
❸ ranging between ～	～の間の範囲である	131

❶ $G\beta\gamma$ dimers containing $G\beta_{1\text{-}4}$ complexed with γ_2 stimulated P-Rex1 activity with EC_{50} values ranging from 10 to 20 nm. (J Biol Chem. 2006 281:1913)

訳 ～は，10から20 nmの範囲であるEC_{50}値を持つP-Rex1活性を刺激した

❷ Sequencing of the C. reinhardtii Rca gene revealed that it contains 10 exons ranging in size from 18 to 470 bp. (Plant Physiol. 2003 133:1854)

訳 それは，大きさにおいて18から470 bpの範囲である10のエクソンを含む

★ rapid 形 急速な／迅速な 用例数 26,274

◈類義語：prompt, quick, fast

■ rapid ＋［名］

❶ rapid increase	急速な増大	450
❷ rapid degradation	急速な分解	372
❸ rapid amplification	急速な増幅	321
❹ rapid activation	急速な活性化	305
❺ rapid growth	急速な増殖	282
❻ rapid identification	急速な同定	229
❼ rapid induction	急速な誘導	226
❽ rapid onset	急速な開始	219
❾ rapid loss	急速な喪失	211

❶ However, after 5-fluorouracil (5-FU) treatment, **there was a rapid increase in plasma vascular endothelial growth factor A (VEGF-A) levels** and expansion of Tie2-positive neovessels. (Blood. 2005 106:505)

訳 血清の血管内皮増殖因子A（VEGF-A）レベルの急速な増大があった

■［副］＋ rapid

❶ more rapid	より急速な	1,135
❷ very rapid	とても急速な	187

❶ **Molecular approaches are now being developed to provide a more rapid and objective identification** of fungi compared to traditional phenotypic methods. (J Clin Microbiol. 2005 43:2092)

訳 より急速で客観的な同定を提供するために，分子的なアプローチが，今，開発されているところである

R

★ rapidly 副 急速に 用例数 16,720

◈類義語：quickly, briefly

■ be rapidly ＋［過分］

★ be rapidly ～	…は，急速に～される	3,012
❶ be rapidly degraded	…は，急速に分解される	236
❷ be rapidly induced	…は，急速に誘導される	205
❸ be rapidly activated	…は，急速に活性化される	101
❹ be rapidly phosphorylated	…は，急速にリン酸化される	100

❷ IL-21 receptor (IL-21R) mRNA is expressed at a low level in human resting T cells but **is rapidly induced by mitogenic stimulation.** (Mol Cell Biol. 2005 25:9741)

訳 ～は，分裂促進的な刺激によって急速に誘導される

■ rapidly ＋［現分］

❶ rapidly growing	急速に増殖する	343
❷ rapidly evolving	急速に進化する	273

❶ In rapidly growing cells, we show that the major sites of RNA polymerase binding are approximately 90 transcription units that include genes needed for protein synthesis.（Proc Natl Acad Sci USA. 2005 102:17693）

訳 急速に増殖する細胞において，

■［副］＋ rapidly

❶ more rapidly	より急速に	1,315
・more rapidly than ～	～より急速に	495
❷ very rapidly	非常に急速に	173

❶ VC^+ tumors grew more rapidly than mock-transfected tumors and exhibited parallel increases in tumor angiogenesis.（Cancer Res. 2001 61:2404）

訳 VC^+腫瘍は，偽の形質移入された腫瘍より急速に増殖した

rare 形 まれな 用例数 6,949

◆類義語：infrequent, uncommon

■ be rare ＋［前］

★ be rare	…は，まれである	1,316
❶ be rare in ～	…は，～においてまれである	260

❶ Our results showed that somatic BHD mutations are rare in sporadic renal tumors.（Cancer Res. 2003 63:4583）

訳 体細胞性のBHD変異は，孤発性の腎臓腫瘍においてまれである

★★ rate 名 割合／速度／比率，動 見積もる 用例数 98,856

rate	速度／割合／比率／見積もる	61,953
rates	速度／割合／比率／見積もる	34,916
rating	見積もる	1,108
rated	見積もった／見積もられる	879

動詞としても用いられる，名詞の用例が非常に多い．複数形のratesの割合は約40%とかなり高い．

◆類義語：velocity, speed, ratio, proportion, fraction, percentage

■ rate ＋［前］

rate atの前にtheが付く割合は圧倒的に高い．また，rate of/forの前にtheが付く割合もかなり高い．

	the%	a/an%	ø%			
❶	67	30	2	rate of ～	～の割合／～の速度	22,046
				・high rate of ～	高い割合（速度）の～	661
				・higher rate of ～	より高い割合（速度）の～	515
				・increased rate of ～	増大した割合（速度）の～	406
				・rate of change	変化の速度（割合）	218
				・rate of progression	進行の速度（割合）	152
❷	43	26	29	rate in ～	～における速度	1,428
❸	81	7	10	rate at ～	～における速度	1,428
				・rate at which ～	～である速度	455
❹	73	18	9	rate for ～	～の速度	1,061

❶ **This high rate of success suggests that** most PESEs function as ESEs in their natural context.（Mol Cell Biol. 2005 25:7323）

訳 この高い割合の成功は，～ということを示唆する

❶ On follow-up, **there were no differences in the rate of change in AVA** or peak and mean gradients when patients were stratified based on the use of bisphosphonates.（J Am Coll Cardiol. 2012 59:1452）

訳 AVAの変化の速度には違いがなかった

❷ **A decrease of the cleavage rate in 4n preimplantation embryos** compared to diploid（2n）embryos was revealed by real-time imaging, using a histone H2b:eGFP reporter.（Dev Biol. 2005 288:150）

訳 4nの着床前胚における卵割の割合の低下

❸ **The presence of nitric oxide（NO）greatly accelerates the rate at which hydrogen peroxide（H_2O_2）kills Escherichia coli.**（Mol Microbiol. 2003 49:11）

訳 一酸化窒素（NO）の存在は，過酸化水素（H_2O_2）が大腸菌を殺す速度を大きく促進する

■［動］＋ the rate of

★	the rate of ～	～の割合／～の速度	11,243
❶	increase the rate of ～	…は，～の速度（割合）を増大させる	825
❷	reduce the rate of ～	…は，～の速度（割合）を低下させる	441
❸	decrease the rate of ～	…は，～の速度（割合）を低下させる	303
❹	enhance the rate of ～	…は，～の割合（速度）を増強させる	218
❺	affect the rate of ～	…は，～の割合（速度）に影響する	204
❻	slow the rate of ～	…は，～の速度（割合）を遅らせる	177
❼	determine the rate of ～	…は，～の割合（速度）を決定する	172
❽	accelerate the rate of ～	…は，～の速度を加速させる	161

❶ The severity of parental depression, as measured by impairment, **significantly increased the rate of a mood disorder** in these grandchildren（relative risk, 2.44; 95% confidence interval, 1.1-5.5; P = .03）.（Arch Gen Psychiatry. 2005 62:29）

R

訳 ～は，気分障害の割合を有意に増大させた

■ [前] + the rate of

❶ increase in the rate of ～	～の速度（割合）の増大	326
❷ decrease in the rate of ～	～の速度（割合）の低下	156
❸ effect on the rate of ～	～の割合（速度）に対する影響	129
❹ reduction in the rate of ～	～の速度（割合）の低下	119

❶ Also, replacement of His[139] in the second trans-membrane segment with Arg caused a dramatic increase in the rate of copper uptake and a large increase in the K_m value for copper.（J Biol Chem. 2005 280:37159）

訳 ～は，銅取込みの速度の劇的な増大を引き起こした

■ rate + [現分]

❶ rate-limiting step	律速段階	1,240
❷ rate-limiting enzyme	律速酵素	446
❸ rate-determining step	律速段階	286

❶ Plant transformation technology is frequently the rate-limiting step in gene function analysis in non-model plants.（Plant J. 2005 43:449）

訳 ～は，しばしば遺伝子機能解析における律速段階である

■ [名] + rate

❶ heart rate	心拍数	2,203
❷ response rate	奏効率	1,577
❸ growth rate	成長速度	1,527
❹ mortality rate	死亡率	1,398
❺ survival rate	生存率	1,033

★ rather 副 むしろ　用例数 17,352

rather thanの用例が非常に多い．

■ rather than

★ rather than ～	～よりむしろ	12,773
❶ rather than by ～	～によってよりむしろ	569
❷ rather than to ～	～によりむしろ	386
❸ rather than in ～	～においてよりむしろ	262

❶ The degree of G protein activation is determined by the combination of Giα and Gβ subunits rather than by the identity of an individual subunit.（Proc Natl Acad Sci USA. 2006 103:212）

訳 Gタンパク質活性化の程度は，個々のサブユニットの独自性によってよりむしろGiαおよびGβサブユニットの組み合わせによって決定される

❷ The physician's primary responsibility in such emergencies is to the public rather than to the individual patient.（Ann Intern Med. 2005 143:493）

訳 そのような緊急時における医師の第一の責任は，個々の患者に対してよりむしろ公衆に対してである

■ not ～ but rather

❶ not ～ but rather …	～ではなくむしろ…	735

❶ This disease is not characterized by highly proliferative cells but rather by the presence of leukemic cells with significant resistance to apoptosis and, therefore, prolonged survival. (Blood. 2002 100:2973)

訳 この疾患は，高度に増殖性の細胞によってではなくむしろ白血病性細胞の存在によって特徴づけられる

★★ ratio 名 比／割合／比率 用例数 35,219

ratio	速度／割合／比率／見積もる	27,281
ratios	速度／割合／比率／見積もる	7,938

複数形のratiosの割合が，約25%ある可算名詞.

◆類義語：rate

■ ratio ＋［前］

ratio forの前にtheが付く割合は非常に高い．また，ratio ofの前にtheが付く割合もかなり高い.

	the%	a/an%	ø%			
❶	61	28	10	ratio of ～	～の比	5,764
				・ratio of ～ to …	～対…の比	1,657
❷	74	15	10	ratio for ～	～に対する比	1,256

❶ The ratio of males to females in a population is known to influence the behaviour, life histories and demography of animals. (Curr Biol. 2012 22:R684)

訳 ある集団におけるオス対メスとの比は，～に影響を与えると知られている

❷ After propensity matching there was no significant difference in the odds ratio for death between the 2 cities (odds ratio: 1.15; 95% confidence interval: 0.51–2.61).

訳 その２つの都市の間の死に対するオッズ比に有意な違いはなかった

■ ［形／名］＋ ratio

❶ odds ratio	オッズ比	6,791
❷ hazard ratio	ハザード比	3,921
❸ risk ratio	リスク比	565
❹ signal-to-noise ratio	シグナル・ノイズ比	424
❺ molar ratio	モル比	362
❻ likelihood ratio	尤度比	339
❼ sex ratio	性比	261

R

rationale ㊔ 理論的根拠／原理 用例数 1,708

rationale	理論的根拠	1,677
rationales	理論的根拠	31

複数形のrationalesの割合は2%しかないが，単数形には不定冠詞が付くことが非常に多い．

◆類義語：principle, basis

■ rationale ＋［前］

	the%	a/an%	ø%			
❶	29	64	6	rationale for ～	～のための理論的根拠	1,307
				・provide a rationale for ～	…は，～のための理論的根拠を提供する	418
				・rationale for ～ing	～するための理論的根拠	283

❶ These results also **provide a rationale for** the development of glycerol-containing therapeutic moisturizers. (J Invest Dermatol. 2005 125:288)

訳 これらの結果は，また，グリセロールを含む治療用の加湿器の開発のための理論的根拠を提供する

reach ㊍ 達する／到達する，㊔ 範囲 用例数 7,861

reached	～に達した／達せられる	2,846
reach	～に達する／範囲	2,575
reaching	～に達する	1,724
reaches	～に達する／範囲	716

名詞としても使われるが，他動詞の用例が非常に多い．

◆類義語：attain, arrive, range, extend, span

■ reach ＋［名句］

❶	reach a maximum	…は，最大に達する	271
❷	reach a plateau	…は，プラトーに達する	159
❸	did not reach statistical significance	…は，統計的な有意には達しなかった	145
❹	reach a peak	…は，ピークに達する	119

❶ The induction of htrA gradually **reached a maximum** at around 20 min, suggesting that HtrA may be involved in a late competence response. (J Bacteriol. 2005 187:3028)

訳 htrAの誘導は，徐々に最大に達した

❸ Although there was a similar trend in the temporal PCA, the differences **did not reach statistical significance**. (Invest Ophthalmol Vis Sci. 2001 42:3337)

訳 差違は，統計的な有意には達しなかった

react ㊐動 反応する　用例数 5,395

react	反応する	1,855
reacted	反応した	1,740
reacts	反応する	1,227
reacting	反応する	573

自動詞の用例が多いが，他動詞として用いられることもある．react withの用例が非常に多い．

◆類義語：respond

■ react ＋［前］

❶	react with ～	…は，～と反応する	3,397
❷	react to ～	…は，～に反応する	277

❶ The Ub-specific protease isopeptidase T/USP5 is shown to **react with** ISG15-VS.（Mol Cell Biol. 2004 24:84）

訳 ～は，ISG15-VSと反応することが示される

R

★★ reaction ㊔名 反応　用例数 51,665

reaction	反応	37,180
reactions	反応	14,485

複数形のreactionsの割合が，約30％ある可算名詞．

◆類義語：response

■ reaction ＋［前］

	the％	a/an％	ø％			
❶	56	13	30	reaction of ～	～の反応	3,423
				・reaction of ～ with …	～の…との反応	1,311
❷	21	18	53	reaction with ～	～との反応	1,412
❸	50	25	15	reaction in ～	～における反応	1,002

❶ We conclude that the **reaction of** NO **with** hemoglobin under normoxic conditions results in consumption, rather than conservation, of NO.（Proc Natl Acad Sci USA. 2002 99:10341）

訳 正常酸素圧の状態下でのNOのヘモグロビンとの反応は，～という結果になる

❷ The **reaction with** the VIP-CRA proceeded more rapidly than with a hapten CRA devoid of the VIP sequence.（J Biol Chem. 2004 279:7877）

訳 VIP-CRAとの反応は，ハプテンCRAとの反応よりも急速に進行した

❸ Mevalonate kinase catalyzes a key **reaction in** this pathway.（J Bacteriol. 2004 186:61）

訳 メバロン酸キナーゼは，この経路における鍵となる反応を触媒する

■ reaction ＋［名］

❶	reaction conditions	反応条件	568

❷ reaction mechanism	反応機構	498
❸ reaction products	反応産物	417

★ reactive 形 反応性の　　用例数 14,193

◆類義語：responsive

■ reactive ＋［前］

❶ reactive with ～	～と反応性の	553
・be reactive with ～	…は，～と反応性である	95
❷ reactive to ～	～に反応性の	203

❶ Isocyanate groups are well suited to serving as a glass coating for arrays, in that they **are highly reactive with many different types of biological compounds**.（Anal Biochem. 2004 326: 55）

訳 ～は，多くの異なるタイプの生物学的化合物と高度に反応性である

■［副］＋ reactive

❶ highly reactive	高度に反応性の	258
❷ more reactive	より反応性の高い	186

■ reactive ＋［名］

❶ reactive oxygen species	活性酸素種	3,633
❷ reactive nitrogen species	活性窒素種	145

reactivity 名 反応性　　用例数 7,055

reactivity	反応性	6,669
reactivities	反応性	386

複数形のreactivitiesの割合は約５％あるが，単数形は無冠詞で使われることが断然多い．

◆類義語：responsiveness

■ reactivity ＋［前］

reactivity ofの前にtheが付く割合は非常に高い．

	the%	a/an%	ø%			
❶	79	2	19	reactivity of ～	～の反応性	1,384
❷	7	2	77	reactivity to ～	～への反応性	687
❸	2	2	74	reactivity with ～	～との反応性	641

❶ We also show that **the reactivity of albumin lysine residues**, including Lys[525], is affected by the status of Cys[34].（J Biol Chem. 2004 279:10864）

訳 アルブミンのリジン残基の反応性

❷ In conclusion, the addition of IL-12 to lamivudine enhances T-cell reactivity to HBV and IFN-γ production.（Hepatology. 2005 42:1028）
訳 ～は，HBVへのT細胞の反応性を増強する

❸ Only the oxidases show high reactivity with molecular oxygen.（Biochemistry. 2004 43:10692）
訳 ～は，分子状酸素との高い反応性を示す

readily 副 容易に／すぐに　用例数 6,931

◈類義語：easily, immediately, soon

■ be readily ＋［過分］

★ be readily ～	…は，容易に～される	2,817
❶ be readily detected	…は，容易に検出される	279
❷ be readily identified	…は，容易に同定される	86

❶ Induction of known LFY target genes was readily detected in these experiments.（Plant J. 2004 39:273）
訳 ～は，これらの実験において容易に検出された

■ readily ＋［形］

❶ readily available	容易に利用できる／すぐに利用できる	550
❷ readily detectable	容易に検出できる	262
・be readily detectable	…は，容易に検出できる	150
❸ readily accessible	容易に到達できる	154
❹ readily releasable	容易に放出できる	148

❶ Analyses of three markers are readily available in clinical laboratories for improved diagnosis.（Curr Opin Pediatr. 2005 17:563）
訳 3つのマーカーの分析が，臨床検査室において容易に利用できる

❷ Only one of these, the tubular carcinoma, was readily detectable by the standard γ-camera.（J Nucl Med. 2004 45:553）
訳 …は，～によって容易に検出できた

■［副］＋ readily

❶ more readily	より容易に	420

❶ On antigenic stimulation, CD4 T cells generally proliferate more readily than CD8 T cells.（Transplantation. 2002 74:836）
訳 CD4 T細胞は，一般にCD8 T細胞より容易に増殖する

rearrangement 名 再構築／再編成　用例数 8,133

rearrangement	再構築	4,163
rearrangements	再構築	3,970

複数形のrearrangementsの割合は約50%と非常に高い．

R

◆類義語：reorganization, reconstruction

■ rearrangement(s) ＋ [前]

	the%	a/an%	ø%			
❶	35	20	42	rearrangement of ～	～の再構築	1,084
❷	7	-	85	rearrangements in ～	～における再構築	415

❶ **Rearrangement of the cytoskeleton in T cells plays a critical role in** the organization of a complex signaling interface referred to as immunologic synapse (IS). (Blood. 2011 118:2492)
訳 T細胞における細胞骨格の再構築は，～において決定的に重要な役割を果たす

reason 名 理由，動 推論する　用例数 3,862

reasons	理由／～を推論する	1,845
reason	理由／～を推論する	1,369
reasoning	～を推論する	372
reasoned	～を推論した／推論される	276

名詞の用例が多いが，他動詞としても用いられる．複数形のreasonsの割合は約60%と非常に高い．

◆類義語：cause, basis, judge, estimate

■ 名 reasons ＋ [前]

reasons forの前にtheが付く割合はかなり高い．

	the%	a/an%	ø%			
❶	67	-	31	reasons for ～	～の理由	793

❶ **The reasons for this phenotypic switch are unclear.** (J Virol. 2006 80:802)
訳 この表現型の切り替えの理由は不明である

■ 名 for this reason

❶	for this reason,	この理由で，	187

❶ **For this reason, we investigated** calcium dynamics in transgenic mice expressing human WT α-synuclein using two-photon microscopy. (J Neurosci. 2012 32:9992)
訳 この理由で，我々は～を精査した

■ 動 [代名] ＋ reason

❶	we reasoned that ～	われわれは，～ということを推論した	232

❶ **We reasoned that** forced Fas-apoptosis resistance would result in earlier and more aggressive gastric cancer in our mouse model. (Cancer Res. 2005 65:10912)
訳 われわれは，～ということを推論した

★★ receive 動 受ける　　用例数 31,602

received	～を受けた	16,434
receiving	～を受ける	8,502
receive	～を受ける	6,034
receives	～を受ける	632

他動詞．受動態の用例は非常に少ない．

◆類義語：take, undergo

■ ［名］＋ receive

❶ patients received ～	患者が，～を受けた	1,934
❷ rats received ～	ラットが，～を受けた	468
❸ mice received ～	マウスが，～を受けた	346
❹ animals received ～	動物が，～を受けた	331
❺ group received ～	グループが，～を受けた	327
❻ subjects received ～	対象が，～を受けた	237

❶ All patients received tacrolimus-based immunosuppression with mycophenolate mofetil and steroids.（Transplantation. 2004 77:1066）

訳 すべての患者が，タクロリムスをベースにした免疫抑制を受けた

■ to receive

★ to receive ～	～を受ける…	3,550
❶ assigned to receive ～	～を受けるよう割り当てられた	1,095
❷ randomized to receive ～	～を受けるために無作為化された	750
❸ likely to receive ～	～を受けそうである	289
・more likely to receive ～	より～を受けそうである	137
・less likely to receive ～	より～を受けそうにない	129

❶ A total of 229 patients were randomly assigned to receive standard medical therapy, and 229 to receive standard medical therapy plus a single-chamber ICD.（N Engl J Med. 2004 350: 2151）

訳 合計229名の患者が，標準の医学療法を受けるように無作為に割り当てられた

■ ［名］＋ receiving

❶ patients receiving ～	～を受けた患者	2,465
❷ mice receiving ～	～を受けたマウス	523
❸ animals receiving ～	～を受けた動物	347
❹ rats receiving ～	～を受けたラット	203

❶ However, the emergence of imatinib resistance and the incomplete molecular response of a significant number of patients receiving this therapy have led to a search for combinations of drugs that will enhance the efficacy of imatinib.（Blood. 2005 105:3270）

訳 この治療を受けたかなりの数の患者の不完全な分子応答が，～の探索につながっている

★★ recent 形 最近の　　用例数 36,921

◆類義語：previous, prior

■ recent ＋［名］

❶	recent studies	最近の研究	7,352
❷	recent advances	最近の進歩	2,450
❸	recent evidence	最近の証拠	1,733
❹	recent work	最近の研究	1,640
❺	in recent years	近年において	1,286
❻	recent data	最近のデータ	1,137
❼	recent findings	最近の知見	1,137
❽	recent reports	最近の報告	911

❶ **Recent studies have shown that** CNK1 also interacts with RalA and Rho and participates in some aspects of signaling by these GTPases.（J Biol Chem. 2004 279:29247）

訳 最近の研究は，～ということを示している

■［副］＋ recent

❶	the most recent ～	もっとも最近の～	642
❷	more recent ～	より最近の～	616

❷ **More recent studies have implicated TRAP-Mediator as a coactivator** for a broad range of nuclear hormone receptors as well as other classes of transcriptional activators.（J Biol Chem. 2002 277:42852）

訳 より最近の研究は，TRAP-Mediatorをコアクチベーターとして関連づけている

★★ recently 副 最近　　用例数 26,381

過去形の用例が断然多く，完了形にも用いられるが，現在形ではほとんど使われない．

◆類義語：previously

■［代名］＋ recently ＋［動］

★	we recently ～	われわれは，最近～	2,303
❶	we recently reported ～	われわれは，最近～を報告した	569
❷	we have recently showed ～	われわれは，最近～を示した	406
❸	we recently demonstrated ～	われわれは，最近～を実証した	310
❹	we recently identified ～	われわれは，最近～を同定した	298

❶ **We recently reported that** the IpaC N terminus is required for type III secretion and possibly other functions.（Infect Immun. 2003 71:1255）

訳 われわれは，最近～ということを報告した

❷ **We have recently shown that** HCV NS5A activates NF-κB via oxidative stress.（J Biol Chem. 2003 278:40778）

訳 われわれは，最近～ということを示した

■ be recently ＋［過分］

★ be recently ～	…は，最近～される	2,264
❶ be recently shown	…は，最近示される	407
❷ be recently identified	…は，最近同定される	365
❸ be recently reported	…は，最近報告される	235

❶ The chromosomal protein Smchd1 was recently shown to play an important role in DNA methylation of CpG islands (CGIs), a late step in the X inactivation pathway that is required for long-term maintenance of gene silencing. (Dev Cell. 2012 23:265)

訳 染色体タンパク質Smchd1は，～において重要な役割を果たすことが最近示された

■ has recently been ＋［過分］

★ have recently been ～	…が，最近～されてきた	2,841
❶ have recently been shown	…が，最近示されてきた	659

❶ MEK1 has recently been shown to translocate to the nucleus. (J Immunol. 2013 190:159)

訳 MEK1は，核に移動することが示されてきた

R

■ a recently ＋［過分］＋［名］

❶ a recently identified ～	最近同定された～	454
❷ a recently described ～	最近述べられた～	370
❸ a recently developed ～	最近開発された～	332
❹ a recently discovered ～	最近発見された～	283

❶ Here, we have studied a recently identified hematopoietic-specific Rho GTPase, RhoH. (Blood. 2005 105:1467)

訳 われわれは，最近同定された造血に特異的なRho GTP加水分解酵素，RhoH，を研究した

■ Recently,

★ Recently,	最近，	4,944
❶ Recently, we ～	最近，われわれは～	1,078

❶ Recently, we showed that the structure of KaiA is that of a domain-swapped homodimer. (Proc Natl Acad Sci USA. 2004 101:10925)

訳 最近，われわれは～ということを示した

■ ［副］＋ recently

❶ more recently	もっと最近	894
❷ most recently	もっとも最近	262

❶ More recently, it has been established that there is no deficit of linoleic acid in atopic eczema. (Am J Clin Nutr. 2000 71:367S)

訳 もっと最近，

■ ［前］＋ recently

❶	until recently,	最近まで	387

★ recipient ㊔ レシピエント／移植患者 用例数 16,103

recipients	レシピエント／移植患者	11,704
recipient	レシピエント／移植患者	4,399

複数形のrecipientsの割合は，約75%と圧倒的に高い．

■ recipients ＋［前］

	the%	a/an%	ø%			
❶	3	-	78	recipients of ～	～のレシピエント	1,561
❷	1	-	64	recipients with ～	～のレシピエント	738

❶ **Patterns were similar among recipients of living donor** and expanded criteria donor transplants. (Transplantation. 2012 94:241)

訳 パターンは，生体移植のレシピエントの間で類似していた

❷ **We conducted a multicenter case-control study of 52 liver transplant recipients with hepatitis C** to assess the incidence of, risk factors for, and outcomes of PEG-IGD. (Gastroenterology. 2012 142:1132)

訳 われわれは，C型肝炎の52名の生体肝移植レシピエントの多施設症例対照研究を行った

■ ［名］＋ recipients

❶	transplant recipients	移植レシピエント	2,354
❷	allograft recipients	同種移植レシピエント	390

★ recognition ㊔ 認識 用例数 20,658

recognition	認識	20,649
recognitions	認識	9

複数形のrecognitionsの用例はほとんどなく，原則，**不可算名詞**として使われる．

◆類義語：cognition, perception, awareness

■ recognition ＋［前］

	the%	a/an%	ø%			
❶	32	2	62	recognition of ～	～の認識	4,908
				・specific recognition of ～	～の特異的認識	244
❷	4	1	73	recognition by ～	～による認識	1,329

❶ The highly conserved DNA-binding ETS domain defines the family and **is responsible for specific recognition of a common sequence motif**, 5'-GGA(A/T)-3'. (Annu Rev Biochem. 2011 80:437)

訳 ～は，共通配列モチーフの特異的認識に責任がある

❷ Surprisingly, the N-terminal flanking region was also important for recognition by 6 of 10 specific T cell hybridomas. (Arthritis Rheum. 2003 48:2375)

訳 …は，また，～による認識にとって重要であった

■ ［名／形］＋ recognition

❶ substrate recognition	基質認識	787
❷ pattern recognition	パターン認識	617
❸ molecular recognition	分子認識	502
❹ DNA recognition	DNA認識	479

■ recognition ＋ ［名］

❶ recognition site	認識部位	557
❷ recognition sequence	認識配列	463

★recognize 動 認識する　用例数 18,121

recognized	認識される／～を認識した	9,581
recognize	～を認識する	4,679
recognizes	～を認識する	2,403
recognizing	～を認識する	1,458

他動詞.

◆類義語：appreciate, perceive, notice

R

■ recognized ＋ ［前］

❶ recognized by ～	～によって認識される	3,012
・be recognized by ～	…は，～によって認識される	1,249
❷ recognized as ～	～として認識される	1,204
❸ recognized in ～	～において認識される	356

❶ One of these elements mediates the response to environmental conditions within pharyngeal muscles and is recognized by the nuclear hormone receptor (NHR) DAF-12. (Science. 2004 305:1743)

訳 ～は，核内ホルモン受容体（NHR）DAF-12によって認識される

❷ The sphingolipid ceramide has been recognized as an important second messenger implicated in regulating diverse signaling pathways especially for apoptosis. (Cancer Res. 2004 64:4286)

訳 スフィンゴ脂質セラミドは，重要な２次メッセンジャーとして認識されている

■ be recognized ＋ ［that節］

❶ be recognized that ～	～ということが認識される	143

❶ It has long been recognized that members of the protein kinase C (PKC) family of signal transduction molecules undergo down-regulation in response to activation. (J Biol Chem.

2004 279:5788)
訳 ～ということが長い間認識されている

■ [副] + recognized

❶ well recognized	よく認識されている	391
❷ increasingly recognized	ますます認識される	334
❸ previously recognized	以前に認識された	288

❶ Aflatoxins are **well recognized** as a cause of liver cancer, but they have additional important toxic effects. (Am J Clin Nutr. 2004 80:1106)
訳 アフラトキシンは，肝癌の原因としてよく認識されている

■ [副] + been recognized

❶ have long been recognized	…は，長い間認識されている	190

❶ Increased amounts of intrathecally synthesized IgG and oligoclonal bands **have long been recognized** as a hallmark of multiple sclerosis (MS). (J Immunol. 2004 173:649)
訳 …は，長い間～の顕著な特徴として認識されている

■ to recognize

★ to recognize ～	～を認識する…	1,357
❶ ability to recognize ～	～を認識する能力	190

❶ Several retinoid, fatty acid, and bile acid ligands were evaluated for their **ability to recognize** the 9-cis-RA binding site. (Biochemistry. 2002 41:4883)
訳 ～が，9-シスレチノイン酸結合部位を認識するそれらの能力について評価された

■ [副] + recognize

❶ specifically recognize ～	…は，～を特異的に認識する	400

❶ Using a monoclonal antibody that **specifically recognizes** tau truncated at Asp421, we show that tau is proteolytically cleaved at this site in the fibrillar pathologies of AD brain. (Proc Natl Acad Sci USA. 2003 100:10032)
訳 Asp421において切断されたtauを特異的に認識するモノクローナル抗体を使って

★ recombination 名 組換え　用例数 16,416

recombination	組換え	16,302
recombinations	組換え	114

複数形のrecombinationsの割合は0.7%しかなく，原則，**不可算名詞**として使われる．

■ recombination + [前]

	the%	a/an%	ø%			
❶	0	0	92	recombination in ～	～における組換え	1,030

❷	1	5	94	recombination between ～	～の間の組換え	628
❸	13	1	75	recombination of ～	～の組換え	395

❶ To investigate its function, an ac79-knockout bacmid **was generated through homologous recombination in Escherichia coli.** (J Virol. 2012 86:5614)

訳 ～は，大腸菌において相同組換えによって作られた

❷ **One mechanism of LOH is mitotic crossover recombination between homologous chromosomes,** potentially initiated by a double-strand break (DSB). (Proc Natl Acad Sci USA. 2011 108:11971)

訳 ヘテロ接合性欠失の一つの機構は，相同染色体の間の有糸分裂交差型組換えである

■ ［形／名］＋ recombination

❶	homologous recombination	相同組換え	2,714
❷	meiotic recombination	減数分裂組換え	582
❸	class switch recombination	クラススイッチ組換え	336

reconstitute 動 再構築する　用例数 4,921

reconstituted	再構築される／～を再構築した	3,970
reconstitute	～を再構築する	581
reconstituting	～を再構築する	249
reconstitutes	～を再構築する	121

他動詞.

■ reconstituted ＋［前］

❶	reconstituted with ～	～によって再構築される	840
	・mice reconstituted with ～	～によって再構築されたマウス	238
❷	reconstituted in ～	～において再構築される	389
	・be reconstituted in ～	…は，～において再構築される	192
❸	reconstituted into ～	～に再構築される	236

❶ Moreover, **tumor metastasis was substantially inhibited in L-selectin-deficient mice reconstituted with wild-type NK cells.** (J Exp Med. 2005 202:1679)

訳 腫瘍の転移が，野生型のNK細胞によって再構築されたL-セレクチン欠損マウスにおいて実質的に抑制された

★ record 動 記録する，名 記録　用例数 11,848

recorded	記録される／～を記録した	6,904
records	記録／～を記録する	2,969
record	記録／～を記録する	1,975

他動詞および名詞として使われる．複数形のrecordsの割合は約60%と非常に高い．

■ 動 be recorded ＋［前］

★	be recorded	…が，記録される	3,263
❶	be recorded in ～	…が，～において記録される	518
❷	be recorded from ～	…が，～から記録される	414
❸	be recorded at ～	…が，～において記録される	170

❶ High-density EEG（256 electrodes）**was recorded in** 17 normal human subjects during adaptation to a visuo-motor rotation of 60 degrees in four incremental steps of 15°．(J Neurosci. 2011 31:14810)

訳 高密度脳波（256電極）が，17名の正常なヒト被験者において記録された

■ 動［名］＋ be recorded

❶	activity was recorded	活動が，記録された	101
❷	currents were recorded	電流が，記録された	96

❶ Single-cell neural **activity was recorded** from SOA neurons in two monkeys with exotropia as they performed eye movement tasks during monocular viewing.（Invest Ophthalmol Vis Sci. 2012 53:3858)

訳 単一細胞神経活動が，SOAニューロンから記録された

■ 動［副］＋ recorded

❶	simultaneously recorded	同時に記録される	146

■ 名 records ＋［前］

	the%	a/an%	ø%			
❶	25	-	71	records of ～	～の記録	648
❷	2	-	82	records from ～	～からの記録	178

❶ Medical **records of** all the patients were reviewed.（Crit Care Med. 2011 39:2659)

訳 すべての患者の診療記録が，再検討された

■ 名［形／名］＋ record(s)

❶	medical records	診療記録	838
❷	fossil record	化石記録	259

recording 名 記録，-ing 記録する　用例数 6,786

recordings	記録	4,138
recording	記録	2,648

名詞の用例が非常に多い．複数形のrecordingsの割合は約60%と非常に高い．

■ recordings ＋［前］

	the %	a/an %	⌀ %			
❶	0	-	99	recordings from ～	～からの記録	807
❷	2	-	95	recordings of ～	～の記録	518
❸	0	-	100	recordings in ～	～における記録	404

❶ Here, **we performed in vivo whole-cell recordings from pyramidal neurons** in the rat dorsal cochlear nucleus (DCN), where intensity selectivity first emerges along the auditory neuraxis.（J Neurosci. 2012 32:18068）

訳 われわれは，錐体ニューロンからの生体内ホールセル記録を行った

■［名／形］＋ recordings

❶	patch-clamp recordings	パッチクランプ記録	469
❷	whole-cell recordings	ホールセル記録	387
❸	electrophysiological recordings	電気生理学的記録	269
❹	intracellular recordings	細胞内記録	261

R

recover 動 回復する／回収する 用例数 6,041

recovered	回復される／回収される／～を回復した／～を回収した	4,561
recover	～を回復する／～を回収する	1,004
recovering	～を回復する／～を回収する	305
recovers	～を回復する／～を回収する	171

他動詞.

◆類義語：restore, retrieve, revert, regain, ameliorate

■ recovered ＋［前］

❶	recovered from ～	～から回収される	1,468
	・be recovered from ～	…は，～から回収される	538
❷	recovered in ～	～において回復される	355
❸	recovered by ～	～によって回復される	227

❶ In a preliminary screen, 12 of the 27 recombinant proteins induced **a response that reduced the number of bacteria recovered from the spleen** or bloodstream of infected mice.（Infect Immun. 2005 73:6591）

訳 脾臓から回収される細菌の数を低下させた反応

❷ **Expression was recovered in tumour cell lines treated with** 5-aza 2-deoxycytidine.（Oncogene. 2003 22:1580）

訳 発現が，～で処理された腫瘍細胞株において回復された

❸ In the BBY samples with 25 mM chloride, we observed that the inhibition induced by azide **is partly recovered by the addition of bicarbonate.**（Biochemistry. 2005 44:12022）

訳 ～は，炭酸水素塩の添加によって部分的に回復される

★ recovery 名 回復／回収／回収率 用例数 11,453

recovery	回復／回収／回収率	11,215
recoveries	回復／回収／回収率	238

複数形のrecoveriesの割合は2％しかなく，原則，**不可算名詞**として使われる．

◆類義語：restoration, amelioration

■ recovery ＋［前］

	the%	a/an%	ø%			
❶	36	6	56	recovery of ～	～の回収／～の回復	2,918
				・recovery of function	機能の回復	109
❷	11	3	83	recovery from ～	～からの回復	1,437
				・recovery from inactivation	不活性化からの回復	162
❸	5	6	80	recovery in ～	～における回復	457

❶ **We used quantitative PCR assays to measure the recovery of DNA from two important fungal pathogens** subjected to six DNA extraction methods.（J Clin Microbiol. 2005 43:5122）

訳 われわれは，2つの重要な真菌病原体からのDNAの回収を測定するために定量的なPCRアッセイを用いた

❷ **Recovery from inactivation was strongly dependent on** voltage and pulse duration.（Biophys J. 2005 89:3026）

訳 不活性化からの回復は，強く～に依存した

■［形／名］＋ recovery

❶	functional recovery	機能的回復	516
❷	fluorescence recovery	蛍光の回復	381

❶ **This finding may be especially important in understanding functional recovery after brain lesions such as stroke.**（J Physiol. 2012 590:4011）

訳 この知見は，脳卒中のような脳障害のあとの機能的回復を理解する際に特に重要であるかもしれない

■［前］＋ recovery

❶	rate of recovery	回復率／回収率	133

❶ **Little effect was observed on the rate of recovery from desensitization** or on the response to the weakly desensitizing agonist kainate.（Mol Pharmacol. 2003 64:269）

訳 脱感作からの回復率に対する効果はほとんど観察されなかった

recruit 動 動員する／採用する／リクルートする 用例数 9,841

recruited	動員される／採用される／～を動員した	5,133
recruit	～を動員する／～を採用する	2,022
recruits	～を動員する／～を採用する	1,469

recruiting	～を動員する／～を採用する	1,217

他動詞.

◈類義語：mobilize, transfer, transport

■ be recruited ＋［前］

★	be recruited	…が，動員される	3,003
❶	be recruited to ～	…が，～に動員される	1,482
❷	be recruited from ～	…が，～から採用される	313
❸	be recruited into ～	…が，～に動員される	161
❹	be recruited by ～	…が，～によって動員される	160

❶ However, **T cells were not recruited to the placenta**. (Infect Immun. 2005 73:6322)

訳 T細胞は，胎盤には動員されなかった

❷ **Subjects were recruited from 246 women participating in** a longitudinal study of anorexia nervosa and bulimia nervosa, now in its 12th year. (Am J Psychiatry. 2001 158:1461)

訳 対象は，～に参加した246名の女性から採用された

❹ We have shown that **the BHC complex is recruited by a neuronal silencer, REST** (RE1-silencing transcription factor), and mediates the repression of REST-responsive genes. (Nature. 2005 437:432)

訳 BHC複合体は，ニューロンのサイレンサー，REST，によって動員される

R

★recruitment ㊔ 動員 用例数 13,040

recruitment	回復／回収／回収率	13,027
recruitments	回復／回収／回収率	13

複数形のrecruitmentsの用例はほとんどなく，原則，不可算名詞として使われる.

◈類義語：mobilization, supplement

■ recruitment ＋［前］

	the%	a/an%	ø%			
❶	35	3	60	recruitment of ～	～の動員	6,501
				・recruitment of neutrophils	好中球の動員	54
❷	0	0	56	recruitment to ～	～への動員	1,324
❸	1	0	83	recruitment in ～	～における動員	283
❹	1	0	83	recruitment into ～	～への動員	265

❶ **Repression by Glis2 appears to involve the recruitment of both CtBP1 and histone deacetylase 3 (HDAC3)**. (Nucleic Acids Res. 2005 33:6805)

訳 Glis2による抑制は，CtBP1およびヒストンデアセチラーゼ3（HDAC3）の両方の動員を含むように思われる

❷ **Leukocyte recruitment to inflammation sites depends on interactions between** integrins and extracellular matrix (ECM). (Blood. 2005 106:3854)

訳 炎症部位への白血球の動員は，～の間の相互作用に依存する

■ ［名］＋ recruitment

❶	neutrophil recruitment	好中球の動員	510
❷	leukocyte recruitment	白血球の動員	312

■ recruitment and ＋［名］

❶	recruitment and activation	動員と活性化	264

❶ T-cell receptor（TCR）stimulation results in the recruitment and activation of the proteins ZAP70 and Lck.（J Mol Biol. 2005 353:1001）

訳 T細胞受容体（TCR）刺激は，～の動員と活性化という結果になる

recurrence 名 再発 用例数 5,278

recurrence	再発	4,857
recurrences	再発	421

複数形のrecurrencesの割合は約10%あるが，単数形は無冠詞で使われることが圧倒的に多い.

◆類義語：relapse

■ recurrence ＋［前］

	the%	a/an%	ø%			
❶	17	8	74	recurrence of ～	～の再発	591
❷	4	3	88	recurrence in ～	～における再発	194
❸	1	4	89	recurrence after ～	～の後の再発	179

❶ Aggregate data from a multinational effort show that in mothers at high risk of having a child with cardiac-NL, the use of HCQ may protect against recurrence of disease in a subsequent pregnancy.（Circulation. 2012 Jul 3;126:76）

訳 HCQの使用は，疾患の再発から保護するかもしれない

■ ［形／名］＋ recurrence

❶	local recurrence	局所再発	404
❷	disease recurrence	疾患の再発	263
❸	tumor recurrence	腫瘍の再発	203
❹	cancer recurrence	癌の再発	174

★★ reduce 動 低下させる／減少させる 用例数 104,698

reduced	低下させられる／低下した／～を低下させた	73,794
reduce	～を低下させる	13,189
reducing	～を低下させる	8,901

reduces	～を低下させる	8,814

他動詞.

◆類義語：decrease, diminish, lower, down-regulate, decline, fall

■ reduce ＋［名句］

❶	reduce the risk of ～	…は，～のリスクを低下させる	962
❷	reduce the number of ～	…は，～の数を低下させる	838
❸	reduce the incidence of ～	…は，～の発生率を低下させる	539
❹	reduce the level of ～	…は，～のレベルを低下させる	515
❺	reduce the rate of ～	…は，～の割合を低下させる	441
❻	reduce the ability of ～	…は，～の能力を低下させる	335
❼	reduce the expression of ～	…は，～の発現を低下させる	286
❽	reduce the amount of ～	…は，～の量を低下させる	282
❾	reduce the frequency of ～	…は，～の頻度を低下させる	244

❶ A high intake of fish **may reduce the risk of AMD**. (Am J Clin Nutr. 2001 73:209)

訳 ～は，AMDのリスクを低下させるかもしれない

❷ Administered to mice before or after infection, both antibodies provided protection against infection **or substantially reduced the number of spirochetes** in the blood of mice after infection. (Infect Immun. 2001 69:1009)

訳 あるいは，スピロヘータの数をかなり低下させた

■［動／名］＋ to reduce

★	to reduce ～	～を低下させる…	5,024
❶	be shown to reduce ～	…は，～を低下させることが示される	288
❷	ability to reduce ～	～を低下させる能力	143

❶ An expression vector for a fragment of SLB containing the LIM-interaction domain **was shown to reduce expression of** Lhx3-responsive reporter genes. (J Biol Chem. 2000 275:13336)

訳 …は，～の発現を低下させることが示された

■ be reduced ＋［前］

★	be reduced	…が，低下する	12,158
❶	be reduced by ～	…が，～だけ低下する／～によって低下させられる	3,074
	・be reduced by approximately ～	…が，およそ～だけ低下する	231
❷	be reduced in ～	…が，～において低下する	2,692
❸	be reduced to ～	…が，～に低下する	984
❹	be reduced with ～	…が，～によって低下する	206
❺	be reduced from ～ to …	…が，～から…に低下する	128

❶ [^{3}H]Deltorphin I binding **was reduced by approximately 50%** in heterozygous animals. (Brain Res. 2002 945:9)

訳 ～が，およそ50%低下した

❷ However, the ability to take up soluble antigen, as determined by fluorescein isothiocyanate-labeled dextran uptake, was reduced in cells cultured with STF. (Infect Immun. 1999 67:1338)

訳 ～が，STFと共に培養された細胞において低下した

❸ Hippocampal met-enkephalin levels were reduced to 50% only at 12 weeks after inoculation. (Brain Res. 1998 793:119)

訳 ～レベルが，接種後12週においてのみ50%に低下した

❺ Finally, the incidence of latency was reduced from 88% to 13% after intranasal immunization. (J Infect Dis. 1998 177:1451)

訳 ～が，88%から13%に低下した

■ ［名］＋ be reduced

❶ levels were reduced	レベルが，低下した	254
❷ activity was reduced	活性が，低下した	194
❸ expression was reduced	発現が，低下した	164

❶ Hepatic apo B-38.9 mRNA levels were reduced by 40%. (J Biol Chem. 2000 275:32807)

訳 肝臓のapo B-38.9メッセンジャーRNAレベルが，40%低下した

■ be ＋［副］＋ reduced

❶ be significantly reduced	…は，有意に低下する	2,630
❷ be markedly reduced	…は，顕著に低下する	691
❸ be greatly reduced	…は，大きく低下する	557
❹ be dramatically reduced	…は，劇的に低下する	397
❺ be substantially reduced	…は，かなり低下する	290
❻ be severely reduced	…は，激しく低下する	185

❷ Moreover, β-catenin levels are markedly reduced in the brains of Alzheimer's disease patients with presenilin-1 mutations. (Nature. 1998 395:698)

訳 β-カテニンのレベルは，アルツハイマー病患者の脳において顕著に低下する

❸ The mutant frequency was greatly reduced in the ribozyme-expressing cells. (Nucleic Acids Res. 2004 32:5820)

訳 変異頻度は，リボザイムを発現する細胞において大きく低下した

■ reduced ＋［名］

❶ reduced levels of ～	低下したレベルの～／～のレベルを低下させた	1,391
❷ reduced expression of ～	低下した～の発現	1,367
❸ reduced risk	低下したリスク	674
❹ reduced ability	低下した能力	448
❺ reduced activity	低下した活性	431

❶ Thus, absence of a behavioral response before puberty is not associated with reduced levels of steroid receptors. (Behav Neurosci. 2002 116:198)

訳 ～は，低下したレベルのステロイド受容体とは関連しない

■ [前] + reducing

❶ effective in reducing ～	～を低下させる際に効果的である	269

❶ Thus **the gate is** effective in reducing **the binding rate for** a ligand 0.4 A bulkier by three orders of magnitude. (Proc Natl Acad Sci USA. 1998 95:9280)

訳 ゲートは，～の結合速度を低下させる際に効果的である

☆☆ reduction 名 低下／減少／還元 用例数 39,892

reduction	低下／還元	35,027
reductions	低下	4,865

複数形のreductionsの割合が，約10%ある可算名詞．a reduction in（～の低下）の用例が非常に多い．ただし，「還元」の意味では不可算名詞として用いられ，the reduction ofは「～の還元」の意味で使われることが多い．

◆類義語：decrease, decline, attenuation, down-regulation, fall

■ reduction + [前]

	the%	a/an%	ø%			
❶	10	71	14	reduction in ～	～の低下	16,864
				・reduction in the number of ～	～の数の低下	419
				・be associated with a reduction in ～	…は，～の低下と関連する	288
				・reduction in the risk of ～	～のリスクの低下	132
❷	29	24	45	reduction of ～	～の還元／～の低下	9,266
				・catalyze the reduction of ～	…は，～の還元を触媒する	179

❶ Here we show that human narcoleptics **have an 85%-95%** reduction in the number of **Hcrt neurons.** (Neuron. 2000 27:469)

訳 ～は，Hcrtニューロンの数の85%-95%の低下を持つ

❶ **This suppressed response** was associated with a reduction in **the influx of** Ca^{2+}. (J Immunol. 1998 161:6812)

訳 この抑制された反応は，～の流入の低下と関連した

❷ **Glutathione reductase** catalyzes the reduction of **glutathione disulfide by NADPH.** (Biochemistry. 2000 39:4711)

訳 グルタチオン還元酵素は，NADPHによるグルタチオンジスルフィドの還元を触媒する

■ [形] + reduction

❶ significant reduction	有意な（顕著な）低下	2,498
❷ ～% reduction	～%の低下	2,450
❸ marked reduction	顕著な低下	761
❹ ～-fold reduction	～倍の低下	696
❺ risk reduction	リスクの低下	473

⑥ dramatic reduction	劇的な低下	433
⑦ greater reduction	より大きな低下	247
⑧ substantial reduction	かなりの低下	244

❶ PC-3 and DU-145 cells that were treated with VES showed a significant reduction in the levels of MMP-9 in the culture medium. (Oncogene. 2004 23:3080)

訳 ～は，MMP-9のレベルの有意な低下を示した

❹ Addition of inositol to the growth medium resulted in a 2-3-fold reduction in gene expression in wild type cells. (J Biol Chem. 1998 273:11638)

訳 ～は，野生型細胞における遺伝子発現の2-3倍の低下という結果になった

redundant 形 重複性の／重複する 用例数 2,182

■ [副] + redundant

❶ functionally redundant	機能的に重複する	237
❷ partially redundant	部分的に重複する	207

❶ The homologous mammalian rho kinases (ROCK I and II) are assumed to be functionally redundant, based largely on kinase construct overexpression. (J Cell Biol. 2005 170:443)

訳 ～は，機能的に重複すると推定される

refer 動 呼ぶ／参照する／照会する 用例数 4,404

referred	呼ばれる／照会される／～を呼んだ	3,290
refer	呼ぶ	523
refers	呼ぶ	374
referring	参照する	217

refer to ～ as …（～を…と呼ぶ），およびその受動態で，be referred to asの用例が非常に多い．

◆類義語：call, mention

■ referred + [前]

❶ referred to as ～	～と呼ばれる	2,063
・be referred to as ～	…は，～と呼ばれる	327
・also referred to as ～	～とも呼ばれる	157
・commonly referred to as ～	一般に～と呼ばれる	98
❷ referred for ～	～のために照会される	469
・be referred for ～	…は，～のために照会される	180
・patients referred for ～	～のために照会された患者	142

❶ The budding yeast homologs of Cdk1 and Polo kinase are referred to as Cdc28 and Cdc5. (Curr Biol. 2005 15:2033)

訳 Cdk1およびPoloキナーゼの出芽酵母ホモログは，Cdc28およびCdc5と呼ばれる

❶ In this study we utilized small hairpin RNAs (shRNAs), also referred to as small interfering RNAs, to target human uPA and uPAR. (J Biol Chem. 2005 280:36529)
訳 低分子干渉RNAとも呼ばれる

■ [代名／名] + refer to

★ refer to ～	…は，～を呼ぶ／～を照会する	859
❶ ～ we refer to as …	われわれが…と呼ぶ～	147
❷ we refer to ～ as …	われわれは，～を…と呼ぶ	111

❶ Mitochondrial aggregation resulted in gross degradation of DNA, a cell death-related process we refer to as mitochondrial aggregation and genome destruction (MAGD). (Cancer Res. 2005 65:4191)
訳 われわれがミトコンドリア凝集とゲノム破壊（MAGD）と呼ぶ細胞死と関連した過程

❷ We refer to this process as heterogeneous gelation. (Biochemistry. 2005 44:1316)
訳 われわれは，この過程を不均一なゲル化と呼ぶ

★ reflect 動 反映する　用例数 14,071

reflect	～を反映する	5,922
reflects	～を反映する	3,069
reflected	反映される／～を反映した	2,798
reflecting	～を反映する	2,282

他動詞.

■ reflect + [名句]

❶ reflect differences in ～	…は，～の違いを反映する	189
❷ reflect changes in ～	…は，～の変化を反映する	117

❶ Variation in chromatin composition and organization often reflects differences in genome function. (Mol Cell. 2012 47:596)
訳 クロマチンの組成と構成の変動は，しばしばゲノム機能の違いを反映する

■ [副／助] + reflect

❶ may reflect ～	…は，～を反映するかもしれない	1,675
❷ likely reflect ～	…は，おそらく～を反映するであろう	342
❸ probably reflect ～	…は，おそらく～を反映する	189
❹ accurately reflect ～	…は，正確に～を反映する	180

❷ Although increased excitability likely reflects changes in synaptic efficacy, the cellular mechanisms underlying injury-induced synaptic plasticity are poorly understood. (J Neurosci. 2012 32:12431)
訳 増大した興奮性は，おそらくシナプスの効率の変化を反映している

R

■ to reflect

★ to reflect ～	～を反映する…	878
❶ appear to reflect ～	…は，～を反映するように思われる	195
❷ be thought to reflect ～	…は，～を反映すると考えられる	89

❶ Cellular susceptibility to UVA-induced necrotic cell death appears to reflect the intracellular level of LIP. (J Invest Dermatol. 2004 123:771)
訳 ～は，LIPの細胞内レベルを反映するように思われる

■ reflected ＋［前］

❶ reflected in ～	～において反映される	1,022
・be reflected in ～	…は，～において反映される	580
❷ reflected by ～	～によって反映される	696
・as reflected by ～	～によって反映されるように	321

❶ The resultant increase in PTEN activity is reflected in decreased activation of Akt by epidermal growth factor and serum. (J Biol Chem. 2006 281:4816)
訳 ～は，Aktの低下した活性化において反映される

❷ Babesiosis followed its normal course of infection in coinfected mice, without evidence for increased severity, as reflected by percentage of parasitemia, spleen weights, and hematologic and clinical chemistry parameters. (J Infect Dis. 2005 192:1634)
訳 寄生虫血症の割合によって反映されるように

refractory 形 抵抗性の／不応性の　用例数 3,456

◆類義語：resistant, tolerant

■ refractory ＋［前］

❶ refractory to ～	～に抵抗性の	1,057
・be refractory to ～	…は，～に抵抗性である	515

❶ Our results indicate that both PS-341 and Zol are effective treatments for ATLL and HHM, which are refractory to conventional therapy. (Cancer Res. 2007 67:11859)
訳 ～は，通常療法に抵抗性である

regard 名 関心／点，動 みなす／考える　用例数 3,553

regard	関心／点／みなす／考える	2,594
regarded	みなされる／みなした／考えられる	738
regards	関心／点／みなす／考える	221

動詞としても用いられるが，名詞の用例が多い．特に，with regard toの用例が非常に多い．

◆類義語：consider, think, concern

■ 名 [前] + regard to

★	regard to ～	～に関して	1,965
❶	with regard to ～	～に関して	1,678
❷	in regard to ～	～に関して	194

❶ Quantitative immunofluorescence analyses showed no detectable loss of microtubule mass during retraction, even with regard to the most labile microtubules. (J Neurosci. 2002 22:5982)
訳 最も不安定な微小管に関してでさえ

■ 名 in this regard

❶	in this regard	この点では	541

❶ In this regard, we have previously shown that CasBrE-induced disease requires late, rather than early, virus replication events in microglial cells. (J Viol. 1999 73:6841)
訳 この点では，われわれは以前に～ということを示した

■ 動 be regarded + [前]

★	be regarded	…は，みなされる	458
❶	be regarded as ～	…は，～とみなされる	420

❶ This model can be regarded as a general explanation for the activity of hyperthermophilic enzymes. (Proc Natl Acad Sci USA. 2004 101:14379)
訳 このモデルは，～に対する一般的な説明とみなされうる

regardless 形 かかわらず　用例数 4,115

regardless ofの用例が圧倒的に多い．

■ regardless + [前]

❶	regardless of ～	～にかかわらず	4,065
	・regardless of whether ～	～かどうかにかかわらず	430

❶ Gap filling was inhibited regardless of whether the loop was retained or removed. (Nucleic Acids Res. 2004 32:6268)
訳 …は，～かどうかにかかわらず抑制された

regeneration 名 再生　用例数 5,971

複数形の用例はなく，原則，不可算名詞として使われる．

■ regeneration + [前]

	the%	a/an%	ø%			
❶	35	1	63	regeneration of ～	～の再生	860
❷	0	0	100	regeneration in ～	～における再生	495

R

❸	0	0	100	regeneration after ～	～のあとの再生	259

❶ The molecular basis of axonal regeneration of central nervous system (CNS) neurons remains to be fully elucidated. (Brain Res. 2012 1453:8)

訳 中枢神経系(CNS) ニューロンの軸索再生の分子基盤は，完全には解明されないままである

■ [名] ＋ regeneration

❶ liver regeneration	肝再生	610
❷ tissue regeneration	組織再生	291
❸ muscle regeneration	筋再生	255
❹ axon regeneration	軸索再生	233

regimen ㊔ 療法／投与計画　用例数 7,220

regimen	療法／投与計画	3,840
regimens	療法／投与計画	3,380

複数形のregimensの割合は約45%とかなり高い.

◆類義語：therapy

■ regimen ＋ [前]

	the%	a/an%	ø%			
❶	23	71	3	regimen of ～	～の療法	429
❷	33	51	7	regimen for ～	～のための療法	214
❸	28	41	1	regimen in ～	～における療法	162

❶ Thirty-nine patients with active SLE were started on a standard regimen of rituximab with intravenous and oral steroids. (Arthritis Rheum. 2011 63:3038)

訳 39人の活動性全身性エリテマトーデスの患者は，リツキシマブの標準療法を開始された

■ [名] ＋ regimen

❶ treatment regimen	投薬治療計画	206
❷ conditioning regimen	移植前処置	158

★★ region ㊔ 領域　用例数 114,047

region	領域	70,336
regions	領域	43,711

複数形のregionsの割合は約40%とかなり高い.

◆類義語：area, portion, territory, site, locus

■ region ＋［前］

region ofの前にtheが付く割合は非常に高い.

	the%	a/an%	ø%			
❶	76	18	0	region of ～	～の領域	19,841
				・region of the protein	そのタンパク質の領域	345
				・region of the gene	その遺伝子の領域	194
				・region of the genome	そのゲノムの領域	121
❷	36	46	3	region in ～	～における領域	2,411

❶ In this paper, by studying cellular processes in Drosophila melanogaster that require fascin activity, **we identify a regulatory residue within the C-terminal region of the protein** (S289). (J Cell Biol. 2012 197:477)

訳 われわれは，そのタンパク質のC末端領域内に調節性の残基を同定する

❷ **A small, proline-rich region in the C-terminal half of MAP4 bound directly to** a Sept 2:6:7 heterotrimer, and to the Sept2 monomer. (Mol Biol Cell. 2005 16:4648)

訳 MAP4のC末端半分の小さなプロリンに富んだ領域は，～に直接結合した

■ region ＋［動］

❶	region contains ～	領域は，～を含む	447

❶ We observed that the region -3724/-3224 of the MUC5AC promoter is critical for CS-induced gene transcriptional activity and that **this region contains two Sp1 binding sites**. (J Biol Chem. 2012 287:27948)

訳 この領域は，２つのSp1結合部位を含む

■［名／形］＋ region

❶	promoter region	プロモーター領域	2,457
❷	untranslated region	非翻訳領域	2,137
❸	coding region	コーディング領域	1,887
❹	C-terminal region	C末端領域	1,382
❺	binding region	結合領域	1,229
❻	regulatory region	調節領域	925

★★ regulate 動 調節する／制御する 用例数 104,745

regulated	調節される／～を調節した	43,835
regulate	～を調節する	25,029
regulates	～を調節する	19,013
regulating	～を調節する	16,868

他動詞.

◆類義語：control, modulate, adjust, drive

R

■ be regulated ＋［前］

★ be regulated	…は，調節される	11,612
❶ be regulated by ～	…は，～によって調節される	7,519
・be regulated by phosphorylation	…は，リン酸化によって調節される	176
❷ be regulated in ～	…は，～において調節される	1,080
❸ be regulated at ～	…は，～において調節される	411
❹ be regulated through ～	…は，～によって調節される	298
❺ be regulated during ～	…は，～の間に調節される	231

❶ Cyclin D expression **is regulated by** growth factors and is necessary for the induction of mitogenesis.（J Biol Chem. 1998 273:29864）

訳 サイクリンD発現は，増殖因子によって調節される

❷ The novel physiological role of copper as a modulator of protein phosphorylation could be central to understanding how copper transport **is regulated in** mammalian cells.（J Biol Chem. 2001 276:36289）

訳 ～は，哺乳類細胞において銅の輸送がどのように調節されるかを理解することに対してきわめて重要でありうる

■［名］＋ be regulated

❶ expression is regulated	発現は，調節される	497
❷ activity is regulated	活性は，調節される	353
❸ genes are regulated	遺伝子は，調節される	154

■［副］＋ regulated

❶ developmentally regulated	発生的に調節される	1,003
❷ tightly regulated	しっかりと調節される	993
❸ differentially regulated	差動的に調節される	783
❹ negatively regulated	負に調節される	694
❺ highly regulated	高度に調節される	571

❶ By RNA and protein analysis, we have demonstrated that Cypher isoforms are **developmentally regulated** in both skeletal and cardiac muscle.（J Biol Chem. 2003 278:7360）

訳 ～は，骨格筋および心筋の両方において発生的に調節される

❷ This process of T-helper cell differentiation is **tightly regulated** by cytokines.（Nature. 2000 407:916）

訳 ～は，サイトカインによってしっかりと調節される

■ regulate ＋［名句］

❶ regulate the expression of ～	…は，～の発現を調節する	1,727
❷ regulate gene expression	…は，遺伝子発現を調節する	1,040
❸ regulate the activity of ～	…は，～の活性を調節する	670
❹ regulate the transcription of ～	…は，～の転写を調節する	237
❺ regulate the function of ～	…は，～の機能を調節する	235

❻ regulate the development of ～	…は，～の発生を調節する	161
❼ regulate the production of ～	…は，～の産生を調節する	141

❶ In adrenal and gonadal tissues **they regulate the expression of the cytochrome P450 steroid hydroxylase genes**, key mediators of steroidogenesis. (J Invest Dermatol. 2001 117:1559)
訳 それらは，チトクロムP450ステロイド水酸化酵素遺伝子の発現を調節する

■［名］＋ regulate

❶ proteins regulate ～	タンパク質は，～を調節する	312
❷ signaling regulates ～	シグナル伝達は，～を調節する	312
❸ pathway regulates ～	経路は，～を調節する	261
❹ factors regulate ～	因子は，～を調節する	177

■［副］＋ regulate

❶ negatively regulate ～	…は，～を負に調節する	2,174
❷ positively regulate ～	…は，～を正に調節する	609
❸ directly regulate ～	…は，～を直接調節する	585
❹ differentially regulate ～	…は，～を差動的に調節する	443

❶ **The egl-1 gene negatively regulates the ced-9 gene**, which protects against cell death and is a member of the bcl-2 family. (Cell. 1998 93:519)
訳 egl-1遺伝子は，ced-9遺伝子を負に調節する

■［動／名］＋ to regulate

★ to regulate ～	～を調節する…	6,907
❶ be shown to regulate ～	…は，～を調節することが示される	393
❷ ability to regulate ～	～を調節する能力	282
❸ be known to regulate ～	…は，～を調節することが知られている	267
❹ appear to regulate ～	…は，～を調節するように思われる	231

❶ **Exogenous iron levels were shown to regulate** ferric hydroxamate uptake in S. aureus. (J Bacteriol. 2000 182:4394)
訳 外来性の鉄のレベルが，～を調節することが示された

★★ regulation ㊔ 調節／制御 用例数 63,511

regulation	調節／制御	63,148
regulations	調節／制御	363

複数形のregulationsの割合は0.6%しかなく，原則，**不可算名詞**として使われる．regulation ofの用例が圧倒的に多い．

◆類義語：modulation, adjustment

R

■ regulation ＋［前］

	the%	a/an%	ø%			
❶	50	2	43	regulation of ～	～の調節	40,649
				・regulation of ～ in …	…における～の調節	4,133
				・regulation of gene expression	遺伝子発現の調節	826
				・regulation of genes	遺伝子の調節	301
				・regulation of expression of ～	～の発現の調節	301
❷	0	1	72	regulation in ～	～における調節	2,185
❸	3	0	59	regulation by ～	～による調節	2,179

❶ Our findings suggest that **IL-12 and IL-18 have different roles in the regulation of gene expression in** NK and T cells.（J Immunol. 1999 162:5070）

訳 IL-12およびIL-18は，～における遺伝子発現の調節において異なる役割を持つ

■［前］＋ the regulation of

❶	involved in the regulation of ～	～の調節に関与している	1,059
❷	implicated in the regulation of ～	～の調節に関与している	557
❸	contribute to the regulation of ～	…は，～の調節に寄与する	284
❹	participate in the regulation of ～	…は，～の調節に関与する	230
❺	play a role in the regulation of ～	…は，～の調節において役割を果たす	228
❻	mechanism for the regulation of ～	…は，～の調節のための機構	166

❶ Although gammadelta T cells **are involved in the regulation of inflammation after infection**, their precise function is not known.（J Exp Med. 2000 191:2145）

訳 ～は，感染のあとの炎症の調節に関与する

❺ These results suggest that **RINT-1 may play a role in the regulation of cell cycle control** after DNA damage.（J Biol Chem. 2001 276:6105）

訳 RINT-1は，細胞周期制御の調節において役割を果たすかもしれない

❻ We propose that clearance of MMP2-TSP2 complexes by LRP **is an important mechanism for the regulation of extracellular MMP2 levels in fibroblasts**, and perhaps in other cells.（J Biol Chem. 2001 276:8403）

訳 ～は，線維芽細胞における細胞外MMP2レベルの調節のための重要な機構である

■［動］＋ the regulation of

❶	investigate the regulation of ～	…は，～の調節を精査する	219
❷	understand the regulation of ～	…は，～の調節を理解する	197
❸	study the regulation of ～	…は，～の調節を研究する	175
❹	examine the regulation of ～	…は，～の調節を調べる	174

❶ **We have investigated the regulation of σE** through a transcriptional and mutational analysis of sigE and the surrounding genes.（Mol Microbiol. 1999 33:97）

訳 われわれは，σEの調節を精査した

■ [形] + regulation of

❶ transcriptional regulation of ～	～の転写制御	1,109
❷ negative regulation of ～	～の負の制御	614
❸ …-dependent regulation of ～	～の…依存性制御	591
❹ …-mediated regulation of ～	～の…仲介性制御	452
❺ differential regulation of ～	～の差動性制御	426
❻ specific regulation of ～	～の特異的制御	288

★★ relate (動) 関連づける／関連する 用例数 65,334

related	関連する／～を関連づけた／関連した	62,063
relate	～を関連づける／関連する	1,556
relating	～を関連づける／関連する	991
relates	～を関連づける／関連する	724

related toの用例が非常に多い．他動詞として用いられることが多いが，自動詞の用例もある．

◆類義語：associate, couple, link, correlate, connect, concern, attach

R

■ be related + [前]

★ be related	…は，関連する	6,283
❶ be related to ～	…は，～に関連する	5,590

❶ This protective effect of eliprodil **may be related to its reduction** (by 78%) of NMDA-induced currents recorded under patch-clamp recording in these cells. (Invest Ophthalmol Vis Sci. 1999 40:1170)

訳 ～は，それの低下に関連するかもしれない

■ to be related to

★ to be related to ～	～に関連する…	465
❶ appear to be related to ～	…は，～に関連するように思われる	216

❶ **Improved outcomes appear to be related to reductions in** organ damage, infection, and severe acute GVHD. (N Engl J Med. 2010 363:2091)

訳 改善された結果は，～の低下に関連するように思われる

■ [副] + related

❶ closely related	緊密に関連する	4,819
・closely related to ～	～に緊密に関連する	1,554
❷ structurally related	構造的に関連する	850
❸ distantly related	かすかに関連する	734
❹ inversely related to ～	～に逆比例する	552
❺ directly related to ～	～に直接関連する	502

❻ highly related	高度に関連する	403
❼ functionally related	機能的に関連する	395
❽ significantly related to 〜	〜に有意に関連する	321

❶ The Neurofibromatosis-2 tumor suppressor gene encodes Merlin, **a member of the Protein 4.1 superfamily most closely related to Ezrin, Radixin and Moesin.** (Development. 2000 127:1315)

訳 エズリン，ラディキシンおよびモエシンに最も緊密に関連するProtein 4.1スーパーファミリーのメンバー

❷ **The predicted protein RpfH is structurally related to the sensory input domain of RpfC.** (Mol Microbiol. 2000 38:986)

訳 予想されたタンパク質RpfHは，RpfCの感覚性入力ドメインに構造的に関連する

❸ **SVOP and SV2 are more distantly related to** eukaryotic and bacterial phosphate, sugar, and organic acid transporters. (J Neurosci. 1998 18:9269)

訳 SVOPおよびSV2は，〜によりかすかに関連する

❹ An unusual feature of DHFR-PLGA expression is that **accumulation of the protein is inversely related to the level of induction of its mRNA.** (J Bacteriol. 2002 184:494)

訳 そのタンパク質の蓄積は，それのメッセンジャーRNAの誘導のレベルと逆比例する

■ [名]-related

❶ age-related 〜	年齢に関連する〜	2,947
❷ treatment-related 〜	治療に関連する〜	697
❸ event-related 〜	出来事に関連する〜	627
❹ disease-related 〜	疾患に関連する〜	532
❺ dose-related 〜	用量に関連する〜	386

❶ **We investigated age-related changes in the BFCN** that may serve as a substrate for this vulnerability. (J Comp Neurol. 2003 455:249)

訳 われわれは，年齢に関連するBFCNの変化を精査した

■ relate ＋ [名] ＋ to

❶ relate 〜 to …	〜を…に関連づける	1,755

❶ **We related these findings to the results of** real-time reverse transcription-PCR quantification of treponemal and cytokine mRNA levels. (A Infect Immun. 2007 75:3021)

訳 われわれは，これらの知見を〜の結果に関連づけた

■ relate to

❶ relate to 〜	…は，〜に関連する	1,858

❶ **These differing effects of GH may relate to the progressive increase of** LV fibrosis in the CM hamster. (Circulation. 1999 100:1734)

訳 これらの異なるGHの影響は，〜の進行性の増大に関連するかもしれない

relation 名 関連 用例数 7,105

relation	関連	5,730
relations	関連	1,375

複数形のrelationsの割合が，約20%ある可算名詞．in relation toの用例が非常に多い．

◆類義語：relationship, association, relevance, link, connection, correlation, implication

■ relation ＋［前］

relation ofの前にtheが付く割合は圧倒的に高い．また，relation betweenの前にtheが付く割合も非常に高い．

	the%	a/an%	ø%			
❶	74	21	0	relation between ～	～の間の関連	1,812
				・examine the relation between ～	…は，～の間の関連を調べる	240
				・investigate the relation between ～	…は，～の間の関連を精査する	99
				・assess the relation between ～	…は，～の間の関連を評価する	88
❷	96	0	3	relation of ～	～の関連	464
				・examine the relation of ～	…は，～の間の関連を調べる	84

❶ In this study, **the authors examined the relation between HDL cholesterol levels and the risk of stroke** in elderly men.（Am J Epidemiol. 2004 160:150）

訳 著者らは，HDLコレステロールレベルと脳卒中のリスクとの間の関連を調べた

■ in relation to

★ in relation to ～	～に関して	2,068
❶ discussed in relation to ～	～に関して論じられる	233

❶ **The results are discussed in relation to cereal genome organization.**（Proc Natl Acad Sci USA. 2002 99:850）

訳 それらの結果は，穀類ゲノムの組織化に関連して論じられる

★ relationship 名 関連性／関係／相関 用例数 24,176

relationship	関連性／関係／相関	17,150
relationships	関連性／関係／相関	7,026

複数形のrelationshipsの割合が，約30%ある可算名詞．relationship betweenの用例が非常に多い．

◆類義語：relation, association, relevance, link, connection, correlation, implication

R

■ relationship(s) + [前]

relationship ofの前にtheが付く割合は圧倒的に高い．また，relationship betweenの前にtheが付く割合も非常に高い．

	the%	a/an%	ø%			
❶	70	26	0	relationship between ～	～の間の関連性	9,641
❷	87	10	3	relationship of ～	～の関連性	1,474
❸	6	9	17	relationship to ～	～との関連性	1,086
❹	14	43	4	relationship with ～	～との関連性	654
❺	38	-	58	relationships among ～	～の間の関連性	638

❶ Here, **we investigate the relationship between matrix metalloproteinase-8 and disease severity** in children with septic shock.（Crit Care Med. 2012 40:379）

訳 われわれは，マトリックスメタロプロテアーゼ8と疾患重症度の間の関連性を精査した

❸ **This prompted us to look at the origin of KCBP and its relationship to SpKinC.**（Plant Physiol. 2005 138:1711）

訳 これが，われわれにKCBPの起源およびSpKinCとそれの関連性を調べるように促した

■ [動] + the relationship between/of

❶	examine the relationship between ～	…は，～の間の関連性を調べる	650
❷	investigate the relationship between ～	…は，～の間の関連性を精査する	484
❸	determine the relationship between ～	…は，～の間の関連性を決定する	233
❹	explore the relationship between ～	…は，～の間の関連性を探索する	226
❺	understand the relationship between ～	…は，～の間の関連性を理解する	195
❻	study the relationship between ～	…は，～の間の関連性を研究する	176
❼	examine the relationship of ～	…は，～の関連性を調べる	119

❶ In the present study, **we examined the relationship between salivary flow rates and maximal bite force** in a community-based sample of men and women 35 years of age or older.（J Dent Res. 2000 79:1560）

訳 われわれは，唾液流速と最大咬合力との間の関連性を調べた

■ [名／形] + relationship(s)

❶	structure-activity relationships	構造－活性相関	592
❷	structure-function relationships	構造－機能相関	413
❸	inverse relationship	逆相関	410
❹	linear relationship	線形相関	318
❺	functional relationship	機能相関	279
❻	causal relationship	因果関係	241
❼	dose-response relationship	用量反応相関	200

❶ **The structure-activity relationship of the cyclic cyanoguanidines is compared with that of** the corresponding cyclic urea analogues.（J Med Chem. 1998 41:1446）

訳 サイクリック・シアノグアニジンの構造－活性相関が，～のそれと比較される

★★ relative　㊖ 比べて／相対的な，㊔ 親族　用例数 36,478

relative	比べて／相対的な／親族	34,445
relatives	類縁体／親族	2,033

形容詞の用例が多いが，名詞としても用いられる．名詞としては，複数形の用例がほとんどである．relative toの用例が非常に多い．

◆類義語：relevant

■ ㊖ relative to

★ relative to ～	～に比べて／～に関連する	13,841
❶ relative to wild-type ～	野生型～に比べて	627
❷ relative to that of ～	～のそれに比べて	432
❸ relative to controls	コントロール群に比べて	317
❹ …-fold relative to ～	～に比べて…倍	201
❺ reduced relative to ～	～に比べて減少した	121

❶ By 4 and 8 wk of age, **the level of type X collagen was increased in growth plate cartilage of transgenic mice relative to wild-type controls.** (J Cell Biol. 1997 139:541)

訳 X型コラーゲンのレベルが，野生型の対照群と比べてトランスジェニックマウスの成長板軟骨において増大した

❹ In this system, **the expression level of ICP0 was reduced more than 1,000-fold relative to the level of expression from HSV-1 vectors.** (J Virol. 2001 75:3391)

訳 ICP0の発現レベルは，HSV-1ベクターからの発現のレベルに比べて1,000倍以上低下した

■ ㊖ relative ＋［名］

❶ relative risk	相対危険度	3,322
❷ relative contributions	相対的な寄与	639
❸ relative importance	相対的な重要性	615
❹ relative abundance	相対的な存在量	443
❺ relative levels	相対的なレベル	295
❻ relative expression	相対的な発現	253

❶ **The relative risk of death decreased by 1.8% per year** over the 1983 to 1999 observation period. (J Clin Oncol. 2005 23:5019)

訳 死の相対危険度は，年に1.8%低下した

■ ㊔ relatives

❶ relatives of ～	～の親族	540
❷ first-degree relatives	一親等親族	368
❸ close relatives	類縁体	122

❶ Compared with healthy controls, **relatives of patients with gastric cancer had a higher prevalence of hypochlorhydria** (27% vs. 3%) but a similar prevalence of H. pylori infection

R

(63% vs. 64%). (Gastroenterology. 2000 118:22)
訳 胃癌の患者の親族は，より高い低酸症の有病率を持っていた

★ relatively 副 比較的／相対的に 用例数 14,855

◆類義語：comparatively, comparably

■ relatively ＋［形］

❶	relatively low	比較的低い	1,212
❷	relatively high	比較的高い	1,147
❸	relatively small	比較的小さい	887
❹	relatively little	比較的わずかな	874
	・relatively little is known about ～	～について比較的わずかしか知られていない	400
❺	relatively large	比較的大きい	483
❻	relatively few	比較的わずかな	463
❼	relatively stable	比較的安定な	332
❽	relatively simple	比較的単純な	306
❾	relatively short	比較的短い	305

❶ These results suggest that LP T cells are differentiated effector cells that respond at high levels **when activated with relatively low levels of Ag- and B7-mediated costimulation** in vivo. (J Immunol. 1999 163:5937)
訳 比較的低いレベルのAgおよびB7に仲介される同時刺激によって活性化されたとき

❷ At E13, **Isl1 is maintained at relatively high levels** in the sensory primordium while down-regulated in the other regions of the cochlear duct. (J Comp Neurol. 2004 477:412)
訳 Isl1は，比較的高いレベルで維持される

❹ Despite intensive study of synaptic plasticity, **relatively little is known about the magnitude and duration of calcium accumulation** caused by unitary events at individual synapses. (Proc Natl Acad Sci USA. 2000 97:901)
訳 カルシウム蓄積の大きさと持続期間について比較的わずかしか知られていない

★★ release 名 放出／遊離，動 放出する 用例数 46,590

release	放出／～を放出する	35,774
released	放出される／～を放出した	7,111
releasing	～を放出する	2,627
releases	放出／～を放出する	1,078

名詞の用例が多いが，他動詞としても用いられる．複数形のreleasesの割合は2％しかなく，原則，**不可算名詞**として使われる．
◆類義語：emission, egress, liberate

■ 名 release ＋［前］

	the%	a/an%	ø%			
❶	46	6	48	release of ～	～の放出	9,889

				・release of ～ from …	～の…からの放出	2,117
				・release of cytochrome c	シトクロムcの放出	459
				・release of Ca^{2+}	Ca^{2+}の放出	283
				・…-induced release of ～	…誘導性の～の放出	211
				・…-dependent release of ～	…依存性の～の放出	143
				・…-mediated release of ～	…仲介性の～の放出	126
❷	2	0	71	release from ～	～からの放出	2,926
				・Ca^{2+} release from ～	～からのCa^{2+}放出	436
❸	1	1	73	release in ～	～における放出	1,574

❶ Release of Ca^{2+} from intracellular Ca^{2+} stores (ICS) is involved in age-related changes in the induction of long-term potentiation. (Brain Res. 2005 1031:125)
訳 細胞内Ca^{2+}の貯蔵（ICS）からのCa^{2+}の放出は，～の加齢性変化に関与する

❶ Based on immunochemical analysis, Mn (II)-induced apoptosis does not lead to the release of cytochrome c into the cytosol. (Proc Natl Acad Sci USA. 2001 98:9505)
訳 ～は，細胞質へのシトクロムcの放出につながらない

❷ Pretreating rat intestinal epithelial cells expressing CCK_2R with EGF increased the level of G17-stimulated Ca^{2+} release from intracellular stores. (J Biol Chem. 2004 279:1853)
訳 ～は，細胞内貯蔵からのG17に刺激されるCa^{2+}放出のレベルを増大させた

❸ Compounds 64 and 65 were potent inhibitors of TNF-α release in the mouse at 100 mg/kg po. (J Med Chem. 2004 47:6255)
訳 ～は，マウスにおけるTNF-αの放出の強力な抑制剤であった

■ 名 [名] ＋ release

❶	Ca^{2+} release	Ca^{2+}放出	2,770
❷	neurotransmitter release	神経伝達物質放出	1,012
❸	cytochrome c release	シトクロムc放出	862
❹	calcium release	カルシウム放出	604
❺	glutamate release	グルタミン酸放出	561
❻	transmitter release	伝達物質放出	552
❼	dopamine release	ドパミン放出	511
❽	insulin release	インスリン放出	383

■ 動 released ＋ [前]

❶	released from ～	～から放出される	1,956
	・be released from ～	…は，～から放出される	849
❷	released by ～	～によって放出される	864
❸	released into ～	～に放出される	435
❹	released in ～	～において放出される	356
❺	released during ～	～の間に放出される	254

❶ The suSLBP is concentrated in the egg pronucleus and is released from the nucleus only when cells enter the first mitosis. (Nucleic Acids Res. 2004 32:811)
訳 細胞が最初の有糸分裂に入るときのみ，～は核から放出される

❷ After 90-99% of the population dies, a small mutant subpopulation uses the nutrients released

R

by dead cells to grow. (J Cell Biol. 2004 166:1055)
訳 ～は，死細胞によって放出された栄養分を利用する

relevance 名 関連／関連性 用例数 3,855

複数形の用例は全くなく，原則，**不可算名詞**として使われる.
◆類義語：relation, relationship, association, link, connection, correlation, implication

■ relevance ＋［前］

relevance ofの前にtheが付く割合は圧倒的に高い.

	the %	a/an %	ø %			
❶	96	1	3	relevance of ～	～の関連	1,898
				・relevance of these findings	これらの知見の関連	114
❷	3	1	77	relevance to ～	～への関連	758
❸	7	2	79	relevance for ～	～に対する関連	249
❹	2	1	58	relevance in ～	～における関連	193

❶ The clinical relevance of these findings may be germane to the regulation of paracrine glucocorticoid formation in disturbed nutritional states such as obesity. (J Biol Chem. 2006 281:341)
訳 これらの知見の臨床的関連は，～の調節に適切であるかもしれない

❷ The synergistic interactions of TLR ligands and antigen might have relevance to the exacerbation of IgE-mediated allergic diseases by infectious agents. (Blood. 2006 107:610)
訳 TLRリガンドおよび抗原の相乗的相互作用は，～の増悪に関連があるかもしれない

■［形］＋ relevance

❶	clinical relevance	臨床的関連性	414
❷	functional relevance	機能的関連性	300
❸	physiological relevance	生理学的関連性	268
❹	biological relevance	生物学的関連性	235

❸ To determine the physiological relevance of these interactions, we examined the effect of mutations in the ARH on LDLR location and function in polarized hepatocytes (WIF-B). (J Biol Chem. 2005 280:40996)
訳 これらの相互作用の生理学的関連を決定するために

★ relevant 形 関連する／実際的な重要性を持つ 用例数 11,659

◆類義語：relative, pertinent, related

■ relevant ＋［前］

❶	relevant to ～	～に関連する	2,527
	・be relevant to ～	…は，～に関連する	996
❷	relevant for ～	～に対して実際的な重要性を持つ	476

❸ relevant in ～	～において実際的な重要性を持つ	268

❶ This model is highly relevant to clinical manifestations of CRAO and is an ideal animal model for research.（Invest Ophthalmol Vis Sci. 2005 46:2133）

訳 このモデルは，CRAOの臨床症状に高度に関連する

❷ These findings are relevant for the mechanism of post-termination complex disassembly.（Biochemistry. 2004 43:12728）

訳 これらの知見は，～の機構にとって実際的な重要性を持つ

■［副］＋ relevant

❶ clinically relevant	臨床的に関連する（重要な）	1,440
❷ physiologically relevant	生理学的に関連する（重要な）	1,228
❸ biologically relevant	生物学的に関連する（重要な）	665
❹ functionally relevant	機能的に関連する（重要な）	276

❶ Furthermore, PP2A may be a clinically relevant drug target for CF, which should be considered in future studies.（J Biol Chem. 2005 280:41512）

訳 PP2Aは，臨床的に重要な薬剤の標的であるかもしれない

❷ Here, we provide further evidence that s-ShhNp is the physiologically relevant form of Shh.（J Biol Chem. 2006 281:4087）

訳 ～は，Shhの生理学的に関連する型である

R

rely 動 頼る／信頼する　用例数 4,863

relies	頼る	2,029
rely	頼る	1,922
relied	頼った	591
relying	頼る	321

自動詞．rely onの用例が非常に多い．

◈類義語：depend

■ rely ＋［前］

❶ rely on ～	…は，～に頼る	3,979
・rely heavily on ～	…は，～に大いに頼る	127
❷ rely upon ～	…は，～に頼る	271

❶ Downregulation of surface CD4 by Nef relies on the ability of this viral protein to redirect the endocytic machinery to CD4.（J Virol. 2000 74:9396）

訳 Nefによる表面CD4の下方制御は，このウイルスタンパク質の能力に依存する

★★ remain 動 ～のままである／残る　用例数 51,685

remains	～のままである	20,247
remain	～のままである	14,147

remained	～のままであった	11,859
remaining	～のままである	5,432

第2文型(S+V+C)の自動詞.

■ remain ＋（[副] ＋）[形／過分]

❶ remain unclear	…は，はっきりしないままである	4,960
❷ remain unknown	…は，知られていないままである	3,274
・remain largely unknown	…は，ほとんど知られていないままである	910
❸ remain elusive	…は，とらえどころのないままである	1,842
❹ remain poorly understood	…は，あまり理解されないままである	1,662
❺ remain controversial	…は，論争の余地があるままである	1,157
❻ remain unchanged	…は，不変のままである	949
❼ remain uncertain	…は，不確かなままである	689
❽ remain stable	…は，安定なままである	597
❾ remain intact	…は，無傷のままである	547

❶ However, **the precise mechanism by which TGase 2 promotes inflammation remains unclear.** (J Biol Chem. 2004 279:53725)

訳 トランスグルタミナーゼ2が炎症を促進する正確な機構は，はっきりしないままである

❷ Yet, the substrates of specificity in complex neuropil **remain largely unknown.** (Neuron. 2004 43:251)

訳 ～は，ほとんど知られていないままである

❹ However, the biological mechanisms by which Ras contributes to metastasis **remain poorly understood.** (Cancer Res. 2002 62:887)

訳 ～は，あまり理解されないままである

❺ **The nature of these B cells remains controversial.** (J Immunol. 2001 166:377)

訳 これらのB細胞の性質は，論争の余地があるままである

❻ Many CD34-positive endothelial cells in GCV-treated tumors showed only weak or marginal LAT1 staining, **whereas CD98 staining remained unchanged.** (J Nucl Med. 2003 44:1845)

訳 一方，CD98染色は不変のままであった

■ remains to *do*

★ remain to ～	…は，まだ～のままである	2,866
❶ remain to be ～	…は，まだ～のままである	2,828
・remain to be determined	…は，まだ決定されないままである	695
・remain to be elucidated	…は，まだ解明されないままである	473
・remain to be established	…は，まだ確立されないままである	224
・remain to be defined	…は，まだ定義されないままである	193
・remain to be identified	…は，まだ同定されないままである	172

❶ The role of this regulated phosphorylation in the adaptation to ischemia **remains to be determined.** (Circ Res. 2004 95:726)

訳 ～は，まだ決定されないままである

■ [名] + remain

❶	mechanisms remain ～	機構は，～のままである	448
❷	questions remain ～	疑問は，～のままである	334
❸	function remains ～	機能は，～のままである	318
❹	levels remained ～	レベルは，～のままであった	244
❺	disease remains ～	疾患は，～のままである	239

❶ However, the underlying **mechanisms remain** unclear. (J Immunol. 2012 188:1961)
訳 根底にある機構は，はっきりしないままである

remarkably 副 著しく　　用例数 4,746

◆類義語：markedly, strikingly, dramatically, prominently, notably, extremely, unusually, significantly

■ remarkably + [形]

❶	remarkably similar	著しく似ている	460
	・be remarkably similar	…は，著しく似ている	309
❷	remarkably high	著しく高い	119
❸	remarkably stable	著しく安定である	106

❶ The genomic architecture of protocadherin (Pcdh) gene clusters **is remarkably similar** to that of the immunoglobulin and T cell receptor gene clusters, and can potentially provide significant molecular diversity. (Genes Dev. 2002 16:1890)
訳 ～は，免疫グロブリンおよびT細胞受容体遺伝子クラスターのそれに著しく似ている

R

removal 名 除去　　用例数 8,631

removal	除去	8,615
removals	除去	16

複数形のremovalsの用例はほとんどなく，原則，**不可算名詞**として使われる．
◆類義語：elimination, abstraction, obliteration, deletion

■ removal + [前]

	the%	a/an%	ø%			
❶	30	1	69	removal of ～	～の除去	6,245
				・by removal of ～	～の除去によって	343
				・after removal of ～	～の除去のあと	319
				・removal of extracellular ～	細胞外～の除去	154
				・upon removal of ～	～を除去するとすぐ	148
❷	1	0	64	removal from ～	～からの除去	233
❸	2	0	67	removal by ～	～による除去	124

❶ In addition, UV-irradiation-induced SEK/JNK activation **was unaffected by removal of extracellular free** Ca^{2+} with EGTA. (Invest Ophthalmol Vis Sci. 2003 44:5102)

訳 ～は，細胞外遊離Ca^{2+}の除去によって影響を受けなかった

❶ **This differentiation was restored after removal of tumor-derived factors.** (J Immunol. 2004 172: 464)

訳 この分化は，腫瘍由来因子の除去のあとに回復された

remove 動 除去する　用例数 8,521

removed	除去される／～を除去した	4,264
remove	～を除去する	1,849
removing	～を除去する	1,468
removes	～を除去する	940

他動詞.

◆類義語：eliminate, exclude, obviate, scavenge, delete, abolish, deprive, withdraw

■ be removed ＋［前］

★ be removed	…が，除去される	2,493
❶ be removed from ～	…が，～から除去される	613
❷ be removed by ～	…が，～によって除去される	374

❶ Both exorbital lacrimal glands **were removed from male Sprague-Dawley rats.** (Invest Ophthalmol Vis Sci. 2003 44:1075)

訳 ～が，雄のSDラットから除去された

❷ When **the stress is removed by a temperature downshift,** σE activity is strongly repressed and then slowly returns to levels seen in unstressed cells. (J Bacteriol. 2003 185:2512)

訳 ストレスが，温度のシフトダウンによって除去される

render 動 ～にする／与える　用例数 4,439

rendered	～にされる／～にした	1,561
renders	～にする	1,240
render	～にする	965
rendering	～にする	673

第5文型(S+V+O+C)の他動詞．補語（C）には，通常，形容詞が用いられる.

◆類義語：give, confer, supply, provide

■ render ＋［名］＋［形］

❶ render ～ resistant to …	‥は，～を…に抵抗性にする	267
❷ render ～ susceptible to …	‥は，～を…に感受性にする	226
❸ render cells ～	…は，細胞を～にする	203
❹ render ～ sensitive to …	‥は，～を…に感受性にする	158

❶ Oncogenic p21^{ras} decreases surface Fas antigen expression and **renders fibroblasts resistant to Fas mediated apoptosis.** (Am J Pathol. 2004 164:1471)
訳 ～は，線維芽細胞をFasに仲介されるアポトーシスに抵抗性にする

❷ In this study, we investigated whether **expression of the IG20 can render cells susceptible to γ-irradiation.** (Cancer Res. 2003 63:8768)
訳 IG20の発現は，細胞をγ-線照射に感受性にしうる

★ repair 名 修復，動 修復する　用例数 23,198

repair	修復／～を修復する	21,692
repaired	修復される／～を修復した	945
repairing	～を修復する	289
repairs	修復／～を修復する	272

名詞の用例が多いが動詞としても用いられる．複数形のrepairsの割合は1％しかなく，原則，不可算名詞として使われる．
◆類義語：restore

■ 名 repair ＋［前］

	the%	a/an%	ø%			
❶	42	1	53	repair of ～	～の修復	2,859
				・repair of DNA double-strand breaks	DNA二重鎖切断の修復	183
				・repair of DNA damage	DNA損傷の修復	114
❷	1	1	70	repair in ～	～における修復	836
❸	0	0	65	repair by ～	～による修復	338

❶ Chk1 both arrests replication forks and **enhances repair of DNA damage** by phosphorylating downstream effectors. (Oncogene. 2012 31:4245)
訳 ～は，DNA損傷の修復を増強する

❷ The promiscuous nature of the archaeal primases suggests that **these proteins might have additional roles in DNA repair in the archaea.** (Trends Genet. 2005 21:568)
訳 これらのタンパク質は，古細菌におけるDNA修復において付加的な役割を担っているかもしれない

■ 名 ［名］＋ repair

❶	DNA repair	DNA修復	4,073
❷	excision repair	除去修復	1,753
❸	mismatch repair	ミスマッチ修復	1,717

■ 名 ［名］＋ and repair

❶	replication and repair	複製および修復	248
❷	recombination and repair	組換えおよび修復	141

❶ Very little is known about protozoan replication protein A (RPA), **a heterotrimeric complex**

R

critical for DNA replication and repair. (J Biol Chem. 2005 280:31460)
訳 DNA複製および修復にとって決定的に重要であるヘテロ三量体複合体

■ 動 be repaired ＋［前］

★ be repaired	…は，修復される	471
❶ be repaired by ～	…は，～によって修復される	155

❶ Double-strand DNA breaks can be repaired by any of several alternative mechanisms that differ greatly in the nature of the final repaired products. (Genetics. 2006 172:1055)
訳 二本鎖DNA切断は，いくつかの代替機構のどれによっても修復されうる

replace 動 置換する／置き換える／交換する　用例数 8,135

replaced	置換される／～を置換した	4,715
replacing	～を置換する	1,662
replace	～を置換する	1,344
replaces	～を置換する	414

他動詞.〈replaced by〉〈replaced with〉の用例が多い.
◆類義語：substitute, exchange, displace, interchange, swap

■ be replaced ＋［前］

★ be replaced	…が，置換される	3,187
❶ be replaced by ～	…が，～によって置換される	1,668
❷ be replaced with ～	…が，～と置換される	1,271

❶ At $\alpha1\beta2/\alpha2\beta1$-interface the β93 cysteine was replaced by alanine (βC93A), and at the $\alpha1\beta1/\alpha2\beta2$-interface the β112 cysteine was replaced by glycine (βC112G). (Biophys J. 1999 76:88)
訳 β93システインが，アラニンによって置換された

❷ Significant reduction in fusion activity was observed, however, when two of the four middle heptadic leucine or isoleucine residues were replaced with alanine. (J Virol. 1999 73:8152)
訳 ～が，アラニンと置換された

■ replace ＋［名］＋ with

❶ replace ～ with …	～を…と置換する	513

❶ This hypothesis was tested by replacing Trp179 with Phe in the APX3M background. (Biochemistry. 2005 44:14062)
訳 この仮説が，Trp179をPheと置換することによってテストされた

replacement 名 置換／交換　用例数 8,517

replacement	置換	7,679
replacements	置換	838

複数形のreplacementsの割合は約10%あるが，単数形は無冠詞で使われることが非常に多い．

◆類義語：substitution, displacement, exchange, interchange, change

■ replacement ＋［前］

	the%	a/an%	ø%			
❶	17	1	81	replacement of ～	～の置換	2,888
				・replacement of ～ with …	～の…との置換	531
				・replacement of ～ by …	～の…による置換	254
❷	5	11	80	replacement in ～	～における置換	242
❸	5	4	77	replacement with ～	～との置換	237

❶ **Replacement of Ser-107 with alanine** yielded an active enzyme with kinetic characteristics similar to those of wild-type APS kinase.（J Biol Chem. 1998 273:28583）

訳 セリン-107のアラニンとの置換

❶ **Replacement of GTP by 7-deaza-GTP** completely abolishes this transition and G-ladder synthesis continues with a constant efficiency of elongation beyond the limit of detection.（Nucleic Acids Res. 2001 29:2601）

訳 GTPの7-deaza-GTPによる置換

❸ **In trials of myeloablative BMT designed to yield total marrow replacement with donor stem cells**, a subset of patients developed mixed chimerism.（Blood. 2002 99:1840）

訳 ドナーの幹細胞との完全な骨髄置換を生じるように設計された骨髄破壊的な骨髄移植の治験において

replicate 動 増幅する／複製する／再現する　用例数 6,717

replicate	増幅する／～を複製する／～を再現する	2,319
replicated	増幅した／再現される／～を再現した	2,230
replicating	増幅する	1,221
replicates	増幅する／～を複製する／～を再現する	947

自動詞として使われることが多いが，他動詞の用例もある．

◆類義語：amplify, proliferate, propagate, grow

■ replicate ＋［前］

❶	replicate in ～	…は，～において増幅する	1,239
	・replicate efficiently in ～	…は，～において効率的に増幅する	107
❷	replicate to ～	…は，～に増幅する	181
❸	replicate within ～	…は，～内で増幅する	163

❶ Under identical conditions, **the parental MHV strain A59 failed to replicate in BHK cells** expressing human Bgp or CEA.（J Virol. 1999 73:638）

訳 親株のMHV A59株は，BHK細胞において増幅することができなかった

R

■ be replicated ＋［前］

★ be replicated	…が，再現される	545
❶ be replicated in ～	…が，～において再現される	230

❶ These results **were replicated in** an independent sample of additional participants.（Curr Biol. 2011 21:677）
訳 これらの結果が，追加の患者の独立した試料において再現された

★★ replication 名 複製／増幅 用例数 35,466

replication	複製／増幅	35,417
replications	複製／増幅	49

複数形のreplicationsの割合は0.1%しかなく，原則，**不可算名詞**として使われる．
◆類義語：amplification, proliferation, growth, duplication, reproduction

■ replication ＋［前］

	the%	a/an%	ø%			
❶	0	0	56	replication in ～	～における複製（増幅）	3,245
				・replication in ～ cells	～細胞における複製（増幅）	393
❷	33	0	65	replication of ～	～の複製（増幅）	2,606
				・replication of ～ in …	…における～の複製（増幅）	305

❶ Both CP-96,345 and anti-SP antibody **inhibited SP-enhanced HIV replication in monocyte-derived macrophages (MDM)**.（Proc Natl Acad Sci USA. 2001 98:3970）
訳 ～が，単球由来マクロファージ（MDM）におけるSPに増強されたHIVの増幅を抑制した

❷ The absence of the viral capsid protein in the tumor cells **excludes productive replication of the virus in neoplastic cells**.（Cancer Res. 2002 62:7093）
訳 ～は，新生物細胞におけるウイルスの増殖性複製を排除する

■［名／形］＋ replication

❶ DNA replication	DNA複製	6,354
❷ viral replication	ウイルスの複製（増幅）	2,923
❸ virus replication	ウイルス複製（増幅）	1,535
❹ RNA replication	RNA複製	899

■［前］＋ replication

❶ origin of replication	複製開始点	306
❷ required for replication	複製（増幅）のために必要とされる	111

■ replication-［形］

❶ replication-competent ～	複製能のある～	534

❷ replication-defective ～	複製欠陥のある～	422

❶ The Phase I study described here represents the first gene therapy trial in which **a replication-competent virus was used to deliver** a therapeutic gene to humans. (Cancer Res. 2002 62:4968)

訳 複製能のあるウイルスが，～を送達するために使われた

■ replication and ＋［名］

❶ replication and repair	複製と修復	230
❷ replication and transcription	複製と転写	198

❶ Proliferating cell nuclear antigen (PCNA), a processivity factor for DNA polymerases δ and ε, **is essential for both DNA replication and repair**. (Nucleic Acids Res. 1999 27:4476)

訳 ～は，DNA複製と修復の両方にとって必須である

★★ report 動 報告する，名 報告　　用例数 96,006

report	～を報告する／報告	48,570
reported	～を報告した／報告される	37,245
reports	～を報告する／報告	8,020
reporting	～を報告する	2,171

動詞および名詞として用いられる．we reportの用例が非常に多い．複数形のreportsの割合が，約５％ある可算名詞．

◆類義語：state, mention, note, describe, document, description

■ 動 report ＋［that節／名句］

❶ report that ～	…は，～ということを報告する	17,303
❷ report the identification of ～	…は，～の同定を報告する	922
❸ report the crystal structure of ～	…は，～の結晶構造を報告する	589
❹ report the results of ～	…は，～の結果を報告する	427
❺ report the development of ～	…は，～の開発を報告する	369
❻ report the discovery of ～	…は，～の発見を報告する	289
❼ report the use of ～	…は，～の使用を報告する	273
❽ report the characterization of ～	…は，～の特徴づけを報告する	269
❾ report the isolation of ～	…は，～の単離を報告する	259

❷ In this study, **we report the identification of** a new gene, ireA (iron-responsive element). (Infect Immun. 2001 69:6209)

訳 われわれは，～の同定を報告する

❹ **We report the results of AFM experiments** on the forced unfolding of barnase in a chimeric construct with I27. (Biophys J. 2001 81:2344)

訳 われわれは，AFM実験の結果を報告する

■ 動 [代名／名] ＋ report

❶ we report ～	われわれは，～を報告する	32,502
・Here we report ～	ここに，われわれは～を報告する	9,314
・Here, we report ～	ここに，われわれは～を報告する	6,028
・we report here ～	われわれは，～をここに報告する	4,379
・we now report ～	われわれは，今，～を報告する	1,750
❷ study reports ～	研究は，～を報告する	315

❶ **Here we report that** EETs have additional fibrinolytic properties.（J Biol Chem. 2001 276:15983）

訳 ここに，われわれは～ということを報告する

❶ **We report here that** the absence of either or both COX isoforms in mice does not result in premature closure of the DA in utero.（Proc Natl Acad Sci USA. 2001 98:1059）

訳 われわれは，～ということをここに報告する

❶ **We now report that** human cell proliferation in NOD-scid mice increased after in vivo depletion of NK cells.（J Immunol. 2000 165:518）

訳 われわれは，今，～ということを報告する

■ 動 be reported ＋ [前／that節]

★ be reported	…は，報告される	11,425
❶ be reported to *do*	…は，～すると報告される	2,927
・have been reported to *do*	…は，～すると報告されている	1,777
・be reported to be ～	…は，～であると報告される	843
❷ be reported in ～	…は，～において報告される	1,624
❸ be reported for ～	…は，～に関して報告される	691
❹ it has been reported that ～	～ということが報告されてきた	291

❶ Vitamin D-binding protein（DBP）**has been reported to contribute to innate immunity**.（J Immunol. 2002 168:869）

訳 ～は，自然免疫に寄与すると報告されている

❷ **No malignancies have been reported in five of the followed-up patients**, and two have had continued enlargement of asymmetric tissue.（Radiology. 1999 211:111）

訳 悪性病変は，フォローアップされた患者のうちの５人において報告されていない

■ 動 [副] ＋ reported

❶ previously reported ～	以前に報告された～／～を以前に報告した	5,964
❷ recently reported ～	最近～を報告した／最近報告された～	1,424
❸ self-reported ～	自己申告される～	1,283

❶ **None of the previously reported mutations was detected**, but four novel mutations（V387M, N435S, V447A, and L452M）were found in the kinase domain in recurrent tumors.（Cancer Res. 2001 61:482）

訳 以前に報告された変異のどれも検出されなかった

❷ **We recently reported that** cytosine arabinoside（AraC）-induced apoptosis of cerebellar neurons involves the overexpression of glyceraldehyde-3-phosphate dehydrogenase

(GAPDH). (J Neurosci. 1999 19:9654)
訳 われわれは，～ということを最近報告した

■ 名 this/first report

❶ in this report	この報告において	4,615
・in this report, we show ～	この報告において，われわれは～を示す	536
❷ the first report	最初の報告	1,417
・the first report of ～	～の最初の報告	838

❶ In this report, we show that SOS induces FGF-dependent dimerization of FGF receptors (FGFRs). (Mol Cell Biol. 2002 22:7184)
訳 この報告において，われわれは～ということを示す

❷ This is the first report of a bacterial two-component signal transduction system that controls gene expression through a heme-responsive mechanism. (J Bacteriol. 1999 181:5330)
訳 これは，～の最初の報告である

★ represent 動 示す／相当する／表す 用例数 29,947

represent	～を示す／～に相当する／～を表す	12,553
represents	～を示す／～に相当する／～を表す	9,541
representing	～を示す／～に相当する／～を表す	4,297
represented	表される／～を示した	3,556

他動詞．能動態の用例が非常に多い．

◆類義語：show, present, exhibit, display, indicate, coincide, match, resemble

■ represent ＋［名句］

❶ represent a novel ～	…は，新規の～に相当する	1,418
❷ represent the first ～	…は，最初の～に相当する	1,401
❸ represent a new ～	…は，新しい～に相当する	886
❹ represent an important ～	…は，重要な～に相当する	711
❺ represent a potential ～	…は，潜在的な～に相当する	382
❻ represent a major ～	…は，主要な～に相当する	363

❶ This discovery represents a novel mechanism for proteasome pathway inhibition in intact cells. (J Biol Chem. 2001 276:30366)
訳 この発見は，～の新規の機構に相当する

❹ In these contexts, PEG-3 may represent an important target molecule for developing cancer therapeutics and inhibitors of angiogenesis. (Proc Natl Acad Sci USA. 1999 96:15115)
訳 PEG-3は，～の重要な標的分子に相当するかもしれない

■［名］＋ represent

❶ study represents ～	研究は，～を示す	229
❷ results represent ～	結果は，～を示す	208
❸ cells represent ～	細胞は，～を示す	198

R

❹ data represent ～	データは，～を示す	160
❺ findings represent ～	知見は，～を示す	147
❻ work represents ～	研究は，～を示す	132

❷ These results represent important steps in characterizing KasA and KasB as targets for antimycobacterial drug discovery. (J Biol Chem. 2001 276:47029)
訳 これらの結果は，～を特徴づける際の重要なステップを示す

■ [動／形] ＋ to represent

❶ appear to represent ～	…は，～に相当すると思われる	184
❷ thought to represent ～	～に相当する考えられる	109
❸ likely to represent ～	おそらく～に相当するであろう	100

❸ The V3Rs are likely to represent a new large family of pheromone receptors in mammals. (Neuron. 2000 28:835)
訳 V3Rは，おそらく新しい大きなファミリーのフェロモン受容体に相当するであろう

■ represented ＋ [前]

❶ represented by ～	～によって表される	993
・be represented by ～	…は，～によって表される	396
❷ represented in ～	～において表される	699

❶ This family of exported fibronectin-binding proteins consists of members Ag85A, Ag85B, and Ag85C and is most prominently represented by 85A and 85B. (Infect Immun. 2000 68:767)
訳 ～は，85Aおよび85Bによって最も顕著に表される

❷ Here we investigate the fundamental question of how multiple languages are represented in a human brain. (Nature. 1997 388:171)
訳 われわれは，どのように複数の言語がヒトの脳において表されるかという基本的な疑問を精査する

■ [名] ＋ representing

❶ peptides representing ～	～に相当するペプチド	119

representation 名 表現／表示 用例数 4,501

representation	表現／表示	2,838
representations	表現／表示	1,663

複数形のrepresentationsの割合が，約35%ある可算名詞.

◆類義語：presentation, expression

■ representation ＋ [前]

	the%	a/an%	ø%			
❶	58	27	14	representation of ～	～の表現	1,520

❷	34	13	40	representation in ～	～における表現	193

❶ This is the first quantitative endogenous perfusion MRI study of the cerebral representation of ongoing, persistent pain due to OA. (Arthritis Rheum. 2012 64:3936)
訳 これは，進行中の持続性疼痛の脳内表現の最初の定量的内因性灌流MRI研究である

representative 形 代表的な／典型的な／特徴を表す，名 代表／典型 用例数 3,545

representative	代表的な／特徴を表す／代表／典型	3,108
representatives	代表／典型	437

形容詞として使われることが多いが，名詞の用例もある．名詞としては，複数形の用例が非常に多い．

■ 形 representative ＋［前］

❶ representative of ～	～を代表する／～の特徴を表す	800
・be representative of ～	…は，～を代表する	303

❶ Health-related information collected in psychological laboratories may not be representative of people's everyday health. (Annu Rev Psychol. 2013 64:471)
訳 心理学の研究室で集められた健康に関連する情報は，人々の日常の健康を代表しないかもしれない

■ 形［副］＋ representative

❶ nationally representative ～	全国を代表する～	297

❶ The purpose of this study was to assess knowledge and awareness about stroke in a nationally representative sample of women. (Circulation. 2005 111:1321)
訳 この研究の目的は，全国を代表する女性のサンプルにおいて脳卒中に関する知識と認識を評価することであった

■ 形 representative ＋［名］

❶ representative sample	代表的なサンプル	243

■ 名 representatives ＋［前］

	the%	a/an%	ø%			
❶	6	-	80	representatives of ～	～の代表	273

❶ Lamprey and hagfish are surviving representatives of the most ancient vertebrates. (Curr Opin Immunol. 2011 23:156)
訳 ヤツメウナギとメクラウナギは，最古の脊椎動物の生き残った代表である

R

repress ㊍ 抑制する 用例数 9,428

repressed	抑制される／～を抑制した	3,627
repress	～を抑制する	2,514
represses	～を抑制する	2,296
repressing	～を抑制する	991

他動詞.

◆類義語：suppress, attenuate, weaken, inhibit, depress, decrease, reduce, constrain

■ repress ＋［名句］

❶	repress transcription	…は，転写を抑制する	857
❷	repress the expression of ～	…は，～の発現を抑制する	256

❶ Chromatin is thought to **repress transcription** by limiting access of the DNA to transcription factors.（Cell. 2001 105:403）

訳 クロマチンは，DNAの転写因子への接近を制限することによって転写を抑制すると考えられる

■ be repressed ＋［前］

★	be repressed	…は，抑制される	1,129
❶	be repressed by ～	…は，～によって抑制される	518
❷	be repressed in ～	…は，～において抑制される	210

❶ We further show that transcription from TnI SURE **is repressed by** GTF3 when overexpressed in electroporated adult soleus muscles.（Mol Cell Biol. 2001 21:8490）

訳 TnI SUREからの転写は，GTF3によって抑制される

❷ Myogenic transcription **is repressed in** myoblasts by serum-activated cyclin-dependent kinases, such as cdk2 and cdk4.（Oncogene. 2002 21:4137）

訳 筋原性の転写は，筋芽細胞において～によって抑制される

★repression ㊔ 抑制 用例数 10,153

repression	抑制	10,149
repressions	抑制	4

複数形のrepressionsの用例はほとんどなく，原則，**不可算名詞**として使われる.

◆類義語：suppression, depression, inhibition, attenuation, decrease, reduction

◆反意語：activation, induction, facilitation, promotion, acceleration

■ repression ＋［前］

	the%	a/an%	ø%			
❶	26	3	67	repression of ～	～の抑制	3,892
				・repression of ～ expression	～発現の抑制	193
				・repression of transcription	転写の抑制	111

❷	5	0	82	repression by ～	～による抑制	704
❸	0	0	70	repression in ～	～における抑制	394

❶ The repression of basal transcription by HSF-4a occurs through interaction with the basal transcription factor TFIIF.（J Biol Chem. 2001 276:14685）

訳 HSF-4aによる基本転写の抑制は，～との相互作用によって起こる

❷ We have developed a yeast model system to address transcriptional repression by the retinoblastoma protein (pRB).（Proc Natl Acad Sci USA. 2001 98:8720）

訳 われわれは，網膜芽細胞腫タンパク質（pRB）による転写抑制に取り組むために酵母モデル系を開発した

■［前］＋ repression

❶	required for repression	抑制のために必要とされる	103

❶ In addition, the glucose induction of the HXT genes is required for repression of gene expression by glucose.（J Biol Chem. 2002 277:46993）

訳 HXT遺伝子のグルコース誘導は，グルコースによる遺伝子発現の抑制のために必要とされる

■［形／過分］＋ repression

❶	transcriptional repression	転写抑制	1,391
❷	～-mediated repression	～に仲介される抑制	718
❸	translational repression	翻訳抑制	319

■［名］＋ and repression

❶	activation and repression	活性化と抑制	207

❶ Elements involved in activation and repression of the DAN/TIR genes were defined in this study, using the DAN1 promoter as a model.（Nucleic Acids Res. 2001 29:799）

訳 DAN/TIR遺伝子の活性化と抑制に関与する要素が，この研究において明らかにされた

★★ require　動 必要とする　用例数 110,869

required	必要とされる／～を必要とした	71,118
requires	～を必要とする	22,770
require	～を必要とする	13,115
requiring	～を必要とする	3,866

他動詞．required forの用例が非常に多い．

◈類義語：need, demand, request, necessitate

■ be required ＋［前］

★	be required	…は，必要とされる	40,172
❶	be required for ～	…は，～に必要とされる	30,502
	・be required for normal ～	…は，正常な～に必要とされる	940

R

・be required for efficient ～	…は，効率的な～に必要とされる	830
・be required for proper ～	…は，適切な～に必要とされる	510
・be required for optimal ～	…は，最適な～に必要とされる	436
・be required for the development of ～	…は，～の発達に必要とされる	247
・be required for the formation of ～	…は，～の形成に必要とされる	232
・be required for activation of ～	…は，～の活性化に必要とされる	226
・be required for the expression of ～	…は，～の発現に必要とされる	161
・be required for the maintenance of ～	…は，～の維持に必要とされる	153
・be required for maintaining ～	…は，～を維持するために必要とされる	115
❷ be required to *do*	…は，～するのに必要とされる	5,323
・be required to maintain ～	…は，～を維持するのに必要とされる	451
・be required to determine ～	…は，～を決定するのに必要とされる	178
・be required to achieve ～	…は，～を達成するのに必要とされる	145
❸ be required in ～	…は，～において必要とされる	1,024

❶ These data indicate that **binding to calmodulin is required for normal transcriptional function of ERα**. (J Biol Chem. 2005 280:13097)

訳 カルモジュリンへの結合は，ERαの正常な転写機能に必要とされる

❶ Activity assays with pure proteins and cell extracts reveal that O_2 (or superoxide) **is required for activation of SOD1 by CCS**. (Proc Natl Acad Sci USA. 2004 101:5518)

訳 ～は，CCS によるSOD1の活性化に必要とされる

❷ **EBNA1 is required to maintain the viral genome** but is not recognized by cytotoxic T cells. (Proc Natl Acad Sci USA. 2004 101:239)

訳 EBNA1は，ウイルスゲノムを維持するのに必要とされる

■ ［名］＋ required for

★ required for ～	～に必要とされる	48,264
❶ genes required for ～	～に必要とされる遺伝子	757
❷ protein required for ～	～に必要とされるタンパク質	531
❸ factors required for ～	～に必要とされる因子	318

❶ To identify **genes required for the degradation of this protein**, A1PiZ degradation-deficient (add) yeast mutants were isolated. (Mol Biol Cell. 2006 17:203)

訳 このタンパク質の分解に必要とされる遺伝子を同定するために

■ ［副］＋ required for

❶ absolutely required for ～	～に絶対的に必要とされる	454
❷ specifically required for ～	～に特異的に必要とされる	276

❶ **Gol is absolutely required for Lit activation**. (J Biol Chem. 2005 280:112)

訳 Golは，Lit活性化に絶対的に必要とされる

■ require ＋［名句］

❶ require both ～	…は，両方の～を必要とする	1,088
❷ require the presence of ～	…は，～の存在を必要とする	862

❸	require activation of ～	…は，～の活性化を必要とする	358

❷ Rather, it also requires the presence of sequences within the carboxyl-terminal E/F domain of ERRα1. (J Biol Chem. 2002 277:24826)

訳 それは，また，～内の配列の存在を必要とする

■ to require

★	to require ～	～を必要とする…	1,084
❶	appear to require ～	…は，～を必要とするように思われる	289
❷	be shown to require ～	…は，～を必要とすることが示される	119

❶ Interestingly, it also appears to require the simultaneous recognition of two Lys-9-methylated histone H3 molecules. (EMBO J. 2004 23:489)

訳 それは，また，2つのLys-9-メチル化ヒストンH3分子の同時認識を必要とするように思われる

★ requirement　名 要求性／必要性／必要条件　用例数 12,869

requirement	要求性／必要性	7,872
requirements	要求性／必要性	4,997

複数形のrequirementsの割合は約40%とかなり高い.

◆類義語：need, demand, request, claim, necessity, prerequisite

■ requirement ＋［前］

requirement ofの前にtheが付く割合は非常に高い.

	the%	a/an%	ø%			
❶	46	45	4	requirement for ～	～に対する要求性（必要性）	5,465
❷	76	21	3	requirement of ～	～の必要性（要求性）	1,126

❶ Thus, we demonstrate the requirement for the U12 spliceosome in the development of a metazoan organism. (Mol Cell. 2002 9:439)

訳 われわれは，～の発生におけるU12スプライソソームに対する要求性を実証する

❷ We previously showed the requirement of both T cells and γ interferon (IFN-γ)-producing non-T cells for the genetic resistance of BALB/c mice to the development of toxoplasmic encephalitis (TE). (Infect Immun. 2004 72:4432)

訳 われわれは，以前にT細胞およびγインターフェロン（IFN-γ）産生非T細胞の両方の要求性を示した

■［形］＋ requirement(s)

❶	structural requirements	構造上の要求性	351
❷	absolute requirement	絶対的な要求性	317

❶ Ecto-AK presented an absolute requirement for magnesium and adenine-based nucleotides. (J Biol Chem. 2003 278:11256)

訳 ～は，マグネシウムに対する絶対的な要求性を示した

★ rescue 動 救出する 用例数 10,646

rescue	～を救出する／救出	4,594
rescued	救出される／～を救出した	4,516
rescues	～を救出する／救出	1,251
rescuing	～を救出する	285

名詞としても使われるが，他動詞の用例が多い．名詞としては，複数形のrescuesの用例はほとんどなく，**不可算名詞**として使われることが多い．

◆類義語：restore, recover

■ 動 be rescued ＋［前］

★ be rescued	…は，救出される	1,837
❶ be rescued by ～	…は，～によって救出される	1,334
❷ be rescued in ～	…は，～において救出される	114

❶ Importantly, this defect could **be rescued by** re-introduction of WT Ocrl1 in both patient and Ocrl1 knock-down cells. (Hum Mol Genet. 2012 21:1835)

訳 この欠損は，野生型Ocrl1の再導入によって救出されうる

■ 動［副］＋ rescued

❶ partially rescued	部分的に救出される	336
❷ completely rescued	完全に救出される	87

■ 動［動／形］＋ to rescue

★ to rescue ～	～を救出する…	965
❶ fail to rescue ～	…は，～を救出できない	168
❷ sufficient to rescue ～	～を救出するのに十分な	145

❶ Enforced expression of an immunoglobulin transgene **failed to rescue** B cell development. (Immunity. 2012 36:769)

訳 免疫グロブリン導入遺伝子の強制発現は，B細胞発生を救出できなかった

■ 名 rescue ＋［前］

	the%	a/an%	ø%			
❶	19	12	68	rescue of ～	～の救出	746

❶ Genetic **rescue of** cell death fails to restore columnar organization and branching patterns, indicating these defects are independent of neuronal loss. (Nat Neurosci. 2012 15:1636)

訳 細胞死の遺伝的救出は，～を回復することができない

★ research (名) 研究／調査 用例数 18,694

research	研究／調査	18,680
researches	研究／調査	14

複数形のresearchesの用例はほとんどなく，原則，**不可算名詞**として使われる．

◆類義語：study, test, examination, investigation, survey, work, trial

■ research ＋［前］

	the%	a/an%	ø%			
❶	5	0	89	research on ～	～に関する研究	1,187
❷	4	0	94	research in ～	～における研究	901
❸	0	0	85	research into ～	～への研究	490

❶ **We review research on the relation between** stigma and the overall quality of life of people with epilepsy. (Lancet Neurol. 2005 4:171)

訳 われわれは，～の間の関連に関する研究を概説する

❷ Although in its infancy, **research in this area potentially unifies** several pathophysiological processes underpinning abnormal pain processing and opens up a different avenue for the development of novel analgesics. (Neuron. 2012 73:435)

訳 この分野における研究は，～を潜在的に統合する

■ research ＋［動］

❶	research is needed to *do*	研究が，～するために必要とされる	347

❶ However, **further research is needed to test whether** viral replication in the periodontium precedes the GBS symptoms. (J Periodontol. 2005 76:2306)

訳 さらなる研究が，～かどうかをテストするために必要とされる

■ ［形］＋ research

❶	future research	将来の研究	698
❷	further research	さらなる研究	661
❸	recent research	最近の研究	607
❹	clinical research	臨床研究	528
❺	previous research	以前の研究	415

■ ［前］＋ research

❶	area of research	研究の分野	137

resemble (動) 似ている 用例数 6,399

resembles	～に似ている	2,079
resemble	～に似ている	1,834
resembling	～に似ている	1,561

R

resembled	～に似ていた	925

他動詞．通常，受動態では用いられない．
◆類義語：match, represent, be similar to

■［副］＋ resemble

❶ closely resemble ～	…は，～に密接に似ている	975
・more closely resemble ～	…は，～により密接に似ている	164
❷ strongly resemble ～	…は，～にとても似ている	79

❶ The cstC gene encodes a 406-amino-acid protein that closely resembles the E. coli ArgD protein, which is involved in arginine biosynthesis. (J Bacteriol. 1998 180:4287)
訳 cstC遺伝子は，大腸菌のArgDタンパク質に密接に似ている406アミノ酸のタンパク質をコードする

■ resemble ＋［名句］

❶ resemble that of ～	…は，～のそれに似ている	524

❶ The Fv-4 gene encodes an envelope (Env) protein whose putative receptor-binding domain resembles that of ecotropic MuLV Env protein. (J Virol. 2001 75:11244)
訳 推定される受容体結合ドメインは，エコトロピックなMuLV Envタンパク質のそれに似ている

reside 動 存在する／住む／属する　用例数 4,417

reside	存在する／住む	1,743
resides	存在する／住む	1,611
residing	存在する／住む	766
resided	存在した／住んだ	297

自動詞．
◆類義語：exist

■ reside ＋［前］

❶ reside in ～	…は，～に存在する	2,381
❷ reside within ～	…は，～内に存在する	544
❸ reside on ～	…は，～に存在する	342
❸ reside at ～	…は，～に存在する	251

❶ The tkv mutation was found to reside in the ACL5 gene, which encodes a spermine synthase and whose expression is specific to provascular cells. (Plant Physiol. 2005 138:767)
訳 tkv変異が，ACL5遺伝子に存在することが見つけられた

❷ Platelets are formed and released into the bloodstream by precursor cells called megakaryocytes that reside within the bone marrow. (J Clin Invest. 2005 115:3348)
訳 骨髄内に存在する巨核球と呼ばれる前駆細胞

■ [名] + reside

❶	cells reside	細胞が，存在する	86
❷	activity resides	活性が，存在する	70

☆ resistance 名 抵抗性／抵抗／耐性 用例数 35,141

resistance	抵抗性／抵抗／耐性	34,973
resistances	抵抗性／抵抗／耐性	168

複数形のresistancesの割合は0.5%しかなく，原則，**不可算名詞**として使われる．resistance toの用例が多い．

◈類義語：tolerance

■ resistance + [前]

resistance ofの前にtheが付く割合はかなり高い．

	the%	a/an%	ø%			
❶	4	1	92	resistance to ～	～に対する抵抗性	9,518
				・confer resistance to ～	…は，～に対する抵抗性を与える	976
				・increased resistance to ～	～に対する増大した抵抗性	351
				・resistance to apoptosis	アポトーシスに対する抵抗性	293
				・enhanced resistance to ～	～に対する増強された抵抗性	154
❷	3	0	87	resistance in ～	～における抵抗性	2,419
❸	62	1	37	resistance of ～	～の抵抗性	1,562
				・resistance of ～ to …	～の…に対する抵抗性	544

❶ These data illustrate that a single molecule **can confer resistance to humoral and cellular immune attack**. (Transplantation. 2003 75:542)

訳 ～は，液性および細胞性の免疫攻撃に対する抵抗性を与えうる

❶ **Increased resistance to apoptosis promotes lymphomagenesis** with aberrant expression of cell survival proteins such as BCL-2 and c-MYC occurring in distinct lymphoma subtypes. (Am J Pathol. 2004 164:893)

訳 アポトーシスへの増大した抵抗性は，リンパ腫発生を促進する

❶ When the activity of the phosphatidylinositol 3-kinase pathway was inhibited, **RGDS-dependent resistance to apoptosis was eliminated**. (J Biol Chem. 2005 280:1733)

訳 RGDS依存性のアポトーシスへの抵抗性が，除去された

❸ **The resistance of tumor cells to chemotherapeutic agents**, such as cisplatin, is an important problem to be solved in cancer chemotherapy. (Cancer Res. 2002 62:4899)

訳 腫瘍細胞の化学療法剤に対する抵抗性

R

■ [前] + resistance

❶ mechanism of resistance	抵抗性の機構	262
❷ development of resistance	抵抗性の発達	141

❶ Pathogenic fungi have many complex mechanisms of resistance to antifungal drugs. (Lancet. 2002 359:1135)
訳 病原性の真菌は，抗真菌剤への多くの複雑な抵抗性の機構を持つ

■ [名／形] + resistance

❶ insulin resistance	インスリン抵抗性	3,985
❷ drug resistance	薬剤耐性	2,014
❸ multidrug resistance	多剤耐性	851
❹ vascular resistance	血管抵抗	739
❺ antibiotic resistance	抗生物質抵抗性	641
❻ disease resistance	疾患抵抗性	611

❶ In high-risk subjects, the earliest detectable abnormality is insulin resistance in skeletal muscle. (Proc Natl Acad Sci USA. 2003 100:8466)
訳 最も初期の検出可能な異常は，骨格筋におけるインスリン抵抗性である

★ resistant 形 抵抗性の／耐性の　用例数 21,079

resistant toの用例が非常に多い.
◆類義語：tolerant, refractory

■ resistant + [前]

❶ resistant to ～	～に抵抗性の	7,088
・resistant to apoptosis	アポトーシスに抵抗性の	178
・resistant to inhibition	阻害に抵抗性の	108
・resistant to infection	感染に抵抗性の	107

❶ Interestingly, despite increased levels of p53, p21-null cells were resistant to apoptosis, suggesting a proapoptotic role of p21 and implying that p53 is a necessary but not sufficient condition for noscapine-mediated apoptosis. (Cancer Res. 2007 67:3862)
訳 p21欠損細胞は，アポトーシスに抵抗性であった

■ [名] + be resistant to

★ be resistant	…は，抵抗性である	3,345
❶ be resistant to ～	…は，～に抵抗性である	3,169
・mice were resistant to ～	マウスは，～に抵抗性であった	160
・cells were resistant to ～	細胞は，～に抵抗性であった	119

❶ We also found that MMP-knockout mice were resistant to $CaCl_2$-mediated aortic injury and did not develop elastin degeneration and calcification. (Circulation. 2004 110:3480)

訳 MMPノックアウトマウスは，$CaCl_2$に仲介された大動脈の損傷に抵抗性であった

■ [副] ＋ resistant to

❶	more resistant to ～	～により抵抗性である	727
❷	highly resistant to ～	～に高度に抵抗性である	267
❸	relatively resistant to ～	～に比較的抵抗性である	248
❹	completely resistant to ～	～に完全に抵抗性である	157

❶ Spores of Bacillus subtilis are significantly more resistant to wet heat than are their vegetative cell counterparts.（J Bacteriol. 2001 183:779）

訳 枯草菌の胞子は，それらの栄養細胞対応物より湿性温熱に有意により抵抗性である

■ [名]-resistant

❶	drug-resistant ～	薬剤耐性の～	1,234
❷	insulin-resistant ～	インスリン抵抗性の～	568
❸	multidrug-resistant ～	多剤耐性の～	568

❶ Lastly, the EGFR-DNR can partially reverse cisplatin resistance in drug-resistant cells.（Cancer Res. 2005 65:3243）

訳 EGFR-DNRは，薬剤耐性細胞においてシスプラチン耐性を部分的に逆戻りさせうる

R

■ resistant ＋ [名]

❶	resistant cells	耐性細胞	547
❷	resistant strains	耐性株	404

★ resolution 名 解像度／分解能／消散 用例数 18,327

resolution	解像度／消散	17,955
resolutions	解像度／消散	372

複数形のresolutions の割合は２％しかないが，単数形には不定冠詞が付く用例は多い．

■ resolution ＋ [前]

	the%	a/an%	ø%			
❶	40	21	39	resolution of ～	～の解像度／～の消散	3,383
❷	1	1	85	resolution in ～	～における解像度	431

❶ Here we report the crystal structure of the PP5 catalytic domain（PP5c）at a resolution of 1.6 A.（J Biol Chem. 2004 279:33992）

訳 われわれは，1.6 A の解像度でPP5触媒ドメイン（PP5c）の結晶構造を報告する

❶ These data demonstrate that T cell-independent influenza-specific Ab promotes the resolution of primary influenza infection and helps to prevent reinfection.（J Immunol. 2005 175:5827）

訳 T細胞非依存性インフルエンザ特異的抗体は，一次インフルエンザ感染の消散を促進する

resolve 動 分離する／解決する／回復させる　用例数 7,582

resolved	分離される／解決される／回復する	4,546
resolve	～を分離する／～を解決する	1,828
resolving	～を分離する／～を解決する	868
resolves	～を分離する／～を解決する	340

他動詞.

◆類義語：solve, unravel, recover, separate, analyze

■ resolved ＋［前］

❶ resolved by ～	～によって分離される／～によって解決される	430
・be resolved by ～	…は，～によって分離される／～によって解決される	281
❷ resolved in ～	～において回復する	215
❸ resolved into ～	～に分離される	107

❶ Kidney proteins **were resolved by** two-dimensional-PAGE and were identified by MALDI-MS.（J Biol Chem. 2002 277:34708）

訳 腎臓のタンパク質が，二次元ポリアクリルアミドゲル電気泳動によって分離された

❷ Diabetes was completely **resolved in** 76.8% of patients and resolved or improved in 86.0%.（JAMA. 2004 292:1724）

訳 糖尿病は，患者の76.8%において完全に回復された

■ to resolve

★ to resolve ～	～を解決する…	1,096
❶ To resolve ～	～を解決するために	194

❶ **To resolve** this issue, we generated BAFF-R-null mice.（J Immunol. 2004 173:2245）

訳 この問題を解決するために，

resource 名 情報源／資源　用例数 5,951

resource	情報源／資源	2,998
resources	情報源／資源	2,953

複数形のresourcesの割合は約50%と非常に高い.

■ resource ＋［前］

	the%	a/an%	ø%			
❶	3	97	0	resource for ～	～のための情報源	683
				・resource for ～ing	～するための情報源	163

❶ Urinary cells may be a good **resource for** the noninvasive diagnosis of CAN.（Transplantation. 2005 80:1686）

訳 尿中の細胞は，～の非侵襲性の診断のためのよい情報源であるかもしれない

respect 名 関心／関係 用例数 6,992

respect	関心／関係	6,661
respects	関心／関係	331

with respect toの用例が圧倒的に多い.

◆類義語：concern, regard, relation, connection, reference, view

■ [動] ＋ with respect to

★ with respect to ～	～に関して	6,369
❶ be discussed with respect to ～	…は，～に関して議論される	119
❷ differ with respect to ～	…は，～に関して異なる	105

❶ These results are discussed with respect to their potential involvement in the pathogenesis of AD.（Brain Res. 2005 1044:206）

訳 これらの結果が，～に関して議論される

★★ respectively 副 それぞれ 用例数 48,780

前にカンマが入ることが，圧倒的に多い.

◆類義語：individually

■ respectively

❶ , respectively	それぞれ，…	47,828
・～-fold, respectively	それぞれ，～倍	771

❶ DNA microarray analysis showed that PG0893 and PG2213 were upregulated 1.4- and 2-fold, respectively, in cells exposed to NO.（J Bacteriol. 2012 194:1582）

訳 PG0893とPG2213は，それぞれ，1.4倍と２倍上昇制御された

R

★ respond 動 応答する／反応する 用例数 14,166

respond	応答する／反応する	7,269
responded	応答した／反応した	3,067
responding	応答する／反応する	2,472
responds	応答する／反応する	1,358

自動詞．respond toの用例が非常に多い.

◆類義語：react

■ respond ＋ [前]

❶ respond to ～	…は，～に応答する	9,258
・cells respond to ～	細胞は，～に応答する	354
・respond to changes in ～	…は，～の変化に応答する	167
・respond to treatment	…は，治療に応答する	138

	・respond differently to ～	…は，～に異なって応答する	149
	・respond normally to ～	…は，～に正常に応答する	120
	・respond poorly to ～	…は，～によく応答しない	119
	・respond well to ～	…は，～によく応答する	105
❷	respond with ～	…は，～と反応する	274
❸	respond in ～	…は，～において反応する	245

❶ **Eukaryotic cells respond to DNA damage** and S phase replication blocks by arresting cell-cycle progression through the DNA structure checkpoint pathways. (EMBO J. 1998 17: 7239)

訳 真核細胞は，DNA損傷に応答する

■［動／名］＋ to respond to

★	to respond to ～	～に応答する…	545
❶	fail to respond to ～	…は，～に応答することができない	263
❷	ability to respond to ～	～に応答する能力	190

❶ Resistant hypertension was cured in two (4.2%) of 48 patients, had improved in 38 (79.1%), and **had failed to respond to treatment in eight** (16.7%). (Radiology. 2003 226:821)

訳 ～は，8名の患者において治療に応答することができなかった

★★ response ㊔ 応答／反応　　用例数 180,367

response	応答／反応	117,412
responses	応答／反応	62,955

複数形のresponsesの割合が，約35%ある可算名詞．in response toの用例が非常に多い．

◆類義語：reaction

■ response(s) ＋［前］

response ofの前にtheが付く割合は圧倒的に高い．

	the%	a/an%	ø%			
❶	20	9	67	response to ～	～に対する応答	47,019
❷	3	-	89	responses in ～	～における応答	5,785
❸	91	6	3	response of ～	～の応答	4,831
				・the response of ～ to …	～の…への応答	1,134

❶ **Molecular mimicry is characterized by an immune response to** an environmental agent that cross-reacts with a host antigen, resulting in disease. (Nat Med. 2002 8:509)

訳 分子擬態は，～への免疫応答によって特徴づけられる

❷ cAMP-dependent protein kinase (PKA) **regulates a broad range of cellular responses in the cardiac myocyte.** (Circulation. 2000 101:1459)

訳 ～は，心筋細胞における広い範囲の細胞応答を調節する

❸ **The response of white adipocytes to hypoxia required HIF-1α**, but its presence alone was incapable of inducing target gene expression under normoxic conditions. (J Biol Chem. 2012

287:18351)

訳 白色脂肪細胞の低酸素へ応答は，HIF-1α を必要とした

■ in response to ＋［名句］

★ in response to ～	～に応答して	27,512
❶ in response to DNA damage	DNA損傷に応答して	599
❷ in response to changes in ～	～の変化に応答して	287
❸ in response to stimulation	刺激に応答して	283
❹ in response to stress	ストレスに応答して	249

❶ The Brca genes are involved in multiple cellular processes in response to DNA damage including checkpoint activation, gene transcription, and DNA repair.（J Biol Chem. 2000 275:23899）

訳 Brca遺伝子は，DNA損傷に応答して複数の細胞過程に関与している

■［動／過分］＋ in response to

❶ activated in response to ～	～に応答して活性化される	411
❷ occur in response to ～	…は，～に応答して起こる	408
❸ regulated in response to ～	～に応答して調節される	354
❹ induced in response to ～	～に応答して誘導される	266
❺ increased in response to ～	～に応答して増大した	246

❶ Although both p38 and p44/p42 MAPK are activated in response to mitogens, they have divergent effects on anchorage-independent growth.（Hepatology. 2001 33:43）

訳 ～は，マイトジェンに応答して活性化される

■［名／形］＋ response(s)

❶ immune responses	免疫応答	8,353
❷ cell responses	細胞応答	5,167
❸ inflammatory response	炎症性応答	3,027
❹ stress response	ストレス反応	2,047
❺ cellular responses	細胞応答	1,545

★ responsible 形 責任のある／原因である 用例数 20,735

responsible forの用例が圧倒的に多い．

◆類義語：due, causative

■ be ＋（［副］＋）responsible

★ be responsible	…は，責任がある	9,454
❶ be responsible for ～	…は，～に責任がある	9,173
・be primarily responsible for ～	…は，～に主に責任のある	373
・be largely responsible for ～	…は，～に主に責任のある	205

R

❶ The lipid bilayers of cell membranes are primarily responsible for the low passive transport of nonelectrolytes across cell membranes, and for the pronounced size selectivity of such transport. (Biophys J. 1999 77:1268)

訳 …は，～の低い受動輸送に主に責任がある

■［名］＋ responsible for

★ responsible for ～	～に責任のある／～の原因である	19,878
❶ mechanisms responsible for ～	～に責任のある機構	1,079
❷ enzyme responsible for ～	～に責任のある酵素	348
❸ genes responsible for ～	～に責任のある遺伝子	311
❹ factors responsible for ～	～に責任のある因子	243

❶ However, little is known about mechanisms responsible for formation of the collateral circulation in healthy tissues. (Circ Res. 2012 111:1539)

訳 ～の形成に責任のある機構については，ほとんど知られていない

■ responsible for ＋［名句］

❶ responsible for binding	結合することに責任のある	127
❷ responsible for regulating ～	～を調節することに責任のある	120
❸ responsible for maintaining ～	～を維持することに責任のある	119
❹ responsible for the majority of ～	～の大多数に責任のある	114
❺ responsible for the production of ～	～の産生に責任のある	91

❶ Each of these ATP-grasp domains contains an active site responsible for binding one molecule of MgATP. (J Biol Chem. 2002 277:39722)

訳 ～は，1 分子のMgATPに結合することに責任のある活性部位を含む

responsive 形 応答性の　　用例数 9,406

responsive toの用例が非常に多い.

◆類義語：reactive

■ responsive to

★ responsive to ～	～に応答性である	1,916
❶ be responsive to ～	…は，～に応答性である	588
❷ more responsive to ～	～により応答性である	153
❸ less responsive to ～	～により応答性のない	134

❶ Transcription of the cucumber hpr-A gene is responsive to cytokinin and light. (Plant Mol Biol. 1998 38:713)

訳 キュウリのhpr-A遺伝子の転写は，サイトカイニンと光に応答性である

❷ Tyrosine kinase activity and activation of MAPK was more responsive to epidermal growth factor stimulation in papilloma cells than in uninfected primary laryngeal cells. (Cancer Res. 1999 59:968)

訳 ～は，非感染性の初代喉頭細胞においてより乳頭腫細胞において上皮成長因子刺激により

応答性であった

❸ The cocultured T cells were also less responsive to IL-12 as assessed by reduced phosphorylation of STAT4 and limited IFN-γ secretion. (J Immunol. 2000 165:3324)

訳 共培養されたT細胞は，また，IL-12により応答性でなかった

■ [名]-responsive

❶	stress-responsive ～	ストレス応答性の～	306
❷	estrogen -responsive ～	エストロゲン応答性の～	186

❶ Heat-shock protein 70 (hsp70) is a stress-responsive gene important for cell survival; induction of hsp70 appears to be mediated, in part, by the prostaglandin pathway. (Gastroenterology. 1998 115:1454)

訳 ヒートショックプロテイン70（hsp70）は，細胞生存のために重要なストレス応答性遺伝子である

■ [名]-responsive ＋ [名]

❶	～-responsive genes	～応答性遺伝子	947
❷	～-responsive cells	～応答性細胞	198

R

responsiveness 名 応答性　用例数 5,033

複数形の用例は全くなく，原則，不可算名詞として使われる．

◆類義語：reactivity

■ responsiveness ＋ [前]

responsiveness ofの前にtheが付く割合はかなり高い．

	the%	a/an%	ø%			
❶	5	1	80	responsiveness to ～	～への応答性	1,661
❷	64	0	30	responsiveness of ～	～の応答性	889
				・responsiveness of ～ to …	～の…への応答性	307
❸	2	0	78	responsiveness in ～	～における応答性	382

❶ We generated human embryonic kidney cells (HEK293) that stably express TLR4 (HEK-TLR4) and examined their responsiveness to LPS by measuring NF-κB activity and production of interleukin-8 (IL-8). (J Biol Chem. 2000 275:20861)

訳 …は，～を測定することによってLPSへのそれらの応答性を調べた

❷ These findings support a model where neuronal activity and Wnts increase the responsiveness of neurons to Wnt signalling by recruiting Fz5 receptor at synaptic sites. (Development. 2010 137:2215)

訳 ～は，Wntシグナル伝達へのニューロンの応答性を増大させる

rest 名 残り，動 静止する／次第である　用例数 9,504

resting	静止した	5,474
rest	残り／静止する／次第である	3,830

rests	残り／静止する／次第である	137
rested	静止した／次第であった	63

名詞および自動詞として用いられる．複数形のrestsの割合は1％しかなく，原則，不可算名詞として使われる．

◆類義語：arrest, remainder, ease

■ 名 rest ＋［前］

rest ofの前にtheが付く割合は，100％と圧倒的に高い．

	the％	a/an％	ø％			
❶	100	0	0	rest of ～	～の残りの部分	864

❶ It was weakly coupled to the rest of the cytoskeleton and promoted the random protrusion and retraction of the leading edge.（Science. 2004 305:1782）

訳 それは，細胞骨格の残りの部分に弱く結合した

■ 名［前］＋ rest

❶	at rest	安静時に	965

❶ Forearm blood flow was measured at rest and during reactive hyperemia by venous air plethysmography.（Crit Care Med. 2000 28:1290）

訳 前腕血流が，安静時と反応性充血の間に測定された

■ 動 rest ＋［前］

❶	rest on ～	…は，～次第である	120

■ 動 resting ＋［名］

❶	resting state	静止状態	509
❷	resting cells	静止細胞	208

❷ Previous studies with GSK-3 inhibitors suggest that GSK-3β is a C/EBPβ kinase in resting cells.（J Biol Chem. 2005 280:32683）

訳 GSK-3βは，静止細胞においてC/EBPβキナーゼである

restoration 名 回復／修復　用例数 2,783

restoration	回復／修復	2,677
restorations	回復／修復	106

複数形のrestorationsの割合は4％あるが，単数形は無冠詞で使われることが非常に多い．

◆類義語：repair, recovery

■ restoration ＋［前］

	the%	a/an%	ø%			
❶	32	1	61	restoration of ～	～の回復	2,095

❶ Restoration of p53 function through the disruption of the MDM2-p53 protein complex is a promising strategy for the treatment of various types of cancer.（J Am Chem Soc. 2012 134: 17059）

訳 MDM2-p53複合体の破壊によるp53機能の回復は，様々なタイプの癌の治療のための有望な戦略である

★ restore 動 回復する 用例数 13,336

restored	回復される／回復する／～を回復した	7,635
restore	～を回復する	2,773
restores	～を回復する	1,884
restoring	～を回復する	1,044

他動詞．

◆類義語：recover, return, regain, reconstruct, ameliorate, repair

R

■ be restored ＋［前］

★ be restored	…は，回復される	2,298
❶ be restored by ～	…は，～によって回復される	806
❷ be restored to ～	…は，～に回復される	361
❸ be restored in ～	…は，～において回復される	330

❷ Maternal haematocrit was restored to normal levels only in animals given supplements for at least 2 weeks.（J Physiol. 2004 561:195）

訳 母のヘマトクリットは，正常なレベルに回復された

■［副］＋ restored

❶ partially restored ～	部分的に回復される／～を部分的に回復した	431
❷ fully restored ～	完全に回復される／～を完全に回復した	187

❶ The ESP EPR signal was partially restored by removal of the chemical reductants.（Proc Natl Acad Sci USA. 1991 88:9895）

訳 ～は，化学的な還元体の除去によって部分的に回復された

■ restore ＋［名句］

❶ restore normal ～	…は，正常な～を回復する	392
❷ restore the ability of ～	…は，～の能力を回復する	192

❷ In both cases, the gene restored the ability of the mutant to gyrate its cell ends and enabled colony spreading in agarose.（Infect Immun. 2004 72:5493）

訳 その遺伝子は，変異体の能力を回復した

★ restrict 動 限定する／制限する 用例数 15,487

restricted	限定される／制限される／～を限定した	12,763
restrict	～を限定する	1,166
restricting	～を限定する	867
restricts	～を限定する	691

他動詞．受動態，特にbe restricted toの用例が非常に多い．

◆類義語：limit, restrain, confine

■ be restricted ＋［前］

★ be restricted	…は，限定される	3,213
❶ be restricted to ～	…は，～に限定される	2,570
・expression is restricted to ～	発現は，～に限定されている	215
❷ be restricted by ～	…は，～によって限定される	193
❸ be restricted in ～	…は，～において限定される	111

❶ In adult mice, the expression is restricted to renal tubules.（J Am Soc Nephrol. 2002 13:1837）

訳 その発現は，腎尿細管に限定されている

■［副］＋ restricted

❶ spatially restricted	空間的に限定される	250
❷ more restricted	より限定される	238
❸ highly restricted	高度に限定される	232

❶ The N termini of Aβ40 pentamers are more spatially restricted than Aβ42 pentamers.（Proc Natl Acad Sci USA. 2004 101:17345）

訳 Aβ40五量体のN末端は，Aβ42五量体より空間的に限定されている

❷ The FAR2 mRNA was more restricted in distribution and most abundant in the eyelid, which contains wax-laden meibomian glands.（J Biol Chem. 2004 279:37789）

訳 FAR2メッセンジャーRNAは，分布においてより限定されていた

■ restricted ＋［名］

❶ restricted expression	限定された発現	329
❷ restricted epitopes	限定されたエピトープ	137

❶ Northern blot and PCR analyses demonstrated a restricted expression of Alt-d in fetal liver, bone marrow, and adult reticulocytes.（Genomics. 2004 84:431）

訳 ノーザンブロットおよびPCR解析は，Alt-dの限定された発現を実証した

restriction 名 制限 用例数 6,709

restriction	制限	6,279
restrictions	制限	430

複数形のrestrictionsの割合は約5％あるが，単数形は無冠詞で使われることが非常に多い．

■ restriction ＋［前］

	the%	a/an%	ø%			
❶	33	5	57	restriction of ～	～の制限	685
❷	6	10	66	restriction in ～	～における制限	174

❶ Translational repression during mRNA transport is essential for spatial restriction of protein production.（Genes Dev. 2008 22:1037）

訳 ～は，タンパク質産生の空間的制限のために必須である

■ restriction ＋［名］

❶	restriction fragment	制限酵素断片	621
❷	restriction enzyme	制限酵素	501

R

★ result 名 結果，動 結果になる 用例数 260,503

results	結果／結果になる	171,507
resulted	結果になった	33,415
result	結果／結果になる	30,624
resulting	結果になる	24,957

名詞の用例の方が多いが，自動詞の用例もかなりある．自動詞としては，主にresult inとresult fromが用いられる．名詞としては，複数形の割合が非常に高い．

◆類義語：consequence, outcome, data, attribute, ascribe, cause, lead to, give rise to, produce

■ 名 results ＋［動］

❶	results suggest ～	結果は，～を示唆する	26,785
❷	results indicate ～	結果は，～ということを示す	16,677
❸	results demonstrate ～	結果は，～を実証する	11,005
❹	results show ～	結果は，～を示す	7,972
❺	results provide ～	結果は，～を提供する	5,493
❻	results support ～	結果は，～を支持する	3,639
❼	results reveal ～	結果は，～を明らかにする	2,110
❽	results identify ～	結果は，～を同定する	1,500
❾	results establish ～	結果は，～を確立する	1,180

❶ The **results suggest** that the spatial pattern of cone system losses in this disease differs from the spatial pattern of rod system losses. (Invest Ophthalmol Vis Sci. 2002 43:2364)

訳 それらの結果は，～ということを示唆する

❷ Our **results indicate** that human aging is associated with a reduction in forearm postjunctional α-adrenergic responsiveness to endogenous NE release and that this might be specific to α1-adrenergic receptors. (Circulation. 2002 106:1349)

訳 われわれの結果は，～ということを示す

❸ Our **results demonstrate** that carnosine can rescue neurons from zinc- and copper-mediated neurotoxicity and suggest that one function of carnosine may be as an endogenous neuroprotective agent. (Brain Res. 2000 852:56)

訳 われわれの結果は，～ということを実証する

■ 名 results ＋［前］

results ofの前にtheが付く割合は非常に高い．

	the%	a/an%	ø%			
❶	75	-	25	results of ～	～の結果	8,311
❷	15	-	85	results from ～	～からの結果	3,900

❶ The **results of** this study are consistent with the hypothesis that the ovarian hormones that drive the menstrual cycle influence genital tract immunity in female primates. (Infect Immun. 1999 67:6321)

訳 この研究の結果は，～と一致する

❷ Recent **results from** animal studies suggest that stem cells may be able to home to sites of myocardial injury to assist in tissue regeneration. (Circulation. 2005 112:1451)

訳 動物研究からの最近の結果は，～ということを示唆する

■ 名 as a result

★	as a result	結果として	6,072
❶	as a result of ～	～の結果として	4,271
	・occur as a result of ～	…は，～の結果として起こる	239
❷	As a result,	結果として，	1,431

❶ In most animal cells studied, chromosome segregation **occurs as a result of** kMT shortening, which causes chromosomes to move toward the spindle poles (anaphase A). (Curr Biol. 2012 22:437)

訳 染色体分離は，kMT短縮の結果として起こる

❷ **As a result,** precise control of the bandgap between these two states is not currently achievable. (J Am Chem Soc. 2012 134:11774)

訳 結果として，これらの２つの状態の間のバンドギャップの正確な制御は，現在，達成可能でない

■ 名 results ＋［過分］

❶	results obtained	得られた結果	1,245
❷	results presented	示される結果	633

❶ In contrast to the **results obtained** with HeLa cells, rwt virus induced apoptosis more slowly than did rM51R-M virus in BHK cells. (J Virol. 2001 75:12169)

訳 HeLa細胞を使って得られた結果とは対照的に

■ 名［前］＋ these results

❶ based on these results	これらの結果に基づいて	532
❷ on the basis of these results	これらの結果に基づいて	271

❶ Based on these results, we propose that α_2(Arg-274) and α_2(Leu-277) are crucial to the efficient transduction of agonist binding into channel gating at the $GABA_A$A-R. (J Biol Chem. 2000 275:22764)

訳 これらの結果に基づいて，われわれは～ということを提唱する

■ 名［形］＋ results

❶ similar results	類似の結果	1,540
❷ experimental results	実験結果	1,022
❸ present results	現在の結果	685
❹ previous results	以前の結果	600

■ 動 result in ＋［名句］

★ result in ～	…は，～という結果になる	81,332
❶ result in increased ～	…は，増大した～という結果になる	2,805
❷ result in the formation of ～	…は，～の形成という結果になる	1,043
❸ result in an increase in ～	…は，～の増大という結果になる	814
❹ result in loss of ～	…は，～の喪失という結果になる	797
❺ result in a decrease in ～	…は，～の低下という結果になる	505
❻ result in activation of ～	…は，～の活性化という結果になる	400
❼ result in inhibition of ～	…は，～の阻害という結果になる	395
❽ result in a reduction	…は，低下という結果になる	385
❾ result in the accumulation of ～	…は，～の蓄積という結果になる	311

❸ V12-ras overexpression resulted in an increase in hypoxia-induced HIF-1α and HIF-2α expression. (Cancer Res. 2001 61:7349)

訳 V12-rasの過剰発現は，～の増大という結果になった

■ 動［名］＋ result in

❶ treatment resulted in ～	処置は，～という結果になった	607
❷ mutation results in ～	変異は，～という結果になる	249
❸ deficiency results in ～	欠損は，～という結果になる	200
❹ infection results in ～	感染は，～という結果になる	145
❺ activation results in ～	活性化は，～という結果になる	114

❶ CD40L treatment resulted in a dramatic decrease in $t_{1/2}$ (< 5 min) for both IκB molecules, which was inhibited by addition of Z-LLF-CHO. (J Immunol. 1998 160:4398)

訳 CD40L処理は，～の劇的な低下という結果になった

■ ㊍ resulting in

❶ , resulting in ～	そして，（それは）～という結果になる	10,318

❶ These results demonstrate that NO availability in the kidney is decreased in SHR, resulting in increased oxygen consumption.（J Am Soc Nephrol. 2002 13:1788）
訳 そして，（それは）増大した酸素消費という結果になる

■ ㊍ result from

★ result from ～	…は，～に起因する	17,286
❶ appear to result from ～	…は，～に起因すると思われる	248
❷ result from mutations in ～	…は，～における変異に起因する	227
❸ likely result from ～	…は，おそらく～に起因する	214
❹ be thought to result from ～	…は，～に起因すると考えられる	126
❺ result from loss of ～	…は，～の喪失に起因する	117
❻ probably result from ～	…は，おそらく～に起因する	105

❹ Breast cancer is thought to result from excessive cumulative exposure to ovarian hormones.（N Engl J Med. 2003 348:2313）
訳 乳癌は，卵巣ホルモンへの過剰な累積性の曝露に起因すると考えられる

★ retain ㊍ 保持する 用例数 10,799

retained	保持される／～を保持した	5,561
retain	～を保持する	2,595
retains	～を保持する	1,616
retaining	～を保持する	1,027

他動詞.
◆類義語：hold, keep, maintain, sustain

■ retain ＋［名句］

❶ retain the ability to *do*	…は，～する能力を保持する	658
❷ retain ～ activity	…は，～活性を保持する	390

❶ Here, we show that memory CD8 T cells retain the ability to respond to dendritic cell-mediated stimulation after adoptive transfer into either $TAP^{-/-}$ (MHC class I-deficient) or wild-type mice.（J Immunol. 2005 175:4829）
訳 記憶CD8 T細胞は，～に応答する能力を保持する

■［名］＋ retain

❶ cells retain ～	細胞は，～を保持する	111

■ retained ＋［前］

❶ retained in ～	～において保持される	966
・be retained in ～	…は，～において保持される	696
❷ retained by ～	～によって保持される	134

❶ The clp mutation affects Lgi4 mRNA splicing, resulting in **a mutant protein that is retained in the cell.** (Nat Neurosci. 2006 9:76)

訳 その細胞において保持されている変異タンパク質

retention ㊔ 貯留／保持　用例数 5,726

restriction	貯留／保持	5,717
restrictions	貯留／保持	9

複数形のrestrictionsの用例はほとんどなく，原則，**不可算名詞**として使われる．

◆類義語：accumulation

■ restriction ＋［前］

	the %	a/an %	ø %			
❶	32	1	61	retention of ～	～の貯留	2,176
❷	2	1	77	retention in ～	～における貯留	375

❶ Corresponding blood, tissue, and bile studies in LEC rats showed incorporation of radiocopper in the liver but without copper excretion in bile, **leading to hepatic retention of the radiotracer.** (J Nucl Med. 2012 53:961)

訳 放射性トレーサの肝貯留につながる

return ㊍ 戻る／回復する，㊔ 復帰／回復　用例数 4,231

returned	戻った／戻される	1,797
return	戻る／復帰	1,746
returning	戻る	391
returns	戻る／復帰	297

自動詞の用例が多いが，名詞としても用いられる．複数形のreturnsの割合が，約10%ある可算名詞．

◆類義語：recover revert, restore, retrieve, regain

■ ㊍ return to

★ return to ～	…は，～に戻る	1,477
❶ return to baseline	…は，ベースラインに戻る	350

❶ After implementation of corrective actions, **the rate of peritonitis returned to baseline.** (Lancet. 2005 365:588)

訳 腹膜炎の比率がベースラインに戻った

■ 名 return ＋［前］

	the %	a/an %	ø %			
❶	4	33	55	return to ～	～への復帰	1,000
❷	31	6	60	return of ～	～の復帰	429

❶ Although **a return to baseline levels was noted** after this period, a distinct rise in iHSP70 occurred again during terminal DC maturation. (J Immunol. 2009 183:388)

訳 ベースライン・レベルへの復帰が認められた

☆☆ reveal 動 明らかにする　用例数 76,943

revealed	～を明らかにした／明らかにされる	47,541
reveal	～を明らかにする	16,416
reveals	～を明らかにする	10,289
revealing	～を明らかにする	2,697

他動詞．revealed thatの用例が非常に多い．

◆類義語：elucidate, clarify, disclose, uncover, manifest, define

■ reveal ＋［that節／名句］

❶	reveal that ～	…は，～ということを明らかにする	30,039
❷	reveal the presence of ～	…は，～の存在を明らかにする	1,264
❸	reveal a novel ～	…は，新規の～を明らかにする	1,138
❹	reveal significant ～	…は，有意な～を明らかにする	616
❺	reveal the existence of ～	…は，～の存在を明らかにする	268
❻	reveal the importance of ～	…は，～の重要性を明らかにする	162
❼	reveal an increase in ～	…は，～の増大を明らかにする	111

❶ **These studies revealed that** the ER-to-Golgi transport of gB requires a nine-amino-acid region (YMTLVSAAE) within its cytoplasmic domain. (J Virol. 2000 74:9421)

訳 これらの研究は，～ということを明らかにした

❷ **Genotyping revealed the presence of** 11 unique strains within the herd. (J Clin Microbiol. 2004 42:5381)

訳 遺伝子型同定は，～の存在を明らかにした

■［名］＋ reveal

❶	analysis revealed ～	分析は，～を明らかにした	4,792
❷	studies revealed ～	研究は，～を明らかにした	2,127
❸	results reveal ～	結果は，～を明らかにする	2,110
❹	data reveal ～	データは，～を明らかにする	1,334
❺	findings reveal ～	知見は，～を明らかにする	1,316
❻	experiments revealed ～	実験は，～を明らかにした	932

❼ microscopy revealed ～	顕微鏡観察法は，～を明らかにした	701
❽ assays revealed ～	アッセイは，～を明らかにした	694
❾ structure reveals ～	構造は，～を明らかにする	662

❶ Sequence **analysis revealed** that the novel mouse and human sulfotransferases display nearly 98% identity in their amino acid sequences.（Gene. 2002 285:39）

訳 配列分析は，～ということを明らかにした

■ to reveal

★ to reveal ～	～を明らかにする…	1,336
❶ fail to reveal ～	…は，～を明らかにすることができない	156

❶ Previous analysis of Slit1; Slit2 double mutant spinal cords **failed to reveal** a defect in commissural axon guidance.（Neuron. 2004 42:213）

訳 …は，～の欠損を明らかにすることができなかった

■ revealed ＋［前］

❶ revealed by ～	～によって明らかにされる	1,372
・as revealed by ～	～によって明らかにされたように	651
・be revealed by ～	…は，～によって明らかにされる	366
❷ revealed in ～	～において明らかにされる	310

❶ This MYB, Pinus taeda MYB4 (PtMYB4), is expressed in cells undergoing lignification, **as revealed by** in situ RT-PCR.（Plant J. 2003 36:743）

訳 in situ RT-PCRによって明らかにされたように

R

★reverse ㊍ 逆転させる／逆戻りさせる，㊎ 逆の　用例数 22,114

reverse	～を逆転させる／逆の	13,586
reversed	逆転される／～を逆転した	7,029
reverses	～を逆転させる	904
reversing	～を逆転させる	595

他動詞および形容詞として用いられる．

◆類義語：inverse, opposite, backward

■ ㊍ be reversed ＋［前］

★ be reversed	…は，逆転される	2,689
❶ be reversed by ～	…は，～によって逆転される	1,791
❷ be reversed in ～	…は，～において逆転される	163
❸ be reversed with ～	…は，～によって逆転される	104

❶ This effect **was reversed by** the addition of the cyclooxygenase inhibitor, ibuprofen, or a DP1 receptor antagonist (MK0524).（J Invest Dermatol. 2010 130:2448）

訳 この効果は，シクロオキシゲナーゼ阻害剤の添加によって逆転された

■ 動［副］＋ reversed

❶ partially reversed	部分的に逆転される／部分的に逆転させた	342
❷ completely reversed	完全に逆転される	261

❶ The inhibition of T cell activation was partially reversed by blocking IL-10.（J Immunol. 2005 175:3225）

訳 T細胞活性化の抑制は，IL-10をブロックすることによって部分的に逆転された

❷ The effects of forskolin were completely reversed by the protein kinase A inhibitor H89, whereas H89 alone increased transport rates.（J Biol Chem. 2006 281:2053）

訳 フォルスコリンの効果は，プロテインキナーゼA阻害剤H89によって完全に逆転された

■ 動 reverse ＋［名句］

❶ reverse the effects of ～	…は，～の効果を逆転させる	147
❷ reverse the inhibition	…は，阻害を逆転させる	87

■ 形 reverse ＋［名］

❶ reverse transcription	逆転写	4,100
❷ reverse transcriptase	逆転写酵素	3,752
❸ reverse phase	逆相	400
❹ reverse genetics	逆遺伝学	355

★review 名 総説／再検討，動 概説する／再検討する 用例数 26,442

review	総説／再検討／～を概説する／～を再検討する	18,872
reviewed	再検討される／～を再検討した	5,239
reviews	総説／再検討／～を概説する／～を再検討する	2,078
reviewing	～を概説する／～を再検討する	253

名詞として用いられることの方が多いが，他動詞の用例もかなりある．複数形のreviewsの割合が，約10%ある可算名詞．

◆類義語：criticize

■ 名 review ＋［動］

❶ review focuses on ～	総説は，～に焦点を当てる	768
❷ review summarizes ～	総説は，～を要約する	725
❸ review discusses ～	総説は，～を議論する	502
❹ review highlights ～	総説は，～を強調する	393
❺ review describes ～	総説は，～を述べる	344
❻ review examines ～	総説は，～を調べる	281
❼ review provides ～	総説は，～を提供する	221

❶ This review focuses on the biochemical functions of Yops, the signaling pathways they modulate, and the role of these proteins in Yersinia virulence.（Annu Rev Microbiol. 2005 59:

69)
訳 この総説は，Yopの生化学的な機能に焦点を当てる

■ 名 [前] ＋ this review

❶ in this review	この総説において	2,392
・In this review,	この総説において，	1,571
❷ the purpose of this review is to *do*	この総説の目的は，～することである	426

❶ In this review, we address the role of costimulatory pathways in allograft rejection and tolerance. (Transplantation. 2005 80:555)
訳 この総説において，われわれは～における共刺激経路の役割に取り組む

❷ The purpose of this review is to summarize recent advances in the diagnosis and treatment of several severe skin diseases seen in children. (Curr Opin Pediatr. 2011 23:403)
訳 この総説の目的は，～における最近の進歩を要約することである

■ 名 review ＋ [前]

	the %	a/an %	ø %			
❶	2	65	25	review of ～	～の再検討	1,966

❶ We undertook a systematic review of 404 published cases of Castleman's disease to identify the role of the surgeon beyond assistance in tissue-based diagnosis. (Ann Surg. 2012 255:677)
訳 われわれは，キャッスルマン病の404例の発表症例の体系的な再検討を行った

■ 名 [形／名] ＋ review

❶ systematic review	体系的な再検討	371
❷ retrospective review	遡及的な再検討	353
❸ the present review	この総説	218
❹ literature review	文献の再検討	170

❷ Clinical base estimates were obtained from retrospective review of all patients receiving ICDs between June 1997 and July 2001 at a single university hospital. (J Am Coll Cardiol. 2005 46:850)
訳 臨床に基づく推定が，すべての患者の遡及的な再検討から得られた

■ 動 be reviewed ＋ [前]

★ be reviewed	…は，再検討される／概説される	2,871
❶ be reviewed for ～	…は，～に対して再検討される	239
❷ be reviewed to *do*	…は，～するために再検討される	206
❸ be reviewed in ～	…は，～において再検討される	169
❹ be reviewed by ～	…は，～によって再検討される	124

❶ Records were retrospectively reviewed for 194 children diagnosed from 1985 to 1999 at St Jude Children's Research Hospital (Memphis, TN). (J Clin Oncol. 2005 23:7152)
訳 記録は，194人の子供に対して遡及的に再検討された

R

■ 動［副］＋ reviewed

❶ retrospectively reviewed	遡及的に再検討される／遡及的に再検討した	351

■ 動［代名／名］＋ review

❶ we review ～	われわれは，～を概説する	2,651
❷ this article reviews ～	この記事は，～を概説する	660
❸ this paper reviews ～	この論文は，～を概説する	189

❶ Here, we review the biology and clinical results of these five species of viruses and discuss lessons learned and challenges for the future.（Oncogene. 2005 24:7802）

訳 ここに，われわれは～の生物学と臨床結果を概説する

★ rich 形 富んだ　　用例数 14,555

◆類義語：abundant

■ rich in

❶ rich in ～	～に富んだ	1,123
・be rich in ～	…は，～に富んでいる	319

❶ The C-terminal dimerization domain is rich in α-helices and shows domain swapping.（J Mol Biol. 2005 354:91）

訳 C末端二量体形成ドメインは，α-ヘリックスに富んでいる

■ -rich ＋［名］

❶ ～-rich region	～に富んだ領域	806
❷ ～-rich domain	～に富んだドメイン	615
❸ ～-rich sequences	～に富んだ配列	370

❶ A small, proline-rich region in the C-terminal half of MAP4 bound directly to a Sept 2:6:7 heterotrimer, and to the Sept2 monomer.（Mol Biol Cell. 2005 16:4648）

訳 MAP4のC末端半分内の小さなプロリンに富んだ領域は，～に直接結合した

■［名］-rich

❶ proline-rich ～	プロリンに富んだ～	1,234
❷ leucine-rich ～	ロイシンに富んだ～	952
❸ cysteine-rich ～	システインに富んだ～	842
❹ GC-rich ～	GCに富んだ～	527

★ rise ㊔ 上昇, ㊍ 上昇する　　用例数 10,334

rise	上昇／上昇する	7,859
rose	上昇した	1,017
rising	上昇する	853
rises	上昇／上昇する	508
risen	上昇した	97

名詞の用例が多いが，動詞としても用いられる．複数形のrisesの割合が，約5％ある可算名詞．give rise toの用例も非常に多い．

◈類義語：elevation, augmentation, elevate, produce, increase, raise

■ ㊔ rise ＋［前］

rise ofの前にtheが付く割合は非常に高い．

	the％	a/an％	ø％			
❶	36	55	5	rise in ～	～の上昇	1,875
❷	72	16	12	rise of ～	～の上昇	306

❶ **A striking feature is a delayed rise in intracellular free** Zn^{2+} **in** CA1 neurons just before the onset of histologically detectable cell death. (Proc Natl Acad Sci USA. 2005 102:12230)

訳 著しい特徴は，～における細胞内遊離Zn^{2+}の遅れた上昇である

■ ㊔ give rise to

❶	give rise to ～	…は，～を生じる	4,206

❶ In zebrafish, **most somitic cells give rise to long muscle fibers** that are anchored to intersegmental boundaries. (Dev Biol. 2005 287:346)

訳 ほとんどの体節細胞は，長い筋肉繊維を生じる

■ ㊍ rise ＋［前］

❶	rise from ～	…は，～から上昇する	215
❷	rise to ～	…は，～へ上昇する	193

❶ **Serum creatinine rose from 1.4+/-0.41 to 2.45+/-1.7 mg/dL** in group 1 (P=0.004) and from 2.1+/-0.45 to 2.62+/-1.2 mg/dL (P=NS) in group 2. (Transplantation. 2007 83:277)

訳 血清クレアチニンは，1.4+/-0.41 mg/dLから2.45+/-1.7 mg/dLへ上昇した

★★ risk ㊔ リスク／危険　　用例数 75,074

risk	リスク／危険	70,740
risks	リスク／危険	4,334

複数形のrisksの割合が，約5％ある可算名詞．

◈類義語：hazard

R

■ risk ＋［前］

	the%	a/an%	ø%			
❶	40	26	32	risk of ～	～のリスク	20,788
				・risk of death	死のリスク	1,183
				・risk of developing ～	～を発症するリスク	1,101
				・risk of breast cancer	乳癌のリスク	454
				・risk of mortality	死のリスク	314
				・risk of stroke	脳卒中のリスク	290
❷	17	24	58	risk for ～	～のリスク	7,342
				・at risk for ～	～のリスクがある	1,063
				・at high risk for ～	～の高いリスクがある	627
				・at increased risk for ～	～の増大したリスクがある	395
				・be at risk for ～	…は，～のリスクがある	347
				・risk for developing ～	～を発症するリスク	309
❸	12	6	65	risk in ～	～のリスク	1,404

❶ The risk of death from natural causes remained lower among Gulf veterans compared with non-Gulf veterans.（Am J Epidemiol. 2001 154:399）

訳 自然の原因による死のリスクは，～の間でより低いままであった

❷ It may be possible to apply this approach to large studies of β-cell function designed to identify changes in islet function in subjects at risk for diabetes.（Diabetes. 2000 49:373）

訳 糖尿病のリスクのある対象における膵島の機能の変化を同定するために設計されたβ細胞機能の大規模な研究

■［動］＋ the risk of

❶	increase the risk of ～	…は，～のリスクを増大させる	1,144
❷	reduce the risk of ～	…は，～のリスクを低下させる	962
❸	decrease the risk of ～	…は，～のリスクを低下させる	185

❶ Declines in the ability to engage in recreational activities and social interactions appear to significantly increase the risk of new depressive symptoms.（Arthritis Rheum. 2001 44:1194）

訳 ～は，新しいうつ症状のリスクを有意に増大させるように思われる

■［前］＋ the risk of

❶	increase in the risk of ～	～のリスクの増大	182
❷	reduction in the risk of ～	～のリスクの低下	146

■［形／過分］＋ risk

❶	increased risk	増大したリスク	6,016
	・be associated with increased risk of ～	…は，～の増大したリスクと関連する	446
❷	high risk	高いリスク	5,637
❸	relative risk	相対リスク	3,322

❹ cancer risk	癌リスク	2,172
❺ higher risk	より高いリスク	1,401
❻ low risk	低いリスク	1,386
❼ cardiovascular risk	心血管リスク	1,106
❽ lower risk	より低いリスク	1,075

❶ However, whether elevations of TNF-α in the stable phase after myocardial ischemia (MI) are associated with increased risk of recurrent coronary events is unknown. (Circulation. 2000 101:2149)

訳 ～は，再発する冠血管イベントの増大したリスクと関連する

■ risk ＋［名］

❶ risk factors	危険因子	8,881
❷ risk stratification	リスク層別化	652
❸ risk assessment	リスク評価	490
❹ risk reduction	リスク軽減	473

❶ The use of antithymocyte globulin and prolonged exposure to ganciclovir are risk factors for the development of ganciclovir resistance. (Transplantation. 1999 68:1272)

訳 ～は，ガンシクロビル耐性の発生の危険因子である

R

robust 形 強固な／頑強な／強い／左右されない　用例数 8,252

◆類義語：strong, potent, powerful, intensive, intense

■ robust ＋［前］

❶ robust to ～	～に対して強固な／～に左右されない	330
・be robust to ～	…は，～に対して強固である／～に左右されない	186
❷ robust in ～	～において強固な	214

❶ Results are robust to variation in unconstrained parameters. (Science. 2001 292:1893)

訳 結果は，～の変動に左右されない

❷ In contrast to these observations, the substance P-immunoreactive innervation of the dorsolateral nucleus remained robust in aged animals and was not significantly different from young adults. (Brain Res. 2005 1036:139)

訳 ～は，老齢の動物において強固のままであった

■ ［副］＋ robust

❶ more robust	より強固な	495
❷ most robust	最も強固な	148

❶ The complex organisms are more robust than the simple ones with respect to the average effects of single mutations.. (Nature. 1999 400:661)

訳 複雑な生物は，～に関して単純な生物より強固である

★★ role ㊔ 役割　　用例数 155,425

role	役割	132,329
roles	役割	23,096

複数形のrolesの割合が，約15%ある可算名詞．〈a role in〉〈the role of〉〈a role for〉の用例が多い．

◆類義語：part

■ role ＋［前］

	the %	a/an %	ø %			
❶	2	85	0	role in ～	～において役割を	56,740
				・role in regulating ～	～を調節する際に役割を	2,365
				・role in the regulation of ～	～の調節において役割を	1,333
				・role in the pathogenesis of ～	～の発病において役割を	1,141
				・role in the development of ～	～の発生において役割を	920
				・role in mediating ～	～を仲介する際に役割を	843
❷	83	14	1	role of ～	～の役割	43,737
				・role of ～ in …	…の際の～の役割	19,117
❸	3	95	0	role for ～	～の役割	20,788
				・role for ～ in …	…の際の～の役割	10,754

❶ These data suggest that **down regulation of DOC-2/hDab2 may play an important role in the development of** gestational trophoblastic diseases.（Oncogene. 1998 17:419）

訳 DOC-2/hDab2の下方制御は，～の発症において重要な役割を果たすかもしれない

❷ **These studies underscore the critical role of chromatin structure in regulating HO gene expression.**（Mol Cell Biol. 2000 20:2350）

訳 これらの研究は，HO遺伝子発現を調節する際のクロマチン構造の決定的に重要な役割を強調する

❸ Together, **these data suggest a novel role for p38 MAP kinase in regulating adhesion of** breast cancer cells to collagen type Ⅳ.（J Biol Chem. 2000 275:11284）

訳 これらのデータは，～の接着を調節する際のp38 MAPキナーゼの新規の役割を示唆する

■［動／前］＋ a role in

★	a role in ～	～において役割を	12,430
❶	play a role in ～	…は，～において役割を果たす	9,875
❷	have a role in ～	…は，～において役割を担う	1,480
❸	consistent with a role in ～	～における役割と一致して	224

❶ The current minimal genetic region contains multiple candidate genes for PPH, **including a locus thought to play a role in lung cancer.**（Am J Respir Crit Care Med. 2000 161:1055）

訳 そして，（それは）肺癌において役割を果たすと考えられる部位を含んでいる

❷ Soy isoflavones exhibit a number of biological effects, suggesting that **they may have a role in**

cancer prevention. (Cancer Res. 1998 58:5231)
訳 それらは，癌の予防において役割を担っているかもしれない

■［動／前］＋ the role of

★ the role of ～	～の役割	30,135
❶ investigate the role of ～	…は，～の役割を精査する	3,885
❷ examine the role of ～	…は，～の役割を調べる	2,752
❸ study the role of ～	…は，～の役割を研究する	1,346
❹ determine the role of ～	…は，～の役割を決定する	1,342
❺ understand the role of ～	…は，～の役割を理解する	996
❻ assess the role of ～	…は，～の役割を査定する	806
❼ insight into the role of ～	～の役割への洞察	471
❽ understanding of the role of ～	～の役割の理解	346
❾ focus on the role of ～	…は，～の役割に焦点を当てる	287

❶ We investigated the role of the mannose-sensitive hemagglutinin (MSHA) type IV pilus as a receptor in phage 493 infection. (Infect Immun. 1998 66:2535)
訳 われわれは，～の役割を精査した

❷ In this study, we examined the role of the N-terminal finger (Nf) of GATA-3 in Th2 cell development. (J Immunol. 2002 168:4538)
訳 われわれは，～の役割を調べた

R

■［動／前］＋ a role for

★ a role for ～	～の役割	8,341
❶ suggest a role for ～	…は，～の役割を示唆する	2,334
❷ support a role for ～	…は，～の役割を支持する	1,147
❸ consistent with a role for ～	～の役割と一致して	562
❹ demonstrate a role for ～	…は，～の役割を実証する	527
❺ indicate a role for ～	…は，～の役割を示す	510
❻ establish a role for ～	…は，～の役割を確立する	244
❼ reveal a role for ～	…は，～の役割を明らかにする	218
❽ identify a role for ～	…は，～の役割を同定する	205

❶ These results suggest a role for an additional, mannose-specific, ER lectin in targeting secretory proteins to the proteasome for destruction. (Biochemistry. 2000 39:8993)
訳 これらの結果は，付加的でマンノース特異的なERレクチンの役割を示唆する

■［形］＋ role

❶ important role	重要な役割	12,899
❷ critical role	決定的な役割	7,053
❸ key role	鍵となる役割	3,928
❹ essential role	必須の役割	3,178
❺ central role	中心的な役割	3,154
❻ major role	主な役割	2,304

❼ potential role	潜在的な役割	2,143
❽ crucial role	決定的に重要な役割	1,971
❾ functional role	機能的な役割	1,829

❶ These results support an important role for the E protein in determining YF virus viscerotropism.（J Virol. 2003 77:1462）

訳 これらの結果は，～におけるEタンパク質の重要な役割を支持する

❼ Because glial tumors are highly invasive and in view of the role of TAX-1 in neurite outgrowth, we investigated the potential role of TAX-1 in glioma cell migration.（Cancer Res. 2001 61:2162）

訳 われわれは，グリオーマ細胞遊走におけるTAX-1の潜在的な役割を精査した

route ㊔ 経路／ルート　用例数 5,155

route	経路／ルート／経路を決める	3,571
routes	経路／ルート／経路を決める	1,413
routing	経路を決める	105
routed	経路を決めた	66

名詞の用例がほとんどだが，動詞としても用いられる．複数形のroutesの割合が，約30%ある可算名詞．

◆類義語：pathway

■ route ＋［前］

	the%	a/an%	ø%			
❶	6	74	0	route to ～	～への経路	1,070
❷	56	32	7	route of ～	～の経路	945
❸	13	83	0	route for ～	～のための経路	373

❶ Finally, a route to optically active material is provided.（J Org Chem. 2005 70:10619）

訳 光学活性物質への経路が提供される

❸ The ubiquitin/26S proteasome pathway is a major route for selectively degrading cytoplasmic and nuclear proteins in eukaryotes.（Plant J. 2001 27:393）

訳 ユビキチン/26Sプロテオソーム経路は，細胞タンパク質および核タンパク質を選択的に分解するための主要な経路である

S

⁑ same 形 同じ 用例数 37,311

〈the same ～〉の用例が圧倒的に多い．

◆類義語：identical, equal, equivalent, comparable

■ the same ＋［名］

★ the same ～	同じ～	35,030
❶ the same ～ as …	…と同じ～	3,505

❶ KA2 Thr-675 mutant subunits were able to co-assemble with GluR5 and GluR6 subunits and were degraded at the same rate as wild-type KA2 subunit protein.（J Biol Chem. 2005 280:6085）

訳 ～は，野生型KA2サブユニットタンパク質と同じ速度で分解された

■ the same ＋［前］

❶ the same as ～	～と同じ	779
・be the same as ～	…は，～と同じである	506
❷ the same in ～	～において同じ	408

❶ This alignment is the same as that observed between collagen fibrils and hydroxyapatite crystals in bone.（Radiology. 2004 233:129）

訳 このアライメントは，コラーゲン原繊維とヒドロキシアパタイト結晶との間に観察されるそれと同じである

❷ This footprint is the same in the presence and absence of RNA.（J Mol Biol. 2004 336:1035）

訳 このフットプリントは，RNAの存在および非存在下において同じである

■［副］＋ the same

❶ essentially the same ～	本質的に同じ～	242
❷ approximately the same ～	おおよそ同じ～	239

❶ Both types of SIRT1 mutant mice and cells had essentially the same phenotypes.（Proc Natl Acad Sci USA. 2003 100:10794）

訳 ～は，本質的に同じ表現型を持っていた

■ the same ＋［名］

❶ at the same time	同時に	1,269
❷ the same conditions	同じ条件	444
❸ to the same extent	同じ程度で	410
❹ the same region	同じ領域	406
❺ the same gene	同じ遺伝子	314
❻ the same cells	同じ細胞	313
❼ the same site	同じ部位	298
❽ the same rate	同じ速度	254

⑨ the same pathway	同じ経路	237

❶ At the same time, the long-time current response at the interfaces surrounded by a thin glass wall of the pipets is enhanced by diffusion of the species from behind the pipet tip. (Anal Chem. 2004 76:5570)

訳 同時に,

❸ Total cellular proteins synthesis was inhibited to the same extent as BBM protein synthesis. (J Am Soc Nephrol. 2001 12:114)

訳 全体の細胞タンパク質合成は, BBMタンパク質合成と同じ程度抑制された

★★ sample 名 サンプル／標本／試料／検体, 動 試料採取する／標本抽出する

用例数 38,267

samples	サンプル／標本／～を試料採取する	23,313
sample	サンプル／標本／～を試料採取する	13,549
sampled	試料採取される／～を試料採取した／～を標本抽出した	1,405

動詞としても用いられるが, 名詞の用例が非常に多い. 複数形のsamplesの割合は, 約65%と圧倒的に高い.

◆類義語：specimen, preparation

■ 名 sample(s) ＋［前］

sample ofの前には, a/anが付く割合が非常に高い.

	the%	a/an%	ø%			
❶	1	-	90	samples from ～	～からのサンプル	3,115
				・samples from patients	患者からのサンプル	342
❷	4	88	2	sample of ～	～のサンプル	2,339
				・random sample of ～	～の無作為標本	276
				・representative sample of ～	～の代表的なサンプル	218
❸	4	-	71	samples with ～	～を持つサンプル	614

❶ Here, we examine samples from patients with IM by use of a new Epstein-Barr nuclear antigen 2 HTA alongside the established latent membrane protein 1 HTA. (J Infect Dis. 2006 193:287)

訳 われわれは, IMの患者からのサンプルを調べる

❷ The HAQ DI (range of scores 0-3) was measured in a random sample of 1,530 adults in the Central Finland District. (Arthritis Rheum. 2004 50:953)

訳 ～が, 1,530名の成人の無作為標本において測定された

■ 名 samples ＋［動］

❶ samples were collected	サンプルが, 集められた	442
❷ samples were obtained	サンプルが, 得られた	402
❸ samples were analyzed	サンプルが, 分析された	193
❹ samples were taken	サンプルが, 採取された	136
❺ samples were tested	サンプルが, テストされた	113

❶ Blood samples were collected at stipulated time intervals to evaluate tracer clearance and metabolism.（J Nucl Med. 2005 46:1916）

訳 血液サンプルが，集められた

❷ Muscle samples were obtained from vastus lateralis, cultured, and differentiated into myotubes.（J Clin Invest. 2005 115:1934）

訳 筋肉サンプルが，外側広筋から得られた

■ 名 samples ＋［過分］

❶ samples obtained	得られたサンプル	416
❷ samples collected	集められたサンプル	385
❸ samples tested	テストされたサンプル	137
❹ samples taken	採取されたサンプル	130

❷ Antibody development was investigated with 154 samples collected from 84 donors 1 to 21 days after their RNA-positive, antibody-negative, index donation.（J Clin Microbiol. 2005 43:4316）

訳 ～が，84名のドナーから集められた154のサンプルで精査された

■ 名［名］＋ samples

❶ blood samples	血液サンプル	1,266
❷ serum samples	血清サンプル	1,066
❸ tissue samples	組織サンプル	713
❹ plasma samples	血漿サンプル	581
❺ DNA samples	DNAサンプル	500
❻ tumor samples	腫瘍サンプル	460
❼ clinical samples	臨床サンプル	399
❽ biopsy samples	生検サンプル	389
❾ urine samples	尿サンプル	388

❷ We also examined serum samples from 31 HIV-positive individuals that contained Tat binding antibodies; 23 of the 31 sera recognized the amino terminus peptide.（J Virol. 2004 78:13190）

訳 われわれは，また，31名のHIV陽性の個々人からの血清サンプルを調べた

■ 動 sampled ＋［前］

❶ sampled from ～	～から標本抽出される／～から試料採取される	190

❶ We have sequenced the MC1R gene in 121 individuals sampled from world populations with an emphasis on Asian populations.（Genetics. 1999 151:1547）

訳 世界中の人種から標本抽出された121の個々人のMC1R遺伝子

★ scale ㊔ スケール／尺度／規模, ㊐ 一定の基準で決める　用例数 17,434

scale	スケール／規模	14,343
scales	スケール／規模	2,659
scaled	一定の基準で決められる	432

動詞としても使われるが，名詞の用例が圧倒的に多い．複数形のscalesの割合が，約15%ある可算名詞．

◆類義語：magnitude, size, amplitude

■ scale ＋［前］

scale ofの前にtheが付く割合は非常に高い．

	the%	a/an%	ø%			
❶	72	26	0	scale of ～	～のスケール	417

❶ Bead regeneration, column repacking, and repetitive measurements **are achieved on the time scale of several minutes.** (Anal Biochem. 2004 331:161)

訳 ～は，数分の時間スケールで達成される

■［形／名］＋ scale

❶	large-scale ～	大規模な～	3,849
❷	time scale	時間スケール	1,412
❸	genome-scale ～	ゲノムスケールの～	447
❹	small-scale ～	小規模な～	261

❶ **Large-scale analysis of the ligand binding specificity of the AhR requires** the use of a high-throughput AhR bioassay system for chemical screening. (Biochemistry. 2002 41:861)

訳 AhRのリガンド結合特異性の大規模な解析は，～を必要とする

★ screen ㊔ 選別／スクリーン, ㊐ 選別する／スクリーニングする　用例数 13,319

screen	選別／スクリーン／選別する	7,592
screened	選別した／選別される／スクリーニングされる	4,091
screens	選別／スクリーン／選別する	1,636

名詞の用例が多いが，動詞としても用いられる．複数形のscreensの割合が，約20%ある可算名詞．動詞としては他動詞の用例が多いが，自動詞としても用いられる．screeningは別項参照．

◆類義語：sort, filter, examine

■ 名 screen ＋［前］

screen ofの前には，a/anが付く割合が非常に高い．

	the%	a/an%	ø%			
❶	2	92	3	screen for ～	～の選別	1,375
				genetic screen for ～	～の遺伝的選別	257
❷	2	95	2	screen of ～	～の選別	552

❶ To identify more negative regulators of MEN, **we carried out a genetic screen for genes that are toxic to cdc5-1 mutants** when overexpressed.（Mol Biol Cell. 2006 17:80）

訳 われわれは，cdc5-1変異体に毒性のある遺伝子の遺伝的な選別を行った

❷ **A systematic screen of a whole-genome microRNA library revealed that** the let-7 and miR-18 families increase mesoderm at the expense of endoderm in mouse embryonic stem cells.（Genes Dev. 2012 26:2567）

訳 全ゲノム・マイクロRNAライブラリの系統的な選別は，～ということを明らかにした

■ 動 be screened ＋［前］

★	be screened	…が，スクリーニングされる	1,625
❶	be screened for ～	…が，～に関してスクリーニングされる	748
❷	be screened by ～	…が，～によってスクリーニングされる	143
❸	be screened with ～	…が，～を使ってスクリーニングされる	108

❶ **Cocultured tumor cells were screened for the expression of 22 genes** associated with inflammation and invasion that also contained an AP-1 and NF-κB binding site.（J Immunol. 2005 175:1197）

訳 共培養された腫瘍細胞が，22の遺伝子の発現に関してスクリーニングされた

■ 動 to screen

★	to screen	スクリーニングする…	1,312
❶	to screen for ～	～をスクリーニングする…	433
❷	be used to screen	…が，スクリーニングするために使われる	299

❶ Thus, **our assay can be used effectively to screen for genes regulating** Wingless distribution or transport.（Genetics. 2005 170:749）

訳 われわれのアッセイは，～を調節する遺伝子をスクリーニングするために効果的に使われうる

❷ **An M13 bacteriophage library was used to screen 10^9 different 12-mer peptide inserts** against PPyCl.（Nat Mater. 2005 4:496）

訳 M13バクテリオファージライブラリーが，10^9の異なる12アミノ酸長ペプチドの挿入をスクリーニングするために使われた

■ 動［代名］＋ screen

❶	we screened	われわれは，スクリーニングした	1,078
	・we screened for ～	われわれは，～を求めてスクリーニングした	204

❶ To test these hypotheses, **we screened patients with PHS and GCPS for GLI3 mutations.**

(Am J Hum Genet. 2005 76:609)

訳 われわれは，GLI3変異に関してPHSおよびGCPSの患者をスクリーニングした

❶ To identify additional proteins that function with syntaxin to control neurotransmitter release and VA action, **we screened for suppressors of the phenotypes** produced by unc-64 reduction of function. (Genetics. 2004 168:831)

訳 われわれは，表現型の抑制因子を求めてスクリーニングした

★ screening 名 スクリーニング／選別／検診, -ing スクリーニングする　用例数 13,340

screening	スクリーニング／選別／検診	13,257
screenings	スクリーニング／選別／検診	83

名詞として用いられることが非常に多い．複数形のscreeningsの割合は約１％しかなく，**不可算名詞**として使われることが多い．

◆類義語：screen, examination

■ screening ＋［前］

	the%	a/an%	ø%			
❶	13	9	75	screening of ～	～のスクリーニング	1,741
❷	2	0	95	screening for ～	～のスクリーニング	1,154

❶ **Screening of colonies by fluorescence microscopy revealed numerous mutants** that exhibited interesting patterns of porin expression. (J Bacteriol. 2005 187:5723)

訳 蛍光顕微鏡によるコロニーのスクリーニングは，たくさんの変異体を明らかにした

❷ **This assay can be adapted for high-throughput screening for potential prenyltransferase substrates and inhibitors.** (Anal Biochem. 2005 345:302)

訳 このアッセイは，潜在的なプレニルトランスフェラーゼ基質と阻害剤のハイスループットスクリーニングに適応しうる

■［名］＋ screening

❶	high-throughput screening	ハイスループットスクリーニング	752
❷	cancer screening	癌検診	436

search 名 探索／検索，動 探索する／検索する　用例数 9,880

search	探索／探索する／検索する	6,221
searches	探索／探索する／検索する	1,455
searching	探索する／検索する	1,139
searched	探索した／検索した／探索される	1,065

名詞および動詞として用いられる．複数形のsearchesの割合が，約20%ある可算名詞．動詞としては自動詞の用例が多いが，他動詞としても用いられる．search forの用例が多い．

◆類義語：investigation exploration, survey, research, explore, investigate, seek

■ 名 search ＋［前］

	the%	a/an%	ø%			
❶	52	37	6	search for ～	～の探索	2,221
❷	6	48	46	search of ～	～の探索	612

❶ Progress in the search for effective therapeutic strategies that can halt this degenerative process remains limited. (FASEB J. 2005 19:489)

訳 この変性過程を停止させうる効果的な治療方針の探索における進展は，限られたままである

■ 動［代名］＋ search

❶ we searched	われわれは，探索した	447
・we searched for ～	われわれは，～を探索した	247

❶ In this systematic review, we searched Medline (1951-2011), Embase (1974-2011), Cochrane Library (2011), and Scisearch (1974-2011) for relevant citations with no language restriction. (Lancet. 2012 379:2459)

訳 われわれは，Medline (1951-2011)，Embase (1974-2011)，Cochrane Library (2011)，およびScisearch (1974-2011) を検索した

❶ Using a mouse model for Turner syndrome, we searched for locus-specific imprinting of X-linked genes in developing brain. (Nat Genet. 2005 37:620)

訳 われわれは，X連鎖遺伝子の座位特異的なインプリンティングを探索した

■ 動 to search

★ to search	探索する…	794
❶ to search for ～	～を探索する…	569
❷ used to search	探索するために使われる	110

❶ This method can be used to search for new biomarkers of organophosphorus agent exposure. (Anal Biochem. 2005 345:122)

訳 この方法は，～の新しい生物マーカーを探索するために使われうる

■ 動 be searched

★ be searched	…は，探索される	351
❶ be searched for ～	…は，～に関して探索される	147

★ secondary 形 二次の／二次的な／続発する　用例数 18,950

◆類義語：second

■ secondary ＋［前］

❶ secondary to ～	～に続発する	1,527
・be secondary to ～	…は，～に続発する	393

S

❶ Thus, loss of function is secondary to a reduction in the in vivo abundance of the expanded protein likely due to degradation. (Hum Mol Genet. 2004 13:2841)
訳 機能の喪失は，～の低下に続発する

■ secondary ＋［名］

❶	secondary structure	二次構造	3,555
❷	secondary end points	二次エンドポイント	381
❸	secondary outcomes	二次結果	327
❹	secondary prevention	二次予防	294

❶ Predictions of secondary structure suggest that these 55 amino acids form a basic, amphipathic helical hairpin. (Biochemistry. 1998 37:10211)
訳 二次構造の予測は，～ということを示唆する

★secrete 動 分泌する　用例数 13,772

secreted	分泌される／～を分泌した	9,538
secrete	～を分泌する	2,105
secreting	～を分泌する	1,807
secretes	～を分泌する	322

他動詞.
◆類義語：discharge, release

■ secreted ＋［前］

❶	secreted by ～	～によって分泌される	1,349
	・be secreted by ～	…は，～によって分泌される	350
❷	secreted from ～	～から分泌される	452
❸	secreted into ～	～の中に分泌される	353

❶ Proteins secreted by Neisseria meningitidis are thought to play important roles in the pathogenesis of meningococcal disease. (Infect Immun. 2005 73:5554)
訳 髄膜炎菌によって分泌されるタンパク質は，～の病因において重要な役割を果たすと考えられる

❷ At Drosophila synapses, Wingless is secreted from presynaptic terminals and is required for synaptic growth and differentiation. (Science. 2005 310:1344)
訳 Winglessは，シナプス前終末から分泌される

❸ In this system, recombinant E3-19K is secreted into the culture medium. (J Virol. 2005 79:13317)
訳 組換え体E3-19Kが，培養液の中に分泌される

■ secreted ＋［名］

❶	secreted protein	分泌タンパク質	599
❷	secreted factors	分泌因子	134

★ secretion 名 分泌　　用例数 20,880

secretion	分泌	20,109
secretions	分泌	771

複数形のsecretionsの割合は４%あるが，単数形は無冠詞で使われることが圧倒的に多い．

◆類義語：discharge, release

■ secretion ＋［前］

	the%	a/an%	ø%			
❶	35	0	62	secretion of ～	～の分泌	4,205
❷	0	0	72	secretion in ～	～における分泌	1,354
❸	0	0	70	secretion by ～	～による分泌	849
❹	0	0	70	secretion from ～	～からの分泌	615

❶ To define the function of IA-2, **we studied the secretion of insulin in** a single cell type, MIN-6, by overexpressing and knocking down IA-2.（Proc Natl Acad Sci USA. 2005 102:8704）

訳 われわれは，～におけるインスリンの分泌を研究した

❷ As determined by ELISA, **endothelin-1（ET-1）induces CXCL1 and CXCL8 secretion in three human melanoma cell lines** in a concentration-dependent fashion.（J Invest Dermatol. 2005 125:307）

訳 エンドセリン-1（ET-1）は，３つのヒトメラノーマ細胞株におけるCXCL1およびCXCL8分泌を誘導する

❸ **The β-cell ATP-sensitive potassium（KATP）channel controls insulin secretion by** linking glucose metabolism to membrane excitability.（J Biol Chem. 2006 281:3006）

訳 β細胞ATP感受性カリウム（KATP）チャネルは，～によるインスリン分泌を制御する

❹ **IFN-γ- and IL-17-stimulated cytokine secretion from mouse peritoneal macrophages was inhibited by** CD200R engagement.（J Immunol. 2006 176:191）

訳 IFN-γおよびIL-17に刺激されるマウスの腹膜マクロファージからのサイトカイン分泌が，～によって抑制された

■［名］＋ secretion

❶	insulin secretion	インスリン分泌	1,937
❷	cytokine secretion	サイトカイン分泌	602
❸	protein secretion	タンパク質分泌	598
❹	fluid secretion	液分泌	277

section 名 切片／セクション　　用例数 6,133

sections	切片／セクション	4,244
section	切片／セクション	1,889

複数形のsectionsの割合は，約70%と圧倒的に高い．

◆類義語：slice

S

■ sections ＋［前］

	the%	a/an%	ø%			
❶	5	-	86	sections of ～	～の切片	870
❷	2	-	94	sections from ～	～からの切片	414

❶ We used tissue sections of surgical lung biopsies from patients with IPF to localize expression of PTEN and α-smooth muscle actin（α-SMA）.（Am J Respir Crit Care Med. 2006 173:112）

訳 われわれは，～の患者からの外科的肺生検の組織切片を使用した

❷ Histopathologic lung sections from 5 patients with Eisenmenger syndrome and from 3 patients with acyanotic PAH were reviewed.（Circulation. 2005 112:2778）

訳 アイゼンメンゲル症候群の５名の患者からの病理組織肺切片

■［名］＋ sections

❶	tissue sections	組織切片	547
❷	cross sections	横断切片	308
❸	brain sections	脳切片	224

★ see 動 見る　用例数 21,444

seen	見られる	19,151
see	～を見る	1,761
saw	～を見た	331
seeing	～を見る	178
sees	～を見る	23

他動詞．seen inの用例が多い．

◆類義語：observe, discover, identify, find, look

■ be seen ＋［前］

★	be seen	…は，見られる	8,055
❶	be seen in ～	…は，～において見られる	3,409
❷	be seen with ～	…は，～に見られる	582
❸	be seen for ～	…は，～に対して見られる	385
❹	be seen at ～	…は，～において見られる	338

❶ No effect was seen in patients with lymphoid disease.（Blood. 2004 103:1521）

訳 リンパ性疾患の患者において効果は見られなかった

❷ Greater acute toxicity was seen with AP.（J Clin Oncol. 2006 24:36）

訳 より大きな急性毒性が，APに見られた

as seen

★ as seen	見られるように	594
❶ as seen in ～	～において見られるように	310

❶ These results could be accounted for by kinetic simulation based on a model in which halothane causes flickering block of open channels, **as seen in muscle nAChRs.** (Mol Pharmacol. 2001 59:732)

訳 筋肉のnAChRにおいて見られるように

[名] ＋ seen

❶ changes seen	見られた変化	138
❷ levels seen	見られたレベル	137

❷ Mirk was stably overexpressed in two colon carcinoma cell lines to **attain levels seen in colon cancers.** (Cancer Res. 2000 60:3631)

訳 大腸癌において見られたレベルを達成する

seek 動 努める／探求する　用例数 6,722

sought	～しようと努めた／～を探求した	5,907
seek	～しようと努める／～を探求する	609
seeks	～しようと努める／～を探求する	206

他動詞．過去形のsought to *do*の用例が圧倒的に多い．受動態で用いられることはあまりない．

◆類義語：try, aim, attempt, address, search, pursue

seek to *do*

★ seek to *do*	…は，～しようと努める	5,491
❶ seek to determine ～	…は，～を決定しようと努める	1,881
❷ seek to identify ～	…は，～を同定しようと努める	549
❸ seek to evaluate ～	…は，～を評価しようと努める	259
❹ seek to characterize ～	…は，～を特徴づけようと努める	179
❺ seek to examine ～	…は，～を調べようと努める	178
❻ seek to investigate ～	…は，～を精査しようと努める	173
❼ seek to assess ～	…は，～を評価しようと努める	167
❽ seek to define ～	…は，～を定義しようと努める	148
❾ seek to test ～	…は，～をテストしようと努める	130

❶ **We sought to determine whether** CDX2 contributes to tumorigenic potential in established gastric cancer. (Oncogene. 2006 25:2048)

訳 われわれは，～かどうかを決定しようと努めた

❸ **This study sought to evaluate the clinical results of** a percutaneous approach to mitral valve repair for mitral regurgitation (MR). (J Am Coll Cardiol. 2005 46:2134)

訳 この研究は，～の臨床結果を評価しようと努めた

■［代名／名］＋ seek to *do*

❶ we sought to *do*	われわれは，～しようと努めた	3,612
❷ study sought to *do*	研究は，～しようと努めた	682
❸ the authors sought to *do*	著者らは，～しようと努めた	176

❶ To overcome this problem, **we sought to identify** constrained peptidomimetic inhibitors that would provide potential new drug leads. (Biochemistry. 2005 44:12362)

訳 われわれは，～を同定しようと努めた

seem 動 思われる　用例数 6,837

seems	思われる	3,560
seem	思われる	2,440
seemed	思われた	796
seeming	思われる	41

自動詞．seem to *do*の用例が非常に多い．第２文型（S+V+C）の自動詞としても用いられる．

◆類義語：appear, think, believe, consider, regard, suspect, know, assume

■ seem to *do*

★ seem to *do*	…は，～するように思われる	5,073
❶ seem to be ～	…は，～であるように思われる	2,501
❷ seem to have ～	…は，～を持つように思われる	225
❸ seem to play ～	…は，～を果たすように思われる	147

❶ In addition, **post-transcriptional regulation seems to be another important mechanism controlling** PNMT expression. (Mol Pharmacol. 2003 64:1180)

訳 転写後調節は，～を制御するもう１つの重要な機構であるように思われる

■ seem ＋［that節］

★ seem that ～	～であるように思われる	201
❶ it seems that ～	～であるように思われる	158
・it seems likely that ～	おそらく，～であるように思われる	130

❶ Thus, **it seems that** the dimeric structure is a major determinant of nNOS stability and proteolysis. (Mol Pharmacol. 2004 66:964)

訳 ～であるように思われる

★ select 動 選択する，形 選ばれた　用例数 17,303

selected	選択される／選択した	13,025
select	選択する／選ばれた	2,781

selecting	選択する	1,128
selects	選択する	369

他動詞の用例が多いが，自動詞としても用いられる．形容詞の用例もある．
◆類義語：choose, sort, screen, prefer

■ be selected ＋［前］

★	be selected	…が，選択される	3,043
❶	be selected for ～	…が，～のために選択される	798
❷	be selected from ～	…は，～から選択される	351
❸	be selected to *do*	…は，～するために選択される	237
❹	be selected by ～	…は，～によって選択される	190
❺	be selected as ～	…は，～として選択される	190
❻	be selected in ～	…は，～において選択される	124

❶ On the basis of its excellent in vivo efficacy and pharmacokinetic profile, **compound 31 was selected for further evaluation** as a clinical candidate and was designated AMG 511.（J Med Chem. 2012 55:7796）
訳 化合物31が，さらなる評価のために選択された

❷ **Controls were randomly selected from women with** no laparoscopic evidence of or history of fibroids（n = 1,268）.（Am J Epidemiol. 2001 153:20）
訳 対照群は，～の女性達から無作為に選択された

■［副］＋ selected

❶	randomly selected	無作為に選択される	777
❷	positively selected	正に選択される	199

■ to select

★	to select ～	～を選択する…	992
❶	be used to select ～	…は，～を選択するために使われる	150

❶ **The same library was used to select shRNAs** that inhibit breast carcinoma cell growth by targeting potential oncogenes.（Proc Natl Acad Sci USA. 2010 107:7377）
訳 同じライブラリーが，shRNAを選択するために使われた

■ select for

★	select for ～	…は，～を選択する	471
❶	to select for ～	～を選択する…	122

❶ In addition, **new screening methods have been designed to select for specific functional genes** within metagenomic libraries.（Curr Opin Biotechnol. 2006 17:236）
訳 新しいスクリーニング法が，特異的な機能遺伝子を選択するために設計されてきた

S

★ selection ㊔ 選択／淘汰 用例数 19,349

selection	選択	19,141
selections	選択	208

複数形のselectionsの割合は1%しかなく，原則，不可算名詞として使われる．

◆類義語：choice, option, screening, sorting, filtering

■ selection ＋［前］

	the%	a/an%	ø%			
❶	44	6	49	selection of ～	～の選択	3,082
❷	3	4	90	selection for ～	～に対する選択	1,132

❶ Therefore, these key residues may have strongly contributed to the selection of these important functions over plant evolution. (Proc Natl Acad Sci USA. 2005 102:7748)

訳 これらの鍵となる残基は，これらの重要な機能の選択に強く寄与したかもしれない

❷ Transport defective strains were generated by selection for resistance to the lethal thiamine analog, pyrithiamine. (Gene. 1997 199:111)

訳 輸送欠陥のある株が，～への抵抗性に対する選択によって作製された

■［形］＋ selection

❶ positive selection	正の選択	1,547
❷ natural selection	自然淘汰	983
❸ negative selection	負の選択	825
❹ site selection	部位の選択	459

❶ Thus, the same MHC molecule that mediates positive selection of 6C5 T cells is also required for HP. (J Immunol. 2004 173:6065)

訳 6C5 T細胞の正の選択を仲介する同じMHC分子は，また，HPのために必要とされる

★ selective ㊢ 選択的な／選択性の 用例数 25,731

◆類義語：elective

■ selective ＋［前］

❶ selective for ～	～に対して選択的な	1,066
・be selective for ～	…は，～に対して選択的である	410

❶ These results suggest that stargazin action is highly selective for AMPA receptors. (Mol Pharmacol. 2003 64:703)

訳 ～は，AMPA受容体に対して高度に選択的である

■［副］＋ selective

❶ highly selective	高度に選択的な	871
❷ more selective	より選択的な	284

■ selective ＋［名］

❶ selective inhibitor	選択的阻害剤	656
❷ selective inhibition	選択的阻害	451
❸ selective pressure	選択的圧力	441
❹ selective advantage	選択優位性	316
❺ selective antagonist	選択的競合剤	302

❶ The latter results were replicated with a newly introduced, **highly selective inhibitor of SK2 channels.** (J Neurosci. 2004 24:5151)

訳 SK2チャネルの高度に選択的な阻害剤

■［形］＋ and selective

❶ potent and selective	強力で選択的な	709

❶ These compounds present a novel platform for **the development of potent and selective PTP1B inhibitors.** (Biochemistry. 2003 42:11451)

訳 強力で選択的なPTP1B阻害剤の開発

★ selectively 副 選択的に　　用例数 11,101

■ selectively ＋［動］

❶ selectively inhibit ～	…は，～を選択的に阻害にする	455
❷ selectively bind	…は，選択的に結合する	314
・selectively bind to ～	…は，～に選択的に結合する	109
❸ selectively target ～	…は，～を選択的に標的にする	270
❹ selectively activate ～	…は，～を選択的に活性化する	215
❺ selectively block ～	…は，～を選択的にブロックする	166
❻ selectively induce ～	…は，～を選択的に誘導する	141

❶ In-gel kinase assays showed that **SB 203580 selectively inhibited a small group of protein kinases** in the photoreceptor cells. (J Biol Chem. 2004 279:22738)

訳 SB 203580は，小集団のタンパク質リン酸化酵素を選択的に抑制した

■［動］＋ selectively ＋［前］

❶ bind selectively to ～	…は，～に選択的に結合する	125

❶ **PSRP1 binds selectively to** 25-nucleotide single-stranded RNA species. (Plant Cell. 2004 16:1979)

訳 PSRP1は，～に選択的に結合する

■ be selectively ＋［過分］

★ be selectively ～	…は，選択的に～される	2,260

❶	be selectively expressed	…は，選択的に発現される	287
❷	be selectively activated	…は，選択的に活性化される	83

❶ PMCA4 was found to be selectively expressed in both synaptic layers.（J Comp Neurol. 2002 451:1）

訳 PMCA4は，両方のシナプス層において選択的に発現されることが見つけられた

selectivity ㊔ 選択性　　用例数 9,433

selectivity	選択性	9,021
selectivities	選択性	412

複数形のselectivitiesの割合は４％あるが，単数形は無冠詞で使われることが多い．

◆類義語：selection

■ selectivity ＋［前］

selectivity ofの前にtheが付く割合は圧倒的に高い．

	the%	a/an%	ø%			
❶	84	3	11	selectivity of ～	～の選択性	1,510
❷	9	4	75	selectivity for ～	～に対する選択性	1,273
❸	14	4	73	selectivity in ～	～における選択性	469

❶ This suggests that human cyclopean acuity for disparity modulations is limited by the selectivity of V1 neurons.（J Neurosci. 2004 24:2065）

訳 ～は，V1ニューロンの選択性によって制限される

❷ Cross-reactivity tests showed high selectivity for heat-shocked C. parvum.（Anal Chem. 2003 75:3890）

訳 交差反応性テストが，～に対する高い選択性を示した

■［形／名］＋ selectivity

❶	high selectivity	高い選択性	337
❷	～-fold selectivity	～倍の選択性	231
❸	substrate selectivity	基質選択性	160

■［名］＋ and selectivity

❶	potency and selectivity	能力と選択性	186
❷	affinity and selectivity	親和性と選択性	175

❷ Endomorphins are opioid tetrapeptides that have high affinity and selectivity for mu-opioid receptors（μORs）.（J Comp Neurol. 2002 448:268）

訳 エンドモルフィンは，μオピオイド受容体（μOR）に対する高い親和性と選択性を持つオピオイドテトラペプチドである

★ sensitive 形 感受性の／鋭敏な 用例数 30,177

sensitive toの用例が多い.

◈類義語：susceptible

■ sensitive ＋［前］

❶ sensitive to ～	～に対して感受性である	8,849
・be sensitive to ～	…は，～に対して感受性である	3,302
・sensitive to changes in ～	～の変化に対して感受性である	164
・sensitive to inhibition by ～	～による抑制に対して感受性である	148

❶ However, only APC/C isolated from mitotic cells was sensitive to inhibition by MCC.（J Cell Biol. 2001 154:925）

訳 有糸分裂細胞から単離されたAPC/Cのみが，MCCによる抑制に感受性であった

■ ［副］＋ sensitive

❶ more sensitive	より感受性である	2,419
・more sensitive to ～	～に対してより感受性である	1,342
・more sensitive than ～	～よりも感受性である	550
・more sensitive than ～ to …	…に対して～よりも感受性である	107
❷ highly sensitive	高度に感受性である	1,379
・highly sensitive to ～	～に対して高度に感受性である	561
❸ less sensitive	より感受性でない	682
❹ most sensitive	最も感受性である	453
❺ very sensitive	とても感受性である	327

❶ These cells are 15-20-fold more sensitive to UV radiation than cells with wild-type pRb.（Oncogene. 2002 21:4481）

訳 これらの細胞は，野生型pRbを持つ細胞より紫外線照射に15-20倍感受性である

❶ ADPR-cyclase in VSMC membranes was more sensitive than CD38 HL-60 ADPR-cyclase to inactivation by N-endoglycosidase F and to thermal inactivation at 45 degrees C.（Circ Res. 2000 86:1153）

訳 ～は，N-エンドグリコシダーゼFによる不活性化に対してCD38 HL-60 ADPRシクラーゼより感受性であった

❷ These studies indicate that after an overnight fast, basal HGP（glycogenolysis）is highly sensitive to the hepatic sinusoidal insulin level.（Diabetes. 1998 47:523）

訳 ～は，肝類洞インスリンレベルに対して高度に感受性である

■ temperature-sensitive ＋［名］

★ temperature-sensitive ～	温度感受性の～	1,811
❶ temperature-sensitive mutant	温度感受性変異体	170

S

★ sensitivity 名 感受性　　用例数 29,857

sensitivity	感受性	28,546
sensitivities	感受性	1,311

複数形のsensitivitiesの割合は４％あるが，単数形は無可算名詞で使われることが多い．
◆類義語：susceptibility

■ sensitivity ＋［前］

selectivity ofの前にtheが付く割合はかなり高い．

	the％	a/an％	ø％			
❶	8	2	77	sensitivity to ～	～に対する感受性	6,201
				・increased sensitivity to ～	～に対する増大した感受性	744
				・enhanced sensitivity to ～	～に対する増強された感受性	223
				・reduced sensitivity to ～	～に対する低下した感受性	155
❷	61	17	18	sensitivity of ～	～の感受性	6,130
				・sensitivity of ～ to …	～の…に対する感受性	1,663
				・have a sensitivity of ～	…は，～の感受性を持っている	190
❸	1	0	92	sensitivity in ～	～における感受性	890
❹	22	5	61	sensitivity for ～	～に対する感受性	614

❶ Consistent with this loss of SOD1 activity, $CCS^{-/-}$ mice **showed** increased sensitivity to **paraquat** and reduced female fertility, phenotypes that are characteristic of SOD1-deficient mice. (Proc Natl Acad Sci USA. 2000 97:2886)
訳 ～は，パラコートに対する増大した感受性を示した

❷ We conclude that the lack of β_1 integrins **decreases** the sensitivity of **ES cells** to **soluble factors** that induce keratinocyte differentiation. (Dev Biol. 2001 231:321)
訳 ～は，ES細胞の可溶性因子に対する感受性を低下させる

❷ **MR imaging** had a sensitivity of **86%** (six of seven), specificity of 100% (14 of 14), positive predictive value of 100% (six of six), and negative predictive value of 93% (14 of 15) for diagnosis of pilonidal sinus disease. (Radiology. 2003 226:662)
訳 磁気共鳴画像法は，86%の感受性を持っていた

❸ Almond-enriched diets **do not alter insulin** sensitivity in **healthy adults** or glycemia in patients with diabetes. (Am J Clin Nutr. 2002 76:1000)
訳 ～は，健康な成人におけるインスリン感受性を変化させない

■［前］＋ sensitivity

❶ increase in sensitivity	感受性の増大	187

❶ The RII transfectants (MIA PaCa-2/RII) **showed a significant** increase in sensitivity **to radiation** when compared with MIA PaCa-2/vector cells. (J Biol Chem. 2002 277:2234)
訳 ～は，放射線に対する感受性の有意な増大を示した

■ sensitivity and ＋［名］

❶ sensitivity and specificity	感受性と特異性	1,474

❶ We confirmed the sensitivity and specificity of this method by demonstration of vigorous CTL responses in a simian-HIV (SHIV)-infected rhesus macaque. (J Virol. 2001 75:73)
訳 われわれは，この方法の感受性と特異性を確認した

■［名／形］＋ sensitivity

❶ insulin sensitivity	インスリン感受性	1,619
❷ high sensitivity	高い感受性	1,158
❸ greater sensitivity	より高い感受性	256

sensitize 動 感作する／感受性を増加させる 用例数 4,089

sensitized	感作される／～を感作した	2,389
sensitizes	～を感作する	587
sensitize	～を感作する	575
sensitizing	～を感作する	538

他動詞.

■ sensitized ＋［前］

❶ sensitized to ～	～に対する感受性が増す／～に対して感作される	262
・be sensitized to ～	…は，～に対する感受性が増す／～に対して感作される	135
❷ sensitized with ～	～によって感作される	101

❶ Colon carcinoma cell lines could be further sensitized to TRAIL-induced apoptosis in vitro by the addition of the chemotherapeutic agent camptothecin. (Cancer Res. 1999 59:6153)
訳 大腸癌細胞株は，TRAILに誘導されるアポトーシスに対してさらに感受性が増しうる

❷ Multiantigen cocktails were evaluated by skin testing guinea pigs sensitized with M. bovis BCG. (Infect Immun. 1998 66:3606)
訳 ウシ型結核菌BCGによって感作されたモルモット

■ sensitized ＋［名］

❶ sensitized mice	感作されたマウス	193
❷ sensitized patients	感作された患者	130

★ separate 形 別々の／別個の／離れた，動 分ける／分離する 用例数 14,380

separate	別々の／別個の／離れた／～を分ける／～を分離する	8,096
separated	分けられる／～を分けた／分離される	5,137
separating	～を分ける／～を分離する	751

S

separates	～を分ける／～を分離する	396

形容詞および他動詞として用いられる.

◆類義語：discrete, distinct, different, divide, interrupt, segregate, resolve, dissociate, isolate

■ 形 separate from

❶ separate from ～	～から離れた／～とは別個の	334
・be separate from ～	…は，～から離れている／～とは別個である	78

❶ These sites **are separate from** the TGFβ receptor phosphorylation sites that activate Smad nuclear translocation. (Genes Dev. 1999 13:804)

訳 これらの部位は，TGFβ受容体リン酸化部位とは別個である

■ 形 separate ＋［名］

❶ separate experiments	別々の実験	185
❷ separate group	別個のグループ	119

❶ In **separate experiments**, we verified that the singlet excited state of NDI-pyr does, indeed, react intermolecularly with acetate, alanine, and glycine. (J Am Chem Soc. 2004 126:4293)

訳 別々の実験において，

■ 動 separated ＋［前］

❶ separated by ～	～によって分離される	2,114
・be separated by ～	…は，～によって分離される	755
❷ separated from ～	～から分離される	747
・separated from ～ by …	…によって～から分離される	175
❸ separated into ～	～に分けられる	280

❶ Analytes **are separated by** using gas chromatography and quantified by using chemical-ionization mass spectrometry that produces predominantly $[M+H]^+$ parent ions. (Proc Natl Acad Sci USA. 2003 100:10552)

訳 分析物は，ガスクロマトグラフィーを使うことによって分離される

❷ The autolysis site of the enzyme is **separated from** its catalytic site in vesicles **by** the lipid bilayer, resulting in a dramatic decrease of the autolysis rate. (Biochemistry. 2004 43:265)

訳 酵素の自己分解部位は，脂質二重層によって小胞内のそれの触媒部位から分離されている

separately 副 別々に／別個に　用例数 2,034

◆類義語：differently, independently, individually

■［動］＋ separately

❶ be analyzed separately	…は，別個に分析される	102

❶ Patients managed with ECMO following cardiac surgery **were analyzed separately** from patients not in the postoperative period. (Crit Care Med. 2004 32:1061)

訳 ～が，術後期間ではない患者たちとは別個に分析された

separation 名 分離　　用例数 8,000

separation	分離	7,030
separations	分離	970

複数形のseparationsの割合が，約10％ある可算名詞．

◈類義語：segregation, isolation

■ separation ＋［前］

	the%	a/an%	ø%			
❶	41	11	47	separation of ～	～の分離	2,077
❷	43	23	33	separation between ～	～の間の分離	230
❸	14	4	77	separation in ～	～における分離	207

❶ This tunnel is too small to accommodate double-stranded DNA and **requires the separation of template and nontemplate strands**.（Mol Cell. 2004 16:609）

訳 ～は，鋳型および非鋳型鎖の分離を必要とする

❷ The structure suggests that the coiled-coil CRs **act as a molecular ruler for the separation between two recognized DNA sequences**.（Proc Natl Acad Sci USA. 2005 102:3248）

訳 ～は，２つの認識されたDNA配列の間の分離のための分子定規として働く

S

⁑ sequence 名 配列／シークエンス，動 配列決定する　　用例数 120,066

sequence	配列／シークエンス／～を配列決定する	75,352
sequences	配列／シークエンス／～を配列決定する	39,685
sequenced	配列決定される／～を配列決定する	5,029

名詞の用例が多いが，他動詞としても用いられる．複数形のsequencesの割合が，約35％ある可算名詞．sequencingは別項参照．

■ 名 sequence ＋［前］

sequence ofの前にtheが付く割合は圧倒的に高い．

	the%	a/an%	ø%			
❶	83	15	0	sequence of ～	～の配列	7,380
❷	13	-	72	sequences in ～	～における配列	2,517
❸	8	-	64	sequences from ～	～からの配列	1,605

❶ **This identification was based on the sequence of the probes** and the sequence of the human genome.（Nucleic Acids Res. 2005 33:e31）

訳 この同定は，プローブの配列に基づいていた

❷ **Chromatin immunoprecipitation analysis demonstrated a direct binding of STAT3 to the putative STAT3 binding sequences in the Mcl-1 promoter**.（Hepatology. 2005 42:1329）

訳 クロマチン免疫沈降分析は，Mcl-1プロモーターにおける推定上のSTAT3結合配列へのSTAT3の直接結合を実証した

❸ Comparison of DNA sequences from different species is a fundamental method for identifying functional elements in genomes.（Nucleic Acids Res. 2004 32:W273）

訳 異なる種からのDNA配列の比較は，～を同定するための基本的な方法である

■ ㊔ sequence ＋［名］

❶ sequence analysis	配列分析	3,748
❷ sequence similarity	配列類似性	2,038
❸ sequence identity	配列同一性	1,492
❹ sequence homology	配列相同性	1,395
❺ sequence data	配列データ	1,174
❻ sequence motifs	配列モチーフ	644
❼ sequence alignment	配列アライメント	641
❽ sequence variation	配列多様性	600
❾ sequence conservation	配列保存	552

■ ㊀ sequenced ＋［前］

❶ sequenced from ～	～から配列決定される	198
❷ sequenced in ～	～において配列決定される	167
・be sequenced in ～	…は，～において配列決定される	117
❸ sequenced to *do*	～するために配列決定される	164

❶ Partial pol and env gp41 regions of the HIV genome were directly sequenced from plasma viral RNA for at least one sample from each patient.（J Virol. 2005 79:11981）

訳 ～が，血清ウイルスRNAから直接配列決定された

■ ㊀［副］＋ sequenced

❶ completely sequenced	完全に配列決定された	198
❷ fully sequenced	完全に配列決定された	146

❶ Therefore, we sequenced the soybean chloroplast genome and compared it to the other completely sequenced legumes, Lotus and Medicago.（Plant Mol Biol. 2005 59:309）

訳 われわれは，ダイズ葉緑体ゲノムを配列決定し，そしてそれを他の完全に配列決定されたマメ科植物，ミヤコグサ属およびウマゴヤシ属と比較した

■ ㊀［過分］＋ and sequenced

❶ cloned and sequenced	クローン化されそして配列決定される	601
❷ isolated and sequenced	単離されそして配列決定される	154

❶ Complete genomes of 12 isolates representing the major lineages of HPV16 were cloned and sequenced from cervicovaginal cells.（J Virol. 2005 79:7014）

訳 ～が，頸膣部の細胞からクローン化されそして配列決定された

■ 動［代名］＋ sequence

❶	we sequenced ～	われわれは，～を配列決定した	542

❶ To test this, we sequenced nAChR genes from five tick species and found that instead of the conserved arginine found in insects, a glutamine was present in all the tick sequences. (Biochemistry. 2012 51:4627)

訳 われわれは，nAChR遺伝子を配列決定した

★ sequencing 名 配列決定／シークエンシング， -ing 配列決定する　用例数 11,067

名詞の複数形の用例はなく，原則，不可算名詞として使われる．

■ sequencing ＋［前］

	the%	a/an%	ø%			
❶	10	0	87	sequencing of ～	～の配列決定	2,458

❶ Sequencing of FGFR3 in KMS-11R cells revealed the presence of a heterozygous mutation at the gatekeeper residue, encoding FGFR3(V555M); consistent with this, KMS-11R cells were cross-resistant to AZD4547 and PD173074. (Oncogene. 2013 32:3059)

訳 KMS-11R細胞におけるFGFR3の配列決定は，～の存在を明らかにした

■［名］＋ sequencing

❶	DNA sequencing	DNA配列決定	1,173
❷	genome sequencing	ゲノム配列決定	763

S

★ series 名 一連／シリーズ　用例数 16,349

単複同形．a series ofの用例が非常に多い．

◈類義語：array, set, class, group, line

■ series ＋［前］

	the%	a/an%	ø%			
❶	4	85	2	series of ～	一連の～／～のシリーズ	12,302
				・a series of ～	一連の～	9,593
				・series of experiments	一連の実験	296
				・a novel series of ～	新規の一連の～	204
				・series of compounds	一連の化合物	183
				・a new series of ～	新しい一連の～	114
				・a complex series of ～	複雑な一連の～	101

❶ In this study, we performed a series of experiments to investigate whether substance P (SP) contributes to neurogenic inflammation in the skeletal muscle tissue. (Brain Res. 2005 1047:38)

訳 われわれは，一連の実験を行った

■ [動] + a series of

❶	generate a series of ～	…は，一連の～を作製する	202
❷	construct a series of ～	…は，一連の～を作製する	133
❸	perform a series of ～	…は，一連の～を実行する	108

❶ We generated a series of L. monocytogenes strains expressing B. anthracis anthrolysin O (ALO) and PI-PLC in place of LLO and L. monocytogenes PI-PLC, respectively. (Infect Immun. 2005 73:6639)

訳 われわれは，～を発現する一連のリステリアモノサイトジェネス菌株を作製した

★ serve 動 役立つ／働く 用例数 17,166

serve	役立つ／働く	9,221
serves	役立つ／働く	4,840
served	役立った／働いた	1,861
serving	役立つ／働く	1,244

自動詞の用例が多いが，他動詞として用いられることもある．serve asの用例が非常に多い．

◆類義語：function, act, behave, work, play, help, contribute

■ serve + [前]

❶	serve as ～	…は，～として役立つ／～として働く	12,607
	・serve as a model	…は，モデルとして役立つ／モデルとして働く	347
	・serve as controls	…は，コントロールとして働く	235
	・serve as a substrate	…は，基質として働く	117
	・serve as a scaffold	…は，足場として役立つ／足場として働く	109
❷	serve to *do*	…は，～するように働く	332

❶ Thus, the ferret may serve as a model for renal disease secondary to intestinal infection with STEC. (J Infect Dis. 2002 185:550)

訳 フェレットは，腎臓疾患のモデルとして役立つかもしれない

❷ This dual function may serve to prevent erroneous γδ T cell activation by cross-reactive cell surface determinants. (J Immunol. 2002 169:1236)

訳 この二重の機能は，誤ったγδ T細胞活性化を防ぐように働くかもしれない

★ set 名 セット，動 セットする 用例数 25,439

set	セット／～をセットする／～をセットした／～をセットされる	18,227
sets	セット／～をセットする	7,212

名詞の用例が多いが，他動詞としても用いられる（原形・過去・過去分詞とも同形）．複数形のsetsの割合が，約30%ある可算名詞．settingは別項参照．

◆類義語：series, group, class, pair

■ set ＋［前］

the%	a/an%	ø%			
❶ 14	85	0	set of ～	セットの～	11,024
			・set of genes	セットの遺伝子	542
			・set of proteins	セットのタンパク質	180
			・set of experiments	セットの実験	120

❶ Through a genomic approach, **we characterized a set of genes that are** implicated in cellular adaptation to IR stress.（Genome Res. 2003 13:2092）

訳 われわれは，～である１セットの遺伝子を特徴づけた

■［形］＋ set(s)

❶ a set of ～	1セットの～	4,171
❷ two sets of ～	2セットの～	475
❸ a diverse set of ～	多様なセットの～	300
❹ different sets of ～	異なるセットの～	229
❺ a large set of ～	大きなセットの～	204
❻ the same set of ～	同じセットの～	197
❼ distinct sets of ～	別個のセットの～	165
❽ a small set of ～	小さなセットの～	161
❾ a common set of ～	共通のセットの～	157

❸ For each quantitative trait locus, **the responsible polymorphism is rare among a diverse set of 13 yeast strains**, suggestive of genetic heterogeneity in the control of yeast sporulation.（Nat Genet. 2005 37:1333）

訳 原因である多型性は，多様なセットの13の酵母菌株の間でまれである

setting 名 状態／設定，-ing セットする　用例数 7,691

setting	状態／設定	4,813
settings	状態／設定	2,878

名詞の用例が非常に多い．複数形のsettingsの割合が，約35%ある可算名詞．

■ setting ＋［前］

setting ofの前にtheが付く割合は圧倒的に高い．

the%	a/an%	ø%			
❶ 96	4	0	setting of ～	～の状態	1,428
			・in the setting of ～	～の状態で	1,271

❶ **These effects occur in the setting of autonomic blockade** and therefore are more likely to be due to the effects of digoxin on intracellular calcium than to its vagotonic effects.（Circulation. 2000 102:2503）

訳 これらの効果は，自律神経遮断の状態で起こる

S

☆ several 形 いくつかの，名 いくつか　用例数 55,986

形容詞の用例が多いが，名詞としても使われる．名詞としては，複数形扱いである．

◆類義語：a number of, some

■ 形 several ＋［名］

❶	several genes	いくつかの遺伝子	1,027
❷	several studies	いくつかの研究	1,001
❸	several lines of evidence	いくつかの証拠	600
	・several lines of evidence suggest ～	いくつかの証拠は，～を示唆する	202
	・several lines of evidence indicate ～	いくつかの証拠は，～を示す	110
❹	several proteins	いくつかのタンパク質	551
❺	several types of ～	いくつかのタイプの～	539

❶ **The expression of several genes** that reside outside of the BBL42-BBL28 operon was not affected by MNNG.（J Bacteriol. 2005 187:7985）

訳 いくつかの遺伝子の発現

❸ **Several lines of evidence suggest that** this effect is mediated by the binding of H^+ to a histidine in the first extracellular loop (His40).（Invest Ophthalmol Vis Sci. 2005 46:1393）

訳 いくつかの証拠は，～ということを示唆する

■ 形 several ＋［形］

❶	several other ～	いくつかの他の～	1,702
❷	several different ～	いくつかの異なる～	1,178
❸	several new ～	いくつかの新しい～	511
❹	several recent ～	いくつかの最近の～	453

❶ Endomorphin-1 (EM1), **in contrast to several other mu opioids**, exhibits a threshold for respiratory depression that is well above its threshold for analgesia.（Brain Res. 2005 1059:159）

訳 いくつかの他のμオピオイドとは対照的に

❷ Activation of muscarinic receptors **has been shown to be neuroprotective in several different models of apoptosis**, but the mechanism of this action is unknown.（Brain Res. 2005 1041:112）

訳 ～は，いくつかの異なるアポトーシスのモデルにおいて神経保護的であることが示されている

■ 名 several ＋［前］

	the%	a/an%	ø%			
❶	0	-	100	several of ～	～のいくつか	1,721

★ severe ㊎ 重篤な／重症の／激しい 用例数 21,756

◆類義語：serious, intense

■ [形] + severe

❶ severe disease	重篤な疾患	609
❷ severe sepsis	重篤な敗血症	481
❸ severe defects	重篤な欠損	324
❹ severe reduction	激しい低下	207

❶ Toxoplasma gondii is an important food- and waterborne opportunistic pathogen **that causes severe disease in immunocompromised patients.** (J Clin Microbiol. 2005 43:5881)

訳 免疫不全患者における重篤な疾患を引き起こす～

■ [副] + severe

❶ more severe	より重篤な	1,991
❷ less severe	より重篤でない	609
❸ most severe	最も重篤な	460

❶ **These lipid abnormalities are more severe in patients with disease complications** and in those with a greater degree of anaemia. (Curr Opin Hematol. 2006 13:40)

訳 これらの脂質の異常は，疾患の合併症を持つ患者においてより重篤である

■ [動] + severe

❶ cause severe ～	…は，重篤な～を引き起こす	635
❷ have severe ～	…は，重篤な～を持つ	504
❸ develop severe ～	…は，重篤な～を発症する	437

❸ In response to oral infection with virulent type 1 or avirulent type II strains of T. gondii, **TCR-δ$^{-/-}$ mice rapidly developed severe ileitis.** (J Immunol. 2005 175:8191)

訳 TCR-δ$^{-/-}$マウスは，重篤な回腸炎を急速に発症した

■ [前] + severe

❶ patients with severe ～	重篤な～の患者	972
❷ moderate to severe ～	中程度から重篤な～	641
❸ result in severe ～	…は，重篤な～という結果になる	395
❹ associated with severe ～	重篤な～と関連する	285

❶ **Studies of patients with severe acute respiratory syndrome (SARS) demonstrate that** the respiratory tract is a major site of SARS-coronavirus (CoV) infection and disease morbidity. (J Virol. 2005 79:14614)

訳 重症急性呼吸器症候群（SARS）の患者の研究は，～ということを実証する

S

severely 副 激しく　用例数 4,762

◆類義語：markedly, significantly, dramatically, drastically, strongly

■ severely ＋［過分］

❶ severely impaired	激しく障害される	682
・be severely impaired	…は，激しく損なわれる	380
❷ severely reduced	激しく低下した	468
❸ severely affected	激しく影響される	312
❹ severely compromised	激しく損なわれる	258

❶ During the first postoperative week, **hand function was severely impaired in all monkeys.** (J Comp Neurol. 2005 491:27)

訳 手の機能は，すべてのサルにおいて激しく損なわれた

❷ **Deletion of exon 4 was associated with severely reduced lipid phosphatase activity,** whereas exon 3 skipping resulted in markedly reduced protein phosphatase activity. (Hum Mol Genet. 2005 14:2459)

訳 エクソン4の欠失は，激しく低下した脂質ホスファターゼ活性と関連した

severity 名 重症度　用例数 9,853

severity	重症度	9,804
severities	重症度	49

複数形のseveritiesの割合は0.5%しかなく，原則，**不可算名詞**として使われる.

■ severity ＋［前］

severity ofの前にtheが付く割合はかなり高い.

	the%	a/an%	ø%			
❶	66	0	34	severity of ～	～の重症度	4,881
❷	2	3	88	severity in ～	～における重症度	303

❶ **Q_A and Q_{CV} values correlated highly with the severity of diabetic retinopathy,** but not with the duration of diabetes. (Invest Ophthalmol Vis Sci. 2013 54:9)

訳 Q_A値およびQ_{CV}値は，糖尿病性網膜症の重症度と高度に相関した

■［名］＋ severity

❶ disease severity	疾患重症度	1,126
❷ symptom severity	症状の重症度	268
❸ injury severity	傷害の重症度	221

★ share ㊐ 共有する　用例数 14,200

share	〜を共有する	5,171
shared	共有される／〜を共有した	5,093
shares	〜を共有する	2,236
sharing	〜を共有する	1,700

他動詞.

■ share ＋［名句］

❶ share a common 〜	…は，共通の〜を共有する	773
❷ share the same 〜	…は，同じ〜を共有する	367
❸ share homology	…は，ホモロジーを共有する	234

❶ One of those enzymes is 4-oxalocrotonate tautomerase, with which CaaD **seems to share a common evolutionary origin.** (Proc Natl Acad Sci USA. 2005 102:16199)

訳 〜は，共通の進化的起源を共有するように思われる

❸ **The bZIP domain of Meq shares homology with Jun/Fos**, whereas the transactivation/repressor domain is entirely different. (Proc Natl Acad Sci USA. 2005 102:14831)

訳 MeqのbZIPドメインは，Jun/Fosとホモロジーを共有する

■［名］＋ share

❶ proteins share 〜	タンパク質は，〜を共有する	191
❷ genes share 〜	遺伝子は，〜を共有する	95

❶ **Class I viral fusion proteins share common mechanistic and structural features** but little sequence similarity. (Proc Natl Acad Sci USA. 2005 102:9288)

訳 クラスⅠウイルス融合タンパク質は，共通の機構的および構造的特徴を共有する

■ shared ＋［前］

❶ shared by 〜	〜によって共有される	956
・be shared by 〜	…は，〜によって共有される	309
❷ shared with 〜	〜と共有される	401
❸ shared between 〜	〜の間で共有される	252
❹ shared among 〜	〜の間で共有される	158

❶ This analysis greatly expands the range of diversity of the AEPs and **reveals the unique active site shared by all members of this superfamily.** (Nucleic Acids Res. 2005 33:3875)

訳 〜は，このスーパーファミリーのすべてのメンバーによって共有されるユニークな活性部位を明らかにする

❷ **Many of these genetic interactions are shared with other genes that are involved in** initiation of DNA replication. (Genetics. 2004 167:579)

訳 これらの遺伝的相互作用の多くは，〜に関与する他の遺伝子と共有される

S

★ shift ㊔ 変化／移動／シフト，㊟ 移す／シフトさせる／シフトする 用例数 20,600

shift	変化／移動／シフト／移す／シフトさせる	12,153
shifts	変化／移動／シフト／移す／シフトさせる	4,673
shifted	変化した／変化させた／シフトした	2,736
shifting	シフトさせる／移す	1,038

名詞として用いられることが多いが，動詞の用例もかなりある．複数形のshiftsの割合が，約30%ある可算名詞．shift inの用例が多い．

◆類義語：change, alteration, modification, conversion, transfer, transition, translocation, alter, convert, translocate, move

■ ㊔ shift ＋［前］

	the%	a/an%	ø%			
❶	5	84	4	shift in ～	～の変化／～のシフト	2,196
❷	32	65	2	shift of ～	～の変化／～のシフト	878
❸	21	75	1	shift from ～ to …	～から…へのシフト	419
❹	27	63	5	shift to ～	～へのシフト	410

❶ The presence of ATP caused a shift in the K_D of the active kinase for ATF2 to 1.70 ± 0.25 muM and for c-Jun of 3.50 ± 0.95 muM.（J Biol Chem. 2012 287:13291）

訳 ATPの存在は，～のK_Dの変化を引き起こした

❷ A shift of HIV-2 distribution was demonstrable between day 10 and day 14 after HIV-2 infection.（Am J Pathol. 2000 156:1197）

訳 HIV-2の分布の変化は，10日目と14日目の間で明白であった

❸ PU.1 excision resulted in a shift from B-2 cells to B-1-like cells, which dramatically increased with the age of the mice.（J Exp Med. 2005 202:1411）

訳 PU.1切除は，B-2細胞からB-1様細胞へのシフトという結果になった

❹ These cellular and molecular changes were consistent with a shift to a proinflammatory phenotype in null chimeras.（Am J Pathol. 2005 167:901）

訳 これらの細胞性および分子的な変化は，炎症誘発性の表現型へのシフトと一致した

■ ㊟ shifted ＋［前］

❶	shifted to ～	～へ移される	453
❷	shifted from ～ to …	～から…へ移される	239

❶ Compared with controls, length-tension curves of the mdx mice were shifted to the right.（FASEB J. 2004 18:102）

訳 mdxマウスの長さ－張力曲線は，右にシフトした

❷ Its synthesis was upregulated when cultures were shifted from 30 to 37℃ and downregulated when cultures were shifted from 37 to 30℃.（Infect Immun. 2001 69:7616）

訳 その合成は，培養が30℃から37℃へ移されたとき上方制御され，そして培養が37℃から30℃へ移されたとき下方制御された

★ short 形 短い 用例数 22,083

■ short ＋［名］

❶ short-term ～	短期間の～	4,294
❷ short chain	短鎖	569
❸ short range	短い範囲	465

❶ Short-term treatment with fosinopril significantly reduced PAI-1 compared with amlodipine in a dose-dependent fashion.（Circulation. 2002 105:457）

訳 フォシノプリルによる短期間の処理は，PAI-1を有意に低下させた

■［副］＋ short

❶ very short	とても短い	328
❷ relatively short	比較的短い	305

shortly 副 直ちに 用例数 1,099

◈類義語：immediately, readily, soon

■ shortly ＋［前］

❶ shortly after ～	～の後直ちに	860
・shortly after birth	生後直ちに	156
❷ shortly before ～	～の直前	118

❶ Homozygous mutant mice die shortly after birth.（Genes Dev. 1998 12:3264）

訳 ホモ変異マウスは，生後直ちに死亡する

★★ show 動 示す 用例数 257,631

show	～を示す	110,799
showed	～を示した	71,477
shown	示される	51,979
shows	～を示す	16,172
showing	～を示す	7,204

他動詞．we show thatのパターンで，研究内容を示すときに使われることが非常に多い．また，研究対象が主語となって，表現型や症状などを示すときにも使われる．

◈類義語：indicate, present, exhibit, display, represent, suggest

■ show ＋［名節］

❶ show that ～	…は，～ということを示す	135,592
❷ show how ～	…は，どのように～かを示す	1,531

❷ We show how logic gates can be implemented. (J Am Chem Soc. 2003 125:1056)
訳 われわれは，どのように～かを示す

■ ［代名／名］＋ show ＋［that節］

❶	we show that ～	われわれは，～ということを示す	48,883
	・Here we show that ～	ここにわれわれは，～ということを示す	11,369
	・In this study, we show that ～	この研究で，われわれは，～ということを示す	1,667
	・Finally, we show that ～	最後に，われわれは，～ということを示す	1,153
	・Furthermore, we show that ～	さらに，われわれは，～ということを示す	1,016
	・In addition, we show that ～	加えて，われわれは，～ということを示す	653
❷	results show that ～	結果は，～ということを示す	6,967
❸	data show that ～	データは，～ということを示す	3,403
❹	studies have shown that ～	研究は，～ということを示している	3,144
❺	analysis showed that ～	分析は，～ということを示した	2,379
❻	experiments showed that ～	実験は，～ということを示した	1,071
❼	findings show that ～	知見は，～ということを示す	946
❽	assays showed that ～	アッセイは，～ということを示した	696

❶ We show that similar levels of proliferation and vasculogenesis expand the primary vasculature in XX and XY gonads. (Dev Biol. 2002 244:418)
訳 われわれは，～ということを示す

❶ Here we show that specific prion variants of [PSI$^+$] and [PIN$^+$] disrupt each other's stable inheritance. (Genetics. 2003 165:1675)
訳 ここにわれわれは，～ということを示す

❷ These results show that TIM folding fits the (4+4) model for folding of ($\beta\alpha$) 8-barrel proteins. (J Mol Biol. 2004 336:1251)
訳 これらの結果は，～ということを示す

■ ［副］＋ show ＋［that節］

❶	we also show that ～	われわれは，また～ということを示す	2,826
❷	we have previously shown that ～	われわれは，以前に～ということを示した	1,856
❸	we now show that ～	われわれは，今～ということを示す	1,325
❹	we further show that ～	われわれは，さらに～ということを示す	1,138

❷ We have previously shown that the first step of thymopoiesis is specifically blocked in aging. (J Immunol. 2004 173:4867)
訳 われわれは，以前に～ということを示した

❸ We now show that this bacterial homologue is not an ion channel, but rather a H^+-Cl^- exchange transporter. (Nature. 2004 427:803)
訳 われわれは，今，～ということを示す

■ show ＋［副／副句］＋［that節］

❶	we show here that ～	われわれは，ここに～ということを示す	3,574
❷	we have shown previously that ～	われわれは，以前に～ということを示した	608
❸	show for the first time that ～	…は，～ということを初めて示す	542

❶ We show here that cat-1 gene transcription is also increased by cellular stress. (J Biol Chem. 2003 278:50000)
訳 われわれは，ここに～ということを示す

■ show ＋［名句］

❶ show significant ～	…は，有意な～を示す	2,078
❷ show increased ～	…は，増大した～を示す	1,984
❸ show reduced ～	…は，低下した～を示す	1,172
❹ show evidence of ～	…は，～の証拠を示す	681
❺ show an increase in ～	…は，～の増大を示す	435
❻ show the presence of ～	…は，～の存在を示す	427
❼ show no evidence of ～	…は，～の証拠を示さない	251
❽ show a decrease in ～	…は，～の低下を示す	207
❾ show a reduction in ～	…は，～の低下を示す	151

❷ We observed that nuclear extracts from TNF-α-activated fibroblasts showed increased Ets-binding activity. (J Immunol. 2004 172:1945)
訳 ～は，増大したEts結合活性を示した

■［名］＋ show

❶ mice showed ～	マウスは，～を示した	2,155
❷ patients showed ～	患者は，～を示した	645
❸ mutants showed ～	変異体は，～を示した	553
❹ animals showed ～	動物は，～を示した	356

❶ On the other hand, Prf-deficient mice showed an increase in the number of VACV-specific $CD8^+$ T cells only in the memory phase. (J Virol. 2011 85:12578)
訳 Prf欠損マウスは，～の数の増大を示した

■［動／名］＋ to show

★ to show ～	～を示す…	3,782
❶ fail to show ～	…は，～を示すことができない	366
❷ be used to show ～	…は，～を示すために使われる	324
❸ the first to show ～	～を示す最初である	144

❷ A luciferase reporter system was used to show that each of the two putative TATA boxes contributed to vIL-6 promoter activity. (J Virol. 2002 76:8252)
訳 ルシフェラーゼレポーターシステムが，～ということを示すために使われた

■ be shown ＋［前］

★ be shown	…は，示される	28,191
❶ be shown to *do*	…は，～することが示される	22,265
・be shown to be ～	…は，～であることが示される	7,317
・be shown to have ～	…は，～を持つことが示される	1,004

S

・be shown to play ～	…は，～を果たすことが示される	678
・be shown to bind	…は，結合することが示される	644
・be shown to induce ～	…は，～を誘導することが示される	508
・be shown to inhibit ～	…は，～を抑制することが示される	499
・be shown to regulate ～	…は，～を調節することが示される	393
❷ be shown by ～	…は，～によって示される	653
❸ be shown in ～	…は，～において示される	589

❶ This pseudoknot was also shown to be essential for replication, and it has a conserved counterpart in every group 1 and group 2 coronavirus. (J Virol. 2004 78:669)

訳 ～は，また，複製に必須であることが示された

❶ Notably, the Sal1 protein was shown to bind calcium through two EF-hand motifs located on its amino terminus. (Genetics. 2004 167:607)

訳 Sal1タンパク質は，～によってカルシウムに結合することが示された

■ be shown ＋［that節］

❶ it is shown that ～	～ということが示される	978
❷ it has been shown that ～	～ということが示されている	735
❸ it was shown that ～	～ということが示された	570

❶ It is shown that LNA ODNs remain associated with plasmid DNA after cationic lipid-mediated transfection into mammalian cells. (Nucleic Acids Res. 2003 31:5817)

訳 ～ということが示される

★★ signal ㊔ シグナル／信号，㊀ 信号を送る　用例数 65,722

signal	シグナル／信号	41,242
signals	シグナル／信号	24,083
signaled	信号を送った／信号を送られる	363
signalled	信号を送った／信号を送られる	34

動詞としても用いられるが，名詞の用例が圧倒的に多い．signalingは別項参照．

■ signals ＋［前］

	the%	a/an%	ø%			
❶	6	-	82	signals in ～	～におけるシグナル	1,609
❷	5	-	92	signals from ～	～からのシグナル	1,602

❷ Integrative circuits in the ENS receive and interpret the chemical signals from the mast cells. (Gastroenterology. 2004 127:635)

訳 ～は，肥満細胞からの化学的なシグナルを受けそして解釈する

■ signal ＋［名］

❶ signal transduction	シグナル伝達	9,502
・signal transduction pathways	シグナル伝達経路	1,705

❷ signal transducer	シグナルトランスデューサー	1,131
❸ signal peptide	シグナルペプチド	1,000
❹ signal intensity	シグナル強度	874

❶ In eukaryotes, the Src homology domain 3 (SH3) is a very important motif in signal transduction. (J Bacteriol. 2003 185:4081)

訳 ～は，シグナル伝達における非常に重要なモチーフである

★★ signaling 名 シグナル伝達，-ing 信号を送る　用例数 95,116

signaling	シグナル伝達	88,872
signalling	シグナル伝達	6,231
signalings	シグナル伝達	13

複数形のsignalingsの用例はほとんどなく，原則，不可算名詞として使われる．signallingはイギリス英語．

◆類義語：signal transduction

■ signaling ＋［前］

	the%	a/an%	ø%			
❶	2	0	50	signaling in ～	～におけるシグナル伝達	6,895
				・signaling in regulating ～	～を制御する際のシグナル伝達	76
❷	1	0	66	signaling by ～	～によるシグナル伝達	2,392
❸	0	0	84	signaling through ～	～を経たシグナル伝達	2,158

❶ Recent studies have demonstrated the vital nature of calcium/calcineurin/NFAT signaling in cardiovascular and skeletal muscle development in vertebrates. (Dev Biol. 2004 266:1)

訳 最近の研究は，心血管および骨格筋発生におけるカルシウム/カルシニュリン/NFATシグナル伝達の重要な性質を実証している

❸ Inhibition of signaling through the phosphatidylinositol 3-kinase (PI3K)-AKT pathway may be particularly important. (J Biol Chem. 2005 280:2092)

訳 ホスファチジルイノシトール３キナーゼ（PI3K)-AKT経路を経たシグナル伝達の抑制

■ signaling ＋［名］

❶ signaling pathways	シグナル伝達経路	9,070
❷ signaling molecules	シグナル伝達分子	1,880
❸ signaling events	シグナル伝達イベント	1,613
❹ signaling cascade	シグナル伝達カスケード	1,306

❶ Modifier screens have been powerful genetic tools to define signaling pathways in lower organisms. (Proc Natl Acad Sci USA. 2000 97:6687)

訳 ～は，下等な生物におけるシグナル伝達経路を決定する強力な遺伝的ツールである

S

■ ［過分／形］＋ signaling

❶	～-mediated signaling	～に仲介されるシグナル伝達	1,702
❷	intracellular signaling	細胞内シグナル伝達	1,547
❸	～-dependent signaling	～依存性シグナル伝達	1,182
❹	downstream signaling	下流シグナル伝達	1,086

❶ Hence the effects of PEA-15 on RSK2 represent **a novel mechanism for the regulation of RSK2-mediated signaling.** (J Biol Chem. 2003 278:32367)

訳 RSK2に仲介されるシグナル伝達の調節のための新規の機構

significance ㊔ 重要性／意義／有意性　　用例数 8,615

significance	重要性／有意性	8,609
significances	重要性／有意性	6

複数形のsignificancesの用例はほとんどなく，原則，**不可算名詞**として使われる．

◆類義語：importance, implication

■ significance ＋［前］

significance ofの前にtheが付く割合は圧倒的に高い．

	the%	a/an%	ø%			
❶	92	3	5	significance of ～	～の重要性	4,662
❷	1	1	58	significance in ～	～における重要性	402
❸	3	3	78	significance for ～	～に対する重要性	347

❶ **The structural and functional significance of the additional MSD1 in MRP1 remains elusive.** (Biochemistry. 2002 41:9052)

訳 MRP1における付加的なMSD1の構造的および機能的重要性は，分かりにくいままである

❷ The accurate identification and characterization of lymph nodes by imaging **has important therapeutic and prognostic significance in patients with newly diagnosed cancers.** (J Nucl Med. 2004 45:1509)

訳 ～は，新たに診断された癌の患者における重要な治療的および予後的重要性を持つ

■ ［形］＋ significance

❶	functional significance	機能的重要性	1,338
❷	statistical significance	統計的有意性	821
❸	clinical significance	臨床的重要性	548
❹	biological significance	生物学的重要性	482
❺	prognostic significance	予後的重要性	374
❻	physiological significance	生理学的重要性	320

⁑ significant 形 有意な／重要な／著しい　用例数 72,807

◈類義語：meaningful, important, marked, striking, prominent

■ significant ＋［名］

❶ significant differences	有意な違い	4,033
・no significant differences	有意な違いのない	1,909
❷ significant increase	有意な増大	3,328
❸ significant reduction	有意な低下	2,498
❹ significant decrease	有意な低下	1,776
❺ significant effect	有意な効果	1,764
❻ significant role	重要な役割	1,588
❼ significant changes	有意な変化	1,402
❽ significant association	有意な関連	946
❾ significant improvement	有意な改善	865

❶ Among women using HT, no significant differences in hormones or SHBG were observed among women who developed CVD and controls. (Circulation. 2003 108:1688)
訳 ホルモンあるいはSHBGの有意な違いは，～の間には観察されなかった

❷ Both compounds resulted in a significant increase in the percent of cancer-free animals. (Cancer Res. 2004 64:2347)
訳 両方の化合物は，～の有意な増大という結果になった

❸ Streptozotocin-induced diabetes was associated with a significant reduction in the expression of GALP mRNA, which was reversed by treatment with either insulin or leptin. (Diabetes. 2004 53:1237)
訳 ～は，GALPメッセンジャーRNAの発現の有意な低下と関連した

■［副］＋ significant

❶ statistically significant ～	統計的に有意な～	4,234
・statistically significant differences	統計的に有意な差	431
❷ clinically significant ～	臨床的に有意な～	837
❸ highly significant ～	高度に有意な～	768
❹ most significant ～	最も有意な～	762

❶ There were no statistically significant differences between treatment groups in any of the secondary efficacy end points. (Crit Care Med. 2004 32:332)
訳 処置グループの間に統計的に有意な差はなかった

❷ This treatment leads to clinically significant sleep improvements within 6 weeks and these improvements appear to endure through 6 months of follow-up. (JAMA. 2001 285:1856)
訳 この処置は，6週間以内に臨床的に有意な睡眠改善につながる

■ be significant ＋［前］

★ be significant	…は，有意である	2,968
❶ be significant in ～	…は，～において有意である	229
❷ be significant for ～	…は，～に対して有意である	195

S

❶ In some human tumor cell lines, the enhanced phosphorylation of Stat3 is inhibited by both PI3K and by Tec kinase inhibitors, suggesting that the link between PI3K and Stat3 is significant in human cancer. (Proc Natl Acad Sci USA. 2011 108:13247)

訳 PI3KとStat3の間の関連は，ヒトの癌において有意である

★★ significantly 副 有意に／著しく 用例数 83,014

◆類義語：markedly, remarkably, strikingly, dramatically

■ significantly ＋［形］

❶ significantly higher	有意により高い	6,393
・be significantly higher	…は，有意により高い	2,959
❷ significantly lower	有意により低い	4,420
❸ significantly greater	有意により大きな	3,144
❹ significantly different	有意に異なる	2,910
❺ significantly better	有意によりよい	883

❶ The mean total number of traditional risk factors was significantly higher in patients who developed vascular events than in those who did not (7.1 versus 5.6). (Arthritis Rheum. 2004 50:3947)

訳 従来のリスク因子の平均合計数は，～である患者において有意により高かった

❷ Hypermethylation of MGMT, RASSF1A, and DAPK was significantly lower in primary melanomas (n=20) compared to metastatic melanomas. (Oncogene. 2004 23:4014)

訳 ～は，原発性メラノーマにおいて有意により低かった

■ significantly ＋［過分］

❶ significantly reduced	有意に低下した	7,000
❷ significantly increased	有意に増大した	5,820
❸ significantly decreased	有意に低下した	2,923
❹ significantly associated with ～	有意に～と関連した	2,417
❺ significantly inhibited	有意に抑制された	1,518
❻ significantly enhanced	有意に増強された	1,381
❼ significantly improved	有意に改良された	1,328
❽ significantly elevated	有意に上昇した	1,110
❾ significantly attenuated	有意に減弱された	1,059

❶ Furthermore, Stat-1 expression was significantly reduced in macrophages deficient in NF-κB1, but not c-Rel. (J Immunol. 2003 171:4886)

訳 Stat-1発現は，～を欠損したマクロファージにおいて有意に低下した

❷ Calcium leakage from the sarcoplasmic reticulum (SR) measured using tetracaine was significantly increased in diabetic myocytes. (Diabetes. 2004 53:3201)

訳 ～が，糖尿病の筋細胞において有意に増大した

❹ In addition, diabetes was significantly associated with breast cancer in women (RR = 1.27, 95% CI: 1.11, 1.45). (Am J Epidemiol. 2004 159:1160)

訳 糖尿病は，乳癌と有意に関連した

❺ This suppression by UVB radiation was significantly inhibited by lutein feeding. (J Invest

Dermatol. 2004 122:510)

訳 ～が，ルテイン摂食によって有意に抑制された

■ [動] + significantly

❶ increased significantly	有意に増大した	1,215
❷ differ significantly	…は，有意に異なる	1,147
❸ contribute significantly to ～	…は，～に著しく寄与する	914
❹ decreased significantly	有意に低下した	661
❺ not change significantly	…は，有意には変化しない	361

❶ PGE_2 production also **increased significantly** in the mesenteric lymph nodes following exposure to viable Salmonella, but not after exposure to killed bacteria.（J Immunol. 2004 172:2469）

訳 PGE_2産生は，また，腸間膜リンパ節において有意に増大した

❷ Stimulus-response curves did not **differ significantly** between the vertex and anterior positions.（J Physiol. 2004 560:897）

訳 刺激応答曲線は，～の間で有意には異ならなかった

❸ MIF has been shown to **contribute significantly to** the development of immunopathology in several models of inflammatory disease.（Am J Pathol. 2003 162:47）

訳 MIFは，免疫病理学の発達に著しく寄与することが示されている

■ significantly + [動]

❶ significantly reduce ～	…は，～を有意に低下させる	1,118
❷ significantly affect ～	…は，～に有意に影響する	957
❸ not significantly alter ～	…は，～を有意に変えない	660

❷ The results indicated that a single CTD in the ICR does not **significantly affect** the Kd of TFIIIA.（J Biol Chem. 2003 278:45451）

訳 ～は，TFⅢAのKdに有意には影響しない

S

■ significantly + [副]

❶ significantly more ～	有意により～	3,765
❷ significantly less ～	有意により～でない	2,132

❶ Pregnant patients were **significantly more** likely to have sustained injuries from assault（odds ratio: 2.6, $P < .001$).（Radiology. 2004 233:463）

訳 妊娠中の患者は，～からの傷害をこうむった公算が有意により高かった

■ [副] + significantly

❶ statistically significantly ～	統計的に有意に～	385

silencing 名 サイレンシング　用例数 9,179

複数形の用例はなく，原則，不可算名詞として使われる．

■ silencing ＋［前］

	the%	a/an%	ø%			
❶	8	1	87	silencing of ～	～のサイレンシング	2,206
❷	0	0	74	silencing in ～	～におけるサイレンシング	570

❶ Methylation of promoter DNA contributes to transcriptional silencing of various tumor-suppressor genes in cancer. (FASEB J. 2008 22:1981)
訳 プロモーターDNAのメチル化は，様々な腫瘍抑制遺伝子の転写サイレンシングに寄与する

■［名／形］＋ silencing

❶	gene silencing	遺伝子サイレンシング	1,701
❷	transcriptional silencing	転写サイレンシング	593
❸	RNA silencing	RNAサイレンシング	402

★★ similar 形 類似の／似ている／同様の　用例数 69,692

similar toの用例が非常に多い．
◆類義語：analogous, equivalent, comparable, homologous

■ similar ＋［前］

❶	similar to ～	～に類似している	25,701
	・be similar to ～	…は，～に類似している	7,352
	・similar to that	それに類似している	6,885
	・similar to that of ～	～のそれに類似している	3,570
	・similar to that observed	観察されたそれに類似している	743
	・similar to that seen	見られたそれに類似している	447
	・similar to those	それらに類似している	5,138
	・similar to other ～	他の～に類似している	425
❷	similar in ～	～において類似している	4,046
	・similar in both groups	両方のグループにおいて類似している	252
❸	similar for ～	～に関して類似している	1,147
❹	similar between ～	～の間で類似している	826
❺	similar among ～	～の間で類似している	318

❶ In ypt11Δ mutants, however, the level of mitochondrial motility in buds was similar to that observed in mother cells. (Mol Biol Cell. 2004 15:3994)
訳 …は，～において観察されたそれに類似していた

❷ Adverse events rates were similar in both groups. (J Am Soc Nephrol. 2004 15:3256)
訳 有害事象率は，両方のグループにおいて類似していた

❹ Two-year blood pressure control was similar between groups. (JAMA. 2003 290:2805)

訳 2年の血圧コントロールは，グループの間で類似していた

■ [副] ＋ similar to

❶	very similar to ～	～に非常に類似している	1,070
❷	more similar to ～	～により類似している	371
❸	most similar to ～	～に最も類似している	275
❹	remarkably similar to ～	～に著しく類似している	254

❶ SPPV and GTPV genomes are very similar to that of lumpy skin disease virus (LSDV), sharing 97% nucleotide identity. (J Virol. 2002 76:6054)

訳 ～は，ランピースキン病ウイルス（LSDV）のそれに非常に類似している

❷ K contrast thresholds and gains were more similar to those of M than P cells. (J Physiol. 2001 531:203)

訳 ～は，P細胞よりM細胞のそれらにより類似していた

■ [名] ＋ similar to

❶	in a manner similar to ～	～に類似した様式で	556
❷	properties similar to ～	～に類似した性質	321
❸	levels similar to ～	～に類似したレベル	320
❹	phenotype similar to ～	～に類似した表現型	248

❶ Moreover, neonatal $CD4^+$ cells up-regulated activation markers in a manner similar to adult $CD4^+$ cells. (J Immunol. 2002 169:4998)

訳 新生児の$CD4^+$細胞は，成人の$CD4^+$細胞に類似した様式で活性化マーカーを上方制御した

■ similar ＋ [名]

❶	similar results	類似の結果	1,540
	・similar results were obtained	類似の結果が得られた	624
❷	similar levels	同様のレベル	759
❸	similar effects	類似の効果	708
❹	a similar pattern	類似のパターン	459
❺	similar changes	同様の変化	302
❻	in a similar manner	類似の様式において	297
❼	to a similar extent	同様の程度	254

❶ Similar results were obtained with assay of lactate dehydrogenase in mice receiving apoptotic cells. (Am J Pathol. 2004 164:1751)

訳 類似の結果が，乳酸脱水素酵素のアッセイによって得られた

❷ One clone of each wild-type strain was isolated, and the two clones (S23-luc7 and S22-luc2) were found to express similar levels of luciferase. (Infect Immun. 2005 73:695)

訳 ～が，同様のレベルのルシフェラーゼを発現することが見つけられた

❼ Each of these promoters is also induced to a similar extent within macrophages. (Infect Immun. 2001 69:5777)

訳 これらのプロモーターのそれぞれは，また，マクロファージ内で同様の程度誘導される

★ similarity　名 類似性／類似　　用例数 11,212

similarity	類似性／類似	7,692
similarities	類似性／類似	3,520

複数形のsimilaritiesの割合は約30%あるが，単数形は無冠詞で使われることも多い．similarity toの用例が多い．

◆類似性：analogy, homology

■ similarity ＋［前］

similarity ofの前にtheが付く割合は非常に高い．

	the%	a/an%	ø%			
❶	0	5	81	similarity to ～	～への類似性	2,707
				・with similarity to ～	～への類似性を持つ	161
				・similarity to other ～	他の～への類似性	119
❷	44	10	40	similarity between ～	～の間の類似性	698
❸	79	2	18	similarity of ～	～の類似性	698
❹	5	4	81	similarity with ～	～との類似性	648

❶ **Intron 2 encodes a DNA endonuclease, I-TwoI, with similarity to homing endonucleases of** the HNH family.（Nucleic Acids Res. 2002 30:1935）

訳 イントロン２は，～のホーミングエンドヌクレアーゼへの類似性を持つDNAエンドヌクレアーゼI-TwoIをコードする

❷ **Given the striking sequence similarity between B1 and the VRK enzymes**, we proposed that they might share overlapping substrate specificity.（J Virol. 2004 78:1992）

訳 B1酵素とVRK酵素の間の著しい配列類似性を考慮して

❹ Phylogenetic analysis most closely grouped MsNramp with other teleost Nramp genes and **revealed high sequence similarity with mammalian Nramp2**.（Infect Immun. 2004 72:1626）

訳 ～は，哺乳類のNramp2との高い配列類似性を示した

■［名／形］＋ similarity

❶	sequence similarity	配列類似性	2,038
❷	structural similarity	構造的類似性	555
❸	significant similarity	有意な類似性	200

❶ ComA, however, **has no significant sequence similarity to any known enolase.**（J Biol Chem. 2003 278:45858）

訳 ～は，既知のどのエノラーゼとも有意な配列類似性を持たない

★ similarly　副 同様に　　用例数 9,510

文頭で用いられることが多い．

◆類義語：likewise, correspondingly, equally, equivalently, identically

■ Similarly

❶ Similarly,	同様に,	4,354

❶ Similarly, we found that Cyp26a1 transcription is epigenetically regulated by RARβ2. (Oncogene. 2006 25:1400)
訳 同様に，われわれは～ということを見つけた

■ similarly ＋［前］

❶ similarly to ～	～と同様に	779
❷ similarly in ～	～において同様に	523

❶ Our results demonstrate that LUG encodes a functional homologue of Tup1 and that SEU may function similarly to Ssn6, an adaptor protein of Tup1. (Proc Natl Acad Sci USA. 2004 101:11494)
訳 SEUは，Ssn6と同様に機能するかもしれない

■ similarly ＋［過分］

❶ similarly treated	同様に処理された	142
❷ similarly increased	同様に増大した	96
❸ similarly reduced	同様に低下した	92

❶ Intravital microscopy studies revealed a dramatic increase in adhesion after tumor necrosis factor (TNF)-α treatment of sdc1$^{-/-}$ mice compared with similarly treated wild-type mice. (Invest Ophthalmol Vis Sci. 2002 43:1135)
訳 同様に処理された野生型マウスに比べて

simultaneous 形 同時の　用例数 4,869

◆類義語：concomitant, concurrent, synchronous

■ simultaneous ＋［名］

❶ simultaneous detection	同時検出	142
❷ simultaneous measurement	同時測定	113

❶ We describe a high-density microarray for simultaneous detection of proteins and DNA in a single test. Anal Chem. 2009 81:5777)
訳 われわれは，タンパク質とDNAの同時検出のための高密度マイクロアレイについて述べる

simultaneously 副 同時に　用例数 6,596

過去分詞の前で使われることも，後で使われることもある．

◆類義語：concomitantly, concurrently, synchronously

■ simultaneously ＋ [過分]

❶ simultaneously recorded	同時に記録される／同時に記録した	146

■ [過分] ＋ simultaneously

❶ recorded simultaneously	同時に記録される／同時に記録した	105

❶ Neuronal activity was recorded simultaneously in the nucleus accumbens, cingulate cortex, and basolateral amygdala.（Behav Neurosci. 2007 121:1243）
訳 神経活動が，同時に記録された

■ [接] ＋ simultaneously

❶ and simultaneously ～	そして，同時に～	323
❷ while simultaneously ～	一方，同時に～	312

❷ Data derived from early work supported a role for TNF as a skeletal catabolic agent that stimulates osteoclastogenesis while simultaneously inhibiting osteoblast function.（Gene. 2003 321:1）
訳 一方，同時に骨芽細胞機能を抑制する

★★ site 名 部位／場所　　用例数 148,958

site	部位	88,000
sites	部位	60,958

複数形のsitesの割合は約40％とかなり高い．
◆類義語：area, region, locus, territory

■ site (s) ＋ [前]

site ofの前にtheが付く割合は非常に高い．

	the％	a/an％	ø％			
❶	78	16	5	site of ～	～の部位	10,332
				・site of action	作用の部位	340
				・site of infection	感染の部位	260
				・site of interaction	相互作用の部位	122
				・site of phosphorylation	リン酸化の部位	118
❷	8	-	68	sites in ～	～における部位／～の部位	5,898
❸	42	53	2	site for ～	～のための部位	3,372

❶ β_2-nAChRs are the initial site of action of nicotine and are implicated in various neuropsychiatric disorders.（J Nucl Med. 2005 46:1466）
訳 ～は，ニコチンの作用の開始部位である

❷ These findings suggest that simple selection pressures may have played a predominant role in determining the sequences of ligand-binding and active sites in proteins.（Proc Natl Acad Sci USA. 2005 102:10153）

訳 ～は，タンパク質のリガンド結合部位および活性部位の配列を決定する際に主な役割を果たしたかもしれない

❸ CREB-1 is phosphorylated at serine 133, **a critical site for activity**, in both T cells and Epstein-Barr virus immortalized B cells. (Oncogene. 2006 25:2170)

訳 活性のために決定的に重要な部位

■ site (s) ＋［動］

❶ sites were identified	部位が，同定された	230
❷ site is located	部位は，位置する	194

❷ **The translational start site is located in exon III.** (Gene. 2005 361:89)

訳 翻訳開始部位は，エクソン３に位置する

■［形／名］＋ site (s)

❶ binding site	結合部位	15,501
❷ active site	活性部位	12,979
❸ start site	開始部位	2,053
❹ phosphorylation sites	リン酸化部位	1,774
❺ cleavage site	切断部位	1,629

❶ **Drugs targeting the γ-secretase nucleotide-binding site represent an attractive strategy for** safely treating Alzheimer disease. (J Biol Chem. 2005 280:41987)

訳 γ-セクレターゼのヌクレオチド結合部位を標的とする薬剤は，～のための魅力的な戦略に相当する

★★ size ㊔ サイズ／大きさ，㊍ 大きさによって分ける　用例数 33,544

size	サイズ／大きさ	28,398
sizes	サイズ／大きさ	3,512
sized	～の大きさの／大きさによって分られる	1,634

動詞としても用いられるが，名詞の用例が圧倒的に多い．複数形のsizesの割合は約10%あるが，単数形は無冠詞で使われることも多い．

◆類義語：magnitude, amplitude, scale

■ size ＋［前］

size ofの前にtheが付く割合は圧倒的に高い．

	the%	a/an%	ø%			
❶	85	6	8	size of ～	～の大きさ	4,992
❷	13	2	80	size in ～	～における大きさ	685

❶ Normal mode analysis of virus capsids **has been limited due to the size of these systems**, which often exceed 50,000 residues. (J Mol Biol. 2005 350:528)

訳 ～は，これらのシステムの大きさのせいで限られている

❷ Nasal MOG **did not reduce infarct size in IL-10-deficient mice**. (J Immunol. 2003 171:6549)

S

訳 ～は，IL-10欠損マウスにおいて梗塞サイズを減少させなかった

■ ［前］＋ size

❶ ranging in size from ～ to …	大きさが～から…までにわたる	174
❷ similar in size to ～	大きさが～に似ている	101

❶ Restriction fragment patterns consisted of 14 to 18 fragments ranging in size from 580 to 40 kbp.（J Clin Microbiol. 2005 43:1205）

訳 制限酵素断片のパターンは，大きさが580 bpから40 kbpにわたる14から18断片から成っていた

■ size and ＋［名］

❶ size and shape	大きさと形	392
❷ size and number	大きさと数	158

❶ Mechanisms that regulate the size and shape of bony structures are largely unknown.（Dev Biol. 2005 278:208）

訳 骨性構造の大きさと形を調節する機構は，ほとんど知られていない

slightly 副 わずかに 用例数 5,105

◆類義語：modestly, marginally, weakly

■ slightly ＋［形］

❶ slightly higher	わずかにより高い	374
❷ slightly lower	わずかにより低い	298
❸ slightly different	わずかに異なる	156
❹ slightly greater	わずかにより大きい	127
❺ slightly larger	わずかにより大きい	103

❶ In contrast, fibrinogen adheres more rapidly to both surfaces, having a slightly higher affinity toward the hydrophobic surface.（J Am Chem Soc. 2005 127:8168）

訳 疎水性の表面に対してわずかにより高い親和性を持っているので

■ slightly ＋［副］

❶ slightly more	わずかにより～	354
・be slightly more	…は，わずかにより～である	155
❷ slightly less	わずかにより～でない	189

❶ Both enzymes, however, show similar unfolding behavior in urea, and the Rhodococcus enzyme is only slightly more tolerant to unfolding by guanidine hydrochloride.（J Bacteriol. 2005 187:7222）

訳 ロドコッカス酵素は，～によるアンフォールディングにほんのわずかにより耐性である

■ slightly ＋［過分］

❶ slightly increased	わずかに増大した	238
❷ slightly reduced	わずかに低下した	234
❸ slightly decreased	わずかに低下した	126

■［副］＋ slightly

❶ only slightly	ほんのわずかに	834

★slow 形 遅い，動 遅くする　用例数 12,452

slow	遅い／遅くする	10,187
slowed	遅くされる／遅くした	1,573
slows	遅くする	692

形容詞の用例が非常に多いが，動詞としても用いられる．

◆類義語：late, gradual

■ 形 slow ＋［名］

❶ slow growth	遅い増殖	296
❷ slow inactivation	遅い不活性化	268

❷ Voltage-gated Kv1.1/Kvβ1.1 A-type channels, as a natural complex, **can switch from fast to slow inactivation under oxidation/reduction conditions**.（Proc Natl Acad Sci USA. 2004 101: 15535）

訳 ～は，酸化/還元状態下で速い不活性化から遅い不活性化に切り替わることができる

■ 形［副］＋ slow

❶ very slow	非常に遅い	225
❷ relatively slow	比較的遅い	190

❶ The best enantioselectivity at room temperature was obtained with the newly synthesized phospholane 8c and benzoic anhydride, but **the reaction is very slow**.（J Org Chem. 2004 69: 1389）

訳 その反応は非常に遅い

■ 動 slow ＋［名句］

❶ slow the rate of ～	…は，～の速度を遅くする	177
❷ slow the progression of ～	…は，～の進行を遅くする	113

❶ This study demonstrates that **bicarbonate supplementation slows the rate of progression of renal failure** to ESRD and improves nutritional status among patients with CKD.（J Am Soc Nephrol. 2009 20:2075）

訳 重炭酸補充は，腎不全の進行の速度を遅くする

S

■ 動 be slowed ＋ [前]

★ be slowed	…は，遅くさせられる	338
❶ be slowed by ～	…は，～によって遅くさせられる	113

slower 形 より遅い　用例数 4,247

slowの比較級.

■ [副] ＋ slower

❶ much slower	ずっとより遅い	334
❷ significantly slower	有意により遅い	226

❶ The fragments are subsequently transported to the Golgi compartment, where **their turnover rate is much slower than that of the full-length presenilin** in the ER.（J Biol Chem. 1998 273: 12436）

訳 それらの代謝回転速度は，全長のプレセニリンのそれよりずっと遅い

❷ The rate at which sterols suppress SREBP processing **is significantly slower in SRD-14 cells than wild type CHO-7 cells.**（J Biol Chem. 2004 279:43136）

訳 ～は，野生型CHO-7細胞よりSRD-14細胞において有意により遅い

■ slower than

★ slower than ～	～より遅い	870
❶ be slower than ～	…は，～より遅い	151
❷ …-fold slower than ～	～より…倍遅い	145
❸ … times slower than ～	～より…倍遅い	98

❷ However, **the rate of diffusion of Ku was approximately 100-fold slower than that predicted from its size.**（J Immunol. 2002 168:2348）

訳 Kuの拡散速度は，その大きさから予想されるそれよりおよそ100倍遅かった

■ slower ＋ [名]

❶ slower rate	より遅い速度	274

slowly 副 遅く／ゆっくりと　用例数 3,368

◆類義語：gradually

■ [副] ＋ slowly

❶ more slowly	より遅く	956
・more slowly than ～	～より遅く	382
❷ very slowly	非常に遅く	143

❶ At high light intensity, **the R214H mutant grew approximately 2.5-fold more slowly than the**

wild type. (J Biol Chem. 2005 280:10395)
訳 R214H変異体は，野生型よりおよそ2.5倍遅く増殖した

★★ small 形 小さい／少ない 用例数 51,124

◆類義語：subtle, slight, modest, insignificant, weak, few, minor, little

■ small + [名]

❶ small molecule	小分子	3,584
❷ small number of ～	少数の～	1,848
❸ small intestine	小腸	1,244
❹ small bowel	小腸	655
❺ a small fraction of ～	わずかな～	502
・only a small fraction of ～	ほんのわずかな～だけ	220
❻ small amounts of ～	少量の～	485
❼ small animal	小動物	483
❽ small subset of ～	小さなサブセットの～	414
❾ small changes in ～	～の小さな変化	381

❷ Surprisingly, however, microarray analysis revealed that **expression of only a small number of genes is affected.** (Cell. 2005 123:1199)
訳 ほんのわずかな数の遺伝子の発現が影響を受ける

❺ Here, we demonstrate that **only a small fraction of the ZmBs repeats interacts with CENH3,** the histone H3 variant specific to centromeres. (Plant Cell. 2005 17:1412)
訳 ほんのわずかなZmBリピートだけがCENH3と相互作用する

❾ Total activity of Lck increased only fourfold, **reflecting small changes in both the amount of protein and specific activity.** (Transplantation. 2005 80:1112)
訳 タンパク質の量と比活性の両方の小さな変化を反映して

■ [副] + (a) small

❶ only a small ～	ほんのわずかな～だけ	1,049
❷ relatively small	比較的小さな	887
❸ very small	非常に小さな	693

❶ **A relatively small number of differences were found within a group of proteins that function as** both RNA binding proteins and transcription factors. (Brain. 2003 126:2052)
訳 比較的少数の違いが，～として機能するひとつのグループのタンパク質の中に見つけられた

smaller 形 より小さい 用例数 8,931

smallの比較級.

■ smaller than/in

❶ smaller than ～	～より小さい	1,406
・be smaller than ～	…は，～より小さい	414
❷ smaller in ～	～においてより小さい	460

❶ The tumor multiplicity in AT2-null mice (1.9 ± 0.3) was significantly smaller than that in wild-type mice (4.1 ± 0.9). (Cancer Res. 2005 65:7660)

訳 ～は，野生型マウスにおけるそれより有意に小さかった

■ [副] ＋ smaller

❶ significantly smaller	有意により小さい	586
・significantly smaller than ～	～より有意に小さい	153
・significantly smaller in ～	～において有意により小さい	104
❷ much smaller	ずっとより小さい	499

❶ Adenoid and tonsil volume was significantly smaller in the subjects with Down syndrome. (Am J Respir Crit Care Med. 2001 163:731)

訳 アデノイドと扁桃腺の体積が，ダウン症候群の患者において有意により小さかった

solely 副 もっぱら／単独で　　用例数 2,406

◆類義語：exclusively, simply, merely, only

■ solely ＋ [前]

❶ based solely on ～	もっぱら～に基づいて	175

❶ Although these findings for retinal degeneration are based solely on self-reported disease, they are consistent with those reported for farmer pesticide applicators. (Am J Epidemiol. 2005 161:1020)

訳 レチナール分解に関するこれらの知見は，もっぱら自己申告される疾患に基づいている

■ solely ＋ [形]

❶ solely responsible for ～	～に単独で責任がある	133

❶ The viral glycoprotein (GP) is solely responsible for virus-host membrane fusion, but how it does so remains elusive. (Proc Natl Acad Sci USA. 2011 108:11211)

訳 ウイルス糖タンパク質(GP)は，ウイルス-宿主膜融合に単独で責任がある

★ solution 名 溶液／解決　　用例数 20,168

solution	溶液／解決	15,868
solutions	溶液／解決	4,300

複数形のsolutionsの割合が，約20%ある可算名詞.

◆類義語：resolution

■ [前] + solution

❶	in solution	溶液中で	4,480
❷	from solution	溶液から	213

❶ These studies show quite clearly that **the heterodimer and heterotetramer complexes do not behave in solution as dimeric structures.** (Biochemistry. 2002 41:13133)
訳 そのヘテロ二量体およびヘテロ四量体複合体は，二量体構造として溶液中では挙動しない

■ solution + [前]

	the%	a/an%	ø%			
❶	15	60	18	solution to ～	～に対する解決	604
❷	20	74	5	solution of ～	～の溶液	568

❶ Here, **we describe a solution to this problem.** (Proc Natl Acad Sci USA. 2005 102:10841)
訳 われわれは，この問題に対する解決を述べる

solve 動 解く／解決する　　用例数 2,892

solved	解かれる／～を解いた	1,790
solve	～を解く／～を解決する	531
solving	～を解く／～を解決する	503
solves	～を解く／～を解決する	68

他動詞.「(結晶構造) を解く」という意味に使われることが多い.
◆類義語：resolve, unravel, reveal, analyze

■ solve + [名句]

❶	solve the crystal structure of ～	…は，～の結晶構造を解く	160
❷	solve this problem	…は，この問題を解決する	85

❶ **We have solved the crystal structure of SMUG1** complexed with DNA and base-excision products. (Mol Cell. 2003 11:1647)
訳 われわれは，SMUG1の結晶構造を解いた

■ 動 be solved + [前]

★	be solved	…が，解かれる／解決される	890
❶	be solved by ～	…が，～によって解かれる／解決される	198
❷	be solved to ～	…が，～の解像度まで解かれる	116
❸	be solved at ～	…が，～の解像度で解かれる	102

❸ **The structure was solved at 1.8A resolution** by isomorphous phasing with a previously solved X-ray crystal structure of the rPGI dimer containing 6-phosphogluconate in its active site. (J Mol Biol. 2002 323:77)
訳 その構造が，1.8Aの解像度で解かれた

S

⁑ some （代名）いくらか，（形）いくらかの／いくつかの／ある 用例数 43,293

◆類義語：several , certain, moderate

■ （形） some ＋［名］

❶ in some cases	いくつかのケースで	1,734
❷ in some patients	幾人かの患者において	537
❸ to some extent	ある程度	248

❶ This suggests that, in some cases, the mutation may subtly alter the entire protein-RNA interface.（J Mol Biol. 2006 356:613）
訳 いくつかのケースで

❸ The differentiation of murine trophoblast giant cells（TGCs）is well characterised at the molecular level and, to some extent, the cellular level.（Dev Biol. 2006 290:13）
訳 ～は，分子レベルで，そしてある程度，細胞レベルでよく特徴づけられている

■ （代名） some of

★ some of ～	～のいくらか	8,412
❶ , some of which ～	それらのいくらかは～	920
❷ at least some of ～	少なくとも～のいくらか	238
❸ some of the most ～	もっとも～なうちのいくらか	175
❹ explain some of ～	…は，～のいくらかを説明する	158
❺ discuss some of ～	…は，～のいくらかを議論する	134
❻ highlight some of ～	…は，～のいくらかを強調する	122
❼ account for some of ～	…は，～のいくらかを説明する	113

❶ A number of these genes contain octamer sequences in their immediate 5' regulatory regions, some of which are conserved in human.（Cancer Res. 2005 65:10750）
訳 それらのいくらかはヒトにおいて保存されている

❷ We found evidence that ligands for multiple $\gamma\delta$ TCRs may be simultaneously expressed on a single cell line, and that at least some of the putative ligands are protease sensitive.（J Immunol. 2004 172:4167）
訳 推定上のリガンドの少なくともいくらかは，プロテアーゼ感受性である

❹ These abnormalities may explain some of the early anomalies in visual function induced by diabetes.（Invest Ophthalmol Vis Sci. 2008 49:2635）
訳 これらの異常は，視覚機能の早期の異常のいくらかを説明するかもしれない

■ （代名）（形） some but

❶ some, but not all,	いくつか，しかしすべてではない	299
❷ some but not all	いくつか，しかしすべてではない	240

❶ Thus, it appears that some, but not all, of the residues in both the N and C termini of CooA play a critical role in the intermolecular interactions of the major pilin with the other structural and assembly proteins.（J Bacteriol. 2006 188:231）
訳 CooAのNおよびC末端の両方の残基のいくつかは，しかしすべてではないが，～の分子間

相互作用において決定的に重要な役割を果たす

★ source 名 ソース／出所／源　用例数 18,313

source	ソース／源	12,371
sources	ソース／源	5,942

複数形のsourcesの割合が，約30%ある可算名詞.

◆類義語：origin, cause

■ source ＋［前］

	the%	a/an%	ø%			
❶	38	60	0	source of ～	～のソース	6,045
				・major source of ～	～の主要なソース	615
				・important source of ～	～の重要なソース	263
				・primary source of ～	～の主なソース	197
				・potential source of ～	～の潜在的なソース	133
				・sole source of ～	～の唯一のソース	129
❷	32	64	1	source for ～	～のソース	555

❶ Genetic data demonstrate that **NR is the major source of NO in** guard cells in response to ABA-mediated H_2O_2 synthesis. (Plant J. 2006 45:113)

訳 NRは，～におけるNOの主要なソースである

■ ［名］＋ source

❶ carbon source	炭素源	546
❷ open source	オープンソース	358
❸ energy source	エネルギー源	216
❹ nitrogen source	窒素源	202

S

specialize 動 特殊化する／特定化する　用例数 4,344

specialized	特殊化される／特殊化した	4,234
specialize	～を特殊化する／～を特定化する	64
specializing	～を特殊化する	33
specializes	～を特殊化する／～を特定化する	13

他動詞．受動態の用例が圧倒的に多い．

◆類義語：specify

■ specialized ＋［前］

❶ specialized for ～	～のために特殊化される	317
・be specialized for ～	…は，～のために特殊化される	148
❷ specialized to *do*	～するように特殊化される	106

❶ The Rev1 DNA polymerase is highly specialized for the incorporation of C opposite template G.（Science. 2005 309:2219）
訳 Rev1 DNAポリメラーゼは，～の取り込みのために高度に特殊化されている

■［副］＋ specialized

❶	highly specialized	高度に特殊化される	228

☆ species ㊔ 種　　用例数 44,693

単複同形.
◆類義語：family, genus

■ species ＋［前］

	the％	a/an％	ø％			
❶	12	10	37	species of ～	…種の～	2,203
❷	16	6	52	species in ～	～における種	1,619

❶ Here, we review the biology and clinical results of these five species of viruses and discuss lessons learned and challenges for the future.（Oncogene. 2005 24:7802）
訳 われわれは，これらの５つの種のウイルスの生物学および臨床結果を概説する

❷ Lactobacillus species in the rectum may contribute to the maintenance of vaginal microflora.（J Infect Dis. 2005 192:394）
訳 直腸におけるラクトバシラス種は，～に寄与するかもしれない

■［前］＋ species

❶	conserved across species	種を超えて保存されている	105
❷	variety of species	様々な種	103

❶ This NLS is conserved across species, among a subfamily of T-box proteins including Brachyury and Tbx10, and among additional nuclear proteins.（Hum Mol Genet. 2005 14:885）
訳 このNLSは，種を超えて保存されている

■［名／形］＋ species

❶	reactive oxygen species	活性酸素種	3,633
❷	other species	他の種	1,022
❸	bacterial species	細菌種	716
❹	plant species	植物種	705
❺	different species	異なる種	684

☆ specific ㊊ 特異的な　　用例数 139,858

◆類義語：special, particular

■ specific ＋［前］

❶ specific for ～	～に特異的な	5,325
・antibodies specific for ～	～に特異的な抗体	428
・T cells specific for ～	～に特異的なT細胞	413
❷ specific to ～	～に特異的な	2,761
・be specific to ～	…は，～に特異的である	1,156

❶ First, **p14 was also immunostained with antibodies specific for the cone PDEγ isoform**. (Genomics. 2002 79:582)

訳 p14は，また，錐体のPDEγアイソフォームに特異的な抗体を使って免疫染色された

❷ Genomic hybridization studies suggest that **the PPE55 gene is specific to the M. tuberculosis complex** and is present in a majority of clinical isolates tested. (Infect Immun. 2005 73:5004)

訳 PPE55遺伝子は，結核菌複合体に特異的である

■［副］＋ specific

❶ highly specific	高度に特異的な	1,293
❷ more specific	より特異的な	455

❶ **A15 was found to be highly specific for the active site of FⅩⅢa** and was covalently bound to fibrin. (Circulation. 2004 110:170)

訳 A15は，FⅩⅢaの活性部位に高度に特異的であることが見つけられた

■［名］-specific

❶ tissue-specific ～	組織特異的な～	3,504
❷ site-specific ～	部位特異的な～	3,345
❸ cell-specific ～	細胞特異的な～	2,590
❹ antigen-specific ～	抗原特異的な～	2,406
❺ sequence-specific ～	配列特異的な～	2,318
❻ virus-specific ～	ウイルス特異的な～	1,522
❼ cell-type-specific ～	細胞タイプ特異的な～	1,458

❶ **To explore the tissue-specific expression of ABCC6**, we first examined various mouse tissues by RT-PCR. (J Invest Dermatol. 2005 125:900)

訳 ABCC6の組織特異的な発現を探索するために

■ specific ＋［名］

❶ ～-specific expression	～特異的な発現	1,930
❷ ～-specific manner	～特異的な様式	1,787
❸ specific genes	特異的な遺伝子	1,770
❹ specific binding	特異的な結合	1,546
❺ specific antibodies	特異的な抗体	1,510
❻ specific inhibitor	特異的な抑制剤	1,261
❼ specific activity	比活性	1,171

★ specifically 副 特異的に　用例数 23,379

動詞の前で使われることも，後で使われることもある．

◆類義語：particularly, especially

■ specifically +［動］

❶	specifically bind to ～	…は，～に特異的に結合する	547
❷	specifically interact with ～	…は，～と特異的に相互作用する	414
❸	specifically recognize ～	…は，～を特異的に認識する	400
❹	specifically target ～	…は，～を特異的に標的にする	387

❶ Furthermore, by yeast two-hybrid and coimmunoprecipitation analyses, we demonstrate that **p28 specifically binds to p10 and p15**, two coronavirus replicase proteins of unknown function. (J Virol. 2004 78:11551)

訳 p28は，p10およびp15に特異的に結合する

■ be specifically +［過分］

★	be specifically ～	…は，特異的に～される	2,933
❶	be specifically expressed in ～	…は，～において特異的に発現する	261
❷	be specifically required for ～	…は，～に特異的に必要とされる	222
❸	be specifically inhibited	…は，特異的に阻害される	128
❹	be specifically associated with ～	…は，～と特異的に関連している	114

❶ **BMPER is specifically expressed in flk-1-positive cells** and parallels the time course of flk-1 induction in these cells. (Mol Cell Biol. 2003 23:5664)

訳 BMPERは，flk-1陽性細胞において特異的に発現する

■［動］+ specifically +［前］

❶	bind specifically to ～	…は，～に特異的に結合する	1,054
❷	interact specifically with ～	…は，～と特異的に相互作用する	376
❸	be expressed specifically in ～	…は，～において特異的に発現する	164

❶ **The C2 domain binds specifically to phosphoinositides** in vitro and is sufficient for localization to membranes in intact cells. (J Cell Biol. 2004 165:135)

訳 C2ドメインは，ホスホイノシチドに特異的に結合する

❸ **Csm is expressed specifically in the heart**, and its expression in the heart is restricted to cardiac myocytes. (J Biol Chem. 2003 278:28750)

訳 Csmは，心臓において特異的に発現する

specification 名 特異化／特定　用例数 3,536

specification	特異化／特定	3,448
specifications	特異化／特定	88

複数形のspecificationsの割合は2％しかなく，原則，不可算名詞として使われる．

◆類義語：speciality

■ specification ＋［前］

specification ofの前にtheが付く割合はかなり高い．

	the%	a/an%	ø%			
❶	60	0	40	specification of ～	～の特異化	1,205
❷	0	0	99	specification in ～	～における特異化	223

❶ Etv2 is essential for the specification of endothelial and hematopoietic lineages.（Curr Opin Hematol. 2012 19:199）

訳 Etv2 は，内皮および造血系の特異化のために必須である

■［名］＋ specification

❶	cell fate specification	細胞運命特定	287

★ specificity ㊔ 特異性　用例数 26,086

specificity	特異性	23,864
specificities	特異性	2,222

複数形のspecificitiesの割合は約10%あるが，単数形は無冠詞で使われることが多い．

◆類義語：speciality

■ specificity ＋［前］

specificity ofの前にtheが付く割合は非常に高い．

	the%	a/an%	ø%			
❶	72	9	17	specificity of ～	～の特異性	6,653
❷	8	2	77	specificity for ～	～に対する特異性	1,968
❸	8	0	86	specificity in ～	～における特異性	926

❶ The specificity of these assays is high, ranging from 93 to 100%.（Curr Opin Gastroenterol. 2005 21:59）

訳 これらのアッセイの特異性は高い

❷ These metallopeptidases exhibit unique specificity for the substrates and peptide bonds they cleave.（Biochemistry. 2005 44:11758）

訳 これらのメタロペプチダーゼは，基質に対するユニークな特異性を示す

■［名／形］＋ specificity

❶	substrate specificity	基質特異性	2,024
❷	binding specificity	結合特異性	891
❸	high specificity	高い特異性	569
❹	dual specificity	二重の特異性	472
❺	sequence specificity	配列特異性	453
❻	tissue specificity	組織特異性	274

■［名］＋ and specificity

❶	sensitivity and specificity	感受性および特異性	1,472
❷	affinity and specificity	親和性および特異性	333

❶ The sensitivity and specificity of this method were estimated at 98.7% and 99.7%, respectively.（J Clin Microbiol. 2005 43:2148）

訳 この方法の感受性と特異性は，それぞれ98.7%および99.7%であると見積もられた

specimen　名 検体／標本／試料　　用例数 9,094

specimens	検体	7,924
specimen	検体	1,170

複数形のspecimensの割合は，約85%と圧倒的に高い．

◆類義語：sample, preparation

■ specimens ＋［前］

	the%	a/an%	ø%			
❶	1	-	78	specimens from ～	～からの検体	1,078
				・specimens from patients	患者からの検体	174
❷	12	-	78	specimens of ～	～の検体	328

❶ Synovial biopsy specimens from patients with juvenile rheumatoid arthritis（JRA）were used for additional comparison.（Arthritis Rheum. 2005 52:3175）

訳 ～の患者からの滑膜生検検体

■ specimens ＋［動］

❶	specimens were obtained	検体が，得られた	111

■［名／形］＋ specimens

❶	biopsy specimens	生検検体	759
❷	clinical specimens	臨床検体	370
❸	tumor specimens	腫瘍検体	251
❹	cancer specimens	癌検体	223
❺	tissue specimens	組織検体	203

★ spectrum　名 スペクトル／スペクトラム／範囲　　用例数 17,166

spectrum	スペクトル／範囲	8,794
spectra	スペクトル／範囲	8,372

複数形のspectraの割合は約50%と非常に高い．

◆類義語：range

■ spectrum ＋［前］

	the%	a/an%	ø%			
❶	42	53	1	spectrum of ～	～の範囲／…範囲の～	4,111
				・a broad spectrum of ～	広域の～／広範囲の～	471
				・a wide spectrum of ～	広範囲の～	320

❶ Histone deacetylases (HDACs) **are major epigenetic modulators involved in** a broad spectrum of **human diseases** including cancers. (Genes Dev. 2011 25:2610)
訳 ～は，広範囲のヒト疾患に関与している主要な後成的修飾因子である

■［名］＋ spectra

❶	NMR spectra	核磁気共鳴スペクトル	795
❷	mass spectra	質量スペクトル	618
❸	absorption spectra	吸収スペクトル	318
❹	difference spectra	差スペクトル	190

speculate ㊍ 推定する　用例数 1,825

speculate	～を推定する	1,350
speculated	推定される／～を推定した	440
speculates	～を推定する	21
speculating	～を推定する	14

S

他動詞.

◆類義語：assume, presume, estimate, infer, predict, expect, postulate, deduce

■［代名］＋ speculate ＋［that節］

★	speculate that ～	…は，～であると推定する	1,498
❶	we speculate that ～	われわれは，～であると推定する	1,014

❶ Hence, we speculate that these novel regulatory elements in the IL-10 family gene locus function via an intermediate regulatory RNA. (J Immunol. 2005 175:7437)
訳 われわれは，～であると推定する

speed ㊔ 速度／スピード　用例数 4,232

speed	速度／スピード	3,693
speeds	速度／スピード	539

複数形のspeedsの割合が，約15%ある可算名詞.

◆類義語：rate, velocity

■ speed ＋［前］

speed ofの前にtheが付く割合は非常に高い．

	the％	a/an％	⌀％			
❶	71	9	19	speed of ～	～の速度	688

❶ The speed of the analysis depends on computational resources and data volume, but will generally be less than 1 d for most users.（Nat Protoc. 2012 7:508）

訳 その分析の速度は，計算資源に依存する

spontaneously 副 自然に／自発的に　用例数 2,886

◆類義語：naturally

■ spontaneously ＋［動］

❶	spontaneously develop ～	…は，～を自然に発症する	267

❶ C3H/HeJ mice spontaneously develop alopecia areata from 5 mo of age and older in females and later in males.（J Invest Dermatol. 2003 120:771）

訳 C3H/HeJマウスは，円形脱毛症を自然に発症する

■ spontaneously ＋［形］

❶	spontaneously hypertensive	自然発症高血圧の	177
❷	spontaneously active	自発的に活性のある	100

spread 名 伝播／蔓延，動 広がる　用例数 5,440

spread	伝播／蔓延／広がる	5,127
spreads	広がる／伝播／蔓延	313

名詞および動詞として用いられる．動詞の原形・過去・過去分詞はすべて同形．名詞の複数形のspreadsの割合は１％しかなく，原則，**不可算名詞**として使われる．

◆類義語：propagation, prevalence, diffusion, propagate, diffuse, extend, expand

■ 名 spread ＋［前］

spread ofの前にtheが付く割合は非常に高い．

	the％	a/an％	⌀％			
❶	78	2	18	spread of ～	～の伝播	1,629

❶ This could also promote the spread of Wolbachia infection, though here the fitness benefits would also help to spread infection when Wolbachia are rare.（Heredity. 2004 93:379）

訳 これは，また，ウォルバキア感染の伝播を促進しうる

■ 動 spread ＋［前］

❶	spread to ～	…は，～に広がる	243
❷	spread from ～	…は，～から広がる	133

❶ Resistance to this drug has emerged in parts of Southeast Asia and **may spread to other regions of the world.** (JAMA. 1997 278:1767)
訳 ～は，世界の他の地域に広がるかもしれない

★ stability 名 安定性　用例数 19,127

stability	安定性	18,395
stabilities	安定性	732

複数形のstabilitiesの割合は4％あるが，単数形に不定冠詞が用いられることはほとんどない．

■ stability ＋［前］

stability ofの前にtheが付く割合は圧倒的に高い．

	the％	a/an％	ø％			
❶	83	1	16	stability of ～	～の安定性	5,871
❷	1	1	83	stability in ～	～における安定性	795

❶ We find that **the helicase domain of gp4 contributes to the stability of the complex** by binding to the ssDNA template. (Proc Natl Acad Sci USA. 2012 109:9408)
訳 gp4のヘリカーゼドメインは，その複合体の安定性に寄与する

❷ By reverse transcription, **the telomerase RNP maintains telomere length stability in almost all cancer cells.** (Hum Mol Genet. 2001 10:677)
訳 テロメラーゼRNPは，ほとんどすべての癌細胞においてテロメア長の安定性を維持する

■［名／形］＋ stability

❶	protein stability	タンパク質安定性	922
❷	mRNA stability	メッセンジャーRNA安定性	870
❸	thermal stability	熱安定性	696
❹	genomic stability	ゲノム安定性	443
❺	genome stability	ゲノム安定性	384
❻	thermodynamic stability	熱力学的安定性	379

stabilization 名 安定化　用例数 5,569

stabilization	安定化	5,563
stabilizations	安定化	6

複数形のstabilizationsの用例はほとんどなく，原則，不可算名詞として使われる．

■ stabilization ＋［前］

	the%	a/an%	ø%			
❶	23	5	70	stabilization of ～	～の安定化	2,881

❶ Finally, **stabilization of** VEGF mRNA coincided with the accumulation of Hsp70 protein in HL60 promyelocytic leukemia cells recovering from acute thermal stress.（Mol Cell Biol. 2013 33:71）

訳 VEGFメッセンジャーRNAの安定化は，Hsp70タンパク質の蓄積と一致した

★ stabilize 動 安定化させる　用例数 12,583

stabilized	安定化される／～を安定化した	3,920
stabilize	～を安定化する	3,289
stabilizing	～を安定化する	3,170
stabilizes	～を安定化する	2,204

他動詞.

■ stabilize ＋［名句］

❶ stabilize the transition state	…は，遷移状態を安定化する	86
❷ stabilize the protein	…は，そのタンパク質を安定化する	80

❶ The third residue of the catalytic triad, Glu334, is found to be essential in **stabilizing the transition state** through electrostatic interactions.（J Am Chem Soc. 2002 124:10572）

訳 ～は，遷移状態を安定化させる際に必須であることが見つけられる

■ stabilized ＋［前］

❶ stabilized by ～	～によって安定化される	1,327
・be stabilized by ～	…は，～によって安定化される	782
❷ stabilized in ～	～において安定化される	264

❶ P23H **is stabilized by** proteasome inhibitors and by co-expression of a dominant negative form of ubiquitin.（J Biol Chem. 2002 277:34150）

訳 P23Hは，プロテアソーム抑制剤によって安定化される

★ stable 形 安定な　用例数 22,398

■ stable ＋［前］

❶ stable in ～	～において安定な	797
・be stable in ～	…は，～において安定である	358
❷ stable for ～	～の間安定な	305
❸ stable at ～	～において安定な	259

❶ The aerobically purified D. vulgaris hydrogenase is stable in air. (J Am Chem Soc. 2001 123: 2771)
訳 好気的に精製されたD. vulgarisヒドロゲナーゼは，空気中で安定である

■ [動] + stable

❶ be stable	…は，安定である	1,846
❷ remain stable	…は，安定なままである	597

❷ Antiviral antibody responses remained stable between 1-75 years after vaccination, whereas antiviral T-cell responses declined slowly, with a half-life of 8-15 years. (Nat Med. 2003 9: 1131)
訳 抗ウイルス抗体反応は，1-75年の間安定なままであった

■ stable + [名]

❶ stable complex	安定な複合体	528
・form a stable complex	…は，安定な複合体を形成する	277
❷ stable isotope	安定な同位体	528
❸ stable disease	安定している疾患	496
❹ stable expression	安定な発現	462
❺ stable transfection	安定なトランスフェクション	280

❶ Our study reveals that RPL11 forms a stable complex with MDM2 in vitro through direct contact with its zinc finger. (J Biol Chem. 2011 286:38264)
訳 RPL11は，MDM2と安定な複合体を形成する

❹ Stable expression of these Gag and Env proteins was observed for more than 12 months. (J Virol. 2002 76:11434)
訳 これらのGagとEnvタンパク質の安定な発現が，12ヶ月以上の間観察された

■ [副] + stable

❶ more stable	より安定な	1,211
・more stable than ～	～より安定な	409
❷ less stable	より安定でない	532
❸ relatively stable	比較的安定な	332
❹ most stable ～	最も安定な～	306
❺ highly stable	高度に安定な	283

❶ The PRLR (S349A) mutant is resistant to ubiquitination and is more stable than its wild-type counterpart. (Mol Cell Biol. 2004 24:4038)
訳 ～は，それの野生型対応物より安定である

stably 副 安定に　　用例数 4,402

■ stably + [動／過分]

❶ stably transfected	安定にトランスフェクトされる	1,292

S

	・cells stably transfected with ～	～を安定にトランスフェクトされた細胞	341
❷	stably express ～	…は，～を安定に発現する	771
❸	stably maintained	安定に維持される	124
❹	stably associated with ～	～と安定に結合している	103

❶ We showed that **FADD$^{-/-}$ MEF cells stably transfected with TRAIL receptors are resistant to** TRAIL-mediated cell death.（J Biol Chem. 2000 275:25065）

訳 TRAIL 受容体を安定にトランスフェクトされたFADD$^{-/-}$マウス胚性線維芽細胞は，～に耐性である

❷ Study of the responses to agonists in **human embryonic kidney 293 cells stably expressing 5-HT2ARs** demonstrated that each agonist elicits a distinct transcriptome fingerprint.（J Neurosci. 2003 23:8836）

訳 5-HT2ARを安定に発現するヒト胎児由来腎臓293細胞

★★ stage 名 段階／ステージ，動 段階に分ける　用例数 33,156

stage	段階／ステージ／段階に分ける	20,593
stages	段階／ステージ／段階に分ける	12,238
staged	段階に分けられた	325

動詞としても用いられるが，名詞の用例が圧倒的に多い．複数形のstagesの割合は約45%とかなり高い．

◆類義語：step, phase

■ stage (s) ＋［前］

stage forは，set the stage forの用例が圧倒的に多い．

	the%	a/an%	ø%			
❶	41	-	45	stages of ～	～の段階	7,101
				・stage of development	発生の段階	484
				・stage of infection	感染の段階	291
				・stage of disease	疾患の段階	169
				・stage of differentiation	分化の段階	132
❷	32	40	10	stage in ～	～における段階	541
❸	98	1	0	stage for ～	～のお膳立て	351
				・set the stage for ～	…は，～のお膳立てをする	300

❶ The few oocytes that were fertilized **failed to progress beyond the two-cell stage of development**.（Dev Biol. 2005 288:405）

訳 ～は，発生の２細胞期を超えて進行することができなかった

❷ **At an early stage in their development**, they must first commit to either the $\gamma\delta$ or $\alpha\beta$ lineages.（Curr Opin Immunol. 1998 10:360）

訳 それらの発達における早期において，

❸ **Our results set the stage for ongoing investigations of the roles of CaMKII in** the formation and function of the olfactory system.（J Comp Neurol. 2002 443:226）

訳 われわれの結果は，～におけるCaMKⅡの役割についての進行中の研究のお膳立てをする

■ [形／名] + stage

❶ early stage	早期	1,755
❷ end stage	末期	1,445
❸ late stage	後期	932
❹ developmental stage	発生ステージ	461
❺ advanced stage	進行したステージ	420

❶ Our results demonstrate that **hemifusion is dominant at the early stage of the fusion reaction.** (J Biol Chem. 2005 280:30538)

訳 半融合は，融合反応の早期において優勢である

❷ Increases in type 1 phosphatase (PP1) activity **have been observed in end stage human heart failure**, but the role of this enzyme in cardiac function is unknown. (Mol Cell Biol. 2002 22:4124)

訳 ～が，末期のヒト心不全において観察されている

■ stage + [形]

❶ stage-specific ～	ステージ特異的な～	657

❶ **The human β-globin genes are expressed in a developmental stage-specific manner** in erythroid cells. (Mol Cell Biol. 2006 26:6832)

訳 ヒトβ-グロビン遺伝子は，発生ステージ特異的な様式で発現される

■ [前] + stage

❶ patients with stage I ～	ステージI～の患者	121

stain (動) 染色する　　用例数 3,536

stained	染色される／～を染色した	2,428
stain	～を染色する	819
stains	～を染色する	289

他動詞.

◆類義語：dye

■ stained + [前]

❶ stained with ～	～によって染色される	509
・be stained with ～	…は，～によって染色される	175
❷ stained for ～	～に対して染色される	304

❶ **Immobilized BAD-1 was stained with ruthenium red dye**, an indicator of calcium-binding proteins. (J Biol Chem. 2005 280:42156)

訳 固定されたBAD-1が，ルテニウムレッド色素によって染色された

❷ To detect apoptotic $CD8^+$ T cells, **sections were assayed by TUNEL and stained for $CD8^+$ T cells.** (J Virol. 2005 79:9019)

S

訳 切片がTUNEL法によってアッセイされ，そしてCD8^{+} T細胞に対して染色された

★ standard 名 標準／基準／スタンダード, 形 標準的な　用例数 17,787

standard	標準／基準／標準的な	15,782
standards	標準／基準／標準的な	2,005

可算名詞および形容詞として用いられる.

◆類義語：criterion

■ 名 standard ＋［前］

standard forの前にtheが付く割合は非常に高い．また，standard ofの前にtheが付く割合もかなり高い.

	the%	a/an%	ø%			
❶	62	13	19	standard of ～	～の基準／標準	596
				・standard of care	標準治療	88
❷	76	21	1	standard for ～	～のための基準／標準	418
				・standard for ～ing	～するための基準	119

❶ Procedural sedation has become the standard of care for managing pain and anxiety in children in the emergency department.（Curr Opin Pediatr. 2012 24:225）

訳 処置時の鎮静は，～のための標準治療になってきた

❷ Renal arteriography is generally regarded as the gold standard for diagnosing RA FMD.（J Am Coll Cardiol. 2003 41:1305）

訳 腎動脈造影は，RA FMDを診断するための究極の判断基準と一般的に見なされている

■ 名［形／名］＋ standard

❶	gold standard	究極の判断基準	680
❷	reference standard	参照標準	331
❸	internal standard	内部標準	272

■ 形 standard ＋［名］

❶	standard deviation	標準偏差	1,110
❷	standard therapy	標準治療	346

❷ In stable patients, intensive statin therapy provides long-term reduction in clinical events when compared with standard therapy.（J Am Coll Cardiol. 2005 46:1405）

訳 標準治療と比較すると

★ start 名 開始, 動 始まる／始める　用例数 10,682

start	開始／始まる／～を始める	5,415
starting	始まる／～を始める	3,741

started	始まる／～を始める／始められる	1,069
starts	始まる／～を始める	457

動詞としても用いられるが，名詞の用例が非常に多い．複数形のstartsの用例はほとんどなく，原則，**不可算名詞**として使われる．動詞としては，自動詞と他動詞の両方で使われる．

◆類義語：initiation, onset, initiate, begin

■ 名 start ＋［前］

start ofの前にtheが付く割合は圧倒的に高い．

	the %	a/an %	ø %			
❶	95	0	5	start of ～	～の開始	870

❶ **At the start of the study,** 50 adults with EoE underwent esophagogastroduodenoscopies (EGDs), biopsies, and skin-prick tests for food and aeroallergens. (Gastroenterology. 2012 142:1451)

訳 研究の開始時に，

■ 名 start ＋［名］

❶ start site	開始部位	2,053
・transcription start site	転写開始部位	1,307
・transcriptional start site	転写開始部位	581
・translation start site	翻訳開始部位	170
❷ start codon	開始コドン	440

■ 動 start to *do*

❶ start to *do*	…は，～し始める	437

❶ **The patient only started to recover** as her plasma glutamine began to return to normal. (Crit Care Med. 2011 39:2550)

訳 その患者は，回復し始めたに過ぎなかった

■ 動 starting ＋［前］

❶ starting from ～	～から始めて	554
❷ starting at ～	～から始めて	386
❸ starting with ～	～で始めて	360

❶ **Starting from known zinc binding sites in crystal structures,** we predicted 202 putative partial surface zinc binding sites in FH, most of which were in SCR-6. (J Mol Biol. 2011 408:714)

訳 結晶構造における既知の亜鉛結合部位から始めて，

❷ **Starting at 55 years of age,** we followed up 61 585 men and women for 700 000 person-years. (Circulation. 2012 125:37)

訳 55歳の年齢から始めて，

❸ **Starting with yeast as a model system,** we investigated an uncharacterized but highly conserved mitochondrial protein (named here Sdh5). (Science. 2009 325:1139)

S

訳 モデルシステムとしての酵母で始めて,

■ 動 starting ＋［前］

❶ starting point	出発点	437
❷ starting materials	出発物質	206

★★ state 名 状態, 動 述べる 用例数 65,833

state	状態／述べる	45,991
states	状態／述べる	19,609
stated	述べた／述べられる	198
stating	述べる	35

動詞としても用いられるが, 名詞の用例が圧倒的に多い. 複数形のstatesの割合が, 約25%ある可算名詞.

◆類義語：status, condition, situation, mention, describe

■ state ＋［前］

state ofの前にtheが付く割合は非常に高い.

	the%	a/an%	ø%			
❶	76	20	2	state of ～	～の状態	6,721
				・state-of-the-art ～	最先端の～	506
❷	29	55	11	state in ～	～における状態	1,127
				・state in which ～	（そこで）～である状態	167

❶ Changes in the phosphorylation **state of** p53 are important in increasing its half-life and potency as a transcription factor.（Oncogene. 2001 20:1076）

訳 p53のリン酸化状態の変化は, それの半減期を増大させる際に重要である

❷ Charge and geometry measures suggest transition **states in which** these features change synchronously, again in contrast to many proton transfer reactions.（J Am Chem Soc. 2005 127:2324）

訳 これらの特徴が同調的に変化する遷移状態

■ ［形／名］＋ state

❶ steady-state ～	定常状態～	8,212
❷ transition state	遷移状態	3,368
❸ solid-state ～	固体の～	1,647
❹ ground state	基底状態	1,206
❺ native state	天然の状態	954
❻ excited state	励起状態	800
❼ phosphorylation state	リン酸化状態	551

❶ The **steady-state** levels of topA transcripts were similar in the various strains.（J Bacteriol. 2003 185:6883）

訳 定常状態レベルのtopA転写物は，さまざまな株において類似していた

statistical 形 統計学的な　用例数 6,964

◆類義語：statistic

■ statistical ＋［名］

❶ statistical significance	統計学的有意	821
❷ statistical analysis	統計学的解析	804
❸ statistical methods	統計学的方法	402
❹ statistical power	検定力	314
❺ statistical model	統計モデル	203

❶ Patients with the t(11;14)(q13;q32) appeared to have better survival and response to treatment, **although this did not reach statistical significance**. (Blood. 2002 99:3735)
訳 これは統計学的有意に達しなかったけれど

statistically 副 統計学的に　用例数 6,178

■ statistically ＋［形／副］

❶ statistically significant	統計学的に有意な	4,234
・be statistically significant	…は，統計学的に有意である	542
・statistically significant differences	統計学的に有意な違い	431
❷ statistically significantly	統計学的に有意に	385
❸ statistically different	統計学的に異なる	284

❶ **No statistically significant differences were observed** between bonds made in vivo and those made in vitro at any time period. (J Dent Res. 1988 67:467)
訳 統計学的に有意な違いは観察されなかった

❷ **LEF was statistically significantly superior to** MTX in improving physical function as measured by the HAQ DI over 24 months of treatment. (Arthritis Rheum. 2001 44:1984)
訳 LEFは，統計学的に有意に～に勝っていた

❸ Progression-free survival **was statistically different between** responder and nonresponder groups (P <.0001). (J Clin Oncol. 2003 21:3853)
訳 …は，～の間で統計学的に異なっていた

★ status 名 状態／状況／地位　用例数 13,835

status	状態／状況／地位	13,796
statuses	状態／状況／地位	39

複数形のstatusesの割合は0.3%しかなく，原則，**不可算名詞**として使われる．

◆類義語：state, condition, situation, position

S

■ status ＋［前］

status ofの前にtheが付く割合は圧倒的に高い．

	the%	a/an%	ø%			
❶	83	1	16	status of ～	～の状態	2,870
❷	4	3	74	status in ～	～における状態	695

❶ Our studies also suggest that **the methylation status of SNCG gene can be used as a sensitive molecular tool in early detections of tumorigenesis.** (Cancer Res. 2005 65:7635)

訳 SNCG遺伝子のメチル化状態は，腫瘍形成の早期の検出における鋭敏な分子ツールとして使われうる

❷ Here we analyze the genomic structure and alternative splicing of L3mbtl and **assess its imprinting status in mouse.** (Genomics. 2005 86:489)

訳 ～は，マウスにおけるそれのインプリンティング状態を査定する

■［名／形］＋ status

❶	health status	健康状態	723
❷	socioeconomic status	社会経済的な地位	597
❸	functional status	機能的状態	553
❹	smoking status	喫煙状況	474
❺	performance status	活動状態	451
❻	methylation status	メチル化状態	414
❼	nutritional status	栄養状態	364
❽	phosphorylation status	リン酸化状態	343

★★ step ㊔ 段階／ステップ，㊍ 段階を進める　用例数 30,693

step	段階／ステップ／段階を進める	20,729
steps	段階／ステップ／段階を進める	9,528
stepping	段階を進める	311
stepped	段階を進めた／段階を進められた	125

動詞として用いられることもあるが，名詞の用例が圧倒的に多い．複数形のstepsの割合が，約35%ある可算名詞．

◆類義語：stage, phase, grade

■ step ＋［前］

step ofの前にtheが付く割合は非常に高い．

	the%	a/an%	ø%			
❶	37	59	0	step in ～	～における段階	5,699
				・step in understanding ～	～を理解する際の段階	81
❷	74	10	0	step of ～	～の段階	2,018

❶ A conformational change coupled to ADP release **is the rate-limiting step in the pathway.**

(Biochemistry. 2005 44:16633)
訳 ～は，その経路における律速段階である

❷ The splicing factor Prp18 is required for the second step of pre-mRNA splicing. (Proc Natl Acad Sci USA. 2000 97:3022)
訳 ～は，プレメッセンジャーRNAスプライシングの２番目の段階のために必要とされる

■ [形] ＋ step

❶ first step	最初の段階	1,980
❷ two-step ～	2段階～	1,297
❸ rate-limiting step	律速段階	1,239
❹ key step	鍵となる段階	772
❺ critical step	決定的な段階	620
❻ important step	重要な段階	532
❼ second step	2番目の段階	471
❽ initial step	初期段階	417
❾ single step	一段階	415

❶ As a first step toward understanding the principles by which the two isoforms assemble at complex promoters, we examined the energetics of PR-B self-association using sedimentation velocity and sedimentation equilibrium methods. (Biochemistry. 2005 44:9528)
訳 ～である原理の理解への最初の段階として

S

★ still 副 まだ　　用例数 11,153

◆類義語：yet

■ be still ＋ [形]

★ be still ～	…は，まだ～である	6,098
❶ be still unclear	…は，まだ不明である	499
・it is still unclear ～	～は，まだ不明である	113
❷ be still unknown	…は，まだ知られていない	331
❸ be still present	…は，まだ存在する	175

❶ However, it is still unclear how the propagation of Tau misfolding occurs. (J Biol Chem. 2012 287:19440)
訳 どのように～かは，まだ不明である

■ be still ＋ [副]

❶ be still poorly understood	…は，まだあまり理解されていない	316
❷ be still largely unknown	…は，まだほとんど知られていない	113

❶ Although the basic pathway of HIV-1 entry has been extensively studied, the detailed mechanism is still poorly understood. (J Biol Chem. 2013 288:234)
訳 詳細な機構は，まだあまり理解されていない

■ still ＋［動］

❶ still remain ～	…は，まだ～のままである	318

❶ The definitive mechanism of pathogenesis still remains to be elucidated. (Curr Opin Hematol. 2013 20:26)

訳 病因の確定的な機構は，まだ解明されないままである

★ stimulate (動) 刺激する 用例数 45,958

stimulated	刺激される／～を刺激した	25,914
stimulate	～を刺激する	7,711
stimulates	～を刺激する	6,425
stimulating	～を刺激する	5,908

他動詞.

◆類義語：prime, treat

■ stimulate ＋［名句］

❶ stimulate transcription	…は，転写を刺激する	328
❷ stimulate the expression of ～	…は，～の発現を刺激する	225
❸ stimulate proliferation	…は，増殖を刺激する	217
❹ stimulate the activity of ～	…は，～の活性を刺激する	187
❺ stimulate the production of ～	…は，～の産生を刺激する	164
❻ stimulate the growth of ～	…は，～の増殖を刺激する	104
❼ stimulate the formation of ～	…は，～の形成を刺激する	102

❶ This assay also demonstrates that β-actin stimulates transcription by RNA polymerase II. (Nat Cell Biol. 2004 6:1094)

訳 β-アクチンは，RNAポリメラーゼⅡによる転写を刺激する

■ ［名／動］＋ to stimulate

★ to stimulate ～	～を刺激する…	3,446
❶ ability to stimulate ～	～を刺激する能力	381
❷ fail to stimulate ～	…は，～を刺激することができない	237
❸ be shown to stimulate ～	…は，～を刺激することが示される	132

❶ We hypothesized that TCDD-mediated inhibition of apoptosis was due to its ability to stimulate the EGF receptor (EGFR) pathway. (Cancer Res. 2001 61:3314)

訳 TCDDに仲介されたアポトーシスの抑制は，EGF受容体（EGFR）経路を刺激するそれの能力のせいであった

■ stimulated ＋［前］

❶ stimulated by ～	～によって刺激される	3,103

	・be stimulated by ～	…は，～によって刺激される	1,424
❷	stimulated with ～	～によって刺激される	1,549
❸	stimulated in ～	～において刺激される	406

❶ Finally, the hMutSβ complex is shown to specifically bind to psoralen ICLs, and **this binding is stimulated by the addition of PCNA.** (Mol Cell Biol. 2002 22:2388)

訳 この結合は，PCNAの添加によって刺激される

❷ **When T cells were stimulated with dendritic cells** infected with vaccinia vectors expressing EBNA1, 18 of 19 donors secreted IFN-γ, whereas only two of 19 secreted IL-4. (J Clin Invest. 2001 107:121)

訳 T細胞が樹状細胞によって刺激されたとき

■ ［名］＋ be stimulated

★	be stimulated	…が，刺激される	2,860
❶	cells were stimulated	細胞が，刺激された	211

★ stimulation 名 刺激　　用例数 31,912

stimulation	刺激	31,766
stimulations	刺激	146

複数形のstimulationsの割合は0.5%しかなく，原則，**不可算名詞**として使われる．

◆類義語：stimulus, irritation, treatment

■ stimulation ＋ ［前］

	the %	a/an %	ø %			
❶	14	2	84	stimulation of ～	～の刺激	10,227
				・stimulation of ～ by …	…による～の刺激	930
				・stimulation of ～ with …	…による～の刺激	747
❷	1	0	98	stimulation with ～	～による刺激	2,024
❸	0	1	98	stimulation in ～	～における刺激	1,040
❹	2	0	96	stimulation by ～	～による刺激	998

❶ In contrast, **stimulation of DNA synthesis by dda helicase requires** direct gp32-dda protein-protein interactions and is relatively unaffected by mutations in gp32 that destabilize its ssDNA binding activity. (J Biol Chem. 2004 279:19035)

訳 ddaヘリカーゼによるDNA合成の刺激は，～を必要とする

❶ **Following the stimulation of cells with tumor necrosis factor α (TNF-α)**, the IκB kinase (IKK) complex rapidly phosphorylates NF-κB1 p105 on serine 927 in the PEST region. (Mol Cell Biol. 2003 23:402)

訳 腫瘍壊死因子α（TNF-α）による細胞の刺激に続いて

❷ Additionally, **after stimulation with thioglycolate**, lpr/lpr and B6 mice showed equivalent numbers of peritoneal macrophages. (J Immunol. 2004 173:7584)

訳 チオグリコール酸による刺激のあと

❹ **These cells are responsive, however, to stimulation by TNFα**, which enhances the production of specific members of the MMP family. (Invest Ophthalmol Vis Sci. 2002 43:260)

S

訳 しかし，これらの細胞はTNFαによる刺激に応答性である

■ stimulation ＋［動］

❶	stimulation induced ～	刺激は，～を誘導した	129
❷	stimulation resulted in ～	刺激は，～という結果になった	108

❶ An anti-Syk-phospho-346 tyrosine antibody indicated that **antigen stimulation induced only a very minor increase in the phosphorylation of this tyrosine.** (Mol Cell Biol. 2002 22:8144)
訳 抗原刺激は，このチロシンのリン酸化のほんのわずかな増大しか誘導しなかった

■［前］＋ stimulation

❶	after stimulation	刺激のあと	789
❷	upon stimulation	刺激するとすぐに	456
❸	in response to stimulation	刺激に応答して	283

❶ ALX overexpression in Jurkat T cells results in inhibition of IL-2 promoter activation **after stimulation with superantigen.** (J Biol Chem. 2003 278:45128)
訳 スーパー抗原による刺激のあと

❷ **Upon stimulation via the B or T cell receptors,** LAX is rapidly phosphorylated by Src and Syk family tyrosine kinases and interacts with Grb2, Gads, and p85. (J Biol Chem. 2002 277:46151)
訳 B細胞受容体あるいはT細胞受容体を介して刺激するとすぐに

❸ This study was undertaken to investigate whether PI 3-kinase participates in cell cycle regulation **in response to stimulation with FGF-2** in CECs. (Invest Ophthalmol Vis Sci. 2003 44:1521)
訳 FGF-2による刺激に応答して

★ stimulus　名 刺激／刺激物　　用例数 21,280

stimuli	刺激	12,384
stimulus	刺激	8,896

複数形stimuliの割合が，約60%と非常に高い．

◆類義語：stimulation, irritation, treatment

■ stimulus/stimuli ＋［前］

	the%	a/an%	ø%			
❶	3	-	92	stimuli in ～	～における刺激	606
❷	33	63	2	stimulus for ～	～の刺激	296

❶ **Many of its neurons respond to visual stimuli in the space near** the arms or face. (Science. 1997 277:239)
訳 それのニューロンの多くは，～の近くの空間における視覚的な刺激に応答する

❷ **Ultraviolet radiation is a potent stimulus for melanosome transport and transfer.** (J Invest Dermatol. 2001 116:296)
訳 紫外線照射は，メラノソーム輸送と移動の強力な刺激である

■ [形] + stimuli

❶	visual stimuli	視覚的な刺激	446
❷	environmental stimuli	環境刺激	327
❸	inflammatory stimuli	炎症性の刺激	306
❹	apoptotic stimuli	アポトーシス性の刺激	300
❺	mechanical stimuli	機械的刺激	274
❻	extracellular stimuli	細胞外刺激	234
❼	external stimuli	外部刺激	230

storage ㊔ 貯蔵／蓄積 用例数 4,761

複数形の用例はなく，原則，不可算名詞として使われる．

◆類義語：store, accumulation

■ storage + [前]

	the%	a/an%	ø%			
❶	56	0	43	storage of ～	～の貯蔵／蓄積	431
❷	1	0	89	storage in ～	～における貯蔵／蓄積	275

❶ LSDs mainly stem from deficiencies in lysosomal enzymes, but also in some non-enzymatic lysosomal proteins, which **lead to abnormal storage of macromolecular substrates.** (J Cell Biol. 2012 199:723)

訳 ～は，高分子基質の異常な蓄積につながる

S

store ㊔ 貯蔵，㊍ 貯蔵する 用例数 6,068

stores	貯蔵／～を貯蔵する	2,378
stored	貯えられる／～を貯蔵した	1,810
store	貯蔵／～を貯蔵する	1,679
storing	～を貯蔵する	201

名詞の用例が多いが，他動詞としても用いられる．複数形のstoresの割合は約60%と非常に高い．

◆類義語：storage, pool, preservation, reserve, preserve

■ ㊔ stores + [前]

	the%	a/an%	ø%			
❶	2	-	83	stores in ～	～における貯蔵	167
❷	9	-	87	stores of ～	～の貯蔵	111

❶ To test its importance in vivo, we inactivated the murine Cybrd1 gene and **assessed tissue iron stores in Cybrd1-null mice.** (Blood. 2005 106:2879)

訳 ～は，Cybrd1欠損マウスにおける組織鉄貯蔵を査定した

■ 名 [形／名] ＋ stores

❶ intracellular stores	細胞内貯蔵	440
❷ Ca^{2+} stores	Ca^{2+}貯蔵	281
❸ iron stores	鉄貯蔵	164

❷ Previously, **caffeine-sensitive intracellular Ca^{2+} stores were shown to play a role in regulating glutamate release from photoreceptors.** (J Physiol. 2003 547:761)

訳 カフェインに感受性の細胞内Ca^{2+}貯蔵は，光受容器からのグルタミン酸放出を調節する際に役割を果たすことが示された

■ 動 stored ＋ [前]

❶ stored in ～	～において貯えられる	579
・be stored in ～	…が，～において貯えられる	207
❷ stored at ～	～で貯えられる	123

❶ **Insulin was stored in the cytoplasm** and released into the culture medium in a glucose-dependent manner. (Am J Pathol. 2005 166:1781)

訳 インスリンが，細胞質に貯えられた

★ strategy 名 戦略／ストラテジー 用例数 27,875

strategy	戦略／ストラテジー	14,714
strategies	戦略／ストラテジー	13,161

複数形のstrategiesの割合は約45%とかなり高い．

◆分類：methodology, method

■ strategy ＋ [前]

	the%	a/an%	ø%			
❶	5	90	0	strategy for ～	～のための戦略	3,428
				・strategy for ～ing	～するための戦略	1,281
❷	2	94	1	strategy to *do*	～するための戦略	2,211
				・strategy to identify ～	～を同定するための戦略	146
❸	35	53	1	strategy of ～	～の戦略	639
				・strategy of ～ing	～する戦略	186
❹	10	81	4	strategy in ～	～における戦略	622

❶ Thus, blocking C5 **is a potential therapeutic strategy for preventing renal injury** in cryoglobulinemia. (J Immunol. 2005 175:6909)

訳 ～は，腎傷害を防ぐための潜在的な治療上の戦略である

❷ To avoid these collateral effects of COX-2 inhibition, **a strategy to identify and block specific prostanoid-receptor interactions** may be required. (J Biol Chem. 2006 281:3321)

訳 特異的なプロスタノイド受容体相互作用を同定し，そしてブロックするための戦略

■ [形] + strategy

❶	therapeutic strategies	治療上の戦略	1,300
❷	treatment strategies	治療戦略	731
❸	novel strategy	新規の戦略	407
❹	new strategy	新しい戦略	368
❺	effective strategy	効果的な戦略	286
❻	management strategies	管理戦略	260
❼	general strategy	一般的な戦略	258

❹ Trp catabolites and their derivatives **offer a new strategy for treating** T_H1-mediated autoimmune diseases such as MS.（Science. 2005 310:850）

訳 …は，～を治療するための新しい戦略を提供する

strength 名 強度　　用例数 8,795

strength	強度	7,694
strengths	強度	1,101

複数形のstrengthsの割合は約15%あるが，単数形は無冠詞で使われることが多い.

◆分類：intensity

■ strength + [前]

strength ofの前にtheが付く割合は圧倒的に高い.

	the%	a/an%	ø%			
❶	87	4	8	strength of ～	～の強度	2,471
❷	9	2	83	strength in ～	～における強度	223

❶ Odds ratio (OR) and 95% confidence interval (95% CI) **were used to investigate the strength of the association.**（PLoS One. 2012 7:e50966）

訳 ～が，関連の強度を精査するために使われた

■ [形／名] + strength

❶	ionic strength	イオン強度	1,352
❷	synaptic strength	シナプス強度	597
❸	field strength	場の強度	210
❹	muscle strength	筋力	202

★★ stress 名 ストレス，動 ストレスを与える／強調する　用例数 38,038

stress	ストレス／～にストレスを与える／～を強調する	35,136
stresses	ストレス／～にストレスを与える／～を強調する	2,000
stressed	ストレスを与えられた／強調される／～にストレスを与える	857
stressing	～にストレスを与える／～を強調する	45

S

他動詞としても用いられるが，名詞の用例が圧倒的に多い．複数形のstressesの割合は約5％あるが，単数形は無冠詞で使われることが圧倒的に多い．

■ stress ＋［前］

	the%	a/an%	ø%			
❶	0	0	89	stress in ～	～におけるストレス	1,285

❶ KGF was shown to significantly reduce DNA damage and cytoskeletal rearrangement caused by oxidative stress in cultured ARPE-19 cells.（Invest Ophthalmol Vis Sci. 2005 46:3435）

訳 KGFは，培養されたARPE-19細胞における酸化ストレスによって引き起こされるDNAダメージと細胞骨格再編成を有意に低下させることが示された

■ stress-［過分］

❶ stress-induced ～	ストレスに誘導される～	2,071
❷ stress-related ～	ストレスに関連する～	365

❶ Caspase-2 is an initiating caspase required for stress-induced apoptosis in various human cancer cells.（J Biol Chem. 2005 280:38271）

訳 カスパーゼ-2は，さまざまなヒト癌細胞のストレスに誘導されるアポトーシスに必要とされる開始カスパーゼである

■ stress ＋［名］

❶ stress response	ストレス応答	2,047
❷ stress conditions	ストレス状態	579

❶ These studies suggest that accumulation of aged and damaged proteins can lead to cellular toxicity and a cell stress response in C3KO muscles, and that these characteristics are pathological features of LGMD2A.（Hum Mol Genet. 2005 14:2125）

訳 ～は，C3KO筋肉における細胞毒性および細胞ストレス応答につながりうる

■ ［形／名］＋ stress

❶ oxidative stress	酸化ストレス	6,891
❷ shear stress	ずり応力	1,262
❸ osmotic stress	浸透圧性ストレス	544
❹ cellular stress	細胞ストレス	484
❺ genotoxic stress	遺伝毒性ストレス	468
❻ heat stress	熱ストレス	409

■ ［前］＋ stress

❶ in response to stress	ストレスに応答して	249
❷ effects of stress	ストレスの効果	103

❶ Activating transcription factor 2（ATF2）is regulated by JNK/p38 in response to stress.（Mol Cell. 2005 18:577）

訳 ～は，ストレスに応答してJNK/p38によって調節される

striking 形 著しい　　用例数 4,148

◆類義語：marked, prominent, profound, manifest, dramatic, significant

■ striking ＋［名］

❶ striking differences	著しい違い	228
❷ in striking contrast	著しく対照的に	224
❸ a striking increase	著しい増大	141
❹ striking similarities	著しい類似性	137

❶ There were also striking differences in resistance to these antibiotics between non-infectious and infectious B. burgdorferi strains.（Gene. 2003 303:131）

訳 ～の間には，また，これらの抗生物質に対する抵抗性の著しい違いがあった

❷ In striking contrast to the wild-type cells, which were diploid, the αB-crystallin$^{-/-}$ cultures had a high proportion of tetraploid and higher ploidy cells, indicating that the loss of αB-crystallin is associated with an increase in genomic instability.（FASEB J. 2001 15:221）

訳 野生型細胞とは著しく対照的に

■［副］＋ striking

❶ most striking ～	最も著しい～	338

❶ One of the most striking features of this complex is the rotation of thymine 2 (T2) away from the DNA helix and into a pocket within the Tnp.（J Biol Chem. 2002 277:11284）

訳 この複合体の最も著しい特徴の１つは，～である

S

strikingly 副 著しく　　用例数 3,080

◆類義語：remarkably, markedly, dramatically, prominently, notably, extremely, unusually, significantly

■ strikingly ＋［形］

❶ strikingly similar	著しく似ている	356
❷ strikingly different	著しく異なる	246

❶ Mice with homozygous deletion of the coding sequence of KCNE1 have inner ear defects strikingly similar to those seen in the corresponding human condition.（J Physiol. 2003 552:535）

訳 …は，～において見られるそれらに著しく似ている内耳の欠損を持つ

❷ The structure is strikingly different from that of the ryanodine receptor at similar resolution despite molecular similarities between these two calcium release channels.（EMBO J. 2002 21:3575）

訳 その構造は，リアノジン受容体のそれと著しく異なっている

★ strong 形 強い／強力な　用例数 19,443

◆類義語：robust, potent, powerful, intensive, intense

■ strong ＋［名］

❶ strong evidence	強力な証拠	1,290
・provide strong evidence	…は，強力な証拠を提供する	802
・results provide strong evidence	結果は，強力な証拠を提供する	153
❷ strong correlation	強い相関	603
❸ strong association	強い関連	426
❹ strong support	強力な支持	334
❺ strong binding	強い結合	285

❶ These results provide strong evidence that multiply charged organic ions are formed by the charged residue mechanism. (J Am Chem Soc. 2003 125:2319)
訳 これらの結果は，～という強力な証拠を提供する

❷ In contrast, there was not a strong correlation between transcriptional activation and oncogenic transformation. (Oncogene. 2002 21:1611)
訳 転写の活性化と発癌性形質転換の間には強い相関はなかった

★ strongly 副 強く　用例数 18,599

◆類義語：potently, robustly, intensively, intensely

■ strongly ＋［動］

❶ strongly suggest ～	…は，～を強く示唆する	2,250
・results strongly suggest that ～	結果は，～ということを強く示唆する	480
・data strongly suggest that ～	データは，～ということを強く示唆する	307
・findings strongly suggest that ～	知見は，～ということを強く示唆する	136
❷ strongly support ～	…は，～を強く支持する	739
❸ strongly inhibit ～	…は，～を強く抑制する	269
❹ strongly influence ～	…は，～に強く影響する	246
❺ strongly indicate ～	…は，～を強く示す	180
❻ strongly implicate ～	…は，～を強く関連づける	161

❶ These results strongly suggest that genistein alters bilayer mechanical properties, which in turn modulates channel function. (Biochemistry. 2003 42:13646)
訳 これらの結果は，～ということを強く示唆する

❷ These data strongly support our hypothesis that mouse ES cells can be accelerated to differentiate into CM by NO treatment. (Proc Natl Acad Sci USA. 2004 101:12277)
訳 これらのデータは，～というわれわれの仮説を強く支持する

■ strongly ＋［過分／形］

❶ strongly associated with ～	～と強く関連する	1,368
・be strongly associated with ～	…は，～と強く関連する	981

❷ strongly inhibited	強く抑制される	656
❸ strongly correlated with ～	～と強く相関する	487
❹ strongly induced	強く誘導される	379
❺ strongly expressed	強く発現する	372
❻ strongly influenced by ～	～によって強く影響される	336
❼ strongly dependent on ～	～に強く依存する	293
❽ strongly reduced	強く低下する	240
❾ strongly activated	強く活性化される	219

❶ The discontinuation of antiplatelet therapy was strongly associated with the development of ST in this patient population. (Circulation. 2004 109:1930)

訳 抗血小板療法の中断は，STの発症と強く関連した

❷ ABA induction of ABI1 and ABI2, negative regulators of ABA signaling, was strongly inhibited by VP1, revealing a second pathway of feed-forward regulation. (Plant Physiol. 2003 132:1664)

訳 ～は，VP1によって強く抑制された

■ [動] ＋ strongly ＋ [前]

❶ correlate strongly with ～	…は，～と強く相関する	305
❷ bind strongly to ～	…は，～に強く結合する	220
❸ depend strongly on ～	…は，～に強く依存する	193
❹ interact strongly with ～	…は，～と強く相互作用する	177

❶ Low TRECs also correlated strongly with extensive chronic graft-versus-host disease (P <.01). (Blood. 2002 100:2235)

訳 ～は，広範な慢性の移植片対宿主病と強く相関した

■ [副] ＋ strongly

❶ more strongly	より強く	826
❷ most strongly	最も強く	491

❶ Nup2p binds Kap60p more strongly than NLSs and accelerates release of NLSs from Kap60p. (EMBO J. 2003 22:5358)

訳 Nup2pは，NLSsより強くKap60pに結合する

❷ Genes most strongly associated with survival were identified by using the Cox proportional hazards survival analysis. (Gastroenterology. 2004 127:S51)

訳 生存と最も強く関連する遺伝子が，～を使うことによって同定された

★★ structural 形 構造上の／構造的な　用例数 38,072

■ structural ＋ [名]

❶ structural changes	構造上の変化	1,805
❷ structural features	構造上の特徴	1,705
❸ structural basis	構造的な基礎	1,377

S

❹ structural studies	構造研究	962
❺ structural information	構造上の情報	937
❻ structural analysis	構造分析	799

❶ The highly helical Kar3/Cik1 nonmotor region and visible stalk indicate that **dimerization with Cik1 causes structural changes in Kar3**. (EMBO J. 2005 24:3214)

訳 Cik1との二量体形成は，Kar3の構造上の変化を引き起こす

❷ **The folded region of KIX has all the structural features of a globular protein**, including three α-helices, two short 3_{10} helices, and a well-packed hydrophobic core. (Biochemistry. 2003 42:7044)

訳 KIXの折り畳まれた領域は，球状タンパク質の構造上の特徴のすべてを持つ

■ ［形］＋ structural

❶ major structural ～	主要な構造上の～	348
❷ secondary structural ～	二次構造上の～	283
❸ significant structural ～	顕著な構造上の～	255

❷ By contrast, probes of CD at far-UV indicate that **secondary structural changes precede the early expansion phase** reported by SAXS and fluorescence. (J Mol Biol. 2004 340:419)

訳 二次構造上の変化は，初期の拡大相に先行する

■ structural and ＋［形］

❶ structural and functional ～	構造的および機能的な～	1,459

❶ **These structural and functional properties of E2F-7 imply a unique role in** regulating cellular proliferation. (Oncogene. 2004 23:5138)

訳 E2F-7のこれらの構造的および機能的な性質は，～におけるユニークな役割を意味する

structurally 副 構造的に　　用例数 5,711

■ structurally ＋［過分／形］

❶ structurally related	構造的に関連している	850
❷ structurally similar	構造的に似ている	609
・be structurally similar to ～	…は，構造的に～に似ている	182
❸ structurally distinct	構造的に別個の	415
❹ structurally diverse	構造的に多様な	349
❺ structurally characterized	構造的に特徴づけられる	238
❻ structurally unrelated	構造的に無関係の	187
❼ structurally homologous	構造的に相同的な	173

❶ **Netrin-G1 is a lipid-anchored protein that is structurally related to the netrin family** of axon guidance molecules. (Nat Neurosci. 2003 6:1270)

訳 ネトリン-G1は，ネトリンファミリーに構造的に関連する脂質アンカータンパク質である

❷ The lack of this domain in the short RGS9 isoform is compensated by the action of **a G protein effector subunit that is structurally similar to this C-terminal domain**. (Neuron. 2003 38:857)

訳 このC末端ドメインに構造的に類似しているGタンパク質効果器サブユニット

❸ These compounds are structurally distinct from those used to build the model and were discovered by testing only 62 library compounds.（J Med Chem. 2004 47:4875）

訳 これらの化合物は，そのモデルを作るために使われたそれらと構造的に別個である

■ structurally and ＋［副］

❶	structurally and functionally	構造的および機能的に	322

❶ Cells of a multicellular organism are genetically homogeneous but structurally and functionally heterogeneous owing to the differential expression of genes.（Nat Genet. 2003 33:245）

訳 ～は，遺伝的に均一だが構造的および機能的には不均一である

★★ structure ㊔ 構造，㊍ 構造化する　用例数 113,262

structure	構造／構造化する	75,479
structures	構造／構造化する	35,292
structured	構造化された／構造化した	2,302
structuring	構造化する	189

動詞としても用いられるが，名詞の用例が圧倒的に多い．複数形のstructuresの割合が，約35%ある可算名詞．

◆類義語：conformation, architecture

■ structure ＋［前］

structure ofの前にtheが付く割合は圧倒的に高い．

	the%	a/an%	ø%			
❶	86	12	0	structure of ～	～の構造	21,187
				・structure of the complex	複合体の構造	294
❷	19	28	48	structure in ～	～における構造	2,231
				・structure in which ～	～である構造	153

❶ The structure of the complex explains how vitronectin binds to and stabilizes the active conformation of PAI-1.（Nat Struct Biol. 2003 10:541）

訳 その複合体の構造は，どのように～かを説明する

❷ The data are consistent with a structure in which the excess electron is bound to the surface of the cluster.（Science. 2005 307:93）

訳 それらのデータは，過剰の電子が～の表面に結合する構造と一致する

■ structure ＋［動］

❶	structure reveals ～	構造は，～を明らかにする	662
❷	structure shows ～	構造は，～を示す	347
❸	structure provides ～	構造は，～を提供する	239

❶ The structure reveals that the BIR3 domain forms a heterodimer with a caspase-9 monomer.（Mol Cell. 2003 11:519）

訳 その構造は，～ということを明らかにする

■ structure-［過分／形］

❶ structure-based ～	構造に基づく～	1,052
❷ structure-specific ～	構造特異的な～	263

❶ The structural information provides a foundation for structure-based design of new inhibitors against these enzymes.（Mol Cell. 2004 16:881）

訳 その構造上の情報は，これらの酵素に対する新しい阻害剤の構造に基づく設計のための基礎を提供する

■［形／名］＋ structure

❶ crystal structure	結晶構造	7,511
❷ secondary structure	二次構造	3,555
❸ chromatin structure	クロマチン構造	1,741
❹ protein structure	タンパク質構造	1,501
❺ three-dimensional structure	三次元構造	1,138
❻ tertiary structure	三次構造	1,138
❼ solution structure	溶液構造	975
❽ X-ray structure	X線構造	794

❷ Hybridization of fluorescent molecular beacons provides real-time detection of RNA secondary structure with high specificity.（Nucleic Acids Res. 2005 33:5763）

訳 ～は，高い特異性を持つRNA二次構造のリアルタイムの検出を提供する

■ structure ＋［名］

❶ structure-activity relationships	構造－活性相関	574
❷ structure prediction	構造予測	546
❸ structure determination	構造決定	497
❹ structure-function relationships	構造－機能相関	361

❶ Toward this goal, three-dimensional quantitative structure-activity relationship (3D-QSAR) models were generated using in vitro data associated with inhibition of P-gp function.（Mol Pharmacol. 2002 61:964）

訳 三次元定量的構造－活性相関（3D-QSAR）モデルが，～を使って作られた

■ structure and ＋［名］／［名］＋ and structure

❶ structure and function	構造と機能	2,128
❷ structure and dynamics	構造と動力学	315
❸ sequence and structure	配列と構造	229

❶ Understanding the structure and function of PKS will provide clues to the molecular basis of polyketide biosynthesis specificity.（Biochemistry. 2004 43:14529）

訳 PKSの構造と機能を理解することは，～への手がかりを提供するであろう

⁑ study ㊔ 研究，㊍ 研究する　用例数 292,639

study	研究／～を研究する	136,750
studies	研究／～を研究する	121,336
studied	研究される／～を研究した	29,973
studying	～を研究する	4,580

名詞の用例が多いが，他動詞としても使われる．複数形のstudiesの割合は約50%と非常に高い．

◆類義語：investigation, test, examination, research, survey, search, work, trial, investigate, explore, examine, look at, dissect

■ ㊔ studies ＋［前］

	the%	a/an%	ø%			
❶	3	-	85	studies of ～	～の研究	14,421
❷	2	-	89	studies in ～	～における研究	5,397
❸	2	-	89	studies with ～	～による研究	3,111
❹	3	-	90	studies on ～	～に関する研究	3,076

❶ Previous studies of this bacterium have focused on mechanisms of adaptation for growth in alkaline environments. (J Bacteriol. 2004 186:818)

訳 この細菌の以前の研究は，アルカリ性環境における成長のための適応の機構に焦点を当ててきた

❷ Studies in experimental models and preliminary clinical experience suggested a possible therapeutic role for the soluble tumor necrosis factor antagonist etanercept in heart failure. (Circulation. 2004 109:1594)

訳 実験モデルにおける研究および予備的臨床経験は，可溶性腫瘍壊死因子のありうる治療上の役割を示唆した

❸ These $BDEs_{O-H}$ were verified with reactivity studies with substrates having known X–H bond energies (X = C, N, O). (J Am Chem Soc. 2003 125:13234)

訳 ～が，基質を使った反応性研究によって検証された

❹ Previous studies on the transcriptional regulation of this gene have used transiently transfected promoter-reporter constructs. (FASEB J. 2004 18:540)

訳 この遺伝子の転写調節に関する以前の研究は，一過性にトランスフェクトされたプロモーターーレポーターコンストラクトを使ってきた

■ ㊔ this study/these studies ＋［動］

❶	this study demonstrates ～	この研究は，～を実証する	1,576
❷	this study provides ～	この研究は，～を提供する	1,375
❸	these studies suggest ～	これらの研究は，～を示唆する	1,349
❹	this study examined ～	この研究は，～を調べた	963
❺	this study was designed to ～	この研究は，～するために設計された	766
❻	this study shows ～	この研究は，～を示す	695
❼	this study investigated ～	この研究は，～を精査した	667
❽	this study was undertaken to ～	この研究は，～するために着手された	635

❶ This study demonstrates that HIV interacts directly with B cells in both lymphoid tissues and peripheral blood. (J Exp Med. 2000 192:637)

訳 この研究は，～ということを実証する

❸ Thus, these studies suggest the TLR2-mediated cytokine response to HSV-1 is detrimental to the host. (Proc Natl Acad Sci USA. 2004 101:1315)

訳 これらの研究は，～を示唆する

❹ This study examined whether inhaled CO was protective against the development of postoperative ileus. (Gastroenterology. 2003 124:377)

訳 この研究は，～かどうかを調べた

■ 名［前］＋ this study

❶	the purpose of this study was to ～	この研究の目的は，～することであった	3,024
❷	～ of this study was to determine …	この研究の～は，…を決定することであった	2,074
❸	the aim of this study was to ～	この研究の目的は，～することであった	2,021
❹	the objective of this study was to ～	この研究の目的は，～することであった	976
❺	the goal of this study was to ～	この研究の目的は，～することであった	922
❻	～ of this study was to investigate …	この研究の～は，…を精査することであった	665
❼	～ of this study was to examine …	この研究の～は，…を調べることであった	641
❽	～ of this study was to evaluate …	この研究の～は，…を評価することであった	640
❾	～ of this study was to assess …	この研究の～は，…を評価することであった	407

❷ The purpose of this study was to determine the potential contribution of an H1 receptor-mediated vasodilatation to postexercise hypotension. (J Physiol. 2005 563:633)

訳 この研究の目的は，～を決定することであった

❸ The aim of this study was to determine the outcomes of patients in whom β blocker therapy was discontinued. (Hepatology. 2001 34:1096)

訳 この研究の目的は，～を決定することであった

❼ The purpose of this study was to examine levels of cell division throughout the telencephalic VZ of juvenile birds. (J Comp Neurol. 2005 481:70)

訳 この研究の目的は，～を調べることであった

■ 名［形］＋ study/studies

❶	present study	現在の研究	12,305
❷	previous studies	以前の研究	9,082
❸	recent studies	最近の研究	7,352
❹	current study	現在の研究	2,831
❺	further studies	さらなる研究	1,838
❻	in vitro studies	試験管内の研究	1,674

❶ The aim of the present study was to determine the utility of new developments in vascular magnetic resonance (MR) technology in patients with TA. (Arthritis Rheum. 2002 46:1634)

訳 現在の研究の目的は，～の有用性を決定することであった

❷ Previous studies have shown the efficacy of this anti-genomic approach in vitro, targeting pathogenic mtDNA templates with only a single point mutation. (Nucleic Acids Res. 2001 29:3404)

訳 以前の研究は，～の有効性を示している

❸ Recent studies have shown that IL-2 is an essential growth factor for these cells.（J Immunol. 2003 171:3435）

訳 最近の研究は，～ということを示している

❹ In the current study, we used the yeast one-hybrid strategy to identify nuclear factors that bind to these three elements.（J Biol Chem. 2004 279:8684）

訳 現在の研究において，

■ 動 study ＋［名句］

❶ study the effect of ～	…は，～の効果を研究する	1,893
❷ study the role of ～	…は，～の役割を研究する	1,346
❸ study the mechanism	…は，機構を研究する	607
❹ study the interaction	…は，相互作用を研究する	326
❺ study the function of ～	…は，～の機能を研究する	269
❻ study the relationship between ～	…は，～の間の関連性を研究する	176
❼ study the regulation of ～	…は，～の調節を研究する	175
❽ study the expression of ～	…は，～の発現を研究する	166

❶ We studied the effects of saccadic eye movements on visual signaling in the primate lateral geniculate nucleus (LGN), the earliest stage of central visual processing.（Neuron. 2002 35:961）

訳 われわれは，～の効果を研究した

■ 動［代名／名］＋ study

❶ we studied ～	われわれは，～を研究した	6,390
・we have studied ～	われわれは，～を研究した	1,443
❷ the authors studied ～	著者らは，～を研究した	129

■ 動 to study

★ to study ～	～を研究する…	11,116
❶ To study ～	～を研究するために	2,303
❷ be used to study ～	…は，～を研究するために使われる	1,364
❸ model to study ～	～を研究するためのモデル	513
❹ system to study ～	～を研究するためのシステム	413
❺ opportunity to study ～	～を研究するための機会	210
❻ tool to study ～	～を研究するための道具	188
❼ approach to study ～	～を研究するためのアプローチ	161

❶ To study the role of IGF-I receptor signaling on cell cycle events we utilized MCF-7 breast cancer cells.（J Biol Chem. 2003 278:37256）

訳 ～におけるIGF-I受容体シグナル伝達の役割を研究するために

■ 動 be studied ＋［前］

★ be studied	…が，研究される	12,112

S

❶ be studied in ～	…が，～において研究される	2,687
❷ be studied by ～	…が，～によって研究される	1,579
・be studied by using ～	…が，～を使うことによって研究される	148
❸ be studied using～	…が，～を使って研究される	1,015
❹ be studied with ～	…が，～を使って研究される	504
❺ be studied for ～	…が，～に対して研究される	414

❶ The role of IL-18 in CpG-induced immune potentiation was studied in splenocyte cultures from control, LPS-conditioned, or CpG-conditioned mice.（J Immunol. 2004 172:1754）

訳 CpGに誘導される免疫増強におけるIL-18の役割が，脾細胞培養において研究された

❷ The Sp-1/AP-2 site near the transcription start site was studied by electrophoretic mobility shift and reporter gene assays.（Gene. 2003 310:133）

訳 転写開始点付近のSp-1/AP-2部位が，電気泳動移動度シフトおよびレポーター遺伝子アッセイによって研究された

■ 動［副］＋ studied

❶ well studied	よく研究される	1,163
❷ extensively studied	広範に研究される	679
❸ previously studied	以前に研究された	340

❶ Artemisinin and its analogues have been well studied for their antimalarial activity.（J Med Chem. 2003 46:4244）

訳 ～が，それらの抗マラリア活性に関してよく研究されている

■ 動［前］＋ studying

❶ model for studying ～	～を研究するためのモデル	440
❷ system for studying ～	～を研究するのためのシステム	329
❸ tool for studying ～	～を研究するための手段	226
❹ useful for studying ～	～を研究するために有用である	144

subcellular 形 細胞内の／細胞下の 用例数 5,263

◆類義語：intracellular

■ subcellular ＋［名］

❶ subcellular localization	細胞内局在	1,792
❷ subcellular distribution	細胞内分布	612
❸ subcellular fractionation	細胞成分分画	425
❹ subcellular compartments	細胞内区画	246
❺ subcellular location	細胞内位置	228

❶ The subcellular localization of Pto was independent of N-myristoylation, indicating that N-myristoylation is required for some function other than membrane affinity.（Plant J. 2006 45:31）

訳 Ptoの細胞内局在は，N-ミリストイル化に依存しなかった

❷ We examined the subcellular distribution of flotillin-1 in different cell types and found that localization is cell type-specific.（J Biol Chem. 2005 280:16125）

訳 われわれは，異なる細胞タイプにおけるflotillin-1の細胞内分布を調べた

■ ［形］＋ subcellular

❶	different subcellular ～	異なる細胞内～	128
❷	specific subcellular ～	特異的な細胞内～	117

❶ In eukaryotes, enzymes of different subcellular compartments participate in the assembly of membrane lipids.（EMBO J. 2003 22:2370）

訳 異なる細胞内区画の酵素が膜脂質の集合に関与する

★★ subject 名 対象／対象者／被験者，動 受けさせる，形 受けやすい

用例数 42,243

subjects	対象者／被験者／～を受けさせる	32,020
subject	対象者／被験者／～を受けさせる／受けやすい	5,723
subjected	受ける／～を受けさせた	4,354
subjecting	～を受けさせる	146

名詞の用例が多いが，他動詞や形容詞としても用いられる．複数形のsubjectsの割合は，約90%と圧倒的に高い．

◆類義語：object, susceptible, apt, suffer

■ 名 subject(s) ＋［前］

subject ofの前にtheが付く割合は非常に高い．

	the%	a/an%	ø%			
❶	8	-	59	subjects with ～	～のある対象（者）	4,741
❷	73	25	2	subject of ～	～の対象（者）	835
❸	8	-	80	subjects in ～	～における対象（者）	811

❶ In the present study, we evaluated whether therapeutic doses of metformin increase AMPK activity in vivo in subjects with type 2 diabetes.（Diabetes. 2002 51:2074）

訳 われわれは，メトホルミンの治療量が２型糖尿病の対象者において生体内でAMPK活性を増大させるかどうか評価した

❷ While ipRGCs have been the subject of much recent research, less is known about their central targets and how they develop to support specific behavioral functions.（Neuron. 2012 75:648）

訳 ipRGCは，最近の多くの研究の対象であった

❸ After 4 weeks of sleeve wear, subjects in the active treatment group reported a 16% decrease in mean WOMAC pain score relative to baseline (P = 0.001).（Arthritis Rheum. 2004 51:716）

訳 積極的治療群の対象者は，～の16%低下を報告した

■ ㊔ subjects ＋［動］

❶ subjects underwent ～	対象者が，～を受けた	242
❷ subjects received ～	対象者が，～を受けた	237
❸ subjects showed ～	対象者が，～を示した	209
❹ subjects performed ～	対象者が，～を行った	179

❷ Depending on the response after 12 weeks, subjects received an additional 12 weeks of treatment with efalizumab or placebo. (N Engl J Med. 2003 349:2004)
訳 対象者が，追加の12週の治療を受けた

■ ㊔［形／名］＋ subjects

❶ control subjects	対照群被験者	4,180
❷ healthy subjects	健康な被験者	1,271
❸ normal subjects	正常な被験者	1,127
❹ comparison subjects	比較被験者	1,115

❶ The mean distal ileum wall thickness at the first examination was 3.0 mm in patients with rotavirus infection and 2.0 mm in control subjects (P = .037). (J Infect Dis. 2004 189:1382)
訳 ～は，ロタウイルス感染の患者において3.0 mm，そして対照群被験者において2.0 mmであった

■ ㊍ be subjected ＋［前］

★ be subjected	…は，受けさせられる	2,170
❶ be subjected to ～	…は，～を受けさせられる	2,150
・rats were subjected to ～	ラットは，～を受けさせられた	163

❶ Male Sprague-Dawley rats were subjected to 45 minutes of superior mesenteric arterial occlusion followed by 90 minutes of reperfusion. (Am J Pathol. 2004 164:1707)
訳 オスのSDラットは，45分間の上腸間膜動脈閉塞を受けさせられた

■ ㊎ subject to

★ subject to ～	～を受けやすい	6,966
❶ be subject to ～	…は，～を受けやすい	1,747

❶ This apoptosis is subject to regulation by signals from growth factor or by Bcl-x_L. (J Immunol. 2002 169:1372)
訳 このアポトーシスは，成長因子からのシグナルによる調節を受けやすい

★ subsequent ㊎ 引き続いた／あとの　用例数 19,118

◆類義語：next, following

■ subsequent to

❶ subsequent to ～	～に引き続いて	649

❶ Western blot analysis showed a similar increase in MT protein levels, with peak times occurring subsequent to increases in mRNA levels. (J Neurosci. 2004 24:7043)
訳 ウエスタンブロット分析は，メッセンジャーRNAレベルの増大に引き続いて起こるピークタイムを持つMTタンパク質レベルの類似の増大を示した

■ subsequent ＋［名］

❶ subsequent activation of ～	引き続いた～の活性化	290
❷ subsequent development of ～	引き続いた～の発生	245
❸ subsequent studies	引き続いた研究	225
❹ subsequent analysis	引き続いた分析	208
❺ subsequent degradation	引き続いた分解	183

❶ Nrf2 accumulation and subsequent activation of the antioxidant response element is regulated by the proteasomal degradation of Nrf2. (J Immunol. 2004 173:3467)
訳 Nrf2の蓄積と引き続いた抗酸化剤応答要素の活性化は，～によって調節される

★subsequently 副 引き続いて 用例数 7,555

◆類義語：continuously, next

■ subsequently

❶ and subsequently	そして引き続いて	2,000
❷ Subsequently,	引き続いて，	1,148

❶ Rabbits were immunized against α-toxin and subsequently challenged with S. aureus strain Newman. (Invest Ophthalmol Vis Sci. 2002 43:1109)
訳 ウサギは，α毒素に対して免疫され，そして引き続いて黄色ブドウ球菌株Newmanに曝露された

❷ Subsequently, we found this mutation in 11 unrelated individuals of diverse ethnic backgrounds. (Blood. 2004 103:1937)
訳 引き続いて，われわれはこの変異を見つけた

■ subsequently ＋［過分／動］

❶ subsequently developed ～	引き続いて～を発症した	156
❷ subsequently used	引き続いて使われる	118

❶ During 6 years of follow-up through January 31, 2000, 266 men subsequently developed nonfatal MI or fatal coronary heart disease. (JAMA. 2004 291:1730)
訳 266名の男性が引き続いて非致死性の心筋梗塞あるいは致死性の冠動脈心疾患を発症した

S

★ subset ㊔ 部分集団／サブセット／一部　用例数 16,432

subset	部分集団／一部	11,707
subsets	部分集団／一部	4,725

複数形のsubsetsの割合が，約30%ある可算名詞.

◆類義語：part

■ subset ＋［前］

	the%	a/an%	ø%			
❶	12	84	0	subset of ～	～の部分集団／～の一部	9,373
				・a subset of ～	～の部分集団	6,945
				・subset of patients	患者の部分集団	524
				・a small subset of ～	～の小さな部分集団	345
				・subset of genes	遺伝子の部分集団	258
				・subset of cells	細胞の部分集団	164
				・a specific subset of ～	～の特異的な部分集団	140

❶ A number of reports showed that a subset of patients with ALS possess mutations in the TDP-43 (TARDBP) gene. (Hum Mol Genet. 2009 18:R156)

㊋ ALSの患者の部分集団は，TDP-43（TARDBP）遺伝子に変異を持つ

★ substantial ㊒ かなりの／実質的な　用例数 10,629

◆類義語：considerable, appreciable, virtual

■ substantial ＋［名］

❶	a substantial proportion of ～	～のかなりの割合	288
❷	a substantial increase in ～	～のかなりの増大	281
❸	a substantial fraction of ～	～のかなりの割合	241
❹	substantial evidence	実質的な証拠	227
❺	a substantial number of ～	～のかなりの数	220
❻	substantial differences	実質的な違い	215
❼	substantial amounts of ～	かなりの量の～	168
❽	a substantial reduction in ～	～のかなりの低下	153

❷ However, the combined treatment is associated with a substantial increase in adverse effects. (N Engl J Med. 2004 350:1937)

㊋ その併用療法は，有害効果のかなりの増大と関連する

❸ It is concluded that a substantial fraction of the missense variants observed in the general human population are functionally relevant. (Genomics. 2004 83:970)

㊋ 一般的なヒト集団において観察されるミスセンス変異体のかなりの割合は，機能的に関連する

★ substantially 副 大幅に／実質的に 用例数 8,182

◆類義語：considerably, appreciably, fairly, quite

■ substantially ＋［過分／動］

❶ substantially reduced	大幅に低下する／大幅に低下させた	707
・be substantially reduced	…は，大幅に低下する	290
❷ substantially increased	大幅に増大する／大幅に増大させた	405
❸ substantially decreased	大幅に低下する／大幅に低下させた	178

❶ SpHus1 phosphorylation is substantially reduced in SpMYHΔ cells after hydrogen peroxide treatment.（J Biol Chem. 2005 280:408）

訳 SpHus1のリン酸化は，SpMYHΔ細胞において大幅に低下する

❷ In women, the risk for breast cancer was substantially increased, being 32% by age 60 years.（Gastroenterology. 2004 126:1788）

訳 乳癌のリスクは，大幅に増大した

■［動］＋ substantially

❶ differ substantially	…は，大幅に異なる	269
❷ increase substantially	…は，大幅に増大する	240
❸ contribute substantially to ～	…は，～に大いに寄与する	229

❶ These results differ substantially from the epidemiologic profile of cervical HPV infection in women.（J Infect Dis. 2004 190:2070）

訳 これらの結果は，～の疫学的なプロファイルと大幅に異なる

❸ The data show that a strong inflammatory response can contribute substantially to local tumor control when the PDT regimen is suboptimal.（Cancer Res. 2004 64:2120）

訳 強い炎症性の反応は，局所の腫瘍の制御に大いに寄与しうる

■ substantially ＋［形／副］

❶ substantially higher	大幅により高い	371
❷ substantially lower	大幅により低い	310
❸ substantially greater	大幅により大きい	209
❹ substantially different	大幅に異なる	189

❶ Comprehensive extended treatments that combine drug and psychological interventions can produce consistent abstinence rates that are substantially higher than those in the literature.（Am J Psychiatry. 2004 161:2100）

訳 ～は，文献にあるそれらより大幅により高い

■ substantially ＋［副］

❶ substantially more ～	大幅により～	356
❷ substantially less ～	大幅により～でない	220

❶ Similarly, late nef alleles were substantially more active than early nef genes in stimulating HIV-1 replication in high CD4-positive cells, including primary lymphocytes, but not in cells

S

expressing low levels of the CD4 receptor.（J Biol Chem. 2003 278:33912）
訳 後期のnefアレルは，初期のnef遺伝子より大幅に活動的であった

substitute 動 置換する／代わりに用いる／代わりになる

用例数 8,672

substituted	置換される／代わりに用いられる／代わりになった	6,433
substitute	～を置換する／～を代わりに用いる／代わりになる	1,316
substituting	～を置換する／～を代わりに用いる／代わりになる	603
substitutes	～を置換する／～を代わりに用いる／代わりになる	320

他動詞の用例が多いが，自動詞としても用いられる．〈substituted for〉と〈substituted with〉の違いに注意．〈B (be) substituted for A〉と〈A (be) substituted with B〉（A：置換前，B：置換後）がほぼ同じ意味になる．

◆類義語：replace, exchange, displace, interchange, swap

■ substituted ＋［前］

❶ substituted for ～	～の代わりに用いられる	546
・be substituted for ～	…が，～の代わりに用いられる	292
❷ substituted with ～	～と置換される	502
❸ substituted by ～	～によって置換される	226

❶ Nae I , a novel DNA endonuclease, shows topoisomerase and recombinase activities when a Lys residue is substituted for Leu 43.（Nat Struct Biol. 2001 8:665）
訳 リジン残基がロイシン43の代わりに用いられる

❷ The contribution of His-51 to catalysis was studied by characterizing ADH with His-51 substituted with Gln (H51Q).（Biochemistry. 2004 43:3014）
訳 触媒作用へのヒスチジン-51の寄与が，ヒスチジン-51がグルタミンと置換されたADHを特徴づけることによって研究された

❸ We have generated three mutants of ficolin α in which the N-terminal cysteines were substituted by serines (Cys4, Cys24, and Cys4/Cys24).（J Biol Chem. 2004 279:6534）
訳 N-末端のシステインがセリンによって置換された

■ substitute for

❶ substitute for ～	～の代わりになる	1,195

❶ The HYD4 isoenzyme did not substitute for HYD3 in H_2 production.（J Bacteriol. 2004 186:580）
訳 HYD4アイソザイムは，HYD3の代わりにならなかった

★ substitution 名 置換

用例数 18,472

substitution	置換	10,676
substitutions	置換	7,796

複数形のsubstitutionsの割合は約40%とかなり高い.

◆類義語：replacement, displacement, exchange, interchange, change

■ substitution(s) ＋［前］

	the%	a/an%	ø%			
❶	23	2	75	substitution of ～	～の置換	3,098
❷	5	-	81	substitutions in ～	～における置換	1,301
❸	4	-	84	substitutions at ～	～における置換	1,060

❶ The defects in signaling caused by substitution of charged amino acids are not caused by changes in the abundance of receptors at the cell surface.（Biochemistry. 2003 42:3004）

訳 荷電アミノ酸の置換によって引き起こされるシグナル伝達の欠陥は，受容体の存在量の変化によっては引き起こされない

❷ Substitutions in the Site Y element markedly reduced inducible MIP-3α reporter activity.（J Biol Chem. 2003 278:875）

訳 サイトYエレメントにおける置換は，誘導性のMIP-3α受容体活性を顕著に低下させた

❸ Human MOG differs from rat MOG at several residues, including a proline for serine substitution at position 42.（J Immunol. 2003 171:462）

訳（それは）42番目のセリンの置換のプロリンを含む

■［名］＋ substitutions

❶	amino acid substitutions	アミノ酸置換	1,633
❷	alanine substitutions	アラニン置換	422
❸	nucleotide substitutions	ヌクレオチド置換	364
❹	base substitutions	塩基置換	297

❶ Of the 140 polymorphic sites identified, 68 were located in the coding region, of which 28 created amino acid substitutions in Nod2.（Genomics. 2003 81:369）

訳 そのうちの28はNod2におけるアミノ酸置換を引き起こした

■ substitution ＋［名］

❶	substitution mutations	置換変異	227
❷	substitution mutants	置換変異体	223
❸	substitution rates	置換率	218

❶ We found two patients homozygous for substitution mutations in CYP17, the gene encoding P450c17.（Nat Genet. 1997 17:201）

訳 われわれは，２人の患者がCYP17における置換変異に対してホモ接合であることが分かった

success 名 成功　　用例数 4,823

success	成功	4,582
successes	成功	241

S

複数形のsuccessesの割合は約５%あるが，単数形は無冠詞で使われることが多い．
◆類義語：achievement

■ success ＋［前］

success ofの前にtheが付く割合は圧倒的に高い．

	the%	a/an%	ø%			
❶	93	0	0	success of ～	～の成功	1,334
❷	5	1	78	success in ～	～における成功	456
				・success in ～ing	～する際の成功	171

❶ The **success of** this approach is dependent on the structural sensitivity of the fragmentation method.（Anal Chem. 2012 84:6814）
訳 このアプローチの成功は，～の構造的感受性に依存している

■ success ＋［名］

❶	success rate	成功率	413

successful 形 成功した　　用例数 6,681

■ be successful ＋［前］

★	be successful	…は，成功する	756
❶	be successful in ～	…は，～において成功する	407
	・be successful in ～ing	…は，～するのに成功する	180

❶ Angiography **was successful in** all primates.（Transplantation. 2004 78:1025）
訳 血管造影は，すべての霊長類において成功した

■ successful ＋［名］

❶	successful treatment	成功した治療	184

successfully 副 うまく／成功して　　用例数 4,715

■ successfully ＋［過分］

❶	successfully used	うまく使われる	292
	・be successfully used	…は，うまく使われる	222
❷	successfully applied	うまく適用される	225
❸	successfully treated	うまく処理される	181

❶ Finally, the tether-in-a-cone model **is successfully used** to analyze experimental spectra from T4 lysozyme.（Biophys J. 2006 90:340）
訳 …は，～を分析するためにうまく使われる

❷ The SAS method was successfully applied to the cysteine protease cathepsin S, which is implicated in autoimmune diseases.（J Am Chem Soc. 2005 127:15521）

訳 SAS法が，システインプロテアーゼカテプシンSにうまく適用された

■ we（have）successfully

❶	we successfully ～	われわれはうまく～	212
❷	we have successfully ～	われわれはうまく～	142

❷ We have successfully delivered an Alexa 488-labeled avidin protein into human glioblastoma cells.（Anal Biochem. 2005 344:168）

訳 われわれは，Alexa 488ラベルされたアビジンタンパク質をヒトの神経膠芽腫細胞にうまく送達した

★★ such 形 そのような　用例数 82,006

such asの用例が非常に多い.

◆類義語：like

■ such as

★	such as ～	～のような…	44,893
❶	factors such as ～	～のような因子	1,051
	・factors, such as ～	…因子，たとえば～	520
❷	proteins such as ～	～のようなタンパク質	931
	・proteins, such as ～	タンパク質，たとえば～	435
❸	processes such as ～	～のような過程	819

❶ Previous studies have found that nuclear-encoded COX subunit genes are under the control of specific transcription factors, such as nuclear respiratory factor 2（NRF-2）.（Gene. 2005 360:65）

訳 核にコードされるCOXサブユニット遺伝子は，特異的な転写因子，たとえば核呼吸因子２（NRF-2），の制御下にある

❸ The c-Jun/AP-1 transcription complex is associated with diverse cellular processes such as differentiation, proliferation, transformation, and apoptosis.（Mol Cell Biol. 2005 25:3324）

訳 c-Jun/AP-1転写複合体は，分化，増殖，癌化およびアポトーシスのような多様な細胞過程に関連する

■ such that

❶	… such that ～	～するほど…である	3,205

❶ The mouse scaramanga（ska）mutation impairs mammary gland development such that both abrogation and stimulation of gland formation occurs.（Genes Dev. 2005 19:2078）

訳 腺形成の抑止と刺激の両方が起こるほど，マウスのscaramanga（ska）変異は乳腺発達を障害する

S

■ such ＋［名］

❶ such patients	そのような患者	461
❷ such studies	そのような研究	441
❸ such cells	そのような細胞	420

suffer 動 患う／受ける　用例数 2,076

suffer	患う／受ける	869
suffering	患う／受ける	573
suffered	患った／受けた	485
suffers	患う／受ける	149

自動詞の用例が多いが，他動詞としても用いられる．suffer fromの用例が非常に多い．

◆類義語：develop, affect

■ suffer from

❶ suffer from ～	…は，～を患う	1,061
・patients suffering from ～	～を患う患者	169

❶ Metastasis of primary tumors leads to a very poor prognosis for patients suffering from cancer.（Proc Natl Acad Sci USA. 2005 102:15901）

訳 原発腫瘍の転移は，癌を患う患者の非常によくない予後につながる

★ sufficient 形 十分な　用例数 18,393

sufficient to *do*の用例が非常に多く，次いでsufficient forの用例が多い．

◆類義語：adequate

■ be sufficient ＋［前］

★ be sufficient	…は，十分である	9,519
❶ be sufficient to *do*	…は，～するのに十分である	6,957
・be sufficient to induce ～	…は，～を誘導するのに十分である	901
・be sufficient to cause ～	…は，～を引き起こすのに十分である	315
・be sufficient to confer ～	…は，～を与えるのに十分である	299
・be sufficient to activate～	…は，～を活性化するのに十分である	225
・be sufficient to mediate ～	…は，～を仲介するのに十分である	200
・be sufficient to promote ～	…は，～を促進するのに十分である	196
・be sufficient to inhibit ～	…は，～を阻害するのに十分である	177
❷ be sufficient for ～	…は，～のために十分である	2,163
・be sufficient for ～ing	…は，～するために十分である	259

❶ Gastric expression of Cdx2 alone was sufficient to induce intestinal metaplasia in mice.（Gastroenterology. 2002 122:689）

訳 ～は，腸の異形成を誘導するのに十分であった

❷ Viral antigen **was sufficient for** PD-1 upregulation, but induction of PD-L1 was required for impairment. (J Clin Invest. 2012 122:2967)

訳 ウイルス抗原は，PD-1の上方制御のために十分であった

■ ［形］＋ and/but not sufficient

❶ necessary and sufficient	必要かつ十分な	2,040
❷ necessary but not sufficient	必要であるが，十分ではない	394

❶ Inactivation of Ran1 (Pat1) kinase is **necessary and sufficient** for cells to exit the cell cycle and undergo meiosis. (Mol Cell Biol. 2000 20:4016)

訳 Ran1（Pat1）キナーゼの不活性化は，細胞が細胞周期から外れ，そして減数分裂を起こすのに必要かつ十分である

sufficiently 副 十分に　　用例数 1,646

◆類義語：adequately, fully, enough

■ sufficiently ＋［形］

❶ sufficiently high	十分に高い	182
❷ sufficiently large	十分に大きい	104

❶ In the presence of **sufficiently high** concentrations of heparin, the deletion mutants exhibited mitogenic activity equal to wild-type FGF-1. (J Biol Chem. 1996 271:26876)

訳 十分に高い濃度のヘパリンの存在下において

S

★★ suggest 動 示唆する／提案する　　用例数 202,437

suggest	～を示唆する	98,655
suggesting	～を示唆する	50,953
suggests	～を示唆する	33,822
suggested	示唆される／～を示唆した	19,007

他動詞．

◆類義語：imply, mean, indicate, represent, show, exhibit, display

■ ［名／代名］＋ suggest

❶ results suggest ～	結果は，～を示唆する	26,785
❷ data suggest ～	データは，～を示唆する	17,981
❸ findings suggest ～	知見は，～を示唆する	9,823
❹ we suggest ～	われわれは，～を提案する	5,115
❺ studies suggest ～	研究は，～を示唆する	4,925
❻ evidence suggests ～	証拠は，～を示唆する	3,166
❼ observations suggest ～	観察は，～を示唆する	2,400
❽ analysis suggests ～	分析は，～を示唆する	906

❶ These **results suggest** that Bax expression leads to an impairment of mitochondrial respiration, inducing toxicity in cells dependent on oxidative phosphorylation for survival. (Mol Cell Biol. 2000 20:3590)
訳 これらの結果は，～ということを示唆する

❹ **We suggest** that interferon-γ could increase the hPepT1 mediated di-tripeptides uptake in inflamed epithelial cells. (Am J Pathol. 2003 163:1969)
訳 われわれは，～ということを提案する

■ suggest ＋［that節／名句］

❶ suggest that ～	…は，～ということを示唆する	150,951
❷ suggest a role for ～	…は，～の役割を示唆する	2,334
・suggest an important role for ～	…は，～の重要な役割を示唆する	435
❸ suggest a mechanism	…は，機構を示唆する	1,143
・suggest a novel mechanism	…は，新規の機構を示唆する	373
❹ suggest a model	…は，モデルを提案する	1,024
❺ suggest the presence of ～	…は，～の存在を示唆する	936
❻ suggest the existence of ～	…は，～の存在を示唆する	883
❼ suggest the possibility	…は，可能性を示唆する	635
❽ suggest the involvement of ～	…は，～の関与を示唆する	501
❾ suggest the importance of ～	…は，～の重要性を示唆する	247

❶ These data **suggest that** BCSC-1 may exert a tumor suppressor activity and is a likely target of the LOH observed on 11q23-q24 in cancer. (Proc Natl Acad Sci USA. 2003 100:11517)
訳 これらのデータは，～ということを示唆する

❷ Recent studies **suggest a role for** PI(5)P in a variety of cellular events, such as tumor suppression, and in response to bacterial invasion. (J Biol Chem. 2004 279:38590)
訳 最近の研究は，～の役割を示唆する

❹ We **suggest a model** in which Dab1 phosphorylation leads to the recruitment of Nck β to the membrane, where it acts to remodel the actin cytoskeleton. (Mol Cell Biol. 2003 23:7210)
訳 われわれは，～であるモデルを提案する

■［副］＋ suggest

❶ strongly suggest ～	…は，～を強く示唆する	2,250
❷ further suggest ～	…は，～をさらに示唆する	1,002

❶ These results **strongly suggest** that the reproductive number for 1918 pandemic influenza is not large relative to many other infectious diseases. (Nature. 2004 432:904)
訳 これらの結果は，～ということを強く示唆する

■ be suggested ＋［前］

★ be suggested	…が，示唆される	9,519
❶ be suggested to *do*	…が，～することが示唆される	1,397
・have been suggested to *do*	…が，～することが示唆されてきた	926
・be suggested to be ～	…が，～であることが示唆される	423
・be suggested to play ～	…が，～を果たすことが示唆される	179

❷ be suggested by ～	…が，～によって示唆される	540

❶ PhaZ2 is thus suggested to be an intracellular depolymerase. (J Bacteriol. 2003 185:3788)
訳 PhaZ2は，このように細胞内解重合酵素であることが示唆される

❶ Ursodeoxycholic acid (UDCA) has been suggested to be of benefit based on open label clinical studies. (Hepatology. 2004 39:770)
訳 ウルソデオキシコール酸（UDCA）は，～であることが示唆されてきた

■ be suggested ＋［that節］

❶ it has been suggested that ～	～ということが示唆されてきた	999
❷ it is suggested that ～	～ということが示唆される	509

❶ It has been suggested that gene expansion occurs as a result of hairpin formation of long stretches of these sequences on the leading daughter strand synthesized during DNA replication. (Biochemistry. 2004 43:14218)
訳 ～ということが，示唆されてきた

■ suggesting ＋［that節］

❶ , suggesting that ～	そして，（それは）～ということを示唆している	1,285

❶ B. judaicus toxins also increased binding of [^{3}H] ryanodine to the purified RyR1, suggesting that a direct protein-protein interaction mediates the effect of the peptides. (J Biol Chem. 2004 279:26588)
訳 そして，（それは）～ということを示唆している

S

suggestive 形 示唆的な　　用例数 1,689

suggestive ofの用例が非常に多い.

■ suggestive of

❶ suggestive of ～	～を示唆している	1,179
・be suggestive of ～	…は，～を示唆している	284

❶ Our finding of impaired performance of fmr1 KO mice on a passive avoidance task is suggestive of a deficit in learning and memory. (Proc Natl Acad Sci USA. 2002 99:15758)
訳 ～は，学習と記憶における欠陥を示唆している

suitable 形 適した／適合性の　　用例数 3,503

suitable forの用例が非常に多い.
◆類義語：appropriate, proper, adequate, compatible

■ suitable for

❶ suitable for ～	～に適している	1,821
・be suitable for ～	…は，～に適している	640

・suitable for ～ing	～するのに適している	291

❶ We provide data demonstrating that **the CSase assay described here is suitable for the determination of the activities of both classes of enzymes.** (Anal Biochem. 2005 347:42)
訳 ここで述べられるCSaseアッセイは，両方のクラスの酵素の活性の定量に適している

summary 名 要約／サマリー 用例数 4,424

summary	要約／サマリー	4,313
summaries	要約／サマリー	111

文頭で，In summaryの用例が非常に多い.

■ In summary

★ in summary	まとめると	3,199
❶ In summary,	まとめると,	3,164

❶ **In summary, our data indicate that** Toso is a functional IgM receptor that is capable of activating signaling molecules, is regulated by IL-2, and is not inherently an antiapoptotic molecule. (J Immunol. 2012 189:587)
訳 まとめると，われわれのデータは～ということを示す

■ summary ＋［前］

	the%	a/an%	ø%			
❶	3	96	1	summary of ～	～のサマリー	284

superior 形 優れた／上位の 用例数 6,283

◆類義語：predominant, dominant, excellent, good, epistatic

■ superior ＋［前］

❶ superior to ～	～より優れている	1,617
・be superior to ～	…は，～より優れている	1,122

❶ Furthermore, **the protection was superior to that achieved with pooled immune γ globulin from human volunteers inoculated with live vaccinia virus.** (J Virol. 2005 79:13454)
訳 その保護作用は，生きたワクシニアウイルスを接種されたヒトのボランティアからのプールされた免疫γグロブリンによって達成されるそれより優れていた

supplement 動 追加する／補う， 名 補充／サプリメント 用例数 3,658

supplemented	追加される／～を追加した／補われる	1,535
supplements	～を追加する／～を補う／補充／サプリメント	1,079

supplement	～を追加する／～を補う／補充／サプリメント	882
supplementing	～を追加する／～を補う	162

名詞としても用いられるが，動詞の用例が多い．

◆類義語：addition, supplementation, recruitment, add, recruit

■ supplemented ＋［前］

❶ supplemented with ～	～を追加された	820
・medium supplemented with ～	～を追加された培地	127
❷ supplemented by ～	～によって補われた	150

❶ The cells were cultured as a suspension in RPMI 1640 medium supplemented with 15% fetal calf serum at 37 ℃ with 5% CO_2 in air.（Anal Biochem. 2000 287:80）

訳 細胞は，15%ウシ胎児血清を追加されたRPMI 1640培地で浮遊液として培養された

★★ support ㊍ 支持する，㊔ 支持／支援 用例数 50,435

support	～を支持する／支持／支援	31,278
supported	～を支持した	6,936
supports	～を支持する	6,143
supporting	～を支持する	6,078

他動詞および名詞として用いられる．複数形のsupportsの用例はほとんどなく，原則，不可算名詞として使われる．

◆類義語：assistance, adjunct, aid, help, assist

S

■ ㊍ support ＋［名句］

❶ support the hypothesis that ～	…は，～という仮説を支持する	3,338
❷ support a model	…は，モデルを支持する	1,977
・support a model in which ～	…は，～であるモデルを支持する	1,233
❸ support a role for ～	…は，～の役割を支持する	1,147
❹ support the idea that ～	…は，～という考えを支持する	1,007
❺ support the notion that ～	…は，～という概念を支持する	941
❻ support the concept that ～	…は，～という概念を支持する	618
❼ support the view that ～	…は，～という考えを支持する	593
❽ support the conclusion that ～	…は，～という結論を支持する	548

❶ These results support the hypothesis that polycystin-1 is a surface membrane receptor that transduces the signal via changes in ionic currents.（J Biol Chem. 2004 279:25582）

訳 これらの結果は，～という仮説を支持する

❷ These results support a model in which the ratio of bound to unbound Ptc molecules determines the cellular response to Hh.（Nature. 2004 431:76）

訳 これらの結果は，～であるモデルを支持する

❸ Together, these results strongly support a role for both RNI and ROI in the host control of C. burnetii infection.（Infect Immun. 2004 72:6666）

訳 これらの結果は，～におけるRNI およびROIの両方の役割を強く支持する

■ 動 [名] + support

❶ results support ～	結果は，～を支持する	3,639
❷ data support ～	データは，～を支持する	3,444
❸ findings support ～	知見は，～を支持する	1,923
❹ studies support ～	研究は，～を支持する	599
❺ evidence supports ～	証拠は，～を支持する	482
❻ observations support ～	観察は，～を支持する	428

❸ These findings support the hypothesis that NPY contribute to, but not be critical for, the nutritional inhibition of sexual receptivity. (Brain Res. 2004 1007:78)

訳 これらの知見は，～という仮説を支持する

■ 動 [副] + support

❶ strongly support ～	～を強く支持する	739

❶ Empirical biological data strongly support the hypothesis that sexual transmission by acutely infected individuals has a disproportionate effect on the spread of HIV-1 infection. (J Infect Dis. 2004 189:1785)

訳 経験的な生物学的データは，～という仮説を強く支持する

■ 動 [名／形] + to support

★ to support ～	～を支持する…	3,573
❶ evidence to support ～	～を支持する証拠	414
❷ ability to support ～	～を支持する能力	217
❸ sufficient to support ～	～を支持するのに十分な	139

❶ Evidence to support a physical association between Oxa1 and the large ribosomal subunit is presented. (EMBO J. 2003 22:6438)

訳 ～の間の物理的な関連を支持する証拠

■ 動 supported by

★ supported by ～	～によって支持される	3,405
❶ be supported by ～	…は，～によって支持される	2,091
❷ further supported by ～	～によってさらに支持される	449

❶ The interaction between these two proteins was further supported by a co-localization of the proteins within rat brain. (J Biol Chem. 2004 279:46946)

訳 これらの２つのタンパク質の間の相互作用は，それらのタンパク質の共存によってさらに支持された

■ 動 [名] + supporting

❶ evidence supporting ～	～を支持する証拠	645
❷ data supporting ～	～を支持するデータ	177

❶ However, evidence supporting a role for neutrophil-endothelial cell interactions was not observed. (Crit Care Med. 2000 28:1290)
訳 好中球ー内皮細胞相互作用の役割を支持する証拠は，観察されなかった

■ 名 support ＋［前］

	the%	a/an%	ø%			
❶	2	2	91	support for ～	～に対する支持	2,697
				・provide support for ～	…は，～に対する支持を提供する	607
				・provide further support for ～	…は，～に対するさらなる支持を提供する	247
				・provide strong support for ～	…は，～に対する強力な支持を提供する	218
				・provide additional support for ～	…は，～に対する追加の支持を提供する	114
				・support for the hypothesis that ～	～という仮説に対する支持	216
❷	5	0	91	support of ～	～の支持	1,576
				・in support of ～	～を支持するように	1,257
				・in support of this hypothesis	この仮説を支持するように	193
❸	3	4	89	support to ～	～に対する支持	624
				・lend support to ～	…は，～に対する支持を与える	229
				・support to the hypothesis that ～	～という仮説に対する支持	92

❶ Our findings provide support for the hypothesis that the molecular pathogenesis of LCMDCs is distinct from that of most DACs. (Am J Pathol. 2001 159:2239)
訳 われわれの知見は，～という仮説に対する支持を提供する

❶ These results provide further support for striatal models of ADHD pathophysiology. (Am J Psychiatry. 2003 160:1693)
訳 これらの結果は，～に対するさらなる支持を提供する

❷ In support of this hypothesis, we found that, when grown in these nutrient concentrations, nsd cells accumulate guanosine tetraphosphate, the cellular starvation signal. (J Bacteriol. 2004 186:3461)
訳 この仮説を支持するように，

❸ EM studies lend support to the hypothesis that C. koseri uses morphologically different methods of uptake to enter macrophages. (Infect Immun. 2003 71:5871)
訳 EM研究は，～という仮説に対する支持を与える

★ suppress 動 抑制する 用例数 23,525

suppressed	抑制される／～を抑制した	10,715
suppress	～を抑制する	6,069
suppresses	～を抑制する	4,223

S

suppressing	～を抑制する	2,518

他動詞.

◆類義語：inhibit, repress, attenuate, weaken, depress, decrease, reduce, interfere

■ suppress ＋［名句］

❶ suppress apoptosis	…は，アポトーシスを抑制する	238
❷ suppress the expression of ～	…は，～の発現を抑制する	228
❸ suppress the growth of ～	…は，～の増殖を抑制する	169
❹ suppress tumor growth	…は，腫瘍増殖を抑制する	159

❸ THANK also strongly suppressed the growth of tumor cell lines and activated caspase-3.（J Biol Chem. 1999 274:15978）

訳 THANKは，また，腫瘍細胞株の増殖を強く抑制した

■［名／動］＋ to suppress

★ to suppress ～	～を抑制する…	641
❶ ability to suppress ～	～を抑制する能力	254
❷ fail to suppress ～	…は，～を抑制することができない	127
❸ be shown to suppress ～	…は，～を抑制することが示される	106

❶ Radicicol, a macrocyclic anti-fungal antibiotic, has the ability to suppress transformation by diverse oncogenes such as Src, Ras and Mos.（Oncogene. 1998 16:2639）

訳 …は，～のような多様な癌遺伝子による形質転換を抑制する能力を持つ

■ be suppressed ＋［前］

★ be suppressed	…は，抑制される	2,842
❶ be suppressed by ～	…は，～によって抑制される	1,562
❷ be suppressed in ～	…は，～において抑制される	444

❶ Both of these defects can be suppressed by overexpression of the Shk1 modulator, Skb1.（J Biol Chem. 1999 274:36052）

訳 これらの欠陥の両方は，Shk1モジュレーター，Skb1，の過剰発現によって抑制されうる

❷ Although ERK activity was suppressed in cells overexpressing KSR1, ERK inhibition alone was insufficient to upregulate TM expression.（Mol Cell Biol. 2003 23:1786）

訳 ERK活性が，KSR1を過剰発現する細胞において抑制された

■［副］＋ suppressed

❶ significantly suppressed	有意に抑制された／有意に抑制した	400
❷ completely suppressed	完全に抑制された／完全に抑制した	217
❸ partially suppressed	部分的に抑制された／部分的に抑制した	200

❶ Furthermore, treatment of subcutaneous tumors in nude mice with 2-5A-anti-hTR significantly suppressed the tumor growth through induction of apoptosis（$P < 0.001$）.（Oncogene. 2000 19:2205）

訳 ～は，腫瘍の増殖を有意に抑制した

❷ Likewise, VIP-mediated Ca^{2+} current inhibition, which is mediated by cholera toxin-sensitive G-protein, **was also completely suppressed by a number of Gα subunits** overexpressed in neurons.（J Neurosci. 1999 19:4755）

訳 ～は，また，いくつかのGαサブユニットによって完全に抑制された

★ suppression 名 抑制　用例数 13,993

suppression	抑制	13,971
suppressions	抑制	22

複数形のsuppressionsの用例は0.2%しかなく，原則，**不可算名詞**として使われる．

◆類義語：inhibition, repression, depression, attenuation, reduction

■ suppression ＋［前］

	the%	a/an%	ø%			
❶	25	4	62	suppression of ～	～の抑制	7,855
				・suppression of ～ by …	…による～の抑制	752
				・…-mediated suppression of ～	…に仲介される～の抑制	424
				・…-induced suppression of ～	…に誘導される～の抑制	259
				・result in suppression of ～	…は，～の抑制という結果になる	168
				・suppression of apoptosis	アポトーシスの抑制	114
❷	7	0	84	suppression by ～	～による抑制	439
❸	5	5	73	suppression in ～	～における抑制	434

❶ **Ethanol-induced suppression of the TNF-α response occurred at a post-transcriptional level.**（Am J Respir Crit Care Med. 2000 161:135）

訳 TNF-α応答のエタノールに誘導される抑制が，転写後レベルで起こった

❶ Together, our results demonstrate that **suppression of apoptosis by PKC-α correlates with its ability of activating endogenous Akt.**（Oncogene. 1999 18:6564）

訳 PKC-αによるアポトーシスの抑制は，内在性のAktを活性化するそれの能力と相関する

★★ surface 名 表面　用例数 63,795

surface	表面	56,362
surfaces	表面	7,433

複数形のsurfacesの用例が，約10%ある可算名詞．

◆類義語：face

S

■ surface ＋［前］

surface ofの前にtheが付く割合は圧倒的に高い．

	the%	a/an%	ø%			
❶	97	1	0	surface of ～	～の表面	6,828
				・surface of the protein	タンパク質の表面	126

❶ Surprisingly, we find that **these residues are located on the surface of the protein** and not in the hydrophobic core.（J Mol Biol. 2003 334:515）

訳 これらの残基は，そのタンパク質の表面に位置する

■ surface ＋［名］

❶	surface expression	表面発現	2,964
❷	surface area	表面領域	1,693
❸	surface receptors	表面受容体	1,101
❹	surface protein	表面タンパク質	1,095
❺	surface antigen	表面抗原	454

❶ We show that **cell surface expression of this immunoregulatory molecule is restricted to a subpopulation of memory B cells**, most of which lack the classical CD27 marker for memory B cells in humans.（J Exp Med. 2005 202:783）

訳 この免疫調節性分子の細胞表面発現は，記憶B細胞の亜集団に限られる

■［名］＋ surface

❶	cell surface	細胞表面	13,631
	・cell surface expression	細胞表面発現	1,016
	・cell surface receptors	細胞表面受容体	889
❷	membrane surface	膜表面	531
❸	binding surface	結合表面	500

★ surgery 名 手術　　用例数 12,726

surgery	手術	12,484
surgeries	手術	242

複数形のsurgeriesの割合は２％しかなく，原則，**不可算名詞**として使われる．

◆類義語：operation

■ surgery ＋［前］

	the%	a/an%	ø%			
❶	0	0	100	surgery in ～	～における手術	515
❷	1	1	97	surgery for ～	～に対する手術	510

❷ **Colonic mucosa was obtained from 9 patients who received surgery for colorectal cancer.** (Gastroenterology. 2011 140:1241)

訳 結腸粘膜が，結腸直腸癌に対する手術を受けた９名の患者から得られた

■ ［形／名］＋ surgery

❶ cardiac surgery	心臓手術	622
❷ bypass surgery	バイパス手術	416
❸ cataract surgery	白内障手術	327

★ surprisingly 副 驚いたことに　用例数 9,034

文頭の用例が非常に多い．

■ Surprisingly

❶ Surprisingly,	驚いたことに，	6,412

❶ **Surprisingly, we found that** in marked contrast to NS, LS mutants are catalytically defective and act as dominant negative mutations that interfere with growth factor/Erk-mitogen-activated protein kinase-mediated signaling.（J Biol Chem. 2006 281:6785）

訳 驚いたことに，われわれは～ということを見つけた

survey 名 調査，動 調査する　用例数 6,423

survey	調査／調査する	4,170
surveys	調査／調査する	1,323
surveyed	調査された／調査した	852
surveying	調査する	78

名詞の用例が多いが，動詞としても用いられる．複数形のsurveysの割合が，約25%ある可算名詞．

◆類義語：research, investigation

■ 名 survey ＋［前］

	the%	a/an%	ø%			
❶	5	87	3	survey of ～	～の調査	1,015

❶ In addition, **a survey of expression data sets shows that** the CP12 paralogs are differentially regulated.（Plant Physiol. 2013 161:824）

訳 発現データセットの調査は，～ということを示す

■ 動［代名］＋ survey

❶ we surveyed	われわれは，～を調査した	217

S

★★ survival ㊔ 生存 　用例数 51,040

survival	生存	50,800
survivals	生存	240

複数形のsurvivalsの割合は0.5%しかなく，**不可算名詞**として使われることが多い．

■ survival ＋［前］

	the%	a/an%	ø%			
❶	43	8	46	survival of ～	～の生存	5,727
				・survival of patients	患者の生存	296
				・survival of mice	マウスの生存	123
❷	0	2	96	survival in ～	～における生存	3,733

❶ The **survival of patients** with diffuse large-B-cell lymphoma after chemotherapy is influenced by molecular features of the tumors.（N Engl J Med. 2002 346:1937）
訳 化学療法後のびまん性大細胞型B細胞性リンパ腫の患者の生存は，腫瘍の分子的特徴によって影響される

❷ Few studies have identified predictors of long-term **survival in** this patient population.（Ann Surg. 2001 234:215）
訳 この患者集団における長期間生存の予知因子を同定した研究はほとんどない

■［前］＋ survival

❶	required for survival	生存のために必要とされる	150
❷	difference in survival	生存の違い	130
❸	predictor of survival	生存の予知因子	108

■［名／形］＋ survival

❶	cell survival	細胞生存	4,255
❷	～-free survival	～の無い生存	3,304
❸	overall survival	全体の生存	3,292
❹	graft survival	移植片生存	2,070
❺	～-year survival	～年生存	1,236
❻	long-term survival	長期間生存	1,034
❼	patient survival	患者生存	1,003
❽	improved survival	改善された生存	871
❾	allograft survival	同種移植片生存	792

❶ Activation of integrins upon binding to extracellular matrix proteins is believed to be a crucial step for the regulation of **cell survival** and proliferation.（J Cell Biol. 1998 142:587）
訳 ～は，細胞生存と増殖の調節のための決定的に重要なステップであると信じられている

❷ MMP-1 levels increase during melanoma progression where they are associated with shorter disease-**free survival**.（J Biol Chem. 2004 279:33168）
訳 それらは，より短い疾患なしの生存と関連する

■ survival ＋［名］

❶ survival rates	生存率	1,520
❷ survival time	生存時間	805

❶ Overall and recurrence-free survival rates in transplanted patients at 5 years were 44% and 48%, respectively.（Ann Surg. 2002 235:533）

訳 移植された患者における全体および再発なしの５年生存率は，それぞれ44%および48%であった

■［名］＋ and survival／survival and ＋［名］

❶ proliferation and survival	増殖と生存	649
❷ growth and survival	増殖と生存	636
❸ survival and proliferation	生存と増殖	333

❷ The epicardium regulates growth and survival of the underlying myocardium.（Dev Cell. 2005 8:85）

訳 心外膜は，基底をなす心筋の増殖と生存を調節する

survive ㊍ 生存する／生き残る／残存する　用例数 4,624

survive	生存する	2,406
survived	生存した	2,030
survives	生存する	188

S

自動詞の用例が多いが，他動詞としても用いられる．

◆類義語：exist, remain

■ survive ＋［前］

❶ survive in ～	…は，～において生存する	462
❷ survive to ～	…は，～まで生存する	397
❸ survive for ～	…は，～の間生存する	284

❸ The donor cells survived for at least 2 months postnatally, the longest time examined.（J Neurosci. 2004 24:4585）

訳 ドナー細胞は，出生後に少なくとも２ヶ月間生存した

■［名］＋ survive

❶ mice survived	マウスが，生存した	189

❶ Hand1-null mice survived to the nine somite stage at which time they succumbed to numerous developmental defects.（Development. 2004 131:2195）

訳 Hand1欠損マウスは，９体節期まで生存した

■ to survive

★ to survive	生き残る…	993
❶ ability to survive	生き残る能力	106

❶ Forty-one ionizing radiation-sensitive strains of Deinococcus radiodurans were evaluated for their ability to survive 6 weeks of desiccation. (J Bacteriol. 1996 178:633)
訳 ～が，6週間の乾燥を生き残るそれらの能力について評価された

★ susceptibility 名 感受性 用例数 13,595

susceptibility	感受性	13,236
susceptibilities	感受性	359

複数形のsusceptibilitiesの割合は2.5%しかなく，原則，不可算名詞として使われる．
◆類義語：sensitivity, responsivity

■ susceptibility ＋［前］

susceptibility ofの前にtheが付く割合は非常に高い．

	the%	a/an%	ø%			
❶	9	3	80	susceptibility to ～	～への感受性	5,247
				・increased susceptibility to ～	増大した～への感受性	706
				・susceptibility to infection	感染への感受性	204
				・associated with susceptibility to ～	～への感受性と関連している	152
				・susceptibility to apoptosis	アポトーシスへの感受性	110
❷	76	5	18	susceptibility of ～	～の感受性	1,404

❶ Cerebellar granule cultures also exhibited an increased susceptibility to exogenous oxidative stress. (Nucleic Acids Res. 2005 33:4660)
訳 ～は，また，外来性の酸化ストレスに対する増大した感受性を示した

❶ Using an IFN-γ-neutralizing antibody in a murine model, we demonstrated increased susceptibility to infection within 24 h. (Infect Immun. 2005 73:8425)
訳 われわれは，24時間以内の感染に対する増大した感受性を実証した

❶ Using congenic mouse strains, we showed that the $H\text{-}2^s$ haplotype of SJL/J mice is not associated with susceptibility to MAV-1. (J Virol. 2005 79:11517)
訳 SJL/Jマウスの$H\text{-}2^s$ハプロタイプは，MAV-1に対する感受性と関連しない

❷ Strikingly, the susceptibility of proinsulin to fibrillation is increased by scission of the connecting peptide at single sites. (J Biol Chem. 2005 280:42345)
訳 プロインスリンの細動への感受性は，連結ペプチドの切断によって増大する

■［過分／名］＋ susceptibility

❶ increased susceptibility	増大した感受性	926
❷ cancer susceptibility	癌感受性	534

❸ disease susceptibility	疾患感受性	482

■ susceptibility ＋［名］

❶ susceptibility gene	感受性遺伝子	526
❷ susceptibility testing	感受性テスト	452
❸ susceptibility loci	感受性座位	438

susceptible 形 感受性の　用例数 7,513

susceptible toの用例が圧倒的に多い.

◆類義語：sensitive, subject, apt

■ susceptible to

★ susceptible to ～	～に対して感受性の	4,176
❶ be susceptible to ～	…は，～に対して感受性である	1,146
❷ more susceptible to ～	～に対してより感受性の	918
❸ highly susceptible to ～	～に対して高度に感受性の	421
❹ susceptible to infection	感染に対して感受性の	421
❺ less susceptible to ～	～に対してより感受性でない	228

❷ Compared with the wild-type (WT) bacterium, a S. aureus mutant with disrupted carotenoid biosynthesis **is more susceptible to oxidant killing**, has impaired neutrophil survival, and is less pathogenic in a mouse subcutaneous abscess model. (J Exp Med. 2005 202:209)

訳 ～は，酸化剤死滅に対してより感受性である

❸ In this study, we show that similar to IFN-$\gamma^{-/-}$ mice, JNK1$^{-/-}$ mice **are highly susceptible to tumor development** after inoculation of both melanoma cell line B16 and lymphoma cell line EL-4. (J Immunol. 2005 175:5783)

訳 ～は，腫瘍発生に対して高度に感受性である

❹ The nematode Caenorhabditis elegans feeds on bacteria but **is susceptible to infection by pathogenic bacteria** in its natural environment. (Nature. 2005 438:179)

訳 ～は，病原菌による感染に対して感受性である

■ susceptible ＋［名］

❶ susceptible mice	感受性マウス	182
❷ susceptible strains	感受性の株	133

❶ We investigated the ability of Δsag1 parasite to induce a lethal intestinal inflammatory response in **susceptible mice**. (J Immunol. 2004 173:2725)

訳 われわれは，感受性マウスにおける致死的な腸の炎症反応を誘導するΔsag1寄生体の能力を精査した

S

suspect 動 疑う／思う　用例数 2,408

suspected	疑われる／～を疑った／思われる	2,231
suspect	～を疑う／～を思う	128
suspects	～を疑う／～を思う	43
suspecting	～を疑う／～を思う	6

他動詞.

◆類義語：suppose, think, feel, appear, believe, consider

■ suspected ＋［前］

❶ suspected of ～	～が疑われる	280
・suspected of having ～	～を持っていると疑われる	194
❷ suspected to *do*	～すると思われる	208
・be suspected to *do*	…は，～すると思われる	119

❶ We measured BNP levels in 11 patients suspected of having either CP or RCMP.（J Am Coll Cardiol. 2005 45:1900）

訳 われわれは，CPあるいはRCMPのどちらかを持っていると疑われる11人の患者においてBNPレベルを測定した

❷ Respiratory syncytial virus（RSV）infection in early life is suspected to play a role in the development of post-bronchiolitis wheezing and asthma.（J Immunol. 2005 175:1876）

訳 …は，～の発症において役割を果たしていると思われる

■［前］＋ suspected

❶ patients with suspected ～	～が疑われる患者	290

❶ Patients with suspected transient ischaemic attack（TIA）with no symptoms or signs when assessed in the ER were excluded from the analysis.（Lancet Neurol. 2005 4:727）

訳 一過性脳虚血発作（TIA）が疑われる患者

★ sustain 動 維持する／持続させる　用例数 10,827

sustained	持続した／～を維持した	8,693
sustain	～を持続させる／～を維持する	1,255
sustaining	～を持続させる／～を維持する	665
sustains	～を持続させる／～を維持する	214

他動詞.

◆類義語：maintain, keep, hold, retain, last, persist, continue

■ sustained ＋［前］

❶ sustained for ～	～の間維持される	216
・be sustained for ～	…は，～の間維持される	167
❷ sustained in ～	～において維持される	157

❸ sustained by ～	～によって維持される	153

❶ Only a small proportion of episodes of clinical remission off medication were sustained for >5 years. (Arthritis Rheum. 2005 52:3554)

訳 ～が，5年以上の間維持された

■ sustained ＋［名］

❶ sustained activation of ～	持続した～の活性化	320
❷ sustained increase in ～	持続した～の増大	265
❸ sustained release	持続した放出	175
❹ sustained virologic response	持続性ウイルス陰性化	155
❺ sustained expression of ～	持続した～の発現	138

❶ Rho-dependent stress fiber accumulation promotes the sustained activation of ERK and subsequent cyclin D1 expression during G_1-S phase cell cycle progression. (J Biol Chem. 2005 280:23066)

訳 Rho依存的なストレスファイバー蓄積は，ERKの持続した活性化を促進する

switch 名 切り換え，動 切り換える／切り換わる　用例数 9,619

switch	切り換え／～を切り換える／切り換わる	7,496
switches	切り換え／～を切り換える／切り換わる	1,131
switched	切り換えられる／～を切り換えた／切り換わった	992

名詞としての用例が多いが，動詞としても用いられる．

◆類義語：conversion, interconversion, alteration, shift, change, convert, turn

■ 名 switch ＋［前］

	the%	a/an%	ø%			
❶	43	50	5	switch from ～	～からの切り換え	683
				・switch from ～ to …	～から…への切り換え	460
❷	23	64	1	switch in ～	～における切り換え	648
❸	34	51	10	switch to ～	～への切り換え	547
❹	53	43	3	switch between ～	～の間の切り換え	318

❶ The phenotypic switch from R5 to X4 virus occurs at low CD4 counts and is accompanied by a rapid rise in viral load and drop in CD4 count. (J Virol. 2006 80:802)

訳 R5からX4ウイルスへの表現型の切り換えが，～において起こる

❷ These data indicate that differences in pattern and level of individual GnRH neuron firing may reflect the switch in estradiol action and underlie GnRH surge generation. (Proc Natl Acad Sci USA. 2005 102:15682)

訳 ～は，エストラジオール作用における切り換えを反映するかもしれない

❸ The switch to AuFONs not only provides a more stable surface for SAM formation but also yields better chemometric results, with improved calibration and validation over a range of 0.5-44 mM (10-800 mg/dl). (Anal Chem. 2005 77:4013)

訳 AuFONへの切り換えは，SAM形成のためのより安定な表面を提供するだけでなく，より

S

よい計量化学的結果も生じる

■ 動 switched ＋［前］

❶ switched to ～	～へ切り換えられる	234
・be switched to ～	…は，～へ切り換えられる	107
❷ switched from ～	～から切り換えられる	127

❶ Most patients were switched to an every-other-week dosing schedule.（J Clin Oncol. 2006 24: 379）

訳 ほとんどの患者は，1 週おきの投薬スケジュールに切り換えられた

❷ By P15, when vision starts, AQP4 and Kir4. 1 localization coordinately switched from horizontal cells to Muller glial cells.（Invest Ophthalmol Vis Sci. 2005 46:3869）

訳 AQP4およびKir4.1の局在は，協調的に水平細胞からミュラーグリア細胞に切り換わった

★ symptom 名 症状 用例数 15,526

symptoms	症状	12,926
symptom	症状	2,600

複数形のsymptomsの割合が，約85%と圧倒的に高い.

◆類義語：manifestation, sign, indication, presentation

■ symptoms ＋［前］

	the %	a/an %	ø %			
❶	30	-	65	symptoms of ～	～の症状	1,826
❷	3	-	94	symptoms in ～	～における症状	781

❶ A history and/or symptoms of testicular disease also were recorded at the time of examination.（Radiology. 2005 237:550）

訳 精巣疾患の病歴および/あるいは症状は，また，検査の時に記録された

❷ Tryptophan depletion induced a transient return of depressive symptoms in patients with remitted MDD but not in controls ($P<.001$).（Arch Gen Psychiatry. 2004 61:765）

訳 ～は，緩解したMDDの患者におけるうつ症状の一過性再発を誘導した

■［前］＋ symptoms

❶ onset of symptoms	症状の開始	223

■［形］＋ symptoms

❶ depressive symptoms	うつ症状	925
❷ clinical symptoms	臨床症状	474

synergistic 形 相乗的な 用例数 3,082

■ synergistic ＋［名］

❶	synergistic effect	相乗効果	317
❷	synergistic activation	相乗作用	188
❸	synergistic interaction	相乗相互作用	164

❶ A synergistic effect between chiPsi and single-stranded DNA binding protein was observed. (J Biol Chem. 2005 280:40465)

訳 chiPsiと一本鎖DNA結合タンパク質の間の相乗効果が観察された

★★ synthesis 名 合成 用例数 41,234

synthesis	合成	40,350
syntheses	合成	884

複数形のsynthesesの割合は2％しかなく，不可算名詞として使われることが多い．

◆類義語：elaboration, construction, production

■ synthesis ＋［前］

	the%	a/an%	ø%			
❶	63	12	24	synthesis of ～	～の合成	10,720
❷	0	0	63	synthesis in ～	～における合成	2,486
❸	1	0	53	synthesis by ～	～による合成	1,162

❶ Expression of the plastid and the cytosolic ACC genes is each driven by two nested promoters responsible for the synthesis of two transcript types. (Proc Natl Acad Sci USA. 2004 101:1403)

訳 ～は，2つの転写タイプの合成に責任のある2つのネステッドプロモーターによってそれぞれ駆動される

❷ Here we show that DNA synthesis in this system is dependent on the viral polymerase processivity factor (UL42). (J Biol Chem. 2004 279:21957)

訳 このシステムにおけるDNA合成は，～に依存する

■［名／形］＋ synthesis

❶	protein synthesis	タンパク質合成	5,316
❷	DNA synthesis	DNA合成	4,119
❸	RNA synthesis	RNA合成	1,045
❹	total synthesis	全合成	605
❺	NO synthesis	NO合成	493

S

★ synthesize 動 合成する　　用例数 14,314

synthesized	合成された／～を合成した	10,803
synthesize	～を合成する	2,207
synthesizing	～を合成する	853
synthesizes	～を合成する	451

他動詞.

◆類義語：construct, generate, produce, create, yield

■ be synthesized ＋［前］

★ be synthesized	…は，合成される	5,461
❶ be synthesized by ～	…は，～によって合成される	713
❷ be synthesized in ～	…は，～において合成される	691
❸ be synthesized as ～	…は，～として合成される	352
❹ be synthesized from ～	…は，～から合成される	347

❶ Type I interferon (IFN) is synthesized by most nucleated cells following viral infection. (J Biol Chem. 2005 280:18651)

訳 1型インターフェロン（IFN）は，ウイルス感染のあとほとんどの有核細胞によって合成される

❷ YPPs are synthesized in the fat body, the insect analogue of the vertebrate liver. (Proc Natl Acad Sci USA. 2004 101:10626)

訳 YPPが，脂肪体において合成される

❸ Recently, it was reported that IL-16 is synthesized as an approximately 80-kDa precursor molecule, pro-IL-16. (J Immunol. 1998 161:3114)

訳 IL-16は，およそ80-kDaの前駆体分子，pro-IL-16，として合成される

❹ Various pyrimidine and purine L-3'-fluoro-2',3'-unsaturated nucleosides were synthesized from their precursors, L-3',3'-difluoro-2',3'-dideoxy nucleosides, by elimination of hydrogen fluoride. (J Med Chem. 2003 46:3245)

訳 ～は，それらの前駆体から合成された

■［副］＋ synthesized

❶ newly synthesized	新たに合成された	1,343
❷ chemically synthesized	化学的に合成された	191

❶ These findings are consistent with a role of hepatocystin in carbohydrate processing and quality control of newly synthesized glycoproteins in the endoplasmic reticulum. (Gastroenterology. 2004 126:1819)

訳 これらの知見は，新たに合成された糖タンパク質の糖プロセシングおよび品質管理におけるhepatocystinの役割と一致する

■ synthesized and ＋［過分］／［動］＋ and synthesized

❶ synthesized and evaluated	合成されそして評価された	342
❷ designed and synthesized ～	～を設計しそして合成した	273

❶ The second generation of methylenecyclopropane analogues of nucleosides 5a-5i and 6a-6i **was synthesized and evaluated for antiviral activity**. (J Med Chem. 2004 47:566)

訳 ～が，抗ウイルス活性のために合成されそして評価された

❷ For this study, **we designed and synthesized a polyamide to** target the TTCCA-motif repeated in the heterochromatic regions of chromosome 9, Y and 1. (Nucleic Acids Res. 2002 30:2790)

訳 われわれは，～するためにポリアミドを設計しそして合成した

■ [代名] + synthesize

❶	we synthesized ～	われわれは，～を合成した	434

❶ For these studies, **we synthesized DNA-RNA chimeric oligonucleotides** with RNA residues in defined positions. (Biochemistry. 2001 40:8749)

訳 われわれは，DNA-RNAキメラオリゴヌクレオチドを合成した

■ to synthesize

★	to synthesize ～	～を合成する…	1,173
❶	ability to synthesize ～	～を合成する能力	126

❶ Mutant yeast lacking any mIF2 **retained the ability to synthesize low levels of a subset of mitochondrially encoded proteins**. (J Biol Chem. 2003 278:31774)

訳 ～は，低いレベルのミトコンドリアにコードされるタンパク質のサブセットを合成する能力を保持した

S

★★ system 名 システム／系　用例数 94,868

system	システム／系	69,454
systems	システム／系	25,414

複数形のsystemsの割合が，約25%ある可算名詞.

◆類義語：method, technique

■ system + [前]

	the%	a/an%	ø%			
❶	41	51	1	system to *do*	～するためのシステム	3,283
				・system to study ～	～を研究するためのシステム	413
				・system to identify ～	～を同定するためのシステム	150
				・system to investigate ～	～を精査するためのシステム	129
❷	10	89	0	system for ～	～のためのシステム	3,279
				・system for studying ～	～を研究するのためのシステム	329
❸	65	28	2	system in ～	～におけるシステム	3,034
				・system in which ～	～であるシステム	626

❹	59	37	1	system of ～	～のシステム	2,028

❶ Mecp2-mutant mice have been used as a model system to study the disease mechanism. (Proc Natl Acad Sci USA. 2005 102:12560)

訳 Mecp2変異マウスは，その疾患の機構を研究するためのモデルシステムとして使われてきた

❷ The zebrafish is an excellent model system for studying the molecular basis of inner ear development and function. (Annu Rev Genet. 2005 39:9)

訳 ゼブラフィッシュは，内耳の発生と機能の分子的基盤を研究するための優れたモデルシステムである

■ ［名／形］＋ system

❶	nervous system	神経系	9,644
❷	immune system	免疫系	4,550
❸	model system	モデルシステム	3,163
❹	expression system	発現システム	1,102
❺	secretion system	分泌系	1,025
❻	visual system	視覚系	843
❼	culture system	培養系	692
❽	regulatory system	制御系	513

❹ Using a simple inducible gene expression system, we found that induced RNA levels within a single bacterium of Escherichia coli exhibited a pulsating profile in response to a steady input of inducer. (Proc Natl Acad Sci USA. 2005 102:9160)

訳 単純な誘導性遺伝子発現システムを用いて

■ system ＋［現分］

❶	system using ～	～を使うシステム	295
❷	system including ～	～を含むシステム	271
❸	system consisting of ～	～から成るシステム	181

★ systemic 形 全身性の　用例数 13,744

◆反意語：local

■ systemic ＋［名］

❶	systemic administration	全身投与	642
❷	systemic inflammation	全身性の炎症	473
❸	systemic infection	全身性の感染	360

❶ In contrast to hepatic TG expression of rIL-10, systemic administration of rIL-10 had only a modest effect on tolerance. (J Immunol. 2005 175:3577)

訳 rIL-10の全身投与は，耐性に対してわずかな効果しか持たなかった

T

★ tag ㊍ タグをつける，㊔ タグ　　用例数 10,602

tagged	タグをつけられる／～にタグをつけた	4,803
tag	タグ／～にタグをつける	3,332
tags	タグ／～にタグをつける	1,595
tagging	～にタグをつける	872

名詞としても用いられるが，他動詞の用例が多い．

◈類義語：label, mark

■ tagged ＋［前］

❶ tagged with ～	～でタグをつけられた	335
・tagged with green fluorescent protein	緑色蛍光タンパク質でタグをつけられた	52

❶ N175I was subsequently introduced into an **ldnt2-D389N construct tagged with green fluorescent protein** and transfected into a Δldnt1/Δldnt2 Leishmania donovani knockout.（J Biol Chem. 2005 280:2213）

訳 緑色蛍光タンパク質でタグをつけられたldnt2-D389Nコンストラクト

■ ［名］-tagged

❶ epitope-tagged ～	エピトープタグをつけられた～	756
❷ GFP-tagged ～	GFPタグをつけられた～	339
❸ His-tagged ～	Hisタグをつけられた～	319
❹ green fluorescent protein-tagged ～	緑色蛍光タンパク質タグをつけられた～	195

❸ N-terminal amino acid sequencing of the recombinant protein yielded **the sequence corresponding to the N terminus of His-tagged gpIFN-γ**.（Infect Immun. 2006 74:213）

訳 HisタグをつけられたgpIFN-γのN末端に対応する配列

★ take ㊍ 取る／服用する　　用例数 24,441

taken	取られる	14,796
take	～を取る／～を服用する	3,248
taking	～を取る／～を服用する	2,935
takes	～を取る／～を服用する	2,270
took	～を取った／～を服用した	1,192

他動詞の用例が多いが，自動詞としても用いられる．taken togetherなどの熟語表現が非常に多い．

◈類義語：receive

■ taken together

★ taken together	まとめると	10,355

❶ Taken together,	まとめると,	9,458
❷ taken together with ～	～とまとめると	384

❶ Taken together, these results suggest that γHV-68-induced IL-12 contributes to the pathophysiology of viral infection while also functioning to limit viral burden. (J Immunol. 2004 172:516)

訳 まとめると，これらの結果は～ということを示唆する

❷ Taken together with previous observations, these results show that the G protein α, β, and γ subunits all play roles in targeting each other. (J Biol Chem. 2004 279:30279)

訳 以前の観察とまとめると

■ taken ＋［前／副］

❶ taken from ～	～から得られた	663
❷ taken up	取り込まれる	560
❸ taken into ～	～に入れられる	494
・be taken into account	…は，考慮に入れられる	323
・be taken into consideration	…は，考慮に入れられる	95
❹ taken to *do*	～するように取られる	325
❺ taken at ～	～において取られる	234

❶ Cells taken from patients exhibit spontaneous chromosomal breaks and rearrangements. (Am J Hum Genet. 2000 66:1540)

訳 患者から得られた細胞は，～を示す

❷ 3TC and FTC are taken up by cells and converted into 3TCTP and FTCTP. (J Mol Biol. 2000 300:403)

訳 3TCおよびFTCが細胞によって取り込まれる

❸ Vitamin D-calcium interdependencies must be taken into account. (Am J Clin Nutr. 2004 80:1735S)

訳 ビタミンDとカルシウムの相互依存が考慮に入れられねばならない

❹ If clinic measurements are used, care should be taken to ensure that these measurements are taken under optimal conditions. (Transplantation. 2003 76:1643)

訳 ～ということを確かめるように注意が払われるべきである

■ take ＋［名／前／副］

❶ takes place	…が，起こる	2,242
❷ take advantage of ～	…は，～を利用する	1,311
❸ take into account ～	…は，～を考慮に入れる	714
❹ take up ～	…は，～を取り込む	474

❶ Our studies provide a deeper understanding of how the CFTR assembly takes place in native cell membrane. (J Biol Chem. 2004 279:24673)

訳 われわれの研究は，どのようにCFTR構築が自然な細胞膜において起こるかについてより深い理解を提供する

❸ Deterministic dual-cell models have been available for sometime; these models take into account the effects of the resultant cell heterogeneity. (J Theor Biol. 2003 221:205)

訳 これらのモデルは，結果として生じる細胞不均一性の影響を考慮に入れる

❹ Here we show that influenza virus may enter and infect **HeLa cells that are unable to take up ligands** by clathrin-mediated endocytosis. (J Virol. 2002 76:10455)

訳 リガンドを取り込むことができないHeLa細胞

■ [名] + taking

❶	patients taking ～	～を服用した患者	280

❶ **Eight (67%) of 12 patients taking pramipexole and two (20%) of 10 taking placebo had an improvement of** at least 50% in their Hamilton depression scale scores. (Am J Psychiatry. 2004 161:564)

訳 プラミペキソールを服用した12名の患者のうち８名（67%）および偽薬を服用した10名の患者のうち２名（20%）が，～の改善を示した

☆ target ㊔ 標的，㊍ 標的にする　用例数 85,935

target	標的／～を標的にする	49,506
targets	標的／～を標的にする	20,073
targeted	標的にされる／狙いを定めた	16,356

動詞としても使われるが，名詞の用例の方が多い．複数形のtargetsの割合が，約30%ある可算名詞．targetingは別項参照．

■ ㊔ target + [前]

	the%	a/an%	ø%			
❶	4	95	0	target for ～	～のための標的	5,893
				・target for ～ing	～するのための標的	499
				・target for the treatment of ～	～の治療のための標的	248
				・target for therapeutic intervention	治療介入のための標的	228
				・target for the development of ～	～の開発のための標的	217
❷	30	62	6	target of ～	～の標的	5,464
❸	6	91	0	target in ～	～における標的	1,054

❶ The protein specified by srtA, sortase, **may be a useful target for the development of new antimicrobial drugs.** (Science. 1999 285:760)

訳 ～は，新しい抗菌剤の開発のための有用な標的であるかもしれない

❷ It seems likely that **CFA synthase is the target of** an unidentified energy-independent heat shock regulon protease. (J Bacteriol. 2000 182:4288)

訳 CFA合成酵素は，～の標的である

■ ㊔ target + [名]

❶	target genes	標的遺伝子	4,825
❷	target cells	標的細胞	2,178

❸ target site	標的部位	860
❹ target proteins	標的タンパク質	725
❺ target sequence	標的配列	393

❶ In the absence of hormone, **several nuclear receptors actively repress transcription of target genes** via interactions with the nuclear receptor corepressors SMRT and NCoR. (Gene. 2000 245:1)

訳 いくつかの核内受容体は，標的遺伝子の転写を活発に抑制する

■ 名 [形／名] ＋ target

❶ therapeutic target	治療標的	2,004
❷ mammalian target	哺乳類の標的	1,158
❸ downstream target	下流の標的	957
❹ potential target	潜在的な標的	811
❺ direct target	直接の標的	492
❻ molecular target	分子標的	458
❼ drug target	薬剤標的	425
❽ novel target	新規の標的	423
❾ specific target	特異的標的	408

❶ **We demonstrate the potential for caveolin-1 as a therapeutic target for this important malignancy.** (Cancer Res. 2001 61:3882)

訳 われわれは，この重要な悪性腫瘍の治療標的としてのカベオリン-1の潜在能を実証する

❸ Furthermore, overexpression of the E2F-1 transcription factor, **a downstream target of Rb,** induced extensive apoptosis and IL-1α release. (Oncogene. 1998 17:1195)

訳 Rbの下流の標的

■ 動 to target

★ to target ～	～を標的にする…	2,886
❶ designed to target ～	～を標的にするように設計される	113

❶ Two hammerhead ribozymes, HRz35 and HRz42, **were designed to target the PDEγ gene** in wild-type C57BL/6 mice. (Invest Ophthalmol Vis Sci. 2005 46:3836)

訳 ～が，PDEγ遺伝子を標的にするように設計された

■ 動 targeted ＋ [前]

❶ targeted to ～	～に向けられる	2,288
・be targeted to ～	…は，～に向けられる	1,108
❷ targeted by ～	～によって標的にされる	968
❸ targeted for ～	～に向けられる	661

❶ We show that **a SHY::eGFP fusion protein is targeted to the cell wall.** (Plant J. 2004 39:643)

訳 SHY::eGFP融合タンパク質は，細胞壁に向けられている

❷ **This critical process for cell division is targeted by G2/M checkpoint.** (J Biol Chem. 2005 280:42994)

訳 細胞分裂にとって決定的に重要なこの過程は，G2/Mチェックポイントによって標的にされる

❸ When oxygen levels are high, the HIF-1α subunit is hydroxylated and is targeted for degradation by the von Hippel-Lindau tumor suppressor protein (VHL). (J Biol Chem. 2005 280:20580)

訳 HIF-1αサブユニットは水酸化され，そしてフォンヒッペル・リンダウ腫瘍抑制タンパク質（VHL）による分解に向けられる

■ 動［副］＋ targeted

❶ specifically targeted	特異的に標的にされる	176
❷ molecularly targeted	分子的に標的にされる	112

■ 動 targeted ＋［名］

❶ targeted disruption of ～	狙いを定めた～の破壊	660
❷ targeted deletion of ～	狙いを定めた～の欠失	525
❸ targeted therapies	狙いを定めた治療法	485

❶ However, mice harboring a targeted disruption of the decorin gene do not develop spontaneous tumors. (Proc Natl Acad Sci USA. 1999 96:3092)

訳 狙いを定めたデコリン遺伝子の破壊を持つマウスは，～を発症しない

★ targeting 名 ターゲッティング，-ing 標的にする　用例数 16,976

名詞の複数形の用例はなく，不可算名詞として使われることが多い．

■ targeting ＋［前］

	the%	a/an%	ø%			
❶	15	1	79	targeting of ～	～のターゲッティング	2,923
❷	4	0	66	targeting to ～	～するためにターゲッティングを／～へのターゲッティング	676
❸	2	0	77	targeting in ～	～におけるターゲッティング	375

❶ In the inactive (GDP-bound) conformation, accessory factors facilitate the targeting of Rab GTPases to intracellular compartments. (Nature. 2005 436:415)

訳 アクセサリー因子は，Rab GTP加水分解酵素の細胞内区画へのターゲッティングを促進する

❷ We have used gene targeting to study the function of the Mus81-Eme1 endonuclease in mammalian cells. (Mol Cell Biol. 2005 25:7569)

訳 われわれは，Mus81-Eme1エンドヌクレアーゼの機能を研究するために遺伝子ターゲッティングを用いた

■［名／形］＋ targeting

❶ gene targeting	遺伝子ターゲッティング	1,184
❷ membrane targeting	膜ターゲッティング	427

T

❸ specific targeting	特異的ターゲッティング	226

★ task ㊔ 課題／作業課題 用例数 11,261

task	課題	8,152
tasks	課題	3,109

複数形のtasksの割合が，約30%ある可算名詞.

■ task ＋［前］

task ofの前にtheが付く割合は圧倒的に高い.

	the%	a/an%	ø%			
❶	12	78	7	task in ～	～における課題	379
				・task in which ～	～である課題	180
❷	83	12	1	task of ～	～の課題	235
				・task of ～ing	～する課題	136

❶ To address this issue **we trained rats to perform a task in which** the size of the predicted reward was signaled before the instrumental response was instructed.（J Neurosci. 2012 32: 2027）

訳 われわれは，～である課題を実行するようにラットを訓練した

■［名］＋ task

❶ memory task	記憶課題	259
❷ discrimination task	弁別課題	245

■ task ＋［名］

❶ task performance	作業能力	259
❷ task force	タスクフォース	236

★ technique ㊔ テクニック／技術 用例数 24,920

techniques	テクニック／技術	13,374
technique	テクニック／技術	11,546

複数形のtechniquesの割合は約55%と非常に高い.

◆類義語：method, skill, technology

■ technique(s) ＋［前］

technique ofの前にtheが付く割合は圧倒的に高い.

	the%	a/an%	ø%			
❶	4	-	90	techniques to *do*	～するためのテクニック	1,246
❷	10	85	1	technique for ～	～のためのテクニック	1,210

				・technique for ～ing	～するためのテクニック	482
❸	85	9	4	technique of ～	～のテクニック	425
❹	5	-	81	techniques in ～	～におけるテクニック	405

❶ Here, **we report the use of single-molecule imaging techniques to study the interactions between** β-amyloid (1-40) peptides and supported synthetic model anionic lipid membranes. (Biophys J. 2012 103:1500)

訳 われわれは，～の間の相互作用を研究するために単分子イメージングテクニックの使用を報告する

❷ **We describe a new technique for the identification of** peptides covalently modified with the maleimide cross-linker o-phenylenebismaleimide (OPBM). (Anal Biochem. 2002 302:230)

訳 われわれは，～の同定のための新しいテクニックを述べる

❸ **The most advanced technique of amyloid typing is laser microdissection** followed by mass spectrometry. (Blood. 2012 120:3206)

訳 アミロイドタイピングの最も進歩したテクニックは，レーザーマイクロダイセクションである

■ technique(s) + [動]

❶	techniques were used to *do*	テクニックが，～するために使われた	344
❷	technique provides ～	テクニックは，～を提供する	101
❸	technique allows ～	テクニックは，～を可能にする	94

❶ **Modeling techniques were used to determine the relationships among** survival, baseline clinical variables, and time-dependent variables. (Circulation. 2004 109:1509)

訳 モデル化テクニックが，～の間の関連性を決定するために使われた

■ [形／名] + technique(s)

❶	imaging techniques	イメージングテクニック	584
❷	new technique	新しいテクニック	257
❸	surgical techniques	外科的テクニック	229

T

technology 名 テクノロジー／科学技術 用例数 8,999

technology	テクノロジー／科学技術	6,005
technologies	テクノロジー／科学技術	2,994

複数形のtechnologiesの割合が，約35%ある可算名詞.

◆類義語：technique, method

■ technology + [前]

	the%	a/an%	ø%			
❶	14	12	65	technology to *do*	～するためにテクノロジーを／～するためのテクノロジー	547
❷	9	53	25	technology for ～	～のためのテクノロジー	358

❶ We employed microarray technology to identify SA target genes. (Infect Immun. 2005 73:5319)

訳 われわれは，SA標的遺伝子を同定するためにマイクロアレイテクノロジーを使用した

❷ We report a new platform technology for visualizing transgene expression in living subjects using magnetic resonance imaging (MRI). (Nat Med. 2005 11:450)

訳 われわれは，導入遺伝子発現を可視化するための新しいプラットフォームテクノロジーを報告する

★ temperature 名 温度 用例数 25,363

temperature	温度	20,230
temperatures	温度	5,133

複数形のtemperaturesの割合が，約20%ある可算名詞.

■ temperature +［前］

temperature ofの前にtheが付く割合はかなり高い.

	the%	a/an%	ø%			
❶	68	24	6	temperature of ～	～の温度	721
❷	8	1	91	temperature in ～	～における温度	330

❶ A cis content as high as 96% could be obtained by lowering the temperature of the polymerization. (J Am Chem Soc. 2012 134:2040)

訳 ～は，重合の温度を下げることによって得られうる

■ temperature-［形］

❶ temperature-sensitive ～	温度感受性の～	605
❷ temperature-dependent ～	温度依存性の～	250

❶ The wbbL of the temperature-sensitive mutant contained a single-base change that converted what was a proline in mc^2 155 to a serine residue. (J Biol Chem. 2004 279:43540)

訳 温度感受性変異体のwbbLは，1塩基変化を含んでいた

■ temperature +［名］

❶ temperature dependence	温度依存性	215
❷ temperature range	温度範囲	139

■［名／形］+ temperature

❶ room temperature	室温	467
・at room temperature	室温において	321
❷ low temperature	低い温度	432
❸ high temperature	高い温度	227
❹ body temperature	体温	140

❶ However, force field and semiempirical calculations on the energy difference between the two

isomers have suggested that t-CHP should be stable at room temperature. (J Am Chem Soc. 2005 127:15983)
訳 t-CHPは，室温において安定なはずである

★ temporal 形 時間的な／側頭の 用例数 12,880

■ temporal ＋［名］

❶	temporal lobe	側頭葉	1,201
❷	temporal resolution	時間的な分解能	434
❸	temporal expression	時間的な発現	357
❹	temporal patterns	時間的なパターン	283

❸ However, plant AGPases differ in several parameters, including spatial and temporal expression, allosteric regulation, and heat stability. (Plant Physiol. 2005 139:1625)
訳 そして，（それは）空間的および時間的な発現を含む

■［形］＋ and temporal／temporal and ＋［形］

❶	spatial and temporal	空間的および時間的な	1,126
❷	temporal and spatial	時間的および空間的な	692

tend 動 傾向がある／しやすい 用例数 3,350

tend	傾向がある	1,486
tended	傾向がある	1,337
tends	傾向がある	473
tending	傾向がある	54

自動詞．tend to *do*の用例が圧倒的に多い．

■ tend to *do*

★	tend to *do*	…は，～する傾向がある	3,252
❶	tend to be ～	…は，～である傾向がある	1,031
❷	tend to have ～	…は，～を持つ傾向がある	330
❸	tend to increase	…は，増大する傾向がある	127

❶ Although SP-A levels tended to be lower in CF compared with non-CF, this was only significant in the presence of bacterial infection. (Am J Respir Crit Care Med. 2003 168:685)
訳 SP-Aレベルは，非CFに比べてCFにおいてより低い傾向があった

❸ In contrast, IFN-γ and TNF-α production tended to increase in HIV-seropositive women with increasing hemozoin levels. (Infect Immun. 2004 72:7022)
訳 IFN-γ およびTNF-α の産生は，～において増大する傾向があった

T

★★ term ㊔ 期間，動 名付ける　用例数 39,273

term	期間／～と名付ける	25,934
terms	期間／～と名付ける	6,960
termed	～と名付けた／名付けられる	6,379

名詞の用例が多いが，他動詞としても用いられる．複数形のtermsの割合が，約20%ある可算名詞．〈long-term〉〈in terms of〉の用例が非常に多い．動詞としては，〈term A B（AをBと名付ける）〉の第５文型で用いられるが，受動態の用例が多い．

◆類義語：period, duration, designate, call, name

■ ㊔［形］＋ term

❶ long-term ～	長期～	18,499
・long-term potentiation	長期増強	1,284
・long-term survival	長期生存	1,035
・long-term memory	長期記憶	506
❷ short-term ～	短期～	4,294
❸ longer term	より長い期間	422

❶ Data suggest that **liver transplantation could offer long-term survival in** selected patients when combined with neoadjuvant chemoradiotherapy. (Lancet. 2005 366:1303)

訳 肝移植は，～における長期生存を提供しうる

■ ㊔ in terms of

★ in terms of ～	～の点から	5,580
❶ be discussed in terms of ～	…が，～の点から議論される	421
・results are discussed in terms of ～	結果が，～の点から議論される	160
❷ be interpreted in terms of ～	…が，～の点から解釈される	213

❶ **The structure of the Cu/Ni complexes is discussed in terms of the implications for prion protein function and disease.** (J Mol Biol. 2005 346:1393)

訳 Cu/Ni複合体の構造が，プリオンタンパク質機能と疾患にとっての意味の点から議論される

■ 動［名］＋ termed

❶ protein termed ～	～と名付けられたタンパク質	260
❷ process termed ～	～と名付けられた過程	201

❶ Hydrophobic agent resistance mediated by the mtr system is also inducible, **which results from an AraC-like protein termed MtrA.** (Mol Microbiol. 2004 54:731)

訳 そして（それは）MtrAと名付けられたAraC様タンパク質に由来する

■ 動［代名］＋ term

❶ we term ～	われわれは，～を…と名付ける	383
・, which we term ～	（そしてそれを），われわれは～と名付ける	174

❶ Using next-generation sequencing, we characterize a phenomenon, **which we term chromothripsis**, whereby tens to hundreds of genomic rearrangements occur in a one-off cellular crisis.（Cell. 2011 144:27）

訳 そしてそれを，われわれはクロモスリプシスと名付ける

★ terminus 名 末端／終端 用例数 19,126

terminus	末端	15,878
termini	末端	3,248

複数形のterminiの割合が，約15%ある可算名詞．C terminusあるいはN terminusの用例が非常に多い．

■ terminus ＋［前］

terminus of/inの前にtheが付く割合は圧倒的に高い．

	the %	a/an %	ø %			
❶	100	0	0	terminus of ～	～の末端	6,695
				・C terminus of ～	～のC末端	2,499
				・N terminus of ～	～のN末端	1,992
				・terminus of the protein	そのタンパク質の末端	239
❷	82	3	0	terminus in ～	～における末端	211

❶ Binding was mediated by the **C terminus of** RhoA and was independent of nucleotide.（J Biol Chem. 2012 287:17176）

訳 結合は，RhoAのC末端によって仲介された

■［名／過分］＋ at ＋ @ ＋ terminus

❶	domain at ～ terminus	～末端のドメイン	229
❷	residues at ～ terminus	～末端の残基	173
❸	located at ～ terminus	～末端に位置する	139
❹	amino acids at ～ terminus	～末端のアミノ酸	104

★★ test 動 検証する／検査する／試す／テストする，名 検証／テスト 用例数 64,411

tested	検証される／検証した／検査した	28,348
test	検証する／検査する／検査	27,299
tests	検証する／検査する／検査	8,764

動詞の用例が多いが，名詞としても用いられる．動詞としては他動詞の用例が多いが，自動詞（test for）の用例もある．複数形のtestsの割合が，約30%ある可算名詞．testingは別項参照．

◆類義語：examine, investigate, survey, analyze, examination, assessment, trial, inspection, testing

T

■ 動 test ＋［名句／whether節］

❶	test the hypothesis that ～	…は，～という仮説を検証する	4,769
❷	test whether ～	…は，～かどうかを検証する	3,193
❸	test this hypothesis	…は，この仮説を検証する	2,310
❹	test the effect of ～	…は，～の効果を検証する	662
❺	test the role of ～	…は，～の役割を検証する	451
❻	test the ability of ～	…は，～の能力を検証する	384
❼	test the efficacy of ～	…は，～の有効性を検証する	158
❽	test the possibility that ～	…は，～という可能性を検証する	111
❾	test the importance of ～	…は，～の重要性を検証する	101

❶ We tested the hypothesis that β-cat is capable of immortalizing and transforming cultured epithelial cells that represent the precursors to colon cancer.（Cancer Res. 2001 61:2097）

訳 われわれは，～という仮説を検証した

■ 動［代名／名］＋ test

❶	we tested ～	われわれは，～を検証した	6,450
❷	study tested ～	研究は，～を検証した	507

❶ In this study we tested whether the NK1R-ir neurons of the VRG are glutamatergic.（J Neurosci. 2002 22:3806）

訳 われわれは，～かどうかを検証した

■ 動 to test

★	to test ～	～を検証する…	10,440
❶	To test ～	～を検証するために	5,441
❷	be used to test ～	…は，～を検証するために使われる	562
❸	be to test ～	…は，～を検証することである	548
❹	be designed to test ～	…は，～を検証するために設計される	211

❶ To test this hypothesis, we have characterized melanopsin following heterologous expression in COS cells.（Biochemistry. 2003 42:12734）

訳 この仮説を検証するために，

■ 動 be tested ＋［前］

★	be tested	…が，検証される	7,469
❶	be tested for ～	…が，～について検証される／～を検証した	1,770
	・be tested for their ability to *do*	…が，～するそれらの能力について検証される	224
❷	be tested in ～	…が，～において検証される	1,397
❸	be tested by ～	…が，～によって検証される	903
❹	be tested on ～	…が，～において検証される	369
❺	be tested with ～	…が，～によって検証される	351

❶ Following 4 rounds of antigen stimulation, **the CTLs were tested for their ability to kill autologous targets** in an Sp17-dependent and HLA-class I-restricted manner in standard cytotoxicity assays. (Blood. 2002 100:961)

訳 CTLが，自己の標的を殺すそれらの能力について検証された

❷ **These compounds were tested in MIC assays** and found to be highly potent against Gram-positive and Gram-negative organisms. (J Med Chem. 2003 46:3655)

訳 これらの化合物が，MICアッセイにおいて検証された

■ 動 [名] + be tested

❶ hypothesis was tested	仮説が，検証された	149
❷ samples were tested	試料が，検査された	113

❶ **This hypothesis was tested by using an antagonist, CJ-12,255 (Pfizer)**, that blocks the binding of SP to the neurokinin 1 receptor (NK-1R). (Proc Natl Acad Sci USA. 2004 101:9115)

訳 この仮説が，拮抗剤CJ-12,255（Pfizer）を使うことによって検証された

■ 動 when tested

❶ when tested	検証されたとき	580

❶ **When tested in the context of** the respective full-length RNAs, the same mutations abolished BMV RNA synthesis in transfected barley protoplasts. (J Virol. 2004 78:13420)

訳 ～に関して検証されたとき

■ 動 test for

★ test for ～	…は，～を検査する	1,584
❶ to test for ～	～を検査するために	604
・be used to test for ～	…が，～を検査するために使われる	116
❷ we tested for ～	われわれは，～を検査した	155

❶ Analysis of variance and logistic regression models **were used to test for changes in outcomes** by genotype. (Am J Respir Crit Care Med. 2006 173:379)

訳 ～が，結果の変化を検査するために使われた

■ 名 test + [前]

test ofの前にtheが付く割合は非常に高い．

	the%	a/an%	ø%			
❶	21	72	4	test of ～	～の検証	799

❶ **As a test of this model,** module 3 of the 6-deoxyerythronolide B synthase has been reengineered to catalyze two successive rounds of chain elongation. (Proc Natl Acad Sci USA. 2012 109:4110)

訳 このモデルの検証として，

T

★ testing 名 検証／試験／テスト，-ing テストする　用例数 11,248

名詞としての用例が多い．複数形の用例はなく，原則，不可算名詞として使われる．

◆類義語：test

■ testing ＋［前］

	the%	a/an%	ø%			
❶	17	1	74	testing of ～	～の検証	1,242
❷	0	0	94	testing for ～	～の検証	731
❸	0	0	90	testing in ～	～における検証	475

❶ This Phase I trial and preclinical studies **support additional testing of bortezomib** in combination with radiation or chemotherapy for androgen-independent prostate cancer. (Cancer Res. 2004 64:5036)

訳 ～は，ボルテゾミブの付加的な検証を支持する

❷ **Testing for antibody to HPV-16 was performed by** capture enzyme-linked immunosorbent assay (ELISA) using viruslike particles. (J Infect Dis. 2004 190:1563)

訳 HPV-16に対する抗体の検証が，～によって行われた

tether 動 つなぎ止める　用例数 3,106

tethered	つなぎ止められる／～をつなぎ止めた	1,779
tether	～をつなぎ止める	903
tethers	～をつなぎ止める	424

他動詞．

◆類義語：anchor

■ tethered ＋［前］

❶ tethered to ～	～につなぎ止められる	562
・be tethered to ～	…は，～につなぎ止められる	194

❶ **The BAG2/Hsp70 complex is tethered to the microtubule** and this complex can capture and deliver Tau to the proteasome for ubiquitin-independent degradation. (J Neurosci. 2009 29:2151)

訳 BAG2/Hsp70複合体は，微小管につなぎ止められている

■［名］/when ＋ tethered

❶ membrane tethered	つなぎ止められた膜	117
❷ when tethered	つなぎ止められたとき	107

❷ MBD3L2 acts as a transcriptional repressor **when tethered to a GAL4-DNA binding domain.** (Genetics. 2005 171:503)

訳 GAL4-DNA結合ドメインにつなぎ止められたとき

theory 名 理論 用例数 7,965

theory	理論	6,739
theories	理論	1,226

複数形のtheoriesの割合が約15%ある可算名詞．ただし，単数形が無冠詞で使われることも多い．

■ theory ＋［前］

theory ofの前にtheが付く割合はかなり高い．

	the%	a/an%	ø%			
❶	60	24	5	theory of ～	～の理論	809
❷	10	5	78	theory to *do*	～する理論	260
❸	27	47	20	theory for ～	～のための理論	226

❶ Fitness landscapes are central in the theory of adaptation.（J Theor Biol. 2013 317:1）
訳 適応度地形は，適応の理論において中心的である

■ theory ＋［動］

❶	theory predicts ～	理論は，～を予想する	198

❶ Theory predicts that stress is a key factor in explaining the evolutionary role of sex in facultatively sexual organisms, including microorganisms.（J Theor Biol. 2013 317:1）
訳 理論は，～ということを予想する

★ therapeutic 形 治療上の／治療の 用例数 23,948

T

■ therapeutic ＋［名］

❶	therapeutic target	治療上の標的	2,004
❷	therapeutic strategies	治療上の戦略	1,300
❸	therapeutic potential	治療可能性	1,090
❹	therapeutic intervention	治療介入	1,076
❺	therapeutic agents	治療薬	1,059
❻	therapeutic approaches	治療方法	816
❼	therapeutic efficacy	治療効果	610
❽	therapeutic options	治療選択肢	393

❶ Thus, GCPII is a potential therapeutic target for the reduction of excitotoxic levels of glutamate and enhancement of extracellular NAAG.（J Med Chem. 2002 45:4140）
訳 GCPⅡは，～の低下に対する潜在的な治療上の標的である

❷ Development of new diagnostic and therapeutic strategies for MDR-TB are urgently needed.（Am J Respir Crit Care Med. 1996 153:317）
訳 MDR-TBに対する新しい診断上および治療上の戦略の開発が，緊急に必要とされる

■ [形] + therapeutic

❶	potential therapeutic ～	潜在的な治療上の～	1,807
❷	novel therapeutic ～	新規の治療上の～	1,447
❸	new therapeutic ～	新しい治療上の～	1,136
❹	important therapeutic ～	重要な治療上の～	346

❷ Inhibition of calpain activity **may represent a novel therapeutic approach for the therapy of hemorrhagic shock.** (FASEB J. 2001 15:171)

訳 ～は，出血性ショックの治療のための新規の治療上のアプローチに相当するかもしれない

★★ therapy 名 治療／療法　　用例数 51,149

therapy	治療／療法	42,834
therapies	治療／療法	8,315

複数形のtherapiesの割合が，約15%ある可算名詞．

◆類義語：treatment, care, practice, cure

■ therapy + [前]

	the%	a/an%	ø%			
❶	3	23	70	therapy for ～	～のための治療	3,411
				・therapy for patients with ～	～の患者のための治療	198
				・therapy for the treatment of ～	～の治療のための療法	117
❷	2	6	86	therapy in ～	～における治療	2,175
				・therapy in patients with ～	～の患者における治療	247
❸	1	0	95	therapy with ～	～による治療	1,767
❹	32	1	64	therapy of ～	～の治療	1,219

❶ Our findings suggest that treatments based on thyroid hormone and IFN-γ **could become effective agents in therapy for patients with EBS.** (Mol Cell Biol. 2004 24:3168)

訳 ～は，EBSの患者のための治療において効果的な薬剤になりうる

❷ **To provide long-term therapy in patients with** severe toxin-induced hepatic parenchymal damage, donor hepatocytes would need to replicate and replace a large portion of the damaged parenchyma. (Nat Med. 2000 6:320)

訳 ～の患者における長期の治療を提供するために

❸ THR-123 acts specifically through Alk3 signaling, as mice with a targeted deletion for Alk3 in their tubular epithelium **did not respond to therapy with THR-123.** (Nat Med. 2012 18:396)

訳 ～は，THR-123による治療に応答しなかった

■ [前] + therapy

❶	response to therapy	治療に対する応答	388
❷	～ weeks of therapy	～週の治療	122

❶ Response to therapy was determined by changes in the size of the tumor on CT using the response evaluation criteria in solid tumors.（J Nucl Med. 2003 44:1）
訳 治療に対する応答が，～によって決定された

■［名／形］＋ therapy

❶	gene therapy	遺伝子治療	2,856
❷	antiretroviral therapy	抗レトロウイルス治療	1,483
❸	radiation therapy	放射線治療	1,362
❹	cancer therapy	癌治療	1,042
❺	combination therapy	併用療法	1,020
❻	replacement therapy	補充治療	925

★ thereby 副 それによって 用例数 11,648

◈類義語：then, so that

■ thereby

★	, thereby	, それによって	8,090
❶	, thereby providing ～	それによって～を提供する	435
❷	, thereby increasing ～	それによって～を上昇させる	282
❸	, thereby allowing ～	それによって～を可能にする	278
❹	, thereby preventing ～	それによって～を防ぐ	271
❺	, thereby reducing ～	それによって～を低下させる	233
❻	, thereby promoting ～	それによって～を促進する	217
❼	, thereby facilitating ～	それによって～を促進する	215
❽	, thereby inhibiting ～	それによって～を阻害する	205

❶ These findings set the stage for determining whether MPIN can restore PIG-A function in multipotential stem cells, thereby providing a potential new therapeutic option in PNH.（Blood. 2001 97:3004）
訳 それによって可能な新しい治療上の選択肢を提供する

❹ This inhibition occurs when the intracellular concentration of heme declines, thereby preventing the synthesis of globin peptides in excess of heme.（EMBO J. 2001 20:6909）
訳 それによってグロビンペプチドの合成を防ぐ

★★ therefore 副 したがって／それゆえ 用例数 28,902

◈類義語：hence, consequently, accordingly, thus

■ Therefore

★	Therefore,	したがって,	12,165
❶	Therefore, we ～	したがって，われわれは～	2,333
	・Therefore, we examined ～	したがって，われわれは～を調べた	192
	・Therefore, we investigated ～	したがって，われわれは～を精査した	188

T

・Therefore, we conclude that ～	したがって，われわれは～であると結論する	150
❷ Therefore, it is ～	したがって，それは～である	285

❶ Therefore, we conclude that ceramide is not a general second messenger for UV-induced apoptosis.（Oncogene. 2002 21:44）

訳 したがって，われわれは～であると結論づける

■ we therefore ＋［動］

★ we therefore ～	われわれは，それゆえ～	2,483
❶ we therefore investigated ～	われわれは，それゆえ～を精査した	256
❷ we therefore examined ～	われわれは，それゆえ～を調べた	222
❸ we therefore propose ～	われわれは，それゆえ～を提案する	195
❸ we therefore conclude that ～	われわれは，それゆえ～であると結論する	165

❶ We therefore investigated the effects of hyperglycemia and hyperlipidemia on macrophage proliferation in murine atherosclerotic lesions and isolated primary macrophages.（Diabetes. 2004 53:3217）

訳 われわれは，それゆえ～の影響を精査した

■ be therefore

★ be therefore ～	…は，それゆえ～である	3,007
❶ it is therefore ～	それは，それゆえ～である	395

❶ It is therefore important to understand the molecular basis of lens opacity due to this mutation.（J Mol Biol. 2011 412:647）

訳 ～を理解することは，それゆえ重要である

★ think　動 考える　　用例数 13,203

thought	考えられる／考えた／考え	12,983
think	考える	215
thinks	考える	5

他動詞受動態の用例が圧倒的に多い．

◆類義語：consider, regard, suppose, suspect, appear, believe

■ be thought to *do*

★ be thought to ～	…は，～すると考えられる	8,751
❶ be thought to be ～	…は，～であると考えられる	3,155
・be thought to be involved in ～	…は，～に関与していると考えられる	196
・be thought to be important	…は，重要であると考えられる	186
❷ be thought to play ～	…は，～を果たすと考えられる	713
❸ be thought to have ～	…は，～を持つと考えられる	342
❹ be thought to contribute to ～	…は，～に寄与すると考えられる	266
❺ be thought to involve ～	…は，～に関わると考えられる	194

❻ be thought to occur	…は，起こると考えられる	186
❼ be thought to function	…は，機能すると考えられる	178
❽ be thought to act	…は，作用すると考えられる	160
❾ be thought to mediate ～	…は，～を仲介すると考えられる	152

❶ Perirhinal cortex in monkeys **has been thought to be involved in** visual associative learning. (Behav Neurosci. 2003 117:1318)

訳 …は，～に関与していると考えられる

❷ **WASP is thought to play a role in actin cytoskeleton organization** and cell signaling. (Blood. 2004 104:4010)

訳 WASPは，アクチン細胞骨格組織化において役割を果たすと考えられる

❹ Although **oxidized LDL (OxLDL) is thought to contribute to lesion formation** and induces macrophage apoptosis, the mechanisms underlying macrophage lysis have not been well defined. (Circ Res. 2003 92:e20)

訳 酸化低密度リポタンパク質（OxLDL）は，病変形成に寄与すると考えられる

■ be thought ＋［that節］

★ be thought that ～	～ということが考えられる	357
❶ it is thought that ～	～ということが考えられる	239

❶ Increasingly, **it is thought that** the cardiovascular benefits of exercise are significantly influenced by adaptations within skeletal muscle and its vasculature. (J Physiol. 2003 548:401)

訳 ～ということが考えられる

■［副］＋ thought

❶ previously thought	以前に考えられた	795
❷ generally thought	一般に考えられる	234

❶ Additionally, **null mutants were generated in genes previously thought to be essential**, indicating that the H_4MPT pathway is not absolutely required during growth on multicarbon compounds. (J Bacteriol. 2003 185:7160)

訳 無発現変異体が，以前に必須であると考えられた遺伝子において作製された

❷ Alternative mating strategies are common in nature and **are generally thought to increase the intensity of** sexual selection. (Proc Natl Acad Sci USA. 2001 98:9151)

訳 …は，一般に～の強度を増大させると考えられる

T

threshold 名 閾値　　用例数 8,401

threshold	閾値	6,341
thresholds	閾値	2,060

複数形のthresholdsの割合が，約25%ある可算名詞．

■ threshold ＋ [前]

threshold forの前にtheが付く割合はかなり高い.

	the%	a/an%	ø%			
❶	67	28	4	threshold for ～	～のための閾値	820
❷	40	56	4	threshold of ～	～の閾値	611

❶ Thus, **LFA-1/ICAM-1 interaction lowers the threshold for B cell activation** by promoting B cell adhesion and synapse formation. (Immunity. 2004 20:589)

訳 LFA-1/ICAM-1の相互作用は，B細胞活性化のための閾値を低下させる

■ [形] ＋ threshold

❶	low threshold	低い閾値	305
❷	high threshold	高い閾値	132

☆ thus 副 このように／したがって　用例数 64,579

文頭の用例が多い.

◆類義語： therefore, hence, consequently, accordingly

■ thus

❶	Thus,	このように，	39,531
❷	and thus	そしてこのように	6,915

❶ **Thus,** our data suggest that IL-17C plays a critical role in maintaining mucosal barrier integrity. (J Immunol. 2012 189:4226)

訳 このように，われわれのデータは～ということを示唆する

tightly 副 堅固に／しっかりと／密接に　用例数 4,827

◆類義語：closely

■ tightly ＋ [過分]

❶	tightly regulated	堅固に調節される	993
	・be tightly regulated	…は，堅固に調節される	686
❷	tightly controlled	堅固に制御された	403
❸	tightly bound	しっかりと結合した	377
❹	tightly linked	密接に関連する	346
❺	tightly associated	しっかりと結合した	315
❻	tightly coupled	堅固に共役した	245

❶ The protein levels of β-catenin **are tightly regulated** by the ubiquitin/proteasome system. (J Biol Chem. 2002 277:27953)

訳 β-カテニンのタンパク質レベルは，ユビキチンプロテアソーム系によって堅固に調節される

❸ Interestingly, at least one copper atom remains tightly bound to N-WNDP even in the presence of excess apo-Atox1. (J Biol Chem. 2002 277:27953)

訳 少なくとも1つの銅原子は，N-WNDPにしっかりと結合したままである

❹ The life cycle of human papillomaviruses (HPVs) is tightly linked to the differentiation program of the host's stratified epithelia that it infects. (J Virol. 2005 79:13150)

訳 ヒトパピローマウイルス（HPV）の生活環は，～の分化プログラムに密接に関連する

■ [動] ＋ tightly ＋ [前]

❶ bind tightly to ～	…は，～に堅固に結合する	293

❶ RuvA binds tightly to the Holliday junction, and then recruits two RuvB pumps to power branch migration. (J Mol Biol. 2006 355:473)

訳 RuvAは，ホリデイジャンクションに堅固に結合する

★★ time ㊔ 時間　　用例数 90,664

time	時間	74,491
times	時間	15,675
timed	計測される	498

動詞としても使われるが，名詞の用例が圧倒的に多い．複数形のtimesの割合が，約15%ある可算名詞．timingは別項参照．

■ time ＋ [前]

time ofの前にtheが付く割合は非常に高い．

	the%	a/an%	ø%			
❶	76	10	13	time of ～	～の時間	8,234
				・at the time of ～	～の時に	3,489
❷	33	5	60	time to ～	～までの時間	4,276
				・time to progression	腫瘍増殖停止時間	346
❸	29	4	59	time in ～	～における時間	1,682

❶ Plasma was collected at the time of bronchoscopy and analyzed for 28 immunomodulating proteins via multiplex bead array or enzyme-linked immunosorbent assay. (Ann Surg. 2013 257:1137)

訳 血漿が，気管支鏡検査の時に集められた

■ [形／名] ＋ time

❶ for the first time	初めて	5,321
❷ over time	時間を経て	4,317
❸ at the same time	同時に	1,269
❹ survival time	生存時間	805

❶ These data demonstrate for the first time that EA activates preganglionic parasympathetic neurons in the NAmb. (Brain Res. 2012 1442:25)

訳 これらのデータは，～ということを初めて実証する

❷ The species distribution has changed over time; in both Atlanta and Baltimore the proportion of C. albicans isolates decreased, and the proportion of C. glabrata isolates increased, while the proportion of C. parapsilosis isolates increased in Baltimore only. (J Clin Microbiol. 2012 50:3435)

訳 種の分布は，時間を経て変化した

■ time ＋［名／形］

❶ time-dependent ～	時間依存的～	3,013
・time-dependent manner	時間依存的様式	642
❷ time course	時間経過	2,933
❸ time points	時点	2,255
❹ time scale	時間尺度	1,412
❺ time period	時限	640

❶ DvSnf7 suppression was observed in a time-dependent manner with suppression at the mRNA level preceding suppression at the protein level when a 240 bp dsRNA was fed to WCR larvae. (PLoS One. 2012 7:e47534)

訳 DvSnf7抑制が，時間依存的な様式で観察された

■ times ＋［形／副］

❶ times more ～	…倍より～	1,338
❷ times higher	…倍より高い	820
❸ times greater	…倍より大きい	561
❹ times faster	…倍より速い	419

❷ Sex-specific maps were also constructed; the recombination rate for females was 1.6 times higher than that for males. (Genetics. 2009 181:1649)

訳 雌の組換え率は，雄のそれより1.6倍高かった

timing ㊔ タイミング　　用例数 5,549

timing	タイミング	5,518
timings	タイミング	31

複数形のtimingsの割合は0.5％しかなく，原則，不可算名詞として使われる．

■ timing ＋［前］

timing ofの前にtheが付く割合は非常に高い．

	the％	a/an％	ø％			
❶	78	0	22	timing of ～	～のタイミング	2,895

❶ Furthermore, the duration of the refractory period predicts the timing of the next activation of the CPG, which may be minutes into the future. (PLoS One. 2012 7:e42493)

訳 不応期の持続期間は，CPGの次の活性化のタイミングを予測する

★ tolerance 名 寛容／耐性 用例数 11,356

tolerance	寛容／耐性	11,319
tolerances	寛容／耐性	37

複数形のtolerancesの割合は0.3%しかなく，原則，不可算名詞として使われる．
◆類義語：resistance, tolerability

■ tolerance ＋［前］

	the%	a/an%	ø%			
❶	2	5	91	tolerance to ～	～に対する寛容	1,635
❷	0	0	92	tolerance in ～	～における寛容	828
❸	37	4	55	tolerance of ～	～の寛容	449

❶ Complement activation also plays an important role in the maintenance of tolerance to self-antigens. (Curr Opin Rheumatol. 2005 17:538)
訳 ～は，また，自己抗原に対する耐性の維持において重要な役割を果たす

❷ The induction of mixed chimerism (MC) is a powerful and effective means to achieve transplantation tolerance in rodent models. (J Immunol. 2005 175:51)
訳 ～は，げっ歯類モデルにおいて移植免疫寛容を達成するための強力で効果的な手段である

■［名／形］＋ tolerance

❶	glucose tolerance	糖耐性	1,194
❷	immune tolerance	免疫寛容	408
❸	peripheral tolerance	末梢性寛容	354

■ tolerance ＋［名］

❶	tolerance induction	耐性誘導	505
❷	tolerance test	耐性試験	281

tolerate 動 許容する／耐える 用例数 4,295

tolerated	許容される／～を許容した	3,552
tolerate	～を許容する／～を耐える	566
tolerates	～を許容する／～を耐える	132
tolerating	～を許容する／～を耐える	45

他動詞．
◆類義語：permit, allow, endure

■ tolerated ＋［前］

❶	tolerated in ～	～において許容される	328
❷	tolerated with ～	～を持って許容される	208

T

❸	tolerated by ～	～によって許容される	168

❶ IL-10 therapy is safe and well tolerated in patients with chronic hepatitis C. (Gastroenterology. 2000 118:655)

訳 IL-10療法は，C型慢性肝炎の患者において安全でよく許容される

■ well tolerated

★	well tolerated	よく許容される／耐用性がよい	2,183
❶	be well tolerated	…は，よく許容される／耐用性がよい	1,404
❷	generally well tolerated	一般によく許容される	190
❸	safe and well tolerated	安全でよく許容される	139

★ tool 名 手段／ツール　用例数 13,989

tool	手段／ツール	7,712
tools	手段／ツール	6,277

複数形のtoolsの割合は45%とかなり高い.

◆類義語：means, method, procedure, avenue, way, instrument

■ tool ＋［前］

	the%	a/an%	ø%			
❶	2	97	0	tool for ～	～のための手段	3,420
				・tool for studying ～	～を研究するための手段	226
				・tool for identifying ～	～を同定するための手段	117
❷	3	91	1	tool to *do*	～する手段	1,308
				・tool to study ～	～を研究する手段	188
❸	4	94	0	tool in ～	～における手段	621
				・tool in ～ing	～する際の手段	134

❶ Use of homology modeling can significantly increase the extent of our approach, making it a useful tool for studying regulatory pathways in many organisms and cell types. (Nucleic Acids Res. 2005 33:5781)

訳 それを調節経路を研究するための有用な手段にする

❷ This simple assay provides a powerful tool to study mRNA dynamics in vivo. (Dev Biol. 2005 284:292)

訳 この単純なアッセイは，メッセンジャーRNAダイナミクスを研究するための強力な手段を提供する

■ ［形］＋ tool

❶	powerful tool	強力な手段	822
❷	useful tool	有用な手段	692
❸	valuable tool	価値ある手段	378
❹	new tool	新しい手段	311

❶ RNA interference (RNAi) has become a powerful tool for genetic screening in Drosophila. (Nucleic Acids Res. 2006 34:D489)

訳 RNA干渉法 (RNAi) は，遺伝的スクリーニングのための強力な手段になっている

★★ total 名 総計，形 総／全　　用例数 33,084

total	総計／総／全	33,063
totals	総計	21

名詞および形容詞として使われる．名詞としては，a total ofの用例が圧倒的に多い．しかし，複数形のtotalsの用例はほとんどない．

◆類義語：overall, whole, entire, gross, all

■ 名 total ＋ [前]

total ofの前には，aが付く割合は100%と圧倒的に高い．

	the%	a/an%	ø%			
❶	0	100	0	total of ～	総計～	7,505
				・a total of ～ patients	総計～名の患者	1,070
				・a total of ～ subjects	総計～名の被験者	128

❶ A total of 372 patients treated with anti-tumor necrosis factor (TNF) therapies met the inclusion criteria. (Arthritis Rheum. 2006 54:54)

訳 抗腫瘍壊死因子 (TNF) 療法で治療された総計372名の患者

■ 形 total ＋ [名]

❶	total body	全身	1,018
❷	total number of ～	～の総数	1,010
❸	total cholesterol	総コレステロール	703
❹	total synthesis	全合成	605
❺	total protein	総タンパク質量	452

❷ The total number of CARTp-IR cells was significantly reduced (by five-fold) at P14, and this reduced number persisted into adulthood. (J Comp Neurol. 2005 489:501)

訳 CARTp-IR細胞の総数は，有意に低下した

■ 形 [前] ＋ total

❶	increase in total ～	総～の増大	257
❷	levels of total ～	総～のレベル	221
❸	decrease in total ～	総～の低下	123
❹	reduction in total ～	総～の低下	120

❷ There were no significant changes in levels of total cholesterol, HDL-C, triglycerides, or VLDL-C in response to treatment with guggulipid in the intention-to-treat analysis. (JAMA. 2003 290:765)

訳 総コレステロール，HDL-C，トリグリセリドあるいはVLDL-Cのレベルの有意な変化はなかった

toxic 形 有毒な／毒性のある，名 有毒物質　用例数 5,023

toxic	有毒な／毒性のある／有毒物質	5,017
toxics	有毒物質	6

名詞の用例もあるが，形容詞の用例が圧倒的に多い．

◆類義語：virulent, aversive

■ toxic ＋［前］

❶ toxic to ～	～に毒性のある	522
・be toxic to ～	…は，～に毒性のある	251

❶ ZFP1 is targeted to the parasite nucleolus by CCHC motifs and significantly altered expression levels **are toxic to the parasites**. (Infect Immun. 2005 73:6680)

訳 ～は，それらの寄生虫に毒性がある

■［副］＋ toxic

❶ less toxic	より有毒でない	227
❷ potentially toxic	潜在的に有毒な	138
❸ highly toxic	高度に有毒な	128
❹ more toxic	より有毒な	116

❶ Our results also indicate that RD114 pseudotypes **were less toxic than VSV-G pseudotypes** in human MSC progenitor assays. (J Virol. 2004 78:1219)

訳 ～は，VSV-G偽型より毒性がなかった

■ toxic ＋［名］

❶ toxic effects	有毒な影響	599
❷ toxic shock	毒素ショック	211

★ toxicity 名 毒性　用例数 11,975

toxicity	毒性	10,468
toxicities	毒性	1,507

複数形のtoxicitiesの割合は約15%あるが，単数形は無冠詞で使われることが非常に多い．

◆類義語：virulence

■ toxicity ＋［前］

toxicity ofの前にtheが付く割合は非常に高い．

	the%	a/an%	ø%			
❶	71	1	27	toxicity of ～	～の毒性	1,099
❷	0	0	79	toxicity in ～	～における毒性	749

❸	4	1	72	toxicity to ～	～に対する毒性	265

❶ However, loss of uracil-DNA-N-glycosylase activity **does not appear to change the toxicity of thymidine deprivation significantly**. (Nucleic Acids Res. 2005 33:6644)
訳 ～は，チミジン欠乏の毒性を有意には変化させないように思われる

■ [現分／過分] + toxicity

❶	dose-limiting toxicity	用量規制毒性	292
❷	～-induced toxicity	～に誘導される毒性	201

❶ Of 15,000 mouse cDNA fragments studied, metallothionein (Mt)-1 and Mt2 emerged **as candidate genes possibly involved in MDMA-induced toxicity to DA neurons**. (J Neurosci. 2004 24:7043)
訳 MDMAに誘導されるDAニューロンに対する毒性におそらく関与する候補遺伝子として

trafficking 名 輸送／トラフィッキング　用例数 8,989

複数形の用例はなく，原則，不可算名詞として使われる．
◆類義語：transportation, transport

■ trafficking + [前]

	the%	a/an%	ø%			
❶	25	0	73	trafficking of ～	～の輸送	2,096

❶ **It is unclear how intracellular trafficking of the SUMOylation enzymes is regulated** to catalyze SUMOylation in different cellular compartments. (J Biol Chem. 2012 287:42611)
訳 SUMO化酵素の細胞内輸送が，どのように制御されるか明らかではない

■ [形／名] + trafficking

❶	membrane trafficking	膜トラフィッキング	658
❷	intracellular trafficking	細胞内輸送	537
❸	protein trafficking	タンパク質輸送	362
❹	vesicle trafficking	小胞輸送	329

transactivation 名 トランス活性化／トランスアクチベーション　用例数 4,999

transactivation	トランス活性化	4,996
transactivations	トランス活性化	3

複数形のtransactivationsの割合はほとんどなく，原則，不可算名詞として使われる．
◆類義語：activation

T

■ transactivation ＋［前］

	the%	a/an%	ø%			
❶	16	2	71	transactivation of ～	～のトランス活性化	981
❷	7	0	71	transactivation by ～	～によるトランス活性化	316
❸	1	1	61	transactivation in ～	～におけるトランス活性化	203

❶ Both proteins influence transactivation of a target gene, lin-48.（Dev Biol. 2006 289:456）

訳 両方のタンパク質は，標的遺伝子lin-48のトランス活性化に影響を与える

❷ However, deletion of the E boxes did not disrupt the transactivation by E2A, raising the possibility of indirect activation via another transcription factor or binding of E2A to non-E-box DNA elements.（Mol Cell Biol. 2004 24:8790）

訳 Eボックスの欠失は，E2Aによるトランス活性化を消滅させなかった

■［過分／形］＋ transactivation

❶	～-mediated transactivation	～に仲介されるトランス活性化	350
❷	～-dependent transactivation	～依存性のトランス活性化	196

❶ Significantly, we demonstrate that E6 inhibits the RXRα-mediated transactivation of target genes, implying that perturbation of RXR-mediated transactivation by E6 could contribute to HPV oncogenesis.（J Biol Chem. 2002 277:45611）

訳 RXRに仲介されるE6によるトランス活性化の混乱は，HPV発癌に寄与しうる

■ transactivation ＋［名］

❶	transactivation domain	トランス活性化ドメイン	604
❷	transactivation activity	トランス活性化活性	259
❸	transactivation function	トランス活性化機能	199

transcribe 動 転写する　用例数 4,462

transcribed	転写される／～を転写した	3,910
transcribe	～を転写する	250
transcribing	～を転写する	206
transcribes	～を転写する	96

他動詞．受動態の用例が非常に多い．

■ transcribed ＋［前］

❶	transcribed in ～	～において転写される	436
	・be transcribed in ～	…は，～において転写される	257
❷	transcribed from ～	～から転写される	397
❸	transcribed by ～	～によって転写される	220

❶ The tomato lat52 gene encodes an essential cysteine-rich protein preferentially transcribed in the vegetative cell during pollen maturation.（Plant Mol Biol. 1998 37:859）

訳 トマトのlat52遺伝子は，～の間に栄養細胞において優先的に転写される必須のシステインに富むタンパク質をコードする

❷ Here we present evidence that the gene is transcribed from three promoters: p1, p2 and p3. (Mol Microbiol. 2002 43:159)

訳 その遺伝子は，3つのプロモーターp1，p2およびp3から転写される

■ ［副］＋ transcribed

❶ divergently transcribed	異なって転写される	133
❷ actively transcribed	活発に転写される	180

❶ acfA and acfD are divergently transcribed genes required for efficient colonization of the intestine. (J Bacteriol. 2005 187:7890)

訳 acfAとacfDは，異なって転写される遺伝子である

★★ transcription ㊔ 転写　　用例数 81,063

transcription	転写	81,030
transcriptions	転写	33

複数形のtranscriptionsの割合は0.4%しかなく，原則，不可算名詞として使われる．

■ transcription ＋［前］

	the%	a/an%	∅%			
❶	34	1	65	transcription of ～	～の転写	7,327
				・transcription of ～ genes	～遺伝子の転写	1,808
❷	1	1	76	transcription in ～	～における転写	3,010
				・transcription in ～ cells	～細胞における転写	309
❸	0	0	93	transcription from ～	～からの転写	1,605
				・transcription from ～ promoter	～プロモーターからの転写	316
❹	0	0	79	transcription by ～	～によって転写を／～による転写	1,572

❶ Hypoxia-inducible factors (HIF-1 and HIF-2) are two closely related protein complexes that activate transcription of target genes in response to hypoxia. (Cancer Res. 2000 60:7106)

訳 低酸素に応答して標的遺伝子の転写を活性化する～

❷ The genetic reduction in β-cell gene transcription in homozygous animals likely contributes to the development of diabetes in the setting of insulin resistance. (Diabetes. 2001 50:63)

訳 ホモ接合動物におけるβ細胞遺伝子転写の遺伝的低下は，おそらく糖尿病の発症に寄与する

❸ Transcription from the transgenic cyclin A1 promoter was repressed in most organs outside the testis, even when the promoter was not methylated. (Mol Cell Biol. 2000 20:3316)

訳 遺伝子導入されたサイクリンA1プロモーターからの転写は，精巣外のほとんどの臓器において抑制された

■ [動] + transcription of

❶ activate transcription of ～	…は，～の転写を活性化する	524
❷ regulate transcription of ～	…は，～の転写を調節する	319
❸ repress transcription of ～	…は，～の転写を抑制する	240

■ [名／形／過分] + transcription

❶ gene transcription	遺伝子転写	3,815
❷ ～-dependent transcription	～依存的な転写	1,491
❸ ～-specific transcription	～特異的な転写	1,065
❹ ～-mediated transcription	～に仲介される転写	882

■ transcription + [名]

❶ transcription factor	転写因子	17,025
❷ transcription initiation	転写開始	1,449
❸ transcription start site	転写開始部位	944
❹ transcription elongation	転写伸長	750
❺ transcription activation	転写活性化	551

❺ Also, like class I promoters, the C-terminal domain of the α-subunit of RNA polymerase appears to play a role in transcription activation. (Mol Microbiol. 2001 39:1504)
訳 ～は，転写活性化において役割を果たすように思われる

★ transcriptional 形 転写の 用例数 35,383

■ transcriptional + [名]

❶ transcriptional activation	転写活性化	4,101
❷ transcriptional activity	転写活性	3,753
❸ transcriptional regulation	転写調節	2,800
❹ transcriptional repression	転写抑制	1,391
❺ transcriptional activator	転写活性化因子	1,257
❻ transcriptional repressor	転写抑制因子	1,195
❼ transcriptional regulator	転写制御因子	1,080
❽ transcriptional control	転写制御	837
❾ transcriptional level	転写レベル	781

❶ Transcriptional activation by TF II-I was severely reduced by overexpression of HDAC3. (J Biol Chem. 2003 278:1841)
訳 TFII-Iによる転写活性化は，HDAC3の過剰発現によって激しく低下させられた

❷ Despite these remarkable similarities, our model demonstrates how a few critical changes in CAR can dramatically reverse the transcriptional activity of this protein. (Mol Cell Biol. 2002 22:5270)

訳 ～は，このタンパク質の転写活性を劇的に逆行させうる

transcriptionally 副 転写的に／転写レベルで 用例数 2,714

■ transcriptionally ＋［形／過分］

❶ transcriptionally active	転写的に活性のある	674
❷ transcriptionally regulated	転写的に調節される	273
・be transcriptionally regulated	…は，転写的に調節される	180
❸ transcriptionally inactive	転写的に不活性である	162
❹ transcriptionally activated	転写的に活性化される	129
❺ transcriptionally silent	転写的に不活性である	123

❶ GAL4-IRF9 was transcriptionally active in reporter gene assays but not in the absence of cellular STAT1 and STAT2.（J Biol Chem. 2003 278:13033）

訳 GAL4-IRF9は，レポーター遺伝子アッセイにおいて転写的に活性があった

❷ In this study, we show that the gene encoding CBP（CBP1）is transcriptionally regulated.（J Bacteriol. 1998 180:1786）

訳 CBP（CBP1）をコードする遺伝子は，転写的に調節される

transduce 動 形質導入する／伝達する／変換する 用例数 5,846

transduced	形質導入された／～を形質導入した／～を伝達した	3,520
transduce	～を形質導入する／～を伝達する	1,075
transducing	～を形質導入する／～を伝達する	896
transduces	～を形質導入する／～を伝達する	355

他動詞.

◆類義語：transmit, transfer, convey, transfect

■ transduced ＋［前］

❶ transduced with ～	～を形質導入された	796
・cells transduced with ～	～を形質導入された細胞	251
・be transduced with ～	…は，～を形質導入される	184
❷ transduced by ～	～によって形質導入される	241
❸ transduced to ～	～に形質導入される	128
❹ transduced into ～	～に形質導入される	108

❶ In this report we show that cells transduced with a replication-deficient recombinant adenovirus expressing Z（rAd-Z）are resistant to LCMV and LFV infection.（J Virol. 2004 78:2979）

訳 複製欠陥のある組換え型アデノウイルスを形質導入された細胞

❷ Human diploid fibroblasts（HDF）are poorly transduced by adenovirus due to a lack of CAR on the surface.（J Virol. 2002 76:1892）

訳 ヒト二倍体線維芽細胞（HDF）は，アデノウイルスによってあまり形質導入されない

T

■ transduce ＋［名］

❶	transduce signals	…は，シグナルを伝達する	290

★ transduction　名 伝達／形質導入／変換　用例数 13,669

transduction	伝達／形質導入	13,640
transductions	伝達／形質導入	29

複数形のtransductionsの割合は0.2%しかなく，原則，**不可算名詞**として使われる．signal transductionの用例が圧倒的に多い．

◆類義語：transmission, transfer, communication, transfection

■ transduction ＋［前］

	the %	a/an %	ø %			
❶	21	1	78	transduction of ～	～の伝達／～の形質導入	7,327
❷	0	0	99	transduction in ～	～における伝達	597
❸	2	0	95	transduction by ～	～による伝達	305

❶ Furthermore, **transduction of** retrovirus-ICSBP in ICSBP$^{-/-}$ macrophages rescued IFN-γ-induced iNOS gene expression.（J Biol Chem. 2003 278:2271）

訳 ICSBP$^{-/-}$マクロファージにおけるレトロウイルス-ICSBPの形質導入は，～を救出した

❷ We propose that this regulatory mechanism might increase signal **transduction in** memory T cells, while limiting TCR cross-reactivity and autoimmunity.（J Immunol. 2003 170:5455）

訳 この調節機構は，記憶T細胞におけるシグナル伝達を増大させるかもしれない

■［名］＋ transduction

❶	signal transduction	シグナル伝達	9,502
	・signal transduction pathways	シグナル伝達経路	1,705

★ transfect　動 トランスフェクトする／移入する／導入する　用例数 12,157

transfected	トランスフェクトされる／移入される／～をトランスフェクトした	11,774
transfecting	～をトランスフェクトする／～を移入する	279
transfect	～をトランスフェクトする／～を移入する	93
transfects	～をトランスフェクトする／～を移入する	11

他動詞．

◆類義語：transform, introduce, transfer

■ transfected ＋［前］

❶ transfected with ～	～をトランスフェクトされる	3,799
・cells transfected with ～	～をトランスフェクトされた細胞	1,705
・be transfected with ～	…は，～をトランスフェクトされる	551
・cells were transfected with ～	細胞は，～をトランスフェクトされた	260
❷ transfected into ～	～にトランスフェクトされる	792

❶ **PSG17 bound to 293 T cells transfected with wild-type CD9** but not the mutant CD9.（Mol Biol Cell. 2003 14:5098）

訳 PSG17は，野生型CD9をトランスフェクトされた293 T細胞に結合した

❷ **A human RhoGDI small interfering RNA was transfected into HeLa cells** to knock down 90% of the endogenous RhoGDI expression.（J Biol Chem. 2004 279:42936）

訳 ヒトRhoGDIの低分子干渉RNAが，HeLa細胞にトランスフェクトされた

■ transfected ＋［名］

❶ transfected cells	トランスフェクトされた細胞	2,165
・in transfected cells	トランスフェクトされた細胞において	553

❶ **In transfected cells**, expression of cyclin D1 or c-myc was not decreased by rapamycin.（J Biol Chem. 2004 279:2737）

訳 トランスフェクトされた細胞において，

■［副］＋ transfected

❶ stably transfected	安定にトランスフェクトされる	1,292
❷ transiently transfected	一過性にトランスフェクトされる	795

❶ The effect of glycosylation on Kv1.1 potassium channel function **was investigated in mammalian cells stably transfected with Kv1.1** or Kv1.1N207Q.（J Physiol. 2003 550:51）

訳 ～が，Kv1.1を安定にトランスフェクトされた哺乳類細胞において精査された

❷ **NIH 3T3 cells were transiently transfected with these vectors** in various combinations.（FASEB J. 2001 15:635）

訳 NIH 3T3細胞は，これらのベクターを一過性にトランスフェクトされた

transfection ㊔ トランスフェクション／移入　　用例数 7,453

transfection	トランスフェクション／移入	7,075
transfections	トランスフェクション／移入	378

複数形のtransfectionsの割合は約5％あるが，単数形は無冠詞で使われることが圧倒的に多い．

◆類義語：transmission, transduction, transfer, import, transformation

■ transfection ＋［前］

	the％	a/an％	ø％			
❶	21	1	78	transfection of ～	～のトランスフェクション	2,654
				transfection of ～ with …	…による～のトランスフェクション	468
				transfection of ～ into …	…への～のトランスフェクション	200
❷	0	0	97	transfection with ～	～のトランスフェクション／～によるトランスフェクション	720
❸	0	0	90	transfection into ～	～へのトランスフェクション	156

❶ Transient transfection of cells with a constitutively active MEK construct significantly protected them from bortezomib/SAHA-mediated lethality.（Blood. 2003 102:3765）

訳 構成的に活性のあるMEKコンストラクトによる細胞の一過性のトランスフェクションは，～を有意に保護した

❷ Their PTEN protein expression was restored by transfection with vinculin or by inhibition of PTEN degradation.（J Biol Chem. 2005 280:5676）

訳 それらのPTENタンパク質発現は，ビンキュリンのトランスフェクションによって回復された

■［前］＋ transfection

❶	after transfection	トランスフェクションのあと	251
❷	following transfection	トランスフェクションに続いて	138

❶ Wild-type Parkin was homogeneously distributed throughout the cytoplasm with a small amount of protein in the nucleus after transfection into human embryonic kidney cells.（Hum Mol Genet. 2003 12:2957）

訳 ヒト胚性腎細胞へのトランスフェクションのあと

■［形］＋ transfection

❶	transient transfection	一過性トランスフェクション	1,550
❷	stable transfection	安定なトランスフェクション	280

❷ Stable transfection of NHERF-1/EBP50 into OKH cells restored the stimulatory effect of U50,488H upon Na^+/H^+exchange.（J Biol Chem. 2004 279:25002）

訳 OKH細胞へのNHERF-1/EBP50の安定なトランスフェクションは，U50,488Hの刺激性の効果を回復した

■ transfection ＋［名］

❶	transfection assays	トランスフェクションアッセイ	726
❷	transfection experiments	トランスフェクション実験	410
❸	transfection studies	トランスフェクション研究	326

★★ transfer 名 移入／移行／転移，動 移行させる　用例数 33,643

transfer	移入／移行させる	28,309
transferred	移行される／移行させた	3,707
transfers	移入	1,039
transferring	移行させる	588

名詞の用例が多いが，動詞としても用いられる．複数形のtransfersの割合は2％しかなく，原則，**不可算名詞**として使われる．

◆類義語：transmission, transduction, communication, movement, shift, translocation, transfection, transmit, transduce, transfect

■ 名 transfer ＋［前］

	the%	a/an%	ø%			
❶	33	0	67	transfer of ～	～の移入／～の移行	5,417
				・transfer of ～ from …	～の…からの移入	751
				・transfer of ～ from … to ‥	～の…から‥への移入	331
				・transfer of ～ to …	～の…への移入	477
				・catalyze the transfer of ～	…は，～の移行を触媒する	187
❷	2	4	91	transfer to ～	～への移入	1,479
❸	7	6	87	transfer from ～	～からの移入	1,331
				・transfer from ～ to …	～から…への移入	570
❹	2	2	93	transfer in ～	～における移入	743
❺	9	2	83	transfer between ～	～の間の移入	609
❻	0	1	96	transfer into ～	～への移入	420

❶ The **transfer of** ubiquitin **to** substrate is a multistep process.（Mol Cell Biol. 1999 19:1759）
訳 ユビキチンの基質への移行は，多段階過程である

❶ β1,4-galactosyltransferase-I（β4Gal-T1）**catalyzes the transfer of** a galactose from UDP-galactose to N-acetylglucosamine.（Biochemistry. 2003 42:3674）
訳 ～は，UDP-ガラクトースからN-アセチルグルコサミンへのガラクトースの転移を触媒する

❷ We determined that AdV gene **transfer to** the murine airway epithelium was inefficient even in GPI-hCAR transgenic mice but that the gene transfer efficiency improved in the absence of Muc1.（J Virol. 2004 78:13755）
訳 マウス気道上皮へのAdV遺伝子移入は，～においてでさえ非効率的であった

❸ This step precedes electron **transfer from** the HE-TPP radical intermediate **to** an intramolecular [4Fe-4S] cluster.（Biochemistry. 2002 41:9921）
訳 この段階は，HE-TPPラジカル中間体から～への電子伝達に先行する

■ 名［名／形］＋ transfer

❶	electron transfer	電子伝達	3,452
❷	gene transfer	遺伝子移入	3,254

T

❸ energy transfer	エネルギー移動	2,688
❹ adoptive transfer	養子移入	1,769
❺ proton transfer	プロトン移動	1,096
❻ charge transfer	電荷移動	931

■ ㊌ transferred ＋［前］

❶ transferred to ～	～に移される	893
・be transferred to ～	…は，～に移される	461
❷ transferred into ～	～に移入される	442
❸ transferred from ～	～から移入される	353

❶ Subsequently, **SHP-2 is transferred to IGF-I receptor** and regulates the duration of IGF-I receptor phosphorylation.（Mol Biol Cell. 2003 14:3519）

訳 SHP-2は，IGF-I 受容体に移される

❸ **The ybhE gene was transferred from the E. coli chromosome to an expression vector.**（J Bacteriol. 2004 186:8248）

訳 ybhE遺伝子が，大腸菌染色体から発現ベクターに移入された

■ ㊌［副］＋ transferred

❶ adoptively transferred	養子性に移入される	615

❶ In addition, flagellin-specific $CD4^+$ T cells induced severe colitis **when adoptively transferred into naive SCID mice.**（J Clin Invest. 2004 113:1296）

訳 ナイーブなSCIDマウスに養子性に移入されるとき

★transform ㊌ 形質転換させる／導入する／癌化させる

用例数 14,384

transformed	形質転換される／～を形質転換させた	6,033
transforming	～を形質転換させる	5,711
transform	～を形質転換させる	2,135
transforms	～を形質転換させる	505

他動詞.

◆類義語：transduce, transfect

■ transformed ＋［前］

❶ transformed into ～	～に形質転換される	430
・be transformed into ～	…が，～に形質転換される	255
❷ transformed with ～	～を導入する	404
❸ transformed by ～	～によって形質転換される	310
・cells transformed by ～	～によって形質転換された細胞	138
❹ transformed to ～	～に形質転換される	133

❶ The newly constructed plasmids were transformed into E. coli for expression of the fusion proteins.（Anal Chem. 1999 71:4321）

訳 新たに構築されたプラスミドが，大腸菌に形質転換された

❷ Subsequently, this integrant strain is transformed with a plasmid expressing the Cre recombinase.（Gene. 2003 304:133）

訳 ～が，Creリコンビナーゼを発現するプラスミドを導入される

❸ In cells transformed by v-Abl, Janus kinase（JAK）tyrosine kinases are constitutively activated.（Oncogene. 2000 19:2523）

訳 v-Ablによって形質転換された細胞において

■ transformed ＋［名］

❶	transformed cells	形質転換された細胞	1,188
❷	transformed phenotype	形質転換された表現型	246

❶ The activation of telomerase by v-Rel may, therefore, partially protect the transformed cells from apoptosis induced by ROS.（J Virol. 2006 80:281）

訳 ～は，形質転換された細胞をアポトーシスから部分的に保護する

■［名］-transformed

❶	Ras-transformed	Rasで形質転換された	334
❷	EBV-transformed	EBVで形質転換された	170

❶ However, TNFα specifically and dose-dependently decreased the ability of v-Ras-transformed RelA-deficient cells to form colonies in soft agar.（Oncogene. 2005 24:6574）

訳 ～は，v-Rasで形質転換されたRelA欠損細胞の能力を低下させた

T

★ transformation 名 形質転換／トランスフォーメーション

用例数 11,708

transformation	形質転換／トランスフォーメーション	10,674
transformations	形質転換／トランスフォーメーション	1,034

複数形のtransformationsの割合が，約10%ある可算名詞.

◆類義語：transfection, oncogenesis

■ transformation ＋［前］

	the %	a/an %	ø %			
❶	42	5	53	transformation of ～	～の形質転換	2,023
❷	5	8	80	transformation in ～	～における形質転換	545
❸	2	0	92	transformation by ～	～による形質転換	507

❶ Overexpression of a novel oncogene MCT-1（multiple copies in a T cell malignancy）causes malignant transformation of murine fibroblasts.（Cancer Res. 2005 65:10651）

訳 ～は，マウス線維芽細胞の悪性形質転換を引き起こす

❷ Constitutively active components of the MEK signaling cascade can induce oncogenic transformation in many cell systems. (Gastroenterology. 2005 129:577)

訳 ～は，多くの細胞系において癌化を誘導できる

■ [形／名] + transformation

❶	malignant transformation	悪性化	721
❷	cell transformation	細胞形質転換	688
❸	cellular transformation	細胞形質転換	535
❹	oncogenic transformation	癌化	418
❺	neoplastic transformation	新生物形質転換	406

★ transient 形 一過性の／過渡応答 用例数 17,013

transient	一過性の／過渡応答	15,367
transients	過渡応答	1,646

名詞としても用いられるが，形容詞の用例が多い．

■ transient + [名]

❶	transient transfection	一過性のトランスフェクション	1,550
❷	transient expression	一過性の発現	910
❸	transient increase	一過性の増大	480

❶ Transient transfection of HIF-1α expression vector induced transcription from p21 promoter construct in prostate cancer cell lines. (Cancer Res. 2000 60:5630)

訳 HIF-1α発現ベクターの一過性のトランスフェクションは，p21プロモーターコンストラクトからの転写を誘導した

❷ Transient expression of genomic RPS4 driven by the 35S promoter in tobacco leaves induces an AvrRps4-independent hypersensitive response (HR). (Plant J. 2004 40:213)

訳 タバコの葉における35Sプロモーターによって駆動されるゲノム性RPS4の一過性の発現は，～を誘導する

transiently 副 一過性に 用例数 4,193

■ transiently + [過分]

❶	transiently transfected	一過性にトランスフェクトされる	795
	・cells transiently transfected with ～	～を一過性にトランスフェクトされた細胞	113
❷	transiently expressed	一過性に発現させられる	466
	・be transiently expressed	…は，一過性に発現させられる	175
❸	transiently increased	一過性に増大した	121

❶ NIH 3T3 cells were transiently transfected with these vectors in various combinations. (FASEB J. 2001 15:635)

訳 NIH 3T3細胞は，これらのベクターを一過性にトランスフェクトされた

❷ These chimeras were transiently expressed in the human embryonic cell line HEK-293T. (J Biol Chem. 2004 279:12242)

訳 これらのキメラは，ヒト胚性細胞株HEK-293Tにおいて一過性に発現させられた

★ transition 名 移行／遷移／転移／トランジション　用例数 22,042

transition	移行／遷移	18,256
transitions	移行／遷移	3,786

複数形のtransitionsの割合が，約15%ある可算名詞.

◆類義語：shift, translocation, entry, transfer, dissemination, transposition, transmission, transduction

■ transition ＋［前］

transition fromの前にtheが付く割合は非常に高い．また，transition betweenの前にtheが付く割合もかなり高い.

	the%	a/an%	ø%			
❶	72	16	10	transition from ～	～からの移行	1,967
				・transition from ～ to …	～から…への移行	1,262
❷	42	32	21	transition to ～	～への移行	937
❸	67	12	20	transition of ～	～の移行	855
❹	37	40	17	transition in ～	～における移行	778
❺	65	18	14	transition between ～	～の間の移行	468

❶ Here we report that the sequence-specific DNA-binding protein, RFX, previously shown to mediate the transition from an inactive to an active chromatin structure, activates a methylated promoter. (J Biol Chem. 2005 280:38914)

訳 不活性から活性クロマチン構造への移行を仲介することが以前に示された配列特異的なDNA結合タンパク質RFXは，メチル化されたプロモーターを活性化する

❷ To facilitate the transition to recovery, we conducted the Moving Beyond Cancer (MBC) trial, a multisite, randomized, controlled trial of psychoeducational interventions for breast cancer patients. (J Clin Oncol. 2005 23:6009)

訳 回復への移行を促進するために

■ transition ＋［名］

❶	transition state	遷移状態	3,368
❷	transition metal	遷移金属	626

❶ Of interest are the structures formed in the transition state that promote calcium binding. (Biochemistry. 1997 36:4607)

訳 遷移状態において形成される構造は興味深い

■［名］＋ transition

❶	phase transition	相遷移	764
❷	permeability transiticn	透過性遷移	446

T

translocate 動 移行する／移行させる／移動させる 用例数 3,652

translocated	移行した／～を移行させた	1,431
translocate	移行する／～を移行させる	934
translocates	移行する／～を移行させる	804
translocating	移行する／～を移行させる	483

自動詞および他動詞の両方で用いられる．他動詞受動態は，自動詞とほぼ同じ意味になる．

◆類義語：transfer, shift, transport, migrate, move

■ translocate ＋［前］

❶ translocate to ～	…は，～に移行する	596
・translocate to the nucleus	…は，核に移行する	282
❷ translocate into ～	…は，～に移行する	158
❸ translocate from ～	…は，～から移行する	121

❶ Activated ERK translocates to the nucleus and phosphorylates the transcription factor ELK-1, which in turn coordinates the expression of downstream target genes such as DMP1 and dentin sialoprotein (DSP). (J Biol Chem. 2012 287:5211)

訳 活性化されたERKは，核に移行する

■ be translocated ＋［前］

★ be translocated	…は，移行する	553
❶ be translocated to ～	…は，～に移行する	173
❷ be translocated into ～	…は，～に移行する	145

❶ We report that albumin is translocated to the nucleus in response to oxidative stress. (Biochemistry. 2004 43:7443)

訳 アルブミンは，酸化ストレスに応答して核に移行する

❷ In addition, YspM is translocated into host cells via the Ysa T3SS. (J Bacteriol. 2008 190:7315)

訳 YspMは，宿主細胞に移行する

★ translocation 名 移行／転位置／転座 用例数 15,187

translocation	移行／転位置／転座	13,565
translocations	移行／転位置／転座	1,622

複数形のtranslocationsの割合は約10%あるが，単数形は無冠詞で使われることが圧倒的に多い．

◆類義語：shift, transition, entry, transfer, transposition, dissemination, transmission, transduction, migration

■ translocation ＋［前］

	the％	a/an％	ø％			
❶	29	2	69	translocation of ～	～の移行／～の転位置	4,288
				・translocation of ～ to …	～の…への移行	493
				・translocation of ～ from … to ‥	～の…から‥への移行	212
❷	3	2	79	translocation to ～	～への移行	628
❸	6	0	90	translocation in ～	～における移行	463
❹	1	0	52	translocation into ～	～への移行	223
❺	1	4	53	translocation from ～	～からの移行	204
				・translocation from ～ to …	～から…への移行	117
❻	0	3	76	translocation across ～	～を超えた移行	182

❶ PBP-deficient hepatocytes in liver failed to reveal PB-dependent translocation of CAR to the nucleus.（Proc Natl Acad Sci USA. 2005 102:12531）

訳 ～は，PB依存的なCARの核への移行を明らかにすることができなかった

❶ Translocation of the BCL2 gene from chromosome 18 to chromosome 14 results in constitutive expression of the gene.（Genomics. 2001 73:161）

訳 BCL2遺伝子の第18染色体から第14染色体への転位置は，～という結果になる

❷ The mechanisms involved in the biogenesis of these vesicles and their translocation to the cell surface are poorly understood.（J Biol Chem. 2004 279:40062）

訳 これらの小胞の生合成およびそれらの細胞表面への移行に関与する機構は，あまり理解されていない

❸ ITT may be a useful tool for dissecting dynamic translocation in various biological systems.（Mol Cell. 2004 15:153）

訳 ITTは，さまざまな生物学的システムにおける動的な移行を精査するための有用な手段であるかもしれない

■［形］＋ translocation

❶	nuclear translocation	核移行	2,162
❷	chromosomal translocations	染色体転座	433

❶ CC3 also inhibits nuclear translocation of transportin itself.（Mol Cell Biol. 2004 24:7091）

訳 CC3は，また，トランスポーチン自身の核移行を抑制する

★ transmission 名 伝達／伝播／伝染　用例数 13,879

transmission	伝達／伝播	13,714
transmissions	伝達／伝播	165

複数形のtransmissionsの割合は１％しかなく，原則，不可算名詞として使われる．

◆類義語：translocation, transduction, transition, entry, transfer, infection

T

■ transmission ＋［前］

	the%	a/an%	ø%			
❶	38	0	61	transmission of ～	～の伝達／～の伝播	2,359
❷	1	0	84	transmission in ～	～における伝達	959
❸	1	0	78	transmission to ～	～への伝達	332

❶ Herpes simplex virus 2 (HSV-2) infection causes significant morbidity and **is an important cofactor for the transmission of HIV infection**. (Nature. 2006 439:89)

訳 ～は，HIV感染の伝播のための重要な補助因子である

❷ Muscarinic K^+ (KACh) channels **are key determinants of the inhibitory synaptic transmission in the heart**. (Biophys J. 2004 87:3122)

訳 ～は，心臓における抑制性シナプス伝達の鍵となる決定要因である

■［形／名］＋ transmission

❶	synaptic transmission	シナプス伝達	2,315
❷	signal transmission	シグナル伝達	200

transmit 動 伝達する／伝染させる　　用例数 4,444

transmitted	伝達される／～を伝達した／伝染する	2,850
transmit	～を伝達する／伝染させる	900
transmitting	～を伝達する／伝染させる	416
transmits	～を伝達する／伝染させる	278

他動詞.

◆類義語：transport, transfer, shift, convey, migrate, move, translocate

■ transmitted ＋［前］

❶	transmitted to ～	～に伝達される	427
	・be transmitted to ～	…は，～に伝達される	275
❷	transmitted by ～	～によって伝達される／～によって伝染する	257
❸	transmitted through ～	～を経て伝達される	164
❹	transmitted from ～	～から伝達される	142

❶ These changes, which **are clonally transmitted to daughter cells**, may contribute to the development of OA. (Arthritis Rheum. 2005 52:3110)

訳 ～は，娘細胞にクローン性に伝達される

❷ For **diseases transmitted by non-sexual direct contacts**, such as SARS or smallpox, individual variation is difficult to measure empirically, and thus its importance for outbreak dynamics has been unclear. (Nature. 2005 438:355)

訳 性的でない直接接触によって伝染する疾患

■ [副] + transmitted

❶ sexually transmitted	性的に伝染する	586
・sexually transmitted disease	性行為感染症	163
・sexually transmitted infections	性行為感染症	106

★ transplant 名 移植／移植片, 動 移植する　用例数 15,780

transplant	移植／移植片／移植する	8,881
transplanted	移植された／移植した	4,020
transplants	移植／移植片／移植する	2,678
transplanting	移植する	201

名詞および動詞として用いられる．複数形のtransplantsの割合が約25%ある可算名詞．

◆類義語：implant, explant, graft, engraft, implantation, engraftment, transplantation

■ 動 transplanted + [前]

❶ transplanted into ～	～に移植される	730
・be transplanted into ～	…は，～に移植される	310
❷ transplanted with ～	～を移植される	379
❸ transplanted to ～	～に移植される	149

❶ In a second series, donor hearts were transplanted into DDAH-I-transgenic or WT mice and procured 30 days after transplantation (n=7 each). (Circulation. 2005 112:1549)

訳 ドナーの心臓が，DDAH-Iトランスジェニックマウスあるいは野生型マウスに移植された

❷ Mice transplanted with Bcl11a-deficient cells died from T cell leukemia derived from the host. (Nat Immunol. 2003 4:525)

訳 Bcl11a欠損細胞を移植されたマウスは，T細胞白血病で死亡した

■ 名 transplant + [名]

❶ transplant recipients	移植レシピエント	2,354
・renal transplant recipients	腎移植レシピエント	393
❷ transplant patients	移植患者	569
❸ transplant rejection	移植拒絶	232

❶ The diagnosis and treatment of this infection in transplant recipients is discussed. (Transplantation. 1999 67:1495)

訳 移植レシピエントにおけるこの感染の診断と治療が議論される

★ transplantation 名 移植／移植術　用例数 18,494

transplantation	移植／移植術	18,112
transplantations	移植／移植術	382

複数形のtransplantationsの割合は2％しかなく，原則，不可算名詞として使われる．

◈類義語：implantation, transplant, implant, explant, graft, engraft, engraftment

■ transplantation ＋［前］

	the%	a/an%	ø%			
❶	12	1	86	transplantation of ～	～の移植	1,114
❷	0	1	97	transplantation in ～	～における移植	749
❸	0	0	95	transplantation for ～	～に対する移植	381
❹	0	1	97	transplantation with ～	～による移植	303

❶ Transplantation of **bone marrow cells** and umbilical cord blood have been attempted as a means of enzyme replacement and have shown limited success.（Proc Natl Acad Sci USA. 2005 102:18670）

訳 骨髄細胞の移植

❷ This protocol could lead to more equal access to **kidney** transplantation in **blood group B recipients**.（Transplantation. 2005 80:75）

訳 血液型Bのレシピエントにおける腎移植

❸ Islet transplantation for **type 1 diabetic patients** shows promising results with the use of nondiabetogenic immunosuppressive therapy.（Proc Natl Acad Sci USA. 2005 102:12153）

訳 1型糖尿病患者に対する膵島移植

■［前］＋ transplantation

❶	after transplantation	移植の後	2,388
	・～ months after transplantation	移植の後～ヶ月	289
	・～ days after transplantation	移植の後～日	252
❷	before transplantation	移植の前	367
❸	at the time of transplantation	移植の時に	289

❶ EBV-related posttransplant lymphoproliferative disease **was not observed up to 12** months after transplantation.（J Infect Dis. 2005 192:1331）

訳 ～は，移植の後12ヶ月までは観察されなかった

■［名］＋ transplantation

❶	liver transplantation	肝移植	1,817
❷	cell transplantation	細胞移植	1,740
❸	bone marrow transplantation	骨髄移植	1,081
❹	renal transplantation	腎移植	511
❺	kidney transplantation	腎移植	505
❻	organ transplantation	臓器移植	477

★ transport （名）輸送，（動）輸送する　用例数 29,594

transport	輸送／～を輸送する	26,455
transported	輸送される／～を輸送した	2,096
transports	輸送／～を輸送する	553

transporting	輸送／～を輸送する	490

名詞の用例が多いが，他動詞としても用いられる．

◆類義語：transfer, shift, translocate, move, carry, convey, bring, trafficking, transition, transposition, transmission, transduction, migration, translocation

■ 名 transport ＋［前］

	the%	a/an%	ø%			
❶	50	2	46	transport of ～	～の輸送	3,678
				・transport of ～ to …	～の…への輸送	233
				・transport of ～ from …	～の…からの輸送	189
				・transport of ～ in …	～の…における輸送	183
				・transport of ～ into …	～の…への輸送	167
				・transport of ～ across …	～の…を超えた輸送	153
❷	3	0	94	transport in ～	～における輸送	1,379
❸	0	0	81	transport by ～	～による輸送	562
❹	3	0	79	transport to ～	～への輸送	552

❶ **After transport of the PIC into the nucleus**, IN catalyzes the concerted insertion of the two viral DNA ends into the host chromosome.（J Virol. 2005 79:8208）

訳 PICの核への移行のあと

❷ Thus, **the markedly reduced glucose transport in muscle** results in increased glycogen synthase activity due to increased hexokinase II, glucose-6-phosphate, and RGL and PTG levels and enhanced PP1 activity.（Mol Cell Biol. 2005 25:9713）

訳 筋肉において顕著に低下したグルコース輸送

❸ **Bile acid transport by cholangiocyte ASBT** can contribute to hepatobiliary secretion in vivo.（Hepatology. 2005 41:1037）

訳 胆管細胞ASBTによる胆汁酸の輸送

T

■ 名 transport ＋［名］

❶	transport activity	輸送活性	667
❷	transport system	輸送システム	463

■ 名［名／形］＋ transport

❶	electron transport	電子伝達	1,008
❷	glucose transport	グルコース輸送	825
❸	nuclear transport	核移行	423

■ 動 transported ＋［前］

❶	transported to ～	～に輸送される	506
	・be transported to ～	…は，～に輸送される	282
❷	transported into ～	～に輸送される	220
❸	transported by ～	～によって輸送される	209

❹ transported from ～	～から輸送される	126

❶ We show that DFrizzled2 is endocytosed from the postsynaptic membrane and transported to the nucleus. (Science. 2005 310:1344)

訳 DFrizzled2はシナプス後膜から取り込まれ，そして核に輸送される

❷ In vitro import assays in digitonin-permeabilized HeLa cells reveal that ORF29p is transported into the nucleus by a Ran-, karyopherin α- and β-dependent mechanism. (J Virol. 2005 79:13070)

訳 ORF29pは，～によって核に輸送される

★★ treat 動 処理する／処置する／治療する／扱う　用例数 47,444

treated	～を処理した／処置される／治療される	39,138
treat	～を処理する／～を処置する／～を治療する	4,533
treating	～を処理する／～を処置する／～を治療する	3,646
treats	～を処理する／～を処置する／～を治療する	127

他動詞．treated withの用例が非常に多い．

◆分類：manipulate, handle, cure, remedy

■ treated ＋［前］

❶ treated with ～	～によって処理される	15,612
・patients treated with ～	～によって治療された患者	1,862
・cells treated with ～	～によって処理された細胞	1,532
・mice treated with ～	～によって処理されたマウス	1,316
・when treated with ～	～によって処理されたとき	420
・treated with either ～	～のどちらかによって処理される	325
❷ treated for ～	～のために治療される／～の間治療される	714
❸ treated by ～	～によって処理される／～によって治療される	533
❹ treated in ～	～において処理される／～において治療される	520
❺ treated at ～	～において処理される／～において治療される	515
❻ treated as ～	～として扱われる	269

❶ An almost identical expression pattern was observed in cells treated with either Taxol or EpoB. (Cancer Res. 2003 63:7891)

訳 ほとんど同一の発現パターンが，TaxolあるいはEpoBのどちらかによって処理された細胞において観察された

❶ We found that both Jurkat and PBMCs arrested efficiently in mitosis when treated with nocodazole. (J Biol Chem. 2002 277:5187)

訳 ノコダゾールによって処理されたとき

❷ These 64 included 15 of the 157 consecutive patients treated for pyoderma gangrenosum at our institution (10 percent). (Engl J Med. 2002 347:1412)

訳 壊疽性膿皮症のために治療された患者

❸ Subjects were treated by their own community practitioners. (Am J Psychiatry. 2002 159:927)

訳 被検者は，自身の地域の開業医によって治療された

■ [名] + be treated with

★ be treated	…が，処理される	6,855
❶ be treated with ～	…が，～によって処理される	4,643
・cells were treated with ～	細胞が，～によって処理された	453
・mice were treated with ～	マウスが，～によって処理された	403
・patients were treated with ～	患者が，～によって処置された	337

❶ Cells were treated with CHA, and the secretion of matrix metalloproteinase (MMP)-2 or the activation of extracellular signal-regulated kinase (ERK1/2) was determined. (Invest Ophthalmol Vis Sci. 2002 43:3016)

訳 細胞が，CHAによって処理された

■ treated + [名]

❶ treated cells	処理された細胞	691
❷ treated mice	処理されたマウス	522
❸ treated animals	処理された動物	495
❹ treated rats	処理されたラット	388
❺ treated patients	治療された患者	378

❶ Indeed, GSE-treated cells showed a strong and sustained increase in phospho-JNK1/JNK2 levels, JNK activity and phospho-cJun levels. (Oncogene. 2003 22:1302)

訳 GSEで処理された細胞は，～の強くそして持続性の増大を示した

❺ TLI had no significant overall effect on survival in treated patients compared with controls (P = 0.62). (Arthritis Rheum. 2001 44:1525)

訳 TLIは，対照群と比べて，処置された患者において生存に対する有意な全体的な影響を持たなかった

T

■ [名]-treated

❶ vehicle-treated ～	媒体で処置された～	514
❷ placebo-treated ～	偽薬で処置された～	307
❸ saline-treated ～	生理食塩水で処置された～	298

❷ Compared with placebo-treated patients, FTY720 subjects did not show a major increase in adverse events or a change in renal function. (Transplantation. 2003 76:1079)

訳 偽薬で処置された患者と比べて

■ to treat

★ to treat ～	～を治療する…／～を処理する…	3,817
❶ used to treat ～	～を治療するために使われる	707
❷ intention to treat ～	～を治療する意図	574
❸ intent to treat ～	～を治療する意図	301

❶ IFN-α 2b (IFN-α) has been used to treat patients with metastatic malignant melanoma and patients rendered disease-free via surgery but at high risk for recurrence. (J Immunol. 2004 172:7368)

訳 IFN-α 2b（IFN-α）は，転移性の悪性黒色腫の患者を治療するために使われている

❷ Meta-analyses of hernia recurrence and persisting pain were based on intention to treat.（Ann Surg. 2002 235:322）

訳 ヘルニア再発および持続性疼痛のメタ解析は，治療する意図に基づいていた

★★ treatment 名 処置／処理／治療 用例数 105,153

treatment	処置／処理／治療	97,892
treatments	処置／処理／治療	7,261

複数形のtreatmentsの割合は７％あるが，単数形は無冠詞で使われることが多い.

◆類義語：therapy, manipulation, intervention, application, care, practice, cure

■ treatment ＋［前］

	the %	a/an %	ø %			
❶	55	1	43	treatment of ～	～の処理／～の治療	24,402
				・treatment of ～ with …	…による～の処理	4,786
				・treatment of cells with ～	～による細胞の処理	757
				・treatment of patients with ～	～による患者の治療	579
❷	1	2	95	treatment with ～	～による処置	13,490
				・by treatment with ～	～による処置によって	1,537
				・after treatment with ～	～による処置のあと	1,228
❸	12	34	48	treatment for ～	～のための治療	3,089
❹	2	3	69	treatment in ～	～における治療	1,710

❶ Treatment of cells with antioxidants and xenobiotics results in the release of Nrf2 from INrf2.（Oncogene. 2001 20:3906）

訳 抗酸化剤と生体異物による細胞の処理は，～という結果になる

❷ Respiratory failure is a serious consequence of lung cell injury caused by treatment with high inhaled oxygen concentrations.（J Biol Chem. 2004 279:16317）

訳 呼吸不全は，高い吸入酸素濃度による処置によって引き起こされる肺細胞傷害の深刻な結果である

❷ After treatment with Ad-gp91ds-eGFP, 4-HNE generation was normalized.（Circ Res. 2004 95:587）

訳 Ad-gp91ds-eGFPによる処置のあと

❸ Orthotopic liver transplantation（OLT）is an effective treatment for patients with advanced primary biliary cirrhosis（PBC）.（Hepatology. 2001 33:22）

訳 ～は，進行した原発性胆汁性肝硬変（PBC）の患者のための効果的な治療である

■［前］＋ the treatment of

❶	target for the treatment of ～	～の治療のための標的	248
❷	used in the treatment of ～	～の治療において使われる	219
❸	useful in the treatment of ～	～の治療において役に立つ	203
❹	approach for the treatment of ～	～の治療のためのアプローチ	169

❺	agents for the treatment of ～	～の治療のための薬剤	156
❻	effective in the treatment of ～	～の治療において効果的な	156
❼	strategy for the treatment of ～	～の治療のための戦略	151

❸ Thus, R-cadherin antagonists may be useful in the treatment of neovascular diseases in which circulating HSCs contribute to abnormal angiogenesis. (Blood. 2004 103:3420)

訳 R-カドヘリンの拮抗薬は，新生血管の疾患の治療において有用であるかもしれない

❹ Our data indicate that Aurora kinase inhibition provides a new approach for the treatment of multiple human malignancies. (Nat Med. 2004 10:262)

訳 Auroraキナーゼの抑制は，複数のヒトの悪性腫瘍の治療のための新しいアプローチを提供する

■ [形／名] + treatment

❶	effective treatment	効果的な治療	905
❷	cancer treatment	癌治療	675
❸	drug treatment	薬物治療	563

★ trial 名 試験／治験／試み 用例数 28,099

trials	試験／治験	15,031
trial	試験／治験	13,068

複数形のtrialsの割合は約55%と非常に高い.

◆類義語：attempt, effort, study, examination, test

■ trial(s) + [前]

	the%	a/an%	ø%			
❶	14	77	4	trial of ～	～の試験	1,438
❷	2	-	79	trials in ～	～における試験	763
❸	2	-	89	trials with ～	～による試験	506
❹	3	-	93	trials for ～	～のための試験	451

❶ This was a double-blind, randomized controlled trial of 32 women with BRCA1/2 mutations. (J Clin Oncol. 2005 23:9319)

訳 これは，BRCA1/2変異を持つ32名の女性の二重盲検無作為化対照試験であった

❷ Clinical trials in humans indicate it is also possible to target these neurotransmitter systems to enhance cognitive performance in patients with chronic deficits. (Curr Opin Neurol. 2005 18:675)

訳 ヒトにおける臨床試験は，～を示す

❸ Here we present the most recent data from clinical trials with some of the promising inhibitors of angiogenesis. (Curr Opin Oncol. 2005 17:578)

訳 われわれは，血管新生の有望な抑制剤のいくつかによる臨床試験からのもっとも最近のデータを提示する

T

■ [形／過分] + trials

❶ clinical trials	臨床試験	5,734
❷ controlled trial	対照試験	1,246
・placebo-controlled trial	プラセボ対照試験	476
・randomized controlled trial	無作為化対照試験	448
❸ randomized trial	無作為化試験	745

★ trigger 動 誘発する／引き金を引く, 名 トリガー 用例数 16,189

triggered	誘発される／～を誘発した	5,788
trigger	～を誘発する／トリガー	4,446
triggers	～を誘発する／トリガー	3,736
triggering	～を誘発する	2,219

他動詞および名詞として用いられる．複数形のtriggersの割合が，約25%ある可算名詞．

◆類義語：induce, evoke, elicit, cause

■ 動 trigger + [名句]

❶ trigger apoptosis	…は，アポトーシスを誘発する	372
❷ trigger activation of ～	…は，～の活性化を誘発する	97

❶ Inhibition of the survival kinase Akt can **trigger apoptosis**, and also has been found to activate autophagy, which may confound tumor attack. (Cancer Res. 2011 71:2654)

訳 生存キナーゼAktの阻害は，アポトーシスを誘発できる

■ 動 to trigger

★ to trigger ～	～を誘発する…	1,281
❶ be sufficient to trigger ～	…は，～を誘発するのに十分である	140

❶ The altered localization of PELP1 **was sufficient to trigger** the interaction of PELP1 with the p85 subunit of phosphatidylinositol-3-kinase (PI3K), leading to PI3K activation. (Cancer Res. 2005 65:7724)

訳 PELP1の変化した局在は，PELP1の～との相互作用を誘発するのに十分であった

■ 動 triggered + [前]

❶ triggered by ～	～によって誘発される	2,362
・be triggered by ～	…は，～によって誘発される	979
❷ triggered in ～	～において誘発される	122

❶ This response **is triggered by** phosphorylation of Chk1 at Ser-345, a known target site for the upstream activating kinase ATR. (Mol Cell. 2005 19:607)

訳 この応答は，Ser-345でのChk1のリン酸化によって誘発される

■ 名 trigger ＋［前］

	the%	a/an%	ø%			
❶	40	58	1	trigger for ～	～のトリガー	244
❷	30	66	1	trigger of ～	～のトリガー	140

❶ Advanced glycation endproduct (AGE) formation is a trigger for the onset of age-related disease. (Proc Natl Acad Sci USA. 2005 102:11846)
訳 終末糖化産物（AGE）形成は，年齢に関連する疾患の発症のトリガーである

turn 名 回転／ターン／順番／変化，動 回転する／変化する

用例数 9,991

turn	回転／ターン／順番／変化する／回転する	7,984
turns	回転／ターン／順番／変化する／回転する	761
turning	回転する／変化する	641
turned	回転した／変化した	605

名詞の用例が多いが，動詞としても用いられる．in turnの用例が非常に多い．

■ turn

❶	in turn	次には／今度は／順に	5,240
	・, which in turn	そして，（それは）次には	2,358
❷	helix-turn-helix	ヘリックス・ターン・ヘリックス	431

❶ The models predict that Gd activates Snk, which in turn activates Ea. (J Biol Chem. 2003 278:11320)
訳 GdはSnkを活性化し，そして，（それは）次にはEaを活性化する

T

★★ type 名 タイプ／型

用例数 164,973

type	タイプ／型／～を分類する	138,160
types	タイプ／型／～を分類する	26,403
typed	分類される／分類した	410

動詞としても使われるが，名詞の用例が多い圧倒的の多い．複数形のtypes割合が，約15%ある可算名詞．
◆類義語：class, group, form

■ type ＋［前］

	the%	a/an%	ø%			
❶	6	-	53	types of ～	…タイプの～	10,457
				・types of cancer	…タイプの癌	364
				・types of cells	…タイプの細胞	226

■ [形／代名] + type(s) of

❶ different types of ～	異なるタイプの～	1,480
❷ two types of ～	2つのタイプの～	1,444
❸ this types of ～	このタイプの～	895
❹ both types of ～	両方のタイプの～	835
❺ other types of ～	他のタイプの～	686
❻ several types of ～	いくつかのタイプの～	539
❼ many types of ～	多くのタイプの～	491
❽ various types of ～	様々なタイプの～	426
❾ distinct types of ～	別個のタイプの～	301
❿ a new type of ～	新しいタイプの～	283

❶ Cell size varies greatly among different types of cells, but the range in size that a specific cell type can reach is limited.（Proc Natl Acad Sci USA. 2012 109:E2561）

訳 細胞の大きさは，異なるタイプの細胞の間で大きく異なっている

■ [形／名] + type

❶ wild-type ～	野生型の～	60,874
❷ cell types	細胞型	8,911
❸ mating type	接合型	646

❶ In this study, the age-dependent 1H MRS profile of transgenic AD mice was compared to that of wild-type mice.（Proc Natl Acad Sci USA. 2005 102:11906）

訳 ～が，野生型マウスのそれと比較された

typical 形 特有の／典型の　用例数 5,646

◈類義語：inherent

■ typical + [前]

❶ typical of ～	～に特有の	1,407
・be typical of ～	…は，～に特有である	275
❷ typical for ～	～にとって特有の	172

❶ This feature is typical of 11-mer repeat containing sequences that adopt right-handed coiled coil conformations.（J Mol Biol. 2003 329:763）

訳 この特徴は，～を含む11アミノ酸長のリピートに特有である

❷ Sequence analysis showed that the 5' flanking region has a structure, which is typical for an interferon-induced gene promoter.（Eur J Biochem. 2001 268:1577）

訳 ～は，インターフェロンに誘導される遺伝子プロモーターにとって特有である

typically 副 主として／典型的に　用例数 6,184

◈類義語：mostly, primarily, chiefly, largely

■ typically ＋［過分］

❶	typically associated with ～	主として～と関連する	233
❷	typically found	主として見つけられる	135
❸	typically used	主として使われる	124

❶ This explains why podocyte injury is typically associated with nephrotic syndrome.（J Clin Invest. 2004 113:1390）

訳 これは，なぜタコ足細胞傷害が主としてネフローゼ症候群と関連するかを説明する

U

ubiquitously ㊐ 遍在性に／一様に／広範に　用例数 1,462

◆類義語：uniformly, widely, broadly, extensively

■ ubiquitously ＋［過分］

❶ ubiquitously expressed	遍在性に発現する	1,037
・be ubiquitously expressed	…は，遍在性に発現する	386

❶ Mammalian cytosolic PITPs include the ubiquitously expressed PITPα and PITPβ isoforms (269-270 residues).（Biochemistry. 2005 44:14760）

訳 ～は，遍在性に発現するPITPαおよびPITPβアイソフォームを含む

unable ㊎ できない　用例数 6,064

◆類義語：incapable　◆反意語：able

■ be unable to *do*

★ be unable to *do*	…は，～することができない	4,859
❶ be unable to bind	…は，結合することができない	263
❷ be unable to induce ～	…は，～を誘導することができない	176
❸ be unable to form ～	…は，～を形成することができない	145
❹ be unable to detect ～	…は，～を検出することができない	140
❺ be unable to grow	…は，増殖することができない	138
❻ be unable to activate ～	…は，～を活性化することができない	115

❶ HXKII is unable to bind VDAC phosphorylated by GSK3β and dissociates from the mitochondria.（Cancer Res. 2005 65:10545）

訳 HXKIIは，VDACに結合することができない

■［代名／名］＋ be unable to *do*

❶ we were unable to *do*	われわれは，～することができなかった	287
❷ mutant was unable to *do*	変異体は，～することができなかった	81

❶ Surprisingly, we were unable to detect any defects in the Kap95, Kap121, Xpo1, or mRNA transport pathways in cells expressing the mutant FG Nups.（J Cell Biol. 2004 167:583）

訳 われわれは，欠損を何も検出できなかった

unaffected ㊎ 影響されない　用例数 5,843

■［動］＋ unaffected

❶ be unaffected	…は，影響されない	3,761
・be unaffected by ～	…は，～によって影響されない	2,134

・be unaffected in ～	…は，～において影響されない	276
❷ remain unaffected	…は，影響されないままである	186

❶ Latency-associated transcript ORF 73 was unaffected by the presence of TPA or CDV, suggesting that it was constitutively expressed.（J Virol. 2004 78:13637）

訳 ～は，TPAあるいはCDVの存在によって影響されなかった

❶ Induction of differentiation markers was unaffected in the absence of C2α and C2β.（Mol Cell Biol. 2005 25:11122）

訳 分化マーカーの誘導は，C2αおよびC2βの非存在下で影響されなかった

❷ Accordingly, the JNK and COX-2 mRNA levels remained unaffected.（Brain Res. 2004 1016:195）

訳 JNKおよびCOX-2メッセンジャーRNAレベルは，影響されないままであった

■［副］＋ unaffected

❶ largely unaffected	ほとんど影響されない	211
❷ relatively unaffected	比較的影響されない	101

❶ Copulation, sexual motivation, and weight gain were largely unaffected, although some differences were observed in copulatory efficiency.（Behav Neurosci. 2005 119:1227）

訳 ～は，ほとんど影響されなかった

uncertain 形 不確かな　用例数 2,220

◆類義語：unclear, ambiguous

■ uncertain ＋［whether節］

❶ uncertain whether ～	～かどうかは確かではない	148
・it is uncertain whether ～	～かどうかは確かではない	93

❶ However, it is uncertain whether overexpression of FGFR1 is causally linked to the poor prognosis of amplified cancers.（Cancer Res. 2010 70:2085）

訳 ～かどうかは確かではない

■［動］＋ uncertain

❶ be uncertain	…は，確かではない	1,029
❷ remain uncertain	…は，不確かなままである	689

❷ Thus, passive immunotherapy is effective at preventing the buildup of intracellular Tau pathology, neurospheroids, and associated symptoms, although the exact mechanism remains uncertain.（Gesnerus. 1990 47 1:21）

訳 正確な機構は不確かなままである

unchanged 形 不変の／変化していない　用例数 3,558

◆類義語：unaltered, invariant

■ unchanged ＋［前］

❶ unchanged in ～	～において不変である	567
❷ unchanged by ～	～によって不変である	216
❸ unchanged after ～	～のあと不変である	132
❹ unchanged from ～	～から不変である	116

❶ CG methylation as measured by R.HpaII/R.MspI ratios was unchanged in cells expressing the transgene.（Nucleic Acids Res. 2005 33:6124）
訳 ～は，導入遺伝子を発現する細胞において不変であった

■［動］＋ unchanged

❶ be unchanged	…は，不変である	1,770
❷ remain unchanged	…は，不変のままである	949

❷ This subcellular pattern of intranodal receptor distribution was unchanged by treatment with FTY-P.（J Immunol. 2005 175:7151）
訳 ～は，FTY-Pによる処理によっても不変であった

■［副］＋ unchanged

❶ essentially unchanged	本質的に不変である	131

❶ Marginal soft-tissue thickness was increased by 0.40 mm (P < 0.05) for the ADM group, whereas the CPF group remained essentially unchanged.（J Periodontol. 2004 75:44）
訳 CPF群は，本質的に不変のままであった

★ unclear 形 不明な　　用例数 12,225

◆類義語：unknown, uncertain, unidentified, unexplained　◆反意語：clear

■［動］＋ unclear

❶ be unclear	…は，不明である	6,069
・function is unclear	機能は，不明である	82
❷ remain unclear	…は，不明なままである	4,960
・mechanisms remain unclear	機構は，不明なままである	121

❷ In mice, Ca sensitization causes ventricular arrhythmias, but the underlying mechanisms remain unclear.（Circ Res. 2012 111:170）
訳 根底にある機構は，不明なままである

■ it is/remains unclear ＋［名節］

❶ it is unclear whether ～	～かどうかは不明である	874
・however, it is unclear whether ～	しかし，～かどうかは不明である	206
❷ it is unclear how ～	どのように～かは不明である	387
❸ it remains unclear whether ～	～かどうかは不明なままである	298

❹	it remains unclear how ～	どのように～かは不明なままである	173
❺	it is unclear if ～	～かどうかは不明である	101

❶ However, it is unclear whether and how a small number of modules could possibly generate a large number of variants of one behavior.（Curr Biol. 2005 15:1712）

訳 しかし，～かどうかは不明である

❷ However, it is unclear how these populations are maintained under steady-state conditions in nonlymphoid peripheral sites, such as the lung airways.（J Immunol. 2006 176:537）

訳 どのように～かは不明である

■ ［副］＋ unclear

❶	still unclear	まだ不明である	509
❷	currently unclear	現在，不明である	114
❸	largely unclear	大部分は，不明である	103

★★ undergo 動 起こす／受ける／経験する 用例数 32,201

underwent	～を起こした/～を受けた	10,200
undergo	～を起こす/～を受ける/～を経験する	9,956
undergoing	～を起こす/～を受ける	6,458
undergoes	～を起こす/～を受ける	3,893
undergone	～を起こした/～を受けた	1,694

他動詞．受動態が用いられることはほとんどない．

◆類義語：experience, receive

■ undergo ＋［名句］

❶	undergo apoptosis	…は，アポトーシスを起こす	1,203
❷	undergo surgery	…は，外科手術を受ける	347
❸	undergo transplantation	…は，移植を受ける	254
❹	undergo a conformational change	…は，立体構造変化を起こす	159

❶ Both bovine aortic endothelial cells and human umbilical endothelial cells were shown to undergo apoptosis in response to angiostatin.（Cancer Res. 2003 63:4275）

訳 ウシの大動脈内皮細胞とヒトの臍帯内皮細胞の両方は，アンジオスタチンに応答してアポトーシスを起こすことが示された

❸ 6,338 (99.6%) patients who underwent transplantation were followed up at 1 year.（Lancet. 1999 354:1147）

訳 移植を受けた 6,338人（99.6%）の患者が，経過観察された

■ ［名］＋ undergo

❶	patients underwent ～	患者は，～を受けた	1,326
❷	cells undergo ～	細胞は，～を起こす	606
❸	subjects underwent ～	被検者は，～を受けた	242

❹ mice underwent ～	マウスは，～を受けた	134

■ [動／名] ＋ to undergo

★ to undergo ～	～を起こす…	775
❶ fail to undergo ～	…は，～を起こすことができない	238
❷ ability to undergo ～	～を起こす能力	154
❸ be shown to undergo ～	…は，～を起こすことが示される	99

❶ Embryos homozygous for the null allele fail to undergo chorioallantoic fusion and die by 10.5 days post coitus. (Development. 2003 130:2681)
訳 ～は，絨毛尿膜融合を起こすことができない

❷ Mutants lacking CBK1 form large aggregates of round cells under all growth conditions and lack the ability to undergo morphological differentiation. (J Bacteriol. 2002 184:2058)
訳 ～は，形態的分化を起こす能力を欠く

■ [名] ＋ undergoing

❶ patients undergoing ～	～を受ける患者	2,533
❷ cells undergoing ～	～を起こす細胞	500

❶ Bacteremia frequently occurs in patients undergoing hemodialysis with dual-lumen catheters. (Ann Intern Med. 1997 127:275)
訳 菌血症は，血液透析を受ける患者においてしばしば起こる

★underlie 動 根底にある／基礎にある　用例数 22,952

underlying	～の根底にある/～の基礎にある	17,286
underlie	～の根底にある/～の基礎にある	4,102
underlies	～の根底にある/～の基礎にある	1,532
underlain	基礎づけられる	21
underlay	～の根底にあった/～の基礎にあった	11

他動詞．現在分詞underlyingの用例が多い．

■ underlie ＋ [名句]

❶ underlie the development of ～	…は，～の発生の根底にある	127

❶ The mechanisms underlying the development of cardiac fibrosis are incompletely understood. (PLoS One. 2012 7:e40196)
訳 心筋線維症の発症の根底にある機構は，完全には理解されていない

■ [名] ＋ underlying

❶ mechanisms underlying ～	～の根底にある機構	3,966
❷ processes underlying ～	～の根底にある過程	208
❸ events underlying ～	～の根底にある事象	153

❶ The mechanisms underlying the neurotoxicity induced by peroxynitrite are still unclear. (J Neurosci. 2004 24:10616)
訳 パーオキシナイトライトによって誘導される神経毒性の根底にある機構は，まだ不明である

■［名］＋ that underlie

★ that underlie ～	～の根底にある…	1,626
❶ mechanisms that underlie ～	～の根底にある機構	459

❶ We are interested in the mechanisms that underlie cell fate determination in the endosporic male gametophytes of the fern, Marsilea vestita. (Dev Biol. 2004 269:319)
訳 われわれは，細胞運命決定の根底にある機構に興味がある

■［動］＋ to underlie

★ to underlie ～	～の根底にある…	561
❶ thought to underlie ～	～の根底にあると考えられる	121

❶ Tyrosine kinases have been implicated in cellular processes thought to underlie learning and memory. (Neuron. 2003 37:473)
訳 チロシンキナーゼは，学習と記憶の根底にあると考えられる細胞過程に関連づけられてきた

■ underlying ＋［名／形］

❶ underlying mechanisms	根底にある機構	1,230
・underlying molecular mechanisms	根底にある分子機構	250
❷ underlying cause	根底にある原因	273
❸ underlying disease	根底にある疾患	180

❶ Mammalian epidermis is maintained by self-renewal of stem cells, but the underlying mechanisms are unknown. (Science. 2005 309:933)
訳 根底にある機構は知られていない

U

underscore 動 強調する　用例数 2,282

underscore	～を強調する	1,158
underscores	～を強調する	518
underscoring	～を強調する	390
underscored	強調される/～を強調した	216

他動詞.
◆類義語：emphasize, highlight, stress

■ underscore ＋［名句］

❶ underscore the importance of ～	…は，～の重要性を強調する	600
❷ underscore the need for ～	…は，～のための必要性を強調する	144

❶ These results underscore the importance of the La N-terminal amino acids in RNA binding and viral RNA translation. (J Virol. 2004 78:3763)
訳 これらの結果は，LaのN末端アミノ酸の重要性を強調する

■［名］＋ underscore

❶ results underscore ～	結果は，～を強調する	260
❷ findings underscore ～	知見は，～を強調する	179
❸ data underscore ～	データは，～を強調する	102

■ be underscored

★ be underscored	…が，強調される	135
❶ be underscored by ～	…が，～によって強調される	129

★ understand 動 理解する　用例数 26,694

understood	～を理解した/理解される	16,317
understand	～を理解する	10,370
understands	～を理解する	7

他動詞．否定的な意味での受動態の用例が多い．understandingは別項参照．
◆類義語：appreciate, regard, know, learn, think, accept

■ understand ＋［名句／how節］

❶ understand how ～	…は，どのように～かを理解する	2,588
❷ understand the role of ～	…は，～の役割を理解する	996
❸ understand the mechanisms	…は，機構を理解する	788
❹ understand the molecular basis	…は，分子基盤を理解する	324
❺ understand the function of ～	…は，～の機能を理解する	287
❻ understand the regulation of ～	…は，～の調節を理解する	197
❼ understand the relationship between ～	…は，～の間の関連性を理解する	195
❽ understand the pathogenesis of ～	…は，～の病因を理解する	182

❶ It is therefore important to understand how cytotrophoblasts respond to changes in oxygen tension. (Biochemistry. 2001 40:4077)
訳 それゆえ，どのように細胞栄養芽層が酸素圧の変化に応答するかを理解することは重要である

❷ To understand the role of autopolysialylation in PST enzymatic activity, we employed a mutagenesis approach. (J Biol Chem. 2000 275:4484)
訳 PSTの酵素活性における自己ポリシアリル化の役割を理解するために

❺ To further understand the function of FER, we have continued our analyses of the interaction of FER with pp120 and other proteins. (J Biol Chem. 1998 273:23542)
訳 FERの機能をさらに理解するために

■ to understand

★ to understand ～	～を理解するために	8,085
❶ To understand ～	～を理解するために	3,021
❷ in order to understand ～	～を理解するために	331
❸ in an effort to understand ～	～を理解しようとして	168
❹ it is important to understand ～	～を理解することは重要である	145

❶ To understand the mechanism of this effect, we examined the modulation of excitatory synaptic inputs onto layer V PFC pyramidal neurons by D1/D5 receptor stimulation. (Proc Natl Acad Sci USA. 2001 98:301)

訳 この効果の機構を理解するために

■ to ＋［副］＋ understand

❶ to better understand ～	～をよりよく理解するために	2,187
❷ to further understand ～	～をさらに理解するために	379

❶ To better understand the role of ELL proteins in the regulation of transcription by RNA polymerase II, we have initiated a search for proteins related to ELLs. (J Biol Chem. 2000 275: 32052)

訳 RNAポリメラーゼⅡによる転写の調節におけるELLタンパク質の役割をよりよく理解するために

■ ［副］＋ understood

❶ poorly understood	あまり理解されていない	7,239
・be poorly understood	…は，あまり理解されていない	4,511
・remain poorly understood	…は，あまり理解されないままである	1,662
・still poorly understood	まだあまり理解されていない	334
❷ not well understood	よく理解されてはいない	2,775
❸ incompletely understood	完全には理解されていない	1,099
❹ not fully understood	十分には理解されていない	991
❺ not completely understood	完全には理解されていない	344
❻ less well understood	よりよく理解されてはいない	265
❼ not clearly understood	明瞭には理解されていない	132

❶ Mechanisms contributing to the maintenance of heterochromatin in proliferating cells are poorly understood. (Mol Cell. 1999 4:529)

訳 増殖性細胞においてヘテロクロマチンの維持に寄与する機構は，あまり理解されていない

❶ But the molecular details of this process remain poorly understood. (Nature. 2002 417:269)

訳 この過程の分子的な詳細は，あまり理解されないままである

❷ The extent and nature of neuronal differentiation characteristics displayed by primary and metastatic melanoma cells are not well understood. (Am J Pathol. 2001 158:2107)

訳 転移性のメラノーマ細胞は，よく理解されてはいない

❸ Atherosclerotic coronary arteries are prone to constriction but the underlying causes are incompletely understood. (Circulation. 2001 104:1114)

訳 しかし，根底にある原因は完全には理解されていない

❹ Mouse myocyte contractility and the changes induced by pressure overload are not fully understood. (Circ Res. 2000 87:588)
訳 ～は，十分には理解されていない

■ be (not) understood

❶	be not understood	…は，理解されていない	938
❷	be understood	…は，理解されている	842
	・little is understood about ～	～についてほとんど理解されていない	99

★ understanding 名 理解，-ing 理解する　用例数 27,164

understanding	理解	27,113
understandings	理解	51

名詞の用例が多い．複数形のunderstandingsの割合は0.2%しかないが，単数形に不定冠詞を伴う用例は多い．understanding ofの用例が非常に多い．
◆類義語：knowledge, realization, awareness

■ understanding ＋［前］

	the%	a/an%	ø%			
❶	12	32	16	understanding of ～	～についての理解	15,372
				・our understanding of ～	～についてのわれわれの理解	5,450

■［動］＋ our understanding of

❶	improve our understanding of ～	…は，～についてのわれわれの理解を改善する	396
❷	advance our understanding of ～	…は，～についてのわれわれの理解を前進させる	376
❸	further our understanding of ～	…は，～についてのわれわれの理解を発展させる	318
❹	enhance our understanding of ～	…は，～についてのわれわれの理解を増強する	295
❺	expand our understanding of ～	…は，～についてのわれわれの理解を広げる	168
❻	extend our understanding of ～	…は，～についてのわれわれの理解を広げる	139

❶ Lessons learned from C. elegans should improve our understanding of how cells become polarized and divide asymmetrically during development. (Dev Cell. 2002 3:157)
訳 ～は，どのように細胞が極性化するかについてのわれわれの理解を改善するはずである

❸ To further our understanding of the regulatory role played by CcrM, we sought to investigate its biophysical properties. (J Biol Chem. 2001 276:14744)
訳 CcrMによって果たされる調節性の役割についてのわれわれの理解を発展させるために

■［前］＋ our understanding of

❶	advances in our understanding of ～	～に対するわれわれの理解の進歩	475
❷	contribute to our understanding of ～	…は，～についてのわれわれの理解に寄与する	173

❸ implications for our understanding of ～	～についてのわれわれの理解に対する意味	156

❸ Although there is much debate regarding these observations, the implications for our understanding of clot formation and therapeutic intervention may be of major importance. (Blood. 2002 100:743)

訳 血栓形成および治療介入についてのわれわれの理解に対する意味が，非常に重要であるかもしれない

■ ［形／過分］＋ understanding of

❶ a better understanding of ～	～についてのよりよい理解	1,462
・to gain a better understanding of ～	～についてのよりよい理解を得るために	179
❷ current understanding of ～	～についての現在の理解	698
❸ improved understanding of ～	～についての改善された理解	356
❹ detailed understanding of ～	～についての詳細な理解	300
❺ greater understanding of ～	～についてのより大きな理解	256
❻ further understanding of ～	～についてのいっそうの理解	234
❼ complete understanding of ～	～についての完全な理解	230
❽ mechanistic understanding of ～	～についての機構的な理解	217
❾ deeper understanding of ～	～についてのより深い理解	171

❶ To gain a better understanding of the role of these genes during stress, their expression has been studied in the drought-resistant relative of tomato, Lycopersicon pennellii. (Plant Physiol. 1993 103:597)

訳 ストレスの間のこれらの遺伝子の役割についてのよりよい理解を得るために

undertake 動 着手する／企てる／行う　用例数 3,859

undertaken	着手される/～に着手した/～を企てた	2,758
undertook	～に着手した/～を企てた	867
undertake	～に着手する/～を企てる	125
undertaking	～に着手する/～を企てる	102
undertakes	～に着手する/～を企てる	7

他動詞.

◆類義語：attempt, try, address, seek, launch

■ be undertaken ＋［前］

★ be undertaken	…は，着手される	2,301
❶ be undertaken to *do*	…は，～するために着手される	1,558
・be undertaken to determine ～	…は，～を決定するために着手される	383
・be undertaken to investigate ～	…は，～を精査するために着手される	214
・be undertaken to examine ～	…は，～を調べるために着手される	130
❷ be undertaken in ～	…は，～において着手される	154

❶ This study was undertaken to determine whether ACE might be present in aortic sclerosis or stenosis lesions. (Circulation. 2002 106:2224)
訳 この研究は，～かどうかを決定するために着手された

■ [代名] ＋ undertake

❶ we undertook ～	われわれは，～に着手した	765

❶ We undertook the present study to assess the role of Rho A/ROCK and its possible relation to ERK1/ERK2 in 5-HT-induced pulmonary artery SMC proliferation. (Circ Res. 2004 95:579)
訳 われわれは，～の役割を評価するために現在の研究に着手した

unexpectedly 副 予想外に　用例数 3,512

文頭で用いられることが多い．

■ unexpectedly

❶ Unexpectedly,	予想外に，	2,036
・Unexpectedly, we found that ～	予想外に，われわれは～ということを見つけた	129
❷ unexpectedly high	予想外に高い	113

❶ Unexpectedly, we found that HDAC4 does not function as a MEF2 deacetylase. (Mol Cell Biol. 2005 25:8456)
訳 予想外に，われわれは～ということを見つけた

❷ This unexpectedly high variation in metabolic gene expression explains much of the variation in metabolism, suggesting that it is biologically relevant. (Nat Genet. 2005 37:67)
訳 この予想外に高い代謝遺伝子発現の変動は，～を説明する

uniformly 副 一様に　用例数 1,673

◆類義語：homogeneously

■ uniformly ＋ [過分]

❶ uniformly distributed	一様に分布する	191

❶ However, Gln3 is not uniformly distributed in the cytoplasm. (J Biol Chem. 2004 279:19294)
訳 Gln3は，細胞質において一様には分布しない

★unique 形 独特の／ユニークな　用例数 23,803

◆類義語：characteristic, peculiar, rare, uncommon, sole, singular

■ unique ＋ [前]

❶ unique to ～	～に独特である	1,875
❷ unique in ～	～において独特である	646

・be unique in ～	…は，～において独特である	521
・unique in ～ing	～する際に独特である	129
❸ unique among ～	～の間で独特である	493

❶ Nonetheless, a significant number of coat proteins are probably unique to each species. (J Bacteriol. 2003 185:1443)

訳 かなりの数のコートタンパク質は，おそらくそれぞれの種に独特である

❸ The SULT2B1 gene is unique among steroid/sterol sulfotransferase genes in that it encodes for two isoforms as a result of an alternative exon ㊕. (Gene. 2006 367:66)

訳 SULT2B1遺伝子は，ステロイド/ステロールスルホトランスフェラーゼ遺伝子の間で独特である

■ unique ＋［名］

❶ unique opportunity	ユニークな機会	390
・provide a unique opportunity	…は，ユニークな機会を提供する	211
❷ unique role	ユニークな役割	371
❸ unique features	ユニークな特徴	325
❹ unique mechanism	ユニークな機構	299
❺ unique properties	ユニークな性質	279

universally 副 普遍的に／一般に／広く 用例数 682

◆類義語：uniformly, widely, broadly, extensively, ubiquitously

■ universally ＋［過分］

❶ universally conserved	普遍的に保存されている	186

❶ These four residues are universally conserved in all known KSs. (Biochemistry. 2000 39:2088)

訳 これらの４つの残基は，すべての既知のKSにおいて普遍的に保存されている

U

★unknown 形 知られていない／未知の，名 未知のもの 用例数 23,564

unknown	知られていない／未知の	23,512
unknowns	未知のもの	52

名詞としても用いられるが，形容詞の用例が圧倒的に多い．

◆類義語：unclear, uncertain, unidentified, unexplained, obscure

■［動］＋ unknown

❶ be unknown	…は，知られていない	10,234
・it is unknown whether ～	～かどうかは，知られていない	581
・function is unknown	機能は，知られていない	223
・mechanism is unknown	機構は，知られていない	152

・it is unknown how ～	どのように～かは，知られていない	133
❷ remain unknown	…は，知られていないままである	3,274
・it remains unknown whether ～	～かどうかは，知られていないままである	139

❶ However, it is unknown whether these complexes indeed perform distinct functions in neuronal tissue. (Mol Biol Cell. 2005 16:128)

訳 ～かどうかは知られていない

❶ The C terminus is also essential, but its function is unknown. (Mol Cell Biol. 2003 23:2733)

訳 それの機能は，知られていない

❷ However, the mechanism by which $\alpha_M\beta_2$ binds fibrinogen remains unknown. (J Biol Chem. 2004 279:44897)

訳 $\alpha_M\beta_2$がフィブリノーゲンに結合する機構は，知られていないままである

■［副］＋ unknown

❶ largely unknown	ほとんど知られていない	2,535
❷ previously unknown	以前には知られていない	1,232
❸ currently unknown	現在知られていない	400
❹ still unknown	まだ知られていない	357

❶ The mechanisms that determine whether neural stem cells remain in a proliferative state or differentiate into neurons or glia are largely unknown. (Dev Biol. 2004 272:203)

訳 ～は，ほとんど知られていない

❷ The outbreak of severe acute respiratory syndrome (SARS) in 2002 was caused by a previously unknown coronavirus-SARS coronavirus (SARS-CoV). (Lancet. 2004 363:2122)

訳 ～が，以前には知られていないコロナウイルスによって引き起こされた

■ unknown ＋［名］

❶ unknown function	未知の機能	1,297
❷ unknown mechanism	未知の機構	529

❶ One of the hallmarks of Orbivirus infection is the production of large numbers of intracellular tubular structures of unknown function. (J Virol. 2004 78:6649)

訳 オルビウイルス感染の特徴の１つは，未知の機能の多数の細胞内管状構造の産生である

unlike 前 形 接 違って／異なって　　用例数 8,703

文頭の用例が非常に多い．

■ unlike ＋［形］

❶ unlike other ～	他の～と違って	808
❷ unlike most ～	ほとんどの～と違って	250
❸ unlike wild-type ～	野生型～と違って	240

❶ Unlike other aminopeptidases, the activity of ERAP1 depended on the C-terminal residue of the substrate. (Proc Natl Acad Sci USA. 2005 102:17107)

訳 他のアミノペプチダーゼと違って

❷ Unlike most virulent wild-type (wt) influenza B viruses, ca B/Ann Arbor/1/66 is temperature sensitive (ts) at 37 degrees C and attenuated (att) in the ferret model. (J Virol. 2005 79: 11014)
訳 ほとんどの病原性のある野生型(wt)のB型インフルエンザウイルスと違って

unlikely 形 副 ありそうにない／なさそうな 用例数 2,134

◆反意語：likely

■ be unlikely to *do*

★ be unlikely to *do*	…は，おそらく～しそうにない	1,194
❶ be unlikely to be ～	…は，おそらく～でありそうにない	484

❶ This defect is unlikely to be due to lack of LC6, because an LC6 null mutant (oda13) exhibits only a minor swimming abnormality. (Mol Biol Cell. 2005 16:5661)
訳 この欠損は，おそらくLC6の欠如のせいでありそうにない

■ unlikely ＋ [that節]

★ unlikely that ～	おそらく～ということはありそうにない	366
❶ it is unlikely that ～	おそらく～ということはありそうにない	215

❶ We conclude that it is unlikely that common CDX2 variants account for a measurable fraction of susceptibility to colorectal cancer in this population. (Cancer Res. 2005 65:5488)
訳 おそらく～ということはありそうにない

unrelated 形 無関係の／血縁のない 用例数 4,534

◆類義語：irrespective, irrelevant, independent

■ unrelated to

★ unrelated to ～	～に無関係の	1,144
❶ be unrelated to ～	…は，～に無関係である	606

❶ These data demonstrate that activation of EW after EtOH is unrelated to hypothermia or stress. (Brain Res. 2005 1063:132)
訳 ～は，低体温とは無関係である

■ [副] ＋ unrelated

❶ structurally unrelated	構造的に無関係の	187

❶ The hallmark of multidrug resistance is cross-resistance to multiple structurally unrelated compounds. (Cancer Res. 2005 65:11694)
訳 多剤耐性の特徴は，複数の構造的に無関係の化合物に対する交差耐性である

U

untreated ㊋ 未処理の／未処置の／未治療の　用例数 4,521

◆類義語：untreated, intact, naive

■ ［前］＋ untreated

❶ compared with untreated ～	未処理の～と比べて	343

❶ Mice treated with IL-21 reject tumor cells more efficiently, and a higher percentage of mice remain tumor-free compared with untreated controls.（J Immunol. 2005 175:2167）

訳 ～は，未処理の対照群と比べて腫瘍のないままである

■ untreated ＋［名］

❶ untreated cells	未処理の細胞	323
❷ untreated patients	未処置の患者	322
❸ untreated controls	未処理の対照群	255

unusual ㊋ 普通でない／異常な　用例数 5,477

◆類義語：extraordinary, uncommon, abnormal, aberrant, anomalous

■ be unusual ＋［前］

❶ be unusual in ～	…は，～において普通でない	228
・be unusual in that ～	…は，～という点で普通でない	125

❶ Pax-6 binding to these elements is unusual in that it appears to require both its homeo and paired domains.（J Biol Chem. 2004 279:34277）

訳 これらのエレメントへのPax-6の結合は，それがそれのホメオドメインとペアードドメインの両方を必要とするようであるという点で普通でない

unusually ㊙ 異常に／非常に　用例数 1,508

◆類義語：abnormally, aberrantly, extraordinarily, markedly, remarkably, strikingly, extremely, very

■ unusually ＋［形］

❶ unusually high	異常に高い	345
❷ unusually large	異常に大きい	181
❸ unusually long	異常に長い	125

❶ The dimer of dengue C protein has an unusually high net charge, and the structure reveals an asymmetric distribution of basic residues over the surface of the protein.（Proc Natl Acad Sci USA. 2004 101:3414）

訳 デングCタンパク質の二量体は，異常に高い正味荷電を持っている

★ up-regulate/upregulate ㊐ 上方制御する　用例数 12,511

up-regulated	上方制御される/～を上方制御した	5,147
up-regulate	～を上方制御する	892
up-regulates	～を上方制御する	700
up-regulating	～を上方制御する	342
upregulated	上方制御される/～を上方制御した	4,009
upregulate	～を上方制御する	636
upregulates	～を上方制御する	513
upregulating	～を上方制御する	272

他動詞.

◆類義語：elevate, increase, induce

■ be up-regulated ＋［前］

★ be up-regulated	…は，上方制御される	2,291
❶ be up-regulated in ～	…は，～において上方制御される	991
❷ be up-regulated by ～	…は，～によって上方制御される	392
❸ be up-regulated during ～	…は，～の間に上方制御される	125

❶ Membrane proteins were also up-regulated in response to most of the metals tested.（J Bacteriol. 2005 187:8437）

訳 膜タンパク質は，また，テストされた金属のほとんどに応答して上方制御された

❷ An additional 10 genes were up-regulated by hypoxia but minimally activated by HIF-1α or HIF-2α transfection.（Cancer Res. 2005 65:3299）

訳 さらに10の遺伝子が，低酸素によって上方制御された

■［名］＋ be up-regulated

❶ expression is up-regulated	発現が，上方制御される	125
❷ genes were up-regulated	遺伝子が，上方制御された	65

❶ Cysteine-rich 61（CCN1）is a matricellular protein of which expression is up-regulated in cancer and various vascular diseases.（Cancer Res. 2005 65:9705）

訳 ～は，その発現が癌において上方制御されるマトリックス細胞のタンパク質である

■［副］＋ up-regulated

❶ significantly up-regulated	有意に上方制御される	209
❷ highly up-regulated	高度に上方制御される	96
❸ markedly up-regulated	顕著に上方制御される	80
❹ strongly up-regulated	強く上方制御される	75

■ up-regulate ＋［名句］

❶ up-regulate the expression of ～	…は，～の発現を上方制御する	198

★ up-regulation/upregulation ㊔ 上方制御 用例数 10,555

up-regulation	上方制御	6,003
up-regulations	上方制御	6
upregulation	上方制御	4,545
upregulations	上方制御	1

複数形の割合は0.1%しかないが，単数形には不定冠詞が付く用例もかなりある．

◆類義語：increase, elevation, augmentation

■ up-regulation ＋［前］

	the%	a/an%	ø%			
❶	22	12	65	up-regulation of ～	～の上方制御	4,450
				・…-induced up-regulation of ～	…に誘導される～の上方制御	226
				・…-mediated up-regulation of ～	…に仲介される～の上方制御	113
❷	4	16	21	up-regulation in ～	～における上方制御	248

❶ The up-regulation of the PrxⅡ protein in radioresistant cancer cells suggested that human peroxiredoxin plays an important role in eliminating the generation of reactive oxygen species by ionizing radiation. (Cancer Res. 2005 65:10338)

訳 放射線抵抗性の癌細胞におけるPrxⅡタンパク質の上方制御は，～ということを示唆した

❶ The regulatory mechanisms behind nicotine-induced up-regulation of surface nicotinic acetylcholine receptors remain to be determined. (J Biol Chem. 2005 280:34088)

訳 ニコチンに誘導される表面のニコチン性アセチルコリン受容体の上方制御の背後にある調節機構は，決定されないままである

★ upstream ㊒ 上流の，㊖ 上流に／上流で 用例数 12,247

◆反意語：downstream

■ upstream ＋［前］

❶ upstream of ～	～の上流	5,335
・… bp upstream of ～	～の…bp上流に	395
・immediately upstream of ～	～のすぐ上流に	329
・… kb upstream of ～	～の…kb上流に	248
・region upstream of ～	～の上流の領域	207
・upstream of the transcription start site	転写開始部位の上流	156
❷ upstream from ～	～の上流に	475

❶ We studied the identity and function of the 528-bp gene immediately upstream of Legionella pneumophila F2310 ptsP (enzyme I^{Ntr}). (Infect Immun. 2005 73:6567)

訳 レジオネラ・ニューモフィラF2310 ptsPのすぐ上流の528 bpの遺伝子の機能

❶ We demonstrate that core promoter activity is contained within a region extending approximately 300 bp upstream of the ATG codon. (Gene. 2005 358:111)

訳 コアプロモーター活性は，ATGコドンのおよそ300 bp上流に広がる領域内に含まれる

❶ Herein we studied the role of the AT-rich region upstream of -17 in transcription regulation of T7 RNA polymerase. (J Biol Chem. 2005 280:40707)

訳 われわれは，T7 RNAポリメラーゼの転写調節における-17の上流のATリッチな領域の役割を研究した

■ [動／過分] ＋ upstream of

❶	act upstream of ～	…は，～の上流で作用する	297
❷	located upstream of ～	～の上流に位置する	280
❸	be upstream of ～	…は，～の上流にある	178
❹	function upstream of ～	…は，～の上流で機能する	176

❶ Tbx1 acts upstream of Smad7 controlling vascular smooth muscle and extracellular matrix investment of the fourth arch artery. (Circ Res. 2013 112:90)

訳 Tbx1は，Smad7の上流で作用する

❷ A similar AT-rich sequence was identified within the intergenic region located upstream of the Rv1057 gene. (J Bacteriol. 2006 188:150)

訳 類似のATリッチな配列が，Rv1057遺伝子の上流に位置する遺伝子間領域内に同定された

■ upstream ＋ [名]

❶	upstream region	上流領域	349
❷	upstream promoter	上流のプロモーター	231
❸	upstream activator	上流の活性化因子	228

■ upstream and ＋ [形／副]

❶	upstream and downstream	上流および下流の	274

❶ These results confirm that regions upstream and downstream of the CWCV motif participate in IGF-1 binding. (Mol Pharmacol. 2006 69:833)

訳 CWCVモチーフの上流および下流の領域は，IGF-1結合に関与する

U

★uptake 名 取り込み　　用例数 17,913

uptake	取り込み	17,852
uptakes	取り込み	61

複数形のuptakesの割合は0.3%しかなく，原則，不可算名詞として使われる．

◆類義語：incorporation

■ uptake ＋ [前]

	the%	a/an%	ø%			
❶	40	4	53	uptake of ～	～の取り込み	3,594
				・uptake of ～ by …	…による～の取り込み	302
				・uptake of ～ in …	…における～の取り込み	208

				・uptake of ～ into …	…への～の取り込み	136
❷	4	0	94	uptake in ～	～における取り込み	1,516
❸	1	1	95	uptake by ～	～による取り込み	905
❹	3	0	93	uptake into ～	～への取り込み	385

❶ However, **little is known of the heme binding and release mechanisms that facilitate the uptake of heme into the pathogenic organism.** (Biochemistry. 2005 44:13179)

訳 病原性生物へのヘムの取り込みを促進するヘムの結合および遊離の機構は，ほとんど知られていない

❷ In this study, **we investigated regulation of iron uptake in GAS** and the role of a putative transcriptional regulator named MtsR (for Mts repressor) with homology to the DtxR family of metal-dependent regulatory proteins. (Infect Immun. 2005 73:5743)

訳 われわれは，GASにおける鉄の取り込みの調節を精査した

❸ **A domain-separation mechanism is proposed for metal uptake by apo-MnSOD.** (Biophys J. 2006 90:598)

訳 ドメイン分離機構が，アポMnSODによる金属の取り込みに対して提案される

■ ［名］＋ uptake

❶	glucose uptake	グルコースの取り込み	1,287
❷	iron uptake	鉄の取り込み	444

★★ use 動 使う，名 使用／有用性　　用例数 320,971

using	～を使う	146,850
used	使われる/～を使った	107,434
use	～を使う/使用/有用性	61,312
uses	～を使う/使用/有用性	5,375

他動詞の用例の方が多いが，名詞としても用いられる．複数形のusesの割合は2％しかなく，原則，**不可算名詞**として使われる．

◆類義語：utilize, exploit, employ, usage, utilization, exploitation

■ 動 use ＋［名句］

❶	use a combination of ～	…は，～の組み合わせを使う	1,668
❷	use this approach	…は，このアプローチを使う	587
❸	use this method	…は，この方法を使う	494
❹	use a series of ～	…は，一連の～を使う	487
❺	use a variety of ～	…は，さまざまな～を使う	472
❻	use this model	…は，このモデルを使う	301
❼	use a mouse model	…は，マウスモデルを使う	298
❽	use a panel of ～	…は，一団の～を使う	297
❾	use this system	…は，このシステムを使う	285

❶ **Using a combination of immunoprecipitation and microarray analysis**, we have identified a number of mRNAs that are bound by the murine Dazl protein both in vivo and in vitro. (Hum

Mol Genet. 2005 14:3899)
訳 免疫沈降およびマイクロアレイ分析の組み合わせを使って

■ 動 [代名／名] ＋ use

❶ we used ～	われわれは，～を使った	16,166
・we have used ～	われわれは，～を使った	4,385
❷ study used ～	研究は，～を使った	451
❸ the authors used ～	著者らは，～を使った	374
❹ method uses ～	方法は，～を使う	210

❶ **We used a sensitive reverse transcription-PCR assay to study** immunoglobulin VH germ line transcripts in proB lines from RAG-deficient mice. (Mol Cell Biol. 1998 18:6253)
訳 われわれは，～を研究するために鋭敏な逆転写PCRアッセイを使った

■ 動 [名] ＋ be used

★ be used	…が，使われる	57,674
❶ analysis was used	分析が，使われた	821
❷ models were used	モデルが，使われた	601
❸ methods were used	方法が，使われた	416
❹ techniques were used	テクニックが，使われた	384
❺ assay was used	アッセイが，使われた	357
❻ system was used	システムが，使われた	346
❼ logistic regression was used	ロジスティック回帰分析が，使われた	303
❽ data were used	データが，使われた	288
❾ approach was used	アプローチが，使われた	285

❽ **Sequence data are used to identify potential target genes** and the results are used to define a prior distribution on the topology of the regulatory network. (Bioinformatics. 2006 22:739)
訳 配列データが，可能な標的遺伝子を同定するために使われる

■ 動 be used to *do*

★ be used to *do*	…が，～するために使われる	37,255
❶ be used to identify ～	…が，～を同定するために使われる	2,360
❷ be used to determine ～	…が，～を決定するために使われる	2,307
❸ be used to assess ～	…が，～を査定するために使われる	1,468
❹ be used to study ～	…が，～を研究するために使われる	1,364
❺ be used to examine ～	…が，～を調べるために使われる	1,275
❻ be used to measure ～	…が，～を測定するために使われる	1,112
❼ be used to investigate ～	…が，～を精査するために使われる	1,042
❽ be used to evaluate ～	…が，～を評価するために使われる	935
❾ be used to estimate ～	…が，～を推定するために使われる	870

❷ **Functional magnetic resonance imaging (fMRI) was used to determine whether** different kinds of visual attention rely on a common neural substrate. (Neuron. 1999 23:747)

訳 機能的磁気共鳴画像法（fMRI）が，～かどうかを決定するために使われた

■ 動 be used ＋［前］

★ be used	…は，使われる	57,674
❶ be used as ～	…は，～として使われる	5,554
・be used as a model	…は，モデルとして使われる	307
❷ be used for ～	…は，～のために使われる／～に対して使われる	3,971
・be used for ～ing	…は，～するために使われる	452
❸ be used in ～	…は，～において使われる	3,806
・be used in conjunction with ～	…は，～と組み合わせて使われる	200
・be used in combination with ～	…は，～と組み合わせて使われる	131
・be used in this study	…は，この研究において使われる	119

❶ Extant salamanders are often used as a model system to assess fundamental issues of developmental, morphological and biogeographical evolution.（Nature. 2001 410:574）

訳 現存しているサンショウウオは，しばしば～を評価するためのモデルシステムとして使われる

❷ This method can be used for negatively or positively charged analytes.（Anal Chem. 1998 70:4578）

訳 この方法は，負にあるいは正に荷電した分析物に対して使われうる

■ 動［副］＋ used

❶ widely used	広く使われる	2,613
❷ commonly used	一般に使われる	2,202
❸ currently used	現在使われる	488

❶ Although MMF is also being widely used in pediatric transplant patients, data documenting its safety are limited.（Transplantation. 1999 68:83）

訳 MMFは，また，小児の移植患者において広く使われている

■ 動 Using

❶ Using ～	～を使って	33,531

❶ Using this approach, we identified 7 novel targets for aberrant methylation in pancreatic cancer.（Gastroenterology. 2006 130:548）

訳 このアプローチを使って，われわれは 7 つの新規の標的を同定した

■ 動［名］＋ using

❶ studies using ～	～を使う研究	2,509
❷ analysis using ～	～を使う解析	1,445
❸ experiments using ～	～を使う実験	1,297
❹ assays using ～	～を使うアッセイ	792

❶ A phylogenetic tree based on homologue content, relative to LT2, was largely concordant with previous studies using sequence information from several loci.（Proc Natl Acad Sci USA. 2002 99:8956）

訳 …は，～からの配列情報を使う以前の研究とおおむね一致した

■ 動 [過分] ＋ using

❶	performed using ～	～を使って行われる	1,463
❷	determined using ～	～を使って決定される	1,436
❸	measured using ～	～を使って測定される	1,372
❹	investigated using ～	～を使って精査される	1,171
❺	studied using ～	～を使って研究される	1,136
❻	assessed using ～	～を使って評価される	1,119
❼	analyzed using ～	～を使って解析される	1,110
❽	obtained using ～	～を使って得られる	1,043
❾	examined using ～	～を使って調べられる	991

❶ Statistical analysis was performed using the Student's t-test for independent means and the chi-square equation. (Crit Care Med. 1998 26:701)

訳 統計分析が，スチューデントt検定を使って行われた

❷ GCF levels of ICTP and IL-1 were determined using radioimmunoassay and enzyme-linked immunosorbent assay techniques, respectively. (J Periodontol. 2002 73:835)

訳 ～が，ラジオイムノアッセイおよび酵素結合免疫吸着測定法の技術を使って決定された

■ 名 use ＋ [前]

use ofの前にtheが付く割合はかなり高い．

	the%	a/an%	ø%			
❶	69	0	30	use of ～	～の使用	30,275
				・support the use of ～	…は，～の使用を支持する	528
				・describe the use of ～	…は，～の使用を述べる	300
				・report the use of ～	…は，～の使用を報告する	273
				・demonstrate the use of ～	…は，～の使用を実証する	221
❷	0	2	74	use in ～	～における有用性	3,437

❶ These data support the use of these transgenic zebrafish as a model system for studies of glomerular pathogenesis and podocyte regeneration. (J Am Soc Nephrol. 2012 23:1039)

訳 これらのデータは，モデルシステムとしてのトランスジェニック・ゼブラフィッシュの使用を支持する

■ 名 [形／名] ＋ use

❶	clinical use	臨床利用	774
❷	drug use	薬物使用	610
❸	alcohol use	アルコール摂取	528
❹	potential use	潜在的使用	526
❺	widespread use	広範な使用	421

U

★ useful ㊋ 有用な／役に立つ　用例数 15,400

〈useful for〉〈useful in〉の用例が多い.

◆類義語：available, helpful, valuable, beneficial, convenient

■ be useful ＋［前］

★ be useful	…は，有用である	6,361
❶ be useful for ～	…は，～のために有用である	2,454
・be useful for studying ～	…は，～を研究するために有用である	98
・be useful for identifying ～	…は，～を同定するために有用である	86
❷ be useful in ～	…は，～の際に有用である	2,169
・be useful in the treatment of ～	…は，～の治療の際に有用である	155
・be useful in identifying ～	…は，～を同定する際に有用である	83
❸ be useful to *do*	…は，～するために有用である	442
❹ be useful as ～	…は，～として有用である	411

❶ This finding is potentially useful for studying hepatic physiology and may also have applications for cell therapy.（J Nucl Med. 2000 41:474）

訳 この知見は，肝臓の生理学を研究するために潜在的に有用である

■［動］＋ useful

❶ prove useful	…は，有用であると判明する	869
・prove useful in ～	…は，～において有用であると判明する	415
・prove useful for ～	…は，～のために有用であると判明する	305

❶ Therapeutic agents that target this pathway may prove useful in the treatment or possible prevention of NEC.（Ann Surg. 2001 233:835）

訳 この経路を標的にする治療薬は，NECの治療あるいは可能な予防において有用であると判明するかもしれない

■［副］＋ useful

❶ potentially useful	潜在的に有用である	417
❷ clinically useful	臨床的に有用である	351
❸ particularly useful	特に有用である	321
❹ very useful	とても有用である	218
❺ most useful	最も有用である	189
❻ more useful	もっと有用である	130
❼ therapeutically useful	治療上有用である	102

■ useful ＋［名］

❶ useful tool	有用なツール	692
❷ useful model	有用なモデル	460
❸ useful information	有用な情報	229

usually 副 通常／普通は　　用例数 5,041

◈類義語：generally, commonly, ordinarily, normally, universally

■ usually ＋［過分］

❶	usually associated with ～	通常～と関連する	190
	・be usually associated with ～	…は，通常～と関連する	104

❶ The disease is characterized by a degeneration of the optic nerve, which is usually associated with elevated intraocular pressure.（Am J Hum Genet. 2004 74:1314）
訳 ～は，通常上昇した眼圧と関連する

utility 名 有用性／実用性　　用例数 6,054

utility	有用性	5,891
utilities	有用性	163

複数形のutilitiesの割合は約2.5%しかなく，原則，不可算名詞として使われる．utility ofの用例が非常に多い．
◈類義語：usefulness

■ utility ＋［前］

utility ofの前にtheが付く割合は圧倒的に高い．

	the%	a/an%	ø%			
❶	97	1	2	utility of ～	～の有用性	3,761
				・demonstrate the utility of ～	…は，～の有用性を実証する	734
				・illustrate the utility of ～	…は，～の有用性を例証する	123
				・utility of this approach	このアプローチの有用性	110
❷	1	3	63	utility in ～	～における有用性	567
				・utility in ～ing	～する際の有用性	164
❸	2	3	60	utility for ～	～のための有用性	341
❹	2	1	59	utility as ～	～としての有用性	206

❶ These studies demonstrate the utility of chromosome deletions for complex trait analysis.（Genetics. 2000 155:803）
訳 これらの研究は，～のための染色体欠失の有用性を実証する

■［形］＋ utility

❶	clinical utility	臨床的な有用性	477
❷	potential utility	潜在的な有用性	356
❸	therapeutic utility	治療上の有用性	131

❶ This study demonstrates the potential utility of gene therapy for systemic delivery of an

U

antiangiogenic agent targeting an endothelium-specific receptor, Tie2.（Proc Natl Acad Sci USA. 1998 95:8829）
訳 この研究は，～のための遺伝子治療の潜在的な有用性を実証する

utilization ㊔ 利用　　用例数 4,268

utilization	利用	4,263
utilizations	利用	5

複数形のutilizationsの用例はほとんどなく，原則，不可算名詞として使われる．
◆類義語：usage, use, exploitation

■ utilization ＋［前］

	the%	a/an%	ø%			
❶	38	2	58	utilization of ～	～の利用	1,454
❷	1	0	94	utilization in ～	～における利用	242

❶ Utilization of an integrative genomic strategy reveals a single gene, the embryonic endoderm transcription factor GATA6, as the selected target of the amplification.（Proc Natl Acad Sci USA. 2012 109:4251）
訳 統合的なゲノム戦略の利用は，単一の遺伝子を明らかにする

❷ These were evaluated for their ability to enhance glucose utilization in cultured L6 myocytes.（J Med Chem. 1998 41:4556）
訳 これらが，培養L6筋細胞におけるグルコース利用を増強するそれらの能力に関して評価された

★ utilize ㊚ 利用する　　用例数 12,224

utilized	利用される/～を利用した	4,699
utilizing	～を利用する	3,209
utilize	～を利用する	2,414
utilizes	～を利用する	1,902

他動詞．
◆類義語：exploit, take advantage of, use, employ

■ be utilized ＋［前］

★	be utilized	…は，利用される	2,301
❶	be utilized to *do*	…は，～するために利用される	822
❷	be utilized in ～	…は，～において利用される	249
❸	be utilized for ～	…は，～のために利用される	190
❹	be utilized by ～	…は，～によって利用される	132
❺	be utilized as ～	…は，～として利用される	130

❶ The recently developed hepatitis C virus（HCV）subgenomic replicon system was utilized to

evaluate the efficacy of several known antiviral agents. (J Virol. 2003 77:1092)
訳 …が，～の有効性を評価するために利用された

■ ［代名］＋ utilize

❶ we utilized ～	われわれは，～を利用した	846
❷ we have utilized ～	われわれは，～を利用した	295

V

vaccinate 動 ワクチン接種する／接種する／予防接種する

用例数 2,200

vaccinated	ワクチン接種される／接種される／～にワクチン接種した	2,047
vaccinating	～にワクチン接種する／～に接種する	77
vaccinate	～にワクチン接種する／～に接種する	57
vaccinates	～にワクチン接種する／～に接種する	19

他動詞.

■ vaccinated ＋［前］

❶ vaccinated with ～	～を接種された	513
・mice vaccinated with ～	～を接種されたマウス	191
・be vaccinated with ～	…は，～を接種される	111

❶ After mating, **75% (15/20) of the mice vaccinated with MOMP** carried embryos in both uterine horns. (Infect Immun. 2005 73:8153)

訳 MOMPを接種されたマウスの75%（15/20）

■ vaccinated ＋［名］

❶ vaccinated mice	ワクチン接種されたマウス	232
❷ vaccinated animals	ワクチン接種された動物	152

❶ At 1 and 7 months postinfection, the lung bacterial burdens were considerably reduced and the lung pathology **was improved in vaccinated mice compared to naive controls.** (Infect Immun. 2005 73:7727)

訳 ～が，未処置の対照群と比べてワクチン接種されたマウスにおいて改善された

vaccination 名 ワクチン接種／接種

用例数 5,681

vaccination	ワクチン接種／接種	5,386
vaccinations	ワクチン接種／接種	295

複数形のvaccinationsの割合は約５%あるが，単数形は無冠詞で使われることが圧倒的に多い.

■ vaccination ＋［前］

	the%	a/an%	ø%			
❶	1	4	93	vaccination with ～	～の接種	655
❷	12	1	75	vaccination of ～	～のワクチン接種	340
❸	0	3	78	vaccination in ～	～におけるワクチン接種	189
❹	1	0	93	vaccination against ～	～に対するワクチン接種	149

❶ **Vaccination with this mutant protects mice from** challenge with S. pneumoniae. (Infect

Immun. 2006 74:586)

訳 この変異体の接種は，～からマウスを保護する

❷ Evidence is accumulating that **universal vaccination of schoolchildren would reduce the transmission of influenza**. (Am J Epidemiol. 2005 162:686)

訳 学童の全員のワクチン接種は，インフルエンザの感染を減らすであろう

❹ **Vaccination against measles prevents** more cases of SSPE than was originally estimated. (J Infect Dis. 2005 192:1686)

訳 麻疹に対するワクチン接種は，～を予防する

■ [前] ＋ vaccination

❶ after vaccination	ワクチン接種のあと	408

validate 動 確証する／検証する／確認する　用例数 7,091

validated	確証される／～を確証した	4,632
validate	～を確証する	1,800
validating	～を確証する	449
validates	～を確証する	210

他動詞.

◆類義語：verify, confirm, ascertain

■ validate ＋ [名句]

❶ validate the use of ～	…は，～の使用を確証する	100

■ to validate ＋ [名]

★ to validate ～	～を確証する…	838
❶ To validate ～	～を確証するために	201
❷ be used to validate ～	…は，～を確証するために使われる	104

❶ **To validate the use of tissue microarrays for immunophenotyping**, we studied a group of 59 fibroblastic tumors with variable protein expression patterns by immunohistochemistry for Ki-67, p53, and the retinoblastoma protein (pRB). (Am J Pathol. 2001 158:1245)

訳 免疫表現型検査のための組織マイクロアレイの使用を確証するために

■ validated ＋ [前]/using

❶ validated by ～	～によって確証される	748
・be validated by ～	…は，～によって確証される	529
❷ validated in ～	～において確証される	435
❸ validated using ～	～を使って確証される	227

❶ **Results were validated by RT-PCR in individual animals** from an independent replication of the experiment. (Proc Natl Acad Sci USA. 2005 102:11533)

訳 結果は，個々の動物においてRT-PCRによって確証された

❷ Many assessment scales **are validated in elderly people** but not used in elderly patients with

cancer. (Lancet Oncol. 2005 6:790)
訳 ～が，高齢者において確証される

valuable 形 役に立つ／価値ある　用例数 3,550

◆類義語：useful, available, helpful, convenient

■ valuable ＋［前］

❶ valuable for ～	～のために役に立つ	351
・be valuable for ～	…は，～のために役に立つ	221
・valuable for ～ing	～するために役に立つ	141
❷ valuable in ～	～において役に立つ	298
・valuable in ～ing	～する際に役に立つ	136

❶ This assay will be valuable for clinical and research-based studies. (J Clin Microbiol. 2005 43: 6139)
訳 このアッセイは，臨床およびリサーチに基づいた研究のために役に立つであろう

■［動］＋ valuable

❶ prove valuable	…は，役に立つと判明する	125

❶ Ultrasound elastography has proven valuable in discriminating these lesions. (Curr Opin Oncol. 2013 25:1)
訳 超音波エラストグラフィーは，これらの病変を識別する際に役に立つと判明している

■ valuable ＋［名］

❶ valuable tool	役に立つ道具	378
❷ valuable information	役に立つ情報	205

❷ Economic analyses can provide valuable information for health care decision makers. (Ann Intern Med. 2005 142:1073)
訳 経済的な分析は，～のために役に立つ情報を提供できる

★★ value 名 値／価値，動 評価する　用例数 35,279

values	値／～を評価する	21,123
value	値／～を評価する	14,000
valued	評価される／～を評価した	144
valuing	～を評価する	12

動詞としても用いられるが，名詞の用例が圧倒的に多い．複数形のvaluesの割合は約60%と非常に高い．

■ value(s) + [前]

value ofの前にtheが付く割合は非常に高い.

	the%	a/an%	ø%			
❶	70	24	4	value of ～	～の値	5,198
❷	26	-	66	values for ～	～に対する値	2,729

❶ The predictive value of HGD for a final HGD or cancer diagnosis was 73%. (Ann Surg. 2012 256:221)

訳 最終HGDあるいは癌の診断のためのHGDの予測値は, 73%であった

❷ Km values for physiological substrates were unaffected, although Kcat was doubled. (Plant Physiol. 2005 139:1625)

訳 生理的基質に対するKm値は, 影響を受けなかった

■ [名/形] + values

❶	predictive value	予測値	1,547
❷	prognostic value	予後値	540
❸	IC50 values	50%抑制濃度値	367

★ variant 名 変異体/バリアント　用例数 26,885

variants	変異体/バリアント	15,733
variant	変異体/バリアント	11,152

複数形のvariantsの割合は約60%と非常に高い.

◆類義語：mutant

■ variants + [前]

	the%	a/an%	ø%			
❶	4	-	80	variants of ～	～の変異体	2,114
❷	2	-	83	variants in ～	～における変異体	1,375

■ [名/形] + variants

❶	splice variants	スプライスバリアント	925
❷	genetic variants	遺伝的変異体	787
❸	sequence variants	配列変異体	450

❶ Expressed sequence tag (EST) data predicts multiple splice variants of both human and mouse ZC3H14. (Gene. 2009 439:71)

訳 発現配列タグ(EST)データは, ヒトとマウスの両方のZC3H14の複数のスプライスバリアントを予測する

★ variation 名 変動／バリエーション／差異／変化　用例数 20,607

variation	変動／バリエーション	15,534
variations	変動／バリエーション	5,073

複数形のvariationsの割合は約25%あるが，単数形は無冠詞で使われることが非常に多い．variation inの用例が多い．

◆類義語：variance, difference, distinction, discrimination

■ variation ＋［前］

	the%	a/an%	ø%			
❶	14	4	82	variation in ～	～における変動	6,167
❷	28	22	46	variation of ～	～の変動	1,185
❸	4	2	94	variation at ～	～における変動	382
❹	11	2	85	variation among ～	～の間の変動	375

❶ Individual variation in gene expression is important for evolutionary adaptation and susceptibility to diseases and pathologies. (Nat Genet. 2005 37:67)
訳 遺伝子発現における個々の変動は，進化的適応のために重要である

❷ Using lacZ fusions, we showed that these tetranucleotide repeats could mediate phase variation of this gene. (Mol Microbiol. 2005 58:207)
訳 これらの四ヌクレオチドリピートは，この遺伝子の相変動を仲介しうる

■［形／名］＋ variation

❶	genetic variation	遺伝的バリエーション	1,837
❷	sequence variation	配列バリエーション	600
❸	antigenic variation	抗原性の変動	360
❹	phenotypic variation	表現型のバリエーション	350
❺	phase variation	相変動	298

❶ These observations indicate that genetic variation in CYP19 might contribute to variation in the pathophysiology of estrogen-dependent disease. (Cancer Res. 2005 65:11071)
訳 CYP19における遺伝的バリエーションは，～に寄与するかもしれない

★ variety 名 多様性　用例数 18,952

variety	多様性	18,683
varieties	多様性	269

a variety ofの用例が非常に多い．

◆類義語：diversity

■ variety ＋［前］

variety ofの前にa/anが付く割合は，99%と圧倒的に高い．

	the%	a/an%	ø%			
❶	1	99	0	variety of ～	種々の～	18,502
				・a variety of ～	種々の～	15,260
				・a wide variety of ～	非常にさまざまな～	2,452
				・variety of cell types	種々の細胞型	352
				・variety of tissues	種々の組織	239
				・variety of conditions	種々の条件	207
				・variety of organisms	種々の生物	170
				・variety of cellular processes	種々の細胞過程	169
				・variety of mechanisms	種々の機構	159
				・a large variety of ～	非常にさまざまな～	140

❶ Recent studies suggest that **green tea flavonoids may be used for the prevention and treatment of a variety of neurodegenerative diseases.** (J Neurosci. 2005 25:8807)

訳 緑茶のフラボノイドは，種々の神経変性疾患の予防と治療のために使われるかもしれない

❶ **Cyclooxygenase 2 (COX-2) overexpression is found in a wide variety of human cancers** and is linked to all stages of tumorigenesis. (Cancer Res. 2005 65:6275)

訳 シクロオキシゲナーゼ２（COX-2）の過剰発現は，非常にさまざまなヒト癌において見つけられる

❶ **Periostin is expressed in a variety of tissues** and expression is increased in airway epithelial cells from asthmatic patients. (J Immunol. 2011 186:4959)

訳 ペリオスチンは，種々の組織において発現する

★ various　形 種々の／さまざまな　用例数 25,445

◆類義語：diverse, myriad

■ various ＋［名］

❶	various types	種々のタイプ	449
❷	various stages	種々のステージ	360
❸	various aspects	種々の面	349
❹	various tissues	種々の組織	341
❺	various forms	種々の型	326
❻	various combinations	種々の組み合わせ	305

❶ **CXCR4 expression is a prognostic marker in various types of cancer,** such as acute myelogenous leukemia or breast carcinoma. (Blood. 2006 107:1761)

訳 CXCR4発現は，種々のタイプの癌における予後マーカーである

■［前］＋ various

❶	response to various ～	種々の～への応答	253
❷	associated with various ～	種々の～と関連する	181
❸	effects of various ～	種々の～の効果	166

★vary ㊍ 変動する／異なる／さまざまである／変動させる

用例数 16,892

varied	変動した／異なった	5,594
varying	変動する／異なる／さまざまである	5,012
vary	変動する／異なる／さまざまである	4,006
varies	変動する／異なる／さまざまである	2,280

自動詞の用例が多いが，他動詞としても用いられる．

◆類義語：fluctuate, oscillate, differ, change, shift

■ vary ＋［前］

❶	vary in ～	…は，～において変動する／～においてさまざまである	1,237
❷	vary with ～	…は，～で異なる	1,227
❸	vary from ～ to …	…は，～から…まで変動する	818

❶ Human immunodeficiency virus type 1 (HIV-1) isolates **vary in** their ability to infect macrophages. (J Virol. 2005 79:4828)

訳 ヒト免疫不全ウイルス１型（HIV-1）単離体は，マクロファージに感染するそれらの能力においてさまざまである

❷ Optimal discriminatory values of BNP **varied with** age and sex. (Circulation. 2004 109:3176)

訳 ～は，年齢と性別で異った

❸ Dissociation constants determined via NMR **varied from** 0.27 **to** >3 mM. (J Am Chem Soc. 2004 126:4453)

訳 NMRによって決定された解離定数は，0.27 mMから3 mM以上まで変動した

■ vary ＋［副］

❶	vary widely	…は，大きく異なる	558
	・vary widely in ～	…は，～において大きく異なる	111
❷	vary significantly	…は，有意に異なる	447
❸	vary considerably	…は，かなり異なる	313
❹	vary greatly	…は，大きく異なる	299
❺	vary substantially	…は，かなり異なる	241

❶ BDA-labeled projection neurons **varied widely in** the shape and size of their cell somas, with mean cross-sectional areas ranging from 60-340 μm^2. (PLoS One. 2012 7:e49161)

訳 BDA標識された投射ニューロンは，それらの細胞体の形と大きさにおいて大きく異なった

■ varying ＋［名］

❶	varying degrees	さまざまな程度	592
❷	varying levels	さまざまなレベル	235

velocity 名 速度 用例数 5,743

velocity	速度	4,653
velocities	速度	1,090

複数形のvelocitiesの割合は約20%あるが，単数形は無冠詞で使われることも多い．

■ velocity ＋［前］

velocity ofの前にtheが付く割合は非常に高い．

	the%	a/an%	ø%			
❶	73	11	16	velocity of ～	～の速度	564
❷	14	3	79	velocity in ～	～における速度	136

❶ The **velocity of** rolling leukocytes was higher in Tph1$^{-/-}$ mice, indicating fewer selectin-mediated interactions with endothelium.（Blood. 2013 121:1008）

訳 回転する白血球の速度は，Tph1$^{-/-}$マウスにおいてより速かった

verify 動 検証する／確認する 用例数 3,066

verified	検証される／～を検証した／～を確認した	2,125
verify	～を検証する／～を確認する	732
verifying	～を検証する／～を確認する	154
verifies	～を検証する／～を確認する	55

他動詞．

◆類義語：validate, confirm, ascertain

■ be verified ＋［前］

★	be verified	…は，検証される	935
❶	be verified by ～	…は，～によって検証される	498
❷	be verified in ～	…は，～において検証される	98

❶ All results **were verified by** immunohistochemistry, cloning, and sequencing.（Cancer Res. 2005 65:10273）

訳 すべての結果が，免疫組織化学によって検証された

■ verify ＋［that節］

❶	verify that ～	…は，～ということを検証する	403

❶ We **verified that** endogenous ETS1 interacted with the ATXN2 promoter by an electromobility supershift assay and chromatin immunoprecipitation polymerase chain reaction.（Hum Mol Genet. 2012 21:5048）

訳 われわれは，～ということを検証した

■ [代名] ＋ verify

❶	we verified ～	われわれは，～を検証した	137

version 名 バージョン／版　　用例数 3,596

version	バージョン／版	2,489
versions	バージョン／版	1,107

複数形のversionsの割合が，約30%ある可算名詞.

■ version ＋ [前]

version ofの前にa/anが付く割合は非常に高い.

	the%	a/an%	ø%			
❶	16	82	0	version of ～	バージョンの～	1,700
				・modified version of ～	修正されたバージョンの～	139
				・truncated version of ～	切断されたバージョンの～	117

❶ A truncated version of ETR1 that lacks both the His kinase domain and the receiver domain failed to rescue the triple mutant phenotype. (Plant Physiol. 2004 136:2961)
訳 Hisキナーゼドメインおよびレシーバードメインの両方を欠く切断されたバージョンのETR1は，～することができなかった

★★ very 副 非常に／とても　　用例数 22,867

◈類義語：unusually, extremely, extraordinarily, greatly, markedly, profoundly

■ very ＋ [形]

❶	very low	非常に低い	2,917
❷	very similar	非常に類似している	2,143
	・be very similar	…は，非常に類似している	1,180
❸	very high	非常に高い	1,691
❹	very little	非常に少ない	1,247
❺	very different	とても異なっている	1,055
❻	very few	非常にわずかな	719
❼	very small	非常に小さい	693
❽	very large	非常に大きい	663

❶ Extracardiac progenitor cells are capable of repopulating cardiomyocytes at very low levels in the human heart after injury. (Circulation. 2005 112:2951)
訳 心外性の前駆細胞は，非常に低いレベルで心筋細胞を再増殖できる

❷ The crystal structure of Ciona $\beta\gamma$-crystallin is very similar to that of a vertebrate $\beta\gamma$-crystallin domain, except for paired, occupied calcium binding sites. (Curr Biol. 2005 15:1684)

訳 ～は，脊椎動物のβγ-クリスタリンドメインのそれと非常に類似している

viability 名 生存率／生存能／生存度／生存力 用例数 5,656

viability	生存率／生存能／生存度	5,628
viabilities	生存率／生存能／生存度	28

複数形のviabilitiesの割合は約５％あるが，単数形は無冠詞の用例が非常に多い．

■ viability ＋［前］

viability ofの前にtheが付く割合は非常に高い．

	the%	a/an%	ø%			
❶	78	1	19	viability of ～	～の生存率	885
❷	1	0	98	viability in ～	～における生存率	369

❶ The viability of vertebrate cells depends on a complex signaling interplay between survival factors and cell-death effectors.（Curr Biol. 2005 15:1762）

訳 脊椎動物細胞の生存率は，～に依存する

■［前］＋ viability

❶	essential for viability	生存能にとって必須である	222
❷	loss of viability	生存能の喪失	128

❶ Attempts to construct an rpoH null mutant in N. gonorrhoeae were unsuccessful, suggesting that RpoH is essential for viability of N. gonorrhoeae.（Infect Immun. 2005 73:4834）

訳 RpoHは，淋菌の生存能にとって必須である

■［名／形］＋ viability

❶	cell viability	細胞生存（率）	483
	・essential for cell viability	細胞生存にとって必須である	106
❷	myocardial viability	心筋の生存（率）	130

❶ Of the numerous two-component signal transduction systems found in bacteria, only a very few have proven to be essential for cell viability.（J Bacteriol. 2005 187:5419）

訳 ほんのわずかだけが，細胞生存にとって必須であると判明している

V

view 名 見解，動 みなす 用例数 8,213

view	見解／みなす	5,213
viewed	みなした／みなされる	1,410
views	見解／みなす	908
viewing	みなす	682

名詞の用例が多いが，他動詞としても用いられる．複数形のviewsの割合が，約15%ある可算名詞．

◆類義語：notion idea, consider, regard

■ ㊔ view ＋［前／同格that節］

view thatの前にtheが付く割合は圧倒的に高い.

	the%	a/an%	ø%			
❶	19	47	25	view of ～	～についての見解	1,962
				・In view of ～	～を考慮して	371
❷	94	6	0	view that ～	～という見解	1,350
				・support the view that ～	…は，～という見解を支持する	592
				・results support the view that ～	結果は，～という見解を支持する	115
				・consistent with the view that ～	～という見解と一致して	157

❶ **Our studies provide a comprehensive view of miRNP/RISC assembly pathways** in mammals, and our assay provides a versatile platform for further mechanistic dissection of such pathways in mammals.（Mol Cell. 2012 46:507）

訳 われわれの研究は，miRNP/RISC構築経路についての包括的な見解を提供する

❶ **In view of potentially harmful effects of acrosomal enzymes on embryo development**, the removal of acrosomes before ICSI is recommended for animals with large sperm acrosomes.（Proc Natl Acad Sci USA. 2005 102:14209）

訳 胚発生に対する先体酵素の潜在的に有害な影響を考慮して

❷ **These results support the view that** SSRE exhibits a degree of Sertoli specificity and acts synergistically with DSRE in controlling the expression of Amh.（Gene. 2005 363:159）

訳 これらの結果は，～という見解を支持する

■ ㊍ be viewed ＋［前］

★	be viewed	…は，みなされる	716
❶	be viewed as ～	…は，～とみなされる	449

❶ Because NS and LS share several features, **LS has been viewed as an NS variant**.（J Biol Chem. 2006 281:6785）

訳 LSは，NS変異体とみなされている

virtually ㊖ 実質的に　用例数 3,081

◆類義語：substantially

■ virtually ＋［形］

❶	virtually all ～	実質的にすべての～	875
❷	virtually identical	実質的に同一の	393
	・be virtually identical	…は，実質的に同一である	226
❸	virtually every ～	実質的にすべての～	135

❶ **The disease recurs in virtually all renal allografts**, and a high percentage of these ultimately

fail.（J Am Soc Nephrol. 2005 16:1392）
訳 その疾患は，実質的にすべての腎同種移植において再発する

❷ The induction was virtually identical to that observed in -2100-FAS-CAT transgenic mice and to the endogenous FAS mRNA.（J Biol Chem. 2000 275:10121）
訳 その誘導は，～において観察されるそれと実質的に同一であった

virulence 名 病原性　用例数 9,911

virulence	病原性	9,906
virulences	病原性	5

複数形のvirulencesの用例はほとんどなく，原則，不可算名詞として使われる.

■ virulence ＋［前］

virulence ofの前にtheが付く割合は非常に高い.

	the%	a/an%	ø%			
❶	77	2	21	virulence of ～	～の病原性	935
❷	2	0	90	virulence in ～	～における病原性	907

❶ TTSS activity is required for the virulence of many pathogenic Gram-negative bacteria including Escherichia coli , Salmonella spp., Yersinia spp., Chlamydia spp., Vibrio spp., and Pseudomonas spp.（J Am Chem Soc. 2012 134:17797）
訳 TTSS活性は，多くの病原性グラム陰性細菌の病原性のために必要とされる

■ virulence ＋［名］

❶	virulence factors	病原性因子	1,337
❷	virulence genes	病原性遺伝子	466
❸	virulence determinants	病原性決定基	250

vital 形 決定的に重要な／生命の　用例数 2,623

◈類義語：important, key, crucial, critical, significant

■ be vital ＋［前］

★	be vital	…は，決定的に重要である	619
❶	be vital for ～	…は，～にとって決定的に重要である	315
❷	be vital to ～	…は，～に決定的に重要である	187

❶ HIV-1 Tat transactivation is vital for completion of the viral life cycle and has been implicated in determining proviral latency.（Cell. 2005 122:169）
訳 HIV-1 Tatのトランス活性化は，ウイルスの生活環の完了にとって決定的に重要である

❷ Vertebrate mineralized tissues are vital to the adaptive evolution of various traits.（Proc Natl Acad Sci USA. 2005 102:18063）
訳 脊椎動物のミネラル化された組織は，さまざまな形質の適応性の進化に決定的に重要である

V

■ vital ＋［名］

❶ vital role	決定的に重要な役割	353

W

warrant ㊍ 正当化する／保証する　　用例数 2,592

warranted	正当化される／～を正当化した／～を保証した	1,450
warrant	～を正当化する／～を保証する	574
warrants	～を正当化する／～を保証する	491
warranting	～を正当化する／～を保証する	77

他動詞として用いられることが非常に多い．

◈類義語：ensure, guarantee, assure, certify, insure

■ be warranted to *do*

★	be warranted	…は，正当化される	1,288
❶	be warranted to *do*	…は，～するために正当化される	286

❶ Further clinical studies are warranted to determine the effectiveness of ^{131}I-mu81C6 mAb based on a target dose of 44 Gy rather than a fixed administered activity.（J Nucl Med. 2005 46:1042）

訳 さらなる臨床研究は，～の有効性を決定するために正当化される

■ warrant further

★	warrant further ～	…は，さらなる～を正当化する	559
❶	warrant further investigation	…は，さらなる研究を正当化する	235
❷	warrant further study	…は，さらなる研究を正当化する	116

❶ This unique cytotoxicity profile warrants further investigation and supports the evaluation of this agent in Phase I clinical trials for patients with B-CLL.（Cancer Res. 2004 64:6750）

訳 このユニークな細胞毒性プロファイルは，さらなる研究を正当化する

★ way ㊔ 方法／様式　　用例数 10,168

way	方法／様式	7,078
ways	方法／様式	3,090

複数形のwaysの割合が，約30%ある可算名詞．

◈類義語：method, means, procedure, fashion, pattern, mode, manner, modality

■ way ＋［前］

way forの前にtheが付く割合は非常に高い．

	the%	a/an%	ø%			
❶	32	54	7	way to *do*	～する方法	1,680
❷	4	45	44	way of ～	～の道／～の方法	726
				・way of ～ing	～する方法	386
				・by way of ～	～を経由して	312

❸	79	18	2	way for ～	～ための道／～のための方法	460
				・pave the way for ～	…は，～に道を開く	262
❹	78	8	0	way in which ～	～である方法	299

❶ Tandem mass spectrometry provides a more efficient way to identify phosphorylated residues in IRS-1. (Anal Chem. 2005 77:5693)

訳 タンデム質量分析法は，IRS-1においてリン酸化された残基を同定するためのより効率的な方法を提供する

❷ DNA enters the herpes simplex virus capsid by way of a ring-shaped structure called the portal. (J Virol. 2005 79:10540)

訳 DNAは，ポータルと呼ばれるリング型構造を経由して単純ヘルペスウイルスキャプシドに入る

❸ These findings pave the way for improving the design of subtype-specific compounds with therapeutic value for neurological disorders and diseases. (Nature. 2011 475:249)

訳 これらの知見は，～のデザインを改善するための道を開く

❹ Human CD4 binds to HIV gp120 in a manner strikingly similar to the way in which CD4 interacts with pMHCⅡ. (Proc Natl Acad Sci USA. 2001 98:10799)

訳 ヒトのCD4は，CD4 がpMHCⅡと相互作用する方法と非常によく似た様式でHIV gp120に結合する

■ [形] ＋ way(s)

❶	different ways	異なる方法	358
❷	effective way	効果的な方法	183
❸	in the same way	同じように	150

❶ Both compounds bind to the canonical ATP-binding site of the kinase domain, but they do so in different ways. (Cancer Res. 2002 62:4236)

訳 しかし，それらは異なる方法でそれを行う

❸ Tumor and normal cells always process the cyclin E-FLAG protein in the same way as endogenously expressed cyclin E. (Cancer Res. 2000 60:481)

訳 腫瘍および正常細胞は，内在性に発現されたサイクリンEと同じようにサイクリンE-FLAGタンパク質を常に処理する

weakly 副 弱く　　用例数 3,130

◆類義語：modestly, marginally, slightly

■ weakly ＋ [過分／過去]

❶	weakly bound	弱く結合した	141
❷	weakly expressed	弱く発現した	88
❸	weakly associated with ～	～と弱く関連した	73

❶ IFN scores were weakly associated with neurologic manifestations. (Arthritis Rheum. 2006 54:2951)

訳 IFNスコアは，神経症状と弱く関連した

■ weakly ＋［前］

❶	bind weakly to ～	…は，～に弱く結合する	101

❶ A retinoic acid analog that binds weakly to RPE65 is not inhibitory.（Proc Natl Acad Sci USA. 2004 101:10030）

訳 RPE65に弱く結合するレチノイン酸アナログは，抑制的ではない

■［副］＋ weakly

❶	only weakly	ほんの弱く	526
❷	more weakly	より弱く	169

❶ Uptake studies using microvillous（apical）membrane vesicles suggest it is either inactive or only weakly active at this site.（J Physiol. 2003 547:849）

訳 それは，不活性であるかあるいはほんの弱い活性があるかのどちらかである

★weight 名 重量／重み／体重，動 重みを加える 用例数 22,276

weight	重量／体重／重みを加える	18,366
weighted	重みを加えられる／重みを加えた	2,424
weights	重量／体重／重みを加える	1,219
weighting	重みを加える	267

名詞の用例が多いが，動詞としても用いられる．複数形のweightsの割合は約5％あるが，単数形は無冠詞で使われることが多い．

■ 名 weight ＋［前］

	the％	a/an％	ø％			
❶	47	23	28	weight of ～	～の重量	833
❷	17	1	79	weight in ～	～における重量	318

❶ The engO gene consists of 2,172 bp and encodes a protein of 724 amino acids with a molecular weight of 79,474.（J Bacteriol. 2005 187:4884）

訳 ～は，79,474の分子量を持つ724アミノ酸のタンパク質をコードする

❷ Also, inhibitors of FAS can cause reduced food intake and body weight in mice.（J Med Chem. 2005 48:946）

訳 FASの抑制剤は，マウスにおいて低下した食物摂取と体重を引き起こしうる

W

■ 名［形／名］＋ weight

❶	molecular weight	分子量	4,045
❷	body weight	体重	3,120
❸	birth weight	出生時体重	1,417

■ 名 weight ＋［名］

❶ weight loss	体重減少	2,231
❷ weight gain	体重増加	1,447

❶ Weight loss and weight maintenance are common concerns for US men and women.（JAMA. 1999 282:1353）

訳 体重減少と体重維持は，合衆国の男性および女性にとって一般的な関心事である

★★ well 副 よく　　用例数 80,726

as well asの用例が非常に多い．

■ as well as

★ as well as ～	～と同様に	45,247
❶ as well as in ～	～においてと同様に	3,315

❶ In this study, we investigated how radiation affects survivin expression in primary endothelial cells as well as in malignant cell lines.（Cancer Res. 2004 64:2840）

訳 われわれは，放射線照射が悪性細胞株においてと同様に初代内皮細胞においてサバイビン発現にどのように影響するかを精査した

■ well ＋［過分］

❶ well characterized	よく特徴づけられる	3,179
❷ well defined	よく定義される	2,883
❸ not well understood	よく理解されてはいない	2,775
❹ well established	よく確立されている	2,554
❺ well tolerated	よく許容されている	2,183
・be well tolerated	…は，よく許容されている	1,404
❻ well known	よく知られている	2,099
❼ well studied	よく研究される	1,163
❽ well documented	よく述べられる	1,084

❶ The PKA-CREB signaling mechanism has been well characterized in terms of nuclear gene expression.（Proc Natl Acad Sci USA. 2005 102:13915）

訳 PKA-CREBシグナル伝達機構は，～の点からよく特徴づけられている

❸ Active CaMKII can bind to NMDA-Rs, but the physiological role of this interaction is not well understood.（Neuron. 2005 48:289）

訳 この相互作用の生理学的役割は，よく理解されてはいない

❹ Although the association with RNAs is well established, it is still unknown how Fmrp finds and assembles with its RNA cargoes and how these activities are regulated.（Hum Mol Genet. 2006 15:87）

訳 RNAとの結合はよく確立されているけれども

❻ These compounds are well known for their important actions in mammalian physiology and disease.（Annu Rev Entomol. 2006 51:25）

訳 これらの化合物は，哺乳類の生理機能および疾患におけるそれらの重要な作用ゆえによく

知られている

■ [副] + well

❶ less well	あまりよく～でない	890
❷ equally well	同様によく～	387

❶ In contrast, screening for hypertensive kidney disease is less well defined. (J Am Soc Nephrol. 2003 14:S144)
訳 高血圧性腎疾患のスクリーニングは，あまりよく定義されてはいない

★ wide 形 広い　用例数 17,206

◈類義語：broad, extensive, widespread

■ wide + [名]

❶ a wide range of ～	広範囲の～	4,183
❷ a wide variety of ～	非常に種々の～	2,452
❸ a wide spectrum of ～	広範囲の～	320
❹ a wide array of ～	広範囲の～	315

❶ UV can cause a wide range of DNA lesions. (Cancer Res. 2004 64:3009)
訳 紫外線は，広範囲のDNA損傷を引き起こしうる

❷ Bcl-2 is an antiapoptotic protein expressed in a wide variety of cell types. (Oncogene. 2003 22:5515)
訳 Bcl-2は，非常にさまざまな細胞のタイプにおいて発現する抗アポトーシス性タンパク質である

★ widely 副 広く／大きく　用例数 9,735

◈類義語：broadly, extensively, universally, ubiquitously

■ widely + [過分／形]

❶ widely used	広く使われている	2,613
・be widely used	…は，広く使われている	1,177
❷ widely expressed	広く発現する	1,028
❸ widely distributed	広く分布する	755
❹ widely accepted	広く受け入れられている	459
・it is widely accepted that ～	～ということは，広く受け入れられている	109
❺ widely studied	広く研究される	237
❻ widely believed	広く信じられている	196
❼ widely available	広く役に立つ	171
❽ widely applicable	広く適用できる	163

❶ Glucocorticoids are widely used in the therapy of inflammatory, autoimmune, and allergic diseases. (J Immunol. 2000 164:1768)

W

訳 グルココルチコイドは，～の治療において広く使われている

❷ TNF-α is widely expressed in proliferative vitreoretinopathy (PVR) membranes and is present in the vitreous of eyes with PVR. (Invest Ophthalmol Vis Sci. 2004 45:2438)

訳 TNF-αは，増殖性硝子体網膜症（PVR）膜において広く発現する

❸ The echinocandins are widely distributed in the body, and are metabolised by the liver. (Lancet. 2003 362:1142)

訳 ～は，体内において広く分布する

❹ It is widely accepted that a relationship exists between deafness and pigmentation in the dog and also in other animals. (Genetics. 2004 166:1385)

訳 ～ということは，広く受け入れられている

■［副］＋ widely

❶ the most widely ～	最も広く～である…	450
・the most widely used ～	最も広く使われている～	248
❷ more widely ～	より広く～	249

❶ High-resolution gas chromatography of volatile derivatives with chiral reagents is the most widely used method. (Anal Biochem. 2002 303:176)

訳 ～は，最も広く使われている方法である

■［動］＋ widely

❶ vary widely	…は，大きく異なる	558

❶ Patterns of sleep vary widely among species, but the functional and evolutionary principles responsible for this diversity remain unknown. (Curr Biol. 2011 21:671)

訳 睡眠のパターンは，種間で大きく異なる

widespread 形 広範な／広く行きわたっている　用例数 5,039

◆類義語：broad, extensive, wide

■ widespread ＋［名］

❶ widespread use	広範な使用	421
❷ widespread expression	広範な発現	118
❸ widespread distribution	広範な分布	115

❶ The widespread use of the pRF vector is a major problem because this vector, which has Renilla luciferase as the 5' cistron and firefly luciferase as the 3' cistron, has been found to generate spliced transcripts. (Nucleic Acids Res. 2005 33:6593)

訳 pRFベクターの広範な使用は，大きな問題である

■ widespread ＋［前］

❶ widespread in ～	～において広く行きわたっている	379
・be widespread in ～	…は，～において広く行きわたっている	283

❶ The $GABA_A$ receptor is widespread in the mammalian brain, and can be specifically labeled

with the receptor agonist [^{3}H]muscimol. (Brain Res. 2003 992:69)
訳 $GABA_A$受容体は，哺乳類の脳において広く行きわたっている

★ work 名 研究，動 働く 用例数 20,990

work	研究／働く	19,903
works	研究／働く	781
worked	働いた	306

名詞の用例が多いが，動詞としても用いられる．複数形のworksの割合は4％あるが，単数形は無冠詞で使われることが非常に多い．

◆類義語：study, investigation, research, operate, function, serve, drive

■ work ＋［前］

	the%	a/an%	ø%			
❶	7	0	90	work in ～	～における研究	983
❷	14	0	82	work on ～	～に関する研究	697

❶ Prior work in our laboratory established a connection between the PALS1/PATJ/CRB3 and Par6/Par3/aPKC protein complexes at the tight junction of mammalian epithelial cells. (Mol Biol Cell. 2004 15:1981)
訳 われわれの研究室における先行研究は，～の間の結合を確立した

❷ Previous work on the phage repressors showed that the monomeric form of the protein is the target of RecA. (J Bacteriol. 2004 186:1)
訳 ファージリプレッサーに関する以前の研究は，～ということを示した

■［形］＋ work

❶	previous work	以前の研究	2,492
❷	recent work	最近の研究	1,640
❸	the present work	現在の研究	786

❸ In the present work, we have analyzed the role of MAP17 expression during mammary cancer progression. (Oncogene. 2012 31:4447)
訳 現在の研究において，

■ work ＋［動］

❶	work demonstrates ～	研究は，～を実証する	527
❷	work suggests ～	研究は，～を示唆する	516
❸	work provides ～	研究は，～を提供する	443
❹	work shows ～	研究は，～を示す	299
❺	work indicates ～	研究は，～を示す	195
❻	work describes ～	研究は，～を述べる	165
❼	work identifies ～	研究は，～を同定する	135
❽	work represents ～	研究は，～を表す	132

❶ Thus, this work demonstrates that Nkx6 acts in a specific neuronal population to link neuronal subtype identity to neuronal morphology and connectivity.（Development. 2004 131:5233）

訳 この研究は，～ということを実証する

❷ Recent work suggests that inhibition of Nrf2 may also depend upon ubiquitin-mediated proteolysis.（Mol Cell Biol. 2004 24:8477）

訳 最近の研究は，～ということを示唆する

❸ This work provides a consensus view of the retinal conformation in rhodopsin as seen by X-ray diffraction, solid-state NMR spectroscopy, and quantum chemical calculations.（Biochemistry. 2004 43:12819）

訳 この研究は，～を提供する

★ yield　名 収率／産生量, 動 産生する／もたらす　用例数 21,869

yield	収率／産生量／～を産生する／～を生じる	8,811
yielded	産生される／～を産生した／～を生じた	5,696
yields	収率／産生量／～を産生する／～を生じる	5,337
yielding	～を産生する／～を生じる	2,025

名詞および動詞として用いられる．複数形のyieldsの割合が，約5％ある可算名詞．

◆類義語：produce, cause, bring about

■ 名 yield ＋［前］

yield ofの前にtheが付く割合はかなり高い．

	the％	a/an％	ø％			
❶	61	28	9	yield of ～	～の収率	1,108

❶ These conditions shorten the analysis time and substantially **improve the yield of cleavage products**.（Anal Biochem. 2005 346:311）

訳 ～は，切断産物の収率を改善する

■ 名［形］＋ yield

❶	high yield	高収率	514
❷	quantum yield	量子収率	443
❸	overall yield	全収率	383
❹	good yield	よい収率	224

❶ Several forms of the voltage-dependent anion-selective channel （VDAC） **have been expressed at high yield in Escherichia coli**.（J Biol Chem. 1998 273:13794）

訳 ～は，大腸菌において高い収率で発現している

■ 動 yield ＋［名句］

❶	yield similar results	…は，類似の結果を生じる	78

❶ Replacement of Pi with sn-glycerol-3-phosphate and 2-aminoethylphosphonate **yielded similar results**.（J Bacteriol. 2004 186:4492）

訳 ～は，類似の結果を生じた

young　形 若い／若年の　用例数 8,507

◆類義語：juvenile

■ young ＋［名］

❶	young adults	若年成人	835
❷	young children	幼児	599

❸ young mice	若いマウス	304
❹ young women	若い女性	279

❶ Muscle mass and fiber type distributions **were maintained at levels similar to those in young adults**. (Proc Natl Acad Sci USA. 1998 95:15603)

訳 ～が，若年成人におけるそれらに類似したレベルで維持された

■ young and ＋［形］

❶ young and old	若いおよび年老いた	182
❷ young and aged	若いおよび年老いた	130

❶ Here we show that only the aminoacyl-tRNA synthetase cofactor p38 **is upregulated in the ventral midbrain/hindbrain of both young and old parkin null mice**. (J Neurosci. 2005 25:7968)

訳 ～は，若いおよび年老いた両方のパーキン欠損マウスの腹側中脳／後脳において上方制御される

Z

zone ㊔ ゾーン／帯／層　　用例数 7,655

zone	ゾーン／帯／層	5,678
zones	ゾーン／帯／層	1,977

複数形のzonesの割合が，約25%ある可算名詞.

◈類義語：region

■ zone ＋［前］

zone ofの前にtheが付く割合はかなり高い.

	the%	a/an%	ø%			
❶	61	28	9	zone of ～	～のゾーン	1,108

❶ In addition, **reduced mineralization was evident in the hypertrophic zone of $AnxA6^{-/-}$ growth plate cartilage**, although apoptosis was not altered compared with wild type growth plates.（J Biol Chem. 2012 287:14803）

訳 低下した石灰化は，$AnxA6^{-/-}$成長板軟骨の肥大層において明白であった

Column
コラム
冠詞のルールを知ろう！

Ⅰ．名詞の種類と不定冠詞

本書の特徴のひとつは，冠詞の使われる頻度について明示したことである．冠詞の使い方には後述するような一般的なルールがあるが，残念ながらそのルールを知っているだけでは，うまく使いこなすことはできない．ルールを踏まえつつ，個々の単語の用例に立ち返って調べ直すことが必要である．まずは，基本的なことを確認しておこう．

1．名詞の分類

名詞には可算名詞と不可算名詞とがあり，可算名詞には単数形と複数形とがある．したがって，名詞の種類は，可算名詞単数形・可算名詞複数形・不可算名詞の３種類である．可算名詞の単数形と複数形の違いは，通常は，綴りが違うので見ればすぐに分かる．ごく稀に単複同形の名詞があるので，それだけに注意すればよい．一方，可算名詞単数形と不可算名詞とは，綴りでは区別できないので注意が必要である．特に，生命科学論文のように形のはっきりしないものを取り扱うことが多い分野では，同じ単語でも可算名詞と不可算名詞の両方で使われるケースが非常に多い．

2．不定冠詞のルール

不定冠詞（a または an）は，可算名詞単数形に対して用いられる．しかし，可算名詞複数形と不可算名詞には，不定冠詞を用いることはできない．定冠詞や代名詞などが前に付くときにも，不定冠詞を用いることはできない．これらは共存せず，どれかひとつしか使えないからである．また，名詞の前にある名詞は，名詞の形容詞的用法といって，役割としては形容詞であるので，名詞が複数連続するときには，冠詞のルールは最後の名詞にだけ適用される.

3. 無冠詞のルール

可算名詞複数形と不可算名詞は，無冠詞で使う．では，逆に無冠詞で使われている名詞は，複数形でなければ不可算名詞なのであろうか．理論上の答えは，イエスである．ただし，同じ単語でも状況によって扱いが変わってくるのは前述した通りである.

Ⅱ. 冠詞の一般的ルール

冠詞には，不定冠詞（a/an）と定冠詞（the）とがあり，以下のように使い分ける.

1. 不定冠詞の用法：原則として次の条件をすべて満たしているときに使う.
   (1) 可算名詞
   (2) 単数形
   (3) 初出

解説：

①「1 つの」を意味する one と語源が同じであり，「1 つの～」という意味を表す（例：This is a pen.）．ただし，この英文を見てわかるように，通常日本語には訳出しない.

②「とある～」という日本語に相当する意味を伝えることもある（例：I like a Japanese folk song.）．ただし，論文ではあまりこの用法は見られない.

③多数の中から不特定のもの（one of many）を表すことになるので，読み手には，他に同類のものが存在すると連想されることもある．This is a goal of our project. という英文の場合，読み手は「これがプロジェクトの目標なんだ」と解釈した上で，「不定冠詞(a)が goal に付いているから，他にも目標があるのかな」と裏事情を勝手に想像することになる．もし，ゴールが 1 つと想定される場合は，初出・可算・単数名詞であっても，the goal of our project とするのが通常である.

2. 定冠詞の用法：原則として「直後の名詞が特定できる」という条件を満たすときに使う.

解説：

①不定冠詞と違い，可算名詞・不可算名詞の区別なく，また単数・複数いずれの名詞にも使う.

②「その」を意味する that と語源を同じくしており,「その～」という意味を表す.（例：This is a pen. The pen is made in Japan.）

③テキスト内ですでに言及があったため特定できる場合に the を付ける．多くの学習者にとって，馴染みがある基本のルールである．また，直接言及がなくても，付随するものにも the を用いる．例えば，“We visited a shrine in Kyoto. The gate was painted red.” の場合，「門」は初出でありながら，定冠詞がよいと言える.

④後に続く修飾語句によって特定できる場合も the を付ける．先ほど紹介した This is the goal of our project. が，意味の誤解なく通じる英文であるのは，後続の of 以下の内容から goal が特定されるから，つまり，「1 つである」という了解が成立するからである.

⑤テキスト外に「特定できる」状況・場面がある，あるいは対象となる名詞の直後の修飾語句が特定する役割を務めている場合は，定冠詞を使う．この作用については，あまり意識がないと思われる．中学校の英語入門期におそらくよく耳にする Look at the blackboard. は，場面・状況が名詞を特定している代表例である．他にも，Please open the window. なども同じである．ここで非常に重要な点は，相手との了解が成立するということである．つまり，世界に 1 つだけのモノには定冠詞を付けるという約束事が成り立つのは，読み手にも世界に 1 つだけのモノだから，「それ」と特定できるという意味である．あるいは，小さな村であれば，Go to the post office and get some stamps. という英文は誤解さ

れずに相手に伝わるはずである．しかし大きな都市では，この英文は相手に戸惑いを生じさせる．この場合，post office に不定冠詞を付けて，Go to a post office and get some stamps. とすべきであろう．このようなことを前提にすると，研究成果を具体的に記述する生命科学論文の場合には，すでに状況設定が整っていることもあり，定冠詞を用いる場面も多いであろう．

⑥そのほかに，定冠詞の用法として，総称を表す使い方がある（例：The dog is a faithful animal.）．しかし厄介なことに，A dog is a faithful animal. という英文もほぼ同様の意味を表すが，定冠詞は，やや格式ばった学術的な内容で使われることがある，という程度のことを理解しておけばよいだろう．論文で一般論を伝える場合には，出番のある用法かもしれない．

Ⅲ．定冠詞を使うときの注意点

以上が一般的なルールであるが，以下，いくつか注意点を取り上げよう．定冠詞 the を使うときには，理由を考えることが重要である．つまり，基本は，不定冠詞もしくは無冠詞であって，なんらかの理由があるときのみ定冠詞が付くわけである．実際，本書を見れば不定冠詞・無冠詞の用例が多いことがわかるであろう．

本書に示すように，名詞直後の前置詞の種類によって冠詞の傾向が大きく変わることにも注目していただきたい．名詞の後に of や同格の that が来ると，前に the が付く割合が非常に高くなる．ただし，その頻度は名詞によって，まちまちである．このような違いを具体的に提示したことが本書の特徴である．本書の活用法としては，前述した冠詞のルールの適用に迷った場合に，基本となる情報を確認することである．基本は不定冠詞もしくは無冠詞だと述べたが，もし，前に定冠詞が来ることが多い組み合わせの場合は，定冠詞を付けることを基本に考えるべきであろう．もちろん，無冠詞が多い組み合わせの場合は無冠詞を基本に，不定冠詞が多い場合は不定冠詞を基本に考えることとなろう．多くのネイティブが選択している用法は，自分でも選択すべき用法である確率が高いというわけである．単純なことだが，多くの人は，極めて確率の高い状況でも，逆に極めて低い状況でも，同じルールのみを用いて判断しようとする．その結果が，とんでもない間違いの連発である．もちろん，単に頻度の多い少ないが選択の理由になるわけではないが，どの確率が高いかを知った上で判断すれば，冠詞の選択の間違いを大幅に減らすことができるはずである．このような点を意識して，本書を活用していただきたい．

Ⅳ．可算名詞・不可算名詞の区別

可算名詞と不可算名詞の１番の違いは，不可算名詞には複数形が存在しないということである．実際に調べてみても，確かに複数形が全く使われない名詞もある．では，複数形が１例でもあれば，それは可算名詞かといえば，もちろんそういうことではない．可算名詞・不可算名詞の境界線は，おそらく複数形の割合が 1〜2%ぐらいのところであろう．それ以上であれば，主に可算名詞として使われる名詞であり，それ以下であれば，主に不可算名詞として使われる名詞である．

もう１つの基準は，不定冠詞が用いられるかどうかという点である．可算名詞単数形を無冠詞で用いることはできず，定冠詞や代名詞などが付かない場合には，必ず不定冠詞を付けるというのが文法上のルールである．実際に調べてみると，不定冠詞が付く可算名詞であっても無冠詞でも使われる場面は非常に多いという不可解な面はあるが，不定冠詞が付く用例の割合がある程度あれば，それは可算名詞と判断すべきであろう．

このように２つの側面から調べた場合，中には驚くような単語もある．例えば analysis などは，複数形（analyses）の用例が断然多いにも関わらず，単数形は無冠詞で使われることが非常に多い．このような事例は，文法書で学ぶルールだけで判断することが困難であることを如実に示している．それぞれの単語の慣用例になるべく合わせることが必要であろう．意外にも類似の傾向を持つ単語はたくさんあるので，それも無駄な努力ではない．

もちろん同じ単語でも状況によって様々な意味で使われ，可算・不可算の区別もそれによって変わってくる．しかし，幸い論文では，単一の意味で使われる名詞が好まれる傾向にあるので，コーパス解析は非常に有用な情報をもたらすはずである．本書に収録した「複数形の割合」や「定冠詞・不定冠詞・無冠詞」の頻度情報は，このような単語の使い方を判断する上で非常に有益なものになるであろう．

INDEX 索引

か

き

く～こ

さ・し

す～そ

た～つ

て・と

な～の

は・ひ

ふ～ほ

ま〜も

や～よ

ら～ろ

わ

■ 編者略歴

河本　健
（かわもと・たけし）

広島大学ライティングセンター特任教授．広島大学歯学部卒業，大阪大学大学院医学研究科博士課程修了，医学博士．高知医科大学助手，広島大学助手，講師などを経て現職．専門は生化学・分子生物学．長年の研究経験を生かして，広島大学その他で英語論文の執筆支援等を行っている．

大武　博
（おおたけ・ひろし）

福井県立大学名誉教授．福井大学教育学部卒業，国立福井工業高等専門学校助教授，福井県立大学助教授，京都府立医科大学（第一外国語教室）教授などを経て福井県立大学学術教養センター教授．コーパス言語学の研究成果を英語教育に援用することが，近年の研究テーマである．

ライフサイエンス英語表現使い分け辞典
第2版

2007年10月10日　第1版第1刷発行
2014年12月15日　第1版第8刷発行
2016年　5月15日　第2版第1刷発行
2020年　4月10日　第2版第2刷発行

編　集　河本　健，大武　博
監　修　ライフサイエンス辞書プロジェクト
発行人　一戸裕子
発行所　株式会社　羊　土　社
〒101–0052
東京都千代田区神田小川町2–5–1
TEL　03（5282）1211
FAX　03（5282）1212
E-mail　eigyo@yodosha.co.jp
URL　www.yodosha.co.jp/

ISBN978-4-7581-0847-8

印刷所　株式会社アイワード

ライフサイエンス辞書プロジェクトの英語の本

ライフサイエンス
英語表現
使い分け辞典 第2版

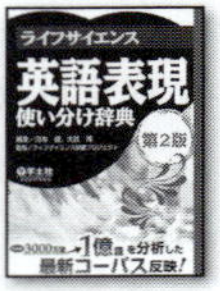

編集／河本　健，大武　博
監修／ライフサイエンス辞書プロジェクト
■ 定価（本体6,900円＋税）
■ B6判　■ 1215頁　■ ISBN978-4-7581-0847-8

9年ぶりに内容刷新

ライフサイエンス英語
動詞 使い分け辞典

動詞の類語がわかれば
アクセプトされる論文が書ける！

著／河本　健，大武　博
監修／ライフサイエンス辞書プロジェクト
■ 定価（本体5,600円＋税）
■ B6判　■ 733頁　■ ISBN978-4-7581-0843-0

「類語～」の動詞情報をさらに発展！

ライフサイエンス
組み合わせ英単語

類語・関連語が一目でわかる

著／河本　健，大武　博
監修／ライフサイエンス辞書プロジェクト
■ 定価（本体4,200円＋税）
■ B6判　■ 360頁　■ ISBN978-4-7581-0841-6

ライフサイエンス　音声データDL
必須
英和・和英辞典
改訂第3版

編著／ライフサイエンス辞書プロジェクト
■ 定価（本体4,800円＋税）
■ B6変型判　■ 660頁　■ ISBN978-4-7581-0839-3

ライフサイエンス
論文を書くための
英作文&用例500

著／河本　健，大武　博
監修／ライフサイエンス辞書プロジェクト
■ 定価（本体3,800円＋税）
■ B5判　■ 229頁　■ ISBN978-4-7581-0838-6

ライフサイエンス　音声データDL
文例で身につける
英単語・熟語

著／河本　健，大武　博
監修／ライフサイエンス辞書プロジェクト
英文校閲・ナレーター／Dan Savage
■ 定価（本体 3,500円＋税）
■ B6変型判　■ 302頁　■ ISBN978-4-7581-0837-9

ライフサイエンス
論文作成のための
英文法

編集／河本　健
監修／ライフサイエンス辞書プロジェクト
■ 定価（本体3,800円＋税）
■ B6判　■ 294頁　■ ISBN978-4-7581-0836-2

ライフサイエンス英語
類語
使い分け辞典

編集／河本　健
監修／ライフサイエンス辞書プロジェクト
■ 定価（本体4,800円＋税）
■ B6判　■ 510頁　■ ISBN978-4-7581-0801-0

発行　羊土社 YODOSHA
〒101-0052　東京都千代田区神田小川町2-5-1　TEL 03(5282)1211　FAX 03(5282)1212
E-mail：eigyo@yodosha.co.jp
URL：www.yodosha.co.jp/
ご注文は最寄りの書店，または小社営業部まで